Ullmann's Renewable Resources

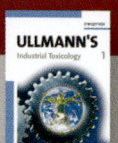

Ullmann's Renewable Resources

WILEY-VCH

WILEY-VCH Verlag GmbH & Co. KGaA

Editor-in-Chief

Dr. Barbara Elvers, Hamburg, Germany

All books published by **Wiley-VCH Verlag GmbH** are carefully produced. Nevertheless, authors, editors, and publisher do not warrant the information contained in these books, including this book, to be free of errors. Readers are advised to keep in mind that statements, data, illustrations, procedural details or other items may inadvertently be inaccurate.

Library of Congress Card No.:
applied for

British Library Cataloguing-in-Publication Data
A catalogue record for this book is available from the British Library.

Bibliographic information published by the Deutsche Nationalbibliothek
The Deutsche Nationalbibliothek lists this publication in the Deutsche Nationalbibliografie; detailed bibliographic data are available on the Internet at http://dnb.d-nb.de.

© 2013 Wiley-VCH Verlag GmbH & Co. KGaA. Published 2013 by Wiley-VCH Verlag GmbH & Co. KGaA

All rights reserved (including those of translation into other languages). No part of this book may be reproduced in any form – by photoprinting, microfilm, or any other means – nor transmitted or translated into a machine language without written permission from the publishers. Registered names, trademarks, etc. used in this book, even when not specifically marked as such, are not to be considered unprotected by law.

Print ISBN: 978-3-527-33369-1

Cover Design Schulz Grafik-Design, Fußgönheim
Typesetting Thomson Digital, Noida, India
Printing and Binding Markono Print Media Pte Ltd, Singapore

Printed on acid-free paper

Preface

This handbook features selected articles from the 7th edition of *ULLMANN'S Encyclopedia of Industrial Chemistry*, including newly written articles that have not been published in a printed edition before. True to the tradition of the ULLMANN'S Encyclopedia, products and processes are addressed from an industrial perspective, including production figures, quality standards and patent protection issues where appropriate. Safety and environmental aspects which are a key concern for modern process industries are likewise considered.

More content on related topics can be found in the complete edition of the ULLMANN'S Encyclopedia.

About ULLMANN'S

ULLMANN'S Encyclopedia is the world's largest reference in applied chemistry, industrial chemistry, and chemical engineering. In its current edition, the Encyclopedia contains more than 30,000 pages, 15,000 tables, 25,000 figures, and innumerable literature sources and cross-references, offering a wealth of comprehensive and well-structured information on all facets of industrial chemistry.

1,100 major articles cover the following main areas:
- Agrochemicals
- Analytical Techniques
- Biochemistry and Biotechnology
- Chemical Reactions
- Dyes and Pigments
- Energy
- Environmental Protection and Industrial Safety
- Fat, Oil, Food and Feed, Cosmetics
- Inorganic Chemicals
- Materials
- Metals and Alloys
- Organic Chemicals
- Pharmaceuticals
- Polymers and Plastics
- Processes and Process Engineering
- Renewable Resources
- Special Topics

First published in 1914 by Professor Fritz Ullmann in Berlin, the *Enzyklopädie der Technischen Chemie* (as the German title read) quickly became the standard reference work in industrial chemistry. Generations of chemists have since relied on ULLMANN'S as their prime reference source. Three further German editions followed in 1928–1932, 1951–1970, and in 1972–1984. From 1985 to 1996, the 5th edition of ULLMANN'S Encyclopedia of Industrial Chemistry was the first edition to be published in English rather than German language. So far, two more complete English editions have been published; the 6th edition of 40 volumes in 2002, and the 7th edition in 2011, again comprising 40 volumes. In addition, a number of smaller topic-oriented editions have been published.

Since 1997, *ULLMANN'S Encyclopedia of Industrial Chemistry* has also been available in electronic format, first in a CD-ROM edition and, since 2000, in an enhanced online edition. Both electronic editions feature powerful search and navigation functions as well as regular content updates.

Contents

Symbols and Units	IX	Chitin and Chitosan	243
Conversion Factors	XI	Cork	255
Abbreviations	XII	Drying Oils and Related Products	275
Country Codes	XVII	Fats and Fatty Oils	291
Periodic Table of Elements	XVIII	Gelatin	363
Raw Materials and Energy	1	Lignin	379
Biorefineries – Industrial Processes and Products	29	Pulp	395
		Polysaccharides	459
Carbohydrates: Occurrence, Structures and Chemistry	59	Starch	517
		Wood	547
Carbohydrates as Organic Raw Materials	89		
Cellulose	123	**Author Index**	601
Cellulose Esters	177		
Cellulose Ethers	225	**Subject Index**	603

Symbols and Units

Symbols and units agree with SI standards (for conversion factors see page XI). The following list gives the most important symbols used in the encyclopedia. Articles with many specific units and symbols have a similar list as front matter.

Symbol	Unit	Physical Quantity
a_B		activity of substance B
A_r		relative atomic mass (atomic weight)
A	m^2	area
c_B	mol/m^3, mol/L (M)	concentration of substance B
C	C/V	electric capacity
c_p, c_v	$J\,kg^{-1}\,K^{-1}$	specific heat capacity
d	cm, m	diameter
d		relative density (ϱ/ϱ_{water})
D	m^2/s	diffusion coefficient
D	Gy (=J/kg)	absorbed dose
e	C	elementary charge
E	J	energy
E	V/m	electric field strength
E	V	electromotive force
E_A	J	activation energy
f		activity coefficient
F	C/mol	Faraday constant
F	N	force
g	m/s^2	acceleration due to gravity
G	J	Gibbs free energy
h	m	height
$\hbar$	$W \cdot s^2$	Planck constant
H	J	enthalpy
I	A	electric current
I	cd	luminous intensity
k	(variable)	rate constant of a chemical reaction
k	J/K	Boltzmann constant
K	(variable)	equilibrium constant
l	m	length
m	g, kg, t	mass
M_r		relative molecular mass (molecular weight)
n_D^{20}		refractive index (sodium D-line, 20 °C)
n	mol	amount of substance
N_A	mol^{-1}	Avogadro constant ($6.023 \times 10^{23}\,mol^{-1}$)
P	Pa, bar*	pressure
Q	J	quantity of heat
r	m	radius
R	$J K^{-1}\,mol^{-1}$	gas constant
R	Ω	electric resistance
S	J/K	entropy
t	s, min, h, d, month, a	time
t	°C	temperature
T	K	absolute temperature
u	m/s	velocity
U	V	electric potential

Symbols and Units (Continued from p. IX)

Symbol	Unit	Physical Quantity
U	J	internal energy
V	m^3, L, mL, µL	volume
w		mass fraction
W	J	work
x_B		mole fraction of substance B
Z		proton number, atomic number
α		cubic expansion coefficient
α	$Wm^{-2}K^{-1}$	heat-transfer coefficient (heat-transfer number)
α		degree of dissociation of electrolyte
$[\alpha]$	$10^{-2} deg\ cm^2 g^{-1}$	specific rotation
η	Pa·s	dynamic viscosity
θ	°C	temperature
$\varkappa$		c_p/c_v
λ	$Wm^{-1}K^{-1}$	thermal conductivity
λ	nm, m	wavelength
μ		chemical potential
ν	Hz, s^{-1}	frequency
ν	m^2/s	kinematic viscosity (η/ϱ)
π	Pa	osmotic pressure
ϱ	g/cm^3	density
σ	N/m	surface tension
τ	Pa (N/m^2)	shear stress
φ		volume fraction
χ	Pa^{-1} (m^2/N)	compressibility

*The official unit of pressure is the pascal (Pa).

Conversion Factors

SI unit	Non-SI unit	From SI to non-SI multiply by
Mass		
kg	pound (avoirdupois)	2.205
kg	ton (long)	9.842×10^{-4}
kg	ton (short)	1.102×10^{-3}
Volume		
m^3	cubic inch	6.102×10^4
m^3	cubic foot	35.315
m^3	gallon (U.S., liquid)	2.642×10^2
m^3	gallon (Imperial)	2.200×10^2
Temperature		
°C	°F	$°C \times 1.8 + 32$
Force		
N	dyne	1.0×10^5
Energy, Work		
J	Btu (int.)	9.480×10^{-4}
J	cal (int.)	2.389×10^{-1}
J	eV	6.242×10^{18}
J	erg	1.0×10^7
J	kW·h	2.778×10^{-7}
J	kp·m	1.020×10^{-1}
Pressure		
MPa	at	10.20
MPa	atm	9.869
MPa	bar	10
kPa	mbar	10
kPa	mm Hg	7.502
kPa	psi	0.145
kPa	torr	7.502

Powers of Ten

E (exa)	10^{18}	d (deci)	10^{-1}
P (peta)	10^{15}	c (centi)	10^{-2}
T (tera)	10^{12}	m (milli)	10^{-3}
G (giga)	10^9	µ (micro)	10^{-6}
M (mega)	10^6	n (nano)	10^{-9}
k (kilo)	10^3	p (pico)	10^{-12}
h (hecto)	10^2	f (femto)	10^{-15}
da (deca)	10	a (atto)	10^{-18}

Abbreviations

The following is a list of the abbreviations used in the text. Common terms, the names of publications and institutions, and legal agreements are included along with their full identities. Other abbreviations will be defined wherever they first occur in an article. For further abbreviations, see page IX, Symbols and Units; page XVI, Frequently Cited Companies (Abbreviations), and page XVII, Country Codes in patent references. The names of periodical publications are abbreviated exactly as done by Chemical Abstracts Service.

abs.	absolute	BGA	Bundesgesundheitsamt (Federal Republic of Germany)
a.c.	alternating current		
ACGIH	American Conference of Governmental Industrial Hygienists	BGBl.	Bundesgesetzblatt (Federal Republic of Germany)
ACS	American Chemical Society	BIOS	British Intelligence Objectives Subcommittee Report (see also FIAT)
ADI	acceptable daily intake		
ADN	accord européen relatif au transport international des marchandises dangereuses par voie de navigation interieure (European agreement concerning the international transportation of dangerous goods by inland waterways)	BOD	biological oxygen demand
		bp	boiling point
		B.P.	British Pharmacopeia
		BS	British Standard
		ca.	circa
		calcd.	calculated
ADNR	ADN par le Rhin (regulation concerning the transportation of dangerous goods on the Rhine and all national waterways of the countries concerned)	CAS	Chemical Abstracts Service
		cat.	catalyst, catalyzed
		CEN	Comité Européen de Normalisation
		cf.	compare
ADP	adenosine 5′-diphosphate	CFR	Code of Federal Regulations (United States)
ADR	accord européen relatif au transport international des marchandises dangereuses par route (European agreement concerning the international transportation of dangerous goods by road)	cfu	colony forming units
		Chap.	chapter
		ChemG	Chemikaliengesetz (Federal Republic of Germany)
AEC	Atomic Energy Commission (United States)	C.I.	Colour Index
		CIOS	Combined Intelligence Objectives Subcommitee Report (see also FIAT)
a.i.	active ingredient		
AIChE	American Institute of Chemical Engineers	CNS	central nervous system
		Co.	Company
AIME	American Institute of Mining, Metallurgical, and Petroleum Engineers	COD	chemical oxygen demand
		conc.	concentrated
ANSI	American National Standards Institute	const.	constant
AMP	adenosine 5′-monophosphate	Corp.	Corporation
APhA	American Pharmaceutical Association	crit.	critical
API	American Petroleum Institute	CTFA	The Cosmetic, Toiletry and Fragrance Association (United States)
ASTM	American Society for Testing and Materials		
		DAB	Deutsches Arzneibuch, Deutscher Apotheker-Verlag, Stuttgart
ATP	adenosine 5′-triphosphate		
BAM	Bundesanstalt für Materialprüfung (Federal Republic of Germany)	d.c.	direct current
		decomp.	decompose, decomposition
BAT	Biologischer Arbeitsstofftoleranzwert (biological tolerance value for a working material, established by MAK Commission, see MAK)	DFG	Deutsche Forschungsgemeinschaft (German Science Foundation)
		dil.	dilute, diluted
		DIN	Deutsche Industrienorm (Federal Republic of Germany)
Beilstein	Beilstein's Handbook of Organic Chemistry, Springer, Berlin – Heidelberg – New York	DMF	dimethylformamide
		DNA	deoxyribonucleic acid
BET	Brunauer – Emmett – Teller	DOE	Department of Energy (United States)

DOT	Department of Transportation – Materials Transportation Bureau (United States)		gefährlicher Güter auf der Straße (regulation in the Federal Republic of Germany concerning the transportation of dangerous goods by road)
DTA	differential thermal analysis		
EC	effective concentration	GGVSee	Verordnung in der Bundesrepublik Deutschland über die Beförderung gefährlicher Güter mit Seeschiffen (regulation in the Federal Republic of Germany concerning the transportation of dangerous goods by sea-going vessels)
EC	European Community		
ed.	editor, edition, edited		
e.g.	for example		
emf	electromotive force		
EmS	Emergency Schedule		
EN	European Standard (European Community)	GLC	gas-liquid chromatography
		Gmelin	Gmelin's Handbook of Inorganic Chemistry, 8th ed., Springer, Berlin – Heidelberg – New York
EPA	Environmental Protection Agency (United States)		
EPR	electron paramagnetic resonance	GRAS	generally recognized as safe
Eq.	equation	Hal	halogen substituent ($-F$, $-Cl$, $-Br$, $-I$)
ESCA	electron spectroscopy for chemical analysis	Houben-Weyl	Methoden der organischen Chemie, 4th ed., Georg Thieme Verlag, Stuttgart
esp.	especially		
ESR	electron spin resonance	HPLC	high performance liquid chromatography
Et	ethyl substituent ($-C_2H_5$)		
et al.	and others	IAEA	International Atomic Energy Agency
etc.	et cetera	IARC	International Agency for Research on Cancer, Lyon, France
EVO	Eisenbahnverkehrsordnung (Federal Republic of Germany)		
		IATA-DGR	International Air Transport Association, Dangerous Goods Regulations
exp (…)	$e^{(\ldots)}$, mathematical exponent		
FAO	Food and Agriculture Organization (United Nations)		
		ICAO	International Civil Aviation Organization
FDA	Food and Drug Administration (United States)	i.e.	that is
		i.m.	intramuscular
FD&C	Food, Drug and Cosmetic Act (United States)	IMDG	International Maritime Dangerous Goods Code
FHSA	Federal Hazardous Substances Act (United States)	IMO	Inter-Governmental Maritime Consultive Organization (in the past: IMCO)
FIAT	Field Information Agency, Technical (United States reports on the chemical industry in Germany, 1945)		
		Inst.	Institute
		i.p.	intraperitoneal
Fig.	figure	IR	infrared
fp	freezing point	ISO	International Organization for Standardization
Friedländer	P. Friedländer, Fortschritte der Teerfarbenfabrikation und verwandter Industriezweige Vol. 1–25, Springer, Berlin 1888–1942		
		IUPAC	International Union of Pure and Applied Chemistry
		i.v.	intravenous
FT	Fourier transform	Kirk-Othmer	Encyclopedia of Chemical Technology, 3rd ed., 1991–1998, 5th ed., 2004–2007, John Wiley & Sons, Hoboken
(g)	gas, gaseous		
GC	gas chromatography		
GefStoffV	Gefahrstoffverordnung (regulations in the Federal Republic of Germany concerning hazardous substances)	(l)	liquid
		Landolt-Börnstein	Zahlenwerte u. Funktionen aus Physik, Chemie, Astronomie, Geophysik u. Technik, Springer, Heidelberg 1950–1980; Zahlenwerte und Funktionen aus Naturwissenschaften und Technik, Neue Serie, Springer, Heidelberg, since 1961
GGVE	Verordnung in der Bundesrepublik Deutschland über die Beförderung gefährlicher Güter mit der Eisenbahn (regulation in the Federal Republic of Germany concerning the transportation of dangerous goods by rail)		
		LC_{50}	lethal concentration for 50 % of the test animals
GGVS	Verordnung in der Bundesrepublik Deutschland über die Beförderung	LCLo	lowest published lethal concentration

LD$_{50}$	lethal dose for 50 % of the test animals	OSHA	Occupational Safety and Health Administration (United States)
LDLo	lowest published lethal dose		
ln	logarithm (base e)	p., pp.	page, pages
LNG	liquefied natural gas	Patty	G.D. Clayton, F.E. Clayton (eds.): Patty's Industrial Hygiene and Toxicology, 3rd ed., Wiley Interscience, New York
log	logarithm (base 10)		
LPG	liquefied petroleum gas		
M	mol/L		
M	metal (in chemical formulas)	PB report	Publication Board Report (U.S. Department of Commerce, Scientific and Industrial Reports)
MAK	Maximale Arbeitsplatzkonzentration (maximum concentration at the workplace in the Federal Republic of Germany); cf. Deutsche Forschungsgemeinschaft (ed.): Maximale Arbeitsplatzkonzentrationen (MAK) und Biologische Arbeitsstofftoleranzwerte (BAT), WILEY-VCH Verlag, Weinheim (published annually)		
		PEL	permitted exposure limit
		Ph	phenyl substituent (—C_6H_5)
		Ph. Eur.	European Pharmacopoeia, Council of Europe, Strasbourg
		phr	part per hundred rubber (resin)
		PNS	peripheral nervous system
		ppm	parts per million
max.	maximum	q.v.	which see (quod vide)
MCA	Manufacturing Chemists Association (United States)	ref.	refer, reference
		resp.	respectively
Me	methyl substituent (—CH_3)	R_f	retention factor (TLC)
Methodicum Chimicum	Methodicum Chimicum, Georg Thieme Verlag, Stuttgart	R.H.	relative humidity
		RID	réglement international concernant le transport des marchandises dangereuses par chemin de fer (international convention concerning the transportation of dangerous goods by rail)
MFAG	Medical First Aid Guide for Use in Accidents Involving Dangerous Goods		
MIK	maximale Immissionskonzentration (maximum immission concentration)		
min.	minimum	RNA	ribonucleic acid
mp	melting point	R phrase (R-Satz)	risk phrase according to ChemG and GefStoffV (Federal Republic of Germany)
MS	mass spectrum, mass spectrometry		
NAS	National Academy of Sciences (United States)		
		rpm	revolutions per minute
NASA	National Aeronautics and Space Administration (United States)	RTECS	Registry of Toxic Effects of Chemical Substances, edited by the National Institute of Occupational Safety and Health (United States)
NBS	National Bureau of Standards (United States)		
NCTC	National Collection of Type Cultures (United States)	(s)	solid
		SAE	Society of Automotive Engineers (United States)
NIH	National Institutes of Health (United States)		
		s.c.	subcutaneous
NIOSH	National Institute for Occupational Safety and Health (United States)	SI	International System of Units
		SIMS	secondary ion mass spectrometry
NMR	nuclear magnetic resonance	S phrase (S-Satz)	safety phrase according to ChemG and GefStoffV (Federal Republic of Germany)
no.	number		
NOEL	no observed effect level		
NRC	Nuclear Regulatory Commission (United States)	STEL	Short Term Exposure Limit (see TLV)
		STP	standard temperature and pressure (0°C, 101.325 kPa)
NRDC	National Research Development Corporation (United States)		
		T_g	glass transition temperature
NSC	National Service Center (United States)	TA Luft	Technische Anleitung zur Reinhaltung der Luft (clean air regulation in Federal Republic of Germany)
NSF	National Science Foundation (United States)		
NTSB	National Transportation Safety Board (United States)	TA Lärm	Technische Anleitung zum Schutz gegen Lärm (low noise regulation in Federal Republic of Germany)
OECD	Organization for Economic Cooperation and Development		
		TDLo	lowest published toxic dose

THF	tetrahydrofuran	UVV	Unfallverhütungsvorschriften der Berufsgenossenschaft (workplace safety regulations in the Federal Republic of Germany)
TLC	thin layer chromatography		
TLV	Threshold Limit Value (TWA and STEL); published annually by the American Conference of Governmental Industrial Hygienists (ACGIH), Cincinnati, Ohio		
		VbF	Verordnung in der Bundesrepublik Deutschland über die Errichtung und den Betrieb von Anlagen zur Lagerung, Abfüllung und Beförderung brennbarer Flüssigkeiten (regulation in the Federal Republic of Germany concerning the construction and operation of plants for storage, filling, and transportation of flammable liquids; classification according to the flash point of liquids, in accordance with the classification in the United States)
TOD	total oxygen demand		
TRK	Technische Richtkonzentration (lowest technically feasible level)		
TSCA	Toxic Substances Control Act (United States)		
TÜV	Technischer Überwachungsverein (Technical Control Board of the Federal Republic of Germany)		
TWA	Time Weighted Average		
UBA	Umweltbundesamt (Federal Environmental Agency)		
		VDE	Verband Deutscher Elektroingenieure (Federal Republic of Germany)
Ullmann	Ullmann's Encyclopedia of Industrial Chemistry, 6th ed., Wiley-VCH, Weinheim 2002; Ullmann's Encyclopedia of Industrial Chemistry, 5th ed., VCH Verlagsgesellschaft, Weinheim 1985–1996; Ullmanns Encyklopädie der Technischen Chemie, 4th ed., Verlag Chemie, Weinheim 1972–1984; 3rd ed., Urban und Schwarzenberg, München 1951–1970	VDI	Verein Deutscher Ingenieure (Federal Republic of Germany)
		vol	volume
		vol.	volume (of a series of books)
		vs.	versus
		WGK	Wassergefährdungsklasse (water hazard class)
		WHO	World Health Organization (United Nations)
		Winnacker-Küchler	Chemische Technologie, 4th ed., Carl Hanser Verlag, München, 1982-1986; Winnacker-Küchler, Chemische Technik: Prozesse und Produkte, Wiley-VCH, Weinheim, 2003–2006
USAEC	United States Atomic Energy Commission		
USAN	United States Adopted Names		
USD	United States Dispensatory		
USDA	United States Department of Agriculture	wt	weight
U.S.P.	United States Pharmacopeia	$	U.S. dollar, unless otherwise stated
UV	ultraviolet		

Frequently Cited Companies (Abbreviations)

Air Products	Air Products and Chemicals	IFP	Institut Français du Pétrole
Akzo	Algemene Koninklijke Zout Organon	INCO	International Nickel Company
		3M	Minnesota Mining and Manufacturing Company
Alcoa	Aluminum Company of America	Mitsubishi Chemical	Mitsubishi Chemical Industries
Allied	Allied Corporation		
Amer. Cyanamid	American Cyanamid Company	Monsanto	Monsanto Company
		Nippon Shokubai	Nippon Shokubai Kagaku Kogyo
BASF	BASF Aktiengesellschaft		
Bayer	Bayer AG	PCUK	Pechiney Ugine Kuhlmann
BP	British Petroleum Company	PPG	Pittsburg Plate Glass Industries
Celanese	Celanese Corporation	Searle	G.D. Searle & Company
Daicel	Daicel Chemical Industries	SKF	Smith Kline & French Laboratories
Dainippon	Dainippon Ink and Chemicals Inc.	SNAM	Societá Nazionale Metandotti
Dow Chemical	The Dow Chemical Company	Sohio	Standard Oil of Ohio
		Stauffer	Stauffer Chemical Company
DSM	Dutch Staats Mijnen	Sumitomo	Sumitomo Chemical Company
Du Pont	E.I. du Pont de Nemours & Company	Toray	Toray Industries Inc.
Exxon	Exxon Corporation	UCB	Union Chimique Belge
FMC	Food Machinery & Chemical Corporation	Union Carbide	Union Carbide Corporation
GAF	General Aniline & Film Corporation	UOP	Universal Oil Products Company
W.R. Grace	W.R. Grace & Company	VEBA	Vereinigte Elektrizitäts- und Bergwerks-AG
Hoechst	Hoechst Aktiengesellschaft	Wacker	Wacker Chemie GmbH
IBM	International Business Machines Corporation		
ICI	Imperial Chemical Industries		

Country Codes

The following list contains a selection of standard country codes used in the patent references.

AT	Austria	IL	Israel
AU	Australia	IT	Italy
BE	Belgium	JP	Japan[*]
BG	Bulgaria	LU	Luxembourg
BR	Brazil	MA	Morocco
CA	Canada	NL	Netherlands[*]
CH	Switzerland	NO	Norway
CS	Czechoslovakia	NZ	New Zealand
DD	German Democratic Republic	PL	Poland
DE	Federal Republic of Germany (and Germany before 1949)[*]	PT	Portugal
		SE	Sweden
DK	Denmark	SU	Soviet Union
ES	Spain	US	United States of America
FI	Finland	YU	Yugoslavia
FR	France	ZA	South Africa
GB	United Kingdom	EP	European Patent Office[*]
GR	Greece	WO	World Intellectual Property Organization
HU	Hungary		
ID	Indonesia		

[*]For Europe, Federal Republic of Germany, Japan, and the Netherlands, the type of patent is specified: EP (patent), EP-A (application), DE (patent), DE-OS (Offenlegungsschrift), DE-AS (Auslegeschrift), JP (patent), JP-Kokai (Kokai tokkyo koho), NL (patent), and NL-A (application).

Periodic Table of Elements

element symbol, atomic number, and relative atomic mass (atomic weight)

1A — "European" group designation and old IUPAC recommendation
1 — group designation to 1986 IUPAC proposal
IA — "American" group designation, also used by the Chemical Abstracts Service until the end of 1986

1A 1 IA	2A 2 IIA	3A 3 IIIB	4A 4 IVB	5A 5 VB	6A 6 VIB	7A 7 VIIB	8 8 VIII	8 9 VIII	8 10 VIII	1B 11 IB	2B 12 IIB	3B 13 IIIA	4B 14 IVA	5B 15 VA	6B 16 VIA	7B 17 VIIA	0 18 VIIIA
1 **H** 1.0079																	2 **He** 4.0026
3 **Li** 6.941	4 **Be** 9.0122											5 **B** 10.811	6 **C** 12.011	7 **N** 14.007	8 **O** 15.999	9 **F** 18.998	10 **Ne** 20.180
11 **Na** 22.990	12 **Mg** 24.305											13 **Al** 26.982	14 **Si** 28.086	15 **P** 30.974	16 **S** 32.066	17 **Cl** 35.453	18 **Ar** 39.948
19 **K** 39.098	20 **Ca** 40.078	21 **Sc** 44.956	22 **Ti** 47.867	23 **V** 50.942	24 **Cr** 51.996	25 **Mn** 54.938	26 **Fe** 55.845	27 **Co** 58.933	28 **Ni** 58.693	29 **Cu** 63.546	30 **Zn** 65.409	31 **Ga** 69.723	32 **Ge** 72.61	33 **As** 74.922	34 **Se** 78.96	35 **Br** 79.904	36 **Kr** 83.80
37 **Rb** 85.468	38 **Sr** 87.62	39 **Y** 88.906	40 **Zr** 91.224	41 **Nb** 92.906	42 **Mo** 95.94	43 **Tc*** 98.906	44 **Ru** 101.07	45 **Rh** 102.91	46 **Pd** 106.42	47 **Ag** 107.87	48 **Cd** 112.41	49 **In** 114.82	50 **Sn** 118.71	51 **Sb** 121.76	52 **Te** 127.60	53 **I** 126.90	54 **Xe** 131.29
55 **Cs** 132.91	56 **Ba** 137.33	57 **La*** 138.91	72 **Hf** 178.49	73 **Ta** 180.95	74 **W** 183.84	75 **Re** 186.21	76 **Os** 190.23	77 **Ir** 192.22	78 **Pt** 195.08	79 **Au** 196.97	80 **Hg** 200.59	81 **Tl** 204.38	82 **Pb** 207.2	83 **Bi** 208.98	84 **Po*** 208.98	85 **At*** 209.99	86 **Rn*** 222.02
87 **Fr*** 223.02	88 **Ra*** 226.03	89 **Ac*** 227.03	104 **Rf*** 261.11	105 **Db*** 262.11	106 **Sg**	107 **Bh**	108 **Hs**	109 **Mt**	110 **Ds**	111 **Rg**	112 **Cn**	113 **Uut**[a]	114 **Fl**	115 **Uup**[a]	116 **Lv**		118 **Uuo**[a]

[a] provisional IUPAC symbol

57 **La*** 138.91	58 **Ce** 140.12	59 **Pr** 140.91	60 **Nd** 144.24	61 **Pm*** 146.92	62 **Sm** 150.36	63 **Eu** 151.97	64 **Gd** 157.25	65 **Tb** 158.93	66 **Dy** 162.50	67 **Ho** 164.93	68 **Er** 167.26	69 **Tm** 168.93	70 **Yb** 173.04	71 **Lu** 174.97
89 **Ac*** 227.03	90 **Th*** 232.04	91 **Pa*** 231.04	92 **U*** 238.03	93 **Np*** 237.05	94 **Pu*** 244.06	95 **Am*** 243.06	96 **Cm*** 247.07	97 **Bk*** 247.07	98 **Cf*** 251.08	99 **Es*** 252.08	100 **Fm*** 257.10	101 **Md*** 258.10	102 **No*** 259.10	103 **Lr*** 260.11

* radioactive element; mass of most important isotope given.

Raw Materials and Energy

ROLF KOLA, B.U.S. AG, Eschborn, Federal Republic of Germany (Chaps. 1, 2)

OTTO VON ELSNER, Wintershall AG, Kassel, Federal Republic of Germany (Sections 3.1–3.3)

WOLFGANG RIEPE, Universität Paderborn, Fachbereich 13, Paderborn, Federal Republic of Germany, Universität Salzburg, Institut Chemie und Biochemie, Salzburg, Austria (Section 3.4)

KNUD REUTER, Bayer AG, Krefeld-Uerdingen, Federal Republic of Germany (Chap. 4)

1.	Introduction	1
2.	Mineral Raw Materials	2
2.1.	Formation of Deposits	2
2.2.	Mineralogy and Raw Material Groups	3
2.3.	Development of Deposits	3
2.4.	Exploitation of Raw Materials	4
2.5.	Availability of Mineral Raw Materials	5
2.6.	Environmental Factors in the Processing of Raw Materials	5
2.7.	Secondary Raw Materials	6
3.	Oil, Gas, and Coal	8
3.1.	Introduction	8
3.2.	Oil	8
3.2.1.	Exploration	8
3.2.2.	Production	8
3.2.3.	Transportation	10
3.2.4.	Loading and Storage	10
3.2.5.	Processing	10
3.2.6.	Uses	11
3.2.7.	Cost	11
3.3.	Gas	11
3.3.1.	Processing	12
3.3.2.	Transportation	12
3.3.3.	Storage	12
3.3.4.	Uses	12
3.3.5.	Cost	12
3.4.	Coal	12
3.4.1.	Mining Engineering	12
3.4.2.	Processing	13
3.4.3.	Utilization	13
4.	Renewable Resources	18
4.1.	Significance	18
4.2.	Important Renewable Resources	19
4.3.	Industrial Use	19
4.3.1.	Natural Fats and Oils	19
4.3.2.	Cellulose and Its Derivatives	21
4.3.2.1.	Cellulose	21
4.3.2.2.	Cellulose Esters	22
4.3.2.3.	Cellulose Ethers	22
4.3.3.	Starch	23
4.3.4.	Low-Molecular Carbohydrates — Sugars	23
4.3.5.	Other Renewable Resources	24
4.3.5.1.	Polyisoprene	24
4.3.5.2.	Natural Resins	24
4.3.5.3.	Plant Polysaccharides ("Gums")	24
4.3.5.4.	Active Substances Based on Renewable Resources	24
4.3.5.5.	Flavors and Fragrances	25
4.4.	Energy Use	25
4.4.1.	Biofuels	25
4.4.2.	Biogas	26
4.4.3.	Heat from Biomass	26
4.5.	Appraisal	26
4.5.1.	Ecobalancing of Renewable Resources	26
4.5.2.	Opportunities and Limits of Use of Renewable Resources	26
	References	26

1. Introduction

"Raw materials" is the name given to those materials whose constituents — after physical or chemical treatment — can be utilized by humans as a whole or in part, or whose stored binding energy can be released in the form of thermal energy.

Naturally occurring materials (e.g., ore, rock, or fossil material) are known as *primary raw materials*. Those materials that have already been in use or subjected to a physico-chemical process, but whose constituents can be recovered and consequently returned to the material cycle, are termed *secondary raw materials*. Examples of these are scrap materials, metallurgical dust and slag, waste glass, and wastepaper.

Renewable resources are raw materials of vegetable or animal origin whose availability can be secured by fresh planting or breeding.

The need for raw materials results from the volume of all commodities and consumer goods that must be produced and from the amount of energy that must be used in the extraction of raw materials, the production of intermediate and end products, and the use of goods during their service life. From this, the important factors affecting raw material consumption derive directly: (1) population density or growth, and (2) standard of living or degree of industrialization.

The extraction of natural raw materials and the production and use of consumer goods are basically associated with effects on the natural environment. The effects can be described in a simplified way as follows:

1. *Extraction of Raw Materials* Impact on the countryside and frequently on the water resources of a region, disturbance of natural biotopes, consumption of energy
2. *Processing of Raw Materials* Production of emissions in air and water, formation of unusable residues, consumption of chemical and physical energy
3. *Manufacture of Consumer Goods* Production of emissions, formation of further residues, consumption of energy
4. *Use of the Goods* Consumption of further materials and energy, creation of waste at the end of the service life, and consumption of disposal site volume

Against the background of a steadily increasing world population and the rise of industrialization and of living standards in the newly industrialized and developing countries, effects on the environment caused by the totality of human activities and questions about the nature and availability of raw materials are becoming increasingly important.

2. Mineral Raw Materials

Mineral is a collective name for all elements and their compounds, mainly inorganic and rarely organic in nature, that occur naturally as constituents of the earth's crust. Solidified minerals, as an aggregate, form the rocks of the earth's crust; 99.4 % of the earth's crust is composed of compounds of only 11 of the 92 natural elements:

SiO_2	59.12 %
Al_2O_3	15.34 %
CaO	5.08 %
Na_2O	3.84 %
FeO	3.81 %
Fe_2O_3	3.08 %
MgO	3.49 %
K_2O	3.13 %
H_2O	1.15 %
TiO_2	1.05 %
P_2O_5	0.30 %

The distribution of chemical elements in the earth's crust is not homogeneous. Dynamic geological changes led to element regroupings (i.e., enrichment and depletion). If a regional element enrichment reaches a mineable concentration, this is called a deposit. Deposits form the natural basis of the raw materials industry.

2.1. Formation of Deposits

The most important deposits are the magmatic deposits. When the magma cooled down slowly, dark, low-quartz or quartz-free rocks were formed in this *early crystallization phase*. These rocks are based mainly on magnesium – iron silicates. Simultaneously, metals such as Cr, Pt, V, Ni, and Cu separated from the liquid magmatic phase and accumulated in the magma chambers (e.g., Ni – Cu – Pt deposits in Sudbury, Canada; → Platinum Group Metals and Compounds, Chap. 3.3.). The subsequent *main crystallization phase* of the magma led to the formation of quartz-bearing eruptive rocks, such as granite. Further cooling and the *late crystallization* occurring at the same time resulted in superheated aqueous solutions (hydrothermal springs), which absorbed a number of the metals important today such as Pb, Zn, Cu, Mo, U, Bi, Hg, Ag, and Au, as well as Ta, Zr, and Nb. Because of temperature and pressure gradients, the solutions diffused into the cavities, veins, and clefts formed during the consolidation stage. Because of appropriate physical and chemical conditions, the metals separated in the form of mixed crystals (e.g., the auriferrous veins of Idaho Springs, United States or the zinc – lead deposits in Trepca, Serbia; → Lead). If, as a result of volcanic processes, hydrothermal springs emerged on the sea bottom, the dissolved metals were precipitated and formed volcanogenic – sedimentary deposits (Zn – Pb – Cu deposits in Kidd Creek, Canada or Pb – Zn deposits in Mount Isa, Australia). If metal-bearing rocks were brought up to the surface by emergence of land and therefore into the weathering zone, they were eroded as a result of physical or chemical processes and redistributed. In this way the so-called sedimentary deposits, in some cases with considerable enrichment, were formed (e.g., the copper belt of Zambia – Zaire; → Copper, Chap. 4.2.; gold placers in South Africa).

Through the settling of rock formations to greater depths, partial melting and new enrichment occurred, and *metamorphic* deposits were formed (e.g., the Zn – Pb deposits in Sulitjelma, Norway or Broken Hill, Australia).

2.2. Mineralogy and Raw Material Groups

In *systematic mineralogy* the minerals are classified by the criteria of chemical composition and crystal structure into nine main groups:

Class I	elements
Class II	sulfides, compounds of As, Sb, Br, Se, and Te
Class III	halide salts
Class IV	oxides, hydroxides
Class V	nitrates, borates, carbonates
Class VI	sulfates, chromates, molybdates, tungstates
Class VII	phosphates, arsenates, vanadates
Class VIII	silicates
Class IX	organic compounds

From the viewpoint of the raw materials industry a grouping such as that shown in Table 1 is appropriate.

2.3. Development of Deposits

Deposit research is concerned with ascertaining the formation of individual deposits and presenting the results in the form of geoscientific (metallogenetic) models. Such a metallogenetic model, for example, led to the discovery of porphyry copper ores on the West Coast of North and South America, in the Southeast Asian island arc, and in the Balkans. About 50 % of present-day copper production originates from such deposits.

Table 1. Mineral raw materials from the viewpoint of the raw materials industry

Raw material group		Examples
1.	Ores	
1.1	Iron ores	hematite, magnetite, pyrite, goethite, siderite
1.2	Nonferrous ores	lead glance, zinc blende, copper pyrite, bauxite, cassiterite
1.3	Ores of alloying metals	molybdenum glance, chromite, vanadinite, cobalt glance, wolframite, rutile
2.	Noble metals	metallic gold, platinum metals
3.	Natural rocks	
3.1	Magmatic rocks	granite, basalt, pumice
3.2	Sedimentary rocks	limestone, sandstone, tufa
3.3	Metamorphic rocks	marble, slate, serpentine
4.	Salt rocks	rock salt, sylvine, Glauber's salt
5.	Quartz minerals	quartz, chalcedony, kieselguhr, flint
6.	Clay minerals	kaolin, montmorillonite, illite, chlorite
7.	Precious and decorative stones	beryl, jade, opal, topaz, turquoise, garnet
8.	Rare earths, reactor fuels	monazite, cerite, thorite, pitchblende
9.	Other minerals	
	Nitrogen minerals	Chile saltpeter, saltpeter
	Phosphate minerals, boron minerals	apatite, phosphorite, borax, colemanite

The *prospecting and development* of raw material deposits is divided into three main stages.

The first stage, *prospecting*, seeks to determine geological structures with potential enrichments in a relatively large target area, and possibly broadly to delimit these structures. During the second stage, *exploration*, the area in which recoverable deposits are suspected is explored in more detail with regard to the concentration, extent, and character of the mineral body. The following phase is the *exploitation* of the deposit. In this, the type of ore or mineral is more closely identified and the content of the entire mineral body calculated. Noteworthy effects on the environment do not appear up to the development of the deposit, if the negligible effects of trial borings are disregarded. This can change, however, drastically when *mining* begins. In many countries, therefore, extensive environmental compatibility tests must be carried out before a mining permit is granted. In a feasibility study, the deposit is evaluated with regard to both technical and economic aspects. This means that the mineral must be minable by the application of available technologies, the necessary energy supply must be guaranteed, and transportation problems must be solved. The value of the ore body must cover the cost of exploration, mining, treatment, and marketing, as well as amortization of the investment within an acceptable period of time. If the profitability study leads to positive conclusions, the deposit is opened up. The time between exploration and opening is between five and ten years. The investment can be as much as $ 2×10^9 under difficult conditions, such as extreme climatic conditions, a lack of infrastructure, and long transportation routes [e.g., in the case of the copper – gold deposit of Ok Tedi (Papua New Guinea)]. Such projects can be handled only by financially strong international consortia.

2.4. Exploitation of Raw Materials

Depending on the structure and position of the mineral body, mining is carried out either two-dimensionally above ground (e.g., in the case of the saltpeter deposits in Atacama, Chile) or in the form of the technically simple and also very economical open-pit mining (e.g., the gigantic iron ore deposits in Minas Gerais, Brazil or Hamersley Range, Australia). In many cases, the position and structure of the ore body require technically complex underground mining.

With the exception of natural rocks such as granite, basalt, limestone, and marble, and of salt minerals such as rock salt, saltpeter, or sylvine, the substance of value in the extracted ore can be assumed to have too low a concentration to make direct metallurgical processing economically possible. In the case of low-content ores — particularly mixed ores with complex interstratification — expensive physicochemical methods are used: crushing, classification, pulverizing, flotation, or the sink-float process. In this way, not only can the gangue be removed, but concentrates of specific metals can also be produced. With decreasing metal content, however, the energy requirements rise exponentially; in many cases, therefore, the availability and price of energy determine the feasibility of a deposit.

The disposal of worthless rock, the creation of cavities or gigantic pits aboveground, the pumping away of mine water and, associated with it, the lowering of groundwater level — aside from intrusion into the landscape and disturbance of natural biotopes — represent the main environmental impacts from the mining of mineral law materials.

The ecological effect profile can therefore be classified as follows:

1. Interference with the natural water distribution in an area and, resulting from this, microclimatic change for plants and ground animals
2. Formation of emissions by deflation of heaps and dumps and from the operation of mechanical processing plants
3. Creation of wastewater through drainage of mine water and operation of flotation plants
4. Formation of residues as a result of overburden and separated gangue
5. Alteration of the natural appearance of the landscape (especially by open-pit mining)

With the exception of emissions into air and water, the remaining effects generally remain regionally limited and can be minimized by appropriate measures.

The depression funnel formed by pumping off mine water is generally many times the actual area of mining. Formation of the depression funnel can be partially compensated by extensive seepage of the water, instead of pumping it off. In many cases, previous treatment of the mine water is necessary. Iron and manganese, for example, must be removed, and in the case of nonferrous deposits, heavy metals (lead, copper, zinc, etc.), and possibly anions such as sulfide and sulfate, must be removed as well.

In open-pit mining, ca. 16×10^9 t of rock is produced per year worldwide, of which about 10×10^9 t is disposed as waste. By refilling, usually after having been stored in bulk for three to eight years, the greater part of a mined area can be recultivated. A mass deficiency remains, however, which can either be covered by filling with other, suitable residues or remain in the form of artificial lakes.

The ground surface must be protected from extensive subsidence, especially in underground mining. Subsidence occurs either through sagging of the overburden rock into the cavities formed by mining or by extensive change of the flows of groundwater and associated subrosive effects (i.e., dissolution of rocks or rock constituents). The effects can be largely limited by a controlled water-distribution system and refilling the galleries and shafts.

2.5. Availability of Mineral Raw Materials

The question of availability and resources arises primarily with reference to the most important commodity metals such as iron, aluminum, copper, zinc, chromium, or nickel.

Salt minerals such as halides, phosphates, nitrates, and sulfates, apart from their recovery by the mining of natural raw materials, can be produced without difficulty by appropriate processes or are obtained to a certain extent as byproducts (e.g., desulfurization gypsum or ammonium sulfate during flue-gas scrubbing).

In structural and civil engineering, *natural rocks* such as granite, basalt, sandstone, and marble have been replaced to a large extent by concrete and steel and are currently used mainly for decorative purposes. Natural rocks based on quartz, clay, and feldspar minerals represent by far the predominant constituent of the earth's crust, and their availability is practically unlimited.

Geochemistry shows that most *metals* occur in the earth's crust with a statistically normal distribution, if the concentration is plotted logarithmically against the quantity, as shown in simplified form in Figure 1. According to this, most metals occur in concentration ranges that are not recoverable under present conditions because of high costs. According to MCKELVEY, by taking into account the actual basic technical and economic conditions, deposit reserves are classified with a constant confidence level of 90 % as follows [7]:

Extent and content of the ore body are known with a limit of error of
$\approx 10\%$: proven reserves
$\approx 20\%$: probable reserves
$\approx 30\%$: possible reserves I
$\approx 50\%$: possible reserves II

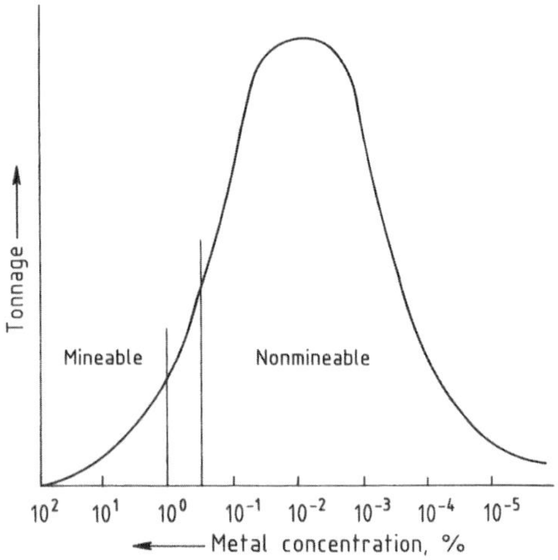

Figure 1. Distribution of metals in the earth's crust

The boundary lines drawn in Figure 2 (according to MCKELVEY) are not to be regarded as definite. More accurate exploration methods widen the field of known reserves; further development of mining methods and preparation processes, energy costs, and raw materials prices influence the boundary line between economic and subeconomic fields. The terms reserves and resources are thus dynamic quantities that are influenced by technoeconomic factors.

Given the reserves known today and the resources presumed with high probability, the possibility of replacing metals with plastics or ceramics, and advancing technical development, the supply of metallic raw materials is ensured for the foreseeable future.

2.6. Environmental Factors in the Processing of Raw Materials

The type of raw material, its recovery method, questions of location, the treatment process, the provision of environmental protection equipment, and other items naturally have a considerable effect on the environment. Table 2 shows the important environmental factors associated with aluminum, for example [8]. The effects of the mining and transportation of bauxite are not considered here. For comparison with the recovery of primary aluminum, the corresponding values for secondary aluminum production are also given.

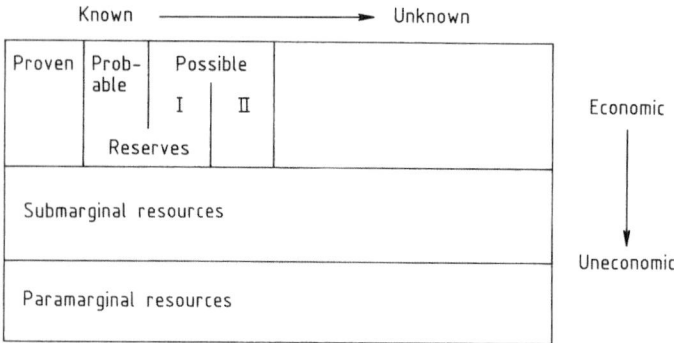

Figure 2. Classification of mineral resources (according to MCKELVEY)

Table 2. Factors with environmental impacts in the production of primary and secondary aluminum

Factor	Primary aluminum	Secondary aluminum
Energy Consumption, MJ/t Al		
Alumina production	27 683	
Scrap preparation		1 752
Melting down, preparation of dross, sweepings, etc.		10 060
Salt slag preparation		3 376
Fused-melt electrolysis	129 618	
Refining	4 473	
Casting	1 956	2 016
Total	163 730	17 204
Residues, kg/t Al		
Red mud	3 175	
Electrolysis residues	328	
Ash		47
Other	147	46
Total	3 650	93
Emissions, kg/t Al		
From primary energy generation	173	
Process specific	31	6
From melt energy consumption		6
Total	204	12

Supplying mineral raw materials to the treatment sites concerned and marketing the products are unavoidably combined with transportation processes. Table 3 lists the environmentally relevant factors of various means of transportation based on a study by the Schweizerisches Bundesamt für Umwelt, Wald und Landschaft (Swiss Federal Office for Environment, Forests and Countryside) [9].

2.7. Secondary Raw Materials [10]

The demand for raw materials and energy has risen steadily with advancing industrialization and growing demand for products and goods, together with the increasing world population. As world population has increased since 1900 from ca. 1.5×10^9 to ca. 6.2×10^9 people in 1980 — that is, by a factor of about 4.1 — energy consumption has risen by a factor of 15; steel production by a factor of 20; the production of typical commodity metals zinc and copper by factors of 15 and 18, respectively; and the production of the rather novel metals nickel and aluminum by factors of 100 and 3000, respectively.

Aside from the production of consumer materials and goods from natural raw materials, the reprocessing of residues and wastes is becoming increasingly important. The return of materials into the material cycle is called recycling, and the substances which are recycled are known as secondary raw materials. The important aspects

Table 3. Specific emissions of various means of transportation (g km^{-1} t^{-1}) and corresponding energy consumption [9]

	Rail	Truck	River boat	Sea-going vessel
Particles	0.006	0.080	0.038	0.011
CO	0.011	0.398	0.220	0.004
HC	0.069	0.199	0.055	0.035
NO_x	0.041	0.995	0.384	0.026
SO_2	0.081	0.080	0.165	0.234
Energy consumption, MJ km^{-1} t^{-1}	0.117	0.850	0.470	0.200

Table 4. Rate of recycling in the Federal Republic of Germany in 1989

	Consumption, 1000 t	From recycled materials[a]	
		1000 t	Percentage
Lead	374	184	49
Zinc	567	216	38
Copper	1 128	553	49
Aluminum	1 979	689	35
Steel[b]	39 700	17 356	44
Plastics	8 100	500	6

[a] Without internal recycling.
[b] Basis: production in 1988 including cast iron.

of recycling are

1. Conservation of natural resources
2. Reduction of total energy consumption
3. Reduction of the disposal site volume required

The reuse of valuable resources has a long tradition, activities being centered on metals. Examples are the remelting of noble metals such as gold and silver, the traditional processing of scrap iron and scrap nonferrous metals, and the processing of high-content metal dross, sweepings, ashes, dusts, etc. The recycling ratios of metals are therefore also distinctly higher than those of other materials, as shown in Table 4, with the Federal Republic of Germany as an example.

Energy saving is a decisive factor, especially in the recycling of metals from scrap, since the expensive energy of reduction has already been provided and only the usually low energy melting must be expended again. To produce a marketable grade, refining or realloying may still be necessary. Figure 3 compares the energy requirements of metals from primary and secondary production.

The essential goal of conventional recycling is the recovery of valuable substances from relatively high-grade residues. The overall "*conventional recycling*" system can be described as follows:

1. The value of the recovered materials or substances exceeds processing costs
2. Recycling is carried out mainly within the process or the industry concerned
3. Considerable quantities of residues of complex composition exist, since the selection of materials and the design of goods have been solely production or application orientated
4. An overall economic – ecological concept is lacking.

Especially in highly industrialized countries, extensive, complex environmental protection legislation requires — aside from the use of efficient waste gas and wastewater purification systems — the conservation of natural resources and, to the greatest possible extent, the avoidance of residues and wastes ($\rightarrow$ Legal Aspects, Chap. 2., $\rightarrow$ Legal Aspects, Chap. 3., $\rightarrow$ Legal Aspects, Chap. 4.). This has led to the development of new technologies especially suitable for processing industrial residues. For a modern raw materials industry with integrated recycling, therefore, some additional key features apply to the so-called *new recycling*:

1. The environment is incorporated in national economic thinking as a resource to be conserved.
2. Residues are also processed whose processing costs exceed the value of the recoverable components.

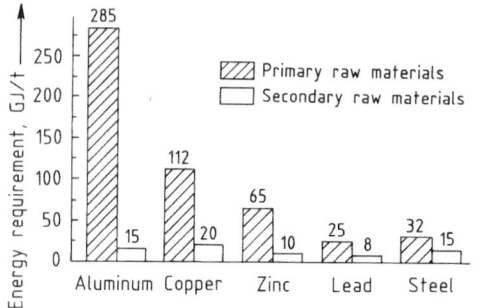

Figure 3. Energy requirements of metal production from primary and secondary raw materials

3. Recycling is not limited to the industry that produces the residue but is also carried out in other or new established sectors of industry.

Examples of this kind of recycling are the processing of salt slag from the secondary alu-minum industry; the processing of zinc-containing filter dust from electrosteel works; the regeneration of used foundry sand; the use of building rubble and slag in civil engineering; the processing of electronic scrap; and the dismantlement of automobiles, household appliances, and other consumer goods. An important precondition for raising the recycling ratios for raw materials from vehicles and appliances is a "recycling-friendly" choice of material and design, a task to which designers and producers are increasingly dedicating themselves.

3. Oil, Gas, and Coal

3.1. Introduction

For millions of years, oil, gas, and coal have been parts of our natural environment. Thus, special environmental protection problems are not expected in connection with the natural occurrence of these substances, nor have they occurred. Only when these materials are used by humans must environmental protection be taken into account: special environmental protection measures are necessary during exploration, production, transportation, storage, processing (→ Oil Refining, Chap. 4.) and use (→ Heating Oil, Chap. 1.6.; see also → Automotive Fuels, Chap. 8.) of these materials and the products obtained from them. Special dangers arise because these materials and their associated substances can form easily flammable mixtures with air. A series of spectacular accidents have shown that at all stages of processing — from the well to the consumer — warning and accident prevention plans must be drawn up in agreement with the authorities and emergency services. Some constituents of these substances and of the products obtained from them are considered environmentally hazardous in the widest sense, and in some cases they are classified as harmful to humans and animals (e.g., carcinogenic [11], mutagenic, or teratogenic) or as substances dangerous in water [12].

For some years, protection of the earth's atmosphere worldwide has been considered one of the greatest ecological challenges. In the scientific discussion, methane [13] present in natural gas and carbon dioxide [14] formed by combustion of oil, gas, and coal play a special part. Within the United Nations, a framework climate convention has been signed by more than 150 governments [15].

3.2. Oil

3.2.1. Exploration

Exploration is carried out by seismic reflection surveying of potential structures after promising areas have been selected by geological studies (→ Oil and Gas). In the surveying, which is carried out with heavy equipment, sometimes even in impassable terrain, the reflections of shock waves are analyzed. The shock waves are generated by vibrators, falling weights, or explosions in 10 to 20 m deep boreholes or by pressure pulses in water. In inhabited areas, the effects of noise and vibration on the population must be minimized. Cultivated areas must be recultivated and the farmers and foresters compensated.

If the result of these geophysical studies is positive, an exploratory well provides final confirmation of the presence of a reservoir; this is then explored further and developed by subsequent wells. If reservoirs must be opened up below towns, military exercise terrain, or areas that have to be protected for environmental reason, this can be done by "directional drilling" from less sensitive locations (→ Oil and Gas). Sometimes one location is sufficient. The same applies to the development of offshore fields from a platform (Fig. 4). Pollution control of seawater in the case of offshore wells in coastal waters is particularly sensitive ecologically. All wastewater, superfluous drilling mud, and drill cuttings must be collected in a barge, transported to land, and processed before the residues are then dumped.

3.2.2. Production

The legal framework governing oil production differs greatly in different countries. Whereas in

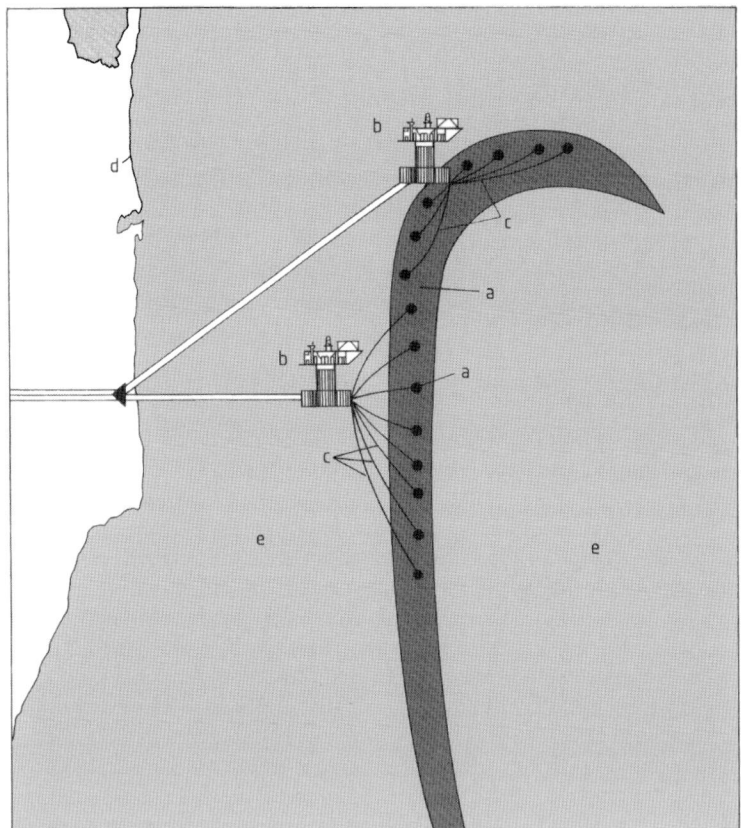

Figure 4. Directional drilling in the Schwedeneck See Oilfield in the German Baltic Sea
a) Oilfield; b) Platform; c) Directional drillings; d) Coast; e) Baltic Sea

highly developed Western countries a multiplicity of laws (on nature protection, soil protection, water pollution control, immission protection, and occupational safety, as well as mining regulations) accord high priority to ecological goals, in the countries of the Third World in the absence of legislation, responsibility for the environment falls to the development companies. The highest environmental standards apply to drilling and production platforms for offshore production, especially near the coast (Gulf of Mexico, North Sea, Baltic Sea) since particularly sensitive ecosystems can be threatened here (Wattenmeer). In particular cases, both feces and rainwater must be completely captured and disposed of on land.

On its way from the earth's surface to the oil field, the well is sunk through a series of strata and groundwater horizons. The oil and its mineral, aqueous, and associated gaseous substances must be guaranteed not to penetrate into these strata and groundwater horizons during extraction. Furthermore, the groundwater horizons themselves must not be brought by the well into contact with other horizons or strata. This is achieved by casing the hole and continuously sealing the outside by injecting cement (→ Oil and Gas, Chap. 3.3.). Production occurs through additional riser pipes in the hole.

On *abandoning production* and shutting down operation, the holes must be filled and the processing site recultivated. These measures are usually carried out on the basis of plugging plans approved by the mine inspectorate. If, at this time, oil contamination of soil or groundwater is observed in the area of the processing site, remedial action must be carried out according to a previously drawn-up plan. The improvement achieved thereby must be followed by analysis of soil and water samples, and completion of the action must be documented.

3.2.3. Transportation

Normally, a series of production locations can be found in an onshore oil field. From these production locations, oil flows in pipelines to collection points. In inhabited areas, these pipelines are laid in the ground and — for safety reasons — covered with at least 1 m of soil and protected against damage by means of concrete slabs or protective tubes if required. Steel lines are coated with plastic and provided with cathodic corrosion protection.

After being processed in central processing plants (see Section 3.2.5), the crude oil is transported further by land (usually by steel pipeline) to the port of loading and unloading or to a refinery near the consumer. Ecological questions have an increasing influence on the route, because both the pipeline and the refilled pipe ditch can affect the microclimate: e.g., the usual sand bed in the pipe ditch causes undesirable drainage of moist biotopes.

In some cases, special environmental protection measures are necessary: e.g., double-walled pipelines in sensitive areas or leading a heated and insulated crude oil line over several thousand kilometers on stilts in order to keep the crude oil liquid and pumpable but not thaw permafrost soils and thereby upset the sensitive balance of nature (Alaska).

Pipelines require regular *monitoring, testing, and maintenance*. During operation, pressures and flows are monitored at several points on the line. If the deviation of the actual value from the target value exceeds defined limits, the line is shut down and depressurized. The line is inspected regularly by patrolling on foot or by flying over with helicopters in order to ensure regular monitoring, for example, if excavation works are carried out in the vicinity of the line. The line is tested at regular fixed intervals by pressure tests and by running through with cleaning scrapers, and condition monitoring by intelligent scrapers developed especially for this purpose. Maintenance of the measurement and control equipment and of the safety and shutoff valves should be carried out according to a fixed maintenance plan.

In individual cases, for smaller fields, transportation can be made by road tankers or railroad tank cars to the refinery or, in the case of offshore fields near the coast, by lighters (double walled for environmental reasons). Sea passages over great distances are carried out in special large tanker vessels (very large crude carriers, VLCCs) between the loading ports of the producing country and the unloading piers of recipient countries. Tanker vessels have single-walled construction: after a series of serious oil accidents at sea, transportation in double-walled vessels has increasingly been demanded since the early 1990s.

3.2.4. Loading and Storage

During the *loading* of processed crude oil into road tank trucks, railroad tank cars, lighters, or oceangoing ships, as well as during the filling of storage tanks, environmental protection measures should be used to protect soil, bodies of water, and air from contaminants: filling operations must be monitored with alarm and cutoff devices; pumps have to be installed in an oil-collecting trough; and air pollution by volatile hydrocarbons must be minimized by gas displacement and gas recycling systems.

The oil can be *stored* in abandoned salt mines, salt caverns, granite caverns (Sweden), and caverns in permafrost soil or in tanks aboveground. For storage in steel tanks, a double tank bottom is increasingly required. The tanks must stand in an oiltight, oil-collecting trough or tank bund or must have a double wall. Floating roof tanks have to be provided with especially effective edge seals or be constructed as fixed roof tanks with internal floating ceilings.

3.2.5. Processing

The crude oil that emerges from the well is contaminated with minerals, water, and gas, which must be separated ($\rightarrow$ Oil and Gas, Chap. 3.4.3.1.). The separated *aqueous constituents* with their dissolved salts are reinjected into the reservoir and in this way disposed of while preserving the environment and maintaining pressure.

Depending on local conditions, the separated *gas* is flared, reinjected into the reservoir, or — after treatment — used for energy generation. Flaring has to be deprecated for environmental reasons. Further processing of crude oil is carried out in oil refineries ($\rightarrow$ Oil Refining), where it is processed to motor fuels, combustibles, and chemical raw materials.

Upon combustion of the refinery products, *sulfur compounds* present in the crude oil will be oxidized to sulfur oxides (mainly SO_2). To protect the environment, these compounds are increasingly being removed from the individual intermediate products by hydrogenation and reduced to hydrogen sulfide. Hydrogen sulfide is processed in Claus plants to elemental sulfur ($\rightarrow$ Sulfur, Chap. 6.3., $\rightarrow$ Sulfur, Chap. 6.4.), which is stored and transported to end users (usually as a liquid). For environmental protection reasons, downstream of the Claus plants, tail gas plants are connected which guarantee that 99 % of the hydrogen sulfide is converted to elemental sulfur ($\rightarrow$ Natural Gas, Chap. 2.7.1., $\rightarrow$ Natural Gas, Chap. 2.7.2.).

Nitrogen-containing compounds present in crude oil are hydrogenated in the aforementioned process to ammonia and fed together with hydrogen sulfide to the Claus plant. Here, ammonia is converted largely to elemental nitrogen.

Organometallic compounds present in crude oil remain in the heavy fuel oil or the petroleum coke. When these fuels are burned the metals remain in the ash — as oxides, sulfates, etc. — or in the flue gas (flue-gas scrubbing is therefore necessary).

3.2.6. Uses

The main constituents of all petroleum products are hydrocarbons, which are used as motor fuels, combustibles, and chemical raw materials. These substances and the products arising from them burn to water and carbon dioxide:

$$C_nH_{2m}+xO_2 \longrightarrow nCO_2+mH_2O$$

The carbon dioxide produced during combustion is a "greenhouse gas" that is judged by some scientists and governments to contribute substantially to global warming. In 1991 a national program for CO_2 reduction was started by the government of Germany. A reduction of 25 – 30 % between 1987 and the year 2005 was fixed as a target, and a catalogue of measures was laid down. From this, a reduction of CO_2 emissions of about 300×10^6 t is expected. The effectiveness of this measure is estimated to be slight, because the building of coal-based power stations in China alone in one year counterbalances German reduction measures. An internationally coordinated policy is therefore being aimed at.

Used *engine and gear oils* were previously disposed of by burning. Today these materials are increasingly reprocessed because legislation in some countries prohibits burning and requires recycling. The reprocessed materials then come onto the market as secondary raffinates.

In the use of *motor fuels,* special emission reduction measures are required for reduction of the environmental pollution caused by road traffic).

3.2.7. Cost

In the case of environmental protection costs, a distinction generally is made between the pure investment for environmental protection and the environmental protection operating costs. A large part of the environmental protection costs results from legal requirements. These regulations require observance of "generally recognized engineering standards" or the "state of the art.". The environmental protection operating costs excluding capital costs of the refineries of the EC states for 1998 have been estimated at about 2 to 10 ECU (European Currency Units) per tonne of product, depending on the state [16]. The environmental protection operating costs of the German petroleum industry, including service of capital, have been determined for 1987 and 1989 as about € 5.6/t in each case [17].

3.3. Gas

Natural gas and petroleum belong to the same class of substances: both are hydrocarbons ($\rightarrow$ Natural Gas) and in almost all cases occur together in nature. Exploration and production are therefore carried out in the same way as in the case of oil (see Sections 3.2.1 and 3.2.2). Special environmental protection and safety measures are necessary in case of sour gas occurrences, since the hydrogen sulfide they contain is toxic to humans, animals, and plants. For production from such wells, subsurface safety valves (SSSV), which are installed from 30 – 100 m below the earth's surface or below the seabed, are a proven technology. In the course of production, soil subsidences may occur, which can make special measures necessary in flat country to protect bodies of water.

3.3.1. Processing

The natural gas produced is processed at the well locations. The undesirable solid constituents, reservoir water, and heavy hydrocarbons are separated and the off-gas is sweetened to remove any hydrogen sulfide and carbon dioxide and dehydrated to avoid corrosion in the process installations connected downstream (→ Natural Gas, Chap. 2.3., → Natural Gas, Chap. 2.4.).

Some gases contain small amounts of mercury, which require further processing since this substance is hazardous to health and therefore must be removed (→ Natural Gas, Chap. 2.3.2.).

3.3.2. Transportation

Natural gas is transported at high pressure in steel pipes laid below ground or in liquid form by special ships as liquefied natural gas (LNG) (→ Natural Gas, Chap. 3.5., → Natural Gas, Chap. 4.1). In ecologically sensitive or especially valuable areas (Wattenmeer, other biotopes, rivers, and roads), the pipelines can be laid by trenchless technology or horizontal drilling. Monitoring, testing, and maintenance are carried out similarly to oil pipelines (see Section 3.2.3).

3.3.3. Storage

Large amounts of gas are stored underground. For this purpose, natural porous reservoirs and artificial salt caverns can be used (→ Natural Gas, Chap. 4.2.). No special environmental considerations apply in the operation of these reservoirs. During creation of the salt caverns by solution mining, environmentally acceptable disposal of the saline brine must be ensured.

3.3.4. Uses

Natural gas consists mainly of methane (→ Natural Gas, Chap. 1.1.). It is used as a source of energy and a chemical raw material. Both methane itself and the carbon dioxide produced during its combustion are climatically active gases, i.e., they absorb and reemit thermal radiation and so contribute to the greenhouse effect. Different calculations exist which [14] characterize the effectiveness of various greenhouse gases. The data available in the literature deviate greatly from each other, because much detailed knowledge is lacking and many effects have still been inadequately explained. Thus, for example, the lifetime in the earth's atmosphere of the gas molecules relevant to climate is subject to considerable uncertainty.

To minimize the greenhouse effect of the gases mentioned, in the case of *methane* the transport, storage, and distribution of natural gas must be as nearly loss-free as possible. *Carbon dioxide emissions* from the combustion of natural gas can be reduced by increasing the overall efficiency of heat production by technical improvements of the firing installations, combined heat and power schemes, and other energy saving measures. To hasten these savings, in various countries reduction programs have been drawn up or the introduction of taxes ("carbon tax") has been considered.

3.3.5. Cost

The costs of a carbon dioxide limitation (stabilization of emissions in the year 2010) are estimated for the United States as ca. $ 10^{12} [13].

3.4. Coal (→ Coal) [18–20]

3.4.1. Mining Engineering

Mining plants need large surface areas for their installation. In planning new mines, hauling shafts and coal preparation installations of existing mines are used as much as possible.

Underground mining of coal causes damage at the surface. Mining damages include subsidence, dislocated inclined planes, stretch, and wrench faults. Subsidence changes the slope of rivers and causes the formation of swamps, fissures in buildings, and thresholds in streets. Mining dam-age may be minimized by application of underground replacement techniques (refilling the cavities with waste rocks or washery tailings).

The *open-cast mining* of coal often requires drainage of the seam and overlying strata. Since open-cast mining may reach depths of 600 m, lowering of the groundwater level in areas adjacent to the mine is also necessary to stabilize the scarp of the mine. For this purpose, wells are installed at the border of the mine and at the entrance area of the groundwater, these wells are arranged to galleries. Modern surface mining techniques demand not only full mechanization with high efficiency (excavators, etc.) but also extensive automation of the sequence of work. The current mining technique using an equipment system consisting of bucket – wheel – excavator, conveyor belt and stacker has a performance of some 100 000 m^3/d. The relation of removed soil to coal (stripping ratio) increases with depth of 700 – 1100 m from about 3 to 5 m^3/t. The removed waste material is used directly for replenishment of the coal-spent area and reclamation.

3.4.2. Processing

Coal normally cannot be mined as clean coal. Run-of-mine coal is generally associated with different amounts of mineral matter. For some purposes, especially power and heat generation, coal may be used directly as so-called ballast coal. However, since mineral matter associated with coal has a considerable negative environmental impact (see Section 3.4.3), preparative measures must be used to separate the mineral matter from the coal. Preparation of coal is thus a primary measure of environmental protection, which separates a major part of harmful components (sulfur, trace elements) from the coal at convenient cost. Preparation lowers the pyrite sulfur level of the run-of-mine coal; separation of organically bound sulfur is not possible with the usual mechanical preparation methods. Pyrite crystals are intergrown with coal to different degrees. To obtain separation, coal must be crushed to grain sizes that correspond to the grain sizes of pyrite. This becomes more difficult the finer the pyrite grains and their distribution within the coal are. An extensive separation of pyrite from coal is connected with a lowering of the amount of the trace elements lead, cadmium, copper, and zinc in coal. The concentration of the trace elements — arsenic, barium, lead, cadmium, chromium, copper, mercury, strontium, and zinc — in the preparation waste increases with increasing ash content of the preparation wastes. Traces of beryllium, cobalt, molybdenum, vanadium, and nickel are bound to the organic matter itself.

Dust emissions resulting from preparation measures are decreased by filtering air and suffocating vapors. The amount of *wastewater* that leaves the preparation installation can be kept low by intermediate cleaning and reuse of part of the water from the preparation. Wastewater from the preparation installation is usually led to clearing ponds. For the elimination of flotation wastes, the deposition of washery tailings on dumps is mainly used. By special construction of the dump, oxidations processes (occurring especially with pyrite) can be suppressed to a great extent, thereby preventing the mobilization of heavy metals.

3.4.3. Utilization

Coal is used as a raw material in coal conversion processes, which can be divided into three main groups (see Fig. 5):

1. Mechanical conversion, including coal preparation and briquetting
2. Processes to obtain secondary fuels, including coking, gasification, liquefaction, and combustion
3. Other processes for the generation of energy (e.g., absorption on active coke and active carbon) including the recovery of byproducts (e.g., benzene, phenols)

Coal is an organic rock, consisting of two essential parts: the organic and the inorganic (mineral) matter. Its components may affect the environment in a typical manner during coal utilization. According to the particular conversion process, different products are formed in different quantities.

The *mineral matter*, which is termed ash or slag, may be used in commercial products. The main portion of trace elements present in coal is

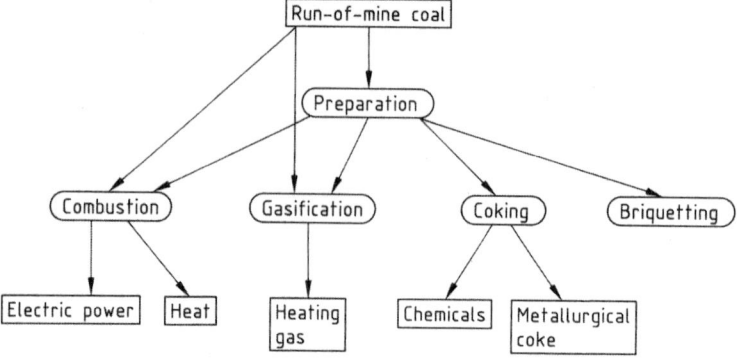

Figure 5. Utilization of coal

bound to the mineral matter, especially the sulfidic components. Care must be taken that these trace elements are not released uncontrolled into the environment.

The aforementioned environmental influences are introduced directly by the coal into the corresponding processes of utilization and valorization. Many other products that affect the environment are formed during conversion itself.

Heat and Energy Production. In the combustion processes used essentially for the production of electric power and heat, CO_2 is a climate-influencing emission, as are SO_2 and NO_x (including N_2O). In unfavorable conditions, toxic compounds may even be formed; a representative example is dioxin.

The development of world coal consumption in petajoules is shown in Figure 6. Consumption is expected to increase only slightly by about 2 % per year in the near future. That means that in the next few years, coal consumption will not exceed 100 000 PJ.

To avoid or reduce environmental problems, fundamental knowledge must be obtained for the development and optimization of technical processes. The main product formed by coal combustion for heat and electricity production is CO_2. As depicted in Figure 7, 3 t of CO_2 results from 1 t of hard coal, that is 1500 m^3 of CO_2. Worldwide coal consumption of 100 000 PJ will result in the generation of 10^{16} m^3 CO_2. Since the CO_2-free combustion of coal is not envisioned, emphasis must be placed on strategies of avoid-

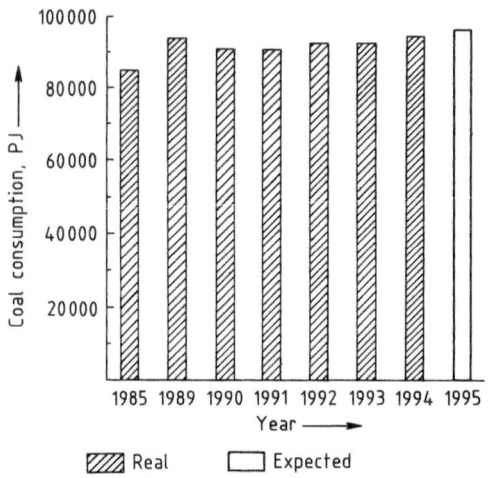

Figure 6. Development of world coal consumption

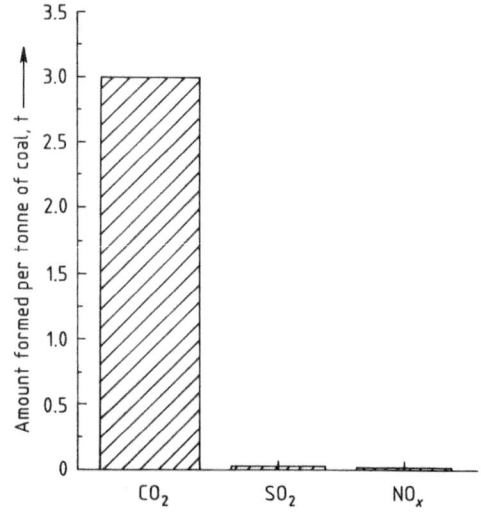

Figure 7. Amount of CO_2, SO_2, and NO_x formed during combustion of 1 t of coal

ance (i.e., economical use of energy ressources, as well as an increase in efficiency, which is demanded of engineering). By gas and steam processes, efficiencies of more than 50 % appear possible. A corresponding reduction of SO_2 and NO_x can be expected from increasing the efficiency of combustion.

The specific coal consumption for energy generation has been reduced from 1200 to 300 g/kW · h today by technical progress, and by more efficient dust precipitation, dust emission has been lowered from 800 to 50 mg per cubic meter of flue gas. At the same time the emission of detrimental matter bound to particles (trace elements, polyaromatic hydrocarbons, etc.) has decreased accordingly.

Emissions from coal combustion can be reduced by primary and secondary measures (Fig. 8). Primary measures include coal preparation, special arrangements at the dust entry (e.g., in a cyclone), improvement of the firing system, lowering of combustion temperature and shortening of combustion time, flue-gas recycling, and eventually the use of additives, which may reduce the emission of dust, SO_2, NO_x, HCl, and HF as well. Yet all of these primary measures together are usually not sufficient, to meet the demands of environmental protection. Accordingly, secondary measures must also be involved in the processes. These are, in terms of the flue gas, mainly the precipitation of dust, but also include gas cleaning by desulfuriza-

tion and DENOX processes ($\rightarrow$ Fixed-Bed Reactors, Chap. 3.). In terms of heat, overall reduction of noxious emissions may be reached by increasing efficiency and better utilization of waste heat.

The schematic of a wet flue-gas desulfurization process, employed for some years in an electric power and heat generating plant in Germany, is shown in Figure 9. Wet washings for desulfurization are accomplished with lime or suspensions of limestone at a temperature near the dew point of the flue gas. The washing liquids are recycled. The best degrees of desulfurization are obtained with hydrate of lime; slaked lime is diluted before entering the absorber (c). The ratio of sprayed liquid and gas to be cleaned is usually between 3 and 10 L/m^3. The absorbent is added in about stoichiometric ratio, because of stoichiometric reaction with the noxious gas (molar ratio Ca: S = 1). Flue gas and washing liquid move countercurrently. The flue gas is thus mixed with finely distributed lime absorbent and sulfur dioxide is dissolved in the washing liquid before it reacts with the absorbent. The contacting time between gas and liquid is ca. 5 min. Part of the washing liquid flow is led to an oxidation stage (f), which is mostly integrated into the washery sump. The calcium sulfite formed is oxidized here to calcium sulfate. From this partial flow the utilizable end product gypsum is brought out, which is crystallized on added crystallization nuclei. Downstream of the absorber, the flue gas

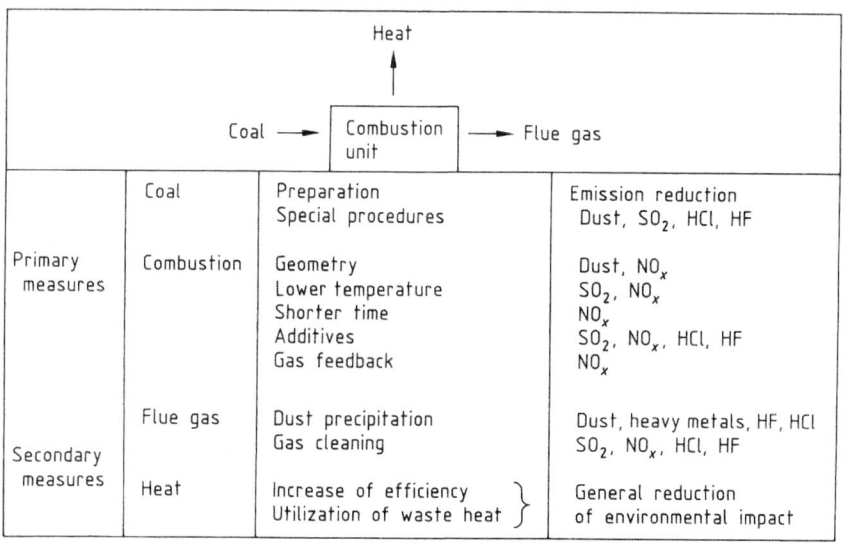

Figure 8. Measures for reduction of emissions formed during combustion of coal

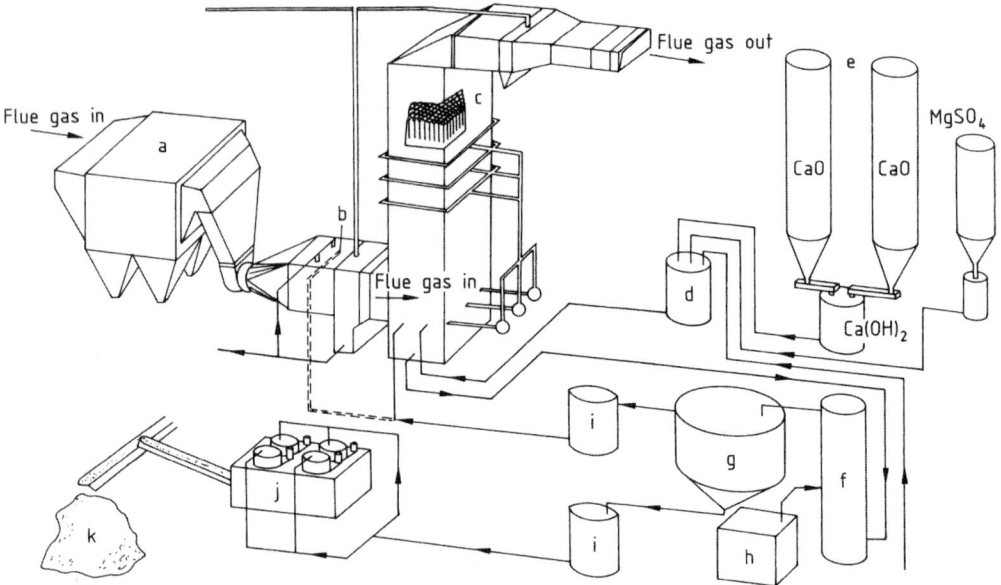

Figure 9. Schematic of a wet flue-gas desulfurization process
a) Electric precipitator; b) Prewasher; c) SO_2 absorber; d) Regenerator; e) Lime preparation station; f) Oxidator; g) Settling tank; h) Compressor; i) Gypsum buffer container; j) Gypsum dryer; k) Gypsum deposit

has to be reheated to 100°C before it is led to the chimney. Elimination degrees for sulfur dioxide of more than 85 % are obtained with this process, i.e., concentrations of <400 mg SO_2 per cubic meter in the cleaned gas.

On a technical scale, NO_x diminution in the flue gas is achieved by selective reaction of NO_x with added ammonia in the presence of a solid catalyst (DENOX process). Nitrogen oxides are reduced thereby to water and nitrogen in the presence of oxygen. The elimination of NO_x occurs at 300 – 500°C. Titanium dioxide or vanadium pentoxide can be used as catalyst (Fig. 10).

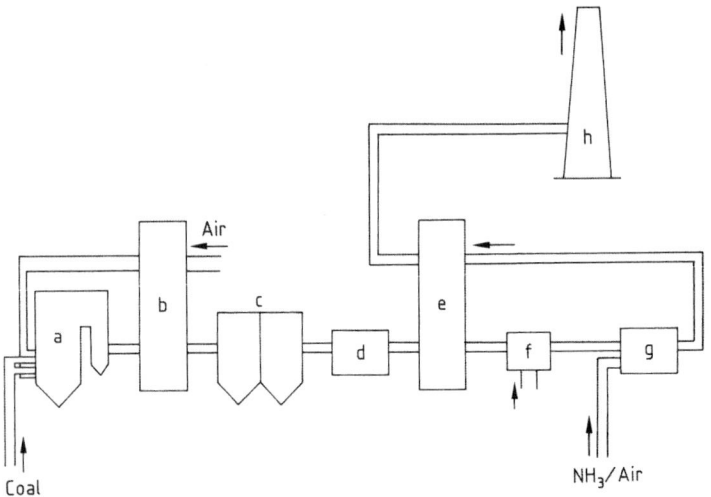

Figure 10. Schematic of a DENOX plant
a) Steam generator; b) Preheater; c) Dust precipitator; d) Flue-gas desulfurization plant; e) Heat exchanger; f) Heater; g) DENOX unit; h) Chimney

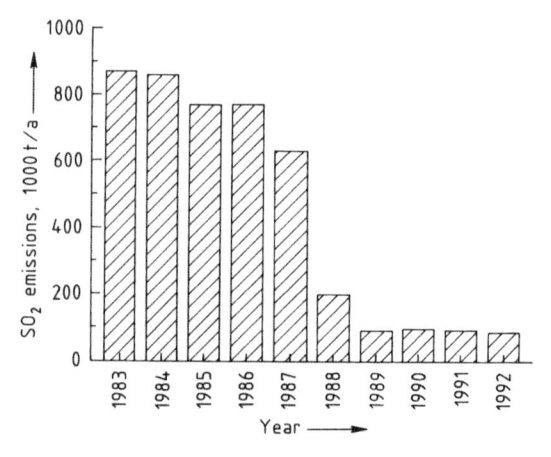

Figure 11. Reduction of SO$_2$ emissions in the Federal Republic of Germany

Another technique for desulfurization and denitrogenation is the so-called activated carbon process. Flue gas is led to an absorber stage where SO$_2$ is absorbed by activated carbon. NO$_x$ is converted to N$_2$ with 80 – 90 % yield.

The success of SO$_2$ reduction due to the introduction of flue-gas desulfurization can be seen in Figure 11. In the Federal Republic of Germany, emission of SO$_2$ was reduced from about 900 000 t/a in 1983 to <100 000 t/a in 1989. The same is true for NO$_x$ emissions, which were reduced by DENOX installations from about 500 000 t/a in 1983 to less than 100 000 t/a in 1992 (Fig. 12). If this reduction could be achieved worldwide when utilizing coal for the generation of electric power and heat, it would be an eminently important contribution to the protection of the environment. The processes are available, but introduction of these techniques often fails because of an unwillingness to invest in environment protection.

Coke Production. The second largest use of coal production after the generation of electric power and heat is in the production of coke. Cokefaction is also connected with environmental impacts, especially gaseous emissions, as well as charging of wastewater. Avoiding emissions from a coking plant is especially difficult because of its large-faced construction. Since the installations stand in the open air and many sources of emissions are distributed over a large area, enclosing them is difficult. Moreover, these diffuse sources (leakages) are related to operations such as filling, pressing, quenching, classifying, and so on. The type and quantity of emissions depend on various influences that occur during cokefaction at various points. Some points and types of emission, as well as limit values and measures that minimize environmental impact, are given below.

Point and type of emission	Limit values and preventive measures
Combustion gas	total sulfur <0.8 g/m^3
Flue gas	NO$_2$ <0.5 g/m^3 (5 % O$_2$)
Filling	dust <25 mg/m^3
Cooling	dust <20 mg/m^3

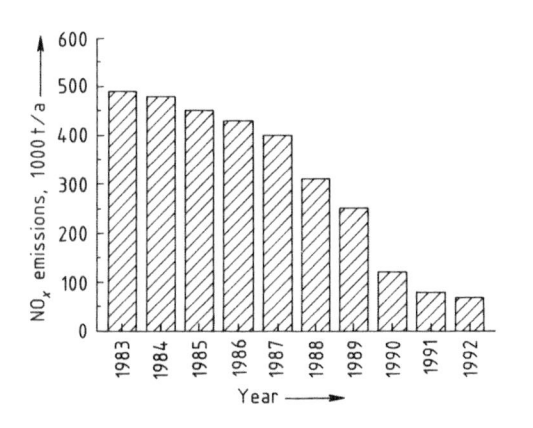

Figure 12. Reduction of NO$_x$ emissions in the Federal Republic of Germany

The wastewater of coking plants with wet quenching is alkaline (pH 8.5 – 9.5); it contains ammonia, phenols, sulfur, sulfanes, and cyan-ides. It must be rid of special contaminants before delivery to the purification plant. Intensively smelling substances such as mecaptans are separated from wastewater. In addition to the elimination of toxic components (phenols and cyanides), oils and tars must be separated.

For the elimination of organic compounds, biochemical processes, extractions, absorption by activated coke, and combustion of wastewater are well suited. Some modern coking plants use the dry coke cooling process in which problems with organic componds no longer occur.

Coal Valorization Processes (Gasification and Liquefaction). Particularly in the case of *gasification* ($\rightarrow$ Gas Production, Chap. 4.), emissions may really affect the environment during sluicing of material streams, especially of residues. Environmental pollution may also originate from flue dust in the raw gas, the raw gas itself, and the process water. The same substances are found here as in combustion processes, and the techniques of diminution are essentially the same. Another problem is that these residues cannot be deposited in an environmentally protective manner without more ado, especially in the case of alkaline catalysts. A solution could be to introduce a stage of inertization before deposition, for example, thermal treatment. The developments required are similar to those for power station byproducts.

In the case of *coal liquefaction* ($\rightarrow$ Coal Liquefaction), e.g., in liquid-phase hydrogenation, damage to the environment can be minimized and easily controlled because of the closed system. The formation of phenols — which formerly was a problem — and the environmental problems caused thereby may be avoided by appropriate operating conditions. The compounds H_2S and NH_3 contained in process water and gas can be eliminated without any problems. Only elimination of the mineral residue of hydrogenation in an environment-protecting manner remains. One may proceed here in the same way as suggested for gasification residues.

The combination of liquefaction and subsequent gasification of the liquefaction residues at 1500°C does not produce more than ca. 10 % ash, which is inert and ready for deposition (Fig. 13).

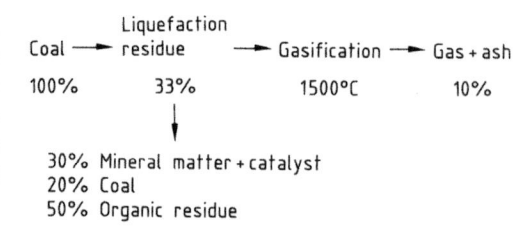

```
                  Liquefaction
Coal    ─→        residue         ─→   Gasification   ─→   Gas + ash
100%              33%                  1500°C              10%
                   │
                   ↓
            30% Mineral matter + catalyst
            20% Coal
            50% Organic residue
```

Figure 13. Reduction of residues formed during coal liquefaction by subsequent gasification of the residues

4. Renewable Resources

Renewable resources are animal and vegetable resources (biomass) that are used as energy carriers or raw materials in the widest sense. The term "renewable" means that the biomass of the agricultural ecosystem on average "survives," i.e., it is used only to the extent that it is regenerated. Agricultural ecosystems are therefore not of themselves lasting; rather, the conditions for their persistence are protection of the ecosystem and abandonment of predatory exploitation.

4.1. Significance

Renewable resources were for centuries virtually the only energy carriers and resource base for chemical products. Although in the 19th and 20th centuries they were driven back, first by coal and later by petroleum, they have never completely lost their importance for chemistry, because many of the structures and properties provided by nature are difficult, uneconomic, or practically impossible to reproduce in a chemical synthesis. Examples are the drying oils, with their importance for paint raw materials, or cel-lulose with its ethers and esters with inorganic and organic acids.

Renewable resources received greater attention again after the oil crises of the 1970s had made clear the dependence on petroleum and the limited nature of oil reserves. In the recent past, environmental reasons have been added as a motivation for renewed interest in renewable resources, since evidence exists that the global CO_2 balance is affected significantly by burning fossil energy carriers. In addition, this discussion also plays a part — although a considerably less important one — for chemical raw materials and their participation in the carbon cycle.

Table 5. Important renewable resources

Cellulosecontaining resources	Starchcontaining resources	Sugarcontaining resources	Oilcontaining resources
Wood	potato	sugarcane	rape
Straw	corn (maize)	sugar beet	flax
Hulls/shells	grain	fodder beet, etc.	soya
Bagasse*	cassava	sweet sorghum	sunflower
Cotton		molasses	oil palm
			coco palm
			castor-oil plant
			peanut
			cotton

*Wastes from cane sugar production.

Not to be underestimated is the fact that many industrialized countries are running short of space for waste disposal. For this reason, the possibilities of waste disposal, especially of plastics and packaging materials, via composting are being discussed increasingly. Biodegradable materials from renewable as well as fossil resources may make an important contribution here.

Finally, agricultural policy reasons exist for using renewable resources. The cultivation of plants that provide raw materials usable for energy production or internal combustion engines ("energy plants") or for the chemical industry ("industrial plants") has been publicized by agricultural interests, especially in countries with excess production, such as the EU countries. The possibility is seen here of cutting back excess production and yet keeping useful agricultural land alive.

The cultivation and use of renewable resources have been encouraged on national and European levels with the aid of subsidies and tax reliefs, and their political importance has thus been emphasized.

Within the framework of the Altener program [21, 22], the proportion of regenerative energy (in which biomass is included) to total energy consumption within the European Union is to be increased from 4 to 8 % by the year 2005, for example, with EU grants.

In Germany renewable resources are encouraged on the national level via the "Fachagentur Nachwachsende Rohstoffe" ("Special Agency for Renewable Resources"). The purpose of this institution is to seek practical opportunities for increased application of renewable resources via improved coordination among the government, science, and industry.

4.2. Important Renewable Resources

Table 5 (modified according to [23]) surveys the quantitatively most important sources of renewable resources.

In the renewable resources used by the chemical industry, natural fats and oils play the most important part. Table 6 shows the estimated consumption of various groups of resources by the chemical industry [24].

4.3. Industrial Use

Industrial application relates above all to use in the chemicotechnical field, where renewable resources are used for the production of intermediate and end products. The most important product lines from renewable resources are based on fats, oils, starch, sugar, and cellulose.

4.3.1. Natural Fats and Oils

World production of natural fats and oils has increased steadily in the last decades and in 1990 reached ca. 80×10^6 t [25, 26]. Vegetable oils, at 60.7×10^6 t, make up the main part, but animal

Table 6. Present use of natural raw materials (in 1000 t) by the chemical industry, estimated

Raw material	Germany (excluding former German Democratic Republic)	EU	World
Oil and fat	700	2700	9500
Cellulose	220	600	5000
Starch	115	390	1750
Sugar	15	65	800

fats (20×10^6 t) also occupy an important place in oleochemistry, especially in the utilization of inedible tallow and fat. In all, probably ca. 14 % of world fat production is used chemically [24, 26], ca. 80 % for human food, and the remaining 6 % as feed material. Castor oil is an exception: because of its unpalatability and special structure it is used exclusively for industrial purposes, mainly for paint raw materials and plastics (e.g., polyamide 11). The 1990 production figures for the most important oils and fats (in 10^6 t) [26] are listed below; for further information also regarding composition, see → Fats and Fatty Acids.

Soybean oil	16.5
Palm oil	10.8
Rapeseed oil (erucic acid free)	8.1
Sunflower oil	8.0
Cottonseed oil	3.8
Peanut oil	3.7
Coconut oil	2.8
Olive oil	1.8
Palmkernel oil	1.3
Linseed oil	0.7
Castor oil	0.4
Others (corn, sesame, etc.)	2.8
Vegetable oils, total	60.7
Tallow, lard	12.3
Butter	6.3
Fish oil	1.4
Animal fats, total	20.0
Total	80.7

Table 7 (after [24]) surveys the most important fields of application of oleochemical products.

Table 7. Most important fields of application of oleochemical products

Product	Application
Fatty acids and derivatives	metallic soaps, detergents, soaps, cosmetics, alkyd resins, paints; textile, leather, and paper industries; rubber, lubricants
Fatty acid methyl esters	cosmetics, detergents
Glycerol and derivatives	cosmetics, toothpaste, pharmaceuticals, foods, paints, plastics, synthetic resins, tobacco, explosives, cellulose processing
Fatty alcohols and derivatives	detergents, cosmetics; textile, leather, and paper industries; duplicator stencils, petroleum additives
Fatty amines and derivatives	fabric softeners, mining, road building, biocides, textile and fiber industries, petroleum additives
Drying oils	paints, varnish, linoleum
Castor oil, ricinoleic acid	polyamide 11, alkyd resins

Alkyd resins (→ Alkyd Resins) are produced on a large scale from natural triglycerides and the fatty acids made from them. Soybean oil is the most significant in this connection as the most important (semi-) drying oil. Linseed oil, which was previously dominant in the production of alkyd resins, is currently used mainly in the manufacture of linoleum.

The most important conversion reactions and products are shown in Figure 14 (according to [26]).

Surfactants. After soaps and petrochemically produced linear alkylbenzenesulfonates, fatty alcohol ethoxylates are the most important surfactants, as measured by consumption in Western industrialized countries (→ Surfactants, Chap. 14.) [27]. Fatty alcohol ethoxylates figured in public debate, being raw materials for fatty alcohol ether sulfates. Fatty alcohol ether sulfates play a large part in cosmetic formulations. Under the acidic conditions of their production by sulfation, traces of dioxane (a suspected carcinogen) can appear as a degradation product of the polyether chain (→ Surfactants). Today, process technology measures enable the dioxane content to be lowered to less than 10 ppm [28].

For nonionic surfactants, greater consideration of exclusively natural structural elements can be expected in the future. Here, the new alkyl polyglucosides, in particular, are interesting for cosmetic applications because of their mucous membrane compatibility, low toxicity, and easy biodegradability [26, 29] (→ Surfactants).

The potential for renewable resources in the surfactant field is increased further by the combination of oleochemical and carbohydrate structural units. A survey of existing carbohydrate surfactants, their significance, and their properties is given in [29]. For a short description of fatty acid esters of sorbitol and sucrose, see → Surfactants, → Surfactants.

Rapeseed oil has recently acquired practical importance as a *noncirculating lubricant*. Mineral-oil-based lubricants from motor saws lead to contamination of forest soil, which can be avoided by the introduction of rapeseed oil products because of their rapid and complete biodegradability [30].

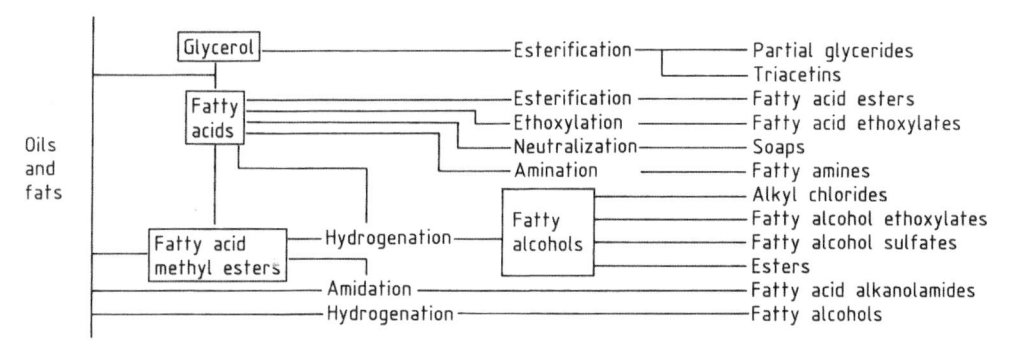

Figure 14. Most important conversion reactions of natural oils and fats and products formed

4.3.2. Cellulose and Its Derivatives

4.3.2.1. Cellulose

The chemical industry worldwide uses ca. 5.1×10^6 t of cellulose (dissolving pulp; EEC: 0.6×10^6 t, Germany: $>0.2 \times 10^6$ t). This constitutes about 3.5 % of the total cellulose production of ca. 150×10^6 t (1990 – 1991), of which the vast majority — more than 98 % — goes into "nonfood use" [31].

Cellulose is the most important product of chemical exploitation of wood ("chemical pulp"); between 30 and 50 % of hardwoods and softwoods consist of cellulose. A particular problem of this wood exploitation is association of the cellulose with lignin and hemicelluloses, each of which makes up 20 – 25 % of the wood.

As a result of the process for producing cellulose from wood, particularly the customary use of sulfur-containing compounds in the sulfite and the kraft (sulfate) processes, lignin and the hemicelluloses must be burned and thus can be utilized only for generation of process heat. A detailed survey of various pulping processes is given in [32] ($\rightarrow$ Paper and Pulp, Chap. 1.3.).

Most of the world chemical pulp production (nearly 90 %) is produced by kraft (sulfate) pulping, in which lignin and hemicelluloses are dissolved out of the wood with an aqueous mixture of sodium sulfide, sodium carbonate, and sodium hydroxide solution ($\rightarrow$ Paper and Pulp, Chap. 1.3.1.) [33]. The complicated bleaching with chlorine – hypochlorite or chlorine dioxide, which is required because of the dark color of the sulfate (kraft) cellulose, can lead to considerable environmental pollution because a number of organic, nondegradable, and unusable chlorine compounds are formed ($\rightarrow$ Paper and Pulp, Chap. 1.6.1., $\rightarrow$ Paper and Pulp, Chap. 1.6.2.). Even though lower wastewater loads are definitely achieved with chlorine dioxide, reservations remain that — aside from the uneconomical size of the operating units — have led to shutdown of the kraft pulping process (e.g., in Germany).

The *sulfite process* [34], which is stagnating worldwide, is less polluting and is used on a large scale in Germany ($\rightarrow$ Paper and Pulp, Chap. 1.3.2.). A few technical drawbacks must be accepted, however, particularly poorer mechanical strength of the (paper) cellulose so prepared; besides, only low-resin woods can be used as raw materials.

In a few processes, an attempt is made to avoid the problems associated with bleaching; examples are the ASAM process as a modification of the sulfite process ($\rightarrow$ Paper and Pulp), and the Organocell ($\rightarrow$ Paper and Pulp) and Acetosolv processes, which both work with organic lignin solvents, the Organocell process requiring ClO_2 [35].

Regenerated cellulose (cellophane), as obtained via alkali cellulose and the xanthate (produced from alkali cellulose by reaction with carbon disulfide), is easily biodegradable [36, 37] and permitted under food laws. Nevertheless, this product is declining as a film material since it is increasingly being replaced by polypropylene which is more resistant to mechanical stress and is considerably cheaper.

Detailed information on fibers from regenerated cellulose (viscose fibers) is given in $\rightarrow$ Cellulose, Chap. 3.1.

Regenerated cellulose, which is made by the copper oxide – ammonia (cuoxam) process in declining production volumes, is an important membrane material for dialyzers ("artificial kidneys") (→ Cellulose, Chap. 3.3.). Copper from the cuoxam process is almost completely recovered with ion exchangers. From an ecological viewpoint, the viscose process is inferior to the cuoxam process. The recovery rate for carbon disulfide used in the viscose process is ≤70 % (→ Cellulose, Chap. 3.4., → Cellulose, Chap. 3.1.4.).

4.3.2.2. Cellulose Esters

Inorganic Cellulose Esters. Cellulose nitrate, the oldest inorganic cellulose ester (nitrocellulose), has great importance as an explosive and a paint raw material, whereas the mixture with camphor is produced still as thermoplastic (celluloid) to a minor extent (e.g., for combs, table tennis balls, spectacle frames). The world capacity for cellulose nitrate amounts to ca. 165 000 t, with declining utilization.

Lacquers made from cellulose nitrate are excellent for open-pore wood (furniture) coatings, but usually contain considerable amounts of organic solvents [38].

For a description of safety aspects in the production, storage, and transport of cellulose nitrate, see → Cellulose Esters, Chap. 1.2.8. The potential hazard of cellulose nitrate depends on the degree of nitration and on the wetting with, e.g., ethanol or n-butanol (at least ca. 20 wt %). Since cellulose nitrate with high nitrogen content is explosive in the dry state, it may only be transported moistened.

Organic Esters. Of the cellulose esters of organic acids, acetic, propionic, and butyric acid esters, as well as the corresponding mixed esters (so-called acetopropionates and acetobutyrates), are especially important technically. The mixed ester of acetic and phthalic acid (monophthalate, from partially hydrolyzed cellulose triacetate and phthalic anhydride) is used to a small extent, especially in galenicals as tablet coatings for regulating drug absorption (dissolving first in the alkaline intestinal medium). For a short description of industrial uses of mixed esters, see → Cellulose Esters, Chap. 2.2.5.

For the production of moldings, thermoplastic cellulose esters of organic acids play a considerably greater part than cellulose nitrate, whose importance has diminished because of its flammability. Cellulose acetate and acetopropionate are important for spectacles (lenses and frames), sunglasses, and sports goggles, see also → Cellulose Esters, Chap. 2.4.9. [39].

During the debate on the biodegradability of plastics, especially packaging materials, thermoplastic cellulose esters have been studied in more detail. Cellulose triacetate proved to be not or not sufficiently rapidly biodegradable [40, 41]. However, the compostability of cellulose esters improves with decreasing degree of substitution. Cellulose-2,5-acetate ("secondary acetate") is apparently biodegraded slowly [41]. Cellulose diacetate, on the other hand, is designated compostable and is available as a commercial product, optionally also for food packaging [42], mixed with, e.g., citrate ester plasticizers that support degradation [43, 44].

Cellulose acetate butyrate (CAB) continues to be important in automobile metallic paints. As a result of the rapid viscosity increase of CAB-containing paints upon evaporation of organic solvents, and despite the very dilute solutions necessary to achieve planar orientation of the aluminum bronze, flawless processing is possible. Because of the high solvent emissions of such paint systems, considerable efforts are being made to replace them by waterborne metallic paints [45].

4.3.2.3. Cellulose Ethers

Cellulose ethers play an important part industrially. In 1987, world consumption — without the state trading countries — was 230 000 t and the EC share 116 000 t [46, 47], with a tendency to rise (for more detailed data see → Cellulose Ethers, Chap. 10.). The most important ethers are carboxymethyl cellulose (or its sodium salt) and methyl cellulose, which together make up more than three-fourths of the world consumption. The hydrophilic character of the cellulose ethers (hydrophilicity depending on the degree of substitution) leads to water-soluble products that are used as thickeners, water absorbers, and adhesives in the building industry; in surface coatings (including methyl cellulose as a wallpaper adhesive); as an auxiliary and additive in detergents (largest use [47]); in foods, cosmetics, and pharmaceuticals; as paper and textile auxiliaries, in petroleum production; etc. Hydroxyethyl cellulose also, as the third largest product of this group, has similar fields of application but is not permitted as a food additive (→ Cellulose Ethers, Chap. 5.1.2.).

4.3.3. Starch

The mixture of amylose (essentially straight chain) and amylopectin (branched), occurring naturally as "starch," has a fluctuating percentage composition depending on its raw material source. The amylopectin fraction usually predominates (ca. 3/4) (→ Starch, Chap. 1.).

Exceptions with a predominant amylose fraction are, e.g., starches from peas and high-amylose corn, grown especially for its high amylose content [48]. Since time immemorial, starch has been used outside the food field (e.g., as a paper and textile auxiliary). In total, ca. 500 000 t of starch are consumed annually in the industrial and technical fields in Germany (excluding the former German Democratic Republic) and more than 2×10^6 t in the EU. Its use in the chemical industry as a chemical raw material in the narrower sense, is considerably less [49], about 600 000 t in the EU [50]. Important applications are as an additive for building materials (e.g., in gypsum wallboards), cosmetics, and detergents, as well as an auxiliary for dyes and pharmaceuticals (→ Starch, Chap. 1.7.).

Table 8 shows the content and applications of starch in selected products in the chemicotechnical field [51].

Starch has recently attracted great interest as a blend raw material for thermoplastics, especially for packing materials, refuse sacks, or mulch films that are claimed to be biodegradable. Such degradation cannot occur in disposal sites, so only composting in appropriate systems is possible [52].

In the composting of blends of polymers customary in the packaging sector with starch, a macroscopically rapidly advancing degradation is initially achieved, based on microbial degradation of the starch contained in the blend. The thermoplastic component of the blend (e.g., polyethylene) is not completely degraded chemically despite the complete disappearance of the film or molding (→ Plastics, Recycling,

Table 8. Uses of starch in different industries

Branch of industry	Products	Starch content*
Paper manufacture	wrapping paper	3.0 – 5.0 %
	newsprint	1.5 – 2.0 %
	graphic paper	1.5 – 2.5 %
Paper processing	corrugated cardboard	10 – 14 g/m^2
	laminated paper	12 – 25 g/m^2
Building materials, etc.	gypsum plasterboard	5 – 7.0 %
	mineral fiberboard	3 – 6.0 %
	granulated coal	2 – 5.0 %
Chemistry, cosmetics, pharmaceuticals	washing starch	15 – 25 %
	washing powder	3 – 5 %
	washing raw materials	30 – 45 %
	toothpaste	6 – 70 %
	tablets, pills	0.5 – 3.0 %

*Based on dry matter.

Chap. 6.2). This is clear from early product examples with degradation curves in which the weight loss is followed and only an asymptotic approach to a maximum value (= starch content) is visible [53, 54]. Thus, a final evaluation of the biological compostability depends also on the second blend component [e.g., polyethylene, ethylene – acrylic acid copolymer [55, 56], partially hydrolyzed poly(vinyl acetate) [57–59]. Increasing degradation of synthetic polymer blend components is aimed at with prooxidants [60]. Composting conditions are decisive for the chronological sequence of degradation. To overcome drawbacks in biodegradability a more recent development uses poly(ε-caprolactone)— instead of, e.g., partially hydrolyzed poly(vinyl acetate) or poly(vinyl alcohol) copolymers — in combination with starch. These blends are totally compostable and have a reduced thermomechanical stability.

4.3.4. Low-Molecular Carbohydrates — Sugars

Products containing both carbohydrate structures and oleochemical building blocks are mentioned in Section 4.3.1.

Mono- and disaccharides, as well as simple derivatives, are produced in considerable amounts. The world production of low-molecular carbohydrates in 1991 – 1992 (in 1000 t/a) is listed below [61]:

D-Sucrose	113 614
D-Glucose	5 000
D-Sorbitol	650
D-Lactose	180
D-Fructose	50
D-Gluconic acid	40
L-Sorbose	25
D-Isomaltulose	20
D-Mannitol	8
D-Maltose	3
Xylitol	3
D-Xylose	1
Dianhydrosorbitol	1

These products are used predominantly in the food sector. Worldwide more than 800 000 t of them go into the non-food area (1991 figures), that is, less than 1 % (EC: 153 000 t; Germany in the chemical industry sector ca. 29 000 t).

In the chemical industry these carbohydrates are used mainly in the pharmaceutical field, as building blocks in polyurethane raw materials (sugar polyethers), and to a considerable extent as a carbon source in biotechnological processes (e.g., for the production of citric acid). To some extent, waste materials, such as molasses and bagasse, are also important here (→ Sugar, Chap. 22.2). Molasses is used to a large extent, e.g., as raw material for the production of citric acid via fermentation by *Aspergillus niger* [62]. The fermentation residues (vinasse), with only a low remaining sugar content, can be used in agriculture as a cattle feed additive or fertilizer, see, e.g., [63] (→ Sugar, Chap. 22.3.).

4.3.5. Other Renewable Resources

4.3.5.1. Polyisoprene

Natural rubber is described in detail in → Rubber, 2. Natural. Aside from natural rubber, which consists essentially of *cis*-1,4-polyisoprene, its *trans*-isomer, gutta percha, is also used industrially — however, to a considerably lower extent [64].

4.3.5.2. Natural Resins

Natural resins are described in → Resins, Natural. For the most important toxicological and legal aspects, see → Resins, Synthetic, Chap. 4. Although the economic importance of natural resins has declined, almost 1×10^6 t/a of rosin is still produced, chiefly in the United States and the People's Republic of China [65, 66] (→ Resins, Natural, Chap. 5.6.).

Important applications are paint and varnish resins and printing inks.

4.3.5.3. Plant Polysaccharides ("Gums")

The polysaccharides that can be used with water for the production of highly viscous gels are grouped under the name gums (see, e.g., [67]).

Some of the most used gums in approximate order of their quantitative importance are guar gum from the seed of the guar plant (→ Polysaccharides); gum arabic (→ Polysaccharides, Chap. 7.), an acacia tree resin; locust bean gum (carob gum) (→ Polysaccharides) from the seed of the carob tree; alginates and carrageenans (→ Polysaccharides, Chap. 4., → Polysaccharides, Chap. 5.) from seaweed, etc.

Aside from their importance in the food industry as emulsifiers, stabilizers, thickeners, etc., gums have been widely used industrially: guar gum (a branched mannose – galactose polysaccharide) in mining, in the textile industry, and above all in the paper industry; gum arabic (alkali – alkaline-earth salts of complex branched polysaccharides) for writing inks, adhesives, and textiles; locust bean gum (branched galactomannan) for textile sizes and dyes, and for the paper industry. Alginates (essentially salts of polysaccharides from guluronic and mannuronic acid units) are used chiefly for foods, pharmaceuticals, dental molding compounds, etc.; and carrageenans (galactans partially esterified with sulfuric acid) are used essentially for foods [67].

4.3.5.4. Active Substances Based on Renewable Resources

The active substances based on renewable resources include the antibiotics produced by fermentation (→ Antibiotics). The most important fermentation raw material is sugar, at ca. 50 000 t/a (EC 1990).

Some vitamins form a second important group of pharmaceutical preparations for which renewable resources and their biotransformation play a part. The most important are Vitamin C (ascorbic acid, worldwide ca. 36 000 t from 60 000 t glucose or sorbitol), Vitamin B_2 (riboflavin, worldwide ca. 2000 t/a), and in considerably smaller amounts Vitamin B_{12}. An important raw material source, among others, is beet molasses.

Finally, very large amounts of L-glutamine (ca. 400 000 t/a worldwide, EU ca. 31 000 t/a) and L-lysine (ca. 40 000 t/a worldwide, EU ca. 10 000 t/a) are produced from agricultural raw materials for use in foods, pharmaceuticals, and feeds.

So-called natural product insecticides play a growing part in public debate. However, they were recommended and used early. Examples are nicotine (from tobacco leaves), pyrethrum (from chrysanthemum flowers), and other plant extracts, in some cases used for centuries ($\rightarrow$ Insect Control, Chap. 3.).

However, unmodified natural substances cannot usually meet the criteria for modern plant protection agents — uniform and controllable action, toxicological acceptability, practicability, and economy. For example, the toxicity of nicotine to warm-blooded animals and the poor stability of pyrethrum make them unable to meet present-day standards. As a rule, therefore, ideas for the use of renewable resources for plant protection begin with modification of the natural model — example: pyrethroids.

The insect repellent azadirachtin occurring in the oil of the neem tree seed is attracting increasing interest. The oil has been used in India from time immemorial for insect control [68].

4.3.5.5. Flavors and Fragrances

Flavors and fragrances, such as those employed in foods, spices, drinks, medicine, toilet articles, cosmetics, and perfume, are often based on natural "renewable resources." In the Federal Republic of Germany, of the 40 000 t of flavors and fragrances produced in 1989 ca. one-third were of natural origin. Ca. two-third were produced synthetically, most of them to natureidentical substances. Fully synthetic aroma substances that do not correspond to natural products are the exception. Aroma substances of both animal and vegetable origin are employed, yet the products of animal origin, e.g., in perfumery (ambergris, musk, civet, etc.), are ever more costly and less used because of their increasing scarcity. A comprehensive survey is given in $\rightarrow$ Flavors and Fragrances.

4.4. Energy Use

The energy use of renewable resources comprises the use of biogas from industrial and agricultural wastes (e.g., for feeding into the public gas grid or for electricity generation); the cultivation of biomass for gasification, fermentation, or direct combustion; and the use of biofuels.

4.4.1. Biofuels

Biodiesel. The use of natural oils such as rapeseed oil or of rapeseed oil fatty acid methyl ester as a fuel for motor vehicles is increasingly discussed. While conventional diesel engines must be converted to use rapeseed oil, rapeseed oil methyl ester ("biodiesel"; produced from rapeseed oil by transesterification with methanol) is suitable directly for diesel engines. Although the ecological and economic evaluation of biodiesel is not yet complete [69, 70], a few aspects can be stressed:

In Germany alone, the area cultivated for rape is about 1×10^6 ha. Of this, up to now, only 42 000 ha (1991) has been used for the non-food area; yet by 1992, ca. 70 000 ha had been contracted for this purpose [70]. The potential is considered to be a maximum of ca. 400 000 ha [69]. Replacement of about 5%, or in the medium term not more than 10%, of diesel fuel is therefore achievable [70]. Environmental advantages then arise from the sulfur-free product and — even on allowing for emission of other greenhouse gases (N_2O, etc.) from the cultivation of the rape — from the reduction of CO_2 emissions due to the virtually closed carbon cycle. If 1 kg of diesel fuel is replaced by the required amount of 1.06 – 1.08 kg [70] rapeseed oil methyl ester, from 0.6 – 2.2 kg [69] to 2.3 – 2.8 kg [70] less CO_2 (including all other climatically active gases) is emitted. In addition savings in crude petroleum use per kilogram diesel fuel of 0.76 – 0.91 kg [70] or 0.55 – 0.94 kg [69] occur.

Ethanol. Molasses as a waste material from cane sugar production is important for the production of ethanol as a motor vehicle fuel ("gasohol"). During fermentation, enormous amounts of slops are formed and must be coped with. An overall estimation of emission reductions as a result of ethanol addition is disputed [71, 72]; while the CO emissions are reduced, NO_x values in the exhaust gas increase [73]. Economically, the high production costs make this alternative fuel, even at only ca. 20% addition, appear uncompetitive ("proalcool" program in Brazil) (→ Ethanol, → Automotive Fuels, Chap. 8.). For efforts in this regard in the United States, see, e.g., [74]. Bagasse (sugarcane residue) is either burned [e.g., for heating purposes (→ Sugar, Chap. 20.5.3.)] or used for the production of high-quality paper or cardboard. For a description of the extremely important fermentative ethanol synthesis, which clearly predominates over the synthesis from ethylene, see → Ethanol, Chap. 5.

4.4.2. Biogas

One possible means of waste utilization is the production of methane (biogas) from organic materials in sewage waters. This anaerobic enzymatic conversion (biomethanation) leads by oxidative and reductive processes to a mixture of ca. 60 mol% methane, ca. 40 vol% carbon dioxide, and small proportions of other gases. Biomethanation can also be carried out with "energy plants" as raw materials. The analogous conversion of organic landfill constituents likewise provides methane – CO_2 (landfill gas). A detailed description, including economic and ecological aspects, can be found in → Methane; → Gas Production, Chap. 1.1.4.

4.4.3. Heat from Biomass

Biomass firing in heating plants (e.g., with waste wood; whole-plant combustion of Miscanthus reeds, for example) plays an increasing role in projects and small decentralized plants [75]. Its profitability is not yet confirmed. However, of the energy sources bioethanol, biodiesel, and whole-plant combustion, the last has the smallest subsidy requirement at the present time [76].

4.5. Appraisal

4.5.1. Ecobalancing of Renewable Resources

Ecobalances seek to follow the life cycle of a product and to register the environmental influences associated with it [77]. Ecobalances in the field of production of renewable resources have already been drawn up [78, 79]. In the course of this, specific balancing problems occur with renewable resources compared to commercial processes. These concern especially the feedstock concept, the description of the emission situation, the land requirement and alternative land use, the dependence of the "production process" on conditions specific to the location (soil, weather, seasons), the description of alternative paths for waste use, and emissions into the soil.

4.5.2. Opportunities and Limits of Use of Renewable Resources

Renewable resoures have opportunities especially in those areas where the product of natural synthesis can be used directly or after simple processing and a lasting, economically advantageous supply of these raw materials is guaranteed. Conversely, renewable resources have their limits in fields of use where the aforementioned conditions are not met.

Thus, the use of renewable resources is reasonable when technical, economic, and ecological advantages over other materials exist.

References

General References

1. Metallgesellschaft Frankfurt: *Metallstatistik* **80** (1993).
2. *Statistisches Jahrbuch der Stahlindustrie 1993*, Verlag Stahleisen mbH, Düsseldorf, 1993.
3. *Metal Bulletin Handbook*, The Metal Bulletin Ltd., London 1993.
4. G. J. S. Govett, M. M. Govett: *World Mineral Supplies-Assessment and Prospectives*, Elsevier, Amsterdam 1976.
5. V. M. Goldschmidt: "Die geochemischen Verteilungsgesetze der Elemente," *Skr. Nor. Vidensk. Akad. Kl. 1 Mat. Naturvidensk. Kl.* 1923.
6. H. Kirsch: *Technische Mineralogie*, Vogel Verlag, Würzburg 1965.

Specific References

7. V. E. McKelvey: *The Mineral Potential of the United States 1975 – 2000*, Univ. Wisconsin Press, Madison 1973.
8. K. Krone et al.: "Ökologische Aspekte der Primär- und Sekundäraluminiumerzugung in der Bundesrepublik Deutschland," *Metall (Berlin)* **44** (1990) no. 6.

9 Bundesamt für Umwelt, Wald und Landschaft (BUWAL): Schriftenreihe Umwelt Nr. 132, Ökobilanzen von Packstoffen Stand 1990, Bern, Feb. 1991.
10 H. Maczek, W. Massion: "Umwelt, Energie und Recycling, " *Erzmetall* **44** (1991) no. 12.
11 World Health Organization (WHO) International Agency for Research on Cancer (IARC); *IARC Monographs on the Evaluation of Carcinogenic Risks to Humans, Occupational Exposures in Petroleum Refining; Crude Oil and Major Petroleum Fuels*, vol. 45, Lyon 1989.
12 Council Directive of December 17, 1979 on the protection of groundwater against pollution caused by certain dangerous substances (80/68/EEC) (List I, no. 7), O.J. no. L 20/43, 26.1.1980.
13 International Petroleum Industry Environmental Conservation Association (IPIECA): *Global Climate Change*, London 1991.
14 DGMK (German Society for Petroleum and Coal Sciences and Technology): Project 448-2 "Approaches and Potentials for Reducing Greenhouse Effects from Fossil Fuels," Hamburg 1992.
15 Conference of the United Nations on Environment and Development (UNCED), Rio de Janeiro 1992.
16 Commission of the European Communities: "Study on the Cost Born by the Refining Industry in the Member States in order to Comply with Environmental Legislation," prepared by Chem. Systems International Ltd., London 1986.
17 Estimation of Mineralölwirtschaftsverband e.V., Hamburg (Petroleum Industry Association in the Federal Republic of Germany, Hamburg).
18 J. F. Unsworth, D. J. Barratt, P. T. Roberts, *Coal Quality and Combustion Performance* in: Coal Science and Techology, vol. **19**, Elsevier, Amsterdam 1991.
19 D. van Velzen (ed.) Sulfur Dioxide and Nitrogen Oxides in Industrial Waste Gases: *Emission, Legislation and Abatement* in: Chemical and Environmental Science, vol. **3**, Kluwer Academic Publishers, Dordrecht 1991.
20 F. H. Friedrich, K. J. Guntermann, M. J. Paersch "Kohle und Umwelt," in: Bergbau, Rohstoffe, Energie, vol. 26, Verlag Glückauf GmbH, Essen 1989.
21 EEC, O.J. no. L 235/41, Sep. 18, 1993.
22 *Ökologische Briefe* **34** (1992) 14 – 15.
23 D. Osteroth: *Von der Kohle zur Biomasse*, Springer Verlag, Berlin 1989.
24 H. Baumann et al., *Angew. Chem.* **100** (1988) 41.
25 H. J. Richtler, J. Knaut, *Fat. Sci. Technol.* **93** (1991) 1.
26 H. Eierdanz in M. Eggersdorfer, S. Warwel, G. Wulff (eds.): *Nachwachsende Rohstoffe — Perspektiven für die Chemie*, VCH Verlagsgesellschaft, Weinheim, Germany 1993, p. 23.
27 H. J. Richtler, J. Knaut, *Seifen Öle Fette Wachse* **117** (1991) 545.
28 H. Baumann, M. Biermann in M. Eggersdorfer, S. Warwel, G. Wulff (eds.): *Nachwachsende Rohstoffe — Perspektiven für die Chemie*, VCH Verlagsgesellschaft, Weinheim, Germany 1993, p. 44.
29 M. Biermann, P. Schulz, *GIT Fachz. Lab.* **35** (1991) 963.
30 *Tribol. Schmierungstech.* **37** (1990) 193.
31 H. U. Woelk, personal communication, see also [48]
32 D. Schliephake et al.: *Nachwachsende Rohstoffe*, Verlag J. Kordt, Bochum 1986.
33 D. Fengel, G. Wegener: *Wood, Chemistry, Ultrastructure, Reactions*, Walter de Gruyter, Berlin 1984.
34 L. Greg et al.: International Bioenergy Directory no. 353, The Bioenergy Council, Washington 1982.
35 R. Patt, *Nachr. Chem. Tech. Lab.* **36** (1988) 26.
36 P. Finch, J. C. Roberts: "Enzymatic Degradation of Cellulose," in: T. P. Nevell, S. H. Zeronian (eds.): *Cellulose Chemistry and its Applications*, Horwood, Chichester 1985, pp. 312 – 343.
37 M. Korn: *Nachwachsende und bioabbaubare Materialien im Verpackungsbereich*, Verl. Roman Kovar, München 1993, pp. 119, 463 – 464.
38 Lacke und Lösemittel, Verlag Chemie, Weinheim – New York 1979, S. 8 – 10.
39 M. Wandel, C. Leuschke, Kunststoffe **74** (1984) 589.
40 in [37] , p. 122.
41 J.-D. Gu, D. T. Eberiel, S. P. McCarthy, R. A. Gross, *J. Environment. Polym. Degradation* **1** (1993) 143.
42 in [37], p. 122.
43 Tubize Plastics: *Biocell 163*, company publication, Tubize 1991.
44 *Chem. Ind. (Düsseldorf)* **114** (1991) no. 3, 64.
45 A. J. Backhouse, *J. Coat. Technol.* **54** (1982) 83.
46 L. Jeromin: *Nachwachsende Rohstoffe für die Chemische Industrie — eine neue verfahrenstechnische Herausforderung?* VDI-Verlag, Düsseldorf 1994.
47 Chemic. Economics Handbook 1989, cited in [46]
48 G. Tegge: *Stärke und Stärkederivate*, 2nd ed., B. Behr's Verlag, Hamburg 1988.
49 H. Baumann, *GIT Fachz. Lab.* **35** (1991) 957.
50 H. Koch, H. Röper, R. Höpke, *Spec. Publ. R. Chem. Soc.* **134** (1993) 157.
51 H. U. Woelk: "Die industrielle Verwendung von Stärke," in: *Stärke im Nichtnahrungsbereich*, Schriftenreihe des BML, Reihe A/Heft 380, Landwirtschaftsverlag, Münster 1990, pp. 5 – 19.
52 V. T. Breslin, *J. Environment. Polym. Degradation* **1** (1993) 127.
53 in [37] , pp. 37 – 55.
54 R. P. Wool, S. M. Goheen, Intern. Symposium on Biodegradable Polymers, Tokyo 1990.
55 F. H. Otey, R. P. Westhoff, C. R. Russel, *Ind. Eng. Chem. Prod. Res. Dev.* **16** (1977) 305.
56 in [36] , pp. 72 – 75.
57 Novamont: "The Biodegradable Nature of Mater-Bi," company publication, Kunstoff-Information ki-Nr. 29558, Bad Homburg 1990.
58 C. Bastioli, V. Bellotti, L. Del Giudice, G. Gilli, *J. Environment. Polym. Degradation* **1** (1993) 181.
59 in [37] , pp. 76 – 85.
60 K. Steinbach, *Umwelt* **23** (1993) 546.
61 F. W. Lichtenthaler, A. Boettcher in : M. Eggersdorfer, S. Warwel and G. Wulff (eds.): *Nachwachsende*

Rohstoffe — Perspektiven für die Chemie, VCH Verlagsgesellschaft, Weinheim, Germany 1993, pp. 151 ff.
62 The Association of American Feed Control Officials, Official Publication, 1977, Section 36, pp. 87 –88.
63 W. Dimmling, *CZ — Chemie Technik* **2** (1973) 425.
64 Winnacker-Küchler: *Chemische Technologie*, 4th ed., vol. **6**, Carl Hanser-Verlag, München 1982, pp. 520 – 521.
65 *Farbe Lack* **95** (1989) 832.
66 *Naturwiss. Rundsch.* **46** (1993) 192.
67 R. L. Whistler: *Industrial Gums*, Academic Press, New York 1973.
68 W. Kraus, A. Klenk, M. Bokel, B. Vogler, *Liebigs Ann. Chem.* 1987, 337.
69 Umweltbundesamt (ed.): Ökologische Bilanz von Rapsöl bzw. Rapsölmethylester als Ersatz für Dieselkraftstoffe (Ökobilanz Rapsöl), Jan. 1993.
70 K. Scharmer, G. Golbs, I. Muschalek, Union zur Förderung von Oel- und Proteinpflanzen e. V. (UFOP)-Studie "Pflanzenölkraftstoffe und ihre Umweltauswirkungen — Argumente und Zahlen zur Umweltbilanz," April 1993.
71 *Naturwiss. Rundsch.* **47** (1994) 76.
72 *New Sci.* **137** (1993) 22.
73 D. Schliephake: "Alkohole für Kraftstoffe," in: D. Schliephake et al. (eds.): *Nachwachsende Rohstoffe*, Verlag J. Kordt, Bochum 1986.
74 *Chem. Ind. (Düsseldorf)* **32** (1980) 84, 748.
75 nawaros — Nachwachsende Rohstoffe/Produktion, Projekte und Politik, no. 2/1993.
76 P. Breloh: "Die Bedeutung nachwachsender Rohstoffe aus Sicht des BML," in: Symposium Nachwachsende Rohstoffe — Perspektiven für die Chemie, Tagungsband, Landwirtschaftsverlag, Münster 1993, p. 19.
77 H. Hulpke, M. Marsmann, *Nachr. Chem. Tech. Lab.* **42** (1994) 11.
78 Dr. Wintzer et al.: Technologiefolgenabschätzung zum Thema "Nachwachsende Rohstoffe," Schriftenreihe des Bundesministers für Ernährung, Landwirtschaft und Forsten, Landwirtschaftsverlag, Münster 1993.
79 G. A. Reinhardt: *Energie- und CO_2-Bilanzierung nachwachsender Rohstoffe — Theoretische Grundlagen und Fallstudie Raps*, Verl. Vieweg, Braunschweig 1993.

Biorefineries – Industrial Processes and Products

BIRGIT KAMM, Research Institute Bioactive Polymer Systems (biopos e.V.), Teltow, Germany

PATRICK R. GRUBER, Outlast Technologies Incorporated, Boulder, Colorado, USA

MICHAEL KAMM, Biorefinery.de GmbH, Potsdam, Germany

1.	Introduction	29	3.4.	Vision and Goals and Plan for Biomass
2.	Historical Outline	30		Technology in the European Union and
2.1.	Historical Technological Outline and			Germany 37
	Industrial Resources	30	4.	Principles of Biorefineries 38
2.2.	The Beginning – A Digest	30	4.1.	Fundamentals 38
2.2.1.	Sugar Production	30	4.2.	Definition of the Term "Biorefinery" 40
2.2.2.	Starch Hydrolysis	30	4.3.	The Role of Biotechnology 41
2.2.3.	Wood Saccharification	31	4.3.1.	Guidelines of Fermentation Section within
2.2.4.	Furfural	31		Glucose-product Family Tree 41
2.2.5.	Cellulose and Pulp	31	4.4.	Building Blocks, Chemicals and Potential
2.2.6.	Levulinic Acid	31		Screening 42
2.2.7.	Lipids	32	5.	Biorefinery Systems and Design 44
2.2.8.	Vanillin from Lignin	32	5.1.	Introduction 44
2.2.9.	Lactic Acid	32	5.2.	Lignocellulosic Feedstock Biorefinery 44
2.3.	The Origins of Integrated Biobased		5.3.	Whole-crop Biorefinery 46
	Production	32	5.4.	Green Biorefinery 48
3.	Situation	35	5.5.	Two-platform Concept and Syngas 49
3.1.	Some Current Aspects of Biorefinery		6.	Biorefinery Economy 50
	Research and Development	35	7.	Outlook and Perspectives 52
3.2.	Raw Material Biomass	35		References 53
3.3.	National Vision and Goals and Plan for			
	Biomass Technology in the United States . . .	36		

1. Introduction

The preservation and management of our diverse resources are fundamental political tasks to foster sustainable development in the 21st century. Sustainable economic growth requires safe and sustainable resources for industrial production, a long-term and confident investment and finance system, ecological safety, and sustainable life and work perspectives for the public. Fossil resources are not regarded as sustainable, however, and their availability is more than questionable in the long-term. Because of the increasing price of fossil resources, moreover, the feasibility of their utilization is declining.

It is, therefore, essential to establish solutions which reduce the rapid consumption of fossil resources, which are not renewable (petroleum, natural gas, coal, minerals). A forward looking approach is the stepwise conversion of large parts of the global economy into a sustainable biobased economy with bioenergy, biofuels, and biobased products as its main pillars (Fig. 1).

Whereas for energy production a variety of alternative raw materials (wind, sun, water, biomass, nuclear fission and fusion) can be established, industry based on conversion of sustainable material, for example the chemical industry, industrial biotechnology, and also the fuel generation, depends on biomass, in particular mainly on plant biomass.

Some change from the today's production of goods and services from fossil to biological raw materials will be essential. The rearrangement of whole economies to implement biological raw materials as a source with increased value

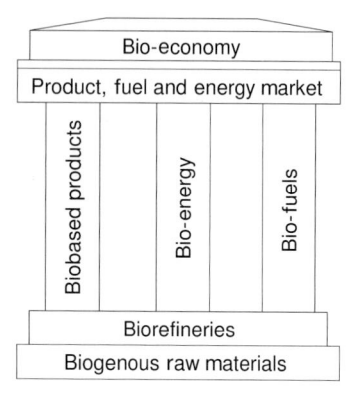

Figure 1. 3-Pillar model of a future biobased economy

requires completely new approaches in research and development. On the one hand, biological and chemical sciences will play a leading role in the generation of future industries in the 21st century. On the other hand, new synergies of biological, physical, chemical, and technical sciences must be elaborated and established. This will be combined with new traffic technology, media- and information technology, and economic and social sciences. Special requirements will be placed on both the substantial converting industry and research and development with regard to raw material and product line efficiency and sustainability.

The development of substance-converting basic product systems and polyproduct systems, for example biorefineries, will be the "key for the access to an integrated production of food, feed, chemicals, materials, goods, and fuels of the future" [4].

2. Historical Outline

2.1. Historical Technological Outline and Industrial Resources

Today's biorefinery technologies are based (1) on the utilization of the whole plant or complex biomass and (2) on integration of traditional and modern processes for utilization of biological raw materials. In the 19th and the beginning of the 20th century large-scale utilization of renewable resources was focused on pulp and paper production from wood, saccharification of wood, nitration of cellulose for guncotton and viscose silk, production of soluble cellulose for fibers, fat curing, and the production of furfural for Nylon. Furthermore, the technology of sugar refining, starch production, and oil milling, the separation of proteins as feed, and the extraction of chlorophyll for industrial use with alfalfa as raw material were of great historical importance. But also processes like wet grinding of crops and biotechnological processes like the production of ethanol, acetic acid, lactic acid, and citric acid used to be fundamental in the 19th and 20th century.

2.2. The Beginning – A Digest

2.2.1. Sugar Production

The history of industrial conversion of renewable resources is longer than 200 years. Utilization of sugar cane has been known in Asia since 6000 BC and imports of cane sugar from oversea plantations have been established since the 15th century. The German scientist A. S. MARGGRAF was a key initiator of the modern sugar industry. In 1748 he published his research on the isolation of crystalline sugar from different roots and beet [5, 6]. Marggraf's student, F. C. ACHARD, was the first to establish a sugar refinery based on sugar beet, in Cunern/Schlesien, Poland, in 1801.

2.2.2. Starch Hydrolysis

In 1811, the German pharmacist G. S. C. KIRCHHOFF found that when potato starch was cooked in dilute acid the starch was converted into "grape sugar" (i.e., d-glucose or dextrose) [7]. This was not only a very important scientific result but also the starting point of the starch industry. In 1806 the French emperor Napoleon Bonaparte introduced an economic continental blockade which considerably limited overseas trade in cane sugar. Thus, starch hydrolysis became of interest for the economy. The first starch sugar plant was established in Weimar, Germany, in 1812, because of a recommendation of J. W. DÖBEREINER to grand duke Carl August von Sachsen-Weimar. Successful development of the sugar beet industry, however, initially obstructed further development of the starch industry [8]. In 1835, the Swedish Professor J. J. BERZELIUS developed enzymatic hydrolyses of starch into sugar and introduced the term "catalysis".

2.2.3. Wood Saccharification

In 1819 the French plant chemist H. BRACONNOT discovered that sugar (glucose) is formed by treatment of wood with concentrated sulfuric acid. 1855, G. F. MELSENS reported that this conversion can be carried out with dilute acid also. Acid hydrolysis can be divided into two general approaches, based on (1) concentrated acid hydrolysis at low temperature and (2) dilute acid hydrolysis at high temperature. Historically, the first commercial processes, named wood saccharification, were developed in 1901 by A. CLASSEN (Ger. Patent 130980), employing sulfuric acid, and in 1909 by M. EWEN and G. TOMLINSON (US Patent 938208), working with dilute sulfuric acid. Several plants were in operation until the end of World War I. Yields of these processes were usually low, in the range 75–130 liter per ton wood dry matter only [9, 10]. Technologically viable processes were, however, developed in the years between World War I and World War II. The German chemist FRIEDRICH BERGIUS was one of the developers. The sugar fractions generated by wood hydrolyses have a broad spectrum of application. An important fermentation product of wood sugar of increasing interest is ethanol. Ethanol can be used as fuel either blended with traditional hydrocarbon fuel or as pure ethanol. Ethanol is also an important platform chemical for further processing [11].

2.2.4. Furfural

DÜBEREINER was the first to report the formation and separation of furfural by distillation of bran with diluted sulfuric acid in 1831. In 1845 the English Chemist G. FOWNES proposed the name "furfurol" (furfur–bran; oleum–oil). Later the suffix "ol" was changed to "al" because of the aldehyde function [12, 13].

Treatment of hemicellulose-rich raw materials with dry steam in the presence of hydrogen chloride gave especially good results [14]. Industrial technology for production of furfural from pentose is based on a development of an Anglo-American company named Quaker Oats. The process was been developed in the nineteen-twenties [E.P. 203691 (1923), F.P. 570531 (1923)]. Since 1922 Quaker Oats Cereal Mill in Cedar Rapids/Iowa, USA, has produced up to 2.5 tons of furfural per day from oat husks. Since 1934 the process had been established as an industrial furfural plant. Furfural was the cheapest aldehyde, at 16–17 cents per lb (lb=pound; 1 metric ton=1000 kg=2204.62442 lb; 1 kg=0.453592 lb) [15]. Until approximately 1960 DuPont used furfural as a precursor of nylon-6.6. Furfural has since been substituted by fossil based precursors.

2.2.5. Cellulose and Pulp

In 1839 the Frenchman A. PAYEN discovered that after treatment of wood with nitric acid and subsequent treatment with a sodium hydroxide solution a residue remained which he called "les cellules", cellulose [16]. In 1854 caustic soda and steam were used by the Frenchman M. A. C. MELLIER to disintegrate cellulose pulp from straw. In 1863 the American B. C. TILGHAM registered the first patent for production of cellulose by use of calcium bisulfite. Together with his brother, Tilgham started the first industrial experiments to produce pulp from wood by treatment with hydrogen sulfite. This was 1866 at the paper mill Harding and Sons, Manayunk, close to Philadelphia. In 1872 the Swedish Engineer C.D. EKMAN was the first to produce sulfite cellulose by using magnesium sulfite as cooking agent [17]. By 1900 approximately 5200 pulp and paper mills existed worldwide, most in the USA, approximately 1300 in Germany, and 512 in France.

2.2.6. Levulinic Acid

In 1840 the Dutch Professor G. J. MULDER (who also introduced the name "protein") synthesized levulinic acid (4-oxopentanoic acid, c-ketovaleric acid) by heating fructose with hydrochloride for the first time. The former term "levulose" for fructose gave the levulinic acid its name [18]. Although levulinic acid has been well known since the 1870s when many of its reactions (e.g., esters) were established, it has never reached commercial use in any significant volume. In the 1940s commercial levulinic acid production was begun in an autoclave in the United States by

A.E. STALEY, Dectur, Illinois [19]. At the same time utilization of hexoses from low-cost cellulose products was examined for the production of levulinic acid [20]. As early as 1956 levulinic acid was regarded as a platform chemical with high potential [21].

2.2.7. Lipids

From 1850 onward European import of tropical plant fats, for example palm-oil and coconut oil, started. Together with the soda process, invented by the French N. LEBLANC in 1791, the industrialization of the soap production began and soap changed from luxury goods into consumer goods. The developing textile industry also demanded fat based products. In 1902 the German chemist W. NORMANN discovered that liquid plant oils are converting into tempered fat by augmentation of hydrogen. Using nickel as catalyst Norman produced tempered stearic acid by catalytic hydration of liquid fatty acids [22]. The so called "fat hardening" led to the use of European plant oils in the food industry (margarine) and other industries.

2.2.8. Vanillin from Lignin

In 1874 the German chemists W. HAARMANN and F. TIEMANN were the first to synthesize vanillin from the cambial juice of coniferous wood. In 1875 the company Haarmann and Reimer was founded. The first precursor for the production of vanillin was coniferin, the glucoside of coniferyl alcohol. This precursor of lignin made from cambial juice of coniferous trees was isolated, oxidized to glucovanillin, and then cleaved into glucose and vanillin [23]. This patented process [24] opened the way to industrial vanillin production. It was also the first industrial utilization of lignin. Besides the perfume industry, the invention was of great interest to the upcoming chocolate industry. Later, however, eugenol (1-allyl-4-hydroxy-3-methoxybenzene), isolated from clove oil, was used to produce vanillin. Today, vanillin production is based on lignosulfonic acid which is a side product of wood pulping. The lignosulfonic acid is oxidized with air under alkaline conditions [25, 26].

2.2.9. Lactic Acid

In 1895 industrial lactic acid fermentation has been developed by the pharmaceutical entrepreneur A. BOEHRINGER. The Swedish pharmacist C. W. SCHEELE had already discovered lactic acid in 1780 and the conversion of carbohydrates into lactic acid had been known for ages in food preservation (e.g. Sauerkraut) or agriculture (silage fermentation). Because of the activity of Boehringer the German company Boehringer-Ingelheim can be regarded as the pioneer of industrial biotechnology. Both the process and the demand for lactic acid by dyeing factories, and the leather, textile, and food industries made the company the leading supplier. In 1932 W. H. CAROTHERS, who was also the inventor of polyamide-6.6, developed, together with van Natta, a polyester made from lactic acid, poly(lactic acid) [27]. In the late 1990s this poly(lactic acid) was commercialized by the company NatureWorks (Cargill, the former Cargill Dow) [28].

2.3. The Origins of Integrated Biobased Production

In the year 1940 the German chemist P. VON WALDEN (noted for his "Inversion of configuration at substitution reactions", the so-called "Walden-Reversion") calculated that in 1940 Germany produced $13 \cdot 10^6$ t of cellulose leaving 5 to 6 million tons of lignin suitable only as wastage. He then formulated the question: How long can national economy tolerate this [29]? Approaches to integrated production during industrial processing of renewable primary products have a long tradition, starting from the time when industrial cellulose production expanded continuously, as also did the related waste-products. Typical examples of this will be mentioned.

As early as 1878 A. MITSCHERLICH, a German chemist, started to improve the sulfite pulp process by fermentation of sugar to ethyl alcohol – it should be mentioned that sugar is a substance in the waste liquor during sulfite pulp production. He also put into practice a procedure to obtain paper glue from the waste liquor. Both processes were implemented in his plant located in Hof, Germany, in the year 1898 [30].

In 1927 the American Marathon Corporation assigned a group of chemists and engineers to the task of developing commercial products from the organic solids in the spent sulfite liquor from the Marathon's Rothschild pulp and paper operations close to Wausau, Wisconsin, USA. The first products to show promise were leather tanning agents. Later, the characteristics of lignin as dispersing agents became evident. By the mid 1930s, with a considerable amount of basic research accomplished, Marathon transferred operations from a research pilot plant to full-scale production [31].

One of the most well known examples is the production of furfural by the Quaker Oats Company since 1922, thus coupling food, i.e. oat flakes, production and chemical products obtained from the waste [13] (see Section 2.2). On the basis of furfural a whole section of chemical production developed – furan chemistry.

Agribusiness, especially, strived to achieve combined production from the very beginning. Modern corn refining started in the middle of the 18th century when T. KINGSFORD commenced operation of his corn refining plant in Oswego, New York [32]. Corn refining is distinguished from corn milling because the refining process separates corn grain into its components, for example starch, fiber, protein and oil, and starch is further processed into a substantial number of products [33].

The extensive usage of green crops has been aim of industry for decades, because there are several advantages. Particularly worthy of mention is the work of OSBORN (1920) and SLADE and BIRKINSHAW (1939) on the extraction of proteins from green crops, for example grass or alfalfa [34].

In 1937 N.W. PIRIE developed the technical separation and extraction methods needed for this use of green crops [35, 36]. By means of sophisticated methods all the botanical material should have been used, both for production of animal feed, isolated proteins for human nutrition, and as raw material for further industrial processes, for example glue production. The residual material, juices rich in nutrients, had initially been used as fertilizer; later they were used for generation of fermentation heat based on biogas production [37, 38].

These developments resulted in market-leading technology, for example the Proxan and Alfaprox procedures, used for generation of protein–xanthophyll concentrates, including utilization of the by-products, however, predominantly in agriculture [39].

In the United States commercial production of chlorophyll and carotene by extraction from alfalfa leaf meal had started in 1930 [40, 41]. For example Strong, Cobb and Company produced 0.5 t chlorophyll per day from alfalfa as early as 1952. The water-soluble chlorophyll, or chlorophyllin, found use as deodorizing agent in toothpastes, soaps, shampoos, candy, deodorants, and pharmaceuticals [42].

A historical important step for today's biorefinery developments was the industry-politics-approach of "Chemurgy", founded in 1925 in the US by the chemist W.J. HALE, son-in-law of H. DOW, the founder of Dow Chemical, and C. H. HERTY, a former President of the American Chemical Society. They soon found prominent support from H. Ford and T. A. Edisons. Chemurgy, an abbreviation of "chemistry" and "ergon", the Greek word for work [43], means by analogy "chemistry from the acre" that is the connection of agriculture with the chemical industry.

Chemurgy was soon shown to have a serious industrial political philosophy – the objective of utilizing agricultural resources, nowadays called renewable resources, in industry. There have been common conferences between agriculture, industry, and science since 1935 with a national council called the "National Farm Chemurgic Council" [44]. The end of Chemurgy started with the flooding of the world market with cheap crude oil after World War II; numerous inventions and production processes remained, however, and are again highly newsworthy. One was a car, introduced by Henry Ford 1941, whose car interior lining and car body consisted 100 % of biosynthetics; to be specific it had been made from a cellulose meal, soy meal, formaldehyde resin composite material in the proportions 70 %: 20 %:10 %, respectively. The alternative fuel for this car was pyrolysis methanol produced from cannabis. Throughout the thirties more than 30 industrial products based on soy bean were created by researchers from the Ford company; this made it necessary to apply complex conversion methods [45]. Hale was a pioneer of ethyl alcohol and hydrocarbon fuel mixture (power alcohol, gasohol) [46]. This fuel mixture, nowadays called E10-fuel, consisting of 10 % bioethanol and 90 % hydrocarbon-based fuel, has been the national standard since the beginning of this millennium in the United States.

Associated with the work of BERGIUS, 1933 [47], and SCHOLLER, 1923 and 1935 [48, 49], wood saccharification was reanimated at the end of World War II. Beside optimization of the process, use of lignocelluloses was of great interest. The continuously growing agribusiness left behind millions of tons of unused straw. Two Americans, OTHMER and KATZEN, were the main pioneers in the field of wood saccharification [50]. Between the years 1935 and 1960 several hydrolysis plants were built in Germany and the United States; in these deal, wood flour, surplus lumber, and also straw were hydrolyzed [51]. One of the most well known plants are those of Scholler/Tornesch located in Tornesch, Germany, with a production rate of 13,000 t/a, in Dessau, Germany, production rate 42,000 t/a, based on wood, in Holzminden, Germany, production rate of 24,000 t/a, also based on wood, in Ems, Switzerland, with a production rate of 35,000 t/a, also wood based, and the plants in Madison and Springfield, United States, and the Bergius plants in Rheinau, Germany (Rheinau I, built 1930, with a production rate of 8000 t/a, based on surplus lumber; Rheinau II, built 1960 with a production rate of 1200 t/a, based on wood) and the plant in Regensburg, Germany, with a production rate of 36,000 t/a [52].

During World War II the plant in Springfield, Oregon, US (using the Scholler-Tornesch process as modified by Katzen) produced 15,000 gallons of ethyl alcohol per day from 300 t wood flour and sawdust, i.e. 50 gallons per ton of wood [53]. The plant in Tornesch, Germany, has been producing approximately 200 liters of ethyl alcohol, purity 100 %, per ton wood and approximately 40 kg yeast per ton of wood. In 1965 there were 14 plants in the former Soviet Union, with a total capacity of 700,000 t/a and an overall annual wood consumption of 4×10^6 t [9].

During the nineteen-sixties wood chemistry had its climax. Projects had been developed, which made it possible to produce nearly all chemical products on the basis of wood. Examples are the complex chemical technological approaches of wood processing from TIMELL 1961 [54], STAMM 1964 [55], JAMES 1969 [56], BRINK and POHLMANN 1972 [57], and the wood-based chemical product trees by OSHIMA 1965 [58]. Although these developments did not make their way into industrial production, they are an outstanding platform for today's lignocellulose conversions, product family trees, and LCF biorefineries (Section 5.2).

Most of the above mentioned technologies and products, some of which were excellent, could not compete with the fossil-based industry and economy; nowadays, however, they are prevailing again. The basis for this revival started in the seventies, when the oil crisis and continuously increasing environmental pollution resulted in a broad awareness that plants could be more than food and animal feed. At the same time the disadvantages of intensive agricultural usage, for example over-fertilization, soil erosion, and the enormous amounts of waste, were revealed. From this situation developed complex concepts, which have been published, in which the aim was, and still is, technological and economical cooperation of agriculture, forestry, the food-production industry, and conventional industry, or at least consideration of integrated utilization of renewable resources.

Typical examples of this thinking were:

- integrated industrial utilization of wood and straw [59];
- industrial utilization of fast growing wood-grass [60, 61];
- complex utilization of green biomass, for example grass and alfalfa, by agriculture and industry [62–64];
- corn wet-grinding procedures with associated biotechnological and chemical product lines [65];
- modern aspects of thermochemical biomass conversion [66];
- discussion of the concept "organic chemicals from biomass" with main focus on biotechnological methods and products (white biotechnology) [67–69] and industrial utilization of biomass [70].

These rich experiences of the industrial utilization of renewable resources, new agricultural technology, biotechnology, and chemistry, and the changes in ecology, economics, and society led inevitably to the topic of complex and integrated substantial and energy utilization of biomass and, finally, to the biorefinery.

3. Situation

3.1. Some Current Aspects of Biorefinery Research and Development

Since the beginning of the 1990s the utilization of renewable resources for production of non-food products has fostered research and development which has received increasing attention from industry and politicians [71–73]. Integrated processes, biomass refinery technology, and biorefinery technology have become object of research and development. Accordingly, the term "biorefinery" was established in the 1990s [4, 74–83]. The respective biorefinery projects are focused on the fabrication of fuels, solvents, chemicals, plastics, and food for human beings. In some countries these biorefinery products are made from waste biomass. At first the main processes in the biorefinery involved ethanol fermentation for fuels (ethanol-oriented biorefineries) [84–88], lactic acid (LA) fermentation [28, 89], propanediol (PDO) fermentation [90], and the lysine fermentation [91] especially for polymer production. The biobased polymers poly(lactic acid) [28], propandiol-derived polymers [92], and polylysine [91] have been completed by polyhydroxyalkanoates [93] and polymerized oils [94].

Many hybrid technologies were developed from different fields, for example bioengineering, polymer chemistry, food science and agriculture. Biorefinery systems based on cereals [95, 96], lignocelluloses [97, 98], and grass and alfalfa [38, 99], and biorefinery optimization tools are currently being developed [100, 101]. The integration of molecular plant genetics to support the raw material supply is currently being discussed intensely [102, 103].

Broin and Associates has begun the development of a second generation of dry mill refineries and E.I. du Pont de Nemours has developed an integrated corn-based biorefinery. In 2001 NatureWorks LCC (to Cargill, former Cargill Dow LLC) started the industrial production of PLA (PLA-oriented Biorefinery) on the basis of maize.

Biorefineries are of interest ecologically [104], economically, and to business, government, and politicians [105–110]. National programs [111, 112], biobased visions [113], and plans [114] have been developed and the international exchange of information is increasing, for example as a result, among others, of series of international congresses and symposia:

- BIO World Congress on Industrial Biotechnology and Bioprocessing [115];
- biomass conferences [116, 117];
- The Green and Sustainable Chemistry Congress [118];
- ACS Symposium Series and [119]
- the Biorefinica symposia series [120, 121].

Currently, biorefinery systems are in the stage of development world-wide. An overview of the main aspects, activity, and discussions is the content of this book. An attempt to systematize the topic "Biorefinery" will be presented below.

3.2. Raw Material Biomass

Nature is a permanently renewing production chain for chemicals, materials, fuels, cosmetics, and pharmaceuticals. Many of the biobased industry products currently used are results of direct physical or chemical treatment and processing of biomass, for example cellulose, starch, oil, protein, lignin, and terpenes. On one hand one must mention that because of the help of biotechnological processes and methods, feedstock chemicals are produced such as ethanol, butanol, acetone, lactic acid, and itaconic acid, as also are amino acids, e.g., glutaminic acid, lysine, tryptophan. On the other hand, only 6×10^9 t of the yearly produced biomass, $1.7–2.0 \times 10^{11}$ t, are currently used, and only 3 to 3.5 % of this amount is used in non-food applications, for example chemistry [122].

The basic reaction of biomass is photosynthesis according to:

Industrial utilization of raw materials from agriculture, forestry, and agriculture is only just beginning. There are several definitions of the term "biomass" [122]:

- the complete living, organic matter in our ecological system (volume/non-specific);
- the plant material constantly produced by photosynthesis with an annual growth of 170 billion tons (marine plants excluded);

- the cell-mass of plants, animals, and microorganism used as raw materials in microbiological processes.

Biomass is defined in a recent US program [111, 112]: "The term "biomass" means any organic matter that is available on a renewable or recurring basis (excluding old-growth timber), including dedicated energy crops and trees, agricultural food and feed crop residues, aquatic plants, wood and wood residues, animal wastes, and other waste materials."

For this reason it is essential to define biomass in the context of the industrial utilization. A suggestion for a definition of "industrial biomass" [111, 112] is: "The term "industrial biomass" means any organic matter that is available on a renewable or recurring basis (excluding old-growth timber), including dedicated energy crops and trees, agricultural food and feed crop residues, aquatic plants, wood and wood residues, animal wastes, and other waste materials usable for industrial purposes (energy, fuels, chemicals, materials) and include wastes and co-wastes of food and feed processing."

Most biological raw material is produced in agriculture and forestry and by microbial systems. Forestry plants are excellent raw materials for the paper and cardboard, construction, and chemical industries. Field fruits are a pool of organic chemicals from which fuels, chemicals, chemical products, and biomaterials are produced (Fig. 2), [72].

Advances in genetics, biotechnology, process chemistry, and engineering are leading to a new manufacturing concept for converting renewable biomass to valuable fuels and products, generally referred to as the biorefinery. The integration of agroenergy crops and biorefinery manufacturing technologies offers the potential for the development of sustainable biopower and biomaterials that will lead to a new manufacturing paradigm [2, 123].

Waste biomass and biomass of nature and agricultural cultivation are valuable organic reservoirs of raw material and must be used in accordance with their organic composition. During the development of biorefinery systems the term "waste biomass" will become obsolete in the medium-term [124].

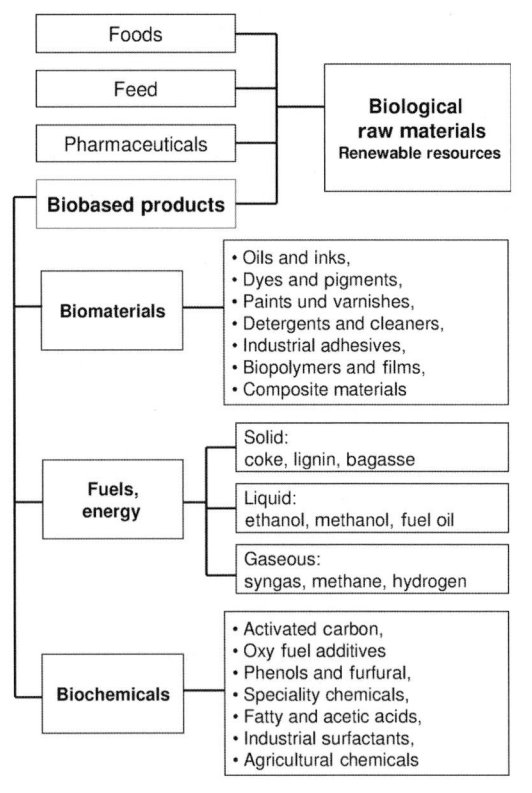

Figure 2. Products and product classes based on biological raw materials [81]

3.3. National Vision and Goals and Plan for Biomass Technology in the United States

Industrial development was pushed by the US President [111] and by the US Congress [112], initially in 2000. In the USA it is intended that by 2020 at least 25 % of organic-carbon-based industrial feedstock chemicals and 10 % of liquid fuels (compared with levels in 1994) will be produced by biobased industry. This would mean that more than 90 % of the consumption of organic chemicals in the US and up to 50 % of liquid fuel needs would be biobased products [4].

The Biomass Technical Advisory Committee (BTAC) of the USA in which leading representatives of industrial companies, for example Dow Chemical, E.I. du Pont de Nemours, Cargill Dow LLC, and Genencor International, and Corn growers associations and the Natural Resources Defense Council are involved and which acts as

Table 1. The US national vision goals for biomass technologies by the Biomass Technical Advisory Committee [113]

Year	2002	2010	2020	2030
BioPower (BioEnergy)Biomass share of electricity and heat demand in utilities and industry	2.8 % (2.7 quad)[a]	4 % (3.2 quad)	5 % (4.0 quad)	5 % (5.0 quad)
BioFuels Biomass share of demand for transportation fuels	0.5 % (0.15 quad)	4 % (1.3 quad)	10 % (4.0 quad)	20 % (9.5 quad)
BioProducts Share of target chemicals that are biobased	5 %	12 %	18 %	25 %

[a] 1 quad=1 quadrillion BTU=1 German billiarde BTU; BTU=British thermal unit; 1 BTU=0.252 kcal, 1 kW=3413 BTU, 1 kcal=4.186 kJ

advisor to the US government, has made a detailed plan with steps toward targets of 2030 with regard to bioenergy, biofuels, and bioproducts (Table 1) [113].

Simultaneously, the plan *Biomass Technology in the United States* has been published [114] in which research, development, and construction of biorefinery demonstration plants are determined. Research and development are necessary to:

1. increase scientific understanding of biomass resources and improve the tailoring of those resources;
2. improve sustainable systems to develop, harvest, and process biomass resources;
3. improve efficiency and performance in conversion and distribution processes and technologies for development of a host of biobased products and
4. create the regulatory and market environment necessary for increased development and use of biobased products.

The Biomass Advisory Committee has established specific research and development objectives for feedstock production research. Target crops should include oil and cellulose-producing crops that can provide optimum energy content and usable plant components. Currently, however, there is a lack of understanding of plant biochemistry and inadequate genomic and metabolic information about many potential crops.

Specific research to produce enhanced enzymes and chemical catalysts could advance biotechnology capabilities.

3.4. Vision and Goals and Plan for Biomass Technology in the European Union and Germany

In Europe there are already regulations about substitution of nonrenewable resources by biomass in the field of biofuels for transportation [125] and the "Renewable energy law" of 2000 [126]. According to the EC Directive "On the promotion of the use of biofuels" the following products are regarded as "biofuels":

(a) "bioethanol", (b) "biodiesel", (c) "biogas", (d) "biomethanol", (e) "biodimethylether",

(f) "bio-ETBE (ethyl tertiary-butyl ether)" on the basis of bioethanol,

(g) "bio-MTBE (methyl tertiary butyl ether)" on the basis of biomethanol, and

(h) "synthetic biofuels", (i) "biohydrogen", (j) pure vegetable oil

Member States of the EU have been requested to define national guidelines for a minimum amounts of biofuels and other renewable fuels (with a reference value of 2 % by 2005 and 5.75 % by 2010 calculated on the basis of energy content of all petrol and diesel fuels for transport purposes). Table 2 summarizes this goal of

Table 2. Targets of the EU and Germany with regard to the introduction of technologies based on renewable resources

Year	2001	2005	2010	2020–2050
Bioenergy Share of wind power, photovoltaics, biomass and geothermal electricity and heat demand in utilities and industry	7.5 %	–	12.5 %	26 % (2030) 58 % (2050)
Biofuels Biomass share of demand in transportation fuels (petrol and diesel fuels)	1.4 %	2.8 %	5.75 %	20 % (2020)
Biobased Products Share of target chemicals that are biobased	8 %			

the EU and also those of Germany with regard to establishment of renewable energy and biofuel [127, 128].

Today there are no guidelines concerning "biobased products" in the European Union and in Germany. After passing directives relating to bioenergy and biofuels, however, such a decision is on the political agenda. The "biofuels" directive already includes ethanol, methanol, dimethyl ether, hydrogen, and biomass pyrolysis which are fundamental product lines of the future bio-based chemical industry.

In the year 2003, an initiative group called "Biobased Industrial Products" consisting of members from industry, small and middle-class businesses, and research, and development facilities met and formulated a strategy paper, called "BioVision 2030" [129]. This strategy paper has been included in the resolution of the German Government (Deutscher Bundestag) on the topic "Accomplish basic conditions for the industrial utilization of renewable resources in Germany" [130]. An advisory committee consisting of members of the chemical industry, related organizations, research facilities, and universities has been established to generate a plan concerning the formulation of the objectives for the third column, bio-products in Europe (Table 2) [131].

A recent vision paper published by the Industrial Biotechnology Section of the European Technology platform for Sustainable Chemistry foresee up to 30 % of raw materials for chemical industry coming from renewable sources by 2025 [132]. Lately European Commission and U S Department of Energy are come to an agreement for cooperation on this field [133].

4. Principles of Biorefineries

4.1. Fundamentals

Biomass, similar to petroleum, has a complex composition. Its primary separation into main groups of substances is appropriate. Subsequent treatment and processing of those substances lead to a whole range of products. Petrochemistry is based on the principle of generating simple to handle and well defined chemically pure products from hydrocarbons in refineries. In efficient product lines, a system based on family trees has been built, in which basic chemicals, intermediate products, and sophisticated products are produced. This principle of petroleum refineries must be transferred to biorefineries. Biomass contains the synthesis performance of the nature and has different C:H:O:N ratio from petroleum. Biotechnological conversion will become, with chemical conversion, a big player in the future (Fig. 3).

Thus biomass can already be modified within the process of genesis in such a way that it is

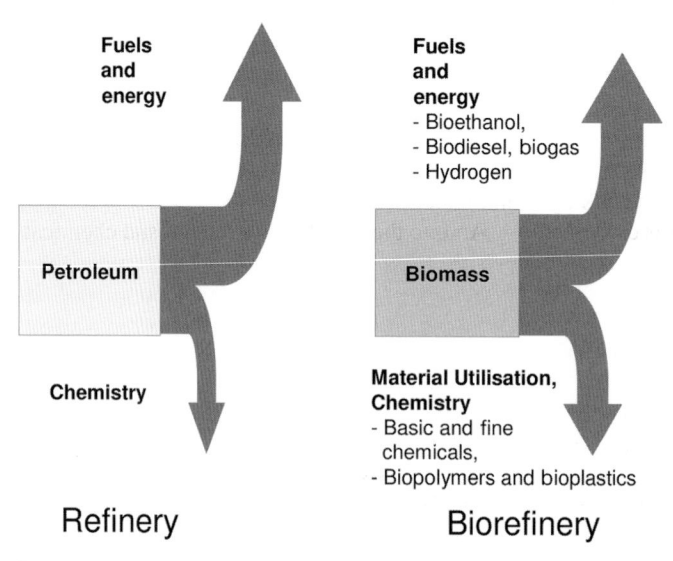

Figure 3. Comparison of the basic-principles of the petroleum refinery and the biorefinery

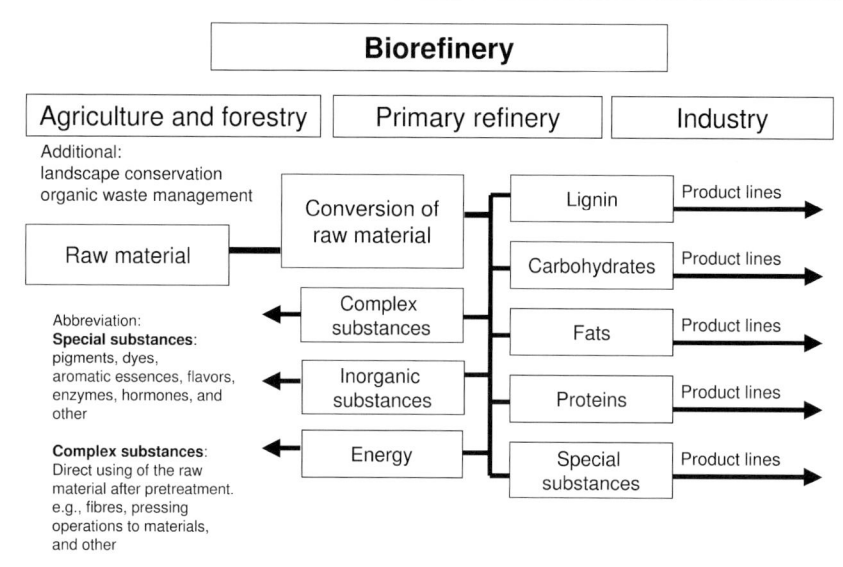

```
                        Biorefinery

 Agriculture and forestry    Primary refinery         Industry
 Additional:
 landscape conservation
 organic waste management   Conversion of    Lignin         Product lines
                            raw material
    Raw material                             Carbohydrates  Product lines

 Abbreviation:              Complex
 Special substances:        substances       Fats           Product lines
 pigments, dyes,
 aromatic essences, flavors, Inorganic
 enzymes, hormones, and     substances       Proteins       Product lines
 other

 Complex substances:        Energy           Special        Product lines
 Direct using of the raw                     substances
 material after pretreatment.
 e.g., fibres, pressing
 operations to materials,
 and other
```

Figure 4. Providing code-defined basic substances (via fractionation) for development of relevant industrial product family trees [81, 82]

adapted to the purpose of subsequent processing, and particular target products have already been formed. For those products the term "precursors" is used. Plant biomass always consists of the basic products carbohydrates, lignin, proteins, and fats, and a variety of substances such as vitamins, dyes, flavors, aromatic essences of very different chemical structure. Biorefineries combine the essential technologies which convert biological raw materials into the industrial intermediates and final products (Fig. 4).

A technically feasible separation operation, which would enable separate use or subsequent processing of all these basic compounds, is currently in its initial stages only. Assuming that of the estimated annual production of biomass by biosynthesis of 170 billion tons 75 % is carbohydrates, mainly in the form of cellulose, starch, and saccharose, 20 % lignin, and only 5 % other natural compounds such as fats (oils), proteins, and other substances [134], the main attention should first be focused on efficient access to carbohydrates, and their subsequent conversion to chemical bulk products and corresponding final products. Glucose, accessible by microbial or chemical methods from starch, sugar, or cellulose, is, among other things, predestined for a key position as a basic chemical, because a broad range of biotechnological or chemical products is accessible from glucose [135]. For starch the advantage of enzymatic compared with chemical hydrolysis is already known [136].

For cellulose this is not yet realized. Cellulose-hydrolyzing enzymes can only act effectively after pretreatment to break up the very stable lignin/cellulose/hemicellulose composites [137]. These treatments are still mostly thermal, thermomechanical, or thermochemical, and require considerable input of energy. The arsenal for microbial conversion of substances from glucose is large, and the reactions are energetically profitable. It is necessary to combine degradation processes via glucose to bulk chemicals with the building processes to their subsequent products and materials (Fig. 5).

Among the variety of microbial and chemical products possibly accessible from glucose, lactic acid, ethanol, acetic acid, and levulinic acid, in particular, are favorable intermediates for generation of industrially relevant product family trees. Here, two potential strategies are considered: first, development of new, possibly biologically degradable products (follow-up products from lactic and levulinic acids) or, second, entry as intermediates into conventional product lines (acrylic acid, 2,3-pentanedione) of petrochemical refineries [81].

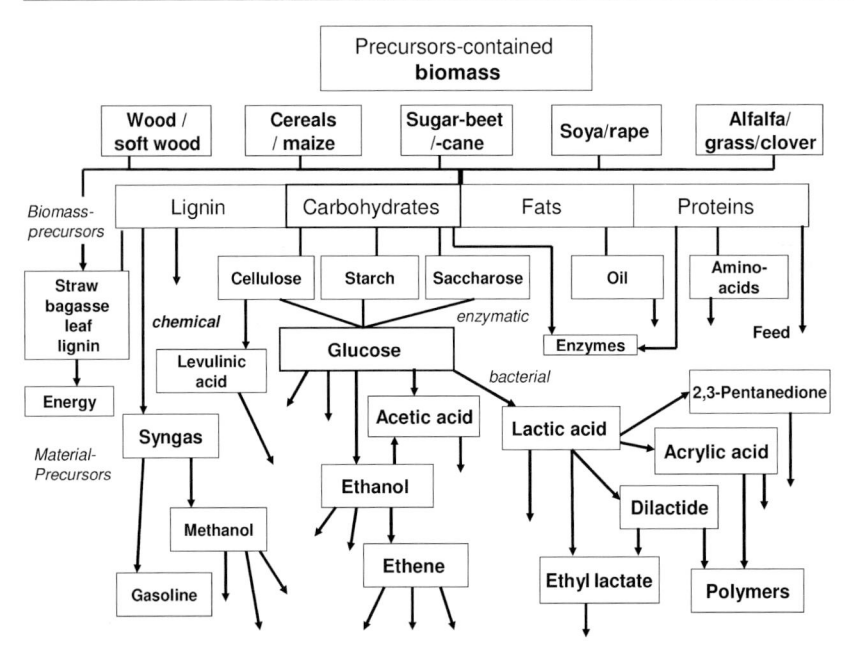

Figure 5. Possible schematic diagram of biorefinery for precursor-containing biomass with preference for carbohydrates [81, 82]

4.2. Definition of the Term "Biorefinery"

The young working field "Biorefinery Systems" in combination with "Biobased Industrial Products" is, in various respects, still an open field of knowledge. This is also reflected in the search for an appropriate description. A selection is given below.

The term "Green Biorefinery" was been defined in the year 1997 as: "Green biorefineries represent complex (to fully integrated) systems of sustainable, environmentally and resource-friendly technologies for the comprehensive (holistic) material and energetic utilization as well as exploitation of biological raw materials in form of green and residue biomass from a targeted sustainable regional land utilization" [76]. The original term used in Germany "complex construction and systems" was substituted by "fully integrated systems". The US Department of Energy (DOE) uses the following definition [138]: "A biorefinery is an overall concept of a processing plant where biomass feedstocks are converted and extracted into a spectrum of valuable products. Based on the petrol-chemical refinery." The American National Renewable Energy Laboratory (NREL) published the definition [139]: "A biorefinery is a facility that integrates biomass conversion processes and equipment to produce fuels, power, and chemicals from biomass. The biorefinery concept is analogous to today's petroleum refineries, which produce multiple fuels and products from petroleum. Industrial biorefineries have been identified as the most promising route to the creation of a new domestic biobased industry."

A general definition is: "Biorefining is the transfer of the efficiency and logic of fossil-based chemistry and substantial converting industry as well as energy production onto the biomass industry".

There is an agreement about the *objective*, which is briefly defined as: "Developed biorefineries, so called "phase III-biorefineries" or "generation III-biorefineries", start with a biomass–feedstock-mix to produce a multiplicity of most various products by a technologies-mix" [77] (Fig. 6).

An example of the type "generation-I biorefinery" is a dry milling ethanol plant. It uses grain as a feedstock, has a fixed processing capability, and produces a fixed amount of ethanol, feed co-products, and carbon dioxide. It has almost no flexibility in processing. Therefore, this type can be used for comparable purposes only.

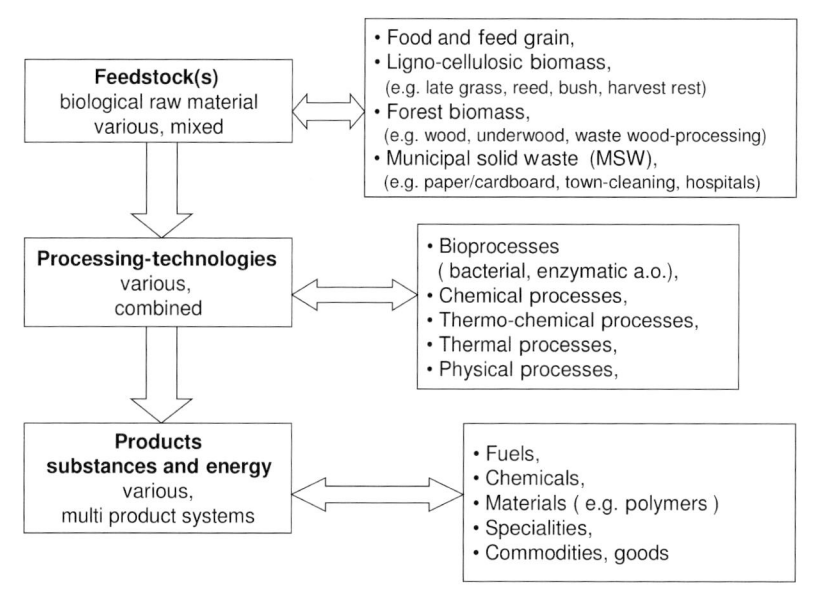

Figure 6. Basic principles of a biorefinery (generation III biorefinery) [81]

An example of a type "generation-II biorefinery" is the current wet milling technology. This technology uses grain feedstock, yet has the capability to produce a variety of end products depending on product demand. Such products include starch, high-fructose corn syrup, ethanol, corn oil, plus corn gluten feed, and meal. This type opens numerous possibilities to connect industrial product lines with existing agricultural production units. "Generation-II biorefineries" are, furthermore, plants like NatureWorks PLA facility [1, 28] (see Chapter 2 and Section 3.1) or ethanol biorefineries, for example Iogen's wheat straw to ethanol plant [147].

Third generation (generation-III) and more advanced biorefineries have not yet been built but will use agricultural or forest biomass to produce multiple products streams, for example ethanol for fuels, chemicals, and plastics [140, 147].

4.3. The Role of Biotechnology

The application of biotechnological methods will be highly important with the development of biorefineries for production of basic chemicals, intermediate chemicals, and polymers [141–143]. The integration of biotechnological methods must be managed intelligently in respect of physical and chemical conversion of the biomass. Therefore the biotechnology cannot remain limited to glucose from sugar plants and starch from starch-producing plants. One main objective is the economic use of biomass containing lignocellulose and provision of glucose in the family tree system. Glucose is a key chemical for microbial processes. The preparation of a large number of family tree-capable basic chemicals is shown in subsequent sections.

4.3.1. Guidelines of Fermentation Section within Glucose-product Family Tree

Among the variety of chemical products, and derivatives of these, accessible microbially from glucose a product family tree can be developed, for example (C-1)-chemicals methane, carbon dioxide, methanol; (C-2)-chemicals ethanol, acetic acid, acetaldehyde, ethylene, (C-3)-chemicals lactic acid, propanediol, propylene, propylene oxide, acetone, acrylic acid, (C-4)-chemicals diethyl ether, acetic acid anhydride, malic acid, vinyl acetate, *n*-butanol, crotonaldehyde, butadiene, 2,3-butanediol, (C-5)-chemicals itaconic acid, 2,3-pentane dione, ethyl lactate, (C-6)-chemicals sorbic acid, parasorbic acid, citric acid, aconitic acid, isoascorbinic acid, kojic acid, maltol, dilactide, (C-8)-chemicals 2-ethyl hexanol (Fig. 7).

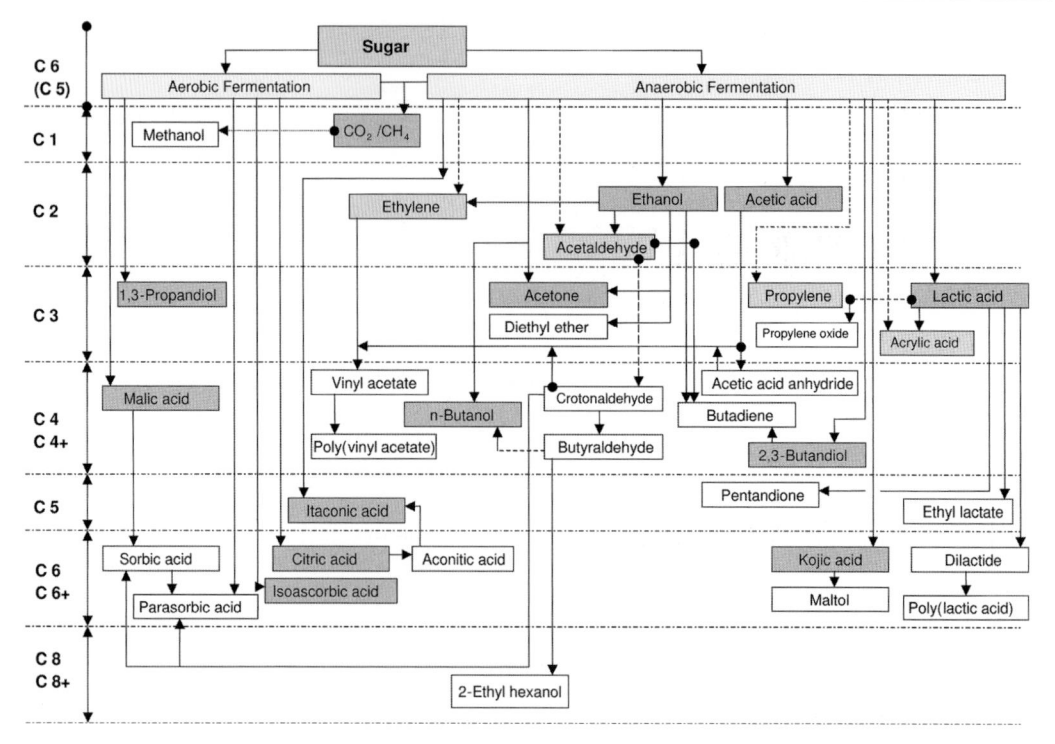

Figure 7. Biotechnological sugar-based product family tree

Guidelines are currently being developed for the fermentation section of a biorefinery. The question of efficient arrangement of the technological design for production of bulk chemicals needs an answer. Considering the manufacture of lactic acid and ethanol, the basic technological operations are very similar. Selection of biotechnologically based products from biorefineries should be done in a way such that they can be produced from the substrates glucose or pentoses. Furthermore the fermentation products should be extracellular. Fermenters should have batch, feed batch, or CSTR design. Preliminary product recovery should require steps like filtration, distillation, or extraction. Final product recovery and purification steps should possibly be product-unique. In addition, biochemical and chemical processing steps should be advantageously connected.

Unresolved questions for the fermentation facility include:

1. whether or not the entire fermentation facility can/should be able to change from one product to another;
2. whether multiple products can be run in parallel, with shared use of common unit operations;
3. how to manage scheduling of unit operations; and
4. how to minimize in-plant inventories, while accommodating necessary change-over between different products in the same piece of equipment [98].

4.4. Building Blocks, Chemicals and Potential Screening

A team from Pacific Northwest National Laboratory (PNNL) and NREL submitted a list of twelve potential biobased chemicals [101]. A key area of the investigation as biomass precursors, platforms, building blocks, secondary chemicals, intermediates, products and uses (Fig. 8).

The final selection of 12 building blocks began with a list of more than 300 candidates. A shorter list of 30 potential candidates was selected by using an iterative review process

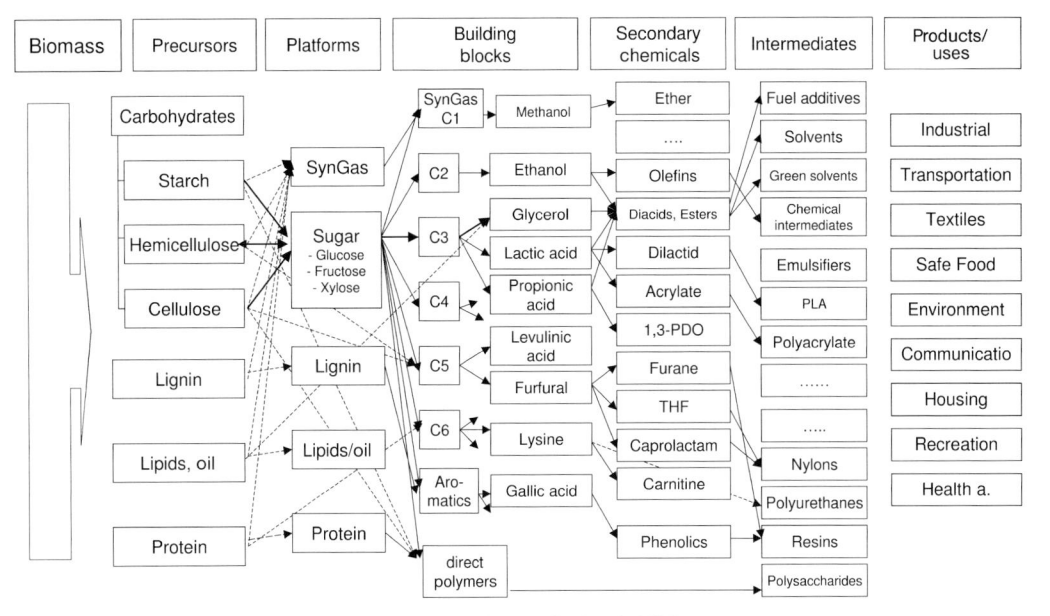

Figure 8. Model of a biobased product flow-chart for biomass feedstock [101]

based on the petrochemical model of building blocks, chemical data, known market data, properties, performance of the potential candidates, and previous industry experience of the team at PNNL and NREL. This list of 30 was ultimately reduced to 12 by examining the potential markets for the building blocks and their derivatives and the technical complexity of the synthetic pathways.

The reported block chemicals can be produced from sugar by biological and chemical conversions. The building blocks can be subsequently converted to several high-value biobased chemicals or materials. Building-block chemicals, as considered for this analysis, are molecules with multiple functional groups with the potential to be transformed into new families of useful molecules. The twelve sugar-based building blocks are 1,4-diacids (succinic, fumaric, and malic), 2,5-furandicarboxylic acid, 3-hydroxypropionic acid, aspartic acid, glucaric acid, glutamic acid, itaconic acid, levulinic acid, 3-hydroxybutyrolactone, glycerol, sorbitol, and xylitol/arabinitol [101].

A second-tier group of building blocks was also identified as viable candidates. These include gluconic acid, lactic acid, malonic acid, propionic acid, the triacids citric and aconitic, xylonic acid, acetoin, furfural, levuglucosan, lysine, serine, and threonine. Recommendations for moving forward include:

- examining top value products from biomass components, for example aromatic compounds, polysaccharides, and oils;
- evaluating technical challenges in more detail in relation to chemical and biological conversion; and
- increasing the suites of potential pathways to these candidates.

The co-production of organic acids, ethanol, and electricity, as well as the opportunities for their integration, will improve overall economics. [132]. Also, the possibilities of catalysis for the production of platform chemicals have not yet sufficiently been exploited [144]. In addition, further investigations on microbial biology and bioprocessing are anticipated [145].

No further down select products from syngas was undertaken. For the purposes of the study from PNNL and NREL hydrogen and methanol comprise the best short-term prospects for biobased commodity chemical production, because obtaining simple alcohols, aldehydes, mixed alcohols, and Fischer-Tropsch liquids from biomass is not economically viable and requires additional development [101].

5. Biorefinery Systems and Design

5.1. Introduction

Biobased products are prepared for a usable economic use by meaningful combination of different methods and processes (physical, chemical, biological, and thermal). It is therefore necessary that basic biorefinery technologies are developed. For this reason profound interdisciplinary cooperation of various disciplines in research and development is inevitable. It seems reasonable, therefore, to refer to the term "biorefinery design", which means: "bringing together well founded scientific and technological basics, with similar technologies, products, and product lines, inside biorefineries". The basic conversions of each biorefinery can be summarized as follows. In the first step, the precursor-containing biomass is separated by physical methods. The main products (M_1–M_n) and the by-products (B_1–B_n) will subsequently be subjected to microbiological or chemical methods. The follow-up products (F_1–F_n) of the main and by-products can also be converted or enter the conventional refinery (Fig. 6).

Currently four complex biorefinery systems are used in research and development:

1. the "lignocellulosic feedstock biorefinery" which use "nature-dry" raw material, for example cellulose-containing biomass and waste;
2. the "whole crop biorefinery" which uses raw material such as cereals or maize;
3. the "green biorefineries" which use "nature-wet" biomass such as green grass, alfalfa, clover, or immature cereal [81, 82]; and
4. the "biorefinery two platforms concept" includes the sugar platform and the syngas platform [101].

5.2. Lignocellulosic Feedstock Biorefinery

Among the potential large-scale industrial biorefineries the lignocellulose feedstock (LCF) biorefinery will most probably be pushed through with the greatest success. On the one side the raw material situation is optimum (straw, reed, grass, wood, paper-waste, etc.), on the other

Lignocellulose + H_2O → Lignin + Cellulose + Hemicellulose
Hemicellulose + H_2O → Xylose
Xylose ($C_5H_{10}O_5$) + acid Catalyst → Furfural ($C_5H_4O_2$) + 3H_2O
Cellulose($C_6H_{10}O_5$) + H_2O → Glucose ($C_6H_{12}O_6$)

Figure 9. A possible general equation for conversion at the LCF biorefinery

side conversion products have a good position on both the traditional petrochemical and future biobased product market [146, 148, 149]. An important point for utilization of biomass as chemical raw material is the cost of raw material. Currently the cost of corn stover or straw is 30 US$/ton and that of corn is 110 US$/ton (3 US$/bushel; US bushel corn =25.4012 kg=56 lb) [150].

Lignocellulose materials consist of three primary chemical fractions or precursors:

- hemicellulose/polyoses, sugar polymers of, predominantly, pentoses;
- cellulose, a glucose polymer; and
- lignin, a polymer of phenols (Fig. 9).

The lignocellulosic biorefinery-regime is distinctly suitable for genealogical compound trees. The main advantages of this method is that the natural structures and structural elements are preserved, the raw materials are inexpensive, and large product varieties are possible (Fig. 10). Nevertheless there is still a demand for development and optimization of these technologies, e.g., in the field of separation of cellulose, hemicellulose and lignin, and utilization of the lignin in the chemical industry.

An overview of potential products of an LCF biorefinery is shown in Figure 11. In particular

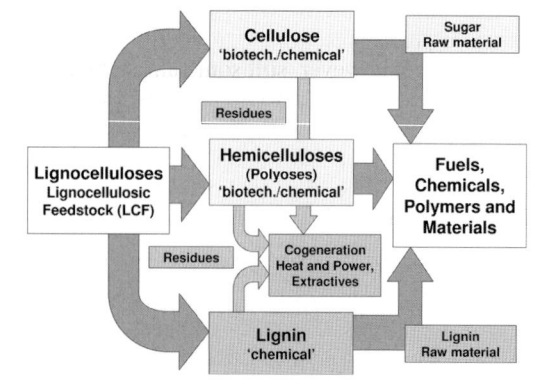

Figure 10. Lignocellulosic feedstock biorefinery

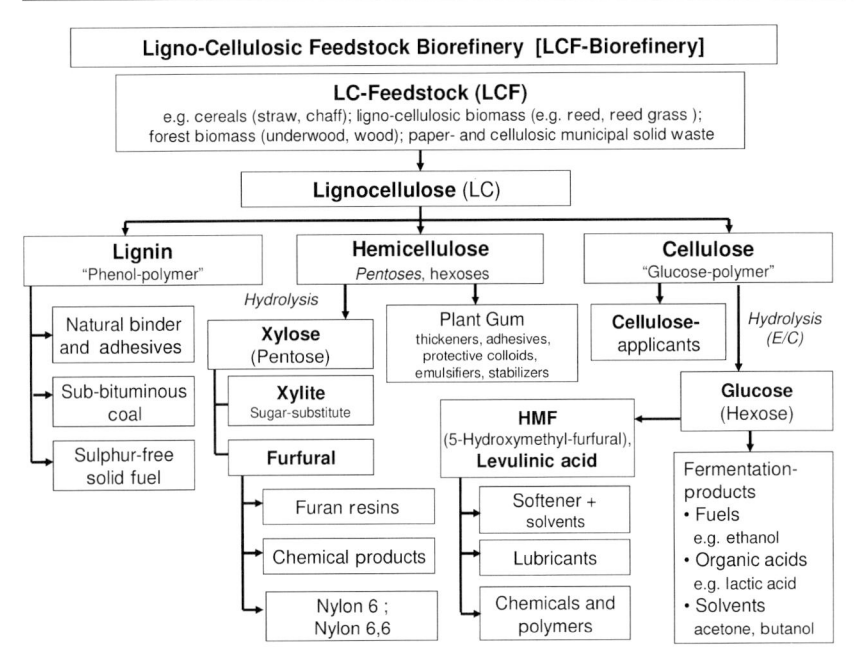

Figure 11. Products of a lignocellulosic feedstock biorefinery (LCF-biorefinery, Phase III) [81, 82, 98]

furfural and hydroxymethylfurfural are interesting products. Furfural is a starting material for production of nylon 6,6 and nylon 6. The original process for production of nylon-6,6 was based on furfural (see also Section 2.2). The last of these production plants was closed in 1961 in the USA, for economic reasons (the artificially low price of petroleum). Nevertheless the market for nylon 6 is huge. Catalytic transformation of these future feedstocks provides new market opportunities. Levulinic acid is one of 12 bioderived feedstocks and it can be obtained directly by cellulose hydrolysis [101]. A number of high value derivatives, such as pyrrolidones, monomers for the synthesis of high Tg amorphous polymers and fuel additives, such as ethyl levulinate can be prepared [123, 151].

There are, however, still some unsatisfactory aspects of the LCF, for example the utilization of lignin as fuel, adhesive, or binder. Unsatisfactory because the lignin scaffold contains substantial amounts of mono-aromatic hydrocarbons, which, if isolated in an economically efficient way, could add a significant increase in value to the primary processes. It should be noticed there are no natural enzymes capable of splitting the naturally formed lignin into basic monomers as easily as is possible for natural polymeric carbohydrates or proteins [152].

An attractive process accompanying the biomass-nylon-process is the already mentioned hydrolysis of the cellulose to glucose and the production of ethanol. Some yeasts cause disproportionation of the glucose molecule during their generation of ethanol from glucose, which shifts almost all its metabolism into ethanol production, making the compound obtainable in 90 % yield (w/w; with regard to the chemical equation for the process).

On the basis of recent technology a plant has been conceived for production of the main products furfural and ethanol from LC feedstock from West Central Missouri (USA). Optimal profitability can be achieved with a daily consumption of approximately 4360 tons of feedstock. The plant produces 47.5 million gallon of ethanol and 323 tons of furfural annually [77].

Ethanol can be used as a fuel additive. It is also a connecting product to the petrochemical refinery, because it can be converted into ethene by chemical methods and it is well-known that ethene is at the start of a series of large-scale technical chemical syntheses for production of important commodities such as polyethylene or poly(vinyl acetate). Other petrochemically produced substances, for example hydrogen, methane, propanol, acetone, butanol, butandiol, itaconic acid, and succinic acid, can also be

5.3. Whole-crop Biorefinery

Raw materials for the "whole crop biorefinery" are cereals such as rye, wheat, triticale, and maize. The first step is mechanical separation into corn and straw, approximately in a 1:1.1–1.3 ratio, respectively, see Fig. 12 [155]. Straw is a mixture of chaff, nodes, ears, and leaves. The straw is an LC feedstock and may further be processed in a LCF biorefinery.

There is the possibility of separation into cellulose, hemicellulose, and lignin and their further conversion in separate product lines which are shown in the LCF biorefinery. The straw is also a starting material for production of syngas by pyrolysis technology. Syngas is the basic material for synthesis of fuels and methanol (Fig. 13).

The corn may either be converted into starch or used directly after grinding to meal. Improved molecular disassembly and depolymerization of grain starch to glucose are key to reducing energy use in the bioconversion of glucose to chemicals, ingredients, and fuels. [156, 157]

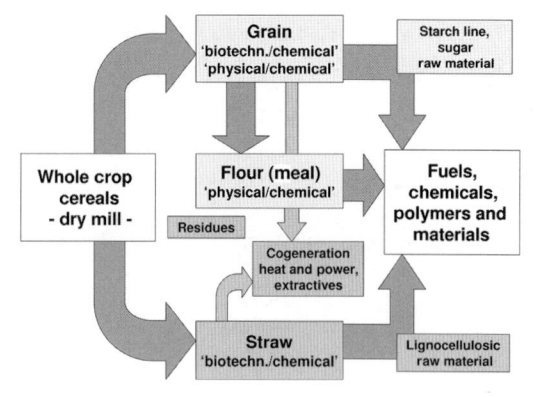

Figure 12. Whole-crop biorefinery based on dry milling

Further processing may be conducted by four processes – breaking up, plasticization, chemical modification, or biotechnological conversion via glucose. The meal can be treated and finished by extrusion into binder, adhesives, and filler. Starch can be finished by plasticization (co- and mix-polymerization, compounding with other polymers), chemical modification (etherification into carboxymethyl starch; esterification and re-esterification into fatty acid esters via acetic starch; splitting reductive amination into ethylenediamine, etc., hydrogenative splitting into sorbitol, ethylene glycol, propylene glycol, and

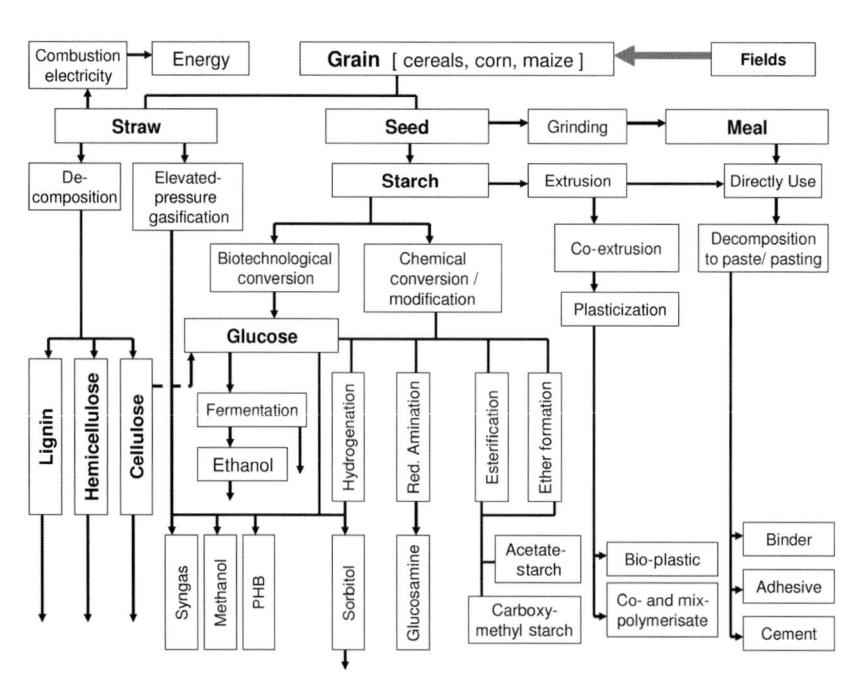

Figure 13. Products from a whole-crop biorefinery [81, 82]

glycerin), and biotechnological conversion into poly-3-hydroxybutyric acid [72, 79, 95, 96, 158, 159]

An alternative to traditional dry fractionation of mature cereals into grains and straw only was been developed by Kockums Construction (Sweden), which later became Scandinavian Farming. In this crop-harvest system whole immature cereal plants are harvested. The whole harvested biomass is conserved or dried for long-term storage. When convenient, it can be processed and fractionated into kernels, straw chips of internodes, and straw meal (leaves, ears, chaff, and nodes) (see also Section 5.4).

Fractions are suitable as raw materials for the starch polymer industry, the feed industry, the cellulose industry, and particle board producers, gluten can be used by the chemical industry and as a solid fuel. Such dry fractionation of the whole crop to optimize the utilization of all botanical components of the biomass has been described [160, 161]. A biorefinery and its profitability has been described elsewhere [162].

One expansion of the product lines in grain processing is the "whole crop wet mill-based biorefinery". The grain is swelled and the grain germ is pressed, releasing high-value oils. The

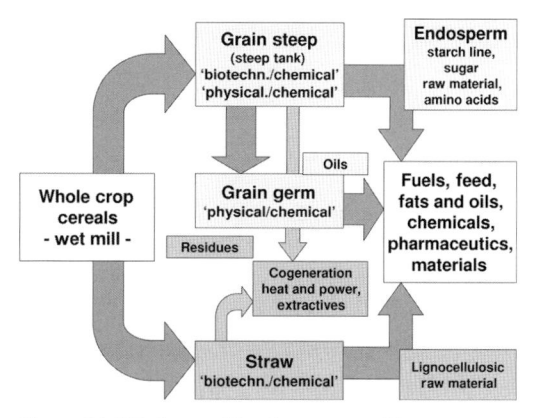

Figure 14. Whole-crop biorefinery, wet-milling

advantages of whole-crop biorefinery based on wet milling are that production of natural structures and structure elements such as starch, cellulose, oil, and amino acids (proteins) are kept high yet well known basic technology and processing lines can still be used. High raw material costs and, for industrial utilization, the necessary costly source technology are the disadvantages. Some of the products formed command high prices in, e.g., the pharmaceutical and cosmetics industries (Figs. 14 and 15). The basic biorefinery

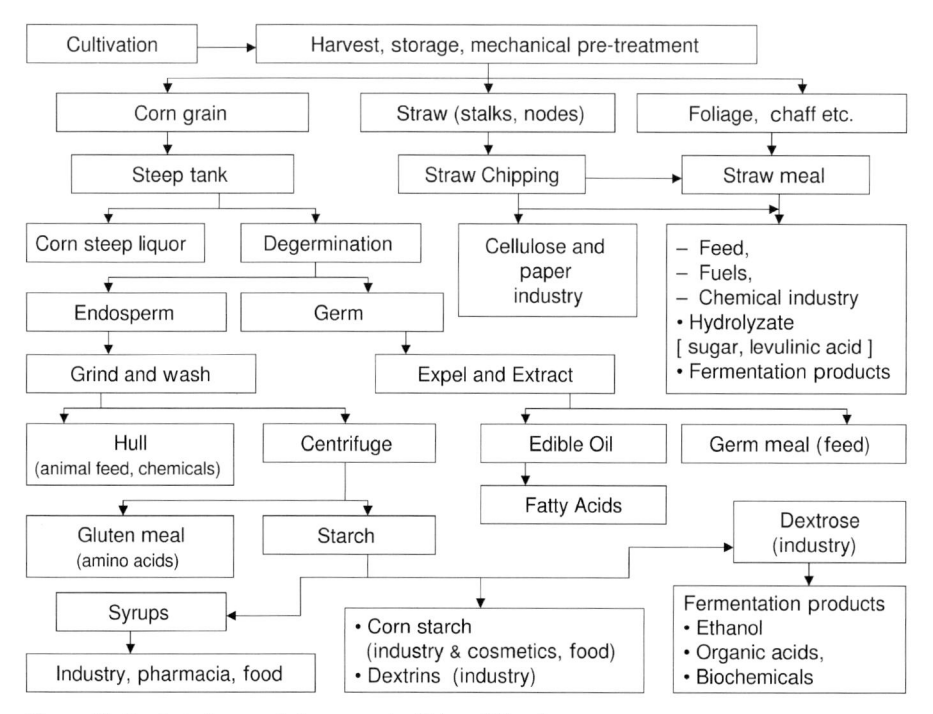

Figure 15. Products from a whole-crop wet mill-based biorefinery

technology of corn wet mills used 11 % of the US corn harvest in 1992, made products worth $7.0 billion, and employed almost 10,000 people [4].

Wet milling of corn yields corn oil, corn fiber, and corn starch. The starch products of the US corn wet milling industry are fuel alcohol (31 %), high-fructose corn syrup (36 %), starch (16 %), and dextrose (17 %). Corn wet milling also generates other products (e.g. gluten meal, gluten feed, oil) [65]. An overview about the product range is shown in Figure 15.

5.4. Green Biorefinery

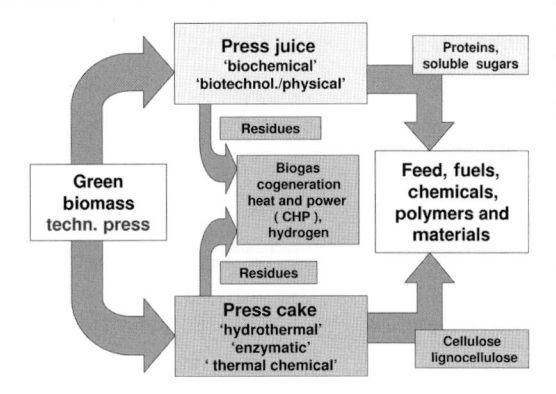

Figure 16. A "green biorefinery" system

Green biorefineries are also multi-product systems and furnish cuts, fractions, and products in accordance with the physiology of the corresponding plant material, which maintains and utilizes the diversity of syntheses achieved by nature. Most green biomass is green crops, for example grass from cultivation of permanent grass land, closure fields, nature reserves, or green crops, such as lucerne, clover, and immature cereals from extensive land cultivation. Green crops are a natural chemical factory and food plant and are primarily used as forage and as a source of leafy vegetables. A process called wet-fractionation of green biomass, green crop fractionation, can be used for simultaneous manufacture of both food and non food items [163].

Scientists in several countries, in Europe and elsewhere, have developed green crop fractionation [164–166]. Green crop fractionation is now studied in approximately 80 countries [167]. Several hundred temperate and tropical plant species have been investigated for green crop fractionation [166, 168, 169] and more than 300,000 higher plants species have still to be investigated. The subject has been covered by several reviews [76, 164–166, 169–173]. Green biorefineries can, by fractionation of green plants, process from a few tonnes of green crops per hour (farm scale process) to more than 100 tonnes per hour (industrial scale commercial process).

Careful wet fractionation technology is used as a first step (primary refinery) to isolate the contents of the green crop (or humid organic waste goods) in their natural form. Thus, they are separated into a fiber-rich press cake (PC) and a nutrient-rich green juice (GJ).

The advantages of the green biorefinery are high biomass profit per hectare, good coupling with agricultural production, and low price of the raw materials. Simple technology can be used and there is good biotechnical and chemical potential for further conversion (Fig. 16). Rapid primary processing or use of preservation methods, for example silage production or drying, are necessary, both for the raw materials and the primary products, although each method of preservation changes the content of the materials.

In addition to cellulose and starch, the press cake contains valuable dyes and pigments, crude drugs and other organic compounds. The green juice contains proteins, free amino acids, organic acids, dyes, enzymes, hormones, other organic substances, and minerals. Application of the methods of biotechnology results in conversion, because the plant water can simultaneously be used for further treatment. In addition, the lignin–cellulose composite is not as intractable as lignocellulose-feedstock materials. Starting from green juice the main focus is directed toward products such as lactic acid and its derivatives, amino acids, ethanol, and proteins. The press cake can be used for production of green feed pellets, as raw material for production of chemicals, for example levulinic acid, and for conversion to syngas and hydrocarbons (synthetic biofuels). The residues of substantial conversion are suitable for production of biogas combined with generation of heat and electricity (Fig. 17). Reviews have been published on the concepts, contents, and goals of the green biorefinery [76, 78, 124, 146].

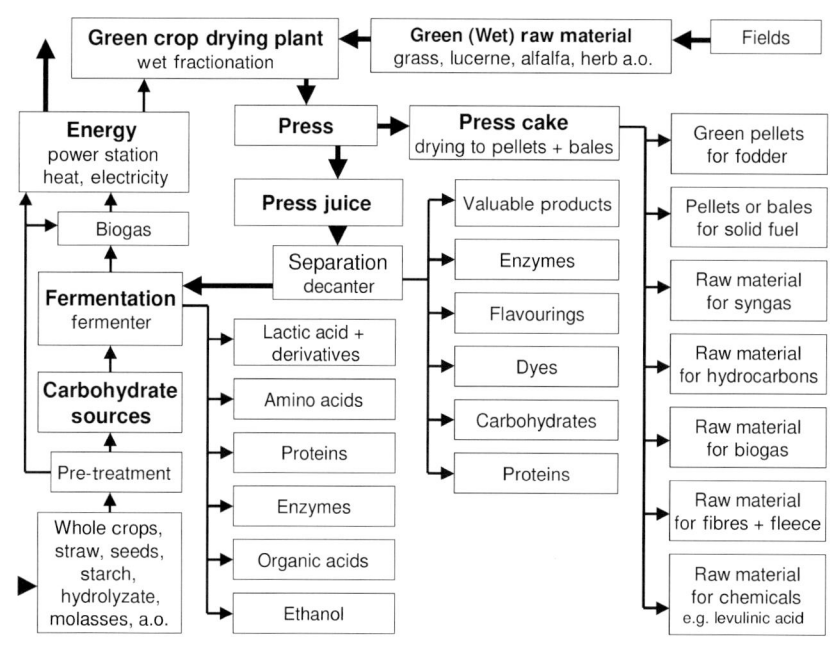

Figure 17. Products from the green biorefinery. In this illustration a green biorefinery has been combined with a green crop-drying plant [81, 82]

5.5. Two-platform Concept and Syngas

The "two-platform concept" is one which uses biomass consisting, on average, of 75 % carbohydrates which can be standardized as a "intermediate sugar platform", as a basis for further conversion, but which can also be converted thermochemically into synthesis gas and the products made from this. The "sugar platform" is based on biochemical conversion processes and focuses on fermentation of sugars extracted from biomass feedstocks. The "syngas platform" is based on thermochemical conversion processes and focuses on the gasification of biomass feedstocks and by-products from conversion processes [66, 101, 139]. In addition to the gasification other thermal and thermochemical biomass conversion methods have also been described – hydrothermolysis, pyrolysis, thermolysis, and burning. The application chosen depends on the water content of biomass [174].

The gasification and other thermochemical conversions concentrate on utilization of the precursor carbohydrates and their intrinsic carbon and hydrogen content. The proteins, lignin, oils and lipids, amino acids, and other nitrogen and sulfur-containing compounds occurring in all biomass are not taken into account (Fig. 18).

The advantage of this concept is that the production of energy, fuels, and biobased products is possible using only slightly complex and low-tech technology, for example saccharification and syngas technology [3]. The sugar platform also enables access to a huge range of family-tree-capable chemicals (Figs. 7 and 8).

In-situ conversion of biomass feedstock into liquid or gas could be one way of using existing infrastructure (developed pipe network), but with

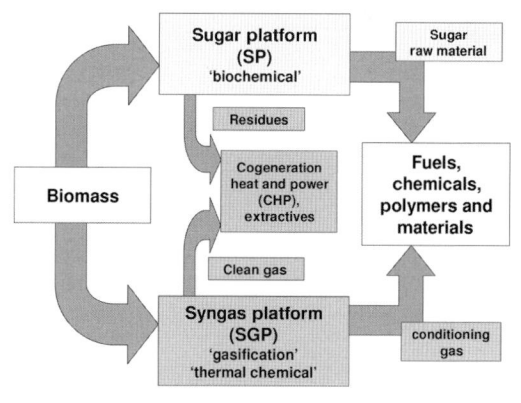

Figure 18. The sugar platform and the syngas platform [139]

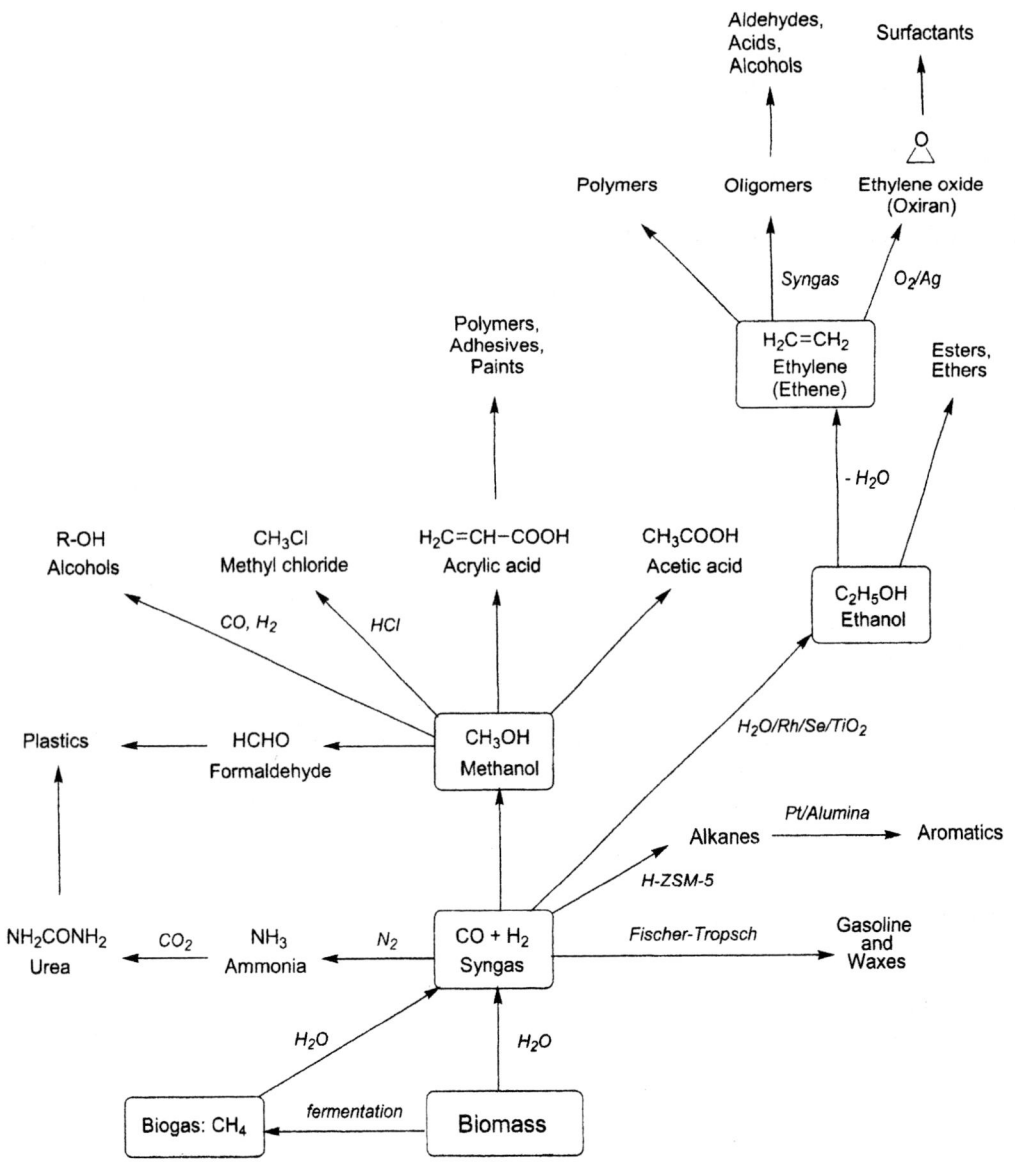

Figure 19. Syngas-based product family tree [175, 190]

the disadvantages of the need to remove heteroatoms (O, N, S) and minerals present in the biomass and the highly endothermic nature of the syngas process [175]. Currently, production of simple alcohols, aldehydes, mixed alcohols, and Fischer-Tropsch liquids from biomass is not economically viable and additional developments are required [101] (Fig. 19).

Recent research approaches for the production of liquid fuels are aqueous-phase catalytic processes for the conversion of biobased building blocks into H and C1–C15 alkanes [176].

6. Biorefinery Economy

Industrial utilization of raw materials from agriculture, forestry, landscape conservation, and organic waste management for large-scale converting industry and chemical industry as well for producing energy is still in its beginnings. The concept of a biorefinery is analogous to a petroleum refinery where a feedstock, crude oil, is converted to fuels, chemical products, and energy. In the case of a biorefinery, plant biomass is used as a feedstock to produce a diverse set of products for all areas of life, such as fuels, chemicals, polymers, lubricants, adhesives, fertilizer, and energy [177, 132, 133].

Although similar to oil refineries, biorefineries exhibit some important differences. First, biorefineries can utilize a variety of feedstocks. Consequently, they require the development of a larger range of processing technologies and biorefinery systems to deal with compositional differences in the feedstock. Second, the biomass feedstock is bulkier (has a lower energy density) than fossil fuels. Therefore, economics dictate decentralized biorefineries closer to feedstock sources. Conversion of lignocellulosic feedstock to the main products ethanol and furfural has been conceived with optimal profitability (see Section 5.2, [77]).

The economics of biorefineries are dependent on the production of co-products, such as protein, chemicals, polymers, and power, to provide revenue streams that offset processing costs. Generation of co-products also results in more efficient utilization of biomass and land along with more effective use of invested capital [178, 179]. Generally, because of low cost, plentiful supply, and amenability to biotechnology, carbohydrates appear likely to be the dominant source of feedstock for biocommodity processing [180].

Biorefineries have the potential to reduce sugar costs from about 8–9 cents per pound in the USA today (ca. 2006) to about 4 cents per pound in three to four years, and even lower as the biorefineries become more integrated. It is even possible to foresee a case in which the net cost of producing sugar in an integrated biorefinery is zero, because byproducts such as lignin and proteins generate the value [181].

Plants like NatureWorks PLA are Generation-II biorefineries (see Section 5.5 [1], [28], [182]). Polylactic acid (PLA) is current produced by using corn as feedstock. The corn-based sugar raw material can account for approximately 25–40 % of the total cost of PLA. Today PLA made from corn is competitive with traditional petrochemical-based polymers, given its combination of performance in applications, costs, and environmental benefits. Lignocellulose biorefineries are important to PLA because of the improvements in both the cost of the raw material and the overall environmental footprint [1].

In a lignocellulosic biorefinery in which multiple value-added co-products are produced and the lignin is used for energy, the cost of PLA would be lower, primarily because of the lower cost of sugars. In this case the cost of sugar raw materials could be as low as 15–25 % of the total cost of PLA production. Importantly, the environmental footprint would improve because of the replacement of petrochemical-based energy by lignin-based renewable energy to drive the manufacturing process. PLA made by a lignocellulosic biorefinery in the future will have competitive advantages.

Biorefineries utilizing lignocellulosic feedstocks hold the promise of lower net feedstock costs combined with an improved environmental footprint. PLA already has an advantage over some traditional petrochemical-based polymers such as poly(ethylene terephthalate), PET. Improvements in cost will make it competitive with polypropylene and polystyrene, too.

Recently (ca. 2006), the goal of the U.S. Department of Agriculture and the U.S. Department of Energy is the additional supply of 10^9 tons of biomass at a price of 35 $ per ton (dry matter) per year for industrial chemical and biotechnological utilization, without the restriction of today's applications of biomass from agriculture and forestry [183]. Annually in the USA, 400×10^6 tons of straw (dry matter) has been produced but not used [184, 146].

Example of Use of Straw for Chemicals and Fuels in Germany. Germany produces on average 50×10^6 t/a (dry matter) of cereal grains. The ratio of grain to straw (chaff, nodes, ears, leaves) is approx. 1:1.3 [155].

That means that approximately 50×10^6 t (dry matter) of straw can be used for industry without reducing today's use in organic fertilizing and natural field cultivation. Furthermore, a life-cycle analysis has shown that the use of straw for a lignocellulosic biorefinery is more ecologically advantageous than other possible applications (field cultivation, production of liquid fuels via Fischer–Tropsch, or energy production) [185].

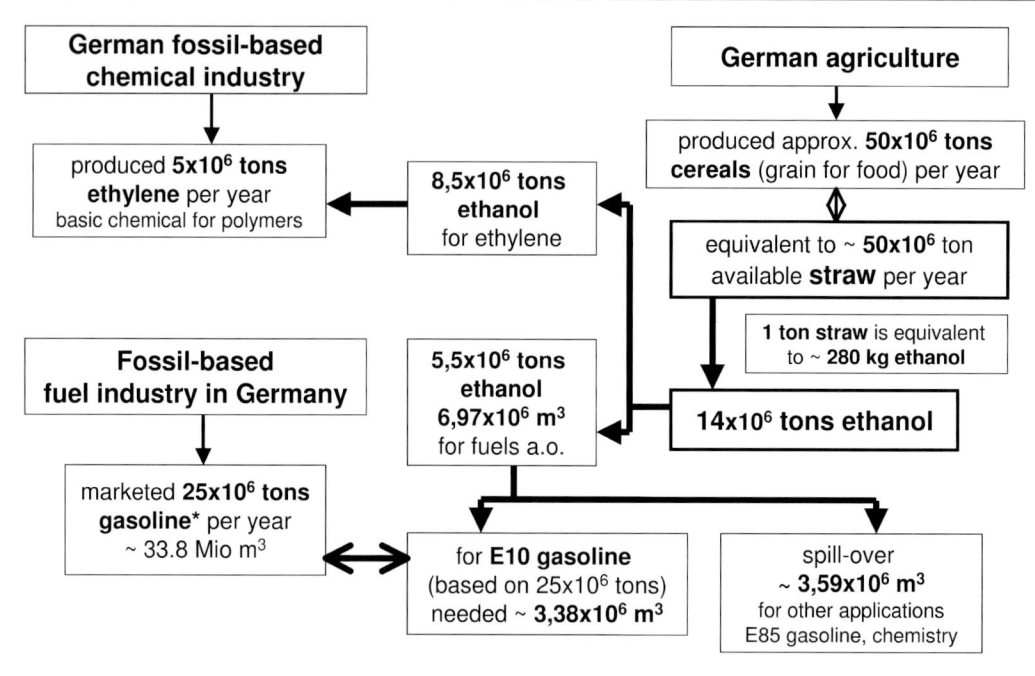

```
┌─────────────────────────────┐                    ┌─────────────────────────────┐
│  German fossil-based        │                    │   German agriculture        │
│  chemical industry          │                    │                             │
└──────────────┬──────────────┘                    └──────────────┬──────────────┘
               ▼                                                  ▼
┌─────────────────────────────┐   ┌──────────────┐   produced approx. 50x10⁶ tons
│ produced 5x10⁶ tons         │◄──│ 8,5x10⁶ tons │◄──cereals (grain for food) per year
│ ethylene per year           │   │ ethanol      │
│ basic chemical for polymers │   │ for ethylene │   equivalent to ~ 50x10⁶ ton
└─────────────────────────────┘   └──────────────┘   available straw per year

                                                     1 ton straw is equivalent
┌─────────────────────────────┐   ┌──────────────┐   to ~ 280 kg ethanol
│  Fossil-based               │   │ 5,5x10⁶ tons │
│  fuel industry in Germany   │   │ ethanol      │
└──────────────┬──────────────┘   │ 6,97x10⁶ m³  │   14x10⁶ tons ethanol
               ▼                  │ for fuels a.o│
┌─────────────────────────────┐   └──────────────┘
│ marketed 25x10⁶ tons        │
│ gasoline* per year          │◄──► for E10 gasoline       spill-over
│ ~ 33.8 Mio m³               │    (based on 25x10⁶ tons)  ~ 3,59x10⁶ m³
│                             │    needed ~ 3,38x10⁶ m³    for other applications
└─────────────────────────────┘                            E85 gasoline, chemistry
```

E10 gasoline is normal gasoline (Otto Petrol*) with 10% ethanol

Figure 20. Exemplary calculation on the use of straw for the production of ethanol for fuels and basic chemicals in Germany

An exemplary calculation for using of this straw for industry (basic chemicals) and fuels (ethanol production) is shown in Figure 20.

The generated ethanol could cover both the demand for ethylene (fossil-based production in Germany: 5×10^6 t) and for E10 fuels (German consumption of gasoline/Otto petrol: 25×10^6 t). The residual ethanol could be applied as a chemical or for E85 fuel (Fig. 20).

7. Outlook and Perspectives

Biorefineries are production plants in which biomass is economically and ecologically converted to chemicals, materials, fuels, and energy. For successful development of "industrial biorefinery technologies" and "biobased products" several problems must be solved. It will be necessary to increase the production of substances (cellulose [186], starch, sugar [187], oil) from basic biogenic raw materials and to promote the introduction and establishment of biorefinery demonstration plants [188, 189]. Ecological transport of biomass must also be developed, for example, utilization of already developed pipelines.

Another important aspect is committing chemists, biotechnologists, and engineers to the concept of biobased products and biorefinery systems and promoting the combined biotechnological and chemical conversion of substances. Last, but not least, the development of systematic approaches to new syntheses and technologies is required to meet the sustainable principles of "ideal synthesis" and "principles of green chemistry and process engineering" [191–193].

The biobased economy and its associated biorefineries will be shaped by many of the same forces that shaped the development of the hydrocarbon economy and its refineries over the past century [194, 182].

Many of the biorefinery elements required for financial success seem to be currently present, for example, high-volume, low-cost applications (ethanol, lactic acid), multiple alternative product streams, ability to shift to different products quickly when required [195], and acceptable cost structure for biomass.

Several financial requirements will enable the biorefinery concept to become a reality in the near future:

- Parts of the overall biorefinery scheme that can operate in a stand-alone manner from a financial perspective, and the ability to evolve from there
- The potential for multiproduct or co-product streams, which may ultimately drive efforts to engineer crops to maximize their potential
- The ability to make use of a variety of feedstocks as hedging strategy and the advantage of pricing opportunities:
- Manageable capital requirements and favorable investment environment
- Government enablement (funding demonstration/development pilot plants, reasonable regulation of biorefinery operations, favorable taxation, strategic market support to give new biorefinery products a head start [196].

References

1 Gruber, P.; Henton, D.E.; Starr, J.; "Polylactic Acid from Renewable Resources", [B. Kamm, P.R. Gruber, M. Kamm (eds.): *Biorefineries – Industrial Processes and Products, Status Quo and Future Directions*] Vol. 2, 381–407, Wiley-VCH, 2006

2 Ragauskas A. J; Williams Ch. K; Davison B. H; Britovsek G.; Cairney J.; Eckert Ch. A; Frederick W. J Jr; Hallett J. P; Leak D. J; Liotta Ch. L; Mielenz J. R; Murphy R.; Templer R.; Tschaplinski T.; The path forward for biofuels and biomaterials. *Science* **311** (5760), (2006) 484–489.

3 Jong, E. de; Ree, R. van; Tuil, R. van; Elbersen, W.; "Biorefineries for the Chemical Industry", [B. Kamm, P.R. Gruber, M. Kamm (eds.): *Biorefineries – Industrial Processes and Products, Status Quo and Future Directions*] Vol. 1, 85–111, Wiley-VCH, 2006

4 National Research Council: *Biobased Industrial Products, Priorities for Research and Commercialization*, National Academic Press, Washington D.C. 2000, a) 74, ISBN 0-309-05392-7.

5 A. S. Marggraf: *Histoire de l'Acadmie Royale des Sciences et Belles Lettres, Ann 1748*, Preußische Akademie, Berlin 1749.

6 A. S. Marggraf: *Chym. Schriften (Chemische Schriften)*, Bd. 2, Berlin 1761–1767.

7 G. S. C. Kirchhoff, *Schweigers Journal für Chemie und Physik*, **4** (1812) 108.

8 C. Graebe: *History of Organic Chemistry*, Bd 1, Springer, Berlin 1920, a) 28, b) 122f.

9 N. Kosaric, F. Vardar-Sukan: "Potential Sources of Energy and Chemical Products" in M. Roehr (ed.): *The Biotechnology of Ethanol*, Wiley-VCH, Weinheim 2001, p. 132, ISBN 3-527-30199-2.

10 S. C. Prescott, C. G. Dunn: *Industrial Microbiology*, 3rd ed., McGraw-Hill, New York 1959.

11 D. Osteroth: *From Coal to Biomass*, Springer, New York 1989, p. 192ff, ISBN 0-387-50712-4.

12 W. J. McKillip: "Furan and Derivatives" in: *Ullmann's Encyclopedia of Industrial Chemistry*, 6th ed., vol. 15, Wiley–VCH, Weinheim 2003, p. 187ff.

13 www.furan.com

14 H. Pringsheim, *Cellulosechemie* **2** (1921) 123.

15 F. Ullmann: *Enzyklopädie der Technischen Chemie*, 2. Aufl., 5. Bd., Urban und Schwarzenberg, Berlin 1930, p. 442ff.

16 A. Payen: *Comptes rendus de l'Acadmie des Sciences (C. r.)* **8** (1839) 51.

17 E. Gruber, T. Krause, J. Schurz: "Cellulose" in: *Ullmann, Enzyklopädie der technischen Chemie*, 4. Aufl. Bd. 9, Verlag Chemie, Weinheim 1975, p. 184–191.

18 G. J. Mulder, *J. Prakt. Chem.* **21** (1840) 219.

19 A. E. Staley, Mfg. Co. A.E. (Decatur Ill.); Levulinic Acid 1942 [C.A. 36, 1612].

20 Kitano *et al.*: "Levulinic Acid, A New Chemical Raw Material – Its Chemistry and Use", *Chemical Economy and Engineering Review* (1975), 25–29.

21 R. H. Leonard: "Levulinic acid as a Basic Chemical Raw Material", *Ind. Eng. Chem.* (1956), 1331–1341.

22 W. Norman: Process for Converting Unsaturated Fatty Acids or their Glycerides into Saturated Compounds, BP 1515, 1903.

23 W. Sandermann: *Grundlagen der Chemie und chemischen Technologie des Holzes*, Akademische Verlagsgesellschaft, Leipzig 1956, p. 147.

24 F. Tiemann, W. Haarmann: "Ueber das Coniferin und seine Umwandlung in das aromatische Princip der Vanille", *Ber. dt. chem. Ges.* **7** (1874) 608–623.

25 M. J. W. Dignum, J. Kerler, R. Verpoorte: "Vanilla Production: Technological, Chemical and Biosynthetic Aspects", *Food Rev. Intern.* **17** (2001) 199–219.

26 L. Vaupel: "Vanille und Vanillin", *Pharm. Zeit.* **38** (2002).

27 H. Carothers, G. L. Dorough, F. J. Van Natta, *J. Am. Chem. Soc.* **54** (1932) 761.

28 P. R. Gruber, M. O'Brien: "Polylactides "Nature Works" PLA" in: Y. Doi, A. Steinbüchel (eds.): *Biopolymers, Polyester III*, Wiley-VCH, Weinheim 2002.

29 P. Walden in: C. Graebe (ed.): *History of Organic Chemistry since 1880*, Bd. 2, Springer, Berlin 1941, p. 686.

30 W. R. Pötsch: *Lexicon of famous Chemists*, Bibliographisches Institut, Leipzig 1988, a) 305, ISBN 3-323-00185-0.

31 Borregaard LignoTech; marathon co.; http://www.ltus.com.

32 B. W. Peckham: "The First Hundred Years of Corn Refining in the United States", *Corn Annual 2000*, Corn Refiners Association, Washington 2000.

33 D. L. Johnson: "The Corn Wet Milling and Corn Dry Milling Industry — A Base for Biorefinery Technology Developments" in B. Kamm, P. R. Gruber, M. Kamm (eds.): *Biorefineries – Industrial Processes and Products, Status Quo and Future Directions*, vol. 1, Wiley-VCH, Weinheim 2006, pp. 345–353.

34 ICI: Improvement in or related to the utilization of grass and other green crops, Brit. Pat. BP 511,525, 1939 (R. E. Slade, J. H. Birkinshaw).
35 N. W. Pirie, *Chem. Ind.* **61** (1942) 45.
36 N. W. Pirie, *Nature* **149** (1942) 251.
37 W. Heier, *Grundlagen der Landtechnik* **33** (1983) 45–55.
38 B. Kamm, M. Kamm: "The Green Biorefinery – Principles, Technologies and Products", *Proceed. 2nd Intern. Symp. Green Biorefinery*, October 13–14, 1999, Feldbach, Austria, 1999, S. 46–69.
39 B. E. Knuckles, E. M. Bickoff, G. O. Kohler, *J. Agric. Food Chem.* **20** (1972) 1055.
40 F. M. Schertz, *Ind. Eng. Chem.* **30** (1938) 1073–1075.
41 W. H. Shearon, O. F. Gee, *Ind. Eng. Chem.* **41** (1949) 218–226.
42 M. A. Judah, E. M. Burdack, R. G. Caroll, *Ind. Eng. Chem.* **46** (1954) 2262–2271.
43 W. J. Hale: *The Farm Chemurgic*, The Stratford Co., Boston 1934.
44 C. Borth: *Pioneers of Plenthy*, Bobbs-Merril Co, Indianapolis, New York 1939.
45 D. L. Lewis: *The Public Image of Henry Ford*, Wayne State University Press, Detroit 1976.
46 E. N. Brandt: *Growth Company – Dow Chemical's First Century*, Michigan State University press, East Lansing 1997.
47 F. Bergius; *Trans. Inst. Chem. Eng. (London)* **11** (1933) 162.
48 H. Scholler: PhD thesis, Technical University of Munich, 1923.
49 H. Scholler: F.P. 777,824, 1935.
50 R. Katzen, G. T. Tsao: "A View of the History of Biochemical Engineering", *Advances in Biochemical Engineering/Biotechnology*, vol. 70, Springer, Berlin 2000.
51 T. F. Conrad: "Holzzuckerbrennerei" in: R. Reiff *et al.* (eds.): *Die Hefen*, vol. II 8F., Verlag Hans Carl, Nürnberg 1962, p. 437–444.
52 S. C. Prescott, C. D. Dunn: *Industrial Microbiology*, 3rd ed., McGraw-Hill, New York 1959.
53 E. E. Harris *et al.*: "Madison Wood Sugar Process", *Ind. Eng. Chem.* **38** (1946) 896–904.
54 T. E. Timell, *Tappi*, **44** (1961) 99.
55 A. J. Stamm: *Wood and Cellulose Science*, Ronald Press, New York 1964.
56 R. L. James: *The Pulping of Wood*, 2nd ed., vol. 1, McGraw-Hill, New York 1969, p. 34.
57 D. L. Brink, A. A. Pohlmann, *Tappi* **55** (1972) 381ff.
58 M. Oshima: *Wood Chemistry – Process Engineering Aspects*, Noyes development Corp., New York 1965.
59 J. Puls, H. H. Dietrichs: "Energy from Biomass", *Proceed. of First European Comm. Inter. Conf. on Biomass*, Brighton, UK, November 1980, p. 348, ISBN: 0-85334-970-3.
60 S. Y. Shen: "Wood Grass Production Systems for Biomass", *Proceed. Midwest Forest Economist Meeting*, Duluth, Minnesota, April 1982.
61 S. Y. Shen: "Biological Engineering for Sustainable Biomass Production" in: E.O. Wilson (ed.): *Biodiversity*, National Academy of Sciences/Smithsonian Institution, National Academic Press, Washington DC 1988; German Edition: *Ende der biologischen Vielfalt?*, 1992, p. 404–416, ISBN 3-89330-661-7.
62 R. Carlsson: "Trends for future applications of green crops", *Forage Protein Conservation and Utilization*, Proceed. of EFC Conf. 1982, Dublin, Ireland, 1982, p. 57–81.
63 R. Carlsson: "Green Biomass of Native Plants and new Cultivated Crops for Multiple Use: Food, Fodder, Fuel, Fibre for Industry, Photochemical Products and Medicine" in: G.E. Wickens, N. Haq, P. Day (eds.): *New Crops for Food and Industry*, Chapman and Hall, London 1989.
64 B. E. Dale: "Biomass refining: protein and ethanol from alfalfa", *Ind. Eng. Product Research and Development* **22** (1983) 446.
65 A. J. Hacking: "The American wet milling industry", *Economic Aspects of Biotechnology*, Cambridge University Press, New York 1986, p. 214–221.
66 D. H. White, D. Wolf: *Research in Thermochemical Biomass*, Elsevier Applied Science, New York 1988.
67 E. S. Lipinsky: "Chemicals from Biomass: petrochemical substitution options", *Science* **212** (1981) 1465–1471.
68 D. L. Wise (ed.): *Organic Chemical from Biomass*, The Benjamin/Cummings Publishing Co., Inc., Menlo Park, California, 1983.
69 J. E. Bailey, D. F. Ollis: *Biochemical Engineering Fundamentals*, 2nd ed., McGraw-Hill, New York 1986.
70 H. H. Szmant: *Industrial Utilization of Renewable Resources*, Technomic Publishing, Lancaster, Pa., 1987.
71 M. Eggersdorfer, J. Meyer, P. Eckes: "Use of renewable resources for non-food materials", *FEMS Microbiol. Rev.* **103** (1992) 355–364.
72 D. J. Morris, I. Ahmed: *The carbohydrate Economy: Making Chemicals and Industrial Materials from Plant Matter*, Institute of Local Self Reliance, Washington D.C. 1992.
73 J. J. Bozell, R. Landucci: *Alternative feedstock program – technical and economic assessment*, US Department of Energy, 1992.
74 L. B. Schilling: "Chemicals from alternative feedstock in the United States", *FEMS Microbiol. Rev.* **16** (1995) 1001–1110.
75 L. B. de la Ross *et al.*: "An integrated process for protein and ethanol from coastal Bermuda grass", *Appl. Biochem. Biotechnol.* **45/46** (1994) 483–497.
76 K. Soyez, B. Kamm, M. Kamm (eds.): "The Green Biorefinery", *Proceedings of 1st International Green Biorefinery Conference, Neuruppin, Germany, 1997*, GÖT, Berlin 1998, ISBN 3-929672-06-5, German and English.
77 D. L. Van Dyne, M. G. Blas, L. D. Clements: "A strategy for returning agriculture and rural America to long-term full employment using biomass refineries" in: J. Janeck

(ed.): *Perspectives on new crops and new uses*, ASHS Press, Alexandria, Va., 1999, p. 114–123.
78. M. Narodoslawsky: "The Green Biorefinery", *Proceedings 2nd Intern. Symp. Green Biorefinery*, Feldbach, Austria, 1999.
79. R. V. Nonato, P. E. Mantellato, C. E. V. Rossel: "Integrated production of biodegradable plastic, sugar and ethanol" *Appl. Microbiol. Biotechnol.* **57** (2001) 1–5.
80. H. Ohara: "Biorefinery", *Appl. Microb. Biotechn. (AMB)* **62** (2003) 474–477.
81. B. Kamm, M. Kamm: "Principles of Biorefineries", *Appl. Microbiol., Biotechnol., (AMB)* **64** (2004) 137–145.
82. B. Kamm, M. Kamm: "Biorefinery-Systems", *Chem. Biochem. Eng. Q.* **18** (2004) 1–6.
83. U.S. Department of Energy (DEO): National Biomass Initiative and Energy, Environmental and Economics (E3) Handbook, www.bioproducts-bioenergy.gov.
84. L. R. Lynd, J. H. Cushman, R. J. Nichols, C. E. Wyman: "Fuel Ethanol from Cellulosic Biomass", *Science* **251** (1991) 1318.
85. F. A. Keller: "Integrated bioprocess development for bioethanol production" in: C.E. Wyman (ed.): *Handbook ob bioethanol: production and utilization*, Taylor and Francis, Bristol 1996, p. 351–379.
86. C. E. Wyman: *Handbook on Bioethanol: Production and Utilization*, Applied Energy Technology Series, Taylor and Francis, Bristol 1996.
87. L. Lynd: "Overview and evaluation of fuel ethanol from cellulosics biomass: technology, economics, the environment, and policy", *Annual Review of Energy and the Environment* **21** (1996) 403–465.
88. M. Galbe, G. Zacci: "A review of the production of ethanol from softwood", *Appl. Microbiol. Biotechnol.* **59** (2002) 618–628.
89. R. Datta *et al.*: "Technological and economics potential of poly (lactic acid) and lactic acid derivatives" *FEMS Microbiol. Rev.* **16** (1995) 221–231.
90. U. Witt, R. J. Müller, H. Widdecke, W.-D. Deckwer: "Synthesis, properties and biodegradability of polyesters based on 1,3-propanediol", *Macrom. Chem. Phys.* **195** (1994) 793–802.
91. T. Yosida, T. Nagasawa: "ε-Poly-L-lysine: microbial production, biodegradation and application potential", *Appl. Microbiol. Biotechnol.* **62** (2003) 21–26.
92. C. Potera: "Genencor and DuPont create "green" polyester", *Genet. Eng. News* **17** (1997) 17.
93. Y. Poirier, C. Nawrath, C. Somerville: "Production of polyhydroxyalkanoates, a family of biodegradable plastics", *Bio/technology* **13** (1995) 142–150.
94. S. Warwel *et al.*: "Polymers and Surfactants on the basis of renewable resources", *Chemosphere* **43** (2001) 39–48.
95. J. J. Bozell: "Alternative Feedstocks for Bioprocessing" in: R.M. Goodman (ed.): *Encyclopedia of Plant and Crop Science*, Dekker, New York 2004, ISBN:0-8247-4268-0.
96. C. Webb, A. A. Koutinas, R. Wang: "Developing a Sustainable Bioprocessing Strategy Based on a Generic Feedstock", *Adv. Biochem Eng./Biotechn.* **87** (2004) 195–268.
97. J. Gravitis, M. Suzuki: "Biomass Refinery – A Way to produce Value Added Products and Base for Agricultural Zero Emissions Systems" in: *Proc. 99 Intern. Conference on Agric. Engineering, Beijing, China 1999*, United Nations University Press, Tokyo 1999, III-9–III-23.
98. D. L. Van Dyne *et al.*: *Estimating the Economic Feasibility of Converting Ligno-Cellulosic Feedstocks to Ethanol and Higher Value Chemicals under the refinery concept: A Phase II Study*, University of Missouri, 1999, OR22072-58.
99. Z. Kurtanjek, *Chemical and Biochemical Engineering Quarterly, Special Issue* **18** (2004) 1–88.
100. J. J. Marano, J. L. Jechura: "Biorefinery Optimization Tools – Development and Validation" in: *25th. Symposium on Biotechnology for Fuels and Chemicals: Program and Abstracts*, National Renewable Energy Laboratory, Golden, CO, No. NREL/BK-510-33708, 2003, p. 104.
101. T. Werpy, G. Petersen (eds.): Top Value Added Chemicals from biomass, U.S. Department of Energy, Office of scientific and technical information, 2004, No.: DOE/GO-102004-1992, www.osti.gov/bridge.
102. O. J. M. Goddijn: "Plants as bioreactors", *Trends Biotechnol.* **13** (1995) 379–387.
103. D. Wilke: "Chemicals from biotechnology: molecular plant genetic will challenge the chemical and the fermentation industry", *Appl. Microbiol. Biotechnol., (AMB)* **52** (1999) 135–145.
104. R. Anex, *Journal of Ind. Ecology, Special Issue* **7** (2003) 1–235.
105. R. G. Ludgar; R. J. Woolsey; The new petroleum. *Foreign Affairs* **78** (1999) 88–102.
106. C. E. Wyman: "Production of Low Cost Sugars from Biomass: Progress, Opportunities, and Challenges" in: R.P. Overend and E. Chornet (eds.): *Biomass: A Growth Opportunity in Green Energy and Value-Added Products*, Pergamon Press, Oxford 1999, p. 867–872.
107. R. Bachmann, E. Bastianelli, J. Riese, W. Schlenzka: "Using plants as plants. Biotechnology will transform the production of chemicals", *The McKinsey Quarterly* **2** (2000) 92–99.
108. J. R. Hettenhaus, B. Wooley: Biomass Commercialization: Prospect in the Next 2 to 5 Years, NREL, Golden Colorado 2000, No. NREL/ACO-9-29-039-01.
109. J. Woolsey: "Hydrocarbons to Carbohydrates, The strategic Dimension", *The Biobased Economy of the 21st Century: Agriculture Expanding into Health, Energy, Chemicals, and Materials*, NABC Report 12, National Agricultural Biotechnology Council, Ithaca, New York 2000, No. 14853.
110. A. Eaglesham, W. F. Brown, R. W. F. Hardy (eds.): *The Biobased Economy of the 21st Century: Agriculture Expanding into Health, Energy, Chemicals, and Materials*, NABC Report 12, National Agricultural Biotechnology Council, Ithaca, New York 2000, No. 14853.

111 US President: Developing and Promoting Biobased Products and Bioenergy, Executive Order 13101/13134, William J. Clinton, The White House, Washington D.C. 1999.
112 US Congress: Biomass Research and Development, Act of 2000, Washington D.C. 2000.
113 Biomass R&D, Technical Advisory Committee; Vision for Bioenergy and Biobased Products in the United States, Washington D.C. Oct. 2002; www.bioproducts-bioenergy.gov/pdfs/BioVision_03_Web.pdf.
114 Biomass R&D, Technical Advisory Committee; Roadmap for Biomass Technologies in the United States, Washington D.C., Dec. 2002; www.bioproducts-bioenergy.gov/pdfs/FinalBiomassRoadmap.pdf.
115 Biotechnology Industrial Organisation: World Congress on Industrial biotechnology and Bioprocessing; http://www.bio.org/worldcongress.
116 Biomass Conferences of the Americas; http://www.nrel.gov/bioam/
117 R. P. Overend, E. Chornet (ed.): "Biomass, a growth opportunity in green energy and value-added products", *Proceedings of the 4th Biomass Conference of the Americas*, Oakland, CA, Aug. 29 – Sept. 2, 1999, Elsevier, ISBN: 0080430198.
118 Green and Sustainable Chemistry Congress; http://www.chemistry.org.
119 J. J. Bozell: "Feedstocks for the future: using technology development as a guide to product identification", *ACS Symp. Series* **921** (2006) 1–12.
120 Biorefinica – International Symposia Biobased Products and Biorefineries; www.biorefinica.de.
121 B. Kamm, M. Hempel, M. Kamm (eds.): "biorefinica 2004, International Symposium Biobased Products and Biorefineries", *Proceedings and Papers*, October, 27 and 28, 2004, biopos, Teltow 2004, ISBN 3-00-015166-4.
122 H. Zoebelin (ed.): *Dictionary of Renewable Resources*, Wiley-VCH, Weinheim 2001.
123 L. E. Manzer: "Biomass derivatives: a sustainable source of chemicals", *ACS Symp. Series* **921** (2006) 40–51.
124 B. Kamm *et al.*: "Green Biorefinery Brandenburg, Article to development of products and of technologies and assessment", *Brandenburgische Umweltberichte* **8** (2000) 260–269.
125 European parliament and Council: Directive 2003/30/EC on the promotion of the use of biofuels or other renewable fuels for transport, Official Journal of the European Union L123/42, 17. 05. 2003, Brussels 2003.
126 Gesetz für den Vorrang erneuerbarer Energien: Erneuerbare Energiegesetz, EEG/EnWGuaÄndG., 29. 03.2000, BGBl, 305, 2000.
127 European parliament and Council; Green Paper "Towards a European strategy for the security of energy supply" KOM2002/321, 26. 06. 2002.
128 Umweltbundesamt: *Klimaschutz durch Nutzung erneuerbarer Energien*, Report 2, Erich Schmidt Verlag, Berlin 2000.
129 BioVision2030-Group: Strategiepapier Industrielle stoffliche Nutzung von Nachwachsenden Rohstoffen in Deutschland, Nov. 2003; www.biorefinica.de/bibliothek.
130 Deutscher Bundestag: Rahmenbedingungen für die industrielle stoffliche Nutzung von Nachwachsenden Rohstoffen in Deutschland schaffen, Antrag 15/4943, Berlin 2005.
131 R. Busch *et al.*: "Biomasse-Industrie – Wie aus "Bio" Chemie wird", *Nachrichten aus der Chemie* **53** (2005) 130–134.
132 European Technology Platform for Sustainable Chemistry, Industrial Biotechnology Section, 2005; www.suschem.org.
133 US Department of Energy (DOE), 1st International International Biorefinery Workshop, July 20/21st 2005, Washington D.C.; www.biorefineryworkshop.com.
134 H. Röper: "Perspektiven der industriellen Nutzung nachwachsender Rohstoffe, insbesondere von Stärke und Zucker", *Mitteilung der Fachgruppe Umweltchemie und Ökotoxikologie der Gesellschaft Deutscher Chemiker* **7** (2001) no 2, 6–12.
135 F. W. Lichtenthaler: "The Key Sugars of Biomass: Availability, Present Non-Food Uses and potential Future Development Lines" in B. Kamm *et al.* (eds.): *Biorefineries – Industrial Processes and Products, Status Quo and Future Directions*, vol. 2, Wiley-VCH, Weinheim 2006, p. 3–59.
136 B. Kamm, M. Kamm, K. Richter: "Entwicklung eines Verfahrens zur Konversion von hexosenhaltigen Rohstoffen zu biogenen Wirk- und Werkstoffen – Polylactid aus fermentiertem Roggenschrot über organische Aluminiumlactate als alternative Kuppler biotechnischer und chemischer Stoffwandlungen" in: P. B. Czedik-Eysenberg (ed.): *Chemie nachwachsender Rohstoffe*, Österreichisches Bundesministerium für Umwelt (BMUJF), Wien 1997, pp. 83–87.
137 B. Kamm, M. Kamm, M. Schmidt, I. Starke, E. Kleinpeter: "Chemical and biochemical generation of carbohydrates from lignocellulose-feedstock (Lupinus nootkatensis), Quantification of glucose", *Chemosphere* **62** (2006) no 1, 97–105.
138 US Department of Energy; http://www.oit.doe.gov/e3handbook.
139 National Renewable Energy Laboratory (NREL); http://www.nrel.gov/biomass/biorefinery.html.
140 S. Fernando, S. Adhikari, C. Chandrapal, N. Murali: "Biorefineries: Current Status, Challenges, and Future Directions", *Energy Fuels*; 2006; **20**(4) 1727–1737.
141 EuropaBio: *White Biotechnology: Gateway to a more sustainable future*, EuropaBio, Lyon, April 2003.
142 BIO Biotechnology Industry Organisation: New Biotech Tools for a cleaner Environment – Industrial Biotechnology for Pollution Prevention, Resource Conservation and Cost Reduction, 2004; http://www.bio.org/ind/pubs/cleaner2004/cleanerReport.pdf.
143 Dti Global Watch Mission Report: Impact of the industrial biotechnology on sustainability of the manufacturing base – the Japanese Perspective, 2004.

144 P. Claus, G. H. Vogel: "Die Rolle der Chemokatalyse bei der Etablierung der Technologieplattform, Nachwachsende Rohstoffe", *Chemie Ingenieur Technik* **78** (2006) no 8, 991–1012 (German).
145 N. Suzuki, H. Yukawa: "Bio-refinery: industrial revolution of the 21st century", *Cellulose Communications* **11** (2004) **4**, 181–187 (Japanese).
146 J. S. Tolan: "Iogen's Demonstration Process for Producing Ethanol from Cellulosic Biomass" in: B. Kamm et al. (eds.): *Biorefineries – Industrial Processes and Products, Status Quo and Future Directions*, vol. 1, Wiley-VCH, Weinheim 2006, pp. 193–208;
147 J. Hettenhaus: "Biomass Commercialization and Agriculture Residue Collection" in: B. Kamm et al. (eds.): *Biorefineries – Industrial Processes and Products, Status Quo and Future Directions*, vol. 1, Wiley-VCH, Weinheim 2006, pp. 317–344.
148 S. Kromus, B. Kamm, M. Kamm, P. Fowler, M. Narodoslawsky: "The Green Biorefinery Concept- Fundamentals and Potential" in: B. Kamm et al. (eds.): *Biorefineries – Industrial Processes and Products, Status Quo and Future Directions*, vol. 1, Wiley-VCH, Weinheim 2006, p. 253–294.
149 B. Kamm, M. Kamm, M. Schmidt, T. Hirth, M. Schulze: "Lignocellulose- based Chemical Products and Product Family Trees" in: B. Kamm et al. (eds.): *Biorefineries – Industrial Processes and Products, Status Quo and Future Directions*, vol. 2, Wiley-VCH, Weinheim 2006, p. 97–149.
150 B. Dale: *Encyclopedia of Physical Science and Technology*, 3rd ed., vol. 2, 2002, p. 141–157.
151 D. J. Hayes, S. Fitzpatrick, M. H. B. Hayes, J. H. R. Ross: "The Biofine Process-Production of levulinic acid from lignocellulosic feedstock" in: B. Kamm et al. (eds.): *Biorefineries – Industrial Processes and Products, Status Quo and Future Directions*, vol. 1, Wiley-VCH, Weinheim 2006, p. 139–164.
152 M. Ringpfeil: Biobased Industrial Products and Biorefinery Systems – Industrielle Zukunft des 21. Jahrhunderts?, 2001; www.biopract.de.
153 J. G. Zeikus, M. K. Jain, P. Elankovan: "Biotechnology of succinic acid production and markets for derived industrial products", *Appl. Microbiol. Biotechnol.* **51** (1999) 545–552.
154 K.-D. Vorlop, Th. Wilke, U. Prüße: "Biocatalytic and catalytic routes for the production of bulk and fine chemicals from renewable resources", in B. Kamm et al. (eds.): *Biorefineries – Industrial Processes and Products, Status Quo and Future Directions*, vol. 1, Wiley-VCH, Weinheim 2006.
155 Die Landwirtschaft 1998 *Lehrbuch für Landwirtschaftsschulen. Band 1. Pflanzliche Erzeugung. 11.* Auflage, München: BLV Verlagsgesellschaft. Münster Hiltrup: Landwirtschaftsverlag. p. 280.
156 G. H. Robertson et al.: "Native or raw starch digestion: a key step in energy efficient biorefining of grain", *Journal of agricultural and food chemistry* **54** (2006) no 2, 353–365.

157 A. A. Koutinas, R. Wang, G. M. Campbell, C. Webb: "A Whole Crop Biorefinery System: A closed System for the Manufacture of Non-food-Products from Cereal" in B. Kamm et al. (eds.): *Biorefineries – Industrial Processes and Products, Status Quo and Future Directions*, vol. 1, Wiley-VCH, Weinheim 2006, p. 165–191.
158 C. E. V. Rossel, P. E. Mantellato, A. M. Agnelli, J. Nascimento: "Sugar Based Biorefinery — Technology for Integrated Production of Poly(3-hydroxybutyrate), Sugar and Ethanol" in: B. Kamm, P. R. Gruber, M. Kamm (eds.): *Biorefineries – Industrial Processes and Products, Status Quo and Future Directions*, vol. 1, Wiley-VCH, 2006, pp. 209–326.
159 A. A. Koutinas, R. Wang, G. M. Campbell, C. Webb: "Developing viable biorefineries for the production of biodegradable microbial plastics and various value-added products", *7th World Congress of Chemical Engineering*, Glasgow, United Kingdom, July 10–14, 2005.
160 F. Rexen: New industrial application possibilities for straw. Documentation of Svebio Phytochemistry Group (Danish), Fytokemi i Norden, Stockholm, Sweden, 1986, p. 12.
161 J. Coombs, K. Hall: "The potential of cereals as industrial raw materials: Legal technical, commercial considerations" in: G. M. Campbell, C. Webb, S. L. McKee (eds.): *Cereals – Novel Uses And Processes*, Plenum Publ. Corp., New York 1997, p. 1–12.
162 E. Audsley, J. E. Sells: "Determining the profitability of a whole crop biorefinery" in: G. M. Campbell, C. Webb, S. L. McKee (eds.): *Cereals – Novel Uses and Processes*, Plenum Publ. Corp.; New York 1997, p. 191–294.
163 R. Carlsson: "Sustainable primary production – Green crop fractionation: Effects of species, growth conditions, and physiological development" in: M. Pessarakli (ed.): *Handbook of Plant and Crop Physiology*, Marcel Dekker, New York 1994, pp. 941–963.
164 N. W. Pirie: *Leaf Protein – Its agronomy, preparation, quality, and use*, Blackwell Scientific Publications, Oxford/Cambridge, UK, 1971.
165 N. W. Pirie: *Leaf Protein and Its By-Products in Human and Animal Nutrition*, Cambridge Univ. Press, UK, 1987.
166 R. Carlsson: "Status quo of the utilization of green biomass" in: S. Soyez, B. Kamm, M. Kamm (eds.): *The Green Biorefinery, Proceedings of 1st International Green Biorefinery Conference*, Neuruppin, Germany, 1997, Verlag GÖT, Berlin 1998, ISBN 3-929672-06-5.
167 R. Carlsson: "Food and non-food uses of immature cereals" in: G.M. Campbell, C. Webb, S.L. McKee (eds): *Cereals – Novel Uses and Processes*, Plenum Publ. Corp., New York, USA, 1997, pp. 159–167.
168 R. Carlsson: "Leaf protein concentrate from plant sources in temperate climates" in: L. Telek, H.D. Graham (eds.): *Leaf Protein Concentrates*, AVI Publ. Co., Inc., Westport, Conn., USA, 1983, pp. 52–80.
169 L. Telek, H. D. Graham (eds.): *Leaf Protein Concentrates*, AVI Publ., Co., Inc., Westport, Conn., USA, 1983.

170 R. J. Wilkins: *Green Crop Fractionation*, The British Grassland Society, c/o Grassland Research Institute, Hurley, Maidenhead, SL6 5LR, UK, 1977.
171 I. Tasaki (ed.): "Recent Advances in Leaf Protein Research", *Proc. 2nd Int. Leaf Protein Res. Conf.*, Nagoya, Japan, 1985.
172 P. Fantozzi (ed.): *Proc. 3rd Int. Leaf Protein Res. Conf.*, Pisa-Perugia-Viterbo, Italy, 1989.
173 N. Singh (ed.): *Green Vegetation Fractionation Technology*, Science Publ. Inc., Lebanon, NH 03767, USA, 1996.
174 C. Okkerse, H. van Bekkum: "From fossil to green", *Green Chemistry* **4** (1999) 107–114.
175 M. Lancaster: "The Syngas Economy" in: *Green Chemistry*, The Royal Society of Chemistry, Cambridge, UK, 2002, p. 205, ISBN: 0-85404-620-8.
176 G. W. Huber, J. A. Dumesic: "An overview of aqueous-phase catalytic processes for production of hydrogen and alkanes in a biorefinery", *Catalysis Today* **111** (2006) 1–2, 119–132.
177 B. Kamm et al.: *Biorefineries — Industrial Processes and Products, Status Quo and Future Directions*, vols. 1 and 2, Wiley-VCH, Weinheim 2006.
178 Biotechnology Industrial Organization: Development of Biorefineries, BioCycle, April, 2005.
179 C. E. Wyman: "Economics of a biorefinery for coproduction of succinic acid, ethanol, and electricity", *Abstracts of Papers, 221st ACS National Meeting*, San Diego, CA, United States, April 1-5, 2001, BIOT-072, American Chemical Society.
180 L. R. Lynd, C. E. Wyman, T. U. Gerngross: "Biocommodity Engineering", *Biotechnol. Prog.* **15** (1999) 777–793.
181 R. Bachmann, J. Riese: "Industrial Biotech — Setting Conditions to Capitalize on the Economic Potential" in: B. Kamm et al. (eds.): *Biorefineries — Industrial Processes and Products, Status Quo and Future Directions*, vol. 2, Wiley-VCH, Weinheim 2006, pp. 445–462.
182 E. T. H. Vink et al.: "The sustainability of Nature Works polylactide polymers and Ingeo polylactide fibers: An update of the future", *Macromolecular Bioscience* **4** (2004) **60**, 551–564.
183 U.S. Department of Agriculture (USDA) and U.S. Department of Energy (DOE) (eds.): *Biomass as Feedstock for a Bioenergy and Bioproducts Industry: The Technical Feasibility of a Billion-Ton Annual Supply*, U.S. Department of Energy, Office of Scientific and Technical Information, P.O. Box 62, Oak Ridge, TN, 2005.
184 R. Katzen, G. T. Tsao: "A View of the History of Biochemical Engineering", *Advances in Biochemical Engineering/Biotechnology*, vol. 70, Springer, Berlin 2000.
185 Office of Technology Assessement at the German Parliament, Monitoring: Industrial conversion of biomass (in print substantial German Parliament Berlin).
186 W. E. F. Mabee, D. G. Evan, P. N. McFarlane, J. N. Saddler: "Canadian biomass reserves for biorefining", *Applied Biochemistry and Biotechnology* **129-132** (2006) 22–40.
187 L. A. Edye, W. O. S. Doherty, J. A. Blinco, G. E. Bullock: "The sugarcane biorefinery: energy crops and processes for the production of liquid fuels and renewable commodity chemicals", *International Sugar Journal* **108** (2006) no 1285, 19–20, 22–27.
188 T. Eggeman, D. Verser: "The importance of utility systems in today's biorefineries and a vision for tomorrow", *Applied Biochemistry and Biotechnology* **129-132** (2006), 361–381.
189 E. Kendall Pye, "Biorefining; a major opportunity for the sugar cane industry", *International Sugar Journal* **107** 1276, 222–224, 226, 228, 230, 253 (2005).
190 A. S. Matlack: "The Use of Synthesis Gas from Biomass", in: *Introduction to Green Chemistry*, Marcel Dekker, New York 2001, p. 369, ISBN: 0824704118.
191 J. H. Clark: "Green Chemistry. Challenges and opportunities", *Green Chemistry* **1** (1999) 1–8.
192 M. Lancaster: "The Biorefinery" in: *Green Chemistry*, The Royal Society of Chemistry, Cambridge, UK, 2002, p. 207, ISBN: 0-85404-620-8.
193 P. T. Anastas, J. C. Warner: *Green Chemistry. Theory and Practice*, Oxford University Press, New York 1998.
194 B. E. Dale, S. Kim: "Biomass Refining Global Impact – The Biobased Economy of the 21st Century" in: B. Kamm et al. (eds.): *Biorefineries – Industrial Processes and Products, Status Quo and Future Directions*, vol. 1, Wiley-VCH, Weinheim 2006, p. 41–66.
195 V. Kochergin, M. Kearney: "Existing biorefinery operations that benefit from fractal-based process intensification", *Applied Biochemistry and Biotechnology* **129-132** (2006) 349–360.
196 B. Dean, T. Dodge, F. Valle, G. Chotani: "Development of Biorefineries- Technical and Economical Consideration" in: B. Kamm et al. (eds.): *Biorefineries – Industrial Processes and Products, Status Quo and Future Directions*, vol. 1, Wiley-VCH, Weinheim 2006, pp. 67–83.

Further Reading

J. H. Clark, F. E. I. Deswarte: *Introduction to Chemicals From Biomass*, Wiley, Chichester 2008.
A. Demirbas: *Biorefineries. For Biomass Upgrading Facilities.* http://dx.doi.org/10.1007/978-1-84882-721-9.
B. Kamm P. R. Gruber, M. Kamm: *Biorefineries - Industrial Processses and Products*, Wiley-VCH, Weinheim 2006.
W. Soetaert E. J. Vandamme: *Biofuels*, Wiley, Chichester 2009.

Carbohydrates: Occurrence, Structures and Chemistry

FRIEDER W. LICHTENTHALER, Clemens-Schöpf-Institut für Organische Chemie und Biochemie, Technische Universität Darmstadt, Darmstadt, Germany

1.	Introduction............................	59
2.	Monosaccharides......................	60
2.1.	Structure and Configuration	60
2.2.	Ring Forms of Sugars: Cyclic Hemiacetals	61
2.3.	Conformation of Pyranoses and Furanoses	62
2.4.	Structural Variations of Monosaccharides	64
3.	Oligosaccharides.....................	65
3.1.	Common Disaccharides...............	65
3.1.1.	Sucrose	65
3.1.2.	α,α-Trehalose and Raffinose	66
3.1.3.	Lactose, Cellobiose, and Maltose	66
3.1.4.	Isomaltulose and Lactulose	67
3.1.5.	Other Heterooligosaccharides..........	67
3.2.	Cyclodextrins	68
4.	Polysaccharides......................	68
4.1.	Cellulose	69
4.2.	Chitin	70
4.3.	Starches	70
4.3.1.	Amylose............................	70
4.3.2.	Amylopectin	71
4.4.	Dextrans	71
4.5.	Inulin	72
4.6.	Other Polysaccharides	72
5.	Nomenclature	73
6.	General Reactions	73
6.1.	Hydrolysis	73
6.2.	Dehydration.........................	74
6.3.	Isomerization........................	75
6.4.	Decomposition	75
7.	Reactions at the Carbonyl Group	75
7.1.	Glycosides	75
7.2.	Thioacetals and Thioglycosides	76
7.3.	Glycosylamines, Hydrazones, and Osazones	77
7.4.	Chain Extension.....................	77
7.5.	Chain Degradation...................	78
7.6.	Reductions to Alditols	79
7.6.1.	D-Glucitol	79
7.6.2.	D-Mannitol	79
7.6.3.	Other Sugar Alcohols	80
7.7.	Oxidation	80
8.	Reactions at the Hydroxyl Groups.......	81
8.1.	Ethers	81
8.2.	Esters of Inorganic Acids..............	81
8.3.	Esters of Organic Acids...............	82
8.4.	Acylated Glycosyl Halides	83
8.5.	Acetals..............................	83
	References	84

1. Introduction

Terrestrial biomass constitutes a multifaceted conglomeration of low and high molecular mass products, exemplified by sugars, hydroxy and amino acids, lipids, and biopolymers such as cellulose, hemicelluloses, chitin, starch, lignin and proteins. By far the most abundant group of these organic products and materials, in fact about two thirds of the annually renewable biomass, are carbohydrates, i.e., a single class of natural products. As the term 'carbohydrate' (German '*Kohlenhydrate*'; French '*hydrates de carbone*') implies, they were originally considered to consist solely of carbon and water in a 1:1 ratio, in recognition of the fact that the empirical composition of monosaccharides can be expressed as $C_n(H_2O)_n$. Today, however, the term is used generically in a much wider sense, not only comprising polysaccharides, oligosaccharides, and monosaccharides, but substances derived thereof by reduction of the carbonyl group (alditols), by oxidation of one or more terminal groups to carboxylic acids, or by replacement of one or more hydroxyl group(s) by a hydrogen atom, an amino group, a thiol group, or similar heteroatomic groups. A similarly broad meaning applies to the word 'sugar', which is often used as a synonym for 'monosaccharide', but may also be applied to simple compounds containing more than one monosaccharide unit. Indeed, in everyday usage 'sugar' signifies table sugar, which is sucrose (German '*Saccharose*'; French '*sucrose*' or '*saccharose*'), a disaccharide composed of the two monosaccharides D-glucose and D-fructose.

Carbohydrates appear at an early stage in the conversion of carbon dioxide into organic compounds by plants, which build up carbohydrates from carbon dioxide and water by photosynthesis. Animals have no way of synthesizing carbohydrates from carbon dioxide and rely on plants for their supply. The carbohydrates are then converted into other organic materials by a variety of biosynthetic pathways.

Carbohydrates serve as sources (sugars) and stores of energy (starch and glycogen); they also form a major portion of the supporting tissue of plants (cellulose) and of some animals (chitin in crustacea and insects); they play a basic role as part of the nucleic acids DNA and RNA. Other carbohydrates are found as components of a variety of natural products, such as antibiotics, bacterial cell walls, blood group substances, glycolipids, and glycoproteins, the latter, due to their multifaceted carbohydrate-based recognition phenomena, forming the basis of glycobiology.

At the turn of the millenium, a large collection of books on carbohydrate chemistry and biochemistry have appeared, ranging from comparatively brief introductions [1–3] to more elaborate monographs [4–7] and *multivolume* comprehensive treatises [8, 9]. They are recommended as more profound sources of information.

2. Monosaccharides

The generic term 'monosaccharide' denotes a single sugar unit without glycosidic connection to other such units. Chemically, monosaccharides are either polyhydroxyaldehydes or *aldoses* (e.g., glucose) or polyhydroxyketones or *ketoses* (e.g., fructose), the ending 'ose' being the suffix to denote a sugar. Monosaccharides are classified according to the number of carbon atoms they contain, i.e., hexoses and ketohexoses (or hexuloses) of the general formula $C_6H_{12}O_6$ or pentoses and pentuloses ($C_5H_{10}O_5$). Subdivisions are made according to functional groups which may also be present, for example, aminohexoses ($C_6H_{13}O_5N$), deoxyhexoses ($C_6H_{12}O_5$), and hexuronic acids ($C_6H_{10}O_7$). Monosaccharides with fewer (trioses, tetroses) or more carbon atoms (heptoses, octoses, etc.) are rare.

A large variety of monosaccharides occur in nature [10], the most common being D-*Glucose* (→ Glucose and Glucose-Containing Syrups, Chap. 1.) [50-99-7], also known as dextrose, blood sugar, or grape sugar ('*Traubenzucker*' in German), is a pentahydroxyhexanal, hence belonging to the class of aldohexoses (see Section 2.1). Glucose can be considered the parent compound of the monosaccharide family, because it is not only the most abundant monosaccharide in nature but also the one most extensively studied. It occurs as such in many fruits and plants, in concentrations of 0.08 – 0.1 % in human blood, and constitutes the basic building unit of starch, cellulose, and glycogen. Other ubiquitous aldohexoses are D-*mannose* [3458-28-4], occurring naturally mainly in polysaccharides ('mannans', e.g., from ivory nut) and D-*galactose* [59-23-4], a frequent constituent of oligosaccharides, notably lactose and raffinose, and of primary cell wall polysaccharides (pectins, galactans, arabinogalactans). A corresponding isomeric 2-ketohexose is D-*fructose* [57-48-7] (→ Fructose, Section 2.1.), the sweetest natural sugar, which occurs in many fruits and in honey, and, glycosidically linked, in sucrose and the polysaccharide inulin (→ Inulin) a reserve carbohydrate for many plants (chicory, Jerusalem artichoke). Other important natural sugars are the aldopentose D-*ribose* [50-69-1], which constitutes a building block of the ribonucleic acids, L-*arabinose*, widely distributed in bacterial polysaccharides, gums and pectic materials, and D-*xylose* [58-86-6], of widespread occurrence in pentosans ('xylans') that accumulate as agricultural wastes (cottonseed hulls, corn cobs).

2.1. Structure and Configuration

D-Glucose, the most abundant monosaccharide, has the molecular formula $C_6H_{12}O_6$ as shown by elemental analysis and molecular mass determination. As evidenced from ensuing reactions (see below) this is consistent with a six-carbon, straight-chain pentahydroxyaldehyde of the following structural formula, an aldohexose in carbohydrate notation (Fig. 1).

This structure contains four asymmetric centers, thus $2^4 = 16$ stereoisomers exist, which can

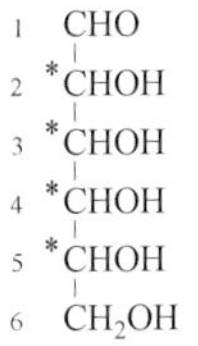

```
1  CHO
2 *CHOH
3 *CHOH
4 *CHOH
5 *CHOH
6  CH₂OH
```

Figure 1. Structural formula of aldohexoses, of which [due to the four chiral centers (marked by *)] 16 stereoisomers are possible

be grouped into eight pairs of enantiomers, and classified as D- and L-sugars. In the D-sugars, the highest numbered asymmetric hydroxyl group (C-5 in glucose) has the same configuration as the asymmetric center in D-glyceraldehyde and, likewise, all L-sugars are configurationally derived from L-glyceraldehyde. A convenient way to show configurational relationships was introduced by EMIL FISCHER in 1891 [11, 12], now termed Fischer projection formula (Fig. 2), as it – literally – projects tetrahedral space relationships into a plane. The resulting formulas are simple to write and easy to visualize, yet they require the setting up of conventions: The carbon chain of a sugar is oriented vertically and to the rear with the aldehyde group at the top; hydrogen atoms and hydroxyl groups at the asymmetric carbon atoms stand out in front. The resulting three-dimensional model is then imagined to be flattened and the groups are laid on the plane of the paper. If the lower-most asymmetric center (C-5 in glucose) has the hydroxyl group to the right, it is considered to have the D-configuration. FISCHER'S decision to place the hydroxyl group of natural glucose to the right, hence D-glucose, was purely arbitrary, yet proved to be a fortunate one, since much later, in 1951, it was proven by special X-ray structural analysis [13] that he had made the right choice.

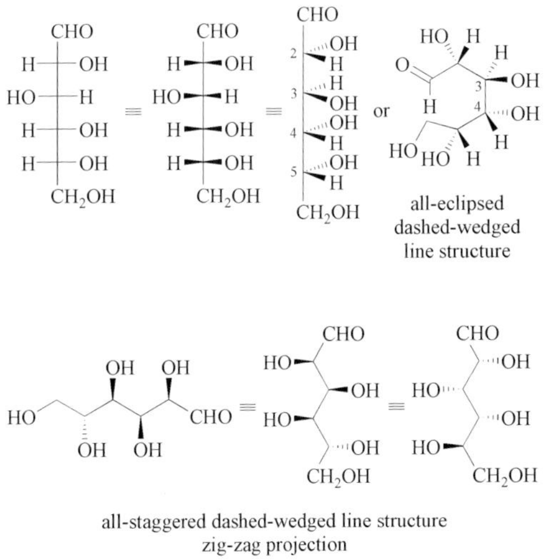

The D-aldose family tree is shown in Figure 3, comprising five of the most important monosaccharides, the aldopentoses D-ribose and D-xylose, and the hexoses D-glucose, D-mannose, and D-galactose, each having the hydroxyl group at the highest-numbered stereocenter (at the bottom) pointing to the right. Likewise, all L-aldoses are configurationally derived from L-glyceraldehyde, entailing a family tree with the lowest hydroxyl group to the left; the respective projection formulas being, in essence, mirror images to those in Figure 3.

A similar system is used to build up the series of ketohexoses or hexuloses, i.e., monosaccharides with a keto group at C-2, which therefore contain one asymmetric carbon atom less (Fig. 4).

2.2. Ring Forms of Sugars: Cyclic Hemiacetals

In the solid state and in solution monosaccharides exist in a cyclic hemiacetal form, ring closure corresponding to reaction between the aldehyde group and either the C-4-OH or C-5-OH. Cyclization involving O-4 results in a five-membered ring structurally related to furan and therefore designated as a *furanose*, whilst hemiacetal formation with O-5 gives rise to an essentially strain-free, hence sterically more favored, six-membered ring, a derivative of pyran, hence termed a *pyranose*. Either ring formation generates a new asymmetric carbon atom at C-1, the anomeric center, thereby giving rise to diastereomeric hemiacetals which are called and labeled α and β. For visualization of the cyclic hemiacetal forms of sugars, HAWORTH, in 1928 [14], introduced his projection formula, in which the rings are derived from the open-chain form and drawn as lying perpendicular to the paper with the ring oxygen away from the viewer. To facilitate this

Figure 2. Configurational representations of the linear (acyclic) form of D-glucose: Traditional Fischer projection formula (top left) and its transformation into the more realistic dashed-wedged line depictions with the six-carbon chain in zigzag arrangement.

```
                        CHO
                        ├─OH
                        CH₂OH
                    D-Glyceraldehyde
```

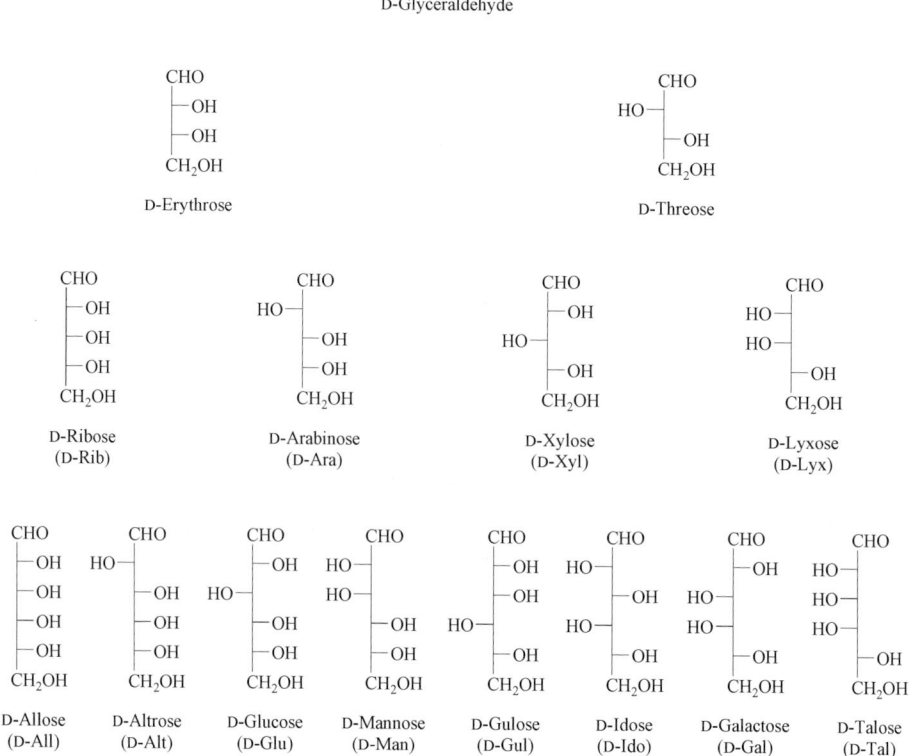

```
         CHO                                    CHO
         ├─OH                              HO─┤
         ├─OH                                  ├─OH
         CH₂OH                                 CH₂OH
       D-Erythrose                          D-Threose
```

```
   CHO           CHO            CHO            CHO
   ├─OH     HO─┤              ├─OH         HO─┤
   ├─OH         ├─OH      HO─┤             HO─┤
   ├─OH         ├─OH          ├─OH             ├─OH
   CH₂OH        CH₂OH         CH₂OH            CH₂OH
  D-Ribose    D-Arabinose    D-Xylose        D-Lyxose
  (D-Rib)     (D-Ara)        (D-Xyl)         (D-Lyx)
```

```
 CHO      CHO       CHO      CHO       CHO      CHO       CHO      CHO
├─OH    HO─┤       ├─OH    HO─┤       ├─OH    HO─┤       ├─OH    HO─┤
├─OH      ├─OH   HO─┤      HO─┤       ├─OH    HO─┤       ├─OH    HO─┤
├─OH      ├─OH     ├─OH      ├─OH   HO─┤      HO─┤     HO─┤      HO─┤
├─OH      ├─OH     ├─OH      ├─OH     ├─OH      ├─OH     ├─OH      ├─OH
CH₂OH    CH₂OH     CH₂OH     CH₂OH    CH₂OH     CH₂OH    CH₂OH     CH₂OH
D-Allose D-Altrose D-Glucose D-Mannose D-Gulose D-Idose  D-Galactose D-Talose
(D-All)  (D-Alt)   (D-Glu)   (D-Man)   (D-Gul)  (D-Ido)  (D-Gal)    (D-Tal)
```

Figure 3. The D-aldose family tree (up to aldohexoses) in their acyclic forms: Common names and Fischer projection formulas, with secondary hydrogen atoms omitted for clarity

mode of viewing, the front part is usually accentuated by wedges as shown in Figure 5 for the β-anomers of pyranose and furanose forms of D-glucose. The projections devised by MILLS in 1954 [15], corresponding to those customary for terpenes and steroids, are also very useful for revealing the stereochemistry of sugars in their cyclic hemiacetal forms: the ring is placed in the plane of the paper with solid or broken wedge-shaped lines to show the orientation of substituents (i.e., OH and CH₂OH groups).

2.3. Conformation of Pyranoses and Furanoses

The concepts of conformation are fundamental to a proper understanding of the structure–property relationships of carbohydrates, most notably of the regio- and stereoselectivities of their reactions. The conformational analysis of monosaccharides is based on the assumption that the geometry of the pyranose ring is essentially the same as that of cyclohexane and, analogously, that of furanoses the same as that of cyclopentane – a realistic view, since a ring oxygen causes only a slight change in molecular geometry. Hence, the rhombus-shaped Haworth formulas which imply a planar ring, and the equally flat dashed-wedged line configurational depictions by Mills (Fig. 5) are inadequate to represent the actual three-dimensional shape of the rings and the steric orientation of the ring substituents (OH and CH₂OH groups). For the six-membered pyranose ring a number of recognized conformers exist [16]: two chairs

CH$_2$OH
|=O
CH$_2$OH

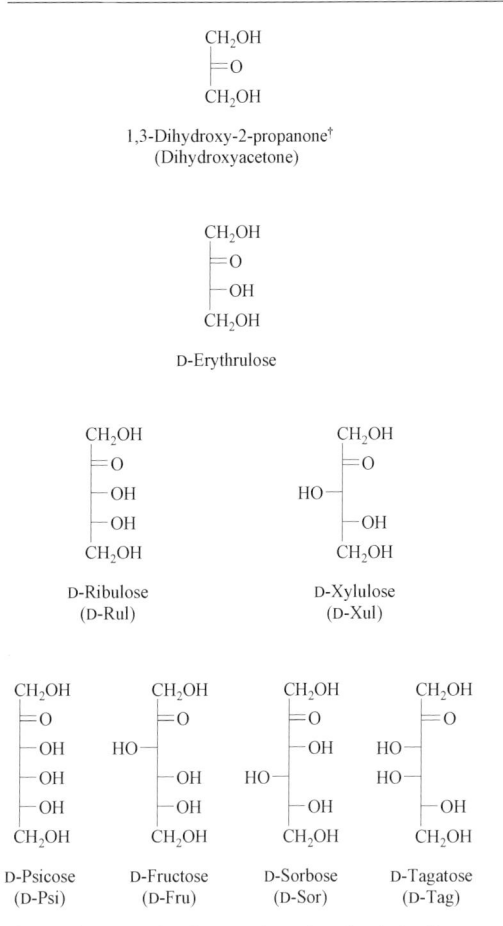

1,3-Dihydroxy-2-propanone†
(Dihydroxyacetone)

CH$_2$OH
|=O
|—OH
CH$_2$OH

D-Erythrulose

CH$_2$OH   CH$_2$OH
|=O |=O
|—OH   HO—|
|—OH |—OH
CH$_2$OH   CH$_2$OH

D-Ribulose   D-Xylulose
(D-Rul)   (D-Xul)

CH$_2$OH   CH$_2$OH   CH$_2$OH   CH$_2$OH
|=O |=O |=O |=O
|—OH   HO—| |—OH   HO—|
|—OH |—OH   HO—| HO—|
|—OH |—OH |—OH |—OH
CH$_2$OH   CH$_2$OH   CH$_2$OH   CH$_2$OH

D-Psicose   D-Fructose   D-Sorbose   D-Tagatose
(D-Psi)   (D-Fru)   (D-Sor)   (D-Tag)

Figure 4. The D-ketohexose (or D-hexulose) family tree: Trivial names, systematic designation (in brackets) and Fischer projection formulas
† Not regarded as being a sugar, due to absence of an asymmetric carbon atom.

β-D-Glucopyranose   β-D-Glucofuranose
β-D-Glc$p$ β-D-Glc$f$

Haworth
projection

Mills
projection

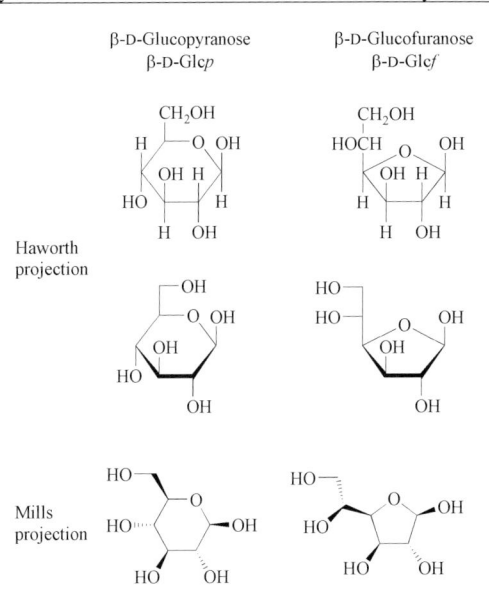

Figure 5. Haworth and Mills projection formulas for the β-anomers of D-glucopyranose and D-glucofuranose (in the formula at the center and at the bottom, the carbon and C-hydrogen atoms are omitted for clarity)

glucose; the other seven D-aldohexoses contain one or more axial substituents.

The hexulose counterpart to the conformational forms of D-glucose is the D-fructose isomerization scheme depicted in Figure 8. Whilst the crystalline product is the β-D-fructopyranose in the $^2C_5$ chair conformation as evidenced by X-ray analysis [17], on dissolution in water, equilibration is essentially instantaneous

($^1C_4$, $^4C_1$), six boats (e.g., $^{1,4}B$ and $B_{1,4}$ in Fig. 6), six skews and twelve half-chairs (e.g., $^OS_2$ and $^5H_4$ forms).

Although there are exceptions, most aldohexoses adopt the chair conformation that places the bulky hydroxymethyl group at the C-5 terminus in the equatorial position. Hence, β-D-hexopyranosides are predominantly in the $^4C_1$ chair conformation, since each of the alternative forms outlined in Figure 6, most notably the $^1C_4$ chair, are energetically less favored. For glucose, this preference means that, in the α-form, four of the five substituents are equatorial, and one is forced to lie axial; in the β-form, all substituents are equatorial (Fig. 7). This situation is unique for

$^4C_1$ $^{1,4}B$ $^2S_0$

$^1C_4$ $B_{1,4}$ $^0H_5$

Figure 6. Conformational forms of pyranose rings: chair (C), boat (B), skew (S) and half-chair (H).
To designate each form, the ring atom numeral lying above the plane of reference appears as a superscript preceding the letter, those below the plane are written as subscripts and follow the letter

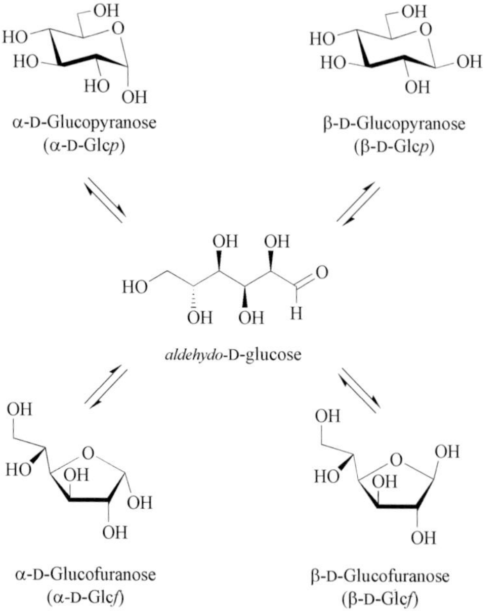

Figure 7. Cyclic hemiacetal forms of D-glucose in configurational representation. In solution, these forms rapidly interconvert through the energetically unfavorable acyclic form; in water at 25 °C the two pyranoid forms are nearly exclusively adopted, the equilibrium mixture amounting to 62 % of the β-*p* and 38 % of the α-*p* anomers. From water, D-glucose crystallizes in the α-pyranose form
The six-membered (pyranose) ring is denoted by the symbol *p* after the three-letter symbol for the monosaccharide (for example, Glc*p*), the five-membered (furanose) ring correspondingly is signated by an *f* (e.g., Glc*f*)

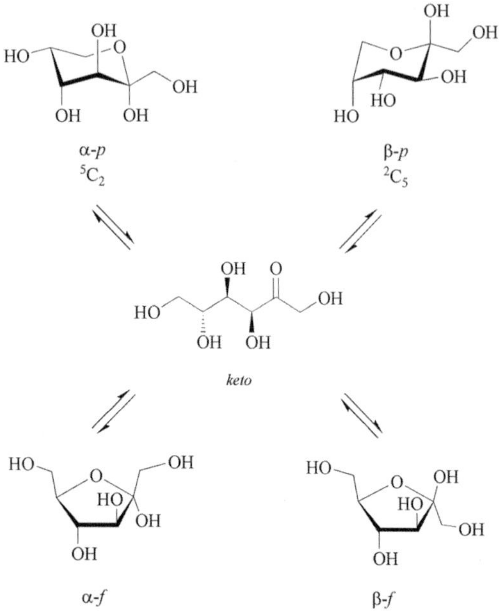

Figure 8. Forms of D-fructose in solution. In water, the major conformers are the β-pyranose (β-*p*, 73 % at 25 °C) and β-furanose (β-*f*, 20 %) forms [18]. On crystallization from water, D-fructose adopts the $^2C_5$ chair conformation in the crystal lattice as evidenced by X-ray analysis [17]

to yield a mixture mainly containing the β-*p*-form (73 % at 25 °C, the only sweet one in fact), together with the β-*f*- (20 %), α-*f*- (5 %) and α-*p*-forms (2 %) [18]. The acyclic form through which equilibration occurs is present only to a minute extent.

The principal conformations of the furanose ring are the *envelope* (*E*) – one atom lying above or below a plane formed by the other four ring atoms – or the *twist* (*T*) arrangement, in which three ring atoms are in a plane and the other two above and below, respectively. As energy differences between the various *E* and *T* conformations are small, the form actually adopted depends on the type of ring substitution (hexoses, hexuloses, pentoses), their configuration, their solvation and the type of intra- or intermolecular hydrogen bonding present. Accordingly, the exact conformation of an individual furanose is usually not known – except for the crystalline state when an X-ray structural analysis is available. Thus, the planar Haworth and Mills projection formulas are the preferred way of drawing furanose forms (Fig. 9).

2.4. Structural Variations of Monosaccharides

Sugars may possess functionalities other than hydroxyl groups. Amino sugars are aldoses, which have a hydroxyl group replaced by an amino functionality, e.g., D-*glucosamine* (2-amino-2-deoxy-D-glucose), which is one of the most abundant sugars. In its *N*-acetylated form (*N*-acetyl-D-glucosamine), it is a constituent of the polysaccharide chitin (→ Chitin and Chitosan), that forms the hard shells of crustaceans and other anthropods, but also appears in mammalian glycoproteins and links the sugar chain to the protein. Monosaccharides lacking a hydroxyl group at the terminal C-6, i.e., 6-deoxy-sugars, are likewise of wide occurrence, for example, L-*rhamnose* (6-deoxy-D-mannose) is found in

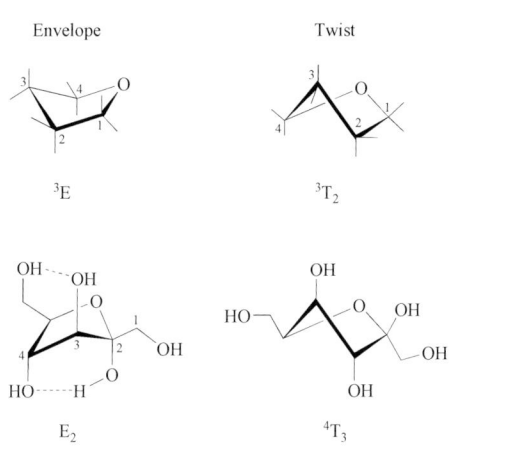

Figure 9. The envelope conformation (top left) is the ^3E form as defined by the C-3 atom lying above the plane formed by the other ring atoms. The defined plane for the twist form (top right) is the triangle given by C-1, C-4, and O-4, entailing the conformational description $^3T_2$. In aprotic solvents (dimethyl sulfoxide) D-fructose populates the E_2 envelope conformation to a substantial extent [18], whilst in crystalline sucrose, the β-D-fructofuranose portion adopts the $^4T_3$ twist form [19, 20] (bottom entries)

plant and bacterial polysaccharides whereas L-*fucose* (6-deoxy-D-galactose) is present in combined form in animals, plants, and microorganisms. 2-Deoxy-D-*erythro*-pentose (2-deoxy-D-ribose) is the exceedingly important sugar component of DNA, various mono-, di- and trideoxy sugars are constituents of many antibiotics, bacterial polysaccharides, and cardiac glycosides.

The uronic acids are aldoses that contain a carboxylic acid chain terminating function, and occur in nature as important constituents of many polysaccharides. The D-*gluco* compound, D-*glucuronic acid*, was first isolated from urine (hence the name), in which it occurs in the form of glycosides and glycosyl esters of toxic substances that the body detoxifies in this way.

N-Acetyl-D-glucosamine
2-acetamido-2-deoxy-D-glucopyranose (D-GlcNAc*p*)

2-Deoxy-D-ribose
2-deoxy-D-*erythro*-pentafuranose (D-dRib*f*)

Branched-chain sugars, i.e., saccharides with a nonlinear carbon chain, are comparatively uncommon, the more widely occurring being D-*apiose* (3-*C*-hydroxymethyl-D-*glycero*-tetrose), abundant in polysaccharides of parsley and duckweed [21], and D-*hamamelose* (2-*C*-hydroxymethyl-D-ribose), a component of the bark of witchhazel [22].

3. Oligosaccharides

Oligosaccharides are compounds in which monosaccharide units are joined by glycosidic linkages, i.e., simple polymers containing between two and ten monosaccharide residues. Accordingly, there are disaccharides – a disaccharide composed of two hexopyranoses can have 5120 distinguishable isomeric forms – trisaccharides, tetrasaccharides, etc. They may be further subdivided into homo- (consisting of only one type of sugar) and hetero-oligosaccharides, and into those that are reducing (presence of a free hemiacetal group) or nonreducing. A comprehensive listing of the di-, tri-, and higher oligosaccharides known up to 1990 is available [23].

3.1. Common Disaccharides

3.1.1. Sucrose

Sucrose, affectionately called "the royal carbohydrate" [24], is a nonreducing disaccharide

3.1.2. α,α-Trehalose and Raffinose

α,α-Trehalose, a nonreducing D-glucosyl D-glucoside, occurs extensively in the lower species of the plant kingdom (fungi, young mushrooms, yeasts, lichens, and algae). In bakers' yeast it accounts for as much as 15 % of the dry mass, in the metabolic cycle of insects it circulates like glucose does in the mammalian cycle. Similarly nonreducing, due to being a galactosylated sucrose, is the trisaccharide *raffinose*, distributed almost as widely in the plant kingdom as sucrose, yet in lower concentration (e.g., less than 0.05 % in sugar beet).

Sucrose [57-50-1]
β-D-Fructofuranosyl α-D-glucopyranoside
β-D-Fru*f*-(2↔1)-α-D-Glc*p*

α,α-Trehalose [99-20-7]
α-D-glucopyranosyl
α-D-glucopyranoside
(α-D-Glc*p*[1↔1]α-D-Glc*p*)

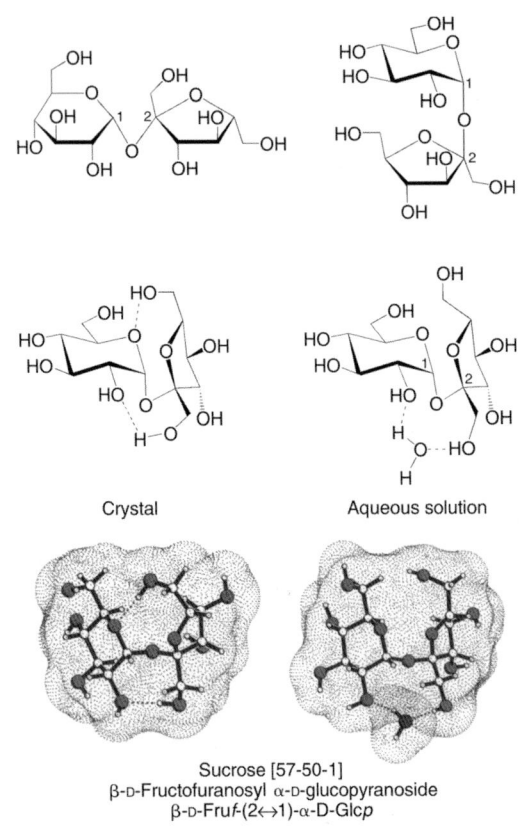

Raffinose [512-69-6]
α-D-galactopyranosyl-(1→6)-sucrose
(α-D-Gal*p*-(1→6)-α-D-Glc*p*(1↔2)-β-D-Fru*f*)

Figure 10. Common structural representations of sucrose (top entries), the molecular geometry realized in the crystal featuring two intramolecular hydrogen bonds between the glucose and fructose portion [19, 20] (bottom left), and the sterically similar disposition of the two sugar units towards each other in aqueous solution form, caused by hydrogen bonding through a 'water bridge' [27]. The bottom entries show the solvent-accessible surfaces (dotted areas) of the crystal form (left) and the form adopted in water [27] (right), clearly demonstrating that sucrose has an unusually compact overall shape, more so than any other disaccharide

because its component sugars, D-glucose and D-fructose, are glycosidically linked through their anomeric carbon atoms: Sucrose is a-β-D-fructofuranosyl α-D-glucopyranoside (see Fig. 10). It is widely distributed throughout the plant kingdom, is the main carbohydrate reserve and energy source and an indispensable dietary material for humans (→ Sugar). For centuries, sucrose has been the world's most plentiful produced organic compound of low molecular mass, the present (2008) annual production from sugarcane and sugar beet being an impressive 169×10^6 t [25]. Its chemistry is fairly well developed [26].

3.1.3. Lactose, Cellobiose, and Maltose

There are only very few naturally occurring oligosaccharides with a free anomeric hydroxyl group, which therefore possess reducing properties. The most important example is *lactose* (milk sugar, → Lactose and Derivatives), an ingredient of the milk of mammals (up to 5 % in cows). As it is produced on an industrial scale, from whey, it represents the only large-scale available sugar derived from animal rather than plant sources. Uses include human food, pharmaceuticals, and animal feeds. The reducing gluco-disaccharides *cellobiose* and *maltose* (malt sugar) are chemical or enzymatic hydrolysis products of the polysaccharides cellulose and starch, respectively, and, hence are not regarded as native oligosaccharides.

Lactose [63-42-3]
β-D-galactopyranosyl-(1→4)-D-glucopyranose
(β-D-Gal*p*-[1→4]-D-Glc*p*)

Cellobiose [528-50-7]
β-D-glucopyranosyl-(1→4)-D-glucopyranose
(β-D-Glc*p*-(1→4)-D-Glc*p*)

Maltose [69-79-4]
α-D-glucopyranosyl-(1→4)-D-glucopyranose
(α-D-Glc*p*-(1→4)-D-Glc*p*)

3.1.4. Isomaltulose and Lactulose

Isomaltulose (palatinose, → Sugar Alcohols, Section 5.1.) and lactulose are both produced in fairly large amounts from sucrose and lactose, respectively, and constitute 6-*O*-glucosyl- and 4-*O*-galactosyl-fructoses. The sucrose → isomaltulose transformation, industrially realized presently (2009) at an estimated 8×10^4 t/a-scale, is effected by a *Protaminobacter rubrum*-induced glucosyl shift from the anomeric fructosyl oxygen to its O-6 position, taking place in a mostly intramolecular fashion via a closed-shell intermediate [28], whilst the generation of lactulose from lactose, presently running at a 12×10^3 t/a level, comprises a base-promoted C-1 → C-2 carbonyl shift. Most of the isomaltulose produced is subsequently hydrogenated to isomalt (→ Sugar Alcohols), a low-calorie sweetener with the same taste profile as sucrose. Lactulose (→ Lactose and Derivatives) has medical and pharmaceutical applications, mainly for treating intestinal disorders.

Isomaltulose [13718-94-0]
α-D-glucopyranosyl-(1→6)-D-fructofuranose
(α-D-Glc*p*-(1→6)-D-Fru*f*)

Lactulose [4618-18-2]
β-D-galactopyranosyl-(1→4)-D-fructopyranose
(D-Gal*p*-β(1→4)-D-Fru*f*)

3.1.5. Other Hererooligosaccharides

Heterooligosaccharides of considerably higher complexity occur in large variety in plants, animals and microorganisms where they are covalently bound to proteins ('glycoproteins') and lipids ('glycolipids') or other hydrophobic entities, and, as such, are implicated in a range of key biological processes: cell-cell recognition, fertilization, embryogenesis, neuronal development, hormone activities, the proliferation of cells and their organization into specific tissues, viral and bacterial infection and tumor cell metastasis [29–31]. Red blood cells, for example, carry carbohydrate antigens which determine blood group types in humans: type A people have the tetrasaccharide in Figure 11 with R = NHAc (i.e., a GalNAc residue) as a key antigen linked by lipid components to the surfaces of red blood cells; in type B blood, the tetrasaccharide determinant is exceedingly similar – formal replacement of NHAc by OH, i.e., GalNAc by Gal – yet on mixing with type A blood leads to clumping and precipitation [32].

All *N*-glycoproteins (*N*-glycans) share the peptide-linked pentasaccharide fragment in Figure 12, consisting of three mannose units in a branched arrangement and two GlcNAc

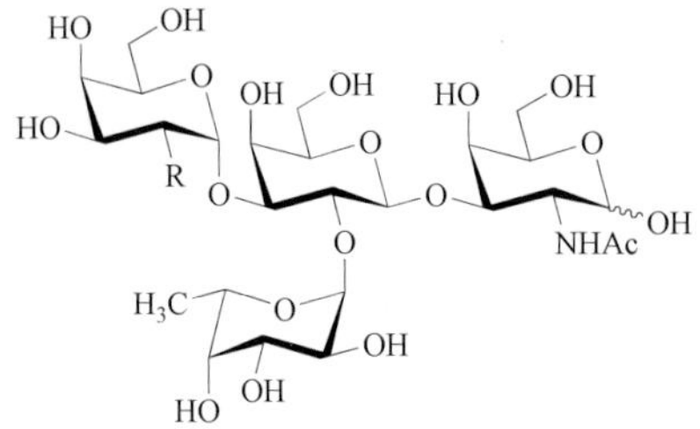

R = NHAc: α-D-GalpNAc-(1→3)-β-D-Galp-(1→3)-D-GalpNAc
                                            2
                                            ↑
                                            1
                                        β-L-Fuc

R = OH: α-D-Galp-(1→3)-β-D-Galp-(1→3)-D-GalpNAc
                                            2
                                            ↑
                                            1
                                        β-L-Fuc

Figure 11. Human blood groups determinants: Differentiation between type A (R = NHAc) and B (R = OH) is effected by relatively simple changes within a branched tetrasaccharide linked to lipid components on the surface of red blood cells

residues, of which the terminal one is N-glycosidically linked to an asparagine moiety of the protein. Branching out from this uniform core region are monosaccharides and oligosaccharide chains of high structural diversity leading to multiple types of branched and unbranched glycoproteins [33].

3.2. Cyclodextrins (→ Cyclodextrins)

Although discovered more than 100 years ago, the cyclic glucooligosaccharides termed cyclodextrins (based on dextrose, which is an old name for glucose) remained laboratory curiosities until the 1970s when they began to be used commercially [34]. Their large-scale production is based upon the degradation of starch by enzymes elaborated by *Bacillus macerans* ('CGTases'), involving excision and reconnection of single turns from the helical α-(1→4)-glucan (amylose) chain (cf. Fig. 13) to provide cyclic α-(1→4)-linked glucooligosaccharides with six, seven and eight glucose units. They are named α-, β- and γ-cyclodextrin, respectively.

Cyclodextrins are truncated cones with well-defined cavities. All the secondary hydroxyl groups are located at the wider rim of the cone leaving the primary CH_2OH groups to protrude from the narrower opening. The respective cavities, as exemplified by that of α-cyclodextrin with its six glucose units (Fig. 14 [35, 36]), are distinctly hydrophobic in character, and show an amazing propensity to form stable complexes with a large variety of equally hydrophobic, sterically fitting guest molecules by incorporating them into their cavities [34, 35], thereby changing the physical and chemical properties of the included guest. The features and properties of the resulting cyclodextrin inclusion compounds has led to the exploitation of cyclodextrins for a wide variety of purposes: as drug carriers [37, 38], as stationary phases for the separation of enantiomers [39, 40], as building blocks for supramolecular structures [41], and as enzyme models [42].

4. Polysaccharides

The bulk of the annually renewable carbohydrate-biomass are polysaccharides (glycans), such as cellulose, hemicelluloses, chitin, starch, and inulin. Invariably composed of

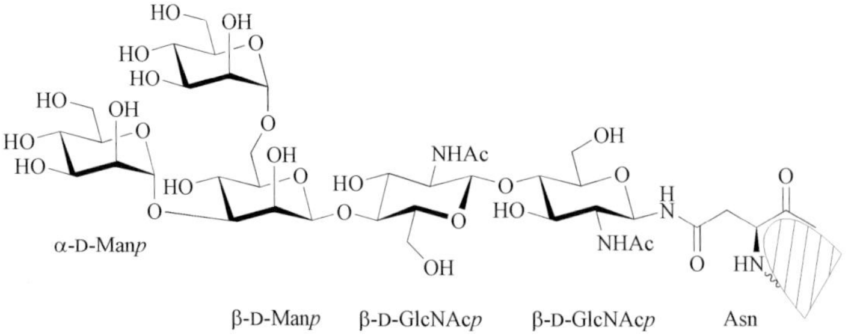

Figure 12. Central core region common to all N-glycoproteins is a pentasaccharide, N-glycosidically linked to the carbamido nitrogen of an asparagine moiety (Asn) within the peptide chain

monosaccharide units, they have high molecular masses and, hence, differ significantly in their physical properties. The majority of naturally occurring polysaccharides contain 80 – 100 units, although a few are made up of considerably more.

4.1. Cellulose (→ Cellulose)

Cellulose [*9004-34-6*] is an unbranched glucan composed of β-(1→4)-linked D-glucopyranosyl units (see Fig. 15) with an average molecular mass equivalent to about 5000 units. It is the most abundant organic material found in the plant kingdom, forming the principal constituent of the cell walls of higher plants and providing them with their structural strength. Cotton wool is almost pure cellulose, but in wood, the other chief source of the polymer, cellulose is found in close association with other polysaccharides (mainly hemicelluloses) and lignin. X-ray analysis and electron microscopy indicate that these long chains lie side by side in bundles, held together by a multiplicity of hydrogen bonds between the numerous neighboring OH groups. These bundles are twisted together to form rope-like structures, which themselves are grouped to

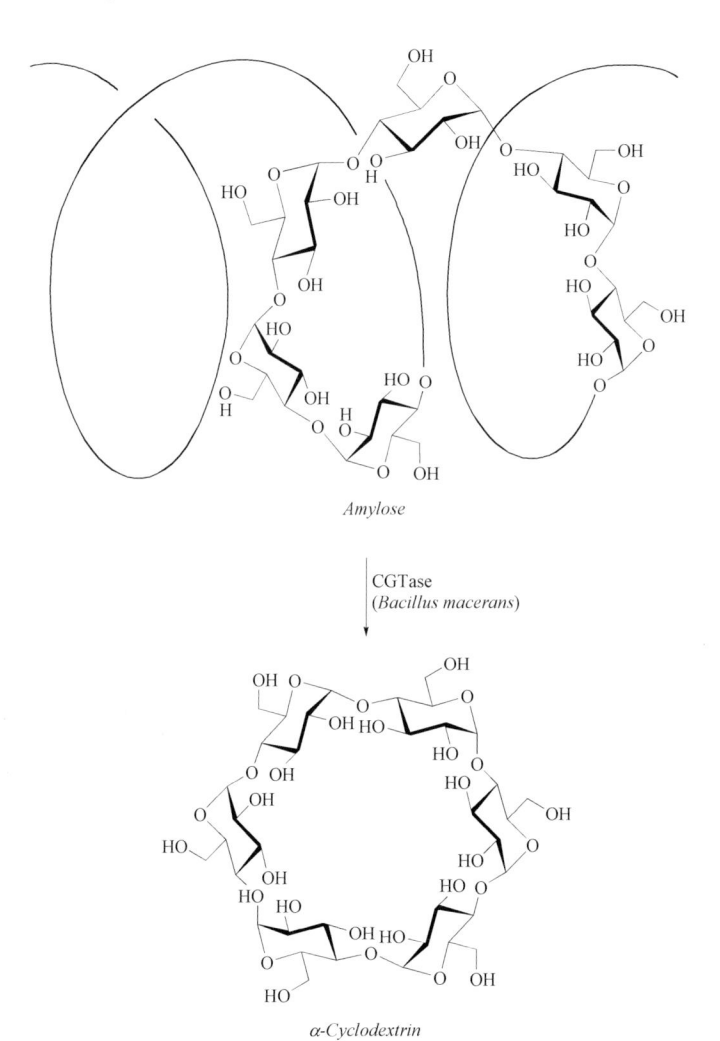

Figure 13. Sketch representation of a left-handed, single-stranded helix of V_H-amylose (top), and of α-cyclodextrin (bottom), which de facto represents a single turn of the amylose helix excised and re-connected by *Bacillus macerans*-derived enzymes (CGTases). The close analogy allows V_H-amylose to be considered as a tubular analogue of α-cyclodextrin.

pyranosyl residues (cf. Fig. 15), of which about one out of every six is not acetylated. Chitin is the major organic component of the exoskeleton (shells) of insects, crabs, lobsters, etc. and, hence, an abundant byproduct of the fishing industries.

Chitosan, a related water-soluble polysaccharide in which the vast majority of residues is not acetylated (i.e., a β-(1→4)-linked chain of 2-amino-2-deoxy-D-glucose residues), can be obtained from chitin by deacetylation in concentrated sodium hydroxide solution.

4.3. Starches (→ Starch)

The principal food-reserve polysaccharides in the plant kingdom are starches. They form the major source of carbohydrates in the human diet and are therefore of great economic importance, being isolated on an industrial scale from many sources. The two components, amylose and amylopectin, vary in relative amounts among the different sources; from less than 2 % of amylose in waxy maize to about 80 % of amylose in amylomaize (both corn starches), but the majority of starches contain between 15 and 35 % amylose.

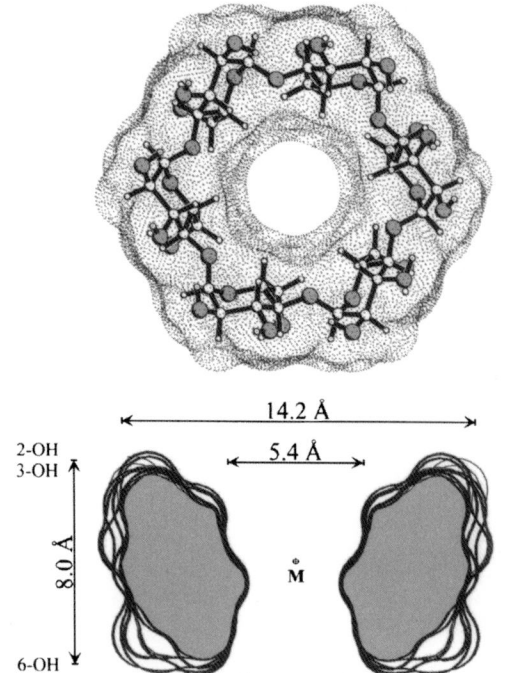

Figure 14. Top: Ball-and-stick model representations of the X-ray-derived solid-state structure of α-cyclodextrin, together with its solvent-accessible surface, shown as a dotted pattern
Bottom: Cross section contour of a plane perpendicular to the macrocycle's mean plane with approximate molecular dimensions [35, 36]

4.3.1. Amylose

Amylose [9005-82-7] is made up of long chains, each containing 100 or more α-(1→4)-linked glucopyranosyl units which, due to the kink in every α-glycosidic linkage, tend to coil to helical segments with six glucose units forming one turn (see sketch in Fig. 13). Amylose is the fraction of starch that gives the intense blue color with iodine; this color arises because iodine molecules become trapped within the hydrophobic channel of the helical segments of the polysaccharide (Fig. 16).

form the fibers that can be seen. In wood (→ Wood, Chap. 1.) these cellulose "ropes" are embedded in lignin to give a structure that has been likened to reinforced concrete.

4.2. Chitin (→ Chitin and Chitosan)

Chitin is a polysaccharide composed of β-(1→4)-linked 2-acetamido-2-deoxy-D-gluco-

Figure 15. Structural representations of segments of cellulose (R = OH), chitin (R = NHAc), and chitosan (R = NH_2)

An illustration of a left-handed, single stranded helix of V_H-amylose forms the top sketch of Figure 16, whereas a more detailed representation of its molecular geometry, based on X-ray diffraction data [43] and calculation of the solvent-accessible contact surfaces [44], is indicated by dots with ball-and-stick models superimposed and presented as the central diagram within the same figure. The channel generated by the helical arrangement of the α-D-glucose residues,

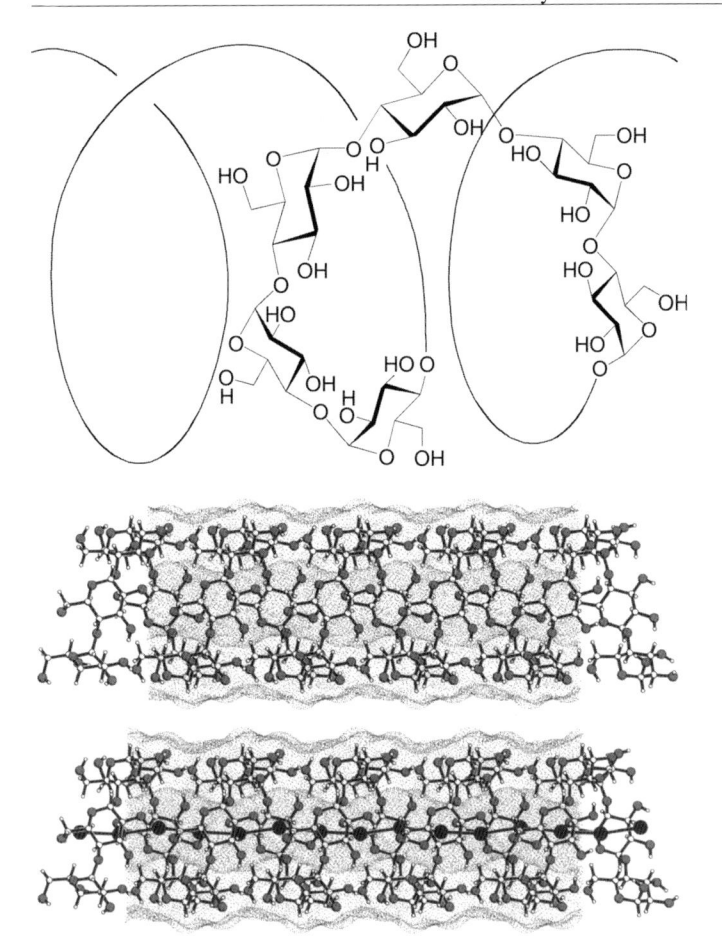

Figure 16. Top: single stranded helix of V_H-amylose; Center: ball-and-stick model of the architecture of the V_H-amylose helical array; Bottom: the complexation of iodine within the central channel of the helical array [43, 44]

of dimensions corresponding to those of the cavity of α-cyclodextrin, is clearly apparent. The outside surface area of V_H-type amylose is uniformly hydrophilic (in conformity with its solubility in water) whereas the central channel is distinctly hydrophobic – making it predestined to incorporate equally hydrophobic guests such as iodine or fatty acids [44]. Thus, in the case of iodine a linear polyiodide chain becomes embedded into this channel (see bottom diagram of Fig. 16), producing an intensely blue-colored starch-iodine complex [44].

4.3.2. Amylopectin

Amylopectin [9037-22-3] is also an α-(1→4)-glucan, yet the molecule is branched via *O*-6 at about every 25 units. The molecular size of amylopectin is of the order of 10^6 D-glucose residues, making it one of the largest naturally occurring molecules. The secondary structure is characterized by several hundred linear chains of about 20 – 25 glucose units each, which are connected in a variety of arrangements to give clusters for which the tassel-on-a-string model (see Fig. 17) is proposed. With iodine, amylopectin produces only a dull-red color, indicating that the short linear chain portions cannot coil effectively to provide the helices required for formation of inclusion complexes.

4.4. Dextrans (→ Dextran)

Dextrans are linear water-soluble α-(1→6)-glucans with only occasional branches via O-2, O-3

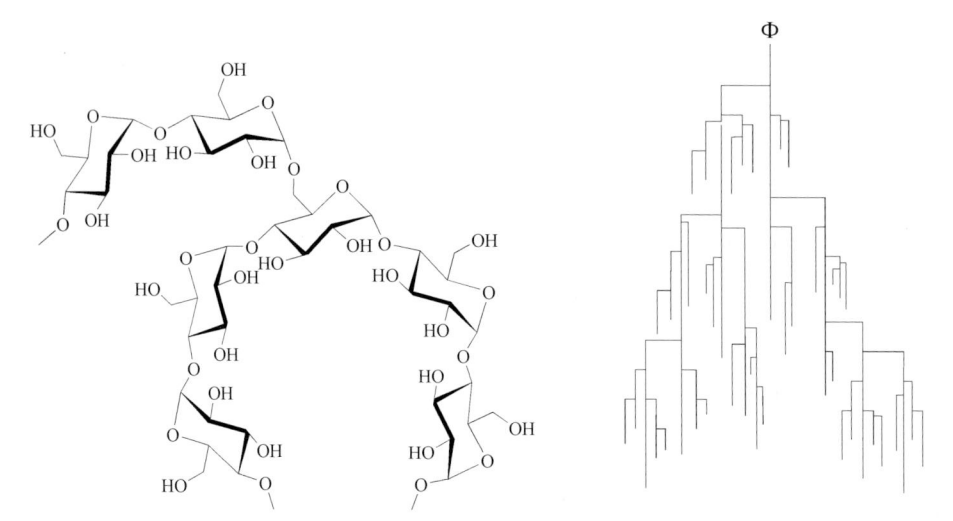

Figure 17. Schematic representation of a section of amylopectin with an α(1→6)-branch of a helical chain of α(1→4)-glucopyranosyl residues (left), with the tassel-on-string model of its higher level structure (φ = reducing end)

or O-4. They are generated from sucrose by a large number of organisms, of which *Leuconostoc mesenteroides* is used to produce the slightly branched commercial dextran, used clinically as a plasma volume expander.

4.5. Inulin (→ Inulin)

Inulin is a polysaccharide composed of β(1→2)-linked D-fructofuranose units with varying chain length of about 15 – 30 units. It is present to the extent of 30 % or more in various plants such as dahlias or Jerusalem artichokes where it replaces starch either partially or completely as the food storage carbohydrate [45, 46]. The structure of inulin is unique in leaving no 'reducing end', as this is glycosidically blocked by an α-D-glucopyranose residue – a sucrose unit in fact (Fig. 18).

Commercial inulins, e.g., those isolated from chicory, have a degree of polymerization far below that found in other polysaccharides, their molecular sizes ranging from around 5 to 30 units [44].

4.6. Other Polysaccharides
(→ Polysaccharides, Section 3.4.1.)

A plethora of other homo- and heteropolysaccharides are found in nature, most notably D-xylans (hemicelluloses with linear chains of β-(1→4)-D-xylopyranosyl units), *pectins* (principal constituent D-galacturonic acid), plant gums (building blocks D-galactose, L-arabinose, L-rhamnose) and various algal and microbial polysaccharides with, in part, unusual sugar units: L-guluronic and D-mannuronic acids in alginates, glucuronic acid and pyruvate acetals in agar, sulfated galactosyl residues in carrageenans, or ribitol phosphates in *teichoic acids*. Excellent accounts on this subject have been given [47, 48].

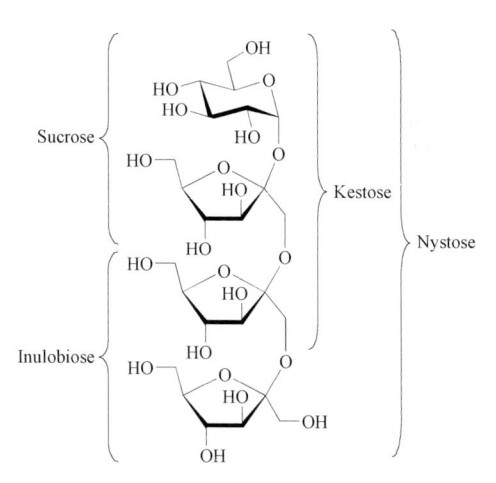

Figure 18. Nystose fragment of inulin, showing sub-fragments corresponding to sucrose, inulobiose and 1-kestose [44]

5. Nomenclature

According to common practice, trivial names are used for monosaccharides and for many naturally occurring oligosaccharides. With the development of carbohydrate chemistry, however, and ever-increasing numbers of newly defined compounds, it has become necessary to introduce a semi-systematic nomenclature which has been approved by the joint commission of IUPAC (International Union of Pure and Applied Chemistry) and IUB (International Union of Biochemistry) [49]. This nomenclature is based on the classical names for monosaccharides which appear, written in italics, as a "configurational prefix". For example, D-*xylo*, L-*arabino* and D-*gluco* refer to the distribution of asymmetric carbon atoms along a carbon chain of any length, designating the configuration of the corresponding monosaccharide.

Monosaccharides with an aldehydic carbonyl or potential aldehydic carbonyl group are called aldoses; those with a ketonic or potential ketonic carbonyl group, ketoses, with the chain length given by the root, such as pentose, hexose, or heptose and pentulose, hexulose, etc. In ketoses the position of the keto group is indicated by the position number. D-Fructose is systematically named D-*arabino*-2-hexulose.

Replacement of a hydroxyl group by hydrogen is indicated by the prefix deoxy, for example, L-rhamnose is a 6-deoxy-L-aldohexose of *manno*-configuration. Replacement of a hydroxyl group by any other substituent is formally regarded as taking place via the deoxysugar. Thus, a sugar with an amino group instead of hydroxyl is called an amino-deoxysugar. Formation of ether groups, most commonly methyl, is indicated by adding '*O*-methyl-' to the front of the name preceded by the number of the carbon atom whose hydroxyl group has been etherified. Esters, for example, acetates are designated by adding either '*O*-acetyl-' before the name or 'acetate' after it, in each case again preceding the descriptor with the appropriate carbon number.

The ring size is indicated by a suffix: pyranose for six-membered rings, furanose for five-membered rings, and pyranulose for six-membered ketose rings. The six-membered cyclic hemiacetal of D-fructose is named D-*arabino*-2-hexopyranosulose. The symbol α or β for the anomeric configuration is always written together with the configurational symbol D or L (α-D, β-D, α-L, β-L).

Names of oligosaccharides are formed by combining the monosaccharide names, usually the trivial names. The nonreducing disaccharide sucrose is a β-D-fructofuranosyl α-D-glucopyranoside. The endings "yl" and "ide" describe the fructose part as the aglycone and the glucose part as the glycone in this "glycoside". It is thus clearly indicated that both sugars are glycosidically linked by their anomeric hydroxyl groups. In reducing oligosaccharides the reducing monosaccharide is the root, and all attached monosaccharide units are named as substituents. The disaccharide lactose is therefore named β-D-galactopyranosyl-(1→4)-D-glucopyranose. Position numbers and arrows indicate a β-configurated glycosidic bond between the anomeric hydroxyl group (carbon atom 1) of D-galactose (glyconic part) and the hydroxyl group at C-4 of D-glucose (aglyconic part).

For a description of more complex oligosaccharides an abbreviation system has come into use – as for oligopeptides and oligonucleotides – that is unambiguous and practical; thus, each monosaccharide is abbreviated by a three-letter symbol, comprising the first three letters of its trivial name, i.e.:

Glucose	Glc	Xylose	Xyl
Fructose	Fru	Arabinose	Ara
Galactose	Gal	Ribose	Rib
Mannose	Man	Deoxyribose	dRib
Fucose	Fuc	Glucosamine	GlcN
Rhamnose	Rha	*N*-Acetylglucosamine	GlcNAc

The anomeric configuration and D- or L- affiliation is written before the three-letter acronym, the ring size (*p* for pyranose, *f* for furanose) is added to the end, followed by the intersaccharidic linkage position in the case of oligosaccharides. Sucrose (see Fig. 10), accordingly, is β-D-Fru*f*-(2→1)-α-D-Glc*p*, lactose β-D-Gal*p*-(1→4)-D-Glc*p*.

6. General Reactions

6.1. Hydrolysis

The hydrolysis of disaccharides such as sucrose and lactose, or polysaccharides like starch and cellulosic materials to their free component sugars is of great importance not only in the food and fermentation industries, but increasingly so in the chemical industry as well toward the generation of bulk chemicals from polysaccharidic waste materials (→ Carbohydrates: Occurence, Structures and Chemistry). This release of the component sugars from di-, oligo- or polysaccharides can be effected by enzymes called glycosidases or chemically by acid treatment.

Enzymatic hydrolysis proceeds with high specificity towards both the sugar and the configuration at the anomeric center. Maltase, an α-D-glucosidase obtainable from barley malt, catalyzes the hydrolysis of α-linked di-, oligo- and polysaccharides (sucrose, maltose, starch, dextrans), whereas the almond emulsin-derived enzyme is a β-glucosidase cleaving only β-linked glycosides.

Acid-induced hydrolysis of glycosides requires comparatively harsh conditions, standard techniques requiring 1 M sulfuric acid at 100 °C for 4 h for hexose-containing polysaccharides and 0.25 M H_2SO_4 at 70 °C for hemicellulose pentosans [50, 51]. The acid hydrolysis of starch, for example, is performed industrially on a 10^6 t/a basis, the resulting D-glucose being used in the liquid form (corn syrup) as a sweetener (→ Glucose and Glucose-Containing Syrups, Chap. 4.).

6.2. Dehydration

Under the rigid acidic conditions required for polysaccharide hydrolysis to the constituent sugars, their partial degradation can usually not be avoided, as strong acid induces elimination of water in various ways. When selecting special conditions though, these dehydrations can be crafted into generating furfural (2-furaldehyde) from the hemicellulose pentosans contained in agricultural and forestry wastes, the pentoses initially formed then being dehydrated. This process is industrially realized on a 2×10^5 t/a level [52] (→ Furan and Derivatives, Chap. 2.), thus furfural is one of the very few biomass-derived large-volume organic chemicals.

Another key biomass-derived chemical of high industrial potential is 5-hydroxymethylfurfural (HMF) readily accessible from fructose or inulin hydrolysates by acid-induced elimination of three moles of water [53]. Developments towards its production from polysaccharides contained in agricultural and forestry waste materials appears to be well advanced [54] (→ Carbohydrates: Occurence, Structures and Chemistry).

When using nonaqueous conditions (e.g., DMSO as the solvent and a strongly acidic resin) the fructose part of disaccharides such as isomaltulose can similarly be converted into the respective, glucosylated HMF-derivative (GMF) without cleaving the acid-sensitive glycosidic linkage [55].

The trisaccharide *raffinose*, a storage carbohydrate in many plants, can be cleaved enzymatically with α-galactosidase into sucrose and galactose. This reaction is used in the beet sugar industry to increase the yield of sucrose, as well as to improve the digestibility of food from leguminous plants. Raffinose can also be fermented by bakers' yeast to form melibiose (α-D-Gal$p$-(1→6)-D-Glc$p$).

6.3. Isomerization

Under basic conditions aldoses isomerize to their C-2 epimers and the corresponding ketoses. Specific conditions may be applied for the preparation of particular products. In 0.035 % aqueous sodium hydroxide at 35 °C for 100 h, for example, any one of the three sugars D-glucose, D-fructose, or D-mannose is converted into an equilibrium mixture containing D-glucose (57 %), D-fructose (28 %), and D-mannose (3 %). This interconversion is known as the *Lobry de Bruyn – van Ekenstein rearrangement* [56], which occurs by enolization of either sugar to the 1,2-enediolate – the mechanism being best visualized in the Fischer projection formulae (Fig. 19). In favorable cases alkali-promoted stereoisomerizations can be of preparative use, especially when the starting sugar is relatively abundant and when structural features minimize competing reactions. Thus, lactulose can be satisfactorily made by the epimerization of lactose (→ Lactose and Derivatives), or maltulose from maltose [57, 58], using either sodium hydroxide alone or with borate or aluminate as coreagents.

Alternatively, C-2-epimerization without ketose involvement can be induced by use of molybdate under mildly acidic conditions. This remarkable transformation (*Bı́lik reaction*) [59] involves a C-1/C-2 interchange within the carbon skeleton.

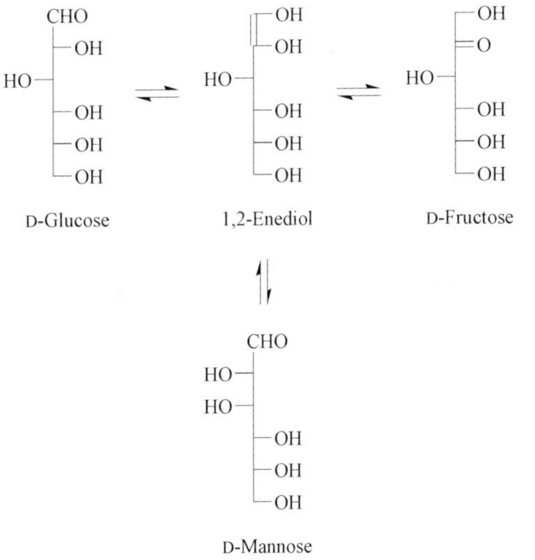

6.4. Decomposition

Exposure of carbohydrates to high temperatures leads to decomposition (dehydration) with darkening (caramelization). This can be used to produce a caramel color (e.g., that of cola beverages). Thermal decomposition in the presence of amino acids (*Maillard reaction* [60]) is responsible for many color- and flavor-forming reactions, such as in the baking of bread and roasting of meat or coffee. The highly complex Maillard reaction, elicited during cooking or the preservation of food, involves condensations, Amadori-type rearrangements of glycosylamine intermediates, and degradations. The dark-colored products formed are responsible for the nonenzymic browning observed with various foodstuffs.

7. Reactions at the Carbonyl Group

In solution, reducing sugars establish an equilibrium between their pyranoid and furanoid hemiacetal forms via the open-chain carbonyl species. Although the latter is present only to a very minor extent, equilibrium between the different forms is fast, so that reducing sugars undergo the typical carbonyl reactions with O-, N-, S-, and C-nucleophiles.

7.1. Glycosides

With alcohols in the presence of acid catalysts reducing sugars give the respective full acetals, called glycosides (Fischer glycosidation) [61].

Figure 19. Lobry de Bruyn – van Ekenstein rearrangement

Depending on the distribution of furanoid and pyranoid tautomeric forms in the reaction mixture, not only glycosides with different ring sizes, i.e., glycopyranosides and glycofuranosides, can result, but also the corresponding α- and β-anomers. Thus, when D-glucose is heated with methanol in the presence of anhydrous hydrogen chloride, pure crystalline methyl α-D-glucopyranoside can be isolated in 90 % yield, whilst the same reaction with D-galactose yields a mixture of the two furanoid and pyranoid methyl galactosides, from which the methyl α-D-galactopyranoside can be obtained in crystalline form but in only 41 % yield.

Although the Fischer glycosidation presents one of the easiest means for preparing glycosides, the synthesis of more complex members of this series, particularly the construction of the biologically important heterooligosaccharides widely distributed in nature, requires the use of more sophisticated methodologies. These preparative techniques generally involve the coupling of suitably OH-group protected glycosyl donors (i.e., glycosides with an anomeric leaving group) with an alcohol component – usually a mono-, di-, or oligosaccharide in which the hydroxyls carry protecting groups [62] except for the one to be glycosylated ('the glycosyl acceptor'). Effective glycosyl donors, derived from D-glucose, are listed in Table 1. They represent the presently most suitable donors for achieving glycosidic bond-forming reactions with high stereocontrol: glycosyl bromides [63], and iodides [64], 2-oxoglycosyl bromides [65–67], anomeric phosphates [68] and trichloracetimidates [69], thioglycosides [70], glycosyl sulfoxides [71], and 1,2-anhydrides [72], some of these methodologies being amenable to combinatorial and solid phase synthesis [73].

For further details on this subject, presently under intense further exploration, some recent general treatments [30, 74, 75–78] are recommended.

Table 1. Established glycosyl donors for the stereoselective synthesis of oligosaccharides

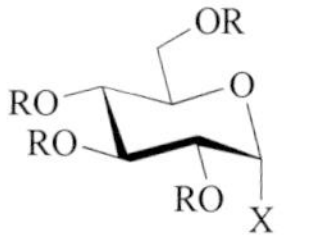

Glycosyl halides, X = Cl, Br

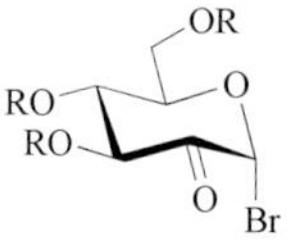

Ulosyl bromides

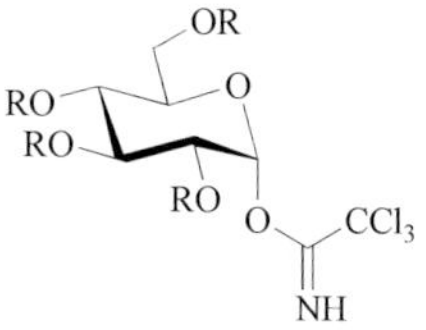

Glycosyl trichloroacetimidates

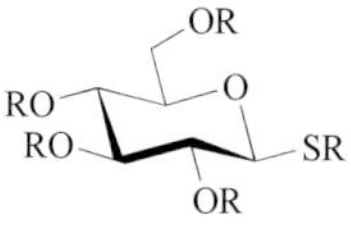

Thioglycosides

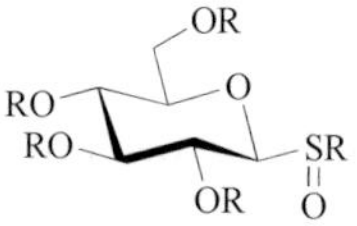

Glycosyl sulfoxides
(R = acetyl, benzoyl, benzyl)

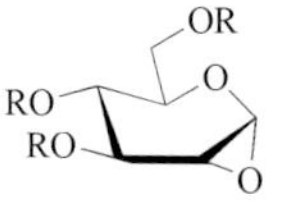

1,2-Anhydro-sugars

tion. These open-chain compounds can be used to prepare monosaccharide derivatives with a free carbonyl group, such as 2,3,4,5,6-penta-O-acetyl-D-glucose:

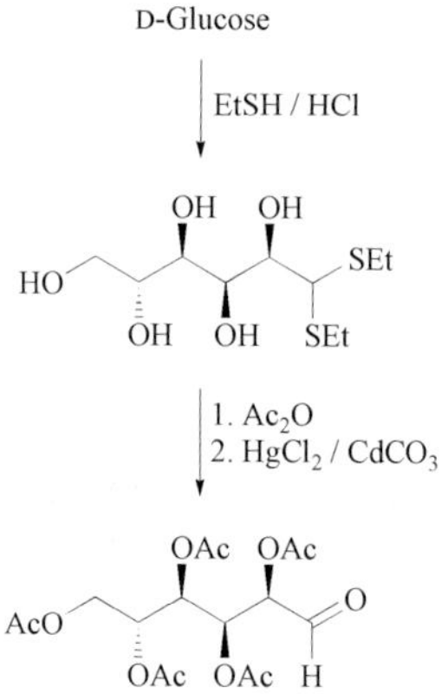

7.2. Thioacetals and Thioglycosides

Sugars react rapidly with alkanethiols in the presence of acid catalysts at room temperature to give acyclic dialkyl dithioacetals as the main products [79], and therefore the reaction is markedly different from the Fischer glycosida-

1-Thioglycosides, established glycosyl donors in oligosaccharide syntheses (upon activation with methyl trifluoromethanesulfonate or other promoters) [70], have to be prepared indirectly, e.g., from peracylated pyranoses (or their 1-halides) by exposure to thiols in the presence of BF_3 etherate or zinc chloride:

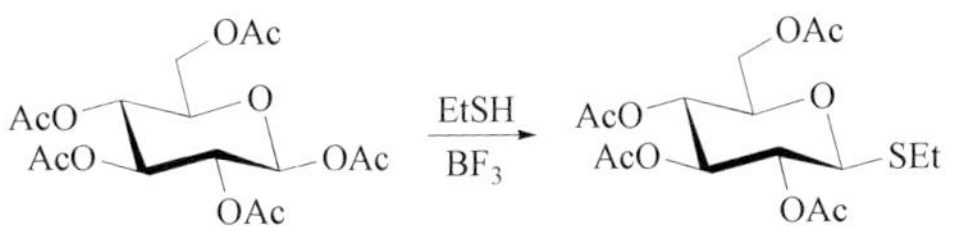

7.3. Glycosylamines, Hydrazones, and Osazones

Aldoses condense with ammonia and with primary and secondary amines upon loss of water – reactions that are analogous to the Fischer glycosidation. The initial condensation products appear to be the open-chain aldimines, which then cyclize to the *glycosylamines* – also called *N*-glycosides. The pyranose forms of the products are preferentially adopted as these are thermodynamically more stable. Accordingly, D-glucose reacts with aniline in methanol to give the α- and β-*N*-glucopyranosides:

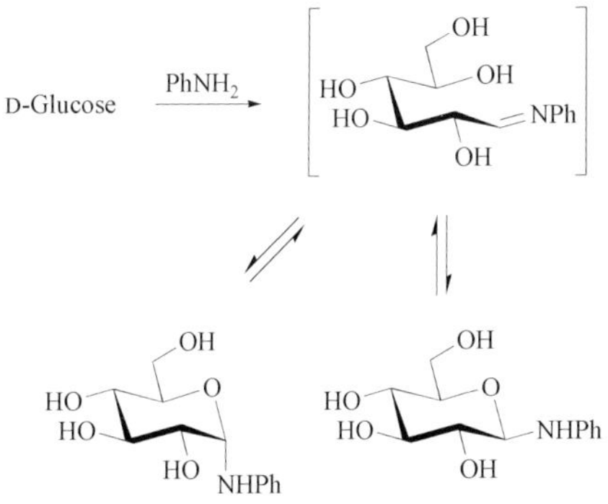

Acids also catalyze a transformation called the *Amadori rearrangement* [80] which often accompanies attempts to prepare glycosylamines from aldoses and amines. This reaction is related to the *Lobry de Bruyn – van Ekenstein* reaction of aldoses and involves the rearrangement of *N*-alkylamino-D-glucopyranosides into 1-alkylamino-1-deoxy-D-fructoses (Fig. 20).

Glycosylamine derivatives are probably involved in the complex Maillard reaction [60], whereby sugars, amines, and amino acids (proteins) condense, rearrange, and degrade during cooking or the preservation of food. Hydrazones and osazones result when aldoses or ketoses are reacted with hydrazine or arylhydrazines, the product depending on the conditions used [81]. With hydrazine acetate in highly acidic medium in the cold, hydrazones are formed, which initially adopt the acyclic structure, but tautomerize in aqueous solution to the cyclic glycosylhydrazine forms. However, when free sugars are treated with an excess of phenylhydrazine, the reaction proceeds further to give – in a formal oxidation of the vicinal C-2-OH group – the highly crystalline, water-insoluble phenylosazones that contain two phenylhydrazine residues per molecule, with a third phenylhydrazine molecule being converted into aniline and ammonia. As C-2 of a sugar is involved in this process, D-glucose, D-mannose, and D-fructose yield the same product:

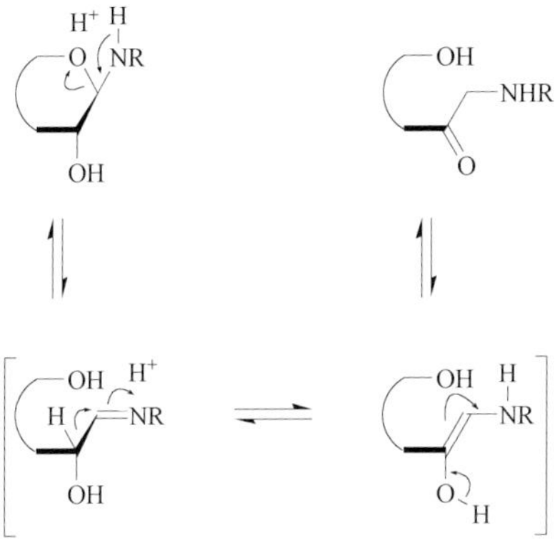

Figure 20. Amadori rearrangement of glycosyl amines induced by acid catalysis: D-Glucopyranosylamine is converted into a 1-alkylamino-D-fructose [79]

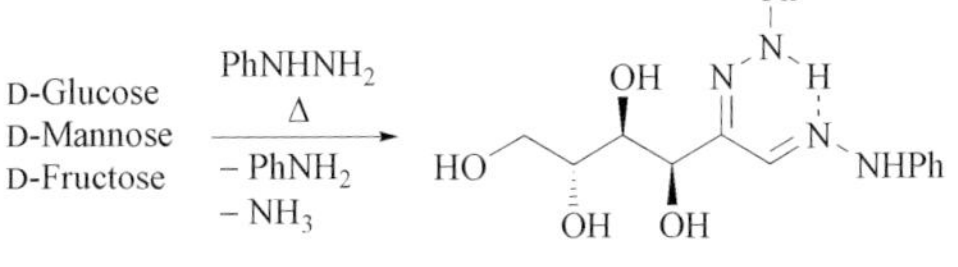

This result played a fundamental role in EMIL FISCHER'S elucidation of the configurational interrelationships of the sugars, eventually leading [11, 12] to the sugar family trees depicted in Figures 3 and 4.

7.4. Chain Extension

The carbonyl group offers excellent opportunities for extension of the sugar chains and the formation of 'higher' sugars. However, only a few carbon nucleophiles can be applied directly to the free sugars, i.e., without protection of the hydroxyl groups. The classical methods comprise the addition of cyanide ion (*Kiliani – Fischer extension*) [82] and of nitromethane under suitable alkaline conditions [83]. In either case, the cyano and nitromethylene group newly introduced can be converted into an aldehyde functionality by hydrolysis of the diastereomeric cyanohydrins to aldonic acids, lactonization and subsequent reduction, or by applying the *Nef reaction* to the nitroalditols; thus, providing methods for the ascent of the sugar series. For example, the rare sugar D-allose can be readily prepared from D-ribose [84] (Fig. 21), whereas the nitromethane addition approach allows the acquisition of the equally scarce hexoses L-glucose and L-mannose from L-arabinose [85] (Fig. 22).

A special case of chain extension by nitromethane is the cyclization of sugar-derived dialdehydes – readily and quantitatively obtained from anomerically blocked glycopyranosides or furanosides by periodate oxidation – to give 3-nitrosugars [86, 87]. As exemplified for methyl β-D-glucopyranoside (Fig. 23), a mixture of 3-nitrohexosides is primarily obtained from which the major product, the D-*gluco* isomer crystallizes out. Subsequent catalytic hydrogenation then provides the 3-amino-3-deoxy-D-glucoside [88].

This nitromethane cyclization sequence can be extended to nitroalkanes (e.g., nitroethane or even nitroacetate [89]), thus providing a ready access – upon hydrogenation of the nitro group – to 3-methyl- or 3–carboxy-branched 3-aminosugars.

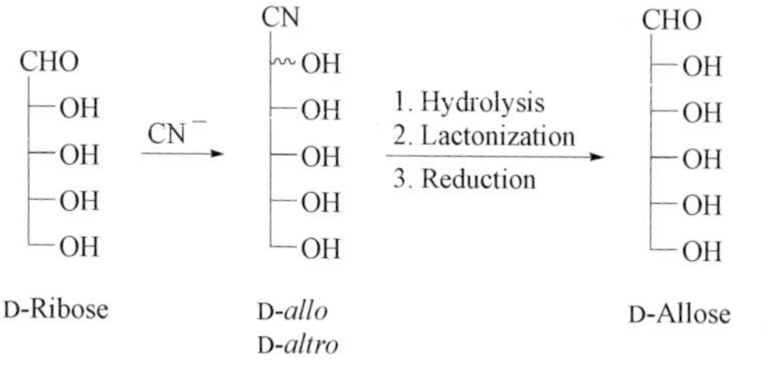

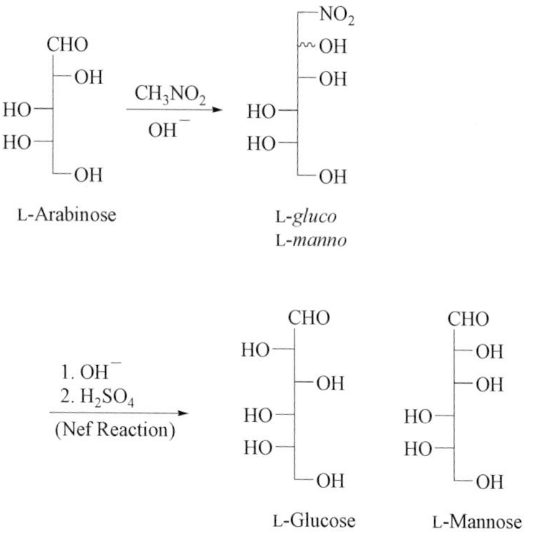

Figure 22. Nitromethane addition to L-arabinose in alkaline medium generates a mixture of the 2-epimeric L-nitroalditols (of L-*gluco* and L-*manno* configuration) which, upon separation, are subjected to the Nef reaction [83]

7.5. Chain Degradation

The removal of a terminal carbon atom from a sugar or sugar derivative to leave an aldehyde group is realizable in a variety of ways, but the yields are often poor. The most practical approach involves conversion of an aldose into the corresponding dialkyl dithioacetal (mercaptal) by reaction with an alkanethiol, then oxidation to the bissulfone with a peracid. Treatment of the bissulfone with dilute ammonia causes expulsion

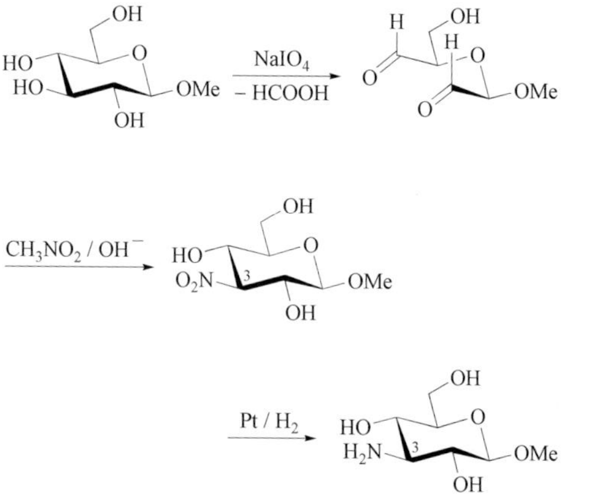

Figure 21. The Kiliani – Fischer cyanohydrin synthesis with D-ribose: the approximate 1:1 mixture of the 2-epimeric D-*allo*- and D-*altro*-cyanohydrins can be separated at the aldonic acid stage, subsequent reduction of the D-allonic acid in the form of its 1,4-lactone then providing D-allose (34 % overall yield)

Figure 23. Conversion of methyl β-D-glucoside into its 3-amino-3-deoxy derivative via the dialdehyde – nitromethane cyclization approach [86]

of the stabilized bis(ethylsulfonyl)methyl carbanion and gives the aldose with one carbon atom less. This three-step protocol smoothly converts D-glucose, for example, into D-arabinose [90]:

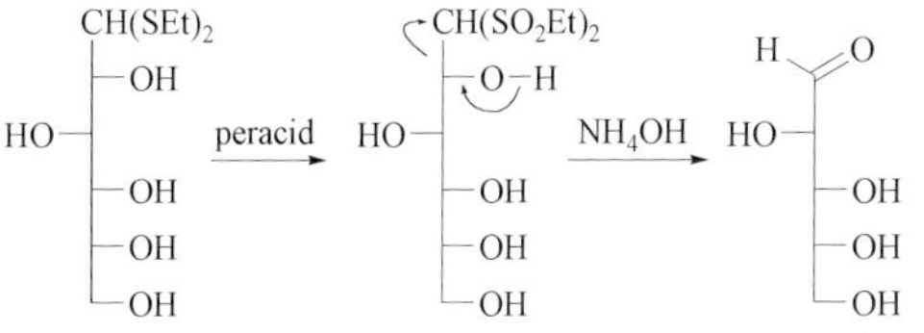

Table 2. Low-caloric, noncariogenic sugar alcohol sweeteners, obtained by catalytic hydrogenation of the parent aldoses. Recommended nomenclature [49] for sorbitol is D-glucitol (hence D-Glc-ol). Being a meso compound, xylitol requires no D- or L-prefix

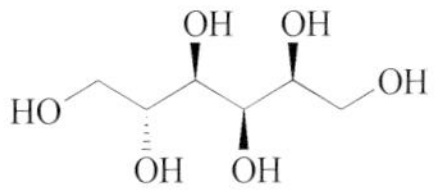

D-Glucitol (sorbitol) (D-Glc-ol)

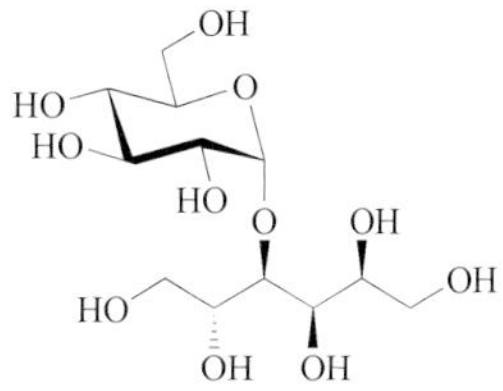

Maltitol (α-D-Glc-(1→4)-D-Glc-ol)

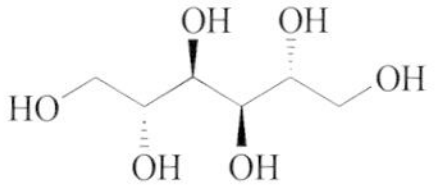

D-Mannitol (D-Man-ol)

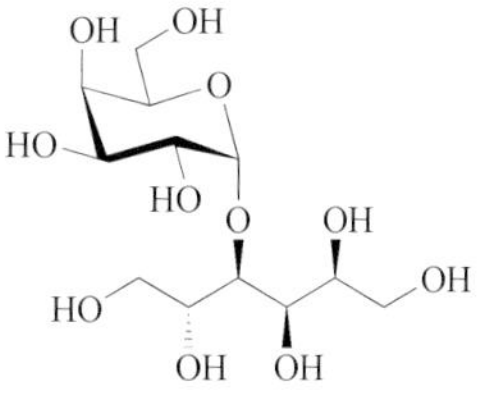

Lactitol (α-D-Gal-(1→4)-D-Glc-ol)

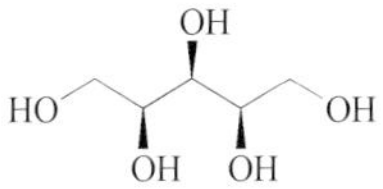

Xylitol (Xyl-ol)

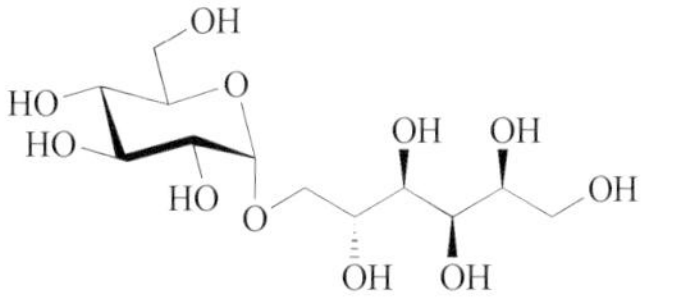

Isomalt (α-D-Glc-(1→6)-D-Glc-ol (α-D-Glc-(1→4)-D-Man-ol)

7.6. Reductions to Alditols

Aldoses and ketoses can readily be reduced to alditols (Table 2) with the generation of a new alcoholic group from the carbonyl functions. The names of the reduced products are derived from the respective aldoses by replacing the 'ose' suffix with 'itol'. Thus, reduction of D-glucose gives D-glucitol. Originally, sodium amalgam was the reducing agent most commonly used for these reductions, but now it has been superseded by others, particularly sodium borohydride in aqueous solution or, for alkali-sensitive sugars, by sodium cyanoborohydride in acetic acid. High pressure hydrogenation of aldoses and ketoses over rare metal catalysts, especially nickel is used for the commercial preparation of alditols (→ Sugar Alcohols).

7.6.1. D-Glucitol

D-Glucitol [50-70-4] (→ Sugar Alcohols, Chap. 3.), common name sorbitol, produced at a level of 9×10^5 t/a worldwide, has a sweet taste and is used in foods for diabetics [91, 92]. It is also the synthetic precursor for ascorbic acid (vitamin C), with about 20 % of the annual production sorbitol going to this use. Sorbitol is used as a humectant in cosmetic and pharmaceutical formulations and in foods. It is also applied as an alcoholic component in the preparation of rigid polyurethane foams. Fatty acid esters of monoanhydrosorbitol (1,4-sorbitan) are widely used as emulsifiers and nonionic surfactants. The mono- and dinitrate esters of 1,4:3,6-dianhydrosorbitol (isosorbide [652-67-5]) are coronary vasodilators.

7.6.2. D-Mannitol

D-Mannitol [87-78-5] (→ Sugar Alcohols, Chap. 4.), is prepared by hydrogenation of the fructose portion of invert sugar [91–93], which yields a mixture of mannitol and sorbitol. In contrast to sorbitol, mannitol is not hygroscopic; world production in 2007 was approximately 3×10^4 t. Mannitol is used in the manufacture of dry electrolytic condensers and synthetic resins; in the pharmaceutical industry as a diluent for solids and liquids and in the preparation of the vasodilator mannitol hexanitrate; in the food industry as

7.6.3. Other Sugar Alcohols

Other sugar alcohols, mostly used as sweeteners, are *xylitol*, obtained by catalytic hydrogenation of D-xylose, which in turn is acquired from wood xylans or maize cobs by acid hydrolysis, and a series of disaccharide alcohols, each manufactured analogously from the respective parent disaccharide: *maltitol* (from partially hydrolyzed starch syrup; Lycasin [93] contains a high proportion of this sugar alcohol), *lactitol* and, most notably, *isomalt*, which due to its mild, pleasant sweetness, ready crystallizability and excellent thermal stability appears presently the most prevalent. Isomalt, also called "palatinit", consists of an approximate 1:1 mixture of α-D-glucosyl-(1→6)-D-sorbitol and α-D-glucosyl-(1→1)-D-mannitol (see Table 2). The latter forms a dihydrate, the two water molecules being attached to the mannitol portion in a hydrogen-bonded water bridge [94]. Isomalt is produced from sucrose through *Protaminobacter rubrum*-induced isomerization to isomaltulose ('palatinose') and subsequent catalytic high-pressure hydrogenation (→ Sugar Alcohols, Section 5.2.).

7.7. Oxidation [95, 96]

Controlled stoichiometric oxidations of carbohydrates to yield glyconic acids or their derivatives are limited to aldoses. Such oxidations can be carried out almost quantitatively either enzymatically by dehydrogenases or oxidases, or chemically with bromine or iodine in buffered solution. Under these conditions, D-glucose – through its pyranose form prevailing in solution – is directly converted into the 1,5-lactone of D-gluconic acid (i.e., the internal ester rather than the free acid), which on addition of base is converted into the salt (in open-chain form). However, by crystallization from aqueous solution it is possible to obtain the free acid or the 1,4-lactone. For different aldonic acids the amounts of each form present at equilibrium vary with structure and with the pH of the solution; in contrast to the free sugars, the five-membered ring lactones are relatively favored.

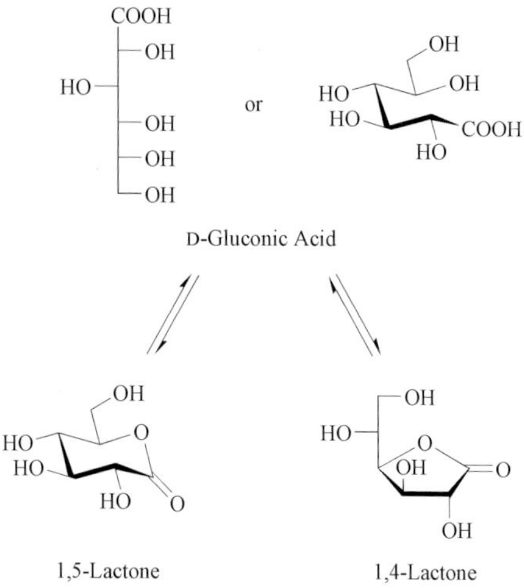

D-Gluconic Acid

1,5-Lactone   1,4-Lactone

Strong nitric acid appears to be one of the few oxidants that is able to oxidize the terminal primary hydroxyl group of aldoses but leave the secondary hydroxyl groups unchanged. D-Glucose treated with this reagent gives D-*glucaric acid* [97], its name being derived by replacing the ending 'ose' in the sugar by 'aric acid'. Aldaric acids can form mono- or dilactones, in the case of D-glucaric acid, the well crystallizing form is the furanoid 1,4-lactone:

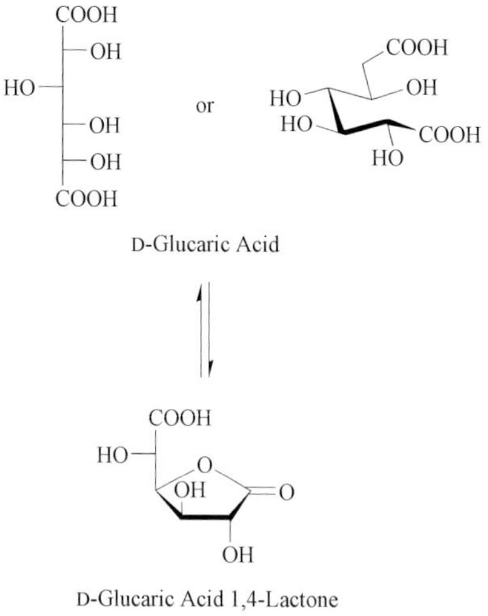

D-Glucaric Acid

D-Glucaric Acid 1,4-Lactone

Under the influence of very strong oxidizing agents such as potassium dichromate or permanganate, sugars suffer oxidative degradation. Hence, these oxidants have no preparative value.

8. Reactions at the Hydroxyl Groups

8.1. Ethers

The most simple compounds of this type are *methyl ethers* which occur in a range of natural carbohydrates. Methyl ethers belong to the most stable *O*-substituted sugar derivatives, such that per-*O*-methylated hexoses can even be distilled. Traditionally, the labeling of free OH- groups in polysaccharides is effected by methylation, structural analysis being then based on the *O*-methyl sugars obtained on hydrolysis. Methyl ethers are conveniently prepared using methyl bromide, iodide, or sulfate in polar aprotic solvents such as dimethylformamide or dimethyl sulfoxide. Agents for deprotonation of the hydroxyl group and for binding the mineral acids liberated include alkali hydroxides or hydrides and barium or silver oxide. With high molecular mass carbohydrates, quantitative deprotonation is best carried out with sodium or potassium methylsulfinyl methanide (the conjugate base of dimethyl sulfoxide) [98].

Benzyl ethers are amongst the most commonly used protecting groups in carbohydrate chemistry [99] as the *O*-benzyl moiety is easily removed by hydrogenolysis (Pd/C, H_2) to yield the respective alcohol and toluene [62]. For the preparation of these ethers, traditional methods involve such reagents as benzyl halides in combination with sodium hydroxide, sodium hydride, or silver oxide.

Triphenylmethyl (trityl) ethers are used mainly for the temporary substitution of primary hydroxyl groups and are usually prepared using trityl chloride in pyridine. Trityl ethers are readily cleaved under mildly acidic conditions; for example, with acetic acid or boron trifluoride in methanol.

Trimethylsilyl ethers, although extremely sensitive both to base- and acid-catalyzed hydrolysis, are often used in analytical and preparative carbohydrate chemistry. The per-*O*-trimethylsilylated monosaccharides and small oligosaccharides are relatively volatile, highly lipophilic, and thermostable, and therefore, ideal derivatives for gas chromatographic analysis. The trimethylsilyl ethers are rapidly formed in pyridine solution by using a mixture of hexamethyldisilazane and trimethylchlorosilane as reagents [100].

Cellulose ethers (→ Cellulose Ethers) are generally manufactured by the Williamson synthesis: namely the reaction of sodium cellulose (prepared by treating cellulose with 20 to > 50 % sodium hydroxide) with an organic halide such as chloromethane or sodium monochloroacetate. The latter reagent produces sodium carboxymethyl cellulose (NaCMC), which is widely used, for example, as a thickening agent in foods. Worldwide production of NaCMC is in the range of several hundred thousand tons per year.

8.2. Esters of Inorganic Acids

Phosphoric acid esters of sugars play vital roles in such fundamental processes as the biosynthesis and metabolism of sugars and, hence, are present in every organism; the most important esters being D-glucose 1-phosphate [59-56-3], D-glucose 6-phosphate [56-73-5], and D-fructose 1,6-diphosphate [488-69-7]. In addition, phosphates of D-ribose and its 2-deoxy derivative form fundamental components of ribonucleic acid (RNA) and deoxyribonucleic acid (DNA), and also of various coenzymes (→ Nucleic Acids, Chap. 7.).

α-D-Glucose-1-phosphate

D-Glucose-6-phosphate

D-Fructose-1,6-diphosphate

D-Ribose-5-phosphate

2-Deoxy-D-ribose-5-phosphate

Adenosine-5-phosphate (AMP)

Both chemical and enzymic methods are available for the synthesis of specific phosphates. Chemically, anomeric phosphate esters are usually prepared either from glycosyl halides or other glycosyl donors by reaction with silver dibenzyl phosphate, whilst phosphorylation of nonanomeric hydroxyl groups is effected with specifically blocked sugar derivatives and diphenyl or dibenzyl phosphorochloridate [101]. Biochemically, phosphates are produced by the action of phosphatases on provided substrates [102].

Sulfate Esters. Sulfate groups are present in many biologically important polysaccharides such as heparin and chondroitin sulfate. Sulfated monosaccharides can be prepared from suitable monosaccharide derivatives by reaction with chlorosulfuric acid in pyridine [103].

Nitrate esters of carbohydrates [104] are not found in nature, yet a large variety ranging from monoesters to peresters have been prepared; favorable conditions being cold nitric acid/acetic anhydride for nonanomeric hydroxyl groups, whilst anomeric nitrate esters are accessible via reactions of acyl glycosyl halides with silver nitrate. Sugar mono- and dinitrates are stable crystalline compounds; examples are adenosine mononitrate (AMN), the nitrate analogue to AMP [105], and the dinitrate of 1,4:3,6-dianhydro-D-glucitol ('isosorbide dinitrate'). This last compound is in broad pharmaceutical use as a coronary vasodilator [106]:

Adenosine 5-nitrate (AMN)   Isosorbide dinitrate

More highly substituted derivatives are heat and shock sensitive, and include mannitol hexanitrate or nitrate esters of cellulose (nitrocellulose). These contain as many as three ONO_2 groups per glucose unit. The product with about 13% nitrogen is the well-known guncotton (→ Cellulose Esters), whereas celluloid is nitrocellulose containing about 10% nitrogen, plasticized with camphor. Celluloid is one of the oldest known plastics and was once the principal photographic and movie film, but has since been replaced by other films because of its high flammability.

8.3. Esters of Organic Acids

For the esterification of the hydroxyl group of free or partially otherwise blocked sugars, acyl halides or acid anhydrides are usually used; for example, acetic anhydride/sodium acetate or zinc chloride, or acetic anhydride/pyridine readily yield the respective *peracetates*.

Perbenzoylation can be effected with benzoyl chloride in pyridine or benzoyl cyanide in acetonitrile with triethylamine as the catalyst. However, the formation of tertiary hydroxyl groups, present in ketoses or branched-chain sugars, usually require the addition of 4-(dimethylamino)pyridine as a coreagent.

Whereas peracetates and perbenzoates of simple sugars are important intermediates for the preparation of the respective glycosyl halides and, hence, acylated glycals and hydroxyglycals (see Section 8.4), those of some polysaccharides are of industrial relevance. Acetate esters of cellulose are manufactured on a large scale, whereby the degree of acetylation determines their solubility and use: the triacetate (3.0-acetate) is soluble in chloroform, the 2.5-acetate in acetone, and the 0.7-acetate in water. These esters, as well as mixed cellulose acetate/propionate and acetate/butyrate are widely used in the production of lacquers, films, and plastics (→ Cellulose Esters).

Polysaccharide esters in which the carbohydrate portion is the acid component occur in the plant kingdom in fruits, roots, and leaves. For example, *pectins* are high molecular mass polygalacturonic acids joined by α-(1→4)-glycosidic links, in which some of the carboxylic acid groups are esterified with methanol (→ Polysaccharides). In the production of fruit juices the formation of methanol, which can be liberated through the action of pectinesterases, should be avoided. Pectins in which 55 – 80% of the carboxyl groups are esterified are called highmethoxyl pectins (HM-pectins), and have the important property of gelling at very low concentrations ($\approx$ 0.5%) in water in the presence of sugars and acid. Low-methoxyl (LM, < 50% of the carboxyl groups esterified) pectins form gels with divalent cations such as a Ca^{2+}; 0.5%

of a low-methoxyl pectin can bind 99.5 % of the water in the gel matrix. These pectins can be used as gelling agents in the production of jellies from fruit juices.

8.4. Acylated Glycosyl Halides

Per-*O*-acylated monosaccharides can be converted smoothly into glycosyl halides by dissolving them in cold solutions of the hydrogen halide in glacial acetic acid (acetates of acid-sensitive oligosaccharides may undergo cleavage of glycosidic bonds). Because of the dominance of the anomeric effect in the pyranosyl cases, the anomer with axial halide is substantially preferred. Accordingly, on acylation and subsequent HBr-treatment, usually performed as a one pot operation, D-glucose yields the 2,3,4,6-tetra-*O*-acetyl-α-D-glucopyranosyl bromide ('acetobromoglucose', R = Ac in Fig. 24) or its benzoylated, pivaloylated (R = *tert*-BuCO) or benzylated (R = $C_6H_5CH_2$) analogues [63]. These halides are commonly used directly for glycosylation reactions, which is the basis of the traditional Koenigs – Knorr procedure [107], or converted into more elaborate glycosyl donors.

Glycosyl halides are also of significance in terms of the use of monosaccharides as inexpensive enantiopure starting materials for the construction of complex, non carbohydrate natural products [108–110], which usually require the reduction of the number of chiral centers paired with the introduction of olefinic or carbonyl unsaturation. Treatment of glycosyl bromides with zinc/acetic acid [111] – or, preparatively more efficient by zinc/1-methylimidazole in ethyl acetate under reflux [112] – results in reductive elimination to give the glycal (this is illustrated in Figure 24 by the formation of tri-*O*-acetyl-D-glucal). Simple 1,2-elimination of hydrogen bromide using diethylamine in acetonitrile in the presence of tetrabutylammonium bromide or by 1,8-diazabicyclo[5.4.0]undec-7-ene (DBU) in DMF [113, 114] yields the respective 2-hydroxyglycal esters. The D-glucose-derived benzoylated example in Figure 24 is an ester of the enol form of 1,5-anhydro-D-fructose, a naturally occurring ketosugar, which may be released from its tetrabenzoate by low-temperature deblocking with sodium methoxide/methanol [115].

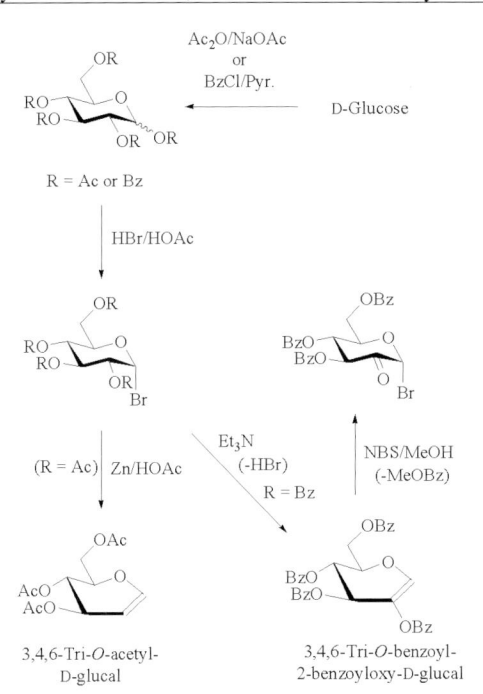

Figure 24. Formation of glycosyl and 2-oxoglycosyl ('ulosyl') bromides from peracylated monosaccharides, as exemplified for the D-glucose case, and their conversions into glucal and 2-hydroxyglucal esters

Endowed with high crystallinity and shelf stability, the hydroxyglycal esters are of considerable preparative interest not only for the generation of a plethora of other unsaturated compounds, e.g., pyranoid enones [116] and enolones [109, 117], but also as precursors for the highly versatile 2-oxoglycosyl ('ulosyl') bromides, produced in high yields simply by exposure to *N*-bromosuccinimide or bromine in the presence of ethanol [118, 119]. The utility of these ulosyl bromides as glycosyl donors in the straightforward synthesis of β-D-mannosides has been amply demonstrated [65–67, 118–120].

8.5. Acetals

Acetals are generally derived from the reaction of an aldehyde or ketone – benzaldehyde and acetone being the most common – with a geometrically suitable diol grouping, of which there is a large variety in free sugars, glycosides, and alditols [121–123]. The reactions are normally carried out in the reagent aldehyde or ketone as solvent with an electrophilic catalyst (H_2SO_4 or

ZnCl$_2$). Acetal formation under these conditions is thermodynamically controlled and usually very specific. Ketones such as acetone or cyclohexanone predominantly bridge vicinal diols to form five-membered cyclic products (1,3-dioxolanes) as exemplified by the di-*O*-isopropylidene derivatives of D-glucose ('diacetone-glucose'), D-galactose and D-mannitol:

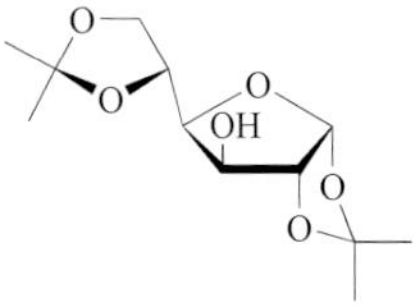

1,2:5,6-Di-*O*-isopropylidene-α-D-glucofuranose

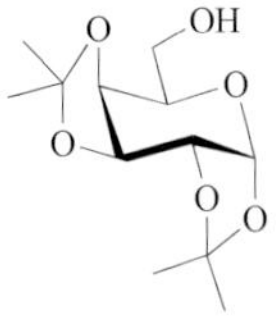

1,2:3,4-Di-*O*-isopropylidene-α-D-galactopyranose

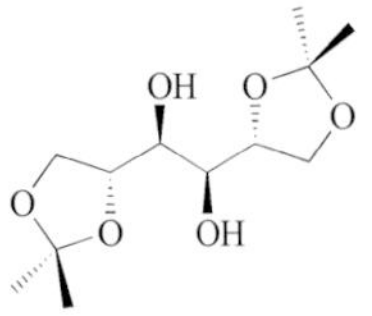

1,2:5,6-Di-*O*-isopropylidene- D-mannitol

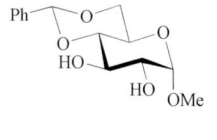

Methyl 4,6-*O*-benzylidene-α-D-glucopyranoside

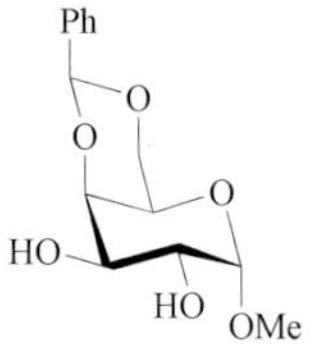

Methyl 4,6-*O*-benzylidene-α-D-galactopyranoside

Aldehydes, however, show a distinct preference for 1,3-diols, as illustrated by the six-membered 4,6-*O*-benzylidene acetals of methyl D-glucoside and D-galactoside.

Introduction of cyclic acetal groups into sugars is simple and satisfactory in terms of yields. As cyclic acetals are stable towards alkali, the entire armory of organic reactions requiring basic conditions can be applied, and due to their ready removal with mild acid (e.g., 90% aqueous trifluoroacetic acid at room temperature), they provide indispensable intermediates in preparative carbohydrate chemistry.

References

1 J. Lehmann: *Kohlenhydrate, Chemie und Biologie*, Thieme, Stuttgart/New York 1996, pp. 372; *Carbohydrates, Chemistry and Biology*, Thieme, Stuttgart/New York 1998, pp. 274.
2 T.K. Lindhorst: *Essentials of Carbohydrate Chemistry and Biochemistry*, 3rd ed., Wiley-VCH, Weinheim 2007, pp. 318.
3 R.V. Stick: *Carbohydrates: The Sweet Molecules of Life*, Acad. Press, San Diego 2001, pp. 256.
4 P. M. Collins, R. J. Ferrier: *Monosaccharides, their Chemistry and their Roles in Natural Products*, Wiley, Chichester/New York 1995, pp. 574.
5 J.F. Kennedy (ed.): *Carbohydrate Chemistry*, Clarendon Press, Oxford 1988, pp. 678.
6 G.-J. Boons (ed.): *Carbohydrate Chemistry*, Blackie Academic, London/Weinheim 1998, pp. 508.
7 D.E. Levy, P. Fügedi (eds.): *The Organic Chemistry of Sugars*, CRC/Taylor and Francis Boca Raton 2006, pp. 880.
8 B. Ernst, G. W. Hart, P. Sinay (eds.): *Carbohydrates in Chemistry and Biology*, vol. 1–4, Wiley-VCH, Weinheim 2000, pp. 2340.
9 B. Fraser-Reid, K. Tatsuta, J. Thiem (eds.): *Glycoscience, Chemistry and Chemical Biology*, vol. I-III, Springer, Heidelberg 2001, pp. 2850.
10 R.M. De Lederkremer, C. Gallo-Rodriguez, "Naturally Occuring Monosaccharides: Properties and Synthesis", *Adv. Carbohydr. Chem. Biochem.* **59** (2004) 9 – 69.
11 E. Fischer, "Ueber die Configuration des Traubenzuckers und seiner Isomeren", *Ber. Dtsch. Chem. Ges.* **24** (1891) 2683 – 2687.
12 F.W. Lichtenthaler, "Emil Fischer's Proof of the Configuration of Sugars: A Centennial Tribute," *Angew. Chem.* **104** (1992) 1577 – 1593; *Angew. Chem. Int. Ed. Engl.* **31** (1992) 1541 – 1556.
13 J.M. Bijvoet, A.F. Peerdemann, A.J. van Bommel, "Determination of the Absolute Configuration of Optically Active Compounds by Means of X-Rays. *Nature* **168** (1951) 271 – 272.
14 W.N. Haworth: *The Constitution of Sugars*, Arnold, London 1929, pp. 104.
15 J. A. Mills, "The Stereochemistry of Cyclic Derivatives of Carbohydrates," *Adv. Carbohydr. Chem.* **10** (1955) 1 – 53.
16 IUPAC-IUB Joint Commission (JCBN): "Conformational Nomenclature for five- and six-membered Ring Forms of Monosaccharides and their Derivatives," *Pure Appl. Chem.* **53** (1981) 1901 – 1905; *Carbohydr. Res.* **297** (1997) 20 – 23.
17 J.A. Kanters, G. Roelofsen, B.P. Alblas, I. Meinders: "The Crystal Structure of β-D-Fructose with Emphasis on the Anomeric Effect," *Acta Cryst.* **B33** (1977) 665 – 672.
18 B. Schneider, F.W. Lichtenthaler, G. Steinle, H. Schiweck: "Distribution of Furanoid and Pyranoid Tautomers of D-Fructose in Dimethylsulfoxide, Water and Pyridine via Anomeric Hydroxyl Proton NMR Intensitites," *Liebigs Ann. Chem.* (1985) 2454 – 2464.
19 G.M. Brown, H.A. Levy: "Sucrose: Precise Determination of Crystal and Molecular Structure by Neutron

19. Diffraction," *Science* **141** (1963) 921 – 923; *Acta Cryst.* **B29** (1973) 790 – 797.
20. J.C. Hanson, L.C. Sieker, L.H. Jensen: "Sucrose: X-Ray Refinement and Comparison with Neutron Refinement," *Acta Crystallogr.* **B29** (1973) 797 – 808.
21. R.R. Watson, N.S. Orenstein: "Chemistry and Biochemistry of Apiose," *Adv. Carbohydr. Chem. Biochem.* **31** (1975) 135 – 184.
22. J. Yoshimura: "Synthesis of Branched-chain Sugars," *Adv. Carbohydr. Chem. Biochem.* **42** (1984) 69 – 134.
23. A. Lipták, P. Fügedi, Z. Szurmai, J. Harangi: *Handbook of Oligosaccharides*, CRC Press, Boca Raton/Boston 1990, vol. I, Disaccharides, pp. 474; vol. II, Trisaccharides, pp. 267; vol. III, Higher Oligosaccharides, pp. 179.
24. A. Hugill: *Introductory Dedicational Metaphor to Sugar and All That. A History of Tate & Lyle*, Gentry Books, London 1978.
25. UN Food & Agriculture Organization. *World Sugar Production 2007/08*. http://www.fao.org/docrep/011/ai474e08.htm (accessed 15 October 2009).
26. Y. Queneau, S. Jarosz, B. Lewandowski, J. Fitreman: "Sucrose Chemistry and Application of Sucrochemicals", *Adv. Carbohydr. Chem. Biochem.* **61** (2007) 218 – 292.
27. S. Immel, F.W. Lichtenthaler: "The Conformation of Sucrose in Water: A Molecular Dynamics Approach," *Liebigs Ann. Chem.* (1995) 1925 – 1937.
28. F.W. Lichtenthaler, P. Pokinskyj, S. Immel: "Sucrose as a Renewable Raw Material. New Reactions via Computer Simulation of Hydroxyl Group Reactivities," *Zuckerind. (Berlin)* **121** (1996) 174 – 190.
29. A. Varki: "Biological Roles of Oligosaccharides," *Glycobiology* **3** (1993) 97 – 130.
30. H.M.I. Osborn, T.H. Khan: *Oligosaccharides. Their Synthesis and Biological Roles*, Oxford University Press, Oxford 2000, pp. 112.
31. H.-J. Gabius, S. Gabius (eds.): *Glycosciences—Status and Perspectives*, Chapman & Hall, London/Weinheim 1997, pp. 631.
32. J. Kopitz: "Glycolipids: Structure and Function", in [31], pp. 163 – 189.
33. N. Sharon, H. Lis: "Glycoproteins: Structure and Function", in [31], pp 133 – 162.
34. J. Szejtli: "Introduction and General Overview of Cyclodextrin Chemistry," *Chem. Rev.* **98** (1998) 1743 – 1753.
35. F.W. Lichtenthaler, S. Immel: "Towards Understanding Formation and Stability of Cyclodextrin Inclusion Complexes. Visualization of their Lipophilicity Patterns," *Starch/Stärke* **48** (1996) 145 – 154.
36. All 3D-structures can be viewed on the www at: http://csi.chemie.tu-darmstadt.de/ak/immel/; (accessed Febuary 25, 2010).
37. J. Szejtli: "Medical Applications of Cyclodextrins," *Med. Res. Rev.* **14** (1994) 353 – 386.
38. J.K. Uekama, F. Hitayama, T. Ivie: "Cyclodextrin Drug Carrier Systems," *Chem. Rev.* **98** (1998) 2045 – 2076.
39. V. Schurig, H.P. Novotny: "Gaschromatographic Separation of Enantiomers on Cyclodextrin Derivatives," *Angew. Chem.* **102** (1990) 969 – 986; *Angew. Chem. Int. Ed. Engl.* **29** (1990) 939 – 958.
40. S. Li, W.C. Purdy: "Cyclodextrins and their Applications in Analytical Chemistry," *Chem. Rev.* **92** (1992) 1457 – 1470.
41. G. Wenz: "Cyclodextrins as Building Blocks for Supramolecular Structures and Functional Units," *Angew. Chem.* **106** (1994) 851 – 870; *Angew. Chem. Int. Ed. Engl.* **33** (1994) 803 – 822.
42. I. Tabushi: "Design Synthesis of Artificial Enzymes," *Tetrahedron* **40** (1984) 269 – 292.
43. T.L. Bluhm, P. Zugenmaier: "Detailed Structure of the V_H-Amylose-Iodine Complex: A Linear Polyiodide Chain," *Carbohydr. Res.* **89** (1981) 1 – 10.
44. S. Immel, F.W. Lichtenthaler: "The Hydrophobic Topographies of Amylose and its Blue Iodine Complex," *Starch/Stärke* **52** (2000) 1 – 8.
45. A. Fuchs (ed.): *Inulin and Inulin-containing Crops*, Elsevier, Amsterdam/New York 1993, pp. 417.
46. A. Fuchs: "Potentials for Non-food Utilization of Fructose and Inulin," *Starch/Stärke* **39** (1987) 335 – 343.
47. J.F. Kennedy, C.A. White: "*The Plant, Algal and Microbial Polysaccharides*", in J.F. Kennedy (ed.): *Carbohydrate Chemistry*, Oxford Science Publ., Oxford 1988, pp. 220 – 262.
48. R.J. Sturgeon: "The Glycoproteins and Glycogen", in J.F. Kennedy (ed.): *Carbohydrate Chemistry*, Oxford Science Publ., Oxford 1988, pp. 263 – 302.
49. IUPAC-IUB Joint Commission: "Nomenclature of Carbohydrates (1996 Recommendations)," *Pure Appl. Chem.* **68** (1996) 1919 – 2008, or *Carbohydr. Res.* **297** (1997) 1 – 92, or *Adv. Carbohydr. Chem. Biochem.* **52** (1997) 47 – 177.
50. J. Szejtli: *Säurehydrolyse glycosidischer Bindungen. Einfluß von Struktur und Reaktionsbedingungen auf die Säurespaltung von Glykosiden, Disacchariden, Oligo- und Polysacchariden*, VEB Fachbuchverlag, Leipzig 1976, pp. 399.
51. C.J. Biermann: "Hydrolysis and Other Cleavages of Glycosidic Linkages in Polysaccharides," *Adv. Carbohydr. Chem. Biochem.* **46** (1988) 251 – 271.
52. O. Theander, D.A. Nelson: "Aqueous, High-Temperature Transformation of Carbohydrates towards Biomass Utilization," *Adv. Carbohydr. Chem. Biochem.* **46** (1988) 273 – 332.
53. (a) B.F.M. Kuster: "Manufacture of 5-Hydroxymethylfurfural", *Starch/Stärke* **42** (1990) 314 – 312. (b) J. Lewkowski, "Synthesis, Chemistry and Applications of 5-Hydroxymethylfurfural and its Derivatives", *ARKIVOC* 2001, 17 – 54.
54. J.N. Chheda, Y. Román-Leshkov, J.A. Dumesic: "Production of 5-Hydroxymethylfurfural and Furfural by Dehydration of Biomass-derived Mono- and Polysaccharides", *Green Chemistry* **9** (2007) 342 – 350.
55. F.W. Lichtenthaler, D. Martin, T. Weber, H. Schiweck: "5-(α-D-Glucosyloxymethyl)furfural: Preparation from

Isomaltulose and Exploration of its Ensuing Chemistry", *Liebigs Ann. Chem.* (1993) 967 – 974.
56   S.J. Angyal: "The Lobry de Bruyn – Alberda van Ekenstein Transformation and Related Reactions," *Topics Current Chem.* **215** (2001) 1 – 14.
57   Nikken Chem. Co., JP 48049938, 1971 (I. Machida, J. Kanaeda, H. Miki, S. Kubomura, H. Toda, T. Shiroishi); *Chem. Abstr.* **79** (1973) 126744v.
58   F.W. Lichtenthaler, S. Rönninger: "α-D-Glucopyranosyl-D-fructoses. Distribution of Furanoid and Pyranoid Tautomers in Water, DMSO and Pyridine," *J. Chem. Soc. Perkin Trans.* **2** (1990) 1489 – 1497.
59   L. Petruš, M. Petrušová, Z. Hricovíniová: "The Bílik Reaction," *Topics Current Chem.* **215** (2001) 16 – 41.
60   F. Ledl, E. Schleicher: "Die Maillard-Reaktion in Lebensmitteln und im menschlichen Körper," *Angew. Chem.* **102** (1990) 597 – 626; *Angew. Chem. Int. Ed. Engl.* **29** (1990) 565 – 594.
61   A.F. Bochkov, G.E. Zaikov: *The Chemistry of the O-Glycosidic Bond*, Pergamon Press, Oxford 1979, pp. 210.
62   T. Ziegler: "Protecting Group Strategies for Carbohydrates", in G.-J. Boons (ed.): *Carbohydrate Chemistry*, Blackie Academic, London/Weinheim 1998, pp. 21 – 45.
63   M. Nitz, D.R. Bundle: "Glycosyl Halides in Oligosaccharide Synthesis", in B. Fraser-Reid, K. Tatsuta, J. Thiem (eds.): *Glycoscience,* vol. II, Springer, Heidelberg 2002, pp. 1497 – 1544.
64   P.J. Meloncelli, A.D. Martin, T.L. Lowary: "Glycosyl Iodides. History and Recent Advances", *Carbohydr. Res.* **341** (2009) 1110 – 1122.
65   F.W. Lichtenthaler, E. Kaji, T. Schneider-Adams: "2-Oxo- and 2-Oximino-glycosyl Halides. Versatile Glycosyl Donors for the Construction of β-D-Mannose- and β-D-Mannosamine-containing Oligosaccharides," *Trends Glycosci. Glycotechnol.* **5** (1993) 121 – 142; *J. Org. Chem.* **59** (1994) 6728 – 6734 and 6735 – 6738; *Carbohydr. Res.* **305** (1998) 293 – 303.
66   M. Nitz, D.R. Bundle: "Synthesis of Di- to Hexasaccharide 1,2-Linked β-Mannopyranan Oligomers and a Terminal S-Linked Tetrasaccharide Congener," *J. Org. Chem.* **66** (2001) 8411 – 8423.
67   F.W. Lichtenthaler: "The Ulosyl Donor Approach to Oligosaccharides with β-D-Man and β-D-ManNAc units", *Chem. Rev.* **110** (2010) in press.
68   E.R. Palmacci, O.J. Plante, P.H. Seeberger: "Oligosaccharide Synthesis in Solution and on Solid Support with Glycosyl Phosphates," *Eur. J. Org. Chem.* (2002) 595 – 606.
69   R.R. Schmidt, W. Kinzy: "Anomeric-oxygen Activation for Glycoside Synthesis: The Trichloroacetimidate Method," *Adv. Carbohydr. Chem. Biochem.* **50** (1994) 21 – 124.
70   P.J. Garegg: "Thioglycosides as Glycosyl Donors in Oligosaccharide Synthesis," *Adv. Carbohydr. Chem. Biochem.* **52** (1997) 179 – 205.

71   (a) D. Crich, L.B.L. Lim: "Glycosylation with Sulfoxides and Sulfinates as Donors or Promoters", *Org. Reactions* **64** (2004) 115 – 251.
(b) L. Yan, D. Kahne: "Generalizing Glycosylation. Synthesis of Blood Group Antigens Le^a, Le^b, and Le^x," *J. Am. Chem. Soc.* **118** (1996) 9239 – 9248.
72   S.J. Danishefsky, M.T. Bilodean: "Glycals in Organic Synthesis. Comprehensive Strategies for the Assembly of Oligosaccharides," *Angew. Chem.* **108** (1996) 1482 – 1522; *Angew. Chem. Int. Ed. Engl.* **35** (1996) 1380 – 1419.
73   K. Fucase: "Combinatorial and Solid Phase Methods in Oligosaccharide Syntheses", in B. Fraser-Reid, K. Tatsuta, J. Thiem (eds.): *Glycoscience*, vol. II, Springer, Heidelberg – New York 2001, pp. 1661 – 1694.
74   G.-J. Boons: "Strategies and Tactics in Oligosaccharide Synthesis", in G.-J. Boons (ed.): *Carbohydrate Chemistry*, Blackie Academic, London/Weinheim 1998, pp. 175 – 222.
75   Ref. [8], vol. 1, pp. 1 – 583.
76   G. Magnusson, U.J. Nilson: "Regio- and Stereoselective Methods of Glycosylation", in B. Fraser-Reid, K. Tatsuta, J. Thiem (eds.): *Glycoscience*, vol. II, Springer, Heidelberg – New York 2001, pp. 1543 – 1588.
77   X. Zhu, R.R. Schmidt: "New Principles of Glycoside-bond Formation", *Angew. Chem. Int. Ed.* **48** (2009) 1900 – 1934.
78   J.T. Smoot, A.V. Demchenko: "Oligosaccharide Synthesis: From Conventional Methods to Modern Expeditious Strategies", *Adv. Carbohydr. Chem. Biochem* **62** (2009) 162 – 250.
79   J.D. Wander, D. Horton: "Dithioacetals of Sugars," *Adv. Carbohydr. Chem. Biochem.* **32** (1976) 15 – 123.
80   T.J. Wrodnigg, B. Eder: "The Amadori and Heyns Rearrangements," *Topics Curr. Chem.* **215** (2001) 115 – 152.
81   H.S. El Khadem, A.F. Fatiadi: "Hydrazine Derivatives of Carbohydrates and Related Compounds," *Adv. Carbohydr. Chem. Biochem.* **55** (1999) 175 – 263.
82   C.S. Hudson: "The Kiliani – Fischer Cyanohydrin Synthesis," *Adv. Carbohydr. Chem.* **1** (1945) 1 – 36.
83   J.C. Sowden: "The Nitromethane and 2-Nitroethanol Syntheses," *Adv. Carbohydr. Chem.* **6** (1951) 291 – 318.
84   F.L. Humoller: "β-D-Allose," *Methods Carbohydr. Chem.* **1** (1962) 102 – 104.
85   J.C. Sowden: "L-Glucose and L-Mannose," *Methods Carbohydr. Chem.* **1** (1962) 132 – 135.
86   F.W. Lichtenthaler: "Cyclizations of Dialdehydes with Nitromethane," *Angew. Chem.* **76** (1964) 84 – 97; *Angew. Chem. Int. Ed. Engl.* **3** (1964) 135 – 148.
87   H.H. Baer: "The Nitro Sugars," *Adv. Carbohydr. Chem. Biochem.* **24** (1969) 67 – 138.
88   F.W. Lichtenthaler: "Amino Sugars and Amino Cyclitols via Cyclization of Dialdehydes with Nitromethane," *Methods Carbohydr. Chem.* **6** (1972) 250 – 260.

89 F.W. Lichtenthaler: "Branched-chain Amino Sugars via Dialdehyde-Nitroalkane-Cyclization," *Fortschr. Chem. Forschg.* **14** (1970) 556 – 577.

90 D.L. MacDonald, H.O.L. Fischer: "The Degradation of Sugars by Means of their Disulfones," *J. Am. Chem. Soc.* **74** (1952) 2087 – 2090.

91 P.J. Sicard: "Hydrogenated Glucose Syrups, Sorbitol, Mannitol and Xylitol", in G. G. Birch, K.J. Parker (eds.): *Nutritive Sweeteners*, Applied Science Publ., London/ New Jersey 1982, pp. 145 – 170.

92 C.K. Lee: "The Chemistry and Biochemistry of the Sweetness of Sugars," *Adv. Carbohydr. Chem. Biochem.* **45** (1990) 199 – 351.

93 P.J. Sicard, P. Leroy: "Mannitol, Sorbitol and Lycasin; Properties and Food Applications", in T.H. Grenby, K.J. Parker, M.G. Lindley (eds.): *Developments in Sweeteners-2*, Appl. Science Publ., London – New York 1983, pp. 1 – 26.

94 H.J. Lindner, F.W. Lichtenthaler: "Extended Zig-zag Conformation of 1-*O*-α-D-Glucopyranosyl-D-mannitol," *Carbohydr. Res.* **93** (1981) 135 – 140.

95 R.M. De Lederkremer, C. Marino: "Acids and other Oxidation Products of Sugars", *Adv. Carbohydr. Chem. Biochem.* **58** (2003) 200 – 306.

96 O. Varela: "Oxidative Reactions and Degradation of Sugars and Polysaccharides", *Adv. Carbohydr. Chem. Biochem.* **58** (2003) 308 – 370.

97 C.L. Mehltretter: "D-Glucaric Acid," *Methods Carbohydr. Chem.* **2** (1963) 46 – 48.

98 H.E. Conrad: "Methylation of Carbohydrates with Methylsulfinyl Anion and Methyl Iodide in DMSO," *Methods Carbohydr. Chem.* **6** (1972) 361 – 364.

99 C.M. McCloskey: "Benzyl Ethers of Sugars," *Adv. Carbohydr. Chem.* **12** (1957) 137 – 156.

100 C.C. Sweely, R. Bentley, M. Makita, W.W. Wells: "Gas-Liquid-Chromatography of Trimethylsilyl Derivatives of Sugars," *J. Am. Chem. Soc.* **85** (1963) 2497 – 2507.

101 C.E. Ballou, D.L. MacDonald: "Phosporylation with Diphenyl Phosphorochloridate," *Methods Carbohydr. Chem.* **2** (1963) 270 – 272; 273 – 276; 277 – 281; 282 – 288.

102 C.-H. Wong, G.M. Whitesides: *Enzymes in Synthetic Organic Chemistry*, Elsevier Science, Oxford, 1994.

103 R.L. Whistler, W.W. Spencer, J.N. BeMiller: "Sulfates," *Methods Carbohydr. Chem.* **2** (1963) 298 – 303.

104 J. Honeyman, J.W.W. Morgan: "Sugar Nitrates," *Adv. Carbohydr. Chem.* **12** (1957) 117 – 135.

105 F.W. Lichtenthaler, H.J. Müller: "Preparation of Nucleoside 5′-Nitrates," *Synthesis* **1974** 199 – 201.

106 L.A. Silvieri, N.J. DeAngelis: "Isosorbide dinitrate," *Profiles of Drug Substances* **4** (1975) 225 – 244.

107 K. Igarashi: "The Koenigs – Knorr Reaction," *Adv. Carbohydr. Chem. Biochem.* **34** (1977) 243 – 283.

108 S. Hanessian: *Total Synthesis of Natural Products: The Chiron Approach*, Pergamon, Oxford/New York 1983.

109 F.W. Lichtenthaler: "Building Blocks from Sugars and their Use in Natural Product Synthesis. A Review with Procedures" in R. Scheffold (ed.): *Modern Synthetic Methods*, vol. 6, VCH Verlagsgesellschaft, Weinheim 1992, pp. 273 – 376.

110 G.J. Boons, K.J. Hale: *Organic Synthesis with Carbohydrates*, Sheffield Acad. Press, Sheffield 2000, pp. 336.

111 E. Fischer: "Ueber neue Reduktionsprodukte des Traubenzuckers: Glucal und Hydroglucal," *Ber. Dtsch. Chem. Ges.* **47** (1914) 196 – 210; *Methods Carbohydr. Chem.* **2** (1963) 405 – 408.

112 L. Somsák, I. Nemeth: "A Simple Method for the Synthesis of Acylated Pyranoid Glycals under Aprotic Conditions," *J. Carbohydr. Chem.* **12** (1993) 679 – 684.

113 M.G. Blair: "The 2-Hydroxyglycals," *Adv. Carbohydr. Chem.* **9** (1954) 97 – 130; *Methods Carbohydr. Chem.* **2** (1963) 411 – 414.

114 R.J. Ferrier: "Modified Synthesis of 2-Hydroxyglycal Esters," *Methods Carbohydr. Chem.* **6** (1972) 307 – 311.

115 M. Brehm, V.H. Göckel, P. Jarglis, F.W. Lichtenthaler:"Expedient Conversions of D-Glucose into 1,5-Anhydro-D-fructose and into Single-stereogenic-center Dihydropyranones, Suitable Six-carbon Scaffolds for the Concise Syntheses of the Soft-coral Constituents Bissetone and Palythazine", *Tetrahedron: Asymmetry* **19** (2008) 358 – 373.

116 N.L. Holder: "The Chemistry of Hexuloses," *Chem. Rev.* **82** (1982) 287 – 33.

117 F.W. Lichtenthaler: "Sugar Enolones — Synthesis, Reactions of Preparative Interest, and γ-Pyrone Formation," *Pure Appl. Chem.* **50** (1978) 1343 – 1362.

118 F.W. Lichtenthaler, U. Kläres, M. Lergenmüller, S. Schwidetzky: "Glycosyl Donors with a Keto or Oximino Function next to the Anomeric Center: Practical Preparation and Evaluation of their Selectivities in Glycosidation," *Synthesis* **1992**, 179 – 184.

119 F.W. Lichtenthaler, T. Schneider-Adams: "3,4,6-Tri-*O*-benzyl-α-D-*arabino*-hexopyranos2-ulosyl Bromide—a Glycosyl Donor for the Efficient Generation of β-D-Mannosidic Linkages," *J. Org. Chem.* **59** (1994) 6728 – 6734.

120 F.W. Lichtenthaler, T. Schneider-Adams, S. Immel: "A Practical Synthesis of β-D-Xyl-(1→2)-β-D-Man-(1→4)-D-GlcOMe, a Trisaccharide Component of the Glycosphingolipid of *Hyriopsis schlegelii*," *J. Org. Chem.* **59** (1994) 6735 – 6738.

121 R.F. Brady: "Cyclic Acetals of Ketoses," *Adv. Carbohydr. Chem. Biochem.* **26** (1971) 197 – 278.

122 A.N. deBelder: "Cyclic Acetals of the Aldoses and Aldosides," *Adv. Carbohydr. Chem. Biochem.* **34** (1977) 179 – 241.

123 D.M. Clode: "Carbohydrate Cyclic Acetal Formation and Migration," *Chem. Rev.* **79** (1979) 491 – 513.

Further Reading

J. N. BeMiller: *Carbohydrates*, "Kirk Othmer Encyclopedia of Chemical Technology", vol. 4, p. 696–733, John Wiley & Sons, Hoboken, 2004, online: DOI: 10.1002/0471238961.0301180202051309.a01.pub2 (January 2004).

A. V. Demchenko (ed.): *Handbook of Chemical Glycosylation*, Wiley-VCH, Weinheim 2008.

H.-J. Gabius (ed.): *The Sugar Code*, Wiley-VCH, Weinheim 2009.

T. K. Lindhorst: *Essentials of Carbohydrate Chemistry and Biochemistry*, 3rd ed., Wiley-VCH, Weinheim 12007.

M. Miljkovic: *Carbohydrates*, Springer, New York, NY 2009.

R. V. Stick, S. Williams: *Carbohydrates*, 2nd ed., Elsevier, Amsterdam 2009.

Carbohydrates as Organic Raw Materials

FRIEDER W. LICHTENTHALER, Clemens-Schöpf-Institut für Organische Chemie und Biochemie, Technische Universität Darmstadt, Darmstadt, Germany

1.	Introduction	89
2.	Availability of Carbohydrates	90
3.	Current Nonfood Industrial Products from Sugars	92
3.1.	Ethanol	92
3.2.	Furfural	93
3.3.	D-Sorbitol (≡ D-Glucitol)	94
3.4.	Lactic Acid and Polylactic Acid (PLA)	94
3.5.	Sugar-Based Surfactants	95
3.5.1.	"Sorbitan" Esters	95
3.5.2.	N-Methyl-N-acyl-glucamides (NMGA)	95
3.5.3.	Alkylpolyglucosides (APG)	96
3.5.4.	Sucrose Fatty Acid Monoesters	96
3.6.	Pharmaceuticals and Vitamins	96
4.	Toward Further Sugar-based Chemicals: Potential Development Lines	97
4.1.	Furan Compounds	98
4.1.1.	5-Hydroxymethylfurfural (HMF)	98
4.1.2.	2,5-Dimethylfuran (DMF)	99
4.1.3.	Furans with a Tetrahydroxybutyl Side Chain	99
4.2.	Pyrones and Dihydropyranones	99
4.3.	Sugar-Derived Unsaturated Nitrogen Heterocycles	101
4.3.1.	Pyrroles	101
4.3.2.	Pyrazoles	102
4.3.3.	Imidazoles	103
4.3.4.	3-Pyridinols	104
4.4.	Toward Sugar-Based Aromatic Chemicals	104
4.5.	Microbial Conversion of Six-Carbon Sugars into Simple Carboxylic Acids and Alcohols	106
4.5.1.	Carboxylic Acids	107
4.5.2.	Potential Sugar-Based Alcohol Commodities by Microbial Conversions	109
4.6.	Chemical Conversion of Sugars into Carboxylic Acids	110
4.7.	Biopolymers from Polymerizable Sugar-Derivatives	112
4.7.1.	Synthetic Biopolyesters	112
4.7.2.	Microbial Polyesters	113
4.7.3.	Polyamides	116
5.	Outlook	117
	References	118

1. Introduction

Because our fossil raw materials, derived from prehistoric organic matter, are irrevocably decreasing and because the pressure on our environment is intensifying, the progressive changeover of the chemical industry to renewable feedstocks for its raw materials emerges as an inevitable necessity [1–3]; i.e. it will have to proceed increasingly to utilize raw materials that prevailed before natural gas and oil outpaced all other resources.

Historically, such raw materials were substantially renewable, as depicted in Figure 1, with the utilization of biomass and coal being almost equal a century ago [4]. In the 1920s, however, coal tar-based materials had taken the lead reaching a maximum around 1930, but thereafter fossil gas and oil irresistibly took over, nearly eliminating coal completely and reducing renewable feedstocks to very modest levels.

This over-reliance of the chemical industry on fossil resources has its foreseeable limits as they are depleting and are irreplaceable, the only question being: When will affordable fossil fuels be exhausted? Or, stated more appropriately: When will fossil-based raw materials have become so expensive that biofeedstocks are an economically competitive alternative?

Fossil oil will be around for a long time, even if it has to be isolated eventually from oliferous rocks or shale. So the prevailing issue is: When will be the end of cheap oil? Experts realistically predict this to occur within the next two to three decades [5–7]. Accordingly, the curve for the utilization of biofeedstocks in Figure 1 will have to rise such that it meets that of fossil raw materials somewhere around 2030–2040.

Ullmann's Renewable Resources
© 2013 Wiley-VCH Verlag GmbH & Co. KGaA. Published 2013 by Wiley-VCH Verlag GmbH & Co. KGaA
ISBN: 978-3-527-33369-1 / DOI: 10.1002/14356007.n05_n07

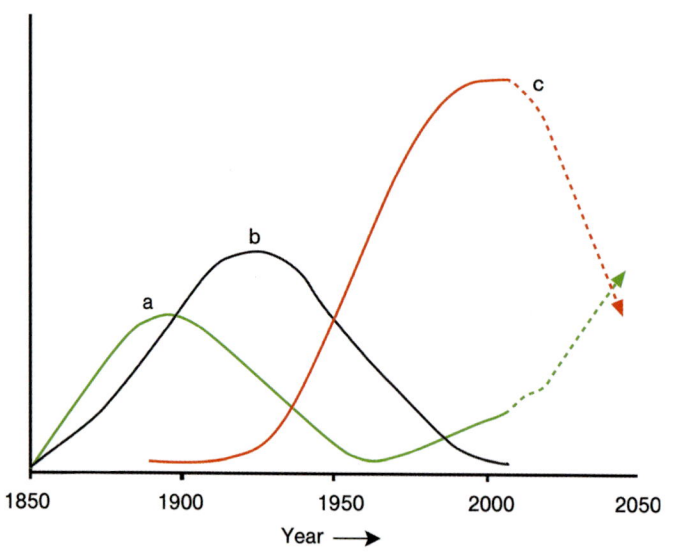

Figure 1. Raw materials used by the chemical industry in historical perspective
a) renewable feedstocks; b) coal; c) natural gas, oil

The terrestrial biomass to be utilized for the transition from fossil to renewable raw materials, however, is – compared to fossil resources – an exceedingly complex array of low and high molecular mass products, exemplified by sugars, hydroxy and amino acids, lipids, and biopolymers such as cellulose, hemicelluloses, chitin, starch, lignin, and proteins. By far the most important class of organic compounds in terms of volume produced are *carbohydrates* as they represent roughly 75% of the annually renewable biomass of about 180×10^9 t (Fig. 2).

Of these, only a minor fraction (ca. 4%) is used by man, the rest decays and recycles along natural pathways.

Thus, carbohydrates, a single class of natural products – aside from their traditional uses for food, lumber, paper, and heat – are the major biofeedstocks from which to develop industrially and economically viable organic chemicals that are to replace those derived from petrochemical sources.

2. Availability of Carbohydrates

The bulk of the annually renewable carbohydrate biomass is made up of polysaccharides, yet their nonfood utilization is confined to the textile, paper, and coating industries, either as such or in the form of simple esters and ethers. Organic commodity chemicals, however, are usually of low molecular mass, so they are more expediently obtained from low molecular mass carbohydrates than from polysaccharides. Accordingly, the constituent repeating units of these polysaccharides – *glucose* (cellulose, starch), *fructose* (inulin), *xylose* (hemicelluloses), etc., or disaccharide versions thereof – are the actual carbohydrate raw materials for organic chemicals with tailor-made industrial applications. They are inexpensive, ton-scale accessible, and provide an ensuing chemistry better worked out and more variable than that of their polymers.

Renewable biomass
($18 \cdot 10^{10}$ t/a)

Carbohydrates 75%

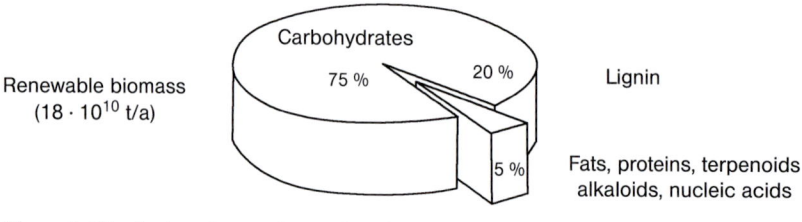

Lignin 20%

Fats, proteins, terpenoids alkaloids, nucleic acids 5%

Figure 2. Distribution of types of natural products in biomass

Table 1 lists the availability and bulk-quantity prices of the eight least expensive sugars, some sugar alcohols and sugar-derived acids – all well below € 10/kg – compared with some basic chemicals from petrochemical sources. The result is stunning, since the five cheapest sugars, their alcohols, and some important sugar-derived acids are within the same price range as basic organic bulk chemicals such as naphtha, ethylene, acetaldehyde, or aniline. Actually, the first three of these sugars, sucrose, glucose, and lactose, are in the price range of some of the standard organic solvents.

The uniqueness of this situation becomes even more imposing when looking at the availability of these sugars. *Sucrose*, "the royal carbohydrate" [8], has for centuries been the world's most abundantly produced organic compound, annual production being an impressive 169×10^6 t in 2007/08 [9]. Similarly bulk scale-accessible are its component sugars D-*glucose*, produced by hydrolysis of starch ($\rightarrow$ Glucose and Glucose-Containing Syrups), and D-*fructose*, generated either from glucose by base-induced isomerization or from inulin or sucrose by hydrolysis ($\rightarrow$ Fructose). *Isomaltulose*, an $\alpha(1\rightarrow6)$-isomer of sucrose, has become accessible on an industrial scale through enzymatic transglucosylation ($\rightarrow$ Sugar Alcohols, Chap. 5.), *lactose* and *maltose* are available in large quantities from whey ($\rightarrow$ Lactose and Derivatives) and starch, D-*xylose*, the cheapest pentose, from wood- or straw-derived xylans. L-*Sorbose* is the cheapest, large-scale accessible L-sugar, because of its production from D-sorbitol (= D-glucitol) in the Vitamin C fabrication process ($\rightarrow$ Vitamins, 7. Vitamin C (l-Ascorbic Acid)). The sugar alcohols D-*sorbitol*, *erythritol* [10], D-*xylitol*, and D-*mannitol* ($\rightarrow$ Sugar Alcohols, Chaps. 2–4), each of comparatively high yearly production via hydrogenation of their parent aldoses, are mainly used as food ingredients because of their sweetening properties, yet also have potential as inexpensive raw materials for broad-scale preparative purposes. The same holds for D-gluconic acid ($\rightarrow$ Gluconic Acid) and the other sugar-derived acids listed.

Despite their large-scale accessibility at comparatively low cost, it seems surprising that the chemical industry currently utilizes these mono- and disaccharides to a minor extent as feedstock for organic chemicals, despite intense efforts within the last decade [11–20] to boost the acquisition of organic chemicals from sugars in Table 1.

There are a variety of reasons for this. Current use of fossil raw materials is more economic and, as important, the process technology for the conversion of petrochemical raw materials into organic chemicals is exceedingly well developed and basically different from that required for transforming carbohydrates into products with industrial application profiles. This situation originates from the inherently different chemical structures of the two types of raw material, the

Table 1. Annual production volume and prices of simple sugars, sugar-derived alcohols and acids as compared to some petrochemically derived basic chemicals and solvents

		World production[a], t/a	Price[b], €/kg
Sugars	Sucrose	169 000 000	0.20
	D-Glucose	30 000 000	0.30
	Lactose	295 000	0.60
	D-Fructose	60 000	1.00
	Isomaltulose	70 000	2.00
	Maltose	3000	3.00
	D-Xylose	25 000	4.50
	L-Sorbose	60 000	7.50
Sugar alcohols	D-Sorbitol	650 000	1.80
	Erythritol	30 000	2.25
	Xylitol	30 000	5.00
	D-Mannitol	30 000	8.00
Sugar-derived acids	Citric acid	1 500 000	1.00
	D-Gluconic acid	60 000	1.40
	L-Lactic acid	150 000	1.75
	L-Tartaric acid	35 000	6.00
	L-Ascorbic acid	80 000	8.00
	L-Glutamic acid	1 500 000	1.20
	L-Lysine	740 000	2.00
Petrochemicals	Ethylene	90 000 000	0.40
	Propylene	45 000 000	0.35
	Benzene	23 000 000	0.40
	Terephthalic acid	12 000 000	0.70
	Aniline	1 300 000	0.95
	Acetaldehyde	900 000	1.10
	Adipic acid	1 500 000	1.70
Solvents	Methanol	25 000 000	0.15
	Toluene	6 500 000	0.25
	Acetone	3 200 000	0.55

[a] Reliable data are available for the world production of sucrose only, the figure given referring to the crop cycle 2007/2008 [9]. All other data are average values based on estimates from producers and/or suppliers as the production volume of many products are not publicly available.

[b] Prices given are those attainable in early 2005 for bulk delivery of crystalline material (where applicable) based on pricing information from sugar industry (sugars) and the Chemical Market Reporter 2009 (acids, basic chemicals, and solvents). The listings are intended as a benchmark rather than as a basis for negotiations between producers and customers. Quotations for less pure products are, in part, markedly lower, e.g. the commercial sweetener "high fructose syrup", which contains up to 95% fructose, may readily be used for large-scale preparative purposes.

Fossil Resources:	Renewable Resources:
HYDRO-CARBONS C_nH_{2n+2}	CARBO-HYDRATES $C_n(H_2O)_n$
oxygen-free, lacking functional groups	overfunctionalized with hydroxyl groups
n-Hexane	D-Glucose
Refinery ($CO_2 \uparrow$)	Biorefinery ($CO_2 \downarrow$)

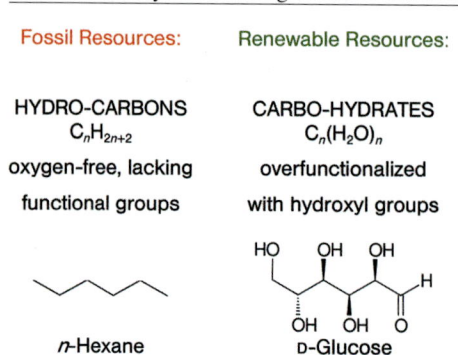

Figure 3. Hydrocarbons versus carbohydrates: more than a play on words, as their names, taken literally, reveal the basic differences in their utilization as organic raw materials

essence of which is manifested in their structure-based names (Fig. 3). Our fossil resources are hydrocarbons, distinctly hydrophobic, oxygen-free and devoid of functionality, thus, organic functional groups such as hydroxyl, amino, aldehyde, acid, ester or halo functionalities have to be introduced – usually into olefinic hydrocarbons such as ethylene, propylene and butane – to obtain the industrially important intermediate chemicals. By contrast *carbohydrates* are overfunctionalized with hydroxyl groups and markedly hydrophilic in nature. Needless to say, that the methods required for converting carbohydrates into viable industrial chemicals – reduction of oxygen content with introduction of C=C and C=O unsaturation – are diametrically opposed to those prevalent in the petrochemical industry.

As higher oil prices, environmental issues, and regulations begin to adversely affect the manufacture of chemicals from fossil raw materials, the inevitable transition to a biobased production system urgently necessitates the systematic elaboration of appropriate chemical and microbial process methods to convert carbohydrates into industrially viable products, be it bulk, intermediate and fine chemicals, pharmaceuticals, agrochemicals, high-value-added speciality chemicals, or simply enantiomerically pure building blocks for organic synthesis.

3. Current Nonfood Industrial Products from Sugars

Current utilization of carbohydrates as a feedstock for the chemical industry – whether for bulk, commodity, intermediate, fine or high-value-added specialty chemicals – is modest when considering their ready availability at low cost and the huge as yet unexploited potential. These examples are outlined briefly that are presently realized on an industrial scale.

3.1. Ethanol

With a production of about 52×10^6 t in 2008 (17335×10^6 gallons [21]), fermentation ethanol ("bioethanol") is the largest volume biobased chemical utilized today. The principal organism for fermentation is *Saccharomyces cerevisiae*, an ascomycetous yeast that can grow on a wide variety of carbohydrate feedstocks – sugar crops, and sugar-containing byproducts such as sugar cane, sugar beet, sorghum, molasses, and (after hydrolysis to glucose) starchy crops such as corn, potatoes and grains, or cellulosic materials; e.g. wood pulping sludges from pulp and paper mills [22]. Developments starting in 2000 [23] replace the conventional yeast by bacteria (*Zymomonas nobilis*) and/or genetically engineered organisms, which seem to improve productivity significantly.

The enormous growth in production of industrial-grade fermentation ethanol in the early 2000s is less due to its use as a solvent or as a starting material for follow-up chemicals, such as acetaldehyde, ethyl esters (e.g. EtOAc) and ethers (Et_2O) as these mostly result from ethylene-based processing lines; rather it is its use as fuel additive, directly mixed to standard gasoline at a level of 5–10%. There is considerable debate [24] about how useful bioethanol is in replacing fossil fuels, arguments going as far as considering this a giant misguided effort to reduce oil insecurity. First of all, present production predominantly relies on grain (wheat, corn) thereby diverting large areas of land for nonfood purposes, which indubitably will result in food price inflation. Furthermore, the production process is not CO_2 neutral as fermentation yields as much CO_2 as ethanol:

$$\text{D-Glucose } (C_6H_{12}O_6) \xrightarrow{\text{fermentation}} 2\,C_2H_5OH + 2\,CO_2$$

Ethanol is economically only viable, at least in Europe, with hefty government subsidies (exemption of gasoline tax) and – a further disadvantage – it cannot be used in diesel engines due to its comparatively low boiling point. Although developments with cellulosic ethanol production [23] and commercialization may allay some of these flaws, the transition to more sustainable and considerably more effective second-generation biofuels such as 2,5-dimethylfuran (see Section 4.1.2) or 1-butanol [24], generated from cellulosic biomass in agricultural and forestry wastes and the like, are imperative.

3.2. Furfural

With an annual production of approximately 25×10^4 t, furfural (2-furfuraldehyde) seems to be the only unsaturated large-volume organic chemical prepared from carbohydrate sources. The technical process involves exposure of agricultural or forestry wastes – hemicellulose up to 25% D-xylose polysaccharides (xylosans) – to aqueous acid and fairly high temperatures; the xylosans are first hydrolyzed, and then undergo acid-induced cyclodehydration (→ Furfural and Derivatives) [26b, 26c].

The chemistry of furfural is well developed, providing a host of versatile industrial chemicals by simple straightforward operations (Scheme 1): furfuryl alcohol (**2**) and its tetrahydro derivative **1** (hydrogenation), furfurylamine (**3**) (reductive amination), furoic acid (**4**) (oxidation) and furanacrylic acid (**5**) (via a Perkin reaction), or furylidene ketones **6** (aldol condensations). Furfural is also the key chemical for the commercial production of furan (through catalytic decarbonylation) and tetrahydrofuran (**8**) (hydrogenation), thereby providing a biomass-based alternative to its petrochemical production by dehydration of 1,4-butanediol (→ Furfural and Derivatives) [26b, 26c]. Further importance of these furan chemicals stems from their ring-cleavage chemistry [27], which has led to a variety of other established chemicals, for example fumaric, maleic and levulinic acid, the last a byproduct of the production of furfural (→ Oxocarboxylic Acids).

The susceptibility of the furan ring in these compounds for electrophilic substitution at C-5 has been widely exploited. Mineral acid-promoted condensations with aldehydes or ketones convert **3** into the difurfural diamine **9** [28], whereas esters of 2-furoic acid afford the respective difurfuryl dicarboxylates (**10** → **11**). Both – the latter on saponification – are relevant monomer components for the generation of polyesters and polyamides [29].

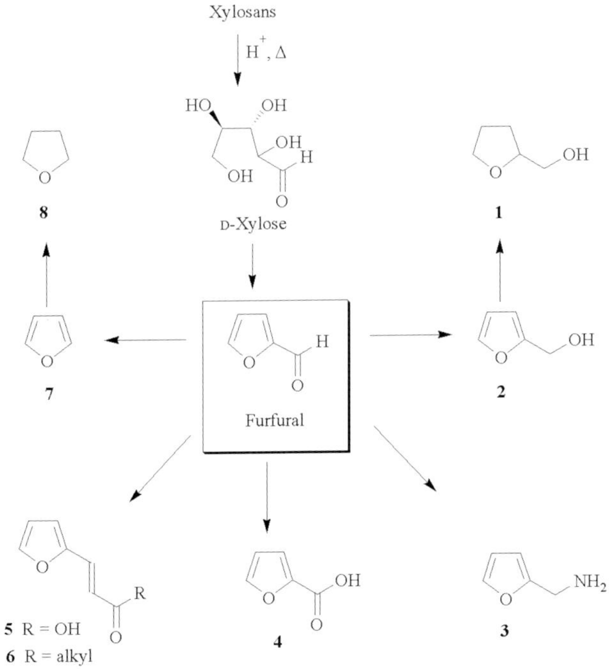

Scheme 1. Furanic commodity chemicals derived from pentosans in agricultural wastes (corn cobs, oat hulls, bagasse, wood chips)

Most of the furfural currently produced is used as a selective solvent in the refining of lubricating oil, and, together with furfuryl alcohol in condensations with formaldehyde, phenol, acetone or urea to yield resins of complex, ill-defined structures, yet excellent thermosetting properties, most notably exhibiting high corrosion resistance, low fire hazard and extreme physical strength (→ Furfural and Derivatives, → Resins, Synthetic, Section 3.1.) [26a, 26b]; they are extensively used in the foundry industry as cores for high-quality castings.

3.3. D-Sorbitol (≡ D-Glucitol)

Although D-sorbitol has a sizable annual production (Table 1), the main consumer is the food industry, primarily as a noncaloric sweetening agent and as a key intermediate in the production of ascorbic acid (vitamin C). However, D-sorbitol has important nonfood applications because of its moisture conditioning, softening and plastifying properties (→ Sugar Alcohols, Chap. 3.). These result in its use in adhesives, paper, printing, textiles, cellulose-based films and pharmaceutical formulations. Other nonfood applications of D-sorbitol result from etherification and polycondensation reactions which provide biodegradable polyetherpolyols used for soft polyurethane foams and melamine–formaldehyde or phenolic resins [28]. Sizable amounts of D-sorbitol are also used for production of the sorbitan ester surfactants (cf. Section 3.5.1).

3.4. Lactic Acid and Polylactic Acid (PLA)

Large amounts of D-glucose – in the crude form obtainable from corn, potatoes or molasses by acid hydrolysis – enter into industrial fermentation processes towards the production of lactic acid (see Scheme 2), citric acid and different amino acids; for example, L-lysine or L-glutamic acid. Although the major use of these products is in food and related industries, nonfood exploitation of lactic acid has made it a large scale, organic commodity chemical. Most of it is subsequently polymerized via its cyclic dimer (lactide) to polylactic acid [30], a high-molecular mass polyester.

Because of its high strength, PLA can be fabricated into fibers, films, and rods that are fully biodegradable (formation of lactic acid, CO_2) and compostable, degrading within 45–60 days. Accordingly, PLA and copolymers of lactic and glycolic acid are of particular significance for food packaging and in agricultural or gardening applications, but they are also are highly suitable materials for surgical implants and sutures, as they are bioresorbable [31].

Since 1989 Cargill has invested some \$ 750×10^6 to develop and commercialize polylactic acid (tradename INGEO by NatureWorks), its Nebraska plant with an annual capacity of 14×10^4 t opened in 2002 [32]. Thus, polylactides, because they combine favorable economics with green sustainability, are poised to compete in large-volume markets that are now the domain of thermoplastic polymers derived from petrochemical sources.

Another green development based on lactic acid is its ethyl ester which has been marketed as

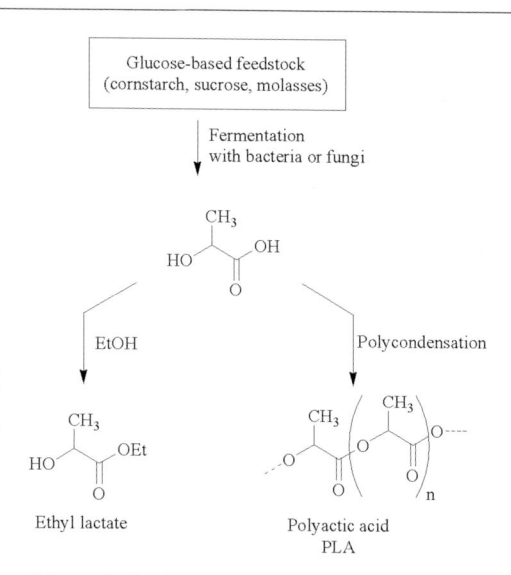

Scheme 2. Production and uses of lactic acid

ELSOL by Vertec Biosolvents for applications in specialty coatings, inks, and directly for cleaning because of its high performance and versatility [33]. As a most benign solvent –green, readily biodegradable, and with excellent low toxicology – this chemical has the potential to displace a variety of petrochemically-based solvents such as acetone, DMF, toluene or N-methylpyrrolidone in industrial processes.

3.5. Sugar-Based Surfactants

Utilization of cheap, bulk-scale accessible sugars as the hydrophilic component and fatty acids or fatty alcohol as the lipophilic part provides nonionic surfactants which are nontoxic, nonskin-irritating and fully biodegradable. Typical examples of such industrially relevant surfactants are fatty acid esters of sorbitol (sorbitan esters) and of sucrose, fatty acid amides of 1-methylamino-1-deoxy-D-glucitol (NMGA) and, most pronounced in terms of volume produced, fatty alcohol glucosides, the so-called alkyl polyglucosides (APGs) [34, 35].

3.5.1. "Sorbitan" Esters

Bulk-scale accessible D-sorbitol (cf. Table 1) readily undergoes dehydration on exposure to mineral acid at fairly high temperatures to give anhydrosorbitol or sorbitan, de facto a mixture of sorbitol and its 1,4-anhydro and 1,4:3,6-dianhydro derivatives, the exact composition depending on the conditions employed [35] (Scheme 3). Esterification of this mixture with C_{16}/C_{18} fatty acid chlorides/base or transesterification with their methyl esters leads to either sorbitan monoesters (SMS for sorbitan monostearate), or di- and triesters ("SMS").

Because of their favorable hydrophilic/hydrophobic balance (HLB) values, sorbitan esters find use as nonionic surfactants and as solubilizers and emulsifiers in cosmetics, pharmaceuticals, textile processing, and a variety of other formulations [40]. Having been commercially available since the 1940s, de facto they constitute one of the first fully green synthetic surfactants, presently produced at an estimated 2×10^4 t/a.

3.5.2. N-Methyl-N-acyl-glucamides (NMGA)

Reductive amination of D-glucose with methylamine smoothly generates an aminoalditol, 1-methylamino-1-deoxy-D-glucitol, which, on amidation with fatty acids gives the corresponding fatty acid amides, carrying a methyl group and a pentahydroxylated six-carbon chain at the amido nitrogen.

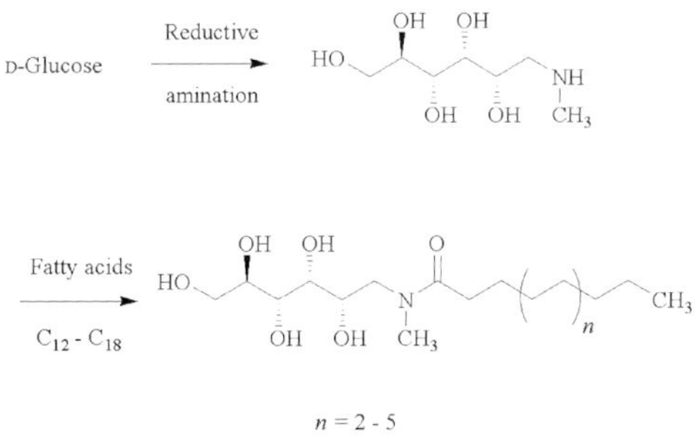

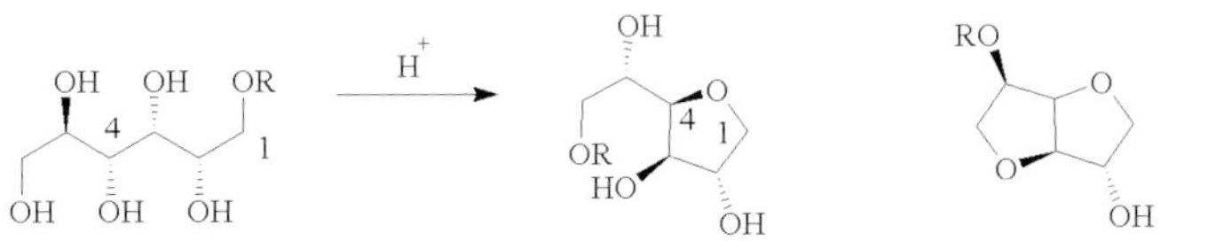

D-Glucitol (R = H) (sorbitol)

1,4-Anhydro-D-Glucitol (R = H)

1,4:3,6-Dianhydro-D-Glucitol (R = H) (isosorbitol)

"Sorbitol monoester" (SMS) : R = C_{16} - C_{18}-acyl

Scheme 3. Dehydration of D-sorbitol to "sorbitan", giving the 'sorbitan monoester' surfactant on esterification with C_{16}/C_{18}-fatty acids

The NMGA's have highly advantageous ecological and toxicological properties which allow their use as surfactants and cleansing agents, and also in cosmetic applications [35, 36].

3.5.3. Alkylpolyglucosides (APG)

APGs are commercially produced by several companies – most notably Cognis, with a capacity in the 50×10^3 t/a range – Kao, Seppic, and ICI. They are by far the most important nonionic surfactants and comprise fatty alcohol glucosides with an alcohol chain length normally between C_8 and C_{14}. Their industrial synthesis entails either direct acid-catalyzed Fischer glycosidation of glucose (in the form of a syrupy starch hydrolysate) or starch itself. The alternate process consists of two stages, the first being Fischer glycosidation with *n*-butanol to butyl glycosides which are subsequently subjected to acid-promoted transacetalization [34].

The resulting product mixtures predominantly contain the α-D-glucosides, as designated in the general formula (Scheme 4) and are marketed as such. APGs are not skin-irritating, have good foaming properties, and are completely biodegradable; hence they are widely used in manual dishwashing detergents and in the formulation of shampoos, hair conditioners and other personal care products [34].

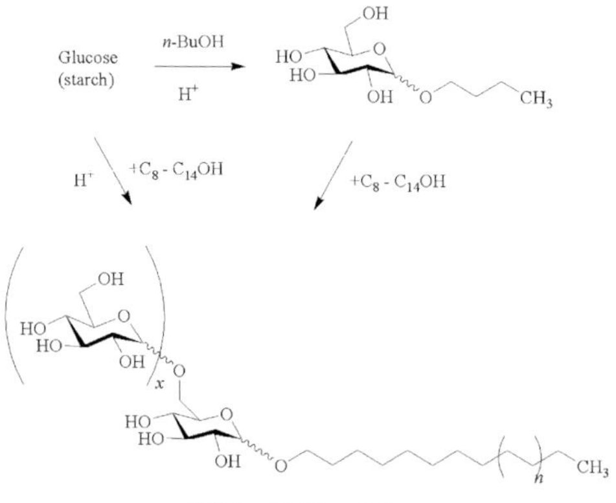

APG's ($x = 0.3$-0.7; $n = 2$-5)

Scheme 4. Synthesis of alkyl polyglucosides (APG's)

3.5.4. Sucrose Fatty Acid Monoesters

These compounds are currently produced at an approximate 4×10^3 t/a only, and are mostly used in cosmetic and personal care formulations because of their attractive dermatological properties. Produced by transesterification of fatty acid methyl esters or with fats, the resulting sucrose monoesters (if 1:1 molar ratios have been used in the process) are not defined products, acylated exclusively at the primary glucose-6-OH as indicated in the formula, but also at the other primary and some of the secondary OH groups [37].

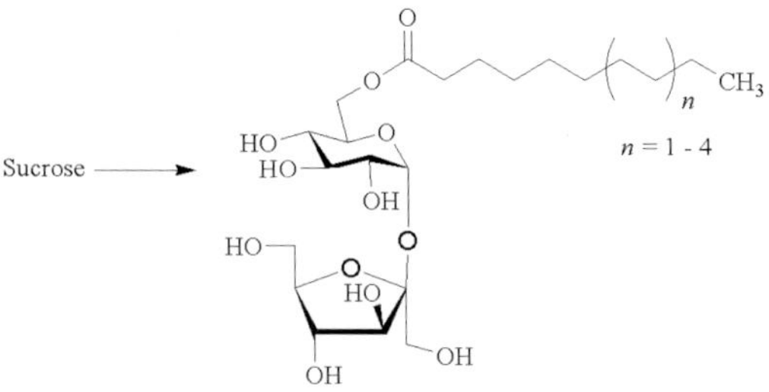

$n = 1$ - 4

3.6. Pharmaceuticals and Vitamins

Alongside the enormous amount of sugars, mostly glucose and sucrose, that flows into the fermentative production of amino and hydroxy acids (see Table 1) (a substantial part of which is for food use), a significant volume of these sugars is used in fermentation processes furnishing high-value-added products – antibiotics and vitamins, much too complex in their structures as to be generated by chemical synthesis. Figure 4 lists several representative examples – penicillins and cephalosporins with an estimated world production in the 7×10^3 t/a range, the aminoglycoside antibiotics of the kanamycin and spectinomycin type, or the optimized bioprocesses for the bulk-scale production of vitamins C and B_6.

Some sugar-derived drugs obtained by chemical means have also reached some importance, e.g. ranitidine (Zantac), an inhibitor of gastric acid secretion – one of the top 30 drugs based on sales (→ Pharmaceuticals, General Survey) – isosorbide dinitrate, a coronary vasodilatator (→ Cardiovascular Drugs, Chap. 15.) [38], or topiramate, a fructose-derived anticolvulsant drug with high antiepileptic efficacy [39].

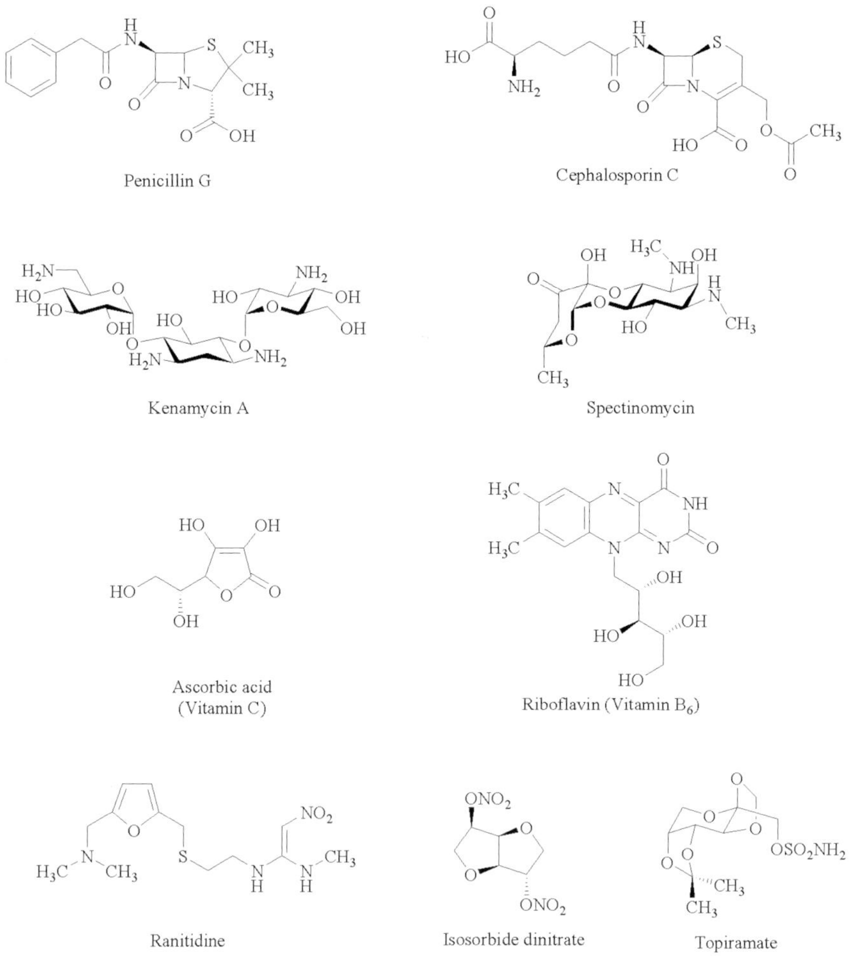

Figure 4. Sugar-derived high-value-added products – antibiotics, vitamins, and pharmaceuticals

4. Toward Further Sugar-based Chemicals: Potential Development Lines

Considering the large-scale, low-cost availability of the basic biomass-sugars listed in Table 1, most notably sucrose, glucose and fructose, their present nonfood utilization by chemical industry is modest indeed; i.e. the huge potential as the raw material for further viable industrial chemicals and materials is largely untapped. In view of the need for the chemical industry to somehow bring about the changeover from fossil raw materials to biofeedstocks, most notably carbohydrates from agricultural crops and waste materials, their further exploitation to produce industrially viable products is one of the major "green" challenges. The attempt to trace those sugar-based development lines along which the further use of the key sugars of biomass is likely to proceed, implies an assessment of many imponderables, particularly with regard to current dynamics in exploiting genetically engineered enzymes and the products resulting from them. Nevertheless, an "inventory" based on the present status may be expedient for focusing efforts on those areas where useful methods leading to promising products already exist yet await further development.

4.1. Furan Compounds

In addition to furfural, an established sugar-based five-carbon commodity with versatile industrial applications (see Section 3.2), several other furan compounds, readily prepared from sugars, hold high promise as industrial intermediate chemicals, albeit – for purely economic reasons – they are not (yet) produced on an industrial scale.

4.1.1. 5-Hydroxymethylfurfural (HMF)

Like many petroleum-derived basic chemicals, e.g., adipic acid or hexamethylenediamine, HMF is a six-carbon compound with broad application potential, inasmuch as it is readily accessible from fructose or inulin hydrolysates by acid-induced elimination of three molecular equivalents of water [40]. Even a pilot-plant-size process has been elaborated [41].

HMF has been used for the manufacture of special phenolic resins of type **12**, because acid catalysis induces its aldehyde and hydroxymethyl group to react with phenol [42].

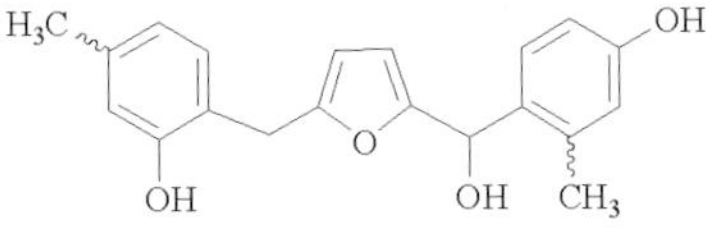

12

Of high industrial potential as key biomass-derived chemicals are various HMF-derived products for which well worked-out, large scale-adaptable production protocols are available (Scheme 5). Of these, 5-hydroxymethylfuroic acid (**13**), the 2,5-dicarboxylic acid **19**, the 1,6-diamine **15**, and the respective 1,6-diol **17** are most versatile intermediate chemicals of high industrial potential, because they are six-carbon analogues of adipic acid, of alkyldiols, or of hexamethylenediamine in the production of polyamides and polyesters. Indeed, an impressive series of furan polyesters and polyamides has been prepared [26, 28] in which the furandicarboxylic acid **19** replaces terephthalic and isophthalic acids in current industrial products (see

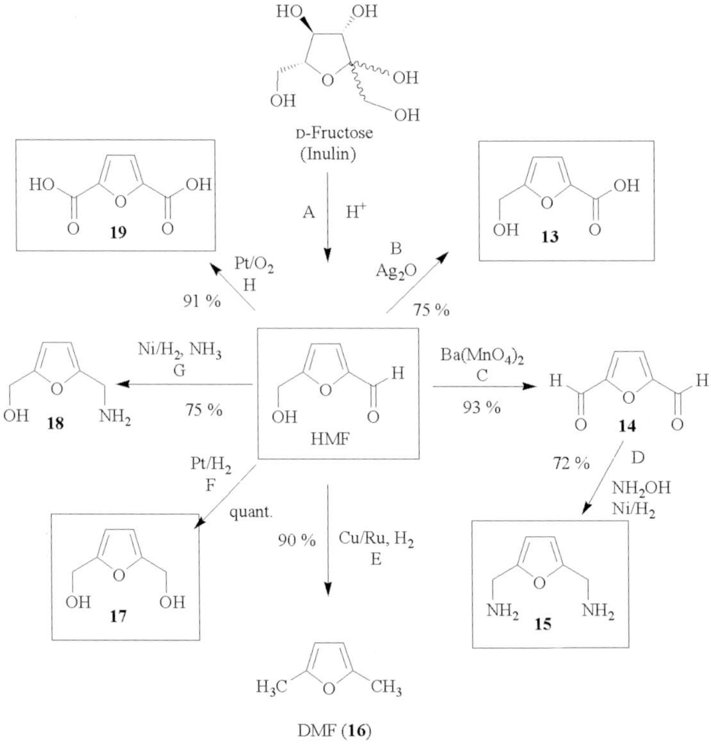

Scheme 5. Versatile intermediate chemicals from HMF.
A [40]; B [43]; C [43]; D [43]; E [44]; F [45]; G [46]; H [47]

Section 4.7.3). However, none has yet proved economically competitive with existing products.

4.1.2. 2,5-Dimethylfuran (DMF)

Hydrogenation of HMF over a copper-ruthenium catalyst affords DMF (**16**) (Scheme 5) in high yield [44]. DMF is considered to be a second-generation biofuel as it has vital advantages over ethanol such as full utilization of the six carbons of the sugar precursors; moreover, its energy content is similar to that of gasoline, it has a higher boiling point (by 16 °C) and it is immiscible with water.

To become a bulk-scale industrial product, however, the raw materials basis for its generation will have to change from D-fructose (Scheme 5) or D-glucose (rearrangement by base to D-fructose) to nonfood raw materials, i.e. cellulosic biomass from agriculture and forestry waste or switch grass and the like. Indeed, there are some promising developments in this direction. Exposure of cellulose to treatment with dimethylacetamide containing lithium chloride in the presence of $CrCl_2$ at 140 °C affords HMF in 54% yield [48] (cf. Scheme 6), whilst untreated lignocellulosic biomass such as corn stover as the raw material, when subjected to these conditions except for using $CrCl_3$ as promoter, delivers HMF in 48% yield [48]. These promising results, upon further development along these veins, are likely to eventually turn HMF and its ensuing products into key biomass-derived chemicals; most notably DMF as a biofuel.

4.1.3. Furans with a Tetrahydroxybutyl Side Chain

A very simple, one-step entry from hexoses to more highly substituted furans is their $ZnCl_2$-mediated reactions with 1,3-dicarbonyl compounds such as 2,4-pentenedione or ethyl acetoacetate. Because only the first two sugar carbons contribute to the formation of the furan, a distinctly hydrophilic tetrahydroxybutyl side-chain is produced. Thus, D-glucose smoothly provides furans **28** and **29** with the D-*arabino* configuration in the polyol fragment [50]; these can be shortened oxidatively to the dicarboxylic acid (**29** → **31**) or a variety of other furan building blocks (Scheme 7).

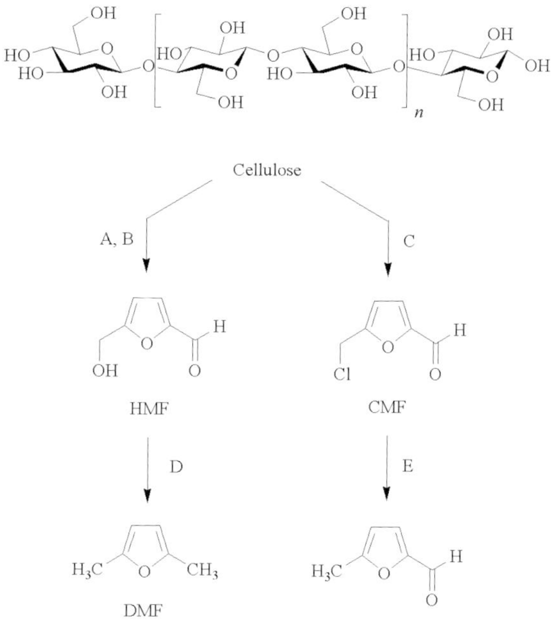

Scheme 6. Reactions and conditions
A, dimethylacetamide/LiCl (10%), $CrCl_2$, [EMIM]Cl, 2 h, 140 °C, 54% from cellulose [48].
B, dimethylacetamide/LiCl (10%), $CrCl_3$, [EMIM]Cl, 140 °C, 48% from corn stover [48].
C, LiCl/HCl in $(CH_2Cl)_2$, 65 °C, 71% [49].
D, Cu/Ru/C, H_2 in *n*-butanol, 120 °C, 79% [44].
E, $PdCl_2/H_2$ in *N,N*-dimethylformamide, 3 h, 40 °C, 88 [49]

In contrast, under mildly basic conditions (aqueous bicarbonate at 85 °C), D-glucose reacts with pentane-2,4-dione in an entirely different manner, producing, via *C*-addition and subsequent retroaldo-type elimination of OAc^-, the 2-*C*-glucosylpropanone **30** [51]. Because this conversion can be performed with the unprotected sugar and in aqueous solution with simple reagents, it may legitimately be referred to as a prototype of green and/or sustainable sugar transformations. The procedure is equally feasible with other aldohexoses and with D-fructose [52]; thus, it is one of the cleanest and most efficient preparative entries into the area of *C*-glycosides, which, as stable "mimics" of the usual *O*-glycosides, command major interest as glycosidase inhibitors [53].

4.2. Pyrones and Dihydropyranones

The bulk-scale-accessible mono- and disaccharides of Table 1 invariably adopt pyran

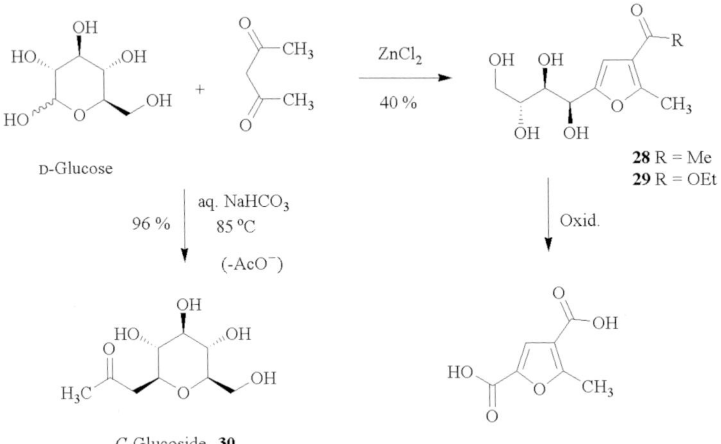

Scheme 7. One-pot conversions of D-glucose into hydrophilic furans [50] or, alternatively, into C-glucosides by reactions with acetylacetone [51]

cyclohemiacetal forms, from which well-elaborated, efficient reaction channels lead to an unusually large variety of unsaturated pyran building blocks; for example, pyrones, dihydropyrans, and dihydropyranones, of which the last two have the additional advantage of being enantiomerically pure. They are treated only cursorily in this context, because their potential as ideally functionalized six-carbon building blocks, particularly for the preparation of pharmaceutical targets, has not been utilized comprehensively.

Pyrones of type **32** (kojic acid) are readily obtained from D-glucose, the former either enzymatically by growing *Aspergillus oryzae* on steamed rice [54] or chemically via pyran 3,2-enolones [55, 56], or the γ-pyrone **33** by oxidation to D-gluconic acid and acetylation [57]. Both compounds, at present, are of little significance as six-carbon building blocks, despite a surprisingly effective route from **33** to cyclopentanoid products of type **34** [58] which is surmised to have industrial potential.

into glycal and hydroxyglycal esters, as depicted for D-glucose in Scheme 8. These esters not only have the oxygen content of D-glucose reduced – a precondition for elaboration of industrially viable products – but carry olefinic unsaturation in the pyran ring.

Despite the ready accessibility of these glucal and hydroxyglucal esters, and their well-developed ensuing chemistry, their exploitation as industrial intermediates is exceedingly modest. Nevertheless, to emphasize their potential toward industrial intermediates, whether as enantiomerically pure building blocks for the synthesis of noncarbohydrate natural products [56, 60] or for agrochemicals and/or high-value-added pharmaceuticals, a particularly versatile array of *six-carbon dihydropyranones* is listed in Figure 5 all are accessible from D-glucose (via the glucal and hydroxyglucal esters) in no more than three to five straight-forward steps.

A bicyclic dihydropyranone, levoglucosenone, is accessible even more directly by vacuum pyrolysis of waste paper [70]. Although the yield achievable is relatively low – levoglucosan is also formed, the proportion depends on the exact conditions (Scheme 9), – relatively large

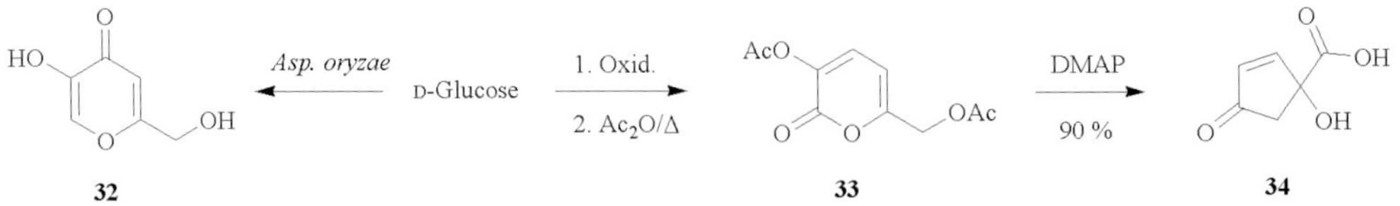

Other highly useful derivatizations of the common sugars, particularly towards the generation of enantiopure six-carbon building blocks, comprise their three-step, one-pot conversion

4.3. Sugar-Derived Unsaturated Nitrogen Heterocycles

Although transformation of sugars into trace amounts of nitrogen heterocycles occurs extensively on exposure of foodstuffs to heat (Maillard reaction [72]), and although a variety of nitrogen heterocycles have been generated from saccharide derivatives [73] procedures meeting preparative standards are exceedingly scarce. Improvements of existing procedures in the early 2000s and the development of new methods have led to the more ready acquisition of various nitrogen heterocycles from carbohydrates, e.g. imidazoles, pyrroles, pyrazoles, pyridines and quinoxalines which, because of their sugar derivation, have hydrophilic side chains – a favorable asset particularly in pharmaceutical applications [17].

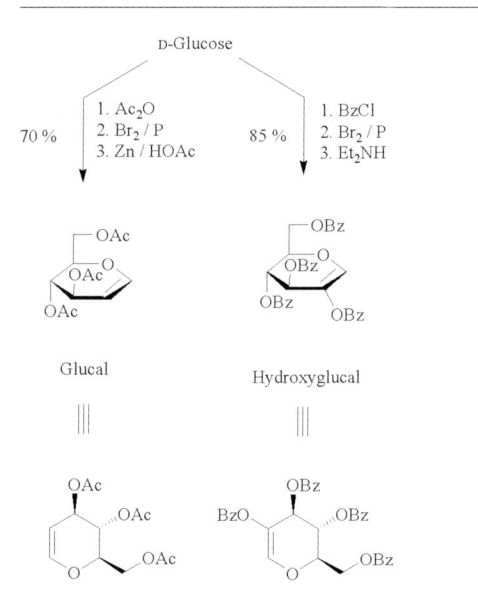

Scheme 8. Readily accessible pyran derivatives from D-glucose [59]

amounts can be amassed quickly. Levoglucosenone has been used for the synthesis of a diverse variety of natural products in enantiomerically pure forms [71].

4.3.1. Pyrroles

The generation of pyrroles by heating a glycerol solution of lactose-derived ammonium salt of

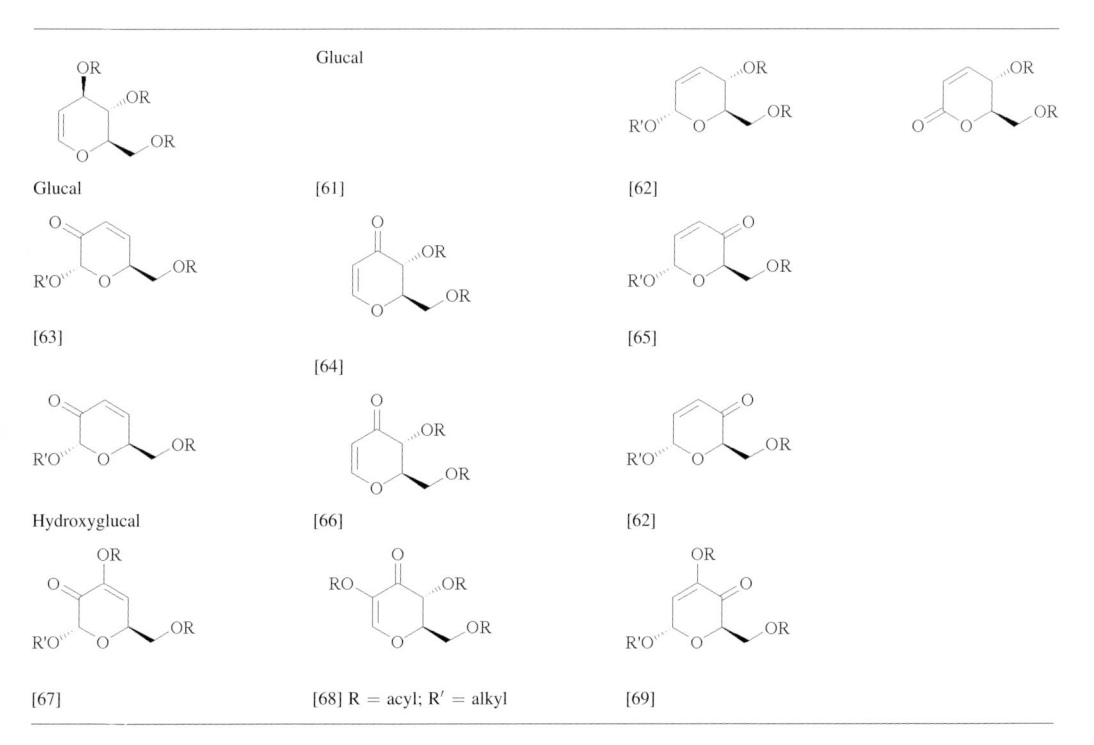

Figure 5. Pyranoid six-carbon building blocks accessible from D-glucose via glucal (upper half) or hydroxyglucal esters (lower entries). All products require no more than three to five straightforward steps from D-glucose [61–69]

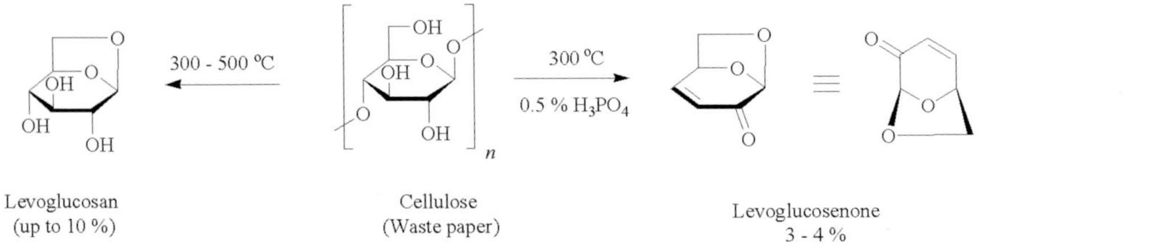

Scheme 9. High-vacuum pyrolysis of cellulose [70]

galactaric acid over a free flame [74] seems to be the highest-yielding acquisition (40%) from a carbohydrate source – a process that, in this or modified form, does not seem to be utilized industrially.

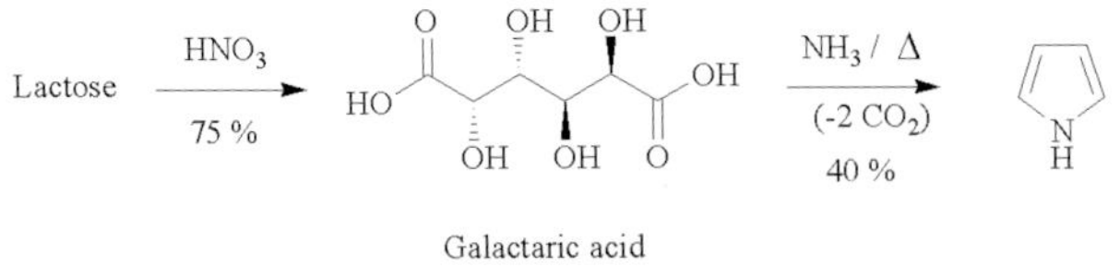

Pyrroles with a hydrophilic tetrahydroxybutyl substituent, e.g. **39**, are available from D-glucosamine by exposure to acetylactone or ethyl acetoacetate under mildly basic conditions [75] or in a one-pot reaction from D-fructose by heating with acetylacetone and ammonium carbonate in DMSO [76].

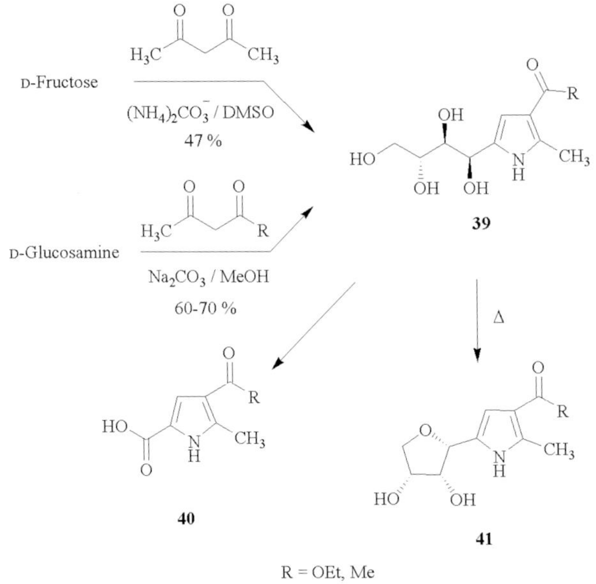

The hydroxylated side chain can, of course, be oxidatively shortened to give a variety of simple pyrrole building blocks, for example the carboxylic acid **40**, or cyclized to a furanoid ring (**39 → 41**) [75], compounds that may be regarded as C-nucleosides.

4.3.2. Pyrazoles

Expeditious four-step approaches to 1-phenyl-pyrazole-3-carboxaldehydes with 5-hydroxymethyl, 5-dihydroxyethyl, or 5-glucosyloxymethyl substituents have been elaborated starting from D-xylose [77], D-glucose, and isomaltulose [78], respectively.

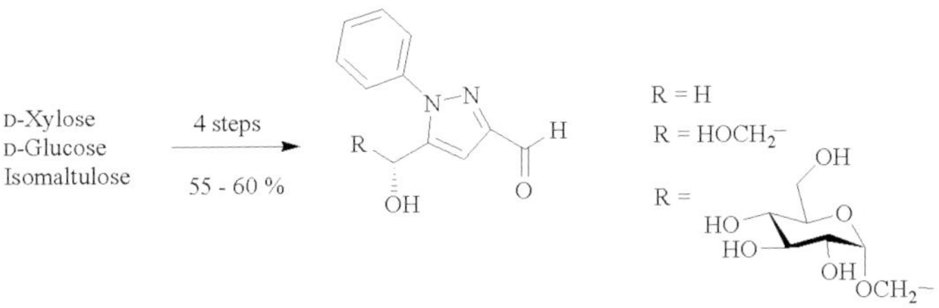

As illustrated for D-xylose, its osazone, formed almost quantitatively on reaction with phenylhydrazine, readily gives the pyrazole when added to acetic anhydride under reflux. Subsqent removal of the N-acetylphenylhydrazone residue with formaldehyde/acetic acid and de-O-acetylation provides a pyrazole aldehyde (57% overall yield from D-xylose), a versatile heterocyclic building block, useful for preparation of pharmaceuticals or monomers for the generation of polyamides and polyesters, e.g. the diamino and diol derivatives [77] (Scheme 10).

4.3.3. Imidazoles

A variety of imidazoles carrying hydrophilic substituents in the 4-position are readily accessible in one-pot procedures from standard monosaccharides. Of those, the formation of 4-hydroxymethylimidazole by a Cu(II)-promoted reaction with formaldehyde and conc. Ammonia [79] is rather unique, because obviously retro-aldolization to glyceraldehyde and dihydroxyacetone is involved (Scheme 11). The retroaldol fission can be partially suppressed, however, on heating D-fructose with formamidinium acetate in liquid ammonia in a pressure vessel [80] or with formamidinium acetate in the presence of boric acid and hydrazine, obviously proceeding via a boric acid complex of the bis(hydrazone) of D-glucosone.

These conditions can be readily applied to pentoses or disaccharides with acceptable yields, as exemplified when using D-xylose [80a] and isomaltulose [80b] in one-pot procedures.

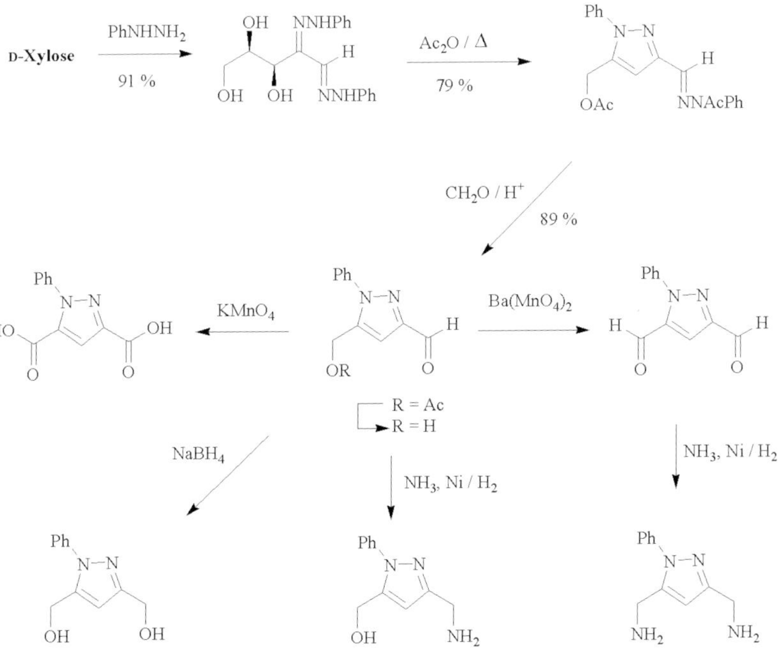

Scheme 10.

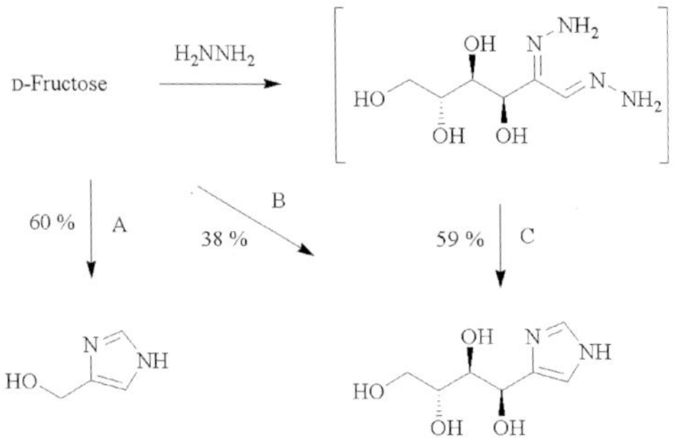

Scheme 11. Fructose-derived hydrophilic imidazoles
A: CH$_2$O, aq. NH$_3$, CuCO$_3$/Cu(OH)$_2$, 2 h, 100 °C [79]; B: formamidine (HOAc/liq. NH$_3$, 15 h, 75 °C [80a]; C: N$_2$H$_4$/ formamidine, HOAc, H$_3$BO$_3$, 3 h reflux [80b]

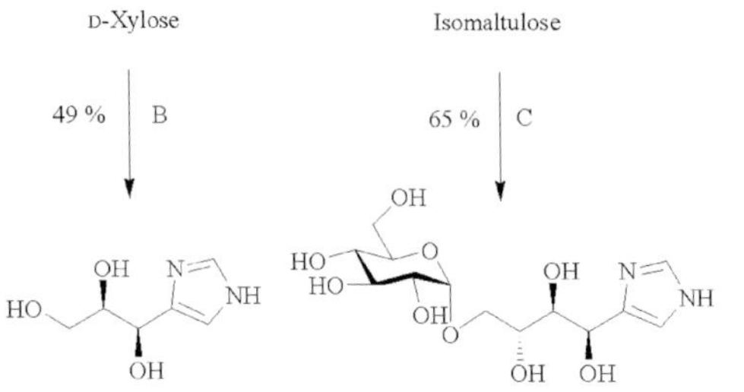

4.3.4. 3-Pyridinols

The conversion of pentosans or pentoses into 3-pyridinol can be effected in a practical three-step sequence, involving acid-induced dehydration to furfural, reductive amination to furfurylamine, and subsequent oxidation with hydrogen peroxide [81], the last step conceivably proceeding through the stage of a 2,5-dihydroxy-2,5-dihydrofurfurylamine, which forms the pyridine nucleus by dehydration to a 5-aminopentenal intermediate and intramolecular aldimine formation. The pyridinol is a prominent intermediate chemical for the preparation of herbicides and insecticides [82], and cholinergic drugs of the pyridostigmine type.

For the conversions of furfurylamines with readily oxidizable hydroxyl groups, e.g. those derived from fructose via HMF/bromine in water/methanol, the entire multistep process to the hydroxymethylpyridinol takes place in a one-pot procedure [83] (Scheme 12).

4.4. Toward Sugar-Based Aromatic Chemicals

On the basis of a 1995 compilation [11], twenty of the 100 major organic chemicals in the US were aromatic compounds, invariably manufactured from fossil raw materials, mostly from the BTX (benzene–toluene–xylene) fraction derived from naphthas in the refineries. There are very few alternatives. The direct thermochemical conversion of biomass into an equivalent BTX product is not realistically feasible, because only small amounts of monocyclic aromatic hydrocarbons – phenols of the catechol and pyrogallol series – are formed on pyrolysis or thermal cracking of woody feedstocks. The same is true for exposure of simple sugars, for example D-xylose, D-glucose or D-fructose, to either basic or slightly acidic aqueous conditions at 100–160 °C [84]. Vanillin, however, is a byproduct of the manufacture of cellulose pulp by the action of alkali on calcium lignosulfonate and may be isolated in yields of up to 25%.

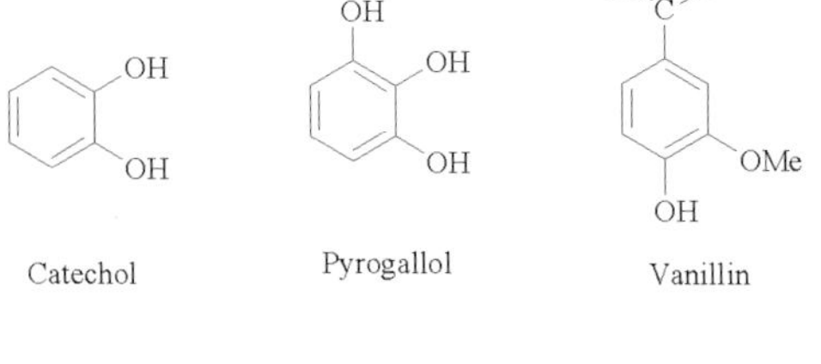

An entirely different, highly-promising approach from sugars to industrially relevant aromatic compounds is based on microbial conversions along the shikimic acid pathway using genetically modified biocatalysts. By incorporation

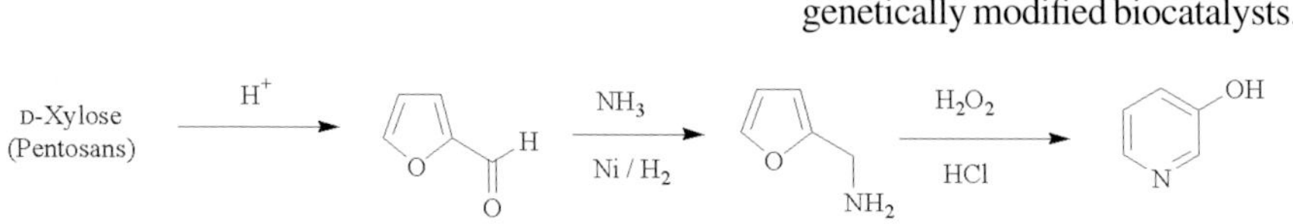

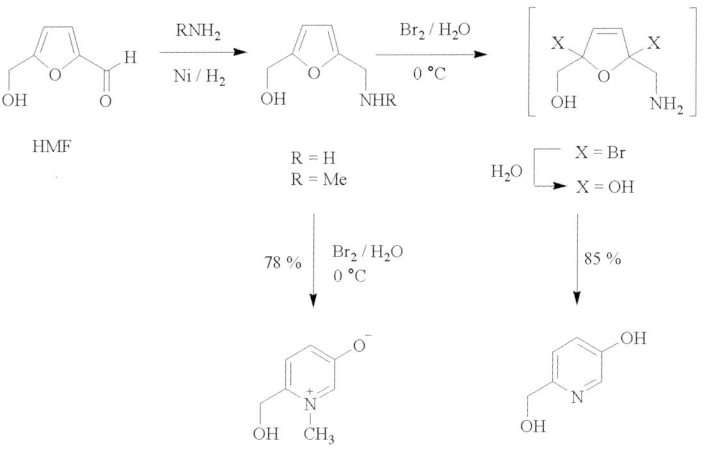

Scheme 12.

of the genomic portion encoding the synthesis of 3-dehydroquinic and 3-dehydroshikimic acid into *Escherichia coli* constructs, the carbon flow is channeled into the accumulation of large amounts of either quinic or shikimic acid [85] (Scheme 13); thus, rendering their availability independent of an often difficult isolation from plant sources. These improvements are likely to lead to pronounced expansion of the synthetic utilization of these enantiomerically pure carbocycles, not only in the pharmaceutical industry, where quinic acid already is already used as the starting material for the synthesis of the anti-influenza drug Tamiflu (oseltamir phosphate [86]), but for the production of bulk scale commodity chemicals such as hydroquinone [87] or phenol [88] by application of simple chemical transformations.

The powerful potential of metabolic engineering is similarly manifested in the *E. coli* biocatalyst-promoted conversion of D-glucose into protocatechuic acid, which can be readily decarboxylated to catechol [89]. This two-step process (see Scheme 13), feasible in a 24% overall yield, may replace the present process used to manufacture of this 25 000 t/a petrochemical commodity (→ Phenol Derivatives, Chap. 2.). Of similar significance seems to be the genetically modified microbe-catalyzed conversion of D-glucose into gallic acid and pyrogallol [90]; the accessibility of these chemicals currently relies on their isolation from plant sources, despite a wide range of uses, particularly as starting materials for pharmaceuticals (→ Phenol Derivatives).

The recent unraveling of the biosynthesis of phloroglucinol may also pave the way to its production from glucose. Thus, by expressing in *Escherichia coli* the *Pseudomonas fluorescens* enzyme (that assembles three molecules of malonyl coenzeme A into an activated diketoheptanethioate) phloroglucinol can be generated from glucose in yields of up to 10 g/L. In this process the initially formed diketoheptanethioate subsequently undergoes cyclization and conceivably spontaneous aromatization into the trihydric phenol [91].

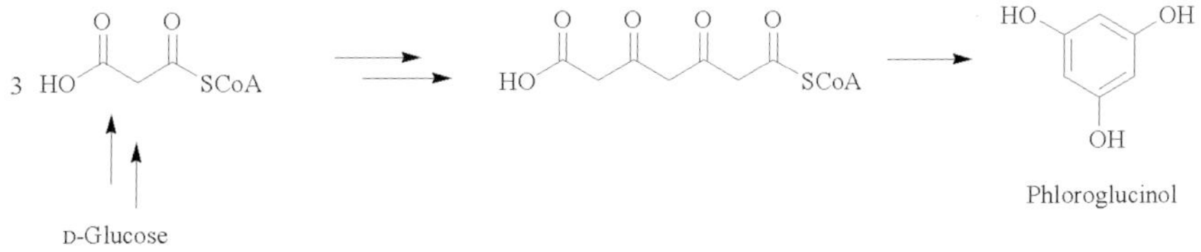

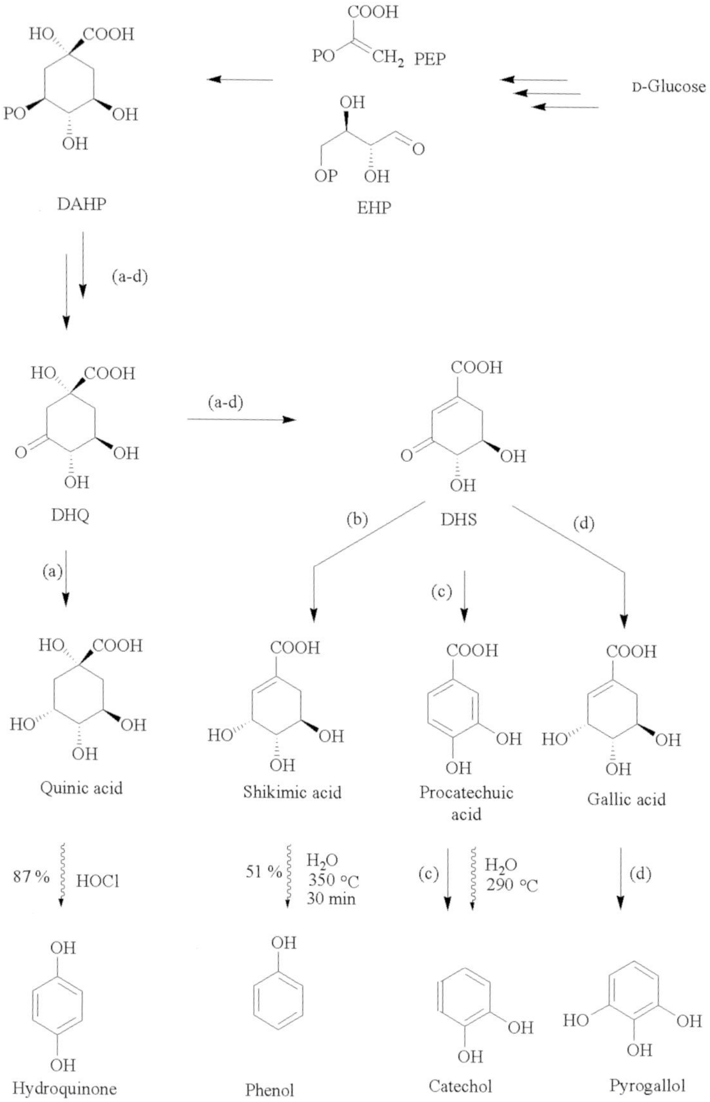

Scheme 13. Metabolic engineering of the shikimic acid pathway intermediates toward aromatic chemicals: → chemical transformations; ⇒ bioconversions with biocatalysts
(a) *E. coli* QP1.1/pKD12.138 [85], (b) *E. coli* SP1.1PTS/pSC6.090B [85], (c) *E. coli* KL3/pWL2.46B [87], (d) *E. coli* KL7/pSK6.161 [88]
Abbreviations: PEP = phosphoenolpyruvic acid, EHP = erythrose-4-phosphate, DAHP = 3-deoxy-D-arabinoheptulosonic acid 7-phosphate, DHQ = 3-dehydroquinic acid, DHS = 3-dehydroshikimic acid

4.5. Microbial Conversion of Six-Carbon Sugars into Simple Carboxylic Acids and Alcohols

Major future development lines towards the economically viable generation of industrial chemicals from carbohydrates lie in the targeted utilization of genetically engineered bioprocesses, most notably, those that involve the bioconversion of sugars into industrially relevant bioalcohols (other than ethanol) and into simple C_3–C_5- carboxylic acids other than those already exploited (i.e. lactic, citric and tartaric acids, and a variety of amino acids, see Table 1). Table 2 lists a variety of acids and alcohols that can be

Table 2. C_3–C_5 carboxylic acids and alcohols producible by microbial fermentation[a]

	Product[b]	Substrate	Microorganism
Acids	Propionic acid	various sugars	Clostridium sp.
			Propionibacterium shermanii
	Pyruvic acid	glucose	Pseudomonas aeruginora
	3-Hydroxypropionic acid	glucose	E. coli constructs
	Butyric acid		Clostridium butyricum
	3-Hydroxybutyric acid	glucose	Alcaligenes eutrophus
	Succinic acid	various sugars	Actinobacillus succinogenes Mannheimia succiniciproducens
	Fumaric acid	various sugars	Rhizopus nigricans Rhizopus arrhizus
	L-Malic acid	various sugars	Parcolobactrum sp. Brevibacterium sp.
	Itaconic acid	various sugars	Aspergillus terreus Aspergillus itaconicus
	2-Oxoglutaric acid	glucose	Pseudomonas fluorescens
Alcohols	n-Propanol	glucose	Clostridium fallax
	Isopropanol	glucose	Clostridium sp.
	1,2-Propandiol	glucose	E. coli constructs
	1,3-Propanediol	glucose, glycerol	Clostridium pasteurianum E. coli mutants
	Glycerol	glucose	yeasts
	n-Butanol	various sugars	Clostridium butylicum
	2,3-Butanediol	glucose, xylose	Klebsiella pneumoniae Bacillus polymyxa
	1,2,4-Butanetriol	xylose (xylonic acid)	E. coli construct

[a] Compiled from [92].
[b] Acids already exploited industrially, i.e. lactic, tartaric and citric acid (cf. Table 1) are not listed here.

obtained by microbial production and substantial further activities in this field are to be expected.

However, which of these products are likely to enter industrial production will be determined by a variety of factors – demand, availability of a genetically engineered biocatalyst, and, not least, by economics, with rising oil prices increasingly providing more favorable conditions.

The products with significant industrial potential marked in bold in Table 2 are being briefly discussed below with respect to the current status of their microbial production and future prospects.

4.5.1. Carboxylic Acids

In 2004 the US Department of Energy (DOE) published a list of twelve sugar-derived chemicals worthy of industrial exploitation [93a–h], of which five are simple carboxylic acids accessible by biotransformations of sugars, mostly glucose:

- 3-Hydroxypropionic acid (3-HPA)
- Four-carbon 1,4-diacids (malic, fumaric, succinic acids)
- Itaconic acid (IA)
- Aspartic acid (Asp)
- Glutamic acid (Glu)

L-Glutamic acid   L-Aspartic acid   Fumaric acid

Each of these are considered to have high "building block potential" either as monomers for the production of novel polyesters and polyamides [94] or as a starting material for a variety of commodity chemicals, currently produced along petrochemical pathways. Accordingly, these five carboxylic acids are to become economically viable products if low-cost fermentation routes can be developed and implemented on an industrial level. Even for *glutamic acid* – for which several fermentation processes have been industrially realized to meet the need for its sodium salt as a flavor enhancer – the productivity of the organism and the final fermentation titer must be improved if it is to become an

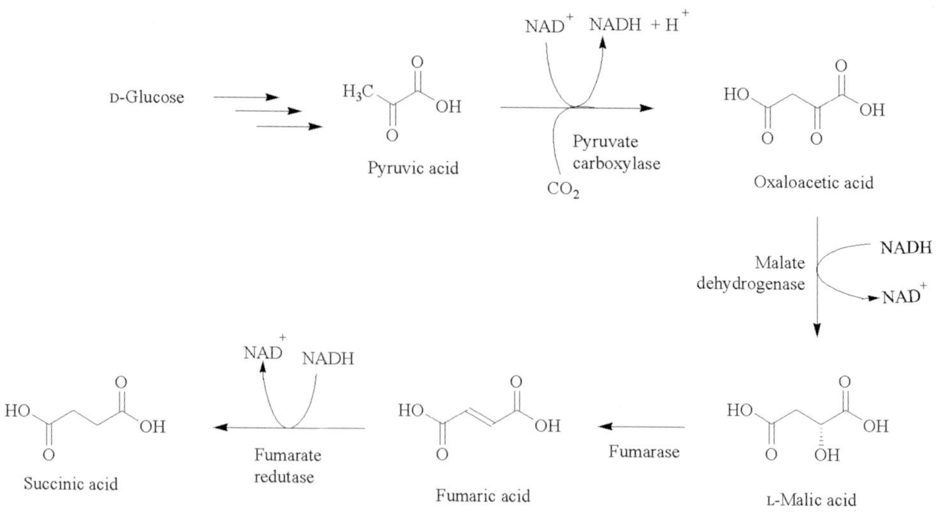

Scheme 14. Glycolytic pathway leading to L-malic, fumaric, and succinic acids [95]

attractive candidate for chemical and microbial follow-up transformations [93a]. The same is true for *aspartic acid*, for which the tricarboxylic acid cycle of biocatalytic organisms can be engineered to overproduce oxaloacetate (cf. Scheme 14) to provide a product competitive with the (racemic) one obtained from petro-derived fumaric acid by amination [93b].

3-Hydroxypropionic Acid (3-HPA) represents a three-carbon building block with the potential to become a key intermediate for a variety of high-volume chemicals, most notably acrylic acid, presently produced on a 3×10^6 t/a level by oxidation of propylene or 1,3-propanediol [93c]. Cargill and Novozymes are jointly developing a low-cost fermentation route from sugars by metabolic engineering of the microbial biocatalyst that produces 3-HPA under anaerobic conditions, yet the hurdles still to be overcome are significant, delaying the immediate introduction of this technology [96].

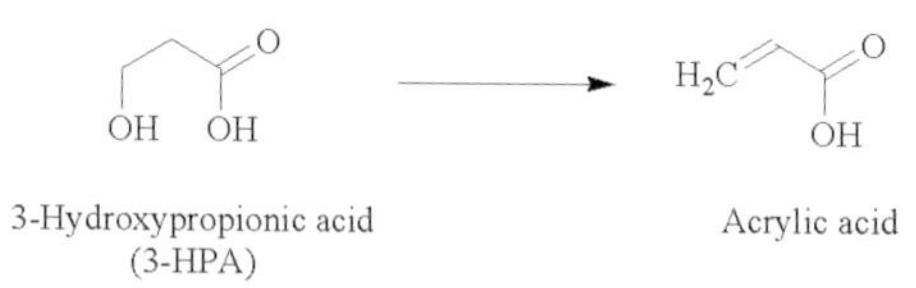

Fumaric Acid is a metabolite of many fungi, lichens, mosses and some plants; it is mainly used as the diacid component in alkyd resins (→ Maleic and Fumaric Acids) and is produced commercially, to some extent, by the fermentation of glucose in *Rhizopus arrhizus* [97]. It seems, however, that improvements in productivity are essential before this approach can be considered an option for replacing the current petrochemical method that requires the catalytic isomerization of maleic acid.

Malic Acid. Most of the malic acid produced, approximately 10^3 t/a, is racemic, because it is derived from petrochemically produced fumaric acid. The L-form can also be produced from fumaric acid by hydration with immobilized cells of *Brevibacterium* or *Corynebacterium* species.

Succinic Acid is used to produce food and pharmaceutical products, surfactants and detergents, biodegradable solvents and plastics, and ingredients to stimulate animal and plant growth. Although it is a common metabolite formed by plants, animals and microorganisms, its current commercial production of about 15×10^3 t/a is from petroleum; i.e. by hydrogenation of malic acid. The major technical hurdles for succinic acid as green, renewable bulk scale commodity chemical – 1,4-butanediol, THF, γ-butyrolactone or pyrrolidones are industrially relevant products – entail the development of very low-cost fermentation routes from sugar feedstocks. Currently available anaerobic fermentations of glucose include the use of an organism genetically cloned from *Aspergillus succinoproducens*, an engineered *E. coli* strain developed by DOE laboratories [93d], and several others [98]. The processes are currently under active development [99]. Production costs are to be at or below $ 0.25/lb to match those from fossil raw materials [93d].

Itaconic Acid (IA) is an α-substituted methacrylic acid; providing a C_5 building block with significant market opportunities. It is currently produced by fungal fermentation at approximately 10^3 t/a [100] and mainly used as a specialty comonomer in acrylic or methacrylic resins, as incorporation of small amounts of itaconic acid into polyacrylonitrile significantly improves its dyeability.

Itaconic acid

To become recognized as a commodity chemical, productivity improvements with the currently used fungi *Aspergillus terrous* and *Aspergillus itaconicus* are required. Progress seems to be encouraging [100]; however, to become competitive with analogous commodities the crucial production price of about $ 0.25/ lb has to be reached [93e] – a significant technical challenge yet to be solved.

4.5.2. Potential Sugar-Based Alcohol Commodities by Microbial Conversions

1,3-Propanediol and terephthalic acid, the dicarboxylic acid component of poly(trimethylene terephthalate) – a high performance polyester fiber with extensive applications in clothing textiles and carpeting – are currently manufactured from petrochemical raw materials by Shell ("Corterra") and DuPont ("Sorona") [101].

Poly(timethylene terephthalate) ('PTT')

For the diol portion of the polyester, 1,3-propanediol, however, biobased alternatives have been developed which rely on a microbial conversion of glycerol [102], a byproduct of biodiesel production, or of corn-derived glucose [103]. For the latter conversion, DuPont has developed a biocatalyst, engineered by incorporating genes from bakers' yeast and *Klebsiella pneumoniae* into *E. coli*, which efficiently converts corn-derived glucose into 1,3-propanediol [103]. The bioprocess is implemented on an industrial scale in a Tennessee manufacturing plant by a DuPont/Tate & Lyle joint venture ("BioPDO").

D-Glucose →(*E. coli* fermentation)→ HO~~~OH

1,2-Propanediol, in its racemic form a petroleum-based high-volume-chemical with an annual production of over 5×10^5 t, is mostly used for the manufacture of unsaturated polyester resins, yet it also has excellent antifreeze properties. Enantiomerically pure (*R*)-1,2-propanediol accumulates along two different pathways via DAHP (cf. Scheme 13) or methylglyoxal, which is then reduced with either hydroxyacetone or lactaldehyde as the intermediates. Both routes have been examined for microbial production from glucose by means of genetically engineered biocatalysts, obtained either by expressing glycerol dehydrogenase genes or by overexpressing the methylglyoxal synthase gene in *E. coli* [104]. This work provides a basis for further strain and process improvements. Another approach entails inoculating silos containing chopped whole-crop maize with *Lactobacillus buchneri*. After storage for four months, yields of 50 g/kg were reported [105]; thus, prospects for elaborating an economically sound bioprocess look promising.

1,2,4-Butanetriol. Used as an intermediate chemical for alkyd resins and rocket fuels, 1,2,4-butanetriol is presently prepared commercially from malic acid by high-pressure hydrogenation or hydride reduction of its methyl ester. In a novel environmentally benign route to this chemical, wood-derived D-xylose is microbially oxidized to D-xylonic acid, followed by a multistep conversion to the product by use of a biocatalyst specially engineered by inserting *Pseudomonas putida* plasmids into *E. coli* [106]. Although further metabolic engineering is required to increase product concentration and yields, microbial generation of

D-Xylose —*Pseudomonas fragi*→ D-Xylonic acid —*Escherichia coli*→ 1,2,4-Butanetriol

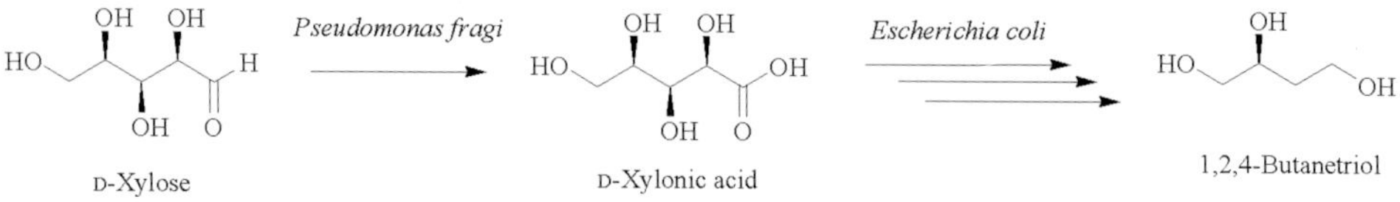

4.6. Chemical Conversion of Sugars into Carboxylic Acids

The sugar-derived carboxylic acids listed in Table 1, i.e. gluconic, citric, lactic, tartaric, and ascorbic acids, are accessible in bulk by fermentation processes and may be considered (and used as) commodity chemicals despite being mostly used for food purposes. In addition to these compounds, however, there are several industrially attractive carboxylic acids obtainable from sugars by chemical means which have high potential as versatile building blocks.

In the 2004 DOE report [93], four of these sugar-derived carboxylic acids have already been singled out as suitable candidates for further development:

- Furan-2,5-dicarboxylic acid
- Glucaric acid
- Levulinic acid, and
- 3-Hydroxybutyrolactone

yet there are a many others that equally merit the development and implementation of low-cost preparative procedures to become competitive products. They are addressed briefly here.

Furan-2,5-Dicarboxylic Acid. The high industrial potential of furan-2,5-dicarboxylic acid (**19**) (cf. Scheme 5), has already been emphasized, because **19** could replace petroleum-derived diacids, such as adipic or terephthalic acid, in the production of polyesters and polyamides.

Aldaric Acids. The aldaric acids of the key hexoses and pentoses as highly hydrophilic diacids also have much potential in industrial use, similar to that of the sugar platform. The most important are D-glucaric acid, a direct nitric acid oxidation product of glucose or starch [107], usually isolated as its 1,4-lactone; galactaric and xylaric acid, accessible from lactose [108] and from D-xylose or hemicellulosic xylans. The technical barriers to their large-scale production are mainly development of efficient and selective oxidation technology for these sugars to eliminate the need for nitric acid as the oxidant. Investment in the Pt- or Au-catalyzed oxidation with oxygen appears to be a promising approach.

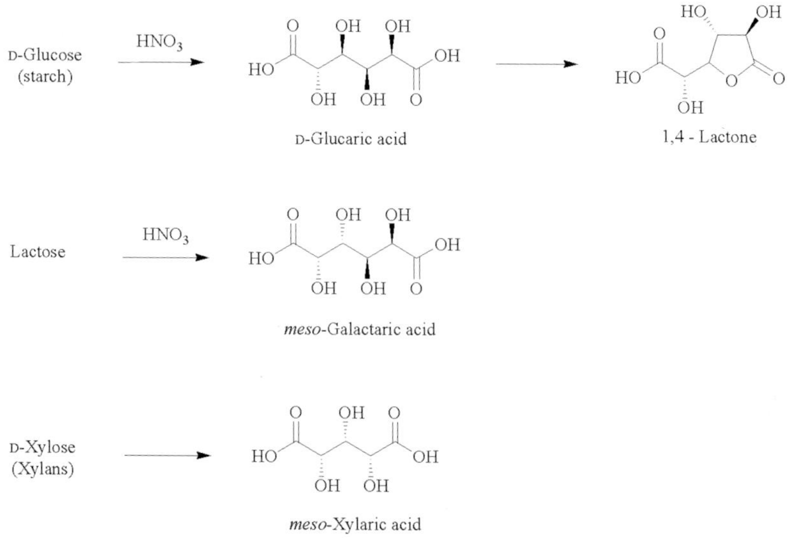

D-Glucose (starch) —HNO_3→ D-Glucaric acid → 1,4-Lactone

Lactose —HNO_3→ *meso*-Galactaric acid

D-Xylose (Xylans) → *meso*-Xylaric acid

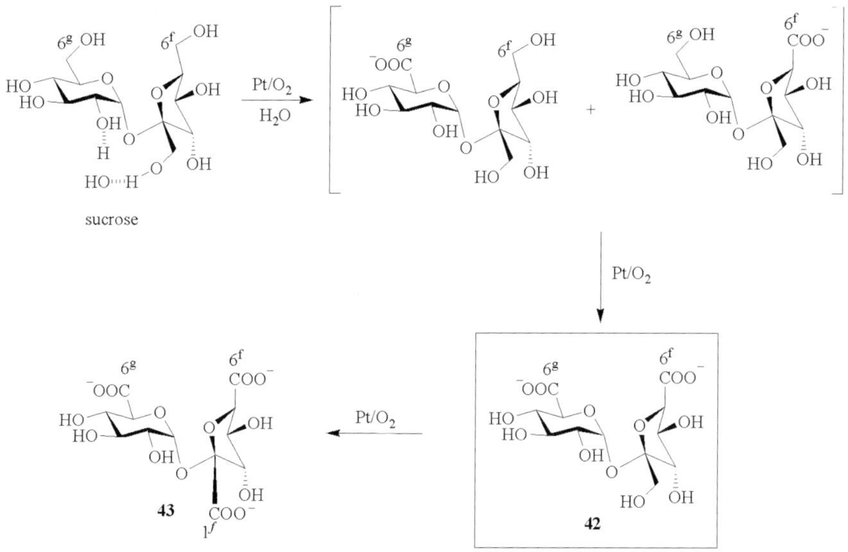

Scheme 15. Catalytic oxidation of sucrose [112, 113]

Because sucrose is cheaper than its component sugars D-glucose and D-fructose, and available in large quantities (cf. Table 1) carboxylic acids resulting from selective oxidation of its primary hydroxyl groups are likely to be of even higher industrial relevance in the future. Through the persistence of an intersaccharidic water-bridge of the 2^g-HO$\cdots$H$_2$O$\cdots$HO-1^f type in aqueous solution [109], oxidation of sucrose with air in the presence of 0.5% Pt/C at 35 °C gives a 9:9:1 ratio of the 6^g-, 6^f- and 1^f-saccharonic acids [110] (Scheme 15). On further oxidation, particularly when using large amounts of the Pt catalyst and higher temperatures (80–100 °C), the formation of sucrose-6^g,6^f-dicarboxylic acid **42** has been observed [111], yet a preparatively useful procedure for its acquisition was developed only end of the 1990s [112] by combining Pt/air-oxidation with continuous electrodialytic removal of **42**, thereby protecting it from further oxidation. Otherwise, on letting the reaction proceed, the sucrose tricarboxylate **43** is obtained [113].

Levulinic Acid ("LA") and *formic acid* are end products of the acidic and thermal decomposition of lignocellulosic material, their multistep formation from the hexoses contained therein proceeding through HMF as the key intermediate, the hemicellulosic part, mostly xylans, furnishes furfural [26a, 26b]. A commercially viable fractionation technology for the specific, high-yield acquisition of LA has been developed (the "Biofine Process" [114]), rendering it an attractive option as an important biorefinery platform chemical [93g].

Levulinic acid is a starting material for a large number of higher-value products, because it can be converted by established procedures into products such as acrylic and succinic acids, pyrrolidines, and diphenolic acid (**44**), which has the potential of replacing bisphenol A in the manufacture of polycarbonate resins. Another derivative is 5-aminolevulinic acid (**45**), applied in agriculture as a herbicide and as a growth-promoting factor for plants. Another asset is that levulinic acid may be converted into 2-methyltetrahydrofuran (**46**), which is used as a liquid fuel extender (Scheme 16). The esters of LA, e.g., ethyl levulinate, have similar industrial potential as oxygenated additives to diesel fuels.

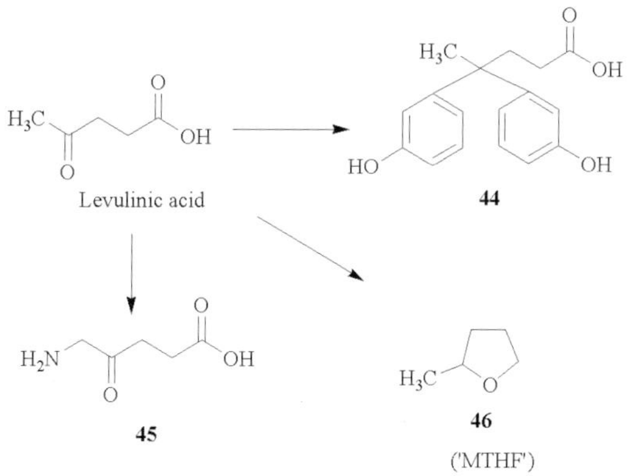

Scheme 16.

3-Hydroxybutyrolactone (3-HBL) is now a specialty chemical for fairly high-value pharmaceuticals. The 2004 DOE report [93h] places it in the list of the top twelve sugar-based candidates worthy of industrial exploitation. Its generation from starch by oxidative (H_2O_2) degradation is regarded as "messy", because it involves multiple steps and results in a variety of side-products. Thus, this process must be improved, or alternatives found – one such alternative is reduction and cyclization of microbially produced L-malic acid (see Section Fumaric Acid, Malic Acid, Succinic Acid) – to fully utilize the potential of this four-carbon building block for the production of a variety of industrially important tetrahydrofuranoid derivatives.

boxylic acids shown in Tables 3 and 4 – only those which are reasonably accessible are listed – the number of possible polyesters is immense, and not all conceivable combinations have been implemented and evaluated for their application profiles. The only one of industrial relevance today is Cargill's polylactic acid (PLA), used as a benign, biodegradable material for packaging, for disposable single use items, and for medical devices (vide supra). Nevertheless on the verge of becoming an industrial bioproduct is Du Ponts's poly(trimethylene terephthalate) – its high-performance fiber Sorona. In this case one of the raw materials, presently petroleum-derived 1,3-propanediol, is being replaced the same component obtained from glucose by microbial fermentation.

Starch $\xrightarrow{H_2O_2/H^+}$ 3-Hydroxybutyrolactone $\xleftarrow{\text{red.}}$ L-Malic acid

4.7. Biopolymers from Polymerizable Sugar-Derivatives

Today, biocompatibility and biodegradability are key functional requirements in the design of new polymeric materials, whether polyesters, polyamides, or polyurethanes ($\rightarrow$ Polymers, Biodegradable). If composed of sugar-derived monomer components, such polymers are nontoxic and biodegradable and, hence, have minimal impact on waste management. They can be safely incinerated and, by composting, can be returned to the ecosystem harmlessly in a carbon dioxide-neutral process.

4.7.1. Synthetic Biopolyesters

The polyester production worldwide is estimated to be approximately 2×10^7 t/a, of which only a small fraction is based on renewable monomers, such as polyols, dicarboxylic acids, or hydroxyalkanoic acids, despite the fact that a huge variety of these building blocks are amenable to either chemical or biotechnological production. As amply illustrated by the large variety of sugar-derived di- and polyols, hydroxyacids, and dicar-

Of the vast number of polyesters prepared from the monomers listed in Tables 3 and 4, those containing furan residues have attracted particular interest [27, 29]. 5-Hydroxymethyl-2-furoic acid, for example, gave a mixture of linear (**47**) and cyclic products (**48**) on polycondensation [115, 116], whereas the 2-furoylacrylic acid analogue afforded the polyester **49**; 2,5-furandicarboxylic acid has been polyesterified with a series of aliphatic diols ($\rightarrow$ **50** [116]), with dianhydrosorbitol ($\rightarrow$ **52** [117]), or with bisphenols ($\rightarrow$ **53** [118]) (cf. Fig. 6). Even the all-furan polyester **54** has been successfully prepared from its respective monomeric components [27] – like polyesters **47–50**, **52** and **54**, a "fully green", naturally resourced product. The same applies to polyester **51**, composed of 1,3-propanediol and the furan-2,5-diacid. This polyester is in fact an analogue of Du Pont's Sorona wherein the terephthalic acid portion is replaced by a bio-counterpart. Given the same fiber properties it would rightfully deserve the "clothing from a cornfield" attribute.

Despite the versatile application profiles of these polyesters – and a vast number of others that have been synthesized – they have been prepared only on the laboratory scale and used

Table 3. Sugar-based alcohols and diamines suitable as monomers for polyesters, polyamides, or polyurethanes

Compound		Derivation
	1,2-Propanediol	D-glucose
	1,3-Propanediol	D-glucose
	Glycerol	fats (D-glucose)
	2,5-Bis(hydroxymethyl) furan	D-fructose
	1,6-Dianhydro-D-sorbitol	D-glucose
	3,5-Bis(hydroxymethyl) pyrazole	D-xylose
	Xylitol	D-xylose
	D-Sorbitol	D-glucose
	1,6-Diamino-1,6-dideoxy-D-glucitol	D-glucose
	2,5-Diamino-1,4:3,6-dianhydrosorbitol	D-glucose
	2,5-Bis(aminomethyl)furan	D-fructose

in experimental applications and so currently are "academic curiosities". They will remain so, however, only as long as the economics are in favor of the production of the monomeric components from fossil raw materials – a situation that will change within the near future [5].

4.7.2. Microbial Polyesters

Microbial polyesters, or PHA for poly(hydroxyalkanoates) [119, 120], constitute a large and versatile family of polyesters produced by various bacteria in which they are deposited

Table 4. Sugar-derived carboxylic acids for potential use as monomers for polyesters and polyamides

Compound		Derivation
	Lactic acid	D-glucose
	3-Hydroxybutyric acid	D-glucose
	5-Hydroxymethylfuroic acid	D-fructose
	3-Hydroxyvaleric acid	D-glucose
	D-Xylaric acid	D-xylose
	D-Glucaric acid	D-glucose
	D-Galactaric acid	lactose
	Furan-2,5-dicarboxylic acid	D-fructose
	Pyrazole-3,5-dicarboxylic acid	D-xylose
	Sucrose-6,6′-dicarboxylic acid	sucrose
	GMF-6,6′-dicarboxylic acid	isomaltulose (sucrose)

Table 4 (Continued)

Compound	Derivation
5-Aminomethylfuroic acid	D-fructose
6-Amino-D-gluconic acid lactam (R = H, CH₃)	D-glucose

within the bacterial cell wall as a storage polymer. Industrial processes have been developed that deliver poly(3-hydroxybutyrate) [poly(3HB)] and a polymer consisting of 3-hydroxybutyric and 3-hydroxyvalerianic acid [poly(3HB)-*co*-3(HV), trade name Biopol]. Both polyesters have outstanding properties in respect to thermoplastic behavior, biocompatibility, and biodegradability, and hence have wide applications in cosmetics, hygiene and agricultural materials, in drug delivery systems, and in medical surgery.

Poly (3HB)

Poly (3HB-*co*-3HV) (Biopol)

Other representative PHA's synthesized by microorganisms contain 3-hydroxyhexanoic, 3-hydroxyoctanoic and malic acid as repeating units. It is to be expected that improvement of fermentation strategies, e.g. by recombinant *E. coli* harboring the microbial PHA biosynthesis

47

48 $n = 3 - 6$

49

50 $x = 6$
51 $x = 3$

52

53

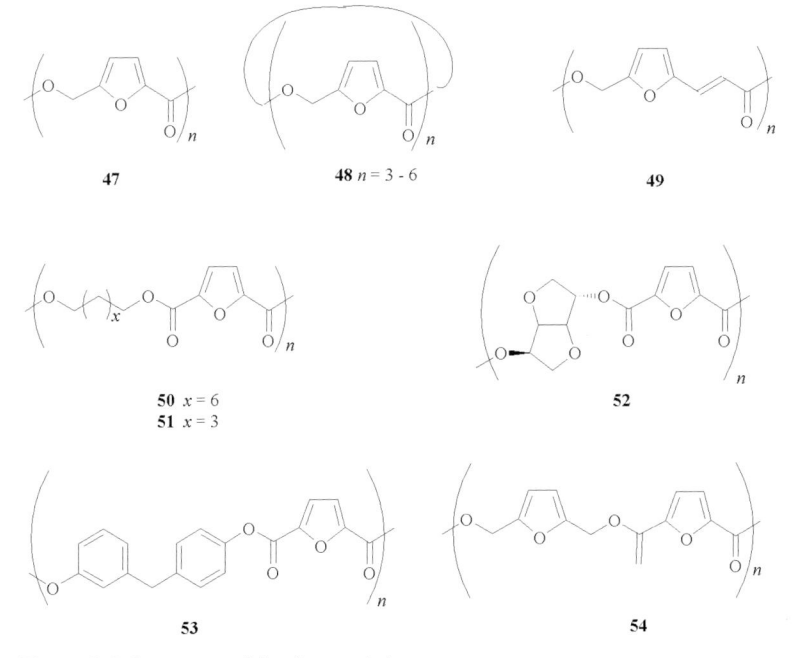

54

Figure 6. Polyesters containing furan moieties

genes, will enhance the economic viability of PHA production, currently on an approximate 3×10^3 t/a level, as they have the potential to replace numerous chemosynthetic polymers in many applications.

4.7.3. Polyamides

More than 90% of the polyamides produced worldwide, amounting to approximately 5.8×16^3 t in 1998 [121], are based on six-carbon monomers, i.e. caprolactam (Nylon 6), and adipic acid/hexamethylenediamine (Nylon 66), the manufacture of which is based exclusively in petroleum-based pathways [121].

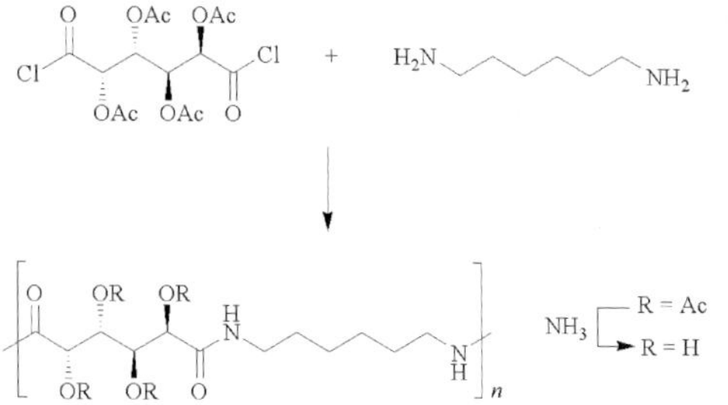

Use of the lactone monomethyl ester of D-glucaric acid proves advantageous in the generation of stereoregular polyglucaramides, leading to an impressive array of aliphatic and aromatic diamines [124].

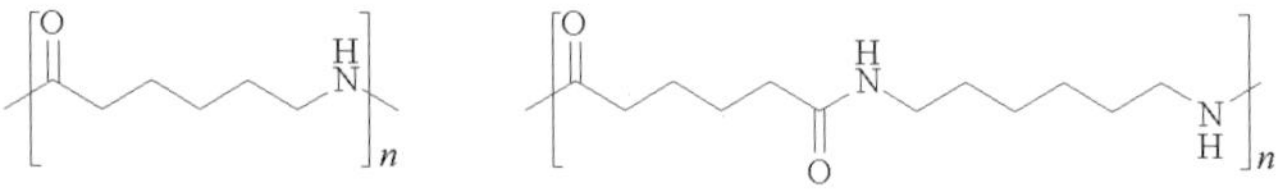

Nylon 6

Nylon 66

When considering the large variety of aminocarboxylic acids, dicarboxylic acids and diamines reasonably accessible from the common six-carbon sugars [122, 123] – expedient examples are listed in Tables 3 and 4 – substitution of the petroleum-based monomers of these polyamides by those derived from sugars seems particularly obvious and promising. Of the myriad of possible combinations of these sugar-derived monomers either with themselves or with the common, petrochemically-derived diamines and dicarboxylic acids, an immense number have been realized. Here, only a few of these polyamides are covered as examples.

Solution or interfacial polycondensation of galactaric acid dichloride in its acetylated form with a variety of aliphatic and aromatic diamines yield a series of polyamides [124], that resulting from 1,6-diaminohexane, resembles a Nylon 6,6 in which half of the methylene hydrogen atoms of the usual adipic acid are substituted by acetoxy groups (R = Ac). These groups can be deacylated with aqueous ammonia to give the tetrahydroxylated Nylon 66.

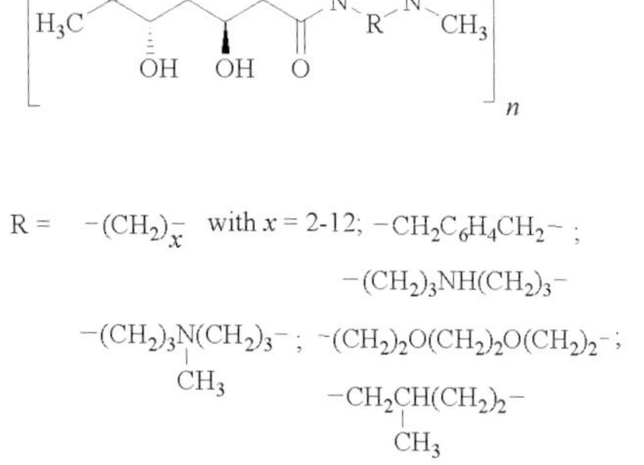

$R = -(CH_2)_x-$ with $x = 2\text{-}12$; $-CH_2C_6H_4CH_2-$;

$-(CH_2)_3NH(CH_2)_3-$

$-(CH_2)_3N(CH_2)_3-$; $-(CH_2)_2O(CH_2)_2O(CH_2)_2-$;
        |
        CH_3

$-CH_2CH(CH_2)_2-$
        |
        CH_3

Of similar practical utility is sucrose-6,6'-dicarboxylic acid (**42**), readily obtained from sucrose by Pt-catalyzed oxidation with oxygen in aqueous medium (cf. Scheme 15), which on amidation of its dimethyl ester with fat amines provides surface-active diamines of type **55** with remarkable tensidometric properties, whereas a reaction with hexamethylenediamine furnishes the interesting, highly hydrophilic polyamide **56** [125].

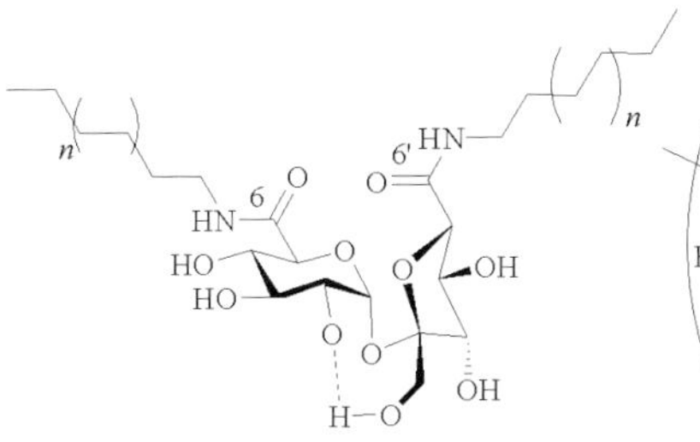

are still cheaper, on average by a factor of five. Eventually, however, with the end of cheap oil in sight [5], and the pressure on our environment increasing, this untoward situation for products from carbohydrate feedstocks will change.

55 (n = 1-4)

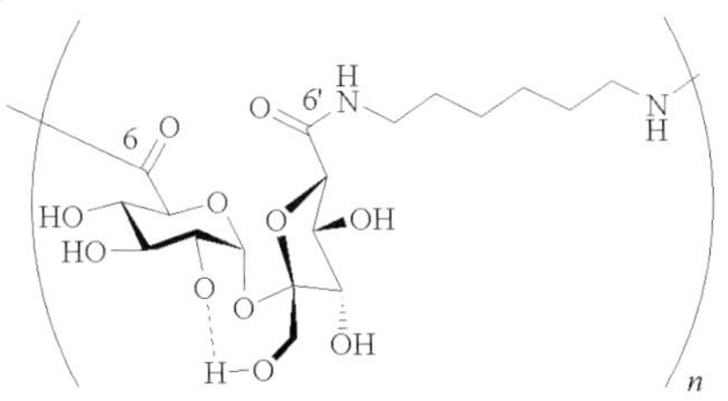

56

Sugar-based "quasi-aromatic" monomers for polyamides, i.e. the furan-2,5-dicarboxylic acid, seem particularly relevant because they have the potential to replace petrochemically derived terephthalic or isophthalic acid in current industrial products. The furan-1,6-diamine has similar potential as a substitute for p-phenylenediamine. Indeed, a series of such furan polyamides has been prepared [126] using the dicarboxylic acid and aliphatic and aromatic diamines. Of these, the polyamide resulting from condensation with p-phenylenediamine, which de facto is an analog of the commercially introduced Kevlar, has particularly promising decomposition and glass temperature properties [127] distinctly better than those of the all-furan polyamides:

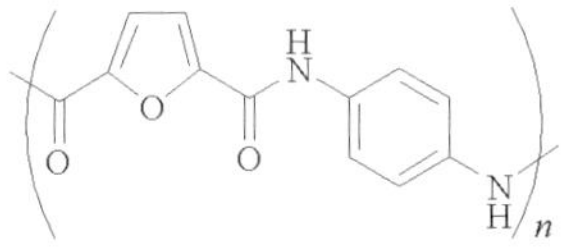

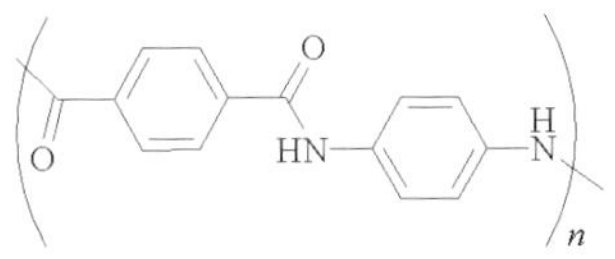

5. Outlook

The unusually diverse stock of readily accessible products described within this account, which covers a wide range of industrial application profiles, lies mostly unexploited – mainly for economic reasons, because equivalent products based on petrochemical raw materials are distinctly cheaper. Notwithstanding this fact, a basic change in the current situation is clearly foreseeable. As the depletion of our fossil raw materials progresses, chemical products derived from them will inevitably increase in price, such that biobased products will become competitive. Realistic estimates [5–7] expect this to occur by 2040.

Kevlar

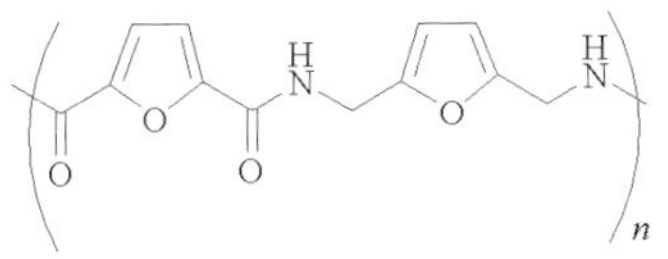

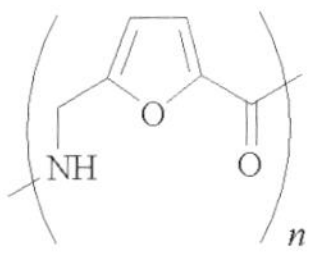

Despite the impressive array of highly useful products – their application profiles compare favorably with those of well-known commercial polyamides – none of these sugar-derived polyamides is currently produced on an industrial scale; the reasons are purely economic, because the products derived from fossil raw material sources

In the meantime, it is imperative that carbohydrates are systematically further exploited leading toward efficient, environmentally benign, and economical process methods for their large-scale conversion into industrially viable products, whether bulk, intermediate, or fine chemicals, pharmaceuticals, or polymeric. General conceptual formulations toward this goal are available [2, 3, 128], yet their straightforward implementation is exceedingly slow. To enhance this, it is essential that national and supranational funding institutions – in Europe the corresponding EU bodies – recognize this. Thus, any promising, innovative research project, irrespective of involving mission-oriented, applied investigations or non-predefined basic explorations, should receive generous support either by funding institutions or by the chemical industry and/or both. Economically sound biobased alternatives to petrochemicals – various potential examples are contained in this account – will then become available as a matter of course.

References

1. C. Okkerse, H. van Bekkum: "From Fossil to Green", *Green Chem.* **1999**, 107–114.
2. The Roadmap for Biomass Technologies in the U.S., Biomass R&D Technical Advisory Committee, US Department of Energy, Accession No. ADA 436527, 2002.
3. R.D. Perlack, L.L. Wright, A. Turhollow, R.L. Graham, B. Stokes, D.C. Erbach: *Biomass as Feedstock for a Bioenergy and Bioproducts Industry: The Technical Feasibility of a Billion-Ton Annual Supply*, Report No. DOE//GO-102995-2135; Oak Ridge National Laboratory, Oak Ridge, TN, 2005. Available electronically at http://www.osti.gov/bridge.
4. W. Umbach, in H. Eierdanz (ed.): *Perspektiven Nachwachsender Rohstoffe in der Chemie*, VCH Publ., Weinheim 1996, pp. IXXX–XLI.
5. (a) C.J. Campbell, J.H. Laherrére: "The End of Cheap Oil", *Sci. Am.* **3** (1998) 60–65. (b) J. Attarian: "The Coming End of Cheap Oil: Hubbert's peak and beyond", *The Social Contracts* **12** (2002) 276–286. (c) C. J. Campbell: "Industry Urged to Watch for Regular Oil Production Peaks, Depletion Signals", *Oil & Gas J.* **101** (2003) 38–45.
6. D.H. Klass: "Fossil Fuel Reserves and Depletion", *Biomass for Renewable Energy, Fuels, and Chemicals*, Acad. Press, San Diego 1998; pp. 10–19.
7. R.H. Hirsch, R. Bezdek, R. Wendling: "Peaking of World Oil Production: Impacts, Mitigation, and Risk Management", US National Energy Technology Laboratory Report, Febr. 2005, 91 pp; http://www.oilcrisis.com/US/NETL/OilPeaking.pdf.
8. A. Hugill: *Sugar and All That–A History of Tate & Lyle*, Introductory Dedicational Metaphor, Entry Books, London 1978.
9. UN Food & Agriculture Organization: World Sugar Production, 2007/08. http://www.fao.org/es/esc/en/20953/21032/highlight_25853en.html (accessed 15 October 2009).
10. (http://www.eridex.com/html (accessed 1 October 2009). http://www.micchem.com/products/Erythritol.htm (accessed 1 October 2009). P. De Cock, C.-L. Bechert: "Erythritol in Non-caloric Functional Beverages", *Pure Appl. Chem.* **74** (2002) 1281–1289.
11. D.L. Klass: "Organic Commodity Chemicals from Biomass", [6], pp. 495–546.
12. F. W. Lichtenthaler (ed.): *Carbohydrates as Organic Raw Materials*, vol. I, VCH Publ., Weinheim 1991, p. 365; G. Descotes (ed.): vol. II, VCH Publ., Weinheim 1993, p. 278; H. van Bekkum, H. Röper, A.G.J. Voragen (eds.): vol. III, VCH Publ., Weinheim 1996, p. 358; W. Praznik (ed.): vol. IV, Wiener Univ. Verlag, Vienna 1998, p. 292.
13. F.W. Lichtenthaler, S. Mondel: "Perspectives in the Use of Low-molecular-weight Carbohydrates as Organic Raw Materials", *Pure Appl. Chem.* **69** (1997) 1853–1866.
14. F.W. Lichtenthaler: "Towards Improving the Utility of Ketoses as Organic Raw Materials", *Carbohydr. Res.* **313** (1998) 69–89.
15. J.J. Bozell (ed.): "Chemicals and Materials from Renewable Resources", *ACS Symposium Series* **784** (2001) 226.
16. F.W. Lichtenthaler: "Unsaturated O- and N-Heterocycles from Carbohydrate Feedstocks", *Acc. Chem. Res.* **35** (2002) 728–737.
17. F.W. Lichtenthaler, in Z.J. Witczak, K. Tatsuta (eds.): "Carbohydrate Synthons in Natural Product Chemistry", *ACS Symposium Series*, **841** (2003) 47–83.
18. F.W. Lichtenthaler, S. Peters: "Carbohydrates as Green Raw Materials for the Chemical Industry", *Comptes Rendue Chimie*, **7** (2004) 65–90.
19. J.J. Bozell, M.K. Patel (eds.): *Feedstocks for the Future: Renewables for the Production of Chemicals and Materials*, Oxford University Press, New York 2006.
20. A. Corma, S. Iborra, A. Velty: "Chemical Routes for the Transformation of Biomass into Chemicals", *Chem. Rev.* **107** (2007) 2411–2502.
21. F.O. Licht: "2008 World Fuel Ethanol Production". Renewable Fuels Association. http://www.ethanolrfa.org/industry/statistics/#E. (accessed 1 October 2009).
22. M.E. Himmel, W.A. Adney, J.O. Baker, R. Elander, J.D. McMillan, R.A. Nieves, J.J. Sheehan, S.R. Thomas, T.B. Vinzant, M. Zhang: "Advanced Bioethanol Production Technologies", in B.C. Saha, J. Woodward (eds.): "Fuels and Chemicals from Biomass", *ACS Symposium Series* **666** (1997) 2–45.
23. H.G. Lawford, J.D. Rousseau: "Cellulosic Fuel Ethanol–Alternative Fermentation Process Design with Wild-type and Recombinant Zymomonas mobilis", *Appl. Biochem. Biotechnol.* **105** (2003) 457–469.
24. J.A. Matthews: "Is Growing Biofuel Crops a Crime against Humanity?", *Biofuels, Biocrops and Biorefining* **2** (2008) 97–99. J. A. Matthews, H. Tan: "Biofuels and Indirect Land Use Change Effects: The Debate Con-

tinues", *Biofuels, Bioproducts, and Biorefining* **3** (2009) 305–317.

25 N. Qureshi, T.C. Ezeji: "Butanol, a Superior Biofuel Production from Agricultural Residues", *Biofuels, Biocrops, and Biorefining* **2** (2008) 319–330.

26 (a) A.S. Mamman, J.-M. Lee, Y.-C. Kim. I.T. Hwang, N.-J. Park, Y.K. Hwang, J.-S. Chang, J.-S. Hwang: "Furfural: Hemicellulose/xylose-derived Biochemical", *Biofuels, Bioproducts, Biorefineries* **2** (2008) 438–454. (b) K.J. Zeitsch: *The Chemistry and Technology of Furfural and its many Byproducts*, Elsevier, Amsterdam 2000, p. 374.

27 A. Gandini, M.N. Belgacem: "Furans in Polymer Chemistry", *Prog: Polym. Sci.* **22** (1997) 1203–1379.

28 M.S. Holfinger, A.H. Conner, D.R. Holm, C.G. Gill, Jr.: "Difurfuryl Diamines by the Acidic Condensation of Furfurylamine with Aldehydes", *J. Org. Chem.* **60** (1995) 1595–1598.

29 C. Moreau, M.N. Belgacem, A. Gandini: "Substituted Furans from Carbohydrates and Ensuing Polymers", *Topics in Catalysis* **27** (2004) 11–30.

30 P. Gruber, D.E. Henton, J. Starr: "Polylactic Acid from Renewable Resources", in *Biorefineries–Industrial Processes and Products*, vol. 2, Wiley-VCH, Weinheim 2006, pp. 381–407.

31 http://www.natureworksllc.com (accessed 30 September 2009).

32 M. McCoy: "Seeking Biomaterials", *Chem. Eng. News* **81** (2003) no. 8, 18; **81** (2003) no. 45, 17–18. S.K. Ritter: "Green Chemistry Progress Report", *Chem. Eng. News* **80** (2002) no. 47, 20.

33 http://www.vertecbiosolvents.com/ (accessed 15 October 2009).

34 W. von Rybinski, K. Hill: "Alkyl Polyglycosides–Properties and Applications of a New Class of Surfactants", *Angew. Chem.* **110** (1998) 1394–1412; *Angew. Chem. Int. Ed.* **37** (1998) 1328–1345.

35 K.-H. Hill, O. Rhode: "Sugar-based Surfactants for Consumer Products and Technical Applications", *Fett/Lipid* **101** (1999) 27–33.

36 P. Jürgens, A. Turowski: "Vergleichende Untersuchung von Zuckerestern, *N*-Methylglucamiden und Glycosiden am Beispiel von Reinigungsprodukten", in *Perspektiven Nachwachsender Rohstoffe in der Chemie*, VCH, Weinheim 1996, pp. 61–70.

37 N.B. Desai: "Esters of Sucrose and Glucose as Cosmetic Materials", *Cosmetics & Toiletries* **105** (1990) 99–107. Mitsubishi-Kagaku Foods Corp.: "Sugar Esters"; http://www.mfc.co.jp/english/whatsse.htm (accessed 20 February 2010).

38 L.A. Silviri, N.J. DeAngelis: "Isosorbide Dinitrate", *Anal. Profiles Drug Subs.* **4** (1975) 225–244.

39 B.E. Maryanoff, S.O. Nortey, J.F. Gardocki, R.P. Shank, S.P. Dodgson: "Anticonvulsant Sulfamates. 2,3:4,5-Di-*O*-isopropylidene-β-D-fructopyranose Sulfamate. *J. Med. Chem.* **30** (1987) 880–887.

40 For pertinent reviews, see: J.N. Chheda, Y. Roman-Leshkov, J.A. Dumesic: "Production of 5-Hydroxymethylfurfural and Furfural by Dehydration of Biomass-derived Mono- and Polysaccharides", *Green Chemistry* **9** (2007) 342–350. J. Lewkowski: "Synthesis, Chemistry and Applications of 5-Hydroxymethylfurfural and its Derivatives", *ARKIVOC* (2001) 17–54. B.F.M. Kuster: "Manufacture of 5-Hydroxymethylfurfural", *Starch/Stärke* **42** (1990) 314–321.

41 H. Schiweck, M. Munir, K. Rapp, M. Voge: "New Developments in the Use of Sucrose as an Industrial Bulk Chemical", in F. W. Lichtenthaler (ed.): *Carbohydrates as Organic Raw Materials*, VCH, Weinheim 1991, pp. 57–94; *Zuckerind. (Berlin)* **115** (1990) 555–565.

42 H. Koch, J. Pein: "Condensations between 5-Hydroxymethylfurfural, Phenol, and Formaldehyde", *Polym. Bull. (Berlin)* **13** (1985) 525–532; *Starch/Stärke* **35** (1983) 304–313.

43 T. El Haj, A. Masroua, J.C. Martin, G. Descotes: "5-Hydroxymethylfurfural and Derivatives by Acid Treatment of Sugars on Ion-exchange Resins", *Bull. Soc. Chim. Fr.* **1987**, 855–860.

44 Y. Roman-Leshkov, C.J. Barret, Z.Y. Liu, J.A. Dumesic: "Production of Dimethylfuran for Liquid Fuels from Biomass-derived Carbohydrates", *Nature* **447** (2007) 982–985.

45 N. Schiavo, G. Descotes, J. Mentech: "Catalytic Hydrogenation of 5-Hydroxymethylfurfural in aqueous Medium", *Bull. Chem. Soc. Chim. Fr.* **128** (1991) 704–711.

46 N. Elming, N. Clauson-Kaas: "6-Methyl-3-pyridinol from 2-Hydroxymethyl-5-aminomethylfuran", *Acta Chem. Scand.* **10** (1956) 1603–1605.

47 Hoechst AG: "Catalytic oxidation of 5-Hydroxymethylfurfural", DE 3826073, 1988 (E.I. Leupold, M. Wiesner, M. Schlingmann, K. Rapp). M.L. Ribeiro, U. Schuchardt: "Cooperative Effect of Cobalt Acetylacetonate and Silica in the Catalytic Cyclization and Oxidation of Fructose to 2,5-Furandicarboxylic acid", *Catalysis Commun.* **4** (2003) 83–86. O. Casanova, S. Iborra, A. Corma: "One-pot, Base-free Oxidative Esterification of 2-Hydroxymethylfurfural into 2,5-Dimethylfuroate", *J. Catalysis* **265** (2009) 109–116.

48 J.B. Binder, R.T. Raines: "Simple Chemical Transformation of Lignocellulosic Biomass into Furans for Fuels and Chemicals", *J. Am. Chem. Soc.* **131** (2009) 1979–1985.

49 M. Mascal, E.B. Nikitin: "Direct, High-yield Conversion of Cellulose into Biofuel", *Angew. Chem. Int. Ed.* **47** (2008) 7924–7926.

50 F. García-Gonzáles: "Reactions of Monosaccharides with β-Ketoesters", *Adv. Carbohydr. Chem.* **11** (1956) 97–143.

51 F. Rodrigues, Y. Canac, A. Lubineau: "A Convenient, One-step Synthesis of β-*C*-Glycosidic Ketones in Aqueous Media", *Chem. Commun.* **2000**, 2049–2050. I. Riemann, M.A. Papadopoulos, M. Knorst, W.-D. Fessner: "*C*-Glycosides by Aqueous Condensation of β-Dicarbonyl Compounds with Unprotected Sugars", *Aust. J. Chem.* **55** (2002) 147–154.

52 S. Peters, F.W. Lichtenthaler, H.J. Lindner: "A *C*-Fructosyl-propanone Locked in a 2,7-Dioxabicyclo[3,2,1] octane Framework", *Tetrahedron: Asymmetry* **14** (2003) 2475–2479.
53 Y. Chapleur (ed.): *Carbohydrate Mimics*, Wiley-VCH, Weinheim, New York 1998, p. 604ff. and references cited therein.
54 A. Beélik: "Kojic Acid", *Adv. Carbohydr. Chem.* **11** (1956) 145–183.
55 F.W. Lichtenthaler: "Sugar Enolones and γ-Pyrone-Formation", *Pure Appl. Chem.* **50** (1978) 1343–1362.
56 F.W. Lichtenthaler: "Building Blocks from Sugars and their Use in Natural Product Synthesis", in R. Scheffold (ed.): *Modern Synthetic Methods*, vol. 6, VCH, Weinheim 1992, pp. 273–376.
57 C. Nelson, J. Gratzl: "Conversion of D-Glucono-1,5-lactone into an α-Pyrone", *Carbohydr. Res.* **60** (1978) 267–273.
58 K. Tajima: "Cyclopentenones from 3-Acetoxy-6-acetoxymethyl-2-pyrone", *Chem. Lett.* **1987**, 1319–1322.
59 For useful preparative procedures, see: *Methods Carbohydr. Chem.* **2** (1963) 318, 326, 405, 427.
60 F.W. Lichtenthaler: "Sugar-derived Building Blocks for the Synthesis of Non-carbohydrate Natural Products", in S. J. Witczak, K. Tatsuta (eds.): "Carbohydrate Synthons in Natural Product Chemistry", *ACS Symposium Series* **841** (2003) 47–83. S. Hanessian: *Total Synthesis of Natural Products: The Chiron Approach*, Pergamon, Oxford 1983.
61 R.J. Ferrier, N. Prasad: "Synthesis of 2,3-Dideoxy-α-D-*erythro*-hex-2-enopyranosides from Tri-*O*-acetyl-D-glucal", *J. Chem. Soc. C* **1969**, 570–575.
62 F.W. Lichtenthaler, S. Rönninger, P. Jarglis: "Expedient Approach to Pyranoid Ene and Enol Lactones", *Liebigs Ann.* **1989**, 1153–1161.
63 S. Hanessian, A.M. Faucher, S. Leger: "Total Synthesis of Meroquiene", *Tetrahedron* **46** (1990) 231–234. N. Ohyabu, T. Nishikawa, M. Isobe: "First Asymmetric Total Synthesis of Tetrodotoxin", *J. Am. Chem. Soc.* **125** (2003) 8798–8805.
64 S. Czernickí, K. Víjayakuraman, G. Ville: "Convenient Synthesis of Hex-1-enopyran-3-uloses: Selective Oxidation of Allylic Alcohols using Pyridinium Dichromate", *J. Org. Chem.* **51** (1986) 5472–5475.
65 B. Fraser-Reid, A. McLean, E.W. Usherwood, M. Yunker: "Synthesis and Properties of some Alkyl 2,3-Dideoxy-2-enopyranosid-4-uloses", *Can. J. Chem.* **48** (1970) 2877–2884.
66 R.J. Ferrier: "Modified Synthesis of 2-Hydroxyglycalesters", *Methods Carbohydr. Chem.* **6** (1972) 307–311.
67 F.W. Lichtenthaler, U. Kraska: "Preparation of Sugar Enolones", *Carbohydr. Res.* **58** (1977) 363–377.
68 F.W. Lichtenthaler, S. Nishiyama, T. Weimer: "2,3-Dihydropyranones with Contiguous Chiral Centers", *Liebigs Ann.* **1989**, 1163–1170.
69 F.W. Lichtenthaler, S. Ogawa, P. Heidel: "Synthesis of Unsaturated Hexopyranosid-4-uloses", *Chem. Ber.* **110** (1977) 3324–3332.
70 J. Gravitis, N. Vedernikon, J. Zandersons, A. Kokorevics: "Furfural and Levoglucosan Production from Deciduous Wood and Agricultural Wastes", in J. J. Bozell (ed.): "Chemicals and Materials from Renewable Resources", *ACS Symposium Series* **784** (2001) 110–122. R.C. Brown, D. Radlein, J. Piskorz: "Pretreatment Processes to increase Pyrolytic Yield of Levoglucosan from Herbaceous Feedstocks", in J. J. Bozell (ed.): "Chemicals and Materials from Renewable Resources", *ACS Symposium Series* **784** (2001) 123–132. F. Shafizadeh, R. Furneaux, T. Stevenson: "Reactions of Levoglucosenone", *Carbohydr. Res.* **71** (1979) 169–191.
71 Z. J. Witczak: "Synthesis of Natural Products from Levoglucosenone", *Pure Appl. Chem.* **66** (1994) 2189–2192.
72 F. Ledl, E. Schleicher: "The Maillard Reaction in Foodstuffs and the Human Body", *Angew. Chem.* **102** (1990) 597–626; *Angew. Chem. Int. Ed.* **29** (1990) 565–594.
73 (a) H. El Khadem: "*N*-Heterocycles from Saccharide Derivatives", *Adv. Carbohydr. Chem.* **25** (1970) 351–405. (b) E.S.H. El Ashry, N. Rashad: "Carbohydrate Hydrazones and Osazones as Raw Materials for Nucleosides and *N*-Heterocycles", *Curr. Org. Chem.* **4** (2000) 609–657.
74 S.M. McElvain, K.M. Bolliger: "Pyrrole", *Org. Synth., Coll. Vol. 1* **1941**, 473–474.
75 F. García-Gonzáles, A. Gómez Sánchez: "Reactions of Amino Sugars with β-Dicarbonyl Compounds", *Adv. Carbohydr. Chem.* **20** (1965) 303–355.
76 A. Rozanski, K. Bielawski, J. Boltryk, D. Bartulewicz: "Simple Synthesis of 3-Acetyl-5-(tetrahydroxybutyl)-2-methylpyrrole", *Akad. Med. Juliana Marchlewskiego Bialymstoku* **1991**, 35–36, 57–63; *Chem. Abstr.* **118** (1992) 22471m.
77 V. Diehl, E. Cuny, F.W. Lichtenthaler: "Conversion of D-Xylose into Hydrophilically Functionalized Pyrazoles", *Heterocycles* **48** (1998) 1193–1201.
78 M. Oikawa, C. Müller, M. Kunz, F.W. Lichtenthaler: "Hydrophilic Pyrazoles from Sugars", *Carbohydr. Res.* **309** (1998) 269–279.
79 R. Weidenhagen, R. Hermann: "4-Hydroxymethylimidazole", *Ber. Dtsch. Chem. Ges.* **70** (1937) 570–583; *Org. Synth., Coll. Vol. III* **1955**, 460–462.
80 (a) J. Streith, A. Boiron, A. Frankowski, D. LeNouen, H. Rudyk, T. Tschamber: "One-pot Synthesis of Imidazolosugars", *Synthesis* **1995**, 944–946. (b) F.W. Lichtenthaler, A. Brust, TU Darmstadt, unpublished results.
81 Sadolin & Holmblad A/S: "Improvement in Preparing 3-Pyridinols", GB 862581, 1961 (N. Elming, S.V. Carlsten, B. Lennart, I. Ohlsson); *Chem. Abstr.* **56** (1962) 11574g.
82 Hoechst AG: "Heterocyclic Phenoxypyridines as Herbicidal Agents and Insecticides", EP 227045, 227046, 1987 (V. Koch, L. Willms, A. Fuss, K. Bauer, H. Bieringer, H. Buerstell); *Chem. Abstr.* **107** (1987) 175892, 134217.
83 C. Müller, V. Diehl, F.W. Lichtenthaler: "3-Pyridinols from Fructose and Isomaltulose", *Tetrahedron* **54** (1998) 10703–10712.

84. I. Forsskahl, T. Popoff, O. Theander: "Reactions of D-Xylose and D-Glucose in alcaline Aqueous Solutions", *Carbohydr. Res.* **48** (1976) 13–21. T. Popoff, O. Theander: "Formation of Aromatic Compounds from D-Glucose and D-Fructose in slightly Acidic Aqueous Solution", *Acta Chem. Scand.* **B30** (1976) 397–402.
85. K.M. Drahts, D.R. Knop, J.W. Frost: "Shikimic acid and Quinic acid: Replacing Isolation from Plant Sources with Recombinant Microbial Catalysis", *J. Am. Chem. Soc.* **121** (1999) 1603–1604.
86. J.C. Rolloff, K.M. Kent, M.J. Postich, M.W. Beeker, H. H. Chapman, D.E. Kelly, W. Lew, M.S. Louie, L.R. McGee, E.J. Prisbe, L.M. Schultze, R.H. Yu, L. Zhang: "Practical Total Synthesis of the Anti-influenza Drug GS-4104", *J. Org. Chem.* **63** (1998) 4545–4550.
87. N. Ran, D.R. Knop, K.M. Drahts, J.W. Frost: "Benzene-free Synthesis of Hydroquinone", *J. Am. Chem. Soc.* **123** (2001) 10927–10934.
88. J.M. Gibson, P.S. Thomas, J.D. Thomas, J.L. Barker, S. S. Chandran, M.K. Harrup, K.M. Drahts, J.W. Frost: "Benzene-free Synthesis of Phenol", *Angew. Chem. Int. Ed.* **40** (2001) 1945–1948.
89. K.M. Drahts, J.W. Frost: "Environmentally Compatible Synthesis of Catechol from D-Glucose", *J. Am. Chem. Soc.* **117** (1975) 2395–2400. W. Li, D. Xie, J.W. Frost: "Benzene-free Synthesis of Catechol: Interfacing Microbial and Chemical Synthesis", *J. Am. Chem. Soc.* **127** (2005) 2874–2882.
90. S. Kambourakis, K.M. Drahts, J.W. Frost: "Synthesis of Gallic Acid and Pyrogallol from Glucose: Replacing Natural Product Isolation with Microbial Catalysis", *J. Am. Chem. Soc.* **122** (2002) 9042–9043.
91. J. Achkar, M. Xian, H. Zhao, J.W. Frost: "Biosynthesis of Phloroglucinol", *J. Am. Chem. Soc.* **127** (2005) 5332–5333.
92. P. Präve, U. Faust, W. Sittig, D.A. Sukatsch (eds.): *Fundamentals of Biotechnology*, VCH, Weinheim 1987.
93. T. Werpy, G. Petersen (eds.): "Top Value Added Chemicals from Biomass vol I–Results of Screening for Potential Candidates from Sugars and Synthesis Gas", Report No. NREL/TP-510-35523, National Renewable Energy Laboratory, Golden, CO, 2004, p. 67. Available electronically at http://www.osti.gov/bridge. (a) "Glutamic Acids", pp. 39–42; (b) "Aspartic Acid", pp. 31–35; (c) "3-Hydroxypropionic Acid (3-HPA)", pp. 29–31; (d) "Four-carbon 1,4-Diacids", pp. 22–25; (e) "Itaconic Acid", pp. 42–44; (f) "Glucaric Acid", pp. 36–38; (g) "Levulinic Acid", pp. 45–48; (h) "3-Hydroxybutyrolactone", pp. 49–51.
94. S.Y. Lee, S.H. Park, S.H. Hong, Y. Lee, S.H. Lee: "Fermentative Production of Building Blocks for Chemical Synthesis of Polyesters", *Biopolymers*, Wiley-VCH, Weinheim 2001, vol. **3b**, chap. 10, p. 265ff.
95. For pertinent reviews, see: (a) S.Y. Lee, S.H. Hong, S.H. Lee, S.J. Park: "Fermentative Production of Chemicals that can be used for Polymer Synthesis", *Macromol. Biosci.* **4** (2004) 157–164. G.T. Tsao, N.J. Cao, J. Du, C. S. Gong: "Production of Multifunctional Organic Acids from Renewable Resources", *Adv. Biochim. Eng. Biotechnol.* **65** (1999) 243–280.
96. Press release by Cargill and Novozymes in January 2008.
97. F.S. Carta, C.R. Soccol, L.P. Ramos, J.D. Fontana: "Production of Fumaric Acid by Fermentation of Enzymic Hydrolyzates Derived From Cassava Bagasse", *Bioresour. Technol.* **68** (1999) 23–28.
98. G.N. Vemuri, M.A. Eiteman, E. Altman: "Succinate Production in Dual-phase *E. coli* - Fermentations depends on the Time of Transition from Aerobic to Anaerobic Conditions", *J. Ind. Microbiol. Biotechnol.* **28** (2002) 325–332. G.N. Vemuri, M.A. Eiteman, E. Altman: "Effects of Growth Mode and Pyruvate Carboxylase on Succinic Acid Production by Metabolically Engineered *E. coli* Strains", *Appl. Environ. Microbiol.* **68** (2002) 1715–1727. R.R. Gokarn, M.A. Eiterman, J. Sridhar: "Production of Succinate by Anaerobic Microorganisms", in B. D. Saha, J. Woodard (eds.): "Fuels and Chemicals from Biomass", *ACS Symposium Series* **666** (1997) 237–279.
99. A. Cukalovic, C.V. Stevens: "Feasibility of Production Methods for Succinc Acid Derivatives", *Biofules, Bioproducts and Biorefining* **2** (2008) 505–529.
100. M. Okabe, D. Lies, S. Kanamasa, E.Y. Park: "Biotechnological Production of Itaconic Acid and its Biosynthesis from *Aspergillus terreus*", *Appl. Microbiol. Biotechnol.* **84** (2009) 597–606. C.S. Reddy, R.P. Singh: "Enhanced Production of Itaconic Acid from Corn Starch and Market Refuse Fruits by Genetically Manipulated *Aspergillus terreus* SKR 10", *Bioresour. Technol.* **85** (2002) 69–71. T. Wilke, K. Welter, K.D. Vorlop: "Biotechnical production of itaconic acid from sugar", *Appl. Microbiol. Biotechnol.* **56** (2001) 289–295.
101. DuPont PTT (Sorona®): http://www2.dupont.com/Sorona/en_US/. (accessed 20 February 2010)
102. M.M. Zhu, P.D. Lawman, D.C. Cameron: "Improving 1,3-Propanediol Production from Glycerol in a Metabollically Engineered *E. coli* by Reducing Accumulation of Glycerol-3-phosphate", *Biotechnol. Prog.* **18** (2002) 694–699.
103. For an informative review, see: A.N. Zeng, H. Biebl: "Bulk Chemicals from Biomass: Ease of 1,3-Propanediol Production and New Trends", *Adv. Biochem. Engineering/Biotech.* **74** (2002) 239–259.
104. N.E. Altaras, D.C. Cameron: "Metabolic Engineering of a 1,2-Propanediol Pathway in *E. coli*", *Appl. Environ. Microbiol.* **65** (1999) 1180–1185. N.E. Altaras, D.C. Cameron: "Enhanced Production of (R)-1,2-Propanediol by Metabolically Engineered *E. coli*", *Biotechnol. Progr.* **16** (2000) 940–946. N.E. Altaras, M.R. Etzel, D. C. Cameron: "Conversion of Sugars to 1,2-Propanediol by *Thermoanaerobacterium thermosaccharolyticum*", *Biotechnol. Progr.* **17** (2001) 52–56.
105. N. Nishino, M. Yochida, H. Shiota, E. Sakaguchi: "Accommodation of 1,2-Propanediol and Enhancement of Aerobic Stability in Whole Crop Maize Silage Inoculated with *Lactobacillus buchneri*", *J. Appl. Microbiol.* **94** (2003) 800–807.

106 W. Niu, M.N. Molefe, J.W. Frost: "Microbial synthesis of the energetic material precursor 1,2,4-butanetriol", *J. Am. Chem. Soc.* **125** (2003) 12998–12999.
107 C.L. Mehltretter: "D-Glucaric Acid", *Methods Carbohydr. Chem.* **2** (1963) 46–48.
108 B.A. Lewis, F. Smith, A.M. Stephen: "Galactaric Acid and its Derivatives", *Methods Carbohydr. Chem.* **2** (1963) 38–46.
109 S. Immel, F.W. Lichtenthaler: "The Conformation of Sucrose in Water: A Molecular Dynamics Approach", *Liebigs Ann. Chem.* **1995**, 1938–1947.
110 Südzucker AG: "Process and Apparatus for the Preparation of Mono-oxidized Products from Carbohydrates", DE 4307388, 1994 (M. Kunz, H. Puke, C. Recker, L. Scheiwe, J. Kowalczyk); *Chem. Abstr.* **122** (1995) 56411.
111 L.A. Edye, G.V. Meehan, G.N. Richards: "Platinum-catalyzed Oxidation of Sucrose", *J. Carbohydr. Chem.* **10** (1991) 11–23; **13** (1994) 273–283.
112 Südzucker AG: "Process for Continuous Manufacture of Di- and Higher-oxidized Carboxylic Acids from Carbohydrates", DE 19542287, 1996 (M. Kunz, A. Schwarz, J. Kowalczyk); *Chem. Abstr.* **127** (1997) 52504.
113 Hoechst AG: "Preparation of Sucrose Tricarboxylic Acid", DE 3535720, 1987 (W. Fritsche-Lang, E.I. Leupold, M. Schlingmann); *Chem. Abstr.* **107** (1987) 59408.
114 Biofine Inc.: "Production of Levulinic Acid by Hydrolysis of Carbohydrate-containing Materials", US 475630, 1996 (S.W. Fitzpatrick); *Chem. Abstr.* **126** (1996) 117739. D.J. Hayes, S. Fitzpatrick, M.H.B. Hayes, J.R.H. Ross: "The Biofine Process–Production of Levulinic Acid, Furfural and Formic Acid from Lignocellulose Feedstocks"", in B. Kamm, P. R. Gruber (eds.): *Biorefineries–Industrial Processes and Products*, vol. **1**, Wiley-VCH, Weinheim 2006, pp. 113–164.
115 J.A. Moore, J.E. Kelly: "Poly(hydroxymethylfuroate)", *J. Polym. Sci., Polym. Chem. Ed.* **22** (1984) 863–864.
116 H. Hirai: "Synthesis of Macrocyclic Oligoesters from 5-Hydroxymethyl-2-furancarboxylic Acid", *J. Macromol. Sci., Chem. Part A* **21** (1984) 1165–1179.
117 R. Storbeck, M. Ballauf: "Synthesis and Thermal Analysis of Copolyesters Derived from 1,4:3,6-Dianhydrosorbitol", *J. Appl. Polymer Sci.* **59** (1996) 1199–1202.
118 J.A. Moore, J.E. Kelly: "Polymerization of Furandicarbonyl Chloride with Bisphenol A", *Polymer* **20** (1979) 627–628.
119 "Polyesters–Properties and chemical Synthesis" in Y. Doi, A. Steinbüchel (eds.): *Biopolymers*, vols. **3a, 3b**, Wiley-VCH, Weinheim 2001, 468 pp. "Polyesters–Applications and Commercial Products"" in A. Steinbüchel, Y. Doi (eds.): *Biopolymers*, vol. 4, Wiley-VCH, Weinheim 2002, 398 pp.
120 Y.B. Kim, R.W. Lenz: "Polyesters from Microorganisms", *Adv. Biochem. Engineering/Biotechnol.* **71** (2001) 51–79.
121 H.-P. Weiss, W. Sauerer: "Polyamides", *Kunststoffe* **89** (1999) 68–74.
122 J. Thiem, F. Bachmann: "Carbohydrate-derived Polyamides", *Trends Polym. Sci.* **2** (1994), 425–432. E.M.E. Mansur, S.H. Kandil, H.H.A. M. Hassan, M.E.E. Shaban: "Synthesis of Carbohydrate-containing Polyamides and Study of their Properties", *Eur. Polym. J.* **26** (1990) 267–276.
123 O. Varela, H.A. Orgueira: "Synthesis of chiral Polyamides from Carbohydrate-derived Monomers", *Adv. Carbohydr. Chem. Biochem.* **55** (1999) 137–174.
124 L. Chen, E. Kiely: "Synthesis of Stereoregular Head/Tail Hydroxylated Nylons Derived from D-Glucose", *J. Org. Chem.* **61** (1996) 5847–5851; US 5 329 044, 1994 (L. Chen, D.E. Kiely); *Chem.. Abstr.* **122** (1994) 56785. D. E. Kiely: "Carbohydrate Diacids: Potential as Commercial Chemicals and Hydrophobic Polyamide Precursors", in J. J. Bozell (ed.): "Chemicals and Materials from Renewable Resources", *ACS Symposium Series* **784** (2001) 64–80.
125 E. Cuny, S. Mondel, F.W. Lichtenthaler: "Novel Polyamides from Disaccharide-derived Dicarboxylic Acids", *Internat. Carbohydr. Symp.*, Whistler 2006, http://csi.chemie.tu-darmstadt.de/ak/fwlicht/PAPERS/paper292.pdf (accessed 20 Febuary 2010).
126 A. Gandini, M.N. Belgacem: "Furanic Polyamides Chemistry", *Prog. Polym. Sci.* **22** (1977) 1238–1246.
127 A. Mitiakoudis, A. Gandini: "Poly(*p*-phenylene)-2,5-furandicarbonamide and Conversion into Filaments and Films", *Macromolecules* **24** (1991) 830–835.
128 M. Eissen, J.O. Metzger, E. Schmidt: U. Schneidewind, "Concepts on the Contribution of Chemistry to a Sustainable Development", *Angew. Chem.* **114** (2002) 402–424; *Angew. Chem. Int. Ed.* **41** (2002) 414–436.

Further Reading

D. S. Argyropoulos (ed.): *Materials, Chemicals and Energy from Forest Biomass*, American Chemical Society Publ., Washington, DC 2007.

M. N. Belgacem, A. Gandini (eds.): *Monomers, Polymers and Composites from Renewable Resources*, Elsevier, Amsterdam 2008.

G. Centi, R. A. Van Santen (eds.): *Catalysis for Renewables*, Wiley-VCH, Weinheim 2007.

J. Clark, F. Deswarte (eds.): *Introduction to Chemicals from Biomass*, Wiley, Chichester 2008.

M. A. Curran: *Biobased Materials*, "Kirk Othmer Encyclopedia of Chemical Technology", 5th edition, John Wiley & Sons, Hoboken, NJ, online DOI: 10.1002/0471238961.biobcurr.a01.

J. Dewulf, H. Van Langenhove (eds.): *Renewables-Based Technology*, Wiley, Chichester 2006.

P. Ranalli (ed.): *Improvement of Crop Plants for Industrial End Uses*, Springer, Berlin 2007.

Cellulose

Hans Krässig, Seewalchen, Austria

Josef Schurz, University of Graz, Graz, Austria

Robert G. Steadman, La Trobe University, Bundoora, Victoria, Australia

Karl Schliefer, Textilforschungsanstalt Krefeld, Krefeld, Federal Republic of Germany

Wilhelm Albrecht, Wuppertal, Federal Republic of Germany

Marc Mohring, J. Rettenmaier & Söhne GmbH + Co, Rosenberg, Germany

Harald Schlosser, J. Rettenmaier & Söhne GmbH + Co, Rosenberg, Germany

1.	**Cellulose**	123	2.7.	Coir	150
1.1.	**Properties**	124	2.8.	Ramie	150
1.1.1.	Molecular Structure	124	2.9.	Economic Aspects	150
1.1.2.	Supermolecular Structure (Texture)	128	2.10.	Occupational Health	151
1.1.3.	Physical Properties	130	3.	**Regenerate Cellulose**	151
1.1.4.	Chemical Properties	131	3.1.	**Viscose Fibers**	152
1.2.	**Occurrence**	138	3.1.1.	Principle of the Viscose Process	152
1.3.	**Production**	139	3.1.2.	Viscose Preparation	153
1.4.	**Quality Testing**	139	3.1.3.	Viscose Fiber Spinning	156
1.5.	**Applications**	140	3.1.4.	Fiber Types	158
2.	**Natural Cellulosic Fibers**	140	3.1.5.	Modified Viscose Fibers	160
2.1.	**Cotton**	141	3.1.6.	Fiber Properties	161
2.1.1.	Molecular Arrangement, Morphology, and Fine Structure	141	3.1.7.	Uses	163
2.1.2.	Properties	143	3.1.8.	Economic Aspects	164
2.1.3.	Production	146	3.2.	**Lyocell Fibers**	165
2.1.3.1.	Harvesting	146	3.3.1.	Principles of the Lyocell Process	165
2.1.3.2.	Ginning	146	3.3.2.	Process Description	166
2.1.3.3.	Byproducts	146	3.3.3.	Fiber Properties	166
2.1.3.4.	Processing	147	3.3.	**Cuprammonium Fibers**	167
2.1.3.5.	Finishing	147	3.4.	**Tentative Cellulose Fiber Production by Other Processes – Outlook**	170
2.1.3.6.	Special Finishes	147	4.	**Ground Cellulose/Powdered Cellulose**	170
2.2.	**Bast (Soft) Fibers**	148	4.1.	Production	170
2.3.	**Jute**	149	4.2.	Properties	171
2.4.	**Flax**	149	4.3.	Uses	171
2.5.	**Hemp**	150		References	173
2.6.	**Leaf (Hard) Fibers**	150			

1. Cellulose

Cellulose [9004-34-6] deserves a special position among the industrially used raw materials for two general reasons. First, cellulose belongs to the natural products which, when used carefully, are inexhaustible since it is regularly regenerated by nature in relatively short time periods. As long as we ensure that the primary sources of cellulose, forests and cotton plantations, are not damaged by destructive lumbering or overcropping, we can expect regular and significant natural annual reproduction.

According to reference [8], the annual yield of cellulosic matter resulting from photoinitiated biosynthesis amounts to approximately 1.3×10^9 metric tons. A tree produces an average of 13.7 g of cellulose daily. If they were lined up, the cellulose chain molecules formed each day would result in a string of 2.62×10^{10} km in length, or 175 times the distance between the sun and the earth.

In wood, cellulose is part of an ingeniously constructed fiber-reinforced composite in which long, stiff cellulose chain molecules organized in thin fibrils constitute the plant reticulum material held together and protected by hydrophobic lignin acting as binder and encasement.

To isolate cellulose from wood for industrial applications, the wooden composite must be broken up by so-called pulping processes. In these treatments, other wood constituents, such as lignin and hemicelluloses, are to a large extent degraded and dissolved. Thus far, these byproducts have found only limited use. In most cases, wood pulp manufacturers concentrate the waste pulping liquors to concentrates consisting of ca. 50 % solids. The organic matter is used as fuel to produce steam and electric power, while the inorganic pulping chemicals (soda, magnesium, or ammonium base and sulfur dioxide) are simultaneously recovered. These recovery processes have practically solved the long-standing environmental problems of the wood pulp industry.

Both cellulose and lignin are biologically degradable and, thus, ecologically beneficial. They will decompose in the open. Cellulose products, such as paper or cellulosic textiles, will decompose and eventually form valuable humus. In industrial use, environmental problems are not caused by cellulose or lignin but by the chemicals used in the isolation or in subsequent chemical processing and transformation into cellulose derivatives, films, or fibers. Therefore, the long-term task of modern cellulose research will be the development of novel processes which yield no or only a few ecologically harmful byproducts. If these efforts are successful, cellulose will surely maintain and strengthen its position as a renewable and environmentally beneficial, industrially important raw material competing with synthetically produced polymers.

1.1. Properties

1.1.1. Molecular Structure

Cellulose is an isotactic β-1,4-polyacetal of cellobiose (4-*O*-β-D-glucopyranosyl-D-glucose).

The actual base unit, the cellobiose, consists of two molecules of glucose. For this reason, cellulose can also be considered as a (syndiotactic) polyacetal of glucose.

Basic Structure. The basic chemical formula of cellulose is the following:

$$C_{6P}H_{10P+2}O_{5P+1} \approx (C_6H_{10}O_5)_P \text{ or } (C_6H_{10}O_5)_n$$

where P = the degree of polymerization; n = the number of units in the chain.

The elemental composition of 44.4 % C, 6.2 % H, and 49.4 % O was already known to PAYEN in 1842 [9]. The molecular mass of the glucose base unit is $m_0 = 162$, and the molecular mass of the cellulose polymer is

$$M_r = m_0 P + 18 \approx 162 P$$

Constitutional Formula. HAWORTH [10] first discovered the covalent bonds inside and between the glucose units while STAUDINGER [11] found the final proof for the macromolecular nature of the cellulose molecule.

Conformational Formula. The glucopyranosic ring adopts a 4_{C_1} chair conformation, as revealed by X-ray crystallography and nuclear magnetic resonance studies [12], [13] with glucose. The chair formation in comparison to the tray conformation exhibits a free stabilization enthalpy of $G_s = 20.05$ kJ/mol [14]. In this conformation, the three hydroxyl groups are positioned in the ring plane while the hydrogen atoms are in a vertical position. It seems only natural to assume that the same conformation also exists in the cellulose molecule.

Structural Anomalies. As a naturally occurring polymer, cellulose always contains small amounts of other constituents in addition to glucose (over 99 %). These may already be partially built into or onto the cellulose molecules during biosynthesis, such as lignin – cellulose complexes [15]. Most of the changes in the molecular structure, however, result from secondary reactions, i.e., hydrolysis or oxidation, during isolation from natural sources. For morphological reasons, such chemical changes occur preferably in the accessible interlinking regions between the crystallites of the elementary fibrils or their aggregations. The glucosidic links in these accessible areas, especially if oxidized sites are also present, split 1 000 – 5 000 times faster than glucosidic linkages inside the well-ordered crystallites. The existence of weak links, as proposed in reference [16], is hard to determine. In homogeneous acid hydrolysis, all glucosidic linkages split at the same rate [17].

Cellulose always contains carboxyl groups: In wood pulp, one –COOH group per 100 – 1000 anhydroglucose units (AHG) exists; in cotton, one –COOH group per 100 – 500 AHG units.

Molecular Size. The molecular size of a polymer can be defined by its average molecular mass ($\bar{M}_r$) or its average degree of polymerization ($\bar{P}$); whereby $\bar{M}_r = \bar{P} \, m_0$ ($m_0 =$ molecular mass of the base unit, i.e., of glucose in the case of cellulose).

By investigation of certain physical properties of cellulose or polyhomologous cellulose derivative solutions, the average degree of polymerization can be determined. Table 1 lists the number average degree of polymerization of a number of celluloses of various origin.

Table 1. Degree of polymerization of celluloses of different origin [18]

Type of cellulose	$\bar{P}_n$
Cotton, raw	7 000
Cotton, raw (according to Russian work)	14 000
Cotton, purified	1 500 – 300
Cotton linters	6 500
Flax	8 000
Ramie	6 500
α-Cellulose (isolated from wood fibers)	1 100 – 800
Spruce, pulped	3 300
Beech, pulped	3 050
Aspen	2 500
Fir	2 500
Bacterial cellulose	2 700
Acetobacter cellulose	600

Table 2. Values for the constants a, K_m, and K_p [19], [20] (concentration in g/mL)

Solvent system		T, °C	a	K_m, ×10^5	K_p, ×10^3
Cellulose	cuoxam	20	0.9	5.54	5.40
Cellulose	cuen	25	0.9	12.5	11.07
Cellulose	cadoxen	20	1.0	3.14	5.09
Cellulose	iron – sodium tartrate	20	1.0	4.08	6.61
CTN	acetone	25	1.0	2.8	4.54
CTN	acetone	25	0.93	7.1	8.06
CTN	butyl acetate	25	1.0	2.8	4.54
CTN	ethyl acetate	25	0.76	37.1	60.10
CTN	ethylene chlorohydrine	25	0.83	15.5	10.57
CTC [20]	acetone	25	0.91	143	146.6
CTC [20]	dioxane	25	0.97	81.3	113.1

For such physical investigations, cellulose solutions in aqueous copper(II)tetrammonium hydroxide (Schweitzer's reagent; Cuoxam), copper(II)ethylenediamine hydroxide (Cuen), alkaline solutions of the ethylenediamine complexes of cadmium or nickel can be used. Cellulose trinitrate (CTN) or cellulose tricarbanilate (CTC) solutions in appropriate solvents are also suitable for such studies (see Table 2). In the latter, it should be kept in mind that chain degradation often occurs in substitution reactions performed under unfavorable conditions.

Light scattering studies performed on dilute solutions of cellulose or cellulose derivatives will yield the weight average ($\bar{M}_w$) and osmotic measurements the number average ($\bar{M}_n$) of the molecular mass (or the corresponding average degrees of polymerization: $\bar{P}_w$ or $\bar{P}_n$). Sedimentation experiments in an ultracentrifuge enable the determination of a higher order average molecular mass, the so-called "Z-average" ($\bar{M}_z$). These various quantities are defined as follows:

$$\bar{M}_w = \frac{\sum N_i \cdot M_i}{\sum N_i} \quad \text{or} \quad \bar{P}_w = \frac{\sum N_i \cdot P_i}{\sum N_i} \quad \text{(weight average)}$$

$$\bar{M}_n = \frac{\sum N_i \cdot M_i^2}{\sum N_i \cdot M_i} \quad \text{or} \quad \bar{P}_n = \frac{\sum N_i \cdot P_i^2}{\sum N_i \cdot P_i} \quad \text{(number average)}$$

$$\bar{M}_z = \frac{\sum N_i \cdot M_i^3}{\sum N_i \cdot M_i^2} \quad \text{or} \quad \bar{P}_z = \frac{\sum N_i \cdot P_i^3}{\sum N_i \cdot P_i^2} \quad \text{(Z-average)},$$

whereby $M = 162\, P$; $i =$ fraction $1, 2, 3 \ldots i$; $N =$ number of molecules with $M_1, M_2 \ldots M_i$ or $P_1, P_2 \ldots P_i$.

The simplest and most widely applied practical method for the determination of the degree of polymerization is based on measuring the "intrinsic viscosity η" (Staudinger index). The intrinsic viscosity expresses the reduced viscosity of a solution at an infinitely small concentration. The latter can be derived from the relative viscosity, which is the ratio of the flow time of the dilute polymer solution of a given concentration (t_{ps}) and that of the solvent (t_s) in a capillary viscometer:

$$\eta_{rel} = t_{ps}/t_s$$
$$\eta_{red} = \frac{\eta_{rel} - 1}{c}$$

wherein c = concentration of the cellulose or its derivative in the solution.

The degree of polymerization can be calculated from [η] by using the formula:

$$[\eta] = \lim_{c \to 0} \eta_{red} = K_p P_v^a \text{ (or } K_m M_v^a)$$

The definition of $\bar{P}_v$ (viscosity average of the degree of polymerization) is as follows

$$\bar{P}_v = \left\{ \frac{w_i P_i^{1+a}}{w_i P_i} \right\}^{1/a}$$

wherein w_i = weight fraction of a molecularly uniform fraction with a degree of polymerization of P_i; $\bar{P}_v$ = viscosity average of the degree of polymerization which for cellulose or cellulose derivative solutions closely resembles the weight average $\bar{P}_w$; and $\bar{M}_v$ = viscosity average of the molecular mass.

Table 2 summarizes some of the more important a, K_m, and K_p values. These values have been obtained by calibration to one of the above-mentioned absolute methods for the determination of the degree of polymerization.

Quite often one speaks simply of the "average degree of polymerization" (DP), which is misleading, however, unless the method of determination is properly stated ($\bar{P}_n$, $\bar{P}_w$, $\bar{P}_z$, or $\bar{P}_v$).

Polymolecularity. Cellulose isolated from its native sources is always polydisperse; i.e., it consists, as do all polymers, of a mixture of molecules with the same basic composition and chemical constitution but differing widely in their chain length or degree of polymerization, respectively. The relative amounts of molecules of various lengths present in a given cellulose substrate can be specified by the so-called differential mass distribution curves. Unlike most synthetic polymers, cellulose substrates have complicated mass distribution functions. Figure 1 shows typical mass distributions of various cellulose samples.

Knowledge of the molecular mass distribution is important for many applications. However, most of the methods applied in the past, such as fractionation by precipitation or selected dissolution, consume a great deal of time and are often subject to objection. In recent years, gel-permeation chromatography, especially in combination with low-angle laser light scattering, was

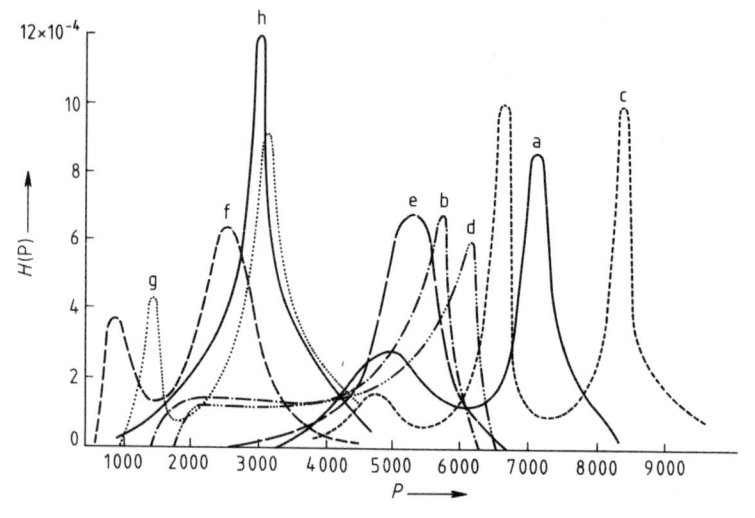

Figure 1. Differential mass distribution curves of various celluloses [18] a) Cotton; b) Cotton; c) China grass (ramie); d) Flax; e) Ramie; f) Balsam; g) White fir; h) Birch

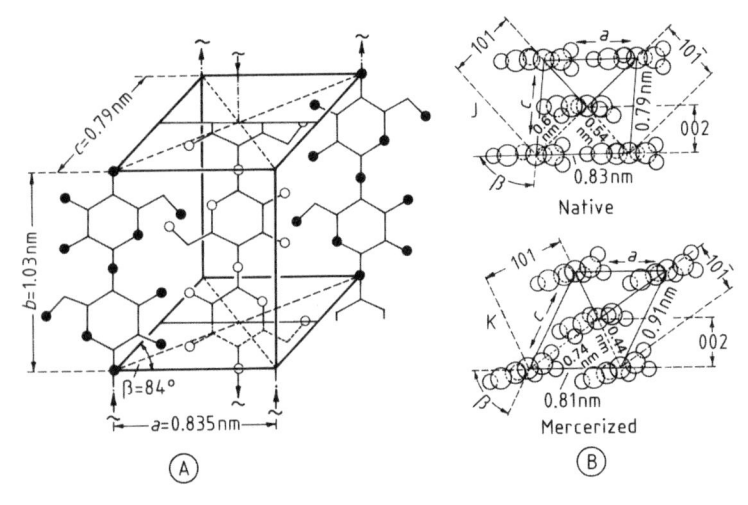

Figure 2. A) Unit cell of the crystal lattice of cellulose I B) Comparison of the unit cell cross-sections of the native cellulose I and mercerized cellulose II

developed to such a state as to provide a fast and highly reproducible method for this purpose [21].

As a convenient measure of the broadness of the mass distribution, the so-called nonuniformity factor (NU) is often used:

$$NU = P_w/P_n - 1$$

A polymer with a normal (most probable) molecular mass distribution will have a nonuniformity factor of 1. In most cases, cellulose substrates show much higher molecular nonuniformities.

Secondary Structure. In solution, the cellulose molecules exist in form of largely expanded coils. In addition to isolated and solvated molecules, cellulose and cellulose derivative solutions frequently also contain supermolecular gel particles, so-called micelles [22].

In solid cellulose, high-order microcrystalline structures ("crystalline regions") alternate with those of a distinctly lower order ("amorphous regions"). Cellulose is polymorph; i.e., depending on the origin or the conditions during isolation or conversion, cellulose will have or can adopt various crystal lattice structures.

Lattice Structure. Native celluloses all show the so-called cellulose I lattice structure. Each unit cell houses two countercurrently arranged cellulose molecules. The lattice of cellulose I is of the monoclinic sphenolitic type. The cellulose chains are in line with the b axis of the unit cell. Figure 2 shows a schematic of the unit cell of the cellulose I modification.

The cellulose molecules are aligned in the fibrillar axis and form the b axis of the unit cell. The length of the b axis is 1.03 nm (10.3 Å), somewhat shorter than the extended length of a cellobiose unit, which suggests a slight helical twist in the cellulose chains along the b axis [23–25]. This twist is caused by intramolecular hydrogen bonding primarily between the hydroxyl groups on the carbon atom C–3 of one glucose unit and the pyranose ring oxygen in the adjacent glucose unit of the same chain molecule. This intramolecular secondary valence bond is also responsible for the relative rigidity of the cellulose molecule [26]. Some authors [27] suggest a second intramolecular hydrogen bond involving the hydroxyl groups on C–6 and C–2 of adjacent glucose units in the same molecule.

In more recent years, some researchers suggested a unit cell for cellulose I in which the a and c axis of the Meyer–Mark–Misch model are doubled [28], [29]. However, these newer interpretations of X-ray and electron diffraction results are more or less closely related to the Meyer–Mark–Misch lattice structure. In reference [30], it is claimed that there is a closer agreement with the observed diffraction intensity data for parallel arrangement of the cellulose molecules in the unit cell. However, this is still controversial [31].

Table 3. Lattice parameters of the unit cells of the cellulose polymorphs

Type	Source	Dimensions, nm			β, degree
		a	b	c	
Cellulose I	cotton	0.821	1.030	0.790	83.3
Cellulose II	cotton, mercerized	0.802	1.036	0.903	62.8
	viscose fiber	0.801	1.036	0.904	62.9
Cellulose III		0.774	1.030	0.990	58.0
Cellulose IV		0.812	1.030	0.799	90.0

The internal cohesion of the cellulose molecules in the unit cells and crystalline domains is due to intermolecular secondary valences – partly hydrogen bonds and partly van der Waal's forces. These bonds can act either between molecules situated in the same crystal lattice plane (intraplanar bonds) or between molecules located in neighboring lattice planes (interplanar bonds). The intraplanar hydrogen bonds are formed primarily between adjacent cellulose molecules in the same 002 lattice planes giving a sheetlike structure. The 002 sheets are then bonded to one another by hydrogen bridges involving the hydroxyl groups on C–6 and the glucosidic ring oxygen atoms of cellulose molecules favorably located in neighboring 002 planes, or by van der Waal's forces acting between neighboring glucopyranose rings.

The unit cells of the other polymorphic structures of cellulose – the most important one being the so-called cellulose II – differ basically in the lengths of their a and c axis and the angle of inclination β. The cellulose II modification is formed as the thermodynamically most stable polymorph when cellulose fibers are treated with concentrated sodium hydroxide solution ($> 14\%$) or precipitated (regenerated) from solution.

In addition to the cellulose I and cellulose II modifications, two other polymorphic lattice structure are known, the cellulose III and cellulose IV crystal modifications. The cellulose III structure is formed when the reaction product of native cellulose fibers is decomposed with liquid ammonia. This modification has a lattice structure closely related to that of cellulose II. The cellulose IV modification is obtained by treating regenerated cellulose fibers in hot baths under stretch. The lattice of this polymorph is closely related to that of cellulose I. Some distinct differences in their infrared absorption spectra seem to indicate their existence. Some researchers however doubt their actual existence [32].

Table 3 lists the lattice parameters of the unit cells of these four polymorphic crystal structures.

Crystallites. The ability of hydroxyl groups to form secondary valence hydrogen bonds is – together with the stiff and straight chain nature of the cellulose molecule – the cause for the high tendency to organize into crystallites in parallel arrangement and crystallite strands (elementary fibrils), the basic elements of the supermolecular structure of cellulose fibers.

The dimensions of the elementary crystallites differ only slightly for native or regenerated cellulose fibers. Their length ranges between 12 and 20 nm ($= 24 - 40$ glucose units) and their width between 2.5 and 4.0 nm. The often observed larger "micro- or macrofibrils" (or fragments thereof) are aggregations of elementary fibrils.

Two questions concerning the crystal structure are still under dispute. The first deals with the antiparallel or parallel arrangement of the cellulose molecules in the crystal lattice as previously mentioned. The second question (still open) concerns the existence or nonexistence of folded chains in the lattice [33–35]. While a folded cellulose chain position in the lattice seems unlikely to most experts, the parallel molecule arrangement in the cellulose I lattice is principally acceptable, under the condition that two cellulose II lattice structures exist, one for heterogeneously mercerized native celluloses with parallel arrangement of the molecules and the other for regenerated cellulose substrates with antiparallel molecule arrangement.

1.1.2. Supermolecular Structure (Texture)

The basic structural element of cellulose fibers is the so-called elementary fibril. It can be seen with the electron microscope, as illustrated in Figure 3.

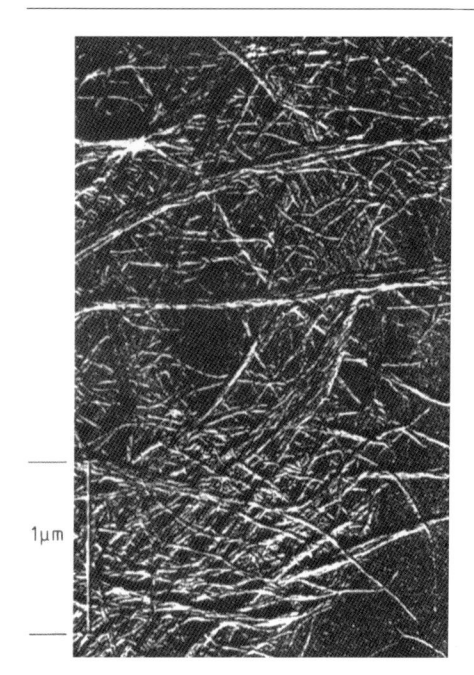

Figure 3. Electron micrograph of the fibrillar nature of cellulose fibers

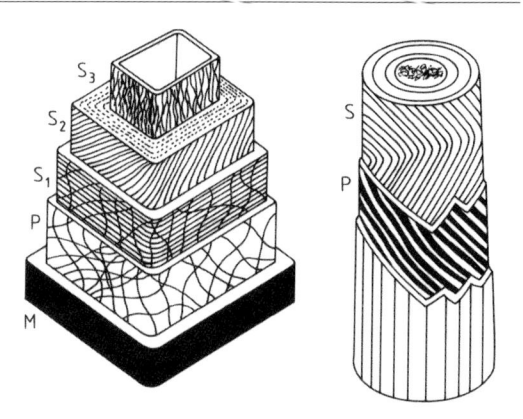

Figure 5. Positioning of the cellulose fibrils in wood (left) and cotton fibers (right) Wood fibers: M) Middle lamella (lignin and hemicelluloses); P) Primary wall (fibril position unarranged); S_1) Secondary wall I (two or more fibrillar layers crossing one another and positioned spirally along the fiber axis); S_2) Secondary wall II (fibrils wound spirally around the fiber axis; S_3) Secondary wall III (fibrils tightly interlaced) Cotton fibers: P) Primary wall (interlaced fibrils); S) Secondary wall (fibrils wound spirally around the fiber axis; in distinct distances along the fiber axis the spiral reverses direction)

The cross-dimensions of the elementary fibrils correspond with those of the elementary crystallites. The elementary fibril is a strand of elementary crystals linked together by segments of long cellulose molecules. The lateral order in the interlinking regions is distinctly less pronounced (amorphous). This structure is schematically shown in Figure 4 [36–39].

Several elementary fibrils associate to form larger aggregations of so-called microfibrils and macrofibrils, which can also be seen with a light microscope.

The elementary fibrils and their aggregations are determined by nature in such native fibers as cotton or wood pulp fibers and are laid down in various cell wall layers in a typical manner [18], [19]. Figure 5 shows the structural organisation of wood pulp and cotton fibers.

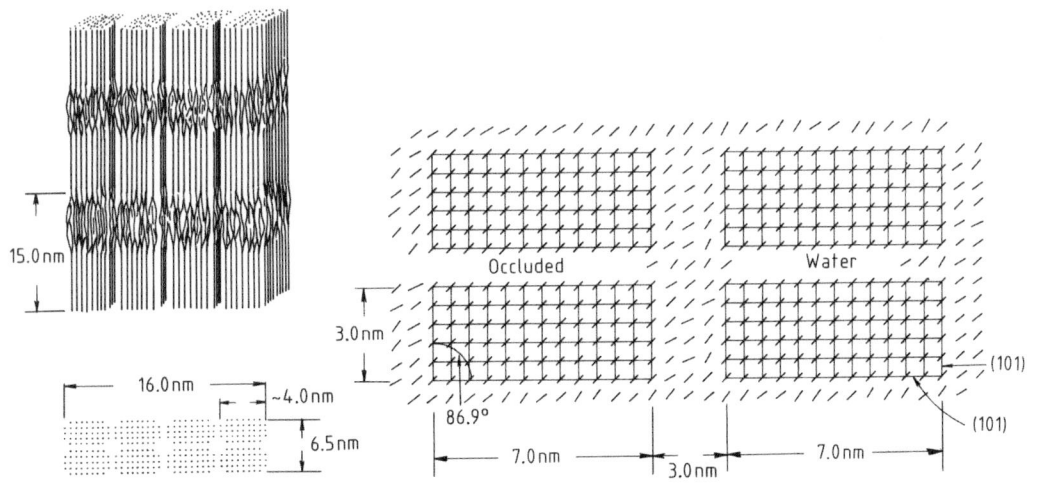

Figure 4. The architecture of elementary fibrils and microfibrils of native celluloses

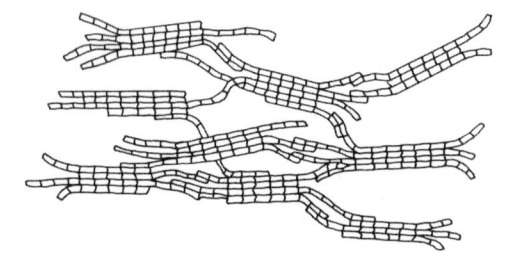

Figure 6. Fringe fibrillar model of fiber structure

Synthetic cellulose fibers, such as viscose, do not have a native morphology. Their supermolecular structure can be described as a network of elementary fibrils and their more or less random associations. This is called a "fringe fibrillar" structure [40], which is shown in Figure 6.

Structure Characterization. The methods used to characterize the molecular and fine structure of native and synthetic cellulose fibers include the following [41]:

1. determination of the average degree of polymerization ($\bar{P}_n$) by the osmotic method;
2. determination of the average crystallite length by meridional X-ray low-angle scattering on slightly hydrolyzed fiber samples or by measurement of the band width of the meridional 040 X-ray wide-angle reflection at half-maximum intensity;
3. determination of the degree of order ("crystallinity," CrI) with a method for separating overlapping equatorial X-ray diffractions [42] and deriving from the band width at half-maximum intensity the average cross-dimensions of the crystalline regions; furthermore, this analysis yields information on the lattice structure, polymorphic composition, and accessibility;
4. determination of the degree of orientation by measuring the azimuthal intensity distribution of major equatorial X-ray diffraction arcs or by IR dichroism.

Structure and Properties. Physicomechanical properties of cellulose fibers such as tenacities, elongations, or moduli in the conditioned or wet state are determined by the following structural parameters [43]: 1) the average length of the fiber-forming molecules ($\bar{P}_n$); 2) the average length of the elementary crystallites ($\bar{P}_{nL}$ = number average "limiting" degree of polymerization); 3) the degree of lateral order (crystallinity, CrI); 4) the degree of orientation (f_r) with respect to the fiber axis; and 5) the presence of heterogeneities (natural defects, incorporated gel or sand particles, etc.).

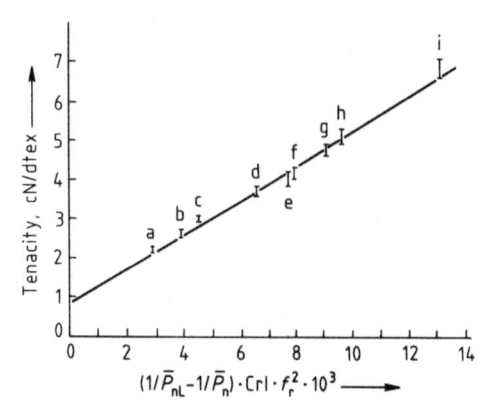

Figure 7. Relation between structure parameters and tenacities of regenerated cellulose fibers a) Regular viscose fiber; b) Medium-strength viscose fiber; c) High-strength viscose fiber; d) High-wet-modulus viscose fiber; e) Polynosic type viscose fiber; f) Medium-strength viscose tire cord; g) High-strength viscose tire cord; h) Meryl fiber; i) Fortisan fiber

This may be illustrated by the following examples: As shown by Figure 7, the tenacity of the conditioned fibers is determined by the length of the molecules in relation to the length of the elementary crystallites building the elementary fibrils ($1/\bar{P}_{nL} - 1/\bar{P}_n$), by the degree of order (CrI), and by the degree of orientation (f_r).

The elongation at break in the conditioned state is mainly dependent on the degree of orientation. Simple geometric considerations give the parameter ($1/\cos \alpha - 1$) in which the angle α derived from the orientation factor (f_r) is the mean angle of deviation of the basic structure elements from the fiber axis. Figure 8 illustrates the relation of this parameter to the breaking elongation of a number of cellulosic fibers.

The wet moduli of the various fibers show a close relation to the product of the length of the elementary crystallites ($\bar{P}_{nL}$), the degree of order (CrI), and the square of the orientation factor ($f_r^{\ 2}$). This is demonstrated in Figure 9.

1.1.3. Physical Properties

Cellulose is relatively hygroscopic. Under normal atmospheric conditions (20 °C, 60 %

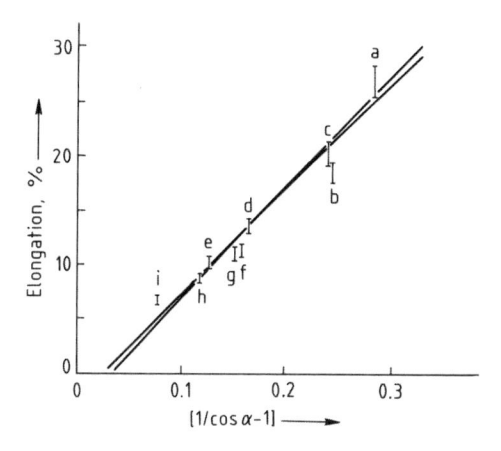

Figure 8. Relation between the degree of orientation and breaking elongation of regenerated cellulose fibers (α = average deviation of structure units from the fiber axis in degree) a) Regular viscose fiber; b) Medium-strength viscose fiber; c) High-strength viscose fiber; d) High-wet-modulus viscose fiber; e) Polynosic type viscose fiber; f) Medium-strength viscose tire cord; g) High-strength viscose tire cord; h) Meryl fiber; i) Fortisan fiber

Table 4. Swelling of various celluloses in water

Type	Swelling, %
Cotton	18
Viscose, continuous filament	74
Cuprammonium rayon	86
Cellulose triacetate	10
Cellulose tripropionate	2.5
Cellulose tributyrate	1.8
Cellulose trivalerate	1.6
Cellulose tristearate	1.0

relative humidity), it adsorbs ca. 8 – 14 % water. Cellulose swells in water (see Table 4). It is, however, insoluble in water or dilute acids. In concentrated acids, solution can be achieved under severe degradation. Caustic solutions cause extensive swelling and dissolution of low molecular mass portions ($P \leq 200$). Solvents for cellulose are listed in Table 2).

Cellulose is nonmelting; thermal decomposition starts at 180 °C; the ignition point is > 290 °C. With chlorine and zinc iodide, cellulose takes on a red-violet to blue color; with phloroglucinol–hydrochloric acid, pure cellulose should not take on a red color (test for residual lignin). Additional data:

Density: 1.52 – 1.59 g/cm³
Refractive index:
  1.62 parallel to the fiber axis
  1.54 perpendicular to the fiber axis

Dielectric constant: 2.2 – 7.2 (at 50 Hz) Highly dependent on humidity conditions
Insulation resistance: $10^{14} - 10^{17}$ Ω cm Highly dependent on humidity conditions
Electric strength: 500 kV/cm
Heat of combustion: 17.46 J/g
Heat of crystallization: 18.7 – 21.8 kJ/mol of glucose
Specific heat: 1.00 – 1.21 J g^{-1} K^{-1}
Coefficient of thermal conductivity:
  0.255 kJ m^{-1} h^{-1} K^{-1} (loosely packed) to
  0.920 kJ m^{-1} h^{-1} K^{-1} (compressed)

Specific internal surface: 10 – 200 m²/g

1.1.4. Chemical Properties

The chemical reactivity of cellulose is determined to a large extent by the supermolecular structure of its solid state. Most of the reactions on cellulose fibers are heterogeneous in nature. The reaction medium acts on a two-phase solid system: (a) the less-ordered (amorphous) regions which are mainly located on the surface of the elementary fibrils or their aggregations and in the interlinking regions between the elementary crystallites in the fibrils, and (b) the well-ordered elementary crystallites or fused associations of the elementary fibrils. Any reaction will first start on the less-ordered surface of the elementary

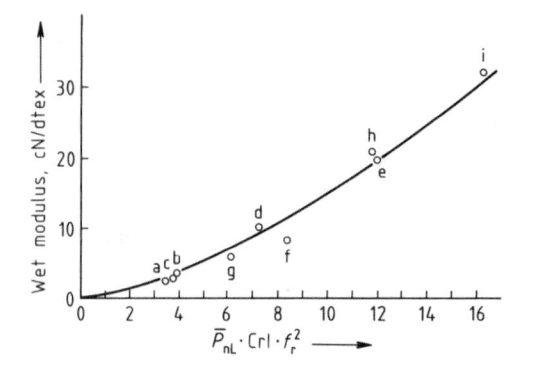

Figure 9. Relation between structure parameters and wet modulus of regenerated cellulose fibers a) Regular viscose fiber; b) Medium-strength viscose fiber; c) High-strength viscose fiber; d) High-wet-modulus viscose fiber; e) Polynosic type viscose fiber; f) Medium-strength viscose tire cord; g) High-strength viscose tire cord; h) Meryl fiber; i) Fortisan fiber

fibrils or their aggregations (topochemical reaction) and then, under favorable conditions, proceed into the interlinking regions between the elementary crystallites to penetrate from both ends into the crystallites. Therefore, as long as the reaction is limited to the accessible surface of the fibrils or fibrillar aggregations and the regions interlinking the elementary crystallites (i.e., up to degrees of substitution (DS) of 1.3 – 1.7), there is no visible effect in the crystalline structure. At increased degrees of substitution (to ca. DS = 2.5), the X-ray diffractogram shows overlapping diffraction bands of the original cellulose I structure and the cellulose derivative. At still higher degrees of substitution, the pure diffraction pattern of the derivative will finally result. This course of reaction implies that partially substituted cellulose derivatives are always a mixture of completely substituted cellulose, partially substituted portions (of block-polymer nature), and unsubstituted cellulose.

A quasi-homogeneous reaction can be achieved when the fiber structure is loosened by swelling treatments to such an extent that all cellulose molecules can react simultaneously. A real homogeneous reaction can, however, only be achieved by bringing the cellulose into a molecularly dispersed solution.

The reactivity of cellulose substrates can be greatly enhanced by activation treatments, such as swelling, solvent exchange, inclusion of structure-loosening additives, degradation, or mechanical grinding, which enlarge accessible surfaces by opening fibrillar aggregations. These treatments also restore, in most cases, the loss of reactivity due to so-called hornification, which occurs when water is removed from cellulose by drying under severe conditions.

Swelling with water or other polar liquids is the most frequently applied activation treatment. It exclusively opens the interfibrillar interstices and swells the less-ordered surface and interlinking regions of the fibrillar elements. The solvent exchange technique is a special kind of activation from the water-swollen state. It allows the introduction of media being inert in subsequent reactions that are unable to swell the cellulose substrate thus maintaining the reactive water-swollen state. An interesting variation of the solvent exchange treatment is the so-called inclusion technique [44]. Inert liquids, such as cyclohexane or benzene, are introduced into the cellulose substrate by solvent exchange from the water-swollen state. During drying, they are permanently incorporated into the interfibrillar interstices or voids, thus preventing fusion of fibrils, i.e., the hornification. Such inclusion celluloses are very reactive, as shown in Table 5.

Another very effective way of activating cellulose fibers is to enhance the accessibility of fibrillar surfaces and to open the less-ordered regions interlinking the crystallites in the fibrils by treatment with systems causing not only interfibrillar, but also intracrystalline, swelling. Some inorganic acids, various salt solutions, and especially certain inorganic and organic bases

Table 5. Acetylation of vacuum-dried native and mercerized cotton after inclusion of benzene using acetic anhydride – sulfuric acid at 60°C

	Native		Mercerized	
Time of acetylation	P 2500 % Acetyl	P 340	P 2200	P 290
Acetylated after vacuum drying:				
24 h	2.0	5.6	0.9	1.3
48 h	3.8	8.3	1.9	2.9
96 h	6.8	12.5	2.5	3.6
192 h	14.2	21.9	3.0	4.8
408 h	37.8	39.9	11.0	14.5
Acetylated after benzene inclusion, benzene wet:				
1 h	19.4	19.1	25.7	24.3
24 h	42.5	44.1	45.0	44.7
Acetylated after benzene inclusion and drying:				
1 h	17.3	16.8	23.9	23.0
24 h	41.5	42.1	43.2	43.1

Table 6. Effect of swelling agents on the (101) interplanar spacing of cellulose [45]

Swelling agent	Observed (101) distance, nm
Liquid ammonia	1.03 – 1.06
Methylamine	1.467
Ethylamine	1.572
n-Propylamine	1.848
n-Butylamine	1.973
n-Amylamine	2.192
n-Hexylamine	2.485
n-Heptylamine	2.874
Hydrazine	1.03
Ethylenediamine	1.226
Tetramethylenediamine	1.465
Tetramethylammonium hydroxide	1.30
Ethyltrimethylammonium hydroxide	1.30
Benzyltrimethylammonium hydroxide	1.65
Dibenzyldimethylammonium hydroxide	1.65

achieve this at distinct concentrations. They apparently penetrate the fiber through existing capillaries and pores by opening the fibrillar interstices and entering the interlinking regions between the crystallites. From there, they enter the elementary crystallites from both ends and force them open. At suitable concentrations and temperatures, they ultimately cause crystal lattice transfer, particularly with respect to opening the 101-plane distances (see Table 6).

So-called mercerization is a frequently used practical method of activation, i.e., the treatment of native cellulose substrates with 10 – 20 % sodium hydroxide solutions at moderate temperature ($< 20\ °C$). In this treatment, sodium cellulose I is formed in which the 101-plane distance is increased from 0.61 nm (6.1 Å) in native cellulose to 1.22 nm (12.2 Å). In this lattice transition, the glucopyranose rings are dislocated and aligned into the 101 lattice plane. The hydroxyl groups on the C–2 and C–6 carbon atoms are thus freely exposed and jut into the widened space between the 101 lattice planes, making them accessible for reactions.

The chemical character of cellulose is determined by the sensitivity of the β-glucosidic linkages between the glucose repeating units to hydrolytic attack and the presence of one primary and two secondary reactive hydroxyl groups in each of the glucopyranose units. These reactive hydroxyl groups are able to undergo exchange, oxidation, and substitution reactions, such as esterification and etherification.

Sorption and Exchange Reactions. Cellulose undergoes sorption and exchange reactions with water and deuterium oxide. These reactions are of special interest since they give a good indication of the accessibility of the cellulose substrate.

The cross-dimensions of the well-ordered regions can be derived from the width of the equatorial X-ray wide-angle diffractions. If it is assumed that the crystallites are surrounded by one layer of disturbed ("amorphous") unit cells and also that the molecules in the next outer layer of the well-ordered crystalline core of the crystallites are in addition accessible for reactants such as water or deuterium oxide, an explanation for the extent of water adsorption and deuterium exchange should be possible, see Figure 10 [46].

Table 7 compares the experimentally obtained water sorption and deuterium exchange values with those predicted from structure investigations according to the above outlined concept.

The internal surface data calculated from crystallite cross-dimensions are also in good agreement with those determined by gas adsorption.

Degradation by Acid Hydrolysis. Degradation in acidic medium is based on the hydrolysis of the β-glucosidic linkages between the glucose base units. The reaction depends strongly on pH and already proceeds at a remarkable rate at low pH and temperatures well under 100 °C. Initially the acetal oxygen of the glucosidic linkage is protonated. Through heterolysis, an intermediate carbonium ion is formed, causing chain-splitting. The carbonium ion finally reacts with water, which reforms the proton. The following reaction scheme illustrates the course of this reaction.

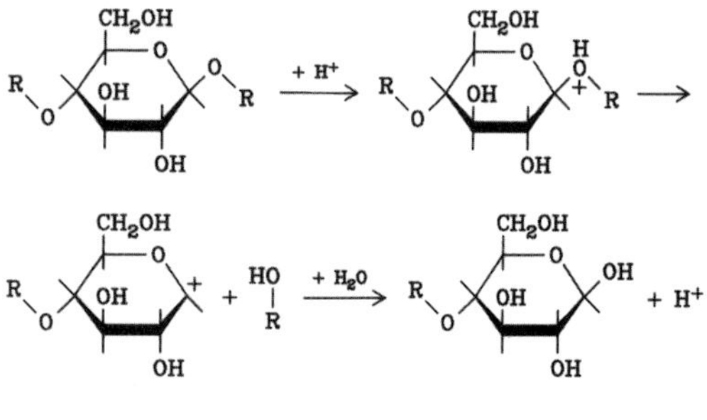

Homogeneous and heterogeneous hydrolysis of cellulose are both first-order reactions. The

134 Cellulose

⊥ A_0=101: 8.3 nm (83Å)   ⊥ A_3=101: 6.8 nm (68Å)

A_4=002: 8.6 nm (86Å)

In crystalline regions: 226 molecules (o)
In amorphous regions: 124 molecules (●)
% crystallinity = 226·100/350 = 64.6

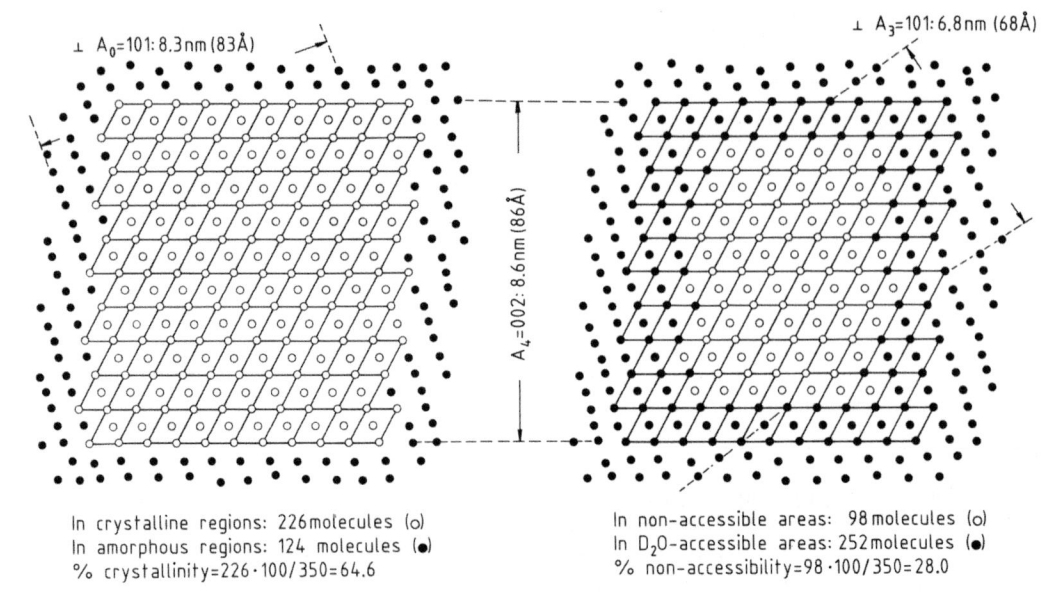

In non-accessible areas: 98 molecules (o)
In D_2O-accessible areas: 252 molecules (●)
% non-accessibility = 98·100/350 = 28.0

Figure 10. Model of crystallinity (left) and accessibility (right) of fibrils in a high-wet-modulus viscose staple fiber [46]

reaction speed is strongly dependent on the acid and the cellulosic material. The course of homogeneous and heterogeneous hydrolysis also reveals a basic difference. In homogeneous hydrolysis degradation proceeds at a constant rate until all of the cellulose is degraded to cellobiose or glucose, respectively. In heterogeneous hydrolysis, the rate decreases continuously and degradation stops almost completely when the number-average degree of polymerization reaches 25 – 100. This corresponds to the length of the elementary crystallites ("level-off degree of polymerization"). Hydrolysis proceeds thereafter at a very slow rate. It is interesting to note that during the course of homogeneous hydrolysis, the degraded and isolated residue adopts an increasingly normal molecular mass distribution as indicated when the ratio of its weight to the number average degree of polymerization approaches a value of 2. This is indicative of a statistical degradation (see Table 8).

In contrast, the residue in heterogeneous hydrolysis tends toward a value of 1 for the ratio of its weight to number-average molecular mass. This is a strong indication that the course of heterogeneous degradation is not determined solely by the sensitivity of the β-glucosidic

Table 7. Crystallinity and accessibility of cellulose substrates from X-ray diffraction in relation to H_2O adsorption and deuterium exchange

Sample	Relative dimensions [a]				% molecules in:			Deuterium exchange, %	H_2O adsorption	
	101	101	002		Ordered regions	Amorphous regions	Accessible regions		Theoretical [b], %	Experimental, %
Cotton	200	155	230	83	17	47		7.8	8.1	
Fortisan	140	100	160	80	20	54	53.5	9.0	9.6	
HWM-rayon	80	70	85	65	35	69	72.0	11.5	12.4	
Polynosic	95	80	100	70	30	60	61.5	10.0	10.9	
Regular rayon	90	75	95	67	33	65	66.5	10.8	11.6	

[a] Derived from X-ray wide-angle diffractograms corrected for the band-widening effects of sample size and beam dimension.
[b] Calculated by assuming that accessible glucose units bind 1.5 mol of H_2O each.

Table 8. Comparison of the changes in molecular non-uniformity during the course of homogeneous and heterogeneous degradation of cellulose [47], [48]

Time, min	$\bar{P}_w$	$\bar{P}_n$	$\bar{P}_w/\bar{P}_n$
Homogeneous hydrolysis in H_3PO_4 at 25 °C			
0	1600	1245	1.29
60	1055	785	1.34
210	610	410	1.50
750	260	150	1.73
2100	100	55	1.82
Heterogeneous hydrolysis with 1 N HCl at 50 °C			
0	1750	1200	1.46
180	800	560	1.43
540	440	310	1.41
1620	250	180	1.38
3240	120	95	1.26

linkages, but primarily by morphological aspects [49]. Hydrolytic attack is almost completely limited to the molecules situated on the surface of the fibrillar strands and the accessible molecule segments connecting the crystallites. It is also important to note that mass loss in heterogeneous hydrolysis performed under moderate conditions is relatively small in the initial fast reaction.

Degradation in Alkaline Media. Hydrolysis of the β-glucosidic linkages in alkaline media occurs at a significant rate only at temperatures above 150 °C. It is most probable that chain-splitting proceeds by way of the 1,2-anhydro configuration [50].

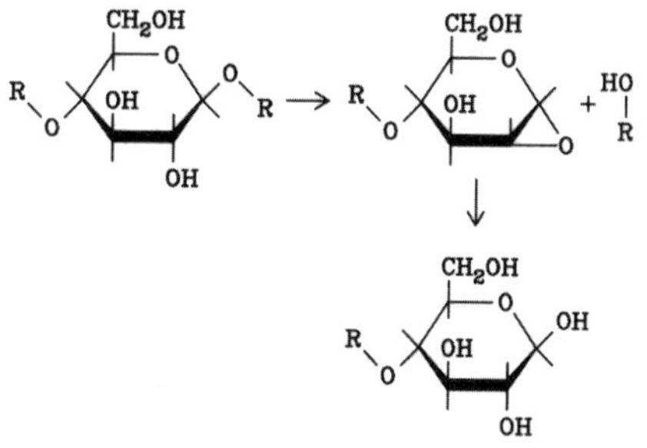

Acidic as well as alkaline hydrolysis of the glucosidic bond is remarkably enhanced (β-elimination) by oxidative changes at the C–2, C–3, or C–6 carbons leading to carbonyl groups. An example of this chain-splitting reaction, which can even occur at moderate temperatures, is [51]:

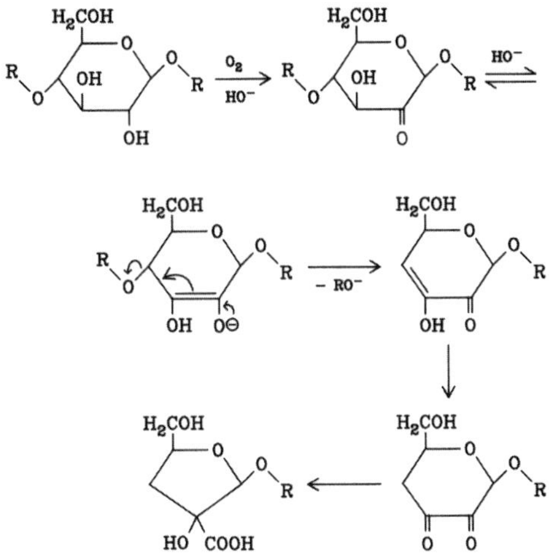

An additional degradation reaction taking place in alkaline media is the so-called peeling reaction. In the course of this reaction, which even takes place at temperatures well below 100 °C, the cellulose chain molecules are degraded step-by-step, beginning at the reducing end and proceeding in a "zipper-like" reaction. The terminal glucose unit is first transformed into the 1,2-enediol, which isomerizes to the corresponding ketose and splits off the chain. The ketose is transformed further into the alkali-stable isosaccharinic acid [52].

The newly formed aldehydic end of the cellulosic chain will repeat the same reaction. When ca. 50 – 60 glucose units are split off under these conditions, the reaction normally stops due to the interference of a chain-stopping reaction. This termination reaction leads by way of the 2,3-enediol to an alkali-stable metasaccharinic acid end group, which stabilizes the cellulose molecules against further degradation.

The degradation in acidic as well as in alkaline media is of great importance in the manufacture of pulp from wood and other plants and in the processing of cellulose derivatives, regenerated fibers, and films.

The microbiological degradation of cellulose should also be mentioned in this connection. This degradation takes place through enzymatic hydrolysis of the β-glucosidic linkages and is of interest in connection with the use of plant biomass [53].

Oxidation Reactions. The hydroxyl groups and the aldehydic end groups participate in the oxidation reactions of cellulose. These reactions form aldehyde, ketone, and carboxyl groups. Extensively oxidized and degraded products are designated as oxycelluloses. Some oxidizing agents show specific action. They attack only specific functional groups, forming defined oxidation products. Other oxidants react nonspecifically with all types of oxidizable groups in the cellulose molecules. Under special conditions hypoiodite and chlorite attack only the aldehyde end group on C–1, oxidizing it to form a carboxylic group. Another oxidant with specific action is periodate, which attacks the glycol configuration on the carbon atoms C–2 and C–3, thus causing ring-splitting and forming a dialdehyde structure.

Nitrogen dioxide (dinitrogen tetroxide) reacts not quite as specifically. It oxidizes, however, with a certain preference the hydroxyl group on C–6 to a carboxyl group and to a lesser extent hydroxyl groups on C–2 and C–3 to ketone groups.

Nonspecific oxidants are chlorine, hypochlorite and chromic acid. They also oxidize all accessible hydroxyl groups to aldehyde, ketone, and carboxyl groups. The oxidative action of chlorine and hypochlorite is extensively used as a bleaching agent in the pulp industry. However, one must keep in mind that the introduction of carbonyl groups on C–2, C–3, and C–6 causes alkali instability of glucosidic linkages in the β-position which initiates degradation under alkaline conditions.

Esterification and Etherification. Both of these substitution reactions are used in industry for the manufacture of widely used products (→ Cellulose Esters, → Cellulose Ethers). The acetate and nitrate esters and the methyl and carboxymethyl ethers of cellulose, and to a lesser extent the ethyl and hydroxyethyl ethers, have acquired practical significance.

To perform the esterification or the etherification reaction properly, the hydroxyl groups in the cellulose substrate must be made accessible for the reaction. The supermolecular structure of the cellulose must be activated before or in the course of the substitution reaction. For this purpose the cellulose substrate is treated with strong acids or alkali. In these treatments addition compounds are formed between the cellulose, acid, or alkali and the water present in the system. During the opening of hydrogen bonds between the molecules in the cellulose substrate, more or less defined addition compounds are formed in the crystalline regions. Their formation is often accompanied by changes in the lattice structure, resulting mostly in an increased distance between the 101 planes. This exposes the hydroxyl groups on C–2 and C–6, making them accessible. Hydrated cations or anions of the reactant, respectively, are incorporated into the widened interplanar space, where they initiate and facilitate the reactions. The addition compounds thus formed are only stable in equilibrium with the reactant and will decompose readily when the system is diluted with water.

The basic principle of the esterification and etherification reaction is quite similar. The first step is a nucleophilic substitution or addition and the formation of an oxonium ion on the carbon atom carrying the reactive hydroxyl group. A surplus of esterification or etherification reactant will lead to the formation of the corresponding ester or ether:

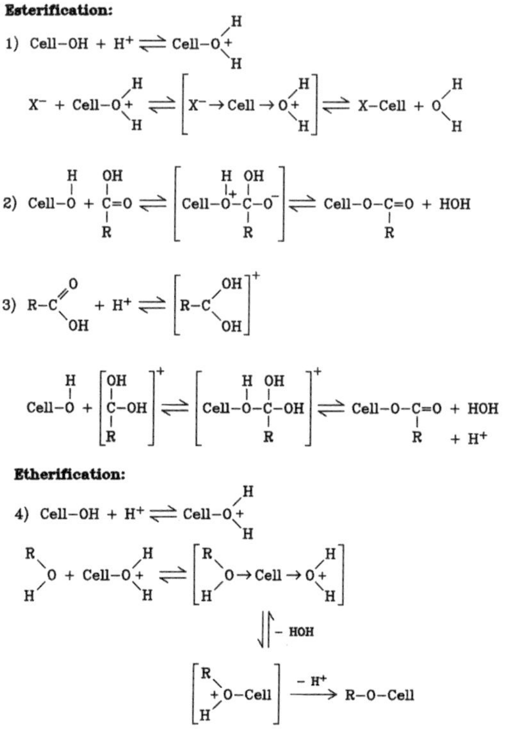

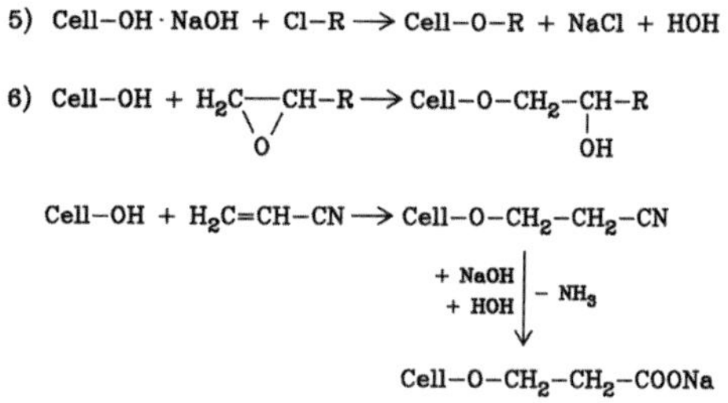

Esterification:

1) Cell–OH + H$^+$ ⇌ Cell–O$^+$H$_2$

   X$^-$ + Cell–OH$_2^+$ ⇌ [X → Cell → OH$_2^+$] ⇌ X–Cell + OH$_2$

2) Cell–OH + C(=O)R(OH) ⇌ [Cell–O–C(OH)(R)–O$^-$] ⇌ Cell–O–C(=O)–R + HOH

3) R–C(=O)OH + H$^+$ ⇌ [R–C(OH)$_2$]$^+$

   Cell–OH + [C(OH)(R)]$^+$ ⇌ [Cell–O–C(OH)(R)–OH]$^+$ ⇌ Cell–O–C(=O)–R + HOH + H$^+$

Etherification:

4) Cell–OH + H$^+$ ⇌ Cell–OH$_2^+$

   R–OH + Cell–OH$_2^+$ ⇌ [R–O(H) → Cell → OH$_2^+$]

   ⇌ [R–O(H)$^+$–Cell] $\xrightarrow{-H^+}$ R–O–Cell

5) Cell–OH · NaOH + Cl–R → Cell–O–R + NaCl + HOH

6) Cell–OH + H$_2$C—CH–R (epoxide) → Cell–O–CH$_2$–CH(OH)–R

   Cell–OH + H$_2$C=CH–CN → Cell–O–CH$_2$–CH$_2$–CN

   $\xrightarrow[+ HOH]{+ NaOH}$ –NH$_3$

   Cell–O–CH$_2$–CH$_2$–COONa

1. General esterification mechanism (formation of an oxonium ion; inorganic acid esterification)
2. Esterification with organic acids (nucleophilic addition)
3. Acid-catalyzed esterification
4. General etherification mechanism (formation of an oxonium ion; alcohol excess leads to ether formation)
5. Etherification of cellulose with alkali consumption
6. Etherification of cellulose without alkali consumption

The esterification reaction is promoted by water-binding catalysts. In etherification, prior swelling of the cellulose substrate with alkali or the transfer to alkaline cellulose is essential for the reaction.

The presence of three hydroxyl groups in each glucose unit allows the formation of mono-, di-, and triesters or ethers, respectively. Contrary to the reaction behavior of primary and secondary hydroxyl groups in low molecular mass alcohols, where the primary hydroxyl group always shows a higher reactivity, the secondary hydroxyl group on C–2 quite often shows preferred reactivity in heterogeneous esterification or etherification of cellulose. Table 9 illustrates this for the relative reaction rate of the hydroxyl groups on C–2, C–3, and C–6 in various etherification reactions [54].

Table 9. Relative etherification velocities at the OH-groups on C–2, C–3, and C–6

Derivative	K_2	K_3	K_6
Methylcellulose	5	1	2
Ethylcellulose	4.5	1	2
Carboxymethylcellulose	2	1	2.5
Hydroxyethylcellulose	3	1	10
Cyanoethylcellulose	3	1	3

The normally expected preferred reactivity of primary hydroxyl groups on C–6 can be observed only in substitution reactions on cellulose in solution. The fact that most substitution reactions on cellulose are performed in heterogeneous systems also has the consequence that most partially substituted cellulose derivatives are actually mixtures of fully substituted, irregularly substituted (block substitution), and unsubstituted cellulose molecules.

The extent of substitution in cellulose derivatives is described by the so-called "degree of substitution" (DS). This value states the average number of substituents linked to each glucose base unit in the cellulose molecules. Since substitution normally occurs irregularly along the cellulose chains, the DS can assume any value between 0 and 3.

Industrially produced cellulose esters and ethers find practical use as fibers, films, lacquers, explosives, adhesives, and as auxiliaries in paper, textile, and food industries. Their properties and, consequently, their applications are primarily dependent on the nature of the substituents, the degree of substitution, the distribution of substituents along the cellulose molecules and from molecule to molecule, and their degree of polymerization.

In particular, the xanthation of cellulose is of great economic importance, i.e., the esterification with dithiocarbonic acid. This reaction is the basis for the so-called viscose process, which is widely used for the manufacture of regenerated cellulose fibers and films (see Chap. 3). The cellulose xanthate is obtained in the reaction of alkali cellulose with carbon disulfide and is soluble even in dilute sodium hydroxide solutions at relatively low degrees of substitution (DS < 1).

Graft Copolymerization. During the last two decades, another way of modifying cellulose, the grafting reaction, has found substantial interest. This method allows the attachment of chemically different side-chains to a given polymer molecule. The reaction mechanisms are principally the same as in the synthesis of polymers [55], [56]. The most frequently used method is to initiate grafting with radical catalysts. To achieve graft copolymerization, the cellulose substrate must be in the presence of peroxides or redox systems (i.e., hydrogen per-

Table 10. Cellulose content of various wood species

Type of wood	Cellulose content, %
Spruce (*Picea abies*)	43
Pine (*Pinus silvestris*)	44
Birch (*Betula verrucosa*)	40
Beech (*Fagus silvatica*)	43
Poplar (*Populus tremuloides*)	53

oxide, hydrogen peroxide – iron(III) ions, Ce (IV) ions) and as such is able to undergo polymerization with such compounds as styrene, acrylic acid, acrylic ester, etc. The introduction of substituents that are able to enhance radical transfer, such as thiol or xanthate groups, promotes the grafting reaction. Industrial applications for graft-modified cellulose fibers are thus far very limited.

1.2. Occurrence

Cellulose is one of the main cell wall constituents of all major plants. It occurs there in varying amounts. Cellulose is found in nonlignified (such as cotton) and lignified (such as wood) secondary plant cells and constitutes as such the major portion of all chemical cell components. Table 10 lists the cellulose content of some wood species.

The cell walls of green algae also contain cellulose. Furthermore, cellulose is also found in the membranes of most fungi and in the cell walls of some flagellates. So-called bacterial cellulose is synthesized by *Acetobacter xylinum* on nutrient media containing glucose. In the animal kingdom one finds cellulose in the tunics of the Tunicatae.

The raw material sources for industrially used cellulose are almost exclusively the fiber cells of more fully developed plants, especially cotton, bast, and leaf fibers for the textile industry, and wood and some other graminaceous plant fibers (i.e., grain straws, bamboo canes, and sugar cane wastes) for the paper industry. Wood consists of up to approx. 40 – 50 % cellulose. Additional wood constituents are lignin, hemicelluloses (i.e., lower molecular mass polysaccharides containing other sugars as monomer units in addition to glucose), extracts (such as resins, gums, fats, waxes, terpenes, etc.), and numerous other native organic compounds in smaller amounts.

1.3. Production

Cotton fibers and wood are the primary raw materials for the production of industrially used cellulose.

Raw cotton fibers and linters contain small amounts of proteins, waxes, pectins, and inorganic impurities in addition to approximately 95 % pure cellulose. These constituents can be removed if they interfere with the intended further processing, with relatively simple procedures. Treatments with hot alkali and subsequent bleaching with hypochlorite, peroxides, or chlorine dioxide are in use.

A sequence of processing steps is necessary for the isolation of relatively pure cellulose (pulp) from wood. To remove most of the lignin and of some of the other wood constituents, a pulping procedure has to be performed. This can either be done with sulfurous acid containing solutions of hydrogen sulfites (i.e., magnesium, calcium, sodium, or ammonium hydrogen sulfite in an excess of sulfurous acid) or with alkaline sodium hydroxide – sodium sulfate solutions at ca. 130 – 180 °C. These treatments reduce wood chips to fibers by using a slight mechanical action. To further purify and brighten the raw pulp stock for the manufacture of paper and for chemical applications, a multistage refining procedure must follow. These refining steps consist of alkali treatments in the presence of oxidizing agents (chlorine, chlorine dioxide, hydrogen peroxide, oxygen, or air) to remove residual lignin followed by bleach sequences with hypochlorite or hydrogen peroxide. In the production of dissolving pulps for the manufacture of cellulose films, fibers, or cellulose derivatives, extractions with cold or hot alkali are used in addition to remove low molecular mass polysaccharides (hemicelluloses) and to increase the pure cellulose (α-cellulose) content. Depending on the number of refining steps and conditions, wood pulps containing up to ca. 99 % pure cellulose can be obtained. The yield from wood, however, will be ca. 70 % for the production of low cellulose content paper pulps and only 30 % for the production of pulps with high cellulose content. Cellulose content higher than 99 % is practically unachievable since the cellulose would undergo extensive degradation in such an attempt and thus be of no further use. Table 11 lists the cellulose contents (α-cellulose) of some major paper and dissolving pulps.

Table 11. α-Cellulose content of various paper and dissolving pulps

Type	α-Cellulose content, %
Paper pulps, bleached:	
Spruce, sulfite	89
Beech, sulfite	89
Spruce, sulfate	82
Birch, sulfate	72
Sulfite dissolving pulps for:	
Regular rayon staple	89 – 91
High-wet-modulus rayon staple	91 – 93
Acetate filament yarn	94 – 95
Sulfate dissolving pulps for:	
Regular rayon staple	93 – 95
High-wet-modulus rayon staple	95 – 98
Rayon tire cord (super II – III)	96 – 98
Acetate filament yarn	98

1.4. Quality Testing

The determination of the residual impurities and the extent of molecular degradation or other chemical changes of the cellulose is, in most cases, not essential for cotton fibers in the textile field or wood pulps in the paper field.

However, celluloses (cotton linters or wood pulp) which will be used either for the manufacture of regenerated cellulose and cellulose acetate fibers or for cellulose derivatives must undergo careful quality testing.

Because the first step in the further processing quite often consists of a treatment with sodium hydroxide solutions, the testing for alkali solubility is very important. The solubility in 5, 10, and 18 % sodium hydroxide solution can give a preliminary indication of the content of noncellulosic low molecular mass polysaccharides (also including low molecular mass cellulose). A more detailed determination of cellulose and noncellulosic polysaccharide content is possible after complete hydrolytic degradation to the corresponding monomeric sugars and the use of chromatographic methods.

One of the most important characteristics for the chemical processing of cellulose is the degree of polymerization (see Molecular Size). For industrial testing practice, viscosimetric methods using standardized flow or ball-fall viscometers are quite satisfactory.

Further test methods for dissolving pulps are the determination of residual lignin, of extractables soluble in organic solvents, and of the composition of inorganic ingredients (ash). Standardized methods for all of these tests are described by ISO (International Standards Organization), ASTM (American Society for Testing and Materials), TAPPI (Technical Association of Pulp and Paper Industry, United States), DIN (German Industrial Norms), or Zellcheming (German Association of Pulp and Paper Chemists and Engineers), etc.

1.5. Applications

The greatest portion of industrially used wood cellulose by far (ca. 150×10^6 t/a) is used after partial removal of the noncellulosic constituents in its original fiber form for the production of paper, board, and nonwovens. The same applies to most of the long-haired cotton fibers (ca. 15×10^6 t/a) and their use in the manufacture of textiles. Only a minor portion (ca. 7×10^6 t/a) is used in form of high-α-cellulose wood pulps or cotton linters as starting material for the production of synthetic cellulosic fibers, primarily viscose and acetate fibers, or regenerated cellulose films, and cellulose derivatives, especially esters and ethers. Small amounts of pure cellulose are further used in the form of microcrystalline cellulose as powder or colloidal suspensions, additives in low-calorie foodstuffs, binder for forming pharmaceuticals into tablets, and for a number of industrial uses.

2. Natural Cellulosic Fibers

All plants produce cellulose, a natural polymer of the plant sugar glucose, as they mature from the "sweet" to the "stringy" phase. This stringy component, located in various parts of the plant, has long been exploited for textile and related uses, with more than 2 000 species having been spun into some kind of yarn or cord at some time or place.

Only a few of these plants are presently cultivated solely for their fibers, cotton being predominant. Few plant fibers lend themselves to home consumption. They are sold mostly as cash crops. All are obtained in impure form and require on-site processing to remove other plant components even before being shipped to spinning mills for further cleaning.

The breakdown of dead plants by bacteria, fungi, and fire is an essential stage in the life cycle of natural flora. Cellulose is thus susceptible to rot, mildew, and fire, three major shortcomings of these fibers in textiles. These problems can be overcome by the application of topical finishes.

Although cellulose predominates, other constituents also influence the properties of these fibers (Table 12). Polymers of plant sugars other than glucose are collectively termed hemicellulose. Lignins appear in most fibers except cotton. Lignified fibers are identified by the phloroglucinol test (1,3,5-benzenetriol [108-73-6]) that is used to distinguish between mechanical and chemical wood pulp ($\rightarrow$ Pulp).

Useful fibers may be obtained from various plant parts:

- the stringy bast component of the stem – jute, flax, hemp, etc.
- leaves – agaves, notably sisal
- leaf petioles – abaca (manila)
- fruit – coconut fiber (coir)
- seed – cotton, kapok

Although few plants have hairy seeds, one genus, *Gossypium,* has been cultivated and bred for its fiber with such success that cotton now

Table 12. Composition (in %) of natural cellulose fibers at 10 % moisture content [57]

	Cotton	Jute	Flax	Hemp	Ramie	Sisal	Abaca
Cellulose	92.7	64.4	62.1	67.0	68.8	65.8	63.2
Hemicellulose	5.7	12.0	16.7	16.1	13.1	12.0	19.6
Pectin	0	0.2	1.8	0.8	1.9	0.8	0.5
Lignin	0	11.8	2.0	3.3	0.6	9.9	5.1
Water solubles	1.0	1.1	3.9	2.1	5.5	1.2	1.4
Wax	0.6	0.5	1.5	0.7	0.3	0.3	0.2

Figure 11. World production of natural cellulosic fibers, 1995 c) Cotton; j) Jute and kenaf; s) Sisal; f) Flax; a) Combined abaca and ramie
Each letter represents $ 400 \times 10^6$ production

accounts for 90 % by value of all plant fibers and, if regenerated fibers are included, 78 % of all cellulosic fibers [58]. Figure 11 gives an idea of the worldwide cultivation of cotton and its economic importance relative to that of other plant fibers.

2.1. Cotton

Table 13 describes the properties of the primary commercial species of cotton, of which American Upland (*Gossypium hirsutum*) is the most important, while Sea Island types account for only a few hundred bales annually. The species *G. hirsutum* has gained importance only in the last 200 years because of the lint's characteristically firm connection to the seed. This lint needed to be separated by hand until the invention of the saw-type gin – generally ascribed to ELI WHITNEY – improved the daily output of lint from 0.5 to 20 kg per person in 1 800, a rate now greatly exceeded.

2.1.1. Molecular Arrangement, Morphology, and Fine Structure

Ginned cotton lint consists of ca. 91 % cellulose (dry mass). The remainder is made up of fibrous impurities, chiefly nitrogen compounds (proteins), wax, and pectins, which are found primarily on the fiber surface. (For the molecular structure and behavior of cellulose, see Chap. 1.)

The unicellular cotton fibers grow on the seeds, whereby a tubular outer layer, the cuticle,

Table 13. Properties of chief species of cotton

	G. barbadense	G. hirsutum	G. herbaceum
Common names	Egyptian American Pima	American Upland	Indian
Number of haploid chromosomes, designation	26, "new-world"	26, "new-world"	13, "old-world"
Usual ginning	roller	saw	roller or saw
Typical properties:			
Staple length, mm	35	26	16
Bundle strength, 3 mm gauge, g/tex[*]	30	24	24
Linear density, mtex[*]	120	200	350
Micronaire	3	4	7
Color	buff	white	tan

[*] 1 mtex = 1 mg/km

is initially formed. When a normal fiber is fully developed, an empty cell channel, the lumen, which contains the residual dead protoplasm, remains.

The originally round fiber takes on the form of a flat or ribbon-like tube twisted in such a way as to have alternating S- and Z-shaped twists irregularly arranged along the length of the fiber. The cell wall of normal mature cotton is 2 – 5 μm thick. The fiber diameter is between 12 and 22 μm and is greatest (between 20 and 40 μm) at the base or seed end. Immature cotton has a thin wall, is only slightly twisted, and has a lumen filled with protoplasm. Prematurely withered ("dead") fibers have an extremely thin cell wall (only ca. 0.5 μm thick) with conspicuous folds.

The cuticle is ca. 0.5 μm thick and covered with a layer of wax and pectins which form wrinkled overlapping rings when they are made to swell, e.g., with cuprammonium hydroxide. The primary wall, 0.5 – 1 μm thick, consisting of ca. 50 % cellulose with pectins, wax, and albumin, follows. The cuticle and primary wall account for ca. 3 % of the fiber's mass. The secondary wall consists of ca. 95 % cellulose and is arranged in numerous concentric layers. The layers denoted by S_1, S_2, and S_3 in Figure 5, consist of tightly packed parallel bundles of fibrils with helical coils along the fiber axis.

The fibrils are considerably less densely packed on the concave side, i.e., the side pushed into the interior of the fiber, than on the convex side, which has a more regular arrangement and a greater packing density [59]. This results from the collapsing of the fiber during its initial drying in the field. As a result, there are internal stresses, which determine the fiber's mechanical properties and, because of differing accessibility, its chemical properties.

The fibrils can be isolated by means of mechanical or chemical degradation (hydrolysis) of the fibers and can be identified with an electron microscope [60]. The major units are the macrofibrils with prevalent diameters of 300, 120, and 60 nm. They consist of microfibrils with a diameter of 10 – 40 nm and a length of 50 – 60 nm. The smallest fibril units are the elementary fibrils consisting of densely packed bundles of single linear polymeric cellulose chains with a diameter of 3.5 – 10 nm [61]. In the longitudinal direction the bundles of fibrils, connected by hydrogen bonds, alternate between highly organized (crystalline) and less-organized regions. This periodicity in the arrangement of the elementary fibrils reappears in the microfibrils in intervals of ca. 50 nm. Figure 4, shows a diagrammatic structural model of the microfibrils, which are constructed of elementary fibrils.

The review of ZAHN [62] summarized progress in elucidating the complex structure of the cotton fiber. Japanese research on the cultured algae *Valonia macrophysa* showed that the algae have regular cell walls. This indicates that the microfibrils are crystals containing 1 200 to 1 400 perfectly arranged cellulose chains [63], see Figure 12. Some of the earlier stress on the "amorphous" parts of the fiber or even of the molecular chains has shifted to assessing the pore-size distribution [64], because pores between 1 and 3 nm are most important for the uptake of dyes and many other finishes. The authors showed an increase in the proportion of pores in that range when cotton is boiled off, and a further increase when mercerized.

Attempts in recent decades to explain fiber strength, extensibility, and maturity in terms of molecular structure, crystallinity and the distribution of degree of polymerization, whether using X-ray, near infrared, chemical, or other techniques that do not discriminate between

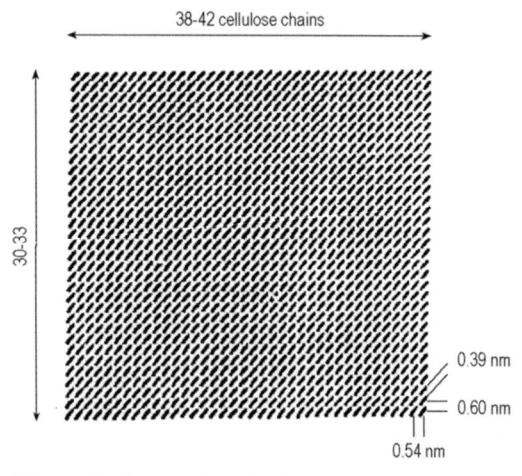

Figure 12. Suggested model for the ac projection of a cellulose crystal from the cell wall of *Valonia macrophysa* No crystalline substructures exist, one single microfibril being one single crystal containing 1 200 – 1 400 perfectly arranged cellulose chains

primary and secondary walls, have commonly yielded only modest correlations, of limited practical value. There is increasing recognition of the difference between the properties of the primary wall, which forms first, and the secondary wall, which usually accounts for most of the fiber's mass [65].

One study of the secondary wall confirmed its close relationship with maturity. The development of the secondary wall was manipulated by changing temperature and light intensity to obtain vast differences in cell wall growth and secondary-wall development [66]. Comparative examination showed that the primary and secondary walls polymerize according to different mechanisms, with higher molecular mass, crystallinity, and crystallite size in the secondary (inner) wall [67].

The transverse dimensions of the cotton fiber are less obvious than in most other fibers, because most commercial measurements confound linear density with maturity and because of the peculiar and variable shapes; even within one boll on the plant, there is great variation in fiber development. Attempts to analyze measured fineness into linear density and maturity, by a variety of techniques, are becoming increasingly succesful. Automated image analysis, commonly longitudinal, but preferably of cross-sections, provides several measures that shed light on corresponding commercial properties:

Fiber Thickness (the inappropriate term diameter is often used) is estimated from the Martin radius, such as the average of four diameters through the cross-section's center of gravity.

Linear Density, in millitex or micrograms per meter, is directly proportional to the cell (i.e., total minus lumen) cross-sectional area, though calibration against known linear densities is needed at present.

Maturity is expressed as degree of thickening, which may be defined as the ratio of the above cross-sectional area to that of a circle having the same perimeter as the fiber.

Varietal Fineness (UK: *Standard hair weight*) has defied reproducible measurement until recently. Each variety and species of cotton has a potential linear density reached when "fully"

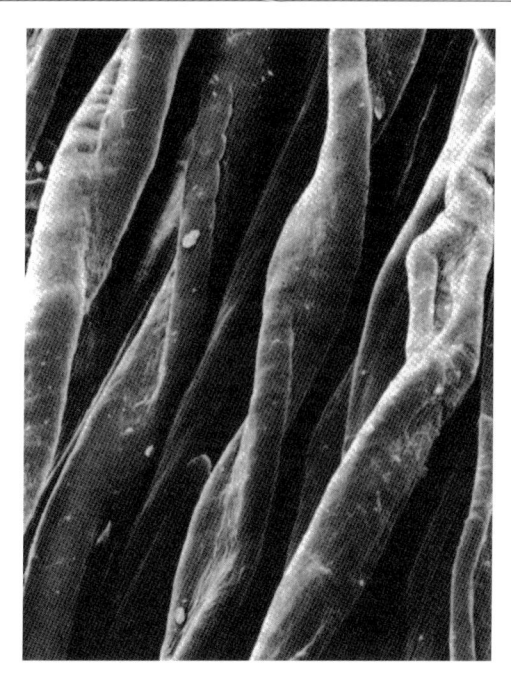

Figure 13. Cotton fiber, longitudinal view, magnification 1 100× Photomicrograph: Shirley Institute, Manchester, England

mature; experience with image analysis suggests that the average fiber perimeter is likely to distinguish best between varieties and to provide breeders with the best means of monitoring intrinsic fineness.

Convolutions with some sharp folds along the cotton fiber are shown longitudinally in Figure 13. Figure 14 is a cross-sectional view at the same magnification (1 100×), showing the lumens. These two figures indicate why cotton fibers are cohesive and easily spun in spite of their relative shortness.

Surface waxes are not strictly part of fiber structure, as they can easily be stripped off. Along with convolutions, they play a major part in cotton's processability, and may be regarded as a natural spin finish and the reason why cotton is the only fiber normally carded and spun without the addition of finishes to alter friction and cohesion.

2.1.2. Properties

Spinning Properties. The chief requirements for spinning textile fibers, and the extent

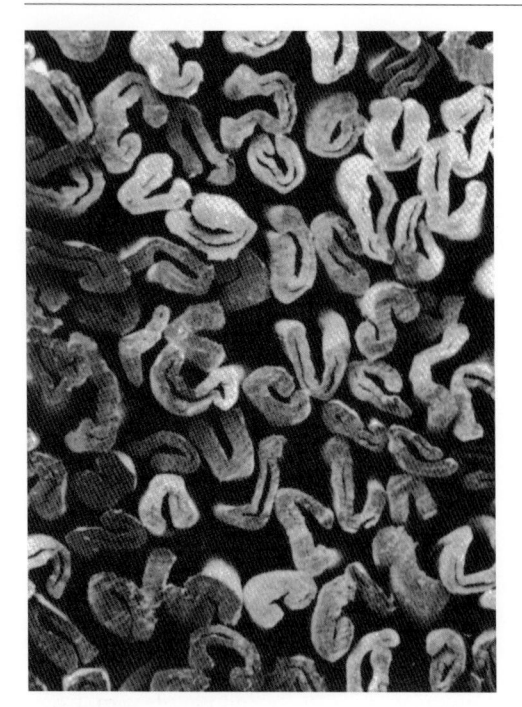

Figure 14. Cotton fiber, cross-sectional view, magnification 1 100×

Table 14. Grades of American Upland cotton, white 1-in. staple length, 1984

Grade number	Traditional grade	Laboratory nonlint content, %	Cotton content after carding, %	Approximate commercial value, US cents/kg
1	good middling			170
2	strict middling	1.9	93.6	169
3	middling	2.3	93.0	167
4	strict low middling	3.1	91.8	163
5	low middling	4.3	90.2	156
6	strict good ordinary	5.5	89.0	135
7	good ordinary	7.8	85.0	121

to which they are met by cotton, are the following:

Fiber Length. Because a longer fiber can be spun into a better yarn with less twist, mills pay a substantial premium for "staple length." This term is important, yet poorly defined. It describes a fiber that is longer than the mean but shorter than the longest fiber. Because of this premium, displacement of cotton by synthetic fibers has been greater at the long end of the cotton spectrum. This is because the price of synthetic staple fibers, which are cut or broken from continuous tow, is independent of fiber length. Cotton fibers vary considerably in length, even on individual seeds. Uniform lengths are coming into increasing demand.

Length Uniformity. Excessive numbers of short fibers increase processing waste and impair spinning. For the production of fine yarn, they must be removed by an expensive combing process.

Impurities and Grade. Cotton lint contains visible particles of leaf, seed, bark, and boll, known collectively as "trash." Although grade is measured optically, either by the classifier or by colorimetry, it is largely a measure of the absence of trash in the cotton. Table 14 illustrates the correlation between grades and other characteristics. Examination of the last two columns shows that the price of cotton increases at ca. 3 times the rate that would be expected from its clean content. This demonstrates the importance of removing trash and dust early.

The steep rise in value as impurities are removed shows, in common commercial grades, a price sensitivity of 3 : 1, i.e., each percentage increase in clean cotton content raises the unit price by 3 %. This provides an incentive for ginners to remove trash in rural areas before bales are shipped to the mills, where waste removal is more expensive. The controversial practice of lint cleaning degrades most fiber properties while maximizing the cotton grower's total revenue [68].

Traditional assessment of grade by classers includes a close look at "leaf", because of the nuisance that it causes, especially to the extent that it is shattered into pepper trash in ginning and in the opening and carding processes at the mill. Colorimeters that earlier made assessment more reproducible failed to distinguish between lint that is generally discolored and white lint that contains abundant trash. This difficulty has been resolved by image analysis using digital trash meters, calibration of which enables reliable "gravimetric" estimation of trash content, measurement of particle size distribution, and conversion between laboratory and classers' measurements ([69] and Table 15).

Table 15. Effect of visible trash content on leaf grade (USDA, 1993)

Classer's leaf grade	Trash measurement, mass %
1	0.08
2	0.12
3	0.18
4	0.34
5	0.55
6	0.86
7	1.56

Further instrumental developments are approaching the classer's skill in distinguishing between different forms of trash, with bark and seed-coat fragments being the worst because, in different ways, they behave in processing much like discrete fibers, and are more likely to cause yarn breaks and unsightly fabric defects.

Color. Discoloration due to weather, excessive heat in drying at the gin, bacetria or insects may indicate sugary secretions or general degradation, and usually causes strength loss which results in price discounts. Colorimetry shows that grade and color correspond two-for-two with regard to reflectance and blueness/yellowness, but colorimetry also shows that greenness/redness has no influence on classification. Depending on the degree of yellowness appropriate for each grade, lint is classified as "white," "light spotted," "spotted," "tinged," or "yellow stained." Excessively "white" cotton may exhibit bacterial damage within the fiber and is discounted as "gray." In Egyptian cotton, a creamy tint is natural and considered desirable.

Micronaire. A complex measure of maturity (ca. one-third) and coarseness (ca. two-thirds), micronaire is determined by airflow instruments in micrograms per inch. The linear density or hair weight is, in fact, usually higher than this reading would indicate. Whatever the theoretical limitations of micronaire, it has proved to be a valid guide for assessing processing performance when the species and variety of the cotton – hence, the full potential of the micronaire – is known. American cottons are discounted if the micronaire is below 3.5, suggesting immaturity, or above 4.9, suggesting coarseness. Low maturity implies insufficient development of the secondary wall, with a tendency to form neps, more yarn breaks, and difficulty in dyeing.

Strength. Fiber strength is rarely a factor in the pricing of a bale of cotton. Not only is a stronger fiber likely to make a stronger yarn and cause fewer processing problems, but as suggested by the trend in fabric weights since 1950, stronger fiber can be used more sparingly, thus making a lighter-weight fabric. Development of hybrid cottons, with strengths almost double that of Upland varieties, promises to widen the scope of cotton's applications.

Elongation and Stress – Strain Behavior. Cotton fiber stretches only ca. 7 % at break, less than most competitive fibers, but well above the 2 – 3 % of most of the other vegetable fibers described in this article. Cotton stretches in response to load in much the same way as polyester. This compatibility of mechanical properties enables the two fibers to be successfully blended and helps to explain why polyester/cotton is the most common fiber blend.

Cohesiveness. The spinning of yarn from discontinuous fibers requires a degree of surface cohesion. This is all the more important if, as in cotton, the fibers are short. Broken cotton yarns usually show evidence that only about half of the fibers have ruptured at the breaking point, while the other half have slipped. In cotton much of the cohesion is due to the natural convolutions in the ribbon-like fiber (see Chap. 1), as can be shown by mercerization, which makes the fibers become rodlike and incapable of being spun. As yet, no commercial method for measuring this property in cotton is available.

Other Properties. *Moisture Uptake.* Cotton has a moisture regain of 8 % at 65 % relative humidity and of 15 % at saturation. Although this is exceeded by wool, jute, and rayon, the lack of a thick cuticle or sheath on the cotton fiber enables the fabric, when freed of natural waxes during finishing, to absorb moisture rapidly. Thus, cotton accounts for 94 % of the United States towel and diaper market [70], and although most printing is now done with pigments, cotton's rapid uptake of water-based dyes is an advantage to the finisher and printer. On the other hand, drying of cotton garments made of such water-retentive fibers after they have been washed adds to energy costs in comparison to those of the less-absorbent noncellulose fibers [71].

Secondary effects of moisture uptake are conspicuous in cotton because it is associated not only with the heat of sorption but with considerable *swelling*. In densely woven fabrics, such as canvas, this swelling blocks large pores, making the fabric impermeable to rain. At the same time, fabric thickening forces the tightly interlaced yarns into a more crimped configuration, which makes the fabric shrink. When the cotton dries, the restoring force is much weaker than the shrinking force, which leaves relaxation shrinkage.

2.1.3. Production

Cotton was originally a tropical perennial, but is now cultivated as a summer annual and with a ratoon crop in some tropical countries. It requires a longer growing season than any other annual crop. As an annual, it can be grown in cold-winter regions provided that there are at least ca. 1 200 degree-days above 15 °C during the growing season (one degree-day is recorded for each degree by which each daily mean temperature exceeds 15 °C). Thus, cotton is cultivated in latitudes as high as 45° N, with about 70 % of the world's cotton now being grown in temperate zones such as Uzbekistan, Northern China, and much of the American cotton belt (Fig. 11). The winter in these areas is cold enough to break the continuity of disease and insect problems.

Particularly in North America, the Mexican boll weevil has influenced the geography of cotton growing. Insects such as the heliothis, including the pink boll worm, and spider mite cause expensive problems. Honeydew that is secreted by aphids, whitefly, and other insects produces sticky cotton, which causes processing difficulties, especially when the fiber is passed through drawing rolls. In analysis, this "entomological" sugar must be distinguished from "metabolic" sugar produced by the maturing plant.

2.1.3.1. Harvesting

In developing countries, seed cotton is harvested by hand. It is pulled from the bolls, usually in two harvests, starting near the bottom of the plants as they ripen and open. Because strong winds may cause losses, stormproof varieties are planted in windy regions. These are short plants which resist lodging and have bolls that only partially open.

In high-wage countries, mechanical harvesting is predominant. Since this method is less discriminating than human fingers, the plants must first be defoliated either by frost or chemically. The seed cotton is still more contaminated than that obtained by hand. There is normally a single passage with the spindle harvester, as the bolls ripen. Some cottons, including all stormproof types, are harvested in a single operation by stripping the entire bolls, which gives a less-expensive, lower-grade lint. Stripping tends to remove strips of bark, which, because they are fibrous, are less easily removed than other impurities.

2.1.3.2. Ginning

Ginning is essentially the separation of lint from seed. Upland cotton is separated by rows of circular saws, which pass through slots that admit lint but not seed. Modern gins perform such functions as drying, blending, and cleaning seed cotton, lint cleaning, packing into bales of 220 kg and 0.5 m^3, invoicing, or marketing. Cottonseed, which accounts for ca. 65 % by weight and 15 % by value of the harvest, is sold separately.

When machine harvesting replaced hand harvesting, a more impure material was delivered to the gins, which resulted in the introduction of lint cleaners. Two passages through lint cleaners, at much higher speeds than in textile mills, restore the lint to the degree of purity that prevailed in the days of hand harvesting and maximize the growers' return. Lint cleaning is particularly useful for stripped cotton. However, it causes many fibers to break and invariably leads to more end breaks in spinning and, thus, a weaker yarn [72]. This weakness can usually not be detected at the fiber-testing stage.

2.1.3.3. Byproducts

Cottonseed directly from the gin is covered with short, coarse fibers called linters. These are cut off in one or two stages and used in wadding or as a raw material for cellulose plastics, such as ethyl cellulose. They were formerly an important raw material in rayon manufacture. (Seed for planting may be completely delinted by acid.)

The seed is then stripped of its hull, leaving a valuable "meat." This is crushed, often with the help of solvent extraction, to obtain the oil which may be purified for human and animal consumption. With an iodine value of 106, it is classified as a semidrying oil. Primary components are linoleic, oleic, and palmitic acid. The remaining meal, with a shelf life too short for retail trade, is used in commercial baking and as animal feed. The cotton seed contains gossypol, which is poisonous to nonruminants and must be removed for nonindustrial uses by hexane extraction.

2.1.3.4. Processing

On arrival at the mill, bales are selected for the desired yarn properties and blended, sometimes with synthetic fibers, in lots of about 30 bales. The first stages of processing greatly increase the bulk and remove much of the trash and dust. Further cleaning is achieved through carding and blending through drawing.

Generally, a roving is spun, which is then drawn and twisted into yarn by a ring and traveler during spinning. The yarn may be plied or finished before being made into a fabric. The newer process of rotor spinning, an open-end method, is excessively expensive for longer fibers, but cotton lends itself well to this technique, especially since rotor spinning is less sensitive to short or immature fibers than ring spinning.

Even though cotton yarn is easily knitted or woven, little is tufted into carpets. To withstand repeated abrasion during weaving, cotton warp yarn that has not been plied is treated with size. Increased speeds and stresses in weaving require yarn of uniform strength, which is largely true for rotor yarn.

2.1.3.5. Finishing

Because of its excellent stability in cool alkali and most bleaches, cotton is a particularly washable fiber, provided that shrinkage is controlled. Wet processing normally begins only after the yarn has been spun, usually after making the "gray" (unbleached) fabric.

Natural waxes inhibit the uptake of dye. The first stages are called "preparation," i.e., for dyeing and printing, and include the following:

1. *Scouring.* Treatment in a hot detergent solution to remove natural impurities from the fiber surface as well as any oil and dirt from processing.
2. *Bleaching.* Removal of any remaining colored impurities through immersion into an oxidizing agent, usually cold hydrogen peroxide; this destroys trash particles or renders them invisible.
3. *Desizing.* Removal of sizes that were applied to the warp before weaving, either by acid hydrolysis or by the use of enzymes to remove natural starches.
4. *Mercerizing.* Originally a specialized process, mercerization is routinely used to increase dye uptake in cotton and polyester/cotton blends. It includes treatment with a cool, concentrated sodium hydroxide or ammonium hydroxide solution, which alters the crystalline structure from cellulose I to cellulose II, thereby improving the accessibility to water and, hence, to water-based dyes. If the yarns are under tension, the luster is enhanced and yarn strength increases by ca. 20 %. This process was formerly used to give cotton fabrics the appearance of the much more expensive silk, as well as to increase strength.

The fabric can then be dyed, usually by direct, vat, or reactive – occasionally basic, sulfur, or azoic – dyes. Pigment printing is popular, partly because of its excellent colorfastness, and because it can be applied to produce a solid shade in imitation of dyeing.

2.1.3.6. Special Finishes

Permanent Press. Because interchain bonds in cellulose are weak, cotton is easily deformed. Crease-resistant finishes, originally developed to reduce the severe wrinkling of rayon, are now widely applied to cotton. Several cross-linking agents, notably dimethylolethylenurea (1,3-bis(hydroxymethyl)-2-imidazolidinone [*140-95-4*]), are in use. These are applied to the fabric, dried, and then reacted by using heat (curing). The harsh texture that is thus imparted is mitigated by applying softeners. The introduction of polyester/treated cotton blends in 1964 proved successful, as the strength of the unaffected polyester enabled the application of resin to provide improved wrinkle resistance.

Flameproofing. Cellulose burns readily in air.

$C_6H_{12}O_6 + 6\,O_2 \rightarrow 6\,CO_2 + 6\,H_2O + \Delta H$

Carbon monoxide is produced in confined spaces:

$C_6H_{12}O_6 + 3\,O_2 \rightarrow 6\,CO + 6\,H_2O + \Delta H$

In the presence of a Lewis acid, only charring occurs:

$C_6H_{12}O_6 \rightarrow 6\,C + 6\,H_2O$

Most flame retardants inhibit the first two reactions. They include simple soluble substances such as borax or alum and insoluble chemicals formed in the fiber by reaction, such as stannous oxychloride or antimony oxide.

Most introduce problems of handle, or "feel," toxicity, or poor durability and have been largely replaced by a variety of organophosphorus compounds or by substitution with synthetic fibers in which an insoluble flame retardant is dispersed before the fiber emerges from the spinneret.

Shrinkproofing. In bottomweight fabrics (those > 140 g/m^2 and used for trousers, workwear, etc.), compressive shrinkage is used. A rubber blanket is moved in opposition to the fabric through the finishing machinery, so that the filling yarns are forced together and the progressive shrinkage that would otherwise occur during repeated laundering is forestalled.

In topweight fabrics, such as shirtings and knits, satisfactory results are usually obtained as a side effect of the permanent-press resin.

Chemical Stonewashing. In many cotton-rich fabrics, especially in woven outerwear, a strong demand for moleskin and stonewashed finishes exists. Mild treatments include emersing, in which moderate abrasion of this fine fiber produces a soft finish. Increasingly, and especially when a "stonewashed" appearance is desired on denim, literal stonewashing is being superseded by controlled enzyme treatment with selected cellulases. Despite a weight loss of about 3 %, the finish does less damage to the fabric, and may be combined with other finishes such as crushing. It is applied to solvent-spun rayons of the Tencel type, the only type of regenerated cellulose fiber whose consumption is not in decline. Some celluloses are beginning to be used for the reduction or bleaching stage of vat dyeing, especially in denim.

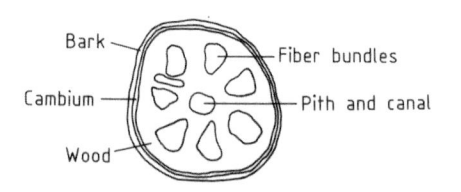

Figure 15. Cross-section of plant stem with bast fiber

2.2. Bast (Soft) Fibers

Plants with a stem containing a fibrous bast component may be commercially processed if the fiber yield and the ease of separation warrant commercial use [73]. The cross-section of the stem, shown in Figure 15, shows several undesired components which must be removed mechanically or chemically, a process that is frequently so odorous that the retting process must be carried out in less-populated regions.

Although the process used varies according to the fiber, region, and available machinery, all bast fibers are generally subjected to the following processes after harvesting [73]:

mechanical or manual removal of seeds and leaves (rippling)
mechanical removal of bark (decortication)
biological removal of woody components through decomposition (retting)

At this stage the fiber, with some residual impurities, is usually baled and shipped to the mill for further processing.

The bast consists of long fiber bundles, which may be as long as the stem. These are made up of ultimate fibers containing lumens and are held together by woody gums; unlike most textile fibers, bast and leaf fibers are bundled. The elementary fibers, even though suitable for pulping, are too short for spinning, so that processing requires mechanical rupture of the bundles into manageable lengths. This is often achieved with a lubricant which softens the gums and allows the elementary particles to separate gently.

In addition to the fibers described here, other bast fibers of commercial importance include urena, sunn hemp, and kenaf.

2.3. Jute

Jute ranks second only to cotton among the natural cellulosic fibers. The net production of jute is greater than that of wool. Declining prices, about 40 ¢/kg in 1995, have reduced its importance since 1970. Since the early 19th century, jute has been intensively grown in the Ganges delta, and except for the substantial increase in jute cultivation in China since 1950, it has seldom been successfully cultivated anywhere else. Most jute-growing countries process jute to produce cloth locally, with only Bangladesh being a major exporter of the fiber [74].

Two varieties are used in commercial production. *Corchorus capsularis* (white jute) is tolerant of waterlogging and is grown on small delta farms subject to flooding during the summer monsoon. Harvesting of the 4-m stems with knives is sometimes done by divers. Good grades, such as Bangla (formerly Dacca) White A and B, are sometimes used for inexpensive apparel. Much bleached white jute in apparel weights appears in world markets as "hemp". The two fibers are not easily distinguished, but jute is identified by turning yellow in sulfuric acid and iodine, not blue.

C. olitorius (tossa or upland jute) is grown on the wet slopes of India, Bangladesh, and Nepal. It is brown and contains noncellulosic impurities which cannot be completely bleached because of resultant excessive fiber damage. Some lowland areas produce "daisee" jute, which is gray or black because of iron salts in the soil.

The stems are retted in rivers, sometimes ponds, giving a fiber yield of only ca. 6 % (i.e., 18 t of "green" jute must be harvested to obtain 1 t of retted fiber). If the prices are high, the stumps (jute butts) are salvaged, though the middle of the stem produces the best fiber.

At the mill, the fiber undergoes blending and spreading with addition of ca. 5 % processing oil before carding. Because the fiber bundles are uneven in both length and diameter, individual yarns are irregular, giving the cloth its characteristic appearance. It is variously known as burlap, hessian, gunny, or sackcloth.

Jute has the shortcomings of cotton, with an even greater tendency to rot, but unlike cotton, it is usually too heavy and cheap to justify the cost of finishing treatments. It is normally not even scoured, formerly leaving the characteristic smell of old vegetable or whale oils on the cloth. Used mostly in carpet backing, sacks and bale wrapping, jute has been largely replaced in the developed world by bulk loading and the lighter, inexpensive polypropylene made from plastic film. Even organizations that advocate the consumption of their own natural fibers have urged the replacement of jute bales with synthetics. In primary carpet backing, polypropylene has also for the most part replaced jute, but it has the drawback of being thermoplastic. A roll of carpet that has just been stored in a cold warehouse may, therefore, be too stiff to lay. Partly for this reason, jute is still common in the secondary backings which are applied to the underside, where its low extensibility helps stabilize the carpet.

2.4. Flax

Unlike cotton, flax (*Linum usitatissimum*) is grown less commonly for its fiber than for the seed, which is crushed to yield linseed oil. Over half of the world's flax is grown in countries that used to be centrally planned. With the sudden transition from subsidized farming to direct exposure to world markets, flax production for fiber has declined, e.g., from 2 000 to 800 t in two years in Estonia, but it remains a fashion fiber and is widely imitated.

The plant is much smaller than jute and is harvested by pulling out of the ground. It is retted on the ground ("dew" retting) or, increasingly, under controlled conditions in tanks. With a fiber yield of only ca. 11 %, an experimental process for retting in the field before harvest is receiving great attention.

At the mill, the fiber is processed more thoroughly than jute, often with "hackling," similar to combing, to separate long ("line") from short ("tow") fibers. About half of the linen yarns are moistened during spinning to lubricate the fiber ultimates which at about 10 mm length are twice as long as jute. Wet-spun flax yarn may be very fine and strong, but its low elasticity limits its ability to blend with other fibers.

The fact that flax has long been the standard against which other fibers have been measured, is expressed in terms such as "bed linen" and "table linen," even though flax fabrics are less durable than those of cotton or blends. The high absorbency of these fabrics destines it for use in towels but this is also combined with severe wrinkling and requires large amounts of crease-resistant resins.

2.5. Hemp

Hemp (Indian hemp) is rarely produced outside of the former Communist countries. The plant (*Cannabis sativa*) is widely cultivated for its leaves, known as marijuana, but the potential for salvaging the fiber in the stems has hardly been exploited. Yields per acre are lower than those for jute, which has similar properties. When grown for fiber, the plants are more closely spaced than for the drug crop, as leaf development is not needed. To distinguish the two uses and to encourage government support for hemp growing, many people have renamed the fiber "industrial hemp".

Because of restrictions on the drug, governments have discouraged production of hemp. World production of true hemp, insufficient to appear on the map (Fig. 11), is probably even lower than official statistics would indicate, because of widespread use of the misnomers "sisal hemp" and "manila hemp."

2.6. Leaf (Hard) Fibers

Fibers obtained from the leaves of plants such as *Yucca filamentosa* and agaves are extracted from the harvested, dried leaves – usually mechanically, since the leaves have spikes at the end and, in some species, along the sides.

These fibers are coarser, but more resistant to rot than the soft fibers and find application in cordage [75]. They have met with increasing competition from synthetic fibers. Common species are *Agave sisalana* (sisal) and *A. fourcroydes* (henequen).

Some banana trees, notably *Musa textilis,* have very fibrous leaf petioles. When separated from the rest of the leaf (tuxying), this abaca fiber can be used for high-quality ropes, often called manila. The natural tan color is sometimes imitated in other fibers through dyeing. Almost as elastic as cotton, manila ropes absorb shock loads better than most vegetable fibers.

2.7. Coir

The fibrous outer layer of the coconut, called coir, may be considered a seed fiber, like cotton, but is usually classified as a fruit fiber. The coarse fibers are commonly processed in India, Sri Lanka, and the West Indies and are woven into coverings for floors and playing surfaces.

2.8. Ramie

Ramie is a bast fiber taken from the stingless nettle, grown in the Philippines and the People's Republic of China. It rivals cotton in purity, whiteness, and strength. A minor fiber, it was omitted from the GATT multifiber agreement. Then it appeared abundantly, often in blends with cotton or polyester, in international trade, unhindered by quotas.

2.9. Economic Aspects

All of the natural cellulose fibers have been affected by the introduction of synthetic substitutes, first as regenerated cellulose (whose total worldwide production is less than one seventh that of cotton), later as fibers of fully synthetic polymers. Because rayon shares many of the problems of natural cellulose – and has a few of its own as well – most of the inroads into markets for plant fibers have been made by such fibers as polyester, nylon, and polypropylene. The replacement of jute and the leaf fibers by polypropylene has been particularly harsh. These newer fibers offer a higher strength : weight ratio than cellulose, as well as greater resistance to rot, mildew, burning, and wrinkling. On the other hand, some plant fibers, especially cotton, have maintained their position, world cotton production being at record levels. Of all the fibers used in apparel, cotton's share in all textiles has remained almost unchanged at near 50% since 1974. Table 16 shows production trends.

A good annual yield in most parts of the world is considered to be about 500 kg/ha (approximately, 1 kg/ha = 1 lb/acre) of cotton or 2 000 kg/ha of jute. Due to a growing world population, food production has been intensified and has caused some land to be diverted from the growth of cotton and jute to food crops. Table 17 shows production

Table 16. Proportions (in %) of the chief apparel-type fibers worldwide [76], [77]

Year	Cotton	Wool	Regenerated cellulose	Noncellulosic synthetics
1930	82	14	3	0
1940	75	12	12	0
1950	71	11	17	1
1960	67	10	17	5
1971	52	7	14	27
1984	49	5	9	37
1995	46	4	6	44

of the chief natural cellulose fibers. In interpreting these data, the increasing proportion of leaf fibers being made into pulp, rather than ropes or other textiles, should be kept in mind.

Many of the fibers in this group, notably cotton and flax, are stronger when wet. Conversely, they are weak when warm and dry, a factor that played a part in reducing cotton's once-dominant position in tire cord.

2.10. Occupational Health

These fibers contain endotoxins which are shaken loose in processing if, as is usually done, the fibers are worked dry. Exposure to these airborne particles over decades may cause byssinosis in textile workers, especially in smokers. Regulations governing this problem [80] have caused profound changes in mills for processing cotton and have led to the redesign of machinery to reduce airborne dust. These changes are responsible for important secondary improvements in the processing behavior of cotton.

3. Regenerate Cellulose

Terms and Definitions. Man-made fibers are divided into two groups – cellulosics, and synthetics.

Table 17. World production of natural cellulose fibers ($\times 10^3$ t) [78], [79]

Year	Cotton	Jute	Flax	Hemp	Sisal	Abaca
1950	6 600	1 417	768	948		
1960	10 100	2 420	650	340		720
1971	11 600	3 446	670	269		821
1983	14 700	4 057	669	230	384	230
1994	18 900	3 530	610	120	380	76

In both cases, the fibers are spun under conditions that are oriented to the end use in the form of continuous filaments. These filaments can be used as they are, textured, combined to heteroyarns, or cut into a staple of defined length.

In Germany, cellulose-based man-made fibers are produced according to DIN 60 001 from cellulose [81]. At present, the following three methods are used:

- the viscose process – by far the most important (see Section 3.1)
- the Lyocell process (see Section 3.2)
- the cuprammonium process (see Section 3.3)
- the acetate process (see → Cellulose Esters, Section 1.3.7.)

Other production methods are discussed in Section 3.4.

While the first three of these processes result in regenerated cellulose fibers, the fourth gives cellulose ester fibers. The viscose process has been widely modified over time. One special kind of regenerated cellulose fibers is the so-called Modal fiber, which is available in two subgroups: the polynosics (= polymer non synthétique), and HWM fiber (= high wet modulus). In the Lyocell process, cellulose is regenerated from a solution in an organic solvent. Cellulose ester fibers are sub-divided into acetate and triacetate fibers (see → Cellulose Esters, Section 2.3.). Table 18 gives a listing of fiber designations according to ISO/TC 38.

In addition, the main processing and application properties of the fibers are indicated by figures, descriptions, and short references such as:

number of single filaments
staple cut in mm
titer in dtex (1 dtex = 1 g/10 000 m)
bright (no dulling agent added)

Table 18. Fiber designation according to ISO/TC 38

Fiber	Generic name	Code
Cuprammonium	cupro	CUP
Lyocell	lyocell	CLY
Modal	modal	CMD
Viscose	viscose	CV
Acetate	acetate	CA
Triacetate	triacetate	CTA

dull/superdull (containing 0.3 – 4.0 % of a dulling agent, generally anatase)
hollow (hollow fibers)
ribbon (tapiform fibers)
crimped/high-crimped fibers (90 – 140 waves/100 mm)

Other descriptions may describe color, moisture retention, flammability, degree of optical brightening, suitability for medical applications, X-ray contrast, recommended uses, etc.

History. Regenerated cellulose fibers are the first artificial fibers ever made. Processes capable of dissolving the cellulose derived from wood or cotton linters were first discovered by SCHÖNBEIN (1845, nitrocellulose soluble in organic solvents), SCHWEIZER (1857, cellulose in cuprammonium solution), CROSS, BEVAN, BEADLE (1885, cellulose sulfidized in sodium hydroxide; 1894, cellulose triacetate in chloroform). The production of threads from cellulose (derivative) solutions goes back to AUDEMARS (1855), who drew filaments from cellulose nitrate submerged in alcohol ether, and SWAN (1883), who transformed nitrocellulose filaments into hydrated cellulose. Members of the Swan family used the threads, which were intended for the manufacture of light bulbs, for their craft work as well. SWAN called his discovery "artificial silk." This early developmental work also set the trend for subsequent uses with regenerated cellulose fibers. In 1891, Count HILAIRE DE CHARDONNET built the first industrial plant for the production of artificial silk according to the nitro process in Besançon, France (daily output ca. 110 kg), and MAX FREMERY established the "Rheinische Glühlampenfabrik" for the production of cellulosic fibers suitable for carbon filament bulbs in Oberbruch (which developed into the Akzo Fibres Group). The spinning techniques were gradually improved: H. PAULY (1897) was granted a patent on a process in which "solutions of cellulose in copper oxide ammoniac are passed through fine holes into a liquid capable of decomposing these solutions" and published for the first time operable instructions. Other patents include TOPHAM'S first spinning centrifuge (1900) and spin pump (1901); THIELE'S stretch-spinning method for cuprammonium silk (1901); MÜLLER'S spin bath for the viscose process (1905); and BOOS' continuous production of artificial silk (1906). A new phase was initiated in 1925, when H. STAUDINGER began describing macromolecules. The introduction of modifiers by COX in 1950 and the development of high-wet-strength fibers initiated by TACHIKAWA in 1951 again increased the variety of cellulosic man-mades. Commercial production of acetate fibers started in 1919; in 1955 they were joined by triacetate (see → Cellulose Esters). When during the 1960s synthetic fibers swept the textile market and water/air-pollution problems began darkening the image of cellulosic man-mades, they seemed doomed. However, after years of fierce competition, cellulosic fibers still maintain their characteristic place among fibers, doubtlessly aided by new developments in raw materials and in fiber manufacture [82]. The search for more ecological and more productive production methods led to the development of the Lyocell process in 1972 and in the course of time to a better adjustment of the raw material to improve the viscose process.

3.1. Viscose Fibers

3.1.1. Principle of the Viscose Process

The starting material for the production of filaments and staple fibers is pulp, mainly produced from wood and linters, but also from annual plants (→ Pulp). Since pulp fibers in their natural state are much too short to be spun into yarns, they are dissolved and regenerated in filament form. Part of these filaments is cut into fibers of the required staple length.

The spinnable solution is prepared by dissolving the raw material "pulp" in sodium hydroxide to obtain alkaline cellulose (first step), which is, if necessary, allowed to age (preripen) until a certain degree of depolymerization is reached (second step). Carbon disulfide is then added (third step). The resulting xanthate is dissolved in dilute sodium hydroxide (fourth step). With modern equipment, these four steps are combined into one.

Subsequent ripening (maturing) or a previously set degree of polymerization (DP) guarantee that the fiber can be perfectly spun and meet the requirements of the textile end use. During the process, the viscose is filtered, deaerated, and finally pressed through the holes of the spinnerets into the regeneration bath, where filaments are formed and drawn off at high speed (Fig. 16).

The cellulose must fulfill certain requirements with regard to processibility (unit weight, swelling behavior, hemicellulose content, inorganic substances, degree of polymerization (DP)) and the properties of the resulting fibers. Many efforts were made to "activate" cellulose by thermal treatment, radiation, introduction of spacers, and surfactants [82]. It has been found that a suitable irradiation with β-rays is the most useful method [83].

Alkalization. When exposed to sodium hydroxide of 20 % or more, the pulp absorbs water and alkali, releases heat (ca. 12 kJ/mol anhydroglucose), and loosens the cellulose lattice. This results in sodium cellulose I (Section 1.1.4), most of which changes into cellulose II during regeneration.

Although alkaline cellulose is an important intermediate product in cotton mercerization and in the preparation of viscose as well as various cellulose ethers, its nature has not yet been fully explored. Cotton mercerization is primarily intended to change the interfibril structure, which is also possible with other swelling agents. Alkaline treatment during viscose preparation is, apart from opening the molecular structure, aimed at transforming as much cellulose as possible into sodium cellulose I. This explains why several different definitions are suggested for the alkaline phase of the cellulose, such as "pseudostoichiometric combination ($C_6H_{10}O_5$) · (NaOH) · $(H_2O)_{3-5}$," or "a more or less loose addition of water molecules and hydroxyl groups of the cellulose to an Na^+OH^- dipole." Besides, some authors mention mesomerizable onium compounds with a cation of limited mobility as well as the fact that some of the cellulose has the form of a hydrated alcoholate Cell-O^-Na^+ · H^+OH^- attached to it [97–99].

Commercial treatment of the pulp starts with the addition of 18 – 22 % sodium hydroxide of 17 – 45 °C. Sometimes depolymerization catalysts such as manganese salts are added. An alkaline cellulose content of ca. 30 % is obtained by pressing. If necessary, the alkaline cellulose is shredded and left to age (generally a continuous process). At this stage, it is important to control the temperature and the surface of the crumbs must not be allowed to dry out. This is particularly important in shortened ripening methods. The degree of polymerization for normal textile

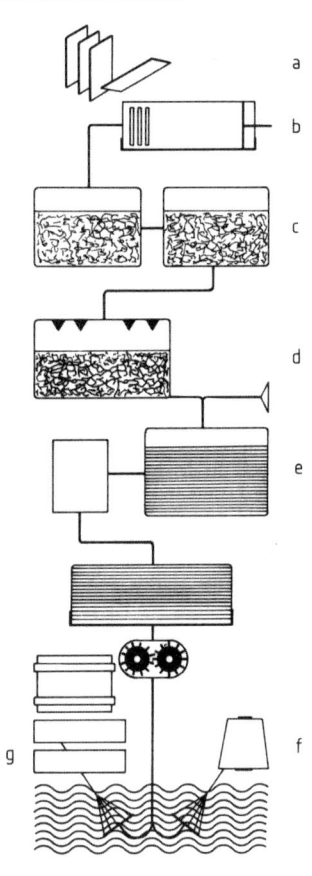

Figure 16. Flow diagram of viscose fiber production a) Pulp; b) NaOH treatment and pressing; c) Shredding, aging, if necessary; d) Carbon disulfide treatment, xanthate crumbs; e) Dissolving xanthate in NaOH, viscose maturing; deaeration, filtration; f) Filament yarn; g) Staple

Fiber properties can be widely varied by changing the processing conditions, such as the compositions of the viscose and the regeneration bath, chemical additives, and modified spinning technology (spinning bath temperature and spinning speed, drawing ratio and geometry, number and nature of the baths, etc.) [86–90].

3.1.2. Viscose Preparation

For textile filaments and staple, the preferred raw material is bleached wood pulp with an α-cellulose content of 89 – 93 %; it comes in sheets, rolls, or crumbs. High-tenacity filament yarns and modal fibers require even higher grades of pulp ($\geq$ 94 % α-cellulose). Cotton linters are used mainly in cotton-producing countries.

fibers is set at ca. 300 and for high-tenacity yarns at ca. 450 and higher.

It has long been known that during the third stage of the viscose process, hemicellulose readily attracts the carbon disulfide, and numerous efforts have been made to reduce the hemicellulose content of the alkaline cellulose. One possibility would be tandem steeping, i.e., two alkaline treatments. The first serves to wash out part of the hemicellulose that is already present in pulp and is formed during swelling and alkaline cellulose formation. After the second alkaline treatment the alkaline cellulose has a low hemicellulose content and distinctly less carbon disulfide is needed for xanthation (20 – 30 % less).

The composition and condition of the alkaline cellulose are governing factors for its processing properties (grinding, preaging, xanthation, filtration). Therefore, it is necessary to check the temperature and the composition of the alkaline solution (hemicellulose, sodium carbonate, alkaline- earth metals, and heavy-metal ions). Sodium carbonate is generally present in percentages between 0.09 and 0.3 %. Higher concentrations may create difficulties in viscose filtration. Varying heavy-metal ion concentrations cause differences in the depolymerization reactions during ripening [100], [101].

Shredding. If the alkaline treatment starts with pulp sheets or rolls, it is followed by shredding. At the same time, the alkaline viscose is adjusted to the ripening temperature. Higher temperatures (up to 45 °C) accelerate the process, lower temperatures (around 20 °C) prolong it. Shredded alkaline cellulose consists of white crumbs. The density by volume is held constant.

Preripening. The pulp and, hence, the alkaline cellulose now usually have a degree of polymerization (DP) which is much higher than desired for fiber formation. The delicate viscose process requires a viscose with an optimal combination of concentration and viscosity. Therefore, the DP must be reduced to ca. 300 (for regular fibers) or 450 or slightly higher (for high-tenacity fibers). The preferred method is to store the alkaline cellulose for a certain period of time at 20 – 45 °C. Care must be taken to avoid drying of the crumbs, which would result in carbonate formation, particularly at higher temperatures. For this reason, the alkaline cellulose is stored in closed containers or is constantly agitated (in continuous operations). Depolymerization is an exothermal process and can be accelerated by suitable catalysts. The results depend on such factors as time, temperature, the nature of the pulp, and the composition of the alkaline cellulose (NaOH, H_2O, hemicellulose, and heavy metals) [102]. This ripening process is either discontinuous (i.e., in containers) or continuous (i.e., in special cabinets provided with conveyor belts or in rotating drums).

It is generally believed that cellulose fiber producers should themselves adjust the preferred DP by ripening the alkaline cellulose. Currently, this is the most economical method. This principle is questionable, since new studies have been published and attractive uses for byproducts from pulp production are thinkable. Preripening can be avoided if pulp is used as feed material, whose degree of polymerization has been adjusted to the desired level by irradiation with β-rays. This treatment also leads to a favorable leveling of the DP.

Xanthation. This term means treatment with carbon disulfide with which the alkaline cellulose reacts and generates heat. The reaction is very rapid and begins on the outer surface of the crumbs – hemicellulose included – and slowly proceeds into the crystalline structure. This homogenating process continues until the viscose is ready for spinning. In commercial production, based on unirradiated pulp an average substitution degree of 0.5 – 0.6 is obtained, reflecting a γ-value (number of xanthate groups per 100 glucose units) of 50 – 60. γ-Values of ca. 50 are sufficient for a readily filterable viscose free from fibrous material. High-tenacity yarns require higher γ-values. Under special conditions, it is even possible to obtain γ-values of up to 300. Xanthation is not uniformly distributed over the cellulose chain, and even with a substitution degree of 1, not all of the glucose groups will have one substituent, let alone one in a specific place. Alkaline cellulose made of irradiated pulp can be reacted with distinctly less carbon disulfide to yield a good soluble xanthate. In practical operation γ-values of 30 – 35 for textile viscose fibers are used.

Table 19 shows the distribution of the xanthate groups over C-2, C-3, and C-6. There are indications that the speed with which the primary

Table 19. Distribution of xanthate groups over C-2, C-3, and C-6, in % [103–105]

Hydroxyl groups occupied	Acc. to [103]	Acc. to [104]	Acc. to [105]	
			fresh γ = 60	ripened viscose γ ≪ 49
C-2	43	38	a	b
C-3	20	28	a	b
C-6	37	34	31	60 – 100

$^a \sum$ (C-2 + C-3) = 69 %;
$^b \sum$ (C-2 + C-3) = 0 – 40 %

and secondary xanthates are formed and destroyed must be different. The hydroxyl groups in position C-2 are the fastest to react with xanthate, but are less stable than those attached to primary C-6 hydroxyl groups.

Xanthation speed depends on temperature, the condition and composition of the alkaline cellulose, the amount of carbon disulfide added, and the pressure of the carbon disulfide vapors formed during the process [103–106]. Pulp that has been adjusted to a desired DP by β-irradiation shows a more favorable behavior during xanthation than untreated pulp because of its narrower DP distribution. A much lesser amount of carbon disulfide is required for reaction and the reaction time is reduced. In addition, the spinning properties of the viscose obtained are even improved, when certain limiting values are complied with.

Side Reactions. Xanthation of alkaline cellulose is accompanied by a great number of side reactions. They are responsible for the characteristic orange color of the xanthate and the viscose made from it.

Carbon Disulfide. Depending on the type of fiber and the alkaline process used, between 150 and 400 kg of carbon disulfide is needed to produce 1000 kg of fiber. CS_2 is a clear, colorless liquid that is to distilled at 46 – 47 °C up to 98.5 %; it should contain < 5 ppm of hydrogen sulfide and be free from carbonyl sulfide and acid. The nonvolatile residue content should be less than 0.01 %, and the density should be 1.26 – 1.27 g/cm³.

Process Techniques. Older equipment may still contain hexagonal drums (heated jacket, teflon coated, vacuum tight). They are filled with alkaline cellulose with adjusted DP and evacuated. The required amount of CS_2 is added via the perforated axle of the drum (15 – 55 %, in the classic process 28 – 35 %). Pressure develops and replaces the vacuum; gradual normalization is aided by cooling. Xanthation takes 1.5 – 2.5 h, after which nitrogen is injected; the remaining carbon disulfide is removed and the orange xanthate crumbs are dumped into the dissolving tank. This early process was work-intensive and exposed the workers to noxious carbon disulfide vapors. During the 1950s, deep-reaching changes took place. Kneaders were introduced which held much larger quantities of alkaline cellulose for xanthation and allowed the first phase of xanthate dissolution to take place. The kneaders were followed by automatic sulfidizing and dissolving equipment.

Processes involving carbon disulfide are dangerous. Sparks may trigger explosions. Safety instructions must be carefully observed.

Viscose Solution. The xanthate crumbs are dissolved in sodium hydroxide either in batches or continuously to give a sticky substance; hence, the name viscose. Dissolution is accelerated by stirring and trituration. The honey-colored viscose contains 7 – 12 % cellulose and 5 – 8 % caustic, as well as various sulfur compounds and carbonate. Viscosity varies from 3.5 to over 10 Pa · s and depends on the concentration of the xanthate, the degree of polymerization of the cellulose, the degree of dissolution, which in turn is a function of dispersion, the degree of substitution and its distribution, and on the purity and concentration of the dissolving sodium hydroxide.

To avoid such common interferences as short fibers and gels, certain chemical auxiliaries are added directly either to the pulp or to the viscose. These auxiliaries are surfactants such as fatty amines or polyglycol compounds in fractions of 1 ppm of viscose.

Maturing. Maturing is another delay period, during which sulfidation continues and substitution becomes more uniform. At the same time, the γ-value is reduced and the amount of byproducts increased, while the ability of the viscose to coagulate also increases (chemocolloidal maturity).

Chemocolloidal maturity can be described by the Hottenroth index (°Ho). It is determined by thorough mixing of 20 g of viscose 3 times with 10 mL of distilled water. A 10% solution of ammonium chloride is added at 20 °C. When "maturity titration" is completed, the viscose changes into a semisolid gel which should adhere to an horizontal stick for at least 20 s. The Hottenroth index corresponds with the number of cubic centimeters of ammonium chloride required to obtain that result. At the start of the maturing process, the Hottenroth index reaches a maximum [107].

This manual maturity test can be easily automated. With the help of suitable apparatus, the maturing process of the viscose can be made much more uniform [108].

The decomposition speed of the sodium xanthate depends on the temperature and concentration of the sodium hydroxide, cellulose, and carbon disulfide [109]. The nature of the reaction has not yet been fully determined.

With older equipment, the maturing process at ca. 18 °C takes several days (50 – 80 h). It can be adjusted to the type of fiber to be spun and the spinning method to be used. Maturing is accompanied by filtration and deaeration.

These three operations have been accelerated and automated by increasing the viscose temperature, the development of various kinds of filters, and the introduction of vacuum evaporation.

3.1.3. Viscose Fiber Spinning

Fiber formation, i.e., regeneration of the cellulose from the viscose with the help of the spin bath and treatment outside the spin bath, includes coagulation of the viscose (sol – gel), deswelling of the filaments, regeneration of the cellulose, neutralization, and stretching of the yarn. The chronological sequence of these steps (which overlap and run parallel) determines the structure and, hence, the properties of the resulting fibers.

Composition and Effect of the Spin Bath. The formation of the thread in the sulfuric acid bath involves a number of chemical reactions. The most important is the decomposition of xanthate into cellulose and carbon disulfide. Besides, the thiocarbonates formed in side reactions during sulfidation and maturation disintegrate. These two basic reactions can be described in a simplified form as follows:

$$2\,\text{Cellulose}-\text{O}-\text{CSSNa} + \text{H}_2\text{SO}_4 \rightarrow 2\,\text{Cellulose}-\text{OH} + 2\,\text{CS}_2 + \text{Na}_2\text{SO}_4$$

$$\text{Na}_2\text{CS}_3* + \text{H}_2\text{SO}_4 \rightarrow \text{Na}_2\text{SO}_4 + \text{H}_2\text{S} + \text{CS}_2$$

* Representative for a number of inorganic thiocompounds.

In addition, carbon dioxide is formed when the caustic lye reacts with the carbon disulfide, and finally, colloidal sulfur is released.

Sulfates – mainly Na_2SO_4, but also MgSO_4 and rarely $(\text{NH}_4)_2\text{SO}_4$ – allow the viscose to coagulate, i.e., macromolecules and associates join into larger aggregates; acids alone initiate rapid disintegration of the viscose, and sulfuric acid of higher concentrations than 55% gives high-tenacity, low-elongation Lilienfeld Silk at low bath temperatures.

The composition of the spin bath is adjusted to the composition of the viscose and the type of fiber to be spun (i.e., titer, tenacity, and crimp). Spin bath temperature, spinning speed, the amount of viscose injected, and the construction of the spinneret are further factors. In general, a spin bath will contain 7 – 12% sulfuric acid, 12 – 24% sodium sulfate, and 0.5 – 3% zinc sulfate. In exceptional cases, part of the sodium and zinc sulfate can be replaced by magnesium sulfate (ca. 5%) so that the swelling of the filament is reduced before winding.

In practical spinning operations, the matured viscose from the tank is injected by precision pumps through fabric filters inside the spinneret support and then through the holes of the spinneret. The body of the spinnerets is made of gold or platinum alloys, with or without rhodium, or of platinum/iridium alloys. Some tantalum and glass spinnerets can still be found. However, since it became clear that the spinnerets are a very important machine part in viscose fiber production, more attention is being paid to high quality. This also includes the finish and form of the spinning holes. Seen in the flow direction, these holes have a conical inlet followed by a cylindrical channel. The cone is about double the length of the channel, but must be adjusted to the thickness of the plate. Depending on the intended fibers and their thickness, the channels have a diameter between 50 and 250 μm. In filament yarn spinning, the number of holes is equal to the desired number of single filaments in the yarn (for example from 18 to 1 000); in staple fiber production, it may surpass 90 000. Spinnerets of this kind are composed of several smaller spinnerets combined into one spinning point. For relatively wide ribbon fibers spinnerets with oblong holes are employed.

When the viscose emerges from the submerged spinneret and enters the spin bath, it begins to solidify. Special "marching-off" devices ensure regeneration of the cellulose in filament form and completion of cellulose regeneration. Spinning and pumping speed are precisely adjusted. Filament yarns of 100 dtex are wound at up to 160 m/min; filament tow for staple production is spun at between 35 and 120 m/min, depending on the fiber type and the number of filaments.

The regenerated cellulose fibers gain more tenacity and less elongation with a higher final degree of orientation. The process is supported by keeping sulfuric acid in the spin bath at a minimum, by adding certain chemicals, and by increasing the zinc content to decelerate the decomposition of the xanthate. Polynosic fibers are spun into a low-temperature bath at lower speeds.

Depending on the composition of the viscose, the spin bath and the flow diagram of the viscose leaving the cylindrical part of the holes starts filament formation in differently structured layers. This is due to the differences in the diffusion speeds of the hydrogen and zinc ions and the reversal of the acceleration diagram of the flowing viscose. On their way out, the degradation products that result from reaction with the faster hydrogen ions meet the slower zinc ions, with which they combine and precipitate ($ZnCS_3$ and ZnS_2). At the same time, two distinct layers are formed in the fiber; the outer one being less-swollen zinc xanthate and the inner one highly swollen sodium xanthate (skin/core). Both layers are differently structured, e.g., regarding the size and total volume of cavities [110–115]. The outer layer differs from the core in the following respects:

smaller crystallites (sometimes cellulose IV residues)
increased strength and elongation
greater resistance to fatigue
less lateral order and density by volume
lower refractory index
lower wet modulus
less accessible to large low-polar molecules
lower defibrillating tendency

The relation between skin and core is about 1 : 3 in standard viscose fibers and 1 : 0.5 or less in high-tenacity fibers.

The cross-sectional shape of the viscose fibers (see Fig. 20) depends on spinning conditions (deswelling, shrinking, drawing, and drying). Addition of varying amounts of zinc sulfate, adjustment of viscose and spin bath composition, variation in maturity, and high byproduct sulfur content may increase the thickness of the skin to almost full fiber thickness (see Section 3.1.4). The amount of crystalline matter (28 – 50%) and degree of orientation, lateral order and size of the crystallites are determined by stretching the filaments at different formation points [90]. In standard viscose yarn production, stretching occurs late in the decomposition phase; high-tenacity yarns are stretched in a less decomposed condition. Stretching of the plastic gel filaments has no measurable effect.

Stretching is generally done in several steps. Little stretch is applied between the spinneret and marching-off device, while a higher stretch is exerted between the drawing rolls on the spinning machine. High-tenacity fibers are drawn on separate roller arrangements. Continuous stretching between rolls mounted at an angle or on godets does not necessarily seem more advantageous than sectional drawing. The total amount of stretch applied ranges between 20% for standard fibers and up to 150% for high-tenacity fibers [116–120].

The sulfuric acid for the coagulation bath (ca. 1 kg of 96% H_2SO_4/kg of fiber) should be highly concentrated (iron storing tanks), of high purity (colorless and free of odor), and free from arsenic. The Fe content should be < 100 ppm, and Cu content < 1 ppm.

The zinc sulfate should not present dispensing problems (hydrated crystals or zinc oxide). Purity requirements are as follows: Cu < 10 ppm, Pb < 30 ppm, Fe/Mn/Ca < 100 ppm each, Mg < 500 ppm, chloride < 0.3%.

During the spinning process, sodium sulfate is produced in the form of Glauber's salt ($Na_2SO_4 \cdot 10\ H_2O$, ca. 1.2 kg/kg of fiber). It is generally calcined and sold to the glass and detergent industries.

Processing water should be soft and low in heavy metals (total and carbonate hardness ≈ 0, Fe < 0.05 mg/L, Mn < 0.03 mg/L, permanganate index ca. 2.5 mg/L).

Surfactant auxiliaries (modifiers, additions to the spin bath, aftertreatment agents, and finishes) are used at various stages of the viscose process. They must be selected for compatibility. Ethylene oxide additives with or without nitrogen and polyalkylene glycols will, for example, promote xanthation, dispersion, and alkalization, hence, the filter properties and spinnability. Surfactants may be added to the pulp or the alkaline cellulose. For high-tenacity fibers, special additives are used as modifiers [121], [122]. Spin bath additions make for a clearer spin bath and reduce deposits on the spinnerets.

Recovery of Chemicals. The steeping sodium hydroxide pressed from the alkaline cellulose is partially cleaned by dialysis and reused. It dissolves the cellulose xanthate and serves as a washing bath for the regenerated fibers.

Spin bath solution is filtered, vacuum-deaerated (removal of H_2S), reduced, and reused. The Glauber's salt generated in the process is removed and calcined.

The carbon disulfide released during the fiber-making process is collected, freed from hydrogen sulfide (if any), and adsorbed on activated carbon. In this way, up to 70 % of the initial amount can be recovered and distilled for reuse.

Part of the hydrogen sulfide is changed into sodium sulfide (for desulfation baths) and oxidized to pure sulfur or sulfuric acid.

3.1.4. Fiber Types

Regular Fibers. Classic spinnable xanthate solution (7 – 10 % cellulose, 5 – 7 % sodium hydroxide, 25 – 35 % carbon disulfide, maturity 10 – 15 °Ho) is spun into an acid salt bath (≥ 80 g/L sulfuric acid, 150 – 300 g/L sodium sulfate, 10 – 20 g/L zinc sulfate) at 45 – 55 °C. The first regeneration trough is 20 – 80 cm long and the fibers are drawn off at a speed of 80 – 160 m/min. The prestretched filaments are washed with cold and hot water and collected on rolls, spools, or cakes for deacidification.

Other treatments have the purpose of removing sulfur, bleaching, application of lubricant, and drying.

The process can also be arranged for continuous operation by using wide machines for collective treatment of yarn sheets or single filament treatment. Collective treatment gives larger throughput, more uniform quality, and larger bobbins. Insulation of the machines is, however, more difficult than in single filament treatment, which is less productive, supplies smaller bobbins, and can result in differences within the yarns (between spinning points, as well as over the length of the threads). The threads are normally spun by immersing the spinnerets into the spin bath and drawing off the filaments upwards. This involves the formation of a "jacket" of varying size between the filaments and around the fiber bundle. The size of this jacket depends on the take-off speed. This jacket, which is stripped off, limits the production speed. That is why more recent production systems involve spinning from top to bottom, similar to the cupro process. In principle, a production speed of ca. 350 m/min is feasible for fine single filaments, this speed is currently limited by the drying operation.

Regular viscose fibers have a rather low wet modulus; wet strength is about 60 % of dry strength, and moisture absorption is about 100 %, this is 2.5 times higher than that of cotton. These facts needed to be considered in textile processing and led to the search for fibers with a higher wet modulus, higher wet strength, and reduced moisture absorption. The successful result of these efforts was the so-called modal fibers (staple) and high-tenacity (industrial) filaments.

Crimped viscose fibers are turned into functional fabrics either alone or in blends with wool, polyester, or acrylics. They are based on a relatively "unripe" grade of viscose (15 – 20 °Ho). Xanthation uses the same amount of carbon disulfide as previously described, but spinning conditions are different. The spin bath is less acid, whereas the second bath has a higher acid content (up to 30 g/L sulfuric acid). When the filament passes the second bath, it is stretched to a higher degree; the initial skin breaks up under the tension, and a new irregular skin is formed. This in turn creates tensions over the cross-section of the yarn which deform the fiber into waves and crimps (Fig. 17). Their number per 100 mm can

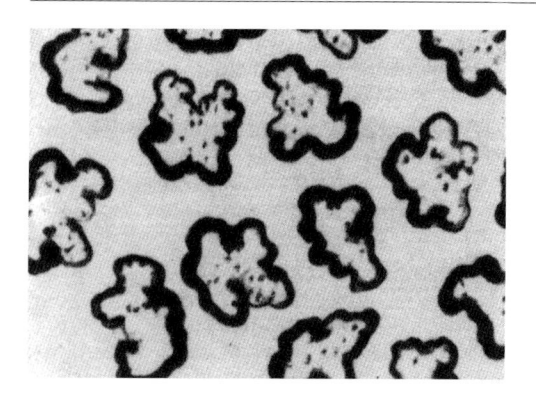

Figure 17. Skin – core structure of a crimped viscose fiber

be widely varied, but processing requirements set certain limits (in a 3-dtex staple fiber, the crimp varies with fiber type from 100 to 125 waves/ 100 mm) [123].

In cellulose fibers, the crimp durability is inferior to the crimp in synthetic fibers, particularly in the wet state.

Industrial (High-Tenacity) Yarns. When filaments are spun from top-quality viscose (ca. 7.5 % cellulose from high-quality pulp, 7 % sodium hydroxide, 38 – 40 % carbon disulfide) and highly stretched during regeneration (up to 100 %), the result is a *high-tenacity filament* or *cord yarn*.

Supercord yarns, i.e., fibers with a homogeneous full-skin cross-sectional structure, require a starting pulp with a high α-cellulose content and a uniformly high degree of polymerization. Viscose ripening is lowered, and carbon disulfide content is increased. The spin bath consists of less sulfuric acid, but a larger amount of zinc sulfate. Spinning speed is reduced, no stretch is applied in the regeneration bath, but very high stretch ($\gg$ 100 %) is applied in the second bath. Modifiers are added to the viscose and bath, e.g., monoamines, polyethylene glycols, oxyethylated ammonium derivatives, and polyglycols condensed with various organic radicals.

Extra-high-modulus fibers (EHM) are obtained by adding formaldehyde to the spin bath, sometimes also to the viscose itself. This will further increase fiber strength and reduce elongation to < 10 %. The fibers show a certain tendency to split into fibrids.

Medium- and high-wet-strength viscose staple is made under much the same conditions as cord and supercord yarns. Compared to the viscose used as regular staple, the viscose for higher tenacity fibers contains more sodium hydroxide, has a lower degree of maturity, and is spun into a regeneration bath containing up to 40 g/L zinc sulfate and more. Spinning speed is lower, but the drawing ratio higher. To ensure optimal uniformity of coagulation and decomposition of the many thousands of single filaments, it is not unusual to use a tube arrangement as in supercord production. This also allows a certain reduction in the amount of spin bath required per spinning point.

Modal Fibers. Modal fibers are also spun from viscose and come in two varieties: *polynosics* and *HWM* (*high wet modulus*). Both are based on higher quality viscose than regular staple (6 – 8 % cellulose, 6.5 – 8.5 % sodium hydroxide, 40 – 50 % carbon disulfide; small amounts of a modifier and some zinc may also be added; the viscose for polynosics can also contain traces of formaldehyde). Modal fibers – they are always cut to staple lengths – are spun into a slightly acid bath of low temperature and with a strong coagulating effect (20 – 30 g/L sulfuric acid, $\geq$ 25 g/L sodium sulfate, < 10 g/L zinc sulfate), which is sometimes difficult to recover. The stretch applied in the second bath may exceed 150 %.

HWM fibers can generally be spun on conventional viscose spinning equipment, while polynosics require essential machine modifications. If formaldehyde is added, special precautions in spin bath treatment are necessary.

Cellulose xanthate for polynosics contains relatively long molecules and is spun at lower speeds than regular viscose. Due to the lower acidity of the bath, the number of nuclei formed when the mixture enters the bath is reduced. They grow into long, tapiform crystallites which lend themselves to easy orientation. The result is a network of long, solid crystalline bundles in whose inner regions even the noncrystalline matter shows a certain degree of orientation and, hence, resistance to penetration. As regeneration speed is also reduced, the nascent filaments remain deformable for a longer time, which allows absorption at higher drawing rates than with HWM fibers. This gives almost complete orientation of the cellulose molecules in the direction of the fiber axis.

In polynosic fibers, the cellulose crystallites are made up of alkali-resistant cellulose II. This is the only type of fiber made from cellulose xanthate which displays the fine fibrils seen in cotton.

In HWM fibers spun in the presence of modifiers, the length of the crystallites and their orientation are somewhat less than those with the polynosics. Moreover, the crystallites contain not only cellulose II, but also orthorhombic cellulose IV [99, p. 90]. Sodium lye of concentrations greater than 4 % causes the lattice of cellulose IV to change into cellulose II. If not subjected to tension during this process, the cellulose II remains unoriented and acts as an instable disturbance within the pattern of ordered macromolecules [124], [125]. This in turn jeopardizes the mercerizing properties and gives a higher elongation than found in polynosics. Practical experience has shown that dry elongation of more than 15 % can result in fiber damage during mercerization.

Addition of formaldehyde or formaldehyde-releasing compounds in the course of the polynosic fiber spinning process will promote gel orientation. Methylene chains will attach to the cellulose chains and strengthen the network to an extent that sometimes allows drawing ratios of 500 – 600 %, a mixed blessing.

3.1.5. Modified Viscose Fibers

Using weakly acidic baths (ca. 15 g/L sulfuric acid) containing more than 100 g/L ammonium sulfate, it is possible to spin so-called coagulation fibers (see Section 3.1.6) from ripe viscose (< 10 °Ho). They have a round cross-section and a scarred surface, are easy to dye, and offer good dimensional stability. Their structure results in reduced strength and, thus, makes production of fine fibers (< 3 dtex) difficult.

With the help of inorganic pigments which are added to the viscose, colorfast spun-dyed fibers are produced. Addition of some 40 % barium sulfate (based on the weight of the fiber substance) gives the fibers X-ray contrast mainly required for medical textiles.

Optical brighteners added to the viscose give a brilliant white. These fibers, however, should not be used for medical purposes. It is also possible to include flame retardants (halogen, phosphorus, and nitrogen compounds). Certain sequestering agents may be used to modify the dye affinity of the fiber so that it accepts acid dyes. Blends of regular and acid-dyeing fibers then allow attractive differential dyeing.

The wet spinning process is well-suited for the addition of substances to the spinning solution. In connection with the viscose process, the additives must, however, be selected for resistance to the strongly alkaline and acid process conditions.

Another fiber modification involves *grafting*. Grafting can be carried out during fiber production as well as in textile finishing. In process grafting, the method should not require extremely drastic changes in the normal schedule and avoid reduction of the spinning speed, which is usually very difficult to manage [126–129].

Depending on the type of monomer or polymer to be grafted (e.g., styrene or acrylonitrile) the grafting process can be initiated by ions ($Ce^{4+/3+}$, $Fe^{3+/2+}$) or free radicals. It is possible to produce radicals on the cellulose by UV radiation in the presence of photosensitive agents [130]. In addition, high-energy rays from γ-ray sources and electron accelerators are used [131–133]. These processes differ in the chronological order of the monomer bath, drying, and source of radiation. The choice of the method is a function of the desired effect, process economics, and the properties of the monomer and polymer. The radiation dose reaches an approximate maximum of 10 J/kg. Yield is affected by the monomer concentration, type of solvent, swelling degree of the textile, and the radiation dose [134]. From an economic standpoint, low-energy radiation, e.g., argon plasma, which as a highly ionized gas emits positive ions and electrons, may also have importance in the future [135].

Grafted fibers are still in a very early phase of commercial use in spite of years of strenuous research efforts. On a pilot scale, some carpet fibers (CV/PAC) as well as flame-retardant, bacteria-resistant, water- and oil-repellent fibers have been spun in the Russia; also included in this research program were acid-protective, ion- and electron-exchanging, complex-forming, antimicrobic, and hemostatic fibers [129].

For more information on fiber modification, → Textile Auxiliaries, 4. Pretreatment Auxiliaries.

3.1.6. Fiber Properties

Structure and Properties. As with cotton, the physical properties and the reaction of the fibers to chemical treatment is governed by the fiber buildup – the macrostructure (morphology) and the fine structure. Compared with cotton, the structure of regenerated cellulose fibers is much simpler [136], [137]. Unlike natural cellulose fibers with an advanced fibrid system in their secondary walls, made up of bundles of chain molecules (elementary fibrids and crystallites) and higher configurations (micro- and macrofibrids), the regenerated fibers rarely show such bundles. Rather, they have orderly and noncrystalline portions which, however, have flowing boundaries. In regular viscose fibers, formation of fibrids is negligible; in modifed fibers intentionally suppressed. The polynosics, on the other hand, have a distinct fibrid structure in the form of relatively stiff ropes 10 – 20 μm across. With them, the primary and secondary bundles are comparable with natural cellulose fibers.

The crystallinity, size of the morphological units, and degree of orientation in regenerated fibers are functions of the coagulation and regeneration conditions and subsequent drawing [90]. Regarding the crosswise order, it is assumed, that straight and folded chains exist side by side, with the number of straightened chains increasing with increased stretch.

Physical Properties. The structural characteristics listed in Table 20 are the outcome of process conditions and result in certain physical properties as shown in Table 21 for three types of regenerated cellulose fiber. The regular viscose fiber data in Table 20 indicate high swelling, low wet modulus, and the typical shrinking behavior of this fiber.

Table 21 indicates that the modal fibers are superior to regular viscose staple with regard to dimensional stability, i.e., the interaction of water retention, relaxation, and wet modulus. The decisive characteristic is the improved wet modulus. Figure 18 illustrates the characteristic steepness of the stress – strain curve of both wet and dry polynosics. This gives increased dimensional stability to woven fabrics even after repeated washing, as indicated in Figure 19. Their utility in certain end uses did not help the polynosics achieve economic importance as had been expected, possibly due to their fibrillating tendency.

Figure 20 shows the lengthwise sections and cross-sections of some regenerated cellulose fibers.

The main properties of these cellulose fibers are [141]:

Viscose fibers
Regular viscose:
    skin – core structure

Table 20. Structural data of regular viscose staple, polynosic fibers, and cotton [138] (Measuring units and methods in parentheses)

	Regular viscose staple	Polynosic fibers	Cotton
Crystallinity, % (X-ray diffraction)	33 – 36	40 – 47	50 – 52
Length of crystalline portions, DP (marginal DP)	60 – 80	110 – 140	60
Thickness of crystalline portions, nm (wide-angle X-ray)	5 – 7	8 – 10	10
Accessibility of OH groups, % (H-D exchange)	60 – 70	50 – 55	50
Chain orientation, % (X-ray diffraction)	70 – 80	80 – 90	
Total chain length, DP (viscosity)	300 – 450	300 – 500	2000
Porosity, cm^3/g [139]	0.016	0.07	0.087

Table 21. Textile data of various viscose fibers

	Regular viscose staple	Modal fibers	
		Polynosic fibers	HWM fibers
Titer, dtex	0.9 – 100	0.9 – 4.2	1.3 – 3.0
Breaking strength, conditioned, cN/dtex	1.8 – 3.3	4.0 – 6.5	4.0 – 5.5
Breaking strength, wet (5 % NaOH, 20 °C), cN/dtex	1.3	3.2	2.5
Elongation at break, conditioned, %	18 – 40	6 – 12	14 – 18
Density, g/cm^3 (65 % R.H., 20 °C)	1.50 – 1.52	1.50 – 1.52	1.50 – 1.52
Water retention, %	90 – 120	55 – 70	66 – 90
Moisture regain, % (65 % R.H., 20 °C)	13	11.5 – 12.5	11 – 12.5
Relaxation shrinkage, %	12 – 15	3 – 5	5 – 7
Wet modulus, cN/dtex	5 – 10	14 – 18	10 – 12
Mercerization	not possible	very good	under certain conditions

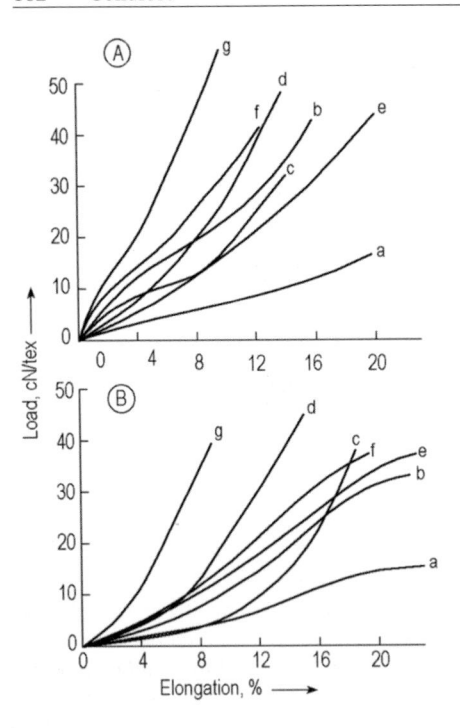

Figure 18. Stress – strain diagrams of various fibers [140] A) Conditioned; B) Wet a) Viscose textile grade; b) Viscose industrial grade; c) Texas cotton; d) Egyptian cotton; e) High-wet fiber; f) High-wet modulus fiber; g) Polynosic

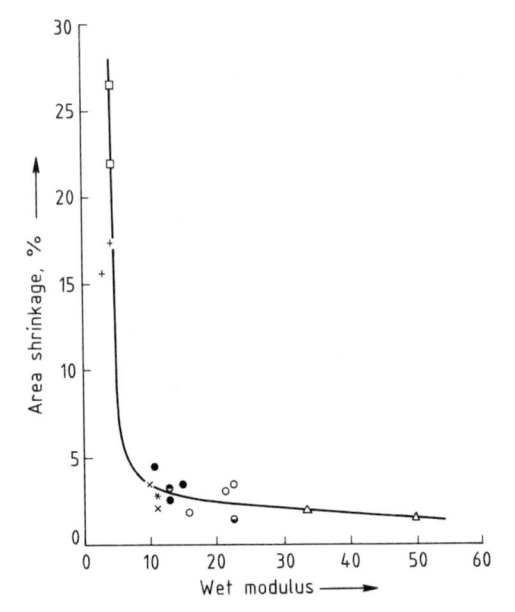

Figure 19. Changes in woven fabric dimensions from 1st to 50th washing [124] + Regular viscose staple □ High-tenacity viscose staple • HWM fiber, commercial ⊖ HWM fiber, experimental ○ Polynosic, commercial ⊖ Polynosic, experimental △ Polynosic with formaldehyde × American cotton I * Egyptian cotton

   medium tenacity
   medium elongation
   1.3 – 50 dtex (standard program)
   bright, dull, extra dull
   optically bleached, spun dyed
Crimped fibers:
   thick/thin skin – core structure
   medium tenacity
   medium elongation
   good crimp (dry)
   2.2 – 25 dtex (standard program)
   bright, dull, spun dyed
High-wet-strength fibers:
   all-skin structure
   high tenacity
   high elongation
   1.3 – 3.1 dtex (standard program)
   bright
Coagulated fibers:
   all-core structure
   low tenacity
   higher elongation
   irregular good crimp
   3.6 – 25 dtex
   bright, dull, spun dyed
Modal fibers
   Polynosics:
      high tenacity
      low elongation
      dimensional stability
      1.3 – 3.6 dtex
      bright
      suitable for mercerization
   HWM:
      high tenacity
      low elongation (but higher than polynosics)
      fair dimensional stability
      1.3 – 3.6 dtex
      bright, optically bleached
      mercerizable under special conditions

Figure 20 illustrates the data shown in Tables 20 and 21. The dark areas of the cross-sections reflect areas of higher orientation and density. Increased thickness of the skin – up to an

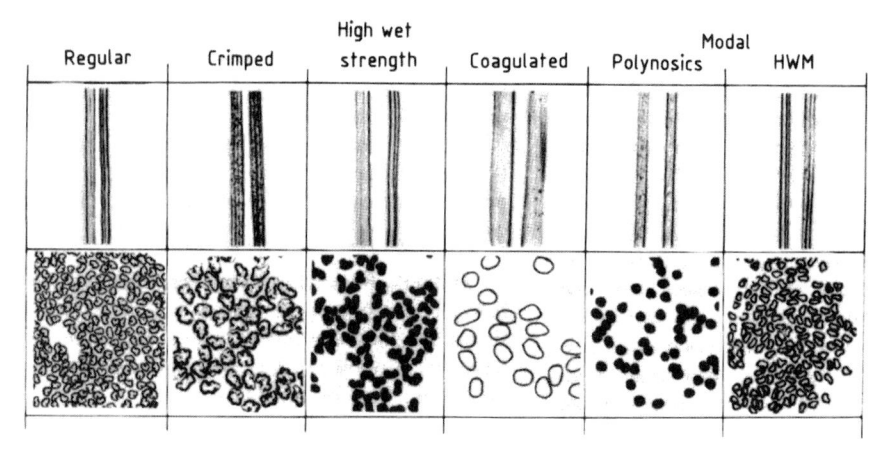

Figure 20. Lengthwise sections and cross-sections of some regenerated cellulose fibers [141]

all-skin structure – is coupled with distinct increases in fiber tenacity.

Chemical Properties. In several respects, viscose fibers behave in the same way as cotton. All these fibers are attacked by acids and oxidizing agents. If properly executed, hypochlorite and peroxide bleaching is possible. Different fibers react differently to dissolution in sodium hydroxide. Regular fibers show a dissolving maximum in ca. 10 % sodium hydroxide, even quicker as the temperature decreases, and will completely dissolve at − 5 °C. The all-core fibers are more easily dissolved in alkali than skin–core fibers.

For these reasons, polynosic and HWM fibers show different strength losses in sodium hydroxide treatments (Table 22). The hydroxyl groups can be utilized for numerous conversions.

Dyeing Properties. Similar to natural cellulosics, regenerated fibers can be dyed with vat dyes, Naphtol-AS, sulfuric, direct, and reactive dyes ($\rightarrow$ Textile Dyeing). Particularly good fastness is obtained with reactive dyes.

Regenerated cellulose fibers absorb much more dye than natural cellulosics. The relationship between production conditions and dyeing properties of regenerated cellulose fibers is extensively described in [88]. It is said that the dye affinity of the regenerated fibers depends on the accessibility, i.e., crystallinity and degree of orientation. HWM fibers are very similar to regular viscose, while the fiber structure of the polynosics puts them nearer to cotton in this respect.

The possibility of making spun-dyed fibers offers some advantages (fastness, cross-dyeing, and low water pollution), but also some disadvantages (minimum order size, variety, and sales disposition).

Similar to dyes also other chemicals can be included in the fiber during spinning if they are resistant to the alkaline and acid processing conditions.

3.1.7. Uses

The varied range of natural and man-made fibers is nowadays used by the textile industry with a keener eye on the specific fiber properties. In this way, the cellulosic man-mades have acquired a leading position in certain applications and are blended or combined with other fibers to add their advantageous properties to the textiles.

Viscose fibers are used in many ways for many textile purposes: clothing, home textiles, and industrial textiles. Regular viscose staple is primarily used on its own for outerwear and nonwoven fabrics. Crimped fibers are used alone or in blends with polyester, wool, or acrylics for woven and

Table 22. Strength loss (in %) of modal fibers in sodium hydroxide of various concentrations (relaxed)

	5 % NaOH	23.5 % NaOH
HWM fibers	40	60
Polynosics	15	30

Table 23. World production of man-made fibers 1970 – 1996 [160]

	1970		1980		1990		1996	
	10^3 t	%	10^3 t	%	10^3 t	%	10^3 t	%
Cellulosics	3 585	43	3 581	25	3 145	16	2 900	12
Synthetics	4 809	57	10 703	75	16 006	84	21 200	88
Man-mades total	8 394	100	14 284	100	19 151	100	24 100	100

knitted outerwear. The most successful use of HWM fibers is in blends with cotton and/or polyester in cotton-type fabrics for shirts, blouses, dresses, raincoats, bed linen, and the like. In blends with cotton, the HWM fibers act as a spin carrier and will sustain mercerization under suitable conditions. In polyester blends, they take the place of cotton; their comparatively high cross-strength makes for fabrics with good wearing properties. High-wet-strength fibers have high work-absorbing capacity (high strength at relatively high elongation) and, therefore, lend themselves as industrial textiles and nonwoven interlinings.

Viscose filament yarns are used mostly in lining fabrics as well as in ladies' outerwear and home furnishings. High-tenacity yarns continue to play an important role in industrial textiles such as automobile tires, conveyor belts, and coated fabrics.

All end uses profit by the following viscose properties: easy processing in the textile and finishing stages, hydrophilic behavior, and attractive cost/benefit ratio. Disadvantages derive from the following: relatively high density in comparison with synthetic fibers and low strength, particularly in the wet state, as well as moderate dimensional stability, which, due to the hydrophilic nature of the fiber, may be unsatisfactory for certain uses. The cost/strength ratio is also rather low. The lack in dimensional stability can often be made up by suitable after-treatment of the fabrics. Care must, however, be taken not to excessively damage the physical properties of the textile.

Viscose fibers hold a secure niche in the textile field. Since raw materials for viscose fiber production are secure, viscose can be expected to continue on an upward trend.

Outlook. Regenerated cellulose fibers hold a firm place among our textile raw materials. Solution of production problems would, therefore, be a sensible proposition. One of the aims should be an optimal cost/benefit ratio.

Regenerated cellulose fibers are popular materials for clothing, home textiles, and industrial textiles for a number of reasons. This also applies to nonwovens for household, personal care, and the wide field of medical applications, where absorption and cleanliness are primary considerations. Possible substitution by fluff has not yet been fully explored.

In the area of industrial fibers the unfavorable cost/tenacity ratio of cellulosic in comparison with synthetic fibers will result in a shift in the quantities produced in favor of synthetic fibers. This is already discernible from the figures given in Table 23. The ecological problems connected with the production of viscose fibers promote investigations in the improvement of process conditions and the search for sulfur-free processes (see Sections 3.3 and 3.4).

3.1.8. Economic Aspects

Growth of the world population and higher standards of living also caused an increase in the demand for textile fibers as illustrated in Table 23. It also indicates that the demand for cellulosic man-made fibers has markedly decreased and lost importance relative to the synthetics.

The fact that production of staple fibers kept growing slightly, while filament yarns lost more and more ground is not shown. This was a consequence of development in the nonwoven field, in garment making, and in the tire yarn industry.

Regional differences in cellulosic man-made fiber production are also interesting. Approximately 22 % is made in Western Europe, ca. 8 % in the United States, 10 % in Japan, and the remainder of more than 60 % is distributed among the other regions. Production sites are being shifted to the Far East to a certain extent (see Table 24). Of the

Table 24. Production development of cellulosic fibers in different parts of the world (in 10^3 t) [160]

	1985	1990	1995
World	3 217.6	3 146.3	3 040.4
Western Europe	709.0	700.1	684.0
Eastern Europe	994.4	833.0	272.6
USA and Canada	287.7	268.9	240.1
Japan	387.7	275.8	297.1
People's Republic of China	180.0	191.7	462.4
India	142.0	216.7	262.2
Indonesia	40.0	60.0	177.0

production methods, the viscose process is by far the most important.

Table 24 gives only general trends because of differences in the computing methods and includes cigarette filter tow (acetate) as well as material for medical use (Cupro).

Depending on fiber type (titer, textile, or industrial grade), viscose filament production units should have a capacity of 10 – 30 000 t. Staple fiber units, on the other hand, should be of an even larger size, i.e., $\geq$ 50 000 t and more. The lower limits are governed by economic considerations and the upper limits by ecological aspects.

The future prospects of cellulose fiber production are not clear. The industry is faced with increasingly stricter environmental regulations; most production equipment is old and the market seems convinced of the advantages of synthetics. Positive aspects of cellulosics are their hydrophilic properties, quantitative importance and the long-term availability of raw materials. These are important reasons for finding solutions to the problems so that adequate textile supplies may be secured for many years to come. Thus, in the long run, the importance of the regenerated cellulose fibers will undoubtedly increase once again.

3.2. Lyocell Fibers

In view of the problems associated with the viscose process, attempts have been made for decades to find a "shorter, simpler, sulfur-free process" for the production of cellulosic man-made fibers (see also Section 3.4). Of the various possible approaches, the NMMO process has now been developed to the commercial stage. The process involves dissolution of cellulose in aqueous N-methyl morpholine N-oxide (NMMO) solutions and subsequent regeneration from this solvent. This process is distinguished from other methods of producing cellulosic man-made fibers by the use of an organic solvent. This also determines the specific property profile of the fibers which justifies that they are classified as separate category within the group of cellulosic man-made fibers.

The basic principles of this process were developed around 1976 by AEC (American Enka Corp., United States) and ARLO (Akzo Research Lab. Obernburg, Federal Republic of Germany) and patented in 1980 [91]. After having discontinued the production of cellulosic staple fibers in 1983, Akzo Fibres granted licences to Lenzing in 1987 and to Courtaulds in 1990 and concentrated their development activities on filament yarns [92].

Courtaulds produced some 70 000 t/a of staple fibers under the trade name Tencel in Mobile, United States and Grimsby, United Kingdom. Lenzing has started production of staple fibers with a production volume of 12 000 t/a in Heiligenkreuz, Austria in 1997, while Akzo Nobel Fibres are preparing for commercial production of filament yarn, following successful conclusion of a market survey based on products from their Obernburg pilot plant. The technology is now also being offered to interested patries by the Thüringisches Institut für Textil- und Kunststoff-Forschung (TITK, Thuringia Institute for Textile and Plastics Research) in Rudolstadt, Federal Republic of Germany.

3.2.1. Principles of the Lyocell Process

The cellulose briefly described in Section 3.1.1, NMMO, and water are mixed, after adding a stabilizer, to give a homogeneous solution [85], [92–94]. The solution is extruded into a regeneration bath consisting of an aqueous NMMO solution to obtain the desired cellulosic man-made fibers (or films). The NMMO used can be recovered at > 99 % and recycled.

A schematic of the process is depicted in Figure 21. As Figure 21 shows, the process – like the production process for synthetic fibers – is short and the regeneration bath can be recycled almost completely. As in the viscose process, certain substances can be added to the solution to impart specific properties to the fibers.

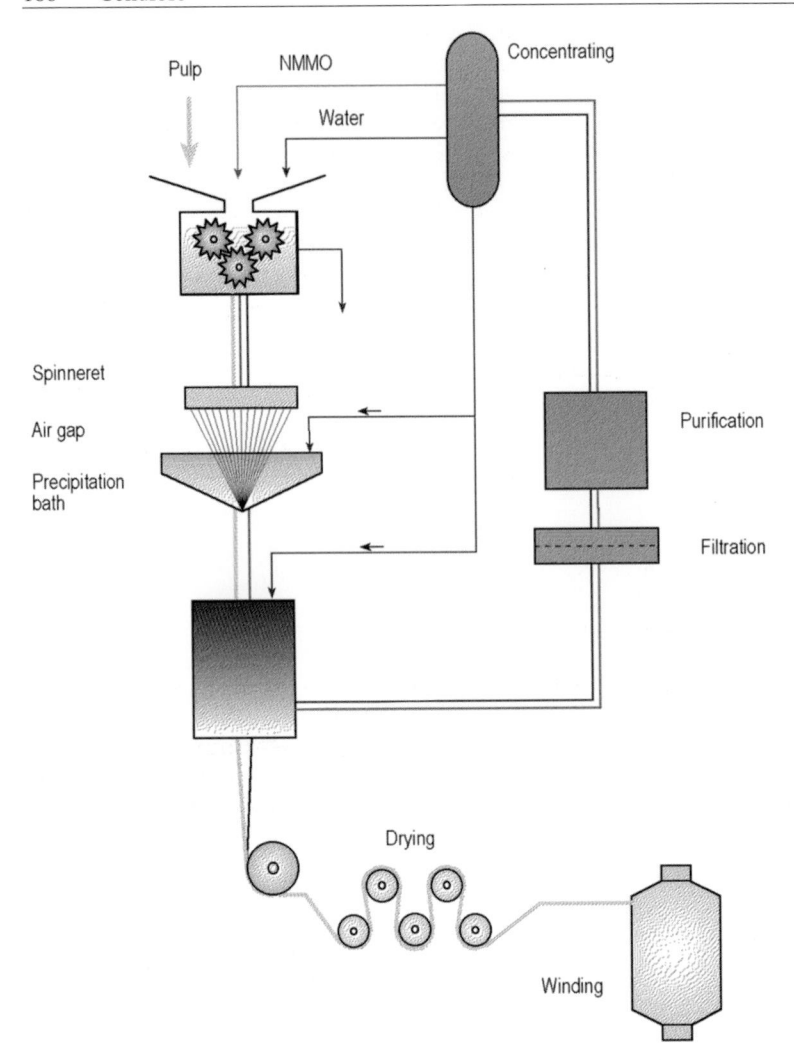

Figure 21. Lyocell process

3.2.2. Process Description

The cellulose is introduced into a mixture of NMMO and water. The slurry (which should be as homogeneous as possible) of, e.g., 13 % cellulose, 20 % water, 67 % NMMO, and some stabilizer is then adjusted to 14 % cellulose, 10 % water, 76 % NMMO, and stabilizer by hydroextraction, to dissolve the cellulose (see phase diagram, Fig. 22). The solution of this composition is extruded at temperatures slightly higher than 100 °C into an aqueous NMMO bath whose composition is beyond the solution range for cellulose. The fibers or films thus obtained are washed and dried. The regeneration bath is purified, concentrated and reused [92].

3.2.3. Fiber Properties

Although lyocell fibers also consist of regenerated cellulose II, their properties differ substantially from those of other cellulosic man-made fibers. This difference can best be demonstrated on finished textiles. That is why Table 25 compares the requirements for four textiles typically made from viscose filament yarn with the performance of traditional viscose and lyocell yarns.

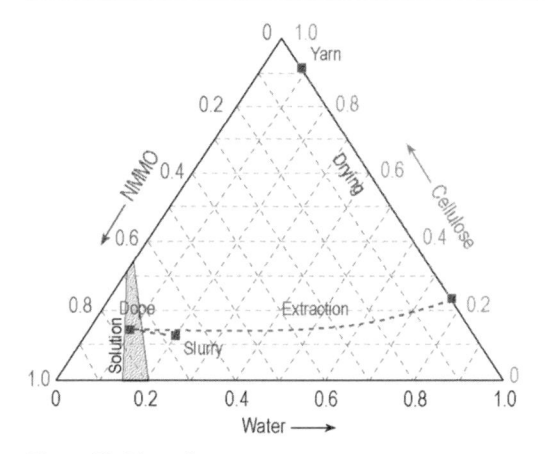

Figure 22. Phase diagram

Table 25 shows that the requirement profiles of the textiles are met by conventional yarns, except for three types of embroidery yarns. It is also evident from Table 25 that with lyocell fibers comparable but lighter textiles with equal good use properties can be obtained which correspond to fashion trends and customers' wishes [92].

However, lyocell fibers have a special property that must be taken into account in processing and use. When subjected to a rubbing treatment, they will – especially in the wet state – form fibrils influencing the feel and appearance of the fabric. This is of advantage for producing a soft handle and/or a peach-skin fabric surface. If this is not desired, various possibilities exist to eliminate this property partially or totally (either during fiber production or aftertreatment of the textiles). These examples and the consequences for fabric design demonstrate that lyocell fibers are no substitutes for viscose fibers, but a fiber group of its own which, when put to appropriate use, opens up new possibilities for textile design without causing the kind of problems associated with viscose fibers [95], [96].

The characteristic of lyocell fibers can also be advantageously utilized in hetero-yarns, specialty textiles, and nonwoven fabrics with specific properties.

3.3. Cuprammonium Fibers

Present Situation. Production of cellulosic fibers after the copper oxide ammonia process (cuoxam process) began to decline in the early 1960s. The reasons were the outdated production equipment, the limited possibilities to increase

Table 25. Comparison of requirement for four typical viscose yarns with performance properties of classical viscose and lyocell yarns, according to [92]

	Requirement for ladies' outerwear					Performance	
	General requirements	L	S	C	E	Enka Viscose (bobbin)	NewCell (current performance)
Total count, dtex	50 – 600	=	=	=	=	67 – 660	55 – 110
Single filament count, dtex	1.5 – 8	=	=	=	=	2.7 – 7.5	1.1 – 2.2
Cross section		=	=	=	=	lobal	round
Uniformity	high	=	=	=	=	+	+
Degree of luster	bright-extra dull	=	=	=	=	+	bright-dull
Tenacity							
dry, cN/tex	≥ 16	=	=	=	>	16 – 20	34 – 40
wet, cN/tex	= 10	=	=	=	>	7 – 10	22 – 27
Elongation							
dry, %		=	=	=	=	16 – 21	6 – 12
wet, %		=	=	=	=	20 – 26	8 – 14
Knot strength, cN/tex	13	=	=	=	>	13 – 16	23 – 27
Loop strength, cN/tex	13	=	=	=	>	13 – 16	18 – 23
Shrinkage at the boil, %	≤ 2	=	=	=	=	0.5 – 2.4	0.5 – 2.0
Moisture gain, %	> 10	=	=	=	=	11 – 14	11 – 13
Water inhibition, %		=	=	=	=	95	60 – 70
Natural shade (whiteness)	natural white	=	=	=	=	+	(+)

L = Lining fabrics; S = Ladies' outerwear from flat yarn; C = Ladies' outerwear from crêpe yarn; E = Embroidering yarn; NewCell = Trade name of lyocell filament yarn from Akzo Nobel; Enka Viscose = Trade name of viscose filament yarns from Akzo Nobel.
= Can be produced without problems, no discernible differences in properties for consumer; > Production advantages; + Better than reqired.

the variety of fiber types available, and the sweeping success of the synthetics. At that time, virtually no one would have predicted the comeback of cuprammonium fibers. Moreover, elimination of the Cu ions from wastewaters seemed an unsurmountable obstacle (hazard for biological water reprocessing). In addition, the fibers were suitable for only a limited range of textiles that was waning under pressure from the synthetics. On the other hand, the largest producer of this type of fiber, Bemberg AG (now Akzo Nobel Fibres AG), diverted its research and development efforts from 1965 on to membranes and used the cuoxam solution for the manufacture of flat and tubular membranes and hollow fibers, currently the standard material worldwide for blood dialysis (Cuprophan). In artificial kidneys, they serve hundreds of thousands of kidney patients as the first artificial organ ever made [142–145]. Membrana – a business unit of Akzo Nobel Fibres – is the only producer of these hollow fibers in the Federal Republic of Germany. Cuprammonium fibers are also made in other countries: in Italy by Bemberg S.p.A., Gozzano (Novara) (filament); in Russia by Chemical Fiber Combine Kalinin (staple); and in Japan by Asahi, Nobeoka (filament and staple).

In view of the decreased importance of cuprammonium filaments and staple, we shall limit ourselves to a short description of the process principles [146–148].

Production. There are no essential differences between filament and staple production. Linters or cellulose pulp, or a mixture of them is opened and mixed with copper hydroxide or a basic copper salt in the presence of a highly concentrated solution of ammonia in water. This results in a viscous spinning solution, which is filtered and deaerated. The method is simpler than the viscose process. The filaments are spun after Thiele's spin-drawing method by extrusion through relatively large spinning holes into a vertical water stream and high stretching. This is followed by copper extraction, neutralization, finish application, and drying. Currently, the process is continuous from start to bobbin/section beam, but some "historical" apparatus is still in use for separate spinning/winding, scouring, drying, and spooling.

Raw Materials. Starting material is generally bleached linters, but a mixture of wood pulp and linters is also occasionally used. It is customary to prepare the linters in a 1 % cuoxam solution at 20 °C to a 25 mPa · s viscosity; wood pulp is adjusted to 16 – 24 mPa · s viscosity at 20 °C, reflecting a polymerization degree (DP) of 1 000 – 1 200 and 800 – 1 000, respectively. Peculiar in cuprammonium fiber spinning from pulp is the fact that the cellulose is not treated with sodium hydroxide, with the result that none of the hemicellulose is washed out. For this reason, the pulp should have a low hemicellulose content ($< 4\%$).

The cellulose is dissolved in basic copper sulfate – obtained from recycling – in complex compounds with ammonia.

Spinning Solution. As a rule, the cellulose is moistened and immersed in the dissolving liquid made up of ammonium hydroxide and copper salt (basic copper sulfate or copper chloride). Since heat is released during the reaction (ca. 40 kJ/kg spinning solution), cooling is recommended so that the temperature does not exceed 25 °C. Heating causes discoloration due to the reaction of copper hydroxide with ammonia, cellulose with copper(II) aminehydroxide hydrate, and the basic copper salt with sodium hydroxide giving ammonium copper(II) hydroxide hydrate. Starting as a highly viscous mass, agitation requires very strong straight or helical mixing blades. The spinning solution is then filtered through presses equipped with noncorroding metal fabric packs (nickel or stainless steel), mesh size 40 – 70 μm. Finally, the filtered spinning solution is deaerated to remove air bubbles; at the same time some of the excess ammonia is eliminated. Depending on the spinning conditions, the solution will then contain 4 – 11 % cellulose, 4 – 6 % Cu, and 6 – 10 % NH_3, and have a viscosity of ca. 200 Pa s at 20 °C. When spun-dyed or delustered fibers are desired, inorganic and/or organic pigment dyes are added.

The solution is very stable. Stored in air-tight vessels, its composition will not change and the γ-value remains constant, ensuring unchanged coagulating properties.

Spinning Methods [148], [149]. Cuoxam cellulose solutions are always spun according to the wet process (→ Fibers, 3. General Production Technology), with one or two bath passages. Strong coagulants are, for example, concentrated sodium hydroxide [150] or 3 – 30 % sulfuric acid. Single-bath wet spinning in NaOH is used for coarse filaments such as upholstery stuffing and hollow fibers [151], [152]. In this case, a bath like that in viscose fiber production is used. A commercial single-bath process using acid and salt is under consideration.

Thus far, however, the Thiele spin-drawing method in a virtually neutral bath is the rule. Vertical spin-draw funnels of glass or transparent plastic are used [153]. The spinning speed of about 150 m/min is very slow compared with the speeds in synthetic fiber spinning. Trials were made to adapt the Thiele funnel spinning method for higher speeds as well.

In several patent applications, Asahi Kasei Kogyo [154] describes methods in which the threads coming from conventional spinning funnels are led over longer drawing and coagulation distances to be finally deposited in a random layer on a perforated conveyor belt for transportation through the subsequent production stages (copper extraction, washing, finish application, and drying). This method is claimed to allow spinning speeds of up to 800 m/min.

Aftertreatment. The plastic and, therefore, mechanically sensitive yarns made up of 20 – 230 single filaments are freed from copper and aftertreated in one continuous operation.

The sheet of parallel threads spun by batteries of several hundred spinning points in staggered rows (thread spacing = 2 – 4 mm) meanders through a bath of 2 – 6 % sulfuric acid and is rinsed in two or three water baths to remove the acid. Finally, the yarn sheet is immersed for finishing and dried on heated drums or in drying chambers to a residual moisture content of ca. 11 % for winding onto beams or bobbins.

American Bemberg Corp. (Beaunit Mills Inc.) introduced a continuous spinning method, in which the funnels are arranged at an angle in a staggered row and the freshly spun yarns are led as a sheet on a straight path through shallow troughs. After the yarns have dried, they are wound onto bobbins [155]. This method is still used in Italy and Japan.

Cuprammonium Nonwovens. In [156], an apparatus for the production of nonwovens from cuprammonium filaments is described. It follows the funnel spinning and drawing principle, although with a rectangular spinneret arrangement from which the threads (ca. 90 000) cascade onto a partially permeable conveyor belt which traverses and causes crosswise entanglement of the yarns. As cuprammonium filaments will securely bake together in an ammoniacal medium, a nonwoven fabric is formed whose thickness and stability can be varied via crossing points and fusion. Production speed is claimed to be ca. 35 m/min.

Cuprammonium Staple. The spinning solution is almost the same as in filament spinning, except that a 1 – 2 % lower ammonia concentration is used to reduce the tendency of the filaments to stick together. Sticking may help in filament spinning, but is disturbing in staple production.

The dimensions of the funnels are much larger. The spinnerets are 80 – 100 mm across and have 2 000 – 3 000 holes of 0.5 – 1 mm across. The filaments emerging from 50 – 100 funnels are collected into thick tows. Aftertreatment takes place in the form of filament tow or cut to staple lengths.

For conversion into fiber laps, 5 – 10 tows are laid parallel and led through troughs of 50 – 100 mm width and 10 – 20 m length and are rinsed in the same way as filament yarns. After drying, the moisture content is readjusted to 11 %. Alternating application of strain and relaxation during wet treatment gives the filaments a regular crimp [157].

Staple fibers with a coarse irregular crimp and a rough surface are obtained if the freshly spun plastic threads are cut into 40 – 150 mm lengths on special machinery [158] and deposited on screen belts for tensionless aftertreatment [146, p. 273], [159].

Recovery of Chemicals. The copper from the ammonia baths is bound by ion exchangers. The "spinning acids" obtained during acid removal and regeneration of the "blue threads" are used to regenerate the ion exchangers. In this way, the acids concentrate around the copper. Neutralization with alkaline water gives basic copper sulfate, which is filtered off and reused to set the spinning solution.

Lately, electrolytic methods have been devised. Copper recovery is $\geq$ 99 %. In addition, part of the ammonia – some 40 – 50 % – is recovered by distillation of the spin bath in special columns.

From the ecological point of view, the cuoxam process no longer poses any problems. At present, it is superior to the viscose process in this respect.

Properties. The most striking properties of cuprammonium fibers are fineness of the filaments, silklike appearance, and silky touch. In this respect, they are far superior to viscose yarns. Moreover, their inner structure makes hollow fibers particularly suitable for blood dialysis. In other respects, they are similar to viscose filaments without, however, reaching the high strength level of the industrial yarns in commercial production and high crimp properties.

3.4. Tentative Cellulose Fiber Production by Other Processes – Outlook

As mentioned earlier, the image of the cellulosics had come under pressure. This first resulted in resignation, as may be inferred from the fewer publications, which are a measure for research activity. After the first oil crisis in November 1973, public thinking changed. In the cellulose fiber field, various activities were initiated. Work was resumed to improve the viscose process with the intention of reducing air and water pollution; other investigators looked for ways of eliminating sulfur from the dissolving process. Assistance came from the plant breeding side in the form of higher cellulose yield per acre.

Air and water pollution was attacked from three sides: first, the reduction of chemicals to limit the problem; second, collection of chemicals where they are generated, concentration where possible, and transportation to ancillary treatment equipment; and third, improvements in the waste treatment plants. For details on waste air treatment see [82]. Suitable custom-tailored water treatment plants are available.

Besides, it must be remembered that the viscose process has set standards of cost, quality, and variety, which any new process must at least equal, if not surpass. Research which shows the possibility that cellulosics may be produced by more efficient methods gives reason for hope [82].

Successful research using NMMO as solvent for cellulose led to the start of staple fiber production (Courtaulds in the United States and the United Kingdom and Lenzing in Austria; see Section 3.2). It is anticipated that also filament yarns will be produced by this process (Akzo Nobel Fibres). The limited variations in the process limit the variety of fiber types produced and their area of application. Experiments with cellulose carbamate [82], [84], zinc chloride – water, sodium hydroxide – water [85], and DMA/LiCl, DMF/LiCl, or DMSO/LiCl [82] as solvents have not been successful thus far. The fibers were either strong enough but too low in maximum elongation, or they were unsatisfactory as far as all physical properties were concerned. In addition, problems arose with optimizing certain production parameters (cellulose concentration, temperature, production conditions, utilization of byproducts, recycling of auxiliaries, etc.).

The insight gained from these experiments and the decades of experience with viscose processing, combined with progress in the petrochemical field, will hopefully lead to useful regenerated cellulose fibers without the help of sulfur compounds. It was intended that the new sulfur-free process should also produce a large variety of types of cellulosic fibers. According to the experience gained, this variety can probably best be obtained by modifications in the regeneration process of dissolved cellulose derivatives.

4. Ground Cellulose/Powdered Cellulose

4.1. Production

Powdered cellulose is obtained from pulp cellulose by milling and fractionation (Figure 23).

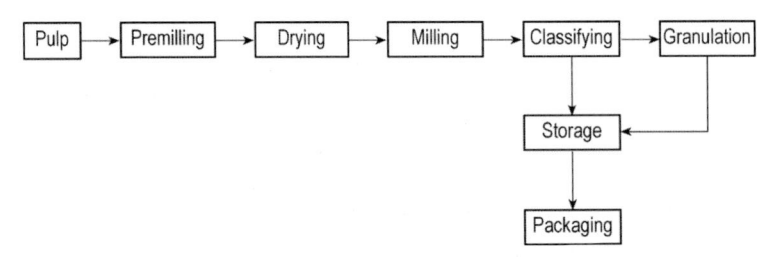

Pulp → Premilling → Drying → Milling → Classifying → Granulation
Classifying → Storage → Packaging

Figure 23. Production of powdered cellulose

Because cellulose, hemicellulose, pectin, and lignin are all incorporated into pulp, chemical reprocessing (chemical pulping) is required for the preparation of pure cellulose. The Kraft (sulfate) or sulfite process is usually used (→ Pulp). The preferred raw material for the economical production of cellulose is wood followed by annual plants such as straw.

Powdered cellulose (Figure 24, top) is obtained from cellulose in several milling and fractionating steps. Size reduction can be performed with jet, pinned-disk, beater, cutting mill, or hammer mills, or in pan grinders. Fibrillated powdered cellulose (Figure 24, middle) is obtained by wet grinding of the cellulose. Products with a defined particle-size distribution (Figure 25) are obtained by sieving and classifying.

To minimize dust formation, improve flowability, and reduce the volume for transport, processes for compacting, granulating, pelletizing, and extrusion of cellulose have been developed.

Microcrystalline cellulose (MCC; Figure 24, bottom) is produced from highly pure cellulose. The more easily accessible amorphous components are released by partial hydrolysis with dilute mineral acids. In this way the crystalline fraction is increased, while the degree of polymerization is reduced by acid treatment to an average DP value (degree of polymerization) of 300. In contrast, native cellulose has a DP of ca. 3000.

4.2. Properties

The physical properties of cellulose are determined by the raw material and the milling technique used. The milling process not only regulates the fiber length and fiber thickness, it also influences fiber properties such as water-binding capacity.

Selected properties of some powdered celluloses (Arbocel) and of a microcrystalline cellulose (Vivapur) are listed in Table 26.

Fiber lengths from 18 to 2200 µm can be produced. The individual fractions have a production-dependent particle-size distribution.

The *fiber diameter* is set by nature and differs corresponding to the type of cellulose.

Whiteness is determined with a spectrometer at 461 nm. Highly pure cellulose has a degree of whiteness of over 90 %.

The *bulk density* is dependent on the fiber length. By using a compacting process, the bulk density can be increased up to 600 g/L. Bulk density is determined according to DIN 53468.

For determination of the *ash content*, the sample is heated at 850 °C in a muffle furnace. Highly pure cellulose has an ash content of max. 0.3 %.

The *pH value* is determined according to NF XVII.

The *water-binding capacity* is determined according to the AACC method. The water-binding capacity depends on the fiber length and the degree of fibrillation. Cellulose fiber can bind up to 15 times its own weight of water.

4.3. Uses

The large range of applications for powdered cellulose can be divided into two main categories:

1. Cellulose functions as a process fiber during the production of semi-finished and end products; it helps to speed up processing and to reduce manufacturing costs.
2. The properties of intermediate or end products can be improved by the use of cellulose: increasing viscosity or improving flowability of dustlike products by converting to resoluble granules.

Cellulose has been used since the mid-1990s as a *disintegration agent* for laundry detergent tablets. Laundry detergent tablets are mainly used in Europe. They have a market share of ca. 10%. Cellulose fiber is also used as a disintegration agent for pigment granulates. Swelling pressure is built up by the capillaries of the cellulose. The fibers increase their volume several times and thus lead to a quick and complete disintegration of granulates, pellets, and tablets.

In spice mixes and grated cheese, cellulose fiber is used as a *flow agent* and *anticaking aid*. It acts as spacers between the individual particles and prevents clumping of the particles by its water-binding capacity.

Cellulose acts as a *transport medium for liquids*: it extracts water and dries but can also bring liquids into a product. Because of their coiled shape, the fibers forms drainage channels

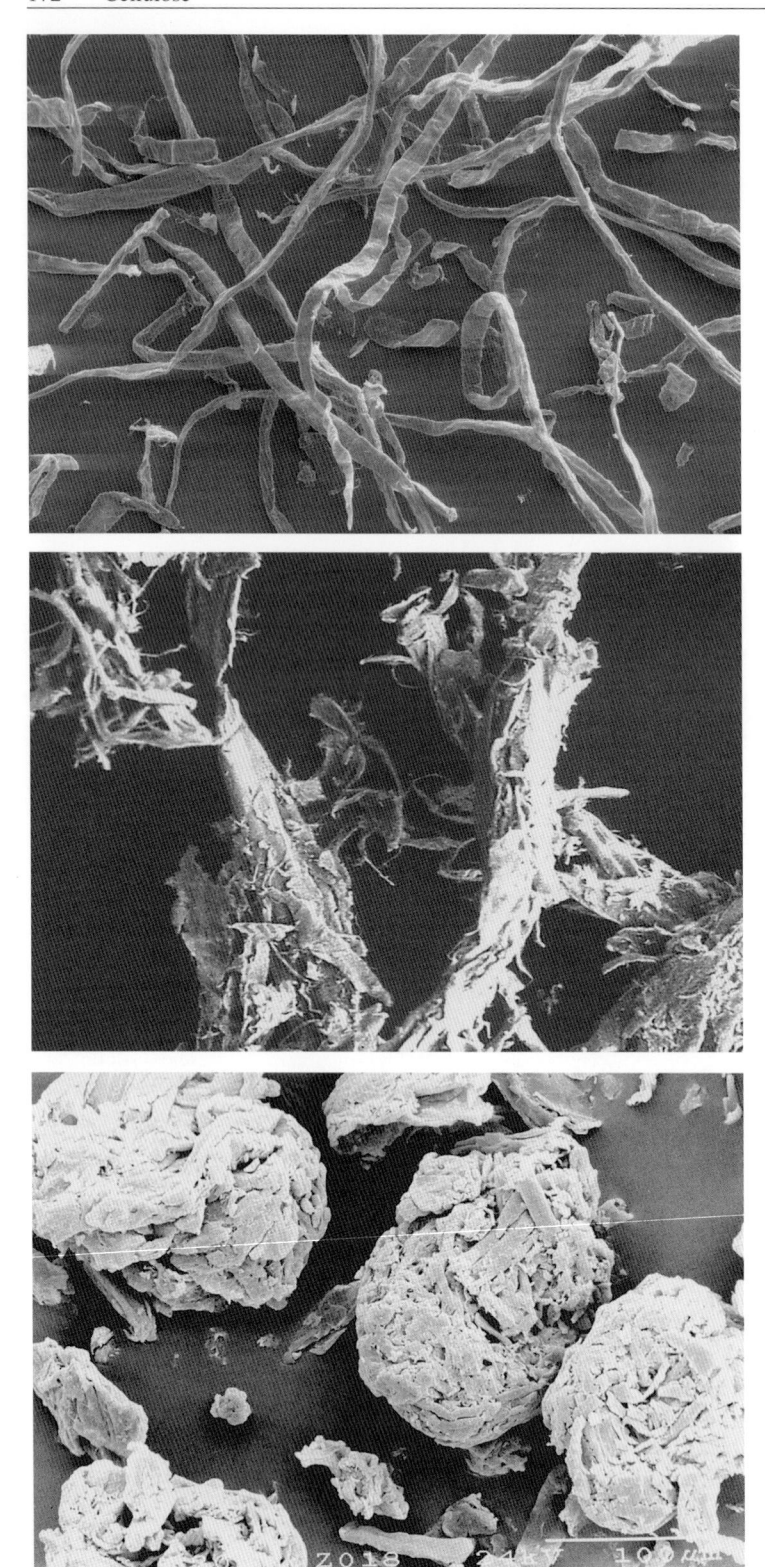

Figure 24. SEM images (×300) of powdered cellulose and microcrystalline cellulose Top: powdered cellulose; Middle: powdered cellulose, highly fibrillated; Bottom: microcrystalline cellulose

Table 26. Selected properties of some powdered celluloses (Arbocel) and a microcrystalline cellulose (Vivapur)

Parameter	Arbocel BE 600–10 TG	Arbocel BE 00	Arbocel BC 1000	Arbocel FIF 400	Vivapur 101
Fiber length, μm	18	120	700	2000	50
Fiber diameter, μm	15	20	20	35	50
Whiteness, %	85 ± 5	86 ± 5	86 ± 5	82 ± 5	86 ± 3
Shape	fiber	fiber	fiber	fiber	particle
Bulk density, g/L	230 – 300	150 – 180	30 – 45	10 – 25	320
Ash content, %	0.3	0.3	0.3	0.3	0.05
pH value	6 ± 1	6 ± 1	6 ± 1	6.5 ± 1	6.5 ± 1
Water-binding capacity, %	400	520	1200	1350	300

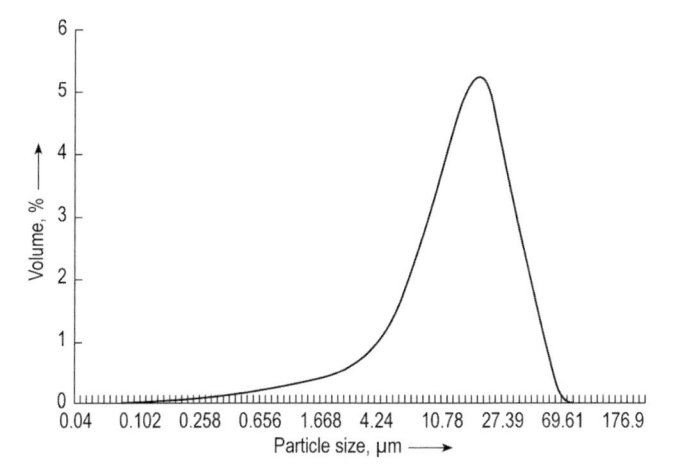

Figure 25. Particle-size distribution of a powdered cellulose (Arbocel BE 600-10 TG)

and thus speed up the transport of liquids. This transport step is not directionally fixed, but rather is determined by a concentration gradient. During drying water must be transported from the interior to the surface. Thus, for example, in the manufacture of cardboard articles, the drying time can be shortened and energy costs reduced. When manufacturing synthetic leather, cellulose is used to speed up the rinsing out of the organic solvents and thus optimize the precipitation process.

Cellulose fibers are used as a *filter aid*. They form a filter cake in suspensions with only a small amount of solids. Slimy solids form dense, compact filter cakes during separation which are difficult for the filtrate to penetrate. The fibers ensure a loose, porous filter cake with good permeability. Some examples of applications are filtration of titanium dioxide (sulfate process), beverages, dextrin (starch gum), and glucose.

In pharmaceuticals and in technical applications such as dishwasher detergent tablets, cellulose is used as a *tabletting aid*. Cellulose makes it possible to reduce the pressing power but maintain the same tablet hardness, increasing the running efficiency and output of the tabletting press.

Due to the high length-to-width ratio (up to 100:1) and the high degree of fibrillation, the fibers form a three-dimensional network. This reinforcing function of the fibers can be used to lend food products an improved texture. In the construction industry (e.g., cement tile adhesive, concrete spacers) the fibers provide improved structural stability.

References

General References

1. E. Treiber (ed.): *Die Chemie der Pflanzenzellwand*, Springer Verlag, Berlin-Göttingen-Heidelberg 1957.
2. A. Frey-Wyssling: *Die pflanzliche Zellwand*, Springer Verlag, Berlin-Göttingen-Heidelberg 1959.
3. S. A. Rydholm: *Pulping Processes*, Interscience Publ., New York 1965.

4 N. I. Nikitin: The Chemistry of Cellulose and Wood, Academy of Science of the USSR, Institute of High Molecular Compounds (Engl. Translation: Israel Program of Scientific Translation, Jerusalem 1966).
5 E. Ott, H. M. Spurlin: *Cellulose and Cellulose Derivatives*, vols. I – III, Interscience Publ., New York 1954 – 1955.
6 N. M. Bikales, L. Segal: *Cellulose and Cellulose Derivatives*, vols. IV and V, Wiley-Interscience, New York 1971.
7 J. C. Arthur (ed.): "Cellulose Chemistry and Technology," *ACS Symp. Ser.* 1977, no. 48.

Specific References

8 W. Sanderman, *Holz Roh Werkst.* **31** (1973) 11.
9 A. Payen: *Troisième mémoire sur le développement des Végétaux*, Extrait des Mémoires de l'Academie Royale des Sciences, Tome III des Savants Etrangères, Imprimerie Royale, Paris 1842.
10 W. N. Haworth, *Helv. Chim. Acta* **11** (1928) 534; *Ber. Dtsch. Chem. Ges. A* **65** (1932) 43.
11 H. Staudinger: *Die hochmolekularen organischen Verbindungen – Kautschuk und Cellulose*, Springer Verlag, 1932 (reprinted 1960).
12 A. J. Michell, H. G. Higgins, *Tetrahedron* **21** (1965) 1109.
13 O. Ellefsen, B. A. Tonnesen in N. M. Bikales, L. Segal (eds.): *Cellulose and Cellulose Derivatives*, Interscience, New York 1971, part IV, p. 151.
14 V. S. R. Rao, P. R. Sundararajan, C. Ramakrishnan, G. N. Ramachandran in G. N. Ramachandran (ed.): *Conformation of Biopolymers*, Academic Press, London-New York 1957, p. 721.
15 D. Fengel, G. Wegener: *Wood – Chemistry, Ultrastructure, Reactions*, Chap. 6.5, De Gruyter, Berlin-New York 1984, p. 167.
16 G. V. Schulz, E. Husemann, *Z. Naturforsch.* **1** (1946) 268.
17 K. Freudenberg, C. Blomquist, *Ber. Dtsch. Chem. Ges B* **68** (1935) 2070.
18 E. Treiber (ed.): *Die Chemie der Pflanzenzellwand*, Springer Verlag, Berlin-Göttingen-Heidelberg 1957, p. 142.
19 J. Schurz: "Theoretische Grundlagen der Viskoseverfahren," in K. Götze (ed.): *Chemiefasern*, vol. I, Springer Verlag, Berlin-Heidelberg-New York 1957.
20 V. P. Shanbhag, *Ark. Kemi* **29** (1968) 1.
21 H. Krässig, *Papier (Darmstadt)* **26** (1971) no. 12, 841.
22 E. Gruber: "Microgel Particles in Solutions of Cellulose and Cellulose Derivatives," *Cellul. Chem. Technol.* **13** (1979) 259.
23 K. H. Meyer, H. F. Mark, *Z. Phys. Chem. Abt. B* **2** (1929) 115.
24 K. H. Meyer, L. Misch, *Ber. Dtsch. Chem. Ges. B* **70** (1937) 266; *Helv. Chim. Acta* **20** (1937) 232.
25 A. Viswanathan, S. G. Shenouda, *J. Appl. Polym. Sci.* **11** (1967) 659.
26 R. H. Marchessault, C. Y. Liang, *J. Polym. Sci.* **43** (1960) 71.

27 J. Blackwell, F. J. Kolpak, K. H. Gardner: "Structure of Native and Regenerated Celluloses," in J. C. Arthur (ed.): "Cellulose Chemistry and Technology," *ACS Symp. Ser.* **1977**, no. 48, 42.
28 G. Honje, W. Watanabe, *Nature (London)* **181** (1958) 326.
29 K. C. Ellis, J. O. Warwicker, *J. Polym. Sci.* **56** (1962) 339.
30 K. H. Gardner, J. Blackwell, *Biopolymers* **13** (1974) 1975.
31 A. Viswanathan, S. G. Shenouda, *J. Appl. Polym. Sci.* **15** (1971) 519.
32 R. H. Atalla in T. E. Timell (ed.): "Proc. VIIth Cellulose Conference, Syracuse 1975," *Polym. Sci. Symp.* 1976, no. 28, 659.
33 S. Watanabe, J. Hayashi, T. Akahori, *J. Polym. Sci.* **12** (1974) 1065.
34 A. Nissan, according to D. W. Jones in N. M. Bikales, L. Segal (eds.): *Cellulose and Cellulose Derivatives*, Wiley-Interscience, New York-London-Sidney-Toronto 1971, p. 118.
35 R. S. J. Manley, *J. Polym. Sci. Polym. Phys. Ed.* **9** (1971) 1025.
36 K. Hess, H. Mahl, E. Gütter, *Kolloid Z.* **155** (1957) 1; **158** (1958) 115.
37 O. Kratky, G. Porod in H. A. Stuart (ed.): *Physik der Hochpolymeren*, vol. 3, Springer Verlag, Berlin-Göttingen-Heidelberg 1955.
38 A. Frey-Wyssling: *Submikroskopische Morphologie des Protoplasmas und seiner Derivate*, Gebr. Bornträger, Berlin 1938.
39 A. Frey-Wyssling: *Submicroscopic Morphology of Protoplasma*, Elsevier, Amsterdam 1949.
40 J. W. S. Hearle in J. W. S. Hearle, R. H. Peters (eds.): *Fiber Structure*, Butterworth – The Textile Institute, London-Manchester 1963.
41 H. Krässig, *Colloid Polym. Sci.* **259** (1981) 1.
42 H. Krässig, *Papier (Darmstadt)* **23** (1969) 881.
43 H. Krässig, *Appl. Polym. Symp.* 1976, no. 28, 777.
44 H. Staudinger, K.-H. In den Birken, M. Staudinger, *Makromol. Chem* **9** (1953) 148.
45 J. A. Howsman, W. A. Sisson in E. Ott, H. W. Spurlin, M. W. Grafflin (eds.): *Cellulose and Cellulose Derivatives*, Interscience Publ., New York-London 1954, part I, Chap. IVB, p. 326.
46 H. Krässig, *Tappi* **61** (1978) no. 3, 94.
47 L. Jörgensen: *Studies on the Partial Hydrolysis of Cellulose*, E. Moestue, Oslo 1950.
48 H. Krässig, *Papier (Darmstadt)* **38** (1984) 571.
49 H. Krässig, *Makromol. Chem.* **26** (1958) 17.
50 O. Samuelson: "Some Undesirable Carbohydrate Reactions During Alkaline Cooking and Bleaching," in: *The Ekman Days 1981*, vol. 2, Int. Symp. Wood Pulp Chem., Stockholm 1981, p. 78.
51 Y. Z. Lai: "Kinetics of Base Catalyzed Cleavage of Glucosidic Linkages," in: *The Ekman Days 1981*, vol. 2, Int. Symp. Wood Pulp Chem., Stockholm 1981, p. 26.
52 B. Lindberg, *Sven. Papperstidn.* **59** (1956) 531.

53 H. Esterbauer, M. Hayn, G. Jungschaffer, E. Taufratzhofer, J. Schurz: "Enzymatic Conversion of Lignocellulose Materials to Sugars," *J. Wood Chem. Technol.* **3** (1983) no. 3, 261.
54 I. Jullander, *Papier (Darmstadt)* **19** (1965) 166, 224.
55 H. Krässig, V. Stannett, *Fortschr. Hochpolym. Forsch.* **4** (1965) 111.
56 J. Schurz: "Chemical Combinations of Natural and Synthetic Polymers," in A. Varmavuori (ed.): *Proc. IUPAC-27th Int. Congress of Pure and Applied Chemistry,* Pergamon Press, Oxford-New York 1980, p. 307.
57 A. J. Turner, *J. Text. Inst.* **40** (1949) 973.
58 Food and Agriculture Organization: *1982 Handbook of Trade,* United Nations, Rome 1983.
59 P. Kassenbeck, *Text. Res. J.* **40** (1970) 330.
60 A. Frey-Wyssling, K. Muhlethaler, R. W. G. Wyckhoff, *Experientia* **4** (1948) 475.
61 H. U. H. Dolmetsch, *Text. Res. J.* **39** (1969) 568.
62 H. Zahn, "Latest Findings on the Micro-Structure of Cotton," *Internat. Cotton Conf.,* Bremen 1988, p. 7.
63 J. Sugiyama, H. Harada, Y. Fujiyoshi, N. Uyeda, *Planta* **166** (1985) 161 – 168.
64 A. A. Saafan, A. M. Habib, *Melliand Textilber.* **68** (1987) 687 – 680.
65 R. G. Steadman: *Cotton Testing,* Textile Inst., Manchester 1996, pp. 73.
66 W. R. Goynes, B. F. Ingber, B. A. Triplett, "Cotton Fiber Secondary Wall Development — Time versus Thickness," *Tex. Res. J.* **65** (1995) 489 – 494.
67 E. K. Boylston, "The Primary Wall of Cotton Fibers," *Tex. Res. J.* **65** (1995) 429 – 431.
68 R. G. Steadman: *Cotton Testing,* vol. 27 (1) Textile Inst., Manchester 1997, pp. 63.
69 United States Department of Agriculture, Agricultural Marketing Service (ed.): *The Classification of Cotton,* Agricultural Handbook no. 566, Memphis TN, 1993, pp. 24.
70 National Cotton Council: *Cotton Counts its Customers,* N.C.C., Memphis 1983. International Institute for Cotton: "Cotton," *Textiles* **11** (1982) 58 – 64.
71 T. L. Van Winkle, J. Edeleanu, E. A. Prosser, C. A. Walker: "Cotton versus Polyester," *Am. Sci.* **66** (1978) 280.
72 J. B. Cocke, R. A. Wesley, I. W. Kirk, *USDA Market Res. Rep.* **1065** (1977) 16 pp.
73 J. E. Ford: "Jute and Other Vegetable Fibers," *Textiles* **4** (1975) 58 – 62.
74 R. H. Kirby: *Vegetable Fibers,* Leonard Hill, London 1963.
75 R. Himmefarb: *Technology of Cordage Fibers and Rope,* Interscience, New York 1957.
76 M. L. Joseph: *Introductory Textile Science,* 4th ed., Holt, Rinehart & Winston, New York 1980.
77 *Textile Organon* **55** (1984) 62.
78 H. L. Roder, *Text. Ind. (Mönchen-Gladbach, Ger.)* **74** (1972) no. 22, 383.
79 Food and Agriculture Organization: *1982 Handbook of Production,* United Nations, Rome 1983.
80 U.S. Department of Labor, Occupational Safety and Health Administration, "Occupational Exposure to Cotton Dust," *Fed. Reg.* **43** (1978) no. 122, 27 349 – 27 418.
81 DIN 60 001, Textile Faserstoffe, August 1970.
82 Meeting report "Cellulosefaserforschung," Dornbirn 1984, *Lenzinger Ber.* 1985, no. 59.
83 W. Albrecht, M. Reintjes, B. Wulfhorst, Fiber Table Appendix 1997, *Chemical Fibers International* **47** (1997) 282.
84 A. Urbanowski, *CTJ* **46** (1996) 260.
85 W. Berger, *CTJ* **44** (1994) 747.
86 K. Götze: *Chemiefasern nach dem Viskoseverfahren,* 3rd ed., Springer Verlag, Berlin 1967.
87 H. F. Mark et al.: *Man-Made Fibres,* vol. 2, Interscience, New York 1968.
88 W. Albrecht, *Melliand Textilber. Int.* **51** (1970) 1487.
89 R. W. Moncrieff: *Man-Made Fibres,* Heywood Books, London 1970, p. 152.
90 C. V. Nikonovich et al., *J. Polym. Sci. Polym. Symp.* 1973, no. 42, 1625.
91 American Enka, US 4 145 523, 1979, US 4 196 282, 1980, US 4 246 221, 1981. Akzo Fibres, DE 30 34 685 C2, 1984.
92 R. Krüger, *CTJ* **44** (1994) 24 – 27.
93 H. Frigo, M. Eibl, D. Eichinger, Lenzinger Berichte (company brochure), vol. 75, 1996, pp. 47.
94 I. Marini, F. Brauneis, *Textilveredlung* **31** (1996) 182 – 187.
95 F. Brauneis, Lenzinger Berichte (company brochure), vol. 75, 1996, p. 105 – 111.
96 H. Nemec, Lenzinger Berichte (company brochure), vol. 9, 1994, pp. 69.
97 J. Chedin, A. Marsaudon, *Makromol. Chem.* **15** (1955) 115.
98 A. Dietl et al., *Papier (Darmstadt)* **20** (1966) 609.
99 J. O. Warwicker et al.: *A Review of the Literature on the Effect of Caustic Soda and Other Swelling Agents on the Fine Structure of Cotton,* Shirley Inst., Manchester 1966, Pamphlet no. 93, p. 75.
100 E. Treiber, *Faserforsch. Textiltech.* **22** (1971) 62.
101 E. Treiber in J. Schurz, *Papier (Darmstadt)* **20** (1966) 66.
102 P. Barthel, B. Philipp, *Faserforsch. Textiltech.* **18** (1967) 266.
103 J. J. Willard, E. Pacsu, *J. Am. Chem. Soc.* **82** (1960) 4350.
104 C. Y. Chen, R. E. Montana, C. S. Grove, *Tappi* **34** (1951) 420.
105 B. Philipp, Ke Tsing Liu, *Faserforsch. Textiltech.* **10** (1959) 555.
106 B. Philipp, *Faserforsch. Textiltech.* **8** (1957) 45.
107 A. Lyselius, O. Samuelson, *Sven. Papperstidn.* **64** (1961) 145.
108 M. E. Schwab, R. Kloss, lecture 23. Int. Chem. Fasertagung, Dornbirn 1984.
109 B. Philipp et al., *Faserforsch. Textiltech.* **21** (1970) 279.
110 A. T. Sorkov et al., *Khim. Volokna* 1971, no. 4, 32; Sowj. Beitrag *Faserforsch. Textiltech.* **8** (1971) 581.
111 H. Klare, A. Gröbe, *Faserforsch. Textiltech.* **8** (1957) 310; **9** (1958) 262; **10** (1959) 155.

112 A. Gröbe, H. J. Gensrich, *Faserforsch. Textiltech.* **21** (1970) 470.
113 P. H. Hermans: *Physics and Chemistry of Cellulose Fibres*, Elsevier, Amsterdam 1948, pp. 172, 365.
114 J. Schurz in [5], pp. 544 – 545.
115 A. Künschner, *Chemiefasern* **15** (1965) 662 – 668, 783 – 788.
116 W. Bandel in [5], p. 671.
117 A. Gröbe, B. Philipp, H. Klare, *Chemiefasern* **15** (1965) 502.
118 D. Vermaas, *Text. Res. J.* **32** (1962) 353.
119 P. H. Hermans: *Physics and Chemistry of Cellulose*, Elsevier, Amsterdam 1949.
120 R. J. E. Cumberbirch, *Rep. Prog. Appl. Chem.* **46** (1961) 233.
121 H. Klare, A. Gröbe, *Österr. Chem. Ztg.* **65** (1964) 218.
122 I. C. Witkamp, W. R. Saxton, *Tappi* **45** (1962) 650.
123 H. Hampe, B. Philipp, *Cellul. Chem. Technol.* **6** (1972) 447.
124 L. Szegö, *Faserforsch. Textiltech.* **21** (1970) 422.
125 V. Jančařik, L. Kuniak, *Faserforsch. Textiltech.* **20** (1969) 491.
126 H. Krässig, *Papier (Darmstadt)* **24** (1970) 926.
127 V. Stannett, H. Hopfenberg, *High Polym.* **5** (1971) 907.
128 S. A. Rogowin, *Lenzinger Ber.* 1970, no. 30, 16.
129 Z. A. Rogowin, L. S. Galbraich: *Die chemische Behandlung und Modifizierung der Zellulose*, Thieme Verlag, Stuttgart 1983.
130 M. LeGall, *Bull. Sci. Inst. Text. Fr.* **2** (1973) no. 6, 77.
131 L. Wiesner, *Melliand Textilber.* **49** (1968) 99.
132 A. S. Hoffmann, *Isot. Radiat. Technol.* **8** (1970) 84.
133 A. Heger, *Dtsch. Textiltech.* **17** (1967) 307; *Textiltechnik (Leipzig)* **23** (1973) 665.
134 W. Bobeth et al.: *Faserforsch. Textiltech.* **24** (1973) 412; *Lenzinger Ber.* 1975, no. 38.
135 Surface Activation Corp. (SAC), US 3 600 122, 1966.
136 H. Dolmetsch, *Melliand Textilber.* **45** (1964) 12.
137 H. Dolmetsch, *Melliand Textilber.* **51** (1970) 182.
138 H. Mark, *Chemiefasern* **15** (1965) 422.
139 D. Paul, D. Bartsch, *Faserforsch. Textiltech.* **23** (1972) 187.
140 P. Weber, *Melliand Textilber.* **50** (1969) 372.
141 W. Albrecht, *Chemiefasern* **63** (1982) 790.
142 W. Bandel: "Entwicklungen auf dem Cuprophan-, Membran- und Hohlfasergebiet", *Mitt. Klin. Nephrologie* **4** (1975) 29.
143 G. Seyfart, W. Henne, G. Marx, P. Schroeder, H. J. Gurland: "Die Kapillarniere", *Biomedizinische Technik* **19** (1974) 174.
144 W. Henne, G. Dünweg, W. Bandel: "A New Cellulose Membrane Generation for Hemodialysis and Hemofiltration," Proc. 2nd Meeting Int. Soc. Artif. Org. (ISAO) 3 (Suppl.) (1979) 466.
145 N. A. Hoenich, T. Frost, D. N. S. Kerr in W. Drukker, F. M. Parsons, J. F. Maher (eds.): "*Replacement of Renal Function by Dialysis,*" Martinus Nijhoff Publ., Den Haag-Boston-London 1979, p. 80.
146 T. Malkomes, A. Reichle in Ullmann, 3rd ed., vol. 11, pp. 260 – 278 (published 1960, numerous references).
147 R. Jährling: *Die Herstellung der Zellwolle und Kunstseide*, Fachbuchverlag, Leipzig 1957.
148 Z. A. Rogowin in: *Chemiefasern*, Thieme Verlag, Stuttgart 1982, pp. 172 – 181.
149 A. Reichle, *Reyon Zellwolle Andere Chem. Fasern* **32** (1954) 133 – 139.
150 H. A. Schlichter: "Untersuchungen über Gleichgewichtsverhältnisse im System Cuoxam-NaOH-Cellulose," Thesis, TH Aachen 1956.
151 Bayer, DE 961 287, 1957.
152 Asahi Kasei Kogyo KK, DE 2 328 583, 1973.
153 Bayer, DE 1 099 690, 1961.
154 Asahi Kasei Kogyo KK, US 3 049 755, 1960; US 3 131 429, 1961; DE 1 660 144, 1965; DE 2 059 177, 1970.
155 American Bemberg Corp., GB 633 108, 1946.
156 Asahi Kasei Kogyo KK, DE 242 964, 1972.
157 Bayer, DE 886 770, 1953.
158 Bayer, FR 1 116 671, 1954.
159 Fiat Final Report 50.
160 CIRFS statistics 1996.

Further Reading

S. S. Cho, P. Samuel (eds.): *Fiber Ingredients - Food Applications and Health Benefits*, CRC Press, Boca Raton, FL 2009.

A. D. French, N. R. Bertoniere, R. M. Brown, H. Chanzy, D. Gray, K. Hattori, W. Glasser: *Cellulose*, "Kirk Othmer Encyclopedia of Chemical Technology", 5th edition, vol. 5, p. 360–394, John Wiley & Sons, Hoboken, NJ, 2004, online: DOI: 10.1002/0471238961.0305121206180514.a01.pub2.

A. Imeson (ed.): *Food Stabilisers, Thickeners and Gelling Agents*, Wiley-Blackwell, Chichester 2010.

A. Lejeune, T. Deprez: *Cellulose - Structure and Properties, Derivatives and Industrial Uses*, Nova Science Publishers, Hauppauge, NY 2009.

A. K. Mohanty, M. Misra, L. T. Drzal (eds.): *Natural Fibers, Biopolymers, and Biocomposites*, CRC Press, Boca Raton, FL 2005.

J. Müssig (ed.): *Industrial Applications of Natural Fibres*, Wiley, Chichester 2010.

H. Sixta (ed.): *Handbook of Pulp*, Wiley-VCH, Weinheim 2006.

A. Steinbüchel (ed.): *Biopolymers*, Wiley-VCH, Weinheim 2006.

J. Stewart, D. Oosterhuis, J. Heitholt, J. Mauney (eds.): *Physiology of Cotton*, Springer, Berlin 2010.

P. Zugenmaier: *Crystalline Cellulose and Derivatives*, Springer, Berlin 2008.

Cellulose Esters

KLAUS BALSER, Wolff Walsrode AG, Walsrode, Federal Republic of Germany

LUTZ HOPPE, Wolff Walsrode AG, Walsrode, Federal Republic of Germany

THEO EICHER, Stuttgart, Federal Republic of Germany

MARTIN WANDEL, Bayer AG, Leverkusen, Federal Republic of Germany

HANS-JOACHIM ASTHEIMER, Rhodia AG, Freiburg, Federal Republic of Germany

HANS STEINMEIER, Rhodia Acetow AG, Freiburg, Federal Republic of Germany

JOHN M. ALLEN, Eastman Chemical Company, Kingsport, TN 37662, USA

1.	Inorganic Cellulose Esters	177	2.1.3.3.	Hydrolysis	202
1.2.	Esterification	179	2.1.3.4.	Precipitation and Processing	203
1.3.	Cellulose Nitrate	180	2.1.4.	Recovery of Reactants	204
1.3.1.	Physical Properties	180	2.1.5.	Properties	204
1.3.2.	Chemical Properties	181	2.1.6.	Analysis and Quality Control	205
1.3.3.	Raw Materials	183	2.1.7.	Uses	206
1.3.4.	Production	184	2.2.	Cellulose Mixed Esters	206
1.3.4.1.	Cellulose Preparation	185	2.2.1.	Production	207
1.3.4.2.	Nitration	186	2.2.2.	Composition	207
1.3.4.3.	Stabilization and Viscosity Adjustment	187	2.2.3.	Properties	207
1.3.4.4.	Displacement and Gelatinization	188	2.2.4.	Other Organic Mixed Esters	208
1.3.4.5.	Acid Disposal and Environmental Problems	188	2.2.5.	Uses	208
1.3.4.6.	Other Nitrating Systems	188	2.3.	Cellulose Acetate Fibers	209
1.3.5.	Commercial Types and Grades	190	2.3.1.	Properties	209
1.3.6.	Analysis and Quality Control	190	2.3.2.	Raw Materials	210
1.3.7.	Uses	192	2.3.3.	Production	210
1.3.8.	Legal Provisions	194	2.3.4.	Uses	211
1.4.	Other Inorganic Cellulose Esters	194	2.3.5.	Economic Aspects	211
1.4.1.	Cellulose Sulfates	194	2.4.	Cellulose Ester Molding Compounds	211
1.4.2.	Cellulose Phosphate and Cellulose Phosphite	195	2.4.1.	Physical Properties of Cellulose Ester Plastics	212
1.4.3.	Cellulose Halogenides	196	2.4.2.	Polymer Modified Cellulose Mixed Esters	214
1.4.4.	Cellulose Borates	196	2.4.3.	Chemical Properties	215
1.4.5.	Cellulose Titanate	196	2.4.4.	Raw Materials	215
1.4.6.	Cellulose Nitrite	196	2.4.5.	Production	217
1.4.7.	Cellulose Xanthate	197	2.4.6.	Trade Names	218
2.	Organic Esters	197	2.4.7.	Quality Requirements and Quality Testing	218
2.1.	Cellulose Acetate	198	2.4.8.	Storage and Transportation	218
2.1.1.	Chemistry	198	2.4.9.	Uses	218
2.1.2.	Raw Materials	199	2.4.10.	Toxicology and Occupational Health	220
2.1.3.	Industrial Processes	199		References	220
2.1.3.1.	Pretreatment	200			
2.1.3.2.	Esterification	201			

1. Inorganic Cellulose Esters

Definition. Cellulose esters are cellulose derivatives which result by the esterification of the free hydroxyl groups of the cellulose with one or more acids, whereby cellulose reacts as a trivalent polymeric alcohol. Esterification can be carried out by using mineral acids as well as organic acids or their anhydrides with the aid of dehydrating substances. Cellulose nitrate [9004-70-0] is the most important and only industrially produced inorganic cellulose ester (abbreviation

CN, according to DIN 7728, T 1, 1978). A comprehensive bibliography on inorganic cellulose esters may be found in [12–19].

Historical Aspects [20]. The nitric acid ester of cellulose is the oldest known cellulose derivative and is still the most important inorganic cellulose ester. The term "nitrocellulose" is still used, but it is not the precise scientific term for cellulose nitrate. Cellulose esters were first described and industrially used at a time when the structure of esters was unknown and information on the polymeric primary material cellulose was not yet available.

The nitration of polysaccharides with concentrated nitric acid had already been described in 1832. H. BRACONNOT obtained a white and easily inflammable powder when he transformed starch with nitric acid. The product obtained was xyloidine. TH.-J. PELOUZE treated paper with nitric acid and obtained an insoluble product containing ca. 6 % nitrogen which he called pyroxiline and which he provided for military use.

C. F. SCHÖNBEIN and R. BÖTTGER are considered to be the inventors of so-called gun cotton (1845). They transformed cotton with a mixture of nitric and sulfuric acid into a highly nitrated product that could serve as a substitute for black powder. Production on an industrial scale was stopped in 1847 because its extremely rapid catalytic decomposition was the cause of numerous plant explosions. Production was legally prohibited in 1865.

The use of cellulose nitrate as an explosive brought new momentum to its further industrial and scientific development, as well as to its economic significance. F. ABEL made a basic breakthrough in 1865 when he succeeded in developing a safe method of handling. He was able to achieve a better washing of the adhering nitrating acid and a hydrolytic decomposition of the unstable sulfuric acid ester by grinding the nitrated fibers in water. This process allowed this product to attain military importance for its use as gunpowder.

In 1875, A. NOBEL phlegmatized nitroglycerine by mixing with cellulose nitrate and discovered blasting gelatin. In the 1880s, smokeless gunpowders were developed. VIEILLE developed Poudre B (blanche) and NOBEL developed Ballistit, the first dibasic gunpowder from cellulose nitrate and nitrogycerol. ABEL and DEWAR developed a similar gunpowder called Cordit.

The discovery that fibrous products could be modified by, for example, dissolution in an alcohol/ether mixture (film for wound protection) or by gelatinization with softeners brought additional uses. Films made from camphor and castor oil were used in collodium photography as carriers for light-sensitive materials (ARCHER, 1851). Nitrofilms found increasing use in photography and cinematography until they were replaced by nonflammable cellulose acetate films.

The year 1869 is considered to be the beginning of the age of plastics. J. W. HYATT discovered celluloid, the first thermoplastic synthetic material. It was originally used as a substitute for ivory in the production of billiard balls.

The practical use of cellulose nitrate as a raw material for lacquers began in 1882, when STEVENS suggested amyl acetate as a highly volatile solvent (Zaponlack). The nitro lacquers achieved importance at first after World War I (FLAHERTY, 1921), when new applications were being sought as a result of the sharply declining demand for gunpowder. Only after the possibility of depolymerization of cellulose nitrate by pressure boiling during production had become known during the 1930s it was possible to use cellulose nitrate in protective and pigmented lacquers. Thus, it became possible to use nitro lacquers for painting automobiles on the assembly line.

Cellulose xanthate [9032-37-5] is a cellulose ester obtained with the inorganic acid dithiocarbonic acid. CH. F. CROSS discovered this important alkali-soluble cellulose ester in 1891 while he was reacting cellulose with alkali and carbon disulfide. It represents the base of the viscose process introduced in 1894 by BEVAN and BEADLE for producing man-made cellulosic fibers (rayon, rayon staple) and cellophane.

Other cellulose esters with inorganic acids are presently only of theoretical interest and have not attained any industrial or economic importance.

Present Significance. Cellulose nitrate is still important, 150 years after its discovery. It is industrially produced in large quantities for diversified applications. The reasons for this are the relatively simple production process with high yields, its solubility in organic solvent systems and its excellent film-forming properties from such solutions (collodion cotton as a raw material for lacquers), compatibility and gelatinability with softeners and other polymers (thermoplastics), as well as inflammability (gun cotton for explosives). Cellulose nitrate has maintained its importance as a raw material for the manufacture of protective and coating lacquers as well as blasting agents and explosives.

Densified products, colored or pigmented chips kneaded with softeners, as well as aqueous dispersion systems with a low solvent content, are available today. They facilitate transport and processing, secure existing application forms, open up new ones, and are becoming increasingly nonpolluting.

The viscose process (see → Cellulose) with its essential intermediate cellulose xanthate will remain, because of the availability of a constantly regrowing raw material supply, an important source of textile fibers for years to come. Alternative processes are intensively being sought to reduce pollution by sulfurous decomposition gases resulting during manufacturing. Cellulose nitrite, the cellulose ester of nitrous acid, is the most prominent example.

Cellulose esters with other inorganic acids have been frequently described and investigated. Cellulose sulfate [9032-43-3] was of some interest because of its solubility in water, but never achieved any practical importance. Cellulose phosphate [9015-14-9], borate, and titanate show interesting properties such as fire retardation, but are not yet of any industrial significance.

1.2. Esterification

Mechanism. The alcoholic hydroxyl groups of cellulose are polar and can be substituted by nucleophilic groups in strongly acid solutions. The mechanism of esterification assumes the formation of a cellulose oxonium ion followed by the nucleophilic substitution of an acid residue and the splitting off of water. Esterification is in equilibrium with the reverse reaction; saponification can be inhibited largely by binding the resulting water.

$$\text{Cell–OH} + \text{H}^+ \longrightarrow \text{Cell–O}^+\!\!\begin{array}{c}\text{H}\\\text{H}\end{array}$$

$$\text{X}^- + \text{Cell–O}^+\!\!\begin{array}{c}\text{H}\\\text{H}\end{array} \longrightarrow \left[\text{X}\!\rightarrow\!\text{Cell}\!\rightarrow\!\text{O}^+\!\!\begin{array}{c}\text{H}\\\text{H}\end{array}\right] \longrightarrow \text{X–Cell} + \text{O}\!\!\begin{array}{c}\text{H}\\\text{H}\end{array}$$

Course of Reaction. The three functional hydroxyl groups on each anhydroglucose unit of cellulose are blocked by intermolecular and intramolecular hydrogen bonds and, therefore, are not freely accessible for the reaction partners. The supermolecular arrangement and microstructure within the cellulose fiber, whose intensity depends on the origin and previous history of the cellulose material, is determined by these hydrogen bonds. The accessibility to the reaction partners and the reactivity of the alcohol groups also depend on this structure.

Due to the fact that cellulose is insoluble in all common solvents, reactions to form derivatives are usually carried out in heterogeneous systems. As the reaction proceeds, new reactive centers are created so that ultimately almost all parts of the cellulose fibers are included and in special cases yield soluble derivatives which react to completion in a homogeneous phase.

Little information is available on the esterification process. The following two reaction types are under discussion:

- An intermicellar reaction, which initially consists of the penetration of the reaction partner into the so-called amorphous regions between the highly organized cellulose micelles and proceeds during the course of esterification from the surface to the innermost regions of the micelles. The reaction speed is determined by diffusion.
- An intramicellar or permutoid reaction, in which the reagent penetrates all regions including the micelles so that practically all cellulose molecules react almost simultaneously. The reaction speed is specified by adjustment of the esterification equilibrium.

Arguments for both mechanisms are based on X-ray analyses. The possibility exists that both reaction types occur and ultimately merge. This depends on the reaction conditions, especially the esterification mixture and the temperature.

The hydrogen bonds between the cellulose molecules are almost completely broken down during esterification. The introduction of ester groups separates the cellulose chains so completely that the fiber structure is either altered or completely destroyed. Whether the cellulose ester is soluble in a solvent or in water depends on the types of substituents added.

Substitution. The esterification reactions do not necessarily proceed stoichiometrically because of equilibrium adjustment. The maximal attainable substitution with a mean degree of substitution (DS) of 3 is generally not reached. A triester can only be obtained under carefully controlled conditions. The primary hydroxyl group on the C-6 atom reacts most readily, while the neighboring hydroxyl groups on the C-2 and C-3 atoms of the anhydroglucose ring react considerably slower due to steric hindrance.

Basically, esterification is possible with all inorganic acids. Limiting factors are the type and the size of the acid residue as well as the varying degree of acid-catalyzed hydrolysis, which can lead to a complete cleavage of the cellulose molecule as the result of statistical chain splitting.

1.3. Cellulose Nitrate

Summary monographs on cellulose nitrate in addition to those in the Reference list can be found in [21] and [22].

1.3.1. Physical Properties

Cellulose nitrate (CN) is a white, odorless, and tasteless substance. Its characteristics are dependent on the degree of substitution.

Density The density of cellulose nitrate is dependent on its nitrogen content and, therefore, on the degree of substitution (Table 1).

The bulk density of commercially available CN types is between 0.25 and 0.60 kg/L for moistened CN cotton, 0.15–0.40 kg/L when converted to dry mass.

Cellulose nitrate chips, which contain at least 18 % dibutyl phthalate in addition to cellulose nitrate, have a density of 1.45 g/cm^3 (measured at 20 °C in an air-comparison pycnometer). The bulk density is 0.3–0.65 kg/L.

Table 1. The density of cellulose nitrate in relation to its nitrogen content (degree of substitution)

Nitrogen content, %	Degree of substitution (DS)	Density at 20 °C, g/cm^3
11.5	2.1	1.54
12.6	2.45	1.65
13.3	2.7	1.71

Table 2. Thermodynamic properties of some cellulose nitrates

Heat of formation	trinitrate	− 2.19 kJ/g
	dinitrate	− 2.99 kJ/g
	cellulose	− 5.95 kJ/g
Heat of combustion	trinitrate	− 9.13 kJ/g
	dinitrate	− 10.91 kJ/g
	cellulose	− 17.43 kJ/g
Specific heat	celluloid film (70 % CN and 30 % camphor)	1.26 – 1.76 J g^{-1} K^{-1}
Thermal conductivity	celluloid film (70 % CN and 30 % camphor)	0.84 kJ m^{-1} h^{-1} K^{-1}
Heat of solution in acetone	CN with 11.5 % N content	− 73.25 J/g
	CN with 14.0 % N content	− 81.64 J/g

Specific Surface The laboratory apparatus described by S. Rossin [23] is best suited for the determination of the specific surface of cellulose nitrate, which is 1 850–4 700 cm^2/g, depending on the fineness of the cellulose nitrate.

The determination of the inner surface according to the BET method showed dependence on the molar mass (i.e., an inner surface area of 1.44 m^2/g would correspond to a molar mass of 180 000 g and a surface area of 2.41 m^2/g would correspond to a molar mass of 400 000 g).

It must, however, be noted that the degassing temperature was lowered from the usual 200 °C to 60 °C due to the fact that cellulose nitrate deflagrates at 180 °C. It is possible that complete desorption did not take place under these conditions.

Thermodynamic Properties see [24, pp. 137–154]. The most important thermodynamic properties are listed in Table 2.

Electrical Properties [24, p. 136]. The following electrical properties were measured on cellulose nitrate containing 30 wt % camphor (celluloid):

Dielectric constant	
at 50–60 Hz	7.0–7.5
10^6 Hz	6.0–6.5
Dissipation factor (tan δ)	
at 50–60 Hz	0.09–0.12
10^6 Hz	0.06–0.09
Specific resistance	10^{11}–10^{12} Ω · cm

Table 3. Mechanical properties of CN lacquer films

Type[a]	Elongation, %	Tensile strength, N/mm^2
E 4	24 – 30	98 – 103
E 6	23 – 28	98 – 103
E 9	23 – 28	88 – 98
E 13	20 – 25	88 – 98
E 15	18 – 23	78 – 98
E 21	12 – 18	78 – 88
E 22	10 – 15	74 – 84
E 24	8 – 12	69 – 78
E 27	5 – 10	59 – 69
E 32	<5	39 – 49
E 34	<3	29 – 49

[a] According to DIN 53179: The E-type designation specifies the CN concentration (% in dry condition) in acetone which gives a viscosity of 400 ± 25 mPa·s.

Mechanical Properties [25]. The stress–strain diagram of cellulose nitrate films shows the elongation and tensile strength to be dependent on the size of the molecule (expressed as a term of viscosity).

The higher the molecular mass of a CN, the more elastic is the film made from it. Films become more brittle and their tensile strength declines with decreasing molecular mass (see Table 3).

Optical Properties Cellulose nitrate films are optically anisotropic because of their microcrystalline structure. The colors change in polarized light in relation to the nitrogen content of the CN:

11.4 % N   weakly red
11.5 – 11.8 % N   yellow
12.0 – 12.6 % N   blue to green

The index of refraction is 1.51, and the maximal light transmission is achieved at 313 nm.

Light Stability Exposure to sunlight, and especially to ultraviolet light, has a detrimental effect on cellulose nitrate film by causing it to become yellowish and brittle. Solvents, softeners, and resins can either promote or hinder yellowing.

1.3.2. Chemical Properties

The three hydroxyl groups of cellulose can be completely or partially esterified by nitrating acid. The varying degrees of nitration can be related to the following theoretical nitrogen contents:

Cellulose mononitrate, $C_6H_7O_2(OH)_2(ONO_2)$· 6.75 % N
Cellulose dinitrate, $C_6H_7O_2(OH)(ONO_2)_2$· 11.11 % N
Cellulose trinitrate, $C_6H_7O_2(ONO_2)_3$· 14.14 % N

Cellulose nitrate with a nitrogen content between 10.8 and 12.6 % is a suitable raw material for lacquers, and CN with > 12.3 % N is suitable for explosives exclusively.

Degree of Substitution – Nitrogen Content – Solubility The degree of substitution can be calculated from the nitrogen content of the various CN types (Fig. 1). The degree of substitution determines the solubility of cellulose nitrate in organic solvents. CN for lacquers can be classified according to its solubility in organic solvents as follows:

alcohol-soluble CN (A types)
   nitrogen content: approx. 10.9 – 11.3 %
   readily soluble in alcohols, esters, and ketones

moderately soluble CN (AM types)
   nitrogen content: approx. 11.4 – 11.7 %
   soluble in esters, ketones, and glycol ethers with excellent blendability or compatibility with alcohol

CN soluble in esters (E types)
   nitrogen content: 11.8 – 12.2 % for lacquer cotton, up to 13.7 % for guncotton

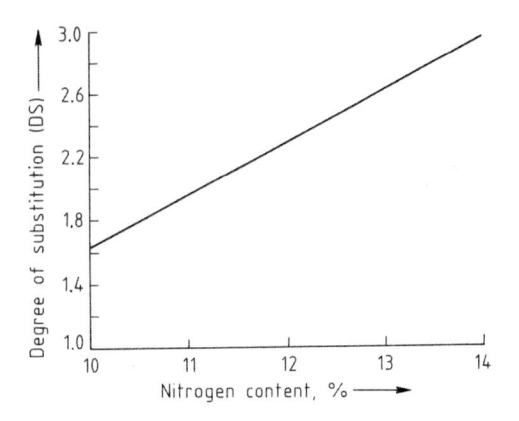

Figure 1. Variation of the degree of substitution with the nitrogen content of cellulose nitrate

readily soluble in esters, ketones, and glycol ethers

Intrinsic Viscosity – Degree of Polymerization [24, pp. 85–121] The mean number of anhydroglucose units in cellulose nitrate molecules is designated as the mean degree of polymerization (DP). The viscosity of the solution (at the same concentration in the same solvent) is generally considered to be a relative measure of the molecular mass. The molecular mass can be mathematically expressed as a function of the intrinsic viscosity (Staudinger – Mark – Houwink equation). For further information → Plastics, Analysis.

Distribution of the Molecular Mass The starting material of cellulose nitrate is natural cellulose, the quality of which is subjected to annual growth cycles. It is, therefore, of great importance to have polymolecular data, such as the mean degree of polymerization and the distribution of the molecular mass, available in addition to viscosity, solubility behavior, and nitrogen values. These values are important, for example, in assessing the mechanical properties and aging processes of polymer products.

The isolation of the polymers according to their molecular mass can be achieved elegantly by gel permeation chromatography (GPC).

Chemical Compatibility An everyday use of cellulose nitrate is in nitro lacquers, where it is dissolved in organic solvents. In this solution, cellulose nitrate is extremely compatible with essential substances in the lacquer formulation such as alkyd resins, maleic resins, ketone resins, urea resins, and polyacrylates. A large number of softeners, such as adipates, phthalates, phosphates, and raw and saturated vegetable oils are compatible with cellulose nitrate.

Chemical and Thermal Stability Cellulose nitrate, as a solid or in solution, should not be brought into contact with strong acids (degradation), bases (denitration), or organic amines (decomposition) since they all induce a destruction of cellulose nitrate. This may proceed very rapidly and lead to deflagration of the cellulose nitrate.

The ester bonds of cellulose nitrate, which can be broken by saponifying agents or by catalysis, are responsible for its physicochemical instability. This substance-specific property is dependent on the temperature, the specimen, and whether catalytically active decomposition products remain or are removed from the sample.

Another basic instability of cellulose nitrate is observed during the production process. Mixed sulfuric acid esters transmit a chemical instability to the nitrocellulose molecule. These mixed esters are destroyed in weakly acid water during the stabilization phase of production. The long reaction time required by this procedure can be considerably shortened by increasing the reaction temperature. The time required can be reduced to only a few hours by raising the temperature to 60–110 °C. Under these conditions the nitrate ester remains stable; the glucosidic bond of cellulose nitrate, on the other hand, is attacked. This property is used to advantage to specifically reduce the degree of polymerization of the cellulose nitrate.

Thermogravimetry, IR spectroscopy, and electron spectroscopy (ESCA) [26], [27] have been used to determine the extent of thermally induced and light-induced decomposition of cellulose nitrate. The reaction proceeds as follows:

$$Cell-O-NO_2 \rightarrow Cell-O\cdot + \cdot NO_2$$

It is proceeded by a series of extremely exothermic oxidation reactions triggered by the NO_2 radical, which often leads to spontaneous deflagration. NO_2 is reduced to NO and in the presence of air NO_2 is reformed, thus initiating an autocatalytic chain reaction, at the end of which the gaseous reaction products CO_x, NO_x, N_2, H_2O, and HCHO are found.

By adding stabilizers such as weak organic bases (diphenylamine) or acids (phosphoric acid, citric acid, or tartaric acid), intermediary nitric oxides can be bound and the autocatalytic decomposition prevented.

Thermal decomposition does not occur at temperatures below 100 °C. The temperature (according to [28]) at which cellulose nitrate spontaneously deflagrates is used as a measure of its thermal stability. A well-stabilized lacquer cotton has a deflagration temperature of ≥ 180 °C. The deflagration temperature of plasticized cellulose nitrate chips with at least 18 wt % softener (i.e., dibutyl or dioctyl phthalate) is ≥ 170 °C.

The Bergmann–Junk test [29] and the warm storage test are additional methods for determining the stability of cellulose nitrate.

1.3.3. Raw Materials

Cellulose. Until the beginning of World War I, the only raw material available for nitration was cellulose obtained from cotton in the form of bleached linters (as flakes or crape). This was due to the high degree of purity (α-cellulose > 98 %), which allowed a high yield and products with good clarity and little yellowing.

Especially in times when linters were scarce, it was possible to produce gunpowder from wood celluloses, even unbleached, other cellulose fibers (annual plants), and even from wood if attention was given to the adequate disintegration of the raw materials. Lacquer types obtained from wood celluloses, especially from hardwood, gave dull and mat films and lacquers with inferior mechanical properties. This is due to the high content of pentosans, which is also nitrated but is easily split by hydrolysis in conventional nitrating acid systems and thus becomes insoluble.

The development of highly purified chemical-grade wood pulp by refinement with hot and cold alkali having R_{18} values of 92 – 95 % (see Table 16) allows this type of raw material to be used in the same manner as were linters, which currently are used only for the production of special and highly viscous CN types. The highly refined prehydrolyzed sulfate pulps with R_{18} values of above 96 % are especially well-suited for nitration. The viscosity range of CN products can be adjusted in advance by choosing an initial cellulose with an adequate DP. A low ash content, and above all a low calcium content, of the cellulose is important in preventing calcium sulfate precipitation during industrial nitration.

A comparative study on the nitrating behavior of linters and wood pulps [30] shows the morphological factors of the fibers (fiber length and distribution, cross-section form and thickness of the secondary wall, and fine structure including packing density, degree of crystallization, and lateral arrangement of the fibrils), the chemical composition of the cellulose (DP and polydispersibility) as well as the type, quantity, and topographic distribution of the accompanying hemicelluloses and lignin to be responsible for the nitrating capability of celluloses. These factors determine the swelling properties and thereby the uniformity of nitration, as well as the compressibility and the relaxation capacity of the fibers, which in turn influence the retention capacity of the fiber mass. Linters with a lower acid retention capacity of 110 – 130 % are definitely superior to wood pulp (acid retention capacity of prehydrolyzed sulfate pulps up to 230 %, of sulfite pulps up to 300 %) in this respect. The suitability of a raw material for nitration can be tested by a specially developed machine that measures the compression and relaxation characteristics of a cellulose fiber pile.

Approximately 150 000 t (3.4 %) of the annual worldwide production of 4.4×10^6 t of chemical-grade pulps are used for the production of cellulose nitrate.

Industrial Nitrating Agents. The so-called nitrating acid as developed by SCHöNBEIN, the nitric acid/sulfuric acid/water system, is still the nitrating agent of choice for industrial purposes. The highest attainable degree of substitution using this system is at DS 2.7 ≙ 13.4 % N. This is achieved only when the nitric acid used is not hydrated and the molar ratio of nitric acid to sulfuric acid monohydrate is 1 : 2. The optimal nitrating mixture is as follows:

	HNO_3	H_2SO_4	H_2O
molar ratio	1	2	2
wt %	21.36	66.44	12.20

Water plays a special role as far as the attainable degree of nitration is concerned. Below 12 % water there is no increase in substitution, but a higher water content results in a drastic decline in the degree of nitration (Fig. 2).

It is assumed that increasingly hydrated nitric acid causes increased swelling and gelatinization of the cellulose so that the nitrating acid is no longer able to penetrate into the inner structures of the micelles.

The desired degrees of esterification can be adjusted by varying the nitrating acid mixture according to the CN types (Table 4), whereby in

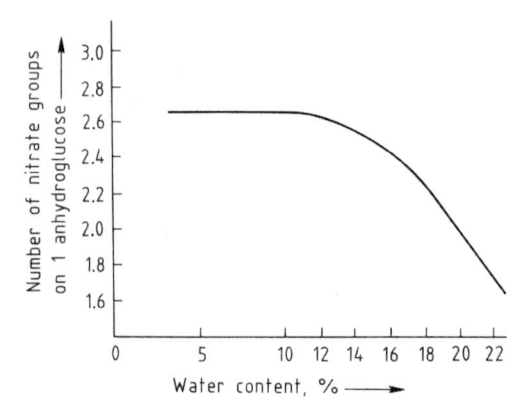

Figure 2. Dependence of the degree of esterification (DS) on the water content of the optimal nitrating mixture (HNO$_3$: H$_2$SO$_4$ = 1 : 2)

industrial processes the nitric acid content is kept nearly constant at 25 – 26 %.

The ternary system HNO$_3$/H$_2$SO$_4$/H$_2$O has been extensively investigated. The results are summarized in Figure 3.

The curves of the same degrees of substitution (% N) in relation to the nitrating acid composition are presented here. The cross-hatched band identifies those areas in which the cellulose material is extremely swollen and gelatinized. Three zones can therefore be differentiated in the phase diagram:

1. Area of technical nitration:

   nitric acid   15 – 100 %
   sulfuric acid 0 – 80 %
   water   0 – 20 %

   In this range, nitric acid is present in a non-hydrated form and induces true nitration (N content 10 %). The industrially used range with 20 – 30 % HNO$_3$, 55 – 65 % H$_2$SO$_4$, and 8 – 20 % water is also included in this area.

2. Area of solution:

   nitric acid   0 – 10 %
   sulfuric acid 60 – 100 %
   water   0 – 40 %

   Little or no nitration takes place in this range. Cellulose is degraded to the point of complete dissolution in concentrated sulfuric acid.

3. Area of swelling: Nitric acid is increasingly hydrated in this range of increasing water content. Nitration decreases rapidly.

A process developed in the United States, but less important, uses magnesium nitrate instead of sulfuric acid as a dehydrating agent [31]. Magnesium nitrate can bind water as its hexahydrate. The nitrating mixture consists of 45 – 94% nitric acid, 3.3 – 34% magnesium nitrate, and 2.7 – 21% water; the ratio of magnesium nitrate : water is 1.2 – 2.2 : 1. A cellulose nitrate with an N content of 11.9% was obtained, for example, with 64.5% HNO$_3$, 19.5% Mg(NO$_3$)$_2$, and 16% H$_2$O. This nitrating system is appropriate for a continuous process, in which waste and washing acid are reprocessed in ion exchangers and the magnesium nitrate is recycled. Thus, acid and sulfate no longer pose a waste disposal problem.

1.3.4. Production

The flow diagram (Fig. 4) shows the industrial production of CN according to the mixed acid process. The viscosity of the end product is determined by the choice of the initial cellulose, and the degree of nitration is determined by the composition of the mixing acid. The final viscosity adjustment follows during the pressure boiling step (see Section 1.3.4.21.3.4.2).

Table 4. Industrially used nitrating acid solutions

CN type	Nitrating acid			N content,%	DS
	% HNO$_3$	% H$_2$SO$_4$	% H$_2$O		
Lacquer cotton A	25	55.7	19.3	10.75	1.90
Celluloid cotton	25	55.8	19.2	10.90	1.95
Lacquer cotton AM	25	56.6	18.4	11.30	2.05
Dynamite cotton	25	59.0	16.0	12.10	2.30
Lacquer cotton E	25	59.5	15.5	12.30	2.35
Powder cotton	25	59.8	15.2	12.60	2.45
Gun cotton	25	66.5	8.5	13.40	2.70

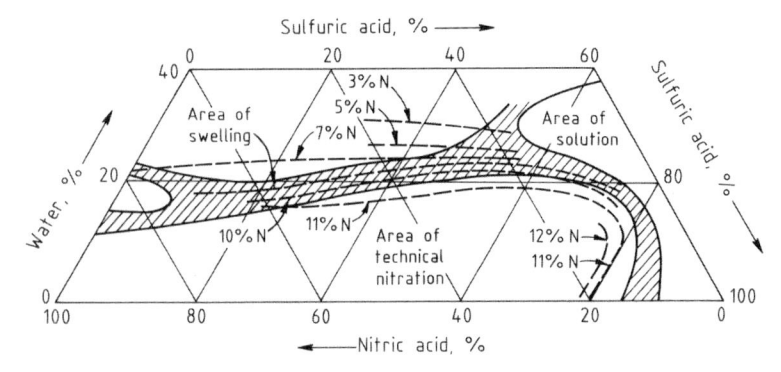

Figure 3. Composition of the nitrating mixtures and attainable N contents of cellulose nitrates

1.3.4.1. Cellulose Preparation

Cotton linters with a moisture content of up to 7% are mechanically disintegrated homogeneously. Pressed pulp sheets must be appropriately shredded to obtain rapid and uniform nitration. Spruce or beech celluloses, preactivated with 20% sodium hydroxide (mercerization), were formerly

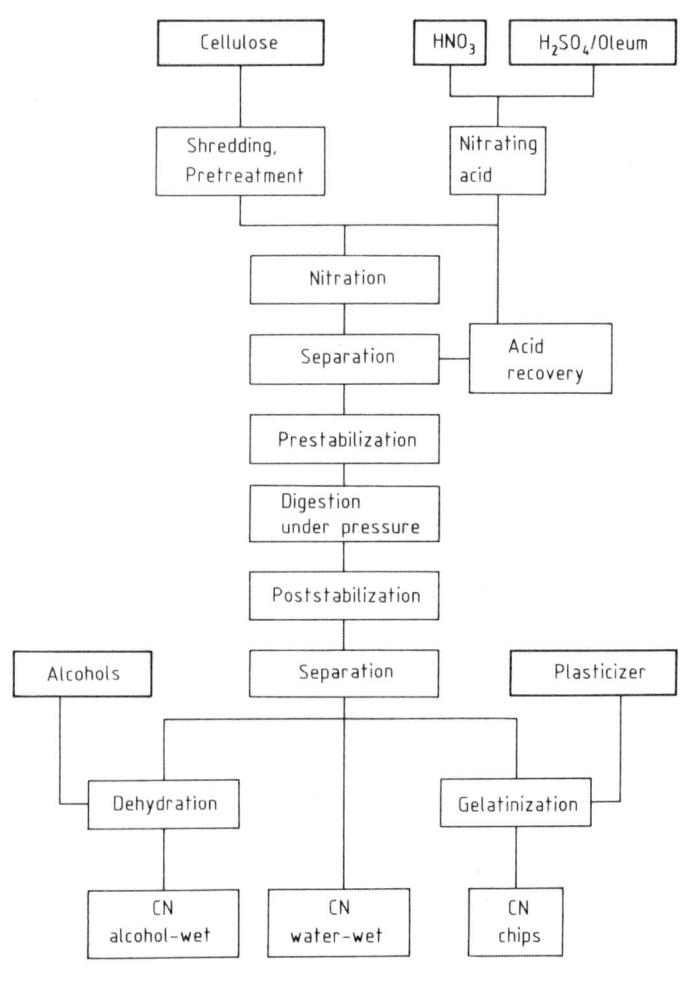

Figure 4. Flow diagram of cellulose nitrate production

used for this purpose in the form of crape papers with a mass per unit area of ca. 25 g/m^3. To avoid the costly transformation of the cellulose to paper sheets, a process was attempted to obtain a loose product resembling linters by direct disintegration of pulps to fibers. A moisture content of 50 % proved to be optimal for nitration and washing out the acid. The required drying of the cellulose flakes before nitration proved to be disadvantageous.

The Stern shredder [32], in which the pulp sheets are torn rather than being cut into small elongated shreds to avoid compression at the edges, was a definite improvement. Currently, cellulose for nitration is used in the form of fluff, shreds, or chips. The packing density and compression behavior of the cellulose fibers in the fiber pile are decisive factors for the swelling and nitration kinetics, as well as the acid retention capability [30].

1.3.4.2. Nitration

Nitration on an industrial scale is still frequently carried out according to a batch process that was developed from a process described by DuPont in 1922. The equipment is constructed of stainless steel. The adjusted and preheated nitrating acid reaches the stirring reactor that is charged with cellulose by means of a measuring system; a large excess of acid (1 : 20 to 1 : 50) is added to retain the ability of the reaction mixture to be stirred and to ensure that heat is carried off. The nitrating temperature is between 10 °C (dynamite type) and 36 °C (celluloid type). The total heat of reaction is estimated to be over 200 kJ per kg of CN, of which the enthalpy of formation of CN is about one-third.

Even though the reaction is nearly complete after ca. 5 min, the mixture remains in the reactor for about 30 min. The temperature must remain constant (cooling), since hydrolytic degradation processes that lead to considerable losses in yield begin at temperatures as low as 40 °C.

The theoretical yield of commonly used industrial types with a DS of 1.8 – 2.7 ($\triangleq$ 10.4 – 13.4 % N) is between 150 and 176 % with respect to cellulose. The practical yield, however, is up to 15 % lower and depends on the type and purity of the cellulose, as well as on the temperature and duration of nitration. Losses arise from the inevitable complete decomposition of cellulose to oxalic acid by way of oligo- and monosaccharides, whereby the nitric acid is reduced to nitrogen oxides, NO$_x$. In addition, mechanical losses during the subsequent separation process, due particularly to short fibers (cellulose from hard wood), must also be taken into consideration.

The reaction mixture is drained from the reactor into the centrifuge, where the excess acid is separated and removed at high speed and reprocessed for recycling. The mixture must remain moist so that it does not ignite and deflagrate.

The degree with which the product retains acid after separation is of economic importance because of the acid loss and the expense of the ensuing washings. Linters with an acid retention of 100 – 130 % clearly surpass wood celluloses in this respect which, depending on the wood and cellulose type as well as its processing, can retain up to 3 times more adhering acid relative to CN [30].

The still acid-moist product is immediately placed into a great excess of water (consistency 1 %) so that the adhering acid is displaced as rapidly as possible and the saponification of the CN is prevented.

Continuous nitrating processes, which are more economical, were developed in the 1960s [33], [34]; they ensure a more uniform product quality and are safer to handle. The nitrating system consists of two or more consecutively arranged straight-run vats or tube systems containing conveyers (screw conveyer or turbulence stirring apparatus) which forward the reaction mixture. The prepared cellulose is directed into this cascading equipment from storage bunkers over automatic weighing scales and continuously mixed with the added nitrating acid. It is important that the cellulose is rapidly added and immediately covered with acid. There it remains for 30 – 55 min. A newer process using a continuous loop-formed pressure reactor [35] requires the cellulose to remain only for 6 – 12 min. The reactant is then sent into a continuously operating special centrifuge, where the excess acid is separated and simultaneously taken up with water. The fact that the reactant remains only a few seconds reduces the risk of spontaneous deflagration and saponification.

Figure 5 shows schematically the continuous process according to Hercules [31].

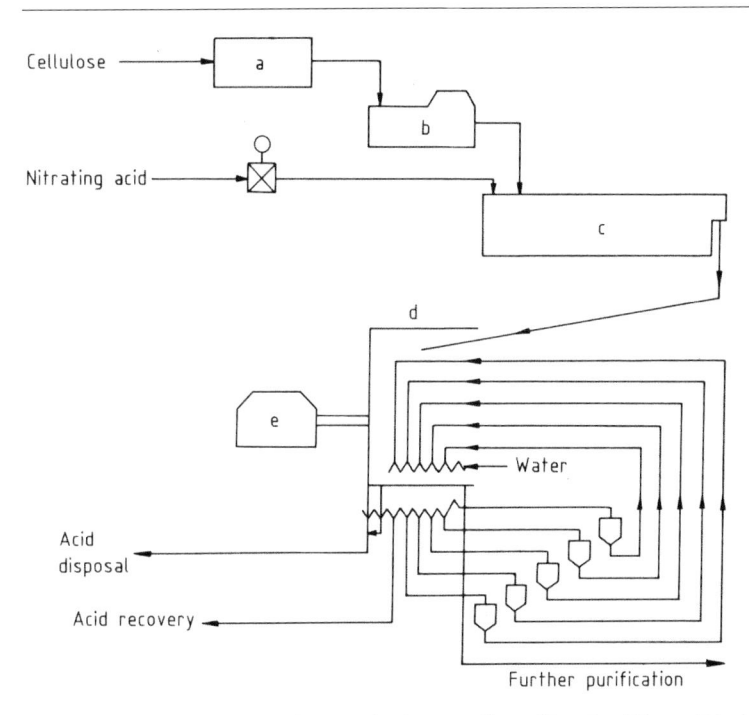

Figure 5. Continuous cellulose nitrate production according to Hercules a) Preconditioning; b) Auto-matic scale; c) Reactor; d) Washing zone; e) Centrifuge

The broken-up and preconditioned cellulose (a) is brought by way of the automatic scales (b) to the continuous reactor (c). The reaction product is centrifuged in a washing zone (d) and simultaneously washed by zones with water in a countercurrent. The product leaves the centrifuge almost free of acid, and the washed out acid can be recycled and reused almost without loss [36].

1.3.4.3. Stabilization and Viscosity Adjustment

The prestabilization step following the prewashing further purifies the product by means of repeated washing and boiling with water that contains 0.5 – 1 % acid residue. The batch method requires large amounts of space, water, and energy; the required boiling time varies between 6 (celluloid type) and 40 h (guncotton). Automatic continuous processes have been developed in this case as well [37].

Most of the remaining sulfuric acid is removed during prestabilization, since it would promote the catalytic self-decomposition of CN. The sulfuric acid is bound by adsorption and esterified. A total sulfate content of 1 – 3 % was found in weakly nitrated CN, of which 70 – 85 % is in the form of the acidic sulfuric acid semiester, while highly nitrated CN contains only 0.2 – 0.5 % total sulfate, of which 15 – 40 % is thought to exist as an ester. Semiesters can be easily saponified and washed out by boiling with water. It is not yet certain whether the so-called resistant sulfate content exists in the form of the neutral sulfuric acid ester or the physically adsorbed sulfuric acid.

The desired final viscosity of the CN is adjusted in the following process step, which is pressure boiling (digestion under pressure) in a consistency of 6 – 8 % at 130 – 150 °C, by means of specific degradation of the degree of polymerization. The remaining extremely low sulfuric acid content induces hydrolytic decomposition at this temperature and under pressure. The viscosity can, for example, be reduced to 1/10 of the initial viscosity within 3 h at 132 °C by using this process. This process made the development of high solid coating and protective nitro lacquers possible. The stabilization process of guncottons is accelerated by pressure boiling; dynamite wools are usually not pressure boiled.

Further product losses are due to chain degradation ranging from soluble cleavage products to oxalic acid. Nitrous gases (NO_x) are released by the reduction of nitric acid, which must be continuously drawn off to avoid decomposition of CN.

Pressure boiling can be achieved batchwise in autoclaves, as well as continuously in a tube reactor of 1 500 m in length and 100 mm in diameter, e.g., with direct steam. A one-pot process in which prestabilization, pressure boiling, and poststabilization are carried out in one operation is described in [38].

During the stabilization process, the remaining sulfuric acid is almost completely removed by additional washing and boiling. While celluloid and lacquer types are finished in flaky, fibrous form, guncotton must be ground. This is done in grinding hollander engines at 12 – 15 % consistency or continuously in a series of cone refiners, whereby the material is gradually concentrated from 3 to 10 % between the various grinding steps. Sorting steps are inserted by hydrocyclones during the final washing processes. The last acid remnants in the fiber capillaries are removed during the grinding process by means of diffusion against water. Weak bases, sodium carbonate, or chalk are used to maintain a pH of 7. Stabilizers (organic acids) may be added during this step.

1.3.4.4. Displacement and Gelatinization

A water-wet CN cotton with a water content of 25 – 35 % remains in the centrifuge after the final separation and is then packed into drums or PE sacks.

Water contained in celluloid and lacquer types is displaced by alcohols specified by the processors (ethanol, 2-propanol, *n*-butanol) in displacement presses or displacement centrifuges. Continuous processes prevail here also [39]. The resulting aqueous alcohols must be distilled to remove the water.

The water-wet CN cotton can be gelatinized with softeners such as phthalates in kneading aggregates and dried on drum or band driers for the production of CN chips [40], [41]. Colored chips are obtained by adding carbon black or pigments from which colored enamels can be produced without the use of ball mills or roller mills.

1.3.4.5. Acid Disposal and Environmental Problems

The nitrous gases formed during the nitrating and stabilizing side reactions are drawn off and washed out in trickling towers. The lower nitrogen oxides are regained after oxidation as 50 – 60 % nitric acid.

The waste acids resulting from the first separation contain 2 – 3 % more water and 3 – 4 % less nitric acid than the initial mixture. They are circulated in a closed system and constantly regenerated with nitric acid and oleum. The acid that adheres to the product must also be replaced.

The proportion of adhering acid depends on the initial cellulose and the CN type. It ranges between 80 % (guncotton) and up to 200 – 300 % (lacquer types) with regard to CN and is removed with the water used for washing and boiling.

Aside from the economic aspects of acid loss in wastewater, environmental considerations are beginning to play an increasingly important role. While older manufacturing facilities using simple centrifuging to remove waste acid produced 300 m^3 of water per ton of CN, containing 0.5 % acid and with a pH of 1, it was possible to reduce the volume of wastewaters to a fraction of its previous volume by almost completely closing the cycles.

Before proceeding into the draining ditch, the wastewater must be separated from the hardly decomposable sludge consisting of cellulose and CN, and then be neutralized. The sulfate proportion can be reduced by calcium sulfate precipitation, while the nitrate proportion remains completely in the wastewater. Organic matter of communal sewage, for example, can be biologically decomposed without additional oxygen, whereby nitrates disappear almost completely as a result of biological denitrification.

The salt/acid process with magnesium nitrate (see Section 1.3.4.1) is more favorable with regard to the wastewater problem. Sulfates are completely absent, and magnesium nitrate is recycled and, therefore, causes no water pollution problems. The amount of wastewater can be reduced by 80 % and the nitric acid requirement by 83 % in comparison to the formerly used discontinuous processes.

1.3.4.6. Other Nitrating Systems

Numerous attempts have been made to improve nitration by the introduction of other nitrating

Table 5. Nitrating systems

Nitrating system	Max. N content, %	Comments
$HNO_3/H_2SO_4/H_2O$	13.4	Industrial nitration
$HNO_3 < 75\%$		"Knecht compound," unstable
78 – 85 %	8	Dissolution in the nitrating acid
85 – 89 %	10	Gelatinization
90 – 100 %	13.3	No swelling
HNO_3 + nitrates, sulfates, phosphates	13.9	
HNO_3 vapor	13.75	Slow reaction, stable nitrate
HNO_3 vapor + nitrogen oxides	13.8	
N_2O_5	14.12	
N_2O_5 in CCl_4	14.14	Trinitrate
HNO_3 in CH_2Cl_2	14.0	
HNO_3 in nitromethane	14.0	Homogeneous reaction
HNO_3 + H_3PO_4/P_2O_5	14.04	Rapid reaction without decomposition (polymer analogue)
	14.12	After extraction with methanol
HNO_3 + acetic acid/acetic anhydride	14.08	Great stability
	14.14	After extraction with ethanol
HNO_3 + propionic acid/butyric acid	14.0	

systems, or at least to increase the degree of substitution. Further details may be obtained from [12], [14], [17], and [21]. Table 5 gives a summary of alternative nitrating systems, none of which was able to displace the ternary system $HNO_3/H_2SO_4/H_2O$ for industrial nitration.

Nitration with pure nitric acid is possible in principle. Esterification is not possible with acid concentrations below 75 %. Acid concentrations less than 75 % cause the formation of the unstable so-called Knecht compound, which has been described as either a molecular complex or an oxonium salt of the nitric acid. Cellulose nitrates with 5 – 8 % N, which dissolve in excess acid, are formed at acid concentrations of 78 – 85 %. Nitrogen contents of 8 – 10 % are attained at concentrations between 85 and 90 % HNO_3; these products have a strong tendency to gelatinize. Heterogeneous nitration without apparent swelling takes place at a HNO_3 concentration above 89 %, and 13.3 % N can be achieved with 100 % HNO_3. Nitration can be increased to 13.9 % N with 100 % HNO_3 by addition of inorganic salts such as sulfates, acid phosphates, and particularly nitrates, preferably in a 15 % concentration.

The nitric acid/phosphoric acid system is of special interest in a 1:1 ratio with 2.5 % phosphorus pentoxide added, with which an almost completely nitrated product of great stability was achieved. The nitric acid/acetic acid/acetic anhydride system in a ratio of 2:1:1 gives highly nitrated and highly stable products in which the fiber structure remains intact largely. After extraction of these nitrates with water or alcohol, the theoretical degree of substitution of the trinitrate may be attained.

Nitrating systems which achieve a high degree of nitration without degradation of the cellulose chain are of special scientific interest. This process is known as polymer analogous nitration. After a critical examination of all known nitrating mixtures, the nitric acid/acetic acid/acetic anhydride system in a ratio of 43:32:25 at 0 °C [42] was recommended for determining the molecular mass of native celluloses of such solutions by using absolute methods and the intrinsic viscosity number [43]. The system anhydrous nitric acid in dichloromethane also allows the application of such polymer analogous reactions at temperatures between 0 and −30 °C [44]. Other authors [45] prefer the system nitric acid/phosphoric acid/phosphorus pentoxide.

Nitration in the Laboratory. Preparative cellulose nitration with HNO_3/H_2SO_4 nitrating acid to products with whatever N content up to 13.65 % is desired, stabilization and stabilization tests, nitration with the nitric acid/phosphoric acid (< 13.9 % N) and nitric acid acetic anhydride systems up to the trinitrate, denitration with hydrogen sulfide to cellulose II, the analytic determination of the N and sulfate content, and the solution of the CN and the viscosity determination of the solution are extensively described in [46].

1.3.5. Commercial Types and Grades

Cellulose nitrates receive, because of their fluffy structure and cottonlike appearance, the additional designation "cotton."

Two parameters are decisive for the industrial use of cellulose nitrate:

Nitrogen content (including the resulting solubility properties)
Viscosity

As seen in Table 6, cellulose nitrates with differing nitrogen contents have various applications. Cellulose nitrates for lacquers are available in numerous viscosities. It is possible to categorize all stages of viscosity according to the European norm (DIN 53179), but the viscosity of cellulose nitrates is primarily categorized by using the Cochius method and the British or American ball drop method (ASTM D 1343 – 69).

In addition to the so-called cotton types densified CN types are available. These may be obtained by either nitrating compressed cellulose or by subsequently compressing the fluffy cellulose nitrate. It is possible to almost double the bulk density by compression.

For safety reasons, the commercially available CN cotton types must be wetted with at least 25 wt % water or aliphatic alcohols. In addition to water, ethanol, *n*-butanol, and 2-propanol may also be used as wetting agent.

The largest manufacturers of cellulose nitrates are the following:

Hercules Inc.	USA
Wolff Walsrode AG	FR Germany
Hagedorn	FR Germany
WNC Nitrochemie GmbH	FR Germany
Société Nationale des Poudres et Explosifs (SNPE)	France
Imperial Chemical Industries (ICI)	Great Britain
S.I.P.E. Nobel S.p.A.	Italy
Unión de Explosivos Río Tinto S.A.	Spain
Bofors	Sweden
Asahi	Japan
Daicel Chemical Industries, Ltd.	Japan

Many countries in South America, Asia, and Eastern Europe maintain small CN production facilities. The total world capacity may be estimated to 150 000 t/a of dry cellulose nitrate.

Table 6. Cellulose nitrate types

Type	N content, %	Degree of substitution (DS)
Celluloid cotton	10.5 – 11.0	1.82 – 1.97
Alcohol-soluble > lacquer cotton	10.9 – 11.3	1.94 – 2.06
Lacquer cotton moderately soluble in alcohol	11.4 – 11.7	2.08 – 2.17
Ester-soluble > lacquer cotton	11.8 – 12.2	2.20 – 2.32
Powder cotton	12.3 – 12.9	2.55 – 2.57
Gun cotton	13.0 – 13.6	2.58 – 2.76

Other Commercial Types. Also available, in addition to cellulose nitrate cotton types, are so-called cellulose nitrate chips, made from cellulose nitrate plasticized by gelatinizing softeners. For safety reasons, the softener content has been established at a minimum of 18 wt %. Chips are preferred in processes where alcohols interfere in the formulation of lacquers.

The dispersions of cellulose nitrate with softeners or resins in water manufactured by the Wolff Walsrode AG are other available forms. The solvent-free or low-solvent dispersions are not polluting and may be used in all areas in which cellulose nitrate lacquers also are used [55].

1.3.6. Analysis and Quality Control

The most important analytical characteristics relate to the determination of the N content and, thereby, the average degree of substitution (DS), as well as the viscosity of the solution as a measure of the average molecular mass or chain-length.

Analytic Tests. The most commonly used analytic procedures are summarized in [25], [46], [47], and [48].

Dry content is determined by careful drying of a small, thinly layered alcohol or water-wet sample at room temperature for 12 – 16 h, in a weighing glass at 100 – 105 °C for 1 h, or with compressed warm air at 60 – 65 °C for 0.5 – 1 h.

Ash content is determined by decomposing a dried sample with HNO_3 and incinerating the

Table 7. Characterization of cellulose nitrates according to DIN 53179

A types	CN concentration, % absol. dry	AM types	CN concentration, % (absol. dry)	E types	CN concentration, % (absol. dry)
				E 1440	4
				E 1160	7
				E 950	9
				E 840	12
		AM 760	14		
		AM 750	15		
				E 730	15
		AM 700	17		
				E 620	21
				E 560	22
				E 510	24
A 500	27	AM 500	27		
A 400	30				
				E 400	27
				E 375	32
		AM 330	36	E 330	34

residue. Specifications require that the ash content should not be above 0.3 %.

N-content is determined by reducing nitrates according to the following reaction (Schulze-Thiemann):

$NO_3^- + 3FeCl_2 + 4HCl \rightarrow 3FeCl_3 + Cl^- + 2H_2O + NO$

or by the following reaction:

$2NO_3^- + 4H_2SO_4 + 3Hg \rightarrow 3HgSO_4 + SO_4^{2-} + 4H_2O + 2NO$

The resulting NO is collected in a Du Pont nitrometer.

Stability Tests. [25]. *Deflagration Temperatures:* Well-stabilized CN deflagrates at temperatures above 180 °C.

Bergmann – Junk Test [29]: A quantity of 2 g of dried CN is kept at a temperature of 32 °C for 2 h in a special apparatus for the elimination reaction, after which time the amount of the developed nitrous gases (after reduction to NO) is determined. CN is stable according to this test if no more than 2.5 cm^3 of NO per gram is measured.

Warm Storage Test: A quantity of 5 g of dried CN is stored in a glass-stoppered tube at 75 °C. Note is then made when the first nitric oxide (red-brown gas) becomes visible. Well-stabilized CN can be stored at 75 °C for at least 10 days.

ASTM Stability Test [48]: After storage at 134.5 ± 0.5 °C the time is noted in which the nitrous gases discolor methyl violet test paper.

Viscosity. *Viscosity according to DIN 53179:* If CN is dissolved in acetone in the appropriate concentration, CN solutions meeting this requirement show a apparent dynamic viscosity of 400 ± 25 mPa · s in the ball drop viscometer according to Höppler (ball no. 4) at 20 °C (Table 7).

Cochius Viscosity [25]: The viscosity of the various cellulose types is measured in commonly used solvent mixtures:

A and AM types: butanol/ethylene glycol/toluene/ethanol in the following proportions 1:2:3:4

E types: butanol/butyl acetate/toluene in the following proportions 3:4:5

Dried CN is dissolved in varying concentrations depending on the type and the time which an air bubble requires to rise 500 mm between two calibrations in a 7 mm Cochius tube at 18 °C is measured in seconds. The Cochius seconds are converted to absolute viscosity units mPa · s by multiplying with the factor 3.64 mPa.

Ball drop method according to ASTM [49]: Dried CN is dissolved according to its viscosity stage in 12.5, 20.0, or 25.0 % ethanol/toluene/ethyl acetate according to [48]. The drop time of the balls with a diameter between 1/4 and 1/16 in. at 25 °C is given in seconds or converted into Pa · s. Figure 6 shows the relationship between the degree of polymerization and the technical viscosity (fall velocity of the balls in a 17.2 % CN solution in acetone).

Comparative viscosity charts for converting the various viscosities and comparing the various types are found in [25].

Solubility and Color. The color and cloudiness of solutions produced according to [48] are

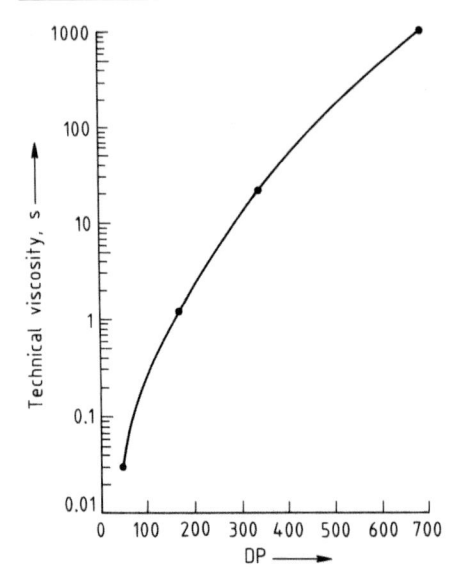

Figure 6. Degree of polymerization (DP) and technical viscosity ("ball drop" in seconds of a 12.2% CN solution with 12% N in acetone). DP = 170 × viscosity

tested visually. Consistency, appearance, and depth of color can be controlled according to [50].

Dilution with Toluene. Toluene is added to a 12.2% CN solution in butyl acetate at 25 °C until CN continuously precipitates. The dilution factor is noted. The dilution ratio of CN solutions with other solvents and blending agents is determined according to [51].

Film Test. The solutions made according to [48] are diluted with an equal volume of butyl acetate and poured as a film onto a glass plate. The dried films are examined for undissolved particles, surface structure, transparency, and gloss.

1.3.7. Uses

Explosives. Explosives may be categorized according to their use:

blasting agents
propellants and shooting agents
detonating agents
igniting agents
pyrotechnical agents

Cellulose nitrates are used primarily as propellants and gun powder, whereby the following distinctions can be made: *monobasic powder*, which is based solely on cellulose nitrate; *dibasic powder*, which contains further energy carriers such as, for example, nitroglycerin or diglycol dinitrate in addition to cellulose nitrate; *tribasic powder*, which contains in addition to the components of the dibasic powder a third agent such as nitroguanidine.

The selection of the cellulose nitrate is of special importance. The types of cellulose nitrates that differ in the degree of nitration were standardized as follows:

CP I (Collodium powder) also known as guncotton, nitrogen content: 13.3 – 13.5%
CP II (Collodium) nitrogen content: 12.0 – 12.7%, mostly 12.6%
PE   (Powder standard) nitrogen content: 11.5 – 12.0%, mostly 11.5%

Aromatic amines, such as diphenylamine, are added to gunpowder as stabilizers. They are capable of binding the nitrous gases generated during the decomposition of the nitric acid ester. A mixture of ca. 80% highly nitrated gunpowder (13.4% N) and ca. 20% less-nitrated collodium cotton (12.5% N) is used for the production of the monobasic propellant powder. Since cellulose nitrate granules are easy to charge electrostatically, they are made conductive with a fine graphite coating.

The multibasic powders usually contain cellulose nitrate CP II. Mixtures of 40% PE cotton and 60% CP I are also used because they have the same energy content as CP II with 12.6% N.

The introduction of a third component to tribasic powder results in a lower heat of combustion in comparison to dibasic powder, thereby lengthening the life of the gun barrel.

Gunpowder is used in small-arms ammunition as well as large-caliber guns and tanks. (For further details → Explosives, Section 7.2.)

Lacquers. Cellulose nitrate lacquers are characterized by the outstanding film-forming properties of the physically drying cellulose nitrate. Moreover, cellulose nitrate is compatible with many other raw materials used in lacquers and can be used advantageously in combination with resins, softeners, pigments, and additives.

In addition to the nonvolatile lacquer components, the composition of the solvent mixture is decisive for the formation of a film.

The most important uses for nitro lacquers are as follows: wood lacquers (especially furniture lacquers), metal lacquers, paper lacquers, foil lacquers (also as hot sealing lacquers, e.g., cellophane, plastic, and metal foils), leather lacquers, adhesive cements, putties, and printing ink (for flexo and gravure printing).

The processes used for applying cellulose nitrate lacquers to substrates are as follows: spraying (compressed-air, airless, and electrostatic spraying), casting (for example, with a curtain coater), rolling (especially for the application of small amounts of lacquer), doctor knife coating, and dipping.

The casting and rolling processes are used for lacquering large, even areas. Irregularly shaped objects are sprayed. The choice of a suitable type of cellulose nitrate (e.g., completely or moderately soluble in alcohol, soluble in esters, degree of viscosity) is dependent on the lacquer type. A highly viscous cellulose nitrate type is used if elastic and thin applications are desired (e.g., leather). However, if hard and thick layers are desired, low-viscosity types are preferred.

The concentration or the degree of viscosity of the cellulose nitrate determine the viscosity of the lacquer solution. However, the formulation of the lacquer must be taken into consideration when the mode of application is chosen. For example, a highly viscous dipping lacquer cannot be sprayed or casted.

Furthermore, the striking differences between ester-soluble and alcohol-soluble types should be taken into consideration when nitro lacquers are formulated (Table 8).

For further information on the formulation of cellulose nitrate lacquers, see [52], [53], and also → Paints and Coatings, 1. Introduction.

Dispersions. Conventional cellulose nitrate lacquers contain between 60 and 90 % organic solvents, which are released during drying. For economic and environmental reasons, it is desirable to substitute organic solvents by water. Aqueous cellulose nitrate/softener dispersions (e.g., Isoderm, Bayer AG; Coreal, BASF; Waloran N, Wolff Walsrode AG) are available for such absorbing substrates as leather [54]. Other aqueous cellulose nitrate dispersions for use on wood, foil, and metal have also been developed (Waloran N, special-types, Wolff Walsrode AG) [55]. The film forming process of water-insoluble cellulose nitrate requires a small amount of coalescents in the dispersion systems.

Celluloid. A special use of cellulose nitrate is in the production of celluloid [56]. Cellulose nitrate with a nitrogen content of 10.5 – 11.0 % is mixed in a kneader with softeners, particularly camphor, and solvents (alcohols).

Normal celluloid contains ca. 25 – 30 % camphor and 70 – 75 % cellulose nitrate. Celluloid that contains 10 – 15 % solvent can be formed into the desired articles in heated piston or screw presses (e.g., tubes and round and profile rods).

In the past decades, celluloid has been widely replaced by synthetic materials and thermoplastics. Celluloid is still of economic importance in the following areas: combs and hair ornaments, toilet articles, office supplies (drafting and

Table 8. CN lacquer cottons

Ester-soluble type	Alcohol-soluble type
Possible use of alcohol in the lacquer formulation	Use of alcohol, especially ethanol, in any desired amount as a solvent
Good dilutability with aliphatic and aromatic hydrocarbons	Good dilutability with aromatic hydrocarbons
Very rapid solvent release	Rapid solvent release
Formation of hard films	Formation of films with thermoplastic properties
Attainment of good mechanical properties as far as the cold-check test, stretch, hardness, and tensile strength are concerned	Attainment of good mechanical properties; some special problems of lacquer production may be solved such as:
	Lacquers which can be diluted with ethanol in any desired manner (wood polishes)
	Odorless lacquers (printing inks)
	Gel dipping lacquers
	Hot sealing waxes (cellophane lacquers and aluminum foil lacquers)

measuring instruments), ping-pong balls, and various special uses.

1.3.8. Legal Provisions

Toxicology and Industrial Safety. Concentrated sulfuric acid, nitric acid, and nitrous gases formed during the production of cellulose nitrate are considered hazardous chemical products [57]:

1. Sulfuric acid
   5 – 15 % EC-No. 016 – 020 – 01 – 5
   above 15 % EC-No. 016 – 020 – 00 – 8

2. Nitric acid
   20 – 70 % EC-No. 007 – 004 – 01 – 9
   above 70 % EC-No. 007 – 004 – 00 – 1

3. Nitrous gases
   EC-No. 007 – 002 – 00 – 0

They are subjected to the Arbeitsstoffverordnung (working substance regulation) [58] and must, therefore, be adequately labeled.

Concentrated nitric acid and mixed nitrating acids are oxidizing when brought into contact with organic materials [59]. The MAK values (maximum working place concentration) are as follows:

nitric acid vapors 10 mL/m^3 (ppm); ≙ 25 mg/m^3
nitrogen oxides (NO$_2$) 5 mL/m^3 (ppm); ≙ 9 mg/m^3

Employees should be examined regularly for obstructive respiratory tract illnesses.

Cellulose nitrate is neither toxic nor hazardous to health [60]. Damping agents in CN and nitrous gases which may be formed during combustion and smoldering processes are potentially hazardous to health if inhaled.

Commercially available phlegmatized cellulose nitrate for the production of lacquer with less than 12.6 % N contains at least 18 % of a gelatinizing softener. According to the first paragraph in [58], cellulose nitrate is a hazardous substance and must be packaged and labeled accordingly. EEC regulations (1982) are similar.

Damping agents such as ethanol and 2-propanol are not subjected to these regulations; butanol belongs to category II d, but is not considered to be hazardous to health in a damped mixture of a maximum 35 % concentration.

Storage and Shipping. Cellulose nitrate, especially guncotton, burns in air with a yellow flame and deflagrates if present in larger quantities, especially after rapid heating. An explosion can be caused by friction or a sharp impact. Dry CN has electrostatic charge. Friction, particularly on metals but also on plastics, can cause sparks which lead to a deflagration. Therefore, cellulose nitrate should be stored in a moist and cool place [60], [61]. Rooms in which cellulose nitrate is processed must be adequately protected according to the guidelines for protection from explosions.

Cellulose nitrate is subjected to the regulations governing explosives [62]. The transportation of phlegmatized cellulose nitrates proceeds according to the most recent versions of the hazardous materials regulation; see [25]. Wetted cellulose nitrate is shipped in thick-walled, galvanized, tightly closing iron or fiber drums which are adequately labeled.

Dried cellulose nitrate may not be shipped under any circumstances.

For further information on the properties, handling, storage, and transportation of hazardous goods, see also [63].

1.4. Other Inorganic Cellulose Esters

Summaries on the esterification products of cellulose with other inorganic acids may be found in [12–19]. For publications on the modification of cellulose, including esterification, see [64].

1.4.1. Cellulose Sulfates

Cellulose sulfates [9032-43-3] are the most frequently investigated of all other inorganic cellulose esters. The ability of concentrated sulfuric acid to dissolve cellulose, particularly in concentrations between 70 and 75 %, has been known since 1819. After precipitation immediately following dissolution, the cellulose contains little or no bound sulfate. An almost homogeneous esterification takes place only if the cellulose is left in a sulfuric acid solution over a longer period of time.

However, the ester yield is very poor. The major portions of the reaction products consist of hydrolytically split decomposition products with a maximum degree of substitution of 1.5.

In their free acidic form, cellulose sulfates are fairly unstable and easily saponified. A semiester was developed in 1953 in the United States [65] in an esterification mixture consisting of 1 mol of cellulose with 20 – 30 % water, 3.5 – 15 mol of sulfuric acid, 0.3 – 1.0 mol of a primary or secondary C_3 – C_5 alcohol, and an inert volatile organic solvent such as toluene or carbon tetrachloride (reaction temperatures between -5 and $-10\,°C$). The product was soluble in hot or cold water, yielded relatively stable, clear, and highly viscous solutions, and was recommended for use as a thickener for aqueous systems (emulsion paints and printing inks, printing pastes for textiles, and food products), as well as for fat- and oil-proof finish, and as paper glue. This product, however, is of no economic importance.

Numerous attempts have been made to find improved preparation methods for water-soluble cellulose sulfates stable to saponification. Known reaction systems are summarized in a tabular overview [65]. The reaction of cellulose with sulfuric acid in organic solvents, especially in lower-mass aliphatic alcohols, gives by way of a heterogeneous reaction fibrous and water-soluble cellulose sulfates with a maximum DS value of 1. More highly substituted products are obtained by reaction with sulfuric acid/acetic acid anhydride (up to a DS value of 2.8) or esterification with chlorosulfuric acid in pyridine or formamide. The reaction with SO_3 only or in various organic systems yields trisubstituted products. The reaction mechanism may be described as the addition of the strongly electrophilic SO_3 to the hydroxyl groups with the succeeding disintegration of the intermediately formed oxonium ion.

$$\text{Cell-OH} + SO_3 \rightarrow \left[\text{Cell}-\overset{H}{\underset{+}{O}}-SO_3^-\right] \rightarrow \text{Cell-O-}SO_3^- + H^+$$

Completely water-soluble, highly viscous sodium cellulose sulfate semiesters are obtained in homogeneous systems by the reaction of cellulose nitrite [67]. The intermediate that is formed and dissolved, cellulose nitrite, is obtained in the N_2O_4/dimethylformamide system and is at the same time transesterified by the SO_3/DMF complex. Uniformly substituted cellulose sulfate with a range of DS values between 0.3 and 2.0 and solution viscosities up to 7000 $mPa \cdot s$ (in 1 % solution) can be obtained by using this process [68]. Such transesterified products can be cross-linked by metal ions to form highly effective thickening agents for aqueous media [69].

Such processes have been further developed [66] and make interesting novel fields of application accessible as a result of the rheological and gel-forming properties of the Na cellulose sulfate semiester.

Mixed esters such as cellulose acetate sulfates, cellulose acetate butyrate sulfates [70], cellulose acetate propionate sulfate [71], and ethyl cellulose sulfates [70], [72] are described in the patent literature.

Being polyelectrolytes, cellulose sulfates form salts and have ion-exchanging properties; thus, they have been recommended for use as cation exchangers [64, p. 65], [73], [74].

1.4.2. Cellulose Phosphate and Cellulose Phosphite

Reaction of cellulose with aqueous phosphoric acid gives the following unstable addition compound: 3 $C_6H_{10}O_5 \cdot H_3PO_4$, from which the cellulose can be regenerated unchanged by reaction with water. Cellulose phosphates [9015-14-9] with a low phosphorus content are obtained by reacting cellulose or linters with phosphoric acid in an urea melt [75]. Higher phosphorus contents and a lower degradation rate of the cellulose may be obtained with excess urea at reduced reaction time (ca. 15 min) and at high temperature (ca. 140 °C). Water-soluble cellulose phosphate with a high degree of substitution may be obtained from a mixture of phosphoric acid and phosphorus pentoxide in an alcoholic medium [76].

Phosphorylated cellulose fibers show increased swelling after partial hydrolytic degradation and transfer into the alkali salt form and were, therefore, suggested for use as adsorbents [77].

Cellulose phosphates with a 17 % phosphorus content (this represents about 3/4 of the maximal possible substitution of triphosphate with 23 % phosphorus) were already produced in 1933 by reacting cellulose with a mixture consisting of concentrated sulfuric acid and phosphoric acid in the presence of a weakly acidic catalyst [78].

Cellulose phosphites [*37264-91-8*] and cellulose phosphonates may be prepared by transesterification with alkyl phosphites. All cellulose esters containing phosphorus have fire-retarding properties [78] and have attracted some interest due to their ion-exchanging effect [74], [79], but are not yet industrially used.

1.4.3. Cellulose Halogenides

Various preparative methods are suitable for the synthesis of halogenated cellulose derivatives [64, p. 64]. Halogenation can be carried out by transesterification of such cellulose esters as tosylate, nitrate, and sulfate with hydrohalic acids [80]. Nucleophilic substitution proceeds considerably faster in homogeneous systems than in heterogeneous aqueous systems.

The Finkelstein transesterification process of cellulose nitrate with sodium iodide in anhydrous acetone leads to deoxyiodo cellulose. The reaction of cellulose with thionyl chloride, $SOCl_2$, in the presence of pyridine produced a monosubstituted, but strongly decomposed and unstable, hydrogen chloride ester.

Halogenation of cellulose improves its water-resistant and flame-resistant properties. Slight fluorination increases oil resistance and lowers the soiling potential of cellulose textiles [64]. Commercial applications are not yet known.

1.4.4. Cellulose Borates

The preparation of cellulose borate succeeded by means of transesterification of methyl and *n*-propyl borate with cellulose [64, p. 7]. The products with a maximum DS value of 2.88 are, however, extremely sensitive to hydrolysis and alcoholysis.

1.4.5. Cellulose Titanate

Cellulose can be reacted to cellulose titanates in a heterogeneous reaction system by reacting it with titanium tetrachloride in DMF or with chlorinated anhydrides, chlorinated ester anhydrides, and esters of the hypothetical orthotitanic acid Ti(OH)$_4$ [81]. Ethyl trichlorotitanate has been shown to be the most reactive. Esters with 16 % titanium content are possible.

Cellulose esters with a titanium content between 3 and 5 % do not burn or smolder. They possess considerable hydrolytic stability in neutral and weakly alkaline media, but are easily hydrolyzed at a low pH.

1.4.6. Cellulose Nitrite

The nitrite of cellulose came of scientific and possibly practical interest as a cellulose derivative in 1974 [67]. It is obtained by reacting cellulose with nitrosyl compounds such as dinitrogen tetroxide, N_2O_4 (corresponding to nitrosyl nitrate), or nitrosyl chloride, NOCl, in dimethylformamide or dimethylacetamide as a proton acceptor and solvent for the resulting ester. The reaction proceeds in a homogeneous phase to the trinitrite.

$$\text{Cell-OH} + \underset{\underset{O}{\|}}{N}\text{-X} \rightarrow \left[\text{Cell-O}\underset{\underset{O}{\|}}{\overset{\overset{H}{|}}{\underset{+}{-N}}}\text{X}^-\right] \rightarrow \text{Cell-O-}\underset{\underset{O}{\|}}{N} + H^+X^-$$

$$X = -O-NO_2 \text{ or } -Cl$$

Cellulose nitrite is extremely sensitive to hydrolysis. Chain degradation to a DP of 200 (level-off DP) was observed to take place within 3 h in the presence of water. The scientific and preparative importance of cellulose nitrite is based on its high reactivity, which may be used to produce many other cellulose esters, also mixed esters, by transesterification in a homogeneous phase [82]. Transesterification to stable cellulose sulfates has already been mentioned [67]. In this manner, water-soluble cellulose nitrates with a DS value of 0.5 – 0.6 may also be obtained [83].

Cellulose solutions produced under cold conditions (up to ca. 5 °C) in a N_2O_4/DMF system are relatively stable to degradation and can be produced, depending on the DP of the cellulose, up to a concentration of 14 %. The cellulose can be regenerated in an unaltered form to cellulose II, with the result that this process has already been considered as an alternative to the environmentally detrimental viscose process [84]. Not only an attempt was made to achieve good mechanical textile properties from the regenerated fibers, but also to recycle the expensive solvent. An economic solution to the competition with the viscose process has not yet been found.

1.4.7. Cellulose Xanthate

Cellulose xanthate [9032-37-5], an important intermediary molecule for the production of regenerated cellulose according to the viscose process (see Cellulose Section 3.2.1), must also be considered as an ester of an inorganic acid, namely the nonexistent thiol – thion carbonic acid.

$$S=C\diagup^{SH}_{OH}$$

The O ester of this compound with organic residues is the xanthic acid and the appropriate salts. Sodium cellulose xanthate is obtained by reacting alkali cellulose with carbon disulfide, which dissolves in dilute sodium hydroxide to an orange-yellowish, highly viscous solution, the so-called viscose.

$$\text{Cell-O}^-\text{Na}^+ + C\diagup^S_S \longrightarrow \text{Cell-O-C}\diagup^{S^-}_S \text{Na}^+$$

The regenerated cellulose is precipitated in the form of fibers (rayon, cord, and rayon staple), foils (cellophane), or tubes in precipitation baths containing sulfuric acid and salts (see → Cellulose, Section 3.1.3.). About 4×10^6 t of regenerated cellulose is presently produced worldwide by using the viscose process.

For further information, see → Cellulose, Chap. 3..

2. Organic Esters

Theoretically, cellulose can be chemically converted into an unlimited number of organic acid esters because each of its anhydroglucose units possesses three reactive hydroxyl groups. Industrial possibilities however, are drastically limited by the complex nature of the cellulose polymer. Highly esterified organic esters therefore are normally produced from a few aliphatic fatty acids with up to four carbon atoms.

Cellulose acetate [9004-35-7], cellulose acetate propionate [9004-39-1], and cellulose acetate butyrate [9004-36-8] have been manufactured for over 70 years and continue to be commercially viable products. Because of their instability, formic acid esters are of no industrial importance.

The only other commercial organic ester of cellulose, cellulose acetate phthalate [9004-38-0], has found use as an enteric coating in pharmaceutical applications. A recently developed commercial cellulose ether-ester, carboxymethyl cellulose acetate butyrate [160047-24-5], has found utility in water-based coating systems. None of the other cellulose esters of organic acids, such as cellulose palmitate, cellulose stearate, esters of unsaturated acids such as crotonic acid, or esters of dicarboxylic acids, are industrially manufactured.

Historical Aspects. Cellulose acetate was first synthesized by P. SCHUTZENBERGER in 1865 by heating cellulose and acetic acid under pressure, whereby a product of very low molecularity was obtained. In 1879, A. P. N. FRANCHIMONT added sulfuric acid to the esterification process, which remains to this day the most frequently used catalyst. The limited solubility of cellulose acetate in less-expensive solvents and poor compatibility with the then-known softeners was a considerable obstacle for its industrial use. The problem was solved in 1904 when F. D. MILES and A. EICHENGRüN simultaneously succeeded in synthesizing an acetone-soluble secondary acetate by partially hydrolyzing a primary triacetate.

During World War I, the less-flammable airplane paints based on cellulose secondary acetate reached considerable importance as a replacement for nitrocellulose. At almost the same time, the manufacture of foils, films, synthetic silk, and plastic masses developed.

An especially high number of publications and patents were achieved between 1920 and 1935. Ultimately, only a few processes proved to be industrially useful, most of which are still used today.

The technology of cellulose ester manufacture has been well established for quite some time. Research continues with respect to rationalization and improvement of production methods to provide products at greater efficiencies and with greater uniformity and improved properties. Development of new applications for cellulose esters continues into the modern era as societal innovations generate uses for their unique properties.

2.1. Cellulose Acetate

2.1.1. Chemistry

The chemistry of cellulose acetate includes two discreet chemical transformations that convert the fibrous cellulose polymer into a commercially functional thermoplastic. The first step involves acetylation of all primary and secondary hydroxyl groups to give a fully esterified cellulose triacetate species that can be isolated or processed further. The relative reactivities of primary versus secondary hydroxyl groups have little consequence in industrial acetylations [83], [86]. The second chemical step involves hydrolysis of a portion of the acetyl groups to regenerate hydroxyl groups and provide a product with the desired level of acetyl substitution per anhydroglucose unit on the cellulose backbone. The level of acetyl substitution is noted as degree of esterification (degree of substitution, DS), where the maximum theoretical DS is three. Other important chemical transformations during the preparation of cellulose acetate include sulfation and desulfation of hydroxyl groups, and molecular weight reduction all as a result of the catalytic effect of the predominant acetylation catalyst, sulfuric acid [87].

Potential cellulose acetylation reagents include acetic acid, acetyl chloride, ketene, and acetic anhydride. Acetic acid reacts very sluggishly with cellulose yielding esters with very low acetyl content [88], [89]. Use of acetyl chloride has been investigated in combination with catalyst, however, no practical commercial process has evolved. Ketene, which would produce no byproduct, has not been demonstrated commercially as an effective acetylation reagent [90]. Only acetic anhydride has achieved commercial significance, where three molecules of acetic anhydride react with three hydroxyl groups in each anhydroglucose unit, yielding three molecules of acetic acid byproduct. The use of acetic anhydride was first demonstrated in 1869 by direct reaction with cellulose in a sealed tube at 180 °C [91]. This elevated temperature most probably would have produced a severely degraded product. The discovery of the potential benefits of catalyst by FRANCHIMONT in 1879 eventually led to opportunities to prepare cellulose ester derivatives at lower temperatures without excessive degradation [92].

A number of acetylation catalysts have been identified [93], however, sulfuric acid is the predominant catalyst of commercial importance. Perchloric acid can be used commercially, however, it presents equipment corrosion issues and safety concerns with respect to the potential instability of its neutralized salts. Other mineral acids are not acidic enough in the acetic acid-acetic anhydride esterification media to be effective. Zinc chloride is no longer used commercially due to the high dosage required (0.5 – 1.0 parts per part cellulose) and recovery costs.

One major advantage of sulfuric acid is that it immediately absorbs onto the cellulose fiber surface during the cellulose pre-swelling stage (pretreatment) prior to acetic anhydride addition. This pretreatment serves to swell the fiber and allows for a more uniform catalyst distribution, which enhances the subsequent reactivity of the cellulose mass. The presence of sulfuric acid during the pretreatment stage also provides a desired level of reduction in chain length through catalyzed hydrolysis of the glycosidic linkages of the cellulose backbone [94], [95].

Upon addition of acetic anhydride (in stoichiometric excess) to the pretreated cellulose mass, the sulfuric acid species immediately bonds to the cellulose hydroxyl groups to form a cellulose sulfate ester acid intermediate [96], [97]. It is further known that sulfuric acid reacts with acetic anhydride to form acetyl sulfuric acid, and therefore both sulfuric acid and acetyl sulfuric acid are believed to play important roles during the esterification reaction [98]. The cellulose sulfate ester acid intermediate reacts with the acetic acid-acetic anhydride medium replacing the sulfate ester group with acetyl. This exothermic esterification must be balanced with the rate of cellulose chain length reduction, via catalyzed acetolysis, in order to meet product molecular weight requirements.

The acetyl hydrolysis reaction is catalyzed by sulfuric acid. It is initiated after the esterification reaction has completed by the addition of water in excess of the amount needed to react with the remaining anhydride. The catalyst and water concentrations are usually adjusted to control the acetyl hydrolysis rate and the level of hydrolytic chain length reduction. Upon hydrolysis to the desired degree of acetyl substitution, the catalyst is neutralized with an acetate salt of calcium, magnesium, or sodium.

As a consequence of the catalytic action of sulfuric acid, a significant amount of the sulfate remains chemically bonded to the hydroxyl groups on the polymer backbone in the form of combined sulfate ester after the acetylation reaction has completed. The major portion of this sulfate ester is hydrolytically cleaved during the water addition. The subsequent hydrolysis reaction also serves to remove additional combined sulfate ester groups [96].

2.1.2. Raw Materials

Cellulose. Post-process performance requirements dictate the use of high purity celluloses for acetate manufacture. The two major naturally occurring cellulose raw material sources for cellulose acetate are cotton and wood.

Cotton linters provide an exceptionally high purity raw material with an α-cellulose content of greater than 99 %. After the long layered spinnable cotton has been freed of the cotton seed by ginning, the remaining shorter fibers on the seed pod are usually removed with two cuts before the seeds go to the oil presses for further processing. The first cut gives about 4 % longer linters relative to the entire cotton flower, which are preferentially processed to medicinal cotton, felt, paper, etc. The second cut gives about 8 % shorter layered linters, which are best suited for further chemical processing.

The raw linters undergo mechanical cleaning by means of screening, pressure boiling in a 3 – 5 % sodium hydroxide solution, and finally acid–alkaline bleaching. Special care should be taken during drying, since local over-drying of cellulose (the water content of which should lie between 3 and 8 %) impairs the reactivity considerably. Table 9 shows analytical values of good linters [99].

Table 9. Analytical values obtained from bleached linters according to [99]

α-Cellulose	99.7 %
β-Cellulose	0.2 %
γ-Cellulose	0.1 %
Carboxyl groups	<0.02 %
Total ash	0.02 %
Degree of polymerization	1000 – 7000

Wood pulp has become the most used raw material for cellulose acetate manufacture. Both softwood (conifer) and hardwood (deciduous) pulps can be purified for this purpose. In early days of cellulose acetate manufacture, wood pulp could only be used for the manufacture of lower-quality products because of the 90–95 % α-cellulose content. Improvements in sulfite and Kraft pulping techniques have allowed for more efficient lignin and hemicellulose removal, providing pulps with greater than 95 % α-cellulose content. Typical pulp properties are provided in Table 16. Higher purity wood pulps have been available since the 1950s and have steadily replaced cotton linters due to cost advantages. Wood pulp-based cellulose acetates are comparable to those produced from linters with respect to tensile strength, color, clarity of the solutions, and light and thermal stability.

Acetic Anhydride. Most manufacturers of cellulose acetate convert the byproduct acetic acid to anhydride directly on the premises and adjust the concentrations as required for their process, generally between 90 and 95 %.

2.1.3. Industrial Processes

Only a few of the proposed industrial processes for the manufacture of cellulose esters have attained industrial significance. Despite the fact that no two manufacturers use identical processes, the following categories can be distinguished:

1. Acetylation in a homogeneous system (solution acetate process)
   Use of glacial acetic acid as a solvent (glacial acetic acid process)
   Use of methylene chloride as a solvent (methylene chloride process)

2. Acetylation in a heterogeneous system (fiber acetate process)

In the solution acetate process, the reaction begins heterogeneous in nature with cellulose fibers dispersed in the reactants and solvents. Upon acetylation to the cellulose triacetate species, the fibers dissolve in the reaction medium to form a homogeneous viscous solution. The dissolved cellulose triester can subsequently be solution hydrolyzed to give the desired level of acetyl and hydroxyl groups in a uniform manner. Direct esterification of cellulose to an acetyl level lower than the triester has not proven feasible.

In the fiber acetate process, a cellulose triacetate fiber is formed in the presence of nonsolvents, similar to the nitration of cellulose. This method does not permit uniform hydrolysis of the acetyl groups.

A flow diagram of the entire glacial acetic acid solution process is shown in Figure 7.

2.1.3.1. Pretreatment

From a traditional chemical perspective, hydroxyl group esterification and ester hydrolysis appear relatively straightforward, however the unique morphology of the cellulose fiber presents challenges that require unique chemical processing techniques. One such step is pretreatment (or activation), which involves swelling the cellulose fibers to allow for efficient diffusion of the acetylation chemicals during the subsequent esterification reaction. A range of swelling agents can theoretically be used, including water, aliphatic acids, alcohols, amines and 10 % aqueous sodium hydroxide solution, however, acetic acid is the most commonly used pretreat reagent for manufacture of cellulose acetate [94]. Non-acetic acid swelling agents need to be removed by exchange with acetic acid prior to acetylation.

The cellulose is generally used with a moisture content between 4 and 7 %. The actual moisture content is heavily dependent on the drying history at the cellulose producer and level of humidity prior to use. With respect to activation, the presence of pulp moisture is beneficial, however it reacts with anhydride during the esterification reaction, driving up the cost of production. Low pulp moistures tend to reduce the activation effectiveness leading to a sluggish esterification reaction.

The ratio of acetic acid to cellulose, activation times, and temperatures are varied based on the manufacturer's process. The activation medium can also contain a small portion of sulfuric acid to further improve the effectiveness of the fiber swelling and to provide time for even distribution of the catalyst prior to acetylation. The catalyst also imparts a desired degree of cellulose chain length reduction through hydrolysis of the glycosidic linkages. The amount of chain length

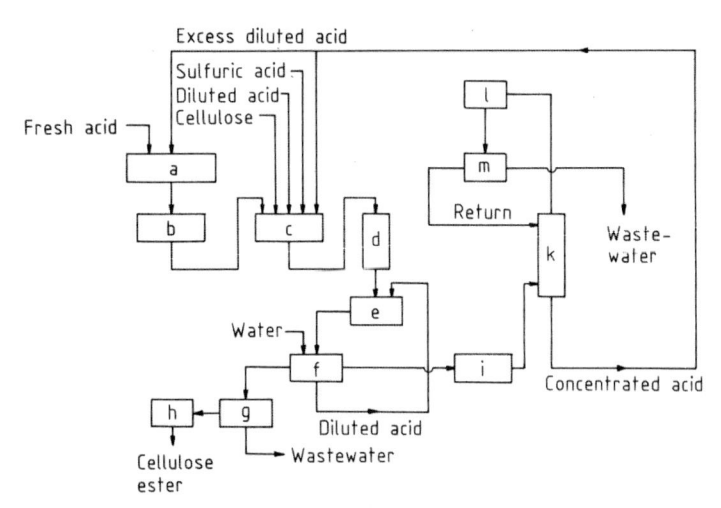

Figure 7. Flow chart for the production of cellulose esters according to [103] a) Acid reconditioning; b) Acidanhydride; c) Esterification; d) Hydrolysis; e) Precipitation; f) Washing; g) Centrifuge; h) Drier; i) Evaporator; k) Azeotropic distillation; l) Cooler; m) Decanter

reduction depends on the temperature, uniformity of the catalyst distribution, and amount of water present during this stage. Higher inherent pulp moisture levels retard the rate of chain length reduction via solvation of the strong acid catalyst [95].

2.1.3.2. Esterification

Acetic Acid Process. The heterogeneous acetylation mixture consists of glacial acetic acid, an excess of 10 – 40 % acetic acid anhydride, the activated cellulose mass, and 2 – 15 % sulfuric acid catalyst based on the cellulose weight. The quantities of each component depend on the manufacturer's acetylation process.

Upon addition of acetic anhydride and catalyst to the activated cellulose mass, the esterification reaction initiates with a rapid exothermic reaction between water contained in the activated cellulose and a portion of the acetic anhydride. The exothermic esterification reaction then proceeds in a heterogeneous fashion whereby the unesterified and partially esterified cellulose fibers are dispersed in a semiliquid mass. The reaction temperature is controlled by reaction vessel cooling and the use of pre-cooled anhydride and acid. Typical reaction temperatures can be up to 50 °C. As the fibers esterify to the cellulose triacetate species, they begin to dissolve in the reaction medium. Upon complete esterification and dissolution of the fibers, a smooth highly viscous solution is formed. A controlled amount of chain length reduction is desirable for purposes of solubility in the reaction medium and final product properties. The amount of chain length reduction is determined by catalyst level, reaction temperature, reaction time, and acid : anhydride ratio [100], [101].

The acetic acid : acetic anhydride reaction medium is actually a poor solvent for the fully substituted triacetylated species. Therefore, the solubilized triacetate species must contain a small portion of chemically combined sulfate ester, which is a reaction intermediate in the sulfuric acid catalyzed esterification. If the reaction is allowed to proceed to the fully substituted cellulose triacetate via replacement of the remaining sulfate ester groups with acetyl groups, the solution viscosity will increase dramatically until the reaction mass becomes gelled. This is known as false viscosity as the molecular mass of the polymer chain actually continues to decrease during this phase. False viscosity is also caused by the presence, type, and quantity of hemicelluloses in the starting cellulose, which become esterified during acetylation [102]. Understanding the ramifications of false viscosity during the esterification reaction is important to providing a final product with the desired molecular weight.

After the reaction solution is free of fibers, the reaction is quenched by adding water or dilute water in acetic acid, which reacts with the excess anhydride and provides water for the subsequent hydrolysis reaction.

Cooled kneaders are suitable reaction vessels in that they allow a rapid and uniform mixture and catalyst distribution through intensive mixing, which is important for controlling the reaction (Fig. 8).

Methylene Chloride Process. Using methylene chloride (*bp* 41 °C) as a solvent presents several advantages over acetic acid. Because methylene chloride is an excellent solvent for cellulose triacetate, lower catalyst concentrations (1 % sulfuric acid) are required at higher esterification temperatures. Furthermore, due to its low boiling point, the heat of reaction can be removed by means of vaporization and return of the cooled methylene chloride. The reaction of highly viscous solutions can, thus, be better controlled. Finally, only a half to a third as much dilute acetic acid must be recycled compared with the glacial acetic acid process.

Table 10 shows typical acetylation recipes for the glacial acetic acid and methylene chloride processes. Figure 9 shows a scene for the production of cellulose acetate according to the methylene chloride process.

Acetylation according to the methylene chloride process is carried out largely in rotating drums (roll vats) or in horizontal containers with shovel-like stirrers on both sides. The problem of corrosion, which arises during esterification and especially during hydrolysis, has only been partially solved by using equipment constructed of bronze, high-alloy steels, or plates containing metals such as silver, titanium, or tantalum.

Fiber Acetate. Cellulose can be esterified maintaining its fiber structure by adding sufficient amounts of nonsolvents during acetylation.

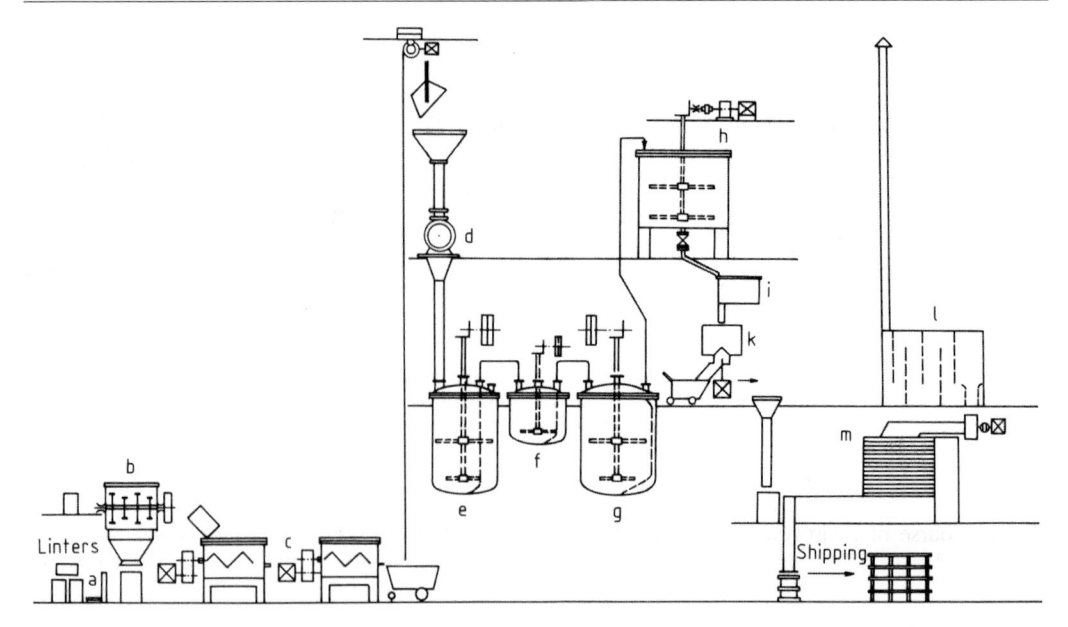

Figure 8. Cellulose acetate production by the kneader method according to [104] a) Weighing scale; b) Sprinkling vat; c) Kneader; d) Mill; e) Rinsing vat; f) Stabilizing vat; g) Bleaching vat; h) Floater; i) Stock pan; k) Centrifuge; l) Dust chamber; m) Drier

Carbon tetrachloride, benzene, or toluene can be used as nonsolvents [106], [107].

Temperature and catalyst concentrations are similar to those required for the solution process. A large amount of nonsolvent is required to keep the loose voluminous cellulose in suspension. Perchloric acid is preferred over sulfuric acid as a catalyst because of the great difficulty in removing the resulting combined sulfuric acid ester in this process.

A uniform hydrolysis of the fibrous acetate to an acetone-soluble material is not possible in a heterogeneous system. Therefore, the use of the fiber acetate process is limited to special applications, such as the manufacture of foils and films from triacetate.

Table 10. Acetylating preparations according to the glacial acetic acid process and the methylene chloride process [104]

	Acetic acid process	Methylene chloride process
Cellulose	700 kg	3 500 kg
Pretreatment	700 kg glacial acetic acid	1 200 kg glacial acetic acid
Acetylation	1 900 kg anhydride 4800 kg glacial acetic acid 50 kg sulfuric acid	10 500 kg anhydride 14000 kg methylene chloride 35 kg sulfuric acid

As shown in Figure 10, fiber acetate is produced by rotation in various directions and at various speeds in a perforated drum enclosed in a metal casing. The shaft of the drum is hollow so that liquid may be added during rotation [108].

2.1.3.3. Hydrolysis

The solution esterification process provides a solubilized cellulose triacetate species, which can be isolated for commercial use. This cellulose triacetate is also known as primary acetate and has an approximate acetyl DS of 2.9. Commercially, the most common hydrolyzed cellulose acetate product is cellulose diacetate, also known as secondary acetate. Cellulose diacetate has an approximate acetyl DS of 2.5.

After the addition of water at the end of esterification, the water content is adjusted to the desired level, generally 5–10%, the concentration of which controls the amount of primary versus secondary hydroxyl groups in the final product [109]. Depending on the manufacturer's process, typical hydrolysis temperatures range between 40 and 80 °C. The acetyl hydrolysis rate is mostly determined by catalyst concentration and temperature. Higher water concentrations can serve to prevent excessive chain length reduction [110].

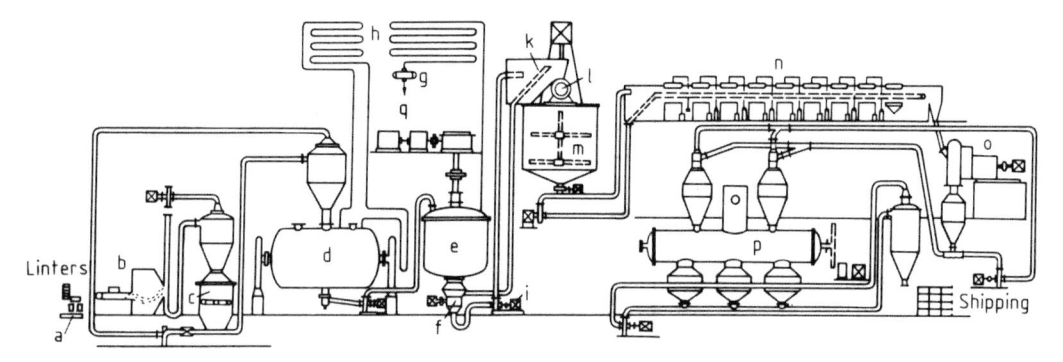

Figure 9. Cellulose acetate production according to the methylene chloride process [105] a) Weighing scale; b) Bale opener; c) Sprinkling vat; d) Acetylator; e) Precipitating vat; f) Prebreaker; g) Vacuum vessel; h) Pipe cooler; i) Pump for viscous substances; k) Filter bath; l) Mill; m) Floater; n) Sprinkling line; o) Centrifuge; p) Vacuum shovel drier; q) To reprocessing of methylene chloride

The course of hydrolysis is constantly monitored by checking the solubility of the secondary acetate. Upon completion of hydrolysis, the sulfuric acid catalyst is neutralized with magnesium, sodium, or calcium acetate, which also serves to neutralize the remaining small portion of bonded sulfate on the polymer backbone for stability purposes during isolation and post-processing steps.

2.1.3.4. Precipitation and Processing

Isolation of the cellulose acetate from the viscous hydrolysis solution involves precipitation of the polymer by either pouring the viscous solution into water, which can also contain a portion of acetic acid, or by adding a dilute solution of acetic acid in water to the stirred solution. The former procedure provides flake particles, whereas the latter procedure provides a powder. The morphology of the precipitate is carefully controlled by the precipitation conditions, which include the acetic acid concentration, degree of agitation, and temperature. Proper control of precipitation conditions provides a particle with an open pore structure, allowing for efficient removal of the acids and salts in the subsequent water washing operation. The methylene chloride process requires that the methylene chloride be completely removed by distillation prior to precipitation. The precipitate can be broken down and thoroughly washed, and the resulting dilute acetic acid can be fed back into the manufacturing cycle.

Water washing is primarily conducted using continuous methods based on the countercurrent principle (Fig. 9). High-quality products for plastic applications are stabilized and bleached. The remaining residual combined sulfate can be removed by either boiling under pressure or heating in 1% mineral acids during stabilization.

After further rinsing and removal of excess water by suctioning, centrifugation, pressing, or by thrust extraction, the product is carefully dried, preferably in a vacuum shovel drier, to less than 1–3% water content. From an efficient process, the cellulose acetate yield is at least 95% of theoretical.

Manufacturers of cellulose acetate include: Eastman Chemical Company, United States; Rhodia Acetow, Germany; Celanese AG, United States; Acordis, United Kingdom; Daicel Chemical Industries, Ltd., Japan; and Acetati SPA, Italy.

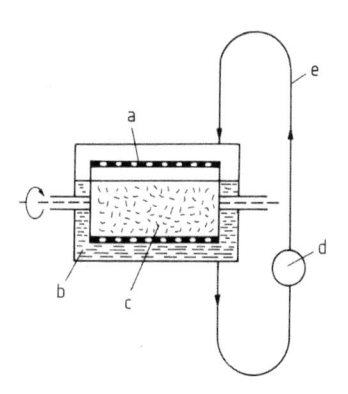

Figure 10. Acetylation equipment for cellulose triacetate fibers [104] a) Perforated drum; b) Reaction solution; c) Cellulose fibers; d) Cooler for acetylating liquid; e) Acetylating liquid circulation

2.1.4. Recovery of Reactants

The recovery of the large amount of acetic acid remaining after product isolation is a critical factor in minimizing costs, which is crucial for the profitability of a cellulose acetate process.

Recovery of Acetic Acid. Depending on the process, 2 – 6 parts of 15 – 25 % dilute acetic acid per part of cellulose remain after isolation of the cellulose ester. This acetic acid is recovered and reprocessed to glacial acetic acid and acetic acid anhydride. For acetic acid recovery, only continuous processes consisting of a combination of extraction and azeotropic distillation are of practical importance. The dilute acid is extracted with ethyl acetate in a countercurrent fashion. This extract is subsequently distilled to remove azeotropic ethyl acetate-water from the top of the column, while yielding 99.8 % glacial acetic acid from the bottom of the column.

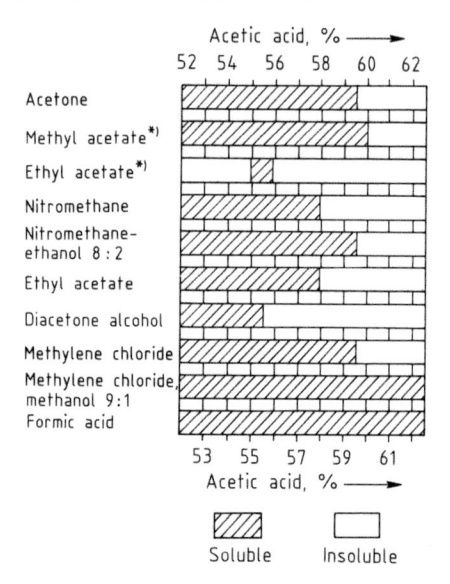

Figure 11. Solubility of cellulose acetate in various solvents (abridged according to [116]) * Technical grade

Recovery of Acetic Anhydride. Since only a portion of the accumulated glacial acetic acid is required for the acetylation process, the remainder must be converted to acetic anhydride.

The ketene process developed by the Wacker Co. (→ Acetic Anhydride and Mixed Fatty Acid Anhydrides, Section 1.3.1.1.) can be used [111]. Pure, almost anhydrous glacial acetic acid is continuously vaporized under a vacuum and converted to ketene in the presence of small amounts of the catalyst, triethyl phosphate. Ketene proceeds to react with glacial acetic acid to form the anhydride (→ Acetic Anhydride and Mixed Fatty Acid Anhydrides, Section 1.3.1.2.).

Recovery of Methylene Chloride. Methylene chloride can be recovered inexpensively because its insolubility in water allows for isolation in almost pure form from the raw solution without further processing steps.

2.1.5. Properties

Properties of cellulose acetates are generally determined by their molecular weight and degree of acetyl substitution. Solution viscosity is used as an indicator for molecular weight, which also correlates to the mechanical properties of the resulting fibers, films, coatings, or plastics. The degree of acetyl substitution primarily determines the solubility and compatibility with solvents, plasticizers, softeners, resins, varnishes, etc., and ultimately also influences the mechanical properties. With decreasing acetyl content, compatibility with plasticizers and solubility in polar solvents increases, while solubility in nonpolar solvents decreases. Figure 11 shows solvent solubility ranges of acetyl substituted celluloses of industrial interest. Also, the span of solubility properties of the full theroretical range of hydrolyzed cellulose acetates is provided in Table 12. This table also provides the relationship between degree of esterification (DS), acetyl content, and bound acetic acid content, which are used interchangeably to describe the degree of acetyl substitution on the polymer backbone.

As described previously, the two commercially produced cellulose acetates are cellulose triacetate (approximate acetyl content of 43.6 %) and cellulose diacetate (approximate acetyl content of 39.5 %). Both cellulose acetates are white, nontoxic, odorless, and tasteless materials, which are commercially available as a powder or flake. They are less flammable than nitrocellulose, resistant to weak acids and largely stable to mineral and fatty oils as well as petroleum. Some physical characteristics of cellulose triacetate and cellulose diacetate are given in Table 11.

Table 11. Physical characteristics of cellulose acetate [99], [104], [115]

Characteristic	Triacetate	Secondary acetate
Density, g/cm^3	1.27 – 1.29	1.28 – 1.32
Thermal stability, °C	>240	ca. 230
Tensile strength of fibers, kg/mm^2	14 – 25	16 – 18
Tensile strength of foils		
longitudinal, kg/mm^2	12 – 14	8.5 – 10
transverse, kg/mm^2	10 – 12	8.5 – 10
Refractive index of fibers toward the fiber axis		
longitudinal	1.469	1.478
transverse	1.472	1.473
Double refraction	−0.003	+0.005
Dielectric constant ε		
50 – 60 Hz	3.0 – 4.5	4.5 – 6.5
10^6 Hz		4.0 – 5.5
Dielectric loss factor tan δ		
50 – 60 Hz	0.01 – 0.02	0.007
10^6 Hz		0.026
Specific resistance, Ω·cm	10^{13} – 10^{15}	10^{11} – 10^{13}
Specific heat, J g^{-1}K^{-1}		1.46 – 1.88
Thermal conduction, J m^{-1}h^{-1}K^{-1}		0.63 – 1.25

Cellulose triacetate is a crystalline polymer that melts with significant decomposition at temperatures close to 300 °C. It has very limited compatibility with plasticizers, making it impractical for melt processing, and is soluble in a very limited number of solvents, the most commercially important being methylene chloride, which is used for film casting applications.

Cellulose diacetate is significantly less crystalline than cellulose triacetate, is soluble in a wider range of solvents, and compatible with a number of commercially useful plasticizers. The most important industrial solvent for cellulose diacetate is acetone, which serves as the basis for the long-standing cellulose acetate fiber business (see Section 2.3). Melt processing applications require use of plasticizers to provide a greater separation between decomposition and melting temperatures and to improve thermoplastic flow properties. Plasticizers also impart desired physical properties in the final product, e.g. they increase the flexibility and toughness of an otherwise brittle polymer. Common compatible cellulose diacetate plasticizers include triacetin, triethyl citrate, diethyl phthalate, and triphenyl phosphate. Careful plasticizer choice is important with respect to permanence in long-term applications [112]. Detailed information on solvent solubility and plasticizer selection can be found in the literature and the manufacturer's informational brochures [112][113][114].

2.1.6. Analysis and Quality Control

Standard test methods for cellulose acetate can be found under ASTM D871–96. The viscosity is determined in practice by usual methods. Along with the relative viscosity, the measurement of 15 – 20 % solutions according to the ball drop method corresponding to ASTM D1343–95 (2000) has been generally accepted.

The acetyl content is generally determined by saponification of the ester with 0.5 N-potassium hydroxide and back-titration of the excess. Historically, free acids from mixed esters (see Section 2.2) are separated from the residue after saponification by means of distillation or extraction and by extraction or azeotropic distillation followed by titration. A comprehensive presentation can be found in [117], [118]. Gas chromatographic methods are routinely used in the modern era for free acid analysis after saponification and acidification.

Determination of the free hydroxyl groups in pure cellulose acetate is not necessary, since a precise analysis of bound acetic acid is possible. Hydroxyl content can be determined by complete esterification with acetic acid anhydride in pyridine and back-titration of the excess [119]. The value is

Table 12. Solubility of cellulose acetate at various degrees of esterification [103]

Degree of esterification	Acetyl content, %	Bound acetic acid, %	Chloroform	Acetone	2-Methoxyethanol	Water
2.8 – 3.0	43 – 44.8 %	60 – 62.5 %	soluble			
2.2 – 2.7	37 – 42 %	51 – 59 %		soluble		
1.2 – 1.8	24 – 32 %	31 – 45 %			soluble	
0.6 – 0.9	14 – 19 %	1.8 – 2.6 %				soluble
<0.6	14	<18 %				

mostly given as a percentage of the hydroxyl or as the hydroxyl value (mg of KOH/g).

Additional quality control methods for unprocessed cellulose esters include: determination of temperature stability by heating to 220 – 240 °C and evaluation of discoloration and melting behavior, determination of free acid as an indicator for the efficacy of the washing process, and determination of the ash content as well as clarity, color, and filterability of the solution.

2.1.7. Uses

Cellulose acetates are used in a broad range of commercial applications including the general areas of films, fibers, plastics, and coatings. Fiber applications of cellulose triacetate and cellulose diacetate are covered in Section 2.3. Plastic applications for cellulose diacetate are described in Section 2.4. For commercial purposes, cellulose diacetate is simply referred to as cellulose acetate, which will be used when discussing applications and uses during the remainder of this text.

Cellulose Triacetate. Cellulose triacetate has been used as a photographic film base since the 1950s. Cellulose triacetate is solution cast from methylene chloride to provide a clear smooth film that is subsequently coated with photographic film emulsions. Important film properties of the solvent cast cellulose triacetate include optical isotropy, high clarity, toughness, scratch and moisture resistance, dimensional stability, and slowness to burn.

A more recent commercial development for cellulose triacetate film has been in the field of liquid crystal displays (LCD) for flat panel televisions, computer monitors, laptops, personal digital assistants, games, automobile navigation screens, view screens on cameras and camcorders, and mobile phones. In this market, cellulose triacetate serves to protect the polarizing films in the display, while maintaining permeability because of its inherent high moisture vapor transmission rate [120].

Cellulose Diacetate. Cellulose acetate's increased compatibility with common solvents and plasticizers versus cellulose triacetate allows it to be utilized in a wider range of commercial film and coating applications. In the film area, it has found utility as a pressure sensitive tape, where its transparency, resistance to "neck-in", and ease of tear properties are valued. Other film applications include clear packaging windows, print protective laminations, and labels. Cellulose acetate can be used in reverse osmosis applications, where it is film cast to achieve desired porosity characteristics for water purification [123].

Cellulose acetate has been used in coating applications since the First World War, where it replaced the more flammable nitrocellulose in airplane coatings. Acetone is a common coating solvent for cellulose acetate and is known for its fast evaporation rate. Other solvents such as ethanol and slower evaporating retarder solvents are commonly incorporated into the formulation to achieve the desired rheological and coated film properties [121]. Plasticizers are often added to the coating formulation to control and generate desired film properties including flexibility, electrical characteristics, flammability, moisture resistance, and weather resistance. Coating applications for cellulose acetate include lacquers for electric insulators, glass, paper/paperboard, release paper, food packaging, plastic, wire, and wood [121].

Cellulose acetate manufactured using cGMP (current Good Manufacturing Practices) techniques is utilized in the pharmaceutical area where it is formulated with the active drug ingredient in osmotic drug and sustained-release systems [122].

In certain environments, cellulose acetate is biodegradable. The biodegradability is strongly influenced by the degree of acetyl substitution, morphology, and choice of plasticizer [123].

2.2. Cellulose Mixed Esters

Apart from the cellulose acetate, cellulose mixed esters of acetic and propionic acid, cellulose acetate propionate [*9004-39-1*], or acetic and butyric acid, cellulose acetate butyrate [*9004-36-8*], are the only other cellulose esters that have attained any notable importance. Pure cellulose propionates [*9004-48-2*] and cellulose butyrates [*9015-12-7*] can also be prepared.

Cellulose esters of higher acids such as isobutyrates, valerates, caprolates, laureates, palmitates, etc., have been prepared, although they are difficult to produce and generally require unique commercially nonviable techniques [124], [125]. The anhydrides of these acids react sluggishly with activated cellulose due to their high degree of steric hindrance. Esters of unsaturated acids and dicarboxylic acids have attained no industrial importance [4].

2.2.1. Production

With respect to the chemistry and processing, the basic description given in Section 2.1) for cellulose acetate is also applicable to mixed cellulose esters, which similarly uses cotton linters and highly purified wood pulp as the cellulose raw material. Mixed ester manufacture uses the homogeneous solution process exclusively (see Section 2.1.3). The esterification chemicals consist of a mixture of anhydrides of acetic acid and propionic acid or of acetic acid and butyric acid with sulfuric acid used as the catalyst.

In comparison to preparation of cellulose acetate, more effective pre-swelling of the cellulose fiber is desirable for the production of cellulose mixed esters as propionic and butyric acids and their anhydrides are slower to penetrate the fiber during the esterification reaction, and the catalyst is less readily absorbed. The reactivity of aliphatic fatty anhydrides decreases very rapidly as the chain length increases. Therefore, the reaction temperature strategies for mixed esters are somewhat different than in the case of cellulose acetate, especially to avoid excessive cellulose chain length reduction during esterification.

Mixed esters consisting of propionic acid and butyric acid or acetic acid, propionic acid and butyric acid are not produced on an industrial scale.

2.2.2. Composition

Cellulose acetate butyrates and propionates can be produced in a wide range of butyryl (or propionyl), acetyl, and hydroxyl contents, as well as a broad range of molecular weights. The ester composition is carefully controlled in the manufacturing process by the ratio of butyryl (or propionyl) to acetyl components in the esterification medium as well as in the subsequent hydrolysis reaction.

Whereas pure cellulose acetates are characterized by their viscosity and acetyl content, mixed esters require data on the individual acids and possibly free hydroxyl groups. Standard test methods for cellulose butyrates and propionates can be found in ASTM D817–96.

Cellulose acetate propionates and cellulose acetate butyrates are manufactured by Eastman Chemical Company in the United States.

2.2.3. Properties

The properties of mixed esters are determined by their molecular weight, molecular weight distribution, degree of substitution of acetyl, butyryl (or propionyl), and hydroxyl content [126]. A wide range of properties is achievable based on the composition of the polymer.

As the degree of substitution changes, the properties of cellulose mixed esters vary over a wide range from pure acetates to pure butyrates, with the propionates occupying a property-profile position between the cellulose mixed esters and pure acetates. For example, if one considers the mixing range from pure cellulose acetate through the various mixing ratios to pure cellulose butyrate for cellulose acetate butyrates (the degree of esterification is adjusted in such a way that the esters in pure cellulose acetate contain 44.8 % acetyl and those in pure cellulose butyrate contain 57.3 % butyryl groups), then the density varies from ca. 1.32 (cellulose acetate) to 1.16 (cellulose butyrate). Similarly, the melting point ranges from ca. 300 °C (cellulose acetate) to 160 °C (cellulose butyrate), while the water absorption at 90 % relative humidity varies between ca.12 % (cellulose acetate) and 1.5 % (cellulose butyrate). Tensile strength, hardness, and stiffness increase with higher acetyl contents, while flexibility increases with higher butyryl content. In general, the glass transition temperature decreases with increased butyryl content. Solubility in ketones, esters, alcohols, glycol ethers, and glycol ether-esters solvents varies over a wide range with the range of usable solvents generally increasing with higher butyryl content [128], [114]. The range of usable plasticizers also increases with higher butyryl content.

Table 13. Characteristic data of Eastman cellulose acetate propionate [114]

	CAP-482-20	CAP-482-0.5	CAP-504-0.2
Acetyl content, %	2.5	2.5	0.6
Propionyl content, %	46	45.0	42.5
Hydroxyl content, %	1.8	2.6	5.0
Viscosity-ASTM D817 formula A and D1343, (s/poise)	20/76	0.40/1.52	0.20/0.76
Melting range, °C	188 – 210	188 – 210	188 – 210
Glass transition temperature, °C	147	142	159

Commercial cellulose acetate butyrates have acetyl contents from 2.0 to 29.5 % and butyryl contents from 17 % to 55 %. For commercial cellulose acetate propionates, the acetyl content varies from 0.6 % to 2.5 %, and the propionyl content from 42.5 to 46.0 %. There are many more grades of cellulose acetate butyrate available than cellulose acetate propionate [114]. Examples of cellulose acetate propionate and cellulose acetate butyrate grades are given in Tables 13 and 14 respectively.

2.2.4. Other Organic Mixed Esters

Cellulose acetate phthalates [9004-38-0] are cellulose mixed esters prepared from hydrolyzed cellulose acetate and phthalic anhydride providing a carboxy-functionalized cellulosic [130], [131]. Cellulose acetate phthalate (C–A–P) is utilized in the controlled release market as an enteric coating for delivery of pharmaceutic. It is applied as a thin film to a tablet formulation, which allows for a slow rate of disintegration in the stomach and a fast rate of disintegration in the small intestine, where the active is released for absorption into the blood stream. Plasticizers are often incorporated to improve the toughness of the coating to protect the tablet during the manufacturing process and time prior to use [123]. This product can also be used as a water- or alkali-soluble textile auxiliary and an antistatic agent in film coating.

2.2.5. Uses

Mixed cellulose esters are compatible with a wider range of plasticizers, soluble in a wider range of solvents, and compatible with a wider range of resins than cellulose acetate. The ability to modify and control polymer structure including acyl content (butyryl, propionyl, or acetyl), hydroxyl content, molecular weight and molecular weight distribution allows one to "dial-in" specific properties to meet unique requirements in a broad range of applications. Many commercial uses have required the development of a special grade to meet the unique performance requirements of a particular application.

Mixed esters based on cellulose acetate butyrate and cellulose acetate propionate are used in the production of molding plastics covered in Section 2.4.

Films. Cellulose acetate butyrate and cellulose acetate propionate can be solvent-cast into clear smooth films. The range of useable solvents is much greater for mixed esters than for cellulose acetates [114]. Cellulose acetate propionate and cellulose acetate butyrate can be cast from many common ketone, ether, and ester ester solvents. Both cellulose esters have been historically used as a base for photographic films. More recent applications of cellulose acetate butyrate film include polarizing films in sunglasses and privacy films for computer screens, which limits

Table 14. Characteristic data of selected Eastman cellulose acetate butyrate [114]

	CAB-171-15	CAB-381-20	CAB-381-0.5	CAB-500-5	CAB-551-0.2
Acetyl content, %	29.5	13.5	13.5	4.0	2.0
Butyryl content, %	17.0	37	38.0	51.0	52.0
Hydroxyl content, %	1.1	1.8	1.3	1.0	1.8
Viscosity-ASTM D817 formula A and D1343, (s/poise)	15/57	20/76	0.50/1.90	5/19	0.20/0.76
Melting range, °C	230 – 240	195 – 205	155 – 165	165 – 175	130 – 140
Glass transition temperature, °C	161	141	130	96	101

the viewing angle of the automatic teller bank machines. Like cellulose acetates, cellulose acetate butryate can be used as an electrical insulating film.

Surface Coatings. Mixed cellulose ester lacquers, with their excellent lightfastness, gloss, low combustibility, and good thermal stability, coupled with their indifference to hydrocarbons, oils, and greases, quickly became established in numerous coating applications upon their development in the 1930's. Mixed cellulose esters are particularly characterized by lower water absorption, good compatibility with extenders, and in the case of the low-viscosity grades, also allow the production of high-solid lacquers.

Of the two categories of mixed esters, cellulose acetate butyrates are predominant in commercial coating applications. A wide range of cellulose acetate butyrate compositions are manufactured ranging from a very high to a low butyryl : acetyl ratio, from low to high molecular weights, and a broad range of hydroxyl content [114]. Cellulose acetate butyrate with lower butyryl content provides coatings with increased hardness and toughness compared to coatings with higher butyryl content. Higher butyryl esters provide softer and more flexible coatings. Increasing the hydroxyl content provides increased solubility in alcohols, hardness, and reactivity with cross-linking agents. Higher molecular weight esters provide greater toughness and better mechanical properties in the final coated film, while lower molecular weight esters allow for higher solids concentration of the coating formulation.

Mixed esters can be used as the major film former, as a modifying resin, or as an additive in a range of coating applications. Mixed esters are utilized in a wide variety of applications including protective and decorative coating systems for wood, plastic, metal, glass, leather, cloth, and paper/paperboard. They are also used in printing inks, nail care, hot melts, and adhesives. Benefits of cellulose esters include reduced drying time, improved flow and leveling, reduced cratering, sag control, color control, improved sprayability, viscosity control, redissolve resistance, metal flake control, pigment dispersibility, reduced blocking, solvent craze resistance, polishability, and gloss control. Other desirable properties such as weatherability, good feel (due to their low specific heat and thermal conductivity), dimensional stability, low temperature impact strength, and color stability upon exposure to UV light are important features provided by mixed esters [114].

Cellulose acetate butyrates are used in conjunction with other coating resins such as thermoplastic or thermosetting acrylics, polyesters, phenolic, melamine, alkyd, crosslinking urea-formaldehyde, and polyisocyanates imparting many of their desired performance benefits described above.

Like cellulose acetate (Section 2.1.7), cellulose acetate butyrate can be manufactured using "cGMP" practices for use in sustained release of pharmaceutical actives in tablet formulations [122].

Carboxymethyl cellulose acetate butyrate (CMCAB) is relatively new mixed ether-ester, which is produced by the esterification of carboxymethyl cellulose [132]. It has found utility in the water-borne coatings area. This new cellulose ester imparts several of the beneficial features of cellulose acetate butyrate to water-based coating systems [133].

2.3. Cellulose Acetate Fibers

Cellulose acetate is the most important cellulose ester. It is primarily used for textile yarn and cigarette filter tow. The cellulose acetate is usually dissolved in a suitable organic solvent and spun by dry spinning (→ Fibers, 3. General Production Technology). Secondary (2.5) acetate with an acetic acid content of 54 – 56 % is normally used, whereas only a small amount of cellulose triacetate is normally produced.

2.3.1. Properties

The viscosity and the filterability of the spinning solution (spinning dope) are particularly important in the production of cellulose acetate fibers. The spinning dope has a high viscosity, which depends on the degree of polymerization. The strength and stretch properties of the fibers also depend on the concentration and the degree of polymerization as well as on the distribution of the acetate groups along the cellulose chain. Because the fibers are produced by extruding the

Table 15. Physical properties of acetate fibers and tow [134]

	Secondary acetate	Triacetate
Strength, cN/dtex	1.0 – 1.5	1.0 – 1.5
Stretch, %	25 – 30	25 – 30
Density, g/cm^3	1.33	1.30
Moisture uptake, % (65 % relative humidity, 20 °C)	6 – 6.5	4 – 4.5
Water retention capability, %	25 – 28	16 – 17
Melting point, °C	225 – 250	decomposition at 310 – 315
DP	300	300

Table 16. Typical properties of acetate wood pulps

Characteristic^a	Sulfite softwood pulp (conifer)	Sulfate hard-wood pulp (deciduous)
R_{10}, %	95	96
R_{18}, %	97	98
Ash, %	0.08	0.08
Silica, %	0.001	0.003
Calcium, %	0.006	0.008
Pentosans, %	1.2	1.2
Moisture content, %	6.5	6.5
Apparent density, g/cm^3	0.45	0.5
DP	2300	1700

^{a} R_{10} and R_{18} are residues in 10 or 18 % sodium hydroxide at 20 °C [135]

spinning dope through minute spinneret holes, insoluble particles must first be removed from the spinning dope by filtration. These particles are primarily composed of very small, incompletely acetylated cellulose fibers or gels, which can obstruct the spinneret holes.

Secondary acetate and triacetate fibers have similar physical properties (Table 15). Their densities are lower than that of viscose rayon fibers and equal to that of wool. For textile yarns, the fibers should be as free of color as possible.

The chemical reactions of cellulose acetate are similar to those of organic esters. Cellulose acetate is hydrolyzed by strong acids and alkali; it is sensitive to strong oxidizing agents but not affected by hypochlorite or peroxide solutions.

Acetate fibers cannot be dyed under the same conditions as viscose rayon fibers because their swelling properties are different. Acetate fibers can only be dyed with water-disperse dyes at the boiling point of the medium usually in the presence of carriers (→ Textile Auxiliaries, 5. Dyeing Auxiliaries). The carriers promote fiber swelling and enhance dye uptake by the fibers. The dyeing process coupled with the textile spinning operation assures color fastness. Triacetate fibers have better wash-and-wear properties than secondary acetate because of better dimensional stability and higher crease resistance.

2.3.2. Raw Materials

Wood pulp produced from various softwood (conifer) or hardwood (deciduous) species is the cellulose source for the production of cellulose acetate fibers. The wood pulps are produced by the sulfite pulping process with hot alkali extraction or by the prehydrolyzed sulfate (Kraft) process with cold caustic extraction (→ Pulp). The lignins and hemicelluloses are removed from the wood to give wood pulps with an α-cellulose content of over 96 % (Table 16). High-purity cotton linters are no longer used in the production of cellulose acetate fibers for economic reasons.

For the production of high-quality cellulose acetate fibers the wood pulp must have good swelling properties for uniform accessibility of the cellulose to the catalyst and the acetylation agent and a uniform reactivity. In addition, it must produce a spinning solution without fibers and gels which can easily be filtered.

2.3.3. Production

The general points discussed in Section 2.2.5 for the production of cellulose acetate also apply here. The sulfuric acid catalyst initially forms the cellulose sulfate ester. The sulfate groups are then replaced by acetyl groups as the acetylation proceeds. The sulfate ester contents is further reduced in the hydrolysis stage. However, any sulfate ester groups remaining at the end of the hydrolysis stage must be neutralized with an appropriate stabilizer, e.g., magnesium salts [136], [137]. Any "free" sulfate ester groups will affect the stability of the acetate because under the influence of heat and humidity they splitt off as sulfuric acid and degrade the fiber [138].

For secondary acetate spinning, acetone is used as the solvent. For triacetate, the solvent is 90 % dichloromethane and 10 % methanol or acetic acid (wet-spinning process). The viscosity of the spinning solution with a cellulose acetate concentration of 20 – 30 % is between 300 and 500 Pa s at 45 – 55 °C. The spinning dope is filtered in one or more steps and is then deaerated in large vessels.

Dry spinning is used almost exclusively; wet spinning is occasionally used for triacetate only. The spinnerets for textile filament have between 20 and 100 holes and those for tow up to 1000. The fibers are formed by evaporating the solvent with a countercurrent of air at 80 – 100 °C in a 4- to 6-m spinning column. The fibers are then stretched while still plastic to increase their strength. Melt spinning of cellulose acetate or triacetate has no commercial importance due to the limited heat stability at the melting point.

A core-skin structure is formed in triacetate fibers. The acetyl groups are distributed very regularly in cellulose triacetate compared to secondary acetate; therefore, crystallization occurs when triacetate fibers are heated at 180 – 200 °C (heat setting) [134], [138], [139]. This heat treatment, which enhances the wash-and-wear properties of triacetate textiles, requires several minutes at 180 °C or several seconds at 220 °C. Heating for shorter periods is not effective and longer heating periods lead to deterioration of the mechanical properties of the textile. Heat-setting reduces water retention to 10 % and water absorption to 2.5 %.

2.3.4. Uses

By blending and twisting of cellulose acetate or triacetate fibers with nylon or polyester a combination of properties is achieved that make them suitable for different end uses in linings. In this way the weaker physical properties of acetate fibers can be compensated for while maintaining the positive characteristics, for instance, the high moisture absorption and the silk-like softness.

Due to the unique hydrophobic – hydrophilic properties, semipermeable membranes made from cellulose acetate fibers have a remarkable potential in desalination (reverse osmosis) of water.

Cellulose acetate hollow fibers are also suitable for gas separation and hemodialysis [140].

For cellulose acetate is non-toxic, biodegradable and the raw material is a renewable natural polymer, it is expected to find application for other uses in the future.

2.3.5. Economic Aspects

Secondary acetate and triacetate fibers for textiles and filter cigarette tow accout for 80 % of all cellulose ester production. The balance is used for plastics and film. Secondary acetate and triacetate textile fibers have a small share (about 1 %) of all textile fiber production.

In the late 1990s synthetic fiber production continued to expand (Table 17), whereas the acetate production was stable at about 850 000 t/a. The acetate fiber production decreased slightly compensated by a slight increase in filter tow production. The five largest manufactures of filter tow are Hoechst-Celanese and Eastman Chemicals in the United States, Rhodia Acetow in Germany, Daicel in Japan, and Acordis in the United Kingdom.

2.4. Cellulose Ester Molding Compounds

Cellulose esters represent a category of plastics that are derived from a natural, renewable, and sustainable resource, cellulose. In the category "plastics made from natural materials", thermoplastics based on cellulose acetate or cellulose mixed esters remain an important category of commercial plastics [142], [143]. As early as 1920, A. EICHENGRüN developed thermoplastic cellulose ester molding compounds as a spraying and molding powder. Cellulose acetate and mixed esters are used in injection molding and extrusion; mixed esters are also used for fluidized-bed dip coating and rotational molding.

Table 17. Worldwide production of textile fibers (1000 t) [141]

Fiber	1993	1995
Man-made fibers[a]	19781	21741
Synthetics	16652	18471
Cellulosics	3129	3270
Cotton	18494	18602
Wool	1687	1767
Silk	68	92

[a] Excl. polyolefin fibers, textile glass fibers, and acetate cigarette filter tow.

The use of inorganic cellulose esters (see Section 1.3.7) is continually decreasing in the plastics sector because of high flammability properties.

2.4.1. Physical Properties of Cellulose Ester Plastics

By themselves, cellulose esters have melting points close to their thermal decomposition temperature and therefore undergo excessive degradation upon melt processing. Addition of a compatible plasticizer reduces the melting range of the cellulose ester, resulting in an easily melt processable polymer. The plasticizer not only significantly improves the melt processability, it also modifies the properties of the polymer to give a softer, tougher, and more flexible plastic when compared to the cellulose ester itself. Common plasticizers for cellulose esters are listed in Section 2.4.4.

The individual cellulose esters generally differ in their mechanical properties and compatibility with plasticizers. Mixed esters are compatible with a wider range of plasticizers than cellulose acetate. As a rule, mixed ester molding compounds have plasticizer loadings ranging from 3 – 25 %, whereas cellulose acetate molding compounds generally contain 15 – 35 % plasticizer. Cellulosic plastic properties vary with ester composition as well as choice and level of plasticizer, resulting in a broad range of conceivable physical properties. Lower plasticizer levels provide a harder surface, greater rigidity, higher heat distortion temperature, higher tensile strength, and better dimensional stability. Impact strength, extensibility, and softness are increased at higher plasticizer levels. Figures 12, 13, 14, 15, 16, 17, 18, 19, 20, 21, 22 show the physical properties as a function of the plasticizer content; the property levels may vary by as much as 15 – 20 % in either direction, depending on the type of plasticizer and the relative viscosity of the cellulose ester. In all of the diagrams and tables featured here, the abbreviations used are as follows:

- CA = Cellulose acetate molding compound (acetic acid content approximately 39 %)
- CP = (CAP) Cellulose acetate propionate molding compound
- CAB = Cellulose acetate butyrate molding compound

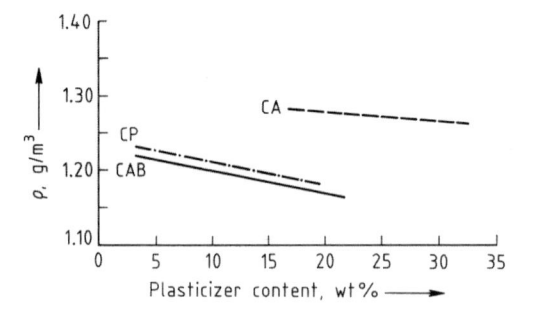

Figure 12. Density of cellulose acetate (CA), cellulose acetate propionate (CP), and cellulose acetate butyrate (CAB) as a function of the plasticizer content (determined in accordance with DIN 53479 or ISO/R 1183)

The indices (e.g. CAB_{10}) give the plasticizer content in percent by weight.

Table 18 shows the electrical properties of medium-hardness cellulose ester molding compounds; their shear moduli and damping properties are given in Figure 23. Figure 24 shows the position of the damping maxima as a function of the plasticizer content.

Long-term properties derived from the tensile creep test are shown in Figures 25–29. Time-to-failure curves, modulus of creep curves and isochronous stress – strain curves of slightly and highly plasticized grades of cellulose acetates, cellulose acetate propionates, and cellulose acetate butyrates are given here.

Figure 30 shows results of the dynamic fatigue test in the tensile pulsating range on a medium-plasticity cellulose acetate, a slightly plasticized cellulose acetate propionate, and a medium-plasticity cellulose acetate butyrate.

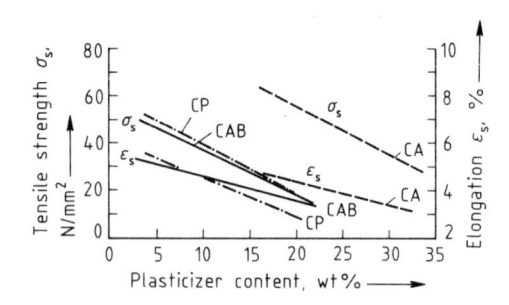

Figure 13. Tensile strength at yield σ_s and elongation ε_s of cellulose acetate, cellulose acetate propionate, and cellulose acetate butyrate as a function of the plasticizer content (determined in accordance with DIN 53455 or ISO/R 527; specimen no. 3, rate of deformation 25 mm/min)

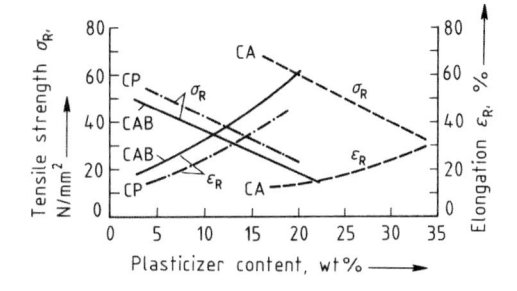

Figure 14. Tensile strength at break σ_R and elongation ε_R of cellulose acetate, cellulose acetate propionate, and cellulose acetate butyrate as a function of the plasticizer content (determined in accordance with DIN 53455 or ISO/R 527; specimen no. 3, rate of deformation 25 mm/min)

Figure 31 shows results of the alternating bending test on slightly and highly plasticized cellulose acetate, cellulose acetate propionate, and cellulose acetate butyrate molding compounds.

Thermoplastic cellulose ester plastics are generally characterized by optical clarity, high mechanical strength, chemical resistance, and toughness. One particularly noteworthy feature is that the material reacts to mechanical stresses by exhibiting cold flow, which helps to eliminate problems with insert molding of metal parts (e.g., stress cracking). Light-stable material is available in a wide range of transparent, translucent, and opaque colors and shades. High surface gloss coupled with antistatic properties (i.e., electrical charges disperse rapidly, not allowing annoying dust patterns to form) ensures that moldings retain their attractive appearance for years. High surface elasticity ensures a good "natural feel" and imparts a

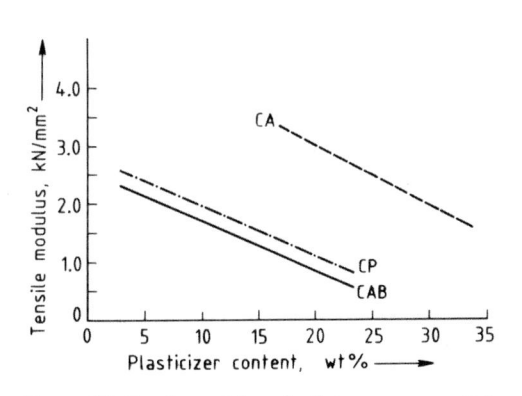

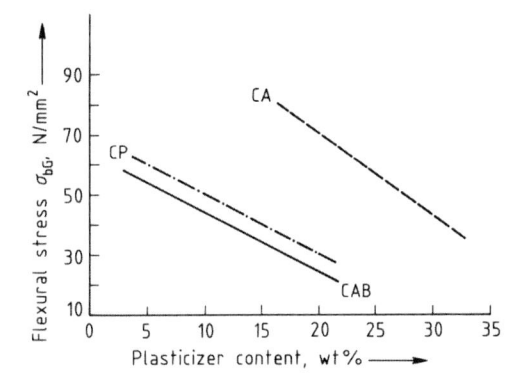

Figure 16. Flexural stress at a given strain σ_{bG} of cellulose acetate, cellulose acetate propionate, and cellulose acetate butyrate as a function of the plasticizer content (determined in accordance with DIN 53452 or ISO/R 178; test specimen 4 × 10 × 80 mm, rate of deformation 2 mm/min)

"repolishing" effect to the material, which means that scratches disappear as the object is used. The relatively low modulus of elasticity gives excellent damping of vibrations, so that the acoustical behavior is not affected by annoying resonance or ambient noise.

Mixed esters absorb considerably less water than cellulose acetates, allowing parts produced from mixed esters to retain their dimensional stability even in humid climates. Cellulose acetate butyrates and (with certain restrictions) cellulose acetate propionates can also be treated with UV inhibitors to ensure serviceability of the moldings during years of outdoor exposure [144]. In principle, cellulose ester molding compounds can be reinforced with glass fibers [146].

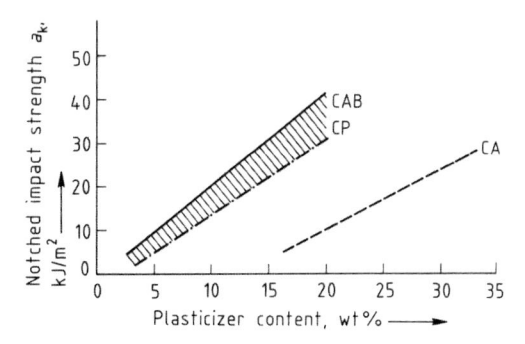

Figure 15. Tensile modulus of cellulose acetate, cellulose acetate propionate, and cellulose acetate butyrate as a function of the plasticizer content (determined in accordance with DIN 53455 or ISO/R 527)

Figure 17. Notched impact strength a_k of cellulose acetate, cellulose acetate propionate, and cellulose acetate butyrate as a function of the plasticizer content (determined in accordance with DIN 53453 or ISO/R 179; specimen no. 2)

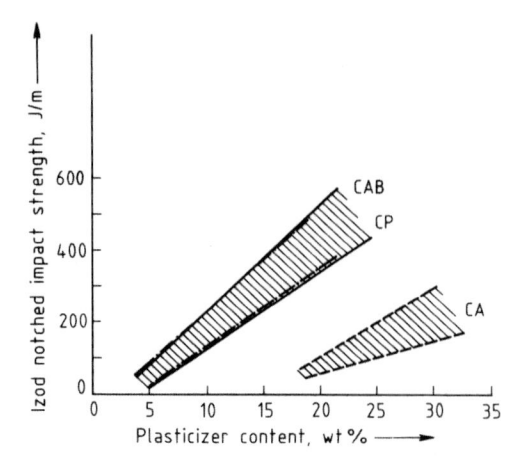

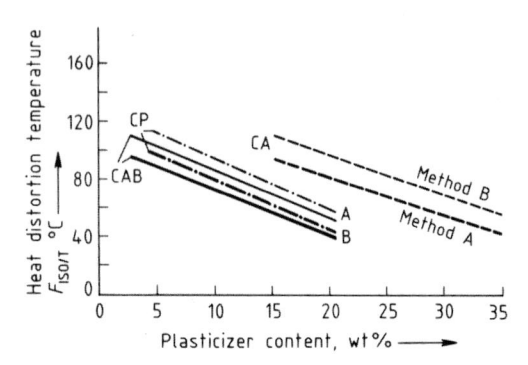

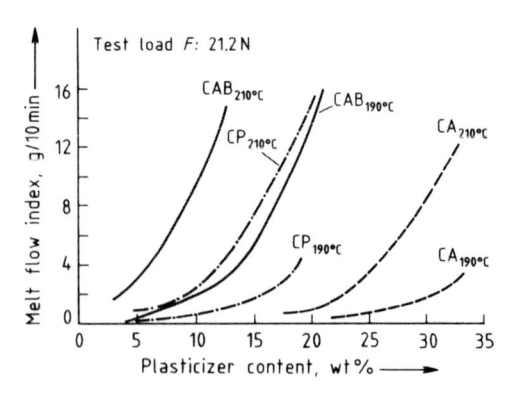

Figure 18. Izod notched impact strength of cellulose acetate, cellulose acetate propionate, and cellulose acetate butyrate as a function of the plasticizer content (determined in accordance with ASTM D 256, Method A, or ISO/R 180; test specimen 63.5 × 12.7 × 3.2 mm)

2.4.2. Polymer Modified Cellulose Mixed Esters

Historically, cellulose acetate butyrates and cellulose acetate propionates have been formulated with polymeric modifiers such as ethylene-vinyl acetate, which provided advantages in heat distortion temperature, creep behavior, stiffness, delayed crazing from long-term outdoor exposure and freedom from potential plasticizer migration [150], [151], [152]. These formulations are no longer commercial available upon Bayer exiting the mixed ester business in the 1980s.

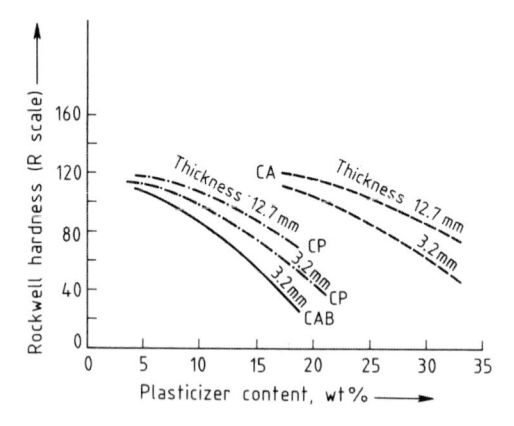

Figure 19. Rockwell hardness (R scale) of cellulose acetate, cellulose acetate propionate, and cellulose acetate butyrate as a function of the plasticizer content (determined in accordance with ASTM D 785)

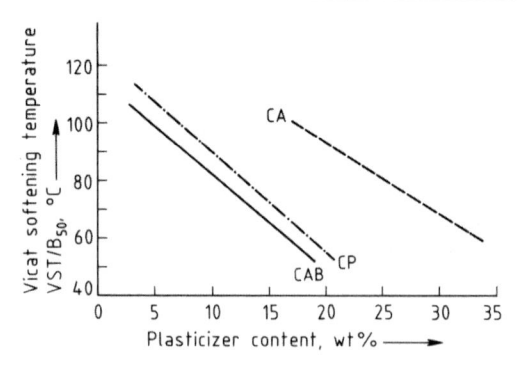

Figure 20. Vicat softening temperature VST/B_{50} of cellulose acetate, cellulose acetate propionate, and cellulose acetate butyrate as a function of the plasticizer content (determined in accordance with DIN 53460/B or ISO/R 306; sheet 10 × 10 × 4 mm)

Figure 21. Heat distortion temperature F_{ISO} of cellulose acetate, cellulose acetate propionate, and cellulose acetate butyrate as a function of the plasticizer content (determined in accordance with ASTM D 648, ISO/R 75, or DIN 53461; test specimen 12.7 × 12.7 × 120 mm)

Figure 22. Melt flow index of cellulose acetate, cellulose acetate propionate, and cellulose acetate butyrate as a function of the plasticizer content (determined in accordance with DIN 53735 or ISO/R 1133)

Table 18. Electrical properties of organic cellulose ester molding compounds

Type of test	Unit	Test specification	Specimen	Cellulose acetate [164]	Cellulose acetate propionate [151]	Cellulose acetate butyrate [151]
Dielectric strength Ed (50 Hz, 0.5 kV/s)						
dry		VDE 0303	Circular	315	355	350
4 days at 80 % rel. humidity	kV/cm	Pt. 2,	discs 95 mm Ø	290	330	330
24 h water immersion		DIN 53481	× 1 mm	280	330	330
Surface resistance R_0						
dry		VDE 0303		8×10^{13}	2×10^{14}	9×10^{13}
4 days at 80 % rel. humidity	Ω	Pt. 3,	$150 \times 15 \times 4$ mm	3×10^{12}	1×10^{13}	9×10^{12}
24 h water immersion		DIN 53482		4×10^{11}	5×10^{12}	9×10^{12}
Insulation resistance R_a						
dry		VDE 0303		5×10^{15}	5×10^{15}	5×10^{15}
4 days at 80 % rel. humidity	Ω	Pt. 3,	$150 \times 5 \times 4$ mm	1×10^{13}	6×10^{13}	5×10^{13}
24 h water immersion		DIN 53482		7×10^{11}	2×10^{13}	2×10^{13}
Volume resistivity ϱ_D						
dry		VDE 0303	Circular	2×10^{15}	1×10^{16}	4×10^{15}
4 days at 80 % rel. humidity	Ω/cm	Pt. 3,	discs 95 mm Ø	2×10^{12}	5×10^{13}	6×10^{13}
24 h water immersion		DIN 53482	× 1 mm	2×10^{11}	1×10^{13}	2×10^{13}
Relative permittivity ε_r, dry						
at 50 Hz		VDE 0303	Circular	5.1	4.1	4.0
at 800 Hz		Pt. 4,	discs 95 and	4.0	3.9	3.8
at 1 MHz		DIN 53483	80 mm Ø × 1 mm	4.1	3.6	3.4
Dissipation factor tan δ, dry						
at 50 Hz		VDE 0303	Circular	0.009	0.005	0.006
at 800 Hz		Pt. 4	discs 95 and	0.019	0.011	0.012
at 1 MHz		DIN 53483	80 mm Ø × 1 mm	0.050	0.026	0.028
Tracking resistance		VDE 030				
KB method		Pt. 1/9.64				
Test solution A		DIN 53480/6	20×15 mm	>600	>600	>600

2.4.3. Chemical Properties

Thermoplastic cellulosic plastics have good resistance to chemically induced stress cracking by typical household, industrial, and medical chemicals. They show good resistance to common everyday chemicals including toothpaste flavorants, aliphatic hydrocarbons, grease, oil, bleach/soaps, ethylene glycol, salt solutions, vegetable and mineral oils, and alcohols [147], [148]. Table 19 provides guide values for resistance to a range of substances, but thorough practical tests are recommended in each case.

2.4.4. Raw Materials

Typical cellulose ester compositions with respect to acetyl, propionyl and butyryl content for production of cellulose ester molding compounds are as follows:

Cellulose acetate propionate molding compounds,
 45 % propionyl acid content
 2.5 % acetyl content

Cellulose acetate butyrate molding compounds,
 37 % butyryl content
 13.5 % acetyl content

Cellulose acetate molding compounds,
 approximately 39.5 % acetyl content
 approximately 37.0 – 38.4 % acetyl (for block acetate only [149])

Of the large number of plasticizers that are compatible with cellulose esters [6], the following have acquired industrial significance, either alone or in combination:

For cellulose acetate propionates and cellulose acetate butyrates:
di-2-ethylhexyl phthalate, dibutyl adipate, di-2-ethylhexyl adipate, dibutyl azelate and dibutyl

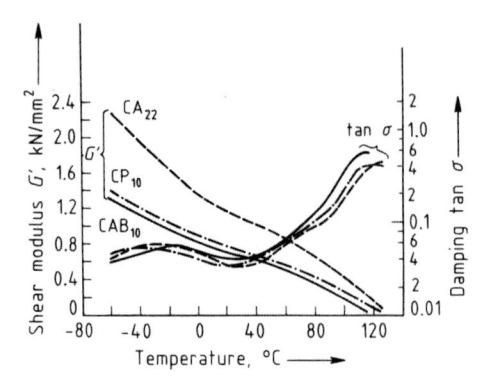

Figure 23. Shear modulus G' and damping tan σ of cellulose acetate$_{22}$, cellulose acetate propionate$_{10}$, and cellulose acetate butyrate$_{10}$ (determined in accordance with DIN 53445 or ISO/R 537)

sebacate, dioctyl azelate, dioctyl sebacate, palmitates, stearates, etc.

For cellulose acetates:

dimethyl, diethyl, dibutyl, di-2-ethylhexyl, and di-2-methoxyethyl phthalate; triphenyl and trichloroethyl phosphate, triacetin, and triethyl citrates.

The *stabilizers* and *antioxidants* used for cellulose ester molding compounds include: alkali salts and alkaline-earth salts of sulfuric, acetic, and carbonic acid, tartaric acid, oxalic acid, citric acid, higher molecular mass epoxides, and phenolic antioxidants. In special cases these stabilizers and antioxidants can be complemented by others [153].

From the range of *ultraviolet absorbers* available, various benzophenones, benzotriazoles, salicylates, and benzoates are recommended for organic cellulose ester molding compounds [154].

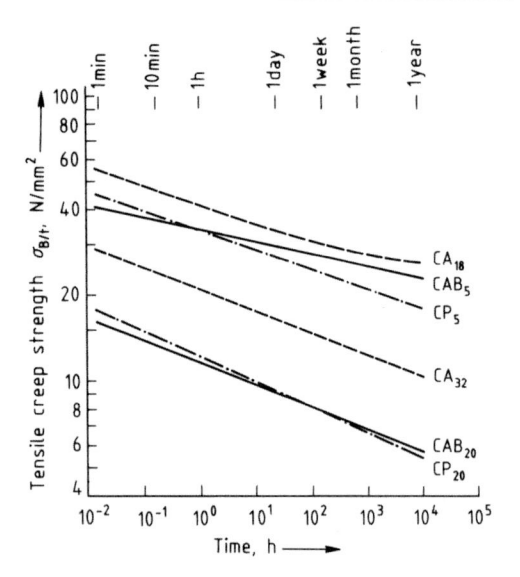

Figure 25. Tensile creep strength $\sigma_{B/t}$ of cellulose acetate, cellulose acetate propionate, and cellulose acetate butyrate (determined in accordance with DIN 53444 or ISO/R 899; test specimen no. 3)

Processing auxiliaries for cellulose ester molding compounds include zinc stearate, butyl stearate, and paraffin oil.

Numerous combinations of *dyes* can be used for coloring cellulose ester molding compounds [155]. The following groups of dyes have proven successful in practice: alkaline, acid, and substantive dyes (provided they are sufficiently soluble in the solvent); Zapon, Sudan, and Ceres

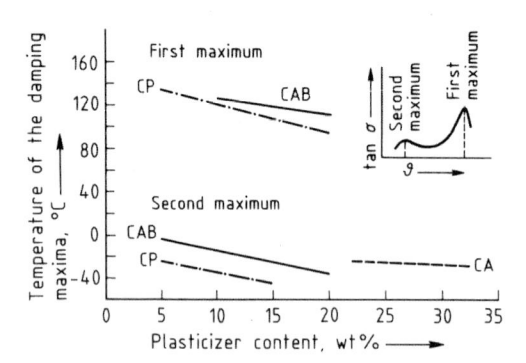

Figure 24. Temperature of the damping maxima of cellulose acetate, cellulose acetate propionate, and cellulose acetate butyrate as a function of the plasticizer content

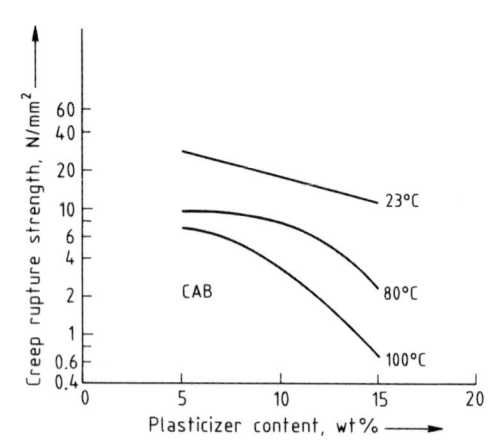

Figure 26. Creep rupture strength $\sigma B10_3$ of cellulose acetate butyrate at 23 °C, 80 °C, and 100 °C as a function of the plasticizer content (determined in accordance with DIN 53444 or ISO/R 899)

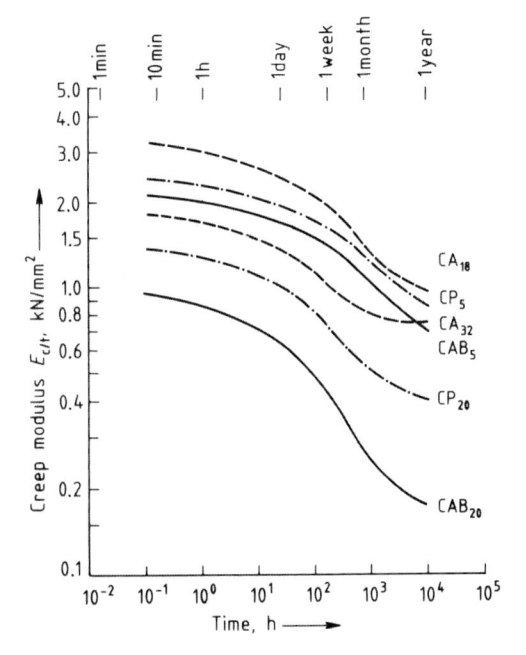

Figure 27. Creep modulus $E_{c/t}$ of cellulose acetate, cellulose acetate propionate, and cellulose acetate butyrate (determined in accordance with DIN 53444 or ISO/R 899)

dyes (provided they are sufficiently resistant to sublimation); and organic and inorganic pigments.

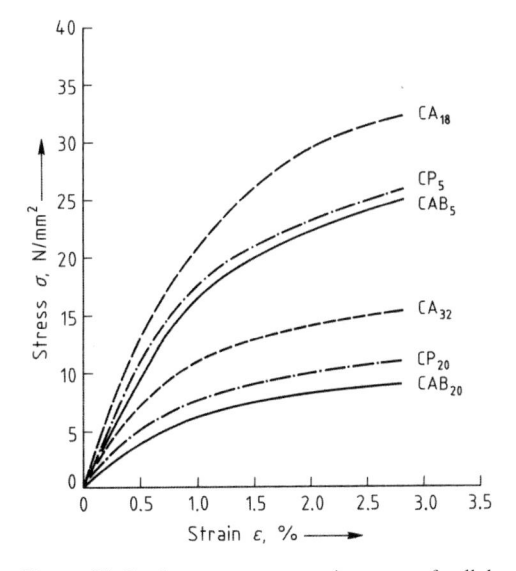

Figure 28. Isochronous stress – strain curves of cellulose acetate, cellulose acetate propionate, and cellulose acetate butyrate for 1 h (determined in accordance with DIN 53444 or ISO/R 899)

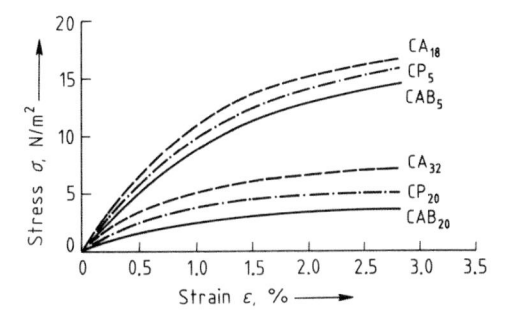

Figure 29. Isochronous stress – strain curves of cellulose acetate, cellulose acetate propionate, and cellulose acetate butyrate for 1000 h (determined in accordance with DIN 53444 or ISO/R 899)

2.4.5. Production

Cellulose acetates, cellulose acetate propionates, and cellulose acetate butyrates are formulated with plasticizers, stabilizers, antioxidants, dyes, and sometimes ultraviolet absorbers and processing aids, utilizing thorough mixing at room temperature. Plastification and homogenization are carried out at higher temperatures (between 150 and 210 °C, depending on the type and the degree of plasticization) in single- or twin-screw kneaders or roll mills. Depending on the type of equipment used, this results in granules in the form of pellets (bulk density 500 – 620 g/L) or cubes (bulk density 400 – 470 g/L).

No wastewater or waste gas problems are associated with the production of thermoplastic

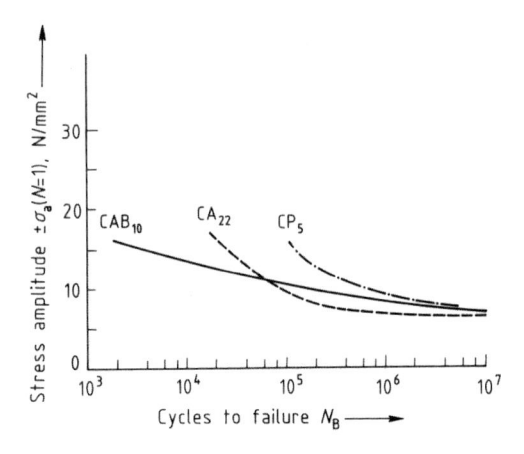

Figure 30. Dynamic fatigue test in the range of pulsating tensile stresses (number of load cycles) of cellulose acetate$_{22}$, cellulose acetate propionate$_5$, and cellulose acetate butyrate$_{10}$ (determined in accordance with DIN 50100; stress amplitude $\pm \sigma_a$ ($N=1$) means stress amplitude under initial loading)

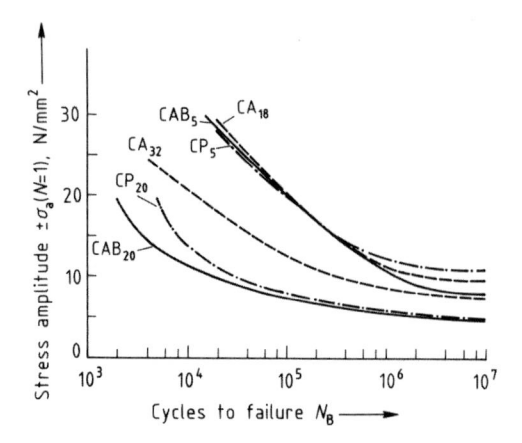

Figure 31. Dynamic fatigue test in the range of alternating flexural stresses (number of load cycles) of cellulose acetate, cellulose acetate propionate, and cellulose acetate butyrate (determined in accordance with DIN 50100; stress amplitude $\pm \sigma_a$ ($N=1$) means stress amplitude under initial loading)

cellulose ester molding compounds. The inevitable plasticizer vapors that occur during processing should be removed by exhaust ventilation.

2.4.6. Trade Names

Trade names of thermoplastic cellulose ester molding compounds are as follows:

Cellulose Acetate Propionates: Tenite Propionate (Eastman Chemical Company, United States), Cellidor CP and Albis (Albis Plastics GmbH, Germany).

Cellulose Acetate Butyrates: Tenite Butyrate (Eastman Chemical Company, Kingsport, TN, United States), Cellidor B and Albis (Albis Plastics GmbH, Germany).

Cellulose Acetate: Tenite Acetate (Eastman Chemical Company, United States), Rotuba Acetate (Rotuba Plastics, United States), Setlithe and Plastiloid (Mazzucchelli, Italy), and Acety (Daicel, Japan).

2.4.7. Quality Requirements and Quality Testing

As has already been stated, the mechanical properties of thermoplastic cellulose ester molding compounds are dependent on the molecular weight properties of the cellulosic polymer as well as the plasticizer selection and content. With respect to cellulose acetate the following tests are all standardized: determination of the viscosity and viscosity ratio in a dilute solution [156], determination of the insoluble constituents [157], viscosity loss during molding [158], light absorption before and after heating [159], and determination of constituents extractable with ethyl ether [160]. The methods described for cellulose acetate are similarly applicable to cellulose mixed esters.

The manufacturers also carry out numerous in-house tests during the course of their quality control programs. These tests include determination of mechanical data, testing of purity, checking thermal stability, colorimetry, determination of flow properties (melt index, Brabender, extrusiometer), etc.

In Germany, cellulose ester molding compounds are standardized in accordance with DIN 7742, Parts 1 and 2. In the United States, standardization is in accordance with ASTM D 706–98 (cellulose acetate), D 707–98 (cellulose acetate butyrate), and D 1562–98 (cellulose acetate propionate).

2.4.8. Storage and Transportation

Storage of thermoplastic cellulose ester molding compounds presents no problems. Even after 10 years in storage, no changes in composition have been detected. It is, however, recommended that thermoplastic cellulose ester molding compounds are pre-dried in accordance with the particular manufacturer's guidelines before processing [161].

Transportation of thermoplastic cellulose esters as a class are not regulated by the DOT (USA), ICAO, and IMDG. They are also not governed by the GGVS/ADR, GGVE/RID, GGVSee/IMDG code or DGR/ICAO regulations for the transportation of hazardous goods.

2.4.9. Uses [162], [143]

Cellulose ester plastics have excellent injection molding, extrusion, and fabrication characteristics. They are known for having a wide processing

Table 19. Typical values[a] for the chemical resistance of organic cellulose ester molding compounds

Solvent	Cellulose acetate (< 55 % acetic acid)	Cellulose acetate (> 55 % acetic acid)	Cellulose acetate propionate	Cellulose acetate butyrate
Water	+	+	+	+
Alcohols	– –	– –	– –	– –
Ethyl acetate	– –	0	0	0
Methylene chloride	– –	0	0	0
Acetone	0	0	0	0
Carbon tetrachloride	+	+	+ –	+ –
Trichloroethylene	+	+ –	– –	– –
Perchloroethylene	+	+	+ –	+ –
Benzene	+	+ –	– –	– –
Xylene	+	+	– –	– –
Petroleum spirit	+	+	+	+
Motor fuel mixture (high octane)	+	+	+ –	+ –
Mineral oil (paraffin)	+	+	+	+
Linseed oil	+	+	+	+
Turpentine oil	+	+	+ –	+ –
Lavender oil	+	+	– –	– –
Ether	+	+	+ –	+ –
Formalin	– –	– –	+ –	+ –
2-Chlorophenol	0	0	0	0
Sulfuric acid, conc.	–	–	–	–
Sulfuric acid, 10 %	+ –	+ –	+ –	+
Hydrochloric acid, conc.	–	–	–	–
Hydrochloric acid, 10 %	–	–	–	–
Nitric acid, conc.	–	–	–	–
Nitric acid, 10 %	–	–	–	–
Caustic potash solution, 50 %	–	–	–	–
Caustic potash solution, 10 %	–	–	+ –	+ –

[a] Key to symbols: + = resistant; + – = resistant, but swells; – – = not resistant; – – – = not resistant, swells; 0 = soluble.

window and excellent processing in secondary operations such as solvent polishing, cutting, cementing, drilling, painting, and decorating [163]. Scrap material generated during processing operations is often easily reprocessed.

In the ophthalmic area, both cellulose acetate and cellulose acetate propionate are used because of their excellent clarity, pleasant feel, colorability, machinability, and chemical resistance. They are also solvent polishable, which imparts a very high gloss finish. Cellulose acetate propionate is more commonly utilized for injection-molded applications, while cellulose acetate is more commonly utilized in extruded sheet applications.

Cellulose acetate propionate has excellent stiffness, good dimensional stability under heat, and low moisture absorptivity, making it an excellent choice for eyeglass frames, high quality frames for sunglasses, protective goggles for industry, and sports goggles [164]]. Due to its high transparency, good impact resistance, and low level of light scattering, cellulose acetate propionate is also used for visor glazings for skiers, drivers, and workers in industry. These applications have been made possible by the surface saponification of the plastic, which ensures permanent antifogging properties (the surface has very good wetting properties and excellent water absorption) [164]. Cellulose acetate propionates with special infrared/ultraviolet-absorbing characteristics can be utilized for welding goggles and certain types of sunglasses [164].

Cellulose acetate is utilized as a sheet material for making high-end ophthalmic frames because of the ease of producing unique color effects. For this purpose, uni- or multicolored plates are either cast or made by extrusion.

Cellulose acetate and cellulose acetate butyrate are easily extruded into transparent profiles for production of tool handles. Benefits in this application include toughness, clarity, ease of post-extrusion machinability, and scuff resistance. Particularly important features include solvent polishability, impact resistance, good feel, and absence of stress cracking. Metal tool parts can easily be insertion molded without stress cracking (the blades can even be driven "cold" into the handles), which brings obvious economic advantages.

Colorability, toughness, and chemical resistant properties make cellulose acetate and cellulose acetate propionate excellent materials for hairbrushes, combs, and toothbrushes. Resistance to stress cracking make cellulose acetate propionate particularly suited for toothbrushes, which requires close spacing of the drill holes and a high bristle density. Cellulose acetate propionate also satisfies the requirement for high tuft pull-out strength.

Cellulose acetate butyrate sheet is used for backlighted sign faces and panels for illuminated advertising signs. Cellulosic sheeting has also been used for machine hoods, lamp covers, outdoor shelters, and dome lights. High light transmission, practically unrestricted choice of colors, antistatic characteristics, easy of processing, ease of joining (by simply gluing), good printing and coating properties, the absence of stress cracking, and finally, excellent mechanical strength are the main factors influencing the choice of this material in this area.

Cellulose esters are used in tubular packaging applications for their clarity, rigidity, scuff resistance, and impact strength properties. Extruded and injection molded cellulose acetate propionate is used in cosmetic packaging because of its post-extrusion machinability, transparency, chemical resistance, and appropriate FDA regulatory status. With their minimal plasticizer migration, cellulose mixed esters are preferred over cellulose acetate for the packaging of toiletries. Other reasons for their use in this sector include the brilliance and depth of color, as well as the ability to produce special color effects.

Decorative trim made of cellulose acetate butyrate combined with aluminum foil [165] has been firmly established in industry for years. An aluminum foil is coated with the cellulose mixed ester and shaped in the crosshead die of an extruder [166]. With its practically unlimited scope for metal and wood effects, elasticity, resistance to detergents, and simple fixing, this combination of materials has been used with great success in the automotive industry [167], as well as in the electrical, audio, and domestic appliance sectors [168].

Extruded cellulose acetate sheet is utilized for high-end playing cards, because of its toughness, durability, dimensional stability, excellent warm feel, and longer lasting life than cards made from other materials.

Other applications of cellulose ester molding compounds include pen barrels, automotive and furniture trim, sporting goods such as fishing lures, health care supplies, medical tubing, appliances for durable goods, and electrical insulation. Further applications include lamp covers, high-quality toys, shoe-string tip films, shoe heels, umbrella handles, curtain rings, toilet seats, tap handles, transparent mouthwash spray attachments, instrument panel covers (glazing), and knife handle grips.

Economic Facts.. With an apparently guaranteed supply of raw materials and a tremendous scope for variation of cellulose ester molding compounds and of their property combinations, this class of plastics should maintain its market significance in special areas of application for years to come.

2.4.10. Toxicology and Occupational Health

In general, components of cellulose ester plastic formulations are listed on the TSCA, EINECS and other worldwide regulatory lists. Components should be reviewed for compliance with a region's appropriate regulatory law.

Cellulose acetate and cellulose propionate molding compounds comply with Recommendation XXVI of the Federal Health Authorities of Germany [169].

There are also various cellulose esters and cellulose ester plasticizers that satisfy requirements of the U. S. Food and Drug Administration. Each should be reviewed for lawful usage prior to use.

References

1 V. Stannet: *Cellulose Acetate Plastics*, Temple Press Ltd., London 1950.
2 K. Thinius: *Analytische Chemie der Plaste*, Springer Verlag, Berlin-Göttingen-Heidelberg 1952.
3 *Kirk-Othmer*, vol. 3, pp. 357 ff.

4 E. Ott, H. M. Spurlin, M. W. Grafflin: *Cellulose and Cellulose Derivatives*, 2nd ed., vol. V, part II, Interscience Publ., New York-London 1954.
5 R. Houwink, A. J. Staverman: *Chemie und Technologie der Kunststoffe*, 4th ed., vol. II/2, Akademische Verlagsgesellschaft Geest & Porting KG, Leipzig 1963.
6 K. Thinius: *Chemie, Physik und Technologie der Weichmacher*, VEB Deutscher Verlag für Grundstoffindustrie, Leipzig 1963.
7 V. E. Yarsley, W. Flarell, P. S. Adamson, N. G. Perkins: *Cellulose Plastics*, Iliffe Book Ltd., London 1964.
8 *Encyclopedia of Polymer Science and Technology*, vol. 3, J. Wiley & Sons, New York 1965.
9 R. Vieweg, E. Becker: *Kunstoff-Handbuch*, vol. III: "Abgewandelte Naturstoffe," Hanser Verlag, München 1965.
10 P. B. Koslov, G. I. Braginski: *Chemistry and Technology of Polymers*, Iskustvo, Moscow 1965 (Russ.).
11 H. Temming, H. Grunert: *Temming-Linters*, 2nd ed., P. Temming, Glückstadt 1972.
12 J. Barsha: "Inorganic Esters," in E. Ott, H. M. Spurlin, M. W. Grafflin (eds.): *Cellulose and Cellulose Derivatives*, 2nd ed., part II, Interscience Publ., New York-London 1954, pp. 713–762.
13 E. D. Klug: "Cellulose Derivatives," in *Kirk-Othmer*, 2nd ed., vol. 4, J. Wiley & Sons, New York-London-Sydney-Toronto 1964, pp. 616–652.
14 G. D. Hiatt, W. J. Rebell: "Esters," in N. M. Bikales, L. Segal (eds.): *Cellulose and Cellulose Derivatives*, part V, Wiley-Interscience, New York-London-Sydney-Toronto 1971, pp. 741–777.
15 J. Honeyman: *Recent Advances in the Chemistry of Cellulose and Starch*, Heywood & Comp., London 1959.
16 E. Sjöström: *Wood Chemistry, Fundamentals and Applications*, Academic Press, New York-London 1981.
17 D. Fengel, G. Wegener: *Wood–Chemistry*, Ultrastructure, Reactions, De Gruyter, Berlin-New York 1983.
18 K. Balser: "Derivate der Cellulose," in W. Burchard (ed.): *Polysaccharide*, Springer Verlag, Berlin-Heidelberg-New York-Tokyo 1985, pp. 84–110.
19 L. C. Wadsworth, D. Daponte: "Cellulose Esters," in T. P. Nevell, S. H. Zeronian (eds.): *Cellulose Chemistry and its Applications*, Ellis Horwood Ltd., Chichester 1985, pp. 344–362. A. Revely: "A Review of Cellulose Derivatives and their Industrial Applications," in *Cellulose Chemistry and its Applications*, pp. 211–225.
20 Company Publication of Wolff Walsrode AG: *100 Jahre Collodiumwolle* (1979).
21 K. Fabel: *Nitrocellulose*, Enke Verlag, Stuttgart 1950.
22 F. D. Miles: *Cellulose Nitrate*, Oliver & Boyd, London 1955.
23 S. Rossin, *Mémorial des Poudres* **40** (1958) 457–471.
24 H. Temming, H. Grunert: *Temming-Linters*, 2nd ed. (engl.), Peter Temming AG, Glückstadt 1973.
25 Company Publication of Wolff Walsrode AG: *Walsroder Nitrocellulose*, no. 500/203/007 (1984).
26 J. Isler, D. Flegier: "The Self-Ignition Mechanism in Nitrocellulose," in J. F. Kennedy, G. O. Phillips, D. J. Wedlock, P. A. Williams (eds.): *Cellulose and its Derivatives*, Ellis Horwood, Chichester 1985, pp. 329–336.
27 A. H. K. Fowler, H. S. Munro: "Some Surface Aspects of the Thermal and X-Ray Induced Degradation of Cellulose Nitrates as Studied by ESCA," in [26] pp. 245–253.
28 Enclosure I of the International Rail Transport Regulations (RID), Appendix 1.
29 E. Berl, *Angew. Chem.* 1904, 982, 1018, 1074. E. Berl-Lunge: *Chemisch-technische Untersuchungsmethoden*, 8th ed., vol. III, Julius Springer, Berlin 1932, p. 1294.
30 P. Kassenbeck: "Faktoren, die das Nitrierverhalten von Baumwoll-Linters und Holzzellstoffen beeinflussen," report 3/83, ICT-Fraunhofer-Institut für Treib- und Explosivstoffe, Pfinztal-Berghausen 1983.
31 Hercules Powder, US 2776965, 1957; US 3063981, 1962.
32 Hercules Powder, US 2028080, 1936.
33 Hercules Powder, US 2950278, 1960.
34 Société Nationale des Poudres et Explosifs, FR 1394779, 1964; DE-OS 1246487, 1965; FR 1566688, 1968; DE-OS 1914673, 1969.
35 Société Nationale des Poudres et Explosifs, DE-OS 2813730, 1977.
36 Hercules Powder, US 2776944, 1957; US 2776964, 1957.
37 Wolff Walsrode, DE-OS 2727553, 1977; DE-OS 2727554, 1977.
38 Wasag Chemie, DE-OS 1771006, 1968.
39 Wolff & Co. Walsrode, DE-OS 1153663, 1961.
40 Wolff & Co. Walsrode, DE-OS 1203652, 1964.
41 Société Nationale des Poudres et Explosifs, DE-OS 2338852, 1973.
42 C. F. Bennett, T. E. Timell, *Sven. Papperstidn.* **58** (1955) 281–286.
43 D. A. J. Goring, T. E. Timell, *Tappi* **45** (1962) 454–460.
44 K. Thinius, W. Thümmler, *Makromol. Chem.* **99** (1966) 117–125.
45 M. Marx-Figini, *Makromol. Chem.* **50** (1961) 196–219.
46 J. W. Green in R. L. Whistler, J. W. Green, J. N. BeMiller (eds.): "Cellulose," *Methods in Carbohydrate Chemistry*, vol. III, Academic Press, New York-London 1963, pp. 213–237.
47 W. J. Alexander, G. C. Gaul: "Cellulose Derivatives," in *Encyclopedia of Industrial Chemical Analysis*, vol. 9, J. Wiley & Sons, New York-London-Sydney-Toronto 1970, pp. 59–94.
48 ASTM D 301–72 (1983): Soluble Cellulose Nitrate.
49 ASTM D 1343–69 (1979): Viscosity of Cellulose Derivatives by Ball-Drop Method.
50 ASTM D 365–79: Soluble Nitrocellulose Base Solutions.
51 ASTM D 1720–79: Dilution Ratio in Cellulose Nitrate Solutions for Active Solvents, Hydrocarbons, Diluents and Cellulose Nitrates.
52 A. Kraus: *Handbuch der Nitrocelluloselacke*, Westliche Berliner Verlagsgesellschaft Heenemann KG, Berlin, vol. 1: 1955, vol. 2: 1963, vol. 3: 1961, vol. 4: 1966.

53 H. Kittel: *Lehrbuch der Lacke und Beschichtungen*, vols. 1–8 Heenemann GmbH, Stuttgart-Berlin 1971–1980.
54 Bayer, DE-OS 2853578, 1978.
55 Wolff Walsrode, DE-OS 3139840, 1981; DE-OS 3407932, 1984.
56 *Ullmann*, 4th ed., **9**, (1975) 179–183.
57 Leaflet M 051 by the German Employers' Liability Insurance Association of the Chemical Industry: *Dangerous Chemical Substances*, Jedermann-Verlag Dr. Otto Pfeffer, Heidelberg 1984.
58 *Verordnung über gefährliche Arbeitsstoffe* 11 Feb. 82 BGBl I, Carl Heymanns, Köln 1982, p. 144.
59 Leaflet M 014 by the German Employers' Liability Insurance Association of the Chemical Industry: *Nitric Acid, Nitrogen Oxides, Nitrous Gases*, Jedermann-Verlag Dr. Otto Pfeffer, Heidelberg 1985.
60 Leaflet M 037 by the German Employers' Liability Insurance Association of the Chemical Industry: *Nitrocellulose*, Jedermann-Verlag Dr. Otto Pfeffer, Heidelberg 1984.
61 E. v. Schwartz: *Handbuch der Feuer- und Explosionsgefahr*, 5th ed., Feuerschutz-Verlag P. L. Jung, München 1958, p. 303.
62 German Explosives Law, publ. 13 Sept. 1976, BGBl I, p. 2737; amended 10 Apr. 1981, BGBl I, Carl Heymanns, Köln 1981, p. 388.
63 H. Dorias: *Gefährliche Güter*, Springer Verlag, Berlin-Heidelberg-New York-Tokyo 1984.
64 Z. A. Rogovin, L. S. Galbraich, W. Albrecht (eds.): *Die chemische Behandlung und Modifizierung der Cellulose*, Thieme Verlag, Stuttgart-New York 1983.
65 Hercules Powder Comp., US 2753337, 1953.
66 B. Philipp, W. Wagenknecht: *Cellul. Chem. Technol.* **17** (1983) 443.
67 R. G. Schweiger, *Tappi* **57** (1974) 86; *ACS Symp. Ser.* **77** (1978) 163; US 4138535, 1977.
68 R. G. Schweiger, *Carbohydr. Res.* **70** (1979) 185.
69 R. G. Schweiger, DE-OS 3025094, 1980.
70 Eastman Kodak, US 3075962, 1963; US 3075963, 1963; US 3075964, 1963.
71 Eastman Kodak, US 3068007, 1963.
72 W. D. Slowig, M. E. Rowley, *Text. Res. J.* **38** (1968) 879.
73 J. Pastyr, L. Kuniak, *Cellul. Chem. Technol.* **6** (1972) 249.
74 J. B. Lawton, G. O. Phillips, *Text. Res. J.* **45** (1975) 4.
75 K. Katsuura, T. Fujinami, *Kogyo Kagaku Zasshi* **71** (1968) 771.
76 Eastman Kodak, US 2759924, 1956.
77 Kimberly-Clark, US 3658790, 1970.
78 National Chemical & Manufacturing, US 1896725, 1933.
79 R. A. A. Muzzarelli, G. Marcotrigiano, C.-S. Liu, A. Frêche, *Anal. Chem.* **39** (1967) 1792.ensp;
80 L. S. Sletkina, A. J. Poljakov, Z. A. Rogovin, *Vysokomol. Soedin* **7** (1965) 199; *Faserforsch. Textiltech.* **2** (1965) 299.
81 D. A. Predvoditelev, M. S. Bakseeva; *Zh. Prikl. Khim. (Leningrad)* **45** (1972) 857; *Faserforsch. Textiltech.* **9** (1972) 361.
82 D. C. Johnson: "Solvents for Cellulose," in *Cellulose Chemistry and its Applications*, Ellis Horwood, Chichester 1985, pp. 181–201.
83 L. P. Clermont, F. Bender, *J. Polym. Sci. Part A-1* **10** (1972) 1669.
84 R. B. Hammer, A. F. Turbak, *ACS Symp. Ser.* **58** (1977) 40. H. L. Hergert, *Tappi* **61** (1978) 63. ITT Rayonier, US 4056675, 1977.
85 C. J. Malm, L. J. Tanghe, B. C. Laird, *J. Am. Chem. Soc.* **72** (1950) 2674.
86 C. J. Malm, L. J. Tanghe, B. C. Laird, G. D. Smith, *J. Am. Chem. Soc.* **75** (1953) 80.
87 C. J. Malm, L. J. Tanghe, *Ind. Eng. Chem.* **47** (1955) 995.
88 Cross, Bevan, Tranquair, *Chem. -Ztg.* **29** (1905) 528.
89 C. J. Malm, H. T. Clarke, *J. Am. Chem. Soc.* **51** (1929) 274.
90 Du Pont, US 1990483, 1935.
91 Schutzenberger, Naudin, *Z. Chem.*, 1869, 264.
92 Franchimont, *Comp. Rend.* **89** (1879) 711.
93 C. J. Malm, L. J. Tanghe, J. T. Schmitt, *Ind. Eng. Chem.* **53** (1961) 363.
94 C. J. Malm, K. T. Barkey, D. C. May, E. A. Lefferts, *Ind. Eng. Chem.* **44** (1952) 2904.
95 C. J. Malm, K. T. Barkey, E. B. Lefferts, R. T. Gielow, *Ind. Eng. Chem.* **50** (1958) 103.
96 C. J. Malm, L. J. Tanghe, B. C. Laird, *Ind. Eng. Chem.* **38** (1946) 77.
97 A. J. Rosenthal, *Pure Appl. Chem.* **14** (1967) 535.
98 L. J. Tanghe, R. J. Brewer, *Anal. Chem.* **40** (1968) 350.
99 H. Temming, H. Grunert: *Temming Linters*, 2nd ed., P. Temming, Glückstadt 1972.
100 C. J. Malm, R. E. Glegg, J. Thompson, L. J. Tanghe, *TAPPI* **47** (1964) 533.
101 C. J. Malm, L. J. Tanghe, *TAPPI* **46** (1963) 629.
102 F. L. Wells, W. C. Schattner, A. Walker, *TAPPI* **46** (1963) 581.
103 B. P. Rousse, *Encycl. Polym. Sci. Technol. 1964–1977*, **3**.
104 *Ullmann*, 3rd ed., **5**, 187 ff.
105 Bios Final Report No. 1600, Item No. 22, H. M. Stationary Office, London 1948.
106 L. Ledderer, DE 200916, 1905.
107 Boehringer, GB 363700, 1930.
108 Bios Final Report No. 1850, Item No. 21, H. M. Stationary Office, London 1948.
109 C. J. Malm, L. J. Tanghe, B. C. Laird, *J. Am. Chem. Soc.* **72** (1950) 2674.
110 C. J. Malm, R. E. Glegg, J. T. Salzer, D. F. Ingerick, L. J. Tanghe, *I&EC Process Design and Development* **5** (1966) 81.
111 Cons. f. Elektrochem. Ind., DE 687065, 1933; US 2108829, 1934.
112 C. R. Fordyce, L. W. Meyer, *Ind. Eng. Chem.* **32** (1940) 1053.

113 Arthur K. Doolittle: *The Technology of Solvents and Plasticizers*, John Wiley & Sons Inc., New York 1954, pp. 203 – 206.
114 Eastman Chemical Company: "Eastman Cellulose Esters" Publication E-325, 2003.
115 Company Publication of Bayer AG, Leverkusen.
116 Company Publication of Hercules Powder Comp., Wilmington, USA.
117 E. Yarsley, W. Flavell, P. S. Adamson, N. G. Perkins: *Cellulosic Plastics*, Iliffe Book Ltd., London 1964.
118 K. Thinius: *Analytische Chemie der Plaste*, Springer Verlag, Berlin-Göttingen-Heidelberg 1952.
119 C. J. Malm, L. B. Genung, R. F. Williams, *Ind. Eng. Chem. Anal. Act.* **14** (1942) 935.
120 J. F. Tremblay, *Chem. Eng. News*, 1999, July 10, 21.
121 Eastman Chemical Company: "Eastman Cellulose Acetate for Coatings" Publication E-305A, January 1996.
122 Eastman Chemical Company: "A Feasibility Study Using Cellulose Acetate and Cellulose Acetate Butyrate", Publication EFC-234, October 2000.
123 K. J. Edgar et al., *Prog. Polym. Sci.* **26** (2001) 1605.
124 C. J. Malm, J. W. Mench, D. L. Kendall, G. D. Hiatt, *Ind. Eng. Chem.* **43** (1951) 684.
125 J. W. Mench, B. Fulkerson, G. D. Hiatt, *I&EC Product Research and Development* **5** (1966) June, 110.
126 C. J. Malm, *Svensk Kemisk Tidskrift* **73** (1961) no. 10, 523.
127 H. Saechtling, W. Zebrowski: *Kunststoff-Taschenbuch*, 18th ed., C. Hanser, München 1971.
128 C. J. Malm, C. R. Pordyce, H. A. Tanner, *Ind. Eng. Chem.* **34** (1942) 430. E. Ott, H. M. Spurlin, M. W. Grafflin: *Cellulose and Cellulose Derivatives*, vol. **V**, Part II, 2nd edn., Interscience Publ. Inc., New York-London 1954. W. Ballas, Internal communication, Bayer AG, Leverkusen.
129 Bayer AG: *Cellit, Cellit BP and PR*, Brochures nos. KL 44172 and KL 44170, Leverkusen 1982.
130 Eastman Chemical Company: "C–A–P Enteric Coating Material for Pharmaceutical Drug Delivery" Publication EFC-205, 1998.
131 C. J. Malm, et al. *Ind. Eng. Chem.* **49** (1957) 84.
132 Eastman Chemical Company., US 5668273, 1997 (J. M. Allen, A. K. Wilson, P. L. Lucas, L. G. Curtis).
133 Eastman Chemical Company., US 5994530, 1999 (J. D. Posey-Dowty, A. K. Wilson, L. G. Curtis, P. M. Swan, K. S. Seo).
134 J. Corbiere, *Faserforsch. Textiltech.* **22** (1971) 71.
135 Technical Association of Paper and Pulp Industry (TAPPI), DIN or ISO-standards.
136 Celanese Corp. of America, US 2597156, 1952 (M. E. Martin, L. G. Reed).
137 Eastman Kodak Co., US 2652340, 1953 (G. D. Hiatt, R. F. Willmans).
138 E. Heim, *Chemiefasern* **16** (1966) 618.
139 *Ullmann*, 4th ed., **9**, 213.
140 W. Pusch: "Membranes from Cellulose and Cellulose Derivatives," in *Wood and Cellulosics*, Ellis Horwood Limited, Chichester 1985, pp. 475–482.
141 *Man-Made Fiber Yearbook*, Fiber Organon, Washington D.C. 1996, p. 12.
142 W. Fischer, *Kunststoffe* **62** (1972) 653.
143 Eastman Chemical Company: "From Trees to Plastics" Publication PPC-100D, 1999.
144 Eastman Chemical Company: "Weathering of Tenite Butyrate" Publication PP-104B, 1999.
145 Eastman Chem. Prod.: Specifications, T 58–3289, Kingsport, Tennessee 1958.
146 W. Fischer, *Kunststoffe* **63** (1973) 292.
147 Eastman Chemical Company: "Tenite Acetate Chemical Resistance" Publication PP-101, 1996
148 Eastman Chemical Company: "Tenite Butyrate Chemical Resistance" Publication PP-102, 2002.
149 *Paist: Cellulosics*, Reinhold Plastics Application Series, Reinhold Publishing Corporation, New York 1958, pp. 45 – 46.
150 W. Fischer, C. Leuschke, H. P. Baasch, *Kunststoffe* **67** (1977) no. 6, 348–252.
151 C. Leuschke, M. Wandel, *Plastverarbeiter* **33** (1982) no. 9, 1095–1098.
152 Bayer, DE-OS 2951800, 1979; DE-OS 2951747, 1979; DE-OS 2951748, 1979. K. Thinius: *Chemie, Physik und Technologie der Weichmacher*, VEB Deutscher Verlag für Grundstoffindustrie, Leipzig 1963.
153 *Mod. Plast. Encycl.* 1970, 1971 840, 870.
154 *Mod. Plast. Encycl.* 1970, 1971, 876.
155 *Mod. Plast. Encycl.* 1970, 1971, 850.
156 ISO/R 1157–1990.
157 ISO/R 1598–1990.
158 ISO/R 1599–1990.
159 ISO/R 1600–1990.
160 ISO/R 1875–1982.
161 Eastman Chemical Company: "Drying Tenite Cellulosic Plastics" Publication PP-105B, 2000.
162 M. Wandel, C. Leuschke, *Kunststoffe* **74** (1984) no. 10, 589 – 592.
163 Eastman Chemical Company: "Secondary Fabrication Techiques for Tenite Cellulosic Plastics" Publication PP-110D, 2000.
164 Bayer AG: *Cellidor für die optische Industrie*, Company Publication KU 40029, Leverkusen 1984.
165 J. Göller, H. Peters, W. Fischer, *Kunststoffe* **65** (1975) no. 6, 326–332.
166 Glas Lab, DE-AS 1226291, 1955 (A. Shanok, V. Shanok, J. Shanok).
167 Bayer AG: *Zierleisten und Profile in der Autoindustrie*, Company Publication KL 40025, Leverkusen 1979.
168 Bayer AG: *Zierleisten in der Elektro- und Möbelindustrie*, Company Publication KL 40011, Leverkusen 1975.
169 *Bundesgesundheitsblatt*, 19th ed., 1976, No. 6, Carl Heymanns Verlag, Köln-Berlin-Bonnhyphen;Muuml;nchen.

Further Reading

J. P. Agrawal: *High Energy Materials*, Wiley-VCH, Weinheim 2010.

L. Bottenbruch, S. Anders (eds.): *Engineering Thermoplastics - Polycarbonates, Polyacetals, Polyesters, Cellulose Esters*, Hanser/Gardner, Cincinnati 1996.

S. Gedon, R. Fengi: *Cellulose Esters, Organic*, "Kirk Othmer Encyclopedia of Chemical Technology", 5th edition, vol. 5, p. 412–439, John Wiley & Sons, Hoboken, NJ, 2004, online: DOI: 10.1002/0471238961.1518070107050415.a01.

J. Müssig (ed.): *Industrial Applications of Natural Fibres*, Wiley, Chichester 2010.

M. C. Shelton: *Cellulose Esters, Inorganic*, "Kirk Othmer Encyclopedia of Chemical Technology", 5th edition, vol. 5, p. 394–412, John Wiley & Sons, Hoboken, NJ, 2004, online: DOI: 10.1002/0471238961.0914151806051407.a01.pub2.

Cellulose Ethers

HEIKO THIELKING, Wolff Cellulosics GmbH & Co. KG, Walsrode, Germany

MARC SCHMIDT, Wolff Cellulosics GmbH & Co. KG, Walsrode, Germany

1.	Introduction... 225	4.2.3.	Processes... 234
2.	Properties... 225	4.2.4.	Characterization... 236
2.1.	Swelling and Dissolving Behavior... 226	4.3.	Hydroxyethyl Cellulose (HEC)... 236
2.2.	Solution State... 226	4.3.1.	Applications / Market... 237
2.3.	Thermally Induced Coagulation... 228	4.3.2.	Synthesis... 237
2.4.	Degree of Substitution... 228	4.3.3.	Processes... 237
2.5.	Molar Mass Distribution... 229	4.3.4.	Characterization... 238
3.	Production... 229	4.4.	Hydroxypropyl Cellulose (HPC)... 238
4.	Product Groups... 231	4.4.1.	Applications/Market... 239
4.1.	Carboxymethyl Cellulose (CMC)... 231	4.4.2.	Synthesis... 239
4.1.1.	Applications / Market... 231	4.4.3.	Processes... 239
4.1.2.	Synthesis... 231	4.4.4.	Characterization... 239
4.1.3.	Processes... 232	5.	Uses... 239
4.1.4.	Characterization... 232	6.	Economic Aspects... 240
4.2.	Methyl and Hydroxyalkyl Methyl Celluloses 233	7.	Toxicology and Occupational Health... 240
4.2.1.	Applications / Market... 234		References... 241
4.2.2.	Synthesis... 234		

1. Introduction

Cellulose ethers are nontoxic, usually water-soluble, white to yellowish powders or granules. There are some cellulose ethers that are not soluble in water, but their share of sales is insignificant as compared with the water-soluble substances. Cellulose ethers are produced by a polymer-analogous reaction of cellulose with low-molecular alkoxylating agents, which can support further functional groups.

The manufacture of cellulose ethers was first published in an article by W. SUIDA [1] in 1905 and the first patents [2] relating to their industrial production were issued as early as 1918. In the decade between 1920 and 1930 carboxymethyl cellulose was the first cellulose ether to gain economic significance, followed by methyl celluloses and hydroxyethyl cellulose some ten years later. These three product categories still dominate the market today.

During the time cellulose ethers have been used, continuous intensive research focusing on production processes and applications has been carried out by both industry and academia. Research was and is imperative to enable this group of substances to be utilized in ever new applications. Despite major progress in the areas of regioselective synthesis and analytics over recent years, the potential of these substances has by no means been fully exploited [3, 4].

Cellulose ethers can be sorted by their economic significance. The most important products in terms of sales are carboxymethylcelluloses (approx. 230 000 tons per annum), methyl- and hydroxyalkylmethylcelluloses (approx. 120 000 t/a), hydroxyethylcelluloses (approx. 60 000 t/a) and hydroxypropylcellulose (less than 10 000 t/a). The demand is served by a small number of manufacturers, which employ proprietary processes and have developed their own areas of expertise.

2. Properties

Cellulose ethers are nontoxic, usually water-soluble, white to yellowish powders or granules.

There are some cellulose ethers, that are not soluble in water, but their share of sales is insignificant as compared with the water-soluble substances.

The bulk density of cellulose ethers is between 300 and 600 g/L. They are lightfast and their storage life is claimed by the suppliers to be between 18 and 36 months. Cellulose ethers turn brown at elevated temperatures (around 200 °C) and show signs of decomposition above 250 °C. Their biodegradability can be influenced by the degree of substitution.

Cellulose ethers are capable of swelling or are colloidally soluble; they increase the viscosity of the solvent and develop a specific rheological profile. Depending on the substitution, cellulose ethers in solution are surface-active (surface tension: water 72 mN/m, carboxymethyl cellulose 70 mN/m, hydroxyethyl cellulose 63 mN/m, methyl cellulose 54 mN/m). The nonionic derivatives are film-forming substances, resulting in transparent films with a high elasticity. Solutions and films obtained from cellulose ethers are compatible with other hydrocolloids.

In many applications, however, where certain properties of an initially dry mixture such as dissolving and solvent binding behavior (water retention) are required, cellulose ethers are not used in ready-made solutions, but are blended with other solids.

These three characteristics – dissolving behavior, the resultant solution structure, and the cellulose ether's capability of binding solvents –

depend on molecular characteristics such as the type, number and distribution of the substituent and the molar mass distribution.

In technical specifications generally only the degree of purity, the viscosity in aqueous solution and in some cases the degree of substitution are stated.

2.1. Swelling and Dissolving Behavior

The different swelling and dissolving behavior of the various cellulose ethers originates from the interplay of the hydrogen bonds between free OH groups on the one hand and hydrophilic and hydrophobic substituent groups on the other. This interplay can be imagined as being the gradual introduction of hydrophobic substituents into native cellulose. Starting from a crystal structure based on hydrogen bridges, the partial imperfection of this structure caused by substituents results in free hydroxyl functions, hence first the substance becomes soluble in alkalies and subsequently in water. Further etherification with hydrophobic substituents results in solubility in organic solvents and ultimately thermoplasticity (intrinsic solubility).

Alongside the hydrophilic-hydrophobic interaction which culminates in "thermodynamic solubility", the particle-size distribution has a crucial impact on the obtainable state of solution. Particle size determines the kinetics of the dissolution process. If the particles are too large, they may not dissolve fast enough; they will swell and release some of their outer layer into the solution; due to the rising viscosity the mobility within the solution will decrease, leaving undissolved particle residues which will no longer dissolve in a reasonable time. Particles, that are too small, on the other hand, may cause formation of lumps during the dissolving process. Again this will ultimately result in undissolved particle residues preventing formation of a homogeneous solution.

To counteract these problems, most suppliers issue recommendations as to which particle size is best for a particular application. Some suppliers also offer specially treated products for enhanced dispersibility. These are mostly powder grades, whose initial dissolving rate is greatly reduced ("delayed or retarded solubility") to ensure that the substance is homogeneously distributed in the solvent before starting to dissolve. This delay can be achieved, for instance, by a reversible cross-linkage with glyoxal. The rate of dissolution is thus pH-dependent.

2.2. Solution State

The state of solution obtained, i.e., the sum of the interaction between the cellulose ether and the solvent and the intramolecular and intermolecular interaction of the cellulose ether molecules, determines the rheological profile of a solution or of an end product.

The standard rheological models of polymer chains dissolved in coil formation can be used to describe the characteristics of cellulose ether solutions, depending on the intrinsic viscosity (also referred to as Staudinger index) and concentration. All known polymeric solution states from the diluted particle solution to the concentrated network solution [5] can thus be described. However, on account of their inclination to form aggregates, cellulose ethers often set scientists impossible tasks to solve.

The viscosity of a cellulose ether solution is generally stated as a characteristic feature. Measured in a 2 wt% aqueous solution, it is sometimes included in the product grade designation for orientation purposes. At first glance this seems to be useful, because many application-related properties are dependent on the same molecular parameters as viscosity. On closer, more discriminating examination, however, this turns out to be misleading, first because the dissolving and measuring methods differ from one supplier to another, which can result in a similar product having twice the viscosity, and second because – although giving a rough guide – the viscosity does not describe the solution structure and seldom the product's behavior in the final application system.

In the range relevant to the applications the viscosity of cellulose ether solutions rises disproportionately with increasing concentration and intrinsic viscosity:

$$\text{viscosity}_{\text{solution}} \propto \text{viscosity}_{\text{solvent}} \cdot (\text{concentration} \cdot \text{intrinsic viscosity})^{\frac{3.4}{a}}$$

($0.5 < a < 1.0$; a: exponent of the Mark–Houwink equation)

Most solutions show a pseudoplastic behavior, which is very important in practical terms.

Many cellulose ether solutions manifest time-dependent rheological functions, so the measurements depend on how the sample has been pretreated. The elasticity of the solutions also increases with rising concentration and intrinsic viscosity, but the exponents are different from those of the viscosity function; therefore the cellulose ether selected can modify the ratio of viscosity to elasticity within certain limits [6]. Some application systems exhibit a solid-like behavior at low shear stress and viscous flow only starts at stress levels above the so-called yield stress (plastic flow behavior).

The intrinsic viscosity describes the volume "filled" by the free molecule. The intrinsic viscosity is often used as a measure of the molar mass or chain length; the correlation is shown by the Mark-Houwink equation. When the intrinsic viscosity is gauged, not only the molar mass is determined but also the quality or performance of the solvent. Therefore effects attributable to pH, temperature and low-molecular salts, which have a substantial impact on the solution structure, are also part of the analysis. Many cellulose ethers can be salted out by "removing" the hydration shell of the molecules. With the ionic products, polyelectrolytic effects such as "repulsion" or "cross-linking" due to the charging of the macromolecules among themselves or with the charges of salts present in the solution also need to be observed.

The maximum possible thickening effect is predetermined by the used cellulose. The actual viscosity achieved by a cellulose ether depends on how well chain scissions can be prevented during the production process and on the type and quantity of substituents introduced into the cellulose molecule. Each substituent represents an increase in molar mass without an increase in the chain length. This reduces the thickening effect. The substituents are, of course, indispensable for solubility and other properties; however, derivatives which predominantly carry substituents with a low molar mass and where the degree of substitution is not too high are particularly effective thickeners. The substituent group with the lowest molar mass is the methyl group which simultaneously cancels out the intermolecular hydrogen bonds very effectively. As pure methyl cellulose manifests a specific temperature-dependent dissolving behavior, which restricts its universal use, mixed cellulose ethers with a low degree of substitution with hydroxyethyl or hydroxypropyl groups are generally used as high-viscosity thickeners, for instance in the construction sector.

The amount of cellulose ether used or its concentration in an aqueous solution has a similarly great effect as that of the chain length. When used as a thickener, the concentration range employed is usually in the range of a moderately concentrated network solution in which the polymer chains are not isolated but looped, forming a temporary physical suspension network.

The temperature dependence (below the flocculation point) of the flow behavior of aqueous cellulose ether solutions follows the principle of temperature-time superposition. At a higher temperature all the molecules move faster. If a flow curve (shear stress as a function of the shear rate) is plotted at a reference temperature, the flow curve determined at a higher temperature follows the same shear stress path, except that it is merely shifted to higher shear rates. Accordingly, at the higher temperature the viscosity values are on a lower level.

Changes in flow behavior are thus predeterminable (as long as there is sufficient distance from the flocculation point). Therefore, when using the cellulose ether as a thickener in aqueous systems, the rheological behavior can be very reliably estimated above and below the reference temperature, if the flow activation energy is known. In the case of many cellulose ethers in aqueous solution the flow activation energy is in the proximity of 40 kJ/mol. Additional substances dissolved in the water may cause a shift in this value, but there is no fundamental change in the correlations.

2.3. Thermally Induced Coagulation

The solubility of cellulose ethers with hydrophobic substituents like methoxy groups decreases as the temperature rises, i.e., the performance of the solvent deteriorates. Cellulose ether dissolved in a cold solution can be made to coagulate by heating the solution. The process is reversible: when the solution cools down, the cellulose ether dissolves again. Depending on the selected type and degree of substitution, the flocculation point of industrial used products is varied between 30 and 100 °C.

This is an important characteristic for industrial production of cellulose ethers because in this way byproducts and salts can be washed out with hot water instead of with mixtures of water and alcohol.

To assess the flocculation point with mixed substituent groups, suitable flocculation diagrams have been published [7].

2.4. Degree of Substitution

The type of substituent and the degree of substitution determine most of the properties of a cellulose ether. With substituent groups where no secondary substitution is possible (methyl, carboxymethyl), the molar substitution (MS), i.e., the absolute number of substituents per anhydroglucose unit, is identical to the degree of substitution (DS), i.e., the number of substituents directly linked to the anhydroglucose unit. In the case of substituents containing hydroxyl groups (hydroxyethyl, hydroxypropyl), for example, further (secondary) reactions with the substituent can occur, i.e., there is no decrease in the number of reactive centers. With these products a distinction has to be made between MS and DS. Comparing MS and DS shows the proportion of side-chain substitution. On account of the simpler analytics involved, frequently only the MS is stated.

To examine the characteristic features more precisely, it is necessary to analyze the substituent distribution. Primary substitution (i.e., prior to any secondary reactions with the substituent) already involves three kinds of distribution:

- Substituent distribution within an anhydroglucose unit: depending on the reagent, the distribution can be between C-2, C-3, and C-6. This distribution can be analyzed by means of nuclear resonance spectroscopy or hydrolysis and subsequent chromatography [8].
- Substituent distribution along the backbone chain: in contrast to the random distribution normally found, a strictly uniform substituent sequence along the chain or a block structure formation are conceivable as theoretical borderline cases, which show up the different effects on solubility and therefore on the properties in general. To gain an impression of the distribution along the chain, various breakdown methods and subsequent liquid chromatography or derivatization and gas chromatography coupled with mass spectroscopy are used to obtain the proportions of non-, mono-, di- and trisubstituted anhydroglucose units. To assess the distribution of these units along the chain, the results of analysis of fragments of various lengths can be compared. Appropriate methods have so far only been developed for particular derivatives and only few research scientists have a good command of them [9, 10].
- Substituent distribution between the individual chains: the substituent density can also vary between the individual chains of the cellulose ether, one known example being the nonsubstituted cellulose fibers in otherwise clear-

soluble products. With ionic products the interchain distribution can be detected by polarographic analysis and an appropriate method for other derivatives is chromatography after specifically marking the free hydroxyl groups. The problem with both methods is that the molar mass distribution has to be known in order to carry out the assessment [11].

In the case of derivatives where a reaction with the substituent is possible, the potential number of different anhydroglucose units and their distribution increases randomly, and the structure becomes enormously complex. At the same time, however, the possibilities for adapting the hydrophilic-hydrophobic interaction to suit individual applications increase as well.

2.5. Molar Mass Distribution

Knowing the molar mass or the molar mass distribution makes it possible in theory to access various material functions. An absolute determination of these variables has so far only been achieved in isolated cases and with an enormous research input. Due to the "poor solubility" and the propensity of most derivatives toward association, producing time-stable molecularly disperse solutions is only rarely possible. Frequently it is a mixture of molecules, associates and aggregates that is analyzed, and the results are often impossible to interpret reliably.

It is precisely because of this complexity that so many attempts to make a breakthrough have been launched in the past, starting with "batch methods" such as static and dynamic light scattering in many different solvents, and "classifying" methods such as ultracentrifugation, size-exclusion chromatography and flow-field-flow fractionation, and also combined methods involving chromatography and various detection systems. All these approaches have in common that, although they give an interesting insight into the solution structure, the results often depend on how the samples have been prepared and how they are influenced by the measuring procedure itself. Taking the measuring conditions into account, some vital assertions can be made relating to existing systems. However, only in rare cases do the results stand up to comparison (between various methods) in respect of an "absolute" molar mass distribution.

3. Production

Cellulose consists of chains of anhydroglucose units (AGUs) linked together glycosidically (β-1-4-bonds) ($\rightarrow$ Cellulose). Accordingly, each AGU still has three free hydroxyl groups which vary in their reactivity and which can be converted to ether groups in polymer-analogous reactions.

The free hydroxyl groups form intermolecular and intramolecular hydrogen bonds, so before the cellulose can be etherified it needs to be 'activated'. For this purpose the cellulose is chemically converted, usually with sodium hydroxide solution (caustic soda), in order to open up its natural structure and to obtain a sufficiently swollen material that is accessible to further reagents. The outcome is known as alkali cellulose or soda cellulose. The degree of swelling is controlled by the ratio of sodium hydroxide solution to cellulose and by the method of adding suspending agents or swelling agents. A minimum quantity of alkali (approx. 0.8 mol per mole of AGU) is required to swell the cellulose and for the reagents. During this activation process several properties of the end product are predetermined, because the degree of swelling and the uniformness of the activation influence the distribution of the substituents introduced in the second stage of cellulose ether production.

After the cellulose has been activated, the etherifying reagent is added. In principle, any etherification reaction known in the context of low-molecular alcohols could be applied, many of which have been studied [3]. Two basic types have become established in the industrial production of cellulose ethers – reactions where sodium hydroxide solution is consumed (e.g., Williamson ether synthesis) and reactions where sodium hydroxide only acts as a catalyst (e.g., alkoxylation). The supernatant caustic solution is neutralized after the reaction.

The kind of cellulose and its quality are crucial with regard to the ultimate properties of the cellulose ether. Therefore, many different types of cellulose obtained from wood pulp or cotton linters are used. The type of cellulose used determines, among others, the maximum chain length, hence the viscosity yield of the final cellulose ether. The chain length is reduced on account of the chemical and thermal stresses imposed during the process. The incidence of chain scissions is inherent in the production process but can be selectively increased. Oxidative, acid and radiation-induced decomposition are some of the known degradation processes which, depending on the product category, are used at various points in the process to obtain a desired low viscosity.

Processes. The processes depend on the synthesis-related reaction stages which can take place in one set of equipment or in several sets arranged in sequence.

A distinction is made between reactions at ambient pressure and reactions carried out at up to 30 bar, depending on the different reagents and on the required weight percentage of cellulose or cellulose ether in the reaction medium (mass fraction). Heterogeneous syntheses are used for industrial-scale production, whereas homogeneous syntheses are so far only found in research. Heterogeneous systems are classified by mass fraction into suspension processes (mass fraction below 10 wt%) and slurry processes (mass fraction approx. 10 – 50 wt%). There is also the option of a gas-phase or vapor-phase process, that works without an inert liquid. The process technology and engineering is adapted to meet the requirements of the mass fraction involved.

The cellulose is usually supplied as rolls of cellulose webs or as cellulose sheets. In the early days of cellulose ether production, the cellulose was treated in an alkali bath (approx. 15 – 20 wt% caustic soda) with the disadvantage that the cellulose had not been ground and was therefore not fully accessible to the activating agent. Consequently, the activation process took 4 h and the ripening time 8 h or more [13]. During the ripening process the alkali cellulose was kept in intermediate tanks where the reaction was allowed to complete at a specific temperature. Prior to the ripening stage any excess activating agent was pressed out and the alkali cellulose was then shredded. Storing it in ripening tanks made handling more difficult and reduced its processability.

To make the cellulose more accessible to the chemicals involved, in modern processes the cellulose is generally ground or milled before the activating and reaction stages. The cellulose is then alkalized with highly concentrated (approx. 30 – 70 wt%) aqueous caustic soda solution. Alkalization usually takes place in mixers or stirred reaction vessels, depending on the mass fraction and the reaction system. The mass fraction can be selected to influence the water balance which affects side reactions or reagent yield and also the quality of the cellulose ethers. The way the chemical reaction is engineered during the alkalization stage also has a significant effect on the quality of the cellulose ethers (see above).

The reaction can be carried out either batchwise, semibatchwise (fed batch) or continuously, and the reaction temperatures vary from room temperature up to 110 °C, depending on the type of etherification. Here again, mixers and stirred vessels are the established equipment. There are some isolated patents describing continuous processes in different equipment set-ups [14–16]. The starting materials are fed as solids or liquids under pressure. In some older processes for the production of methyl or hydroxyalkylmethyl celluloses gaseous reagents are used.

The equipment assembled downstream of the reaction depends on the required degree of purity. In general a distinction is made between technical and nontechnical cellulose ethers which differ in their content of byproducts and neutralization products – predominantly salts. Depending on the washing agents and the type of wash (displacement or dilution-type wash) the equipment used for solid-liquid separation includes suction filters, belt filters, drum filters, centrifuges and decanters, operating either continuously or batchwise. The washing processes are carried out in aqueous or alcoholic media, depending on the product group.

If the product is still wet with solvent after the washing sequence, the inert dissolving or washing agents can be recovered by displacing them with water prior to the drying stage. Suitable equipment for this are, for example, continuous mixers. Auxiliary facilities are used for regenerating the reagents and washing media and for extracting byproducts.

At the end of the process line the products are converted in forming stages from their fibrous form into the required supply form, usually powder grades with a defined particle size distribution and containing small amounts of residual moisture (in most cases < 10 wt%). Various drying and comminuting equipment is used in a number of different combinations. This process step is also utilized for modifying the cellulose ethers so that they offer optimal properties for the individual fields of application. The particle size distribution ranges from granular grades with an average grain diameter of approx. 0.5 mm to very fine powders where the percentage of particles with a diameter of less than 63 µm is higher than 50 wt%.

4. Product Groups

4.1. Carboxymethyl Cellulose (CMC)

In terms of sales volume carboxymethyl celluloses are the largest product group. They are available in various levels of purity from completely "unpurified" to "highly purified" for food-grade applications. The CMC group also includes mixed ethers with hydroxyalkyl substituents, that have a predominantly ionic character. The structure of CMC is shown in Figure 1.

4.1.1. Applications / Market

The market can be divided into two segments: one for "purified" and another for "unpurified" CMC. However, the terms "purified" and "unpurified" are subject to regional and supplier-specific differences in definition. Products with an active content of 55 – 85 % are considered to be "unpurified"; 85 – 95 %: "semi-purified"; around 98 %: "purified"; and products with an active content greater than 99.5 % are classed as "highly purified" (high-purity grades).

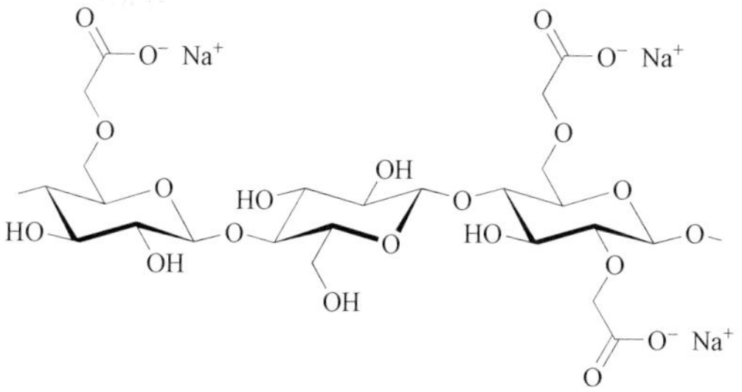

Annual sales worldwide amount to approx. 230 000 t, of which purified CMC accounts for 130 000 t and unpurified CMC totals 100 000 t. It is difficult to accurately apportion figures to the two segments, not only because of varying definitions but also due to the fact that in some applications both purified and unpurified grades of CMC can be used, and many suppliers have both purified and unpurified products in their portfolios.

CMC can be produced with less technological input than the other cellulose ethers, which explains why a comparatively large number of manufacturers (more than 20) exists. The leading producer is CP Kelko with a market share of around 33 %. The largest regional market and the region with the largest production capacity is currently Europe, which still accounts for some 60 % of global capacity.

The initial application fields of CMC were detergents, where it acts as a soil carrier, and in deep-well drilling for oil and water, where it is used as a flotation aid in drilling mud. Unpurified grades were and in some cases still are used in these applications. Purified products are used in surface coatings and various other technical areas such as in the paper industry for coatings and pulp sizing to improve fiber retention, filler/pigment/dye yield as well as strength. CMC also improves the printability and smoothness of paper. Together with gelatin CMC is used as a coacervate to encapsulate ink in the production of noncarbon copy papers. Users of high-purity grades include the cosmetics and pharmaceutical industries, where CMC is used for instance as a fat-free ointment base or as a tablet disintegrant. In food and pet food production CMC's simplest function is to improve consistency and to provide a low-calorie replacement for starch and proteins, or to control more complex functions such as the freeze-thaw stability of deep-frozen products or the creaminess of ice cream.

4.1.2. Synthesis

CMC is produced in a Williamson ether synthesis from alkali cellulose with sodium chloroacetate or with chloroacetic acid itself, which reacts

Figure 1. Structure of CMC

in-situ with caustic soda to form the salt. The process takes place in an aqueous or aqueous-alcoholic medium (slurry). The alcohols used are ethanol, isopropanol, *tert*-butanol and mixtures thereof.

Cell–OH · NaOH+ClCH$_2$COONa

→Cell–O–CH$_2$COONa+NaCl+H$_2$O

Alkalyzation is carried out at room temperature; the reaction takes place between 50 °C and the boiling point of the slurry medium under appropriate system pressure. The reaction is exothermic. With an activation energy of 87.9 kJ/mol it is highly temperature-dependent. The yield from this reaction in relation to chloroacetic acid is between 65 % and 80 %. Hydrolysis of chloroacetic acid occurs as a side reaction, forming glycolate. Commercial CMCs are produced with a degree of substitution between 0.2 and 1.5. Carboxymethylation takes place with a slight preference toward C2- followed by C6- and C3-substitution of the anhydroglucose unit [17].

4.1.3. Processes

In the early days of CMC production the cellulose was activated in alkaline steeping tanks [18]. After the excess caustic soda solution had been pressed out and the alkali cellulose had been shredded to fibers, the reaction was carried out with the addition of sodium chloroacetate in a kneader, where temperatures reached 30-35°C. The reaction was completed in a tumbling drum at 50 °C. The material was then washed with a mixture of methanol and water, filtered, ground and finally dried.

The first continuous process was used by Wyandotte Chem. Corp. (1947), but there was no washing stage, so it was only suitable for unpurified CMC [19]. Sodium hydroxide solution and monochloroacetic acid were added to the ground cellulose in a rotating drum. After a dwell time of 3 h the product was allowed to "ripen" for 10 h in drums.

In contemporary CMC production processes the ground cellulose is processed batchwise either as a suspension in the case of a low mass fraction of cellulose or, if the mass fraction is high, in mixers in the presence of "inert" media (see Fig. 2). Both processes use one or more short-chain alcohol(s) as a mass-transfer and heat-exchanging medium or as a suspending agent. For processes with a high mass fraction ethanol is usually employed a slurry medium, whereas isopropanol is used in suspension-type processes [20, 21]. By placing several sets of the same kind of equipment in parallel or in series it is possible to implement a quasi-continuous operation in the activation and reaction stages.

The starting materials and suspension or slurry media are either solid or liquid at normal room and reaction temperatures. Activation and reaction can therefore be carried out at ambient pressure.

Depending on the required product purity, the product is washed with alcohol-water mixtures, preferably with the same alcohol used in the reaction. Washing with water is not possible because of the water-solubility of the products. To meet product purity requirements, it is necessary to remove the alcohols from the product after the washing stage and prior to the grinding and drying processes.

Suspension agents and washing liquids are collected and reprocessed either by distillation and extraction or membrane processes.

4.1.4. Characterization

Carboxymethyl cellulose becomes water-swellable or alkali-soluble even at a low degree of substitution. The water-solubility of industrially produced CMC begins at a DS of approx. 0.6. The solubility increases up to a rather theoretical DS value of 3.0. Research work has shown that products with a lower degree of substitution (DS < 0.6) can manifest good solubility, the reason being different substituent distribution along the backbone [22].

Aqueous carboxymethyl cellulose solutions have very complex solution structures, because boundary conditions such as pH or salts simultaneously affect various characteristics. The solution state achieved depends, as with other cellulose ethers, on the nature of the dissolution process and the performance of the solvent. The parameter "solvent performance" covers all thermodynamic effects of protons, salts and other low-molecular substances on the interaction between the solvent and the macromolecule.

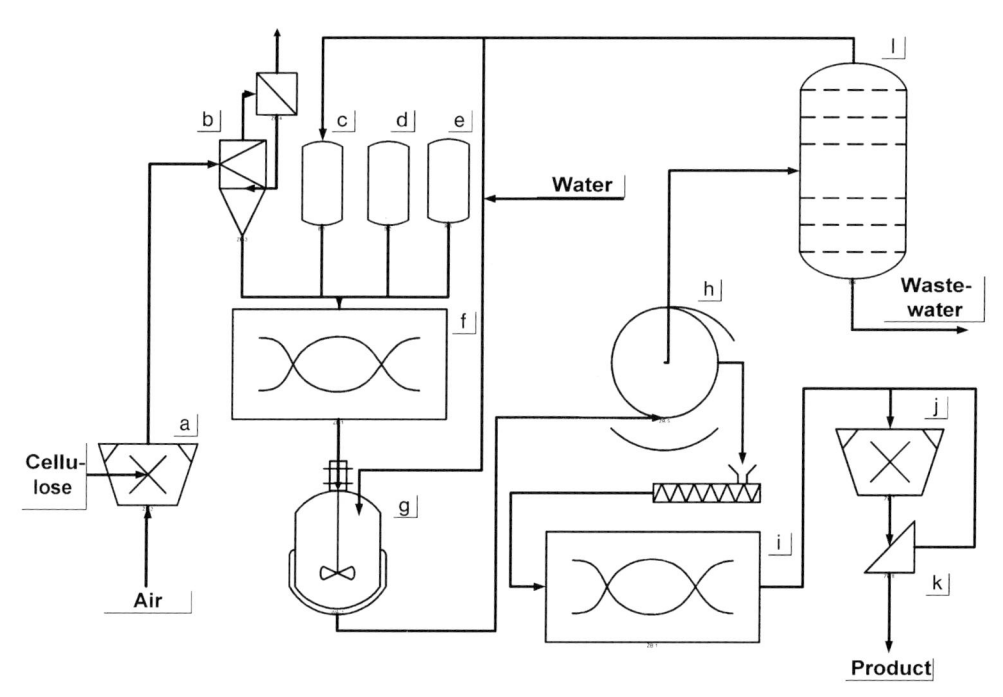

Figure 2. Reaction mixer processes for purified CMC production a) Cellulose shredding; b) Cellulose buffer; c) Slurry medium dosing vessel; d) Caustic soda dosing vessel; e) Chloroacetic acid dosing vessel; f) Reaction mixer; g) Washing vessel; h) Filtration equipment; i) Dryer; j) Comminuting; k) Screening equipment; l) Slurry medium recycling

In addition CMC is a weak carboxylic acid. The proton concentration in the solution not only affects the solvent's performance but simultaneously alters the number of free carboxylate groups. CMC's equivalence point is at a pH of 8.25; its pK_a value is between 4 and 5. In its acid form CMC is insoluble in water. Depending on the degree of substitution CMC can be precipitated from the solution at low pH values. Deprotonation increases with rising pH values and the character of the polyelectrolyte becomes apparent: The repulsion among the individual substituents becomes more distinct and the salt sensitivity increases. When CMC comes into contact with selected divalent or trivalent salts, this leads to complexing and in some cases to interlinking and precipitation of the CMC. With divalent salts these effects depend largely on the ratio of substituent to salt [23].

The many different possibilities of microscopic interaction are also reflected in the macroscopic properties of CMC. From a rheological point of view CMC solutions are inclined to form superstructures. Viscosities of 2 wt% solutions ranging from 3 mPa·s to 100 000 mPa·s are available on the market. Most of the products exhibit pseudoplasticity, some of them a thixotropic behavior. Depending on the selected degree of substitution and the product group, the solutions are either smooth-flowing or highly gelatinous.

Commercial grades are not only categorized according to their purity, their morphology and their mean viscosity, but also according to their degree of substitution. Water-soluble CMC commodities are available with a DS of around 0.7, 0.9 and 1.2. These three product groups do not only differ in their DS, but also in their solution structure, hence their rheological behavior.

4.2. Methyl and Hydroxyalkyl Methyl Celluloses

Pure methyl cellulose (MC) and hydroxyalkyl methyl celluloses (HAMC) are collectively known as methyl cellulose. Pure methylcellulose only accounts for a minor share of the market. The mixed ethers hydroxyethyl methyl cellulose (HEMC, see Fig. 3) and hydroxypropyl methyl

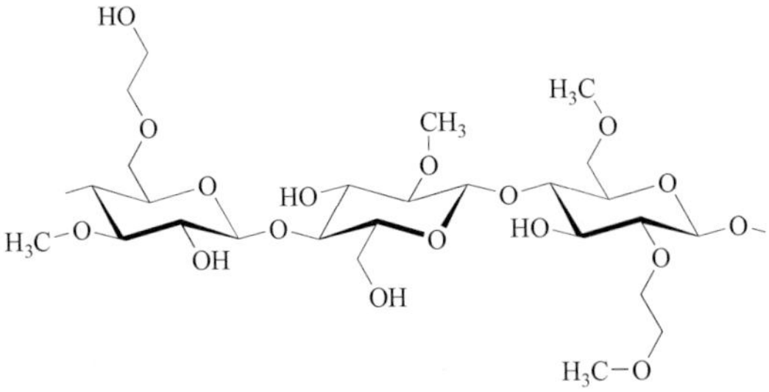

Figure 3. Structure of HEMC

cellulose (HPMC) have established themselves in particular for technical applications.

4.2.1. Applications / Market

The market for methyl cellulose and hydroxyalkyl methyl celluloses was in the range of 120 000 t in 2003, the market leader being Dow Chemical Company, with Shin-Etsu Chemical and Wolff Cellulosics having slightly lower market shares. High-purity grades are marketed for applications in the food, pharmaceutical and cosmetics sectors and purified products are used in building materials and industrial applications. The building material segment is by far the largest field of application.

The applications and therefore also the types of product used in the building materials sector vary from region to region due to different construction methods and traditions (→ Dry Mortars). In the USA joint compounds account for the largest segment; in Europe the main applications of MC and HAMC are plasters, renders, and tile adhesives. Efficient modern application methods would not be possible without the use of methyl celluloses. Properties such as water retention, open time, wet adhesion, initial thickening and setting behavior are regulated using only low additive contents of between 0.01 and 2 wt% (in relation to the building material system). Accordingly, the cellulose ether has to meet high quality requirements.

In the life-science sectors only nonhydroxyalkylated methyl cellulose (modified vegetable gum) and hydroxypropyl methyl cellulose (carbohydrate gum) are used. In the pharmaceutical industry methyl celluloses are employed as tablet base and in coatings of controlled-release drugs. In the food and cosmetics industries it is the thickening and emulsifying properties that are exploited to give the products the desired consistency and texture. The most noteworthy industrial application is the use of highly methylated hydroxypropyl methyl cellulose as a protective colloid in the polymerization of vinyl chloride.

4.2.2. Synthesis

The synthesis of hydroxyalkyl methyl cellulose is a combination of both principles of synthesis in industrial cellulose ether production: methylation is along the lines of Williamson ether synthesis where a stoichiometric amount of NaOH solution is consumed, whereas alkylation only requires catalytic amounts of alkaline solution.

Pure MC is produced from alkali cellulose by a reaction with gaseous or liquid methyl chloride. In side reactions the methyl chloride is hydrolyzed to form methanol, and subsequently the methanol is etherified by methyl chloride, forming dimethyl ether. The reactions take place between 70 and 120 °C. Methylation is an exothermic reaction, the activation energy being 80 kJ/mol.

$$Cell-OH \cdot NaOH + ClCH_3 \rightarrow Cell-O-CH_3 + NaCl + H_2O$$

The grades produced have a degree of substitution of 1.7 and 2.3. In case of pure methylation the substituent is found to prefer C2-, followed by C6- and C3-positions of the anhydroglucose unit [25, 26].

In the course of hydroxyalkyl methyl cellulose production alkoxylation of the cellulose takes place prior to or parallel to its methylation. As both the cellulose and the alkoxy groups formed can be methylated and alkoxylated, the outcome is a potentially indefinite number of different products, allowing the range of properties to be tailored to the requirements of specific applications.

4.2.3. Processes

The development of processes to produce methyl cellulose and hydroxyalkyl methyl celluloses began with methods at ambient pressure, i.e. gas circulation processes, where the cellulose is activated (for example by a steeping process), shredded and subsequently brought into contact with the reagents in a mixer. Nonreacted gaseous feed materials and byproducts are continuously extracted, the byproducts are condensed out and the gas flow is returned to the reactor. The product is then washed, filtered, compressed, ground and dried.

The slurry process (see Fig. 4), operated at pressures of up to 30 bar, is the process in most widespread use today for producing HAMC batchwise or semibatchwise in a mixer. Processes patented in the 1950s suggest the use of pressure-resistant reactors without any additional inert slurry media [27]. More recent patents propose adding inert slurry media, mostly ethers [28]. The slurry medium acts as a heat-transfer medium, which evaporates and is condensed in a dome-shaped condenser.

The reagents are added to the ground cellulose in any sequence, depending on the targeted product characteristics. The slurry medium has to be added prior to etherification on account of the heat generated during the exothermic reactions. Once the reaction has been completed, low-boiling slurry media are extracted by reducing the pressure and are collected for reuse. Byproducts are discharged. High-boiling slurry media are separated during the washing stage and subsequently reprocessed.

Suspension-type processes are mentioned in several patents. Their advantages are their simple process engineering, whereas disadvantages can be seen in the more complex preparation of the suspension agent.

The flocculation point of MC and HAMC allows the products to be washed with water above this flocculation point. To do so, the contents of the reactor are led into a stirred tank filled with hot water or they are suspended in the reaction mixer; the salts formed during the reaction are dissolved and any nonreacted activation reagents are neutralized with acid. The suspension is then separated in a solid-liquid separating unit. Further washing stages may be required, depending on the product specifications, in order to comply reliably with restrictions in content of byproducts.

Again depending on the type of product, the HAMC is then conditioned for the grinding and drying stage. Drying and grinding can be carried out in succession or in combined milling and drying units. The required fineness of grind, i.e.,

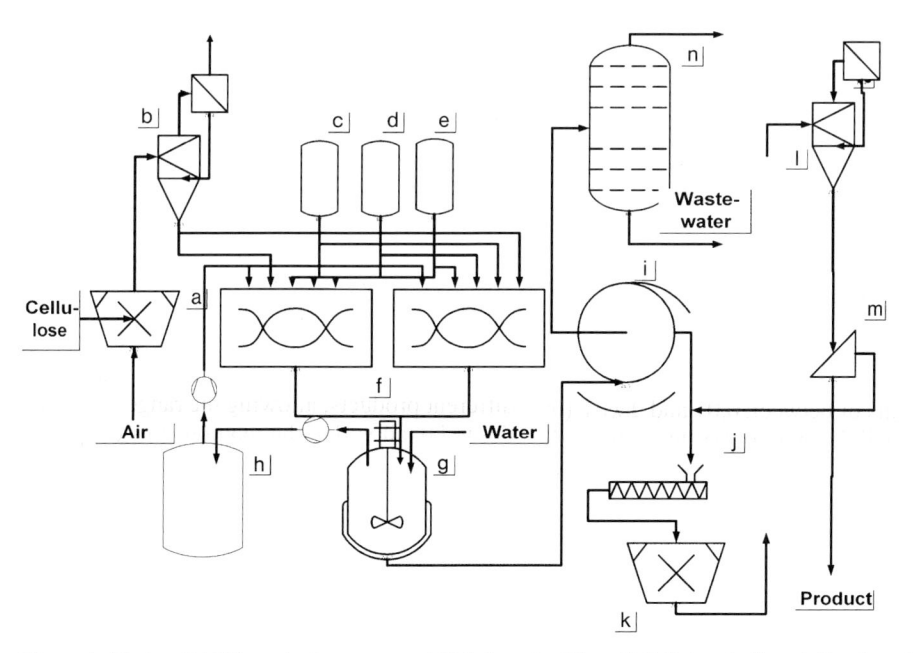

Figure 4. Modern HAMC production process a) Cellulose shredding; b) Cellulose buffer; c) Caustic soda dosing vessel; d) Alkoxylation agent dosing vessel; e) Methyl chloride dosing vessel; f) Reaction mixer; g) Washing vessel; h) Gas buffer; i) Filtration equipment; j) Conditioning equipment; k) Comminuting; l) Stream dryer; m) Screening equipment; n) Slurry medium recycling

the particle size, which in some cases is less than 63 µm in more than 50 wt% of the product, can often only be achieved in screening and sifting stages, after which the coarse-grained material is recycled.

4.2.4. Characterization

With a degree of substitution of around 1, methyl celluloses exhibit good water solubility at room temperature. The higher the degree of substitution, the better the solubility in water. However, it decreases again with further methylation and changes to solubility in organic solvents at a DS of approx. 2.6, because the material becomes increasingly hydrophobic.

Available on the market are pure methyl celluloses with a degree of substitution between 1.7 and 2.3, hydroxyethyl methyl celluloses with methylation levels between 1.3 and 2.0 and ethoxylation levels between 0.1 and 0.4, and hydroxypropyl methyl celluloses with degrees of methylation between 1.3 and 2.1 and of propoxylation between 0.1 and 1.0. The limits of substitution are more or less predetermined by the products' flocculation point, because washing with water is less expensive than with alcohol.

The flocculation point, i.e., the gelation and subsequent coagulation as the temperature rises, is a characteristic found only in nonionic cellulose ethers. By varying the methyl and/or alkyl content of hydroxyalkyl methyl celluloses, it is possible to modify the macromolecule's affinity to its water shell and therefore the flocculation point: with increasing methyl substitution the number of hydrophilic groups is reduced, and the flocculation point drops. Accessibility to the hydrophilic groups increases in proportion to the rise in hydroxyethylation, and the flocculation point shifts to higher temperatures. (The introduction of hydroxypropyl substituents results in opposite characteristics where, depending on the degree of substitution, the flocculation point either rises or drops.) Especially for applications where there is a high ambient temperature, e.g. on building sites in the summer, a methyl cellulose (MHEC or MHPC) with a sufficiently high flocculation point should be used to eliminate any undesired changes in characteristics.

The rheological properties of methyl celluloses and hydroxyalkyl methyl celluloses vary within a broad range, exhibiting some very different features, because there is a wide viscosity range in the grade spectrum available and because the products can have very differing chemical structures.

Products with a particularly low viscosity, such as HPMC for tablet coatings with a typical viscosity of between 3 and 6 mPa·s at a concentration of 2 wt% in water, still show a purely Newtonian (ideal) flow behavior in aqueous solutions with a HPMC concentration of 10 – 20 wt% at room temperature, and the viscosity of these solutions can be as high as 10 000 mPa·s. The Newtonian behavior is crucial for spraying highly concentrated solutions and for good leveling and film formation.

HEMC or HPMC grades with a higher viscosity for use as thickeners, however, exhibit a non-Newtonian flow behavior even with at low concentrations. In building materials applications, such as tile adhesives, the high static shear viscosity guarantees standing strength and resistance to demixing as well as high water retention, while the non-Newtonian properties ensure rapid and easy workability. The thickening characteristics of nonionic cellulose ethers are scarcely affected by the presence of dissolved (even polyvalent) salts or by changes in the pH value, so they are also ideal for use in cement-based building materials in which, after water is added, highly alkaline conditions prevail.

The most effective way to achieve good thickening at a low shear rate is to use small input quantities of high-viscosity grades. In these systems the viscosity soon drops considerably as the shear rate increases. A high thickening effect at high or at least at medium shear rates is desirable for improved workability in certain high-quality building material systems, and it can only be obtained with higher levels of methyl cellulose, which also provide additional benefits, such as long open times and improved adhesion in the case of tile adhesives.

4.3. Hydroxyethyl Cellulose (HEC) [29]

Hydroxyethyl celluloses are the third largest product group in the marketplace. They are nonionic and soluble in both cold and hot water. Their manufacture was first described in 1920 in

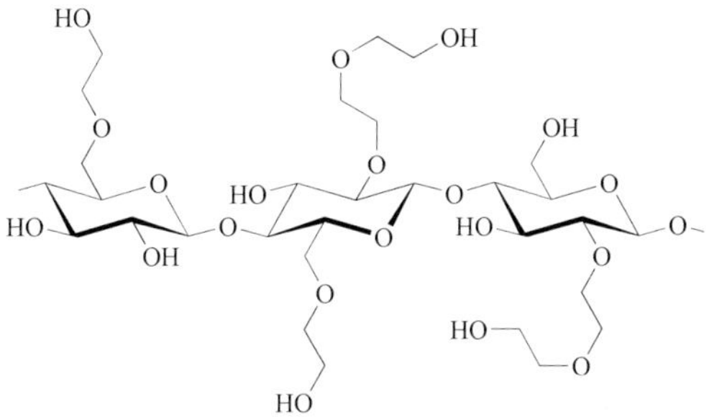

Figure 5. Structure of HEC

a patent granted to Farbenfabriken Bayer. The structure of HEC is shown in Figure 5.

4.3.1. Applications / Market

The hydroxyethyl cellulose market is served mainly by Hercules Incorporated/Aqualon and Dow Chemical Company. Annual sales are around 60 000 t, most of which are used in the construction industry, and specifically in surface coatings. Hydroxyethyl cellulose is also found in cosmetic applications where it competes against HPMC derivatives. Other applications are in drilling fluids and as a protective colloid in emulsion polymerization.

In the paint industry HEC is the cellulose ether product group with the largest market share. Although products based on HEMC, HPMC or CMC have smaller shares, they are firmly established in these applications for technical and cost-related reasons. Beside pure HEC this product group also includes mixed cellulose ethers similar to HEC: ethyl hydroxyethyl cellulose (EHEC) and hydrophobically modified (hm) variants of HEC. The additional hydrophobic properties are imparted to hm-HEC by minor secondary substitution, for instance with long-chain alkyl residues (C_{12} through C_{24}) [30]. The hydrophobically modified grades are, like purely synthetic associative thickeners, capable of providing additional effects by interacting with particle surfaces, in particular those of the binder in latex paints, and by forming micelles and mixed micelles with other surfactants contained in the paint. These options are exploited especially in high-grade paint formulations in order to reduce spattering when applying the paint with a roller, and to obtain improved brush resistance and leveling.

4.3.2. Synthesis

Hydroxyethyl cellulose is formed by reaction of cellulose with ethylene oxide. This ethoxylation only requires catalytic quantities of alkaline solution.

$$Cell-OH + OH^- \rightarrow Cell-O^- + H_2O$$
$$Cell-O^- + CH_2CH_2O \rightarrow Cell-O-CH_2CH_2-O^-$$
$$Cell-O-CH_2CH_2-O^- + H_2O \rightarrow$$
$$Cell-O-CH_2CH_2-OH + OH^-$$

Just as the cellulose alcoholate reacts with the ethylene oxide, also the hydroxyl ions present can react with the ethylene oxide or with the glycolates formed in the process. Large amounts of alkali reduce the yield of the main reaction, but to produce HEC a minimum quantity of caustic soda solution is essential to break down the cellulose.

Ethoxylation takes place in the presence of an "inert" solvent (isopropanol, *tert*-butanol, 1,2-dimethoxyethane or acetone); it starts at a very low temperature ($\approx 30\ °C$) and is highly exothermic. HECs are produced with a molar substitution of 1.5 and 3.5, these degrees being controlled via the quantity of ethylene oxide added. The yield is between 40 and 75 % in relation to ethylene oxide. The reaction of ethylene oxide takes place in the following order of preference (from high to low): alkoxy groups, then C6 positions, then (followed by some distance) C2 and C3 positions [31]. The relationship of the reaction rates can be influenced by varying the alkali concentration, e.g., by partial neutralization [32]. After the reaction has taken place, the alkali quantity added needs to be neutralized.

4.3.3. Processes

For the production of HEC, suspension processes above ambient pressure have become the established methods [33]. Stirred pressure vessels are used for the reaction (see Fig. 6).

The ground cellulose is suspended in the suspending medium and mixed with the alkalizing agent. After alkalization, liquid ethylene oxide is added to the suspension and allowed to react for about 2 h.

Salts and byproducts have to be washed out of the products. As with other cellulose ethers, this

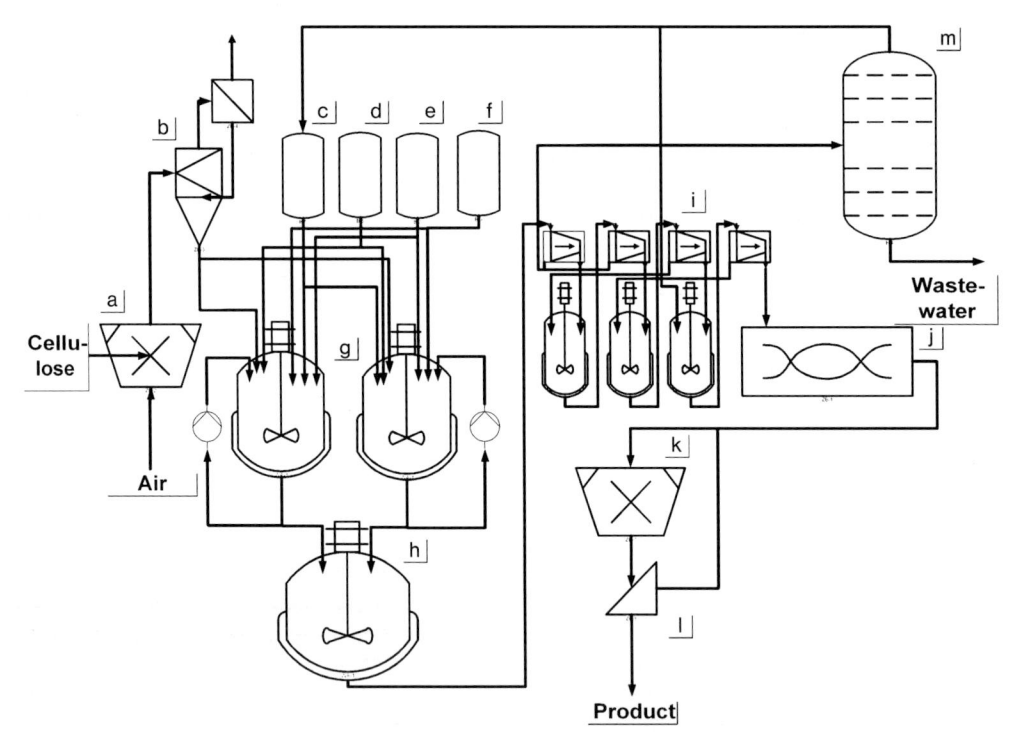

Figure 6. Process for the production of HEC/HPC with stirred tanks a) Cellulose shredding; b) Cellulose buffer; c) Slurry medium dosing vessel; d) Caustic soda dosing vessel; e) Acid dosing vessel; f) Alkoxylation agent dosing vessel; g) Stirred reaction tank; h) Washing vessel; i) Washing and filtration process; j) Stream dryer; k) Comminuting; l) Screening equipment; m) Slurry medium recycling

involves appropriate equipment for washing and solid-liquid separation either in series or in a combined unit. As this cellulose ether does not have a flocculation point, mixtures of water and an organic solvent are used, followed by drying and grinding.

Parallel to this, the suspension medium and any extracting agents have to be regenerated and recycled. Distillation is the method usually selected for this.

4.3.4. Characterization

Hydroxyethyl celluloses are available on the market with viscosities between 10 mPa·s and 100 000 mPa·s and molar substitutions of 1.5 and 3.0 (up to 4.5 in the case of hydrophobically modified grades). They are water-soluble and have no thermal flocculation point. Hydroxyethyl celluloses are also more difficult to salt out than methylated derivatives and are not displaced from water even by larger amounts of organic solvent. The degree of substitution varies in the region of 1 in industrially produced grades, i.e., higher hydroxyethylation leads chiefly to secondary, tertiary, and quaternary substitution of the hydroxyalkyl substituent (formation of ethylene oxide side chains).

The rheological behavior of HECs is comparable with that of carboxymethyl celluloses. By introducing a small number of hydrophobic side chains, grades can be obtained, whose rheological profile in solution is highly modified, particularly when mixed with surfactants. The hydrophobic residues form part of the surfactant micelles and this increases the viscosity, especially at elevated shear rates.

4.4. Hydroxypropyl Cellulose (HPC)

In relation to the quantities of cellulose ethers used worldwide, hydroxypropyl celluloses are the smallest product group. Their properties range from simple swellability, through solubility in

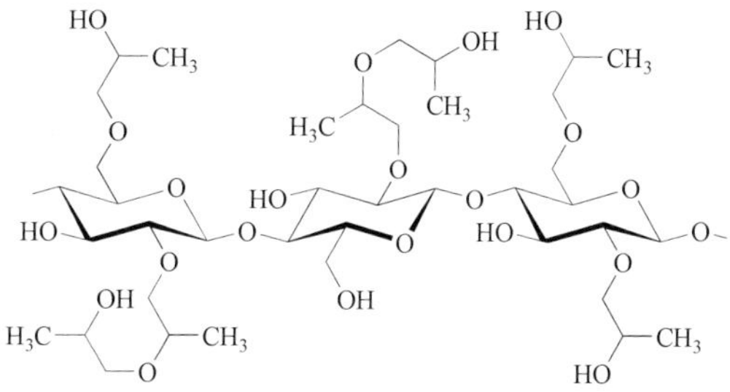

Figure 7. Structure of HPC

cold water to enhanced thermoplasticity, which enables them to be processed in melt extruders. Hydroxypropyl cellulose was first launched on the market in the USA at the end of the 1960s. The structure of hydroxypropyl cellulose is shown in Figure 7.

4.4.1. Applications/Market

The annual global market for hydroxypropyl cellulose is far below 10 000 t. The main fields of application are in the pharmaceutical and cosmetics industries [34]. High-purity grades are available for food and pharmaceutical applications, and purified grades are supplied to the cosmetics industry and for technical applications.

Hydroxypropyl cellulose is supplied by Hercules Incorporated/Aqualon, by Shin-Etsu Chemical and by Nippon Soda.

4.4.2. Synthesis

The synthesis of hydroxypropyl cellulose does not differ from that of hydroxyethyl cellulose. Prior to the reaction with propylene oxide, the cellulose is activated with caustic soda solution. The reaction has a somewhat higher activation energy and requires a higher reaction temperature than that with ethylene oxide. The byproducts are propylene glycol and polypropylene glycols which, like the salts from the neutralization stage, need to be removed in one or several washing stage(s).

4.4.3. Processes

The production processes for HPC are largely the same as for HEC. The reaction is generally carried out in a slurry. The slurry media described in relevant literature are hexane, toluene, tetrahydrofuran and dioxane, and also the common alcohols. Apart from the slurry process, the gas-phase method – which produces high-quality grades due to the high degree of substitution – is also applied. Since highly substituted HPC has a thermal flocculation point, however, washing can be carried out using hot water, which is not possible with HEC [35].

4.4.4. Characterization

Highly substituted hydroxypropyl cellulose is readily soluble in cold water and also soluble in polar organic solvents. The HPCs available on the market have a high molar substitution (3.0 – 4.5) and mean viscosities between 10 and 30 000 mPa·s. HPC solutions are generally smooth-flowing and exhibit scarcely any tendency toward thixotropic behavior. Gelation and coagulation begins at relatively low temperatures (40 – 50 °C). Hydroxypropyl cellulose is surface-active and tends to foam. Highly concentrated HPC solutions can develop liquid-crystalline phases. With a sufficiently high degree of substitution hydroxypropyl cellulose is thermoplastic and can be processed at approx. 150 °C. The extruded moulded parts are water soluble yet, due to their low propensity to absorb water, they are not sticky even when the humidity is high.

HPC with a low molar substitution of between 0.2 and 0.4, on the other hand, swells in water, but is not water-soluble. These grades are used as binders or disintegrants in the manufacture of tablets.

5. Uses

Cellulose ethers are used in a wide variety of fields ranging from oil drilling to industrial applications such as polymerization, from surface coatings and construction materials to the health care sector, cosmetics, food and pharmaceuticals. Not all types of cellulose ethers are equally suitable in all fields of application. Table 1 provides an overview of the great variety of applications and the appropriate types of cellulose ether. A listing of this kind

Table 1. Major fields of application in which the individual product groups are used

Carboxymethyl cellulose	Methyl cellulose, hydroxyalkyl methyl cellulose	Hydroxyethyl cellulose	Hydroxypropyl cellulose
Paper	tile adhesives	latex paints	adhesives
Detergents	plasters/renders	adhesives	ceramics
Drilling for oil and gas	pharma/cosmetics	building materials	cosmetics
Pharma	joint compounds	cosmetics	encapsulation
Cosmetics	wallpaper paste	drilling for oil & gas	food
Textile industry	polymerization	agriculture	household goods
Food	food	paper	printing inks
Coatings	latex paints	synthetic resins	polymerization
Encapsulation	cement extrusion	textile industry	films

can only be incomplete, but indicates the versatility and above all the adaptability of the products to the various fields in which they are to perform.

In many cases the amount of cellulose ether required is only between 0.02 and 2 %. This small level is, however, crucial for properties such as water-binding capacity, film formation and thickening as well as for controlling other rheological properties. Many modern products and processing techniques would not be possible without the use of cellulose ethers.

Each field of application requires specific solutions which in many cases involve modifying the cellulose ether product. This means that, once a suitable product category has been selected, all the parameters relating to the appropriate type of substitution and viscosity range have to be fine-tuned. Besides their chemical structure, characteristics such as product purity and forming, i.e., the physical state in which the products are made available, play a substantial part. Fine-tuning requires a great deal of experience in respect of the properties of these products as well as with regard to the systems and methods of application. Most suppliers provide this expertise and offer additional technical service.

6. Economic Aspects

In 2003 the sales volume of the cellulose ether market worldwide was approximately $\$ 2 \times 10^9$. The largest market is Europe, followed by the USA and Japan. Annual consumption is ca. 420 000 t.

Sorted in the order of their economic significance (by sales volume), carboxymethylcelluloses rank first with approx. 230 000 t/a, while methyl- and hydroxyalkylmethylcelluloses are second with approx. 120 000 t/a, followed by hydroxyethylcelluloses on third place with approx. 60 000 t/a and last by hydroxypropylcellulose with less than 10 000 t/a.

Taking all market segments together, the cellulose ether market has grown continuously in recent years. However, there has been and still is a noticeable shift among the individual markets. As opposed to vigorous growth in the threshold countries and moderate growth in Europe and the USA, the Japanese market is dwindling.

The market is served overall by a small number of suppliers of cellulose derivatives. Most of the manufacturers produce a variety of different products for several market segments. The carboxymethyl cellulose market is spearheaded by CP Kelko, a company with a market share of 33 %. The remaining market is shared by 10 to 15 small to medium-size suppliers. The methyl and hydroxyalkyl methyl cellulose market is served by a few major producers [Dow Chemical Company, Wolff Cellulosics, and Shin-Etsu Chemical (incl. SE Tylose, the former cellulose ether division of Clariant AG)] with a market share totaling 80 %. Hydroxyethyl cellulose is supplied more or less exclusively by Hercules and Dow Chemical Company, whereas HPC is mostly supplied by Hercules and Shin-Etsu Chemical.

7. Toxicology and Occupational Health

Toxicology. Cellulose ethers are generally nontoxic. High-purity grades of most commercial products are approved as food additives and for use in cosmetic compositions.

Handling. Fine powders of cellulose ethers form explosive dusts in air as do natural polysaccharides or sawdust. Dry nonionic ethers undergo electrostatic charging similar to that of other organic polymers. When cellulose ethers are stored and handled, the general precautions concerning powdered organic polymers must be observed. The flammability is similar to that of cellulose. Spilling of solutions forms very slippery films that are difficult to remove.

Ecology. The biodegradation of cellulose ethers by cellulase-producing microorganisms also occurs in wastewater and therefore prevents the accumulation of cellulose ethers. Glucose, glucose ethers, and ether oligomers result from enzymatic hydrolysis; they are further degraded to carbon dioxide and water in slow bioreactions. Biostable or toxic metabolites are not known.

Cellulose ethers have no fish toxicity and are poor nutrients for most microorganisms. Nevertheless, wastewater bacteria may adapt to enhanced cellulose ether degradation after some exposure.

The suppliers provide Safety Data Sheets containing detailed data with regard to handling and the toxicological properties of the respective products.

References

1. W. Suida, *Monatshefte Chemie* **26** (1905) 413–427.
2. E. Jansen, DE 332203, 1918.
3. D. Klemm, et al.: *Comprehensive Cellulose Chemistry*, Vol. 2, WILEY-VCH Verlag GmbH, Weinheim, Germany 1998.
4. J. Schurz, et al.: "Methodische Fortschritte in der Instrumentanalytik", *Das Papier* **12** (1999) 712–764.
5. W. W. Graessley, *Polymer* **21** (1980) 258.
6. W.-M. Kulicke, et al.: "Characterization of aqueous carboxymethylcellulose solutions in terms of their molecular structure and its influence on rheological behaviour", *Polymer* **37/13** (1996) 2723–2731.
7. R. Dönges, *Das Papier* **12** (1997) 653–670.
8. P. Käuper, et al.: "Development and evaluation of methods for determining the pattern of functionalization in sodium carboxymethylcelluloses", *Angew. Makromol. Chem.* **260** (1998) 53–63.
9. P. Mischnick, et al.: "Structure Analysis of 1,4-Glucan Derivatives", *Macromol. Chem. Phys.* **201** (2000) 1985–1995.
10. P. W. Arisz: "Substituent distribution along the cellulose backbone in O-methylcelluloses using GC and FAB-MS for monomer and oligomer analyses", *Carbohydr. Res.* **271** (1995) 1–14.
11. B. Reiche, H. Jehring, H. Dautzenberg, B. Philipp: "Elektrosorptionsuntersuchungen an Carboxymethylcellulosen", *Faserforschung und Textiltechnik* **29/5** (1978) 324–328.
12. "Cellulose", in: *Ullmann*, 6th ed., vol. 6, WILEY-VCH Verlag GmbH, Weinheim, Germany 2002, pp. 593–645.
13. O. Wurz: *Celluloseäther - Herstellung und Anwendung*, Roether, Darmstadt 1961.
14. Kalle AG, DE 1543136, 1969 (F, Eichenseer, S. Janocha, H. Macholdt).
15. Dow Chemical Company, US 4015067, 1975 (G.Y.T Liu, C.P. Stange).
16. Henkel KgaA, DE 2929011, 1981 (W. Willi, H. Leischner, W. Rähse, F.-J. Carduck, N. Kühne).
17. H. M. Spurlin, *J. Am. Chem. Soc.* **61** (1939) 2222.
18. J. Voss: "Celluloseäther", in *Kunststoffhandbuch*, vol. 3, Hanser, München 1965 pp. 349–397.
19. R. N. Halder, W. F. Waldeck. F. W. Smith, *Ind. Eng. Chem.* **44** (1952) 2803.
20. V. Stigsson, G. Kloow, U. Germgård, *PaperAsia* **17/10** (2001) 16–21.
21. R. Dönges, *Das Papier* **12** (1997) 653–670.
22. T. Heinze, T. Liebert, P. Klüfers, F. Meister: "Carboxymethylation of cellulose in unconventional media", *Cellulose* **6** (1999) 153–165.
23. P. Käuper: "Natriumcarboxymethylcellulose: Bestimmung der chemischen Struktur, der Lösungsstruktur und der Elektrolytwechselwirkungen", *Berichte aus der Chemie* (1998), Shaker, Aachen.
24. "Dry Mortars", in: *Ullmann*, 6th ed., vol. 11, WILEY-VCH Verlag GmbH, Weinheim, Germany 2002, pp. 83–108..
25. R. Dönges, *Br. Polym. J.* **23** (1990) 315–326.
26. K. G. Rosell,; *J. Carbohydr. Chem.* **7** (1988) 525–536.
27. Henkel & Cie GmbH, GB 754876, 1953.
28. Britisch Celanese Ltd., GB 909039, 1960 (E. W. Hitchin, H. Bates).
29. G. M. Powell: "Hydroxyethylcellulose", in *Water-Soluble Gums Resins*, McGraw-Hill, New York 1980.
30. L.-M. Zhang, *Carbohydr. Polym.* **45** (2001) 1–10.
31. M. G. Wirick, *J. Polym. Sci.* **6** (1968) 1705–1718.
32. P. W. Arisz, H. T. T. Thai, J. J. Boon, W. G. Salomon, *Cellulose* **3** (1996) 45–61.
33. Hercules Powder Company, US 688486, 1950 (E.D. Dlug, H.G. Tennent).
34. R. W. Butler, E. D. Klug: "Hydroxypropylcellulose", in *Water-Soluble Gums Resins*, McGraw-Hill, New York 1980.
35. Hercules Powder Company, US 3479190, 1969 (A. J. Ganz).

Chitin and Chitosan

SHIGEHIRO HIRANO, Tottori University, Tottori, Japan

1. Introduction 243
2. Molecular Structure and Conformation . . 243
2.1. Chitin. 244
2.2. Chitosan. 245
3. Raw Materials and Production 245
3.1. Isolation of Chitin from Crab and Shrimp Shells 245
3.2. Preparation of Chitosan from Chitin 246
4. Metabolism and Biosynthesis 246
5. Chemical Properties 246
5.1. Reactions on the Amino Group 246
5.1.1. N-Acylation 246
5.1.2. Formation of N-Alkylidene and N-Arylidene Derivatives 247
5.1.3. N-Alkylation and N-Arylation 247
5.2. Reactions at the Hydroxyl Group 248
5.3. Reactions at C-6. 248
5.4. Graft Polymerization on Chitin and Chitosan. 248
6. Application Forms and Formulations 248
7. Uses 249
8. Economic Aspects 250
9. Toxicology and Environmental Aspects . . . 250
References 251

1. Introduction

In 1811 BRACONNOT [1] isolated from a fungus a compound which he called "Fungin". In 1823 ODIER discovered that this compound is a constituent of the exoskeleton of insects and gave it the name "chitin", which means envelope in Greek [2].

In 1859 ROUGET [3] prepared a compound from chitin by treatment with concentrate alkalis. This compound was named "chitosan" by HOPPE-SEILER [4] in 1894.

Both chitin and chitosan are naturally occurring cationic biopolymers, that are obtained from crab and shrimp shells, being waste materials of marine food processing companies. Chitin and chitosan are biodegradable, biocompatible with animal and plant cells and tissues, and almost nontoxic. Both can be processed into various products including hydrogels, beads, membranes, and sponges.

2. Molecular Structure and Conformation

Chitin is a linear (1→4)-linked 2-acetamido-2-deoxy-β-d-glucopyranan (N-acetyl-β-d-glucosaminan) in the chair $^4C_1$ conformation without any branching, and chitosan is a linear (1→4)-linked 2-amino-2-deoxy-β-d-glucopyranan (β-d-glucosaminan) (see below). Between chitin and chitosan, a number of derivatives of partially N-deacylated chitins and partially N-acetylated chitosans exist. Chitins are insoluble in aqueous organic acids (e.g., acetic acid), although chitosans are soluble. Their physical properties vary with the molecular mass, the degree of substitution (d.s.) and the distribution of free amino group on the chain.

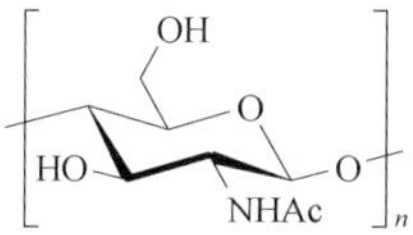

Chitin

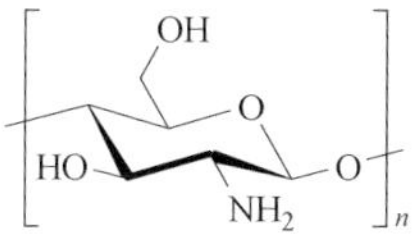

Chitosan

2.1. Chitin

On the basis of X-ray diffraction patterns, chitin occurs in two polymorphs (α- and β-chitin) in the solid state, owing to the orientation of the extended chains and inter- and intra-molecular hydrogen bonds (Fig. 1). In α-chitin the chains have an antiparallel, in β-chitin a parallel orientation [5], [6]. The α-orientation is more stable and more widely distributed in the nature than the β-orientation which is found only in squid pens. The β-orientation transforms into the α-orientation in the solid state by treatment with acids, but the reverse reaction does no occur under these conditions.

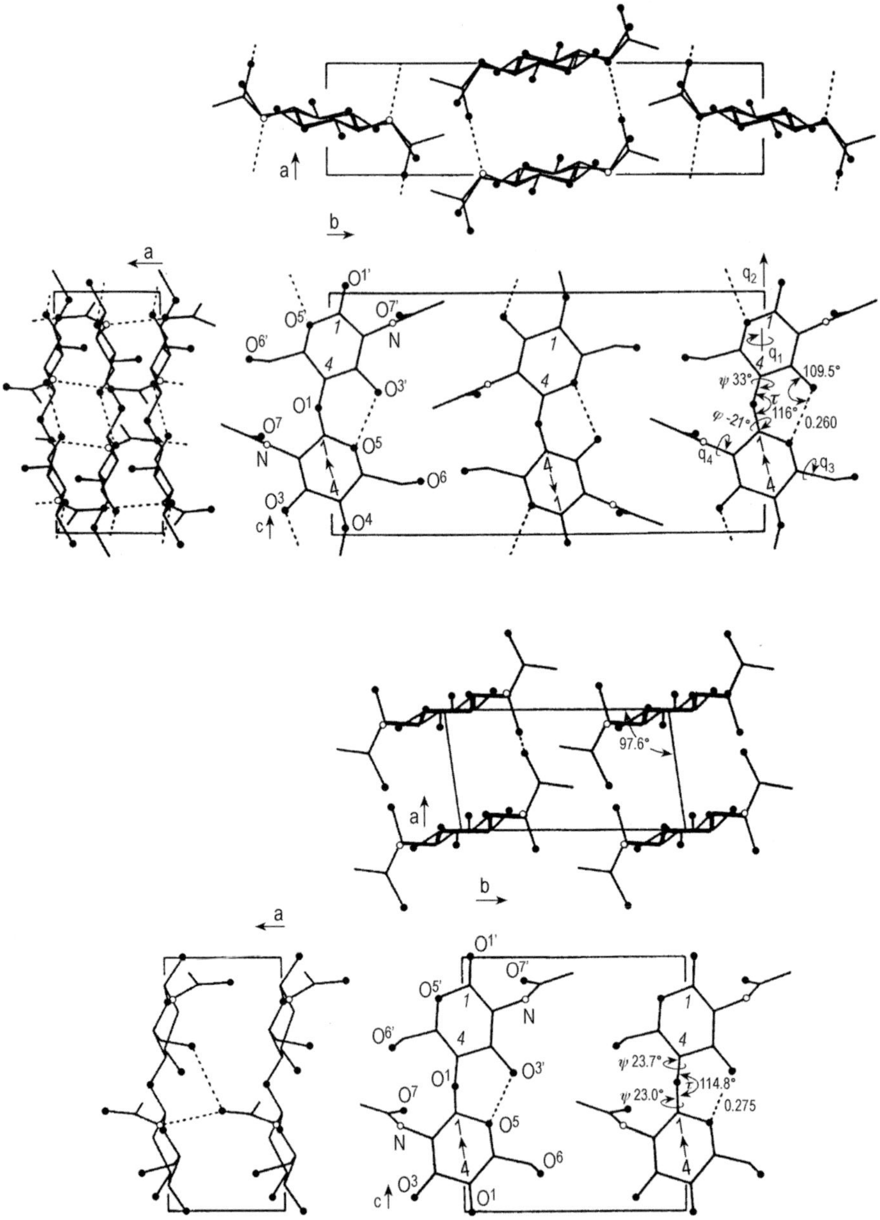

Figure 1. Molecular conformations of chitin in the sold state β-chitin in the parallel conformation (above) and α-chitin in the antiparallel conformation (below)

Table 1. Chitin content of some organisms

Organism	Chitin content, %
Crustacea	
Carcinus (crab)	0.4–3.3[a]
Callinectes (blue crab)	14[a]
Paralithodes (king crab)	10.4[a], 35[b]
Pleuroncodes (red crab)	1.3–1.8[b]
Crangon (shrimp)	5.8[b]
Insects	
Blatella (cockroach)	10[b]
Coleoptera (beetle)	5–15[b]
Grasshopper	2–4[a]
May beetle	16[b]
Mollusks	
Clam shell	6.1[b]
Krill	40.2[b]
Oyster shell	3.6[b]
Squid pen	41[b]
Microorganisms	
Aspergillus niger	42[c]
Lactarius vellereus (mushroom)	19.0[c]
Mucor rouxii	44.5[c]
Penicillium notatum	18.5[c]
Saccharomyces cerevisiae (baker's yeast)	2.9[c]

[a] Wet body weight.
[b] Dry body weight.
[c] Dry weight of the cell wall.

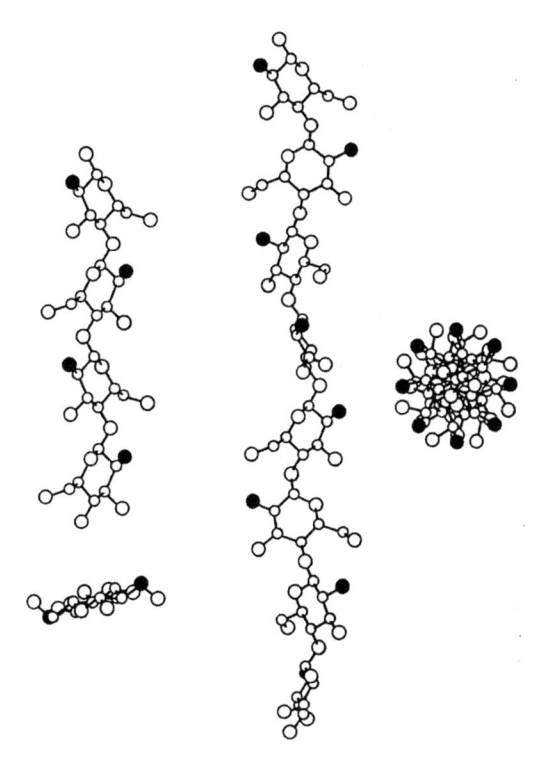

Figure 2. Molecular conformations of chitosan at the solid state. The two-fold helix conformation is shown on the left: a side view (above) and a sectional view (below). The eight-fold helix conformation is shown on the right: a side view (left) and a sectional view (right).

2.2. Chitosan

Chitosan occurs mainly in two molecular conformations: as extended two-fold helix and as extended eight-fold helix (Fig. 2) [7]. The eight-fold helix conformation transforms into the two-fold helix under conditions of high humidity [8]. No ordered conformation is present in the aqueous acidic solution of chitosans. The molecular flexibility increases with increasing *N*-deacetylation, with increasing ionic strength in the solutions, and increasing temperature.

3. Raw Materials and Production

It is estimated that about 100×10^9 t of chitin are produced every year on the earth. As shown in Table 1, chitin is distributed widely in the exoskeleton of crustaceans (crab, shrimp etc.), in the cartilages of mollusks (krill, squid etc.), in the cuticles of insects (cockroach, beetle etc.), and in the cell walls of micro-organisms (fungi, yeasts, etc.) [9]. At present the major industrial source of chitin and chitosan are the shell wastes of crabs and shrimps. The shells mainly consist (on a dry basis) of chitin (20–30 %), proteins (30–40 %), calcium carbonate (30–50 %), lipids, and astaxanthin (less than 1 %).

3.1. Isolation of Chitin from Crab and Shrimp Shells

Chitin is isolated from crab and shrimp shells as follows: (1) Ca_2CO_3 in the shell flakes is dissolved out by treatment with dilute HCl (demineralization), (2) astaxanthin pigments and lipids are extracted with organic solvents, e.g., acetone and ethanol, (decolorization), and (3) proteins are extracted with dilute NaOH or digested enzymatically by proteases or microorganisms [10] (deproteinization). Chitin is obtained as residue in the form of flakes.

Astaxanthin is extracted more efficiently from fresh wet shells than from dry shells [11]. The pigment is easily decomposed by exposing to sunlight in the air-drying step, because the conjugated double bonds in the pigment molecule are very sensitive to ultraviolet light. An oxidative bleaching treatment with H_2O_2 or NaOCl is used at some companies in order to obtain a white-colored product, but this process partially cleaves the glycosidic linkage at the N-deacetylated portions, resulting in the formation of low molecular mass products having a 2,5-anhydro-d-mannose residue at the reducing end group.

3.2. Preparation of Chitosan from Chitin

For preparation of chitosan, chitin flakes are treated in suspension with aqueous 30–60 % NaOH at 80–120 °C with stirring for 4–6 h, and this treatment is repeated once or more times for obtaining highly N-deacetylated products. However, repeated treatment is generally accompanied with depolymerization. Alternatively, the clear homogeneous solution of the sodium salt of chitin (alkaline chitin) in aqueous 14 % NaOH is treated at 25 °C to give rise to the partially N-deacetylated derivatives (d.s. about 0.5 for NAc), which are soluble in water [12]. A random distribution of N-acetyl group is found in these products [13].

Chitosan can be prepared from chitin also enzymatically [14], [15]. A powdered sample of chitin is treated with N-deacetylase (EC 3.5.1.41) (see Chap. 4) or with microbes which secret N-deacetylase. The enzymatic method yields chitosan with low degree of N-deacetylation and low degree of depolymerization.

The degree of acetylation in chitosan may be determined (1) by the C : N ratio in the elemental analyses, by, ^{15}N NMR, or ^{13}C NMR spectroscopy [16–18], by IR spectroscopy [19] by colloidal titration [20], and by pyrolysis-gas chromatography [21].

4. Metabolism and Biosynthesis

Chitinase (EC 3.2.1.14) and lysozyme (EC 3.2.1.17) hydrolyze the internal β-(1→4)-glycosidic linkages of chitin to give chitin oligosaccharides, which are further hydrolyzed by β-d-glucosaminidase (EC 3.2.1.30) to afford N-acetyl d-glucosamine. Chitin is converted to chitosan by N-deacetylase (EC 3.5.1.41) [22], and chitosan is hydrolyzed by chitosanase (EC 3.2.1.132) to give chitosan oligosaccharides, which are hydrolyzed by β-d-glucosaminidase to give rise to d-glucosamine (Fig. 3). Chitinase is widely distributed in plants, insects, fish and microorganisms, and lysozyme is found in animals [23].

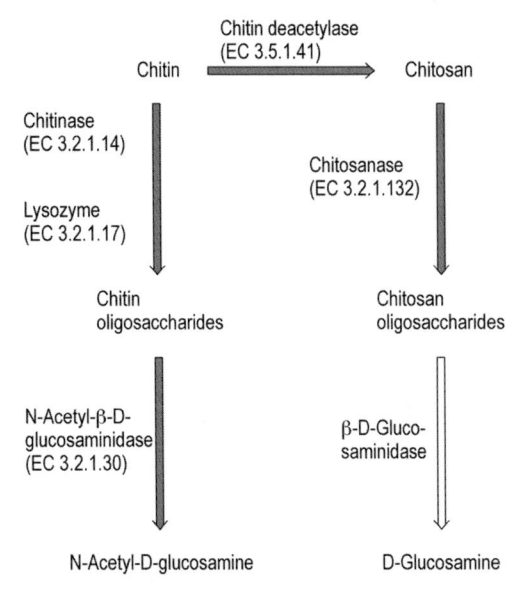

Figure 3. Enzymes for the hydrolyses of chitin and chitosan

5. Chemical Properties

Chitin has two hydroxyl groups at C3 and C6 of the repeating N-acetyl-d-glucosamine moiety. The OH group at C6 is more reactive then that at C3. Chitosan has one reactive amino group at C2 and two OH groups at C3 and C6. The chemical reactions on these groups give rise to various derivatives (Table 2).

5.1. Reactions on the Amino Group

5.1.1. N-Acylation

Chitosan is regioselectively N-acylated by treatment with carboxylic anhydrides [24], [25]. The reactions with low fatty acid anhydrides (C_2–C_{10})

Table 2. Chemical structures of chitin and chitosan derivatives[a]

Chitosan derivatives	R^1	R^2	R^3
Reactions at the amino group			
1) Acyl	–NHC(=O)–R	–OH	–CH$_2$OH
2) Alkylidene + arylidene	–N=CH–R	–OH	–CH$_2$OH
3) Alkyl + aryl	–NH–CH$_2$R	–OH	–CH$_2$OH
4) Deacyl	–NH$_2$	–OH	–CH$_2$OH
5) Imido	–N(C=O)$_2$=R	–OH	–CH$_2$OH
6) Metal chelate	–NH2Me	–OH	–CH$_2$OH
7) Nitro	–NHNO2	–OH	–CH$_2$OH
8) Salts	–NH$_3^{+-}$R	–OH	–CH$_2$OH
9) Sulfate	–NHS(=O)$_2$ONa	–OH	–CH$_2$OH
10) Sulfonyl	–NHS(=O)$_2$R	–OH	–CH$_2$OH
11) (Thiol)thiocarbonyl	–NH(C=S)SH	–OH	–CH$_2$OH
Reactions at the hydroxyl group			
1) Acyl	–NHAc	–OC(=O)–R	–CH$_2$O(C=O)–R
2) Alkyl + aryl	–NHAc	–O–R	–CH$_2$O–R
3) Metal alcoholate	–NHAc	–ONa	–CH$_2$ONa
4) Nitro	–NHAc	–ONO$_2$	–CH$_2$ONO$_2$
5) Phosphate	–NHAc	–OP(=O)(ONa)$_2$	–CH$_2$OP(=O)(ONa)2
6) Sulfate	–NHAc	–OS(=O)$_2$ONa	–CH$_2$OS(=O)$_2$ONa
7) Sulfonyl	–NHAc	–O–S(=O)$_2$–R	–CH$_2$O–S(=O)$_2$–R
8) Siryl	–NHAc	–O–Si(R)$_3$	–CH$_2$–Si(R)$_3$
9) (Thiol)thiocarbonyl	–NHAc	–OC(=S)SR	–OC(=S)SR
Reaction at the carbon chain			
1) 3,6-Anhydro	–NHC(=O)CH$_3$		
2) Azido	–NHAc	–OH	–CH$_2$N$_3$
3) Halo	–NHAc	–OH	–CH$_2$X
4) Mercapto	–NHAc	–OH	–CH$_2$SH
5) Oxidation	–NHAc	–OH	–COOH
6) Oxidative-deamination	–O–C5	–OH	–CH$_2$OH
7) Reduction	–NHAc or –NH$_2$	–OH	–CH$_3$

[a] Ac, acetyl; R, alkyl or aryl group; X, halogen.

yield N-acyl derivatives having a d.s. of 1.0, the reactions with high fatty acid anhydrides (C$_{16}$–C$_{18}$) give rise to the products having d.s. < 1.0.

Chitosan is also N-acetylated by reactions with acyl halides [26], with carboxylic acids in the presence of dehydration agents [e.g., dicyclohexylcarbodiimide (DCC)] [27], and with halocarboxylic anhydrides [28]. Under these conditions not only the amino group but also the hydroxyl groups are acylated.

5.1.2. Formation of N-Alkylidene and N-Arylidene Derivatives

The amino group of chitosan reacts with a series of aldehydes and ketones [29–37], to yield the N-alkylidene, N-arylidene, and ketamine (Schiff's base) derivatives. The reactions of chitosan with long-chain alkyl mono-aldehydes give rise to the corresponding derivative [38]. Reactions with dialdehydes (e.g., glutaraldehyde and glyoxal) [33], [37], [39], [40] give the cross-linked derivatives.

5.1.3. N-Alkylation and N-Arylation

N-Alkyl and N-aryl derivatives of chitosan are produced (1) by direct N-alkylation and N-arylation, and (2) by the reduction of the N-alkylidene and N-arylidene derivatives.

In direct N-alkylation and N-arylation, the amino group of chitosan reacts with alkyl halides [41], [42] in the presence of a base to form the N-mono and N-dialkyl derivatives. The N-dialkyl derivatives react further with alkyl halides to form the N-trialkyl quaternary ammonium slats [36], [41], [43].

The *N*-alkylidene and *N*-arylidene derivatives are reduced by treatment with NaBH₃CN or NaBH₄ to yield the corresponding *N*-alkyl and *N*-aryl derivatives [31], [34], [42], [44–48].

5.2. Reactions at the Hydroxyl Group

Chitin, *N*-acylchitosans, and *N*-alkylidene and *N*-arylidene chitosans are *O*-acylated by treatment with acyl halides [49], [50] and with carboxylic anhydrides in pyridine [51], [52] or methanesulfonic acid [53], [54]. *O*-Acyl chitosan derivatives are produced from the *O*-acyl *N*-arylidene- and *N*-alkylidene derivatives by acid hydrolysis of the *N*-arylidene- and *N*-alkylidene groups [55].

The *O*-alkyl and *O*-aryl derivatives are produced from the sodium salt of chitin by treatment with an alkylating agent (e.g., dialkyl sulfate or an alkyl halide).

3-*O*-(1-carboxyethyl)chitosan (muraminan) is prepared as follows (see below):

N-1-naphthylmethylenechitosan, which is produced from chitosan by treatment with 1-naphthaldehyde, is reacted with trityl chloride to yield the *N*-(1-naphthylmethylene)-6-*O*-trityl derivative. The compound reacts with 2-chloropropionic acid in *N-N*-dimethyl acetamide to form the 3-*O*-(1-carboxyethyl)-*N*-(1-naphthyl-methylene)-6-trityl derivative. Subsequently, both, the Schiff's base and the trityl groups, are removed by treatment with aqueous 1 N HCl to give rise to the muraminan [56].

5.3. Reactions at C-6

6-Carboxychitosan is produced from chitosan by treatment with chromium trioxide in acetic acid [57]. 6-Carboxychitin is prepared from chitin by the regioselective oxidation at C6 with NaOCl at 25 °C in aqueous solution [58].

5.4. Graft Polymerization on Chitin and Chitosan

Chitin-polystyrene copolymers are produced by cationic and radical graft copolymerizations of styrene onto 6-iodo-6-deoxychitin [59] and 6-mercapto-6-deoxychitin [60]. Chitin-poly(*N*-acetylethyleneimine) copolymers are produced by graft copolymerization of 2-methyl-2-oxazoline onto both the 6-tosyl and 6-iodo-6-deoxy derivatives [61].

Onto the amino group of chitosan, the lactams of amino acids (e.g., *l*-alanine *N*-carboxylic anhydride) are graft-copolymerized to give rise to chitosan-poly(amino acids) and chitosan-poly(peptides)copolymers [62], [63]. The grafting of poly(4-vinylpyridine) [64] and poly(2-methyl-2-oxazoline) [65] and poly(isobutyl vinyl ether) [66] on chitosan are also performed.

6. Application Forms and Formulations

Hydrogels Various hydrogels are produced from chitin and chitosan. Chitosan oxalate hydrogel is prepared by dissolution of chitosan in aqueous oxalic acid by heating and subsequent cooling. The hydrogel melts on heating at 80 – 90 °C and solidifies on cooling at room temperature. Chitosan hydrogel can be obtained from the oxalate hydrogel by treatment with aqueous 1 N NaOH [67], [68].

Hydrogels can also be produced from the *N*-acyl, *O*-acyl-*N*-acyl and the alkylidene/arylidene derivatives of chitosan. Chitosan-Ca alginate and

N-acylchitosan-collagen hydrogels are also known.

Chitosan Polyelectrolytes are formed by dissolving chitosan in aqueous organic acids (e.g., acetic, propionic and lactic acids), and reaction of the produced cationic chitosan with polyanions (e.g., chondroitin sulfate, hyaluronate [69], heparin [70], other acidic polysaccharides [71], [72], DNA [73], and some acidic proteins [74]).

Hard Composites of Chitosan and Metals are produced from the corresponding hydrogels (see page 7) by soaking in an aqueous $NaCO_3$, K_2CO_3 or Na_2HCO_3 solution at room temperature for a few days or weeks, [75–81].

Membranes Chitin and chitosan membranes are prepared as follows:

1. Dissolution of the polymers in formic acid [12], [82–85], aqueous acids [86], [87], or organic solvents [26], [44]. The solutions are spread as a thin layer on a glass plate. After removing the solvent the corresponding membrane is formed
2. Air drying of a thin-sliced piece of N-acylchitosan hydrogels or their blend hydrogels (see page 7) [88], [89].
3. Chemically modified membranes are produced by chemical modification of chitin and chitosan membranes in the solid state [90].

N-Acylchitosan membranes are stable in both aqueous acid and aqueous alkaline solutions, but chitosan membranes and N-alkylidene- and N-arylidene-chitosan membranes are unstable in aqueous acid solutions. Flow rates of water through N-acylchitosan membranes are in the range of $10.0-23.6 \times 10^{-3}$ m/cm^2 min under a pressure of 29.4×10^4 Pa, and are almost independent on the membrane thickness (12–60 μm). The flow rate of water through chitosan membrane (30–35 μm in thickness) is 7.1×10^{-4} mL/cm^2 min, and the rate decreases with an increase of the membrane thickness [92].

7. Uses

Chitin and chitosan are utilized as ecologically harmless materials in various fields. These include biotechnology, water-treatment, medicine and veterinary medicine, textiles, membranes, cosmetics, agriculture, and food industry.

Chitin and chitosan hydrogels are used as media for affinity chromatography of enzymes (e.g., chitinase and lysozyme) [91] and lectins (e.g., wheat germ agglutinin, WGA); as media for gel-permeation chromatography [37], [67], [92]; for the isolation of bovine serum albumin [93]; as materials for the preparation of membranes, sponges and sponge sheets; as immobilization media of enzymes [94], [95] and cells [96]; and as wound-dressing materials [97].

Polyelectrolytes are used in paper sizing; as textiles auxiliaries [98]; for immobilization of enzymes [99], [100]; and for wastewater treatment [101], [102].

Metal composites of chitin and chitosan are used as solid electrolytes for secondary lithium cells (chitosan-lithium trifluorate) [103]; and media for the removal of uranium ions [104], [105] and harmful radioisotopes from water [106].

Chitin and chitosan fibers are used as antibacterial, antithrombogenic, and thrombogenic (hemostatic), deodorizing, moisture-controlling, and nonallergenic fibers. Staple fibers and nonwoven fabrics are used as materials for bandages, for wound-dressing in plant and animals, as perfume-releasing fibers, and as textile materials for underwear, sportswear, and socks [107].

Chitin filament (about 5 μm in diameter), bundled up into 16–20 filaments, finds application as surgical suture, which has an enough strength for clinical uses [108]. The suture is digestible in the tissues by lysozyme and chitinase, and its digestion period is controlled by the structure of N-acyl groups and by their d.s.

Chitosan beads (>1 mm in diameter) and microspheres (0.1–10 μm in diameter) chitin beads, and composite beads are used as media for anion-exchange and affinity chromatography [109]; as controlled release carriers of drugs and agrochemicals [110], [111]; as encapsulating materials for mammalian cells, microbes, and drugs [112–125]; and for immobilization of enzymes [126].

Membranes find application as active transport membranes for halogen and organic ions; in protein purification [127], [128]; as affinity membranes for purification of lysozyme [129]; in dialysis, as edible chitosan-gelatin films [130],

[131]; electrochemical luminescence sensors [132]; as enzyme-immobilized membranes [133], [134], anion-exchange, [135–137]; pervaporation [138], [139], potentiometric response membranes [140]; in reverse osmosis and ultrafiltration [92].

Colloidal chitin and powders of chitin and chitosan are used as cosmetic ingredients, solid substrates for chitinase, lysozyme and chitosanase, and emulsifiers in food processing [141].

Aqueous acidic solutions of chitosan, and of O-carboxymethylchitin, O-hydroxyethylchitin (neutral) and 3,6-methoxychitin in water are highly viscous, and have moisture retaining, antielectrostatic, hair-protecting, odor-absorbing, and anti-bacterial properties. They are used as cosmetic ingredients [142], [143]; and coating agents for lumber surface.

Chitosan and chitin have an *antimicrobial action*, [144], [145], which is even more pronounced in the corresponding quaternary ammonium salts (e.g., N-trimethylchitosan hydrochloride). The cationic amino groups of chitosan bind to anionic groups on microbe cell walls, resulting in their growth inhibition. Chitin and chitosan also have an antiviral action in vitro [146]. On supplying chitosan as a feed ingredient in animals, the proportion of the useful microbes (e.g., *Bifidobacteria*) in the intestine increases and the proportion of the harmful microbes decrease [147].

Chitin and chitosan are applied as additives for nonfermented pickles, as anti-microbial agents; antimicrobial food processing additives, and soil amendments in agriculture [148].

Chitosan N-acyl derivatives are used as artificial blood vessels, and materials for contact lenses, chitosan is applied as sealing material for arterial puncture sites [149].

Treatment with chitin and chitosan, induces production of various extracellular bioactive compounds that enhance the self-defense function of plants against pathogen infections and diseases resulting in the increase of plant tissue production [150]. Chitin and chitosan activate mammalian cells [151]. As the result, immune reactions [152], [153] are stimulated by producing lysozyme, interleukins [154], leukotriene B4 and prostaglandin E2. Chitin and chitosan thus find application as agricultural materials; coating materials for plant seeds; medical, and veterinary materials [155].

Because of its *hypocholesterolemic function* [156–160] chitosan is used as an additive for feeds and foods; and as ingredient for health foods.

The coating or spraying of an aqueous solution of chitosan over the surface of fruit and vegetables slows down their respiration of CO_2 and ethylene gases. As the results, the ripening of fruits is delayed, the storage life of fruit and vegetables are prolonged, and their freshness is kept for a long period [161–163].

8. Economic Aspects

The industrial production of chitin and chitosan from crab shells was started in Japan in 1987. At that time, a relatively large amount of crab shells was produced as waste in shellfish processing companies, and easy to collect without any costs. The first industrial application of chitosan was as a flocculating and dewatering agent in municipal wastewater treatment. Since then, the utilization of chitin and chitosan has been expanded into wide fields. The estimated consumption of chitin and chitosan in 1998 in Japan was about 832 t as calculated as chitosan. About 30 % of the total consumption was for wastewater treatment, about 24 % as foods and feeds additives, and about 14 % in agriculture. In the future, the usage in textile, medical, and veterinary materials will increase, that in the wastewater treatment will decrease.

9. Toxicology and Environmental Aspects

Chitin and chitins are almost nontoxic (LD_{50} 16 g per kilogram body weight for rats). No abnormal symptoms are observed with several animals after the oral administration of chitosan for 8 months at a daily dose of 0.7 – 0.8 g per kilogram body weight, and after the intravenous injection of low molecular mass chitosan (M_r ca. 3000) and chitosan oligosaccharides (M_r 304 – 1162) for 11 days at a daily dose of 4.5 mg per kilogram body weight.

Chitin and chitosan are degradable in the biosphere, in the agricultural soil, and in the hydrosphere to produce oligosaccharides.

The biodegradation rate varies with seasons in the soil and hydrosphere, and is controlled by chemical modification [164–170].

References

1. H. Braconnot, *Ann. Chim. Phys.* **79** (1811) 269.
2. A. Odier, *Mem. Soc. Hist. Nat. Paris* **1** (1823) 29.
3. C. Rouget, *Comp. Rend.* **48** (1859) 792.
4. F. Hoppe-Seiler, *Ber.* **27** (1894) 3329.
5. R. Minke, J. Blackwell, *J. Mol. Biol.* **120** (1978) 167.
6. Y Saito, T. Okano, H. Chanzy, J. Sugiyama, *J. Struc. Biol.* **114** (1995) 218.
7. K. Ogawa, *Agric. Biol. Chem.* **55** (1991) 375.
8. P. Cairn, *Carbohydr. Res.* **235** (1992) 23.
9. H. M. Cauchiel, G. Murugan, J. P. Thome, H. J. Dumont, *Hydrobiologia* **359** (1997) 23.
10. S. L. Wang, S. H. Chio, *Enzyme Micorb. Technol.* **22** (1998) 629.
11. Jpn. Kokai Tokkyo Koho, JP 99 49 972, 1999 (M. Matsumara, J. Sato).
12. T. Sannan, K. Kurita, Y. Iwakura, *Makromol. Chem.* **176** (1975) 1191.
13. S. Aiba, *Int. J. Biol. Macromol.* **13** (1991) 40.
14. US Patent 5 739 015, 1998 (V. R. Srinivasan).
15. K. Tokuyasu, M. Ohnishi-Kameyama, K. Hayashi, *Biosci. Biotech. Biochem.* **60** (1996) 1598.
16. K. M. Varum, M. W. Anthonsen, H. Grasdalen, O. Smidsrod, *Carbohydr. Res.* **211** (1991) 17.
17. G. Yu, F. G. Morin, G. A. R. Nobes, R. H. Marchessault, *Macromolecules* **32** (1999) 518.
18. K. M. Varum, M. W. Anthonsen, H. Grasdalen, O. Smidsrod, *Carbohydr. Res.* **217** (1991) 19.
19. J. G. Domszy, G. A. F. Roberts, *Macromol. Chem.* **186** (1985) 1671.
20. H. Terayama, *J. Polym. Sci.* **8** (1952) 243.
21. H. Sato, S. Mizutani, S. Tsuge, H. Ohtani, K. Aoi, A. Takasu, M. Okada, S. Kobayashi, T. Kiyosada, S. Shoda, *Anal. Chem.* **70** (1998) 7.
22. A. Martinou, U. Bowiotis, B. Stokke, K. M. Varumd, *Carbohydr. Res.* **311** (1998) 71.
23. M. Matsumiya, K. Miyauchi, A. Mochizuki, *Fish Sci.* **64** (1998) 166.
24. S. Hirano, Y. Ohe, H. Ono, *Carbohydr. Res.* **47** (1976) 315.
25. L. Vachoud, N. Zydowicz, A. Domard, *Carbohydr. Res.* **302** (1997) 169.
26. S. Fuji-i, H. Kumagai, M. Noda, *Carbohydr. Res.* **83** (1980) 389.
27. K. Kurita, T. Sannan, Y. Iwakura, *Makromol. Chem.* **178** (1977) 2595.
28. S. Hirano, Y. Kondo, *Nippon Kagaku Kaishi* (1982) 1622.
29. J. Dutkiewicz, *J. Macromol. Sci. Chem.* **20** (1983) 877.
30. S. Hirano, N. Matsuda, O. Miura, H. Iwaki, *Carbohydr. Res.* **71** (1979) 339.
31. G. K. Moore, G. A. F. Roberts, *Int. J. Biol. Macromol.* **3** (1981) 337.
32. S. Hirano, T. Osaka, *Agric. Biol. Chem.* **47** (1983) 1389.
33. R. A. A. Muzzarelli, F. Tanfani, M. Emanuelli, *Carbohydr. Res.* **107** (1982) 199.
34. P. Tong, Y. Baba, Y. Adachi, K. Kawazu, *Chem. Lett.* (1991) 1529.
35. R. A. A. Muzzarelli, A. Zattoni, *Int. J. Biol. Macromol.* **8** (1986) 137.
36. L. A. Nud'ga, E. A. Plisko, S. N. Danilov, *Zh. Obshch. Khim.* **43** (1973) 339. *Chem. Abst.* **80** 96251.
37. S. Hirano, N. Matsuda, T. Tanaka, *Carbohydr. Res.* **71** (1979) 344.
38. K. Kurita, M. Ishiguro, T. Kitajima, *Int. J. Biol. Macromol.* **10** (1988) 124.
39. R. A. A. Muzzarelli, G. Barontini, R. Rocchetti, *Biotechnol. Bioeng.* **18** (1976) 1445.
40. R. A. A. Muzzarelli, F. Tanfani, S. Mariotti, E. Emanuelli, *Carbohydr. Res.* **104** (1982) 235.
41. E. Kokufuta, Y. Hirai, I. Nakamura, *Makromol. Chem.* **182** (1981) 1714.
42. I. D. Hall, M. Yalpani, *Biopolymers* **20** (1981) 1413.
43. A. Domard, R. Rinaudo, C. Terassin, *Int. J. Biol. Macromol.* **8** (1986) 105.
44. R. A. A. Muzzarelli, F. Tanfani, M. Emanuelli, S. Mariotti, *J. Membr. Sci.* **16** (1983) 295.
45. R. A. A. Muzzarelli, F. Tanfani, *Carbohydr. Res.* **5** (1985) 297.
46. R. A. A. Muzzarelli, F. Tanfani, S. Mariotti, E. Emanuelli, *Carbohydr. Polym.* **2** (1982) 145.
47. M. Yalpani, I. D. Hall, *Can. J. Chem.* **59** (1981) 2934.
48. J. M. Harris, E. Struck, M. Case, M. S. Paley, M. Yalpani, J. Vanalstine, D. Brooks, *J. Polym. Sci. Part A: Polym. Chem.* **22** (1984) 341.
49. S. Hirano, S. Kondo, Y. Ohe, *Polymer* **16** (1975) 622.
50. K. Kaifu, N. Nishi, T. Komai, *J. Polym. Sci. Polym. Chem. Ed.* **18** (1981) 2361.
51. S. Hirano, Y. Koide, *Carbohydr. Res.* **65** (1978) 166.
52. K. Kurita, S. Ishi, K. Tomita, S. Nishimura, K. Shimoda, *J. Polym. Sci. Part A: Polym. Chem.* **32** (1994) 1027.
53. N. Nishi, J. Noguchi, S. Tokura, H. Shiota, *Polym. J.* **11** (1979) 27.
54. N. Nishi, H. Ohmura, S. Nishimura, O. Somorin, S. Tokura, *Polym. J.* **14** (1982) 919.
55. G. K. Moore, G. A. F. Roberts, *Int. J. Biol. Macromol.* **4** (1982) 246.
56. S. Hirano, Y. Kondo, H. Inui, F. Hirano, K. Nagamura, T. Yoshizumi, *Carbohydr. Polym.* **31** (1996) 29.
57. D. Horton, E. K. Just, *Carbohydr. Res.* **29** (1973) 173.
58. R. A. A. Muzzarelli, C. Muzarelli, A. Cosani, M. Terbojevich, *Carbohydr. Polym.* in press.
59. K. Kurita, H. Yoshino, S. Inoue, K. Yamamura, S. Ishi, S. Nishimura, *Macromolecules* **25** (1992) 3791.
60. K. Kurita, S. Hashimoto, H. Yoshino, S. Ishii, S. Nishimura, *Macromolecules* **29** (1996) 1939.
61. K. Kurita, S. Hashimoto, S. Ishii, S. Nishimura, *Polym. J.* **28** (1996) 686.

62. K. Kurita, A. Yoshida, Y. Koyama, *Macromolecules* **21** (1988) 1579.
63. K. Kurita, S. Iwaki, S. Ishi, S. Nishimura, *J. Polym. Sci. Part A: Polym. Chem.* **30** (1992) 685.
64. H. Caner, H. Hasipoglu, O. Yilmaz, E. Yilmaz, *Eur. Poly. J.* **34** (1998) 493.
65. K. Naka, R. Yamashita, T. Nakamura, I. Ohki, S. Maeda, K. Aoi, A. Takasu, M. Okada, *Int. J. Biol. Macromol.* **23** (1998) 259.
66. S. Yoshikawa, T. Takayama, N. Tsubokawa, *J. Appl. Polym. Sci.* **68** (1998) 1883.
67. S. Hirano, R. Yamaguchi, N. Fukui, M. Iwata, *Carbohydr. Res.* **201** (1990) 145.
68. S. Hirano, R. Yamaguchi, N. Fukui, M. Iwata in C. G. Gebelein (ed.): *Biotechnology and Polymers*, Plenum New York 1991, pp. 181–188.
69. A. Denuziere, D. Ferrier, O. Damour, A. Domard, *Biomaterials* **19** (1998) 1275.
70. Y. Kikuchi, *Makromol. Chem.* **157** (1974) 2209.
71. Y. Kikuchi, H. Fukuda, *Makromol. Chem.* **178** (1977) 2895.
72. S. Hirano, C. Mizutani, R. Yamaguchi, O. Miura, *Biopolymers* **17** (1978) 805.
73. S. C. W. Richardson, V. H. J. Koll, R. Duncan, *Int. J. Pharm.* **178** (1999) 231.
74. R. Yamaguchi, A. Arai, S. Hirano, T. Ito, *Agric. Biol. Chem.* **42** (1978) 1297.
75. S. Hirano, O. Miura, R. Yamaguchi, *Agric. Biol. Chem.* **41** (1977) 1755.
76. S. Hirano, K. Yamamoto, H. Inui, K. I. Draget, K. M. Varum, O. Smidsrod, *Adv. Chitin Sci.* **2** (1997) 1.
77. S. Hirano, K. Yamamoto, H. Inui, K. I. Draget, K. M. Varum, O. Smidsrod, *Studies on Surface and Catalysis* **114** (1998) 621.
78. C. H. Chu, H. Kumagai, T. Sakiyama, S. Ikeda, K. Nakamura, *Biosci. Biotech. Biochem.* **60** (1996) 1627.
79. S. Hirano, Y. Yamada, K. Yamamoto, H. Inui, M. Ji, *Advan. Chitin Sci.* **1** (1996) 137.
80. S. Hirano, K. Yamamoto, H. Inui, M. Ji, M. Zhang, *Macromol. Symp.* **105** (1996) 149.
81. S. Hirano, K. Yamamoto, H. Inui, *Energy Convers. Manage.* **38** (1997) S517.
82. S. Tokura, N. Nishi, J. Noguchi, *Polym. J.* **11** (1979) 781.
83. T. Uragami, Y. Ohsumi, M. Sugihara, *Polymer* **22** (1981) 1155.
84. S. Aiba, M. Izume, N. Minoura, Y. Fujiwara, *British Polymer* **17** (1985) 38.
85. S. Aiba, M. Izume, N. Minoura, Y. Fujiwara, *Carbohydr. Polym.* **5** (1985) 285.
86. S. Hirano, K. Tobetto, Y. Noishii, *J. Biomed. Mater. Res.* **15** (1981) 903.
87. A. L. Andrady, P. Xu, *J. Polym. Sci. Part B: Polym. Phys.* **35** (1997) 517.
88. S. Hirano, *Agric. Biol. Chem.* **42** (1978) 1939.
89. O. W. Achaw, M. F. A. Goosen, *Polym. Prepr. (ACS)* **35** (1994) 75.
90. E. Khor, A. Wan, C. A. Tee, F. Chae, G. W. Hastings, *J. Polym. Sci. Part A: Polym. Chem.* **35** (1997) 2049.
91. S. Hirano, H. Kaneko, M. Kitagawa, *Agric. Biol. Chem.* **55** (1991) 1683.
92. S. Hirano, K. Tobetto, M. Hasegawa, N. Matsuda, *J. Biomed. Mater. Res.* **14** (1980) 477.
93. M. F. A. Goosen, O. W. Achaw in M. F. A. Goosen (ed.): *Applications of Chitin and Chitosan*, Technomic Lancaster 1997, pp. 233–254.
94. H. Tanibe in M. Yabuki (ed.): *Applications of chitin and chitosan*, Gihodo Publishing Co. Tokyo 1990, pp. 123–154.
95. T. Y. Hsien, G. C. Rorrer, *Ind. Eng. Chem. Res.* **36** (1997) 3631.
96. S. Kirkwood, *Ann. Rev. Biochem.* **43** (1974) 401.
97. S. L. Wang, S. H. Chio, *Enzyme Microb. Technol.* **22** (1998) 634.
98. M. R. Julia, J. M. Canal, D. Jocic, *Rev. Quim. Text.* **140** (1988) 58, 62. *Chem. Abst.* **130** 326239.
99. W. S. Ngah, L. K. Wan, *Ind. Eng. Chem. Res.* **38** (1999) 1411.
100. H. Minamisawa, H. Iwanami, N. Arai, T. Okutani, *Anal. Chim. Acta.* **378** (1999) 279.
101. L. Guerrero, F. Omil, R. Mendez, J. M. Lema, *Bioresour. Technol.* **63** (1998) 221.
102. M. Hashimoto, Y. Ishii, Y. Ohi, 55th Annual Conf. Water Pollution Control Federation, St. Louis, Mo. USA, 1982.
103. N. M. Morni, A. K. Arot, *J. Power Sources* **77** (1999) 42.
104. T. Sakaguchi, T. Horikoshi, A. Nakajima, *Nippon Nogei Kagaku Kaishi* **53** (1979) 211.
105. S. Hirano, Y. Kondo, Y. Nakazawa, *Carbohydr. Res.* **100** (1982) 431.
106. R. A. A. Muzzarelli, B. Spalla, *J. Radiational Chem.* **10** (1972) 27.
107. M. Yoshikawa, T. Midorikawa, T. Otsuki, T. Terashi, *Eur. Pat. Appl. EP* **794** (1997) 223.
108. K. M. Nakajima, K. Atsumi, K. Kifune in J. P. Zikakis (ed.): *Chitin, Chitosan and Related Enzymes*, Academic Press London 1984, pp. 407–410.
109. A. Kristane, A. Nysaeter, H. Grasdalen, K. M. Varum, *Carbohydr. Poly.* **38** (1999) 23.
110. C. Aval, J. Akbuga, *Int. J. Pharm.* **168** (1998) 9.
111. F. Bugamelli, M. A. Raggi, I. Orieti, V. Zecchi, *Arch. Pharm. (Weinheim, Ger.)* **331** (1998) 133.
112. P. Giunchedi, L. Genta, B. Conti, R. A. A. Muzzarelli, U. Conte, *Biomaterials* **19** (1998) 157.
113. R. H. Chen, M. L. Tsaih, *J. Appl. Polym. Sci.* **66B** (1997) 161.
114. M. J. B. Miguez, B. C. Rodrigues, M. N. M. Sandrez, M. C. M. Laranjeira, *J. Encapsulation* **14** (1997) 639.
115. M. L. Huguet, E. Dellacherie, *Process Biochem. (Oxford)* **31** (1996) 745.
116. K. Nilsson, W. Scheirer, O. Merten, L. Ostberg, E. Liehl, H. Katinger, K. Mosbach, *Nature* **302** (1983) 629.
117. J. Wang, P. Li, H. Shi, Y. Qian, *J. Environ. Sci. (China)* **9** (1997) 283.
118. A. Javis, T. Springs, W. Chipuru, M. Sullivan, G. Koch, *In Vitro* **18** (1982) 267.

119 L. L. McCormick, D. K. Lichatowich, *J. Polym. Sci., Polym. Lett. Ed.* **17** (1979) 479.
120 Y. Kawashima, T. Handa, A. Kosai, H. Takenaka, S. Lin, Y. Ando, *J. Pharm. Sci.* **74** (1985) 264.
121 T. Yoshioka, R. Hirano, T. Shioya, M. Kako, *Biotechnol. Bioeng.* **35** (1990) 66.
122 K. Aiedeh, E. Gianasi, I. Orienti, V. Zecchi, *J. Microencapsulation* **14** (1997) 567.
123 F. L. Mi, T. B. Wong, S. S. Shyu, *J. Microencapsulation* **14** (1997) 577.
124 N. Tarimci, D. Ermis, *Int. J. Pharm.* **147** (1997) 71.
125 P. Calvo, J. L. Vila-Jato, M. J. Alonso, *Int. J. Pharm.* **153** (1997) 41.
126 A. Illanes, L. Wilson, C. Altamirano, A. Aillapan, *Prog. Biotechnol.* **15** (1998) 27.
127 X. Zeng, E. Ruckenstein, *J. Membr. Sci.* **148** (1998) 195.
128 E. Ruckenstein, X. Zeng, *J. Membr. Sci.* **142** (1998) 13.
129 R. Eli, Z. Xianfnag, *Biotechnol. Bioeng.* **58** (1998) 117.
130 I. S. Arvanitoyannis, A. Nakayama, S. Aiba, *Carbohydr. Polym.* **37** (1998) 371.
131 C. Caner, P. J. Vergano, J. L. Wiles, *J. Food Sci.* **63** (1998) 1049.
132 C. Zhao, N. Egashira, Y. Kurauchi, K. Ohga, *Anal. Sci.* **14** (1998) 439.
133 G. Cho, I. S. Moon, J. S. Lee, *Chem. Lett.* (1997) 577.
134 H. S. Liu, W. H. Chen, J. T. Lai, *Appl. Biochem. Biotechnol.* **66** (1997) 57.
135 J. T. Shieh, R. Y. M. Hung, *J. Membr. Sci.* **127** (1997) 185.
136 A. Ito, M. Sato, T. Anma, *Angew. Makromol. Chem.* **248** (1997) 85.
137 Y. M. Lee, S. Y. Nam, D. J. Woo, *J. Membr. Sci.* **133** (1997) 103.
138 T. Uragami, M. Saito, K. Takigawa, *Makromol. Chem. Rapid Commun.* **9** (1988) 361.
139 J. Jegal, K. H. Lee, *J. Appl. Polym. Sci.* **71** (1999) 671.
140 B. H. Choi, Y. J. Yun, *Anal. Sci. Technol.* **11** (1998) 235.
141 P. C. Schulz, M. S. Rodriguz, L. F. DelBlanco, M. Pistonesi, E. Agulls, *Colloid. Polym. Sci.* **276** (1998) 1159.
142 S. Hirano, K. Hirochi, K. Hayashi, T. Mikami, H. Tachibana in C. G. Gebelein (ed.): *Cosmetic and Pharmaceutical Polymers*, 1991, pp. 95–104.
143 S. Hirano, Y. Akiyama, M. Ogura, Y. Ayaki in S. Tokura, I. Azuma (eds.): *Chitin Derivatives in Life Science*, Jap. Soc. Chitin/Chitosan Sapporo 1992, pp. 115–120.
144 S. Uchida: "Antimicrobial function of chitin and chitosan, and its applications," in M. Yabuki (ed.): *Applications of Chitin and Chitosan*, Ed. Japn Chitin/Chitosan Soc. Gihodo, Tokyo 1990, pp. 71–98.
145 A. Tokoro, M. Kobayashi, N. Tatewaki, K. Suzuki, Y. Okawa, T. Mikami, S. Suzuki, M. Suzuki, *Microbiol. Immun.* **33** (1989) 357.
146 J. Iida, T. Une, C. Ishihara, K. Nishimura, S. Tokura, N. Mizukoshi, I. Azuma, *Vaccine* **5** (1987) 270.
147 K. Suzuki, Y. Okawa, K. Hashimoto, S. Suzuki, M. Suzuki, *Microbiol. Immunol.* **28** (1984) 930.
148 K. Ohta, A. Taniguchi, N. Konishi, T. Hosoki, *Fort. Science* **34** (1999) 233.
149 A. Hoekstra, H. Struszczyk, O. Kivekas, *Biomaterials* **19** (1998) 1467.
150 H. Yamamoto, T. Koga, S. Hayakawa, Y. Ohgata, T. Kurasaki, K. Tohyama, *Nihon Sakumotsugakkai Kiji* **67** (1998) 452.
151 K. Nishimura, S. Nishimura, H. Seno, N. Nishi, I. Saiki, S. Tokura, I. Azuma, *Vaccine* **5** (1987) 136.
152 Y. Shibata, L. A. Foster, M. Kurimoto, H. Okamura, R. Nakamura, K. Kwawajiri, J. P. Justice, M. R. Van Scott, N. Myrvik, *J. Immunol.* **161** (1998) 4283.
153 S. Minami, Y. Suzuki, Y. Okamoto, T. Fujinaga, Y. Shigemasa, *Carbohydr. Polym.* **36** (1998) 151.
154 T. Mori, Y. Irie, S. Nishimura, S. Tokura, M. Matsuura, M. Okumura, T. Kadosawam, T. Fujinaga, *J. Biomed. Mater. Res.* **43** (1998) 469.
155 K. Suzuki, T. Mikami, Y. Okawa, A. Tokoro, S. Suzuki, M. Suzuki, *Carbohydr. Res.* **151** (1986) 403.
156 I. Furda, *A. C. S. Symposium Series* **214** (1983) 105.
157 D. J. Ormrod, C. C. Holmes, T. E. Miller, *Atherosclerosis* **138** (1998) 329.
158 S. Hirano, H. Seino, Y. Akiyama, I. Nonaka in C. G. Gebelein, R. L. Dunn (eds.): *Progress in Biomedical Polymers*, Plenum New York 1990, pp. 283–290.
159 S. Hirano, Y. Akiyama, *J. Sci. Food Agric.* **69** (1995) 91.
160 P. T. Mathew, K. G. R. Nair, *Fish Technol.* **35** (1998) 46.
161 A. E. Ghaouth, J. Arul, R. Ponnampalam, M. Boulet, *J. Food Sci.* **56** (1991) 1618.
162 A. E. Ghaouth, R. Pnnampalm, F. Castaigne, J. Arul, *Host. Sci.* **27** (1992) 1016.
163 A. E. Ghaouth, J. Arul, C. Wilson, N. Benhamou, *Physiol. Mol. Plant Pathol.* **44** (1994) 427.
164 S. Hirano, Y. Yagi, *Carbohydr. Res.* **83** (1980) 103.
165 K. Tomihata, Y. Ikada, Y., *Biomaterials* **18** (1997) 567.
166 S. Hirano, Y. Yagi, *Agric. Biol. Chem.* **44** (1980) 963.
167 S. Hirano, H. Tsuchida, N. Nagao, *Biomaterials* **10** (1989) 574.
168 N. Hutadilok, T. Mochimasu, H. Hisamori, K. Hayashi, H. Tachibana, T. Ishii, S. Hirano in R. A. A. Muzzarelli (ed.): *European Chitin Soc.*, Ancona 1993, pp. 289–294.
169 N. Hutadilok, T. Mochimasu, H. Hisamori, K. Hayashi, H. Tachibana, T. Ishii, S. Hirano, *Carbohydr. Res.* **268** (1995) 143.
170 S. Hirano, Y. Yagi, *Carbohydr. Res.* **92** (1981) 319.

Further Reading

A. Steinbüchel (ed.): *Biopolymers*, Wiley-VCH, Weinheim 2006.

T. Uragami, S. Tokura (eds.): *Material Science of Chitin and Chitosan*, Springer; Kodansha, Tokyo 2006.

Cork

LUIS GIL, Instituto Nacional de Engenharia e Tecnologia Industrial, Lisboa, Portugal

CRISTINA MOITEIRO, Instituto Nacional de Engenharia e Tecnologia Industrial, Lisboa, Portugal

1.	Introduction	255
1.1.	Definition and Origin	255
1.2.	History	256
1.3.	Morphology	256
1.3.1.	Microscopic Aspects	256
1.3.2.	Macroscopic Aspects	257
1.4.	Chemical Composition	258
1.4.1.	Suberin	258
1.4.2.	Lignin	258
1.4.3.	Polysaccharides	260
1.4.4.	Waxes	262
1.4.5.	Tannins	262
1.4.6.	Other Components	262
1.5.	Physical and Mechanical Properties	262
2.	Cork Extraction	265
3.	Production of Cork Based Products	266
3.1.	Disks and Cork Stoppers	266
3.2.	Granulates and Broken	267
3.3.	Composition Cork	267
3.3.1.	Floor and Wall Coverings	267
3.3.2.	Agglomerated Disks and Cork Stoppers	268
3.3.3.	Corkrubber	268
3.4.	Insulation Corkboard	268
3.5.	Other Specific Products	268
4.	Uses	269
4.1.	Stoppers	269
4.2.	Civil Construction	269
4.3.	Industrial Applications	269
4.4.	Automotive Industry	269
4.5.	Other Specific Uses	270
5.	Quality and Standardization	270
5.1.	Standardization and Testing	270
5.2.	Quality in Cork Production	270
6.	Economic Aspects	270
6.1.	Forestall Production	270
6.2.	Personnel and Companies	271
6.3.	Consumption of Products and Markets	271
6.4.	Producers	271
7.	Environmental and Toxicological Aspects	272
7.1.	Industrial Wastes	272
7.2.	Recycling and Treatments	272
7.3.	Emissions	272
8.	Legal Aspects	273
	References	273

1. Introduction

1.1. Definition and Origin

Cork is the suberous covering (suberose parenchyma, or bark) of the species *Quercus Suber* L., commonly known as the cork oak. Cork is composed of an aggregate of cells, about 42 millions per cubic centimeter, which have five layers walls. Cork is a very light material, elastic and flexible, and impermeable to gases or liquids. It is imperishable and a good electric, thermal, sound, and vibration insulator [1].

The cork oak grows mainly in the Mediterranean zone, especially where there is an Atlantic influence, and Portugal produces more than half of the world's cork. The cork tree has a natural reproduction by acorns or, more frequently by producing shoots [2]. The cork oak is a dicotyledon of the family Fagaceae, and it is part of the genus Quercus having dozens of varieties [2].

All the trees have a thin layer of cork in their bark but *Quercus suber* is the only one where the cork forms a several centimeters thick layer around the trunk, with the function of insulating the tree from heat and loss of moisture and protecting from damage by animals and fire. Cork regenerates after each stripping.

Today cork products are used for thermal insulation in refrigerators and rockets, acoustic insulation in submarines and recording studios, seals and joints in woodwind instruments and combustion engines, as energy-absorbing medium in flooring, shoes, and packaging, and of course as stoppers [3]. The utilization and processing of commercial cork depend very much on its structural quality [4].

1.2. History

The cork oak dates from the Oligocene stage of the Tertiary Period. It is thought that it was disseminated from the region covered by the Tyrrhenian Sea and that it migrated to the Iberian region in the Miocene stage [2].

Strangely, the first peoples to put cork to what it became after were not of these regions. In ancient China (3000 B.C.) cork was used for fishing buoys, and amphoras were found closed with cork in excavations in Pompeii and in tombs in Egypt [5]. In ancient Greece the bark of the cork was used to make buoys for fishing nets, sandals, and stoppers for wine and olive oil vessels.

In the early days of Portuguese nation, cork was used in building (including shipbuilding). By the 14th century it was already an economic asset of national importance in Portugal. The great step forward for the generalized use of cork was taken by the French Benedictine monk DOM PÉRIGNON (1639 – 1715), proctor of the Hautvillers' Abbey who developed champagne production. He observed that the wooden stoppers wrapped and soaked in olive oil, that were used for the containers holding sparkling wine often jumped out. So he used cork stoppers with surprisingly good results and soon cork became essential for wine producers [2].

In 1750 in Angullane (Gerona), close to the French border, the first stopper factory opened, marking the beginning of cork industry, which then expanded to Portugal. Around 1770, when Port Wine trade had begun to flourish, the cork stopper became indispensable.

The real boom of the cork industry began in the last quarter of the 19th century. In 1915 Portuguese cork oak forests covered between 300 000 and 470 000 ha and nowadays encompass more than 700 000 ha. In 1927 Portuguese annual cork production was 16 000 t and is now about ten times higher [2].

At the end of the 19th century, an American manufacturer of lifejackets accidentally found that it was possible to produce agglomerated cork. This was the beginning of a new world of opportunities in which waste cork and cork previously thought to be of no commercial value could also be used [6].

In the first half of the 20th century, the world cork market was controlled by a handful of multinational companies. However, their productive investment was minimal, especially in Portugal and Spain. They worked in these countries through small offices that acted as purchasing centers for the raw materials and some manufactured goods, that were channeled to other centers for processing and distribution. In the 1960s, Portuguese companies started to invest in modern factories and today Portuguese cork industry processes more than 2/3 of world's cork [1]. Nowadays Spain is the second world cork producer and transformer.

1.3. Morphology

1.3.1. Microscopic Aspects

Cork has an open structure, which is made up of cells of 20 to 30 millionths of a millimeter in size, forming the suber and the rays. The cell lumen contains mainly nitrogen. A thickness of 1 mm corresponds to about thirty layers of cells, 1 mm^3 has about 20 000 – 40 000 cells [2], [7].

Cork tissue is made up of dead micro-cells, generally 14-sided polyhedrons, slotting in one against the other. The contents of newly formed cells disappear during growth and the subsequent suberization process of the membranes, which imparts the completion of communication with the plant living tissues.

The cells are staked in long rows with thin walls. The radial section also shows five, seven and eight sided shapes [3]. Cells walls are covered with very thin layers of waxes and suberin which make them impervious to air and water and resistant to attack by many acids. The cell walls are conjugated. The cork cells are closed hexagonal prisms (average) stacked in rows so that two cells share the hexagonal faces, and the rows are staggered so that the membranes forming the hexagonal faces are not continuous across rows (see Fig. 1).

The cellular structure of cork is significantly influenced by temperature; upon heating cork cells expand and the cell walls stretch, attaining

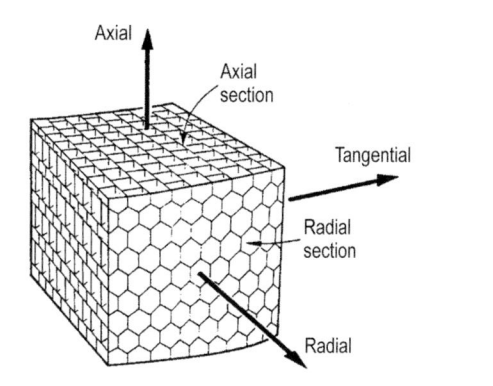

Figure 1. Diagram of cork tree and cork with axis system

their maximum volume (twice the original volume) at 250 °C [8].

The intracellular layers comprise five parts: two made of cellulose, a further two of hard matter with impermeable components (suberin, waxes) and a final ligneous part, whose function is to maintain structure and rigidity [1].

The cell walls have a thickness of about 1 – 2 μm, a height of about 10 – 45 μm and a hexagonal face edge of about 20 μm (see Fig. 2). The density of cork is related to that of the cell wall material, which is close to 1150 kg/m^3 [3], [7]. The thickness of cork cells is lower in spring and greater in autumn. The cell height has also a minimum value in autumn.

The unevenness of membrane thickness, height and diameter of the cells can affect the cork's mechanical and physical properties, namely compressibility and elasticity. The prismatic shape of the cells makes the cork anisotropic [7].

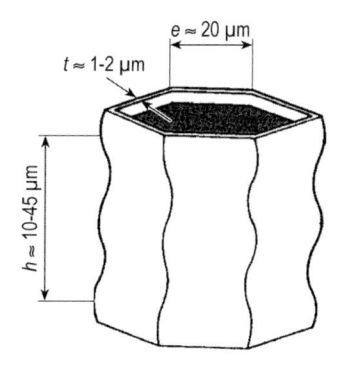

Figure 2. Micrograph of cork cells and corrugated cell showing dimensions

When cork deforms, cell walls bend and buckle, giving resilience. Tensile deformation along the prism axis unfolds the corrugations, straightening prism walls. Compressive deformation folds the corrugations up to a certain point where the cells collapse completely. Axial compression produces almost no lateral expansion, due to the fact that cells fold [3], [7].

1.3.2. Macroscopic Aspects

The cork tree has a remarkable capacity to create a suberous tissue from its inner bark. The external productive layer or phloem grows in the direction of the bark and produces cork [6]. As the phellogen is only active for 6 – 7 months each year, the differences in cell size and the thickness of cell membranes between cork produced in different year seasons leave discernible rings (growing rings), which change from 1.5 to 4 mm in thickness on average, mainly according to the nature of the soil, climatic conditions, and vitality of the tree [6]. Younger trees produce usually very porous and cracked cork and old and aging trees form thin cork.

Natural cork contains lenticels, which are tubular channels that connect the outer surface of the bark to its inner surface allowing oxygen into and carbon dioxide out from the new cells that grow there.

The life cycle of the cork oak produces three qualities of suberose tissue: virgin cork, reproduction cork from the second stripping, and reproduction cork from subsequent stripping. *Virgin cork* is the suberose tissue extracted for the first time (usually when the tree is 25 years old), it is hard and has an irregular structure. *Summer virgin* is the cork extracted from a level above the previous extraction. *Winter virgin* is the virgin cork from the pruned branches, containing fragments of fiber and ligneous tissue. The first *reproduction cork* (taken at least nine years later) is more regular than virgin cork but does not yet have quality enough for cork stopper production [1]. Only the second reproduction cork can be used for cork stoppers. All the other types of cork can be used for agglomerates.

The thickest suberose layer is generally formed in the growing cycle subsequent to cork extraction and then the cork produced diminishes progressively year after year.

Porosity is a consequence of the lenticels, orifices or pores in the tissue. Porosity depends on the area occupied by lenticular channels cut tangentially. If porosity is excessive (> 4 %) the economic value of cork may decrease for some applications [1].

1.4. Chemical Composition

The chemical constitution of cork depends on factors such as geographical origin, climate and soil conditions, genetic origin, tree dimension, age and growth conditions [1], [9], [10].

According to the proposed model [11], the cellular structure walls of cork consists of a thin and lignin-rich middle-lamella primary wall, a thick secondary wall made up from alternating suberin and waxes lamellae and a thin tertiary wall of lignin and polysaccharides. Studies on chemical composition show that the secondary wall is lignified and therefore may not consist exclusively of suberin and waxes [9]. Further information on suberin structure has been obtained from solid-state ^{13}C NMR spectroscopy studies, showing that the aliphatic chain of suberin is spatially separated from lignin and polysaccharides [12].

The main chemical constituents of cork can be divided in five groups:

Suberin	33 – 50 %
Lignin	12.6 – 30 %
Cellulose and polysaccharides	12 – 20 %
Waxes	3.5 – 7.9 %
Tannins and other phenolic substances	6 – 7 %

1.4.1. Suberin

The structure of suberin in cork is not yet fully understood. It has been proposed that suberin consist of a polyester structure composed of long-chain fatty acids, hydroxyfatty acids and phenolic moieties, linked by ester groups, as shown in Figure 3 [13].

Suberin is insoluble in all solvents but can be depolymerized by cleavage of the ester linkages by alkaline hydrolysis, alcoholysis, transesterification (catalyzed by soda, alkoxides or boron trifluoride) or reduction. The aliphatic components of suberin include fatty acids, fatty alcohols (mainly C_{20}, C_{22}, and C_{24} alcohols) and glycerol.

Glycerol (**1**), is present as component of esters of suberinic acids [14–18].

The fatty acids include monoacids, α,ω-diacids and ω-hydroxyacids, which can have different functionalities including double bonds, vicinal hydroxy and epoxide groups. The most important acids are 9,10-dihydroxyoctadecanedioic (phloionic acid) (**2**), 18-hydroxyoctadec-9-enoic (**3**), octadec-9-enedioic (**4**), docosanedioic (phellogenic acid) (**5**), 22-hydroxydocosanoic (phellonic acid) (**6**), and 9,10,18-tridihydroxyoctadecanoic acids (phloionolic acid) (**7**). The composition of suberin monomers is shown in Table 1 [14], [15], [19–27].

Several triterpenes and sterols constituents of waxes were also isolated from suberin [19].

The composition of the aromatic fraction of cork suberin is not completely elucidated. The structure of the phenolic part displays features similar to those of lignins. Alkali-depolymerized cork contains large amounts of phenolic compounds (ca. 60 %) [28]. Aromatic acids, not occuring in lignin, have been identified, such as 4-hydroxy-3-methoxycinnamic (ferulic) (**13**), 3,5-dimethoxy-4-hydroxycinnamic (sinapinic) (**14**), and 4-hydroxy-2-methoxybenzoic acids (**15**), together with 4-hydroxy-3-methoxybenzaldehyde (vanillin) (**16**), and esters of the ferulic acid esterified with fellonic (**17**) and other suberinic acid [14], [17], [29–31].

1.4.2. Lignin (→ Lignin)

Although several attempts have been made to extract and characterize cork lignin its structure has not been fully established. The nature of linkages between lignin and aromatic units of suberin is unclear. It is not possible to isolate a "Milled Wood Lignin" (MWL) by the extraction of ball-milled cork with aqueous dioxane at room temperature, applied for the isolation of lignin from wood [32], indicating that, if any "lignin-like polymers" occur in cork, they must be covalently bound to suberin [33], [46]. Furthermore, the calculation of 40 % "lignin" in cork from the methoxyl content (5.6 %) of whole cork [46], does not take into account the methoxy groups in hemicelluloses, e.g., 4-O-methyl-glucurono-xylan, as well as those in suberin itself,

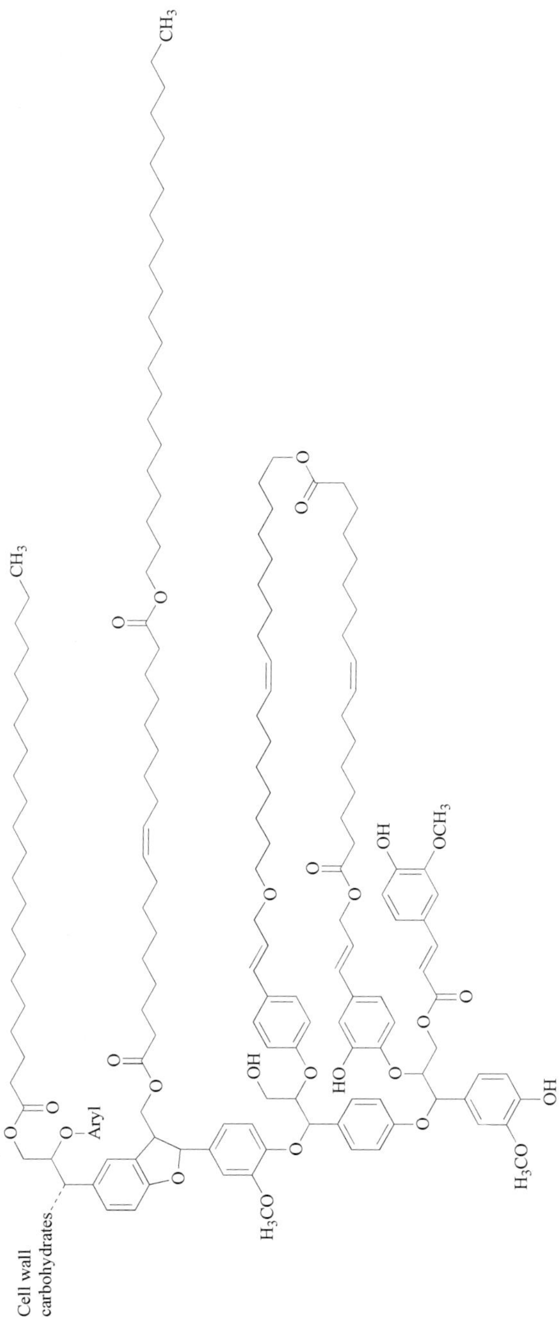

Figure 3. Schematical structure of cork suberin

Table 1. Composition of suberin monomers of cork (%)

Compounds	Structure	Content in cork, %
Aliphatic		
Propane-1,2,3-triol (glycerol) **1** [56-81-5]	HO-CH$_2$-CH-OH-CH$_2$OH	2.0 – 7.0
1-Alkanos (C$_{18}$ – C$_{28}$)	CH$_3$(CH$_2$)$_n$OH	2.3 – 2.8
Alkanoic acid (C$_{16}$ – C$_{28}$)	CH$_3$(CH$_2$)$_n$COOH	1.3 – 10.0
α,ω-Alkanodioic acids (C$_{16}$ – C$_{26}$)	HOOC(CH$_2$)$_n$COOH	6.2 – 12.3
α,ω-Alkenodioic acids (C$_{18}$:1 – C$_{22}$:1)	HOOC(CH$_2$)$_n$CH=CH(CH$_2$)$_m$COOH	2.8 – 25.4
ω-Hydroxyalkanoic acids (C$_{16}$ – C$_{26}$)	HOCH$_2$(CH$_2$)$_n$COOH	10.8 – 33.0
ω-Hydroxyalkenoic acids (C$_{18}$:1 – C$_{22}$:1)	HOCH$_2$(CH$_2$)$_n$CH=CH(CH$_2$)$_m$COOH	11.4 – 25.7
C$_{16}$ Functionality acids		
7,16-Dihydroxyhexadecanoic acid **8**	HOCH$_2$(CH$_2$)$_8$CHOH(CH$_2$)$_5$COOH	2.1 – 4.0
8,16-Dihydroxyhexadecanoic acid **9**	HOCH$_2$(CH$_2$)$_7$CHOH(CH$_2$)$_6$COOH	
C$_{18}$ Functionality acids		
Octadec-9-en-1,18-dioic acid **10** [2027-47-6]	HOOC(CH$_2$)$_7$CH=CH(CH$_2$)$_7$COOH	2.8 – 10.2
9,10-Epoxy-octadecane-1,18-dioic acid **11**	HOOC(CH$_2$)$_7$HC—CH(CH$_2$)$_7$COOH, O (epoxide)	4.0 – 15.9
9,10-Epoxy-18-hydroxy-octadecanoic acid **12** [3233-92-9]	HOH$_2$C(CH$_2$)$_7$HC—CH(CH$_2$)$_7$COOH, O (epoxide)	5.0 – 15.3
9,10,18-Trihydroxy-octadecanoic acid (phloionolic acid) **7** [496-86-6]	HOCH$_2$(CH$_2$)$_7$CHOHCHOH(CH$_2$)$_7$COOH	2.0 – 5.4
9,10-Dihydroxy-octadecanedioic acid (phloionic acid) **2**	HOOC(CH$_2$)$_7$CHOHCHOH(CH$_2$)$_7$COOH	2.0 – 8.8
Aromatic		
4-Hydroxy-3-methoxycinnamic acid (ferulic acid) **13** [537-98-4]	HOC$_6$H$_3$(OCH$_3$)CH=CHCOOH	7.0
3,5-Dimethoxy-4-hydroxycinnamic acid **14** (sinapinic acid) [530-59-6]	HOC$_6$H$_2$(OCH$_3$)$_2$CH=CHCOOH	
4-Hydroxy-2-methoxy-benzoic acid **15**	4-(HO)C$_6$H$_3$ – 2-(OCH$_3$)COOH	
4-Hydroxy-3-methoxybenzaldehyde (vanillin) **16** [121-33-5]	4-(HO)C$_6$H$_3$ – 3-(OCH$_3$)CHO	
Ferulic acid esterified with methyl fellonate **17**	HO-C$_6$H$_3$(OCH$_3$)-CH=CH-C(O)-O-(CH$_2$)$_{21}$-COOCH$_3$	

e.g., ferulic acid esterified with methyl fellonate (**17**, Table 1).

When ground, extracted reproduction cork was heated for 4 h at 160 °C with ethanol/water (1:1, v/v), an "Organosolv Cork Lignin-Like Polymer" (OCL) was obtained in just 2.6 % yield, that still contained "residual aliphatic moieties" [40].

In order to hydrolise the assumed ester bonds between lignin and suberin (Figure 4), ground reproduction cork was first treated with sodium methylate in methanol and the residual cork (34.4 %) then treated by the Björkman procedure, to yield 4.1 % "Saponified Milled Cork Lignin" (MCL$_{sap.}$), which, based on the results of analytical pyrolysis, was claimed to be similar to softwood MWL [42]. However, a comparison of MCL$_{sap.}$ with MWL from spruce wood by thioacidolysis revealed that MCL$_{sap.}$ contained 34 % less β-O-4 linkages and, by permanganate oxidation, gave 21 – 23 % less oxidation products than the spruce MWL [43].

As, β-O-4 linkages are indicative for the discrimination between lignin and bark polyphenols, taking into account the low yields of permanganate oxidation products as well as those of OCL and MCL$_{sap.}$, it may be well concluded that the aromatic moieties in cork, linked to suberin, are to be classified as bark polyphenols rather than lignins.

1.4.3. Polysaccharides

Cellulose represent ca. half of the total polysaccharides of cork (5 – 10 %). Hemicelluloses i.e.,

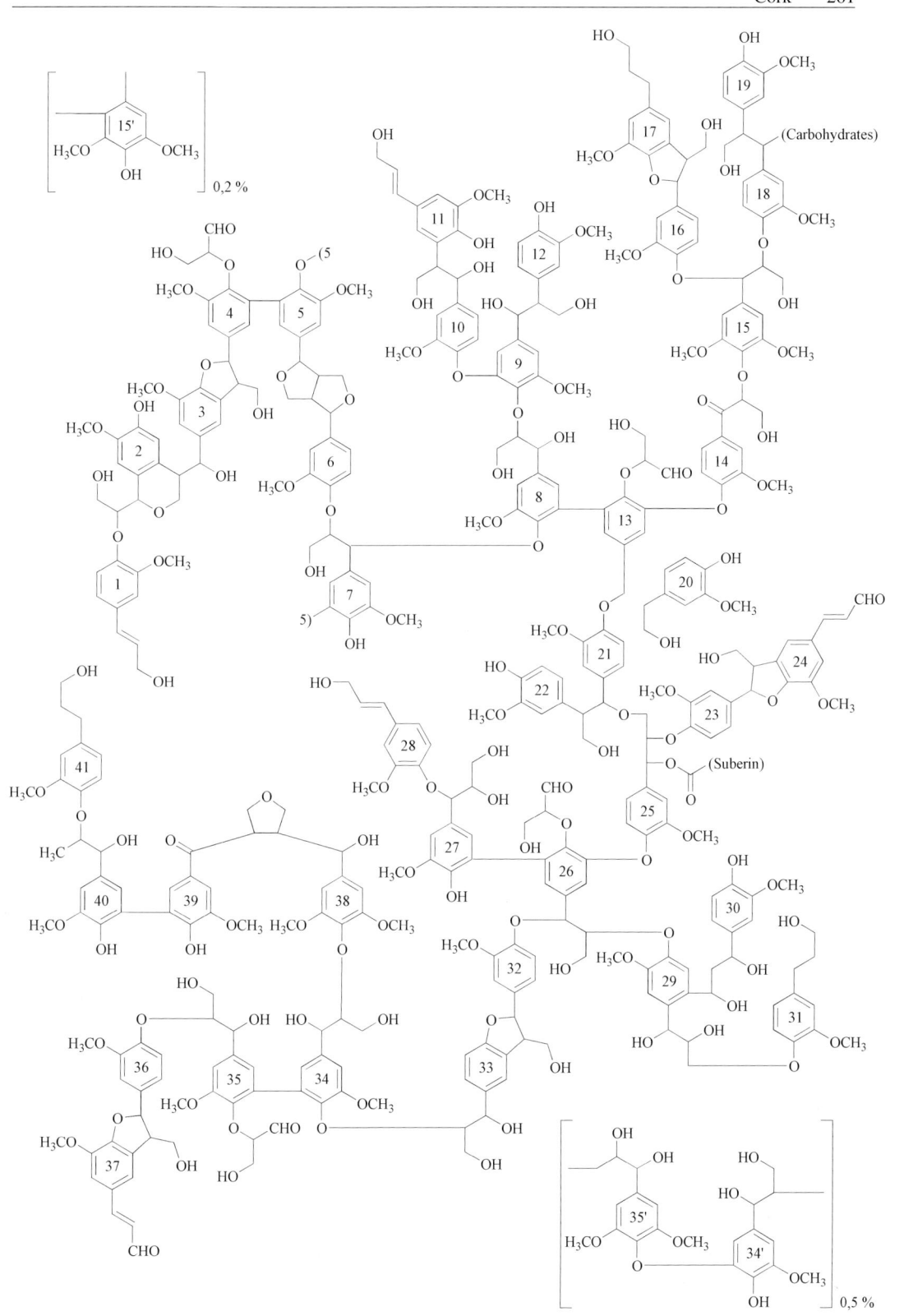

Figure 4. Model of the lignin structure of *Quercus suber*

$\rightarrow$ 4)-β-D-Xyl p-(1$\rightarrow$ 4)
2
↑
1
4-O-Me-α-GlcpA

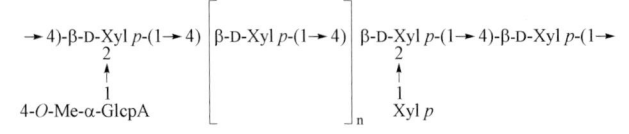

$\left[\text{β-D-Xyl } p\text{-(1} \rightarrow \text{4)} \right]_n$

β-D-Xyl p-(1$\rightarrow$ 4)-β-D-Xyl p-(1$\rightarrow$
2
↑
1
Xyl p

Figure 5. Model of hemicellulose structure of *Quercus suber*

heteropolymers, which include several different sugar moieties (arabinose, xylose, glucose, galactose, etc.) constitute the other half [15].

Three types of hemicelluloses are found in cork: hemicellulose A and two hemicelluloses B (B-1 and B-2). Hemicellulose A contains xylose (95.2 %) and 4-O-methylglucuronic acid (5.9 %) [40]. Hemicellulose B-1 is composed of xylose, 4-O-methylglucuronic acid, arabinose, galactose, mannose, and glucose in the molar ratios 135 : 12 : 7 : 11 : 2 : 30 [41]. Hemicellulose B-2 is more highly branched than hemicelluloses A and B-1 and contains xylose, arabinose, glucose, galactose, 4-O-methylglucuronic acid, and rhamnose in the molar ratio of 17 : 12 : 12 : 6 : 4 : 1 [42]. The model of the hemicellulose structure from cork of *Quercus suber* L. is depicted in Figure 5 [43].

1.4.4. Waxes

The neutral compounds, extracted from cork by apolar solvent, include n-alkanes ($C_{16} - C_{34}$), fatty alcohols (with even numbers of C-atoms from $C_{18} - C_{26}$, containing traces of intermediate alcohols with an odd number of C-atoms and some unsaturated alcohols), glycerol, triglycerides, triterpenes of the friedelane and lupane families, and a very small fraction of monocarboxylic fatty acids (C_{16} and C_{24}). The triterpenes represent ca. 50 % of waxes and 2 – 5 % of cork [10], [15], [44].

The main components of the acid fraction are fatty acids, saturated even numbered $C_{14} - C_{24}$, odd C_{15}, C_{17}, C_{21}, and ω-hydroxyacids, triterpenes, sterols, and C_{20} and C_{24} alcohols. Table 2 shows the chemical composition of the cork waxes, chemical and common names and structures [9], [10], [40–52].

1.4.5. Tannins

The tannins are phenolic and polyphenolic compounds which can be extracted from cork by polar solvents, such as water. The phenolic compounds may be monomeric (phenols, benzoic and cinnamic acids) and polymeric (hydrolyzable and condensed tannins). *Hydrolyzable tannins* are composed mainly of trihydroxybenzoic acid (gallic acid) (**56**) [*149-91-7*] or hexahydroxydifenic acid esterified with glucose (**57**). *Condensed tannins* originate from flavonoid type precursors by polycondensation reactions [44].

The phenolic compounds extractable from cork represent ca. 6 – 7 % of total cork corresponding to 95 – 98 % of tannins.

1.4.6. Other Components

The ash content of cork is ca. 1 – 3 %. Calcium is the most abundant inorganic constituent, in addition cork contains potassium, phosphorus ⋏and magnesium [1], [53].

1.5. Physical and Mechanical Properties

Cork has a low density, which usually varies from 120 – 140 kg/m³ depending on the region, the quality of the tree and the degree of humidity. Another special characteristic of cork is its great elasticity. When cork is rapidly compressed under elastic conditions it returns immediately to about 85 % of its initial volume and in 24 h about 95 % of its initial dimension are reached.

The average properties of cork are listed below [7]:

Density	120 – 140 kg/m³
Young's modulus	4 – 60 MN/m³
Shear modulus	2 – 30 MN/m²
Poisson ratio	0.15 – 0.22
Collapse strength (compression)	0.5 – 2.5 MN/m²
Fracture strength (tension)	0.8 – 3.0 MN/m²
Loss coefficient	0.1 – 0.3
Coefficient of friction	0.2 – 0.4
Thermal conductivity	0.025 – 0.028 J m K^{-1}

Table 2. Chemical composition of cork waxes of *Quercus suber*

Compounds	Structure	Content in cork, %
Fatty alcohols	$H_3(CH_2)_nCH_2OH$	
Octadecanol **18** [*112-92-5*]	n = 16	
Icosan-1-ol (eicosanol) **19** [*629-96-9*]	18	
Heneicosan-1-ol **20**	19	
Docosan-1-ol **21** [*661-19-8*]	20	
Tricosan-1-ol **22** [*3133-01-5*]	21	
Tetracosan-1-ol **23** [*506-51-4*]	22	0.7
Pentacosan-1-ol **24** [*26040-98-2*]	23	
Hexacosan-1-ol **25** [*506-52-5*]	24	
Propane-1,2,3-triol (glycerol **1** [*56-81-5*])	$OHCH_2CH_2OHCH_2OH$	
Triterpenes		
Friedelan-3-one (friedelin) **26** [*559-74-0*]		
3-Hydroxyfriedel-3-en-2-one **27**		
2α-Hydroxyfriedel-3-one (cerin) **28** [*468-67-7*]		

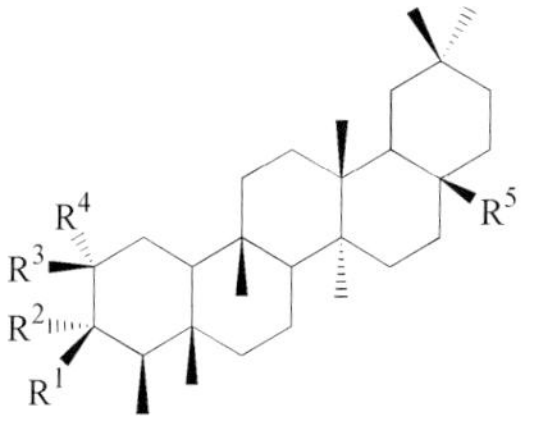

26 $R^1, R^2 = O; R^3, R^4 = H; R^5 = H$
28 $R^1, R^2 = O; R^3 = H; R^4 = OH; R^5 = H$
29 $R^1, R^2 = O; R^3 = H; R^4 = OAc; R^5 = H$
30 $R^1, R^2 = O; R^3 = OAc; R^4 = H; R^5 = H$
31 $R^1 = H; R^2 = OH; R^3, R^4 = O; R^5 = H$
32 $R^1, R^2 = O; R^3, R^4 = H; R^5 = CH_2OH$

2α-Acetoxyfriedel-3-one (cerin acetate) **29**		
2β-Acetoxyfriedel-3-one **30**		
3α-Hydroxyfriedelan-2-one **31**		
28-Hydroxyfriedelan-3-one (canophyllol) **32**		2 – 5

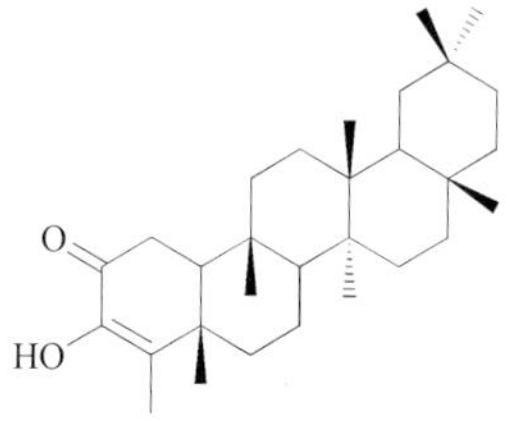

27

Lupane		
20(29)-Lupen-3β-28-diol (betulin) **33** [*473-98-3*]		
3β-Hydroxy-20(29)-lupen-28-oic acid (betulinic acid) **34** [*472-15-1*]		

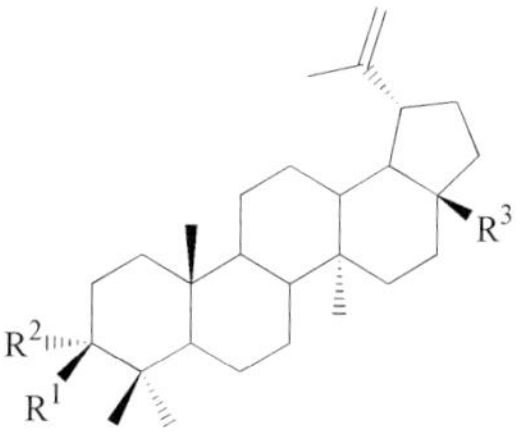

33 $R^1 = OH; R^2 = H; R^3 = CH_2OH$
34 $R^1 = OH; R^2 = H; R^3 = COOH$
35 $R^1, R^2 = O; R^3 = COOH$
36 $R^2 = H; R^3 = CH_2OH; R^1 =$

(*Continued*)

Table 2. (*Continued*)

Compounds	Structure	Content in cork, %
3-Oxo-20(29)-lupen-28-oic acid (betulonic acid) **35** [*4481-62-3*] Lup-20(29)-en-28-ol-3β-yl-caffeate **36**		
Sterols β-Sitosterol **37** [*83-46-5*]	**37** $R^1 = OH; R^2 = CH_2CH_3$ **38** $R^1 = OH; R^2 = CH_3$	
Campesterol **38** [*474-62-4*] Suberindiol **39**	**39**	
Fatty acids Tetradecanoic acid **40** [*544-63-8*] Pentadecanoic acid **41** [*1002-84-2*] Hexadecanoic acid **42** [*57-10-3*] Heptadecanoic acid **43** [*506-12-7*] Octadecanoic acid **44** [*57-11-4*] Icosanoic acid (Eicosanoic acid) **45** [*506-30-9*] Heneicosanoic acid **46** [*2363-71-5*] Docosanoic acid **47** [*112-85-6*] Tetracosanoic acid **48** [*557-59-5*] Heptadec-9-enoic acid **49** [*10136-52-4*] Hexadec-12-enoic acid **50** Heptadec-12-enoic acid **51** Octadec-9-enoic acid [*2027-47-6*] **52** Octadeca-9,12-dienoic acid [*2197-37-7*] **53** 18-Hydroxy-octadeca-9,12-dienoic acid **54** 18-Hydroxy-octadec-9-enoic acid **3** *n*-Alkanes 2,6,10,15,19,23-Hexamethyl-2,6,10,14,18,22-tetra-cosanehexaene (squalene) [*111-02-4*] **55**	$H_3C(CH_2)_nCOOH$ $n = 12$ 13 14 15 16 18 19 20 22 $H_3C(CH_2)_6CH=CH(CH_2)_7COOH$ $H_3C(CH_2)_2CH=CH(CH_2)_{10}COOH$ $H_3C(CH_2)_3CH=CH(CH_2)_{10}COOH$ $H_3C(CH_2)_7CH=CH(CH_2)_7COOH$ $H_3C(CH_2)_4CH=CHCH_2CH=CH(CH_2)_7COOH$ $HOCH_2(CH_2)_4CH=CHCH_2CH=CH(CH_2)_7COOH$ $HOCH_2(CH_2)_7CH=CH(CH_2)_7COOH$	1.1
$C_{20} - C_{34}$	$CH_3(CH_2)_nCH_3$	0.6

Because of its small Poisson coefficient (approx. 0) compression of cork parallel to the prism axis produces little lateral spreading [7], a property useful in gaskets. Cork is almost impermeable to gases and liquids, mainly due to the hydrophobic nature of the cell membranes. Cork sticks well to smooth surfaces even when humid, due to its elasticity, to the suction effect produced by the broken cells in contact with the walls and traces of resinous substances, aspects which are very important in a stopper. The friction coefficient $\mu_{cork-glass}$ is around 0.6.

Figure 6 shows a complete stress – strain curve for cork, depicting all the characteristics of a cellular solid. Cork is linear-elastic in the first stage, then its structure collapses and gives an almost horizontal plateau until the complete collapse of the cells causes the curve to rise steeply.

When a sharp object like a pin is stuck into cork, the deformation is highly localized. A layer of cork cells, occupying a thickness of about a quarter of the diameter of the indenter, collapses as it suffers a large strain. The indenter volume is absorbed by the collapse of the cells.

The presence of suberin renders the cork impermeable, and because of its content of tannins cork has no tendency to rot. A high concentration of oxygen is needed to incinerate cork and when the combustion starts (only at very high temperatures) cork burns on the surface and a layer of carbon forms and prevents the fire to spread. One can say it is a self-extinguishing material.

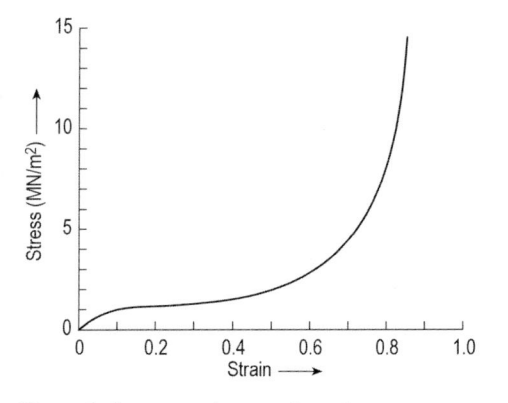

Figure 6. Stress – strain curve for cork

2. Cork Extraction

The first cork extraction (virgin cork) is done after the trunk of a cork oak has reached 0.7 m of perimeter [1].

The stripping when done well does not harm the trees because the first layer of reproduction cork merges with the continuously developing virgin layer in the unpeeled part of the tree. The "mother" (trunk surface after peeling) changes from a pink color to red ochre and afterwards to a reddish brown and finally to a gray crust-like formation [2].

After extraction, the reproduction cork is produced at a rate of 1.5 to 4 mm per year. In Portugal it is required by law to wait at least for 9 years after stripping before a new extraction in order to get a useful harvest (about 2 to 5 cm). This thickness is needed because it is related with stopper diameter. In other locations a cycle of 12 years or even more may be needed. The collection of cork may be repeated every 9 years up to 12 or 15 harvests. Afterward, the tree will no longer be acceptably productive in general. Although yield varies greatly, an adult cork oak, with normal development, can produce from each stripping between 40 to 60 kg of cork [1], although much higher values (> 1 t per strip) can be achieved.

Harvest is between beginning of June and end of August (Fig. 7). The cork is extracted in spring or summer because at this time the cork comes away easily from the trunk since the tree is growing and new cork cells are being generated. It is in the summer that the phelogen is in full production of new cork cells and it is easier to extract the bark by the disruption of these new-formed cells. The cut must be carefully done so as not to affect the mother layer. Care must be taken not to damage the cambium, which unlike the phellogen cannot regenerate itself. When wounded it heals very slowly and wood may start to rot. Even when this is avoided a scar will be seen in future cork and reduce its value. In the first harvest the good practice is not to cut higher than two times the perimeter of the tree at chest level and this height may increase in the following peelings up to three times this perimeter. A special axe is used for the harvest. The blade is used to make the incision while the end of the handle is shaped specially to detach the cork [1]. Besides this manual stripping also a mechanical

Figure 7. Cork tree and cork extraction

process exists, based on small machines, that work using a blade with a sawing motion.

After the harvest "planks" are obtained that are dried in big-formed piles in the forest for approximately 3 – 6 months [1].

3. Production of Cork Based Products

3.1. Disks and Cork Stoppers

After harvest and stabilization cork is taken to the factory where the first operation of boiling takes place [1]. The major structural alteration due to boiling is the attenuation of the cork cells corrugation and an expansion of 10 – 15 % in the radial direction and 5 – 7 % in the other directions [54]. Boiling also decreases porosity by about 50 % [55] (see Fig. 8).

In boiling the cork is immersed in boiling water for approximately 1 h, usually in big open tanks with direct fire underneath. The aim of this operation is to remove dirt, insects, etc. to extract contaminating products and to render the planks softer and more flexible, and thus to facilitate the workability. The boiling water is regularly changed (at least every week) in modern systems.

After boiling to mature cork, it is left to ripen for two to four weeks in cellars (or some days after the new boiling process), sometimes under controlled hygrometric conditions. Very few thermophilic species of microflora, which eventually survived boiling and further recontamination, may appear on the planks. Finally, the edges are cut to make the planks rectangular, and the planks are inspected and selected according to thickness and quality (i.e., flexibility, homogeneity, color, evenness, and natural defects). 40 – 50 % of the original cork weight is lost in this step of cork preparation (separation of defects, drying, edge cutting) [1]. The best cork (about 25 %) goes to stopper production and the remainder is used for agglomerates.

The rectangular planks are cut into strips with a thickness corresponding to the diameter of the cork stopper and a width corresponding to the height of the stopper. The corks are punched with a cutting tube (rotating), the machines being more or less automated [1]. As lenticels lie

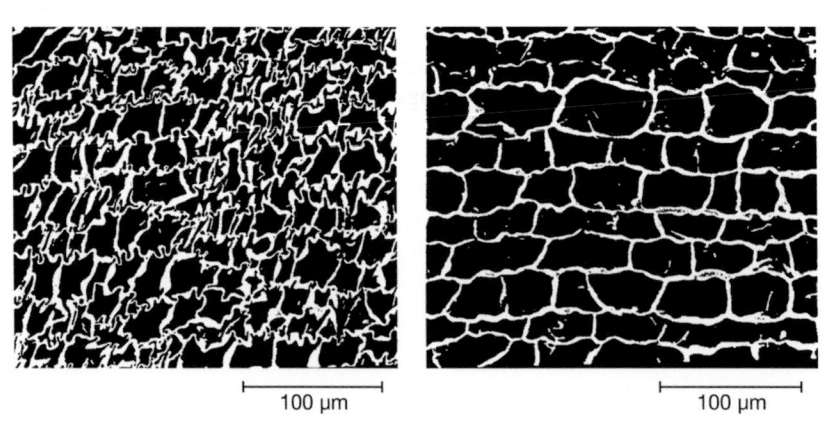

Figure 8. Effect of boiling on the cellular structure of cork

parallel to the cell prism axis, cork cut parallel to this axis will leak. That is why natural cork stoppers are cut with the prism axis (and lenticels) perpendicular to the cork stopper axis [3], [56]. After this, dimensional rectification, such as ends trimming and control of height and diameter, is performed.

The corks are classified in four to six categories, this selection being made by hand by specially qualified workers or by classifying machines (electronic optical reading). A good cork should be light and even in color, the pores being light and few in number. Harder corks may be chosen for wine under pressure [1].

The selection of cork stoppers is one of the most important operations in cork processing. Several parameters are analyzed, and a definition of cork defects for stoppers may be as follows [1]:

- Belly spots – surface depression caused by the dense inner surface of the cork strip;
- Chips – pieces missing from the cork surface (near the ends) due to dry or brittle cork material or faulty corker jaws;
- Cracks – fissures in the cork surface, which can cause chips;
- Dry year – narrow woody growth rings caused by lack of rain;
- Greenwood – undulation in the cork surface caused by moisture;
- Hardwood – hard rough areas on the cork surface caused by boring too close to the bark surface;
- Porosity – high-density pore activity;
- Worm holes – holes caused by worms.

To improve the appearance, the pores or lenticels can optionally be filled in with a mixture of cork powder and glue. Afterward the corks are washed (in tanks or automatic rotating machines), nowadays normally with a peroxide solution. After washing the corks are dried to 6 – 8 % moisture content. They can also be satin finished, i.e., covered by a fine film of paraffin (solid, oil emulsion) or silicone to facilitate cork stopper extraction and are marked (ink or fire) [1], [56]. All the products used in finishing are of food grade.

Cork stoppers can be sold as finished or semifinished products. In the first case marked and treated corks are packed in plastic bags under vacuum and SO_2. In the second case raffia or jute bags are usually used.

To produce *cork disks*, the thinnest good quality cork (not thick enough for cork stopper manufacturing) is used. From these planks the belly and the rasp are cut to obtain a clean slab of cork from which disks are cut and then dimensionally finished.

3.2. Granulates and Broken

Waste cork from the stopper industry, low quality cork (refuse), and possibly virgin cork are used to produce *cork granulates*. The granulates are separated and classified according to density and grain size. These cork granulates can be used as a final product in several applications or used as raw material for the production of composition cork. The finest are used to produce linoleum.

Broken is the name of small pieces of cork obtained from the breaking and first trituration of cork waste in order to reduce volume, e.g., for transportation.

3.3. Composition Cork

Composition cork is made of granules, which have been joined together using different synthetic or natural binding agents [1]. The agglutination is made, e.g., with asphalt, rubber, casein, cement, gypsum, plastics, urethane, melaminic, or phenolic resins [2]. Composition cork is usually produced in sheets, rolls, or blocks, or even extruded or molded.

3.3.1. Floor and Wall Coverings (→ Floor Coverings)

Floor and wall coverings made of cork are usually produced in the form of tiles, with different thickness, densities and finishing: simply polished, waxed, varnished or with a poly(vinyl chloride) layer. This last group may use a decorative sheet between the PVC and the agglomerate underneath [1]. A cork layer associated to a medium-density fiberboard (MDF) basis, for example, constitute a new type of flooring, known as floating floor coverings. The finest cork granules are used to produce linoleum [2].

Wall coverings have a density of 200 – 300 kg/m^3 and floor coverings of 400 – 500 kg/m^3. A mixture of cork granules and glue and/or other additives is filled into a mold which is then closed and heated usually at more than 120 °C for some hours to produce a block which upon cooling (or not) is then sliced into sheets, which are then dimensionally finished. These sheets can be painted, varnished, or covered with layers of other materials.

3.3.2. Agglomerated Disks and Cork Stoppers

For the production of agglomerated cork stoppers cork waste from the cutting stage (until 90 % of the granulated material) is used or material rejected at the stopper sorting stage. Contaminated material should be avoided.

There are simple agglomerated corks and two types of composite stoppers, the stopper for champagne and sparkling wine (head in agglomerate and two or more disks in the bottom) and the "1 + 1" stopper for the other wines (body in agglomerate and one disk in each end).

Champagne corks comprise an upper part of agglomerated cork made be molding, extrusion, or block cutting and usually two or three disks of natural cork, which are in contact with the wine. This special type of corks has thre distinct parts: the "head" that protrudes out of the bottle, the "wheel" which is the part in the neck of the bottle, and the "mirror" which is the part in contact with the wine [6].

3.3.3. Corkrubber

The production of corkrubber is similar to the production of rubber-like products. Rubber and cork granules are mixed in rolls and the mats obtained are introduced into a mold, which is heated for polymerization. The heating process may take from several hours (in a typical oven) to some minutes (in microwave systems).

The most common corkrubber materials include a portion of cork granulates (60 – 70 kg/m^3), ranging from 15 – 260 parts of cork per 100 parts of rubber in weight. The main types of rubbers used are styrene-butadiene, acrylonitrile-butadiene and acrylic rubbers.

3.4. Insulation Corkboard

Insulation corkboard (ICB) is a product made of granules of cork expanded (steambacked) in autoclaves (> 300 °C, ~ 50 kPa) and agglomerated by pressure and temperature without any exogenous glue. Heat expansion results in an increase in cell volume of about 100 % and so the granules are compressed against each other and in the junctions the cells collapse. The blocks formed are then cut into planks which are then finished as required [57], [58]. The main characteristics of ICB are listed in Table 3 [57], [59].

Although such an agglomerate is produced with the lowest quality types of cork it has very favorable properties for some applications. Insulation corkboard is used for three applications: thermal insulation, acoustical absorption, an vibration insulation. The first application is the most important one. Acoustical corkboard, for acoustic insulation is fabricated from specially selected raw materials, using granules of smaller size, lower density, and careful fine finish [59]. For vibration insulation higher densities are used.

Agglomerate for insulation cork board is exclusively made from cork granules (mainly from winter virgin cork) of a larger size (usually 5 – 20 mm) than those used in composition cork.

3.5. Other Specific Products

Most other cork products are made by similar processes as used for floor and wall covering cork products. For example, *filler and expansion joints* are specially designed to neutralize the

Table 3. ICB average characteristics

Characteristic	Value
Density	110 kg/m^3
Thermal conductivity coefficient	0.045 W m^{-1} K^{-1}
Specific heat	1.9 MJ kg^{-1} K^{-1}
Thermal expansion coefficient	40×10^{-6}
Limit of pressure*	10 000 kN/m^2
Permeability	3×10^{-5} m^{-1} h^{-1} Pa^{-1}
Tensile strength	0.05 MPa
Compressive strength at 10 %	0.25 MPa
Bending tension**	1.6 kN/cm^2

* limit of pressure under which the material has an elastic behavior
** force for bending a material with a cross surface

expansion and contraction phenomena that can jeopardize concrete constructions, being an excellent protection against cracks, usually due to temperature variations. These materials can withstand continuous strain, whatever the moisture conditions are.

4. Uses

4.1. Stoppers

Good wines are made to grow and mature and they have to be protected from the environment but have to establish with it a slow and subtle relationship of exchange that only the cellular structure of cork can ensure. The friction coefficient of cork is high providing the ability to adhere glass due to the elastic force and the microscopic suction pads (cut cells).

Agglomerated corks are most commonly used in sparkling wines, but also in a sizable percentage of table wine (fast consumption wine). These corks are of very uniform size and density but are unlikely to provide the aging capabilities of the traditional cork stopper [1]. Colmated corks (pores covered with a mixture of cork powder are glue) are attractive low cost stoppers but are used only in common table wines.

The type of surface treatment applied on the cork stopper is determined by the type of wine and corkage cadence. The recommended moisture content for the stoppers in corkage is 6 – 8 % and the corker machine must press only 1 – 1.5 mm below the entry bore, avoiding compressions of more of 1/3 of the stopper diameter. When filling, an empty space level should be respected. After corking the bottle must be maintained in a stand-up position for some minutes to allow the dimensional recovery of the cork stopper.

In champagne corks, the base, where sealing is most critical, is usually made of two disks cut with the lenticels crossing the disks' thickness. Leakage problem is avoided because the lenticels of two or more disks do not connect [6].

4.2. Civil Construction

In buildings and other civil construction works, cork products may be used for thermal insulation and insulation, acoustic correction, floor covering, wall covering, false ceilings, and expansion joints:

- Thermal insulation – ICB
- Vibration insulation – ICB, corkrubber
- Acoustic correction – ICB, composition cork
- Floor covering – composition cork, cork rubber
- Wall covering – soft composition cork
- Expansion joints – composition cork

Some of these cork products may be used as composites with other construction materials, for example, ICB with polyurethane, composition cork with MDF and wood veneer, or may be covered by waxes, varnishes, and paints.

4.3. Industrial Applications

Some specific applications of cork in the industry include: cork slabs and sheets, pipe insulation, battery mold coating, cold storage insulation, machinery anti-vibration sheets, and storage tank insulation.

As expanded cork agglomerate (ICB) maintains its properties at very low temperatures, it has a wide range of applications in the refrigeration industry. The electrical industry uses corkrubber to stop leaks of transformer fluids. Similar products are also used in the manufacture of electrical switches, lightning apparatus and the like [2]. The ceramic industry uses cork disks to smooth edges and to eliminate defects [2].

High density expanded agglomerate planks are one of the best means of insulating machines and damping vibrations [2], [59].

4.4. Automotive Industry

Cork makes good gaskets because it accommodates large elastic distortion and volume change and its closed cells are impervious to water and oils. Almost all the gaskets used in automobiles are made of corkrubber since this combination is easy to cut and due to its flexibility and resistance it can be used in gaskets with very narrow borders [2], [7].

4.5. Other Specific Uses

Other specific uses include: baseballs, cricket, hockey, and golf balls, golf clubs, plate mats, memoboards, musical instruments, polishing blocks, radioactive isotope capsules, ashtrays, bath mats, beehives, candle holders, glass coasters, cork paper, insulation in solid fuel rocket motor cases, desk pads, cork textiles, fishing rods rings and handles, fishing net floats, handicraft items, hot pads, jar lids, buoys, safety helmets, shoes, shuttlecocks, table-tennis bats, whistles, bungs for gun cartridges, calendars, etc.

5. Quality and Standardization

5.1. Standardization and Testing

Cork standardization was initiated in Portugal in 1957 with the establishment of the Technical Committee (TC-16) aiming at the development of standards for cork and industrial cork products, from raw material and terminology to finished products. Currently about 40 standards exist. ISO/TC-87 working since 1958 has the same aim and is responsible for the production of international standards. Portugal has the Chairmanship and the Secretariat of this TC. Within CEN (Standardization European Committee) there are three TC whose scope is directly related to agglomerated cork applications: CEN/TC-88 (thermal insulation), CEN/TC-99 (wall coverings) and CEN/TC-134 (resilient floor coverings). Several European Standards are now published concerning test methods and product specifications, supporting harmonized standards to be converted at a national level [60].

For a list of European and international standards see [60].

The requirements and *methods of testing* for the different cork products, as well as these products classification are described in both national and international and/or European standards [60].

5.2. Quality in Cork Production

There are a large number of productive units, certified according to the EN ISO 9000 series standards. Furthermore, the recent adoption and implementation of the International Code for Cork Stopper Manufacturing Practice reinforces and adds credibility to the quality guarantee process. There are also accredited laboratories, which specifically perform specific tests for cork products.

The Cork Quality Council has instituted a common grading standard in order to eliminate differences in different companies to categorize cork stoppers (visual characteristics). This standard is designed to be used by buyers. The criteria allow classification of cork stoppers in four grades (A, B, C, and D). The visual grades are designed to describe a range of corks of similar appearance.

6. Economic Aspects

6.1. Forestall Production

The cork oak adapts to soils practically unable of supporting any other economically viable crops, and prevents the degradation of these soils. The value of this tree should not be measured merely by the cork produced but rather by the whole aspects of agriculture, forestry, grazing, hunting, etc.

Because of particular climatic conditions (temperature, light, and rainfall) the cork oak established itself in the Western Mediterranean basin. It is found in Northern Africa (Morocco, Algeria, Tunisia), in Southern France (e.g., Corsica), Italy (Sardegna), Spain and, above all in Portugal. Worldwide cork oak forests cover an area of about 2.2×10^6 ha [1], [61]. A breakdown of cork forest surface and cork production per year worldwide is given in Table 4.

Table 4. World cork wood surface and production per year

Country	Surface, ha	Production, t
Portugal	660 000 – 730 000 (28 – 35 %)	170 000 – 185 000 (54 – 63 %)
Algeria	410 000 – 480 000 (12 – 21 %)	15 000 – 27 000 (4 – 12 %)
Spain	410 000 – 500 000 (18 – 23 %)	80 000 – 89 000 (22 – 29 %)
Morocco	340 000 – 400 000 (9 – 18 %)	13 000 – 18 000 (4 – 7 %)
France	100 000 – 110 000 (1 – 5 %)	5000 – 32 000 (1 – 11 %)
Italy	90 000 – 100 000 (4 – 10 %)	20 000 – 26 000 (5 – 6 %)
Tunisia	90 000 – 100 000 (3 – 6 %)	8000 – 9000 (1 – 2 %)

The world average cork productivity is around 150 kg ha^{-1} a^{-1}. An average tree (in age and size) may provide up to 100 kg per strip, with 7 – 11 kg/m^2 of debarked area. However, in the best cases these values can be much higher [1]. Although prices can vary considerably, as a reference, the price of good quality cork for stoppers in Portugal, in the campaign of 2000, was around 3 – 3.5 €/kg.

The undergrowth in corkwoods is rich in a wide variety of aromatic and medicinal plants and is also home to game of significant economic value. Acorns are used for swine food, wood is used for small pieces but mainly for production of firewood and charcoal [61].

6.2. Personnel and Companies

At present more than 900 companies are operating in the cork business in Portugal, mainly in the north of the country (Santa Maria da Feira). These firm employ about 15 000 persons, 85 % of the factories have less than 20 workers. In Spain about 280 companies are active in this field, mainly in Cataluña (province of Girona), Extremadura and Andaluzia, 120 of these companies produce cork stoppers.

In Portugal about 6000 workers work seasonably in cork extraction and this figure allows an extrapolation for other countries. It is estimated that there are about 150 000 people around the world, which are dependent on the cork sector, including production, processing, and marketing.

6.3. Consumption of Products and Markets

Taking the Portuguese industry as reference, the breakdown of cork consumption according to product is as follows [61]:

Natural cork stoppers	57 %
Agglomerated cork stoppers	11 %
Composition cork board	17 %
Insulation cork board	6 %
Other	8 %
Non finished products	1 %

As a reference (for the year 2000) a first quality cork stopper can cost as much as € 0.35 but a very wide range of prices exists for all types of cork stoppers, starting at less than € 0.05.

One square meter of simple cork floor covering can have a price as low as € 7 – 8, but also here a wide range is found.

Annual world consumption of cork products is given below [2]:

Cork stoppers for wine	13×10^9
Stoppers for champagne and sparkling wine	1.5×10^9
Insulation cork board	150 000 m^3
Floor tiles	10×10^6 m^2

The importance of the main sectors of cork products utilization, taking the Portuguese production as a reference, is as follows [4]:

Wine sector	61 %
Automobile sector	11 %
Civil construction	15 %
Other	13 %

6.4. Producers

Currently the cork industry is spread over different areas of activity:

- The preparer branch: selection and preparation of cork for the manufacturing industry
- The manufacturer branch: production of items by simply cutting the cork planks
- The crusher branch: production of cork granulates and raw materials for the cork board industry
- The blender branch: production of different types of cork boards

Cork products are manufactured both in small family companies and very large companies with thousands of employees. Some of the companies produce only one type of cork products, for example stoppers, and others manufacturing the entire variety of cork products.

Portugal and Spain process around 75 % of the world's cork production; France, Italy, and the North African countries produce together around 10 %, the remaining 15 % is scattered over different countries in Eastern Europe, Japan, Switzerland, Germany, UK, Latin America, India, and USA [1].

The most important producer is the Portuguese Amorim Group, which started its business in 1870. This group has 34 factories in Portugal and other countries with a total of more than 4000 workers and total sales in 1999 of about € 512 million, accounting for about 1/5 of the world cork processing capacity. This group produces cork components for almost all kinds of applications.

The second largest cork producing country is Spain, with the company Corchos de Mérida that produces 200×10^6 of cork stoppers (natural and granulates) with a transaction volume of € 18 million.

7. Environmental and Toxicological Aspects

7.1. Industrial Wastes

Most of the waste materials produced in cork processing are used in other types of cork products and therefore cannot be considered as an industrial waste. However, the cork processing still gives rise to real specific waste materials such as cork dust, cork boiling wastewaters, stoppers washing waste solutions, and steambaking condensates.

Cork Powder Waste (cork dust) is formed in different industrial cork processing operations and from different cork types. Worldwide cork powder production is around 50 000 t/a [62]. For studies on cork dust composition see [63].

By definition, cork dust is the group of cork particles with a size smaller than 0.25 mm [62]. Cork dust and cork fungi (from cork processing and manipulation) may be present in the air, even if extraction systems are installed in the factories. Prolonged occupational contact with these agents, if care with protection systems such as masks is not taken, may be related with an illness called suberosis.

Considerable quantities of cork dust are produced in grinding, granules cleaning and separation, cutting, sanding, and other finishing operations. More than 25 % of processed cork becomes cork dust [62]. Pure cork dust may be used in cork stoppers finishing operations and other specific applications [68], [69]. The rest is incinerated to produce steam (e.g., for ICB backing) or energy (drying, electricity etc.). This waste has a high heating value (15 – 28 MJ/kg). For ICB production 180 kg of cork dust are consumed per cubic meter [1] or 1.45 kg per kilogram ICB [70].

Steambaking Condensates. The ICB steambaking condensates are dark brown substances, which consist mainly of tannins and waxes. The amount of steambaking condensates formed annually is difficult to estimate but according to some calculations is about 1000 t/a.

Wastewater. The wastewater from cork processing amounts to 0.45 m^3 per ton boiled cork [1] (120 – 130 t/a world wide, mostly tannins).

7.2. Recycling and Treatments

Cork materials formed as waste in one operation can often be used in another operation of cork transformation. For example, the waste of cork from the cork stopper industry is used to produce cork granulates and cork powder from different origins is used to fill pores of some stoppers or is used as an energy source in cork factories [62] (see Section 7.1).

Cork grinding waste can be used as a substrate for plants with very good results [71] in germination and plant growing.

Used cork stoppers may be used as a raw material for the production of cork granulates or for the production of utilitarian wares (e.g., cutlery holders, table wares, pullers) according to a patented process [67].

7.3. Emissions

In the production and processing of some cork based products sometimes adhesives, varnishes, paints, glues, coatings, and other additives are used. Substances are formed or liberated and are emitted from the plants and eventually from finished products.

When urea – formaldehyde or phenol – formaldehyde resins are used as adhesives for cork materials, the finished products can potentially emit formaldehyde. The emission depends on the type of binding agent, catalyst, operational conditions, stabilization periods etc. (see →

Formaldehyde; → Phenolic Resins; → Amino Resins). Studies carried out on cork materials showed that the level of formaldehyde emitted from cork board ranged from about 0.02 to 0.2 mg/100 g, formaldehyde emitted by natural cork ranged from 0.03 – 0.04 mg/100 g. The stabilization period (after manufacturing and before marketing) has almost no effect on this [72].

8. Legal Aspects

Protection of acorn production date back as far as the Costumes de Castelo Rodrigo e Castelo Melhor in 1209 in Portugal [2]. Further legislation was enacted during the following centuries to protect the cork tree, setting regulations for thinning, pruning, and stripping the cork. Current legislation covers the fields of forestland, cork oak, cork, and the cork oak farmer. The cork oak is protected from the moment of germination. A special license is required to fell trees. The same law requires the tree owners to take preventive measures, regarding to diseases, fire and damages by animals. Thinning is only permitted if the normal density is disrupted and the owner must notify the authorities. Pruning is only permitted from December to March. The tree may only be stripped for the first time when the trunk has a perimeter of 70 cm and extraction of reproduction cork is expressly prohibited if this cork is less than 9 years old [1], [2].

References

1. L. Gil, "Cortiça — Produção, Tecnologia e Aplicação", Ed INETI, Lisbon, 1998.
2. M. A. Oliveira, L. Oliveira: *The Cork*, Ed. Corticeira Amorim, Rio de Mouro 2000.
3. L. J. Gibson, M. F. Ashby: *Cellular Solids. Structure and Properties*, Pergamon Press, Oxford 1988.
4. W. Liese, H. Gunzerodt, N. Parameswaran: *Biological Alterations of Cork Quality Affecting its Utilization*, Ed. Instituto Produtos Florestais, 1983.
5. L. Gil, "História da Cortiça", Ed. APCOR, Santa Maria de Lamas, 2000.
6. "The art of cork", Ed. ICEP/AIECN/AIEC, Lisboa.
7. M. F. Ashby: *Cork in Concise Encyclopedia of Wood and Wood Based Materials*, Pergamon Press, 1989, pp. 66 – 69.
8. H. Pereira, *Wood Sci. Technol.* **26** (1992) 259 – 269.
9. H. Pereira, *Wood Sci. Technol.* **22** (1988) 211 – 218.
10. E. Conde, M. C. García-Vallejo, E. Cadahía, *Wood Sci. Technol.* **33** (1999) 271 – 283.
11. P. Sitte, *Protoplasma* **54** (1962) 555 – 559.
12. A. M. Gil, M. Lopes, J. Rocha, C. Pascoal Neto, *Int. J. Biol. Macromol.* **20** (1997) 293 – 305.
13. P. E. Kolattukudy, *Recent Adv. Phytochem.* **11** (1977) 185 – 246.
14. M. F. P. V. S. Bento, "Estudo dos Componentes da Cortiça e da Estrutura da Suberina por Técnicas Espectrométricas", Universidade Técnica de Lisboa, ISA, Ph. D. Thesis, 1997.
15. A. J. V. Marques, "Isolamento e Caracterização Estrutural da Lenhina da Cortiça de *Quercus Suber* L.", Ph. D. Thesis, Universidade Técnica de Lisboa, ISA, 1998.
16. N. Parameswaran, W. Liese, H. Günzerodt, *Holzforschung* **35** (1981) 195 – 199.
17. J. Carvalho: *A Text book of course of F.S.E., Técnicos para a Indústria Cortiçeira*, Universidade Técnica de Lisboa, IST Lisboa, 1987.
18. J. Graça, H. Pereira, *Holzforschung* **51** (1997) 225 – 234.
19. E. Seoane, E. Villacorta, M. J. De La Villa, *Anal. Quím.* **67** (1971) no. 3, 329 – 334.
20. B. Rodríguez-Miguéne, I. Ribas Marques, *Anal. Quím.* **68** (1972) no. 11, 1301 – 1306.
21. P. J. Holloway, *Chem. Phys. Lipids* **9** (1972) 158 – 170.
22. C. Agullô, E. Seoane, *Chem. Ind. (London)* **17** (1981) 608 – 609.
23. M. Arno, M. C. Searra, E. Seoane, *Anal. Quím. C* **77** (1981) 82 – 86.
24. P. J. Holloway, *Phytochemistry* **22** (1983) 495 – 502.
25. N. Cordeiro, N. M. Belgacem, A. Gandini, C. Pascoal Neto, *Bioresource Technol.* **63** (1998) 153 – 158.
26. N. Cordeiro, M. N. Belgacem, A. J. D. Silvestre, C. Pascoal Neto, A. Gandini, *Int. J. Biol. Macromol.* **22** (1998) 71 – 80.
27. L. Gil, *Silva Lusitana* **3** (1995) no. 1, 73 – 84.
28. A. Guillemonat, *Ann. Fac. Sci. Marseille* **30** (1960) 43 – 54.
29. A. Guillemonat, J. C. Trainard, *Bull. Soc. Chim. France* (1963) 142 – 144.
30. E. Agullô, E. Seoane, *Anal. Quim.* (1982) 78e, 389 – 393.
31. J. Graça, H. Pereira, *J. Wood Chem. Technol.* **18** (1998) no. 2, 207 – 217.
32. A. Björkman, Svensk, *Papperstid.* **59** (1956) 477 – 485.
33. W. Zimmermann, H. H. Nimz, E. Seemüller, *Holzforschung* **39** (1985) 45 – 49.
34. C. Pascoal Neto, N. Cordeiro, A. Seca, F. Domingues, A. Gandini, D. Robert, *Holzforschung* **50** (1996) 563 – 568.
35. A. J. V. Marques, H. Pereira, D. Maier, O. Faix, *Holzforschung* **48** (1994) 43 – 50.
36. A. J. V. Marques, H. Pereira, D. Maier, O. Faix, *Holzforschung* **50** (1996) 393 – 400.
37. A. J. V. Marques, H. Pereira, D. Maier, O. Faix, *Holzforschung* **53** (1999) 167 – 174.
38. C. Pascoal Neto, D. Evtuguin, D. Robert, *J. Wood Chem. Technol.* **14** (1994) no. 3, 383 – 409.

39 M. Lopes, C. Pascoal Neto, D. Evtuguin, A. J. D. Silvestre, A. Gil, N. Cordeiro, A. Gandini, *Holzforschung* **52** (1998) 146 – 148.
40 A. Asensio, *Carbohydr. Res.* **161** (1987) 167 – 170.
41 A. Asensio, *Carbohydr. Res.* **165** (1987) 134 – 138.
42 A. Asensio, *Can. J. Chem.* **66** (1988) 449 – 450.
43 A. Asensio, *J. Nat. Prod.* **51** (1988) no. 3, 488 – 491.
44 H. Pereira, *Boletim Instituto Produtos Florestais, Cortiça* **483** (1979) 259 – 264.
45 J. L. B. López, G. G. Curbera, I. Ribas Marques, *Anal. Fis. Quím.* **62-B** (1966) 865 – 869.
46 E. Seoane, E. Villacorta, *Anal. Quím.* **66** (1970) 1015 – 1016.
47 E. Seoane, E. Villacorta, M. J. De La Villa, *Anal. Quím.* **67** (1971) 329 – 334.
48 J. Fernandez-Salgado, I. Ribas-Marques, *Acta Científica Compostelana* **IX** (1972) no. 3 – 4, 139 – 144.
49 J. Fernandez-Salgado, I. Ribas-Marques, *Anal. Quím.* **70** (1974) no. 4, 363 – 366.
50 S. K. Talapatra, D. K. Pradhan, B. Talapatra, *Ind. J. Chem.* 16 B (1978) 361 – 365.
51 A. Patra, S. K. Chaudhuri, S. K. Panda, *J. Nat. Prod.* **51** (1988) 217 – 220.
52 M. L. Mata, "Identificação dos Constituintes Químicos do Pó da Cortiça e sua Valorização por Hemissíntese", M.Sc. Thesis, FCT-UNL, Lisboa, 1991.
53 F. Mata, V. Marques, H. Pereira, *Boletim Instituto Produtos Florestais, Cortiça* **569** (1986) 68.
54 M. E. Rosa, H. Pereira, M. A. Fortes, *Wood Fiber Sci.* **22** (1990) no. 2, 149 – 164.
55 F. Cumbre, F. Lopes, H. Pereira, *Wood Fiber Sci.* **32** (2000) no. 1, 125 – 133.
56 K. C. Fugelsang, D. Callaway, T. Toland, C. J. Moller, *Practical Winery & Vineyard* Jan. – Feb. (1997) 50 – 55.
57 H. Pereira, E. Ferreira, *Mater. Sci. Eng. A* **111** (1989) 217 – 225.
58 L. Gil, *Wood Sci. Technol.* **30** (1996) 217 – 223.
59 H. Medeiros, ABC Insulation Corkboard, Ed. JNC, Porto.
60 M. F. Bicho, L. Gil, *"Cortiça — Guia Normativo"*, Ed. IPQ/CTCOR. Lisbon 1999.
61 "The Cork oak and Cork", Ed. DGDR, Lisboa, 2000.
62 L. Gil, *Biomass and Bioenergy* **13** (1997) no. 1/2, 59 – 61.
63 L. Gil, J. Santos, M. I. Florêncio, *Boletim Instituto Produtos Florestais-Cortila* **575** (1986) 255 – 261.
64 E. Moiteino, "Aplilalões de triterperiódes dos Condensados Negios da Cortila em Quimila Firne", M. Se Thesis, FET-UNL, 1992.
65 EP 651 760, 1994 (E. Moiteino, M. R. Tavanes, M. Y. Mareloeunto); *Chem. Abstr.* **122** (1995) 187820.
66 E. Moiteino, F. Justino, R. Tavanes, M. Y. Marelo Curto, M. I. Florêncio, M. S. Y. Naseimento, M. Pedro, F. Cequeira, M. M. Pinto, *J. Nat. Prod.* **64** (2001) 1273 – 1277.
67 Pt 102 013, 1997 (L. Gil).
68 L. Gil, *Ingenium* **54** (1991) 58 – 71.
69 L. Gil, *Ingenium* **48** (1991) 54 – 61.
70 L. Gil, *The European Congress on Renewable Energy Implementation*, Athens (1997) 705 – 706.
71 A. Pes, *Boletim Instituto Produtos Florestais-Cortiça* **471** (1978) 13 – 17.
72 L. Gil, N. Maurício, G. Cáceres, *Holz als und Werkstoff* **58** (2000) 47 – 51.

Drying Oils and Related Products

ULRICH POTH, BASF Coatings AG, Münster, Federal Republic of Germany

1. Definitions 275
2. History............................. 276
3. Chemical Compositions 277
4. Sources and Types 277
4.1. Linseed Oil 277
4.2. Perilla Oil 278
4.3. Tung Oil 278
4.4. Oiticica Oil 278
4.5. Fish Oils 280
4.6. Safflower Oil 280
4.7. Sunflower Oil 280
4.8. Soybean Oil 280
4.9. Cottonseed Oil 280
4.10. Dehydrated Castor Oil 280
4.11. Tall Oil 281
5. Production.......................... 281
6. Testing and Specifications 282
7. Film Formation 282
7.1. Properties of Drying Oils for Film Formation 282
7.2. Cross-Linking by Reaction with Oxygen .. 282
7.3. Influence of Catalysts on Cross-Linking .. 283
7.4. Yellowing........................... 284
8. Modifications 284
8.1. Boiled Oils 285
8.2. Stand Oils 285
8.3. Blown Oils 285
8.4. Factice, Sulfurized Oils 286
8.5. Isomerized Oils 286
8.6. Copolymers 286
8.7. Maleinized Oils 286
8.8. Oil Boiling with Rosin, Copal, and Amber 287
8.9. Oil Boiling with Phenolic Resins 287
8.10. Modifications by Transesterification 287
8.11. Modifications Based on Fatty Acids..... 288
9. Application, Importance, and Future Aspects 288
References 288

1. Definitions

Drying oils are fatty oils (→ Fats and Fatty Oils) from different natural sources, which are able to form coating films by reaction with atmospheric oxygen. The reaction may take place with the original, unmodified oils or after modification. Oils which are able to form films on their own, without being modified, are called drying oils in the narrow sense. Oils which are able to form films if they are modified or mixed with other film-forming components are called *semidrying oils*. *Nondrying oils* are also used to produce binders for coatings formulations, but these oils and there modifications are not able to form films by reaction with atmospheric oxygen.

The reaction with atmospheric oxygen (autoxidation) takes place with unsaturated fatty acids containing more than two double bonds in the molecule. Triglycerides containing a sufficient amount of these fatty acids are drying oils in the narrow sense. Triglycerides with a significant content of fatty acids with two double bonds are semidrying oils. Oils containing predominantly fatty acids with only one double bond are not able to dry.

Drying oils in the narrow sense are linseed oil, perilla oil, tung oil, oiticica oil, and different fish oils. Semidrying oils are cottonseed oil, soybean oil, sunflower oil, safflower oil, hemp oil, poppy oil, walnut oil, sesame oil, maize oil, grape seed oils and the oils of Euphorbia, Coriander, Brassica, and Limanthes species. Peanut oil, castor oil, palm tree oil, olive oil, and almond oil are nondrying oils.

Tall oil is not a triglyceride, but a mixture of mainly unsaturated fatty acids and rosin. Tall oil is formed as a byproduct in kraft (sulfate) pulping (→ Tall Oil). Refined tall oil can be used in the same way as fatty acids from soybean oil in the coating industry.

Dehydrated castor oil is a drying oil. Dehydrated castor oil is prepared from castor oil by dehydration with sulfonic acids or at high temperatures. Dehydrated castor oil contains a rather high amount of 9,11-linoleic acid.

2. History

It was in the Stone Age that humans began to grow plants and the early advanced civilizations developed an agricultural system. Important was the cultivation of flax to produce linen, but linseed was also used for production of linseed oil. First documents which give information on the growing of flax are dated to the third dynasty in the Egyptian "Old Empire" (Pharaoh Snofru, approx. 2650 B.C.). These documents only describe the use of linseed oil as food and for lamps, there is no indication that they were used for painting. But it can be assumed that oils were used as binders for pigments in addition to inorganic binders, albumin, waxes, glues, and blood.

First paintings were artistic paintings for religious objects or for objects representing those in power. In the classic period of ancient Greece paints were also used to protect objects. Wooden ships were painted with varnishes prepared from mixtures of waxes, rosins, and oils (semidrying oils).

The first written document how to prepare oil varnishes dates back to the 11th century. The monk RODGERUS OF HELMARSHAUSEN named THEOPHILUS PRESBYTER described in his "Schedula diversarum artium" [1] how to prepare a varnish by melting amber (lat. glassa) together with linseed oil. In the Middle Ages drying oils were combined with metal oxides (white lead, red lead or minium, brown stone) not only to prepare different colors, but also to improve drying (early 13th century, HERAKLIUS, "De coloribus et artibus Romanorum" [2]). Furthermore, it was found by empirical experiments that the drying properties of oil varnishes could be improved if the oils were exposed to sunlight (stand oil), or cooked (boiled oil). Use of other driers (metal vitriols) in oil varnishes was introduced. The knowledge to prepare advanced oil varnishes was kept as an important secret [3].

In 1298 MARCO POLO reported the use of tung oil as raw material in paints for the ships in the Chinese Empire. But it was not until the late 19th century that tung oil was shipped to Europe for production of advanced oil varnishes.

In the beginning of the 15th century the use of turpentine oil (prepared by the new technique of distillation, developed in the Oriental Empires) was introduced into the preparation of oil varnishes (VAN EYCK brothers in Brugge and Gent). This invention formed the basis for the works of the most famous painters in the Netherlands at that time.

In the 16th century the first Chinese and Japanese painted articles were imported to Europe and the Europeans started to copy those techniques by using oil-based coatings. The demand for painted articles became larger and the first coating manufactories were founded in France and later in Germany. In the 18th century the method to produce sheet metal was developed in England. The coatings for the articles from sheet metal consisted of drying oil varnishes modified with furnace soot and asphalt. At the end of the 18th century the coating technique was also introduced in France and Germany. The development of traffic gave a new impetus for founding and extension of paint industries, because of an increased demand for coatings for horse coaches and ships. In the 19th century the industrial production of coatings and paints started in England, the Netherlands, and in Germany. Raw materials for the oil based coatings were linseed oil, tung oil, amber, copals, and different types of rosin. The English varnish processing plants were famous.

When the natural raw materials became rare (mainly amber and copals), chemists started to look for synthetic replacements. The first synthetic resins which were combined with oils were the phenolic resins. Later so-called oil-reactive phenolic resins were prepared. The alkyd resins formed the most important group of synthetic resins which combined the air drying of oils or oil-derived fatty acids with the physical drying of a polyester molecule. Alkyds became the most important group of resins for paints and coating from the 1930s to the 1980s and nearly totally replaced the older varnishes based on oils combined with amber, copals, and rosins. Currently alkyds are also being replaced by other binder systems.

3. Chemical Compositions [5]

Fatty Acids with Isolated Double Bonds. Important contents of drying oils are the fatty acids with isolated double bonds, i.e., linoleic, linolenic, arachidonic, and clupanodonics acid.

Linoleic Acid [60-33-3], all-*cis*-9,12-octadecadienoic acid, M_r 280.45, mp − 5.0 °C, bp (1333.2 Pa) 224 °C, iodine value 181 %, acid number 200. Various oils, which are used for the production of air-drying coatings, contain a significant amount of linoleic acid: safflower oil, sunflower oil, cottonseed oil, and soybean oil. Linoleic acid is also the main component of tall oil and a constituent of other semidrying oils which are used for specialties or artist paints.

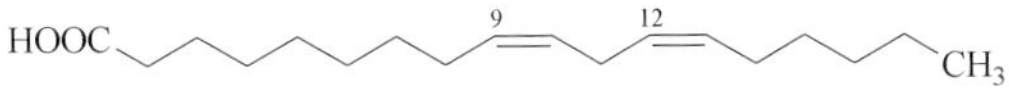

HOOC~~~~~9~~~12~~~CH$_3$

9, 12-Linoleic acid

Linolenic Acid [463-40-1], all-*cis*-9,12,15-octadecatrienoic acid, M_r 278.44, mp − 5.0 °C, bp (1333.2 Pa) 225 °C, iodine value 274 %, acid number 202. Linolenic acid is the main constituent of linseed oil and perilla oil.

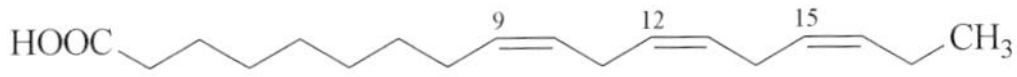

HOOC~~~~9~~~12~~~15~~CH$_3$

9, 12, 15-Linolenic acid

Arachidonic Acid [7771-44-0], all-*cis*-5,8,11,14-eicosatetraenoic acid, M_r 304.47, mp − 49.5 °C, iodine value 334 %, acid number 184, and *clupanodonic acid* [2548-85-8], all-*cis*-4,8,12,15,19-dodecosapentaenoic acid, M_r 330.52, mp − 78.0 °C, bp (666 Pa) 236 °C, iodine value 384 %, acid number 170. These fatty acids are components of fish oils, which have nearly the same properties (for coatings and varnishes) as linseed oil.

Fatty Acids with Conjugated Double Bonds. Also fatty acids with conjugated double bonds are important constituents of industrially used drying oils, i.e., α-eleostearic acid, α-licanic acid, and conjugated linoleic acid.

α-Eleostearic Acid [4337-71-7], 9-*cis*,11-*trans*,13-*trans*-octadecatrienoic acid, M_r 278.44, mp + 49 °C, bp (1560 Pa) 235 °C, iodine value 274 %, acid number 202, is a component of tung oil.

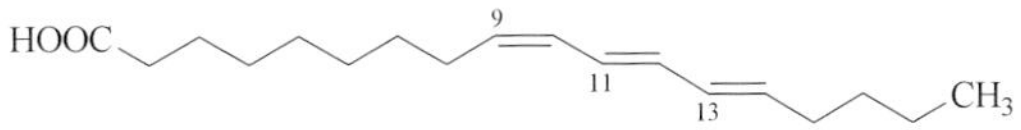

HOOC~~~~~9~~=~~11~~=~~13~~CH$_3$

α-Eleostearic acid

α-Licanic Acid, 9-*cis*,11-*trans*,13-*trans*, trans-4-ketooctadecatrienoic acid, M_r 292.40, mp + 75 °C, iodine value 260 %, acid number 202, is a constituent of oiticica oil.

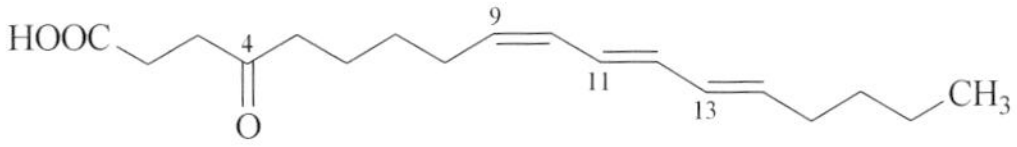

HOOC~~4~~O~~~9~~=~~11~~=~~13~~CH$_3$

α-Licanic acid

Conjugated Linoleic Acid, all-*trans*-9,11-octadecadienoic acid, M_r 280.45, iodine value 260 %, acid number 202, is prepared by isomerization of 9,11-linoleic acid (see Sections 8.5) or by dehydration of ricinolic acid form castor oil (see Section 4.10).

There are other high unsaturated fatty acids in Boleco oil, Isano oil, in the oil of Marigold and in the oils of Balsaminaceae species but until now they did not find any industrial use.

4. Sources and Types [6], [7]

The composition of important drying oils is shown in Table 1. Properties of commercial drying oils are listed in Table 2.

4.1. Linseed Oil [8], [9]

Linseed oil is a drying oil. It is the oil of the seed of flax (*Linum usitatissimum*) which has been cultivated by humans since the Stone Age. Flax is mainly grown to produce fibers (linen) and is cultivated in the Baltic states, Canada, USA, Argentina, India, and since the nineties also (again) in middle Europe. The seed consists of

Table 1. Composition of drying oils (figures are given in %)

Name of oil	14:0	16:0	18:0	20:0	22:0	24:0	16:1	18:1	18:2	18:3	20:1	20:4.5	22:1	22:5.6
Cottonseed oil	0 – 2	17 – 29	1 – 4	0 – 1			0 – 2	13 – 44	33 – 58	0 – 3				
Soybean oil	traces	7 – 11	2 – 6	0 – 3			traces	15 – 33	43 – 56	5 – 11				
Sunflower oil	traces	3 – 7	1 – 5	0 – 1	0 – 1			14 – 33	44 – 68	traces				
Safflower oil	traces	3 – 6	1 – 4			traces		13 – 21	73 – 79	traces				
Linseed oil		4 – 6	2 – 5	0 – 1				12 – 34	17 – 24	35 – 60				
Herring oil	3 – 8	8 – 13	1 – 3				6 – 9	17 – 22	1 – 4	0 – 1	9 – 15	6 – 9	1 – 16	6 – 18
Menhaden oil	7 – 8	17 – 19	3 – 4				7 – 10	13 – 16	0 – 1	0 – 1	1 – 2	13 – 19[a]	0 – 2	11 – 17
Sardine oil	6 – 8	16 – 19	2 – 4				8 – 12	10 – 16	1 – 3	0 – 2	2 – 8	13 – 29[a]	1 – 6	6 – 18
Tung oil		3 – 4	1 – 3					4 – 9	8 – 10	77 – 86[b]				
Oiticica oil		6 – 7	4 – 5					8 – 13	11 – 25	73 – 83[c]				

[a] and C 18:4.
[b] α-Eleostearic acid.
[c] α-Licanic acid.

32 – 43 % oil, the main fatty acids are linoleic acid and linolenic acid. Linseed oil is mainly used as a raw material for paints (varnishes, stand oils, alkyds, combinations with phenolic resins, maleinized oils, urethane oils, styrolized oils) and glaziers putty and to produce linoleum, normally not for foods. Films from linseed oil are stable to water and other agents but they are not sufficiently light-stable and have a tendency for yellowing.

4.2. Perilla Oil

Perilla oil is prepared from the seeds of *Perilla fructescens* (Labiatae species), grown in India and Indochina. The seed consists of 30 – 51 % oil. The composition of Perilla oil is similar to that of linseed oil. The oil is of local importance.

4.3. Tung Oil [10–16]

Tung oil is isolated from the nut-like seeds of the tung tree (*Aleurites fordii*, Euphorbiaceae species) in China and Japan. Tung trees are also cultivated in USA, Malawi, and Argentina. The seed consists of 25 – 45 % oil. 80 % of the oil consists of the triple unsaturated, conjugated α-eleostearic acid. Tung oil is easy to polymerize and is still being used for coatings. Tung oil is used for preparation of stand oils, for alkyds, and in combinations with phenolic resins. Tung oil is often mixed with linseed oil. Oxygen uptake during drying is lower than that of linseed oil. Films from tung oil are more elastic than those from linseed oil and more stable to yellowing. Products based on tung oil and its modifications are used for water-resistant paints (marine varnishes) and corrosion protection. A special use is the preparation of wrinkle finishes. These finishes dry very fast in the presence of driers and form regularly wrinkled surfaces (paints for machine facings).

4.4. Oiticica Oil [17]

Oiticica oil is isolated from the seeds of *Licania rigida* (Chrysobalanaceae species) in Brazil. The seed consists of 50 – 60 % oil, which consists mainly of licanic acid. Like α-eleostatic acid licanic acid contains three conjugated double

Table 2. Properties of drying oils (commercial, refined)

Name of oil	Color Gardner (max.)	Acid number (max.)	Iodine value (Wijs)	Saponification number	Unsaponifiables (max.), %	Viscosity at 20 °C, mPa · s	Titer[a], °C	Refractive index n_{20}^D	Density at 20 °C, g/cm^3	Specifications
Cottonseed oil [8001-29-4]	4	1	110 – 120	188 – 196	1.5	approx. 60	30 – 37		0.918	
Soybean oil [8001-22-7]	2	0.4	120 – 141	188 – 195	1.5	approx. 60	20 – 24	1.472 – 1.477	0.917 – 0.925	DIN 55938 ASTM D 1462 BS 5170
Sunflower oil [8001-21-6]	2	1	125 – 135	188 – 196	1.5	approx. 60	16 – 24	1.475	0.920	
Safflower oil [8001-23-8]	2	1	140 – 150	188 – 196	1.5	approx. 60	16 – 17	1.475 – 1.478	0.923 – 0.926	DIN 55942
Linseed oil [8001-26-1]	4	0.6	160 – 200	188 – 195	1.5	approx. 60	19 – 21	1.4785 – 1.4825	0.926 – 0.933	DIN 55933 DIN 55934 ISO 150 BS242 BS 243
Fish oil [8016-13-5]	6	4	190 – 205	188 – 194	1.5	approx. 50	25 – 31	1.482	0.930	
Tung oil [8001-20-5]	8	5	160 – 180	189 – 195	1.0	200 – 400	36 – 37	1.511 – 1.522	0.930 – 0.940	DIN 55936 ASTM D 12 ISO 277 BS 391
Oiticia oil [8016-35-1]	11	10	145 – 155	186 – 193	1.5		44 – 47	1.510	0.977 – 0.988	ASTM D 601

[a] Temperature range where the oil becomes turbid during cooling.

bonds and has a keto group in addition. That is why oiticica oil has nearly the same properties as tung oil. It is of local interest.

4.5. Fish Oils [18–21]

The various fish oils originate from different organs of fish and differ in their composition. Oils from herring, menhaden, and sardine are of industrial interest. All of these oils contain fatty acids with 18 to 22 carbon atoms and with up to fife isolated double bonds. The drying properties of fish oils are comparable with those of linseed oil. Fish oils have a specific odor, the same holds true for products from fish oil. Fish oils are used for varnishes and alkyds for anticorrosive coatings and primers.

4.6. Safflower Oil [22], [23]

Safflower oil is a semidrying oil. It is obtained from the seeds of Safflower (*Carthamus tinctorius*, Compositae species), which is grown in Egypt, Turkey, and North America. The pure seed consists of 46 – 50 % oil. The oil contains approx. 70 % linoleic acid. Safflower oil is for high-quality foods and in the coating industry mainly for nonyellowing alkyds.

4.7. Sunflower Oil [24], [25]

Sunflower oil is also a semidrying oil. It is isolated from seeds of sunflower (*Helianthus annuus*), which is grown in Eastern Europe, China, North America, Middle America, Australia, and Eastern Africa. The seeds consist of 42 – 63 % oil. The oil contains approx. 60 % linoleic acid. In the coating industry sunflower oil is applied in high quality alkyds (nonyellowing) and in artists' paints.

4.8. Soybean Oil [26], [27]

Soybean oil is a semidrying oil. Soybeans are the fruits of soy (*Glycine max*, Leguminosae species) which is mainly grown in China, Argentina, Brazil, and USA. The beans consist to 18 – 20 % of oil. The oil contains approx. 50 % linoleic acid. Soybean oil is mainly used for foods, the paint industry uses only a small part of the large quantities of soybean oil produced. Fatty acids from soybean oil are used for alkyds. Also stand oil are prepared from soybean oil. Soybean oil is much slower drying than linseed oil. But the films are more durable and stable to yellowing.

4.9. Cottonseed Oil

Cottonseed oil is also a semidrying oil. It is obtained from the seeds of the cotton plant (*Gossypium arboreum, G. hirsutum, G. barbadeuse*, and other species), which are located on the cotton fibers. Cotton is mainly cultivated in Egypt, China, Japan, North, and Middle America. The seed kernels consists of 14 – 25 % oil. The oil contains approx. 40 % linoleic acid. Cottonseed oil is used for foods and a raw material for fatty acids for alkyds (nonyellowing).

4.10. Dehydrated Castor Oil [28–30]

Castor oil consists of a mixture of fatty acids which contain up to 90 wt % ricinoleic acid (12-hydroxy-9-octadecanoeic acid). Castor oil is isolated from the bean-like seeds of the castor plant (*Ricinus communis*), which is cultivated in Brazil, India, the Philippines, South Africa, and Southeast Europe. The seed consists of 50 – 60 wt % oil. Castor oil is a nondrying oil. In the paint industry castor oil is used for polyurethane coatings, alkyds, and for the preparation of dehydrated castor oil and its fatty acids.

Dehydrated castor oil is prepared by dehydration of castor oil with sulfonic acids and removing the crystalline hydrates. Dehydrated castor oil consists mainly of conjugated 9,11-linoleic acid and some 9,12-linoleic acid. During dehydration polymerization may occur as side reaction. Dehydrated castor oils are available with different viscosities. Dehydrated castor oil is a drying oil. The film forming properties are similar to those of tung oil, but the drying velocity of dehydrated castor oil is lower. The films are stable against yellowing, are elastic and water resistant. Dehydrated castor oil is used for oil varnishes – often combined with other oils – for pigment pastes and for printing inks.

4.11. Tall Oil (→ Tall Oil) [31–37]

Tall oil is a byproduct of the kraft (sulfate) pulping process. Crude tall oil consists of a mixture of fatty acids (45 – 55 wt %), rosin (30 – 40 wt %), and unsaponifiable components (10 – 13 wt %). The fatty acids can be separated and used as components for alkyds. Tall oil has a similar fatty acids composition as soybean oil, but normally the content of oleic acid is higher. Furthermore, tail oil fatty acids contain unusual isomers, e.g., octadecadienoic acids with double bonds in the 5-,9-, and 5-,12-positions.

5. Production

The industrial production of oils from vegetable materials (oilseeds) and animal materials (selected fish organs) by pressing (cold and hot) and extraction is described in detail in → Fats and Fatty Oils.

Refining. [6], [38] (→ Fats and Fatty Oils, Chap. 6.). Some of the trace components contained in fatty oils are accepted for use of oils in foods. But when used in coatings the trace components may negatively affect the film forming properties. The impurities extracted together with fatty oils include gums, phosphatides, carbohydrates, proteins, colorants, and waxes.

The classic refining procedure consists of the following sequential, stepwise treatments of the oil: degumming, bleaching, deacidizing, and, if necessary, deodoration and dewaxing.

Degumming is performed by adding approximately 0.1 wt % of phosphoric acid or the equivalent amount of citric acid and subsequent agitation of the oil for approx. 30 min at 80 – 90 °C. The acids coagulate the gums and the coagulate is removed by filtration or by centrifugation. The process can be supported by a treatment with fuller's earth.

Bleaching removes the vegetable colors, such as chlorophyll and carotene, autoxidation products, soaps, traces of heavy metals, residues of phosphorus-containing materials, and odorants. In bleaching solid adsorbents, such as activated fuller's earth and in special cases activated carbon, are added. Bleaching is carried out in closed, agitated tanks under partial vacuum at 80 – 90 °C in the presence of 1 – 5 wt % fuller's earth for up to 30 min. Afterwards, the oil is filtered over filter presses.

Deacidizing Crude oils may contain 1 – 3 wt % of free fatty acids. This content can be lowered to below 0.1 wt % by deacidizing. In the older process equivalent amounts of caustic soda (sodium hydroxide) are added which convert the free fatty acids to sodium soaps at 90 °C. The process is carried out under slow stirring to avoid emulsifying. The sodium soaps are separated in form of an aqueous, concentrated soap solution (soap stock). The soap stock is separated and removed and the oil is dried under vacuum. The process is problematic regarding the residual soap contents in the oil and the oil losses by saponification.

In the newer process the fatty oils are deacidized by distillation (physical deacidizing). The oil, which is free from other impurities, which may form decomposition products during the process, is vacuum steam-distilled at 210 – 270 °C and the fatty acids are distilled off. The final acid content is 0.01 wt %. This process includes also a deodorization step. The process avoids contamination of the oil with soaps and the loss of neutral oil is lower than in the older process.

Deodorization is essentially a vacuum steam-distillation, by which odorous volatile compounds, mainly aldehydes and ketones, are driven from the nonvolatile oils. Other volatiles, such as free acids, sterols and tocopherol, are also removed. Deodorization is carried out with steam under reduced pressure at 1.0 – 2.5 kPa and 180 – 220 °C for 2 – 6 h.

Dewaxing Linseeds and sunflower seeds are covered with a thin layer of wax. Particles of this wax were dissolved during oil pressing and extraction. Wax dissolved in the oil crystallizes during storage, especially during cooling and the oil becomes turbid. The wax contents would result in film defects in the coatings. Dewaxing consists of cooling the oil to 4 – 10 °C and holding it at this temperature until the crystallization process is finished. Then the wax is removed by filtration.

6. Testing and Specifications [39]

The properties of drying oils are tested in specific tests. The methods and target values are defined in national (DIN, DGF, ASTM, BS) and international (ISO) standards and specifications. Oils are characterized by color, acid number, saponification value, amount of unsaponifiable contents, iodine number (amount and activity of double bounds), viscosity, refractive index, and specific density ($\rightarrow$ Fats and Fatty Oils, Section 11.3., see Table 2). The fatty acid composition is determined by gas and liquid chromatography [40].

7. Film Formation

7.1. Properties of Drying Oils for Film Formation

Besides their ability to dry by reaction with atmospheric oxygen, drying oils have other specific properties. Due to the long aliphatic chains of the fatty acids and their double bond content, the coatings based on oils and related products have a low viscosity and excellent wetting properties on different types of substrates (metals, wood) and pigments. That is why drying oils can form the base for "high-solids" paints, corrosion protection paints, pigment pastes, fillers, and artist's paints ($\rightarrow$ Artists' Colors). In addition, drying oils have an excellent solubility in a broad variety of solvents, including simple aliphatic hydrocarbons, which are low cost solvents, odorless, and physiologically safe. Also the oils themselves are no hazardous products. Their molecular structure imparts excellent flow and leveling properties during film formation. When cross-linking is finished the films still contain mainly aliphatic structures, which are hydrophobic and rather stable to water and some water-soluble agents. A disadvantage is their sensitivity to basic agents, which will saponify the ester groups. That is why oils and related products are not the optimum bases of waterborne coatings. But in principle the hydrophobicity of the fatty acids is an advantage also for waterborne coatings, if the drying oils are incorporated well and protected from saponification (e.g., in aqueous polyurethane dispersion based on alkyds with unsaturated fatty acids). Another disadvantage is that drying requires much time and cannot be accelerated sufficiently by higher temperatures and catalysis.

7.2. Cross-Linking by Reaction with Oxygen

The mechanisms of film formation of drying oils have been studied in investigations on oil molecules and appropriate model substances [41]. The process occurs in different overlapping steps and the total reaction sequence and the molecular structure of the film forming polymers are complex [42], [43]. Different reaction mechanisms are followed, depending on the type of double bonds and the use of catalysts.

Oils with Nonconjugated Double Bonds.
The initial autoxidation step in the case of oils with nonconjugated double bonds is the formation of hydroperoxides in allyl position to the double bonds [44–48]. Since the 11-position of linoleic and linolenic acids and – in the case of linolenic acid also the 14-position – are activated by two double bonds, the reactivity of these positions is the highest. First a radical is formed at these positions by dehydrogenation, then oxygen is added and a hydroperoxide is formed by recombination with the hydrogen radical. The reactivity of drying oils is based on the mesomeric stabilization of the radical intermediate: the unpaired electron is delocalized over several carbon atoms along the chain, and less energy is required to eliminate the hydrogen radical [49].

Additionally isomerization may take place, analytical investigation on esters of linoleic acid, for example, showed higher amounts of conjugated unsaturated 9- and 13-hydroperoxides besides hydroperoxides in 8-, 10-, 12- and 14-position [50–53].

Since the next steps of film forming lead to different higher molecular products, the analysis of these products is much more complex. Decomposition products of the film in different states, byproducts of the film-forming reaction, were investigated by mass spectroscopy [54–56]. Time-laps infrared spectroscopy was used to follow the process [57]. The analytical result led to the interpretation that the next important step is the formation of peroxy radicals by decomposition of the hydroperoxide (bimolecular disproportionation at higher concentration). Then a free-radical chain process is started by formation of oxygen bridges between fatty acid molecules and by activation of C-radicals and formation of C–C bridges. The reaction is terminated by recombination of different radicals.

The reaction is strongly influenced by temperature, light, and traces of substances which may act as catalysts or inhibitor. For example, at lower temperatures more C–O–C bonds are produced than at higher temperatures, where formation of C–C bridges is favored [58].

Oils with Conjugated Double Bonds. react differently from those with isolated double bonds. Tung oil for example absorbs less oxygen during drying, 12 wt % instead of 16 wt % in case of linseed oil [49], [59–61]. Analysis shows that the starting reaction is not the formation of hydroperoxides, but of cyclic peroxides formed by direct oxygen attack on the conjugated bond system [62]. Reaction of the peroxides with allylic methylene groups or dissociation leads to radicals. The radicals initiate a radical chain reaction, yielding C–O–C bonds and C–C bridges [63]. Due to the easy 1,4-addition of radicals to conjugated double bonds, formation of C–C bridges is preferred. The reaction is terminated by recombination of radicals or disproportionation.

7.3. Influence of Catalysts on Cross-Linking

All oxidative drying systems are formulated by adding small amounts of catalysts for drying, the so-called driers. Driers (siccatives) are oil-soluble metal salts of organic acids. The soaps with best solubility are obtained from 2-ethylhexanoic acid and naphthenic acids, in earlier times rosin acids and linoleic acid were in use [64], [65].

The soaps are produced from metal oxides or hydroxides and the acids, either by fusion or by precipitation [66]. The driers are usually dissolved in aliphatic or aromatic hydrocarbons, water-emulsifiable versions are available [67], [68].

The driers are classified in two groups, distinguished by the mode of action: active driers and auxiliary driers.

Active Driers. consist of metals occurring in several oxidation states and being capable to undergo redox reactions. The degree of activity may be also influenced by the molecular compatibility of the metal ion with the double bond system. Examples of metals in active driers, in order of decreasing activity, are:

Co > Fe > Mn > Ce > Pb > Zr

The metals are added to drying oils and related products (drying alkyds) at a concentration of 0.005 – 0.2 wt %. The driers speed up two reaction steps of the drying process, the activation of methylene bonds to form hydroperoxides and, more important, the redox reaction in which radicals are formed from the hydroperoxides [69–75]. If the hydroperoxides are transformed faster to radicals, the radical chain polymerization takes place more efficiently. The films are more homogeneous and become stable to dissolution in a shorter time. The metal ions are reoxidized during the process. A simplified illustration of the reaction of a drying oil with oxygen in the presence of cobalt driers is shown in Figure 1.

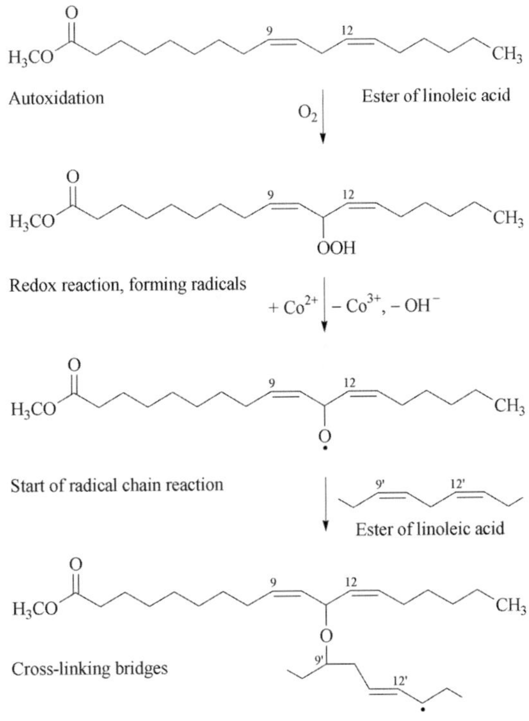

Figure 1. Simplified reaction of a drying oil with oxygen in the presence of a cobalt drier

Suitable combinations of metals allow synergistic improvement of the total drying effect. Thus, in the past most siccatives consisted of combinations of cobalt and lead. Cobalt promotes the formation of hydroperoxides rather fast on the film surface. This is connected with the risk that the surface is cross-linked more quickly than the other coating layers. Thus, oxygen cannot diffuse into the film and wrinkles can be formed. Lead promotes hydroperoxide decomposition and accelerates polymerization and ensures that the film is formed evenly over all coating layers. Also zirconium ions form complex compounds with hydroperoxides that decompose slowly [76], zirconium has the same effect as lead. Since lead is defined as a hazardous component, most of the new drier compositions consist of mixtures of cobalt and zirconium. But also cobalt is in discussion as possible health risk. Because no real alternative to cobalt exist, the newer formulation contain smaller amounts of cobalt driers.

Auxiliary Driers are soaps of metals which form only one oxidation degree but which activate the primary active driers. They do not catalyze autoxidation by themselves. Auxiliary driers are added in higher concentration than the primary driers, up to 0.6 wt % of metal. Examples of metals for auxiliary driers are calcium, barium, and zinc. The effect is explained by the basic reaction of these driers, which mainly compensate the effect of acid groups present the oil, and thus prevent precipitation of the active drier metals [77]. Auxiliary driers can also be absorbed on surfaces, e.g., from pigments instead of the active driers and the active driers are not involved in the process [78]. Barium is more toxic than lead and should no longer be used.

If the total content of metals in a film is rather high, they will be active indefinitely, causing oxidative breakdown of the film. Therefore, the amount of driers used must be carefully balanced to achieve a high degree of cross-linking but to avoid an early film decomposition.

During storage of wet paints reaction with oxygen must be avoided. The addition of oximes (methyl ethyl ketoxime) stabilizes against skin formation by reaction with oxygen in the container [79]. But now also methyl ethyl ketoxime is defined as a toxic substance and has to be substituted.

7.4. Yellowing

Films of highly unsaturated oils and related products with isolated double bonds tend to yellow under specific conditions. Films of linseed oil and related products yellow in the dark, but the yellowing will largely disappear if the film is exposed to light. Irreversible yellowing takes place if films from linseed oil and related products are aged or baked at higher temperatures. The yellowing is caused by formation of colored polyenes or quinone structures, starting from conjugated unsaturated hydroperoxides via ketones as intermediates [80–82]. The formed 1,4-diketones may also react with ammonia (farmhouses, stables, and rest rooms) to yield substituted pyrroles that can form colored products by oxidation [83], [84]. For white and other light colors linseed oil should be replaced by less yellowing products from semidrying oils, or better, alkyds based on these oils.

But also the driers (manganese, cobalt) can form complexes with oximes, which may cause discoloration in films, mainly by aging [85].

8. Modifications

The rather effective cross-linking of drying oils by reaction with atmospheric oxygen suffers from the disadvantages of a low viscosity of the unsaturated triglycerides, and the fact that the drying process has to proceed to a significant extent until a rather hard solvent- and chemicals-resistant coating film is formed. This process requires a lot of time. Early in history the painters tried to improve the drying properties of oils. One way was the combination of the chemical drying of oils with products which form films by physical drying (waxes, rosins, amber). When the compatibility or solubility constituted a problem, the components were heated to form stable mixtures. Furthermore, it was found that addition of particular pigments to the oils improved the drying velocity of the coating films. If oils were stored in glass bottles in the sun light or heated in the absence of oxygen the oil became more viscous. The higher viscosity of these oils results also in a reduction of drying time.

When rosins, copals, and amber became rare or were too expensive, various synthetic resins were developed as combination partners for oils. In the beginning mainly phenolic resins were used. Phenolic resins were later replaced by alkyds, which combine the chemical drying of double bonds of fatty acids with physical drying of a phthalic polyester.

8.1. Boiled Oils

Boiled oils are combinations of drying oils and stand oils with siccatives. To accomplish a homogeneous distribution of the driers in the oil, the oxides or salts of the driers are incorporated by heating the mixture to higher temperatures ($\geq$ 100 °C, boiling). The metal oxides or salts form salts with part of the fatty acids by saponification. The main bases for boiled oils is linseed oil. The driers may consist of 0.1 – 0.3 wt % cobalt or about 0.4 wt % lead, sometimes combined with 0.1 wt % of manganese.

Boiled oils are used for varnishes, mainly lithographic varnishes, and printing inks (pigment pastes).

Currently it is easier to produce varnishes in a cold process, i.e., by mixing oils and stand oils (or their mixtures with other resins) with solutions of metal driers.

8.2. Stand Oils

Stand oils are drying oils with increased viscosity. The viscosity increase results from heating of the oils in the absence of oxygen. During heating an intermolecular polymerization process occurs. The main reaction is a Diels – Alder reaction between unsaturated fatty acids, acting as partly as dienes and partly as dienophiles [86], [87]. The step that controls the reaction rate of oils with *nonconjugated double bonds* (linseed oil) is the isomerization into conjugated double bonds with concurrent trans rearrangement [88–90]. Therefore, oils with conjugated double bonds react very fast. Stand oils consist of esters of polymeric fatty acids, containing a significant amount of cyclic polymeric fatty acids [91]. The reaction may be catalyzed by the addition of small amounts of anthraquinone [92].

The process used for manufacturing of stand oils must be adjusted to the reactivity of the drying oils. Stand oils of linseed oil and soybean oil are produced in agitated stainless steel reactors at 280 – 300 °C in an inert gas stream (carbon dioxide, nitrogen). The reactor may be heated by heat-transfer oils, superheated steam, electrical induction, or resistance heating. The reactor should be equipped with water cooling facilities (half-pipes outside the reactor wall or internal cooling coils). The process is controlled by temperature and time, the progress of reaction is determined by measuring the viscosity and the acid number of samples. A linseed stand oil needs 40 – 45 h at 280 °C to attain viscosities of about 10 – 15 Pa · s. Viscosity increase of soybean oil is much slower. Decomposition products of the production of stand oils are removed by distillation, under vacuum if necessary.

Tung oil is heated in small batches (20 – 50 kg) rapidly to 260 °C, held for 7 – 12 min at that temperature (exothermic reaction takes place) and then flash-cooled with minimum the same amount of cold tung stand oil.

Stand oils are used in varnishes, oil paints, enamel varnishes, and fillers. Stand oils dry more homogeneously than their starting oils and give more elastic and water-resistant coating films. Although stand oils dry more slowly than drying alkyds, stand oils are used in combination with alkyds to raise the solids of paints to reduce the amount of solvent to evaporate. Linseed stand oil has a lower tendency for yellowing than linseed oil. Tung stand oil and soybean stand oil are even more stable to yellowing than linseed oil. Tung stand oil is superior to linseed stand oil in forming hard and water-resistant films, but the films are more brittle. Thus, linseed stand oil is often combined with tung stand oil in varnishes.

8.3. Blown Oils

For production of blown oils, drying oils are blown with air at 90 – 120 °C. During this process, hydroperoxides are formed and decompose into radicals which lead to increase in molecular mass by a radical chain transfer reaction. In addition, keto groups and hydroxy groups are formed. Blown oils have viscosities up to 3 Pa · s and are much more polar than stand oils. The drying properties are reduced.

Blown oils from linseed oil and soybean oil are used as plasticizers for polymers which dry physically, for printing inks, for putties, and sealings. Blown linseed oil is used for core binders for casting.

8.4. Factice, Sulfurized Oils

Factice is a mixture of drying oils with sulfur or disulfur dichloride. Like oxygen, sulfur is able to react with the double bond system in a radical chain reaction. The polymerization occurs very effectively over the whole film layer. Varnishes containing factice oils can be applied in a wet-on-wet process.

8.5. Isomerized Oils

During isomerization isolated double bonds are transferred into conjugated double bonds and cis double bonds are converted into trans double bonds. The process is carried out with metal catalysts at 100 – 140 °C. Linseed oil and safflower oil are isomerized in the presence of a nickel catalyst, finely distributed on diatomaceous earth or active carbon [93–95]. Isomerized oils take up less oxygen during drying than the corresponding starting oils and form elastic films with higher water resistance.

The alkali isomerization process is carried out with the methyl ester of fatty acids to afford fatty acids with conjugated double bonds. They are used in alkyds, mainly for stoving enamels [96–98].

8.6. Copolymers

Copolymers of drying oils are produced by heating unsaturated oils with unsaturated monomers [99], [100], preferably styrene. Additionally, vinyltoluene [101], α-methylstyrene, dicyclopentadiene [102], pentadiene, and indene [103] are used as monomers, also methylmethacrylate is described. Copolymerization is carried out at higher temperatures, catalysts are peroxides, e.g., di-*tert* butyl peroxide [104]. The process may be run in bulk or in solution. A feed process for monomer dosing is preferred for most exothermic reactions to guarantee a reproducible process. Unreacted monomers are distilled off after the process. The content of styrene in styrenated oils is 20 to 45 wt %. During the polymerization of nonconjugated oils and styrene, grafting of styrene molecules to chains onto oil molecules occurs. Starting from the active methylene groups of the oil, polysterene bridges are formed [105], [106]. It is important to avoid the formation of styrene homopolymer, because it is not compatible with the oil. *Conjugated oils* form Diels – Alder adducts connected by short bridges of polystyrene, the viscosity increases more rapidly than the viscosity resulting from the grafting nonconjugated oils [107], [108]. The compatibility of styrenated oils with other polymers is limited. Using vinyltoluene, α-methylstyrene, or dicyclopentadiene gives copolymers with a much better compatibility and solubility. Cyclopentadiene and indene form only Diels – Alder adducts both with conjugated and nonconjugated oils.

Copolymers of oils have improved drying properties, the resulting films are harder and more resistant to water and chemical agents than the films from drying oils and alkyds.

8.7. Maleinized Oils [109–112]

Maleic anhydride is a dienophile and easily reacts with conjugated double bonds in a Diels – Alder reaction to form a substituted cyclohexane dicarboxylic anhydride. The process is carried out at 80 – 120 °C [113], [114]. Nonconjugated unsaturated oils react with maleic anhydride in a ene reaction via the active methylene groups to form substituted succinic anhydride derivatives [115], [116]. But also Diels – Alder reactions may take place if the isolated double bonds are converted into conjugated during the process. The reaction is carried out at 180 – 240 °C.

Maleinized oils with 5 – 8 wt % of maleic anhydride (less than 1 mol per triglyceride) are esterified with glycerol and other polyols to form oil-modified polyesters. Their properties are similar to those of alkyds, and oil-modified polyesters dry faster than the original oils and form harder and more water-resistant films.

Maleinized oils with 16 – 24 wt % maleic anhydride are reacted with water or alcohols to yield maleinized oils with high acid numbers (100 – 200). After neutralization of the acid groups with ammonia or – more preferably – amines, the maleinized oil becomes water-thinnable [117]. The product is combined with phenolic or amino resins [118–120]. These water-thinnable combinations are mainly used as anticorrosive primers for electrodeposition. In electrodeposition the part to be coated is immersed in a tank and direct current is applied. In the process the binder systems (together with pigments and additives) are coagulated on the anode, which is the surface of the substrate to be coated [121–123]. The coagulated film can be rinsed with water and is then stoved at 175 – 180 °C for 20 – 30 min. The process ensures that film formation takes place on all parts which are immersed in the tank. Starting in the 1960s, such systems gained importance as corrosion protective primers in the automotive industry.

In the late 1970s the systems were mainly substituted by a resin system for a cathodic deposition process. But the maleinic oils are still in use for industrial primers.

8.8. Oil Boiling with Rosin, Copal, and Amber

In the past oils – and also drying oils – were used as solvent for different natural resins. The target was to convert the natural resins into a suitable form for application and to combine the film properties of the natural resins with the drying properties of the oil. The higher the melting point of the natural resins, the better the film properties. Copals were better than rosins, and amber was better than copals. To improve the properties of rosins, rosins were transferred into rosin salts with calcium and zinc, into rosin esters with glycerol and pentaerythrol, and maleinized and esterified (→ Resins, Natural, Section 5.5.). It was not easy to melt the high molecular natural resins (copals and amber) and starting in England special know-how was gathered for this process.

The combinations of oils, different natural resins and related products and siccatives formed the basis for the oil varnishes, the first industrially used coatings.

8.9. Oil Boiling with Phenolic Resins

When the natural resins became rare the investigation to prepare synthetic resins began. The first synthetic resins used for industrial coatings were the phenolic resins (→ Phenolic Resins). But the normal phenolic resins (resols) are not compatible with oil. The first step to overcome the problem was the combination of rosin-modified phenolic resins with drying oils [124]. The second step was the preparation of so-called alkylphenolic resins (from alkyl-substituted phenols) [125], [126]. These alkylphenolic resins are able to react with oils, preferable tung oil, to yield adducts between the double bond system and the methylol and OH group of the phenolic resin [127]. The process is carried out at 240 °C, using a mixture of 2 – 4 parts tung oil and 1 part alkylphenolic resin. The reaction products have good drying properties and the films are highly resistant to water and chemicals. The product were used for industrial coatings and especially for yacht varnishes. For this application they were much more suited than alkyd varnishes. These combinations are often replaced by polyurethane coatings, but yacht varnishes are still available.

8.10. Modifications by Transesterification

If drying oils are heated with an excess of polyols, glycerol, pentaerythrol, or dipentaerythrol at 230 to 260 °C, mono- and diglycerides [128–131] are obtained. The alcoholysis (transesterification) is an equilibrium process. Salts of alkali or alkaline-earth metals serve as catalysts. To avoid the formation of incompatibilities of the catalyst in the reaction mixture or the final product, the use of lithium salts is preferred. The reaction is controlled by the added quantity of polyols, the reactivity of the OH groups, the temperature, and the time. The progress of the reaction is determined by measuring the solubility of the reaction mixture in methanol or ethanol. Transesterification products are starting materials for production of alkyds, urethane oils, and urethane alkyds.

Alkyds. (→ Alkyd Resins) are prepared by reaction of the transesterification products and phthalic anhydride. The process was developed in 1927 and was the first important process for production of air-drying alkyds.

Urethane Oils [132], [133] are prepared from the transesterification products and toluylene diisocyanate. Products obtained by reaction of the transesterification products with mixtures of phthalic anhydride (first esterified) and toluylene diisocyanate are called urethane alkyds. Urethane oils and urethane alkyds dry faster than alkyds to give harder and more resistant films, but the films are not weather-resistant. Urethane oils and urethane alkyds are used for wood coatings, preferably for parquet varnishes.

8.11. Modifications Based on Fatty Acids

Fatty acids are produced from drying oils by saponification. The fatty acids are the basis for the production of alkyds (→ Alkyd Resins) and epoxy esters. In contrast to oil varnishes and paints, alkyds and epoxy esters based on unsaturated fatty acids are still important coating materials.

9. Application, Importance, and Future Aspects

Oil varnishes and oil paints have been largely replaced by more modern systems, mainly alkyds, but there are still some oil paint products on the market. For people, who want to use natural raw materials, oil-based paints are still of interest for furniture coatings (mainly for "do-it-yourself"-applications), but there is nearly no industrial use. Products derived from drying oils, i.e., air-dryings alkyds and epoxy esters, and maleinized oils, still play a more important role in the paint industry. In Germany the market volume of solventborne coating materials is estimated to be approximately 300 000 t in 1999. Oil paints and oil varnishes have a share of only 0.6 %, but this is approximately 12 % more than in 1998. Coatings based on air-drying alkyds made up nearly 20 % of this volume, but the amount is decreasing in comparison to 1998 [134].

Advantages of oil-based or fatty acid-based coating materials are: excellent wetting of pigments and surfaces, easy application, good flow and leveling, low health risks, environmental compatibility, and optimum hydrophobicity of films. However, they suffer from two important disadvantages: (1) Oil- or fatty acid-based coating materials need too much time for film formation by air drying to be used in advanced industrial coatings application processes (e.g., automotive coating application) and (2) these coating materials are not easy to transfer into water-based systems – to avoid organic solvents for ecological and economical reasons – due to the sensitivity to react by saponification. On the other hand, since the 1980s there has been growing interest to increase the use of natural renewable raw materials. Drying oils and related products meet this demand. Several institutes started to investigate the possibility to decrease the agricultural food production in Europe and to replace growing of plants for food partly by growing of plants which produce raw materials for industrial use [135–140]. There are new proposals to grow plants which contain drying or semidrying oils or plants which contain fatty acids plus other chemically reactive groups.

References

General References

1 Albert Ilg (ed.): *Presbyter, Theophilus: Rodgerus von Helmarshausen: Schedula diversarium artium*, Verlag Wilh. Braumüller, Wien 1874.
2 Albert Ilg (ed.): *Heraklius: De coloribus et artibus Romanorum*, Verlag Wilh. Braumüller, Wien 1873.
3 K. Herberts: *Die Maltechniken*, Econ-Verlag, Düsseldorf 1957.
4 L. Stryer: *Biochemie*, Spectrum Verlag, Heidelberg 1990.
5 Unichema: Fatty Acid Data Book. 3rd. edition 1992.
6 E. Gulinsky: *Pflanzliche und tierische Fette u. Öle*, C. R. Vincentz Verlag, Hannover 1963.
7 H. Kittel: *Lehrbuch der Lacke und Beschichtungen*. Verlag W. A. Colomb, Stuttgart-Berlin 1971.

Specific References

8 W. Schuster, R. Marquard, *Fette, Seifen, Anstrichm.* **76** (1974) 207.
9 J. D. v. Mikusch, *Farbe + Lack* **58** (1952) 303 – 306.
10 J. D. v. Mikusch, *Farben, Lacke, Anstrichst.* **4** (1950) 149.
11 J. Baltes, *Fette Seifen* **52** (1950) 462.
12 J. Baltes, F. Weghorst, O. Wechmann, *Fette, Seifen, Anstrichm.* **63** (1961) 413.
13 A. E. Rheineck, D. D. Zimmerman, *Farbe + Lack* **70** (1964) 641.

14 C. I. Atherton, A. F. Kertess, *Fette, Seifen, Anstrichm.* **68** (1966) 279.
15 E. Fonrobert: *Das Holzöl*, Berliner Union-Verlag, Stuttgart 1951.
16 T. P. Hilditch, A. Mendelewitz, *J. Sci. Food Agric.* **2** (1951) 548.
17 M. Hassel, W. Lawrence, *Paint Ind.* **74** (1959) no. 7, 10, and no. 8, 15.
18 J. D. Hetchler, *J. Am. Oil Chem. Soc.* **31** (1954) 503.
19 R. J. De Sesa, *Farbe + Lack* **69** (1963) 752.
20 H. W. Chatfield, *Paint Manuf.* **30** (1960) 45.
21 H. A. Bhatt, P. V. Tagdiwala, *Paintindia* **29** (1970) no. 4, 19; no. 7, 19.
22 G. B. Kromer, *Oleagineux* **18** (1963) 21.
23 A. E. Rheineck, L. O. Cummings, *J. Am. Oil Chem. Soc.* **43** (1966) 409.
24 W. Schuster, *Fette, Seifen, Anstrichm.* **74** (1972) 150.
25 A. Müller, *Farbe + Lack* **57** (1951) 240.
26 F. G. Sietz, *Fette, Seifen, Anstrichm.* **67** (1965) 411.
27 N. Minoru et al., *Farbe + Lack* **71** (1965) 839.
28 G. W. Priest, J. D. v. Mikusch, *Ind. Eng. Chem.* **32** (1940) 1314.
29 N. A. Ghanem, Z. H. Abd El-Latif, *J. Paint Technol.* **39** (1967) 144.
30 C. I. Atherton, A. F. Kertess, *J. Oil Colour Chem. Assoc.* **49** (1966) 340.
31 W. Sandermann: *Naturharze, Terpentinöl, Tallöl*, Springer Verlag, Berlin 1960.
32 W. Asche, *Farbe + Lack* **68** (1962) 448 – 453, 518 – 525.
33 H. W. Chatfield, *Paint Oil Colour J.* **141** (1962) 1262.
34 G. Eick, *Farbe + Lack* **69** (1963) 616.
35 W. M. Kraft et al., *J. Am. Oil Chem. Soc.* **42** (1965) 96.
36 K. B. Gilkes, T. Hunt, *J. Oil Colour Chem. Assoc.* **51** (1968) 389.
37 J. T. Geoghegan, H. G. Arlt, C. O. Myatt, *J. Paint Technol.* **40** (1968) 209.
38 R. V. V. Nicholls, W. H. Hoffman, *Off. Dig. Fed. Paint Varn. Prod. Clubs* **24** (1952) no. 4, 245 – 260.
39 H. P. Kaufmann: *Analyse der Fette und Fettprodukte*, Springer Verlag, Berlin 1958.
40 DIN 53 782 and DIN 55 957.
41 G. H. Hutchinson, *J. Oil Colour Chem. Assoc.* **56** (1973) 44.
42 P. O. Powers, *Ind. Eng. Chem.* **41** (1949) 304.
43 J. R. Chipault, E. C. Nickell, W. O. Lundberg, *Off. Dig. Fed. Paint Varn. Prod. Clubs* **23** (1951) no. 11, 740.
44 L. Dulog, *Congr. FATIPEC 8th*, (1966), 17 – 25.
45 H. R. Rawls, P. J. Van Santen, *J. Am. Oil Chem. Soc.* **47** (1970) 121.
46 R. Criegee, *Justus Liebigs Ann. Chem.* **522** (1936) 84.
47 A. Rieche, *Angew. Chem.* **50** (1937) 520.
48 E. H. Farmer, *Trans. Faraday. Soc.*, **38** (1942) 340.
49 H. P. Kaufmann, *Fette, Seifen, Anstrichm.* **59** (1957) 153.
50 F. Hasbeck, W. Grosch, *Lipids* **18** (1983) 706.
51 E. N. Frankel, C. D. Evans, D. G. McConnell, E. Selke, et al., *J. Org. Chem.* **26** (1961), 4663.
52 F. Haslbeck, W. Grosch, J. Firl, *Biochim. Biophys. Acta* **750** (1983) 185.
53 F. Haslbeck, W. Grosch, *Fette, Seifen, Anstrichm.* **86** (1984) 408.
54 A. M. Gaddis, R. Ellis, G. T. Currie, *J. Am. Oil Chem. Soc.* **38** (1961) 371.
55 J. H. Skellon, *J. Oil Colour Chem. Assoc.* **46** (1963) 1001.
56 R. Clement, *Congr. FATIPEC 14th*, 1978, 177 – 182.
57 J. H. Hartshorn, *J. Coat. Technol.* **54** (1982) no. 687, 53 – 61.
58 P. S. Hess, G. A. O'Hare, *Ind. Eng. Chem.* **42** (1950) 1424.
59 M. Dyck, *Farbe + Lack* **67** (1961) 442.
60 H. Wexler, *Chem. Rev.* **64** (1964) 591.
61 W. Treibs, *Fette Seifen* **54** (1952) 3.
62 R. N. Faulkner, *J. Appl. Chem.* **8** (1958), 448.
63 L. A. O'Neill, *Chem. Ind. (London)* 1954, 384.
64 W. J. Stewart, *Off. Dig. Fed. Paint Varn. Prod. Clubs* **26** (1954) no. 6, 413 – 425.
65 DIN 55 901.
66 J. Skalsky, *Prog. Org. Coat.* **4** (1976) 137.
67 R. Hurley, F. Buono, *J. Coat. Technol.* **54** (1982) no. 694, 55 – 61.
68 R. G. Middlemiss, *J. Water Borne Coat.* **8** (1985) no. 11, 3 – 9.
69 C. E. H. Bawn, *Discuss. Faraday Soc.* **14** (1953) 181.
70 E. R. Mueller, *Ind. Eng. Chem.* **46** (1954) 562.
71 N. Uri, *Chem. Ind. (London)* 1956, 515 – 517.
72 C. E. H. Bawn, J. A. Sharp, *J. Chem. Soc.* 1957, 1854.
73 F. W. Heaton, N. Uri, *J. Lipid Res.* **2** (1961) 152.
74 M. Allard, *Double Liaison* **18** (1971) no. 194, 503 – 523.
75 N. S. Baer, N. Indictor, *J. Coat. Technol.* **48** (1976) no. 623, 58 – 62.
76 M. Giesen, *Dtsch. Farben Z.* **20** (1966) 236.
77 E. Krejcar, K. Hájek, O. Kolář, *Farbe + Lack* **74** (1968) 115.
78 G. Rieck: *Congr. FATIPEC 10th*, 1970, 239.
79 M. Giesen, *Congr. FATIPEC 7th*, 1964, 349.
80 F. Franks, B. Roberts, *J. Appl. Chem.* **13** (1963) 302.
81 A. C. Elm, *Off. Dig. Fed. Paint Varn. Prod. Clubs* **29** (1957) no. 4, 351 – 385.
82 O. S. Privett, M. L. Blank, J. B. Covell, W. O. Lundberg, *J. Am. Oil Chem. Assoc.* **38** (1961) 22.
83 L. A. O'Neill, S. M. Rybicka, T. Robey, *Chem. Ind. (London)* 1962, 1796 – 1797.
84 L. A. O'Neill, *Paint Technol.* **27** (1963) 44.
85 G. Link, *Farbe + Lack* **92** (1986) 279.
86 C. Boelhouwer, J. Th. Knegtel, M. Tels, *Fette, Seifen, Anstrichm.* **69** (1967) 432.
87 D. H. Wheeler, *Off. Dig. Fed. Paint Varn. Prod. Clubs* **23** (1951) no. 11, 661 – 668.
88 R. F. Paschke, D. H. Wheeler, *J. Am. Oil Chem. Soc.* **26** (1949) 278.
89 R. F. Paschke, D. H. Wheeler, *J. Am. Oil Chem. Soc.* **31** (1954) 208.

90. A. L. Clingman, D. E. A. Rivett, D. A. Sutton, *J. Chem. Soc.* 1954, 1088 – 1090.
91. J. A. MacDonald, *J. Am. Oil Chem. Soc.* **33** (1956) 394.
92. G. H. Hutchinson, *J. Oil. Colour Chem. Assoc.* **41** (1958) 474.
93. S. B. Radlove, H. M. Teeter, W. H. Bond, J. C. Cowan, J. P. Kass, *Ind. Eng. Chem.* **38** (1946) 997.
94. J. D. von Mikusch, *Farbe + Lack* **57** (1951) 341 – 345.
95. J. D. von Mikusch, *Farbe + Lack* **57** (1951) 393 – 399.
96. J. D. von Mikusch, *Farben, Lacke, Anstrichst.* **4** (1950) 149.
97. P. L. Nichols, S. F. Herb, R. W. Riemenschneider, *J. Am. Chem. Soc.* **73** (1951) 247.
98. J. Baltes, F. Weghorst, O. Wechmann, *Fette, Seifen, Anstrichm.* **63** (1961) 413.
99. BE 13 378, 1900 (A. Kronstein).
100. J. Scheiber, *Farbe + Lack* **63** (1957) 443.
101. S. R. Vranish, R. M. Jackman, R. F. Helmreich, W. A. Henson, *Off. Dig. Fed. Paint Varn. Prod. Clubs* **29** (1957) no. 1, 56 – 74.
102. H. P. Kaufmann, H. Gruber, *Fette, Seifen, Anstrichm.* **62** (1960) 607.
103. H. P. Kaufmann, H. Brüning, *Fette, Seifen, Anstrichm.* **62** (1960) 1146.
104. F. R. Mayo, C. W. Gould, *J. Am. Oil Chem. Soc.* **41** (1964) 25.
105. K. Hamann, O. Mauz, *Fette, Seifen, Anstrichm.* **58** (1956) 528.
106. J. B. Crofts, *J. Appl. Chem.* **5** (1955) 88.
107. E. F. Redknap, *J. Oil Colour Chem. Assoc.* **43** (1960) 260.
108. M. Dyck, *Farbe + Lack* **67** (1961) 148.
109. US 2 033 131, 1931 (Ellis-Foster).
110. US 2 033 132, 1931 (Ellis-Foster).
111. US 2 188 882, 1934 (E. T. Clocker).
112. Springer & Möller, DE 635 926, 1935.
113. W. G. Bickford, E. F. DuPre, C. H. Mack, R. T. O'Connor, *J. Am. Oil Chem. Soc.* **30** (1953) 376.
114. J. Sauer, R. Sustmann, *Angew. Chem.* **92** (1980) 779.
115. C. P. A. Kappelmeier, J. H. van der Neut, *Kunststoffe* **40** (1950) 81.
116. A. E. Rheineck, T. H. Khoe, *Fette, Seifen, Anstrichm.* **71** (1969) 644.
117. A. E. Rheineck, R. A. Heskin, *J. Paint Technol.* **40** (1968) 450.
118. Glasurit Werke, DE-AS 15 19 167, 1965 (G. Tröger, H. Noak).
119. Am. Cyanamid, DE-OS 16 69 593, 1967 (J. N. Koral).
120. PPG, DE-OS 1938 191, 1969 (E. J. Kapalko, W. H. English).
121. Ford Motor Co., DE-AS 15 46 944, 1962 (A. B. Gilchrist).
122. ICI, DE-OS 15 46 995, 1963 (N. McPherson, J. P. Burden).
123. Herberts, DE-AS 2 737 174, 1977 (H.-P. Patzschke, D. Saatweber).
124. G. Dantlo, *Peintures, Pigments, Vernis* **23** (1947) 171.
125. Koller & Co., EP 334 527, 1929; FP 676 456, 1929 (H. Hönel, Beck).
126. V. H. Turkington, I. Allen, *Ind. Eng. Chem.* **33** (1941) 966.
127. H. Hultzsch, *Kunstst.* **37** (1947) 43.
128. K. Hajek, J. Stanek, J. Hires, *Plaste Kautsch.* **15** (1968) 679.
129. R. Schöllner, L. Läbisch, *Fette Seifen Anstrichm.* **69** (1967) 426.
130. N. A. Ghanem, F. F. Abd. El-Mohsen, *J. Oil Colour Chem. Assoc.* **50** (1967) 441.
131. F. Mort, *J. Oil Colour Chem. Assoc.* **47** (1964) 919.
132. A. C. Jolly, *J. Oil Colour Chem. Assoc.* **47** (1964) 919.
133. K. Hajek, J. Kitzler, J. Stanek, E. Krejcar, *Farbe + Lack* **77** (1971) 422.
134. Statistics from Verband der Lackindustrie e.V., Frankfurt, Germany 1999.
135. J. T. P. Derksen, B. G. Muuse, F. P. Cuperus in D. J. Murphy (ed.): *Designer Oil Crops, Breeding, Processing and Biotechnology*, VCH, Weinheim, Germany 1993.
136. J. T. P. Derksen, F. P. Cuperus, *Industrial Crops and Products* **3** (1995) 57.
137. B. G. Muuse, F. P. Cuperus, J. T. P. Derksen, *Industrial Crops and Products* **3** (1992) 57.
138. H. G. Hauthal, "Bunsentage 1992, Nachwachsende Rohstoffe," *Nachr. Chem. Techn. Lab.* **40** (1992) no. 9, 996.
139. E. M. S. Hammersveld, F. P. Cuperus, J. T. P. Derksen: Modified and Novel Veg. Oils in a New Gen. of Emulsion Paints, Agrotechnical Research Institute, Wageningen, Netherlands 1994.
140. Project: VOICI (Vegetable Oils for Innovation in Chem. Ind.) in ECLAIR (Europ. Collaborative Linkage of Agriculture and Industry through Research), Brüsselsss 1993.

Further Reading

Z. W. Wicks Jr.: *Drying Oils*, "Kirk Othmer Encyclopedia of Chemical Technology", 5th edition, vol. 9, p. 142–155, John Wiley & Sons, Hoboken, NJ, 2005, online DOI: 10.1002/0471238961.0418250923090311.a01.pub2

Fats and Fatty Oils

ALFRED THOMAS, Unimills International, Hamburg, Federal Republic of Germany

ULLMANN'S ENCYCLOPEDIA OF INDUSTRIAL CHEMISTRY

1.	Introduction	292	11.3.	Oils and Fats	330
2.	Composition	293	11.3.1.	Physical Methods	330
2.1.	Glycerides	293	11.3.2.	Chemical Methods	332
2.2.	Fatty Acids	293	12.	Storage and Transportation	336
2.3.	Phospholipids	297	13.	Individual Vegetable Oils and Fats	337
2.4.	Waxes	298	13.1.	Fruit Pulp Fats	337
2.5.	Sterols and Sterol Esters	298	13.1.1.	Palm Oil	337
2.6.	Terpenoids	299	13.1.2.	Olive Oil	338
2.7.	Other Minor Constituents	300	13.1.3.	Avocado Oil	339
3.	Physical Properties	300	13.2.	Seed-Kernel Fats	339
3.1.	Melting and Freezing Points	300	13.2.1.	Lauric Acid Oils	339
3.2.	Thermal Properties	301	13.2.1.1.	Coconut Oil	339
3.3.	Density	302	13.2.1.2.	Palm Kernel Oil	340
3.4.	Viscosity	303	13.2.1.3.	Babassu Oil and other Palm Seed Oils	341
3.5.	Solubility and Miscibility	304	13.2.1.4.	Other Sources of Lauric Acid Oils	341
3.6.	Surface and Interfacial Tension	304	13.2.2.	Palmitic – Stearic Acid Oils	342
3.7.	Electrical Properties	304	13.2.2.1.	Cocoa Butter	342
3.8.	Optical Properties	305	13.2.2.2.	Shea Butter, Borneo Tallow, and Related Fats (Vegetable Butters)	342
4.	Chemical Properties	306	13.2.3.	Palmitic Acid Oils	343
4.1.	Hydrolysis	306	13.2.3.1.	Cottonseed Oil	343
4.2.	Interesterification	306	13.2.3.2.	Kapok and Related Oils	344
4.3.	Hydrogenation	307	13.2.3.3.	Pumpkin Seed Oil	344
4.4.	Isomerization	307	13.2.3.4.	Corn (Maize) Oil	345
4.5.	Polymerization	308	13.2.3.5.	Cereal Oils	345
4.6.	Autoxidation	308	13.2.4.	Oleic – Linoleic Acid Oils	346
5.	Manufacture and Processing	309	13.2.4.1.	Sunflower Oil	346
5.1.	Vegetable Oils and Fats	309	13.2.4.2.	Sesame Oil	346
5.1.1.	Storage and Handling of Raw Materials	309	13.2.4.3.	Linseed Oil	347
5.1.2.	Cleaning and Dehulling	309	13.2.4.4.	Perilla Oil	348
5.1.3.	Expelling	310	13.2.4.5.	Hempseed Oil	348
5.1.4.	Extraction	311	13.2.4.6.	Teaseed Oil	348
5.2.	Land-Animal Fats	315	13.2.4.7.	Safflower and Niger Seed Oils	348
5.3.	Marine Oils	316	13.2.4.8.	Grape-Seed Oil	348
5.4.	Synthetic Fats	317	13.2.4.9.	Poppyseed Oil	349
6.	Refining	317	13.2.5.	Leguminous Oils	349
6.1.	Degumming	317	13.2.5.1.	Soybean Oil	349
6.2.	Deacidification (Neutralization)	318	13.2.5.2.	Peanut Oil	350
6.3.	Bleaching	320	13.2.5.3.	Lupine Oil	350
6.4.	Deodorization	322	13.2.6.	Cruciferous Oils	351
7.	Fractionation	324	13.2.6.1.	Rapeseed Oil	351
8.	Hydrogenation	325	13.2.6.2.	Mustard Seed Oil	352
9.	Interesterification	327	13.2.7.	Conjugated Acid Oils	352
10.	Environmental Aspects	329	13.2.7.1.	Tung Oil and Related Oils	352
11.	Standards and Quality Control	329	13.2.7.2.	Oiticica Oil and Related Oils	352
11.1.	Sampling	330	13.2.8.	Substituted Fatty Acid Oils	352
11.2.	Raw Materials	330			

Ullmann's Renewable Resources
© 2013 Wiley-VCH Verlag GmbH & Co. KGaA. Published 2013 by Wiley-VCH Verlag GmbH & Co. KGaA.
ISBN: 978-3-527-33369-1 / DOI: 10.1002/14356007.a10_173

13.2.8.1.	Castor Oil	352	14.1.4.	Horse, Goose, and Chicken Fat 355
13.2.8.2.	Chaulmoogra, Hydnocarpus, and Gorli Oils	353	**14.2.**	**Marine Oils** 356
13.2.8.3.	Vernonia Oil	353	14.2.1.	Whale Oil 356
14.	**Individual Animal Fats**	353	14.2.2.	Fish Oil 356
14.1.	**Land-Animal Fats**	353	**15.**	**Economic Aspects** 357
14.1.1.	Lard	353	**16.**	**Toxicology and Occupational Health** .. 359
14.1.2.	Beef Tallow	354		**References** 359
14.1.3.	Mutton Tallow	355		

1. Introduction

Naturally occurring oils and fats are liquid or solid mixtures consisting primarily of glycerides. Depending on whether they are solid or liquid at ambient temperature, they are referred to as *fats* or *oils*, respectively. Naturally occurring oils and fats always contain minor constituents such as free fatty acids, phospholipids, sterols, hydrocarbons, pigments, waxes, and vitamins. The nomenclature rules for glycerides have been summarized [18].

History. A century ago ecological, religious, and social factors still played a more important role than technology in the choice and utilization of oils and fats.

Primeval humans utilized animal fats, making cheese and butter from goat's milk. Oilseed plants were cultivated during the neolithic period. Poppy seeds have been found in remains of Bronze-Age bread; rapeseed and linseed, together with millstones, have been found in Bronze-Age dwellings.

Linseed, almonds, and sesame seed were part of Egypt's natural flora. Sesame oil had mythical significance. The oil-bearing safflower plant is still grown in Egypt. Olive oil came from Palestine, Syria, and Crete. The Phoenicians and Greek colonists introduced the olive tree to Sicily and Italy. Cotton is one of the oldest cultivated plants; it was grown 2600 years ago in India as a source of both oil and fiber. Soy and hempseed are mentioned as oilseed plants in a Chinese document of 2838 B.C.

The oilseeds were ground with a pestle and mortar or between stones. Simple mills of the type still being used in some developing countries – a concave stone rotating on a convex one – also evolved. The Egyptians developed the sack or expeller press. The Greeks and Romans used a grinding device known as the "trapetum". In North Africa, mechanical presses were used in processing plants that approached the size of modern factories.

The processing of oil fruits and seeds in Central and Northern Europe advanced more slowly. Oilseeds, primarily linseed, hempseed, and rapeseed, were pulped in hollowed stones, and the oil was expelled from the pulp by pressing between two cloth-covered frames. This domestic-type process was practiced up to the 16th century. Industrial oil milling developed primarily in regions where linseed was grown extensively. A wood engraving dated 1568 depicts an oil mill with a horse-drawn vertical millstone on a stone bed. The ground oilseed was heated in a kettle over an open fire and finally "beaten" in wedge presses.

Animal fats were obtained by rendering fatty tissue and by churning cream. Up to the middle of the 19th century, tallow and butter were the most important edible fats in Europe, lard and vegetable oils playing only a minor role.

Toward the end of the 19th century the production of oil by hydraulic pressing and solvent extraction was introduced. This process gave relatively high oil yields but necessitated posttreatment of the oils by neutralization, bleaching, and deodorization.

The invention of margarine in the 1870s gave further impetus to the oil-processing industry. With the discovery of oil hydrogenation (hardening) at the beginning of the 20th century, liquid oils could be converted into spreadable, consistent fats. In the 1930s, interesterification and fractionation were developed as further methods to modify the consistency of oils and fats.

Apart from being used for edible purposes, oils and fats are referred to in the Old Testament as cosmetic products and lamp fuel. Anointing with oil symbolized royal dignity. HOMER and HERODOTUS refer to the use of fats as "processing aids" during weaving. The ability of fats to calm waves was studied by Indian scholars 3000 years ago. The Egyptians supposedly used fats as lubricants to transport stone blocks. They were also familiar with the use of drying oils in varnishes and paints.

The elucidation of the chemical nature of fats was initiated by SCHEELE, who produced glycerol from olive oil around 1780. CHEVREUL subsequently (ca. 1815) recognized that fats were predominantly esters of fatty acids and glycerol.

Modern research and development is focussed on the application of new biotechnological principles to the production and modification of oils and fats [19], as well as on nutritional aspects, problems of trace contaminants, and environmental pollution. The breeding of new plant varieties is an important method for increasing the types of oils and fats. Examples are new varieties of rapeseed (low erucic acid content), safflower, and sunflower (high in oleic acid). New sources of oils and fats are being exploited by cultivation of wild plants such as the jojoba and *Cuphea* shrubs.

2. Composition

2.1. Glycerides

Naturally occurring fats contain about 97 % triglycerides (triacylglycerides), i.e., triesters of glycerol with fatty acids; up to 3 % diglycerides (diacylglycerides); and up to 1 % monoglycerides (monoacylglycerides). Tri-, di-, and monoglycerides consist of 1 mol of glycerol esterified with 3 mol, 2 mol, or 1 mol of fatty acid, respectively. The triglycerides of naturally occurring oils and fats contain at least two different fatty acid groups. The chemical, physical, and biological properties of oils and fats are determined by the type of the fatty acid groups and their distribution over the triglyceride molecules. The melting point generally increases with increasing proportion of long chain fatty acids or decreasing proportion of short chain or unsaturated fatty acids. Milk fat (butterfat) and coconut oil, which contain a high proportion of C_6–C_{12} fatty acids, have lower melting points than fats such as tallow and lard, which contain predominantly C_{16} and C_{18} fatty acids. Vegetable oils are liquid at ambient temperature because of their high proportion of unsaturated fatty acids.

The properties of a triglyceride are also determined by the position of the various fatty acid groups in the triglyceride molecule (i.e., 1-, 2-, or 3-position). The total number N of possible triglycerides (including positional isomers) from x different fatty acids is

$$N = \frac{x^2 + x^3}{2}$$

However, the proportions of different triglycerides in a naturally occurring fat generally do not conform to a statistical distribution. In vegetable oils and fats, unsaturated fatty acids are linked preferentially to the 2-position of the glycerol group, whereas in animal fats they appear primarily in the 1- and 3-positions ("2-random" and "1,3-random" distributions) [20]. Extreme examples of nonrandom distributions of fatty acid groups over the triglyceride molecule are cocoa butter (ca. 40 % 1-palmito-3-stearo-2-olein) and lard (ca. 20 % 2-palmito-1,3-diolein).

2.2. Fatty Acids

The fatty acids that form the triglycerides of naturally occurring oils and fats are predominantly even-numbered, straight-chain, aliphatic monocarboxylic acids with chain lengths ranging from C_4 to C_{24}. Unsaturated fatty acids differ in number and position of double bonds and in configuration (i.e., cis or trans isomers). The more common fatty acids are known by trivial names such as butyric, lauric, palmitic, oleic, stearic, linoleic, linolenic (→ Fatty Acids). Crude oils contain significant amounts of free fatty acids.

The chief fatty acids in some commercial oils and fats are listed in Tables 1 (saturated fatty acids) and 2 (unsaturated fatty acids). The fatty acid composition of most vegetable oils and fats is relatively simple; they consist predominantly of palmitic, oleic, and linoleic acids [20]. The fatty acids of land-animal fats mainly have a chain length of C_{16} or C_{18}. They are formed by biosynthetic conversion of carbohydrates, proteins, or fats, or originate directly from ingested fat.

Ruminant fats contain 5 – 10 % trans fatty acids, which are produced from linoleic and linolenic acid in the rumen. Marine oils contain a high proportion of polyunsaturated fatty acids

Table 1. Saturated fatty acids in various oils and fats, (main sources: Unilever; Food RA, Leatherhead; ITERG[a])

	Saturated fatty acids[b], g/100 g fatty acids							
	C_{10} and lower	C_{12}	C_{14}	C_{16}	C_{18}	C_{20}	C_{22}	C_{24}
Liquid vegetable oils								
Almond			tr	6.5 – 7	1 – 2.5	tr	tr	
Avocado			tr	10 – 26	0.5 – 1			
Corn germ		tr – 0.5	tr – 0.3	9 – 12	1 – 3	ca. 0.5	tr – 0.5	<0.5
Cottonseed		tr	0.5 – 2.0	21 – 27	2 – 3	<0.5	tr	tr
Grape-seed				4 – 11	2 – 5	tr	tr	tr
Linseed		tr	tr	5 – 6	3 – 5	<0.5	tr – 0.2	tr
Olive				7 – 16	2 – 4	ca. 0.5	tr	tr
Peanut (Africa)			tr	7 – 12	1.5 – 5	ca. 1.5	2 – 4	1 – 2
Peanut (South America)			tr	10 – 13	1.5 – 4	ca. 1.5	3 – 4	1.5 – 2
Pumpkinseed				7 – 13	6 – 7	tr		
Rapeseed (high erucic)		tr	tr	2 – 4	1 – 2	0.5 – 1	0.5 – 2.0	0.5
Rapeseed (low erucic)		tr	tr	3 – 6	1 – 2.5	<1	tr – 0.5	tr – 0.2
Ricebran			ca. 0.5	13 – 18	ca. 2	0.5 – 1		ca. 0.5
Safflower		tr	tr	ca. 5	2 – 3	ca. 0.5	ca. 1	ca. 1
Sesame			tr	8 – 10	3 – 6	ca. 0.5		
Soybean			<0.5	8 – 12	3 – 5	<0.5	tr	
Sunflower		tr	tr – 0.1	5.5 – 8	2.5 – 6.5	<0.5	0.5 – 1.0	<0.5
Wheat germ				12 – 14	ca. 1	0.5		tr
Consistent vegetable fats								
Babassu oil	ca. 12	42 – 44	15 – 18	8 – 10	2 – 3	tr		
Coconut oil	ca. 13	41 – 46	18 – 21	9 – 12	2 – 4	tr	tr	
Cocoa butter		tr	tr	23 – 30	32 – 37	<1	ca. 0.5	
Palm kernel oil	ca. 7	41 – 45	15 – 17	7 – 10	2 – 3	tr – 0.3	tr – 0.5	
Palm oil (Africa)		tr	1 – 2	41 – 46	4 – 6.5	ca. 0.5		
Palm oil (Indonesia)		tr – 0.5	ca. 1	41 – 47	4 – 6	ca. 0.5		
Animal fats								
Beef tallow			tr	2 – 4	23 – 29	20 – 35	<0.5	tr
Butterfat	7 – 9	2 – 5	8 – 14	24 – 32	9 – 13	2		
Chicken fat			ca. 1	20 – 24	4 – 7			
Goose fat				20 – 22	4 – 11			
Horse fat	tr		ca. 0.5	3 – 6	20 – 30	6 – 10	tr	
Lard	tr		<0.5	ca. 1.5	24 – 30	12 – 18	ca. 0.5	
Mutton tallow	tr		ca. 0.5	1 – 4	22 – 30	15 – 30	tr	tr
Marine oils								
Fish oils								
Japanese	tr		tr	ca. 6	ca. 16	ca. 3	<0.5	tr	tr – 1
Menhaden	tr		tr	ca. 9	ca. 20	ca. 4	tr – 1	tr	tr – 1
Scandinavian	tr		tr	6 – 8	11 – 15	1 – 3	tr – 0.5	tr	tr – 1
South American	tr		tr	ca. 7	17 – 19	2 – 4	ca. 0.5	tr	tr – 1
Whale oil	tr		ca. 0.5	4 – 10	10 – 18	1 – 3	tr		

[a] ITERG = Institut des Corps Gras, Centre Technique Industrielle, Paris.
[b] tr = traces (<0.05 %).

with a chain length of $C_{20} - C_{24}$. Land-animal fats and marine oils contain numerous odd-numbered and branched fatty acids in trace concentrations. More than 80 different fatty acids have been found in milk fat, and more than 40 in lard. However, most of these fatty acids occur only in traces. They can be of industrial significance for some oils and fats and they can play a role in the identification of fats and their detection in mixtures.

The fatty acid composition of a naturally occurring fat is determined genetically. The fatty acid composition of oleaginous seeds can be changed by developing new varieties. Examples are low-linoleic safflower and sunflower oils and low-erucic rapeseed oil. Environmental factors can influence the fatty acid composition within certain limits. The proportion of unsaturated fatty acids in the glycerides of linseed, soybean, and sunflower oils, for example, generally increases

Table 2. Unsaturated fatty acids in various oils and fats, (main sources: Unilever; Food RA, Leatherhead; ITERG[a])

	Unsaturated fatty acids[b], g/100 g fatty acids									
	$C_{14:1}$	$C_{16:1}$	$C_{18:1}$	$C_{18:2}$	$C_{18:3}$	$C_{20:1}$	$C_{22:1}$	$C_{20:x}^{c}$	$C_{22:x}^{c}$	$C_{24:1}$
Liquid vegetable oils										
Almond		<0.5	65 – 69	21 – 25	tr	tr – 0.1				
Avocado		2 – 12	44 – 76	8 – 25	ca. 1					
Corn germ		<0.5	25 – 35	40 – 60	ca. 1	ca. 0.5	tr – 0.1			
Cottonseed		<1	14 – 21	45 – 58	tr – 0.2	tr				
Grape-seed			12 – 33	45 – 72	1 – 2					
Linseed		tr	18 – 26	14 – 20	51 – 56	<0.5				
Olive		1 – 2	64 – 86	4 – 15	0.5 – 1	0.5				
Peanut (Africa)		<0.5	50 – 70	14 – 30	tr	0.5 – 1.5	tr			
Peanut (South America)		<0.5	35 – 42	39 – 44	tr	0.5 – 1.5	tr			
Pumpkinseed			24 – 41	46 – 57						
Rapeseed (high erucic)		ca. 0.5	11 – 24	10 – 22	7 – 13	ca. 10	41 – 52			
Rapeseed (low erucic)		0.1 – 0.5	52 – 66	17 – 25	8 – 11	1.5 – 3.5	tr – 2.5	tr – 0.1		tr
Ricebran			ca. 44	30 – 40						
Safflower		tr	12 – 20	70 – 80	tr	tr				
Sesame		tr	35 – 46	40 – 48	tr – 0.5	<0.5				
Soybean		tr	18 – 25	49 – 57	6 – 11	<0.5				
Sunflower		<0.5	14 – 34	55 – 73	tr – 0.4	<0.5		tr – 0.3		
Wheat germ			ca. 30	40 – 55	ca. 7					
Consistent vegetable fats										
Babassu oil			14 – 16	1 – 2						
Coconut oil	tr	tr	5 – 9	0.5 – 3	tr	tr				
Cocoa butter		ca. 0.5	30 – 37	2 – 4						
Palm kernel oil			10 – 18	1 – 3	tr – 0.5	tr – 0.5				
Palm oil (Africa)		<0.5	37 – 42	8 – 12	tr – 0.5	tr				
Palm oil (Indonesia)		ca. 0.5	37 – 41	ca. 10	tr – 0.5	tr				
Animal fats										
Beef tallow	ca. 0.5	2 – 4	26 – 45	2 – 6	ca. 1	<0.5		tr	ca. 0.5	
Butterfat	ca. 2	3	19 – 33	1 – 4	2 – 6				ca. 2	
Chicken fat		ca. 7	38 – 44	18 – 23	ca. 1				0.5 – 1	
Goose fat			41 – 74	7 – 19						
Horse fat		3 – 10	36 – 40	6 – 11	4 – 9	tr – 0.5		tr	1 – 2	
Lard	tr	2 – 3	36 – 52	10 – 12	ca. 1	0.5 – 1		<0.5	tr	
Mutton tallow	ca. 0.5	3 – 4	31 – 56	3 – 7	1 – 2				ca. 0.5	
Marine oils										
Fish oils										
Japanese	tr	ca. 7	ca. 14	ca. 2	ca. 1	ca. 7	ca. 6	ca. 15	ca. 12	tr – 1
Menhaden	tr	ca. 11	ca. 13	ca. 2	ca. 1	ca. 2	ca. 1	ca. 14	ca. 11	tr – 1
Scandinavian	tr	6 – 11	12 – 15	1 – 2	0.5 – 1	9 – 16	14 – 20	6 – 10	5 – 11	tr – 1
South American	tr	9 – 11	14 – 15	1 – 2	0.5 – 1	1 – 2	1 – 2	7 – 19	10 – 14	tr – 1
Whale oil	1 – 3	13 – 20	24 – 33	1 – 2	tr	10 – 15	4 – 10	1 – 6	5 – 7	tr

[a] ITERG = Institut des Corps Gras, Centre Technique Industrielle, Paris.
[b] tr = traces (<0.05 %).
[c] x > 1.

as the climate becomes colder and wetter. The subcutaneous fats of marine animals living in the colder parts of the oceans have a particularly high content of unsaturated fatty acids and thus a relatively low melting point. Linoleic acid cannot be synthesized by the animal or human organism and is hence referred to as an "essential" fatty acid.

Naturally occurring oils and fats are distributed homogeneously in varying concentrations in vegetable and animal tissues. In plants they are found predominantly in the seeds and the fruit pulp where they serve as a source of energy.

Biosynthesis. The synthesis of fatty acids in plants and animals generally starts with "activated" acetic acid, i.e., acetyl coenzyme A, which is derived from carbohydrates via pyruvic acid. Animals also synthesize acetyl coenzyme A from amino acids.

A distinction is made between *de novo synthesis* of saturated fatty acids and *elongation* of

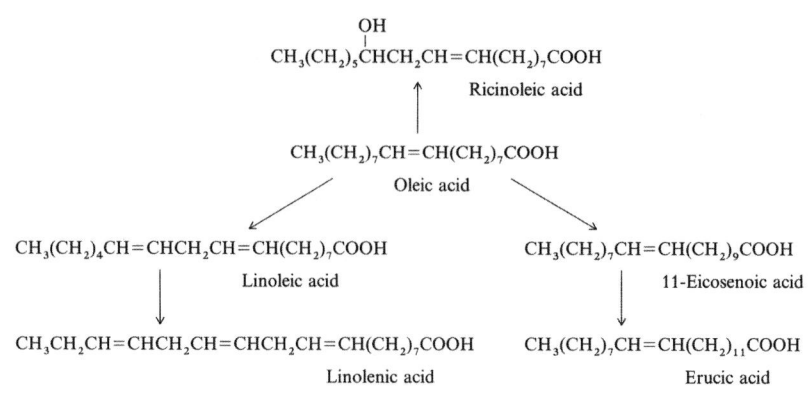

$$CH_3(CH_2)_5\overset{OH}{\underset{|}{CH}}CH_2CH=CH(CH_2)_7COOH$$
Ricinoleic acid

$$CH_3(CH_2)_7CH=CH(CH_2)_7COOH$$
Oleic acid

$$CH_3(CH_2)_4CH=CHCH_2CH=CH(CH_2)_7COOH$$
Linoleic acid

$$CH_3(CH_2)_7CH=CH(CH_2)_9COOH$$
11-Eicosenoic acid

$$CH_3CH_2CH=CHCH_2CH=CHCH_2CH=CH(CH_2)_7COOH$$
Linolenic acid

$$CH_3(CH_2)_7CH=CH(CH_2)_{11}COOH$$
Erucic acid

Figure 1. Unsaturated fatty acids derived from oleic acid

a fatty acid chain. De novo synthesis leads primarily to palmitic acid with smaller amounts of lauric, myristic, and stearic acids.

De novo synthesis:

$$CH_3COCoA \xrightarrow{+CO_2} HOOCCH_2COCoA \xrightarrow[-HSCoA]{+CH_3COCoA}$$
Acetyl CoA　　　　　Malonyl CoA

$$CH_3COCH(COOH)COSCoA \xrightarrow{-CO_2} CH_3COCH_2COSCoA$$

$$\xrightarrow{+H_2} CH_3CH(OH)CH_2COSCoA \xrightarrow{-H_2O}$$

$$CH_3CH=CHCOSCoA \xrightarrow{+H_2}$$

$$CH_3CH_2CH_2COSCoA \xrightarrow[-HSCoA]{+Malonyl\,CoA}$$
Butyryl CoA

$$CH_3CH_2CH_2COCH(COOH)COSCoA\text{ etc.}$$

The conversion of palmitic into stearic acid and the formation of longer chain saturated and unsaturated fatty acids presumably takes place by elongation, which involves addition of acetyl coenzyme A to activated fatty acids (RCOSCoA).

Elongation:

$$RCOOH \xrightarrow[-H_2O]{+HSCoA} RCOSCoA \xrightarrow[-HSCoA]{+CH_3COSCoA}$$

$$RCOCH_2COSCoA \longrightarrow \longrightarrow RCH_2CH_2COOH$$

The synthesis of fatty acids from C_2 units explains the predominance of even-numbered fatty acids in naturally occurring oils and fats. Bacterial degradation of feed in the rumen leads not only to acetic and butyric acids but also to propionic acid (C_3); incorporation of these C_3 units into fat synthesis in the udder explains the occurrence of branched and odd-numbered fatty acids in milk fat.

Oleic acid appears to play a key role in plants. It is probably synthesized from short-chain fatty acids and can be dehydrogenated into more highly unsaturated fatty acids; be converted into substituted acids such as ricinoleic acid; or elongated into erucic acid by specific enzyme systems (Fig. 1).

In the animal organism oleic acid is formed by dehydrogenation of stearic acid. Further dehydrogenation to linoleic and linolenic acid does not take place. The animal organism can, however, further desaturate linoleic and linolenic acid by introducing double bonds between the carboxyl group and the double bond nearest to it. Together with an elongation of the fatty acid chain this leads to the formation of γ-dihomolinolenic acid ($C_{20:3}$), arachidonic acid ($C_{20:4}$), and eicosapentaenoic acid ($C_{20:5}$), which are precursors of prostaglandins.

Three activated fatty acids (RCOSCoA) react successively with a molecule of phosphorylated glycerol to form a triglyceride molecule. The factors that determine the distribution of the fatty acids over the 1-, 2-, and 3-positions of the glycerides are not fully known. Phosphatides are also synthesized by this route.

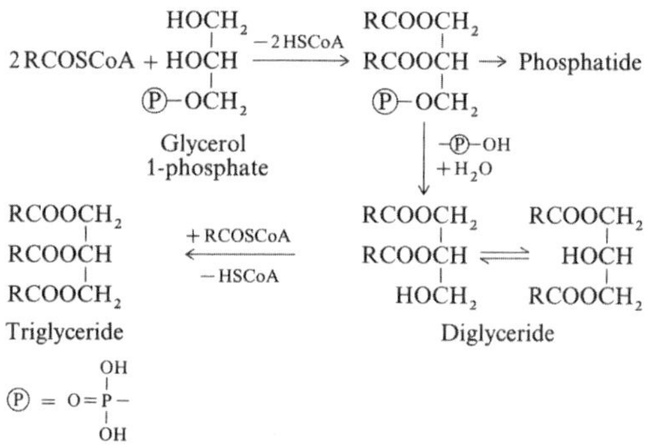

$$2\,RCOSCoA + \underset{\substack{|\\ \text{HOCH}\\ |\\ \textcircled{P}\text{-OCH}_2}}{\text{HOCH}_2} \xrightarrow{-2\,HSCoA} \underset{\substack{|\\ \text{RCOOCH}\\ |\\ \textcircled{P}\text{-OCH}_2}}{\text{RCOOCH}_2} \rightarrow \text{Phosphatide}$$

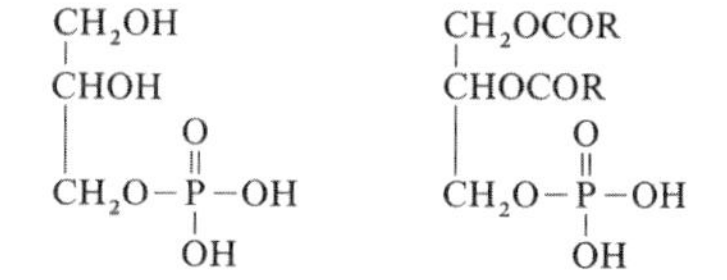

Glycerol 1-phosphate

$$\downarrow \begin{array}{c} -\textcircled{P}\text{-OH} \\ +H_2O \end{array}$$

Triglyceride ⇌ Diglyceride (with +RCOSCoA / −HSCoA)

$\textcircled{P} = \underset{\substack{|\\ OH}}{O=\overset{OH}{\underset{|}{P}}-}$

Biodegradation. During digestion and absorption of fats, triglycerides are successively split into di- and monoglycerides, glycerol, and fatty acids by lipases. The products are absorbed by the intestinal epithelial cells either as water-soluble complexes or as micelles. The fatty acids are biodegraded mainly via β-oxidation:

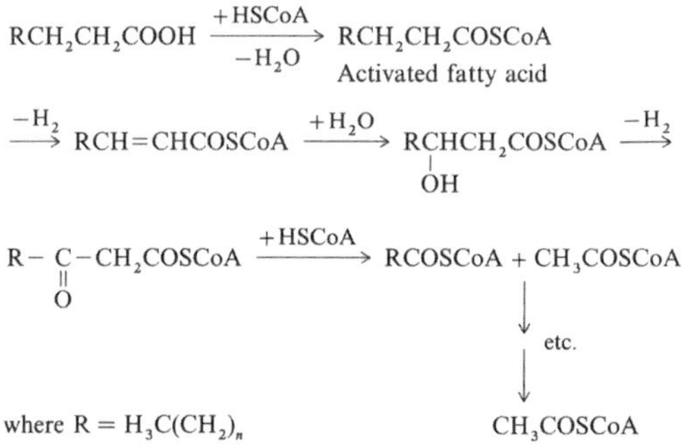

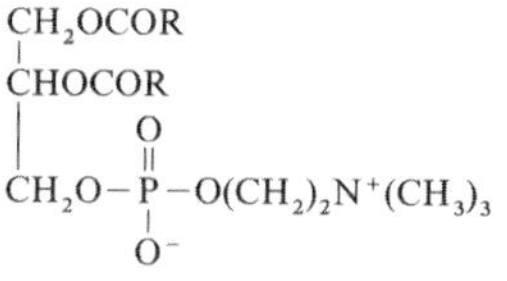

$$RCH_2CH_2COOH \xrightarrow[-H_2O]{+HSCoA} \underset{\text{Activated fatty acid}}{RCH_2CH_2COSCoA}$$

$$\xrightarrow{-H_2} RCH=CHCOSCoA \xrightarrow{+H_2O} RCHCH_2COSCoA \xrightarrow{-H_2}\;|\;OH$$

$$R-\underset{\substack{\|\\ O}}{C}-CH_2COSCoA \xrightarrow{+HSCoA} RCOSCoA + CH_3COSCoA$$

↓ etc.
↓
CH$_3$COSCoA

where R = H$_3$C(CH$_2$)$_n$

The bio-oxidation of unsaturated fatty acids involves additional steps.

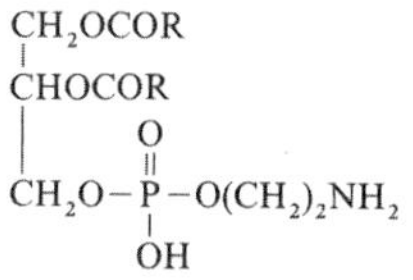

2.3. Phospholipids [20]

Phospholipids are essential constituents of the protoplasm of animal and plant cells; they are mostly present as lipoproteins and lipidcarbohydrate complexes. Oilseeds, cereal germs, egg yolk, and brain are the richest sources of phospholipids. Esters of glycerophosphoric acid (glycerol 1-phosphate) are usually referred to as phosphatides (→ Lecithin).

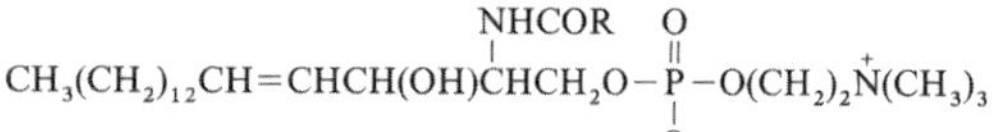

Glycerol 1-phosphate

Phosphatidic acid
R = alkyl

Some important phosphatides are phosphatidylcholine [8002-43-5] (lecithin), phosphatidylethanolamine [5681-36-7] (cephalin), phosphatidylinositol [2816-11-7], and phosphatidylserine.

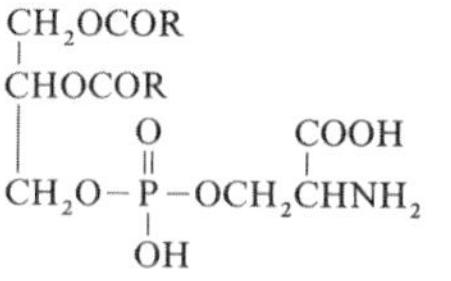

Phosphatidylcholine

Phosphatidylethanolamine

Phosphatidylserine
R = alkyl

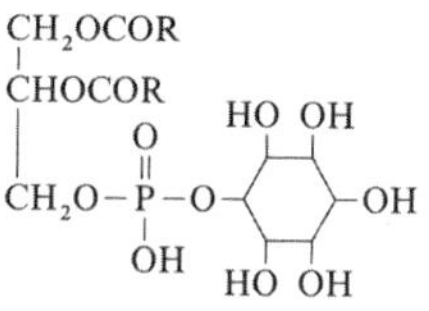

Phosphatidylinositol

The plasmalogens are ethers of fatty alcohols and phosphatidic acid; they occur in animal tissue.

Sphingolipids are derivatives of the amino alcohol sphingosine. For example, sphingomyelins are constituents of the phospholipids in the brain, blood plasma, and erythrocytes.

$$CH_3(CH_2)_{12}CH=CHCH(OH)\overset{NHCOR}{\underset{|}{C}}HCH_2O-\overset{O}{\underset{|}{\overset{\|}{P}}}-O(CH_2)_2\overset{+}{N}(CH_3)_3$$
$\qquad\qquad\qquad$ Sphingomyelin $\qquad\qquad$ O$^-$

The cerebrosides are derivatives of sphingosine and either galactose (galactolipids) or glucose (glycolipids). They are important constituents of the myelin nerve sheath. The gangliosides are based on neuraminic acid and are found in the ganglia cells of the brain.

Lecithin, cephalin, inositol phosphatides, and phosphatidic acid are the principal phospholipid components of plant origin. During pre-refining of crude vegetable oils, especially soybean and rapeseed oil, most of the phosphatides are removed as sludge by hydration with water. Drying of this sludge yields "lecithin," which is used, often after further modification, in the food industry as an emulsifier, antispattering agent, dispersant, or viscosity reducing agent. Lecithin is also used in pharmaceuticals, toiletries, animal feeds, and as a mold release agent and emulsifier – dispersant [21]. The phosphatide contents of crude oils and fats are listed in Table 2; the phosphatides are almost completely removed during refining of oils and fats.

2.4. Waxes

Waxes are esters of fatty alcohols and fatty acids (→ Waxes). Free and esterified fatty alcohols (e.g., cetyl, stearyl, oleyl alcohols) occur in considerable concentrations in marine oils. Ethers of glycerol and fatty alcohols (batyl, chimyl, and selachyl alcohols) are also found in animal tissues. The wax in the seed coat of sunflower seed causes the oil to become cloudy at refrigerator temperatures and is therefore removed by winterization (see Chap. 7).

2.5. Sterols and Sterol Esters

The major part of the nonsaponifiable matter of oils and fats consists of sterols present as such or as fatty acid esters and glycolipids. The most important sterol in animal fats is cholesterol [57-88-5]. β-Sitosterol [83-46-5] is the predominant sterol in vegetable oils and fats, although traces of cholesterol are also present. Total sterol concentrations are shown in Table 3, and sterol compositions in Table 4. Some of the sterols are removed during the deodorization step of refining oils and fats, without, however, changing their relative composition. Sterols are therefore a useful tool in checking authenticity. Vitamin D_2 [50-14-6] (calciferol) is present in milk and butter, and vitamin D_3 [67-97-0] (cholecalciferol) in cod liver oil. Other animal and vegetable oils contain hardly any vitamin D.

Table 3. Minor constituents of crude oils and fats

	Content, wt %		
	Phosphatides	Tocopherols	Sterols
Babassu fat		0.003	
Beef tallow	<0.07	0.001	0.08 – 0.14
Butterfat	<1.4	0.003	0.24 – 0.50
Castor oil			0.5
Cocoa butter	0.1	0.003	0.17 – 0.20
Coconut oil		0.003	0.05 – 0.1
Cod liver oil			0.42 – 0.54
Corn germ oil	1 – 2	0.1 – 0.3	0.8 – 2.2
Cottonseed oil	0.7 – 0.9	0.04 – 0.11	0.27 – 0.6
Fish oil			ca. 0.3
Lard	<0.05	0.003	ca. 0.1
Linseed oil	0.3	0.11	0.37 – 0.42
Olive oil		0.01 – 0.03	0.1 – 0.2
Palm oil	0.05 – 0.1	0.02 – 0.12	0.04 – 0.08
Palm kernel fat			0.08 – 0.12
Peanut oil	0.3 – 0.4	0.02 – 0.07	0.19 – 0.29
Rapeseed oil	2.5	0.07 – 0.08	0.5 – 1.1
Sesame oil	0.1	ca. 0.05	0.4 – 0.6
Soybean oil	1.1 – 3.2	0.09 – 0.12	0.2 – 0.4
Sunflower oil	<1.5	0.07 – 0.1	0.25 – 0.45
Wheat germ oil	0.1 – 2.0	ca. 0.28	1.3 – 1.7

Table 4. Sterol composition[a] in crude oils (main sources: Food RA, Leatherhead; Unilever; ITERG[b])

	Coconut	Corn germ	Cotton-seed	Olive	Palm	Palm kernel	Peanut	Rapeseed	Soybean	Sunflower
Cholesterol	0.6 – 2	0.2 – 0.6	0.7 – 2.3	0 – 0.5	2.2 – 6.7	1 – 3.7	0.6 – 3.8	0.4 – 2	0.6 – 1.4	0.2 – 1.3
Brassicasterol	0 – 0.9	0 – 0.2	0.1 – 0.9			0 – 0.3	0 – 0.2	5 – 13	0 – 0.3	0 – 0.2
Campesterol	7 – 10	18 – 24	7.2 – 8.4	2.3 – 3.6	18.7 – 29.1	8.4 – 12.7	12 – 20	18 – 39	16 – 24	7 – 13
Stigmasterol	12 – 18	4 – 8	1.2 – 1.8	0.6 – 2	8.9 – 13.9	12.3 – 16.1	5 – 13	0 – 0.7	16 – 19	8 – 11
β-Sitosterol	50 – 70	55 – 67	80 – 90	75.6 – 90	50.2 – 62.1	62.6 – 70.4	48 – 65	45 – 58	52 – 58	56 – 63
Δ5-Avenasterol	5 – 16	4 – 8	1.9 – 3.8	3.1 – 14	0 – 2.8	4 – 9	7 – 9	0 – 6.6	2 – 4	2 – 7
Δ7-Stigmastenol	2 – 8	1 – 4	0.7 – 1.4	0 – 4	0.2 – 2.4	0 – 2.1	0 – 5	0 – 5	1.5 – 5	7 – 13
Δ7-Avenasterol	0.6 – 2	1 – 3	1.4 – 3.3		0 – 5.1	0 – 1.4	0 – 5	0 – 0.8	1 – 4.5	3 – 6

[a] Sterols as percentage of total sterol fraction.
[b] ITERG = Institut des Corps Gras, Centre Technique Industrielle, Paris.

2.6. Terpenoids

The nonsaponifiable part of most fats also contains traces of terpenes and terpene alcohols. The triterpene squalene [7683-64-9] occurs in relatively high concentrations (up to 0.5 %) in olive oil. Shea butter contains 2 – 10 % of kariten, a rubber-like hydrocarbon.

Carotenoid pigments occur widely in oils and fats. Approximately 70 different carotenoids, ranging in color from yellow to deep red, are known. The most well-known carotenoids are the isomeric tetraterpenes ($C_{40}H_{56}$) α-carotene [7488-99-5], β-carotene [7235-40-7], and γ-carotene [472-93-5], lycopene [502-65-8], and xanthophyll [127-40-2]. Crude palm oil contains up to 0.2 % α- and β-carotene.

The bulk of the carotenoids are removed during refining, primarily during bleaching and deodorization. β-Carotene (provitamin A) can be oxidized to vitamin A [68-26-8] in the animal organism. Most vegetable oils and fats do not contain significant concentrations of vitamin A; in most countries margarine therefore contains added vitamin A. Vitamin A is present in high concentrations in fishliver oils. Butter contains ca. 0.003 – 0.0015 % vitamin A.

The most important compounds of the vitamin E [59-02-9] group are α-, β-, γ-, and δ-tocopherols:

$$\text{structure: chromanol ring with } R^1, R^2, R^3 \text{ substituents; HO– on ring; CH}_3 \text{ on ring; O in ring with } (CH_2CH_2CH_2CH(CH_3))_3CH_3 \text{ side chain}$$

	R^1	R^2	R^3
α-Tocopherol	CH_3	CH_3	CH_3
β-Tocopherol	CH_3	H	CH_3
γ-Tocopherol	H	CH_3	CH_3
δ-Tocopherol	H	H	CH_3

These compounds act as a vitamin (rat fertility factor) and as antioxidants. α-Tocopherol has the highest biological activity.

Tocopherols occur only in traces in animal fats, whereas vegetable oils contain appreciable concentrations (see Table 5). Refined oils and fats still contain ca. 80 % of the original tocopherols, the main losses occurring during deodorization. Deodorizer distillates are a valuable source of natural tocopherols.

Table 5. Tocopherols (mg/kg) in crude oils (main sources: Food RA, Leatherhead; Unilever; ITERG** [22])

	Coconut	Corn germ	Cotton-seed	Olive	Palm	Palm kernel	Peanut	Rapeseed	Safflower	Sesame	Soybean	Sunflower
α-Tocopherol	0 – 17	270 – 370	140 – 540	100 – 250	80 – 95*	0 – 44	50 – 300	230 – 300	ca. 580	2 – 18	60 – 75	600 – 1000
β-Tocopherol	0 – 14	10 – 16		0 – 4	0 – 5*	0 – 123	100 – 400	5 – 10	ca. 18	<0.5	10 – 20	15 – 35
γ-Tocopherol	0 – 2	640 – 860	160 – 625	8 – 14	5 – 15*	0 – 6		400 – 550	ca. 16	500 – 540	580 – 740	3 – 35
δ-Tocopherol		30 – 50	0 – 17	<10	0 – 5*		0 – 20	15 – 20	<0.5	<0.5	275 – 320	0 – 7
Total	0 – 31	1100 – 1300	410 – 1169	110 – 270	200 – 1200		175 – 700	670 – 800	ca. 610	500 – 550	900 – 1200	650 – 1000

* Tocopherols as percentage of total tocopherol fraction.
** ITERG = Institut des Corps Gras, Centre Technique Industrielle, Paris.

2.7. Other Minor Constituents

Sesame oil contains 0.3 – 0.5 % sesamolin [526-07-8], a glycoside of the phenol sesamol [533-31-3], and 0.5 – 2.0 % sesamin [607-80-7]. These minor constituents give a characteristic color reaction (the basis of the Baudouin test) and impart stability to oxidation.

Gossypol [303-45-7], a toxic polyphenol with pronounced antioxidant activity, is found in crude cottonseed and kapokseed oils (0.5 – 1.5 %). Gossypol is removed during refining in the lye neutralization step.

Crude linseed, rapeseed, soybean, olive, avocado, and many other vegetable oils contain the green pigments chlorophyll and phaeophytin. The chlorophyll content is particularly high in oil from immature seeds and is generally determined by harvesting and climatic conditions. Chlorophyll is generally regarded to be indicative of inferior crude oil quality; it can be removed by treatment with acidic absorbents such as bleaching earth.

Crude oils and fats may contain traces of proteins, the concentration depending on processing conditions. In addition, vegetable oils may contain carbohydrates. These compounds are removed almost completely in refining.

Autoxidation of the fatty acid groups of triglycerides leads to the formation of volatile and nonvolatile oxidation products. The type and quantity of the volatile compounds (ketones, aldehydes, and alcohols) depend on the initial fatty acid composition and the oxidation conditions. These volatile products are responsible for the typical odor and taste of oils and fats. One of the main aims of refining is to remove these odoriferous compounds, which in some cases occur in concentrations of only 10^{-3} ppm. The nonvolatile oxidation products generally have little odor and taste, but they can act as oxidation promoters.

Most crude oils and fats contain traces of pesticide residues and metals (e.g., Fe, Cu, Pb, As, Cd, and Hg) as a consequence of crop treatment and environmental influences. Whereas phosphate-based pesticides used to treat oilseeds decompose with time, chlorinated pesticides are stable and gradually migrate into the oil.

Peanut oil may contain traces of aflatoxins produced by growth of *Aspergillus flavus* on the seed.

Most oils and fats, particularly coconut oil, contain varying concentrations of polycyclic aromatic hydrocarbons. These hydrocarbons are introduced during smoke-drying of the raw materials prior to storage and further processing.

Crude rapeseed oil contains up to 50 ppm of sulfur in the form of elemental sulfur, isothiocyanates, and 5-vinyl-2-oxazolidinethione [500-12-9] (goitrin) derived from glucosinolates (sulfur-containing glucosides) in the seed.

All of these undesirable contaminants are reduced to negligibly low levels in the course of refining, if necessary with the aid of additional steps such as adsorptive treatment with activated charcoal to remove polycyclic aromatic hydrocarbons.

3. Physical Properties [23]

3.1. Melting and Freezing Points

The *melting point* of the even-numbered, saturated fatty acids increases with increasing chain-length, and decreases with increasing degree of unsaturation (see also → Fatty Acids, Section 2.1.). The glycerides show a similar behavior (Table 6). Since the naturally occurring fats are mixtures of glycerides, they melt over a wide range of temperature (for methods of melting point determination, see Section 11.3.1).

Table 6. Melting points of fatty acids and glycerides in their stable polymorphic forms

	mp (slip point), °C			
Fatty acid		1-Monoglyceride	1,3-Diglyceride	Triglyceride
Butyric acid	−7.9			
Caproic acid	−3.9	19.4		−25.0
Caprylic acid	16.3	40.0		8.3
Capric acid	31.3	53.0	44.5	31.5
Lauric acid	44.0	63.0	56.5	46.5
Myristic acid	54.4	70.5	65.5	57.0
Palmitic acid	62.9	77.0	72.5	65.5
Stearic acid	69.6	81.0	78.0	73.0
Arachidic acid	75.4	84.0	(75.0)	(72.0)
Behenic acid	79.9			81.0
Oleic acid	16.3	35.2	21.5	5.5
Elaidic acid	45.0	58.5	55.0	42.0
Erucic acid	34.7	50.0	46.5	30.0
Linoleic acid	−5.0	12.3	−2.6	−13.1
Linolenic acid	−11.0	15.7	−12.3	−24.2
Ricinoleic acid	5.0			

Table 7. Melting points of some polymorphic triglycerides

	mp, °C		
	α	β′	β
Trilaurin	15.0	35.0	46.5
Trimyristin	33.0	46.5	57.0
Tripalmitin	45.0	56.0	65.5
Tristearin	54.5	65.0	73.0
Triolein	−32.0	−12.0	5.5
Trielaidin	15.5	37.0	42.0
1,2-Dicapriolaurin	17.5	26.0	30.0
2-Capriodilaurin	23.0	33.0	38.5
1-Laurodimyristin	37.0	42.0	46.5
1-Laurodipalmitin	45.0	49.5	54.0
2-Laurodipalmitin	47.0	50.0	53.5
1,2-Dicapriostearin	32.0	38.0	41.0
1,3-Dicapriostearin	34.0	40.0	44.5
2-Palmitodistearin	56.0	64.0	68.5
1,3-Dicaprioolein	−10.2	0.6	6.2
1,2-Dilauroolein	−10.0	4.8	16.0
1,2-Dipalmitoolein	ca. 18.0	ca. 31.0	34.5
1,3-Dipalmitoolein	26.5	33.5	38.0
1,2-Distearoolein	ca. 30.0	ca. 40.0	
1,3-Distearoolein	37.0	41.5	44.0
1,3-Dipalmitoelaidin	ca. 40.0	53.0	54.0
1-Stearodibehenin	61.3	71.0	73.5
1-Lauro-2-myristo-3-palmitin	37.0	44.0	49.0
1-Lauro-2-myristo-3-stearin	27.5	45.5	49.5
1-Palmito-3-stearo-2-olein	18.2	33.0	38.0

The melting point also depends on the polymorphic form of the glycerides, i.e., the crystalline structure (Table 7). The packing density and the spatial arrangement of the triglyceride molecules depend on the crystallization conditions. X-ray diffraction patterns (short spacing lines) can distinguish between three different polymorphic forms [25]. The lowest melting and most labile form is designated α. The most stable form is called β and the intermediate one β′. The α form is obtained by rapidly cooling the molten triglycerides, the β′ form by suitable tempering. Polymorphic changes induced by processing conditions are of great practical importance since they can significantly influence the properties of, for example, a margarine (oral melting behavior, sandiness) or chocolate (fat bloom).

The *congeal* or *set point* (point of solidification) is generally lower than the melting point.

The *solids content* of a fat at different temperatures is normally determined by pulsed nuclear magnetic resonance [24]. Fats exhibit an increase in volume (dilatation) on melting that is disproportionately larger than that on heating a liquid fat; an obsolete method of measuring solids content at different temperatures was based on dilatation. *Dilatation* or *solids content curves* can be used to characterize a fat. Figure 2 reflects the large proportion of 1-palmito-3-stearo-2-olein in cocoa butter and the more complex glyceride composition of lard.

The *latent heat of fusion* increases with increasing chain length and increasing degree of saturation (see Table 8). Naturally occurring fats generally have a lower heat of fusion than simple glycerides.

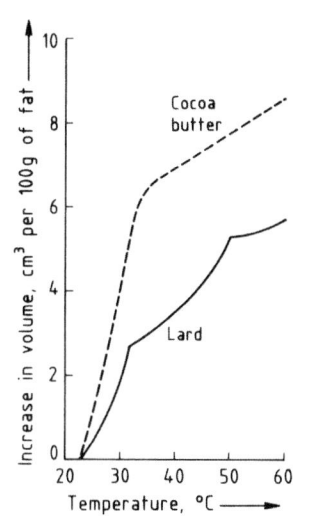

Figure 2. Dilatation curves for cocoa butter and lard

3.2. Thermal Properties

The approximate *heats of combustion* of oils and fats can be calculated from the following

Table 8. Latent heat of fusion of some fats

	Heat of fusion, J/g
Butterfat	81.6
Cottonseed oil	86.0
Fully hardened cottonseed oil[*]	185.0
Peanut oil	90.9
Partially hardened peanut oil[**]	103.4
Trilaurin (β-form)	193.5
Tripalmitin (β-form)	222.0
Tristearin (β-form)	228.0

[*] Iodine value ca. 1.
[**] Iodine value ca. 60.

Table 9. Specific heat of oils and fats

	Specific heat, J/g	Temperature, °C
Trilaurin	2.130	66.0
Trimyristin	2.152	58.4
Tripalmitin	2.173	65.7
Tristearin	2.219	79.0
Soybean oil	2.060	80.4
	2.000	60.0
Linseed oil	2.050	70.7
Cottonseed oil	2.200	90.0
Hardened cottonseed oil*	2.177	79.6
Olive oil	2.300	110
Palm oil	2.400	140
Sunflower oil	2.500	175.0

*Iodine value ca. 6.

formula:

Heat of combustion (J/g)
= 47645 − 4.1868 × iodine value
 − 38.31 × saponification value

Using this equation, values ranging from 37 765 J/g (9020 cal/g) for coconut oil to 40 528 J/g (9680 cal/g) for a high-erucic rapeseed oil have been obtained.

The *specific heat* of liquid oils and fats increases with increasing chain length and degree of saturation (see Table 9); it also increases with temperature.

Table 10. Vapor pressure of triglycerides (temperatures corresponding to a vapor pressure p of 6.7 Pa (0.05 mm Hg) and 0.13 Pa (0.001 mm Hg)

	Temperature, °C	
	$p = 6.7$ Pa	$p = 0.13$ Pa
Tributyrin	91	45
Tricaproin	135	85
Tricaprylin	179	128
Tricaprin	213	159
Trilaurin	244	188
Trimyristin	275	216
Tripalmitin	298	239
Tristearin	313	253
Soybean oil	308	254
Olive oil	308	253
1,3-Distearoolein	315	254
1-Myristo-2-palmito-3-stearin	297	237
1-Palmito-2-lauro-3-stearin	290	232
1-Myristo-2-lauro-3-stearin	282	223
1-Palmito-2-capro-3-stearin	280	223
1-Capro-2-lauro-3-myristin	249	189

Table 11. Boiling points of some monoglycerides

	Pressure, Pa	bp, °C
Monocaprin	133.3	175
Monolaurin	133.3	186
Monomyristin	133.3	199
Monopalmitin	133.3	211
Monostearin	26.7	190
Monoolein	26.7	186

Triglycerides of long-chain fatty acids have extremely low *vapor pressures*; typical data for some triglycerides and oils are shown in Table 10. Monoglycerides have significantly higher vapor pressures (see Table 11). There is relatively little information on the *heats of vaporization* of fats (see Table 12).

Oils and fats are relatively poor thermal conductors. Data for *thermal conductivity* are limited. Most data lie within ± 10 % of the following general relationship:

Thermal conductivity (Wm^{-1}K^{-1}) = 0.181 − 0.00025 t

where t = temperature in °C.

The *smoke, flash,* and *fire points* of oils and fats are measures of their thermal stability when heated in air (Table 13).

3.3. Density

The density of fatty acids and glycerides decreases with increasing molecular mass and degree of saturation (Tables 14 and 15). Oxidation generally leads to higher densities. A high free fatty acid content tends to decrease the density of a crude oil. The following formula can be used to estimate the density of an oil:

d_{15}^{15} = 0.8475 + 0.0003 × saponification value
 + 0.00014 × iodine value

Table 12. Heat of vaporization of some triglycerides calculated for 0.13 – 67 Pa

	Heat of vaporization, J/g
Tricaprylin	247.0
Tricaprin	226.1
Trilaurin	213.5
Trimyristin	205.2
Tripalmitin	201.0
Tristearin	188.4
Soybean oil	209.3

Table 13. Smoke, flash, and fire points of refined oils and fats

	ffa[*], %	Smoke point, °C	Flash point, °C	Fire point, °C
Rapeseed oil	0.08	218	317	344
Peanut oil	0.09	207	315	342
Peanut oil	0.11	198	333	363
Peanut oil	1.0	160	290	–
Cottonseed oil	0.04	223	322	342
Cottonseed oil	0.18	185	318	357
Soybean oil	0.04	213	317	342
Sunflower oil	0.1	209	316	341
Coconut oil	0.1	200	300	
Coconut oil	0.2	194	288	329
Coconut oil	1.0	150	270	
Palm oil	0.06	223	314	341
Hardened peanut oil (mp 32/34 °C)	0.04	226	314	340
Hardened soybean oil (mp 42/44 °C)	0.04	223	318	342
Beef tallow	0.4		316	344
Beef tallow	5.0		266	344

[*] ffa = free fatty acid.

Table 14. Density of triglycerides

	ϱ, g/cm³		
	80 °C	15 °C	25 °C
Tricaprin	0.8913		
Trilaurin	0.8801		
Trimyristin	0.8722		
Tripalmitin	0.8663		
Tristearin	0.8632		
Triolein		0.9162	0.9078
Trilinolein		0.9303	
Trilinolenin		0.9454	

Up to 260 °C the density decreases by about 0.00064 g/cm³ per temperature increase of 1 °C. The following equations apply for the density ϱ (in g/L) of commercial oils and fats (t = temperature in °C):

Soybean oil	933.4 − 0.657 t
Sunflower oil	932.7 − 0.680 t
Sesame oil	933.0 − 0.700 t
Cottonseed oil	931.7 − 0.755 t
Peanut oil	927.0 − 0.642 t
Olive oil	928.5 − 0.700 t
Palm oil	925.0 − 0.655 t
Palm kernel oil	940.0 − 0.740 t
Coconut oil	932.0 − 0.745 t
Rapeseed oil	925.5 − 0.700 t
Tallow	956.8 − 0.898 t
Fish oil	940.0 − 0.700 t

3.4. Viscosity

Oils tend to have a relatively high viscosity because of intermolecular attraction between their fatty acid chains. Generally, viscosity tends to increase slightly with increasing degree of saturation and increasing chain length (see Tables 16 and 17). There is an approximately linear relationship between log viscosity and temperature. The viscosity of oils tends to increase on prolonged heating due to the formation of dimeric and oligomeric fatty acid groups.

Table 15. Density of fats and oils

	ϱ_{15}, g/cm³	d_{25}^{25}
Vegetable fats		
Babassu oil	0.9250	
Castor oil	0.950 – 0.974	0.945 – 0.965
Coconut oil	0.919 – 0.937	0.869 – 0.874[a]
Cocoa butter	0.945 – 0.976	0.856 – 0.864[a]
Corn germ oil	0.920 – 0.928	0.916 – 0.921
Cottonseed oil	0.917 – 0.931	0.916 – 0.918
Grape-seed oil	0.919 – 0.936	
Hempseed oil	0.924 – 0.932	0.923 – 0.925[b]
Linseed oil	0.930 – 0.935	0.931 – 0.936
Mustard seed oil	0.912 – 0.923	
Olive oil	0.914 – 0.925	0.909 – 0.915
Oiticica oil	0.9518 – 0.9694	0.978[b]
Palm oil	0.921 – 0.947	0.898 – 0.901[c]
Palm kernel oil	0.925 – 0.935	0.860 – 0.873[a]
Peanut oil	0.911 – 0.925	0.910 – 0.915
Perilla oil	0.927 – 0.933	min. 0.932[b]
Poppyseed oil	0.923 – 0.926	
Rapeseed oil	0.910 – 0.917	0.906 – 0.910
Ricebran oil		0.916 – 0.921
Safflower oil	0.923 – 0.928	0.919 – 0.924
Sesame oil	0.921 – 0.924	0.914 – 0.919
Shea butter	0.917 – 0.918	
Soybean oil	0.922 – 0.934	0.917 – 0.921
Sunflower oil	0.920 – 0.927	0.915 – 0.919
Tung oil	0.936 – 0.945	0.940 – 0.943[b]
Wheat germ oil		0.925 – 0.933
Animal fats		
Beef tallow	0.936 – 0.952	0.860 – 0.870[a]
Butterfat	0.935 – 0.943	
Herring oil	0.917 – 0.930	
Horse fat	0.915 – 0.932	
Lard	0.914 – 0.943	0.858 – 0.864[a]
Menhaden oil	0.925 – 0.935	
Mutton tallow	0.936 – 0.960	
Sperm oil	0.875 – 0.890	
Whale oil	0.914 – 0.931	0.910 – 0.920

[a] $d_{15.5}^{99}$;
[b] $d_{15.5}^{15.5}$;
[c] $d_{37.8}^{37.8}$.

Table 16. Viscosity of triglycerides at 70°C

	η, mPa · s
Tributyrin	3.0
Tricaproin	5.9
Tricaprylin	8.8
Tricaprin	11.7
Trilaurin	14.6
Trimyristin	17.6
Tripalmitin	20.5
Tristearin	23.4

3.5. Solubility and Miscibility

Nearly all fats and fatty acids are easily soluble in common organic solvents such as hydrocarbons, chlorinated hydrocarbons, ether, and acetone. Castor oil is an exception in that it is only partially soluble in petroleum ether but easily so in ethanol. The solubility of fatty acids in ethanol is greater than that of the corresponding triglycerides. The solubility of fats in organic solvents decreases with increase in molecular mass and increases with degree of unsaturation. The differences in solubility enable categories of glycerides to be separated by fractional crystallization although complete separation is rarely achieved because of mutual solubility effects.

The water solubility of fats is low and decreases with increasing chain length and with decreasing temperature. The solubility of water in cottonseed oil is, for example, 0.14 wt % at 30 °C and 0.07 wt % at 0 °C.

Table 17. Viscosity of naturally occurring fats and oils

	η, mPa · s			
	20 °C	30 °C	40 °C	50 °C
Castor oil	1000	454	232	128
Coconut oil		39	26	19
Cottonseed oil	80	55	38	27
Fish oil	60	43	32	22
Lard			35	25
Linseed oil		48	33	25	18
Olive oil	80	55	40	30
Palm oil			40	28
Palmkernel oil		43	29	20
Peanut oil	78	50	32	23
Poppyseed oil	63			
Rapeseed oil (high erucic)	85	60	40	30
Sesame oil	65			25
Soybean oil	65	45	33	25
Sunflower oil	68	47	35	26
Tallow				25

Table 18. Solubility of gases in fats

	Gas solubility, vol %				
	t, °C	N_2	H_2	O_2	CO_2
Cottonseed oil	30.5	7.11	4.63		
	49.6	7.79	5.40		
	78.2	8.91	6.73		
	101.5	9.76	7.83		
	147.8	11.83	10.24		
Soybean oil	ca. 20.0	ca. 4.95		ca. 2.65	
Lard	41.5	7.65	5.218		
	73.2	8.79	6.58		
	111.3	10.38	8.50		
	147.3	12.06	10.35		
Hardened tallow	64.3		6.14		92.0
(iodine value 1)	67.0	8.44		14.50	
	84.7			15.35	
	88.0				79.1
	139.4	11.68	9.79		61.9

The solubility of gases in oils generally increases with increase in temperature, the reverse holding for carbon dioxide (Table 18).

3.6. Surface and Interfacial Tension

The surface and interfacial tensions of some oils are shown in Table 19. The interfacial tension is markedly reduced by the presence of surface-active agents, e.g., phosphatides, monoglycerides, free fatty acids, and soaps.

3.7. Electrical Properties

Dry oils, fats, and fatty acids are poor conductors of electricity. Recorded values for the specific resistance of stearic acid are 0.6×10^{11} Ω at 100 °C and 22.3×10^{11} Ω at 186 °C, and for oleic acid 2×10^{11} and 83×10^{11} Ω at comparable temperatures. The dielectric constant of most

Table 19. Surface and interfacial tension of some oils

	Surface tension, mN/m			Interfacial tension at 70 °C, mN/m
	20 °C	80 °C	130 °C	
Cottonseed oil	35.4	31.3	27.5	29.8
Coconut oil	33.4	28.4	24.0	
Castor oil	39.0	35.2	33.0	
Peanut oil				29.9
Soybean oil				30.6

oils lies in the range 3.0 – 3.2 at 25 – 30 °C. Castor and oiticica oil have dielectric constants of about 4 in this temperature range because of the hydroxyl and keto groups in the fatty acid chains. In emulsion systems, e.g., butter or margarine, the dielectric constant is affected by both the emulsion structure and the moisture content.

3.8. Optical Properties

The refractive index of oils, fats, and fatty acids generally increases with increasing chain length, number of double bonds, and extent of conjugation (see Tables 20 and 21). The refractive index of fatty acids is much lower than that of the corresponding triglycerides. Prolonged heating leads to an increase in refractive index due to the introduction of polar groups into the fatty acid chain.

There are a number of equations depicting a relationship between refractive index and other data. The following equation has been suggested for fresh, nonhydrogenated oils and fats:

$$n_D^{40} = 1.4643 - 0.00066 \times \text{saponification value}$$
$$- \frac{0.0096 \times \text{acid value}}{\text{saponification value}}$$
$$+ 0.0001711 \times \text{iodine value}$$

Table 20. Refractive indices of glycerides

	n_D^{60}
Tricaprin	1.4370
Trilaurin	1.4402
Trimyristin	1.4428
Tripalmitin	1.4452
Tristearin	1.4471
Triolein	1.4548
Trilinolein	1.4645
1,2-Dilaurostearin	1.4437
1,3-Dilaurostearin	1.4442
1,2-Dilauroolein	1.4456
1,3-Dilaurostearin	1.4459
1,2-Dipalmitoolein	1.4480
1,2-Distearoolein	1.4494
1-Stearodiolein	1.4524
Monocaprin	1.4443
Monolaurin	1.4462
Monomyristin	1.4480
Monopalmitin	1.4499
1-Lauro-3-olein	1.4472
1-Stearo-3-olein	1.4507

Table 21. Refractive indices of naturally occurring oils and fats

	n_D^{20}
Babassu oil	1.449 – 1.450*
Beef tallow	1.454 – 1.459*
Butterfat	1.452 – 1.457*
Castor oil	1.477 – 1.479
Cocoa butter	1.453 – 1.458*
Coconut oil	1.448 – 1.450*
Corn oil	1.474 – 1.476
Cottonseed oil	1.472 – 1.477
Grape seed oil	1.474 – 1.478
Herring oil	1.470 – 1.475*
Lard	1.458 – 1.461*
Linseed oil	1.479 – 1.481
Menhaden oil	1.480
Mustardseed oil	1.470 – 1.474
Mutton tallow	1.455 – 1.458*
Oiticica oil	1.4921 – 1.4945
Olive oil	1.467 – 1.471
Palm oil	1.453 – 1.456*
Palmkernel oil	1.449 – 1.452*
Peanut oil	1.460 – 1.472
Perilla oil	1.481 – 1.483
Rapeseed oil	1.472 – 1.476
Safflower oil	1.4754
Sesame oil	1.473 – 1.476
Soybean oil	1.470 – 1.478
Sunflower oil	1.474 – 1.476
Tung oil	1.517 – 1.526
Whale oil	1.463 – 1.471
Wheat germ oil	1.469 – 1.478*

* n^{40}.

Pure glycerides do not absorb in the visible region of the spectrum (400 – 750 nm). However, naturally occurring oils and fats invariably contain pigments that have characteristic absorption bands (carotene at 450 nm, chlorophyll and phaeophytin at 660 nm). Most of these pigments are removed during refining. Unsaturated oils and fats absorb in the ultraviolet region between 200 and 400 nm. Conjugated double bonds show characteristic maxima at 232 nm (dienes, e.g., 9,11-*trans, trans*-linoleic acid) and at 268 nm (trienes, e.g., β-eleostearic acid). Conjugated tetraenoic acids absorb between 290 and 320 nm.

In the infrared region (0.075 – 1000 μm) chain substituents such as epoxy, hydroxyl, keto, and cyclopropene groups as well as trans double bonds exhibit specific absorption peaks:

monoglycerides	1.43 μm
hydroxy fatty acids	3.20 μm
ester carbonyl group	5.83 μm
trans double bonds	10.0 – 10.35 μm
isolated trans double bonds	10.35 μm

The fingerprint region around 8 μm can be used to determine the fatty acid chain length.

X-ray spectroscopy is employed to characterize the various polymorphic crystal structures of pure glycerides (see Section 3.1).

Nuclear magnetic resonance of hydrogen atoms is used in structural identification and to determine the solids content of fats at different temperatures. Nuclear magnetic resonance measurement of phosphorus and nitrogen atoms can be employed to analyze the composition of phosphatide mixtures.

The optical activity of enantiomorphic trigly-cerides with different fatty acid groups in the 1- and 3-positions is usually too small to be measured [20].

4. Chemical Properties

The chemical reactions of fats are basically those of esters and hydrocarbon chains ($\to$ Esters, Organic). Only those reactions that are primarily relevant to the processing of edible oils and fats are dealt with in this chapter.

4.1. Hydrolysis

Glycerides can be hydrolyzed into fatty acids and glycerol:

$$\begin{array}{l} CH_2OCOR \\ CHOCOR \\ CH_2OCOR \end{array} + 3\,H_2O \rightleftharpoons \begin{array}{l} CH_2OH \\ CHOH \\ CH_2OH \end{array} + 3\,RCOOH$$

The reaction is reversible; in practice the equilibrium can be shifted to the right by using a large excess of water, high temperatures, and high pressures.

Hydrolysis is catalyzed by inorganic and organic acids, e.g., sulfonated hydrocarbons. In the enzymatic hydrolysis of glycerides with pancreatic lipase, the fatty acid groups in the 1- and 3-positions, are split off preferentially.

A fat can also be hydrolyzed with alkali (saponification, see also $\to$ Soaps):

$$\begin{array}{l} CH_2OCOR \\ CHOCOR \\ CH_2OCOR \end{array} + 3\,NaOH \longrightarrow \begin{array}{l} CH_2OH \\ CHOH \\ CH_2OH \end{array} + 3\,RCOONa$$

4.2. Interesterification

Like other esters, glycerides can be transesterified by acidolysis or alcoholysis ($\to$ Esters, Organic). In the presence of an alkaline catalyst and an excess of glycerol, triglycerides form a mixture of mono- and diglycerides (alcoholysis).

The acyl groups of glycerides can also be exchanged inter- and intramolecularly without addition of acids or alcohols (interesterification).

$$\begin{array}{l} CH_2OCOR^1 \\ CHOCOR^1 \\ CH_2OCOR^1 \end{array} + \begin{array}{l} CH_2OCOR^2 \\ CHOCOR^2 \\ CH_2OCOR^2 \end{array} \rightleftharpoons$$

$$\begin{array}{l} CH_2OCOR^2 \\ CHOCOR^1 \\ CH_2OCOR^1 \end{array} + \begin{array}{l} CH_2OCOR^1 \\ CHOCOR^2 \\ CH_2OCOR^2 \end{array} \rightleftharpoons \text{etc.}$$

Even at 200 – 300 °C interesterification proceeds very slowly, but the reaction can be accelerated by using an alkaline catalyst such as a metal alkoxide. With such a catalyst the reaction is complete within one minute at 80 °C. Interesterification is of practical importance since it enables the physical properties of a fat, e.g., melting behavior and consistency, to be modified without changing the fatty acids chemically, as occurs in hydrogenation (hardening).

Interesterification may be either random or directed. *Random interesterification* leads to a random distribution of the fatty acid groups over the triglyceride molecules. This is demonstrated for interesterification of equal proportions of tristearin (S – S – S) and triolein (O – O – O):

$$\begin{array}{cc} S-S-S & + & O-O-O \\ 50\% & & 50\% \end{array}$$
$$\downarrow$$

S–S–S	S–O–S	O–S–S	S–O–O	O–S–O	O–O–O
12.5%	12.5%	25%	25%	12.5%	12.5%

S = stearic acid, O = oleic acid

In *directed interesterification*, the temperature is reduced to such an extent that the highest melting glycerides are continously frozen out of the reaction mixture, in turn continuously shifting the reaction equilibrium. In this way a fat can be separated into higher and lower melting fractions. The higher melting fraction contains the glycerides of saturated fatty acids (stearin fraction), whereas the glycerides of unsaturated fatty acids are found in the lower melting fraction (olein fraction). Directed interesterification of stearodiolein can yield 33.3 % of tristearin and 66.7 % of triolein:

O–S–O
↓
S–S–S   O–O–O
33.3 %   66.7 %

S = stearic acid, O = oleic acid

4.3. Hydrogenation

The double bonds in a fatty acid chain can be wholly or partially saturated by addition of hydrogen in the presence of a suitable catalyst such as nickel, platinum, copper, or palladium. Hydrogenation always leads to an increase in melting point and is therefore also called "hardening". Partial hydrogenation can lead to isomerization of cis double bonds to trans double bonds.

The catalyst, the oil, and the hydrogen must be brought into mutual contact under suitable temperature and pressure conditions. The reaction rate depends on mixing intensity, the type of oil or fat, temperature, catalyst activity, and concentrations of catalyst and dispersed hydrogen. Hydrogenation is an exothermic process. Industrial nickel catalysts are generally obtained by precipitation of nickel hydroxide or carbonate on kieselguhr, silica gel, alumina, or similar carriers, followed by reduction to metallic nickel, or by in situ production of metallic nickel from nickel formate. Such heterogeneous catalysts have a large activated surface. During hydrogenation the double bonds form transient complexes with the active centers of the catalyst. These complexes disintegrate after reaction of the double bonds with hydrogen, leaving the catalyst in its original form [26]. The active centers of the catalyst can be inactivated or poisoned by a number of compounds such as phospholipids, sulfur compounds, organic acids, and oxidized lipids.

A fatty acid with several double bonds, such as linolenic acid ($C_{18:3}$), is hydrogenated more quickly to linoleic acid ($C_{18:2}$) or oleic acid ($C_{18:1}$) than is linoleic acid to oleic acid or oleic acid to stearic acid ($C_{18:0}$). The reaction sequence occurring during hydrogenation can be represented schematically as follows:

$$C_{18:3} \xrightarrow[+H_2]{k_1} C_{18:2} \xrightarrow[+H_2]{k_2} C_{18:1} \xrightarrow[+H_2]{k_3} C_{18:0}$$

The term selectivity is used to indicate which of these reactions is fastest. Selectivity I is defined as the ratio k_2/k_3; it is related to the proportion of saturated glycerides formed and to the melting behavior of the product. Selectivity II, expressed as the ratio k_1/k_2, must be as high as possible if the concentration of linoleic acid in the hydrogenated product is to be maximized [27].

Selectivity can be influenced by the catalyst type (surface area, pore size, etc.) and by altering the reaction conditions. An increase in selectivity, i.e., an increase in partial hydrogenation, promotes isomerization of cis to trans double bonds.

At temperatures above 200 °C and with a low hydrogen concentration, catalytic hydrogenation of polyunsaturated fatty acid groups can lead to the formation of traces of cyclic aromatic compounds [28].

The double bonds of substituted fatty acids such as ricinoleic acid can also be hydrogenated under suitable reaction conditions. Cyclopropane or cyclopropene groups behave as double bonds and lead to branched fatty acids on hydrogenation.

Iron pentacarbonyl and cobalt octacarbonyl are examples of homogeneous hydrogenation catalysts.

Reduction with hydrazine does not lead to isomerization; there is also no selectivity ($k_1 = k_2 = k_3$).

4.4. Isomerization

Naturally occurring fatty acids exist predominantly in the cis form. An equilibrium mixture in which the higher melting trans form predominates can be formed by heating to 100 – 200 °C in the presence of catalysts such as nickel, selenium, sulfur, iodine, nitrogen oxides, or sulfur dioxide.

If selenium or oxides of nitrogen and sulfur are used in the cis – trans isomerization (elaidinization) of oleic acid, there is virtually no positional isomerization. However, cis–trans isomerization of linoleic and linolenic acid leads to conjugated double bonds.

Nonconjugated systems can be isomerized into conjugated systems by heating in an alkaline solution at 200 °C (→ Fatty Acids, Section 2.1.). If reaction times and temperatures are extended, linolenic acid can be converted into cyclohexadiene and benzene derivatives:

$$\text{cyclohexadiene derivative with }(CH_2)_nCH_3\text{ and }(CH_2)_mCO_2H \quad\quad \text{benzene derivative with }(CH_2)_nCH_3\text{ and }(CH_2)_mCO_2H$$

Isomerization can occur if oils and fats are heated at temperatures above 100 °C in the presence of bleaching earth, kieselguhr, or activated charcoal.

4.5. Polymerization

Dimeric, oligomeric, and polymeric compounds are formed by heating unsaturated fatty acids at 200–300 °C [29]. The rate of polymerization increases with increasing degree of unsaturation; saturated fatty acids cannot be polymerized. Thermal polymerization of polyunsaturated fatty acid groups is normally preceded by isomerization and conjugation of double bonds. Thermal polymerization involves formation of new carbon–carbon bonds by combination of acyl radicals and by Diels–Alder reactions, while oxidative polymerization involves formation of C–O–C bonds. Thermal dimerization is catalyzed by Lewis acids such as boron trifluoride; industrial processes for dimerizing oleic acid are based on this principle [30]. Heating of oils during refining or during household use does not lead to a significant increase in dimeric triglycerides. Up to 2 % dimeric triglycerides can be encountered in fresh raffinates; these dimers are not toxic, and are largely excreted as such.

4.6. Autoxidation

Autoxidation, the oxidation of olefins with oxygen, plays a decisive role in the development of rancidity, off-flavors, and reversion flavors in oils and fats during their production and storage. Autoxidation of oil-containing products such as oilseeds and spent bleaching earths can lead to their spontaneous combustion. Autoxidation of drying oils is an important initial stage of polymerization leading to stable surface films (→ Drying Oils and Related Products, Chap. 3.).

Autoxidation involves the formation of a hydroperoxide on a methylene group adjacent to a double bond; this step proceeds via a free-radical mechanism:

$$-CH_2-CH=CH- \xrightarrow{\text{Activation}} -\overset{\bullet}{C}H-CH=CH-$$

Chain reaction:

$$-\overset{\bullet}{C}H-CH=CH- + O_2 \longrightarrow -\overset{|}{\underset{OO^\bullet}{C}H}-CH=CH-$$

$$-\underset{OO^\bullet}{\overset{|}{C}H}-CH=CH- + -CH_2-CH=CH- \longrightarrow$$

$$-\underset{OOH}{\overset{|}{C}H}-CH=CH- + -\overset{\bullet}{C}H-CH=CH-$$

Autoxidation is characterized by an induction period during which free radicals are formed. This phase is triggered by light (photooxygenation), heat, and the presence of compounds that readily form free radicals (e.g., hydroperoxides, peroxides, and transition metals). Photooxygenation, i.e., light-induced oxidation, leads to a particularly fast buildup of radical concentration. The formation of singlet oxygen under the influence of short-wave radiation and a sensitizer such as chlorophyll or erythrosine probably plays a key role in this reaction.

The reactivity of a methylene group in forming a hydroperoxide is enhanced by a second adjacent double bond. Hence linoleic acid oxidizes 10 to 20 times faster than oleic acid. Linolenic acid reacts about three times faster than linoleic acid, since two doubly activated methylene groups are present. The olefin radical formed subsequently isomerizes; in a 1,4-diene such as linoleic acid, isomerization leads to conjugated hydroperoxides:

$$-CH=CH-\underset{13\quad 12\quad 11\quad 10\quad 9}{CH_2-CH=CH-}$$

$$\downarrow$$

$$-CH=CH-\overset{\bullet}{C}H-CH=CH-$$

$$\swarrow \quad\quad \searrow$$

$$-\overset{\bullet}{C}H-CH=CH-CH=CH- \quad -CH=CH-CH=CH-\overset{\bullet}{C}H-$$

$$\downarrow O_2 \text{ etc.} \quad\quad\quad \downarrow O_2 \text{ etc.}$$

$$-\underset{OOH}{\overset{|}{C}H}-CH=CH-CH=CH- \quad -CH=CH-CH=CH-\underset{OOH}{\overset{|}{C}H}-$$

The intermediate hydroperoxides are labile compounds that decompose into a number of different products: epoxides, alcohols, diols, keto compounds, dicarboxylic acids, aldehydes, and isomerization and polymerization products. The volatile carbonyl compounds formed in this process are responsible for the taste and odor of oxidized oils and fats.

When the radical concentration has reached a certain limit, the chain reaction is gradually stopped by mutual combination of radicals.

Antioxidants prolong the induction period by reacting with the intermediate products of the chain reaction, forming inactive radicals. Tocopherols are naturally occurring antioxidants. Butylhydroxyanisole (BHA), butylhydroxytoluene (BHT), and propyl gallate are among the most effective synthetic antioxidants ($\rightarrow$ Antioxidants). Certain organic or inorganic acids (citric, tartaric, ascorbic, phosphoric acids) have a synergistic effect without being true antioxidants; this effect is presumably based on the inactivation of trace metals or reduction of oxidized antioxidants.

5. Manufacture and Processing

Oils and fats are either of vegetable or animal origin. The approximate proportions of the corresponding world production are 55 % vegetable oils, 40 % land-animal fats, and 5 % marine oils.

5.1. Vegetable Oils and Fats

5.1.1. Storage and Handling of Raw Materials

The handling of oleaginous seeds and fruits during transport and storage has a decisive influence on the quality of the crude oils. The oils in fruit pulp (e.g., olive and palm) are very susceptible to enzymatic hydrolysis since they are finely dispersed in moist cell tissue. The oils contained in seeds are more stable but can also be attacked. A high moisture content accelerates the uptake of oxygen and the corresponding release of carbon dioxide due to degradation of starch. The heat generated in this reaction promotes lipolysis, growth of microorganisms (which can lead to the formation of mycotoxins such as aflatoxins), formation of undesirable color and odor, and – in extreme cases – may result in a coagulation of seed in a silo and even spontaneous combustion [31], [32]. Damage of seed cells, mechanically or by pests, also promotes lipolysis and lipoxidation. Free fatty acids, oxidation products, and coloring matter formed by lipolysis, lipoxidation, and degradation of protein and carbohydrate can in general be removed from the oil only with difficulty, with corresponding yield losses.

In order to minimize these effects, oilseeds must be dried before storage, preferably under mild conditions to avoid cellular damage. The critical moisture content roughly correlates with the hygroscopic equilibrium at 75 % relative humidity and varies between 6 and 13 %, depending on the protein and carbohydrate content. In the case of copra, rapeseed, and sunflower seed, the critical moisture content is 7 %; in the case of soybeans 13 %.

Transport and handling in bulk has almost entirely superseded that in bags. Seed is normally stored in concrete silos that are ventilated to avoid local generation of heat and water pockets; such silos can be up to 70 m in height and up to 12 m in diameter. The silos are filled from the top, emptied from the bottom, and normally equipped with special vibrating devices at the bottom to minimize bridging.

5.1.2. Cleaning and Dehulling

On arrival at the oil mill, oilseeds still contain plant residues, dust, sand, wood, pieces of metal, and foreign seed, which must be removed prior to further processing by screening, air classification, and passage over magnets.

Some oilseeds are dehulled (decorticated) before being further processed, especially if only the oil is to be expelled, since the hulls tend to retain part of the oil. Preexpelling and subsequent extraction with solvents is the most common process; a certain proportion of hulls in the material to be extracted can be desirable since they facilitate percolation. An excessively high proportion of hulls can, however, impair the quality of the crude oil since the solvent also dissolves wax from the seed coat. A further reason for dehulling can be the need to produce so-called high protein meal, for example from soybeans.

Dehulling is normally preceded by relatively intense drying of the seed material to help loosen the seed coat from the meats. Dehulling is normally performed by using impact disintegration or passage through rolls to break up the seeds. The hulls are separated by screening and air-classification.

5.1.3. Expelling

Modern processing of oilseeds generally involves a combination of expelling and solvent extraction. Seed is normally preexpelled to a residual fat content of ca. 20 %, and the expeller cake is then extracted to 1 – 2 % fat. Preexpelling is omitted if the fat content of the seed is 20 % or less. High-pressure expelling down to ca. 4 % residual fat is no longer important, mainly because of the higher processing costs.

Most of the fat present in seeds and fruit pulp is in the endosperm and hypocotyl cells, with much less in the seed coat. The processing conditions are largely determined by the size and stability of the oil-containing cells. With palm fruit and olives, mere heating or boiling with water suffices to burst the membranes of the oil-containing cells to liberate the oil. Nearly all oleaginous seeds must, however, first be comminuted and thermally pretreated to isolate the fat in an acceptable yield. Comminution partially destroys the cellular structure, increasing the total internal surface and facilitating access to the oil within the seed.

Coarse materials such as copra are first broken by passage through interlocking rolls. Fluted rolls are used for nearly all other seeds. Smooth or so-called flaking rolls are used to complete seed comminution. These three types of rolls are shown in Figure 3; they can be arranged in parallel, diagonally, or sequentially. In the more modern, parallel arrangement, the rotational speed of the two rolls differs by 5 – 10 % to produce an additional shear effect. The separation of the rolls and their circumferential speed is adjusted to the type of seed and the degree of cell rupture required for the subsequent expelling and extraction.

The comminuted seed is conditioned prior to expelling by moistening and heating in suitable equipment. During this treatment lipoproteins decompose, proteins coagulate, and intracellular oil bodies coalesce. The cell walls themselves are not macerated during conditioning. Coagulation of protein facilitates expelling and subsequent percolation in the extractor. Undesirable enzymes and microorganisms are deactivated at temperatures above 80 °C and in the presence of sufficient moisture [33].

By suitably conditioning cottonseed, the toxic gossypol can be deactivated by association with denatured protein. In rapeseed, suitably severe conditioning deactivates the enzyme myrosinase and hence improves the quality of the meal and the oil. However, excessive heating impairs the nutritive quality of the expeller cake as well as the color and taste of the expeller oil and must be avoided. An excessively high moisture content can reduce the yield of expeller oil. Each type of seed must therefore be adjusted to an optimal moisture content and temperature during conditioning.

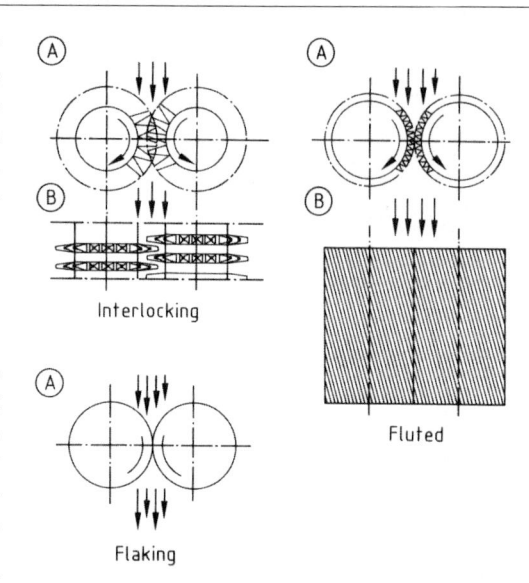

Figure 3. Schematic representation of various rolls used for expelling oilseeds A) Side view; B) Top view

Conditioning is performed in horizontal jacketed tubes (conditioners) or in vertical stack cookers. The stack cooker basically consists of up to five kettles arranged in a tier and through which the seed material successively moves from the top to the bottom; each kettle is equipped with a sweep stirrer and facilities for direct and indirect steam heating (Fig. 4). Generally, wet cooking with direct steam takes place in the top kettle, and drying, with increased temperatures and venting, in the bottom kettles.

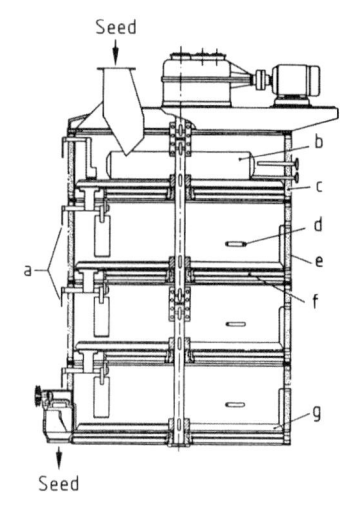

Figure 4. Vertical stack cooker for conditioning oilseeds (Krupp) a) Level control; b) Steam heating; c) Insulation; d) Steam jet; e) Manhole; f) Steam heating; g) Sweep stirrer

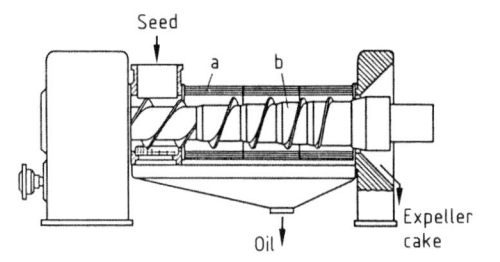

Figure 5. Screw press (Krupp) a) Drainage barrel; b) Worm shaft

Expelling (pressing) separates the oil from the solid phase, the so-called expeller cake, and can be done in continuous or batch presses. Levers, wedges, or screws were used to apply pressure in the more primitive types of batch presses, but modern batch presses are almost invariably operated by a hydraulic system. They can be divided into the open type, in which the seed must be wrapped in press cloths, and the closed type, which dispenses with press cloths and confines the material in a cage-type construction. Open-type presses can be subdivided into plate and box presses, closed types into pot and cage presses.

Continuous expellers or screw presses (e.g., Simon – Rosedowns, Krupp) have generally replaced batch presses. Deoiling in these presses is efficient and uniform since the layer of material to be expelled is relatively thin and is continuously broken up. Several types exist, all with the common principle of a worm shaft rotating in a cylindrical drainage barrel (Fig. 5). The barrel is composed of armored, rectangular bars that fit into the barrel bar frame. The individual bars are separated by bar spacing clips, the specific spacings – ranging from 0.1 to 0.4 mm – depending on the type of seed and degree of expelling required. The required pressure is built up either by a gradual increase in the diameter of the worm shaft or by a gradual restriction of the barrel diameter. Specific pressure profiles and cake thickness can be further controlled by variations in the pitch of the worm and choke mechanisms in the barrel. The expelled oil is discharged through the barrel spacings while the expeller cake emerges at the end of the barrel.

The capacity of screw presses varies widely, being dependent primarily on the cross-sectional ratio and the speed of the screw. The pressure should be built up fairly slowly. Initially, the channels in the expeller cake are sufficiently large for ready discharge of oil; in the latter sections of the screw press, the pressure is increased to about 300 MPa (3000 bar) to expel more tightly bound oil and to burst intracellular oil bodies.

Up to 200 t of seed per day per press is normally preexpelled to a residual fat content of 15 – 25 %. The expeller cake is then comminuted on fluted rolls, and possibly also pelletized, prior to solvent extraction.

Screw presses used to reduce the residual fat content to 3 – 6 % are similar in design to preexpellers but have a larger compression ratio (1 : 25 instead of 1 : 15) and have a lower worm speed (9 – 12 rpm versus 30 – 45 rpm).

A specific worm arrangement normally gives optimal results for only one particular type of seed. Difficulties encountered on changing the type of seed can often be overcome by modifying conditioning and comminution. Preexpellers that do not require seed conditioning of sunflower seed have been developed; however, they do not work as well with unconditioned rapeseed.

The expeller oil still contains so-called foots, which must be removed by filtration.

5.1.4. Extraction [34], [35]

Very much lower residual fat contents can be achieved by solvent extraction than by expelling.

Seed material with an oil content of ca. 20 % and expeller cake with 15 – 25 % fat are usually subjected to solvent extraction. Both must be flaked prior to extraction. This pretreatment creates a large internal surface and yields thin (ca. 0.3 mm), firm flakes, which form a loose layer in the extractor and provide uniform, short diffusion pathways. The flakes must have a minimum moisture content and elasticity to prevent them from crumbling during transport to the extractor, which would lead to too dense a packing and hence poor percolation of solvent. However, a high moisture content also impairs percolation.

Ideally, the extraction solvent should dissolve only glycerides but not undesirable components such as coloring matter, gums, and phospholipids. The solvents must not contain toxic components, and should be recoverable with minimum loss, be safe in handling, and be readily removable from the extracted material. For these reasons aliphatic hydrocarbons, especially hexane, are used almost exclusively. Technical hexane with a boiling point range of 55 – 70 °C has proved to be optimal. Hexane can be readily removed from the oil at temperatures below 100 °C in vacuo and can be stripped from the meal with steam. The solubility of hexane in the condensed water is only 0.1 %.

For special purposes, e.g., the production of heat-labile pharmaceuticals, lower-boiling hydrocarbons such as pentanes are sometimes used. Extraction with propane or carbon dioxide under supercritical conditions is reserved for special products of high intrinsic value because of the high equipment costs involved. Castor oil, being relatively polar, is preferably extracted with the higher-boiling heptane. Alcohols (methanol, ethanol, propanol, and butanol) and furfural are specially suited for the extraction of relatively wet materials. Extraction of oilseeds with alcohols leads to relatively high concentrations of phosphatides, glycolipids, carbohydrates, and similar constituents in the crude oil, although the glycerides can in principle be concentrated by cooling or extraction of the alcoholic solutions. Alcohols are generally not used as primary extraction solvents. However, they are occasionally used as secondary extraction solvents to remove gossypol from cottonseed meal, thioglycosides from rapeseed meal, sugars from soybean meal (to produce protein concentrates), and alkaloids from bitter lupine meal.

Chlorinated hydrocarbons such as trichloroethylene and dichloromethane are of interest because of their safety in handling and high extraction capacity but do not appear to be used for extraction of oilseeds because of the potential toxicity risk from residual solvent in the meal. Processes in which the oil is displaced by hot water are known but have not achieved industrial importance [36].

Oilseeds are generally extracted in a countercurrent process: pure solvent is contacted with material that has already been largely extracted, and the oil-rich solution is contacted with nonextracted material. Such extractions can be performed either continuously or in a batch process.

Batch Processes. In a discontinuous process the solvent successively passes through a battery of up to batch extractors (e.g., from Buss). Each of these extractors is a cylindrical vessel with a large diameter : height ratio and has an opening for filling and emptying, a screen in the bottom part, and facilities for pumping solvent in and out, for direct and indirect steam heating, and a sweep stirrer to move the extracted meal during solvent stripping (desolventization) (Fig. 6).

After filling the extractor with comminuted seed material, the upper opening is closed and solvent pumped in. The displaced air escapes via a circuit; the hexane vapors in the displaced air are recovered. After the extractor has been filled with solvent, the hexane containing the extracted oil is pumped into a second extractor

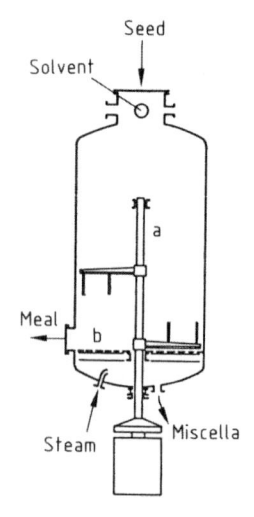

Figure 6. Batch extractor a) Stirrer; b) Screen

filled with seed material. This process is repeated until the hexane solution has been pumped through 3 or 4 extractors. The solvent strongly enriched with oil (the miscella) from the last extractor is filtered to remove seed particles and then distilled. In this type of process, exhaustive extraction can yield oil concentrations of at least 35 % in the final miscella.

When the contents of the first extractor have been extracted sufficiently, the solvent tap is closed; the pure extraction solvent is then passed directly to the second extractor while the fifth extractor becomes part of the battery. After discharging the solvent, remaining solvent is stripped from the extracted seed material by passing in steam. When the vapors leaving the extractor are free of solvent, the extractor is emptied via the lower opening.

The sequence of filling, extraction, desolventization and emptying is repeated for each extractor in a battery in such a way that while one extractor is being filled, the contents of three or four extractors are extracted, residual solvent is stripped in one or two extractors, and one extractor is emptied.

The vapors from the extractors and the distillation are condensed. The resulting hexane – water mixture is separated mechanically and the hexane recycled. Such a battery with 10 extractors of 7 m^3 each can reach a daily throughput of up to 1200 t seed, depending on the type of seed. The quantity of hexane circulated is ca. 100 t. The solvent loss is normally between 0.35 and 0.45 %, based on seed input.

These discontinuous plants can be automated, and have the principal advantage that special treatment of meal (e.g., detoxification of aflatoxin-containing peanut or cottonseed meal with ammonia) can be carried out fairly easily. Their capacities cannot, however, match those of fully continuous plants, which can process up to 200 t seed per hour.

Continuous Processes. Of the many fully continuous processes that have been proposed and developed, countercurrent gravity percolation has become the most popular because of its high capacity. The comminuted seed material is transported through the extractor and sprayed with solvent in the various extraction stations. The solvent becomes increasingly enriched with oil and is passed through the seed bed repeatedly

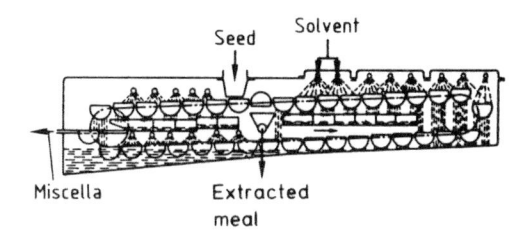

Figure 7. MIAG basket conveyor extractor

according to the number of extraction stations. This arrangement has the advantage that the miscella is continuously cleaned by filtration through the seed bed.

The MIAG basket conveyor extractor (Fig. 7) uses a horizontal arrangement of baskets in which the flaked seeds are first sprayed with solvent or miscella and then passed through a solvent bath. The baskets containing the extracted and drained flakes are automatically inverted and discharged into a hopper from which the meal is conveyed to the desolventizers and dryers.

In modern plants with a capacity of up to 5000 t of seed per day, circular cell and traveling belt horizontal extractors are generally preferred. The Blaw – Knox Rotocel extractor is especially popular because it offers a large capacity within a relatively small space. It uses baskets that travel in a horizontal circle. After countercurrent extraction, the baskets are emptied through hinged sieve bottoms.

The carousel extractor from Extraktionstechnik operates on a similar principle (Fig. 8), but has a fixed discharge point instead of movable basket bottoms.

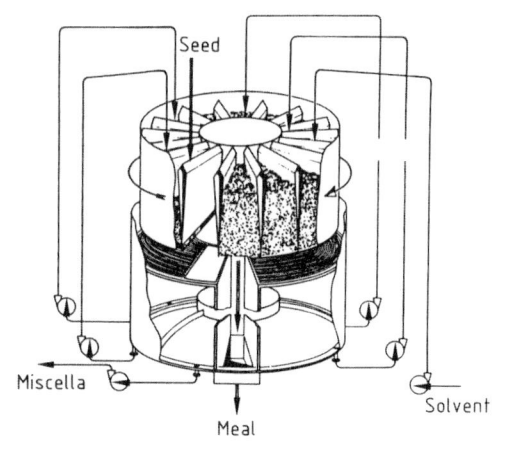

Figure 8. Carousel extractor

In traveling belt extractors, such as the De Smet extractor, the seed material is conveyed on an endless sieve belt, the height of the bed being regulated mechanically. Miscella and solvent are sprayed onto the seed material. Miscella percolates throught the belt, falls into compartments in the bottom of the extractor housing, and is picked up by a series of pumps and recirculated countercurrent to the flakes.

The Lurgi frame belt extractor works on a similar principle [37]. The endless frame belt runs over two bar-sieve belts which form the bottoms of the frames. The flaked oilseeds are filled into the frames of the upper belt, tipped onto the lower belt after half the extraction time has elapsed, and then discharged after completion of extraction and passage through a dripping zone. This separation into distinct extraction steps and intermediate decompaction effects a particularly uniform extraction.

There are numerous other types of extractors which are of less practical importance. In Hildebrandt, Schlotterhose, Ford, Detrex, Adler, and Olier extractors, the comminuted seed material is transported through a solvent bath by a screw; Figure 9 shows the principle involved. In a Kennedy extractor (Fig. 10), the material to be extracted is conveyed to various sections of a trough by means of impeller wheels. In the Sherwin – Williams process the finely comminuted seed material is thoroughly mixed with solvent in three stages and then separated centrifugally from the miscella. The structure of the seed material is not as critical for these extractors as for basket or belt-type extractors. However, they require a far greater ratio of solvent to seed material. Their miscella are also contaminated with fines, which requires a separate filtration step. These types of extractors are therefore virtually obsolete.

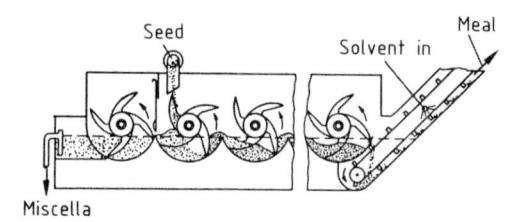

Figure 10. Kennedy extractor

Böhm, Tyca, Allis – Chalmers, Bonotto, and Anderson extractors are essentially columns divided into a number of sections in which the seed material moves from the top to the bottom, countercurrent to a rising flow of solvent. These extractors do not match the performance of the large basket or belt-type extractors.

Several attempts have been made to develop processes obviating preexpelling. In the Filtrex process, which is used to a limited extent, oilseeds such as rapeseed, sunflowerseed, or peanuts can be extracted directly to about 1 % residual fat without preexpelling, provided the seed material has been flaked to about 0.1 mm (the cell structure is probably destroyed) and subsequently conditioned by "crisping" [38]. A drawback is that emersion-type extractors are required. In the Direx process the seed material is only coarsely ground, extracted to 14 – 16 % residual fat with hexane, flaked, and then extracted to ca. 1 % residual fat in a second extraction step. The theory is that flaking the hexane-moist material bursts the oil cells. This process has been applied only on a small scale (ca. 5 t/h).

A more recent development is the Alcon process [39] for soybeans, in which the essential step is to heat the flakes to ca. 100 °C with direct steam, followed by drying. The special flake conditioning prior to extraction deactivates enzymes such as phospholipases and lipoxygenase, and presumably dissociates lipoprotein complexes; this leads to lower residual fat contents in the meal and a more easily deslimable crude oil with improved oxidation characteristics.

Extracted oilseeds can be desolvized in a series of horizontal steam-jacketed tubes. The material is propelled through the tubes by screws and is then passed through toasters, similar to stack cookers, that are heated by both live and indirect steam. Alternatively, desolventization can be performed in one step in desolventizers – toasters. After drying and cooling, the meal is

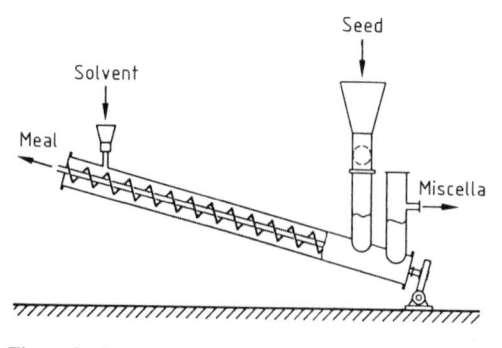

Figure 9. Screw extractor

ready for use as an animal feed component. Toasting not only reduces residual solvent to levels below the lower explosion limit but also deactivates nutritionally undesirable enzymes such as urease and trypsin. In some cases, superheated solvent vapor is used for desolventization. This process and flash desolventization are used when minimal protein denaturation is desirable.

Solvent is recovered from the filtered miscella by distillation in multiple-stage evaporators, which are constructed to minimize thermal damage to the oil. The miscella is first concentrated to 96 % oil and finally stripped with injected steam in a falling-film evaporator. The vapors from the toaster, dryer, cooler, and distillation unit are condensed and separated into water and hexane, which is recycled. Vented air is passed through special adsorbents such as activated charcoal or mineral oil to recover solvent. Total solvent loss is 0.2 – 0.3 %, based on seed input.

5.2. Land-Animal Fats

Animal fats can be readily isolated from tissue by heating since the cell membranes of animal cells are much weaker than those of plant cells. The intracellular fat expands on heating and bursts the membrane. Extraction with solvents is limited to offal. Fatty tissue is particularly susceptible to decomposition so that rapid processing is necessary if refrigeration is not possible. Microbial and autolytic degradation rapidly lead to oxidation, hydrolysis, color deterioration, and unpleasant odors and flavors, which are difficult to remove from the fat. The free fatty acid content therefore serves as a convenient, primary quality criterion.

In *dry rendering*, the comminuted fatty tissue is digested by cooking in steam-jacketed vessels equipped with agitators [40]. Local overheating of fatty tissue must be avoided since this can easily lead to unpleasant off-flavors. In modern automated plants for working up animal byproducts, the raw material is first heated and sterilized in a dry-rendering vessel. The resulting sludge is then separated, possibly after predrying, into crude meal (cracklings) and a liquid phase containing fat and water by a continuous screw press (see Fig. 11, Krupp Kontipress process) or by a decanter, i.e., a centrifuge with a built-in screw conveyor (Alfa–Laval Centrimeal process). The fat is finally separated from the aqueous phase by centrifuging. The 7 – 12 % residual fat content of the meal can be reduced to 1 – 4 % by solvent extraction.

Wet Rendering, which involves treating the fatty tissue with direct steam, tends to give better yields and quality than dry rendering. In modern continuous rendering plants (e.g., as built by Alfa – Laval, Westfalia, and Sharples) selected tissue is coarsely comminuted and heated to 50 – 60 °C. It is then heated quickly to 80 – 90 °C with direct steam to deactivate oxidizing enzymes. The greaves are then separated in a decanter centrifuge, and the fat is clarified by further centrifuging, cooled, possibly texturized, and packed (Fig. 12). Fat obtained by this meth-

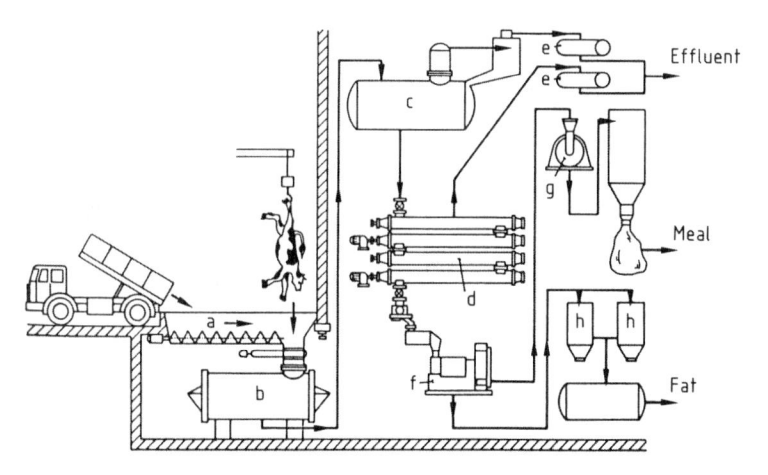

Figure 11. Kontipress system for working up carcasses (Krupp) a) Raw material; b) Cooker – sterilizer; c) Buffer; d) Dryer; e) Condenser aggregate; f) Screw press; g) Meal grinder; h) Buffer

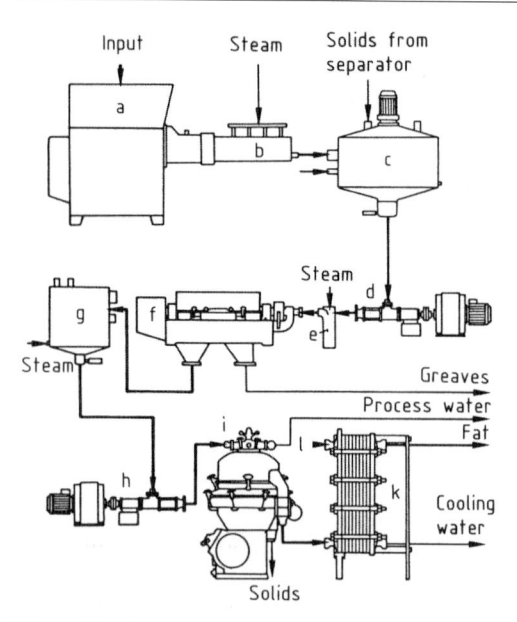

Figure 12. Alfa – Laval Centriflow rendering plant a) Cutter; b) Renderer; c) Buffer; d) Propeller pump; e) Steam heater; f) Decanter; g) Buffer; h) Propeller pump; i) Separator (centrifuge); k) Plate heat exchanger; l) Cooling water

od in up to 99 % yield contains 0.1 – 0.2 % moisture; the content of free fatty acids is virtually identical with that of the raw material. This mild processing technique minimizes the formation of off-flavors. Further refining of lard or tallow for edible purposes is controlled by national legislation.

5.3. Marine Oils

Whale Oil is isolated by rendering. The blubber is cooked with live and indirect steam, and the oil is then separated by centrifugation from insoluble tissue and water-soluble matter.

Similar processes are used to isolate *fish oils* [41]. Crushed fish (herring, menhaden, sardines, pilchards) with a dry matter content of 15 – 22 % is first cooked in closed vessels at relatively low pressure. This operation sterilizes the fish and coagulates the protein to produce a fish mass that can be easily pressed. Subsequent screw pressing separates most of the water and oil from the press cake. The press water is prepurified over screens or with a decanter (desludger) and then separated by centrifugation into crude fish oil and stickwater (or glue water). The press cake (ca. 50 % water) is dried, sometimes also extracted with a solvent, and ground into fish meal. Evaporation of the stickwater gives solubles which are added to the meal. Industrial plants are offered by Alfa – Laval (Centrifish) and Westfalia (see Fig. 13)

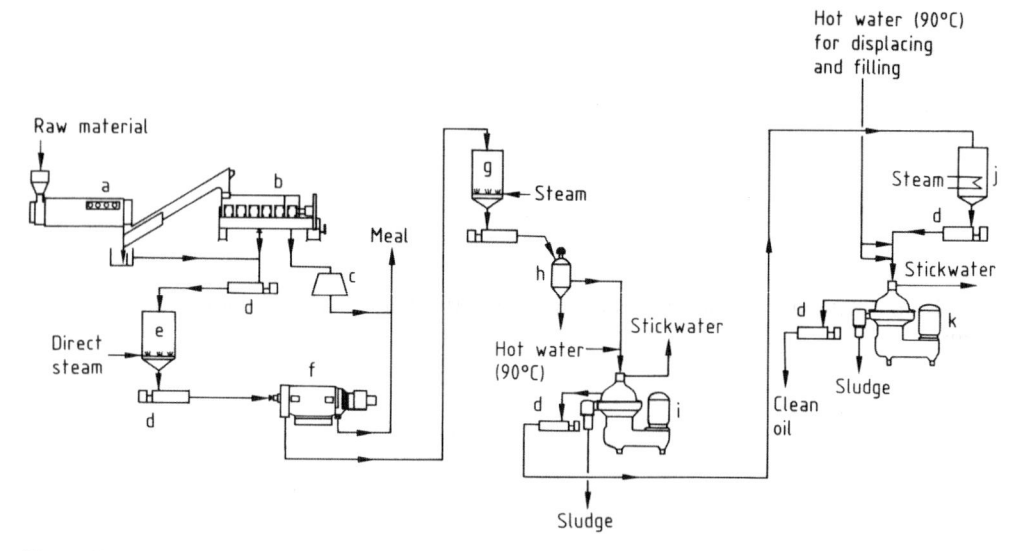

Figure 13. Fish meal and fish oil plant (Westfalia) a) Cooker; b) Power press; c) Crusher; d) Eccentric screw pump; e) Preheater; f) Decanter for clarification of press water; g) Press water preheater (92 °C); h) Rotary brush strainer; i) Separator with self-cleaning bowl (1st stage, press water deoiling); j) Crude oil preheater (92 °C); k) Separator with self-cleaning bowl (2nd stage, oil polishing)

Liver oils are usually obtained on board ship by mild rendering of crushed liver, sometimes assisted by pressure-release disintegration of the oil cells, and subsequent separation in centrifuges. The Solexol process used in South Africa employs liquefied propane as the extraction solvent. This method can also be used to fractionate fishliver oils and to isolate vitamin concentrates.

5.4. Synthetic Fats

Pure synthetic glycerides [20] and phospholipids have few industrial uses. Synthetic fats for edible purposes are of little importance with the exception of specific dietary products such as medium-chain triglycerides of C_8 and C_{10} fatty acids (MCT fats), which are administered in some postsurgical cases [42].

6. Refining

Crude oils and fats obtained by expelling, extraction, or rendering contain trace components which are undesirable for taste, stability, appearance, or further processing. These substances include seed particles, dirt, phosphatides, carbohydrates, proteins, fatty acids, trace metals, pigments, waxes, oxidation products of fatty acids, and toxic components such as polycyclic aromatic hydrocarbons, gossypol, mycotoxins, sulfur compounds (in fish and cruciferous oils), and pesticide residues. The aim of refining edible fats and fats for industrial purposes is to remove these undesirable components as far as possible without significantly affecting the concentration of desirable constituents such as vitamins and polyunsaturated fatty acids, and without significant loss of the major glyceride components. Some oils, such as olive oil, butterfat, and cold-pressed sunflower oil or linseed oil, are not refined in order to preserve their typical flavors.

Refining usually involves the following stages:

1. Precleaning to remove phosphatides (degumming)
2. Neutralization by treatment with lye or by distillation
3. Decolorization by adsorptive treatments (bleaching)
4. Deodorization or stripping in vacuo

6.1. Degumming [43]

Phosphatides, gums, and other complex colloidal compounds can promote hydrolysis of an oil or fat during storage and also interfere with subsequent refining. They are therefore removed by degumming or desliming.

The degumming method depends on the type of the oil and the phosphatide content. In hydration degumming, which involves treating the oil or fat with water or steam, phosphatides absorb water to form a sludge that is insoluble in the oil. This process is used to remove soy lecithin (1 – 3 %) from soybean oil: 2 – 5 % water, based on oil, is intimately mixed with the crude oil at 70 – 80 °C; after a contact time of 1 – 30 min the phosphatide sludge is separated by centrifuging. The degree of hydration can be increased by using acids such as citric or phosphoric, bases, or salts but at the expense of lecithin quality. Such modified hydration processes are therefore confined to post-degumming and to oils whose lecithins have little commercial value. Degumming and neutralization can also be carried out together but usually with increased losses of neutral oil owing to the formation of emulsions. The use of sulfuric acid is generally restricted to industrial oils that are not to be further refined.

Some technical oils such as linseed oil can be degummed by heating to 240 – 280 °C; the precipitation of gums on heating is referred to as oil "breaking."

Precleaning of a crude oil by treatment with bleaching earth or other adsorbents, normally in combination with acids such as phosphoric or citric, can be advantageous.

Oils that are refined by distillation (physical refining, see Distillative Neutralization) should have a very low phosphatide content. This can be achieved by special degumming techniques involving the addition of citric acid followed by hydration at ca. 20 °C for several hours before centrifugal separation [44]. Another "superdegumming" technique involves separating the phosphatide micelles from a solution of the crude oil in hexane by ultrafiltration using special polysulfone or polyacrylonitrile membranes [45].

6.2. Deacidification (Neutralization)

Commercially available crude oils and fats contain on average 1 – 3 % free fatty acids. Whereas a soybean oil can have as little as 0.5 % free fatty acids, some palm and fish oils can contain up to 10 %. The free fatty acid content of refined fats should be below 0.1 %. Although longer chain fatty acids do not generally affect taste, the short-chain acids can impart a soapy, rancid flavor. When the free fatty acids are removed as soaps by treatment with lye, other undesirable constituents such as oxidation products of fatty acids, residual phosphatides and gums, phenols (e.g., gossypol), and aflatoxins are also "washed out." In practice, deacidification is performed mainly by treatment with lye or by distillation. Deacidification by esterification with glycerol, by selective extraction with solvents, or by adsorbents is not industrially important.

Alkali Neutralization [46]. The usual method of alkali neutralization is treatment with weak lye, either batchwise or continuously. The more concentrated the lye, the more readily are undesirable constituents taken up in the soapstock. Dilute alkaline solutions (ca. 1 N) do not saponify the oil, but greater neutral oil losses occur due to occlusion in the soapstock. The losses increase with increasing content of free fatty acids. Concentrated lye (4 – 7 N) yields a relatively concentrated soap which takes up little neutral oil, but it can saponify substantial amounts of the neutral oil, especially in those oils that have a relatively high proportion of short-chain fatty acids. The optimum lye concentration therefore depends on the quality of the crude oil and the desired quality of the refined product. The neutralization process can be further influenced by temperature, residence time, and mode and intensity of mixing the lye and the fat phase.

Batch Neutralization is done in an open or closed vessel (up to 75 t content) with a conical base. The reactor is fitted with a heating mantle, heating coils, and facilities for injecting live steam (Fig. 14). Lye and oil can be mixed without undue emulsification by using special coil and frame stirrers. Soap settles in the conical bottom of the neutralization vessel. Discharge of neutral oil with the soap can be minimized by using ultrasonic or dielectric measurement of the phases. Closed vessels can also be used for bleaching if they can be evacuated. Whereas weak lye is generally sprayed onto the oil at ca. 90 °C without stirring, stronger lye is normally stirred into the oil at 40 – 80 °C. Sodium carbonate is sometimes used instead of caustic soda; it does not saponify but tends to lead to foaming. Oils with a relatively high concentration of free fatty acids are neutralized with strong lye in order to limit the total volume. The required amount of lye, including a slight excess, is based on the free fatty acid content (ffa).

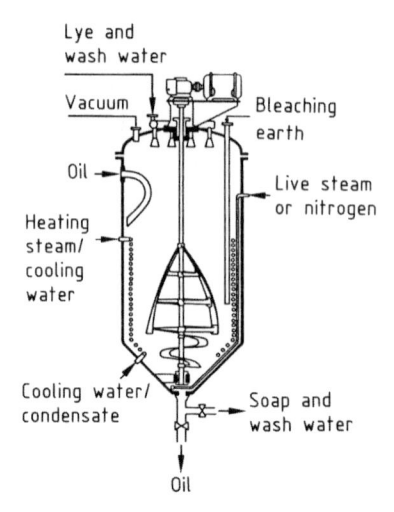

Figure 14. Combined neutralizer – bleacher

After neutralization and discharge of soapstock, the oil is washed with dilute lye (ca. 0.5 N) and then with water in order to reduce the soap content below 500 ppm. Incomplete removal of soap impairs the color of the oil and the efficacy of the following refining steps. Polyunsaturated oils such as soybean oil are sometimes subjected to posttreatment with sodium carbonate – silicate to remove residual traces of oxidized glycerides and phosphatides. Although soapstock can be utilized as such in chicken broiler feed, it is normally converted into "acid oil" (a mixture of fatty acids and neutral oil) by batch or continuous treatment with 30 % sulfuric acid at 70 °C.

The quantity of acid oils is a measure of the refining efficacy. A refining factor of 2 means that for each mass unit of free fatty acid in the crude oil, 2 mass units of acid oil are produced. The difference between acid oil and free fatty acid content is mainly due to occluded neutral oil.

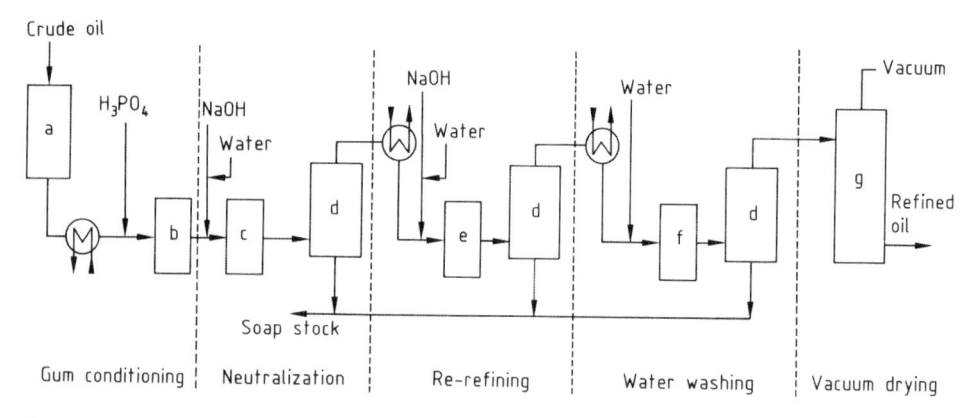

Figure 15. Flow diagram of continuous neutralization of oils and fats a) Crude oil storage tank; b) Gum conditioning mixer; c) Neutralizing mixer; d) Separator (centrifuge); e) Re-refining mixer; f) Water-washing mixer; g) Vacuum dryer

Acid oils are used for technical purposes and as energy carriers in animal feeds [47].

Batch neutralization is more flexible than continuous neutralization but gives lower refining yields.

Continuous Processes achieve the same or greater capacities in a smaller space with better yields, provided long continuous runs can be maintained. Continuous plants from Alfa – Laval, Westfalia, and Sharples are widely used (Fig. 15). In these plants, the oil is generally first conditioned by addition of citric or phosphoric acid in small continuous mixing chambers and then brought into contact with lye. Soap and gums are separated in centrifuges. After re-refining and passage through a washing centrifuge, the oil is practically free of fatty acid and soap. The Sharples plant uses high-speed open centrifuges with a high separation efficiency, while the Alfa – Laval short-mix lines employ completely closed centrifuges with a mean residence time of only a few seconds. The capacity can be increased considerably by using self-discharging centrifuges (e.g., from Westfalia).

The Zenith process employs a different principle: a fine stream of oil rises through a vessel filled with dilute caustic soda and is skimmed off continuously at the top. The spent lye must be replaced. This type of neutralization requires efficient pre-degumming and has the advantage of relatively low losses of neutral oil [48].

Neutralization with aqueous ammonia is an old process which is being reexamined because of environmental pollution problems with lye. The ammonium soaps can be separated by centrifuging and split into fatty acids and ammonia by heating, the ammonia being recycled [49].

Distillative Neutralization. In distillative neutralization (physical refining) the free fatty acids are continuously removed from the crude oil with water vapor in vacuo. Prior to this stripping, gums, phosphatides, and trace metals must be almost completely removed since these compounds impair the oil quality during distillation. They are preferably removed by treatment with acids and bleaching earth.

The economics of distillative neutralization improve with increasing content of free fatty acids. The fatty acid factor is approximately 1.1, thus the neutral oil losses are far lower than with alkali neutralization. An added advantage is the avoidance of soap-stock splitting and the consequent effluent problems. Distillative neutralization was formerly restricted to fairly saturated oils with a high free fatty acid content (e.g., coconut oil, palm oil, and tallow). However, it is now being extended to polyunsaturated oils such as rapeseed, soybean, and sunflower oil because techniques are available for reducing the phosphatide content of the crude oils to sufficiently low levels to obviate lye neutralization [50].

Temperatures between 240 and 270 °C are normally used to achieve free fatty acid contents of less than 0.1 %. At these temperatures decomposition and removal of carotenoid pigments result in a significant bleaching effect, this is particularly noticeable in palm oil. Older plants such as those built by Wecker and Feld & Hahn have now been generally superseded by modern plants (e.g., Lurgi, Girdler, HLS, and EMI), which have the advantage of a smaller spread in residence time and hence better process control (especially in combined deacidification, deodorization, and heat bleaching). The latter plants are very similar to the corresponding deodorizer plants (see Figs. 20 and 21) except for an additional fatty acid condenser between the deodorizer and the vacuum assembly.

Other Processes. *Esterification* of the free fatty acids with glycerol may be economically feasible for highly acid oils. The esterification is carried out at 160 – 210 °C in the presence of catalysts such as magnesium oxide, alkali silicates, or metals (Zn or Sn).

Solvent Extraction of free fatty acids is of interest only for highly acid oils. Olive oil with 22 % ffa can be deacidified to ca. 3 % ffa by liquid – liquid extraction with ethanol. Furfural can be used to dissolve polyunsaturated glycerides and fatty acids.

In the Solexol process, liquefied propane is used as solvent. Saturated glycerides are taken up in the propane phase while nonsaponifiable matter, fatty acids, oxidation products, and polyunsaturated glycerides remain largely undissolved. This process has been used to fractionate fish and fishliver oils and may increase in importance with the advent of supercritical fractionation and extraction techniques.

Various adsorbents such as silica gel and alumina have been proposed for the *selective adsorption* of free fatty acids [51]. Adsorption on strongly basic ion-exchange resins has also been used. Plant-scale application of such processes has, however, not been reported.

6.3. Bleaching

Degumming followed by alkali neutralization generally does not lead to significant decolorization of an oil or fat. A bleaching step with solid adsorbents such as bleaching earth or activated charcoal is normally employed. Bleaching with air or chemicals is not used for edible fats.

The aim of adsorptive treatment is to remove pigments such as carotenoids and chlorophyll but also residues of phosphatides, soaps, trace metals, and oxidation products such as hydroperoxides and their nonvolatile, polar decomposition products. These compounds can have a deleterious effect on the course of further processing (especially hardening and deodorization) and on the quality of the final product. Some of the oxidation products that are removed by adsorption can promote oxidation of the oil. Since some of these undesirable compounds can be removed by suitable degumming and neutralization, there is an interdependence between the pretreatment conditions and the treatment with adsorbents. This interdependence can also involve deodorization (heat bleaching) where, for example, the carotenoid content is reduced.

The adsorption of pigments follows Freundlich's isotherm: the amount of pigment adsorbed decreases with decreasing concentration of the pigment in the oil. Although adsorption should be most efficient if the oil is passed through a layer of adsorbent, in practice the adsorbent is normally stirred into the oil and then filtered off after a certain residence time.

The choice of the adsorption process and the type and concentration of adsorbent is governed by factors such as pretreatment, desired quality of the fully refined product, filtration speed of the oil, and oil retention by the adsorbent. Oil retention can be up to 50 wt % on bleaching earths and nearly 100 wt % on activated charcoal.

Natural and Activated Earths are the most frequently used adsorbents in the refining of oils and fats. Activated earths are obtained by leaching natural earths (predominantly aluminum silicates containing montmorillonite) with hydrochloric acid followed by washing with water, drying, and sizing. This process increases the internal surface of the bleaching earth by partial dissolution of the aluminum and iron oxides; calcium and magnesium ions are partially replaced by hydrogen ions.

The amount of bleaching earth to be used and the optimum bleaching conditions must be determined empirically. Lower concentrations (0.5 – 1.0 %) are required for activated earths than for natural earths. Activated earths catalyze oxidation of the oil. For this reason bleaching should be done below 100 °C in vacuo, and the subsequent filtration should be carried out with limited access of atmospheric oxygen. Bleaching is sometimes performed in two steps: a conditioning and adsorption stage at relatively low temperature (ca. 65 °C), and a second stage at elevated temperature (ca. 100 °C) to fix the adsorbed components onto the adsorbent surface [52].

In physical refining, bleaching is often preceded by addition of citric or phosphoric acid to chelate trace metals and to precipitate hydrated phosphatides. This can also be achieved by using bleaching earths containing added acid.

The moisture content of bleaching earths lies between 5 and 10 %. Except for some types of natural earths, the moisture content does not appear to have a significant influence on bleaching capacity.

Activated Charcoal (0.1 – 0.4 %) is sometimes used in combination with bleaching earth for oils that are difficult to bleach. Activated charcoal has also assumed importance as an adsorbent for removing polycyclic aromatic hydrocarbons from oils and fats [53]. It has been recommended for use in a fixed-bed process for the physical refining of soybean oil [54].

Bleaching and filtration is still mostly done *batchwise*. The oil is first dried in vacuo at ca. 30 kPa (30 mbar) to an optimum moisture content, frequently in the same closed vessels in which the oil was neutralized and washed (Fig. 14). The required amount of adsorbent is then drawn into the oil at 80 – 90 °C, preferably through a pipe reaching into the oil. After stirring for a few minutes to half an hour in vacuo, the oil – earth slurry is pumped over filter presses. Automated frame and plate presses or centrifugally dischargeable, closed-disk filters of the Funda (Fig. 16) and Schenk type are preferred; these also have the advantage of limiting access of air. The filter cake, containing 30 – 50 % oil, is often extracted with hot water (Thomson process) or with hexane to recover the bulk of the adsorbed oil. The quality of the recovered oil depends on the degree of unsaturation, the type of extraction, and the age of the spent earth before extraction. It is normally not economic to regenerate spent bleaching earth.

Continuous Bleaching is especially relevant when the other refining steps are also performed continuously. The degummed or degummed and

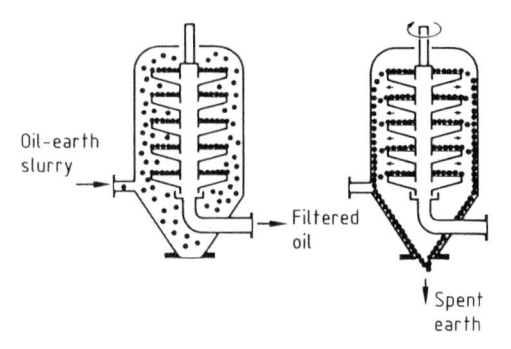

Figure 16. Closed disk filter with centrifugal discharge (Funda)

neutralized oil is normally dried and deaerated in vacuo and charged together with bleaching earth into an evacuated vessel where the oil – earth slurry is fed through various compartments, possibly at different temperatures, and is then filtered in a closed-disk system (Fig. 17).

Bleaching with oxygen or chemicals is not practiced for edible fats and is used only occasionally for technical fats. Oxidizing agents such as hydrogen peroxide, sodium peroxide, benzoyl peroxide, potassium permanganate, chromium salts, hypochlorite, and chlorine dioxide, and reducing agents such as sulfurous acid and sodium dithionite have been proposed. Hydrogen peroxide is used when adsorptive methods do not suffice. Castor, coconut, peanut, and olive oils and certain types of tallow and bone greases can be successfully bleached with 0.05 – 0.1 % of a 20 % aqueous solution of sodium chlorite. Fats such as red palm oil are sometimes bleached by very slight hydrogenation which destroys carotenoid pigments.

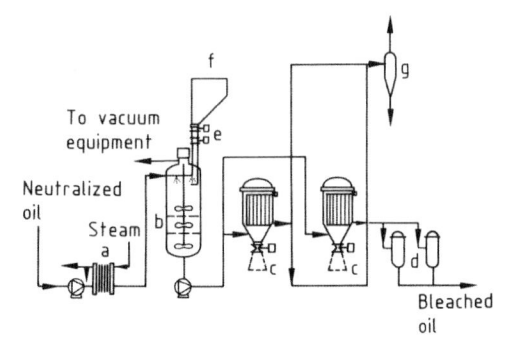

Figure 17. Alfa – Laval bleach system a) Plate heat exchanger; b) Bleacher; c) Filters; d) Polishing filters; e) Bleaching-earth dosage equipment; f) Bleaching-earth storage tank, g) Cyclone

6.4. Deodorization

Deodorization is the last step of the refining sequence, in which odors and flavors are removed from the bleached oil or fat. It is essentially a steam distillation process in which volatile compounds are separated from the nonvolatile glycerides. The odoriferous compounds are primarily aldehydes and ketones formed by autoxidation during handling and storage and can have a flavor threshold value of a few ppm. Other volatile components such as free fatty acids, alcohols, sterols, or tocopherols are also partially removed by deodorization.

Deodorization is not, however, a purely physical process. During deodorization, flavor compounds can be formed by hydrolysis and thermal decomposition, and peroxides are decomposed by heat. These reactions presumably play a major role in the flavor stability improvement achieved by deodorization, especially in vegetable oils. In practice, therefore, the residence times and steam volumes required are significantly greater than those calculated for normal steam distillation. At the relatively high deodorization temperatures used, traces of air can initiate oxidation with severe impairment of taste and flavor stability. Because thermal decomposition products of proteins, carbohydrates, phosphatides, and soaps can also interfere with deodorization, the oil must be well prerefined, i.e., freed from phosphatides and soap.

Deodorization is performed in vacuo because of the low partial vapor pressure of the compounds to be removed. Consumption of stripping steam increases with decreasing vapor pressure of the volatile components and with increasing pressure in the deodorizer, it ranges from 5 to 10 wt % (based on the oil) for batch processes and from 1 to 5 wt % for continuous processes. Deodorization temperatures range from 190 to 270 °C and the pressure from 0.13 to 0.78 kPa (1 – 6 mm Hg). The volume of stripping steam required is 10 – 20 m^3 per kilogram of oil [24]. Carry-over losses of oil are minimized by limiting the velocity of the injected steam, by providing a sufficiently large headspace, and by baffles. The distillate removed during deodorization is about 0.2 % of the oil, and contains very little neutral oil. In a batch process a deodorizing time of at least 2 h is required for saturated or hydrogenated fats; about 4 h is needed for other oils and fats. Although attempts have been made to determine the end point of deodorization by analytical data, in practice this is done organoleptically.

The steam injected into the oil must be dry and oxygen-free. In order to deactivate prooxidative trace metals, aqueous citric acid solution (2 – 5 mg of citric acid per 100 g of oil) is often injected into the oil toward the end of the deodorization.

The deodorizer is normally constructed of stainless steel if temperatures of 190 °C are to be exceeded. Deodorizing temperatures up to 270 °C are employed, for example, to complete removal of pesticide residues [55], to heat-bleach palm oil, and to complete removal of free acids during physical refining. Since dimerization and isomerization of fatty acid groups increase with time and temperature [56], a deodorization at 270 °C should not last more than 30 min.

Batch Deodorizers are still widely used (Fig. 18). They are about twice as high as they are wide and are normally half-filled to provide sufficient head space. Their capacity can vary from 5 to 25 t. The bottom part is fitted with heating coils and a steam distributor through which stripping steam is injected into the oil. The vacuum system (0.6 – 1.2 kPa) comprises a booster and ejector assembly. Barometric condensers can be eliminated by using dry condensing plants. When deodorization is completed, the oil is cooled to ca. 90 °C in the deodorizer and to 40 – 60 °C in a subsequent vacuum cooler; cooling minimizes oxidation that occurs on

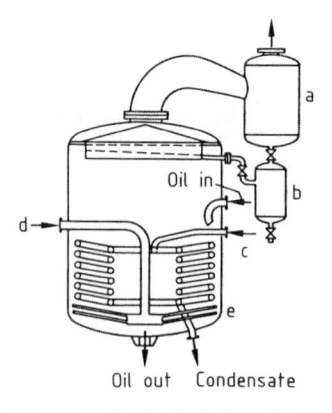

Figure 18. Batch deodorizer a) Barometric condenser; b) Fat trap; c) Heating steam; d) Stripping steam; e) Steam distributor

contact with atmospheric oxygen at elevated temperatures. The total cycle of charging, heating, deodorization, cooling, and discharging requires up to 8 h, the deodorization process itself up to 4 h. The deodorizer distillate has the appearance of a milky emulsion and contains calcium soaps, nonsaponifiable matter, and a small fraction of neutral oil.

Semicontinuous and Continuous Deodorization systems are replacing batch deodorizers because of the savings in steam resulting from more efficient stripping and heat recovery.

The semicontinuous Girdler deodorizer (Fig. 19) consists of an iron shell fitted with stainless steel trays. At defined time intervals (e.g., every 30 min), a certain quantity of oil flows into the top tray of the deodorizer where it is deaerated in vacuo and then heated to ca. 170 °C by exchange with the hot effluent oil. After 30 min the oil is discharged into the second tray where it is heated to the operating temperature (normally 230 – 250 °C). Injection of steam (at 0.6 – 1 kPa) takes place in the 3rd and 4th trays, and cooling in the 5th tray. Operation of the unit is made fully automatic by use of a timing device which opens and closes the oil valves. The semicontinuous Lurgi plant works on a similar principle (Fig. 20); the stripping steam is injected through special orifices and further distributed through the oil by steam-lift pumps. These types of deodorizers have capacities of up to 20 t/h.

Fully continuous deodorizers generally comprise a high cylindrical column fitted with a number of trays with bubble caps or similar

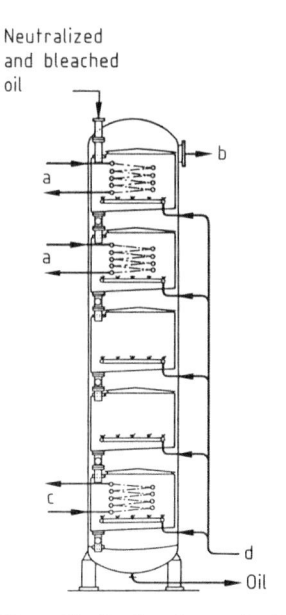

Figure 19. Semicontinuous deodorizer (Girdler) a) Heating steam; b) Vacuum aggregate; c) Cooling water; d) Stripping steam

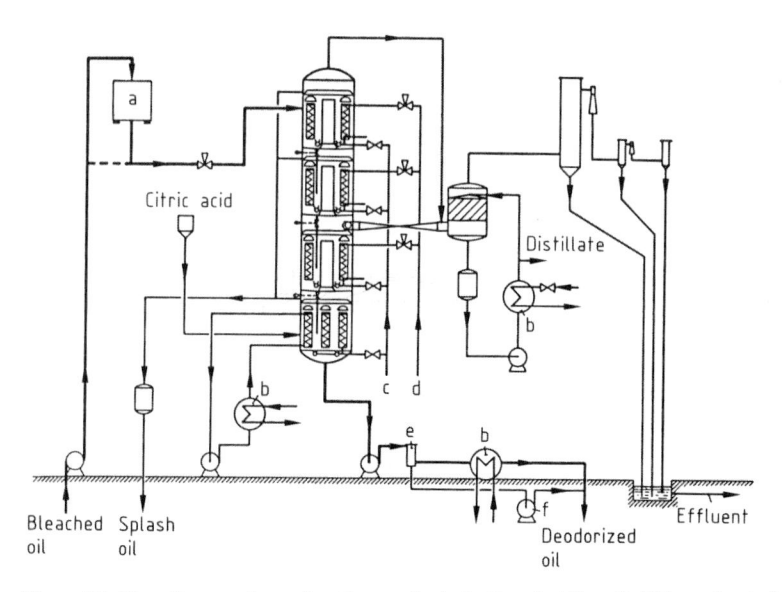

Figure 20. Flow diagram of a semicontinuous deodorization plant (Lurgi) a) Measuring tank; b) Heat exchanger; c) Stripping steam; d) Heating steam; e) Filter; f) Pipe emptying

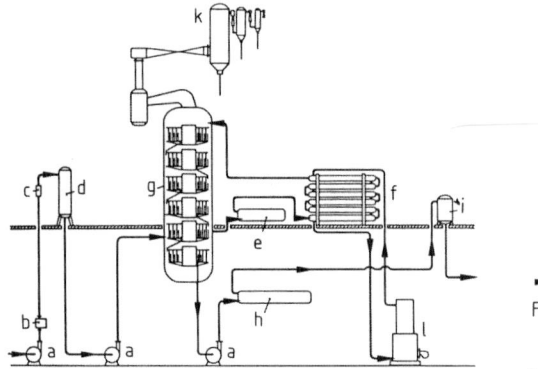

Figure 21. Continuous deodorizer (Krupp) a) Pump; b) Filter; c) Meter; d) Deaerator; e) Preheater (low-pressure steam); f) Preheater (high-pressure steam); g) Kreuzstrom – Konti deodorizer; h) Cooler; i) Filter; k) Vacuum aggregate; l) Steam kettle

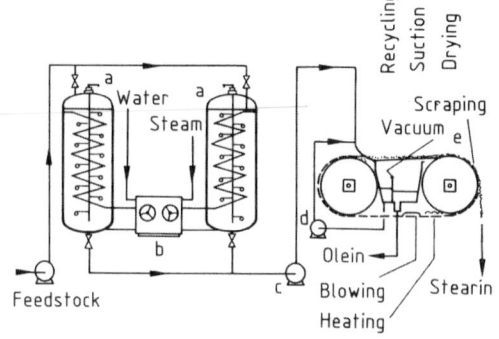

Figure 22. Dry fractionation plant a) Crystallizers; b) Circulating unit; c) Feed pump; d) Recycling pump; e) Florentine filter

devices, down which the oil flows countercurrent to the stripping steam. The oldest continuous deodorizer is probably that built by Foster – Wheeler. Modern versions [57] are made by EMI, Gianazza, Mazzoni, Wurster & Sanger, Votator, de Smet, Ex-Technik, HLS, Lurgi, and Krupp (Fig. 21). An alternative construction is that of EBE in which the oil flows countercurrent to the stripping steam in horizontal tubes. The Campro deodorizer has a continuous tray-in-shell design that combines plug flow with a unique thin-film stripping concept.

7. Fractionation

The aim of fractionating fats is to remove either undesirable components or to isolate desired components with special properties. Edible oils can, for example, be winterized by removing waxes and saturated components; cocoa butter substitutes can be obtained from palm oil. Except for the molecular distillation of monoglycerides, distillative techniques cannot be employed. Industrial fractionation techniques are based on crystallization or liquid – liquid extraction. Separation of glycerides by selective adsorption is only used occasionally for purification purposes. Fractional crystallization can be subdivided into dry processes, processes employing solvents, and processes involving selective wetting of fat crystals. Directed interesterification combines selective fractionation with interesterification (see Section 4.2).

The oldest example of *dry fractionation* is the separation of tallow into stearin and olein by pressing; this separation came into use toward the end of the 19th century in connection with the production of margarine. The lower-melting glycerides (olein) were squeezed out of the crystal network of the higher-melting glycerides by pressing blocks of tallow wrapped in filter cloth.

This method has been superseded by modern processes (Fig. 22) in which the fat is slowly chilled to the required temperature under controlled conditions in tempering vessels fitted with low-speed agitators, or in scraped coolers. Relatively large crystals of the higher-melting glycerides are obtained which can readily be filtered off in filter presses or drum filters. Slow cooling is important to obtain a relatively small number of nuclei which can then grow into large crystals. In addition, slow cooling promotes selective crystallization of the higher-melting glycerides and avoids the formation of mixed crystals containing both high- and low-melting glycerides. The difference in melting point must be at least 10 °C to achieve a sufficiently sharp, reproducible separation.

Winterization of edible oils such as sunflower or cottonseed oil involves removing waxes, which cause turbidity at refrigerator temperatures. After bleaching or deodorization, the oils are kept at ca. 10 °C for at least 1 h and then filtered. An alternative process is to combine dewaxing with superdegumming of the crude oil (see also Section 6.1) [44]. Peanut oil is not normally winterized because its turbidity is caused by small glyceride crystals that are difficult to filter.

The separation of higher- and lower-melting glycerides can be improved by *fractionation using solvents* such as hexane, acetone, or methanol, in which the unsaturated fats are more readily soluble. Costs are much higher than for dry fractionation because of the need for solvent recovery. This relatively sharp fractionation method is generally only economical for preparing specialty products. For example, fractionation of palm oil with acetone gives a fraction (ca. 40 % of the feed) consisting of disaturated-monounsaturated glycerides, primarily 1,3-di-palmito-2-olein.

The separation of relatively small fat crystals can be facilitated by *selective wetting*, for example with sodium lauryl sulfate or decyl sulfate in aqueous solution. The crystal suspension, produced either batchwise or continuously, is mixed with an aqueous solution containing the wetting agent and normally also an electrolyte such as magnesium sulfate. The crystals covered with wetting agent migrate to the aqueous phase as a suspension that can be separated by centrifugation. Alfa – Laval recommends such a process (Lipofrac process) for the fractionation of palm oil and for the winterization of other oils.

Selective Extraction [58] has long been used in refining mineral oil. Liquefied hydrocarbons or furfural are used as selective solvents in the separation of fats and fatty acids. This technique not only permits the separation of fats and fatty acids according to the degree of unsaturation or the molecular mass, but also effects the partial removal of coloring matter, gums, and vitamins.

Liquefied propane preferentially dissolves saturated components (Solexol process; see also Other Processes) [59]. This process has been used to fractionate tallow. The solubility of fats and fatty acids increases with decreasing molecular mass. Furfural preferentially dissolves unsaturated components. It is miscible with fatty acids and monoglycerides at room temperature, but with triglycerides only at elevated temperatures. Depending on the operating temperature, furfural is used alone or together with hexane. The selective solubility behavior is less pronounced for differences in molecular mass [60]. The furfural process has been used to produce high-performance drying oils from soybean and linseed oils.

8. Hydrogenation [61], [62]

Hydrogenation of the carbon – carbon double bonds of unsaturated glycerides leads to an increase in melting point (hardening, see also Section 4.3). The combination of hydrogenation, interesterification, and fractionation offers the opportunity of economically meeting the increasing demand for specialty products for foodstuff and technical applications from available raw materials.

Trans fatty acids produced by isomerization during hydrogenation also occur in animal fats such as butterfat and tallow. Although there are indications that trans fatty acids may be metabolized differently from their cis isomers [63], long-term studies have demonstrated their toxicological safety [64]. As far as their effect on blood lipid levels is concerned, they are comparable with saturated fatty acids.

The selectivity of hydrogenation is determined by the type and activity of the catalyst [65] (controlled poisoning and a large pore size increase selectivity), the operating temperature (both selectivity and cis – trans isomerization increase with increasing temperature), stirring – mixing efficacy (increasing the hydrogen concentration at the catalyst surface by intensive stirring – mixing reduces selectivity), and the hydrogen pressure (increasing the pressure reduces the selectivity).

Neither absolute selectivity nor complete isomerization (or complete suppression of isomerization) can be achieved. However, the process can be steered in a desired direction by suitable choice of the above parameters to produce fats with different degrees of saturation and melting properties from a given oil. For example, peanut oil can be hardened nonselectively to an iodine value of 70 – 90 yielding a soft, semiliquid fat. If the same oil is hardened to the same iodine value under conditions that promote isomerization, a firm, plastic fat with a melting point of ca. 33 °C is obtained. Palm oil is sometimes hydro-bleached, i.e., hydrogenated under very mild conditions that lead to selective hydrogenation of carotenoids.

The oil should be at least degummed and bleached prior to hydrogenation and preferably neutralized and bleached as well. Catalyst poisons include phosphatides, sulfur compounds, soaps, and oxidation products from poor handling and/or bleaching. Free fatty acids do not poison the catalyst but tend to slow down the hydrogenation reaction.

The hydrogen must be pure and dry and can be produced by various routes. The most important industrial process is decomposition of natural gas or propane by passage over a nickel catalyst in the presence of steam at 800 – 900 °C. Electrolysis of water is used to a lesser extent. The hydrogen must be specially purified to remove hydrogen sulfide and carbon monoxide, which are strong catalyst poisons.

Nickel, platinum, and palladium are preferred catalysts. Platinum and palladium are especially suited for laboratory-scale and low-temperature hydrogenations. Only nickel is used on a production scale. Many different types of nickel catalysts are available (→ Heterogeneous Catalysis and Solid Catalysts, 2. Development and Types of Solid Catalysts). They essentially consist of metallic nickel on a carrier such as silica gel, silicic acid, or alumina. The carrier is treated with a nickel salt such as the carbonate, hydroxide, or sulfate, which is converted into nickel oxide by roasting. The oxide is then reduced to metallic nickel in a stream of hydrogen at 300 – 400 °C. Other routes involve decomposition of nickel formate at 240 °C or dissolution of aluminum from a nickel – aluminum alloy with alkali (Raney catalyst). In addition, there are a number of mixed catalyst systems (copper, chromium, cobalt) with special selectivities.

Batch Hydrogenation (Fig. 23) is normally performed on a scale of 5 – 25 t. The dry oil is pumped into the hydrogenation vessel that has been previously flushed with hydrogen, and heated to just below the required operating temperature (100 – 180 °C). The catalyst (0.01 – 0.1 % active nickel) is then introduced and the hydrogen pressure increased. Alternatively, the oil is vacuum-deaerated while it is heated before the catalyst and hydrogen are introduced. Hydrogenation is an exothermic reaction; a decrease in the iodine value of one unit causes a temperature increase of 1.5 – 2.0 °C. The required operating temperature is therefore maintained by controlled cooling. Hydrogen is introduced through a sparge ring at the bottom of the hydrogenation vessel, pumped out at the top, and recirculated. Sparging and fast stirring ensure adequate dispersion of the catalyst in the oil and

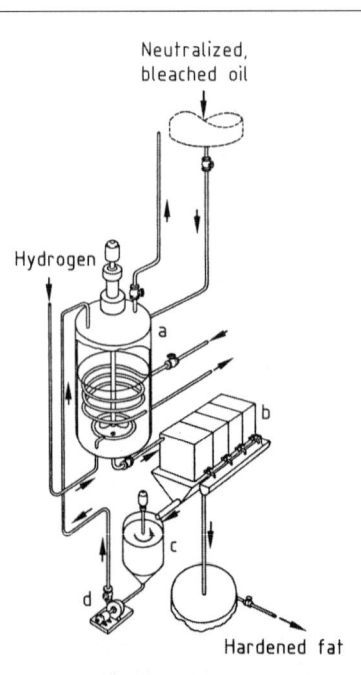

Figure 23. Flow diagram of a batch hardening plant
a) Converter; b) Filter press; c) Catalyst tank; d) Catalyst pump

optimum mass transport. The usual working pressure is 0.15 – 0.3 MPa (1.5 – 3 bar).

In hydrogenation autoclaves that operate without hydrogen recirculation ("dead-end" process), spent hydrogen is replaced at a somewhat higher pressure [66]. The intensive stirring required can be achieved with a turbine agitator or similar device (Fig. 24).

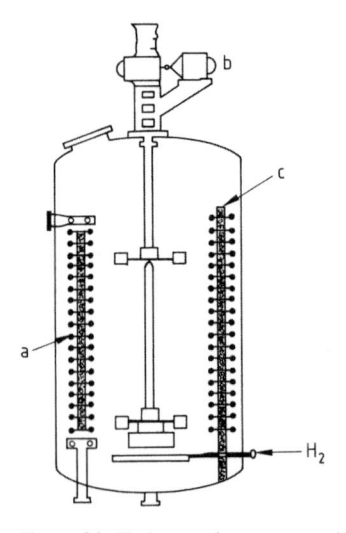

Figure 24. Hydrogenation converter (dead-end design)
a) Heating and cooling coil; b) Turbine agitator; c) Baffle

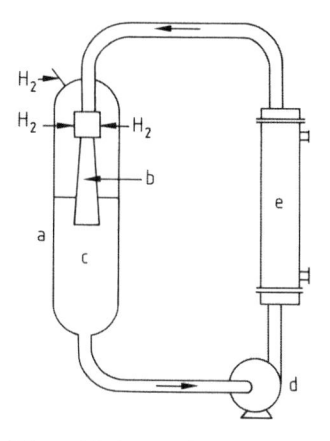

Figure 25. Loop hydrogenation reactor a) Autoclave; b) Mixing and reaction zone; c) Reactant – solvent – catalyst suspension; d) Primary loop recirculation pump; e) Primary loop heat exchanger

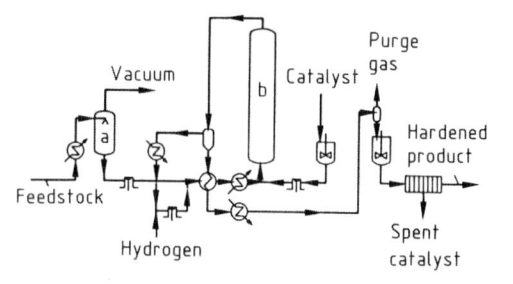

Figure 26. Continuous hydrogenation of fatty acids and fatty oils (Lurgi) a) Dryer; b) Reactor

So-called loop hydrogenation reactors (e.g., Buss, Fig. 25) have become increasingly popular. The oil – catalyst mixture is constantly circulated through an external heat exchanger during heating and cooling. This loop reduces total cycle time; it also gives better temperature control and thus less product variation from batch to batch.

When hydrogenation is complete, the product is cooled to 80 – 90 °C and the catalyst is filtered off. Since the acid value tends to increase slightly during hydrogenation, the crude hardened fat is usually postneutralized with lye and bleached before it is deodorized. Posttreatment with citric or phosphoric acid, followed by bleaching prior to deodorization, is an alternative to lye posttreatment. In either case the residual nickel content is reduced to <0.2 ppm. Deodorization of a hardened fat is essential since hydrogenation leads to typical hardening flavors which must be removed.

Continuous Hydrogenation is primarily suitable for the production of one particular product over a longer period of time. Fully continuous processes are the exception rather than the rule. In semicontinuous processes several reaction vessels or autoclaves are arranged sequentially (e.g., Alfa – Laval system). In fully continuous plants, either the preheated oil and hydrogen are passed over a stationary catalyst bed, or all three components are mixed prior to entering the reactor (see Fig. 26). In the Procter & Gamble process, the reactor contains a special stirrer to ensure intensive mixing in a thin film. The Girdler Corporation developed a process in which three Votator heat exchangers are arranged sequentially; the first vessel serves as preheater, the second as a hydrogenator, and the third as a cooler. Continuous hydrogenation of miscella at low temperatures over a catalyst bed produces little isomerization but has not achieved industrial importance.

The course of hydrogenation is normally monitored by changes in analytical data such as iodine value or refractive index. The hydrogenated products are characterized primarily by melting point and solids content at different temperatures.

9. Interesterification

Prior to interesterification, fats must be neutralized to less than 0.1 % ffa and dried to avoid excessive deactivation of the catalyst, which is normally sodium ethoxide (see also Section 4.2). The catalyst (0.1 – 0.3 %) is finely dispersed in the fat. After completion of the interesterification reaction, which is carried out at 80 – 100 °C, excess catalyst is deactivated by addition of water. The resulting soap is removed by washing with water and bleaching. During interesterification, sodium ethoxide is converted into fatty acid ethyl esters, which are removed during the final deodorization step.

Interesterification can be performed batchwise (Fig. 27), usually in a neutralization–bleaching vessel, or continuously. In the continuous process, fat with suspended catalyst passes through a tubular reactor (Fig. 28) [67]. In directed interesterification (see Section 4.2) the

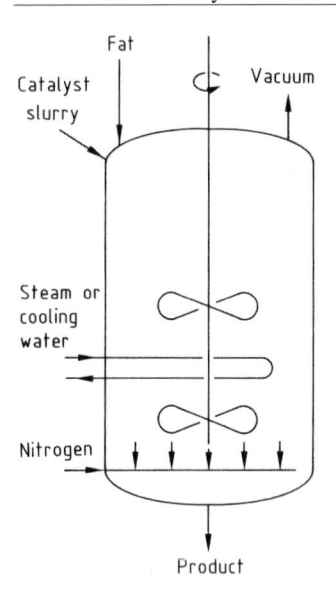

Figure 27. Batch reactor for interesterification

fat is cooled, for example, in a scraped-surface cooler to crystallize the higher-melting glycerides.

New types of fats can be created by suitable choice of interesterification components. Depending on the starting components, the melting point of the fat after interesterification can be lower or higher. Figure 29 shows the solids content (dilatation) curves of native and interesterified palm oil; the glyceride composition is given in Table 22.

Acidolysis. An example of the industrial application of acidolysis is the transesterification of coconut oil with acetic acid and the subsequent esterification of excess acid with glycerol. A mixture of aceto fats (laurodiacetin, myristodiacetin, etc.) is obtained which are used as plasticizers and coatings. Transesterification of coconut oil with higher fatty acids, employing continuous distillation of the liberated lower fatty acids ($C_6 - C_{10}$), yields fats with properties similar to those of cocoa butter. Acidolysis is usually performed in the presence of an acid or base catalyst; no catalyst is required at temperatures of 260 – 300 °C.

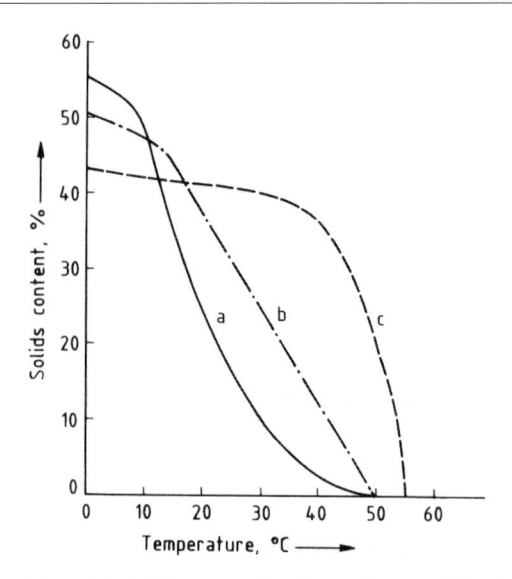

Figure 29. Solids content of native and interesterified palm oil (Alfa – Laval) a) Native palm oil; b) Randomly interesterified palm oil; c) Directedly interesterified palm oil

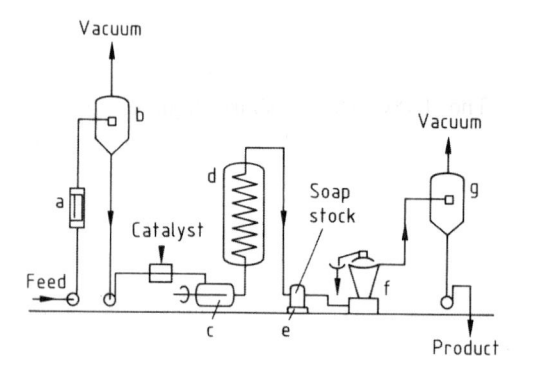

Figure 28. Continuous interesterification a) Heat exchanger; b) Dryer; c) Homogenizer; d) Tubular reactor; e) Mixer; f) Centrifuge; g) Dryer

Table 22. Fatty acid and glyceride composition of native and interesterified palm oil[*]

	Native palm oil	Interesterified palm oil	
		Random interesterification	Directed interesterification
mp, °C	42	47	52
Fatty acid composition, mol %			
S	51	51	51
U	49	49	49
Glyceride composition, mol %			
S_3	7	13	32
S_2U	49	38	13
SU_2	38	37	31
U_3	6	12	24

[*] S = saturated fatty acid. U = unsaturated fatty acid.

Alcoholysis. The replacement of the glycerol moiety of a fat with lower monohydric alcohols proceeds at lower temperatures. The Bradshaw process [68] utilizes transesterification of fats with methanol as the first step in the continuous production of soap. Fat with an acid value of less than 1.5 is stirred with excess methanol for a few minutes in the presence of 0.1 – 0.5 % caustic soda at ca. 80 °C. On subsequent standing, practically dry glycerol settles at the bottom of the reaction vessel. The same principle is employed in the preparation of fatty acid methyl esters for laboratory analyses. If the acid value is higher than 1.5, boron trifluoride is normally used as a catalyst.

The exchange of glycerol with other polyhydric alcohols, such as pentaerythritol, is best done under conditions at which glycerol can be distilled off, i.e., at 200 – 250 °C and 5 – 40 kPa (50 – 400 mbar), in the presence of an alkaline catalyst.

The production of mono- and diglycerides by reacting fats with an excess of glycerol is a special case of transesterification of fatty acid esters with polyhydric alcohols. These glycerides are especially important as emulsifiers. The production process consists of heating a fat with 25 – 40 % glycerol at 205 – 245 °C in the presence of sodium hydroxide or a sodium alkoxide. Part of the catalytic effect is probably based on soap enhancing the solubility of glycerol in fat. Industrial products contain 40 – 50 % monoglycerides. Oxidation and undesirable color effects can be minimized by working in vacuo or under an inert gas, and by using stainless steel reaction vessels. Monoglyceride concentrates can be produced by molecular or high-vacuum distillation.

10. Environmental Aspects

Crude oils and fats can be contaminated with traces of various substances such as metals (e.g., lead, arsenic, cadmium, selenium, mercury), chlorinated organic pesticides (DDT, dieldrin), mycotoxins (e.g., aflatoxins), and sulfur compounds. These trace contaminants are removed in the course of refining [55].

Wastewater from lye neutralization and from the barometric condensers of deodorizers can be discharged into communal sewage systems or surface waters only if certain limits for residual fat, pH, sediment, sulfate, and chemical and biological oxygen demand are met. Otherwise biological pretreatment is required. Since the cost of biological or other pretreatment of wastewater can be considerable, there has been a trend toward physical refining (i.e., refining without lye neutralization) by employing special degumming techniques, bleaching, and distillative deodorization.

Bleaching earth is normally extracted before being deposited or burnt. The extracted fats can be recycled.

Deodorizer distillates [55] can be used in animal feeds, as a source of vitamin E, for technical applications, or simply burnt.

11. Standards and Quality Control

Standardized methods for quality control have been issued by the following organizations:

Association Française de Normalisation (AFNOR), American Oil Chemists' Society (AOCS), Association of Official Agricultural Chemists (AOAC), American Society for Testing and Materials (ASTM), British Standards Institution (BSI), Deutsche Gesellschaft für Fettwissenschaft (DGF), Deutsches Institut für Normung (DIN), International Association of Seed Crushers (IASC), International Union of Pure and Applied Chemistry (IUPAC), International Standards Organization (ISO), the Netherlands Normalisatie Instituut (NNI), and the Federation of Oils, Seeds, and Fats Associations (FOSFA).

There is an increasing trend for the various national and international methods to be uniformly standardized as ISO procedures; these are generally adopted as official methods within the European Community.

The IASC (8 Salisbury Square, London EC 4P 4 AN, UK) has issued a list of associations dealing with or issuing standard forms of contract and/or quality standards for oilseeds, oils, and fats. This association has also drawn up a list of official institutions and associations that deal with or issue methods for grading, sampling, and preparing samples for analysis, as well as methods for analysis that are referred to in standard forms of contract or usually applied in the trade of oilseeds, oils, and fats.

11.1. Sampling

Correct sampling is a difficult process requiring careful attention; it must take into account that separation can occur during storage of fats and fatty products. Hulls, for example, tend to separate from oilseeds because of their lower density. Oils and fats tend to form a sediment of water and dirt; for this reason a representative average sample is often taken in bypass during pumping.

A special zone sampler consisting of a metal cylinder with a valve at both ends is often used for sampling the contents of a tank. The sampler is lowered into the tank, the valves being operated with a cord. Starting from the bottom of the tank, the oil is sampled at various depths up to the surface. Smaller containers (drums) are sampled with a tube or shutter scoop; solid fats in retail packs are sampled with a sampling scoop.

Special sample dividers are used for oilseeds. Sampling apparatus for bags includes sack-type spears or triers, cylindrical and conical samplers, and hand scoops. Shovels, scoops, cylindrical and conical samplers, or mechanical samplers are used for drawing small periodical samples from a flow of oilseeds. Shovels, quartering irons, or riffles are used for mixing and dividing.

Samples for analysis of oilseeds, expeller cakes, and meals should preferably be at least 2 kg, and of oils and fats at least 250 g. The samples should be homogenized immediately prior to analysis. Larger oilseeds are comminuted prior to analysis.

11.2. Raw Materials

Buying – selling contracts for oilseeds, crude oils, and fats normally specify a maximum permissible concentration of impurities. If this level is exceeded the buyer receives a discount. The key parameters in oilseeds are fat and moisture contents.

To determine the *fat content* of oilseeds, expeller cakes, and meals, the comminuted sample is extracted with a solvent and the fat taken up by the solvent is determined refractometrically, gravimetrically, densitometrically, or by the change in dielectric constant. The gravimetric methods are time-consuming but most reliable. Each method gives reproducible but different values, depending on the type of solvent (usually hexane or petroleum ether) and the extraction method. The fat content can also be determined directly without solvent extraction by nuclear magnetic resonance or by near infrared absorption.

According to ISO 659, the oilseeds are extracted with *n*-hexane or petroleum ether in a Bolton extractor. After extraction for several hours, the sample is removed from the extractor thimble, reground, and reextracted. This is repeated a third time. The extract is evaporated and the residual oil dried to constant weight at 103 ± 2 °C.

Unilever has developed a far quicker method which gives identical values to those obtained with the ISO method. The sample is first comminuted in a ballmill in the presence of petroleum ether, and then extracted for 2 h in a Bolton extractor.

When determining the fat content of fruit pulps and animal tissues, predrying of the substrate at 105 °C in vacuo is recommended. When extracting animal tissue polar solvents such as ether – HCl or ethanol – chloroform are used to facilitate liberation of fat from lipoprotein complexes present in aqueous dispersion.

Water and *volatile matter* are generally determined by weighing a sample before and after heating under standardized conditions, usually 103 °C. Crude oils are normally analyzed for water and volatile matter, color, particulate dirt, phosphatides, and free fatty acids. Up to 0.5 % moisture and impurities are generally regarded as technically unavoidable. Inferior oils usually have a dark color and a high content of free fatty acids and oxidized components that is indicated by a relatively high absorption at 232 nm, high anisidine and totox numbers, and a high concentration of oxidized fatty acids. These analytical data give an approximate indication of the expected yield and quality of the fully refined product. Admixture can be detected within certain limits by determining the composition of fatty acids, sterols, and tocopherols. Very severe admixture can be detected by iodine value, refractive index, and saponification value.

11.3. Oils and Fats

11.3.1. Physical Methods

The *density* or *specific gravity* at a given temperature is determined by weighing a known volume, by determining the buoyancy, or by electromagnetic excitation of the natural frequency of the sample.

The *refractive index* is normally determined with an Abbe refractometer at 20 °C for oils which are liquid at this temperature and at 40, 60, or 80 °C for fats. The temperature coefficient is ca. 0.0036 K^{-1}. A correlation exists between refractive index and iodine value (see Section 3.8); for this reason the refractive index is often used to follow the course of hydrogenation.

Most oils and fats rotate *polarized light* by only a few tenths of a degree; higher values are observed with substituted fatty acids such a ricinoleic acid.

In melting a distinction is made between the slip point, the point of incipient fusion, and the point of complete fusion. The *slip point*, the usually cited melting point, is the temperature at which a sample plug placed in an open glass capillary begins to rise when gradually heated in a water bath or by other means. This method can be automated. The slip point value is influenced by the method of cooling used prior to melting. Whereas for conventional fats the sample is normally solidified by chilling the sample and capillary at -10 °C for a few minutes, fats that exhibit distinct polymorphism, e.g., cocoa butter and substitutes, are conditioned at higher temperatures for up to 18 h.

The *point of incipient fusion* is the temperature at which a sample placed in a U-tube begins to flow on being warmed. The *point of complete fusion* is the temperature at which a sample becomes completely clear.

Flow and *drop points* are primarily of importance for technical greases and some plastic fats. They are determined in an Ubbelohde apparatus. Flow point is the temperature at which a drop begins to be clearly formed at the end of a thermometer, and drop point is the temperature at which the drop falls.

Dilatation, i.e., the increase in volume of a fat on passing from the solid to the liquid state at a given temperature, can be used to determine the concentration of solid components at a certain temperature (Solid Fat Index, SFI). This type of measurement has been replaced by the much quicker *nuclear magnetic resonance* method which is based on the difference between the response from the hydrogen nuclei in the solid phase and that from all the hydrogen nuclei in the sample.

Differential Thermal Analysis permits the quantification of changes in the solid phase of a fat on being heated. A sample is placed in a metal block which is heated at a constant rate. The difference between the temperature of the block and that of the sample is characteristic of crystal modifications occurring during melting [69].

On passing from the liquid to the solid state, the temperature remains constant for a certain period of time because of the liberated heat of fusion; with sufficient insulation an increase in temperature can be observed. The *titer value* of a fat is the maximum value to which the temperature rises in the cooling curve. Cooling curves are frequently used to check the purity of fats which consist predominantly of a few triglyceride types, e.g., cocoa butter.

The *cloud point* is the temperature at which an oil or molten fat begins to become turbid when cooled under controlled conditions. The temperature stability of an oil is normally assessed after keeping a carefully dried sample at 0 °C for $5-8$ h.

The smoke, flash, and fire points of oils and fats are measures of their thermal stability when heated in air. Table 13 shows some typical data. The *smoke point* is the temperature at which smoke is first detected in a standard apparatus with specially designed, draft-free illumination. The *flash point* is the temperature at which volatile products are evolved at such a rate that they can be ignited but do not continue to burn. The *fire point* is the temperature at which the production of volatile products is sufficient to support continuous combustion after ignition. The smoke point is important in the assessment of refined oils and fats used for deep frying. It is governed primarily by the free fatty acid content. The flash point is a measure of residual solvent in crude oils, although gas – liquid chromatographic methods are now generally preferred for this determination.

The *viscosity* of an oil can be determined in a flow (Engler), capillary (Ubbelohde), fall (Höppler), or shear (Haake) viscometer.

There are a number of different methods for determining the *color* of an oil or fat. In the Lovibond tintometer the color of the sample in a standardized cell with a length of $1''$ (2.45 cm) or 5 1/4″ (13.3 cm) is matched – either manually or automatically – with the color of standard yellow, red, and blue color slides. Other colorimetric scales are based on standard iodine or potassium dichromate solutions. The most objective way of assessing color is to determine the absorption at specified wavelengths in the range 370 – 780 nm.

The characteristic absorption maxima of specific lipid groups in the infrared and ultraviolet ranges can be used qualitatively and quantitatively (see Section 3.8). Fatty acids with isolated double bonds such as linoleic, linolenic, and arachidonic acids do not exhibit specific UV absorption. After isomerization with alkali they can, however, be determined by the characteristic absorption maxima of the corresponding conjugated fatty acids. Infrared spectroscopy is used primarily for the identification of specific groups such as hydroxyl and trans double bonds.

Optical Rotatory Dispersion Spectroscopy is used only in studies of biosynthesis and metabolism of lipids.

Mass spectroscopy, especially in combination with gas chromatography and other methods of separation, is often used to determine the structure of oxidation products and substituted fatty acids.

11.3.2. Chemical Methods

For preliminary identification and evaluation, determination of chemical data such as the iodine value (see below) and the fatty acid composition often suffice. In many cases, however, it is necessary to test for characteristic minor constituents directly or by specific chemical reactions. Table 23 gives some examples.

The complexity of oils and fats makes the determination of their precise composition difficult. For this reason analytical values are used to express certain properties. Analytical values in this sense are defined as the equivalent amounts of certain reagents which react with specific groups in the lipid molecules. Many of the older analytical values are no longer used because of the introduction of modern analytical techniques such as column, paper, gas – liquid, and high – performance liquid chromatography, and spectrophotometry. However, the principle of determining analytical values is still valid since it can authenticate and assess the quality of an oil or fat relatively quickly with modest laboratory facilities.

The *saponification value* is the number of milligrams of potassium hydroxide required to saponify (hydrolyze) 1 g of fat and is related to the molecular mass of the fat.

The *hydroxyl value* is expressed as the number of milligrams of potassium hydroxide required to neutralize the acetic acid needed to acetylate 1 g of fat.

The *carbonyl value* is defined as the number of milligrams of CO (carbonyl groups) per gram of fat. The basis for this value is the conversion of the carbonyl group into an oxime with hydroxylamine in alcoholic KOH:

$$RCO + NH_2OH \rightarrow RC = NOH + H_2O$$

Hydroperoxides interfere with this determination.

The acidity of an oil or fat arises from free inorganic or organic acids. The *acid value* is defined as the number of milligrams of potassium hydroxide required to neutralize 1 g of fat. The content of free fatty acids (ffa) expresses how many parts of free fatty acids are contained in 100 parts of fat. This content is – depending on the type of oil or fat – expressed as % oleic acid (M_r 282), % palmitic acid (M_r 256), or % lauric acid (M_r 200). If not specified, it is calculated as oleic acid.

Peroxides are determined iodometrically according to Wheeler (cold) or Sully (hot) and are expressed as milliequivalents of active oxygen per 1000 g of fat. The peroxide value can also be expressed in millimoles per kilogram (Lea value, 1 mequiv/kg = 0.5 mmol/kg) or in milligrams per kilogram (1 mequiv/kg = 8 mg/kg). It is important to use fresh reagents and to exclude oxygen as far as possible during the determination.

The reaction with anisidine is often used to determine the carbonyl compounds produced during autoxidation. The *anisidine value* is de-

Table 23. Detection of oils and fats

Oil/Fat	Characteristic constituent	Specific reaction or detection procedure	Limit of detection	Remarks
Animal fats	cholesterol	isolation of sterols from the unsaponifiable matter by TLC or HPLC, and separation by HPLC, TLC, or GLC	1 % in vegetable oils and fats	Vegetable oils and fats contain traces of cholesterol in addition to phytosterols while in animal fats only cholesterol but no phytosterols have been found.
	branched C_{13}–C_{17} fatty acids and odd-numbered, unbranched C_{11}–C_{19} fatty acids	GLC of fatty acid methyl esters	ca. 2 % in vegetable oils and fats	The branched fatty acids may have to be concentrated, e.g., by urea adduct formation.
Lard	palmitic acid in the glyceride 2-position	isolation of the glyceride fraction with 1 saturated and 2 unsaturated fatty acid groups, followed by determination of the fatty acid composition in the glyceride 2-position	1 % lard in tallow	Determination of the fatty acid composition in the 2-position without prior glyceride fractionation gives a significantly higher limit of detection.
Partially hardened and unhardened marine oils	highly unsaturated C_{20}–C_{24} fatty acids	Tortelli – Jaffé reaction (green coloration with bromine in chloroform/acetic acid) GLC of the fatty acid methyl esters	ca. 2 % in other oils and fats	Not strictly specific; highly unsaturated vegetable oils can also react.
Vegetable oils and fats	phytosterols	isolation of sterol esters and separation as sterols by TLC, HPLC, and GLC	ca. 2 % in animal fats	Sterols occur in vegetable oils and fats in the free and esterified form, while in animal fats they are only present in the free form.
Castor oil	ricinoleic acid	TLC; IR spectroscopy	ca. 1 % in other oils and fats	Hydroxy fatty acids can be formed during autoxidation of oils and fats.
Coconut oil and palm kernel oil	lauric acid	GLC of the fatty acid methyl esters	ca. 2 % in other vegetable oils except oils and fats rich in lauric acid, e.g., babassu fat	
Cottonseed oil and other oils of the malvaceae, tiliaceae, and bombaceae families	malvalic acid	Halphen reaction (red coloration on heating with sulfur in CS_2)	depending on extent of refining	Hardened cottonseed oil does not give this reaction; malvalic and sterculic acid are converted into saturated, branched fatty acids during hardening.
Crude palm oil	α- and β-carotene	isolation from nonsaponifiable matter by HPLC		Indicative of distillative or other heat treatment.

(*Continued*)

Table 23 (*Continued*)

Oil/Fat	Characteristic constituent	Specific reaction or detection procedure	Limit of detection	Remarks
Peanut oil	lignoceric and arachidic acid	GLC of the fatty acid methyl esters	4	Hardened marine oils and certain vegetable oils also contain higher molecular mass fatty acids.
Olive oil	α-tocopherol	isolation and separation by HPLC	ca. 5 % other oils in olive oil	
Partially hardened fats	trans fatty acids	IR spectroscopy	ca. 2 % in vegetable oils	Animal fats contain up to 10 % trans fatty acids.
Rapeseed oil	erucic acid	GLC of the fatty acid methyl esters	ca. 1 % of "high erucic acid" rapeseed oil in other vegetable oils	Admixture with "low erucic acid" rapeseed oil cannot be detected satisfactorily this way. Marine oils also contain erucic acid.
	brassicasterol	isolation and separation of sterols by TLC, HPLC, and GLC	ca. 5 % in other vegetable oils	Independent of erucic acid content.
Sesame oil	sesamol and sesamin	Baudouin reaction (red coloration with furfural and HCl)	ca. 0.5 % in other oils and fats	
Soybean oil	linolenic acid	GLC of the fatty acid methyl esters	ca. 10 % in e.g., sunflower oil	
	δ-tocopherol	isolation and separation by HPLC	ca. 5 % in e.g. sunflower oil	
Teaseed oil		Fitelson test (modified Liebermann – Burchard test)	e.g., 5 % in olive oil	
Tung oil	eleostearic acid	GLC of the fatty acid methyl esters	ca. 1 % in other oils and fats	

fined as 100 times the absorption of a 1 % sample solution in a 1-cm cell after reaction with anisidine. This value has been recommended for monitoring the refining of oils and fats. It is also used in combination with the peroxide value to characterize the overall autoxidative state of an oil or fat (*totox value*):

Totox value = 2 × peroxide value + 1 × anisidine value

The *iodine value* denotes the percentage by weight of iodine bound by 100 g of fat and is an approximate measure of the degree of unsaturation. The usual method is that of Wijs. The rate of addition of iodine to double bonds depends on their distance from the carboxyl group. Conjugated unsaturated fatty acids are only partially halogenated; substituted fatty acids such as keto acids and cyclic acids can also react. Trans fatty acids react more slowly than the cis isomers.

The *oxidative stability* of oils and fats is determined by their composition and their previous history. A number of empirical methods have been developed to express oxidative stability [70]. These tests are performed at elevated temperature and do not necessarily correlate with organoleptic stability at ambient temperature. The most frequently practiced methods are the Schaal test and the Swift test (or Active Oxygen Method); automated versions of the latter are popular, e.g., the Rancimat apparatus.

The *moisture content* can be determined by weighing before and after drying or by codistillation with toluene. It is most conveniently determined by the Karl Fischer titration, which is based on the following reaction:

$2 H_2O + SO_2 + I_2 \rightarrow H_2SO_4 + 2 HI$

Impurities such as dirt and seed particles are not dissolved on treating the sample with petroleum ether or hexane and can be collected and weighed as filter residue.

Traces of *residual solvent* are best determined by gas chromatography, either by direct injection or after enrichment by the so-called headspace technique [71].

Nonsaponifiable matter is defined as that material which is a soluble constituent of the fat and which remains insoluble in the aqueous phase after saponification of the fat and can be extracted with petroleum ether or diethyl ether; diethyl ether tends to give higher values. Nonsaponifiable matter consists primarily of sterols, alcohols, hydrocarbons, and vitamins.

Oxidized fatty acids contain one or more hydroperoxide, hydroxyl, or keto groups in the fatty acid chain and can be determined by their reduced solubility in petroleum ether. Only part of the oxidized fatty acids are detected by this type of analysis. Thus, the separation of fats into polar and nonpolar fractions on silica gel tends to be preferred when, for example, characterizing used frying fats [72].

To determine *resin acids* in the presence of fatty acids, the sample is esterified with methanol in the presence of an acid catalyst; resin acids escape esterification.

Inorganic acids can be detected by extracting the sample with hot water and testing the aqueous extract with methyl orange.

Free alkali can similarly be detected by dissolving the sample in ether – ethanol and adding phenolphthalein (red coloration). Traces of soap can be detected down to 10 ppm by dissolving the sample in ether – ethanol, adding bromophenol blue (which gives a green-blue color in the presence of soap), and titrating with very dilute HCl.

Many crude seed oils contain considerable concentrations of *phosphatides*. These can be determined by dry digestion of the sample with magnesium oxide and photometric determination of the resulting phosphate after addition of molybdate [73]. An alternative is the direct determination of phosphatides by atomic absorption or plasma emission spectroscopy.

Methods for determining traces of *mineral oil* in edible oils are based on the fluorescence of mineral oils, their insolubility in acetic anhydride, and the fact that they are not retained on alumina. Headspace gas chromatography is also used.

The gas chromatographic determination of the *fatty acid composition* of an oil or fat has replaced the time-consuming, chemical techniques of fractionating fatty acids into single components or groups. Analytical values such as the Reichert – Meissl, the Polenske, the Kirchner, and the butyric acid number (as measures of the concentration of lower molecular mass or steamvolatile fatty acids) are now virtually obsolete. Prior to gas chromatographic analysis, the fatty acids are converted into their methyl esters either by direct transesterification of the sample with methanolic KOH (acid values <2) or by saponification (acid values >2) followed by esterification of the fatty acids with methanol in the presence of BF_3. Full details of these procedures and of the various types of columns, column packings, and detectors are given in ISO 5508/5509. If a marker fatty acid is used as an internal standard, quantitative determination of the total amount of monomeric, nonoxidized fatty acids is possible; nonsaponifiable matter, oxidized, and polymerized fatty acids are retained on the gas chromatographic column.

Polymeric fatty acids are conveniently determined by special high-performance liquid chromatography (HPLC) or gel permeation chromatography [74].

The *glyceride structure* may have to be determined to authenticate an oil or fat. The methods for determining positional isomers are based on fractional crystallization, selective oxidation, chromatographic separation, or enzymatic hydrolysis of the fatty acid in the 2-position; they are normally too expensive for routine control unless specialty products are involved.

Mono- and diglycerides are formed by hydrolysis of fats and therefore occur in higher concentrations in fats from damaged seeds. Mono-, di-, and triglycerides can be separated by column or thin-layer chromatography. 1-Monoglycerides can be determined directly in fats by oxidative fusion with periodic acid:

$RCOOCH_2CHOHCH_2OH + H_5IO_6$

$\rightarrow HCHO + RCOOCH_2CHO + HIO_3 + 3H_2O$

Monoglycerides can be isomerized into the 1-isomers prior to addition of periodic acid by treating the sample with perchloric acid.

12. Storage and Transportation [75]

Crude Oils and Fats. If crude oils and fats are to be stored for a long time, removal of insoluble impurities by sedimentation, filtration, or centrifugation is advisable. Seed particles and cell fragments contain fat-splitting enzymes, which can gradually hydrolyze glycerides in the presence of moisture and produce an increase in free fatty acid concentration. Proteins are good nutrients for microorganisms.

Further measures for avoiding undue quality deterioration include drying the crude oil to less than 0.2 % moisture and storage without heating; consistent fats should be stored at not more than 10 – 15 °C above the melting point. Frequent pumping should be avoided to prevent saturation of the oil with air and consequent autoxidation. The addition of antioxidants is useful primarily for animal fats; most vegetable oils contain sufficient concentrations of natural antioxidants. Most crude vegetable oils can be stored for some months without significant loss in quality if the above precautionary measures are observed. Animal fats can be stored for only relatively short periods of time without loss in quality. Continual monitoring of the acid value, peroxide value, and other data that characterize the state of oxidation during storage is recommended.

Raffinates. Since natural antioxidants, such as tocopherols, phenols, and synergists, are partially removed during the refining process, semi- and fully refined oils often have lower oxidation stability than the crude oils. This is especially true for oils with a relatively high content of polyunsaturated fatty acids. Therefore, raffinates must be protected as far as possible from light, air, moisture, relatively high temperatures, and trace metal prooxidants if they are to be stored for long periods.

Raffinates are best stored in closed containers with a relatively small ratio of surface area to volume in order to minimize access and infusion of air. Any heating should be uniform without local overheating; the temperature should not be higher than 10 °C above the melting point. Aeration can be further reduced by using a bottom-loading pipe and a nitrogen atmosphere during both storage and transportation. Immediately after deodorization, nitrogen should be introduced into the suction side of the loading pump that feeds the storage or transport vessels.

The storage containers are preferably made of aluminum or stainless steel, although mild steel can be used if access of air is limited and if cooling is adequate. Raffinates can be kept stored in large tanks for several weeks without an increase in peroxide value; however, the peroxide value tends to increase each time the oil is pumped. Raffinates are transported in bulk in ships and trucks, in drums, tins, and bottles; consistent fats can also be transported in block or powder form if the melting point is sufficiently high.

13. Individual Vegetable Oils and Fats

Vegetable oils and fats are normally divided into fruit pulp fats and seed oils. Although there are few fruit pulp fats (e.g., palm oil and olive oil), the number of seed oils is considerable. Seed oils can be conveniently subdivided according to their characteristic fatty acids:

1. Lauric acid oils (e.g., coconut and palm kernel oil)
2. Palmitic – stearic acid oils (e.g., cocoa butter)
3. Palmitic acid oils (e.g., cottonseed and corn oil)
4. Oleic – linoleic acid oils (e.g., sunflower and linseed oil)
5. Leguminous oils (e.g., soybean and peanut oil)
6. Cruciferous oils (e.g., rapeseed oil)
7. Conjugated oils (e.g., tung oil and oiticica oil)
8. Substituted fatty acid oils (e.g., castor oil)

13.1. Fruit Pulp Fats

Palm oil and olive oil are the most important products within this group.

13.1.1. Palm Oil [76]

Palm oil [8002-75-3] is obtained from the fruit pulp of the oil palm (*Elaeis guineensis*). The fruit bunches, weighing 10 – 25 kg, contain 1000 – 2000 plum-sized fruits which are dark red because of their carotene content. The fruits contain 35 – 60 % oil, depending on the moisture content. The oil palm is native to West Africa, and is extensively cultivated in Malaysia, Indonesia, and Central Africa. The oil yield has been increased to 4.5 t of oil per hectare and year by the introduction of insect pollination and by selecting and cloning high-yielding cultivars. Malaysian palm oil production was 4.13×10^6 t in 1985 and is expected to reach 6×10^6 t/a by 1990.

Saponification value	195 – 205
Iodine value	44 – 58
Nonsaponifiable matter	0.5 %
mp	36 – 40 °C
n_D^{40}	1.453 – 1.456
Carotene	500 – 2000 ppm

For fatty acid composition see Tables 1 and 2.

Palm oil contains up to 40 % oleodipalmitin, consisting primarily of 2-oleodipalmitin.

Production. The fresh palm fruits contain very active fat-splitting enzymes (lipases), which must be deactivated as quickly as possible prior to isolating the oil. Since lipases are particularly active in damaged fruits, the whole fruit bunches are sterilized by heating with live steam in autoclaves; this deactivates the enzymes and loosens the fruit from the bunches.

The sterilized fruit bunches are threshed in a rotary drum stripper consisting of longitudinal channel bars; the fruit falls through whereas the empty bunch stem is retained in the drum. The separated, sterilized fruit is then converted into an oily mash by a mechanical stirring process known as digestion. The digested fruit mass is passed through screw presses, and the liquor from the press is clarified by static settling or by using decanter centrifuges. After washing with hot water and drying, the oil can be stored and transported.

High-grade palm oil contains ca. 3 % free fatty acids, inferior grades up to 5 %. Crude plantation oils obtained from palm fruits processed under conditions that minimize autoxidation and lipolysis contain ca. 2 % free fatty acids and can be readily bleached; such oils are traded as super prime bleachable (SPB) grades.

Refining. Lye neutralization of palm oil may entail relatively high losses of neutral oil since inferior crude oils contain high concentrations of mono- and diglycerides, which exert a strong emulsifying effect. Palm oil is therefore often neutralized by distillation at temperatures up to 270 °C and 0.5 – 0.8 kPa (5 – 8 mbar) after pretreatment with phosphoric or citric acid and bleaching earth to remove foreign matter and traces of copper and iron compounds. During distillative neutralization (stripping), the oil is also bleached (i.e., heat-bleached), since carotenes are decomposed and removed. The carotenes in palm oil mainly consist of α-and β-carotene, with lower concentrations of γ-carotene, lycopene, and xanthophylls.

Distillative neutralization can be performed under conditions that simultaneously deodorize the oil. For very light raffinates, distillative neutralization is often followed by posttreatment consisting of a lye wash, bleaching with earth, and deodorization. Color and stability are largely a function of the quality of the crude oil. "White" palm oil, e.g., for biscuit fats, can generally only be made from crude SPB oils and by using fairly large amounts of bleaching earth. Bleaching with oxidizing agents, with air injected at 110 – 115 °C, or with activated earth at 140 – 160 °C is used only for technical applications. Palm oil can be effectively decolorized by mild hydrogenation at temperatures below 100 °C (hydrobleaching).

Palm oil can be separated into a solid and a liquid fraction. Two-stage cooling to 32 – 34 °C and 25 – 27 °C followed by separation over filter presses yields 20 – 25 % palm stearin, mp 50 – 52 °C, and 75 – 80 % liquid palm olein. The liquid glycerides can be separated from the solid fraction by centrifugation after addition of surfactant solutions. Fractionation in the presence of solvents (acetone, hexane) gives 10 % of a stearin melting at ca. 55 °C, 60 % olein, and 25 – 30 % of an intermediate fraction with properties similar to those of cocoa butter. The yield of the intermediate fraction can be increased to ca. 50 % by interesterification of the combined stearin and olein fractions followed by repeated fractionation.

Use. Palm oil is used predominantly for edible purposes. Refined palm oil is used extensively in shortenings and margarines. Carotene concentrates from crude oil have been largely replaced by synthetic carotene as coloring agents for margarines.

Palm stearin can be used in margarine and shortening blends whereas palm olein is used primarily in liquid frying fats and shortenings.

Technical applications of palm oil include tempering of metal and production of soap.

Other Palm Oil Varieties. These are commercially relevant only in their countries of origin. Tucum oil (saponifaction value 200, iodine value 40) is obtained from the fruit pulp of *Astrocaryum vulgare*, a palm growing in Central America.

13.1.2. Olive Oil

Olive oil [68153-21-9] is obtained from the fruit of olive trees (*Olea oleaster, Olea europaea oleaster*) and some related varieties such as *Olea americana* which is grown in South America.

Saponification value	185 – 196
Iodine value	80 – 88
Nonsaponifiable matter	<1.4 %
mp	−3 °C
n_D^{25}	1.466 – 1.468

For fatty acid composition see Tables 1 and 2.

Olive oil has a yellow to greenish-yellow color and becomes cloudy below −5 °C. Varieties with a higher cloud point can be made by winterization. The glycerides consist of 45 – 60 % monosaturated dioleins, 25 – 34 % dioleolinolein, and 4 – 29 % triolein. Saturated triglycerides are present at <1 %. The nonsaponifiable material contains saturated and unsaturated hydrocarbons, including 0.1 – 0.7 % squalene.

Production. The quality of olive oil depends on the ripeness of the olives, the type of harvesting (picking, shaking), intermediate storage, and type of processing. Olives contain 38 – 58 % oil and up to 60 % water. Ripe olives should be processed as quickly as possible since lipases in the pulp cause rapid hydrolysis of the oil, impairing its quality for edible purposes. Top-grade oils are made from fresh, handpicked olives by comminution, pasting, and cold pressing.

Traditionally, olives were ground into a paste with stone mills; today modern milling equipment is used. Milling is followed by mashing, possibly with addition of salt. The pulp is then pressed and the press oil clarified by settling or centrifuging. Open-cage presses are being replaced by continuous screw expellers. The mashed pulp can also be separated in a horizontal decanter, the crude oil being recentrifuged after addition of wash water. An alternative is the use of machines to remove the kernels from the pulp; the residue is separated in self-discharging centrifuges.

Cold pressing, which yields vierge grades (also referred to as virgin, Provence, or Nizza oil), is generally followed by warm pressing at ca. 40 °C, which gives an oil with a less delicate flavor. The yields depend on the equipment used. The press cake (pomace) contains 8 – 15 % of a relatively dark oil, called Sanza or Orujo, which can be extracted with hexane and is used for technical purposes; after refining it is also fit for edible consumption.

Olive Kernel Oil is obtained by pressing and solvent extraction of cleaned kernels. It is similar to olive oil but lacks its typical flavor.

Use. Cold-pressed olive oil is a valuable edible oil. Trade specifications are based primarily on the content of free fatty acids and flavor assessment. In some countries, warm-pressed olive oil with a high acidity is refined by neutralization, bleaching, and deodorization, and flavored by blending with cold-pressed oil.

Testing for Adulteration. Because of its high price, olive oil is sometimes admixed with other oils. Some oils can be detected by specific color reactions: cottonseed oil by the Halphen test, sesame oil by the Baudouin reaction, teaseed oil by the Fitelson test (see Table 23). Gas chromatographic determination of the fatty acid and sterol composition is generally used to quantify the degree of adulteration. Olive-pomace oil and re-esterified olive oil, obtained by esterification of olive oil fatty acids or Sanza or Orujo oil with glycerol, can be detected by the glyceride composition.

13.1.3. Avocado Oil [77]

Avocado oil is isolated from the pulp of the pearlike fruit of the tree *Persea gratissima* L. growing in Southern Europe, South Africa, the Middle East, and Central America. Its composition and properties are similar to those of olive oil (see Tables 1 and 2). The oil is used in cosmetics; in California smaller quantities of avocado oil are used as special salad oil. The avocado fruit weighs up to 1.5 kg and contains 40 – 80 % oil in dry matter. World production of avocado fruit is ca. 1.5×10^6 t/a, but only a small fraction is used for oil production.

13.2. Seed-Kernel Fats

Oilseeds are the major source of oils and fats.

13.2.1. Lauric Acid Oils

Coconut, palm kernel, and babassu oil are the most important lauric acid oils. They contain >40 % of lauric acid and ca. 15 % of myristic acid.

13.2.1.1. Coconut Oil [76]

The coconut palm (*Cocos nucifera* or *Cocos butyracea*) is cultivated in coastal areas around the world within 20 ° of either side of the equator. The Philippines, Indonesia, Southern India, Sri Lanka, Equatorial Africa, and the West Indies are important producers. Around 1820 coconut oil [8001-31-8] was introduced to Britain (vegetable butter).

Saponification value	250 – 262
Iodine value	7 – 10
Nonsaponifiable matter	0.15 – 0.60 %
mp	20 – 28 °C
n_D^{40}	1.448 – 1.450

For fatty acid composition see Tables 1 and 2

The glycerides contain 50 – 60 % caprylo-lauromyristin and up to 20 % myristodilaurin in addition to smaller concentrations of laurodimyristin.

Production. In plantations the coconut palm reaches a height of 30 m and from its 6th to 30th years annually yields 50 – 70 coconuts with a diameter of 10 – 12 cm. The hard shell, covered by a fibrous husk, encloses the white endosperm tissue 1 – 2 cm thick, the copra. To obtain a light, flavor-stable coconut oil, the fresh copra with a water content of 60 – 70 % is dried in the sun or with hot air. This treatment prevents bacterial decomposition and lipolysis of the fat. Dry copra contains 60 – 67 % oil.

The dried copra is processed in an oil mill in two steps. About two-thirds of the oil is first obtained by expelling broken and rolled copra in continuous screw presses. The residual fat content of the expeller cake can be reduced to ca. 5 % by high-pressure expelling and to 2 – 4 % by subsequently extracting the expeller cake with hexane.

In order to obtain an edible oil of good quality, the crude coconut oil must be neutralized, bleached, and deodorized. Normally, the crude oil contains ca. 5 % free fatty acids but it can be lye-neutralized without great loss of neutral oil. The neutralized, washed, and dried oil contains only small amounts of pigments, phosphatides, and other constituents. It is decolorized with 1 – 2 % of bleaching earth and 0.1 – 0.4 % of activated charcoal. Activated charcoal also serves to remove polycyclic aromatic hydrocarbons deposited on the copra by drying with flue gases [53]. Crude coconut oil with a relatively high content of free fatty acids can be advantageously neutralized and deodorized by distillation after pretreatment with phosphoric acid and bleaching earth – activated charcoal.

Saponification value	242 – 254
Iodine value	16 – 19
Nonsaponifiable matter	0.2 – 0.8 %
mp	23 – 30 °C
n_D^{40}	1.449 – 1.452

For fatty acid composition see Tables 1 and 2

The properties of palm kernel fat are very similar to those of coconut oil. However, it contains more oleic acid and only half as much C_8 and C_{10} fatty acids. The glycerides of palm kernel oil consist of 60 – 65 % trisaturated, ca. 25 % disaturated – monounsaturated, and 10 – 15 % monosaturated – diunsaturated components. Its stability is similar to that of coconut oil.

Uses. Coconut oil is used for cooking, and is also an important component of vegetable margarines. Because of its high latent heat of fusion, it produces a pronounced cooling effect in the mouth on melting. This effect also makes it a valuable fat for biscuit filling, confectionery products, and confectionery coatings (couvertures). Coconut oil can undergo enzymatic hydrolysis in foods containing relatively high concentrations of water leading to so-called perfume rancidity. This reaction can be counteracted by incorporating at least 30 % sugar or by drying. Coconut oil has a high resistance to oxidative deterioration.

Coconut stearin, mp 27 – 32 °C, and coconut olein can be obtained by fractional crystallization and pressing of prerefined coconut oil. Coconut stearin is used in couverture for confectionery products.

Hydrogenated coconut oil (iodine value 2 – 4, mp 30 – 32 °C) is produced from prerefined coconut oil with fresh nickel catalyst at 140 – 180 °C. In the summer months it is often used in admixture with unhardened coconut oil to raise the melting point of couvertures and similar products.

Crude coconut oil, meeting certain color specifications, is also used in the manufacture of special soaps.

13.2.1.2. Palm Kernel Oil [76]

Palm kernel oil [*8023-79-8*] is derived from the kernels of the oil palm (*Elaeis guineensis*) (see Section 13.1.1).

Production. The palm nuts are separated by air classifiers from the fibers of the press cake obtained on pressing palm fruit (see Section 13.1.1). After drying in silo dryers, the nuts are cracked with centrifugal crackers, and the shell is separated from the kernel with air and water separation systems. The kernel, 1 – 2 cm in size, is dried from an initial moisture content of 20 – 25 % to about 4 – 7 %; it then contains 44 – 57 % oil. Following cracking and rolling, the palm kernels are passed through a screw expeller. The residual oil in the press cake can be further reduced by high-pressure expelling or, more commonly, by extraction with hexane.

Crude palm kernel oil often has a higher free fatty acid content than coconut oil (up to 15 %, calculated as oleic acid). Palm kernel oil with a free fatty acid content <5 % can be lye-neutralized without significant neutral oil losses. At higher contents of free fatty acids, distillative neutralization (possibly followed by lye postneutralization) is generally preferred. Treatment with 1 – 2 % activated bleaching earth yields a light, yellowish oil. Deodorization generally requires a longer period of time than for coconut oil in order to obtain a stable, bland raffinate.

Use. Top-grade palm kernel raffinates are used in couvertures and margarines. Palm kernel stearin and olein are obtained by fractional crystallization of prerefined palm kernel oil at ca. 25 °C. Palm kernel stearin, mp 30 – 32 °C, is often used as such, or blended with coconut stearin, as a filling mass or couverture for confectionery products. Hydrogenated palm kernel oil (mp 39 °C, iodine value 1) is made in the same way as hydrogenated coconut oil and is mostly used in the confectionery trade. The melting point of hydrogenated palm kernel oil can be reduced from 39 °C to 33 – 35 °C by interesterification. Interesterified, hydrogenated palm kernel oil increases the range of special tailor-made products for application in confectionery products.

13.2.1.3. Babassu Oil and other Palm Seed Oils

The babassu palm (*Orbygnia speciosa, Attalea funifara*) is a native of the great forest regions of the Brazilian states Maranhão, Piauí, Pará, and Minas Geraes. The number of babassu trees in these forests is estimated at $>10^9$. Collecting and transport of the babassu nuts to the oil mills and removal of the extremely hard shell prior to expelling make it difficult to exploit this interesting source of lauric oil. A bunch holds 200 – 600 fruits; 1 t of fruit yields about 125 kernels containing 63 – 70 % fat after drying to ca. 4 % moisture. The fat can be isolated by expelling and subsequent extraction with hexane.

Saponification value	242 – 253
Iodine value	10 – 18
Nonsaponifiable matter	0.2 – 0.8 %
mp	22 – 26 °C
n_D^{40}	1.449 – 1.451

For fatty acid composition see Tables 1 and 2

Use. Refined babassu oil is known in South America as an edible fat. In European countries it is used in place of palm kernel oil in margarine blends. The export of babassu kernels is only of minor economic importance; total production is likely to be around 150 000 t/a.

Other palm seed oils are used as edible fats in their countries of origin and do not play a role in international trade. Some of their properties are listed in Table 24. Cohune oil is obtained from the kernels of the cohune palm (*Attalea cohune*), a native of Mexico and Honduras. Its properties are similar to those of coconut oil. Murumuru oil, somewhat harder than coconut oil, is contained in the kernels of the palm *Astrocaryum murumuru*, which is found in the Northern provinces of Brazil. It contains ca. 40 % lauric acid, 35 % myristic acid, and only little caprylic and capric

Table 24. Properties and analytical data of seed oils of some South American palm varieties

	Cohune oil	Murumuru oil	Ouricuri oil	Tucum oil
Density (60 °C), g/cm³	ca. 0.893	0.893	0.898	0.893
n_D^{60}	1.441	1.445	1.440	1.443
Saponification value	251 – 257	237 – 242	ca. 257	240 – 250
Melting point, °C	18 – 24	32 – 34	18 – 21	30 – 36
Iodine value	10 – 14	11 – 12	14 – 16	10 – 14
Nonsaponifiable matter, %	0.4	0.3	0.3	0.3

acids. Ouricuri oil, from the kernels of the Brazilian palm *Syagrus coronata*, has a lower melting point than murumuru oil. It contains 45 % lauric acid and 10 % myristic acid in addition to 20 % fatty acids with 6, 8, and 10 carbon atoms. The kernels of *Astrocaryum tucuma*, a native of Guyana, Venezuela, and Northern Brazil, yield tucum oil.

13.2.1.4. Other Sources of Lauric Acid Oils

Up to 43 % of lauric acid is contained in the seed oil of the evergreen laurel tree (*Laurus nobilis*), from which the name of this acid was derived. Laurel oil contains about 30 % trilaurin; it has no commercial importance.

The seed oils of *Myristica* species contain relatively high concentrations of myristic acid. Nutmeg butter [8008-45-5] is made by expelling ground and boiled seed kernels of *Myristica officinalis*; it contains 39 – 76 % myristic acid. *Myristica otoba* yields otoba fat. Ucuhuba fat is obtained from *Virola surinamensis*, a native to Brazil. Dika fat is contained in the seed kernels of *Irvingia gabonensis* and *I. barteri*, native to West Africa. None of these fats is involved in international trade.

Cuphea is a herbaceous, annual plant, native to Mexico and also found in Northern Brazil and Nicaragua. The seed oil contains various medium-chain fatty acids which, depending on the species, account for 40 – 80 % of the total fatty acids. Cuphea oil could be a substitute for coconut and palm kernel oil and serve as a natural source of capric acid; agronomic research is directed toward adaptation and yield improvement of *Cuphea* [78].

13.2.2. Palmitic – Stearic Acid Oils

13.2.2.1. Cocoa Butter

Cocoa butter is the seed fat of *Theobroma cacao* L., a tree reaching a height of 9 m, which is cultivated in many tropical countries (→ Chocolate, Section 6.1.). The cocoa tree is grown mainly in West Africa (Ivory Coast, Nigeria) and South Africa (Brazil). The world production of cocoa beans is in the order of 1.3×10^6 t/a.

Saponification value	190 – 200
Iodine value	35 – 40
Nonsaponifiable matter	0.2 – 0.5 %
mp	28 – 36 °C
n_D^{40}	1.453 – 1.458

For fatty acid composition see Tables 1 and 2

Cocoa butter has a pleasant aromatic flavor and keeps well. The cocoa butter glycerides are hard and brittle up to about 28 °C and melt in a range of 4 – 5 °C with a pronounced oral cooling effect. The melting point of cocoa butter depends on the type and length of tempering. After rapidly cooling to 0 – 5 °C, cocoa butter melts at 26 – 30 °C, whereas the melting point increases to 32 – 35 °C after tempering for 40 h at 28 °C. The dilatation curve (plot of solids content at different temperatures) is very steep (see Fig. 2).

The glycerides of cocoa butter consist roughly of 2 % trisaturated compounds, 14 % 2-oleodipalmitin, 40 % 1-palmito-3-stearo-2-olein, 27 % 2-oleodistearin, 8 % palmitodiolein, and 8 % stearodiolein.

Production. Cocoa butter is isolated from cocoa beans by fermentation, roasting, dehulling, grinding, and expelling (→ Chocolate, Section 6.1.). The fat content of dehulled cocoa beans is about 54 – 58 % in dry matter. In some countries the fat obtained from nondehulled cocoa beans by expelling or solvent extraction may also be called cocoa butter.

Use. Cocoa butter is used extensively in the production of chocolate and other confectionery products, and to a lesser extent in the pharmaceutical and cosmetics industries.

Test for Adulteration. Because of its high price, cocoa butter is sometimes admixed with press or extraction fat from damaged cocoa beans or byproducts, animal fats, coconut oil and stearin, palm oil and palm oil fractions, or hydrogenated vegetable or marine oils.

Extraction cocoa butter can be identified by its relatively dark color, by its strong fluorescence under ultraviolet light, by various color reactions, and by its behavior in simple tests such as the "smear test" (extraction fat tends to smear while authentic cocoa butter tends to disintegrate into small, hard pieces). In addition, extraction fat has a higher content of nonsaponifiable matter (cocoa butter contains 0.2 – 0.5 %, extraction fat 1.9 – 2.8 %, fat from cocoa bean hulls about 7.5 % nonsaponifiable matter) and a higher content of linoleic acid (10 – 29 % compared to 2 – 4 % in authentic cocoa butter).

Animal fats such as tallow and hydrogenated fish oils can be detected by gas chromatographic analysis of the fatty acid composition and by their cholesterol content. Hydrogenated vegetable oils can be recognized by their content of trans fatty acids. Coconut oil and similar oils can be detected by their concentrations of C_8, C_{10}, C_{12}, and C_{14} fatty acid groups. Addition of cocoa butter-like fats such as fractions of Borneo tallow, illipé fat, shea butter, and palm oil can sometimes be detected by their concentration of terpene alcohols. Gas chromatographic analysis of chocolate fat triglycerides can detect so-called cocoa butter equivalents in cocoa butter and certain types of chocolate; this technique can be augmented by analyzing the fatty acid composition at the 2-position [79].

13.2.2.2. Shea Butter, Borneo Tallow, and Related Fats (Vegetable Butters)

Shea butter, also known as Karité butter or Galam butter, is obtained from the seeds of a tree growing mainly in West Africa (*Butyrospermum parkii*). Plantations tend to be uneconomic since shea nuts can be harvested only after the tree is about 15 years old.

Shea butter is similar to cocoa butter (see Table 25). The content of nonsaponifiable matter, which contains rubber-like hydrocarbons such as kariten [$(C_3H_8)_n$, mp 63 °C], can be as much as 11 %.

The kernels can be readily separated from the fruit and the thin hulls. In Africa they are

Table 25. Properties and analytical data of some vegetable butters

	Shea butter	Borneo tallow	Sal butter	Illipé butter	Mowrah butter	Katiau fat	Phulwara butter
Myristic acid, %		ca. 15					
Palmitic acid, %	5 – 6	18 – 22	8 – 9	ca. 28	16 – 27	ca. 10	54 – 57
Stearic acid, %	36 – 42	39 – 44	ca. 35	ca. 14	ca. 20	ca. 19	ca. 4
Arachidic acid, %		ca. 1	ca. 12				
Oleic acid, %	49 – 50	38 – 42	ca. 42	ca. 50	41 – 66	ca. 69	ca. 36
Linoleic acid, %	4 – 5	traces	ca. 3	8 – 9	9 – 14	2 – 3	3 – 4
Saponification value	178 – 196	189 – 200		186 – 200	187 – 195	189 – 192	188 – 200
Iodine value	55 – 67	29 – 38		50 – 60	58 – 63	53 – 67	40 – 51
mp, °C	32 – 42	34 – 39		25 – 29	23 – 31	30	38 – 43
n_D^{40}	1.4635 – 1.4668	1.4561 – 1.4573		1.459 – 1.462	1.458 – 1.461	1.461 – 1.462	1.455 – 1.458
Nonsaponifiable matter, %	2 – 11	<2		1.4 – 2.3	1.2 – 2.1	0.4 – 0.5	2 – 2.8
Fat content in seed kernel, %	45 – 55		20	50	50		60 – 65

ground, boiled, and the fat is skimmed off. The commercial production of shea butter by expelling in continuous screw presses is difficult and requires special processing conditions. The green-brown crude fat can be neutralized only with considerable loss of neutral oil.

Refined shea butter can be used as an edible fat. The stearin fraction, primarily 2-oleodistearin, which can be obtained by crystallization and removal of the liquid fraction, is a valuable cocoa butter substitute. The export quantities (from Nigeria, Dahomey, Upper Volta) amount to ca. 50 000 t/a.

Borneo tallow (tengkawang tallow) is derived from *Shorea stenoptera*, a plant growing in the East Indies and Malaysia; there are about 100 different species including *S. gysbertsiana, S. palembassica,* and *S. seminis*. The proportions of component fatty acids in these species are very similar. The size of nut varies considerably from one variety to another, and only a few are worth harvesting. Borneo tallow resembles cocoa butter chemically and physically more closely than any other fat. For this reason it is used in preparing cocoa butter equivalents or extenders.

Sal butter is derived from *Shorea robusta*, commonly known as the sal tree, found extensively in parts of Northern and Central India. Illipé fat is derived from the group of Madhuca seed fats (also known as Bassia seed fat) and is itself produced from the Indian plant *Madhuca longifolia*. Another Indian plant, *Madhuca latifolia,* is the source of mowrah fat which is very similar in composition. Illipé and mowrah fats contain sufficient unsaturated fatty acids to make them resemble shea butter rather than the firmer cocoa butter and Borneo tallow. Illipé and mowrah fats are used principally in Asia for technical purposes but also for edible purposes when fully refined. Both fats are used to prepare cocoa butter extenders. Katiau fat is derived from the plant *Madhuca mottleyana,* and phulwara butter (Indian butter) from *M. butyracea.*

For properties of the various vegetable butters see Table 25 [80].

13.2.3. Palmitic Acid Oils

The oils in this group have a relatively high content of palmitic acid, mostly above 10 %. Cottonseed oil and some cereal oils are the most important representatives.

13.2.3.1. Cottonseed Oil

Cottonseed oil [8001-29-4] is obtained from the seeds of different varieties of cotton (*Gossypium*). The most important producers are the United States, the former Soviet Union, China, Pakistan, Brazil, Egypt, and India. *Gossypium barbadense* and *G. hirsutum* are preferred in the United States, while *G. herbaceum* is native to the Asian countries. The United States produce about one-third of the world's cottonseed oil (see also → Cellulose, Section 2.1.).

Saponification value	190 – 198
Iodine value	100 – 117
Nonsaponifiable matter	0.5 – 1.5 %
mp	ca. 0 °C
n_D^{40}	1.464 – 1.468

For fatty acid composition see Tables 1 and 2

The oil contains 0.4 – 0.6 % of malvalic acid, which presumably is the basis of the Halphen color reaction. The glycerides of cottonseed oil consist of approximately 12 – 14 % disaturated – monounsaturated, 52 – 58 % monosaturated – diunsaturated, and 25 – 30 % triunsaturated compounds.

Production. Egyptian cottonseed (not dehulled) contains 22 – 24 % oil, the American varieties on average 19.5 %. About 50 cottonseeds are contained in a pod, which opens when it is ripe. After removal of the cotton, the seeds are still covered with fine hairs (linters), which are removed with delintering machines that essentially consist of rotating sawtooth discs [81]. The seeds are then dehulled by passage through "dehullers" (essentially rotating knives) and screens – air classifiers. The dehulled seeds (30 – 40 % oil content) are expelled in continuous screw presses and the expeller residue is extracted with solvent.

Gossypol is present in the seed in concentrations of 0.4 – 2.0 %. However, only traces are detectable in the crude oil, because gossypol is partially deactivated (presumably by interaction with protein) during conditioning and expelling the seed. Crude cottonseed oil has a dark color because of dissolved resins and pigments, and must therefore be refined before further use. The crude oil is first clarified by settling or centrifuging and then neutralized with lye, mostly in continuous centrifuge lines. The neutralized oil is free of gossypol, light yellow, stable during storage, and is usually traded in this form. A very stable oil can be obtained by postneutralization, bleaching, and deodorization. Stearin separates out at temperatures between 0 and 8 °C. An oil with a low cloud point is obtained by removing these solid glycerides by cooling and filtration (winterization).

Use. Winterized cottonseed oil is a very stable salad oil. Cottonseed oil is also used in margarine blends without being winterized. Lard-like fats for use in margarine blends and shortenings are obtained by hydrogenation under iomerizing conditions. Such fats have a melting point between 30 and 35 °C and an iodine value of 60 – 80. Cottonseed oil hardened to a melting point of 36 – 37 °C is suitable as a couverture fat for confectionery products. The characteristic Halphen color reaction disappears on hydrogenating the oil.

13.2.3.2. Kapok and Related Oils

Kapok oil is derived from the fruit of a tree belonging to the Bombaceae family that grows in Indonesia, Sri Lanka, the Philippines, and South America.

The seeds in the fruit pods are covered with fine hairs (kapok), which are used in upholstering and contain ca. 25 % oil. Crude kapok oil, produced by preexpelling and solvent extraction, has a reddish-brown color and is similar to cottonseed oil in composition and further processing. It contains little or no gossypol. Kapok oil contains ca. 15 % of a C_{18} cyclopropenoic acid and therefore responds strongly to the Halphen reaction.

Okra oil is obtained from the seeds of *Hibiscus esculentis,* a family related to Gossypium and found in the United States and some Mediterranean countries. Kenafseed oil is contained in the seeds of *Hibiscus canabinus,* a plant native to India where it is grown for its fibers.

The properties of kapok, okra, and kenafseed oils are listed in Table 26.

13.2.3.3. Pumpkin Seed Oil

Pumpkin Seed Oil is popular as a high-quality, edible oil in Southeast Europe. It is obtained

Table 26. Properties of kapok, okra, and kenafseed oils

	Kapok oil	Okra oil	Kenafseed oil
Myristic acid, %	ca. 0.5	0 – 4	
Palmitic acid, %	10 – 16	23 – 33	ca. 14
Stearic acid, %	3 – 8	0.5	ca. 6
Arachidic acid, %	ca. 1	traces	
Oleic acid, %	ca. 50[*]	26 – 42	ca. 45
Linoleic acid, %	ca. 30	30 – 40	ca. 23
n_D^{40}	1.464 – 1.468	1.462 – 1.467	1.465 – 1.466
Saponification value	189 – 197	192 – 199	189 – 195
Iodine value	86 – 110	90 – 100	93 – 105
Nonsaponifiable matter, %	0.5 – 1.8	0.7 – 1.4	0.4 – 3.4

[*] Including 15 % malvalic acid.

from carefully decorticated pumpkin seeds by grinding, conditioning the moistened ground seeds, expelling in hydraulic presses, and clarification by settling in special tanks, mostly on smaller locations. It has a reddish-brown color and a nutty taste. Refining yields an oil with properties similar to those of sunflower oil.

Saponifaction value	185 – 198
Iodine value	117 – 130
Nonsaponifiable matter	0.6 – 1.5 %
Solidification point	ca. −15 °C
n_D^{40}	1.466 – 1.469

For fatty acid composition see Tables 1 and 2

Melon Seed Oil is very similar in composition to pumpkin seed oil [82], the major fatty acids being palmitic (ca. 12 %), stearic (ca. 11 %), oleic (ca. 11 %), and linoleic (ca. 65 %).

13.2.3.4. Corn (Maize) Oil

Corn oil (maize oil) [8001-30-7] is obtained from the germ removed during the processing of corn (*Zea mays* L.) into starch.

Properties. Winterized corn oil becomes cloudy at about −10 °C.

Saponification value	187 – 196
Iodine value	109 – 133
Nonsaponifiable matter	1.3 – 2.0 %
n_D^{40}	1.465 – 1.466

For fatty acid composition see Tables 1 and 2

Corn oil glycerides consist of ca. 2 % disaturated – monounsaturated, 40 % monosaturated – diunsaturated, and 58 % triunsaturated compounds.

Production. Corn contains 3.5 – 5 % oil, about 80 % of which is in the germ and about 20 % in the endosperm. The corn germ itself contains about 36 % oil. The germs can be separated by wet or dry processing. Wet processing tends to be preferred because it gives a higher oil yield. The cleaned corn is first conditioned by soaking in warm water and is then milled and slurried with water. The germs are collected by flotation, washed, and dried. In dry processing, the ground corn is separated into germ and endosperm fractions by screening and air classification; suitable moistening and conditioning facilitate this fractionation. The oil is isolated from the germs by preexpelling followed by extraction with hexane.

The golden yellow, crude corn oil contains up to 3 % free fatty acids. After lye neutralization, bleaching, and deodorization it yields a light, highly stable oil. The content of tocopherols, predominantly γ-tocopherol, is nearly 1000 ppm. Salad oils that remain clear at refrigerator temperatures are obtained by winterization, during which about 500 ppm wax is removed.

Use. Corn oil is a universal oil; it is used especially as a salad oil and in salad dressings.

13.2.3.5. Cereal Oils

Oils are generally not extracted from cereal grains except rice and wheat. The cereal oils are, however, very important as constituents of bread, other bakery products, and animal feeds.

Wheat Germ Oil is extracted from the wheat germ (8 – 11 % oil) obtained as a by-product of milling wheat. It has a very high tocopherol content and is therefore popular as a dietetic oil.

Saponification value	180 – 189
Iodine value	115 – 126
Nonsaponifiable matter	3.5 – 6.0 %
mp	ca. 0 °C
n_D^{40}	1.468 – 1.478

For fatty acid composition see Tables 1 and 2

Rice Bran Oil [68553-81-1] is an important byproduct of rice processing. It is obtained by hexane extraction of the bran (brown outer coating) and the germ (8 – 16 % oil content) removed during grinding, dehulling, and polishing rice. The bran fraction must be extracted immediately, preferably after heat-sterilization, to destroy the very active lipases. Refined rice bran oil has a light yellow color and is a relatively stable edible oil. Rice bran oil hardened to 30 – 35 °C is often used in margarines and shortenings. In Japan and India ca. 100 000 t of rice bran oil are produced annually.

Saponification value	183 – 194
Iodine value	92 – 109
Nonsaponifiable matter (including ca. 330 mg squalene per 100 g oil)	3.5 – 5.0 %
mp	ca. −10 °C
n_D^{40}	1.466 – 1.469

For fatty acid composition see Tables 1 and 2

13.2.4. Oleic – Linoleic Acid Oils

This group encompasses a large number of drying and semidrying oils derived from plants growing in temperate and colder regions (see also → Drying Oils and Related Products). Some of these oils, such as hazelnut oil, poppyseed oil, walnut oil, and teaseed oil, only have regional significance, while sunflower oil, sesame oil, linseed oil, and safflower oil have considerable commercial importance.

13.2.4.1. Sunflower Oil

Sunflower oil [8001-21-6] is the seed oil of the sunflower, *Helianthus annuus,* that originated in America and is now grown extensively in East, West, and South Europe, the United States, Canada, South America, China, India, South Africa, and Australia. The annual plant reaches a height of 1 – 3 m, the short-stem varieties being preferred for ease of harvesting. The air-dry seeds contain about 45 % oil.

Properties. Sunflower oil begins to solidify at −16 °C; the cloud point is 0 – 5 °C.

Saponification value	188 – 194
Iodine value	125 – 144
Nonsaponifiable matter	0.4 – 1.4 %
n_D^{40}	1.466 – 1.468

For fatty acid composition see Tables 1 and 2

The linoleic acid content of sunflower oil is 50 – 70 % and tends to increase with decreasing temperatures during growing [83]. The glycerides contain 35 – 45 % diunsaturated and 56 – 63 % triunsaturated compounds. Linolenic acid is present to the extent of <0.4 %.

Production and Use. Sunflower oil is obtained by preexpelling followed by extraction of the expeller cake with hexane. Decortication is practiced to a lesser degree; the hulls have to be burnt. The crude oil can be readily refined into a light, stable oil by lye neutralization, bleaching, and deodorization.

Cold-pressed sunflower oil is produced by expelling at ≤ 80 °C; posttreatment is limited to filtration and clarification. Such a "natural" oil is often recommended for dietetic purposes because of its relatively high concentration of phosphatides. Trace contaminants such as metals and pesticide residues, however, remain in the oil. Refined sunflower oil is used extensively in salad oils and margarine blends because of its oxidative stability. Winterization (removal of 500 – 2000 ppm of wax) of the deodorized or the neutralized and bleached oil at 10 – 15 °C yields an oil which does not cloud at refrigerator temperatures. Sunflower oil that is hydrogenated to a melting point of 32 – 35 °C (iodine value 70 – 75), under isomerizing conditions, is widely used in "one-oil" margarine blends.

Smaller quantities of a new sunflower variety are being grown in California; these plants yield an oil with an oleic acid content of about 80 % and hence a very high oxidative stability. This oil is recommended especially for salad and frying oils [84].

13.2.4.2. Sesame Oil

Sesame oil [8008-74-0] is contained in the seeds of *Sesamum indicum*, a plant resembling linseed and grown mainly in China, India, Africa, and Mexico. The flat seeds are 2 – 3 mm long and vary in color from white to brown-black; the oil content is 45 – 55 %. Harvesting is relatively difficult because of varying rates of ripening and the ripe seed pods shattering the seed. More uniform varieties with improved harvesting properties are being developed.

Properties. Sesame oil solidifies at −6 to −3 °C.

Saponification value	187 – 193
Iodine value	136 – 138
Nonsaponifiable matter	0.9 – 2.3 %
n_D^{40}	1.4665 – 1.4675

For fatty acid composition see Tables 1 and 2.

Crude sesame oil is yellowish and contains up to 3 % free fatty acids. The glycerides contain about 66 % dioleolinolein and oleodilinolein. Refined oil has a high stability due in part to the content of sesamol, a natural antioxidant formed by hydrolysis of sesamolin. The combined concentration of sesamol and sesamolin is 0.1 – 0.2 %. Sesamol gives a red color with an ethanolic solution of furfural in the presence of hydrochloric acid (color reaction of Baudouin and Villavechia); it also responds positively in color tests with an acidic solution of stannous chloride (Soltsien test), and with 3 % hydrogen peroxide – 75 % sulfuric acid (Kreis test).

Use. Sesame oil is widely used as an edible oil in the countries of origin. A small amount is exported for use in salad oils, salad dressings, and margarines. The addition of sesame oil to margarine is mandatory in some countries to provide a rapid means of identifying margarine by the above color reactions; for this application the oil may only be lightly bleached and deodorized at 150 °C.

13.2.4.3. Linseed Oil

Linseed oil [8001-26-1] is obtained from linseed or flax (*Linum usitatissimum*), which grows best in the temperate regions of Europe, Asia, and America. The flax fiber plant is a different variety of the same species. There is considerable variation in the iodine value of linseed oil from different regions. The iodine value of the oil tends to increase with severity of climate, but genetic and seasonal (e.g., rainfall) variations also have an influence [85]. While 180 – 185 is a typical iodine value range for individual lots, it may vary from 140 to 205.

The main producer countries are Canada, the United States, and Argentina. The brown, flat seeds contain ca. 40 % oil and 6 – 8 % water.

Properties. Fatty acid composition and other properties of linseed oil are given in Table 27 (see also Tables 1 and 2). Most applications require a linseed oil with a linolenic acid content of about 50 % and an iodine value of 170 – 190 [86], [87]. Linseed oil glycerides consist of about 5 % disaturated – monounsaturated, 43 % monosaturated – diunsaturated, and 52 % triunsaturated glycerides, which begin to solidify at −18 to −27 °C. Most of the glycerides contain an average of seven double bonds per molecule, which is the reason for the excellent drying properties of linseed oil (→ Drying Oils and Related Products). The use of linseed oil for edible purposes is limited to smaller quantities of cold-pressed oil (for dietetic purposes) and to hydrogenated linseed oil, which can be used in low concentrations in shortenings, and margarine blends.

Some wild linseed varieties (e.g., *Linum capitatum* and *L. flavum*) have linolenic acid contents as low as 11 %. Modern biotechnological techniques are being used in Australia and Canada in an attempt to transfer this property to *L. usitatissium*. An alternative approach is γ-ray induced mutation of *L. usitatissimum*.

Production. Linseed oil is obtained from the seed by preexpelling followed by hexane extraction of the press cake. After refining it is used predominantly in the technical sector for paints and coatings. A small amount is used in East European countries as edible oil. Linseed oil has a characteristic odor and flavor which cannot be removed entirely by refining. The crude oil is rich in phosphatides and gums which sediment on storage. The crude oil is often pre-deslimed by treatment with hot water followed by centrifugation. Oils virtually free of phosphatides and gums are made by post-desliming with sulfuric or phosphoric acid. Very light drying oils can be

Table 27. Fatty acid composition and analytical data of vegetable oils of the oleic – linoleic group

	Linseed oil	Perilla oil	Hempseed oil
Myristic acid, %	ca. 0.5		
Palmitic acid, %	5 – 7	7 – 8	ca. 6
Stearic acid, %	3 – 5	traces	ca. 2
Arachidic acid, %	ca. 6		ca. 2
Lignoceric acid, %	traces		
Oleic acid, %	16 – 26	ca. 8	6 – 20
Linoleic acid, %	14 – 24	ca. 38	46 – 70
Linolenic acid, %	50 – 65	44 – 50	14 – 28
Saponification value	188 – 196	188 – 196	190 – 194
Iodine value	170 – 204	170 – 204	140 – 170
Solidification point, °C	10 – 21	12 – 17	15 – 17
n_D^{25}	1.4786 – 1.4815	1.4800 – 1.4820	
Nonsaponifiable matter, %	ca. 1.5	<1.5	<1.5

made by a further posttreatment consisting of lye neutralization and earth bleaching. To obtain virtually odorless oils, deodorization is required as a final step. Heating refined linseed oil to 260 – 285 °C increases its viscosity and gives a so-called "boiled" linseed oil, which is used in special paints and coatings.

13.2.4.4. Perilla Oil

The seeds of *Perilla ocymoides*, an Asian plant, yield perilla oil (30 – 50 %), which resembles linseed oil in appearance and application. For properties see Table 27. World production is about 25 000 t/a. Perilla oil is sometimes blended with linseed oil; in Asia it is also used as an edible oil.

Lallemantia oil (Asia) and Chia seed oil (Mexico) are further Labiatae seed oils with similar compositions.

13.2.4.5. Hempseed Oil

The seeds of hemp (*Cannabis sativa* L.) contain 30 – 35 % oil. Hempseed is grown in the Soviet Union, India, China, Japan, and Chile, and yields fibers in addition to the oil. Hashish can also be obtained from Indian hempseed varieties. Hempseed itself can be used as a foodstuff. The oil is normally produced by pressing the seed. For properties see Table 27. Hempseed oil has been largely replaced by linseed oil for technical purposes.

13.2.4.6. Teaseed Oil

Teaseed oils are obtained in China, Japan, India, and Turkey from the hazelnut-like seeds of tea shrub varieties, which contain about 60 % oil. Tea shrubs grown for oil are generally not suited for tealeaf harvesting. Teaseed oil is produced primarily from the seeds of *Thea sasanqua* n. (sasanqua oil) and *Thea japonica* n. (tsubaki oil) by drying, grinding, conditioning, and pressing or solvent extraction. After refining the oil is used for edible purposes, as a special lubricant, and for toiletries.

The principal acids of teaseed oils [88] are palmitic (ca. 16 %), oleic (ca. 60 %), and linoleic acid (ca. 22 %). Teaseed oils can be detected in concentrations above 10 % in olive oil by the Fitelson color reaction. The expeller cake and extraction meal cannot be used as such in animal feeds because of their high saponin contents.

Total world production is ca. 30 000 t, half of which is exported.

13.2.4.7. Safflower and Niger Seed Oils

Safflower Oil [8001-23-8] is the seed oil of the thistle-like safflower plant (*Carthamus tinctorius*), thriving in the West of the United States and in Mexico, North Africa, and India. The plant can be grown under fairly arid conditions, the oil yield increasing with available moisture in the soil. The seeds resemble small sunflowerseed kernels and can be harvested mechanically. The oil content of the seed is 25 – 37 %.

Saponification value	180 – 194
Iodine value	136 – 152
Nonsaponifiable matter	0.3 – 2.0 %
Solidification point	−13 to −25 °C
n_D^{40}	1.467 – 1.469

The fatty acid composition is similar to that of sunflower oil (see Tables 1 and 2).

Production. Modern processing involves grinding the seed followed by conditioning, expelling, and hexane extraction of the expeller cake.

Use. Safflower oil has a high oxidative stability and is being used increasingly in salad oils and dietetic margarines because of its high content of linoleic acid. Some years ago special varieties with oleic and linoleic acid contents of 80 % and 15 % respectively ("high oleic" varieties) were developed and recommended as especially stable frying oils. World production amounts to about 200 000 t/a.

Niger Seed Oil, closely resembling safflower oil in its composition, is derived from the seed of the gingli plant, which is grown fairly extensively in Ethiopia, Togo, and India.

13.2.4.8. Grape-Seed Oil

Grape-seed oil [60-33-3] is a valuable oil obtained from the grapeseed left in winery pomace [89]. The oil content averages 15 % (dry weight basis).

Properties. The flavor of grape-seed oil resembles that of olive oil; however, its fatty acid composition is similar to that of sunflower oil (see Tables 1, 2).

Saponification value	180 – 196
Iodine value	124 – 143
Nonsaponifiable matter	0.3 – 1.6 %
Solidification point	−10 to −20 °C
n_D^{40}	1.464 – 1.471

Use. After suitable refining (removal of chlorophyll can be difficult), grape-seed oil keeps relatively well and is used as a salad and cooking oil and in special margarine blends. In France and Italy, 5000 t and 20 000 t, respectively, of grape-seed oil are produced annually. It has been estimated that the world production of grape seed is 1.4×10^6 t, with a potential oil yield of about 190 000 t [82].

13.2.4.9. Poppyseed Oil

Poppyseed oil is obtained from the seeds of different varieties of *Papaver somniferum*. The oil content of air-dry seeds is 44 – 50 %. Cold-pressed oil is suitable for edible purposes, as is preexpelled and extracted oil after refining. Poppyseed oil is also used in smaller quantities for superior oil paints. The major fatty acids are stearic (ca. 10 %), oleic (ca. 16 %), and linoleic acid (70 – 75 %).

Saponification value	ca. 193
Iodine value	ca. 140
n_D^{40}	ca. 1.468

The world production of poppy seed amounts to about 25 000 t/a.

13.2.5. Leguminous Oils

The commercially most important representatives of this group are peanut oil and soybean oil. Soybean oil [8001-22-7] holds first place in the worldwide production of vegetable oils.

13.2.5.1. Soybean Oil

The cultivation of the soybean (*Glycine maxima*) in the United States and South America has steadily increased. The soybean has been known in China for 5000 years. Main producing countries today are the United States, Brazil, and Argentina. Cultivation in the temperate regions of Europe is possible but the hectare returns are generally not sufficiently attractive. New, higher-yielding varieties are constantly being developed; in the United States they thrive best in the "corn-belt," i.e., Illinois, Minnesota, Iowa, Indiana, Ohio, and Missouri.

The soy pods contain up to four soybeans (5 – 10 mm in length). The oil content varies between 17 and 22 % and the protein content reaches 40 – 45 % in dry matter. Climatic conditions have a considerable influence on the oil and protein content; late sowing and a cool, wet climate tend to lower the oil content and to increase the protein content. Development of low linolenic soybeans in the United States has been hindered by lower seed productivity.

Properties. Refined soybean oil has a light yellow color and a bland flavor.

Saponification value	188 – 195
Iodine value	120 – 136
Nonsaponifiable matter	0.5 – 1.5 %
Solidification point	−15 to −8 °C
n_D^{40}	1.465 – 1.469

For fatty acid composition see Tables 1 and 2.

The oil contains 40 – 60 % triunsaturated, 30 – 35 % diunsaturated, and up to 5 % monounsaturated glycerides. In spite of the high tocopherol content (up to 1200 ppm) the stability of the refined oil is limited, primarily because of the content of linolenic acid. Autoxidation leads to "green", "seedy" off-flavors. Soybean oil can be detected in other oils by its fatty acid, tocopherol, and sterol contents.

Production. Soybeans are normally cracked and flaked prior to extraction with hexane. Preexpelling is not normally performed. Soybean meal is a valuable animal feed. Crude soybean oil contains 0.5 – 1.0 % free fatty acids as well as up to 2.5 % phosphatides, which are removed by hydration with 2 – 3 % water at 70 – 80 °C followed by centrifuging. Drying of the resulting lecithin sludge yields lecithin. Soybean oil is often postdegummed with phosphoric or citric acid prior to neutralization with lye, earth bleaching, and deodorization.

The yield of lecithin can be increased by steam-heating the flakes prior to extraction [39]. With the aid of special degumming techniques [44], the phosphorus content of the crude oil can be reduced to 20 ppm, which facilitates subsequent physical refining (i.e., direct treatment with acids and earth followed by combined stripping of free fatty acids and deodorization).

Use. Soybean oil is used in almost all fatty products, from salad oils to margarines. Hydrogenation with fresh catalyst at ca. 100 °C yields a liquid oil with significantly improved stability due to hydrogenation of linolenic acid. Hydrogenation under isomerizing conditions to melting points ranging from 36 to 43 °C yields products for use in shortenings and margarines. Liquid soybean oil is a so-called semidrying oil; it is used in combination with tung oil and as a component of alkyd resins for coatings.

13.2.5.2. Peanut Oil

Peanut oil (groundnut oil) [8002-03-7] is obtained from the seed kernels of the peanut plant (*Arachis hypogaea*), native to South America. The peanut is now being grown in China, India, South Africa, West Africa, Argentina, and the United States. Senegal and Nigeria are the main African export countries. The annual plant thrives best on light, sandy soils and reaches maturity in 4 – 5 months. Peanuts for further processing are mostly shipped after shelling, whereas peanuts for direct consumption are normally exported in the pods. Peanuts are often designated according to their shipping ports, e.g., Bombay, Casamance, Rufisque, Bissao nuts.

Properties. The iodine value of peanut oil is 84 – 105; Rufisque oil has an iodine value of 84 – 90 and Argentinian oil a value of 100 – 105 in line with the higher content of linoleic acid. The saponification value is 185 – 196, n_D^{40} 1.461 – 1.465. Tocopherols and other antioxidants as well as hydrocarbons and sterols are found in the nonsaponifiable matter (0.5 – 1.0 %). The oil solidifies below 0 °C. For fatty acid composition see Tables 1 and 2. The oil contains 36 – 46 % triunsaturated, 47 – 52 % diunsaturated, and 7 – 11 % monounsaturated glycerides.

Production. The peanut pods contain up to four hazelnut-sized kernels which are covered with a thin, reddish-brown skin and contain 5 – 12 % water, 45 – 50 % oil, and 23 – 35 % protein. The pods are first removed by passage through cracking rolls or disk mills and over screens. This step also loosens the skins, which are removed by aspiration. The kernels are then screened, cracked, and expelled in screw presses. The residual oil in the expelled cake is extracted with hexane. Crude peanut oil contains low concentrations of phosphatides and up to 1.5 % free fatty acids. Lye neutralization, earth bleaching, and deodorization yield a light, slightly yellowish oil with a very good oxidative and flavor stability. Peanut kernels to be processed into oil and meal can contain up to 0.5 ppm of aflatoxins. The bulk of the aflatoxins remain in the meal where they can be deactivated by treatment with, for example, ammonia; traces of aflatoxin that might enter the oil are removed during neutralization with lye.

Use. Peanut oil is used as a superior salad and cooking oil, less so in margarines. In shortenings it is used as such or after hardening (*mp* 31 – 38 °C). On cooling to 6 – 8 °C, peanut oil becomes cloudy due to the crystallization of glycerides containing $C_{20} - C_{24}$ fatty acid groups. The glycerides can be removed by filtration but only with a low yield of clarified oil.

Peanut Butter is made by grinding specially roasted peanuts and homogenizing the mash with addition of liquid and possibly also hydrogenated peanut oil.

13.2.5.3. Lupine Oil

Lupins are attracting attention worldwide as an arable crop for possible use in foods and feeds. *Lupinus albus, L. angustifolius, L. luteus,* and *L. mutabilis* are potential sources of edible oil. Whereas the first three species (sweet varieties) contain only 4 – 9 % oil in the dried seed, the seed oil content of *L. mutabilis* approaches that of soybeans. The main fatty acid components are palmitic (7 – 13 %), stearic (2 – 7 %), oleic (27 – 53 %), linoleic (18 – 47 %), linolenic (2 – 9 %); lower concentrations of $C_{20} - C_{22}$ fatty acids are found [90]. The meal of *L. mutabilis* (bitter lupins) cannot be used as an animal feed without prior extraction of the alkaloids.

13.2.6. Cruciferous Oils

The seed oils of the traditional cruciferous plants are mainly characterized by their content of glucosinolates (sulfur-containing glycosides) and erucic acid. Rapeseed oil is the most important cruciferous oil.

13.2.6.1. Rapeseed Oil [91]

Rapeseed oil [8002-13-9] is the seed oil of different varieties of *Brassica napus* and *B. campestris*. The air-dry seed contains about 40 % oil and 7 – 9 % water. The production of rapeseed, one of the oldest oil crops known, has increased dramatically over the past 10 years. The introduction of so-called single-zero rapeseed varieties, i.e., varieties with an erucic acid content of less than 5 % of the total fatty acids, has made rapeseed oil universally applicable as an edible oil. Generally, it can be used to replace soybean oil. Whereas single-zero, spring-sown rapeseed varieties are grown in Canada, the EEC countries (with the exception of Denmark) generally prefer winterhardy single-zero varieties because of the higher crop yields. Within the European Community the traditional rapeseed varieties with up to 50 % erucic acid in the oil are now Within the European Community only single-zero varieties are eligible for subsidy, and the United States has recently permitted the use of rapeseed oil with max. 2 % erucic acid. Whereas single-zero, spring-sown rapeseed varieties are grown in Canada, the EEC countries (with the exception of Denmark) generally prefer winterhardy single-zero varieties because of the higher crop yielused only for technical and specialty purposes.

The increase in planted acreage has led to greater availability and to processing throughout the year; a significant improvement of the quality of the meal and the oil has been achieved [92]. Canada, the European Community, Poland, Sweden, China, India, and Pakistan are the main producing countries. To facilitate the marketability of meal, double-low varieties, i.e., low in erucic acid and glucosinolates, are grown in Canada (spring-type) and are being introduced in the EEC (winter-type). The winter-sown varieties are larger in size and generally easier to process than the smaller spring-sown varieties.

Properties [92]. For fatty acid compositions see Tables 1 and 2. Analytical data such as iodine value and saponification value are functions of the erucic acid content. The high level of erucic acid in traditional rapeseed oil has been associated with irreversible organ changes in animal tests [92].

Production. Modern rapeseed processing entails flaking and cooking the seed prior to preexpelling in continuous screw presses; the press cake is extracted with hexane. Crude rapeseed oil contains up to 2 % free fatty acids and ca. 2.5 % phosphatides; it is degummed with water or acids, lye-neutralized, bleached with earth, and deodorized, analogously to soybean oil. The stability is also comparable. The phosphatide content of the crude oil can be reduced to 0.1 % (i.e., 20 ppm phosphorus) by special degumming techniques [44], which enable physical refining to be carried out.

The enzymatic degradation products of sulfur-containing glycosides (glucosinolates) in rapeseed can influence the taste of the crude oil and impair the nutritional properties of the meal [92]. The enzyme responsible for glucosinolate degradation, myrosinase, is destroyed by adequate cooking of the seed prior to expelling. Sulfur compounds are removed from the oil during refining; residual traces can affect the course of hardening since they act as catalyst poisons. The residual sulfur compounds in the extracted meal can be deactivated by treatment with alkali and heating.

Use. Traditional rapeseed oil containing up to 50 % erucic acid can be hardened into steep-melting products, *mp* 32 – 34 °C, suitable as partial coconut oil substitutes in shortenings and couvertures. Modern single-zero varieties are used as such, after mild hardening (which eliminates the linolenic acid), and after hardening under isomerizing conditions to raise the melting point to 30 – 43 °C. The reduced complexity in fatty acid composition of the hardened oil leads to a different crystallization behavior, which can be overcome by interesterification. Treatment of rapeseed oil with sulfur or sulfur compounds is the basis of mastic production.

13.2.6.2. Mustard Seed Oil

Mustard seed oil [8007-40-7] is derived from the seeds of black mustard (*Brassica nigra* or *Sinapis nigra*), brown mustard (*Brassica juncea*), and white mustard (*Brassica alba* or *Sinapis alba*). The seeds contain 30–35 % oil and high levels of glucosinolates. The erucic acid content of the oil is 40 – 50 %; it is therefore of interest primarily for technical applications.

13.2.7. Conjugated Acid Oils

13.2.7.1. Tung Oil and Related Oils

Tung oil [8001-20-5] is obtained from the kernels (oil content about 50 %) of the fruit of the tung tree, *Aleuritis fordii* (Chinese tung oil) or *Aleuritis cordata*, syn. *vernica* and *verrucosa* (Japanese tung oil). The oil from *Aleuritis montana* is almost identical in composition. Oils similar to Chinese tung oil are obtained from the nuts of *Aleuritis trisperma* (kekuna oil) and *A. moluccana* or *A. triloba* (lumbang oil). Tung trees are cultivated in Southern China, Southern Russia, the Southern United States, and Argentina. Tung oil contains about 80 % α-eleostearic, 4 % linoleic, 3 % linolenic, 8 % oleic, 1 % stearic, and 4 % palmitic acid. For properties, production, and use, see → Drying Oils and Related Products.

13.2.7.2. Oiticica Oil and Related Oils

Oiticica oil [8016-35-1] is obtained mainly from the nuts of the Brazilian oiticica tree, *Licania rigida*. This evergreen tree reaches a height of 20 m and produces 150 – 900 kg of nuts per year. The kernels contain 55 – 63 % oil, which is generally obtained by pressing. Fresh oiticica oil is yellowish and of lardlike appearance. Oiticica oil for export is heated for 30 min at 210 – 220 °C, after which it remains liquid. Licanic acid, 4-oxo-9,11,13-octadecatrienoic acid, is the principal fatty acid. The oil yields varnishes similar to those made from tung oil (→ Drying Oils and Related Products).

Boleko or isano oil (ongoke oil) is obtained from the nuts of *Ongokea gore* ENGLER, a tree growing in the Congo regions. It contains isanic and isanolic acids, which are fatty acids with conjugated triple bonds.

Parinarium oils are contained in the kernels of various tropical trees of the family Rosaceae and have little commercial significance. *Parinarium laurinum*, a tree growing in Japan and Oceania, yields a seed oil which contains parinaric acid, an acid with 4 conjugated double bonds.

Néou oil, obtained from *P. macrophyllum*, contains about 30 % eleostearic acid. Eleostearic acid, in addition to licanic acid, has been found in po-yoak oil (obtained from *P. sherbroense*).

13.2.8. Substituted Fatty Acid Oils

13.2.8.1. Castor Oil

The seeds of the castor tree, *Ricinus communis*, contain 45 – 50 % castor oil [8001-79-4]. The evergreen castor tree belongs to the family Euphorbiaceae and grows in tropical and subtropical countries. It can reach a height of 10 m.

Properties. The fatty acids of castor oil consist of 87 – 91 % ricinoleic acid, 2 % stearic and palmitic acid, 4 – 5 % oleic acid, 4 – 5 % linoleic acid, and 1 % dihydroxystearic acid. Pure castor oil can be readily identified by its hydroxyl value, viscosity, and specific gravity. The oil is soluble in ethanol at room temperature and in boiling hexane.

d_{25}^{25}	0.945 – 0.965
n_D^{40}	1.466 – 1.473
Solidification point	−12 to −18 °C
Viscosity at 20 °C	935 – 1033 mPa s
$[\alpha]_D$	+7.5 to +9.7 °
Saponification value	177 – 187
Iodine value (Wijs)	82 – 90
Hydroxyl value	161 – 169
Acetyl value	144 – 150
Nonsaponifiable matter	0.2 – 0.3 %

Production. Cold pressing of the spotted, oblong seeds yields a light, viscous oil which is used primarily in pharmaceutical products. Subsequent expelling at elevated temperatures and extraction with hexane yields a yellowish-brown oil which is used primarily in technical applications. The extracted meal contains ricin, a toxic protein, and hence cannot be used as an animal feed.

Use. A drying oil can be made by dehydrating castor oil (dehydrated castor oil, → Drying

Oils and Related Products, Chap. 4.). Pyrolytic cleavage at 300 °C gives undecylenic acid and heptaldehyde, raw materials for polymers and perfumes. Alkali fusion of castor oil yields 2-octanol and sebacic acid.

Hydrogenation yields a fat with a melting point of 84 – 86 °C and waxlike properties, finding application as a special lubricant. Air-blown castor oil is used as plasticizer for varnishes and polymers. The oil is also used in the production of transparent soap, textile processing aids, special lubricants, and toiletries.

13.2.8.2. Chaulmoogra, Hydnocarpus, and Gorli Oils

The seed kernels of *Taractogenos kurzii*, a tree native to Southeast Asia, contain 48 – 55 % chaulmoogra oil, the chief fatty acid components of which are chaulmoogric acid, hydnocarpic acid, gorlic acid, and lower homologues of hydnocarpic, palmitic, and oleic acid. Chaulmoogra oil and related oils such as hydnocarpus oils, obtained from Southeast Asian *Hydnocarpus* varieties, and the African gorli oil have been used in the treatment of leprosy; for properties see Table 28.

13.2.8.3. Vernonia Oil [93]

The seeds of *Vernonia anthelmintica* (of the Compositae family) contain an oil with a high percentage of vernolic acid (*cis*-12,13-epoxy-*cis*-9-octadecenoic acid). This epoxy acid is used in protective coatings, plastics, and other industrial products. Efforts have been made to grow *Vernonia* varieties as new industrial crops in the United States and Asia.

14. Individual Animal Fats

Animal fats are obtained from the milk or the fatty tissue of certain groups of animals (see also → Milk and Dairy Products). The content of depot fat varies from 0 to 60 %. The concentration and composition of the fat depend on type of animal, age, sex, and diet.

14.1. Land-Animal Fats [94]

14.1.1. Lard

Lard [*61789-99-9*] is the fat rendered from fresh, clean, sound fatty tissues from pigs in good health at the time of slaughter. The tissues do not include bones, ears and tails, internal organs, windpipes, or large blood vessels. Rendered pork fat is obtained from the tissues and bones of pigs in good health at the time of slaughter; it may contain fat from bones, skin, ears and tails, and other tissues fit for human consumption. A range of different quality grades is available.

Properties. European lard (iodine value ca. 60) is often harder than America lard (iodine value ca. 70).

Table 28. Properties of some "leprosy oils"

	Chaulmoogra oil (*Taractogenos kurzii*)	Hydnocarpus oil (*Hydnocarpus whigtiana*)	Gorli oil (*Oncoba echinate*)
Palmitic acid, %	ca. 4	1 – 2	7 – 8
Oleic acid, %	13 – 15	5 – 7	2 – 3
Lower homologues of hydnocarpic acid, %	ca. 0.5	3 – 4	
Hydnocarpic acid, %	ca. 35	ca. 50	
Chaulmoogric acid, %	ca. 23	27	ca. 75
Gorlic acid, %	ca. 23	12 – 13	ca. 14
n_D^{40}	1.471 – 1.473	1.472 – 1.473	
$[\alpha]_D^{25}$, °	+49.8	+55	+51.7
mp, °C	22 – 30	22 – 24	42 – 44
Saponification value	198 – 208	200 – 208	190 – 194
Iodine value	98 – 105	93 – 101	94 – 100
Nonsaponifiable matter, %	<0.5	<0.5	1 – 1.5

mp	30 – 40 °C
Solidification point	22 – 32 °C
Saponification value	193 – 202
Nonsaponifiable matter	0.1 – 1.0 %

Lard consists of 4 – 8 % trisaturated, 32 – 40 % disaturated – monounsaturated, 45 – 50 % monosaturated – diunsaturated, and 3 – 10 % triunsaturated triglycerides. The main fatty acid components are shown in Tables 1 and 2. Lard contains low concentrations of arachidonic acid (0.4 – 0.9 %), trans fatty acids, as well as traces of branched and odd-numbered fatty acids such as saturated C_{15}, C_{17}, and C_{19} fatty acids and monounsaturated C_{17} and C_{19} fatty acids. Triunsaturated C_{18}, C_{20}, and C_{22} fatty acids are also present. These typical fatty acids, together with the cholesterol that occurs in relatively high concentrations in animal fats, are the basis for detecting lard and other animal fats in vegetable oils and fats.

Production. Lard is obtained by dry or wet rendering. The best quality lard (acid value <0.8; moisture <0.1 %) is made from selected tissue which has been washed and cooled immediately after slaughter. After comminution, the fatty tissue is wet-rendered by steam heating at 50 – 60 °C (for continuous wet-rendering plants see Section 5.2). Prime steam lard and leaf lard are obtained similarly from nonselected or partially selected tissue. Packer's lard retains a typical flavor; refiner's lard has to be refined. Inferior varieties, made in part from inedible offal, include so-called white, yellow, and brown greases, which are used primarily in technical applications.

Lard can be fractionated into a higher- and lower-melting fraction. In order to prevent the use of inferior raw materials, some European countries do not permit the refining of lard. The addition of antioxidants significantly improves the organoleptic and oxidative stability. Lard can be refined either by neutralization with lye, followed by bleaching with ca. 1 % earth and deodorization at 180 – 240 °C, or by simple pretreatment with phosphoric acid – bleaching earth and subsequent distillative stripping. Interesterification improves the performance of lard as a shortening in terms of creaming and cakemaking [95].

Use. Lard is used extensively as an edible fat (shortening, margarines) and as an energy source in animal feeds.

14.1.2. Beef Tallow

Properties. The main fatty acids of beef tallow [*61789-97-7*] are shown in Tables 1 and 2. In addition, it contains up to 4 % of characteristic branched and odd-numbered fatty acids. These, together with cholesterol, identify tallow in vegetable oils and fats. Tallow also contains 6 – 10 % of trans fatty acids, resulting from bacterial hydrogenation in the rumen. Top-grade tallow is white to greyish-white; in summer it is slightly yellow because of carotenes taken up with fresh roughage. Table 29 shows analytical data for various beef tallow products. Beef tallow contains 14 – 26 % trisaturated, 22 – 34 % disaturated – monounsaturated, and 40 – 64 % monosaturated – diunsaturated glycerides.

The stability of beef tallow and its products is relatively poor in the absence of an added antioxidant. Primary quality criteria are odor and taste, concentration of free fatty acids, and absence of impurities and of refined low-quality tallow (white grease).

Table 29. Properties and analytical data of beef tallow and derived products

	Beef tallow	Stearin	Olein	Bone grease	Neatsfoot oil
Density, g/cm³	0.898 – 0.908 (40 °C)		0.914 – 0.924 (15 °C)		
n_D^{60}	1.451 – 1.454	ca. 1.449		1.451 – 1.452	1.461 – 1.463
mp, °C	40 – 50	50 – 55	23 – 35	44 – 45	
Saponification value	193 – 200	190 – 198	193 – 198	190 – 200	188 – 198
Iodine value	32 – 47	14 – 25	40 – 53	49 – 53	67 – 80
Nonsaponifiable matter, %	0.3 – 0.8	0.3 – 0.8	0.1 – 0.5	0.5 – 0.6	0.1 – 0.6

Production. The choice of fatty tissue determines the quality of beef tallow. Fat rendered from the carcasses of cattle is generally not quite as hard as that obtained from sheep and goats. The softest, most unsaturated fat is found under the skin, and the firmer fat is located near the middle of the animal. The consistency of individual lots of tallow is therefore influenced by the procedure used in trimming the carcass prior to fat rendering.

Edible tallow (dripping, max. 1 % ffa) is obtained from clean, sound, fatty tissues of beef cattle in good health at the time of slaughter. Premier jus (max. 0.5 % ffa) is obtained by rendering the fresh fat (killing fat) of heart, caul, kidney, and mesentery, excluding cutting fats at low temperature (50 – 60 °C). Tallow to be incorporated into margarine blends must be refined. Fractionation at 26 °C yields a lower-melting fraction (olein) and a higher-melting fraction (stearin). Tallow olein is an important component of some margarines and shortenings. Stearin is used in shortenings and in specialty margarines. Bone grease is obtained by expelling and extracting of comminuted bones ; it is used primarily in the technical sector and for animal feeds. Neatsfoot oil [8037-20-5] is a low-melting, inedible fat rendered from the feet of cattle; it is used as a special lubricant and leather dressing aid.

Use. Only a small part of the annually produced beef tallow is used as edible fat. The concentration in margarine is limited because of its poor stability but it is used more widely in shortenings. The technical grades are important raw materials for soap and fatty acid derivatives and for use in animal feeds. Specifications for technical tallows used for soapmaking are shown in Table 30.

14.1.3. Mutton Tallow

Mutton tallow finds applications primarily in the technical sector. Superior grades are similar to corresponding beef tallows.

Saponification value	192 – 198
Iodine value	31 – 47
Nonsaponifiable matter	0.1 – 0.6 %
Density (40 °C)	0.896 – 0.898 g/cm^3
n_D^{40}	1.455 – 1.458
mp	44 – 55 °C

The main fatty acids are palmitic, stearic, and oleic acids (Tables 1 and 2). Up to 4 % of branched and odd-numbered fatty acids and traces of trans fatty acids are also present.

14.1.4. Horse, Goose, and Chicken Fat

Horse Fat is softer than lard or rendered pork fat. The fatty acid composition (Tables 1, 2) is strongly influenced by diet. Branched and odd-numbered fatty acids occur up to about 2 %. The nonsaponifiable matter is <0.5 %, the melting point 29 – 40 °C. Horse fat is occasionally used in meat and sausage products in some countries.

Goose Fat has a relatively low melting point, 25 – 35 °C; its consistency is often improved by addition of lard.

Chicken Fat is also low-melting, mp 23 – 40 °C. The fatty acid composition (see Tables 1 and 2) is strongly influenced by diet.

Table 30. The Society of British Soap Makers specifications for tallows and greases (based on testing methods specified in B.S.3919 : 1976)

Grade	ffa ^{a}, %	Bleached oil colour (red)	Moisture and dirt, %	Nonsaponifiable matter, %	Titer, °C	Iodine value
Tallow 1	max. 3.0	max. 0.5, 5 1/4 inch cell	max. 0.5	max. 0.5	min. 40.0	max. 57
Tallow 2	max. 5.0	max. 1.0, 5 1/4 inch cell	max. 1.0	max. 1.0	min. 40.0	max. 57
Tallow 3	max. 8.0	max. 3.0, 5 1/4 inch cell	max. 1.0	max. 1.0	min. 40.0	max. 57
Tallow 4	max. 12.0	max. 4.0, 1 inch cell	max. 1.0	max. 1.5	min. 40.0	max. 60
Tallow 5	max. 15.0	max. 12.0, 1 inch cell	max. 1.0	max. 1.5	min. 40.0	max. 60
Tallow 6	max. 20.0	no limit	max. 1.0	max. 2.0	min. 40.0	max. 60
Grease	max. 20.0	–	max. 2.0	max. 2.0	36.0 – 40.0	max. 63

* ffa = free fatty acids.

14.2. Marine Oils [96]

Marine oils are the oils obtained from whales and fish. Fish oils have a high content of highly unsaturated (4 – 6 double bonds) C_{20}, C_{22}, and C_{24} fatty acids (cf. Tables 1 and 2). The oils from freshwater fish contain less C_{20} and C_{22} fatty acids and more oleic and linoleic acid than those from sea fish. These differences are caused primarily by differences in the composition of the feed. Only oils obtained from sea fish are commercially important.

14.2.1. Whale Oil

The whale population has been depleted to such an extent that whale oil is now completely overshadowed in commercial importance by the various fish oils. The Antarctic has long been the catching area. Whales are processed in modern factory ships (see also Section 5.3). The color of the oil can be very light, and the concentration of free fatty acids very low. Since 1913, when it was discovered that the oil can be converted into a stable edible fat by hydrogenation, many thousands of tons of hardened whale oil have been produced for margarine, cooking fats, and soapmaking.

Saponification value	185 – 205
Iodine value	110 – 135
Nonsaponifiable matter	<2 %
Slip point	22 °C
n_D^{65}	1.4554 – 1.4579

Whale oil has been graded according to free fatty acid content and color (apart from moisture and dirt):

Grade 0: pale yellow, max. 0.5 % ffa
Grade 1: pale yellow, max. 2.0 % ffa
Grade 2: amber yellow, max. 6.0 % ffa
Grade 3: pale brown, max. 15.0 % ffa

Grade 0 – 2 whale oils are hydrogenated for edible purposes. Prior to hydrogenation the oil is lye-neutralized and bleached. Whale oil can be hydrogenated into soft and hard fats. Soft fats with a melting point between 33 and 38 °C are produced by hydrogenation under isomerizing conditions with a partially inactivated nickel catalyst. Fats with a melting point between 40 and 45 °C are produced with relatively fresh catalyst. The crude hydrogenated fats are filtered to remove catalyst, treated with weak lye, bleached, and deodorized. In some cases, posttreatment can be restricted to treatment with citric acid – bleaching earth followed by distillative stripping. Hydrogenated whale oil is stable for about 6 months. The softer fats are used primarily for shortenings, the higher melting ones for preserves, dry soups, and special margarines.

The main components of sperm oil [8002-24-2] are esters of fatty acids and fatty alcohols (waxes). Sperm oil is a valuable raw material for high-quality cosmetic preparations. It is being replaced by jojoba oil, a seed oil from a shrub (*Simmondsia chinensis*) indigenous to the Sonoran desert and being cultivated in arid zones around the world. Its seeds contain 50 – 60 % of a liquid wax composed of esters of fatty acids and fatty alcohols [97].

14.2.2. Fish Oil [98]

The bulk of fish oils is produced in modern plants by cooking comminuted fish or fish offal, expelling in screw presses, and centrifugation; or by extraction of dried fish meal with hexane (see Section 5.3). Refining and hydrogenation yields fats of edible quality which are used in margarine and shortening blends. For historical reasons fish oils are not permitted as edible oils in the United States. The main crude oil origins are Japanese (iodine value ca. 180), U.S. (menhaden: iodine value 150 – 160), South American (iodine value ca. 200), and Scandinavian (iodine value 130 – 140).

The oil content of fish is subject to seasonal variations and generally lies between 8 and 20 %. Typical properties of various fish oils are shown in Table 31 (for fatty acid composition see Tables 1 and 2). All fish oils contain small concentrations of branched and odd-numbered fatty acids.

Use. Fish oils are processed into hydrogenated fats analogously to whale oil [100]. How-

Table 31. Analytical data for various fish oils [99]

	Herring oil	Sardine oil	Pilchard oil	Menhaden oil	South American oil
n_D^{40}	1.470 – 1.475	1.473 – 1.475	1.473 – 1.476	1.473 – 1.474	1.474 – 1.478
Saponification value	183 – 192	187 – 190	188 – 194	188 – 194	189 – 194
Iodine value	120 – 160	165 – 185	180 – 190	150 – 180	185 – 206
Nonsaponifiable matter, %	0.8 – 1.3	1.0 – 1.6	0.8 – 1.5	0.8 – 1.5	0.8 – 1.5

ever, due to their content of polyunsaturated fatty acids and sulfur compounds, hydrogenation requires more hydrogen and more catalyst. The melting points of hardened fish oils range from 31 to 45 °C. Organoleptic stability generally improves with decreasing iodine value of the starting oil and with increasing melting point of the hardened product. Hardened fish oils are used in shortenings and margarine blends. They have good creaming and cake-making properties but do not perform as well in frying.

Much of the oil of cod, halibut, and shark is stored in the liver. Cod-liver oil and halibut-liver oil are valued for pharmaceutical purposes because of their high content of vitamin A (1500 – 50 000 i.u./g) and vitamin D (40 – 200 i.u./g). Typical properties of cod-liver oil are as follows:

Saturated fatty acids, %	
C_{14}	6
C_{16}	6
C_{18}	8
Unsaturated fatty acids, %	
C_{14}	traces
C_{16}	20
C_{18}	29
C_{20}	26
C_{22}	10
Saponification value	180 – 197
Iodine value	150 – 175
Nonsaponifiable matter, %	ca. 1
n_D^{25}	1.481

15. Economic Aspects

Figures for the world production and exports of important fats and oils are shown in Tables 32–35.

Table 32. World production of oil fruits, oilseeds, oils, and fats according to countries (annual average, expressed as oil/fat in 1000 t)

	1934/38	1960	1970/71	1973/74	1980/81	1981/82	1982/83	1983/84	1984/85[a]
Europe	3328[b]		4790[c]						
EEC		2300		4220	5615	5550	6065	6215	6080
EFTA		1135		710	855	850	930	910	930
Other West European countries		990	805	1015	835	650	1180	750	1285[d]
Soviet Union	2000	3090	5335	6030	no data	4870	5340	5315	5105
Other East European countries	1231	1595	1975	2430	2920	2950	2975	3040	3290
Africa						3015[e]			
West Africa	1132	2010	1845	1685	1380		2970	2800	2790
Other African countries	379	985	1420	1540					
America									
Argentina/Brazil	995	1525	2270	3080	5155	5100	5465	5935	6750
USA	3543	7470	11185	11945	14415	15620	16515	13535	14940
Other American countries	565	1480	3165	2605	3345	3210	3180	3340	3675
Asia									
Sri Lanka/Burma	216	240	320	215	265	290	330	365	375
India/Pakistan	2274	2725	3615	3620	3890	4750	4290	4875	4825
Indonesia/Malaysia	931	780	1425	1815	4510	5380	5355	5530	6800
Philippines	452	840	965	800	1585	1500	1345	1090	1140
Chinese Republic	3341	2905	3100	3070	4250	5685	6325	6160	6650
Other Asian countries	628	960	1460	1645	1105	1045	1045	1090	1120
Australia/Oceania	575	755	955	890	1100	1140	1115	1240	1270
World production	21590	31785	44630	46320	58880	62505	65485	63145	68155

[a] Preliminary data.
[b] $\sum$ EEC, EFTA, and other West European countries.
[c] $\sum$ EEC, EFTA.
[d] Including Spain.
[e] $\sum$ West Africa and other African countries.

Table 33. World production of oil fruits, oilseeds, oils, and fats (annual average, expressed as oil/fat in 1000 t)

	1934/1938	1960	1970/71	1973/74	1980/81	1981/82	1982/83	1983/84	1984/85 [a]
Vegetable oils									
Cottonseed	1453	2315	2450	2825	3035	3280	3025	3040	3770
Peanut [b]	1755	2760	3230	3965	2605	3455	2900	3190	3195
Corn [c]	95	185	325	390	665	695	745	785	805
Olive	950	1285	1605	1550	1980	1495	2025	1575	1785
Rapeseed	1273	1105	2220	2170	3675	4040	4905	4700	5500
Safflower	[d]	[d]	170	220	265	260	245	260	265
Sesame	563	485	755	720	615	730	620	700	695
Soybean	1263	3780	7700	9670	13275	13885	15255	13540	14675
Sunflower	435	1170	3335	4250	4695	5250	6000	5530	6165
Babassu	27	50	105	125					
Coconut	1633	2060	2375	2015	3150	2990	2860	2435	2640
Palm kernel	355	450	490	430	585	700	730	760	870
Palm	644	1100	1745	2260	4550	5410	5410	5645	6720
Linseed	1040	1025	1255	795	660	670	815	670	700
Castor	178	230	340	465	325	350	350	375	420
Tung	121	115	120	95	105	95	100	95	95
Others [e]	220	450	65	75	25	25	25	25	25
Animal fats									
Butterfat [f]	4160	4435	4950	5225	5450	5705	6200	6255	6135
Lard	2913	4370	4365	4180	4760	5050	5130	5185	5210
Tallow	1679	3455	5270	5320	6360	6200	6100	6110	6210
Fish oil	325	470	1170	970	1100 [g]	1300 [g]	1090 [g]	1265 [g]	1280 [g]
Whale oil and sperm oil	507	490	190	140					

[a] Preliminary data.
[b] Until 1960 including 10 000 – 15 000 t teaseed.
[c] Including technical olive oil.
[d] Included under "others".
[e] Including oiticica, mowrah, niger, hempseed, perilla, and other commercially important oils.
[f] Including ghee fat.
[g] $\sum$ Fish oil, whale oil, and sperm oil.

Table 34. World exports of oil fruits, oilseeds, oils, and fats according to countries (annual average, expressed as oil/fat in 1000 t)

	1934/38	1965	1970	1974	1980	1981	1982	1983
Europe								
Western Europe and Iceland		738	1020	1335	1693	1820	1658	2022
Eastern Europe and Soviet Union		315	783	740	532	458	548	466
Africa								
Former French Equatorial/West Africa	298	409	395	341	312	194	313	325
Nigeria	455*	737	392	182	107	75	55	65
Former Portuguese Africa	62	93	111	96				
Sudan		115	100	105	85	91	93	52
Zaire	97	115	173	104	29	25	19	15
Other African countries	212	259	315	335	211	273	209	137
America								
Argentina/Uruguay	577	416	469	224	1426	1079	1243	1548
Canada		320	556	520	1071	1215	1012	1024
United States	100	3356	4479	5319	8246	7976	8534	7603
Other American countries	133	377	579	926	234	213	334	213
Asia								
India/Sri Lanka	589	142	105	123	75	138	166	163
Indonesia	529	260	317	304	610	201	323	449
Malaysia	132	164	435	883	2517	2827	3187	3346

Table 34 (*Continued*)

	1934/38	1965	1970	1974	1980	1981	1982	1983
Philippines	348	797	608	631	1002	1126	1080	1035
China	742	195	153	134	139	246	190	310
Other Asian countries		122	137	263	455	508	593	649
Australia/Oceania	363	540	640	530	755	749	790	804
Antarctic/Arctic	507	350	209	138	18	14	14	11
World exports	5829	8301	9820	13233	20819	20957	21589	21618

*Including Ghana, Gambia, Sierra Leone.

Table 35. World exports of oil fruits, oilseeds, oils, and fats (annual average, expressed as oil/fat in 1000 t)

	1934/38	1965	1970	1974	1980	1981	1982	1983
Vegetable oils								
Cottonseed	189	395	307	369	495	473	522	320
Peanut	826	967	817	677	678	519	641	669
Olive*	136	113	237	262	219	205	162	278
Rapeseed	51	272	481	670	1149	1364	1224	1274
Safflower		70	50					
Sesame	69	70	104					
Soybean	432	1802	2962	3903	6665	6757	7162	6770
Sunflower	26	263	700	629	1431	1535	1558	1832
Babassu	12	12	14					
Coconut	1057	1283	1087	923	1365	1455	1411	1376
Palm kernel	320	371	327	383	431	419	462	479
Palm	447	551	744	1327	2961	2818	3336	3567
Linseed	572	473	421	271	423	420	352	402
Castor	81	206	236	227	203	202	180	184
Tung	80	42	36					
Others	68	121	185	358	680	714	755	759
Animal fats								
Butterfat	500	499	641	704	775	740	671	608
Lard	173	290	347	372	432	439	389	366
Tallow	162	1225	1517	1575	2193	2181	2068	2025
Fish oil	121	445	554	500	701	702	682	698
Whale oil and sperm oil	507	350	209	138	18	14	14	11

*Including olive oil for industrial uses.

16. Toxicology and Occupational Health

With the exception of technical oils and fats such as castor and tung oil, oils and fats are classed as foods or food additives and do not entail toxicological or occupational health hazards.

References

General References

1 E. W. Eckey: *Vegetable Fats and Oils*, Reinhold Publ. Co., New York 1954.
2 H. P. Kaufmann: *Analyse der Fette und Fettprodukte*, vols. I and II, Springer Verlag, Berlin – Göttingen – Heidelberg 1958.
3 V. C. Mehlenbacher: *The Analysis of Fats and Oils*, The Garrard Press, Champaign, Ill., 1960.
4 J. Devine, P. N. Williams: *The Chemistry and Technology of Edible Oils and Fats*, Pergamon Press, Oxford 1961.
5 R. Lüde: *Die Raffination von Fetten und fetten Ölen*, 2nd ed., Th. Steinkopff, Leipzig 1962.
6 A. J. C. Anderson: *Refining of Oils and Fats for Edible Purposes*, 2nd ed., Pergamon Press, London 1962.
7 W. Wachs: *Öle und Fette*, Part II, Paul Parey, Berlin – Hamburg 1964.
8 "Fette und Lipoide," *Handbuch der Lebensmittelchemie*, vol. IV, Springer Verlag, Berlin – Heidelberg – New York 1965.

9. L. V. Cocks, C. V. van Rede: *Laboratory Handbook for Oil and Fat Analysis,* Academic Press, London –New York 1966.
10. F. D. Gunstone: *An Introduction to the Chemistry and Biochemistry of Fatty Acids and their Glycerides,* Chapman and Hall, London 1967.
11. H. A. Boekenoogen: *Analysis and Characterisation of Oils, Fats and Fat Products,* Interscience, New York 1968.
12. A. J. Vergroesen (ed.): *The Role of Fats in Human Nutrition,* Academic Press, London 1975.
13. H. Pardun: *Analysis of Edible Fats,* Paul Parey, Hamburg 1976.
14. D. Swern (ed.): *Bailey's Industrial Oil and Fat Products,* 4th ed., Wiley & Sons, New York 1979.
15. *Kirk-Othmer,* **9**, 795; **21**, 417; **23**, 717.
16. F. B. Padley, J. Podmore (eds.): *The Role of Fat in Human Nutrition,* VCH Verlagsgesellschaft, Weinheim, Germany 1985.
17. R. J. Hamilton, J. B. Rossell (eds.); *Analysis of Oils and Fats,* Elsevier Applied Science Publ., Barking, ngland 1986.

Specific References

18. J. Baltes: "Gewinnung und Verarbeitung von Nahrungsfetten," *Grundlagen und Fortschritte der Lebensmitteluntersuchung,* vol. 17, Paul Parey, Berlin, Hamburg 1975, p. 42.
19. J. B. M. Rattray, *JAOCS J. Am. Oil Chem. Soc.* **61** (1984) 1701.
20. F. D. Gunstone: *An Introduction to the Chemistry and Biochemistry of Fatty Acids and their Glycerides,* Chapman & Hall, London 1967.
21. M. Szuhaj, G. List: *Lecithins,* AOCS monograph, 1985.
22. A. J. Speek et al., *J. Food Sci.* **50** (1985) 121.
23. M. L. Meara: *Leatherhead Food RA, Scientific and Technical Survey no. 110,* July 1978.
24. K. van Putte, J. v. d. Enden, *Fette Seifen Anstrichm.* **76** (1974) 316.
25. E. S. Lutton, *J. Am. Oil Chem. Soc.* **49** (1972) 1.
26. B. G. Linsen, *Fette Seifen Anstrichm.* **73** (1971) 411, 753; **70** (1968) 8.
27. J. W. E. Coenen in I. Morton, D. N. Rhodes (eds.): *The Contribution of Chemistry to Food Supplies,* Butterworths, London 1974.
28. H. Wissebach, *Tenside* **3** (1966) 285.
29. A. K. Sen Gupta, H. Scharmann, *Fette Seifen Anstrichm.* **71** (1969) 873.
30. H. W. G Heynen et al., *Fette Seifen Anstrichm.* **74** (1972) 677.
31. E. W. Trautschold: *Die Lagerung von Sojabohnen unter Qualitätsaspekten,* DGF-paper, München 1970.
32. H. P. Kaufmann et al.: *Neuzeitliche Technologie der Fette und Fettprodukte,* Aschendorffsche Verlagsbuchhandlung, Münster 1956 – 1965.
33. I. E. Liener: *Toxic Constituents of Plant Foodstuffs,* Academic Press, New York – London 1969.
34. "Oilseed Processing Symposium 1976," *J. Am. Oil Chem. Soc.* **54** (1977) 473A.
35. "Deutsche Gesellschaft für Fettwissenschaft": *Gewinnung von Fetten und ölen aus pflanzlichen Rohstoffen durch Extraktion,* Industrieverlag von Hernhaussen KG, Hamburg 1978.
36. S. Skipin, *Chem. Abstr.* **29** (1935) 7682. A. Carter et al., *J. Am. Oil Chem. Soc.* **51** (1974) 137.
37. P. König, *Fette Seifen Anstrichm.* **64** (1962) 23.
38. J. Furman et al., *J. Am. Oil Chem. Soc.* **36** (1959) 454.
39. G. Penk, *Proc. A.S.A. Symp. Soybean Process. 2nd* 1981.
40. W. H. Prokop, *JAOCS J. Am. Oil Chem. Soc.* **62** (1985) 805.
41. S. M. Barlow, M. L. Windsor: *Techn. Bulletin no. 19,* Int. Ass. of Fish Meal Manufacturers, UK, Sept. 1984.
42. Th. Wieske, H.-U. Menz, *Fette Seifen Anstrichm.* **74** (1972) 133. V. K. Babayan, *J. Am. Oil Chem. Soc.* **51** (1974) 260.
43. O. L. Brekke in: *Handbook of Soy Oil Processing and Utilisation,* Amer. Soybean Ass., 1980.
44. H. J. Ringers, J. C. Segers, US 4 049 686, 1977.
45. A. K. Sen Gupta, US 4 062 882, 1977.
46. F. V. K. Young, *Proc. A.S.A. Symp. Soybean Process. 2nd* 1981.
47. A. N. Sagredos, K. Remse, *Fette Seifen Anstrichm.* **85** (1983) 185. P. Röttgermann, *Fette Seifen Anstrichm.* **85** (1983) 190.
48. Y. Hoffmann, *J. Am. Oil Chem. Soc.* **50** (1973) 260.
49. H. Pardun, *Fette Seifen Anstrichm.* **81** (1979) 297.
50. D. C. Tandy, W. J. Macpherson, *JAOCS J. Am. Oil Chem. Soc.* **61** (1984) 1253. A. Forster, A. J. Harper, *J. Food Sci.* **49** (1984) 23.
51. H. P. Kaufmann, D. Schmidt, *Fette Seifen* **47** (1940) 294.
52. M. Kock, *Proc. A.S.A. Symp. Soybean Process. 2nd* 1981.
53. G. Biernoth, H. E. Rost, *Chem. Ind.* 1967, 2002. G. Biernoth, *Fette Seifen Anstrichm.* **70** (1968) 217. A. N. Sagredos, D. Sinha-Roy, *Dtsch. Lebensm. Rundsch.* **75** (1979) 350.
54. D. B. Erskine, W. G. Schuliger, *Chem. Eng. Prog.* **67** (1971) 41.
55. A. Thomas, *Fette Seifen Anstrichm.* **84** (1982) 133.
56. J. B. Rossell et al., *Proc. A.S.A. Symp. Soybean Process. 2nd* 1981 .
57. H. Stage: *Proc. A.S.A. Symp. Soybean Process. 2nd* 1981.
58. A. E. Rheineck, R. T. Holman et al.: *Progress in the Chemistry of Fats and Other Lipids,* vol. 5, Pergamon Press, New York 1958.
59. H. J. Passino, *Ind. Eng. Chem.* **41** (1948) 280.
60. S. W. Gloyer, *J. Am. Oil Chem. Soc.* **26** (1949) 162.
61. "Deutsche Gesellschaft für Fettwissenschaft": *Die Hydrierung von Fetten,* Industrieverlag von Herrnhaussen KG, Hamburg 1979.
62. H. B. W. Patterson: *Hydrogenation of Fats and Oils,* Elsevier Applied Science Publ., Barking, England, 1983.
63. F. A. Kummerow, *J. Am. Oil Chem. Soc.* **51** (1974) 255.
64. E. le Breton, P. le Marchal: *Riv. Ital. Sostanze Grasse* **47** (1970) 231. F. Camurati et al., *Riv. Ital. Sostanze Grasse* **47** (1970) 241.
65. P. v. d. Plank, *Fette Seifen Anstrichm.* **76** (1974) 337.

66 R. C. Hastert, *JAOCS J. Am. Oil Chem. Soc.* **58** (1981) 169.
67 B. Screenivasan, *J. Am. Oil Chem. Soc.* **55** (1978) 803.
68 G. Bradshaw, *Soap Sanit. Chem.* **18** (1941) no. 5, 23, 69.
69 A. J. Haighton, L. Vermaas, *Fette Seifen Anstrichm.* **74** (1972) 615.
70 J. R. Rossell: "Measurement of Rancidity in Oils and Fats," *Leatherhead Food RA, Scientific and Technical Survey no. 140,* Sept. Sept. 1983.
71 M. Arens, E. Kroll, *Fette Seifen Anstrichm.* **85** (1983) 307.
72 G. Guhr et al., *Fette Seifen Anstrichm.* **83** (1981) 373.
73 H. Karstens, *Fette Seifen Anstrichm.* **70** (1968) 400.
74 M. Unbehend et al., *Fette Seifen Anstrichm.* **75** (1973) 689.
75 K. G. Berger: *Recommended Practices for Storage and Transport of Edible Oils and Fats,* Palm Oil Research Institute of Malaysia, 1986.
76 "Proceedings of the World Conference, Kuala Lumpur, 1984, on Palm, Palm Kernel and Coconut Oils," *JAOCS J. Am. Oil Chem. Soc.* **62** (1985) .
77 Y. Lozano et al., *Rev. Fr. Corps Gras* **32** (1985) 377.
78 F. Hirsinger, *JAOCS J. Am. Oil Chem. Soc.* **62** (1985) 76. S. A. Graham, R. Kleiman, *JAOCS J. Am. Oil Chem. Soc.* **62** (1985) 81.
79 D. Gegion, K. Staphylakis, *JAOCS J. Am. Oil Chem. Soc.* **62** (1985) 1047.
80 R. Banerji et al., *Fette Seifen Anstrichm.* **86** (1984) 279.
81 A. E. Bailey: *Cottonseed and Cottonseed Products,* Interscience, New York 1948.
82 B. S. Kamel et al., *JAOCS J. Am. Oil Chem. Soc.* **62** (1985) 881.
83 W. Schuster, *Fette Seifen Anstrichm.* **74** (1972) 150.
84 R. Purdy, *JAOCS J. Am. Oil Chem. Soc.* **62** (1985) 523.
85 R. Marquard et al., *Fette Seifen Anstrichm.* **80** (1978) 213.
86 W. Schuster, R. Marquard, *Fette Seifen Anstrichm.* **76** (1974) 207.
87 J. D. v. Mikusch, *Farbe + Lack* **7** (1952) 303.
88 T. Yaziciovglu et al., *Fette Seifen Anstrichm.* **79** (1977) 115.
89 G. W. Rohne, *Fette Seifen Anstrichm.* **86** (1984) 172.
90 B. J. F. Hudson et al., *J. Plant Foods* **5** (1983) 15.
91 L.-A. Appelquist, R. Ohlson: *Rapeseed,* Elsevier, Amsterdam 1972.
92 A. Thomas, *JAOCS J. Am. Oil Chem. Soc.* **59** (1982) 1.
93 M. Y. Raie et al., *Fette Seifen Anstrichm.* **87** (1985) 325.
94 O. Dahl: *Schlachtfette,* Fleischforschung und Praxis no. 10, Verlag der Rheinhessischen Druckwerkstätte, Alzey 1973.
95 E. S. Lutton et al., *J. Am. Oil Chem. Soc.* **39** (1962) 233.
96 "Fish Oils and Animal Fats," *Proc. Leatherhead Food RA,* Feb. 1986.
97 T. K. Miwa, *JAOCS J. Am. Oil Chem. Soc.* **62** (1985) 377.
98 M. E. Stansby: *Fish Oils, Their Chemistry, Technology, Stability, Nutritional Properties and Uses,* The Avi Publishing Co., Westport, Conn., 1967.
99 F. V. K. Young: *The Chemical and Physical Properties of Crude Fish Oils,* Int. Ass. Fish Meal Manuf., Fish Oil Bulletin no. 18, June 1986.
100 F. V. K. Young: *The Refining and Hydrogenation of Fish Oils,* Int. Ass. Fish Meal Manuf., Fish Oil Bulletin no. 17, June 1986.

Further Reading

G. S. Breck, S. C. Bhatia: *Handbook of Industrial Oil and Fat Products,* CBS Publ., Delhi 2008.
C. K. Chow: *Fatty Acids in Foods and their Health Implications,* 3rd ed., CRC Press, Boca Raton, FL 2008.
A. J. Dijkstra, R. J. Hamilton, W. Hamm (eds.): Trans Fatty Acids, Blackwell, Oxford 2008.
F. D. Gunstone: *Oils and Fats in the Food Industry,* Blackwell, Oxford 2008.
G. L. Hasenhuettl: *Fats and Fatty Oils,* "Kirk Othmer Encyclopedia of Chemical Technology", 5th edition, John Wiley & Sons, Hoboken, NJ, online DOI: 10.1002/0471238961.0601201908011905.a01.pub2.
R. D. O'Brien: *Fats and Oils - Formulating and Processing for Applications,* 3rd ed., CRC Press, Boca Raton, FL 2009.
F. Shahidi, A. E. Bailey (eds.): *Bailey's Industrial Oil & Fat Products,* 6th ed., Wiley-Interscience, New York 2005.

Gelatin

WILFRIED BABEL, Deutsche Gelatine-Fabriken Stoess AG, Eberbach, Federal Republic of Germany

DIETER SCHULZ, Deutsche Gelatine-Fabriken Stoess AG, Eberbach, Federal Republic of Germany

MONIKA GIESEN-WIESE, Deutsche Gelatine-Fabriken Stoess AG, Eberbach, Federal Republic of Germany

UWE SEYBOLD, Deutsche Gelatine-Fabriken Stoess AG, Eberbach, Federal Republic of Germany

HERBERT GAREIS, Deutsche Gelatine-Fabriken Stoess AG, Eberbach, Federal Republic of Germany

EBERHARD DICK, Deutsche Gelatine-Fabriken Stoess AG, Eberbach, Federal Republic of Germany

REINHARD SCHRIEBER, Deutsche Gelatine-Fabriken Stoess AG, Eberbach, Federal Republic of Germany

ANNELORE SCHOTT, Deutsche Gelatine-Fabriken Stoess AG, Eberbach, Federal Republic of Germany

WINFRIED STEIN, Fritz Häcker GmbH & Co., Vaihingen/Enz, Federal Republic of Germany

1.	Introduction	363	4.3.	Pharmaceuticals	372
2.	Structure and Properties	364	4.4.	Cosmetics	373
2.1.	Structure	364	4.5.	Photography	373
2.2.	Physical Properties	365	4.6.	Technical Gelatin and Glue	374
2.3.	Chemical Properties	367	4.6.1.	Matches	374
2.4.	Microbiological Analysis	368	4.6.2.	Coated Abrasives	375
2.5.	Quality Specifications	368	4.6.3.	Paper Sizing	375
3.	Raw Materials and Production	368	4.6.4.	Adhesives	375
3.1.	Raw Materials	368	4.6.5.	Other Uses	376
3.2.	Production	369	5.	Economic Aspects	376
4.	Uses	370		References	376
4.1.	General Aspects	370			
4.2.	Food Industry	371			

1. Introduction

Gelatin is a mixture of high molecular mass polypeptides produced from collagenous animal tissues such as hide splits, pigskin, and bones. Collagen is the most commonly occurring protein in the human and animal body, accounting for about 30 % of the total protein content. Since animal horns and hoofs are not composed of collagen they cannot serve as raw materials, contrary to popular belief. The name gelatin has been used since about 1700 and is derived from the Latin *gelatus* which means frozen [1]. There is evidence that gelatin has been used for at least 4000 years [2]. The great variety of applications of gelatin depend on its solubility in hot water, its availability in a wide range of qualities, and its ability to form thermally reversible gels, which is unique for a protein and is due to the fact that the primary structure of collagen still exists in collagen-derived polypeptides.

2. Structure and Properties

2.1. Structure

Collagen is not a uniform substance but an entire protein family. To date, some 19 polymorphs of collagen have been identified. All types of collagen have a triple-helix structure in which three protein chains are intertwined to form a rigid strand (Fig. 1). The length of the triple helix and the type and position of the nonhelical regions vary from one collagen type to an other.

The "classical" type I collagen, the most important type in the manufacture of gelatin, occurs in skin and bone. Type II collagen occurs almost exclusively in cartilage tissue, and type III collagen occurs besides type I in skin. The concentration of type III collagen is strongly dependent on age: embryonic skin can contain up to 50 %, but older skin only 5 – 10 %. The other collagen types occur only in low concentrations and fulfil mostly organ-specific functions.

Collagen consists of three helical polypeptide chains wound around each other and connected by intermolecular cross-links. The triple helix is formed intercellularly from three individual chains (α-chains). Chain selection occurs via intra- and intermolecular disulfide bridges in the region of the N- and C-terminal peptides, additionally supported by glycopeptide structural elements. While collagen types II and III contains three identical chains (homotrimer $[\alpha_1(II)]_3$), type I collagen consists of two identical and one somewhat differing chain (heterotrimer $[\alpha_1(I)]_2\alpha_2(I)$). The triple helix molecule is 300 nm long and 1.4 nm in cross section. The amino acid sequences of the α_1 and α_2 chains are highly homologous (ca. 90 % correspondence).

Collagen contains large amounts of two unusual amino acids: 4-hydroxyproline and ϵ-hydroxylysine, which are the result of post-translational modification: special enzyme systems facilitate the intracellular hydroxylation of proline and lysine on the growing collagen chain. Other enzymes are responsible for the glycosylation of individual hydroxylysine residues. Type I collagen, for example, contains 0.5 wt % carbohydrate, and type III about 10 %. The carbohydrate portion comprises galactose and a 1,2-linked galactose – glucose disaccharide. The ether-like bond between collagen and the saccharide occurs between the C1 atom of galactose and the ϵ-hydroxyl group of hydroxylysine.

The enzyme-catalyzed modification steps (hydroxylation and glycolyzation) are restricted to growing individual chains; as soon as triple-helix formation is initiated, modification ceases. As stability of the helix is dependent on the hydroxyproline content, helix formation and proline hydroxylation are interdependent and self-regulating.

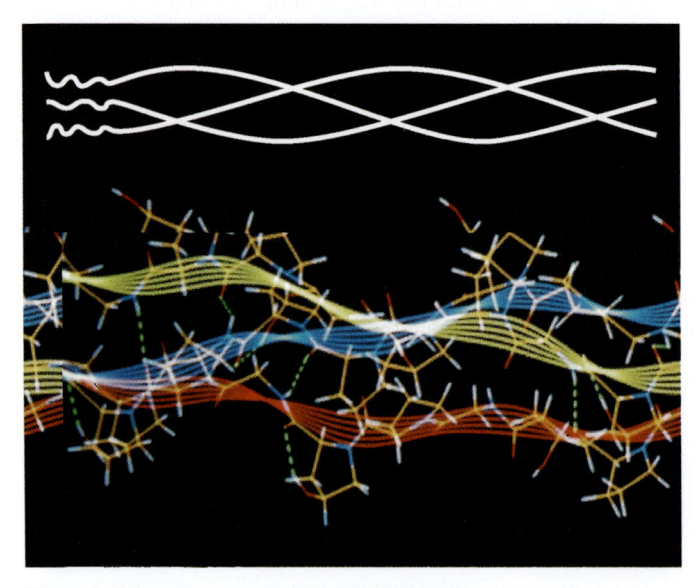

Figure 1. Schematic representation of the triple helix

Table 1. Comparison of the amino acid composition of collagen and gelatin: number of amino acids per 1000 (rounded off)

Amino acid	Type I collagen (bovine)	Type A gelatin[*]	Type B gelatin[**]
Alanine	114	112	117
Arginine	51	49	48
Aspargine	16	16	-
Aspartic acid	29	29	46
Glutamine	48	48	-
Glutamic acid	25	25	72
Glycine	332	330	335
Histidine	4	4	4
4-Hydroxyproline	104	91	93
ε-Hydroxylysine	5	6	4
Isoleucine	11	10	11
Leucine	24	24	24
Lysine	28	27	28
Methionine	6	4	4
Phenylalanine	13	14	14
Proline	115	132	124
Serine	35	35	33
Threonine	17	18	18
Tyrosine	4	3	1
Valine	22	26	22

[*] Type A gelatin: acid-pretreated pigskin gelatin.
[**] Type B gelatin: alkali-pretreated bone gelatin [1]

The helical part of all three collagen types contains 1014 amino acids per chain, and the N- and C-terminal nonhelical extension peptides (telopeptides) are between 9 and 26 amino acids long. The triple helix molecule has a molecular weight of about 290 000, including the short N- and C-terminal globular extension peptides.

Collagen consists of one-third glycine and 22 % of the imino compounds proline and hydroxyproline; the remaining 45 % comprises some 17 other amino acids (Table 1). The content of acidic and basic amino acids is relatively high. About one-third of the acidic amino acids glutamic acid and aspartic acid is present in the amidated form as glutamine and asparagine. Cysteine is completely absent, and of the sulfur-containing amino acids only methionine is present at low concentration (exception: type III which has two cysteine residues per 1000 amino acids).

The "classical collagen" types are characterized by repetitive tripeptide units — (glycine-X-Y-)$_n$ — where proline occurs in the X and Y positions and 4-hydroxyproline exclusively in the Y position. Several other amino acids also have preferred positions, presumably for steric reasons. For example, glutamic acid, phenylalanine, and leucine occur frequently in the X position, and arginine mostly in the Y position. Generally, collagen types from different species are to a high degree homologous (usually about 90 %); and, within the same species, type I collagen from skin, bones, and tendons are identical.

Triple-helix, water-soluble collagen (types I to III) is converted extracellularly by cross-linking and fibril formation into highly associated, water-insoluble collagen. These intra- and intermolecular cross-links are responsible for a dramatic increase in stability of the collagen molecule relative to thermal, mechanical, and enzymatic influence. The type and extent of the covalent cross-links are age-dependent, an aspect that is of considerable importance for the gelatin industry. Process parameters and extraction conditions must hence be adapted to the age of the raw materials (e.g., young pig or calf skin / old cattle hide).

2.2. Physical Properties

Commercial gelatin is a vitreous solid with a faint color; it is almost tasteless and odorless and usually contains 9 – 13 % moisture. It gives typical protein reactions and is hydrolyzed by most proteolytic enzymes to give its peptide or amino acid components.

Depending on the degree of hydrolysis, two different types of gelatin are obtained:

1. Gelatin, gelling type [*9000-70-8*]: When the granules are immersed in cold water they are hydrated to discrete swollen particles. On heating, these swollen particles dissolve to form a solution.
2. Gelatin hydrolysate (also known as collagen hydrolysate), nongelling type [*68410-45-7*]: In this product range the molecular mass is reduced to such an extent (< 15000 D) that no gelation is observed. This type, mostly produced as a fine powder is soluble in cold water. Gelatin and gelatin hydrolysate are insoluble in less polar organic solvents such as benzene, petroleum ether, ethanol, acetone and tetrachloromethane. Gelatin and gelatin hydrolysate are insoluble in less polar organic solvents such as benzene, petroleum ether, ethanol, acetone and tetrachloromethane. Gelatin is amphoteric: in acidic solution

gelatin is positively charged and migrates as a cation in an electric field; in alkaline solution it is negatively charged and migrates as an anion. The intermediate point, where the net charge is zero and no migration occurs, is known as the isoionic or isoelectric point. Type A gelatin, produced by acidic processing of collagenous raw materials, has a broad isoelectric region between pH 6.0 and 9.5. Type B gelatin, produced by the alkaline processing of collagenous raw materials, has an isoelectric point between pH 4.7 and 5.6. The difference in the isoelectric points of type A and type B gelatins is the result of partial desamidation of glutamine and asparagine to the corresponding glutamic acid and aspartic acid. Mixtures of type A and B, as well as gelatins produced by modifications of the above-mentioned processes, may exhibit isoelectric points outside these ranges. Gelatin in solutions containing no ions other than H^+ and OH^- is known as isoionic gelatin [3]. These solutions can be readily prepared by the use of ion-exchange resins.

One of the most important properties of gelatin (gelling type) is the formation of heat-reversible gels in water. Gelatin readily forms gels over a wide range of pH with a variety of solutes. The formation and stability of a gelatin gel is influenced by a number of factors. When an aqueous solution of gelatin with a concentration greater than ca. 1 % is cooled to about 35 – 40 °C it first increases in viscosity and on further cooling it forms a gel. The strength of the gel depends on concentration and the intrinsic strength of gelatin used, which is a function of structure and molecular mass. Gel formation is not fully understood, but is believed to result from hydrogen bonding; the gelatin molecules are arranged in micelles that form a semisolid gel and bind water. Quantitative measurement of this property is important, both for control and to determine the amount of gelatin required for a given purpose because gelatin is used in many products for its gel-forming properties. Modern theory proposes that the first step in gelation is the formation of locally ordered regions caused by the partial random return (renaturation) of gelatin to collagen-like helices (collagen fold). Next, a continuous fibrillar three-dimensional network of fringed micelles forms throughout the system, probably due to nonspecific bond formation between the more ordered segments of the chains. Hydrophobic, hydrogen, and electrostatic bonds may be involved in cross-linking. Since these bonds are disrupted on heating, the gel is thermoreversible. Formation of cross-links is the slowest part of the process, so that under ideal conditions the strength of the gel increases with time as more cross-links are formed. The total effect is a time-dependent increase in average molecular mass and in order. Renaturation involves association between components (peptide chains) that differ in degree of cross-linking, chain length, and chemical composition [4], [5].

Gelatin is stable for long periods of time when suitably stored in sealed containers, at normal ambient temperatures to prevent ingress or loss of moisture. When gelatin is heated above 45 °C in air at relatively high humidity it may lose its ability to swell and dissolve [6].

Gel Strength is determined with a Bloom gelometer [7], [8], [10–13]. A 6.67 % solution of the gelatin sample is prepared in a special wide-mouthed test bottle, which is then cooled to 10.0 ± 0.1 °C and kept for 17 ± 1 h for maturation at this temperature. The firmness of the resulting gel is then measured with a gelometer. This instrument impresses a standard plunger (12.7 mm diameter, plane surface, sharp edges) into the surface of the gel. The force required to depress the plunger 4 mm into the gel is the gel strength or Bloom value of the gelatin. Commercial gelatins vary from ca. 50 to 300 g Bloom.

Viscosity is usually determined by using a calibrated viscosity pipette (Bloom pipette), which measures the efflux time of 100 mL of a 6.67 % solution at 60 °C.

For gelatin hydrolysate, the viscosity of a 10 % or 20 % solution at 25 °C is determined, but other concentrations or temperatures are possible, according to the different uses. For measurement of higher viscosity values, rotation viscometers are also used.

Molecular mass fractions (≥ 200 000 D) have some impact on viscosity, medium molecular mass fractions (55 000 – 300 000 D) have an influence on gel strength. The viscosity of gelatin solutions increases with increasing gelatin concentration and with decreasing temperature; viscosity is at a minimum at the isoionic point.

Gelatin solutions should not be exposed to temperatures above 60 °C for prolonged periods of time because the gel strength and viscosity decrease [14].

Color and Clarity of gelatin and gelatin hydrolysate can be important for certain applications. These parameters can be measured with a spectrophotometer, but visual assessment against standard gelatins is also common practice. The color of gelatin depends on the raw material and the production process. Pigskin gelatins usually are less strongly colored than those made from bone or hide. Turbidity may be due to insoluble or foreign matter in the form of emulsions or dispersions that are stabilized by the protective-colloid action of gelatin, or can be caused by an isoelectric haze. This haze is at a maximum at the isoelectric point in ca. 2 % solution [15].

Gelatin typically contains ca. 9 – 13 % *moisture*, which can be determined by drying a sample at 105 °C for 17 h.

Technical Gelatins are normally specified by their gelling power and viscosity. The test methods employed are generally similar to those used for edible or pharmaceutical gelatins, but different gelatin concentrations may be employed. For lower grades of technical gelatins (and animal glues) 12.5 % solutions (15 g in 105 mL water) are used.

Methods for sampling and testing of animal glues are described in an international standard [16]. In addition to gel strength and viscosity the measurement of moisture content, melting and setting point, foam characteristics, pH and fat content are also described.

In some countries the viscosity of technical gelatin is measured with a cup viscometer, normally the Engler viscometer [17], [18], [19], and using a 17.75 % solution at 40 °C. The viscosity (in degrees Engler, °E) is the flow time of the gelatin solution compared with that of an equal volume of water at 20 °C. Relationships between viscosity measured by capillary and by the Engler method are available [19], [20].

Gelatin also undergoes a phenomenon called coacervation when colloidal particles separate from the liquid phase. Coacervates are formed when two hydrophilic sols carrying opposite charges are mixed in suitable amounts [21]. If gelatin from an acid-treated precursor is mixed with gelatin from an alkali-treated precursor and the resulting gel is prepared at pH 5 – 7, different degrees of turbidity result. This effect depends on the ratio of the two gelatins in the mixture and the pH. This phenomenon arises from the formation of a complex coacervate between the gelatin micelles with opposite charge. Adjustment of the pH above or below the isoelectric range of both gelatins gives a clear gel; both gelatins are now either positively or negatively charged and hence are mutually compatible. A common application of complex coacervation is the use of gelatin and gum arabic to produce oil-containing microcapsules for carbonless paper manufacture [22], [23] (see → Microencapsulation).

By addition of plasticizers such as glycerin, sorbitol and lubricants like stearate it is possible to produce a blend with thermoplastic properties that can be used in extrusion and injection molding processes [24].

2.3. Chemical Properties

Gelatin is formed by the particle hydrolysis of collagen protein resulting in a mixture of protein fragments of varying molecular mass (gelatin 15 000 – 400 000 Dalton, gelatin hydrolysate < 15 000 D). The amino acid composition (see Table 1) corresponds to that of the collagen from which it is derived. Hydroxyproline and hydroxylysine are two unusual amino acids found in gelatin. The determination of the hydroxyproline content can be used for identification purposes [25].

The quantitative analysis of the elements gives about 50.5 % carbon, 25.2 % oxygen, 17.0 % nitrogen and 6.8 % hydrogen. The ash content of commercial gelatin varies with the origin of the raw material and the method of processing. Gelatins derived from pigskin contain small amounts of chlorides and sulfates resulting from acid treatment before extraction; gelatin from bone and hide contains calcium and sodium salts from the lime used in pretreatment.

Dry gelatin stored in airtight containers at room temperature remains unchanged for long periods. Degradation in solution may be caused by extremes of pH, temperature and by proteolytic enzymes such as papain or trypsin.

Gelatin can be chemically modified to change its properties. It undergoes typical protein reactions, including acylation and carbamylation [26]. Such products include succinylated, phthalated, and carbamylated gelatin. These products are used for special pharmaceutical and photographic applications. A new type is methacrylated gelatin, which is suitable for technical applications. Permanent cross-linking of gelatin can be achieved by reaction with aldehydes such as formaldehyde, glyoxal, or glutaraldehyde [26].

2.4. Microbiological Analysis

Microbiological testing is in most cases carried out according to methods described in the European Pharmacopeia (Ph. Eur.) and the United States Pharmacopeia (U.S.P.). Important test parameters are total aerobic count, including molds and yeasts, after an incubation period of up to 5 d at incubation temperatures of 30 – 35 °C and 20 – 25 °C and counts of colony-forming units on tryptic soy agar and Sabouraud agar. Tests for Escherichia coli and Salmonella species include at least one enrichment step and detection of typical growth on selective nutrition media.

2.5. Quality Specifications

Gelatin, as a protein, is subject to contamination by microorganisms; good manufacturing practice must be followed to ensure a clean product. In addition to rendering gelatin unacceptable for human consumption, bacteria can degrade gelatin to a point at which it loses its gel-forming property. Conductivity and pH have a major impact on gelatin quality due to their importance regarding optimal growth conditions for bacteria [27].

Most gelatin manufacturing plants meet the strict requirements of ISO 9000 [28] and ensure consistent high quality by the implementation of HACCP (hazard analysis and critical control points) techniques [29]. The result is one of the purest proteins available to the food, pharmaceutical, and photographic industries. As a pure, natural protein, gelatin is a food, not an additive and, does not carry an E number.

The majority of commercial gelatins contain less than 3000 nonpathogenic bacteria per gram.

Table 2. Gelatin specifications

Parameter	Ph. Eur. limits	U.S. P. limits
pH	3.8 – 7.6	
Total ash, %	2	2
Loss on drying, %	15	
Sulfur dioxide, mg/kg	200	40[*]
Peroxides, mg/kg	100	
Arsenic, mg/kg	1	0.8
Heavy metals, mg/kg	50	50
Phenolic preservatives	0	
Total viable count CFU[**]	1000/g	1000/g
Escherichia coli	0 in 1 g	0 in 10 g
Salmonella	0 in 10 g	0 in 10 g

[*] Gelatin used for capsule manufacture can contain up to 1500 mg/kg.
[**] CFU = colony-forming units.

For pharmaceutical-grade gelatin the pharmacopeia (Ph. Eur., U.S.P.) set a limit of 1000 bacteria per gram, and Salmonella species and Escherichia coli must be absent in 10 g and 1 g, respectively [9], [10]. Physical and chemical characteristics of gelatin are listed in Table 2, together with microbiological specifications for pharmaceutical-grade gelatin.

Up to now there are no official limits for microbiological, physical or chemical parameters for food-grade gelatin. Therefore the Gelatin Manufacturers of Europe (GME) cooperate with official authorities to set microbiological, physical, and chemical limits for food-grade gelatin. It is planned to release a monograph for food-grade gelatin which contains not only the quality limits but describes also the test methods to ensure compliance with those limits.

Since the raw materials are exclusively obtained from animals fit for human consumption, and the production process includes numerous purification steps like washing, filtration, ion exchange and a final UHT sterilization step, gelatin is a safe product with regard to bacteriological, viral, and TSE (transmissible spongiform encephalopathies, e.g., Scrapie and BSE) safety [30], [31] and therefore does not carry any risk for the health of consumers.

3. Raw Materials and Production

3.1. Raw Materials

Gelatin is commercially derived from animal collagen of skins and bones. The principle raw materials worldwide are cattle hides, bones and pigskins. In principle all raw materials are fit for human consumption. Sources from other mammalian species and fish sometimes are also used. Each raw material requires special treatment to remove noncollagenous extraneous substances such as fat and minerals. The resulting pure collagen is then hydrolyzed to gelatin, which is soluble in hot water.

Bones. The so-called green bones are supplied by slaughterhouses and meat packers. The fresh raw material is transported directly to the degreasing plants of the gelatin producer. After preselection the bones are crushed into small particles, degreased, dried and sorted according to size. The mineral components of bones are hydroxyapatite and calcium carbonate. To remove the inorganic components the bones are treated for about a week or more with cold diluted acid. The resulting collagen is called ossein, the starting material for subsequent gelatin extraction.

Cattle Hides. Skins from calves and beef are obtained from animals slaughtered for human use. The subcutaneous, fat-containing tissue is mechanically separated and the skins are split horizontally into two parts: the part previously covered with hair is tanned and subsequently used as a high-quality leather. The parts previously in contact with the flesh are the so called hide splits. These residual thinner sections remain untanned and are dried or preserved with salt or calcium hydroxide to become the raw material used in the production of gelatin.

Pigskin. Pigskin is the most important raw material for production of edible gelatin in Europe and United States. Pigskins are supplied by slaughterhouses and meat processing plants and are frequently cooled or even frozen to prevent deterioration.

3.2. Production

Each raw material requires a special pretreatment of the collagen containing material to render it soluble in hot water. During pretreatment procedures the collagen swells and softens, peptides and cross-links are hydrolyzed, and various substances, regarded as impurities, are extracted. Two main types of gelatin are distinguished: type A from acid pretreated raw material and type B from alkaline processing. The production process is shown in Figure 2.

Acid Process. The acid process is used for pigskin and sometimes for special types of ossein. Washed pigskins are treated with diluted mineral acids for approximately 24 h at low temperature. This pretreatment is sufficient to break acid-labile peptide bonds in pigskin collagen. After partial neutralization the gelatin is then extracted with hot water. Usually, the gelatin is extracted stepwise with successive increase of temperature and time. Gelatins from successive extractions show different physical and chemical properties. The first extract has the highest gel strength and molecular mass and a very low color. Later extracts, obtained at elevated temperatures, contain a higher amount of low molecular weight peptides and therefore have lower gel strength and are more intensive in color and clarity.

Alkali Process. For the production of type B gelatin, cattle hides and ossein are pretreated with alkali (lime or sodium hydroxide) at ambient temperature. The liming time varies from several weeks to several months. The treatment removes impurities and splits cross links and peptides. The liming time is a critical parameter for the further processing.

The dilute solutions obtained by hot-water extraction contain 3 – 10 % gelatin. The liquors are filtered and deionized to remove suspended matter and undesirable amounts of inorganic ions. The clarified, dilute gelatin solution is concentrated to 25 – 35 wt % by vacuum evaporation, filtered, and sterilized. The concentrated liquor is rapidly chilled to a gel, extruded as noodles, which are deposited on a stainless steel net for drying with hot, sterile air. The net passes slowly through a drying chamber, which is divided into several zones with controlled temperature and humidity. Typical temperature range is from about 30 °C at the beginning up to about 70 °C in the final zone. The dried gelatin, having a moisture content of ca. 10 %, is broken and milled. The physical, chemical and microbiological properties of each extract are tested.

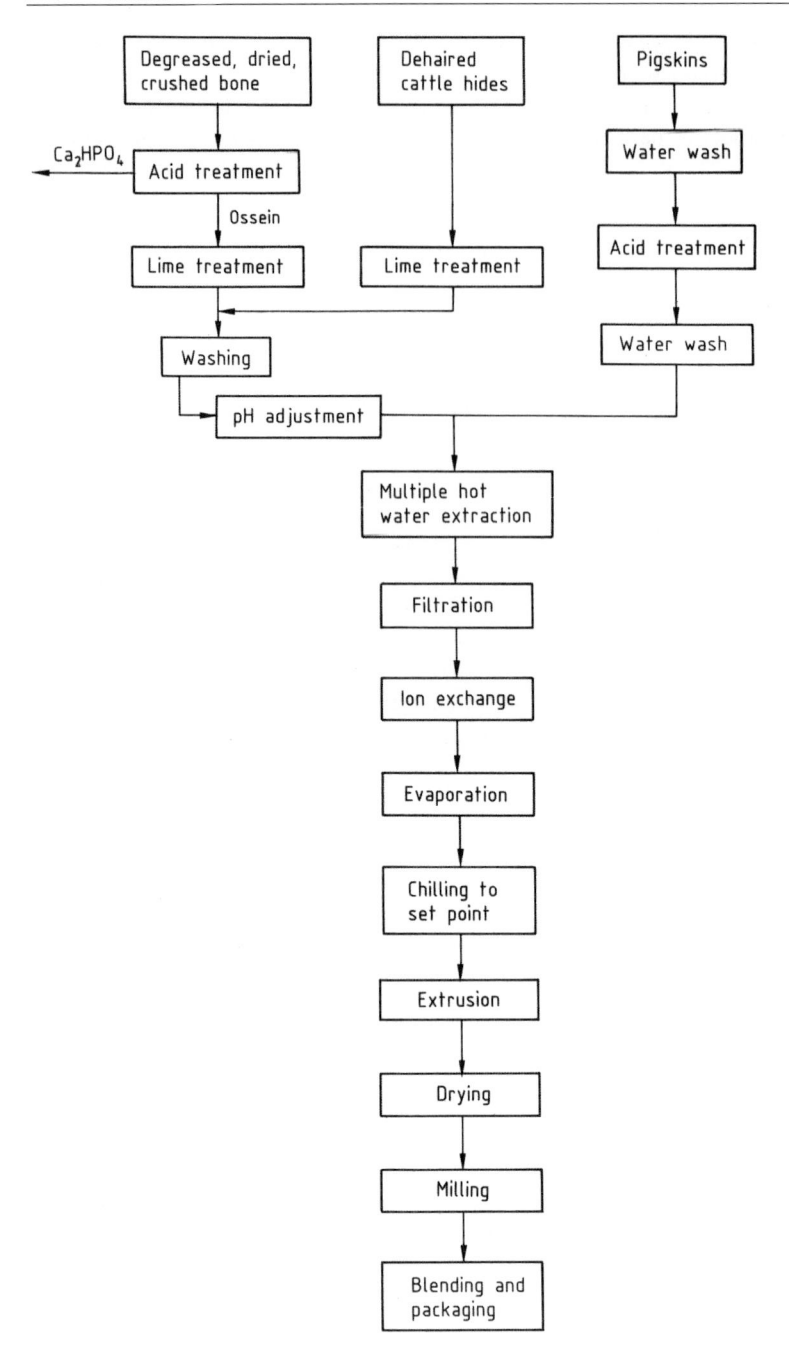

Figure 2. Gelatin production process

4. Uses

4.1. General Aspects

Although basically a food stuff, gelatin has long been valued not only for its nutritional contribution but also for its functional qualities. Therefore, the use of gelatin is not limited to the food industry. Very important applications are found in the pharmaceutical, cosmetic, photographic and technical area.

The use of gelatin in all these application fields depends largely on the structure function relationships that influence the functionality [33]. The main influencing molecular parameters are:

- Amino acid composition and sequence
- Secondary, tertiary, quaternary conformation
- Net charge and charge distribution
- Hydrophobicity/hydrophilicity ratio
- Size and shape
- Secondary reactions (intra- and intermolecular)
    Hydrogen bonding
    Ionic bonding
    Van der Waal's bonding

The functional role of gelatin does not arise from a single physicochemical property, rather it is a manifestation of a complex interaction of multiple properties. In spite of this the amino acid composition and sequence is the main reason for several functional properties. Gelatin is composed of about 65 % of polar or ionized amino acids. This high proportion of these polar amino acids influences hydration properties such as solubility and water-binding capacity. The type of hydrophobic amino acid residues and the hydrophilicity ratio are associated with amphiphilic behavior, which is important for foaming and emulsifying properties.

Due to its molecular structure and its structure – function relationship gelatin, shows the following properties:

- Gel formation
- Water binding
- Thickening
- Emulsion formation and stabilization
- Foam formation and stabilization
- Film formation
- Adhesion/cohesion
- Protective colloid function

4.2. Food Industry

Edible gelatin and gelatin hydrolysate are widely used ingredients in the production of food. Different properties and different applications require the production of suitable gelatin types [32]. Because of the ease of measuring out the required quantity, *leaf gelatin* is sold worldwide for household use. For cream cakes and cream desserts, special instant gelatins are available for preparing foods without heating. In the food industry powdered gelatin and gelatin hydrolysate are used in the processing of confectionery, dairy products, desserts, meat products, beverages, and so on.

Confectionery. For the confectionery industry, gelatin and gelatin hydrolysate are used due to their different properties:

- In the production of gums and jellies, gelatin is responsible for gel formation, texture, and the bright shiny appearance.
- Aerated confectioneries such as marshmallows owe their light and soft character to the ability of gelatin to form and stabilize foams.
- In fruit chews, gelatin or gelatin hydrolysate is used to prevent products becoming sticky.
- The film-forming properties of gelatin are used in the production of dragees.
- In caramel and liquorice sweets, gelatin provides excellent texture and mouthfeel.
- In pastilles, an elastic structure and excellent melting properties are provided by gelatin.
- In lozenges and compressed products, the binding properties of gelatin guarantee retention of shape.

Dairy Products. Gelatin enhances the texture of dairy products in an optimal way. Many product properties can be controlled by using different quantities and types of gelatin:

- Gelatin acts as a protective colloid in yogurts, thus preventing syneresis. The consistency can be adjusted from creamy to almost solid.
- Soft cheese can be adjusted in consistency from creamy to sliceable.
- Cream and toppings can be stabilized to retain their shapes.
- Sour cream retains its spreading and good melting qualities.
- The ability of gelatin to bind water, form emulsions and provide stability is exploited in the production of low-fat dairy products.
- The melting behavior of ice cream is substantially enhanced by the improved emulsion and finer crystal structure made possible by the addition of gelatin. In this application, gelatin is frequently combined with other hydrocolloids.

Desserts. Crystal-clear, firm to soft gels are required for dessert and fruit jellies; these can be fulfilled by selecting the most suitable gelatin type. In the production of whipped mousse and cream desserts, excellent foam formation and stability of the gelatin are important selection criteria. Different types of instant gelatin with different gelling properties can be used in the production of ready-to-eat desserts, both in mass production and in the households.

Meat and Sausage Products. Gelatin is used as gelling agent in the production of crystal-clear aspic meat and sausage products. Depending on the gelatin type the texture of the gel can range from soft to sliceable. In addition gelatin guarantees the brilliance and color of these products. Nongelling gelatin hydrolysates are used in edible dips and in combination with gelatin as adhesives in meat and sausage coatings.

Additionally both gelatin and gelatin hydrolysate are used to optimize the following technological and sensory quality parameters:

- Reduction of jelly and fat residues in canned sausage
- Improvement of spreading, quality, and softness in emulsified sandwich spreads
- Whipping agent for low-calorie sandwich spreads
- Improved homogeneity of binding in cooked sausage
- Protein enrichment in cured meats
- Rapid reduction of a_w values (water activity, see → Foods, 2. Food Technology, Section 1.4.) and shorter maturation times in raw sausage preparations
- Stabilization of emulsions, dispersions and suspension

Beverages. Gelatin and gelatin hydrolysate are used for the clarification of beverages, especially wine, fruit juices and beer. Clarification removes, substances causing turbidity and/or bitterness and have a detrimental effect on optics and taste. Secondly, such pre-clarification can enhance the efficiency of subsequent centrifugation. If necessary, gelatin treatment can be enhanced by adding silicic acid solution or bentonite. This helps to give the wines and juices the desired degree of brilliance.

Bakery Products. In the bakery industry, powder gelatin, leaf gelatin and instant gelatin are primarily used for the binding or gelling of fillings, as well as for the stabilization of creams.

Other Food Applications. Other industrial uses of gelatin in the food area are:

- Binder and gelling agent in canned seafood and seafood aspic products.
- Protein enrichment of food such as in beverages and dietetic products.
- Film former in panned confectionery products such as sugared almonds and hard-panned chewing gum varieties.
- Stabilizer and film former in icings and glazings for bakery products.
- Thickener and emulsifier in soups, sauces, and gravies.
- Emulsifier in half-fat margarines and butter.

4.3. Pharmaceuticals

Gelatin is widely used in the pharmaceutical industry: for the manufacture of soft and hard capsules, as a bulking agent for the coating of water-insoluble vitamins, as a starting material for the production of human plasma expanders, as a primary component for surgical sponges and as a binder in tablets. In addition to its use as an excipient (i.e., auxiliary agent in pharmaceuticals, see above) gelatin is recommended as a food supplement in the case of joint diseases.

The highest demand for pharmaceutical gelatin is for the production of soft and hard capsules. Soft capsules are made by feeding two plasticized gelatin ribbons between two revolving dies. Pharmaceuticals – either as a liquid or dispersed in an oil – are injected into the capsules as they are formed. Subsequently the capsules are dried in a two-step process.

Gelatin types used for the manufacture of soft capsules are 150 – 175 Bloom limed bone or limed hide, 190 – 210 Bloom acid pigskin or acid bone or a blend of 150 – 170 Bloom acid pigskin / limed bone and hide gelatin. Viscosities are in the range of 2.8 – 4.5 mPa · s. Acid-processed hide and bone gelatin types are preferred for water-miscible or hygroscopic contents. Due to their lower viscosity, less water can be used for the gel mass formulation. Glycerin or polyhydric alcohols like sorbitol are frequently used as plasticizers. In general, the gel mass is composed of 40 – 50 % gelatin, 20 – 30 % plasticizer and water.

Hard capsules, consisting of gelatin, dyes or pigments, and sometimes some processing aids, are produced as two separate parts, known as bodies and caps, by dipping stainless steel pins into a preheated 25 – 30 % gelatin solution. Dried and prelocked hard capsules are forwarded to pharmaceutical companies or contract laboratories to be filled mainly with powdered or granulated pharmaceuticals and sealed. Gelatin types used for the manufacture of hard capsules are 220 – 260 Bloom limed bone or blends of 220 – 260 limed bone/acid pigskin gelatin types. The use of gelatin to coat tablets is a growing market. In a process similar to the manufacture of hard capsules, the tablets are dipped into a concentrated gelatin solution and dried. In another process, the tablets are pressed into prefabricated hard capsules and sealed. Gelatin capsules and coatings protect pharmaceuticals from moisture, oxygen, and microbial deterioration.

Low-bloom gelatins (70 – 140 Bloom) are frequently used to coat vitamins A, D, and E for animal feed and human vitamin preparations. Emulsions of oil-soluble vitamins in gelatin solution containing carbohydrates are spray-dried to give free-flowing powders. Vitamins are thus available in a convenient and stable form that offers protection from light and oxidation.

Specially processed gelatin types have been successfully used as plasma extenders capable of maintaining osmotic pressure. The osmotic pressure of gelatin is strongly dependent on the average molecular weight. Thus the retention time of the plasma extender (generally: 24 – 48 h) is influenced by the size of the molecules. Commercial products are manufactured by partial thermal degradation of high-bloom limed bone gelatin, followed by chemical modification with succinic anhydride or cross-linking the resulting polypeptides with glutardialdehyde or isocyanates. Products containing ca. 3 % of degraded and modified gelatin are liquid at room temperature. Hemostatic sponges are manufactured by whipping a sterile gelatin solution into a foam, which is rendered insoluble by adding formaldehyde on drying. Gelatin sponges impregnated with thrombin can be left in place after surgery or extraction of teeth. Proteolytic enzymes completely digest the sponge.

Gelatin can act as a binder in the manufacture of tablets. Pharmaceuticals are dispersed in a gelatin solution and dried. The ground powder is subsequently compressed into tablets. Glycerin/gelatin masses are used for the production of zinc/gelatin dressings and suppositories.

4.4. Cosmetics

Gelatin and more frequently its hydrolysates find wide application in skin and hair-care products. Hydrolyzed collagen is nontoxic and produces no skin irritation, sensitization, or phototoxicity.

In skin-cleansing preparations, gelatin or gelatin hydrolysates reduce the irritation potential of anionic surfactants.

Due to their affinity to hair keratin, gelatin hydrolysates protect the structure of hair. Particularly in preparations for permanent waving, bleaching, or coloring damage of hair structure can be minimized.

4.5. Photography

Gelatin is an important raw material for all silver halide based photographic materials. For this application mostly alkaline-processed bone gelatin types are used. As a hydrocolloid and a polyelectrolyte, gelatin controls growth of silver halide crystals and prevents coagulation [34]. In addition, gelatin contains substances which influence the photographic properties of the light-sensitive material [35]. Gelatin also stabilizes hydrophobic additives in formulations (e.g., dye couplers) [36], [37]. The ability to undergo reversible sol – gel transition combined with the rheological properties of gelatin enable coating of several emulsion layers at the same time on a paper or film base [38], [39]. Gelatin can be cross-linked by a variety of chemical substances [40] to increase mechanical strength and stability of the coated layers to higher temperatures under acid or alkaline conditions without losing the ability to swell. Therefore chemicals can still diffuse through the different layers. This is essential for the development of the latent image and dye formation during processing of the exposed photographic material.

According to their application and their use in photographic materials, gelatin can be classified as follows:

- Emulsion gelatin
- Oil-in-water emulsion gelatin
- Support gelatin

Emulsion Gelatin. During the *precipitation* of the silver halide crystals, gelatin acts as a protective colloid. It stabilizes the silver halide crystals and influences their growth and their crystal habit. For this application highly purified lime-treated bone gelatin is used to prevent contamination of the silver halide by photographically active substances that may otherwise be present in gelatin. Specially modified gelatin may be used as well. By adjusting the pH, the modified gelatin forms a coacervate. This process is reversible and allows washing of the emulsion.

During *chemical ripening*, gelatin is used to control chemical reactions of added components with the silver halide surface in order to achieve desired photographic properties (e.g., sensitivity and contrast). Since gelatin itself could contain sensitizing or restraining substances, a consistent manufacturing process with a reliably high level of purification is required.

Oil in Water Emulsion Gelatin. Gelatin is used to aid formation of stable dispersions of organic substances (e.g., dyes and color couplers) as finely dispersed oil droplets. For this application the surface active properties of gelatin are important.

Support Gelatin. The physical properties of support gelatin are more important in these application areas than the chemical properties. In antihalation and antistatic coatings gelatin serves as a binding medium for light absorbing and electrically conducting materials. Other uses are as follows:

Before a photographic emulsion is applied to a film base, adhesion must be provided between the hydrophobic base material and the hydrophilic gelatin emulsion. For this purpose so-called *scrubbing gelatin*, mixed with organic solvents or acids, is applied to the substrate as an intermediate layer [41]. Gelatin used for this application must be free of any electrolytes that can flocculate with the organic solvents.

Backing Gelatin is applied as an additional layer on the back side of the film base to prevent curling, which is caused by the different swelling and shrinkage behavior of the emulsion coatings and the film base material.

Top-coat Gelatin is an abrasion-resistant layer which protects the emulsion layers against mechanical stress, rupture, and friction. For some applications, a glossy surface may also be desirable.

Baryta Gelatin is used as a nearly colorless binder for a suspension of barium sulfate which is applied to photographic papers in order to provide a highly reflective white appearance.

4.6. Technical Gelatin and Glue

Technical gelatin is a convenient term for gelatin that is used for nonedible purposes. Chemically, like pure gelatin, it is a mixture of proteins obtained by partial hydrolysis of waste collagen from animal connective tissue, bone, and skin. Technical gelatins, however, usually have inferior physical properties (gelling strength and viscosity of standard aqueous solutions, measured under standard conditions) and contain a higher proportion of nonprotein material than pure gelatin. The nonprotein component varies with the origin of the raw material and is usually residual mucopolysaccharides and soluble inorganic salts arising from the animal protein waste from which the gelatins are derived.

Technical gelatin has many applications based on its gelation and surface properties, for example: adhesives, remoistenable adhesive tapes, abrasives, matches, paper impregnation, plaster manufacture, production of stabilizers for emulsions, flocculation agents, and many more.

4.6.1. Matches

Only very clean glues are suitable for the production of matches, where they are applied as a binder for the matchhead composition. The inclusion of very fine air bubbles in the matchhead is very important to ensure the presence of additional oxygen and good ignition behavior.

Therefore the glue must show well-defined foaming behavior as far as stability and structure of the foam is concerned. Special test methods and instruments have been developed to control these important properties. As fat and grease act as undesirable defoamers, they must be carefully eliminated during production. Furthermore, the glue should contain only very small amounts of salts. Especially chlorides lower the setting point of the glue. In countries with high humidity in the air this leads to softening of the matchhead, rendering ignition more difficult or even impossible. Cross-linking with other components of the matchhead may counteract this undesirable performance to a certain extent. Technical gelatin, due to its high gelation power, is also used as an adhesive to produce the slide cover of the matchbox, which is produced at speeds of several thousands of pieces per minute.

4.6.2. Coated Abrasives

In coated abrasives technical gelatin is used as a binder between the paper substrate and the abrasive component. Gelatin-impregnated papers and unglued papers are used as substrates. Usually two coats are applied by two-roller coating devices. As a base coat a concentrated warm gelatin solution is used to apply a thick adhesive film with good spreading behavior and good flat-lying properties. To achieve good spreading properties, gelatin with certain thixotropic properties is required. In some cases starch is added. After the first coat the abrasive material is added, and a high-voltage electric charge is used to align the particles to improve the abrasive performance.

In addition to gelatin fillers, the second coat may contain pigments and additives with a cross-linking effect to toughen the dried film and to reduce moisture sensitivity, thereby increasing the life cycle of the coated abrasive. The coated paper is usually dried in loops in an oven. Compared with abrasive materials coated with phenolic resins, gelatin provides better and more elastic binding to the substrate. A disadvantage is the lower water resistance. The main application for gelatin coated abrasives is therefore the traditional glass or sandpaper used for smoothing wood, leather, and paint surfaces.

4.6.3. Paper Sizing

Technical gelatin is utilized as a size coating in the production of high-quality all-wood, rag – wood, and rag-based papers:

- Currency and other security papers must possess high performance for multiple use, good moisture and abrasion resistance, and good adherence to printing inks.
- In aquarelle papers gelatin provides the desired reduction in surface fluff.
- Most hand-made papers contain gelatin as sizing agent.

4.6.4. Adhesives

Gelatin has many functions in the adhesive field. A rough surface is required to allow mechanical grab of the glue, and at least one substrate must be permeable to water vapor to accomplish proper bonding. Gelatin glues are generally applied as a solution at ca. 60 °C. They have a unique two-step setting behavior. If applied in normal thickness to a substrate at room temperature, gelation takes place within parts of a second. A bond is formed which will allow handling of the joined materials almost immediately. The much higher final strength and the high temperature resistance (up to 190 °C) comes after complete evaporation of the free water.

Modified gelatin glues with special additives to develop the desired properties are used for:

- Bookcase making on fast-running machines, where stiff cover materials with high resilient strength must often be turned in.
- Book-spine reinforcement: Gelatin gives the book spine the necessary stability. It is also suitable for bonding the gauze or other material. The glue has to form an elastic film after drying and should have good adherence to previously applied synthetic dispersion glues.
- Lamination of paper to cardboard: glue based on gelatin exhibits good flat-lying properties and long open pot life, and only very little shrinkage of the paper is observed. It is used for the production of posters, jig saw puzzles and games.
- Box covering: cardboard boxes are covered with decorative cover materials. The gelatin-based glue must have good flat-lying properties and strong gelation power.

- Gummed tapes and papers: here often unmodified warm gelatin solutions are used. It is important to dry the solution with hot air to form a glasslike film which can be activated with cold water.
- Musical instruments and furniture: only for high-quality products gelatin based glues are still used today. In mass production, synthetic dispersion glues have replaced technical gelatin.

4.6.5. Other Uses

- Very fine ground gelatin is combined with gypsum to provide special rheological properties to surfacer compounds.
- Gelatin is employed for partly replacing casein in glues.
- Gelatin films exhibit antifogging behavior and are therefore used in the production of gas masks.
- As gelatin films are nonflammable, they are used for compostable grave lights.
- The resistance of gelatin to most organic solvents opens up applications in paintbrush production and the manufacture of paper gaskets.
- Gelatin is used in rubber mixtures to smooth the surface.

5. Economic Aspects

The gelatin market in total has shown only a moderate growth of about 2 to 3 % per annum over the last 25 years. Nevertheless, large differences could be seen from year to year with regard to different areas of application or geographical regions. Pharmaceutical use of gelatin in the late 1990s increased by more than 10 % per year in North America, South America, and the Far East, but there was practically no growth in Europe. The use of edible gelatin declined in the Far East and showed no growth in Western Europe.

Worldwide production of high-grade gelatin in 1997 was ca. 245 000 t and ca. 50 000 t of technical gelatin (glue). The distribution of the production is listed in Table 3.

The raw material situation is summarized in Table 4.

The worldwide market shares of the different applications were the following:

Table 3. Gelatin production by region (in 10^3 t)

Territory	Gelatin	Technical gelatin (glue)
Western Europe	110	8
North America	42	
Asia/Australia/Africa	43	30
Latin America	42	9
East Europe	8	3

Table 4. Raw materials for gelatin in production

Raw material	Gelatin	Technical gelatin (glue)
Pigskin	38 %	10 %
Bones	33 %	40 %
Hide splits	29 %	50 %

Edible gelatin	55 %
Pharmaceutical gelatin	17 %
Photographic gelatin	11 %
Technical gelatin (glue)	17 %

Regarding production capacity the leading manufacturers rank as follows:

High-grade gelatin
1. Deutsche Gelatine-Fabriken Stoess AG, Germany
2. SKW Biosystems, France
3. Leiner Davis Gelatin International, Australia
4. Tessenderlo B.V., Belgium

Technical gelatin (glue)
1. Fritz Häcker GmbH & Co., Germany
2. Trobas B.V., Netherlands
3. Ind. Cola e Gelatina Campo Belo Ltda., Brazil
4. Colas e Gelatinas Rebière Ltda., Brazil

References

1. P. I. Rose: "Gelatin", in *Encyclopedia of Polymer Science and Engineering*, vol. 7, Wiley, London 1987, pp. 488 – 513.
2. A. Lucas in H. Carter (ed.): *Tut-ench-Amun*, Brockhaus Verlag, Leipzig 1927.
3. A. Veis: *The Macromolecular Chemistry of Gelatin*, Academic Press, New York 1964, pp. 6 – 8.
4. G. Stainsby: "Gelatin Gels," in A. M. Pearson, T. R. Dutson, A. J. Bailey (eds.): *Advances in Meat Research, Collagen as Food*, vol. 4, Van Nostrand Reinhold Company, New York 1987.
5. L. Slade, H. Levine: "Polymer-Chemical Properties of Gelatin in Foods," in [4]

6 R. T. Jones in K. Ridgway (ed.): *Hard Capsules Development and Technology*, The Pharmaceutical Press, London 1987, pp. 41 – 42.
7 Gelatin Manufacturers Institute of America, Inc., *Standard Methods for the Sampling and Testing of Gelatins*, New York 1986.
8 S. Williams (ed.): *Official Methods of Analysis of the Association of Official Analytical Chemists*, 14th ed., 23, AOAC, Inc., Arlington, Virginia 1984, p. 429.
9 National Formulary XXIII, United States Pharmacopeial Convention, Inc., Rockville, Maryland 1997.
10 *European Pharmacopeia*, 3rd ed., European Council, Strasbourg 1997.
11 *Supplementary Standardised Methods for the Testing of Edible Gelatine*, Gelatine Manufactures of Europe, Brussels 1998.
12 O. T. Bloom, US 1 540 979, 1925.
13 British Standard 757: "Methods for sampling and testing gelatine (physical and chemical methods)," 1975.
14 R. Schrieber, "Edible Gelatine: Types, Uses and Application in the Food Industry, " *Gordian* **76** (1976) 356 – 364.
15 Gelatin Manufacturers Institute of America, Inc., *Gelatin*, New York 1986.
16 ISO 9665, Animal glues – Methods for sampling and Testing, 1993.
17 Bulgarian Standard BDS 543–72 "Bone glue", 1972; BDS 1560–63 "Technical gelatin", 1963.
18 USSR Standard GOST 2067–71 "Bone glue", 1971; GOST 3252–46 "Hide glue", 1946.
19 C. A. Finch, S. Ramachandran: *Matchmaking: Science, Technology, and Manufacture*, John Wiley & Sons, Chichester 1983.
20 E. Sauer: *Tierische Leime und Gelatine*, Springer Verlag, Heidelberg 1958.
21 J. E. Vandegaer (ed.): *Microencapsulation: Processes and Applications*, Plenum Press, New York 1974, pp. 21 – 27.
22 National Cash Register, US 2 800 457, 1957 (B. K. Green, L. Schleicher).
23 M. H. Gutcho: *Microcapsules and Microencapsulation*, Noyes Data Corporation, Park Ridge, New Jersey, 1976, pp. 3 – 60.
24 Deutsche Gelatine-Fabriken Stoess AG, EP 0 354 345, 1993 (Deutsche Gelatine-Fabriken Stoess AG).
25 *Standardised Methods for the Testing of Edible Gelatine*, Gelatine Manufactures of Europe, Brussels 1998.
26 R. C. Clark, A. Courts: "The Chemical Reactivity of Gelatin," in A. G. Ward, A. Courts (eds.): The Science and Technology of Gelatin, Academic Press, London, pp. 209 – 241.
27 U. Seybold: "Mikrobiologie von Gelatine und Gelatineprodukten, "in H. Weber (ed.): *Mikrobiologie der Lebensmittel: Fleisch und Fleischerzeugnisse*, Behr's Verlag, Hamburg 1997.
28 International Standard Organization ISO 9000.
29 Codex Alimentarius, Food and Agricultural Organization of the United Nations, Rome.
30 R. Schrieber, U. Seybold: "Gelatin production, the six steps to maximum safety, "in F. Brown (ed.): *Transmissible Spongiform Encephalopathies – Impact on Animal and Human Health*, vol. 80, Karger, Basel 1993.
31 U. Manzke, G. Schaf, R. Poethke, K. Felgenhauer, M. Muder, "On the Nervous Proteins from Materials Used for Gelatin Manufacture During Processing, " *Pharm. Ind.* **58** (1996) 9, 837 – 841.
32 A. Schott, "Speisegelatine, Herstellung, Eigenschaften und Einsatzgebiete," *Food. Technologie Magazin* **7** (1996) 3, 14 – 20.
33 E. Dick, "Gelatine a look at functionality, " *IFI* **21** (1998) 2, 42 – 46.
34 M. Szücs: "Die Rolle der Gelatine während der Fällung und des Wachstums von Silberhalogenidkristallen, " in H. Amman-Brass, J. Pouradier, F. Moll (eds.): *Photographic Gelatin*, IAG, Fribourg 1996, p. 338.
35 T. Ohno et al., *J. Photogr. Sci.* **33** (1985) 207 – 211.
36 R. Schrieber, *Gordian* **73** (1973) 282 – 287.
37 http://www.dgfstoess.com
38 F. J. Moll: "Gelatin in the Photographic Industry, "in H. Amman-Brass, J. Pouradier (eds.): *Photographic Gelatin*, IAG, Fribourg 1994, p. 181.
39 T. Kobayashi et al.: "On the Rates of Gelation and the Setting points of Gelatin Solutions," in M. Austin (ed.): *J. Photogr. Sci.* **40** (1992) p. 181.
40 T. Takahashi: "The Mechanism of the Crosslinking Reaction of Gelatin with Hardener, "in H. Amman-Brass, J. Pouradier, F. Moll (eds.): *Photographic Gelatin*, IAG, Fribourg 1996, p. 593.
41 W. Krafft in K. Keller (ed.): *Science and Technology of Photography*, Verlag Chemie, Weinheim, Germany 1993, p. 66.

Further Reading

A. Imeson (ed.): *Food Stabilisers, Thickeners and Gelling Agents*, Wiley-Blackwell, Ames IA 2010.
T. R. Keenan: *Gelatin*, "Kirk Othmer Encyclopedia of Chemical Technology", 5th edition, John Wiley & Sons, Hoboken, NJ, online DOI: 10.1002/0471238961.0705120111050514.a01.pub2.
R. Schrieber, H. Gareis: *Gelatine Handbook*, Wiley-VCH, Weinheim 2007.
A. Steinbüchel, S. K. Rhee (eds.): *Polysaccharides and Polyamides in the Food Industry*, Wiley-VCH, Weinheim 2005.
N. J. Zuidam, V. Nedovic (eds.): *Encapsulation Technologies for Active Food Ingredients and Food Processing*, Springer, New York 2009.

Lignin

BODO SAAKE, Institut für Holzchemie, Bundesforschungsanstalt für Forst- und Holzwirtschaft, Hamburg, Germany

RALPH LEHNEN, Institut für Holzchemie, Bundesforschungsanstalt für Forst- und Holzwirtschaft, Hamburg, Germany

1.	Occurrence and Functions.............	379	6.	Analysis........................... 389
2.	Structure and Biosynthesis	380	6.1.	Detection and Quantification........... 389
3.	Physical Properties	384	6.2.	Isolation.......................... 389
4.	Chemical Properties	385	6.3.	Spectroscopic Methods................ 390
4.1.	Kraft Pulping......................	385	6.4.	Degradation Techniques 391
4.2.	Sulfite Pulping	386	6.5.	Molar Mass Determination 391
4.3.	Other Pulping Processes	386	7.	Uses.............................. 392
4.4.	Pulp Bleaching.....................	387	8.	Toxicology......................... 392
4.5.	Yellowing Reactions.................	387		References......................... 393
5.	Commercial Lignins.................	387		

1. Occurrence and Functions

Lignin is one of the three major constituents of vascular plants, the other two being cellulose and hemicelluloses. The name lignin is derived from the Latin word *lignum* meaning wood. After cellulose, lignin is the most abundant natural (terrestrial) organic polymer. Its content is higher in softwoods (27–33 %) than in hardwoods (18–25 %) and grasses (17–24 %). The highest amounts of lignin (35–40 %) occur in compression wood on the lower part of branches and leaning stems of conifers [1, 2]. Lignin does not occur in algae, lichens, or mosses [3], whereas the "lignins" of bark differ in their structure from typical wood lignins [4].

Lignin is a randomly branched polyphenol, made up of phenylpropane (C_9) units, rendering it discernible from the other two major wood components by its UV absorption maximum at 280 nm. Figure 1 shows a typical UV microscopic imaging profile on a microtome cut of an individual spruce wood fiber [5]. Novel instruments can as well provide color-coded 2D or 3D plots, showing the lignin distribution over the cross-section of fibers [6]. The highest lignin concentration ($\approx 70\%$) is found between adjacent cell walls (middle lamella) and at the cell corners, while it is much lower ($\approx 20\%$) across the secondary wall. However, due to the much larger volume of the secondary wall, most lignin ($\approx 80\%$) is located in the secondary wall of the wood cells.

In accordance with its distribution, lignin performs three important functions in the xylem tissue that are essential to the life of plants: Due to its lipophilic character, lignin decreases the permeation of water across the cell walls, that consist of cellulose fibers and amorphous hemicelluloses, thus enabling the transport of aqueous solutions of nutrients and metabolites in the conducting xylem tissue. Secondly, lignin imparts rigidity to the cell walls and, in woody parts, together with hemicelluloses, functions as a binder between the cells generating a composite structure with outstanding strength and elasticity. Finally, lignified materials effectively resist attacks by microorganisms by impeding penetration of destructive enzymes into the cell walls.

The above mentioned functions of lignin are impressively demonstrated by the strength of the trunks of giant redwoods (*Sequoia*) in California, growing up to more than 100 m, supporting crown structures of several tons in weight for

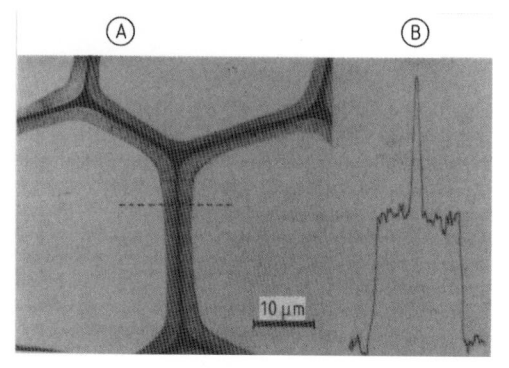

Figure 1. A) Black spruce earlywood
Cross section of tracheids of black spruce earlywood photographed in ultraviolet light ($\lambda = 240$ nm); B) Densitometer tracing across the tracheid wall showing the variation of lignin concentration along the dotted line [5]

thousands of years, unparalleled by human constructions.

2. Structure and Biosynthesis

After cell growth has ceased, lignin is formed by a dehydrogenative polymerization of three *p*-hydroxycinnamyl alcohols (monolignols): i.e., *p*-coumaryl (**1**), coniferyl (**2**), and sinapyl alcohol (**3**) (Fig. 2), which are formed from D-glucose via shikimic acid [7]. It has to be mentioned that in lignin chemistry the designation of the carbon atoms does not follow the IUPAC nomenclature. The aromatic ring is numbered assigning the phenolic OH group to position 4. The propane side chain is labeled with Greek letters, starting from the benzylic carbon, termed α carbon. The phenylpropanoid monomers are oxidized to phenoxy radicals by hydrogen peroxide in the presence of peroxidase. The unpaired electron of the phenoxy radical is delocalized over the aromatic ring and the conjugated olefinic double bond of the side chain, the highest density of the unpaired electron being at the β-carbon atom, C-5, C-1, and O-4, as illustrated for coniferyl alcohol in Figure 3. Random recombination of two monomeric phenoxy radicals leads to dilignols, which are further oxidized to oligolignols and finally to the polymeric lignin. The most important structural units and their designation are depicted in Figure 4.

In order to prove this pathway FREUDENBERG and his group carried out tracer experiments with carbon-14 labeled D-glucose, phenylalanine, coniferyl alcohol, and D-coniferin [7]. In addition they performed in vitro model studies by dehydrogenation of coniferyl alcohol with peroxidase and hydrogen peroxide, obtaining polymers which were not identical but rather similar to lignin [7]. The resulting products were called dehydrogenation polymers (DHP) and have been intensively used as a lignin polymer model by many research groups.

The random polymerization by recombination of phenoxy radicals explains why lignin — in contrast to most other natural polymers such as proteins, polysaccharides, nucleic acids, and natural rubber — has an irregular structure and is optically inactive, though carbon atoms α and β in the side chains of the phenylpropane structural units are asymmetric. However, it has to be

Figure 2. Monomeric lignin precursors

Figure 3. Dehydrogenation of coniferyl alcohol (**2**) yielding phenoxy radicals

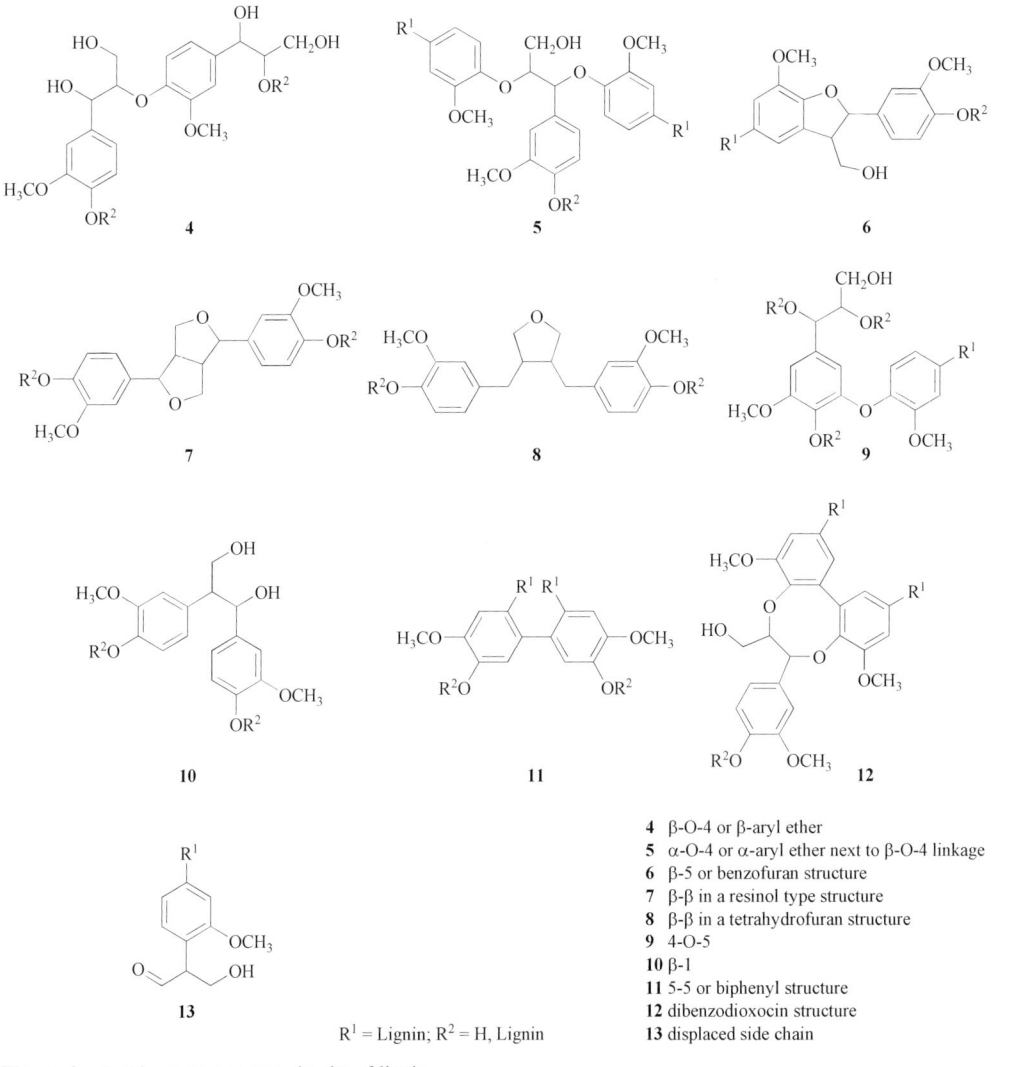

Figure 4. Most important structural units of lignin

4 β-O-4 or β-aryl ether
5 α-O-4 or α-aryl ether next to β-O-4 linkage
6 β-5 or benzofuran structure
7 β-β in a resinol type structure
8 β-β in a tetrahydrofuran structure
9 4-O-5
10 β-1
11 5-5 or biphenyl structure
12 dibenzodioxocin structure
13 displaced side chain

R^1 = Lignin; R^2 = H, Lignin

kept in mind that the last polymerization step takes place within the secondary cell wall of differentiated wood cells in close contact with partly crystalline cellulose fibers, so that the aromatic nuclei in lignin may partly be aligned tangentially to the secondary wall [8]. Recently (in the early 2000s) a new hypothesis has evolved, postulating that the synthesis of lignin might be at least partly controlled by dirigent proteins. Such proteins are involved in the polymerization of lignans, a group of oligomeric products which belong to the extractives of lignocellulosic plants. This hypothesis is discussed rather intensely and summaries of the pros and cons have been published by the main protagonists [9, 10].

The random coupling reaction, which is still the most widely accepted concept, results in a three-dimensional, amorphous polymer without a regular structure or repeating unit. Accordingly, no definite lignin structure can be determined although several schemes have been proposed as a statistical structural model for various plant species. A first concept for spruce lignin was developed by FREUDENBERG [7] based on 18 C_9 units, of which 2.5 (units 2, 3, 5) are coumaryl, one a syringyl (3,5-dimethoxy-4-hydroxyphenyl) (unit 16), and 14.5 guaiacyl (3-methoxy-

4-hydroxyphenyl) units (Fig. 5). A structural scheme of beech lignin was proposed by NIMZ consisting of 25 structural units, of which six (5/6, 9/10, 24/25) are partly replaced by the dilignol units enclosed in brackets (Fig. 6). The concentrations of the latter units are below 4%. The scheme shows a representative section of a ten to twenty times larger beech lignin molecule, in which ten different interunit bond types are randomly distributed. The relative amounts of interunit linkages for the two lignin structures are listed in Table 1 [11]. The data show as well that the calculated C_9 formulas are in good correlation with experimental data for isolated lignins. The most abundant interunit linkages in both schemes are the β-O-4 linkages. It is the concentration of these linkages in which lignins typically differ from other polyphenols occurring also in plant materials. During the lignification process the quinonemethide can also react with hydroxyl or carboxyl groups of polysaccharides forming covalent ether and ester bonds in the α-C position. These covalent bonds make it difficult to obtain lignin without carbohydrate impurities for both analytical and technical purposes. For this close association of lignin and carbohydrates the term lignin carbohydrate complex (LCC) has been coined [12].

Although the structures in Figures 5 and 6 do not include some new features such as the dibenzodioxocin structure (Fig. 3, **12**) discovered in the 1990s [15], they are still good models to describe the fundamental differences of softwood and hardwood lignins. According to the high proportion of guaiacyl units softwood lignins are called guaiacyl lignins (G-lignins), while hardwood lignins are termed guaiacyl–syringyl lignins (GS-lignins). Grass lignins additionally

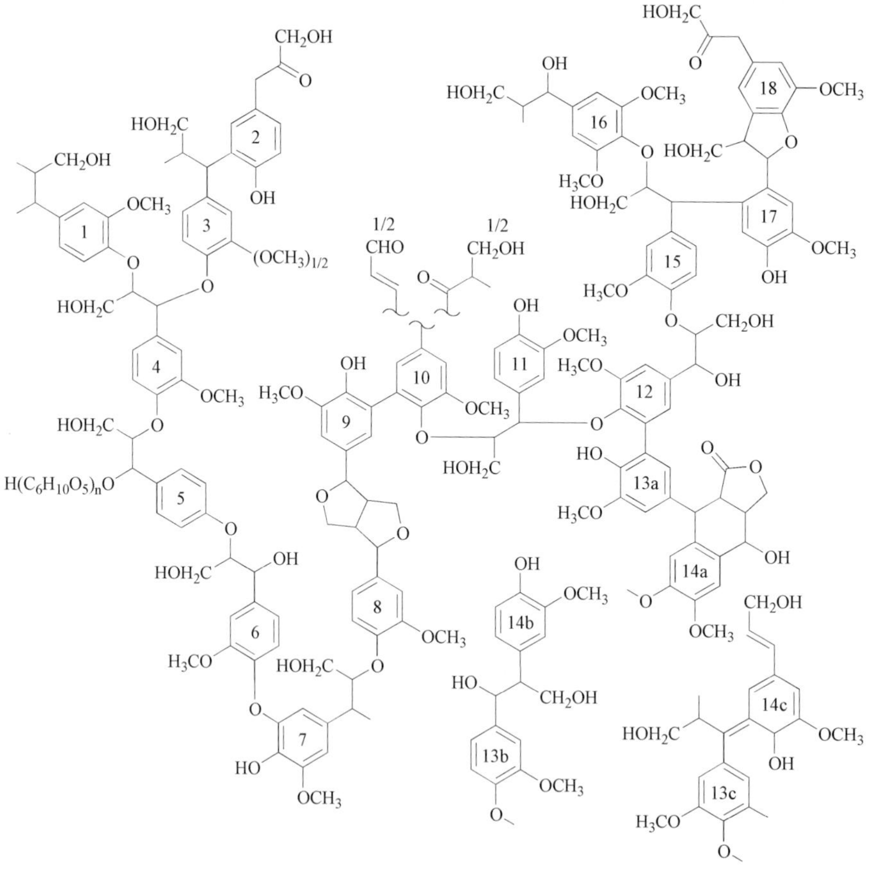

Figure 5. Constitution of spruce lignin after FREUDENBERG [7]

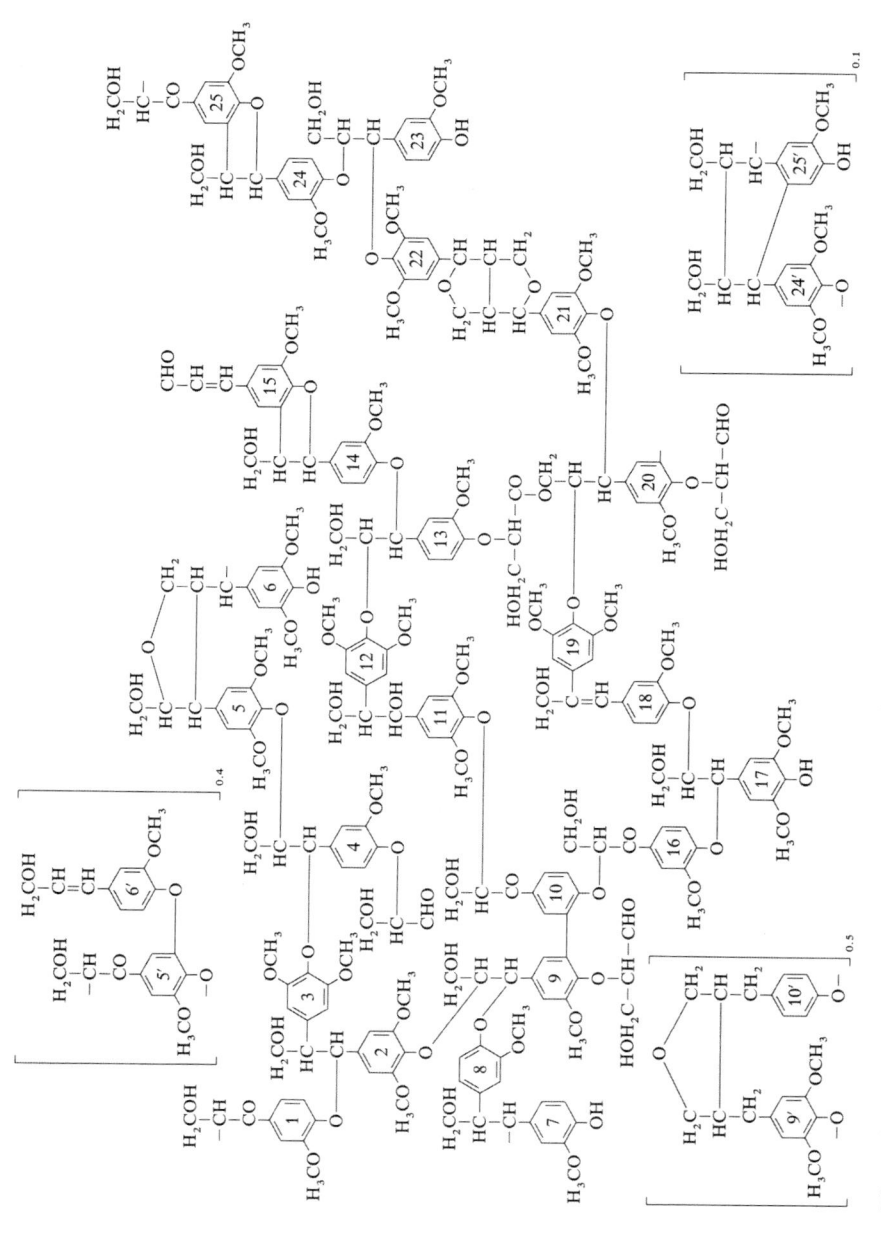

Figure 6. Constitutional scheme of beech lignin after Nimz [11]

contain p-hydroxyphenyl (coumaryl) units (GSH-lignins).

Accordingly, hardwood lignins have the highest methoxy values with 1.21–1.52 per C_9 unit, followed by softwood lignins with 0.80–1.01 and grass lignins with about 0.6–0.9 [16]. However, the low methoxy values of grass lignins may also be caused by substantial amounts of p-coumaric acid, linked to the γ-carbon atoms of grass lignins via ester bonds [17]. The methoxy group content of compression wood MWL from larch (*Larix leptolepis*) was found to be 0.2 per C_9 unit lower than that of normal wood MWL [18]. From the comparison of the carbon-13 NMR spectra of acetylated compression wood MWL with those of acetylated GSH- and H-DHPs (p-hydroxyphenyl dehydrogenation polymers), compression wood lignin was classified as a guaiacyl-p-hydroxyphenyl (GH)-lignin [19].

Variations in the methoxy group contents cause changes in the constitution of the lignins. As shown in Table 1 the proportions of α-O-4 and β-1 interunit linkages in beech lignin are significantly higher than those in spruce lignin, while the content of 5–5 and 4–O–5 linkages is lower, meaning that softwood lignins are more condensed than hardwood lignins and hardwoods may be more easily delignified than softwoods, agreeing with the differences in the chemical properties of hardwood and softwood lignins. Consequently, compression wood lignins have the highest level of condensation with the lowest solubility and degradability. A review with almost 700 references on the history, chemistry, and technology of lignin is given in [20].

3. Physical Properties

Solubility. As a branched polymer, native lignin is insoluble in all neutral solvents at room temperature. At temperatures above 100 °C or in alkaline solvents native lignin is partly dissolved as a result of degradation reactions. For isolated or technical lignin the solution properties vary depending on the isolation or production process and sample purity. Milled wood lignins are soluble in dioxane and acetone, containing 5–10 % water, as well as in dimethylformamide and pyridine. Lignosulfonates can be dissolved in water, while all isolated lignins are soluble in alkali. Most lignins can be solubilized in organic solvents such as dimethyl sulfoxide, dimethylformamide, dioxane, and pyridine at least for analytical purposes.

Molar Mass. The molar mass of lignin in situ is unknown and that of isolated lignins depends on the conditions of isolation. In general the molar mass determination of lignins with various methods show a large deviation (see

Table 1. Proportions (%) of interunit linkages, functional groups and elementary composition per C_9 unit of spruce and beech lignin according to structural schemes in Figures 5 and 6

Bond types, functional groups and elementary composition of C_9 unit	Spruce lignin	Beech lignin
β-O-4	39–48	32–37
α-O-4	11–16	28–32
β-5	6–10	8
β-β	7–10	6.4
5–5	7–9	2
4-O-5	6–7	2
β-1	2	16
α-5	7.2	
α-β		4
Aliphatic OH	92–98	88
Benzylic OH (α-OH)	18	4
Phenolic OH	29.4	16
Ketone groups	13.8	16
Aldehyde groups	2.8	4
Methoxy groups	91.7	136
C_9 formulae of structural schemes	$C_9H_{7.82}O_{2.4}(OCH_3)_{0.92}$ [7]	$C_9H_{7.16}O_{2.44}(OCH_3)_{1.36}$ [11]
C_9 formulae of MWLs*	$C_9H_{7.95}O_{2.4}(OCH_3)_{0.92}$ [13]	$C_9H_{7.10}O_{2.41}(OCH_3)_{1.36}$ [14]

*MWL (milled wood lignin), see Section 6.2

Section 6.5). For spruce milled wood lignin an average molar mass (M) of 11 000 g/mol has been estimated by ultracentrifugation [21], corresponding to a weight-average degree of polymerization of approximately 60 C_9 units. For the most important commercial lignins from the kraft, soda, and sulfite pulping process molar masses in a range between 3 000 and 20 000 g/mol and polydispersities between 2 and 12 have been reported, while the intrinsic viscosities were in a range from 0.04 to 0.08 dL/g. The exponent a from the Mark–Houwink–Sakurada equation varies between 0.1 and 0.5, which is in the range between an Einstein sphere ($a = 0$) and a compact coil ($a = 0.5$) [22, 23].

Thermal Properties. Lignin is an amorphous thermoplastic polymer. The glass transition temperature T_g of lignin is strongly affected by moisture content and hydrogen bonding, a phenomenon of great importance for the wood processing industry. By dynamic mechanical thermal analysis the T_gs of native lignins in wood were calculated resulting in lower values in hardwoods (65-85 °C) than in softwoods (90-105 °C). For isolated lignins the T_g values are higher and show deviations according to the raw material, production process, and measurement procedure: MWL lignins: 110 to 160 °C, kraft lignins: 124–174 °C, steam explosion lignin: 113–139 °C [24]. The effect of moisture was demonstrated for an isolated periodate lignin which had a T_g of 195 °C when dry and a T_g of 90 °C when containing 27 % water [25]. Thermal properties are as well depending on molar mass. On fractionated kraft lignins the fraction of lowest molar mass (M 620 g/mol) had a T_g of only 32 °C while at highest molar mass (M 180 000 g/mol) T_g amounted to 173 °C. A similar tendency was observed for the temperature of start of decomposition which increased with molar mass from 181 to 238 °C [26]. The calorific value of lignin depends on the purity of the sample. The dry matter of the black liquor from the kraft process has a value around 23.4 MJ/kg [1].

4. Chemical Properties

Besides ketone and aldehyde groups, the most reactive functional groups in lignin are the phenolic groups, the benzyl alcohol (α-OH) and noncyclic benzyl ether groups (α-O-4), which under acidic and alkaline conditions are prone to condensation reactions. From an industrial standpoint the reactions of lignin in pulping processes are most important since these reactions influence not only the pulping process itself but the structural features of technical lignins obtained as by products.

4.1. Kraft Pulping

The leading technical pulping process with sodium hydroxide and sodium sulfide is called kraft pulping. The word "kraft" originates from the German word "Kraft" (strength), meaning the strength of the obtained kraft pulps in comparison to that of sulfite pulps.

As a main reaction of the kraft process moieties with free phenolic groups are converted into quinonemethide groups (**15**), which add hydrogen sulfide ions at the α-carbon atoms (Fig. 7).

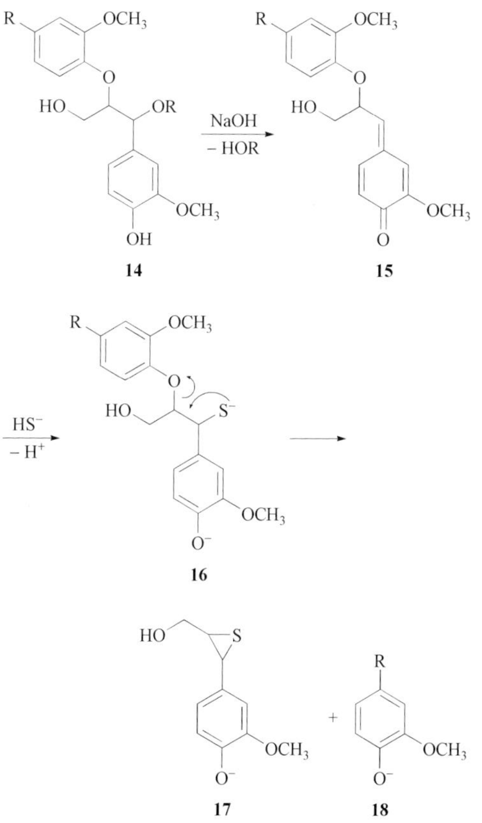

Figure 7. Mechanism of kraft cooking

The thus formed benzylthiolate anion (**16**) loses its β-phenolate anion in a neighboring displacement reaction. The units with free phenolic groups thus created may again form quinone-methides and add hydrogensulfide ions, if their α-carbon atoms are bonded to hydroxyl or non-cyclic ether groups. At the high reaction temperature the sulfur from compound (**17**) is partly split off. Especially in the last stage of the kraft process carbon-carbon bonds can be formed between lignin units, a reaction which is referred to as "condensation". This results in structures which are very difficult to cleave in further process stages. Furthermore, the hydrogensulfide ions lead to demethylation reactions and subsequently to the formation of methyl mercaptan which resulted in former times in odor problems in the mills. An extensive review on the reactions and technology of pulping processes is given in [27].

4.2. Sulfite Pulping

In the technical sulfite pulping processes wood is usually reacted with calcium (pH ≈1-2) or magnesium sulfite (pH ≈3-5) at ca. 125 to 150 °C for 3 to 7 h. The three major reactions occurring simultaneously are sulfonation, hydrolysis, and condensation. The dissolution of lignin in aqueous solution is mainly caused by the introduction of hydrophilic sulfonic acid groups at the α-carbon atoms (Fig. 8). In the *sulfurous acid* process at pH 1–1.6 and in the *acid sulfite* process at pH 1.8–3.1, benzyl alcohol and benzyl ether groups in lignin form benzylium ions (**20**), that preferably add nucleophilic sulfite ions, but to some extent may also lead to condensation products (**21**) [28, 29]. The degree of condensation may be less in the *bisulfite process* at pH 4.5 and the *bisulfite-sulfite* process at pH 7. The β-O-4 bond is rather stable under acidic conditions. Some sulfite processes operate under neutral or alkaline conditions using sodium or ammonia as base. In these processes β-O-4 bonds can be cleaved as well depending on the reaction conditions. The spent sulfite liquor, besides degraded hemicelluloses contains lignosulfonic acids (**22**).

4.3. Other Pulping Processes

Especially in countries with small wood resources the soda process is used for the production of pulp from annual plants applying only sodium hydroxide as a reaction chemical. In this process only α-aryl ether bonds in phenolic lignin units are effectively cleaved [29]. The addition of the redox catalyst system anthrahydroquinone/anthraquinone can improve the delignification when raw materials with higher lignin content are processed. Anthrahydroquinone can catalyze the cleavage of lignin in a similar manner to sulfide, improving especially the cleavage of β-O-4 linkages [30]. It can be applied in all alkaline processes such as kraft or alkaline sulfite pulping.

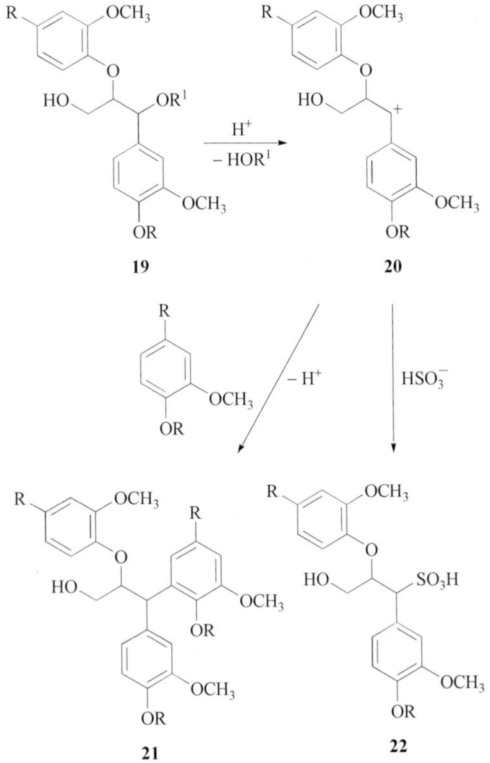

A large number of pulping processes based on organic solvents have been investigated over the last decades and were summarized under the term "organosolv" processes. Some were based on organic solvents (especially methanol and ethanol) with and without addition of catalysts while others were applying organic acids (mainly formic and acetic acid) [31, 32]. Up to now none of these processes has been implemented in industry. Nevertheless, in many papers on new applications of lignin still organosolv lignins from

Figure 8. Mechanism of sulfite cooking

former pilot plant trials are used as a raw material, especially when a sulfur-free lignin is required.

Steam explosion or steam refining is a process route for the production of fibers from wood or straw. These processes are often catalyzed autohydrolytically by organic acids liberated from the hemicelluloses of hardwoods or straw. During steam explosion a part of the lignin ($\approx$10–15 %) is rendered water-soluble while the rest can be extracted with dilute sodium hydroxide solution. During the steam treatment an acid-catalyzed hydrolysis of aryl ether linkages takes place while on the other hand homolytic cleavages occur. At high reaction severity a recondensation of lignin can occur as well. Accordingly steam explosion lignins have higher molar masses than sulfite, kraft, or organosolv lignins. Steam explosion lignins might become interesting in the future, because this process is the first step for saccharification of lignocelluloses for glucose or ethanol production, a process route which is currently under discussion as a source for bioenergy and fuel [31, 33] (the whole issue of [33] deals with steam explosion.)

4.4. Pulp Bleaching

Unbleached pulps still contains between 1.5 and 6 % of lignin. Since the residual lignin is intensively colored it has to be removed in order to produce pulps with high brightness and brightness stability. Previously chlorine and chlorine dioxide were major bleaching chemicals leading to toxic chlorinated phenols in the bleaching effluents (measured as adsorbable organic halogen (AOX). In order to reduce these effluents, chlorine (C) was almost completely and chlorine dioxide (D) partly replaced by alkaline oxygen (O), hydrogen peroxide (P), and ozone (Z), leading to *totally chlorine free* (TCF) or *elemental chlorine free* (ECF) bleached pulps. In the production of ECF pulps chlorine dioxide is still applied, but no gaseous chlorine. In the beginning of these developments it was often assumed that TCF bleaching should be the long-term goal for an environmental-friendly industry. However, it has been shown that a modern ECF sequence can be operated with very low AOX levels and accordingly ECF is now the dominating technology for modern pulp mills [34].

4.5. Yellowing Reactions

In the presence of alkali and oxygen or in daylight lignin shows an intense discoloring, which is observed by the yellowing reaction of paper, paper products, and wood. A major reaction pathway is the absorption of ultraviolet light by the α carbonyl group leading to a phenoxy radical which reacts subsequently with oxygen to form quinoid chromophores (Fig. 9).

The photoyellowing is a rather complex process which was recently summarized in [35]. For applications of lignin the sensitivity to UV light has to be considered being a potential problem for product stability or a positive feature when UV absorbing properties are required.

5. Commercial Lignins

Up to now commercial lignins are exclusively obtained as byproducts from the chemical pulping industry. The kraft process is the dominating technology with about 89 % of the total production capacity while in 2000 the share of the sulfite process had declined to only 3.7 %. Nevertheless, lignosulfonates from sulfite processes are still of major importance for the application of lignin as industrial products. About 10 % of the total pulp production is based on nonwood plant materials like bast and leaf fibers, straw, sugar cane bagasse, or bamboo. These materials are easier to delignify and are mainly produced by the soda process. They are especially important raw materials in countries with small wood resources like India and China [36]. In former times these

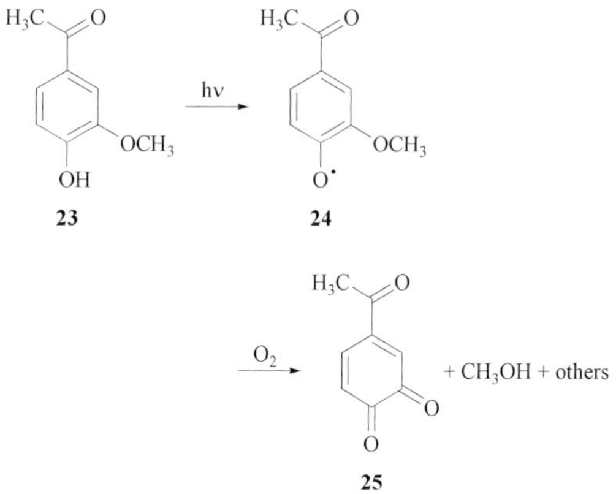

Figure 9. Main reaction pathway of photo yellowing

lignins were not commercially available. Recently a plant for lignin recovery from nonwood soda pulping has been installed which claims an annual production capacity of 10 000 t sulfur-free lignin [37] and accordingly soda lignins might become an alternative in the future. In 1998 some 50×10^6 t of lignin were produced in the western hemisphere as a byproduct of pulp, less than 2 % of this amount was isolated and sold. Kraft lignin, making up some 95 % of the totally produced lignin, is predominantly burnt to cover the energy demands of kraft mills and the recovery of caustic soda in the mill. Most of the commercially used kraft lignin is sulfonated to water-soluble lignosulfonates. Excluding the former Soviet Union, lignosulfonate production capacity in 1998 was estimated to be about 975 000 t/a with some new capacities under construction in South Africa [38].

The lignins are contained in the pulping liquor as a mixture with other wood degradation products with strongly fluctuating composition depending on the process and the raw material. Typical compositions of pulping liquors from hardwoods and softwoods obtained by the sulfite and kraft process are summarized in Tables 2 and 3 [39]. Kraft lignins are preferably obtained by precipitation from "black liquors" with mineral acids or carbon dioxide, whereas crude lignosulfonates are partly used directly as "spent sulfite liquors". The isolation of lignosulfonates can be achieved by addition of excess lime, by treatment with a long-chain alkylamine and extraction, or by ultrafiltration. In order to produce high-purity lignosulfonates a chemical destruction of sugars, or fermentation treatments can be included into the refinement procedure [38]. The sulfur contents of lignosulfonates (4–8 %) are higher than those of kraft lignins (1–1.5 %). Due to the sulfonate groups the lignosulfonates show good

Table 2. Example for the composition of spent sulfite liquors from softwood and hardwood [39]

Component	Percentage of total solids	
	Softwood	Hardwood
Lignosulfonate	55	42
Hexose sugars	14	5
Pentose sugars	6	20
Acetic and formic acid	4	9
Resin and extractives	2	1
Ash	10	10

Table 3. Example for the composition of kraft black liquors from softwood and hardwood [39]

Component	Percentage of total solids	
	Softwood	Hardwood
Kraft lignin	45	38
Xyloisosaccharinic acids	1	5
Glucoisosaccharinic acid	14	4
Hydroxy acid	7	15
Formic acid	6	6
Acetic acid	4	14
Resin and fatty acids	7	6
Turpentine	1	
Others	15	12

solubility in water over the entire pH range but are insoluble in many organic solvents. Kraft lignins are not soluble in water but in alkali at pH > 10.5. They are as well soluble in many organic solvents. The molar masses of lignins are difficult to determine in a reliable way and often results in controversial discussion of data. Furthermore molar masses depend strongly on the pulping conditions. In most cases the molar masses of lignosulfonates are higher compared to kraft lignin and even values up to 60 000 g/mol have been reported [40], although lower data are found for most samples [22, 23]. For Kraft lignins the data are mostly below 10 000 g/mol [22, 23, 39].

Kraft lignins contain larger amounts of phenolic hydroxyl, carboxyl, and catechol groups than lignosulfonates and are more likely to possess some unsaturated side chain structures. For certain application lignins can be further modified in order to improve the dispersing, complexing, or binding properties. These treatments can be sulfonation, sulfoalkylation, desulfonation, oxidation, carboxylation, amination, cross-linking, graft polymerization, and various combinations of the previous methods. Most of the kraft lignins are subjected to sulfonation either with sodium sulfite at 150–200 °C or oxidative with oxygen and sulfite. They can also be sulfomethylated with sulfide and formaldehyde at temperatures around 100 °C. Lignosulfonates are mostly obtained as magnesium, calcium, ammonium, or sodium salts [38–41].

Trade Names: A large variety of trade names and products exists. However, most of these products are marketed by two companies,

MeadWestvaco Speciality Chemicals and Borregaard-LignoTech, where the first company is dominating the kraft lignin and the latter the lignosulfonate market. Some trade names are listed below:

Unsulfonated kraft lignin	Indulin
Sulfonated kraft lignin	Polyfon, Reax
Lignosulfonates	Ameri-Bond, Borresperse, Borresol, Collex, Diwatex, Dustex, Dynasperse, Kelig, LignoBond, LignoSol, Marasperse, Norlig, Ufoxane, Ultrazine, Wafex, Wanin, Zewa
Soda lignin	Protobind, Biosurfact

6. Analysis

Due to its complicated structure the analysis of lignin is very difficult. For the pulp industry the analysis of lignin contents in pulps and raw materials is specified, while for technical lignins no well-defined industrial standards exist. For these reasons it is difficult to obtain reliable, detailed information on technical products. Due to this situation the following information can provide only a small insight into the field of lignin analysis. A review of most classical techniques for lignin analysis is given in [42].

6.1. Detection and Quantification

The easiest methods to detect lignin in plant materials are color reactions. However, those tests can be perturbed by some polyphenolic plant extractives. In the *Wiesner reaction* the material is treated with phloroglucinol in hydrochloric acid, leading to a purple conjugated quinone [43], formed between phloroglucinol and cinnamyl aldehyde groups in lignin. The reaction is positive with all lignins, but is weak with hardwood lignins with high contents of syringyl units.

For the *Mäule test* [44] lignified material is treated with a 1 % potassium permanganate solution, followed by washing, a treatment with 12 % hydrochloric acid, and a final moistening with aqueous ammonia. With hardwoods an intensive purple-red color is produced, while with softwood lignins the color is an indefinite brownish shade. The chromophores are probably chlorinated quinones.

The quantitative determination of lignin in plants most commonly used, the *Klason procedure*, is based on the observation that cellulose and hemicelluloses are hydrolyzed by concentrated sulfuric acid to soluble sugars while lignin is condensed to an insoluble cross-linked polymer which is recovered by filtration on a filter crucible and can be gravimetrically determined. This procedure is a standard test for raw materials in the pulp industry [45].

Another method for the quantitative determination of lignins in wood uses the solubility of wood in a mixture of acetyl bromide and acetic acid [46]. After removal of the reagent, the absorbance of the resulting solution is measured at 280 nm. A detailed overview on all analytical methods for the determination of lignin in raw materials as well as in pulps is given in [47]. A critical comparison of several methods including not only wood and straw but various agricultural residues was presented in [48].

6.2. Isolation

The complete isolation of unchanged lignin from wood is not possible, due to its intimate merging with cellulose and hemicelluloses in the secondary walls of wood cells. Lignin has to be degraded before extraction and accordingly the structure depends on the conditions of isolation. The most common procedure for structural and analytical purposes is production of *milled wood lignin* (MWL), also called *Björkman lignin* [49]. In the isolation procedure preextracted, dry wood meal is milled for, e.g., 30 days in a vibratory ball mill under a nonswelling solvent (toluene). In the subsequent extraction step with dioxane – water (9:1) up to 50 % of the lignin can be isolated from conifer wood which can be further purified in order to reduce carbohydrate impurities. Further methods for the isolation of lignins from pulps and pulping liquors are summarized in [42].

Recently it has been proposed to dissolve ball-milled cell wall material in dimethyl sulfoxide and tetrabutylammonium fluoride or *N*-methylimidazole. The product mixture of lignin and polysaccharides can be directly acetylated in the solution and allows further analyses, e.g., by 2D NMR techniques. Although this is no isolation procedure in the strict sense it can serve the same purposes for analytical techniques. A major advantage is that a higher lignin yield than in traditional isolation procedures is obtained [50].

For pulp fibers a method has been developed which is based on the enzymatic saccharification of polysaccharides and allows recovering of the undissolved lignin residue from the aqueous solution [51].

6.3. Spectroscopic Methods

UV Spectroscopy. The aromatic structure of lignin gives rise to a strong absorption maximum at 280 nm, while wood carbohydrates are transparent in the near-ultraviolet range. As, according to the Lambert – Beer law, the absorption of UV light by dissolved lignin is proportional to its concentration, UV absorption can be used for quantitative determination of lignin in solution or as a detection method in liquid chromatography [52]. In alkaline solution, the UV spectrum of lignin shows a bathochromic shift and hyperchromic effect, due to the formation of phenolate anions. An alkaline ionization difference spectrum (Δ_i) is obtained by subtracting the UV spectrum of the neutral solution from that of the alkaline solution. Δ_i-Spectra are often used to obtain information on the amount of phenolic hydroxyl groups. These spectra show as well specific absorption peaks for conjugated (carbonyl and olefinic) groups in lignin, which thus can be determined semiquantitatively [53, 54]. The UV absorption can be used also in the solid state for UV microscopy on microtome cuts (Fig. 1).

FTIR spectra of lignins are usually measured on finely ground solid samples dispersed in potassium bromide pellets. In solution the lignin spectra are influenced by strong interfering absorption bands of the solvents, i.e., semiquantitative work can be carried out better with KBr pellets using controlled pressing conditions and weight ratio of lignin to potassium bromide. The assignment of infrared absorption bands of lignins has been reviewed in [55] and [56].

NMR Spectroscopy. Of all physical methods, NMR spectroscopy provides by far the most complete information on the chemical structure of lignins. The assignment of some 40 signals in the ^{1}H broad-band decoupled carbon-13 NMR spectra of spruce and beech MWLs and DHPs as well as their acetates has been achieved in the early 1970s by comparison with the spectra of more than 50 lignin model compounds [57]. By now ^{1}H and ^{13}C shifts for a variety of model compounds have been determined and a large database is provided by the U.S. Dairy Forage Research Center in Madison, Wisconsin. This database is continuously updated and available free of charge in the internet [58]. Although solid-state NMR (^{13}C cross-polarization magic angle spinning (CP-MAS) NMR) spectra of lignin have been frequently recorded [59], the solution NMR techniques have by far the higher importance due to their higher resolution. One-dimensional high resolution ^{1}H or ^{13}C spectra of lignin can be obtained directly or after acetylation. The acetylation can improve the resolution of spectra and provide as well information on the primary, secondary, and phenolic OH groups from the carbonyl signals of the corresponding esters. For a good signal-to-noise ratio extreme long measurement times are required for ^{13}C spectra [60]. Here the short measurement times of ^{1}H spectra are advantageously, although the information is more limited. Besides the methoxy signals, a differentiation of phenolic and aliphatic hydroxyl groups, and information on the protons in aromatic and various aliphatic structures can be obtained. As an alternative ^{31}P and ^{19}F NMR spectra can be recorded with good sensitivity and short measurement times after labeling reactions with P and F containing reagents. The resulting spectra cannot provide information on the entire lignin structure, but on different hydroxyl, aldehyde, keto and quinone groups [61, 62].

The large number of signals in the carbon and proton spectra results in an intense overlapping of signals. This can be resolved by heteronuclear single- and multiple-bond shift correlation spectra which provide 2D plots for proton and carbon shifts. Under normal operation conditions 2D spectra are not suitable for quantification. However, various approaches have been developed in the early 2000s in order to enable a quantitative determination of lignin either by a calibration of the 2D spectra [63] or by a remodeling of the pulse sequences [64]. Also 3D spectra of lignin have been recorded [65, 66].

6.4. Degradation Techniques

Many methods for the determination of intermonomer bonds of lignin are based on analytical degradation procedures. The degradation by alkaline solutions of potassium permanganate was used intensively by FREUDENBERG'S group in the 1950s and 1960s and further modified by MIKSCHE and coworkers. Accordingly different variations of the method exist and are reviewed in [61]. The protocol consists of several reaction steps and result in mono- and dimeric aromatic carboxylic methyl esters, which can be analyzed by gas chromatography. This shows not only the proportion of hydroxyphenyl, guaiacyl, and syringyl units but allow calculation of the abundance of different lignin substructures on a molar basis. The procedure is applicable to all kind of lignins but has the disadvantage of a rather complicated multi-step protocol which requires an elaborated data treatment in order to conclude on the original lignin structure [68].

The thioacidolysis of lignin encompasses mainly the cleavage of β-O-4 linkages with boron trifluoride etherate and ethanethiol in anhydrous media. The monomeric degradation products can be analyzed by gas chromatography and normally between 1000 to 3000 μmol/g of mono- and dimeric degradation products are obtained. The protocol is accordingly simpler and allows an evaluation of the amount of alkyl-aryl ethers in lignin samples. A disadvantage is the strong odor of the reagents and the fact that the method is less informative on degraded samples, which is sometimes the case with some technical lignins [68, 69].

Derivatization followed by reductive cleavage (DFRC) was introduced as a new degradation method for lignin analysis [70]. Lignin is derivatized with acetyl bromide, followed by a reductive cleavage with zinc dust in a mixture of dioxane, acetic acid, and water (5:4:1). The resulting products are then acetylated and analyzed by gas chromatography. In the DFRC procedure hydroxyphenyl, guaiacyl, and syringyl units linked by α- and β-aryl ethers are cleaved and transferred into 4-acetocinnamyl acetate, coniferyl diacetate, and sinapyl diacetate.

The pyrolysis of lignin or lignocellulosic materials, i.e., the thermal degradation in the absence of oxygen, leads to the formation of a large variety of volatile components, which contain information on the original structure. In analytical pyrolysis this degradation technique is combined online or offline with other analytical techniques and most importantly with mass spectrometry (Py-MS) or gas chromatography and mass spectrometry Py-GC/MS. In the GC or GC/MS analysis of pyrolysis degradation products more than 100 peaks are found. Accordingly, the evaluation requires a large database and is based on statistical evaluation methods. Information on various pyrolysis techniques is found, e.g., in [71, 72]. A general review on MS techniques in combination with all relevant lignin degradation techniques can be found in [73].

6.5. Molar Mass Determination

Size-exclusion chromatography (SEC) is the most common method to determine molar masses of lignins. Frequently, lignins were acetylated followed by SEC in THF using columns based on polystyrene – divinylbenzene [74]. However, in various round robin tests carried out by the World Energy Association and the European network Eurolignin this approach resulted in a tremendous variation of the calculated molar masses [75, 76]. This problem might be partly due to aggregation phenomena and partly due to interaction with the column material. Furthermore, the acetylation is not suitable for lignins containing carbohydrate impurities and for lignosulfonates, the most important technical lignins. Lignosulfonate samples have been recently investigated using a complex mixture of water, DMSO, sodium phosphate buffer, and sodium dodecyl sulfate [40]. A universal eluent system applicable to all technical lignins is sodium hydroxide, however, the handling is not so easy because most commercial prepacked columns for HPSEC cannot withstand the high alkalinity [74]. Recently DMSO – water – lithium bromide and dimethylacetamide – lithium chloride were applied as eluents, showing a good agreement with alkaline SEC [23]. The determination of absolute molar masses by light scattering is complicated by the fluorescence of lignin, while viscosimetric methods suffer from the low overall viscosity of the samples. All methods are affected by the impurities of technical samples especially by carbohydrates and ash. Recently matrix-assisted laser desorption ionization/time of flight (MALDI/TOF) was applied to lignins after a fractionation of the samples on SEC columns in order to obtain fractions with narrow polydispersity [77].

7. Uses

Most applications of lignin and lignosulfonates are based on their dispersing, binding, complexing, and emulsion-stabilizing properties. About 50 % of all *lignosulfonates* produced worldwide are used for concrete mixtures, usually in the form of calcium or sodium salts. Addition of 0.1-0.3 % lignosulfonates to cement retards the setting or hydration of concrete. The second most important application is as a binder for animal feed pellets, where mainly calcium and ammonium salts are applied in order to improve pellet durability and abrasive resistance. A maximum dosage of 4 % is possible in finished pellets. Primarily chrome and ferrochrome salts of lignosulfonates function in oil well drilling muds as mud thinners, clay conditioners, viscosity-control agents, and fluid-loss additives. Mud systems conditioned with 0.2–0.5 % lignosulfonates, applied in the crude oil industry, perform well at high pressures and at temperatures of up to 175 °C. Another major market, especially for crude lignosulfonate spent liquors, is the dust control application for stabilizing insurfaced roads. In addition to these bulk applications a variety of speciality markets exist. Lignosulfonates are used for the granulation, complexation, or encapsulation of pesticides and as additives for gypsum boards in order to disperse the stucco. Lignosulfonates are used as dispersant in water-based paints and inks. Lignosulfonates and spent sulfite liquors are used for water treatment plants and paper machines to reduce deposits and slime formation or to complex metals such as zinc in cooling-water cycles. They are used in industrial cleaner formulations, for the complexation of nutrients in soil stabilization, for dyes or as additives in the brick industry. Modified lignosulfonates are additives in lead – acid batteries, which extend the service life of the product significantly. As further future uses the stabilization of enzyme formulations, biocide neutralization, and applications exploiting the antiviral and the chelating properties have been targeted [38].

Borregaard LignoTech still produces vanillin from softwood lignosulfonates, while all other producers closed down their production due to lower costs for vanillin production from petrochemical feedstocks. This was also the reason for stopping the production of dimethyl sulfoxide (DMSO) from kraft lignin by heating black liquor with sulfur to above 200 °C to obtain dimethyl sulfide (DMS), which upon oxidation with dinitrogen tetroxide (N_2O_4) yielded DMSO [78].

Up to now, much work has been carried out on new utilizations of lignin (see for instance [79]). Only few applications are economically feasible, taking into account that lignin is currently burnt, covering not only the energy demand of the pulp mills but providing a significant surplus of energy. Sometimes the argument occurs that the utilization of lignin as a chemical feedstock will increase at higher oil prices. However, one has to keep in mind that the economic burning value of lignin is increasing as well when the overall energy costs are rising. Accordingly, the changes for implementing new applications are higher when lignin is used not as a bulk chemical but as a specialty product using its specific properties.

8. Toxicology

Lignosulfonates are considered as nontoxic at the typical use levels and they are approved, e.g., by the U.S. Food and Drug Administration for a variety of applications in food packaging and food production. Examples are the use as a boiler-water additive for the production of steam, which comes into contact with food (21 CFR 173.310), or the approval of calcium lignosulfonates as a "specific usage additive" for the regulations on food additives for direct addition to food for human consumption (21 CFR 172.172). Some data for various lignosulfonates are listed in Table 4 [38]. However, one should keep in mind that the properties of a specific

Table 4. Data on the toxicity of different lignosulfonates [38]

Property	Calcium lignosulfonate	Sodium lignosulfonate	Ammonium lignosulfonate
Acute toxicity LD_{50}, mg/kg	> 2 000	> 10 000	> 10 000
Eye irritant	no	no	no
Skin irritant	no	no	no
Fish toxicity LC_{50}, mg/L	> 1 000	> 1 000	> 1 000
Bacteria toxicity EC_{10}, mg/L	5 000	5 000	343

product are not only influenced by its lignosulfonate component but by its impurities.

References

1. D. Fengel, G. Wegener: *Wood: Chemistry, Ultrastructure, Reactions*, DeGruyter, Berlin-New York 1984.
2. K. V. Sarkanen, C. H. Ludwig (eds.): *Lignin, Occurrence, Formation, Structure*, Wiley-Interscience, New York 1971, pp. 45, 66, 75.
3. H. H. Nimz, R. Tutschek, *Holzforschung* **31** (1977) 101–106.
4. W. Zimmermann, H. H. Nimz, E. Seemüller, *Holzforschung* **39** (1985) 45–49; C. Pascoal Neto, N. Cordeiro, A. Seca, F. Domingues, A. Gandini, D. Robert, *Holzforschung* **50** (1996) 563–568.
5. B. J. Fergus, A. R. Prokter, J. A. N. Scott, D. A. I. Goring, *Wood Sci. Technol.* **3** (1969) 117–138.
6. G. Koch, J. Puls, J. Bauch, *Holzforschung* **57** (2003) 339.
7. K. Freudenberg, A. C. Neish: *Constitution and Biosynthesis of Lignin*, Springer, Berlin-Heidelberg-New York 1968.
8. R. H. Atalla, *Wood Chem. Technol.* **7** (1987) 115.
9. L. B. Davin, N. G. Lewis, *Curr. Opin. Biotechnol.* **16** (2005) 407.
10. J. Ralph et al., *Phytochem. Rev.* **3** (2004) 29.
11. H. H. Nimz, *Angew. Chem.* **86** (1974) 336; *Angew. Chem. Int. Ed. Engl.* **13** (1974) 313.
12. T. Koshijima, T. Watanabe: *Association between Lignin and Carbohydrates in Wood and other Plant Tissues*, Springer, Berlin – Heidelberg 2003.
13. K. Freudenberg, J. M. Harkin, *Holzforschung* **18** (1964) 166.
14. K. Freudenberg, G. S. Sidhu, *Holzforschung* **15** (1961) 3.
15. P. Kanhunen, P. Rummakko, J. Sipila, G. Brunow, I. Kilpelainen *Tetrahatron Lett.* **36** (1995) 169–170.
16. K. V. Sarkanen, C. H. Ludwig: *Lignin, Occurrence, Formation, Structure*, Wiley-Interscience, New York 1971, pp. 64, 56, 75.
17. T. Higuchi, I. Kawamura, *Holzforschung* **20** (1966) 16.
18. S. Yasuda, A. Sakakibara, *Mokuzai Gakk.* **21** (1975) 363.
19. H. H. Nimz, D. Robert, O. Faix, M. Nemr, *Holzforschung* **35** (1981) 16.
20. J. L. McCarthy, A. Islam in W. G. Glasser, R. A. Northey, T. P. Schultz (eds.): "Lignin: Historical, Biological, and Materials Perspectives", *ACS Symp. Ser.* **742** (1999) 2–99.
21. A. Björkman, B. Person, *Svensk Papperstidn.* **60** (1957) 285.
22. W. G. Glasser, V. Davé, C. E. Frazier, *J. Wood Chem. Technol.* **13** (1993) 545.
23. O. Ringena, S. Lebioda, R. Lehnen, B. Saake, *J. Chromatogr. A* **1102** (2006) 154.
24. W. G. Glasser in W. G. Glasser, R. A. Northey, T. P. Schultz (eds.): "Lignin: Historical, Biological, and Materials Perspectives", *ACS Symp. Ser.* **742** (1999) 216–238.
25. D. A. I. Goring, in K. V. Sarkanen, C. H. Ludwig (eds.): *Lignin, Occurrence, Formation, Structure*, Wiley-Interscience, New York 1971, p. 732.
26. H. Yoshida, R. Mörck, K. P. Kringstad, H. Hatakeyama, *Holzforschung* **41** (1987) 171.
27. H. Sixta, A. Potthast, A. W. Krotscheck in H. Sixta (ed.) *Handbook of Pulp*, vol. 1, Wiley-VCH, Weinheim 2006, pp. 109–472.
28. J. Gierer, *Svensk Papperstidn.* **18** (1970) 571.
29. J. Gierer, *Wood Sci. Technol.* **14** (1980) 241.
30. J. Gierer, O. Lindberg, I. Noren, *Holzforschung* **33** (1979) 213.
31. R. A. Young, M. Akhtar (eds.): *Environmental Friendly Technologies for the Pulp and Paper Industry*, John Wiley & Sons, New York 1998, pp. 3–256.
32. "Lignin", in *Ullmann's*, 7th ed., Wiley-VCH Verlagsgesellschaft, Weinheim 2002.
33. G. Glasser, R. S. Wright, *Biomass Bioenergy* **14** (1998) 219.
34. H. Sixta (ed.): *Handbook of Pulp*, Wiley-VCH, Weinheim 2006.
35. C. Heitner, J. C. Scaiano (eds.): "Photochemistry of lignocellulosic materials", *ACS Symp. Ser.* **531** (1993) 1–223.
36. H. Sixta (ed.) *Handbook of Pulp*, vol. 1, Wiley-VCH, Weinheim 2006, pp. 8–9.
37. http://www.asianlignin.com.
38. J. D. Gargulak, S. E. Lebo in W. G. Glasser, R. A. Northey, T. P. Schultz (eds.): "Lignin: Historical, Biological, and Materials Perspectives", *ACS Symp. Ser.* **742** (1999) 304.
39. *Ullmann's*, 5th ed., A15, VCH Verlagsgesellschaft, Weinheim 1990.
40. G. E. Fredheim, S. M. Braaten, B. E. Christensen, *J. Wood Chem. Technol.* **3** (2003) 197.
41. W. G. Glasser, S. Sarkanen (eds.): "Lignin Properties and Materials", *ACS Symp. Ser.* **397** (1989) .
42. S. Y. Lin, C. W. Dence (eds.): *Methods in Lignin Chemistry*, Springer Verlag, Berlin 1992.
43. E. Adler, K. J. Björkquist, S. Häggroth, *Acta Chem. Scand.* **2** (1948) 93; E. Adler, L. R. Ellmer, *Acta Chem. Scand.* **2** (1948), 839.
44. C. Mäule, *Beitr. Wiss. Botanik* **4** (1900) 166.
45. Acid-Insoluble Lignin in Wood and Pulp, TAPPI T 222 OS-74, TAPPI Press, Atlanta.
46. D. B. Johnson, W. E. Moore, L. C. Zank, *Tappi* **44** (1961) 793.
47. S. Y. Lin, C. W. Dence (eds): *Methods in Lignin Chemistry*, Springer Verlag, Berlin 1992, pp. 21–61.
48. R. S. Fukushima, R. D. Hatfield, *J. Agric. Food Chem.* **52** (2004) 3713.
49. A. Björkman, *Svensk Papperstidn.* **59** (1956) , 477; **60** (1957) 158.
50. F. Lu, J. Ralph, *The Plant J.* **35** (2003) 535.
51. B. Hortling, E. Turunen, L. Sundquist, *Nord. Pulp. Pap. Res. J.* **7** (1992) 144.

52 S. Y. Lin in S. Y. Lin, C. W. Dence (eds): *Methods in Lignin Chemistry*, Springer Verlag, Berlin 1992, pp. 217–232.
53 G. Aulin-Erdtman, *Svensk Papperstidn.* **55** (1952) 745.
54 O. Goldschmid, *J. Am. Chem. Soc.* **75** (1953) 3780.
55 H. L. Hergert in K. V. Sarkanen, C. H. Ludwig (eds.): *Lignin, Occurrence, Formation, Structure*, Wiley-Interscience, New York 1971, p. 267.
56 O. Faix, *Holzforschung* **45** (1991) Suppl. 21.
57 H.-D. Lüdemanny, H. H. Nimz, *Biochem. Biophys. Res. Comm.* **52** (1973) 1162; *Angew. Makromol. Chem.* **175** (1974) 2393, 2409.
58 S. A. Ralph, J. Ralph, L. L. Landucci: *NMR Database of Lignin and Cell Wall Model Compounds*, November 2004. Available at URL http://ars.usda.gov/Services/docs.htm?docid=10491.
59 G. J. Leary, R. H. Newman in S. Y. Lin, C. W. Dence (eds): *Methods in Lignin Chemistry*, Springer Verlag, Berlin 1992, pp. 146–160.
60 D. R. Robert, G. Brunow, *Holzforschung* **38** (1984) 85.
61 L. G. Akim *et al.*, *Holzforschung* **55** (2001) 386.
62 R. M. Sevillano, G. Mortha, M. Barrelle, D. Lachenal, *Holzforschung* **55** (2001) 286.
63 L. Zhang, G. Gellerstedt in *Proc. of Sixth European Workshop on Lignocellulosics and Pulp* 2004, pp. 7–10.
64 H. Koskela, I. Kilpelainen, S. Heikkinen, *J. Magn. Res.* **174** (2005) 237.
65 T. M. Liitia, S. L. Maunu, B. Hortling, M. Toikka, *J. Agric. Food Chem.* **51** (2003) 2136.
66 J. Ralph, F. Lu, *Org. Biomol. Chem.* **2** (2004) 2714.
67 G. Gellerstedt in S. Y. Lin, C. W. Dence (eds): *Methods in Lignin Chemistry*, Springer Verlag, Berlin 1992, pp. 322–333.
68 W. G. Glasser in W. G. Glasser, R. A. Northey, T. P. Schulz: "Classification of lignin according to chemical and molecular structure", *ACS Symp. Ser.* **742** (2000) 216–238.
69 C. Rolando, B. Monties, C. Lapierre in S. Y. Lin, C. W. Dence (eds): *Methods in Lignin Chemistry*, Springer Verlag, Berlin 1992, pp. 334–349.
70 F. Lu, J. Ralph, *J. Agric. Food. Chem* **45** (1997) 2590.
71 D. Meier, O. Faix in: S. Y. Lin, C. W. Dence (eds): *Methods in Lignin Chemistry*, Springer Verlag, Berlin 1992, pp. 177–199.
72 P. Bocchini, G. C. Galletti, S. Camarero, A. T. Martinez, *J. Chromatogr.*, A **773** (1997) 227.
73 S. Reale, A. Di Tullio, N. Spreti, F. De Angelis, *Mass Spectrom. Rev.* **23** (2004) 87.
74 M. E. Himmel, J. Mlynár, S. Sarkanen in C. Wu (ed.) *Handbook of Size Exclusion Chromatography*, Marcel Dekker, New York, 1995, pp. 353–379.
75 T. H. Milne, H. L. Chum, F. Agblevor, D. K. Johnson, *Biomass Bioenergy* **2** (1992) 341.
76 A. Abächerli *et al.*, in *Proc. of the 7th International Lignin Institute Forum*, Barcelona, ILI, Lausanne, 2005, p. 119.
77 A. Jacobs, O. Dahlman, *Nord. Pulp Pap. Res. J.* **15** (2000) 120.
78 D. W. Goheen in K. V. Sarkanen, C. H. Ludwig (eds.): *Lignin, Occurrence, Formation, Structure*, Wiley-Interscience, New York 1971, p. 807.
79 T. Q. Hu (ed.): *Chemical Modification, Properties, and Usage of Lignin*, Kluwer Academic/Plenum Publishers, New York 2002.

Further Reading

J. H. Clark, F. E. I. Deswarte (eds.): *Introduction to Chemicals from Biomass*, Wiley, Chichester 2008.
M. Ek, G. Gellerstedt, G. Henriksson: *Pulping Chemistry and Technology*, de Gruyter, Berlin 2009.
R. B. Gupta, A. Demirbas: *Gasoline, Diesel, and Ethanol Biofuels from Grasses and Plants*, Cambridge University Press, New York, NY 2010.
C. A. S. Hill: *Wood Modification*, Wiley, Chichester 2006.
S. E. Lebo Jr., J. D. Gargulak, T. J. McNally: *Lignin, Kirk Othmer Encyclopedia of Chemical Technology*, 5th edition, vol. **15**, p. 1–25, John Wiley & Sons, Hoboken, NJ, 2005, online: DOI: 10.1002/0471238961.12090714120914.a01.pub2 (December 2001)
L. A. Lucia, O. J. Rojas (eds.): *The Nanoscience and Technology of Renewable Biomaterials*, Wiley, Chichester 2009.
D. M. Mousdale: *Biofuels*, CRC Press, Boca Raton, FL 2008.
E. A. Paul: *Soil Microbiology, Ecology, and Biochemistry*, 3rd ed., Academic Press, Amsterdam 2007.
R. P. Wool, X. S. Sun: *Bio-Based Polymers and Composites*, Elsevier / Academic Press, Amsterdam 2005.

Pulp

RUDOLF PATT, University of Hamburg, Hamburg, Germany

OTHAR KORDSACHIA, University of Hamburg, Hamburg, Germany

RICHARD SÜTTINGER, Schongau, Germany

1.	**Chemical Pulp**	395
1.1.	Raw Materials	396
1.2.	**Processing of Raw Materials**	399
1.2.1.	Processing of Wood	399
1.2.2.	Processing of Annual Plants	403
1.3.	**Pulping Processes**	403
1.3.1.	Alkaline Pulping Processes	405
1.3.1.1.	Kraft (Sulfate) Pulping	405
1.3.1.2.	Soda Pulping of Annual Plants	409
1.3.2.	Sulfite Process	410
1.3.2.1.	Reactions in Sulfite Pulping	411
1.3.2.2.	The Various Sulfite Processes	413
1.3.2.3.	Multistage Sulfite Processes	415
1.3.2.4.	Anthraquinone-Catalyzed Sulfite Pulping	416
1.3.3.	Other Pulping Processes	416
1.3.3.1.	Nitric Acid Pulping	416
1.3.3.2.	Pulping with Organic Solvents	417
1.3.3.3.	Pulping with Organic Acids	418
1.4.	**Technology of Pulp Production**	418
1.4.1.	Pulping	418
1.4.2.	Washing and Screening of Pulp	420
1.4.3.	Pulp Drying	424
1.5.	**Waste Liquor Treatment**	425
1.5.1.	Recovery of Useful Products from Waste Liquor	425
1.5.2.	Energy Generation from Waste Liquor	426
1.5.3.	Chemicals Recovery	429
1.6.	**Pulp Bleaching**	432
1.6.1.	Conventional Bleaching	432
1.6.2.	Chlorine-Reduced and Chlorine-Free Pulp Bleaching	436
1.6.3.	Technology of Bleaching	439
1.7.	**Environmental Aspects of Pulp Production**	441
1.8.	**Pulp Properties and Applications**	442
2.	**Mechanical Pulp**	443
2.1.	**Introduction**	443
2.1.1.	The Raw Material, Wood	444
2.1.2.	Characteristics Required of Mechanical Pulps	444
2.2.	**Groundwood**	445
2.2.1.	The Grinder Stone	446
2.2.2.	Groundwood Production	447
2.2.3.	Newer Methods of Groundwood Production	447
2.3.	**Refiner Mechanical Pulps**	448
2.3.1.	Historical Survey	448
2.3.2.	Theory of the Refiner Process	448
2.3.3.	TMP Production	448
2.3.4.	New Methods of Refiner Pulp Production	449
2.4.	**Screening of Mechanical Pulps**	449
2.4.1.	Purpose of Screening	449
2.4.2.	Mechanical Pulp Screening	450
2.5.	**Bleaching of Mechanical Pulps**	451
2.6.	**Comparison of Mechanical Pulps**	452
	References	452

1. Chemical Pulp

Wood and other lignocelluloses, such as straw or other annual plants consist of many different cell types in which cellulose, hemicelluloses, and lignin are the main chemical components. In the chemical digestion of lignocelluloses, lignin, which is embedded predominantly between the wood fibers as a binder and in the outer cell wall layers, and the hemicelluloses are largely dissolved out of the fiber matrix. The cohesion of the structural elements is lost in this process. The fibrous material obtained after digestion, consisting principally of cellulose, is called pulp.

The chemicals used in the digestion cannot quantitatively eliminate lignin from the wood matrix. In addition, unwanted, sometimes colored accompanying substances, called wood extractives, may remain in the cell walls. For most uses, the residual lignin and the extractives must be eliminated in a subsequent bleaching treatment. Moreover, partial or complete removal of hemicelluloses by bleaching and purification is required if the cellulose in the pulp is to be employed as raw material for chemical products.

Pulps that are used to make paper and board are called paper or chemical pulps, and those processed in the chemical industry are known as dissolving pulps. Pulps are also classified according to the type of raw material used (softwood or long-fiber pulps, hardwood or short-fiber pulps, straw pulps, etc.) and the pulping process employed (e.g., kraft, sulfite, and soda pulps).

The pulping and bleaching processes must be adapted to the final use of the pulp. The digestion of the raw material and the bleaching of the pulp can be carried out with varying intensity, or the latter step can be completely omitted. Depending on the intensity of pulping, it is possible to differentiate between semichemical and high-yield pulps and between chemical pulps and dissolving pulps. Depending on the residual lignin content of chemical and dissolving pulps, a difference is also made between harder and softer pulp qualities. Semichemical and high-yield pulps can be used with or without bleaching, but chemical and dissolving pulps must usually be bleached. There are slightly bleached, partially bleached, and fully bleached qualities of pulp. Paper and board production is by far the dominant use; only about 5 % is processed in the chemical industry.

1.1. Raw Materials

Apart from good availability and processibility, the distribution and morphology of the cell types and the chemical composition are important criteria for the suitability of lignocelluloses for pulp production. The proportion of fibrous elements with a favorable length-to-width ratio is crucial for the technological properties of the resultant pulps. The cellulose content as well as the amount of lignin and its structure determine the pulp yield and the conditions to be used for digestion. In particular, acid sulfite digestion, and also the further processing of pulp, can be adversely affected by accompanying substances, such as fats, waxes, phenolic components, or mineral inclusions. The average composition of important raw materials for pulp production is presented in Table 1. Note that considerable deviations in individual analytical data can be observed, depending on the method of analysis and especially on the sample of raw material used. However, the values given do exemplify the essential differences in the chemical composition of conifers (softwoods), deciduous woods (hardwoods), and annual plants.

The most important raw material for the production of pulp is wood, softwoods being preferred. However, hardwoods have gained increasing importance because, in comparison with softwoods, they have some interesting advantages for the production of pulp and paper (Table 2). Only 7 – 8 % of pulp production is based on annual plants, which are mainly employed when wood is not available [16].

The morphology of the lignocelluloses varies considerably (see also → Wood). Conifers have the most homogeneous structure. They consist to more than 90 % of tracheids with an average length of 3 mm and a width of 30 – 50 µm. These fiber cells have a length-to-width ratio

Table 1. Chemical composition of important raw materials for pulp production

Common name	Scientific name	Cellulose, %	Hemi-cellulose, %	Pentosan, %	Lignin, %	Ethanol/benzene extract, %	Ash, %	Silica, %
Scots pine	Pinus sylvestris	44	26	9	28	4	0.4	
European spruce	Picea abies	43	27	9	29	2	0.4	
Douglas fir	Pseudosuga menziesii	47	22	7	30	5	0.3	
Silver fir	Abies alba	43	27	11	29		0.5	
Silver birch	Betula verrucosa	46	36	25	19	2	0.3	
European birch	Fagus sylvatica	45	35	22	22	2	0.3	
Trembling aspen	Populus tremuloides	50	31	17	18	4	0.3	
Blue gum	Eucalyptus globulus	47	27	17	26	1.5	0.3	
Wheat straw	Tritium vulgare	38	36	28	19	4	5	3.5
Rice straw	Oryza sativa	32	36	25	12	4	16	12
Sugarcane bagasse	Saccharinum officinarum	38	36	27	21	3	2	1.5
Bamboo	Dendrocalamus strictus	35	32	18	26	5	3.5	2.5

Table 2. Advantages of softwoods and hardwoods for pulping and papermaking

	Softwoods	Hardwoods
Wood production	usually higher productivity	in some cases extremely high productivity (e.g., eucalyptus, poplar), higher stock stability
Wood	more uniform raw material, especially better stem formation, easier to debark, longer fibers, higher yields in acidic cooking	higher specific weight (e.g., eucalyptus, beech, birch) → lower transport cost → higher digester capacity → higher pulp output lower lignin content → easier to pulp → less energy and chemicals consumption higher yields in alkaline cooking
Pulp	better wet web strength, better drainage properties	requires less beating energy, rapid strength development during beating, better bleachability → fewer bleaching stages → lower chemical demand → less pollution (chlorine-free bleaching possible)
Paper	high strength properties, especially tear strength, better runability	better sheet formation (higher content of hemicelluloses), usually smoother surface, higher bulk, higher opacity → better printability

> 100:1 and are ideally suited to papermaking. On the other hand, the parenchyma cells, which occur only in small amounts in conifers, have unfavorable technological properties. They are very small and have an unfavorable length-to-width ratio of ca. 3 : 1.

The structure of hardwoods is much more heterogeneous. The most important cell type are the libriform (or scleremchyma) fibers, which account for 40–60 % of the tissue of most deciduous trees. The average length of hardwood fibers is usually less than 1 mm. The fiber width varies greatly, in the range 10–40 µm, depending on the type of wood. In addition, hardwoods contain vessels and, in comparison with conifers, a higher content of parenchyma cells. The vessels are very voluminous, with relatively thin, partially perforated walls and a low length-to-width ratio.

The cellular dimensions of annual plants are extremely variable. In comparison with wood, the cells are, in most cases, shorter and have thinner walls, and the parenchyma content is very high. A simplified survey of the cellular composition and morphology of various raw materials for pulp is presented in Figure 1 [17].

In the pulp and paper industry, a morphologically correct differentiation of cell types is not made. All cells are designated as fibers. The suitability of cells for papermaking depends essentially on the length-to-width ratio and the ratio of twice the radial cell-wall thickness to the

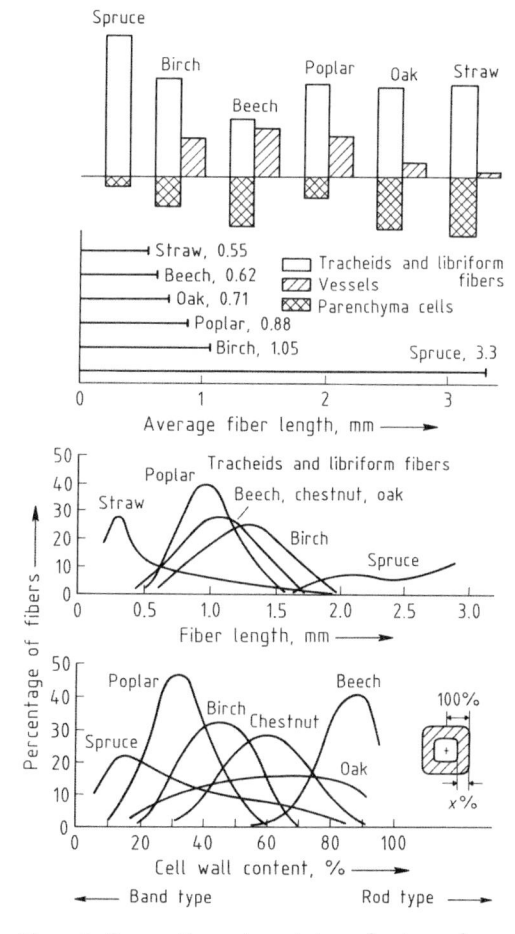

Figure 1. Composition and morphology of various pulp raw materials

lumen diameter. The latter is important for the flexibility or stiffness of a structural element. In this respect, considerable differences can be observed within the same cell type due to the seasonal growth rhythms and to the formation of specific cell structures because of special stresses in certain tissue regions. At the start of the growth period, cells having a large diameter and thin walls (springwood) are formed, while at the end of the vegetation period, thick-walled cells are produced (summerwood). Compressive and tensile stresses can also result in the formation of especially thick-walled cells that are capable of absorbing these stresses. The increase in the wall thickness of spruce tracheids during the growth period is shown in Figure 2.

The structure of the tissue elements influences the technological properties of the resulting paper in the following manner. In papermaking, the fibers suspended in water deposit on the wire of the paper machine to form a fiber web. Hydrogen bonds are responsible for bonding the fibers together. Apart from the presence of water, formation of hydrogen bonds requires hydrophilic groups for contact between the fiber surfaces, and small interfiber distances. The removal of the hydrophobic lignin from the surface of the fiber exposes the cellulose and hemicelluloses with their hydrophilic OH groups. The tensile strength and burst strength of a paper sheet depend primarily on the number of hydrogen bonds per unit volume between the fiber surfaces. Therefore, during paper sheet formation it must be ensured that the interfiber distance required for the formation of hydrogen bonds is attained. Springwood fibers with thin walls and large diameters (i.e., with large volumes) collapse during paper sheet formation because the lignin responsible for the stiffness of the cell wall was dissolved during digestion. Hence, the fiber becomes much more flexible, and its surface can touch neighboring fibers in large areas of the paper web, leading to increased hydrogen bond formation. This effect can be increased still further by mechanical treatment of the pulp in a beating step.

Summerwood fibers, having thick cell walls and being less flexible, hardly collapse and therefore form fewer hydrogen bonds per unit volume. However, they contribute greatly to the tear strength of the fiber web because of their inherent high fiber strength. In addition, they increase the volume of the paper and improve the optical properties by increasing the opacity of the paper. Thus, long, thick-walled fibers are required for high tear resistance of the pulp, while high fiber flexibility and a large surface-to-volume ratio are required for good interfiber bonding and high tensile strength. Therefore, fibers with a good length-to-width ratio have optimal properties for papermaking. This ratio should be at least 100 : 1. Figures 3 and 4 show the embedding of unbeaten and slightly beaten fibers in a paper web.

In Europe, sawmill residues and wood from thinning of forests are used in the production of wood pulp. In countries with good growth conditions, wood for pulp production is predominantly cultivated on plantations. Conifers, particularly spruce and pine, are preferentially

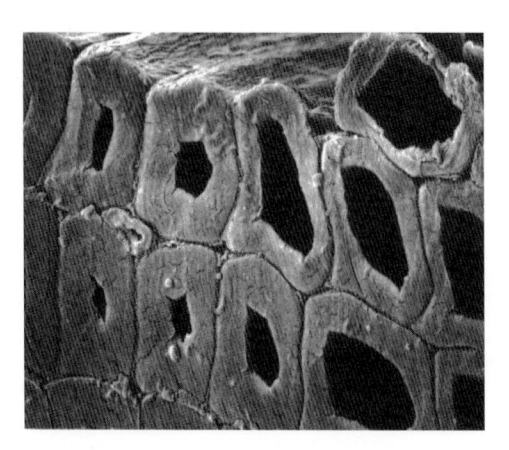

Figure 2. Increase of the cell wall thickness of spruce tracheids during the growth period

Figure 3. Electron micrograph of a test sheet made of unbeaten oak pulp

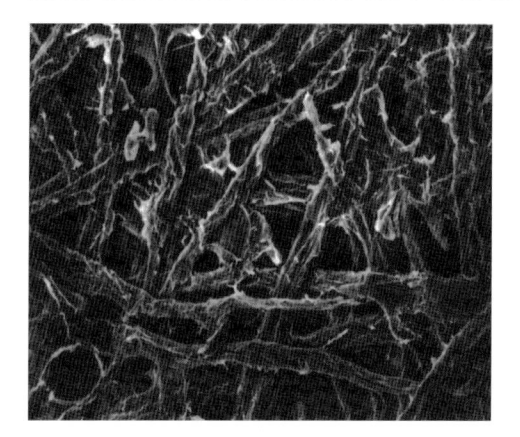

Figure 4. Electron micrograph of a test sheet made of oak pulp after 20 min beating

used because of their favorable morphological structure. However, pulps of good quality can also be made from hardwoods, such as birch, beech, eucalyptus, and poplar (see Tables 1 and 2). In particular, the importance of eucalyptus for the fiber industry should increase steadily because of its high growth rate and the excellent properties of pulps made from this wood [18–20]. Each type of wood is normally processed separately because, to achieve optimal pulp properties, the production process must be adjusted to the specific requirements of that particular type of wood. Mixtures of woods of greatly varying morphology and chemical composition, e.g., mixtures of hardwoods and softwoods or mixtures of hardwoods of widely varying wood density, lead to unsatisfactory digestion results and pulp properties.

1.2. Processing of Raw Materials

1.2.1. Processing of Wood

Wood is the most important raw material for the pulp industry and is obtained as logs, saw-mill residues, and chips. Logs normally enter the pulp mill with bark, but industrial residues should be bark-free.

Storage of Wood. Weeks and months often pass between wood harvesting and its processing in the pulp mill. In this time, biological respiration of living cells, chemical reactions, and attack by microorganisms can degrade the wood. The extent of wood degradation depends on many factors, the most important being the duration of storage, the storage conditions, the climate, and the type of wood.

In the storage of very fresh wood, especially from summer felling, the respiration of nutrients still present in the wood continues, generating heat, especially in the center of wood chip piles. Temperatures of 60 °C can be reached in only one to two weeks and in extreme cases temperatures as high as 90 °C have been observed. Spontaneous ignition can even occur under these conditions. After this increase in temperature, fungi and bacteria attack the wood, reducing the temperature again. Normally a temperature gradient arises, with the highest temperatures in the upper part of the pile.

Apart from microbial oxidation, chemical autooxidative and hydrolytic degradations also occur. In autooxidation, the wood components react with atmospheric oxygen; the final products are frequently organic acids. Substances that primarily undergo autooxidation are wood extractives, especially unsaturated fatty acids, resin acids, and terpenes. Although hydrolysis, particularly of carbohydrates, contributes only to a small extent to the generation of heat, it releases acetic acid from the hemicelluloses, which causes further hydrolysis.

Fungi and bacteria have the greatest destructive effect on wood. Most of the microorganisms that destroy wood have an optimum temperature between 25 and 50 °C; the most damaging microorganisms have an optimum of 30 °C. The optimal moisture content for microbial attack is 24 – 32 %. Degradation is accompanied by a reduction of strength of the wood and of the individual tissue components. Frequently, discoloration of the wood also occurs.

No general statements can be made about the losses of wood incurred during storage because of a great variety of interconnected factors. However, it can be assumed that wood losses of at least 1 % occur for each month of storage. Considerably higher values are observed for hardwoods than for softwoods, and the degradation of sapwood is greater than that of heartwood. The loss in digestion yield resulting from the decay of wood during storage is substantially greater. Microbial attack results not only in the loss of wood substance, but also in the reduction of the degree of polymerization of cellulose. This results in decreased pulp strengths. In addition, condensation reactions occur in lignin which lead to darkening of the wood and reduction of lignin reactivity. Hence, stronger digestion and bleaching conditions are required, which result in an increased degradation of carbohydrates.

The duration of storage of wood should be limited to two to three months to prevent storage damage. Since fungi do not attack highly moist wood, it is of advantage to store wood in a water saturated state. This applies in particular to the storage of chips. In principle, chemical treatment with wood preservatives is possible, but is now hardly considered for economic and environmental reasons.

Wood Debarking. The bark of a tree accounts for 10 – 20 % of the trunk, depending on the type of tree, its age, and the growth conditions. The bark must be removed before the wood is cooked because it interferes with the digestion process and reduces the quality of the pulp. Bark has a higher content of lignin and extractives and a much lower content of cellulose than wood. Up to 25 % of the tissue components of hardwood barks is fiber. However, softwood barks have a fiber content of at most 5 %. The presence of bark leads to an increased consumption of digestion chemicals, and reduces not only pulp yield, but also the technological properties and the brightness of the pulp. In addition, acidic digestion processes can only degrade and dissolve bark to a limited extent. Hence, the resulting pulp contains more impurities and requires stronger bleaching conditions. Bark, especially outer bark, contains considerable amounts of colored substances which usually have a phenolic structure. Bark phenols can cause serious digestion problems in the sulfite process. They react with lignin, preventing its sulfonation and dissolution. The kraft process can cope with the largest proportions of bark because most of the bark components are quite readily soluble in alkali. The higher the quality of pulp required, the lower the content of bark that can be tolerated. The use of bark-containing wood also has the disadvantage that dirt such as sand and stones enters the pulping process, and must subsequently be removed.

The efficiency of debarking depends on the structure of the bark, the strength of bonding between the bark and the wood, and the shape of the trunk. The most important influencing factor is the bonding between the bark and the wood, which depends on the time of felling and storage influences. In most cases, wood felled during the vegetation period can be debarked in the fresh state rather easily. Wood which is felled in summer and allowed to dry in its bark can be debarked only with great difficulty. However, considerable differences are observed between different types of wood. After exceptionally long storage, the bark can be removed from the wood without additional mechanical effort because of microbial degradation. However, overstored wood should not be used for the production of pulp. The debarking of wood during periods of frost causes great problems as frozen bark can hardly be removed from the wood. Therefore, wood of this type must be subjected to steam treatment before debarking. Wood which is suitable for debarking should be as straight as possible and contain only a few depressions and knots. Before debarking, the wood is cut into segments.

In the pulp industry, drum barking is the most frequently used debarking process (Fig. 5). In this process, small short logs are continuously fed into a slightly-inclined, horizontally rotating drum. The bark is sheared off by friction between the logs and between the logs and the inner wall of the drum. The bark falls through slots in the drum wall onto a conveyor belt below the drum. Although the spraying of water improves debarking, this process is usually not employed today because the resulting wastewater is polluted and

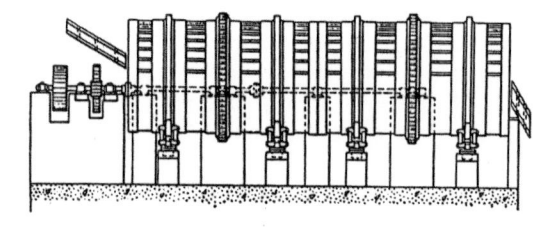

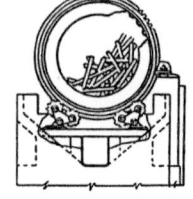

Figure 5. Drum barker

the heating value of the bark is reduced because of the increased moisture content. Drum barkers have diameters of 3 – 5 m, lengths of up to 30 m, and rotate at 5 – 10 rpm. Since the wood is subjected to intense mechanical stress in the drum, losses occur, but these do not normally exceed 3 % for normal wood. However, losses are much higher if the wood has been damaged by fungi or is very thin.

Ring barkers are normally used only for barking large-diameter logs (Fig. 6). In this process, logs are passed through a rotating ring and scrapers are hydraulically pressed against the surface of the trunk. Wood losses of 3 – 5 % can occur and, in addition, the barking efficiency for strongly adhering bark is frequently inadequate.

Knife barking machines and hydraulic barkers are no longer used in the debarking of wood. The output of knife barkers is too low and wood losses too high, while hydraulic barkers require too much energy and cause wastewater problems.

The debarking efficiency should be in the range of 85 to 95 %, depending on the digestion process used.

The bark obtained from the debarking process is usually utilized, either directly or after press dewatering, for the production of power in a bark combustion furnace.

The debarking of chips has been the subject of intensive studies, but the results were unsatisfactory. The debarking of chips is still an unsolved problem.

Production of Chips. In pulp production, the digestion of wood requires homogeneous impregnation with digestion chemicals. Therefore, chipping of wood is necessary. However, in the mechanical production of chips the wood can be damaged in various ways, and the resulting fiber damage influences the pulp quality. The smaller the chips, the more often the fibers are cut. The extent to which chipping reduces the average fiber length also depends on the average native length of the wood fiber. The smaller the average fiber length of the wood, the smaller the chips can be cut without reducing the average fiber length. Therefore, the chip size is always a compromise between the demand for chips that are mechanically as undamaged as possible and the necessity of achieving the quickest and most extensive impregnation with the digestion liquor. Moreover, the chips produced should have as homogeneous a size distribution as possible. Optimally dimensioned chips are 25 – 35 mm long and 3 – 7 mm thick.

The machines used most frequently are disk chippers (Fig. 7). They have a disk diameter of 1.2 – 4.5 m, with chopping knives fastened in holes in the disk. The number of knives per disk varies between 4 and 14 and the speed between 300 and 900 rpm. The wood is transported towards the rotating disk at an angle of about 45 ° to the longitudinal direction of the wood, and chopped into slices. Due to the wedge-shaped form of the knife and the position of the knife with respect to the cutting surface, the knife blade applies a shearing force in the longitudinal direction of the wood which results in a predetermined chip thickness. The length of the chips depends on how far the knife projects out of the chopping disk. The cut slices are flung outwards,

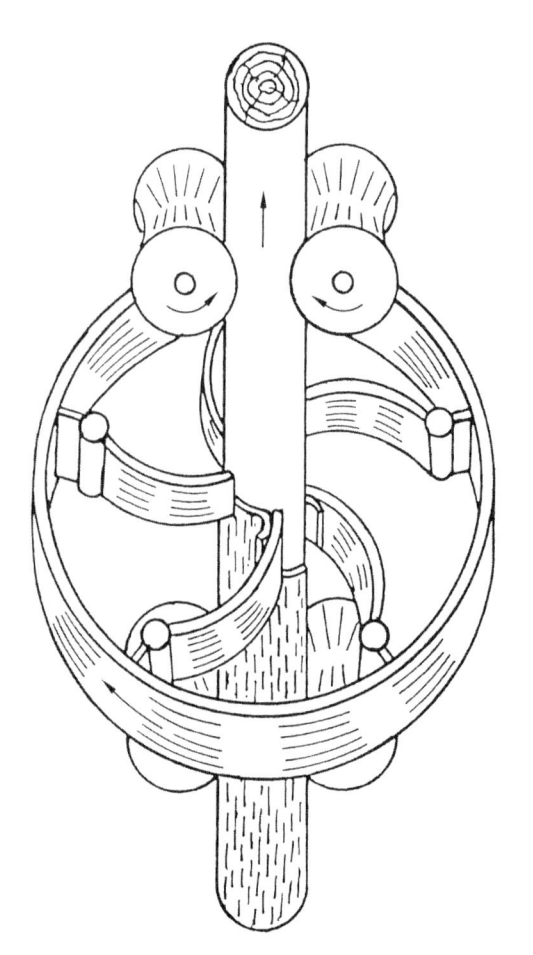

Figure 6. Ring barker

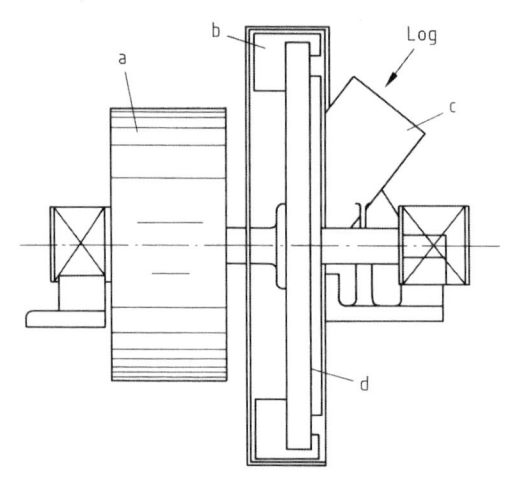

Figure 7. Schematic of a disk chipper
a) Flywheel; b) Crushers; c) Spout; d) Disk with knives

and are then disintegrated into chips by a crusher fastened at the periphery of the chipping disk. Disk chippers have a capacity of more than 100 m³ of wood per hour.

Drum chippers (Fig. 8) are used for chopping round wood and sawmill scraps in particular. The knives of a drum chipper are arranged on a rotating drum. They penetrate the wood from above at an angle of about 40 °; during chopping, the cutting angle changes constantly because of the rotary motion. The wood is transported into the chopper at right angles to the drum axis. Depending on the cutting angle, the chips have surfaces that are compressed to a greater or lesser extent. The proportion of fines is high.

A development that has considerably influenced sawmill techniques in the last years is the headrige cutter (Fig. 9). By using two parallel cutter heads of greatly varying shape, chips can be cut directly out of the trunk surface, resulting in a beam with two smooth cut surfaces. The knives can be arranged on a conical cutter head in

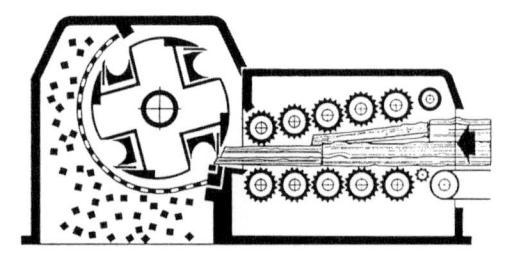

Figure 8. Schematic of a drum chipper (Courtesy of Pallmann Maschinenfabrik, Zweibrücken, Germany)

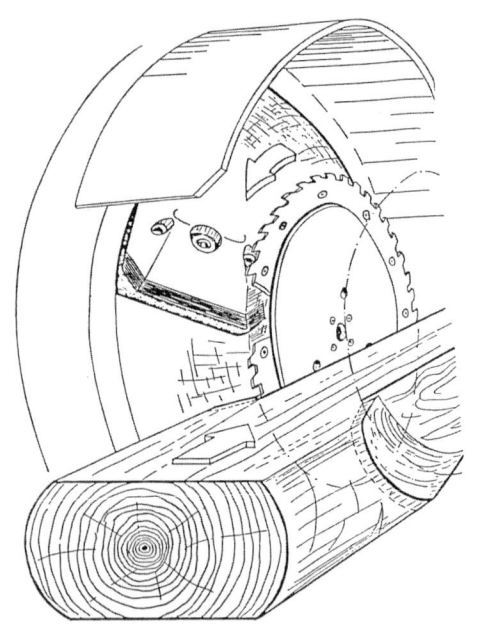

Figure 9. Schematic of a headrige cutter (Courtesy of Linck Holzbearbeitungstechnik, Oberkirch, Germany)

spirals or rings. The space between the cutter heads can be varied, depending on the dimensions of the log to be cut. The knives penetrate the wood at an angle of 35 – 40 °. As in the drum chipper, the cutting angle changes constantly due to the rotation of the cutting heads. In the final phase of cutting, the chips are removed from the wood parallel to the fiber direction, resulting in little fiber damage. The design of the knives determines both the length of the chips and their thickness, so that the production of chips with a homogeneous size distribution is possible.

The CCL chipper functions in a similar manner to the cutting heads of the headrige cutter. The two cutting heads are fitted together to form a V-shaped opening on the surface. The trunk is transported horizontally towards the rotating disk, and the knives cut in the same way as in the headrige cutter, except that the entire log is cut. With well-dimensioned wood and a constant cutting geometry, very homogeneously dimensioned chips having little fiber damage can be produced.

In chip production, not only are fibers shortened due to cutting, but damage also occurs within the chips. This damage includes primarily cracks in the middle lamella, and cracks and breaks in the inner, cellulose-rich cell wall layers

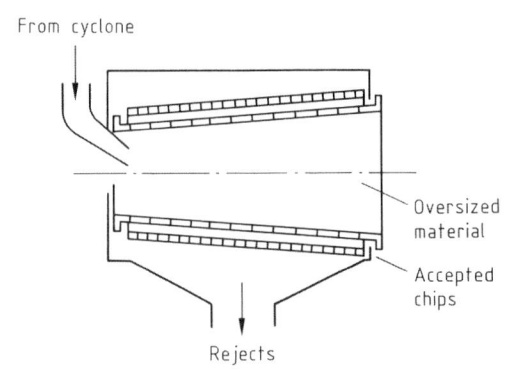

Figure 10. Rotary screen

due to the slipping of the parallel fibril layers. The latter, in particular, represents irreversible damage and allows the digestion chemicals to attack the carbohydrates directly. This results in poorer yields and strength losses. In thin-walled springwood fibers, damage is more pronounced than in the more stable summerwood fibers. This mechanical damage is particularly evident in acid sulfite digestion. It can reduce the strength of sulfite pulps by 10 – 20 %. Fewer negative effects are observed in alkaline digestion.

Chips must be screened after production. Oversized and fine components are normally separated by rotating or vibrating screens (Figs. 10 and 11). Oversized chips are rechipped and the sawdust fraction is used for another purpose, e.g., production of chipboard and fiberboard. The entire course of wood processing is shown in Figure 12.

1.2.2. Processing of Annual Plants

The processing of annual plants for pulp production poses specific problems [21]. Harvesting is limited to a few weeks a year; hence, a pulp mill requires an extensive store capacity to ensure an all-year supply. As a result of their morphological and chemical structure, most annual plants are attacked rapidly and intensively by microorganisms. To minimize degradation, these plants should be stored as dry as possible or in a water-saturated state. Straw and strawlike annual plants are usually stored as bales.

It is essential that the plants be thoroughly washed before processing to remove adhering soil and other impurities. Chopping to lengths of about 4 cm is usually carried out in a chipper.

Plant material of high pith content, especially sugarcane, must be subjected to pith removal because the parenchyma pith cells are unsuitable for papermaking. Pith removal can be achieved either in the moist or in the dry state. Wet pith removal in hammer mills and screens is more intensive, but pollutes the wastewater considerably. In the dry process, the starting material is crushed in disk mills and the pith is removed with screens or by air separation.

1.3. Pulping Processes

At elevated temperatures lignin, a large proportion of the hemicelluloses, and some of the cellulose are dissolved by the action of pulping chemicals on wood. The fiber yield is 45 – 90 %, depending on the intensity of the chemical action. The dissolution of lignin weakens the fiber matrix. The fibers can be separated easily after about 40 – 45 % of the wood substance has been dissolved. This stage is depicted in Figure 13 for the pulping of spruce chips.

The corresponding pulps are called chemical pulps. If the pulping process is continued and if in addition to lignin, the hemicelluloses are also dissolved extensively, pulp yields of less than 45 % are obtained, but the pure cellulose content is high. Pulps of this type are processed further into chemical products. Very mild pulping

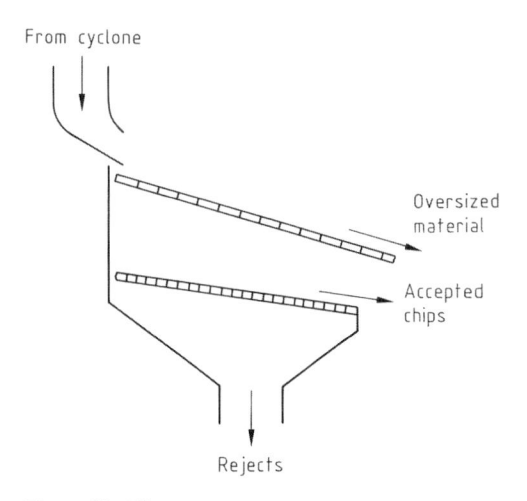

Figure 11. Vibratory screen

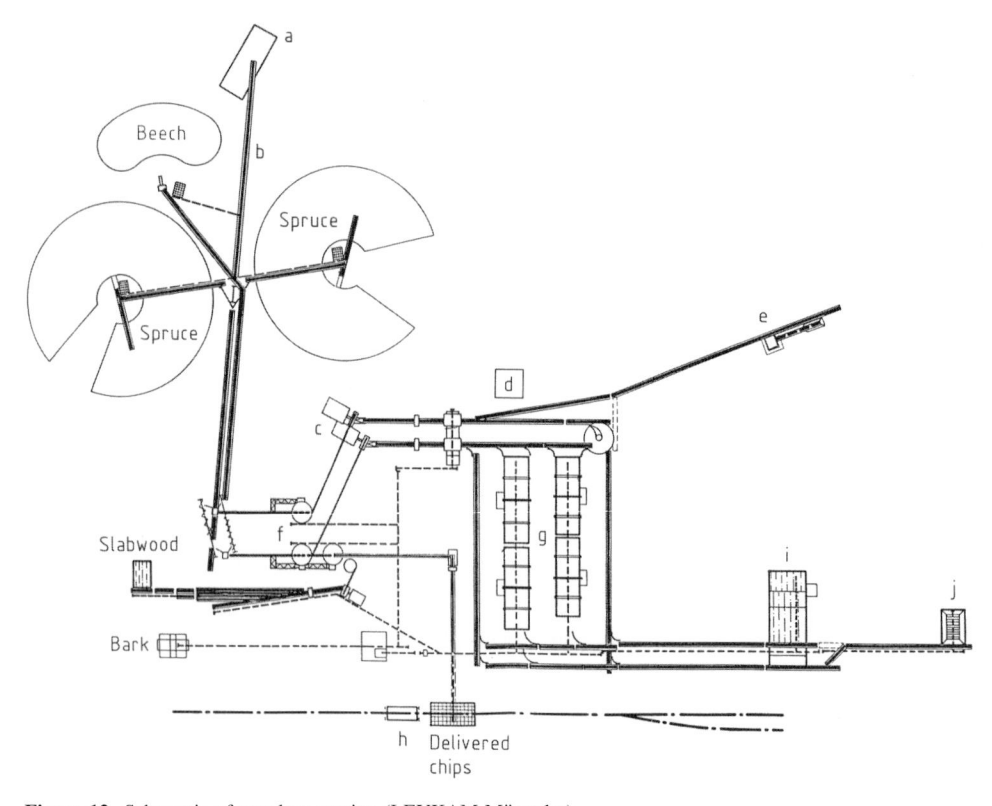

Figure 12. Schematic of wood processing (LEYKAM-Mürztaler)
a) Digester house; b) Chip storage; c) Chipper; d) Control room; e) Intermediate storage for debarked beech wood; f) Screening; g) Debarking; h) Tipping device; i) Saw; j) Log feeder

Figure 13. Electron micrograph of a spruce chip after the defibration point has been reached

conditions result in high yield pulps or semichemical pulps. The yields are in the range of 60 – 90 %, and the defibration point is not reached during digestion, i.e., the bonding substances in the wood are not rendered sufficiently soluble to disrupt the fiber binding and separate the cells from one another. Therefore, mechanical defibration must follow digestion. A survey of the various processes involved in pulp production is given in Table 3.

The most important pulping processes are the sulfate or kraft process, the acid sulfite process, and the soda process. The advantages and disadvantages of the kraft and sulfite processes are listed in Tables 4 and 5, respectively.

The sulfate or kraft process has become increasingly important and is now the principal pulping process, accounting for ca. 80 % of world pulp production. In comparison, the importance of the sulfite process has decreased steadily. Today, only 10 % of the world pulp production is obtained by this method. The soda process is

Table 3. Pulp production processes

Process	Raw material	Yield, %	Brightness, % ISO	Breaking length[a], m	Tear strength[a], cN	Uses
High-yield neutral sulfite	softwood	85–93	60–80	6000–7000	80–100	wood-containing printing paper tissue, fluff
Neutral alkaline sulfite		85–90	60–80	4000	60	writing and printing paper tissue
Neutral sulfite, semichemical	hardwood	70–80	60–80	7000	80	fluting for corrugated board
Sulfate, semichemical	softwood	55–70	30	8000	150	liner for corrugated cardboard
Sulfite, chemical	softwood	48	90	7000	70	wood-containing papers and boards
	hardwood	43	90	5500	60	wood-free papers
Kraft	softwood	43	88	9000	130	wood-free and wood-containing papers and boards
	hardwood	50	90	8000	90	wood-containing and wood-free papers and boards
Soda	straw	43	90	6000	45	wood-free fine papers

[a] Strength values based on a degree of beating of 30 °SR.

Table 4. Advantages and disadvantages of kraft pulping

Advantages	Disadvantages
Universal raw material basis	Small process flexibility
High insensitivity to bark	Low yields for softwoods
Fast pulping process, short cover-to-cover times	High residual lignin content and poor bleachability of the pulps
High yields for hardwoods	High bleaching chemicals demand
Good pulp strength	Indispensable use of chlorine-containing bleaching agents; high water pollution
Low extract content of the pulps	Offensive smell due to volatile reduced sulfur compounds

Table 5. Advantages and disadvantages of sulfite pulping

Advantages	Disadvantages
High process flexibility	Limited raw material basis
High yields for softwoods	High requirements on raw material quality, especially high sensitivity to bark
Good bleachability	Poor pulp strength
Low water pollution	Low pulping yields for hardwoods
Possible utilization of dissolved carbohydrates	Air pollution by SO_2 emission

used primarily for the pulping of annual plants and is responsible for 5 % of the total pulp production. The remaining pulp production is based on other processes.

1.3.1. Alkaline Pulping Processes

1.3.1.1. Kraft (Sulfate) Pulping

The sulfate or kraft process exhibits a higher selectivity of delignification and, for this reason, has almost completely replaced the older soda or sodium hydroxide pulping process. Only in the case of annual plants that are easily cooked has the soda process been able to assert itself because the sulfate process offers no distinct advantages [22].

The predominant position of the kraft process is, above all, due to its applicability to almost all lignocellulose-containing raw materials, its low sensitivity to variations in the quality of raw materials, and the excellent technological properties of the resulting pulps (see also Table 4). Furthermore, the original disadvantages of the kraft process, such as poor bleachability of the pulps and complicated recovery of chemicals, have now largely been eliminated.

The terms "sulfate" and "soda" process are derived from the make-up chemicals, sodium sulfate and sodium carbonate, which are added in the recovery cycle to compensate for chemical losses.

Pulping Conditions. The digestion liquor (white liquor) used in the kraft process consists of sodium hydroxide and sodium sulfide as active pulping chemicals, as well as sodium carbonate as residue from the causticizing step. Furthermore, small amounts of Na_2SO_4, Na_2SO_3, and $Na_2S_2O_3$ from side reactions are present in the pulping liquor. The following definitions are used to characterize white liquor, sodium equivalents expressed as Na_2O or $NaOH$ being used to calculate the chemicals employed.

Titratable alkali	$NaOH + Na_2S + Na_2CO_3$
Active alkali	$NaOH + Na_2S$
Effective alkali	$NaOH + 1/2\, Na_2S$
Sulfidity, %	$100 \times \frac{Na_2S}{NaOH+Na_2S}$
Causticizing efficiency, %	$100 \times \frac{NaOH}{NaOH+Na_2CO_3}$
Degree of reduction	$100 \times \frac{Na_2S}{Na_2S+Na_2SO_4}$

The sodium sulfide added is hydrolyzed almost completely to sodium hydroxide and sodium hydrogen sulfide

$Na_2S + H_2O \rightarrow NaOH + NaSH$

and, therefore, only half of the Na_2S is effective available alkali.

In comparison with the sulfite process (see Table 5), the kraft process is a very uniform method that exhibits little flexibility and is well suited to continuous production of a given quality of pulp. In particular, pH adjustments over a wide range are not possible. As a result of the large amounts of sodium hydroxide used, the pH value at the start of the digestion is between 13 and 14. It decreases continuously during the course of digestion because organic acids are liberated from lignin and carbohydrates during the pulping reaction.

The amount of chemicals required for pulping, their composition and the pulping parameters to be applied depend on the type of raw material used, the quality of pulp desired, and especially on the extent of delignification required. The production of semichemical or high-yield pulps requires between 10 – 15 % of active alkali, chemical pulp based on hardwoods requires 18 – 22 %, and the corresponding softwood pulp 20 – 25 %, calculated in each case as NaOH. The amount of sodium solution used should be calculated such that the pH does not fall below 10, because otherwise the dissolved lignin would reprecipitate on the pulp fibers. Nevertheless, a drop in pH towards the end of the digestion process is, as a rule, desired because then dissolved xylan fragments reprecipitate on the cellulose fibers [23–26].

A very important pulping parameter is the sulfidity of the white liquor. The optimal sulfidity depends on the raw material, the amount of chemicals used, and the pulping temperature. Since the Na_2S used in the kraft process improves the selectivity of pulping by accelerating the delignification reactions without simultaneously enhancing the dissolution of carbohydrates, pulping time can be shortened by maintaining a high sulfidity. This raises the pulp yield, and the higher hemicellulose content increases the tensile strength of the pulp. However, using more sodium sulfide leads to more odors due to the formation of reduced sulfur-containing gases, and increases the corrosiveness of the digestion liquor. A sulfidity of 25 – 40 % is normally applied in softwood pulping, while a lower sulfidity in the range 20 – 30 % is chosen for hardwoods.

Kraft pulping is usually carried out at 165 – 175 °C. The digestion time at the maximum temperature is in the range of 1 – 2 h, depending on the type of wood, the extent of delignification desired, and the digestion temperature. The heating of the digester can be carried out relatively quickly because, as a rule, the impregnation of the chips poses no problem due to the strong swelling of the fiber walls under the strongly alkaline conditions. The liquor to wood ratio influences the intensity of pulping through the chemical concentration. Depending on the chip size and the packing density of the chips in the digester, the pulping process (continuous or batch) etc., the liquor to wood ratio is adjusted to 2.5 – 4.5. A higher ratio favors uniform impregnation of the chips but reduces the delignification rate. As a result of the poor bleachability

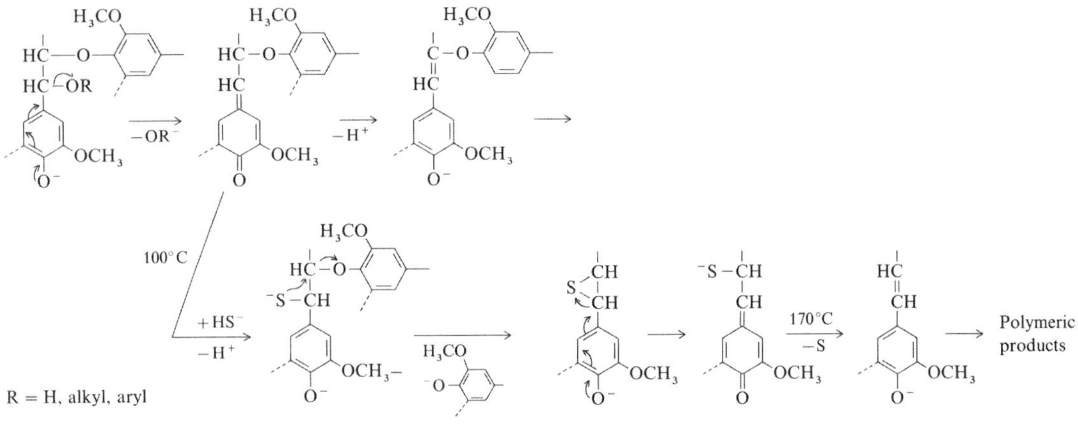

Figure 14. Main reactions of lignin during soda and kraft pulping

of kraft pulps and the pollution caused by pulp bleaching, efforts are being made to continue digestion to give the lowest possible kappa number so as to achieve an easier bleachability of the pulp. The kappa number is a measure of the residual lignin content of the pulp. The amount of lignin that remains in the pulp can be determined approximately by multiplying this number by the factor 0.15 for sulfate pulp and 0.165 for sulfite pulp.

Chemical pulps made from conifers (spruce and pine) are cooked to kappa numbers of 25 – 30, giving yields between 45 and 47 %. In comparison, hardwoods give yields of >50 % when delignified to kappa numbers of 15 – 20.

Pulping Reactions. Lignin is a macromolecule based on phenylpropane units, three-dimensionally cross-linked through various ether and C–C bonds (→ Lignin). The dominant bonds are α- and, above all, β-arylether bonds. A large proportion of these bonds must be cleaved in the pulping process to achieve the fragmentation and subsequent dissolution of lignin. Compared with the soda process, delignification is accelerated in the kraft process by the presence of strongly nucleophilic hydrogen sulfide ions, and the extent of lignin breakdown is increased, especially in the case of softwoods. Phenylpropane units with free phenolic OH groups are very quickly attacked. The initial step is the formation of a quinone methide structure, which already results in lignin degradation by cleavage of α-arylether bonds. The HS^- ion (OH^- ion) can subsequently add to the quinone methide intermediate and, as a strong nucleophile, can initiate the cleavage of an arylether bound to a neighboring β-C atom (Fig. 14). Phenolate anions formed by this cleavage reaction favor the continued breakdown of lignin.

Even β-arylether bonds in nonphenolic lignin units are cleaved, albeit more slowly, if an adjacent ionized hydroxyl group was originally present (OH groups in α- or γ-position) or were formed during the pulping reaction. During kraft pulping, some of the methoxyl groups of the phenolic ring are split off to give methyl mercaptan (CH_3SH), which can be converted subsequently to dimethyl sulfide (CH_3SCH_3) or dimethyl disulfide (CH_3SSCH_3). These compounds are responsible for the unpleasant odor surrounding kraft pulp mills. Apart from the cleavage of arylether bonds, carbon – carbon bonds are also cleaved to some extent. The degradation of lignin by the pulping chemicals competes with condensation reactions caused by internal nucleophiles, especially in the final phase of digestion. These reactions counteract further delignification by forming C–C bonds and greatly reduce the bleachability of the pulp produced [27–36].

The polysaccharides in the raw material are also attacked and degraded under the conditions of alkaline pulping. The carbohydrate losses incurred and the reduction of the degree of polymerization are caused essentially by two reactions: the depolymerization reaction and alkaline hydrolysis. The former is primarily responsible for yield losses, while the latter mainly causes a reduction of the chain length. In particular, the hemicelluloses are subject to degradation reactions because of their amorphous structure and their lower degree of polymerization. Galactoglucomannans and glucomannans are dissolved much faster than the fairly alkali resistant xylans [37, 38]. For this reason, the xylan-rich hardwoods give higher yields in alkaline pulping than the low-xylan conifers.

The acetyl side groups of the xylans of hardwoods and the galactoglucomannans of softwoods are hydrolyzed at the very start of digestion. The primary depolymerization reaction begins above ca. 100 °C. In this process, sugar units are successively cleaved in a β-alkoxy elimination reaction, commencing at the reducing end of the carbohydrate chains [39–42]. An average of 50 – 60 units are cleaved from the cellulose chain molecules until depolymerization is stopped by competing stopping reactions and formation of an alkali-stable carboxyl group at the end of the chain [43–45]. Chain scission reactions occur at temperatures above ca. 150 °C due to alkaline hydrolysis, which lowers the degree of polymerization of cellulose [46–48]. The ends of the chains, exposed by the hydrolytic cleavage of glycosidic bonds, are the starting point for the secondary depolymerization reaction. With increasing digestion time and more extensive lignin dissolution, the carbohydrate degradation reactions dominate and, because of the reduced selectivity of the pulping reaction, digestion must be terminated to avoid excessive losses of yield and pulp strength.

The carbohydrate losses caused by the depolymerization reactions can be limited by reduction or oxidation of the aldehydic end groups of the cellulose to give alkali-stable alcohol or carboxyl groups [49–52]. Although reduction by the addition of 1 – 2 % sodium borohydride, based on o.d. (oven dry) raw material, produces up to 6 % higher pulp yields [53, 54], the high cost make this process uneconomical. In comparison, oxidation by the addition of 3 – 4 % polysulfide is used on an industrial scale in some cases. A 6 % increase in yield can be obtained by polysulfide digestion, which, however, also leads to an increased emission of odorous sulfur compounds [55–59]. The use of anthraquinone (AQ-catalyzed pulping) stabilizes carbohydrates against degradation by oxidizing the end groups.

Anthraquinone-Catalyzed Pulping. Anthraquinone (AQ) which, when used in catalytic amounts of 0.05 – 0.2 % based on o.d. wood in alkaline pulping, causes efficient stabilization of carbohydrates and simultaneous acceleration of the delignification reaction [60, 61]. The addition of 0.05 % AQ to softwood pulping increases pulp yield at a given kappa number by 2 – 3 % [62, 63]. Although various AQ derivatives are also effective catalysts for alkaline pulping, they are less cost effective than anthraquinone [64].

The effect of anthraquinone is based on the AQ/AHQ (AHQ = anthrahydroquinone) redox system. Already in the initial phase of pulping, anthraquinone oxidizes the reducing end groups of the carbohydrates to give aldonic acid end groups and thus stabilizes them towards degradation by the depolymerization reaction. The monoanion of the anthrahydroquinone produced in this reaction can add to the α-C atom of the lignin quinone methide intermediate. AHQ monoanion can initiate the cleavage of β-aryl-ether bonds in the same way as HS^- ions. AHQ is oxidized in this redox reaction, regenerating AQ for further redox cycles [65–70].

Anthraquinone was first used on an industrial scale in 1977. In the meantime it has been employed in more than 70 pulp mills on a temporary or permanent basis [71, 72]. Since anthraquinone is relatively expensive and cannot be recovered because it is decomposed extensively in the pulping process, the decision to use AQ depends on the wood costs, the pulp prices attainable, and on the specific production conditions. In principle, many different objectives can be pursued with the use of AQ in the sulfate process:

1. Increase in pulp yield and therefore enhanced production in the case of limited capacity for evaporation or combustion of waste liquor
2. Reduced wood consumption per tonne of pulp
3. Energy savings by use of lower pulping temperatures or shorter digestion times
4. Reduced pulping chemicals demand per tonne of pulp
5. Reduction of sulfidity in order to reduce the formation of volatile, odorous sulfur compounds
6. Increase in the extent of delignification in order to improve the bleachability and reduce water pollution caused by bleaching effluents

In comparison to kraft pulping, a far greater effect can be achieved by the use of anthraquinone in the soda process. During lignin degradation, AHQ acts in a similar way to hydrogen sulfide ions. Therefore, the rate and extent of delignification is increased to a much greater amount by the addition of AQ to the soda process. In fact, the application of the soda AQ process permits the production of bleachable chemical pulp from softwoods. The same applies to alkaline and semialkaline sulfite pulping (see Section 1.3.2.4) [73–78].

Prehydrolysis Kraft Process. The chemical industry requires dissolving pulps with a high α-cellulose content for processing into cellulose products. If the kraft process is used, the wood hemicelluloses must be degraded by acid hydrolysis prior to digestion because the alkaline pulping conditions stabilize the hemicelluloses remaining in the pulp towards further alkaline attack. In contrast, the hemicelluloses in sulfite dissolving pulps can be removed from the fiber matrix by an alkaline purification step. Prehydrolysis can be carried out by using mineral acids (e.g. 0.25 – 0.5 % H_2SO_4 based on o.d. wood) at 120 – 140 °C, or by steam or water treatment at 160 – 170 °C. In the latter case, acetyl groups are split off from the xylans of hardwoods or the galactoglucomannans of softwoods, resulting in a pH of 3 – 4. Whereas cellulose is very stable to such acid treatment, hemicelluloses are broken down into smaller, water-soluble chains. Bonds between hemicelluloses and cellulose and lignin are hydrolyzed as well. The average degree of polymerization of the hemicelluloses is decreased by about 30 % during prehydrolysis. If softwoods are used, ca. 10 – 15 % of the organic matter is dissolved in this step, and with deciduous woods the losses are 15 – 20 %. In particular when hemicellulose-rich hardwoods are used, the large-scale utilization of the hydrolysate obtained is of interest. Dissolved sugars and their degradation products, such as furfural, hydroxy-methylfurfural, levulinic acid, acetic acid, formic acid, and methanol, are present in the hydrolysate.

The subsequent kraft pulping is carried out analogously to the process used in the production of paper pulps. The digestion conditions depend primarily on the required α-cellulose content of the pulp. The hemicelluloses still present in the wood after prehydrolysis are subject to intensive degradation, especially through the depolymerization reaction that can commence at the numerous carbonyl groups formed during prehydrolysis.

Table 6. Prehydrolysis conditions for hard- and softwood

	Charge, %/o.d. wood	Temperature, °C	Time, h
Prehydrolysis with mineral acids			
Hardwood, H_2SO_4	10 – 30	70 – 80	0.5 – 3
Hardwood, H_2SO_4	0.3 – 0.5	120 – 140	0.5 – 3
Hardwood, HCl	0.5 – 0.7	95	0.5 – 3
Softwood, H_2SO_4	0.25 – 0.5	130 – 140	0.5 – 1
Prehydrolysis with liberated organic acids			
Hardwood		160 – 170	0.5 – 3
Softwood		170 – 175	0.5 – 1

Softwoods, and hardwoods, and low-ash annual plants are suitable raw materials for prehydrolysis kraft pulping. Hardwoods have a higher hemicellulose content than softwoods and usually give lower yields, but they can be much more easily and extensively delignified. Also, the risk of lignin condensations occurring during prehydrolysis is far lower. These condensations impede the subsequent pulping and bleaching steps. The conditions for the prehydrolysis of hardwoods and softwoods are summarized in Table 6 [79–85].

1.3.1.2. Soda Pulping of Annual Plants

The most important annual plants for the pulp industry are agricultural waste materials (bagasse and straw) and naturally occurring or cultivated bagasse and reeds. Cultivated fiber plants, such as kenaf, sisal, jute, hemp, and cotton, are valuable raw materials for the production of specialty pulps [16, 21, 86–88].

For the pulping of annual plants, the soda process is usually employed. The kraft and neutral sulfite processes are used less frequently. The acid sulfite process is not applied because it produces brittle pulps with a high ash content and inadequate strength properties.

Since annual plants are impregnated easily and have a low content of reactive lignin, the amount of chemicals required is lower than in wood pulping. In soda pulping, 10 – 15 % NaOH, based on the raw material, is normally employed at a pulping temperature of 160 –170 °C. The yields attainable are in the range 40 – 55 % and are influenced greatly by the type and quality of the raw material, especially by the lignin content and the proportion of parenchyma cells. A high parenchyma content of the raw material not only reduces the yield, but also increases alkali consumption and diminishes the dewatering properties and strength of the pulp by increasing the formation of fines. For this reason, bagasse is subjected to a depithing process to eliminate parenchyma pith cells before pulping.

While various fiber plants such as hemp, kenaf, jute, and sisal give high strength pulps, sugar-cane bagasse and straw pulps have lower strengths than wood pulps because of their short, thin-walled fibers. However, these pulps are commonly added in the production of fine papers to give the paper certain characteristics, e.g., a smooth, closed surface [89].

1.3.2. Sulfite Process

The sulfite process is characterized by its high flexibility. In principle, the entire pH range can be used for pulping by changing the dosage and composition of the chemicals. In comparison, kraft pulping can be carried out only with highly alkaline digestion liquor. Thus, the use of sulfite pulping permits the production of many different types and qualities of pulps for a broad range of applications (Table 7).

Today, the term sulfite process also refers to the many different variations of this process. Previously, this term was reserved for the acid (calcium) bisulfite process, which was used almost universally. The sulfite process can be subdivided into the following types of pulping:

1. acid bisulfite pulping,
2. bisulfite pulping,
3. neutral sulfite pulping,
4. alkaline sulfite pulping,
5. multistage sulfite pulping, and
6. anthraquinone-catalyzed sulfite pulping.

All these processes are based on the use of aqueous sulfur dioxide and a so-called base, i.e., calcium, magnesium, sodium, or ammonium. In the pulping system, pH-dependent equilibrium reactions occur that depend on the ratio of SO_2 to base. If sodium hydroxide is added to the aqueous sulfur dioxide, first bisulfite and then sulfite are formed:

$H_2O + SO_2 \rightarrow SO_2 \cdot H_2O$

$SO_2 \cdot H_2O + NaOH \rightarrow NaHSO_3 + H_2O$

$NaHSO_3 + NaOH \rightarrow Na_2SO_3 + H_2O$

Table 7. Sulfite pulping processes

Process	pH range	Base	Active agent	Digestion temperature, °C	Digestion time, min	Pulp yield, %	Special applications
Acid (bi)sulfite	1 – 2	Ca^{2+}, Mg^{2+}, Na^+, NH_4^+	$SO_2 \cdot H_2O$, H^+, HSO_3^-	125 – 143	180 – 420	40 – 50	dissolving pulp, tissue, printing paper
Bisulfite	3 – 5	Mg^{2+}, Na^+, NH_4^+	HSO_3^-, H^+	150 – 170	60 – 180	50 – 65	printing paper
Neutral sulfite (NSSC)	5 – 7	Na^+, NH_4^+	HSO_3^-, SO_3^{2-}	160 – 180	25 – 180	75 – 90	corrugated medium
Alkaline sulfite	9 – 13.5	Na^+	SO_3^{2-}, OH^-	160 – 180	180 – 300	45 – 60	kraft-type pulp
Two-stage sulfite Stora							
1st stage	6 – 8	Na^+	HSO_3^-, SO_3^{2-}	135 – 145	126 – 360	50 – 60	greaseproof paper
2nd stage	1 – 2		$SO_2 \cdot H_2O$, H^+, HSO_3^-	125 – 140	120 – 240		
Sivola							
1st stage	3 – 4	Na^+	HSO_3^-, H^+	140 – 150		35 – 45	dissolving pulp
2nd stage	7 – 10		SO_3^{2-}	160 – 180	60 – 180	55 – 65	kraft-type pulp
AQ – sulfite	9 – 13	Na^+	SO_3^{2-}, OH^-, AHQ	170 – 175	150 – 240		
(NS – AQ, AS – AQ)							

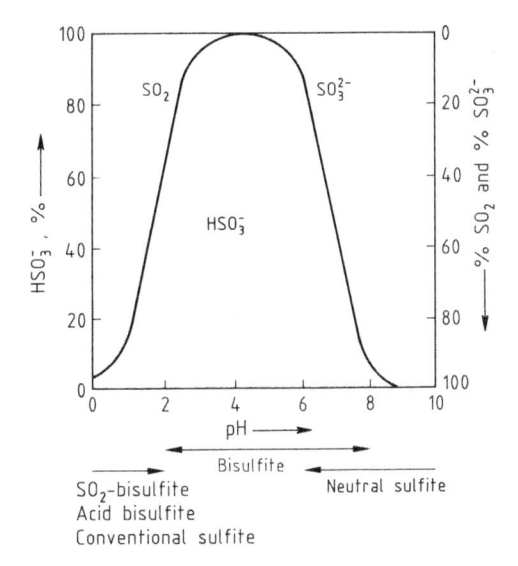

Figure 15. Composition of a dilute SO_2–H_2O solution versus pH at 20 °C

A pure bisulfite solution is present at pH 4.5 and a pure unbuffered sulfite solution at approximately pH 9 (see Fig. 15). At these inflection points, a slight change in the chemical composition leads to a large pH shift [90]. As shown below, practically the entire pH range can be used for sulfite pulping with Na as "soluble" base (the same applies to NH_4).

Chemical composition	pH
$SO_2 \cdot H_2O/NaHSO_3$	1.5 – 4.5
$NaHSO_3$	4.5
$NaHSO_3/Na_2SO_3$	4.5 – 9.0
Na_2SO_3	9.0
Na_2SO_3/Na_2CO_3	9 – 12.5
$Na_2SO_3/Na_2CO_3/NaOH$	10 – 13
$Na_2SO_3/NaOH$	13 – 14

Since calcium sulfite is only slightly soluble, this base can only be used in acid bisulfite pulping, i.e., in the presence of excess SO_2; Ca$(HSO_3)_2$ precipitates at pH values above 2.0 – 2.3. In comparison, magnesium bisulfite solutions are stable up to pH 5 – 6, so that acid bisulfite as well as the bisulfite (magnefite) and multistage pulpings can be carried out. Due to the rise in the pH value with increasing pulping temperatures [91], the pH of the cold pulping liquor must be in the range of 1.5 (acid bisulfite) and 4.0 (bisulfite) to avoid precipitation in the digestion phase.

In practice, apart from the chemical composition and the pH values, the terms free, combined, and total SO_2 are used to characterize sulfite solutions. The content of theoretically base bound bisulfite and sulfite ions is referred to as combined SO_2. The difference between the total SO_2 and the combined SO_2 is then free SO_2. According to this definition, half of a pure bisulfite solution is free and the other half is bound SO_2.

1.3.2.1. Reactions in Sulfite Pulping

In sulfite pulping, the reactivity of lignin is essentially determined by the pH of the pulping liquor. Fundamental differences exist in the mechanism, rate, and extent of delignification under acidic, neutral, and alkaline conditions. Regardless of the pH, the α-position of the quinone methide intermediate formed from the lignin phenylpropane units is always the most reactive site for attack.

Under acidic pulping conditions, aqueous sulfur dioxide is the active pulping chemical that causes extensive sulfonation, especially by addition to the α-C of the propane side chain. Bisulfite ions can also take part in the sulfonation reactions. The initial step involves the protonation and subsequent cleavage of α-hydroxy, α-alkoxy, and α-aroxy groups leading to the formation of resonance stabilized carbonium – oxonium ions. Since these ions are very reactive, the weakly nucleophilic pulping reagents can readily add to the α-C atom (Fig. 16).

Other sulfonation reactions can occur on the γ-C aldehyde groups or on the α-C carbonyl groups of the propane side chain. The solubility of the lignin molecule is increased by the hydrophilic sulfo groups. However, the fragmentation of lignin is restricted to the cleavage of non-cyclic α-arylether bonds between the lignin units. In acid sulfite pulping, the β-arylether bonds are essentially stable due to the absence of strong nucleophiles, and methoxy groups attached to the phenolic ring are also not subject to cleavage [36, 92–94].

Condensation reactions of nucleophilic lignin fragments compete with the sulfonation reaction. Both reactions are favored by low pH values. The danger of lignin fragments undergoing recondensation to give higher molecular, partially degradable units exists, especially if the amount of

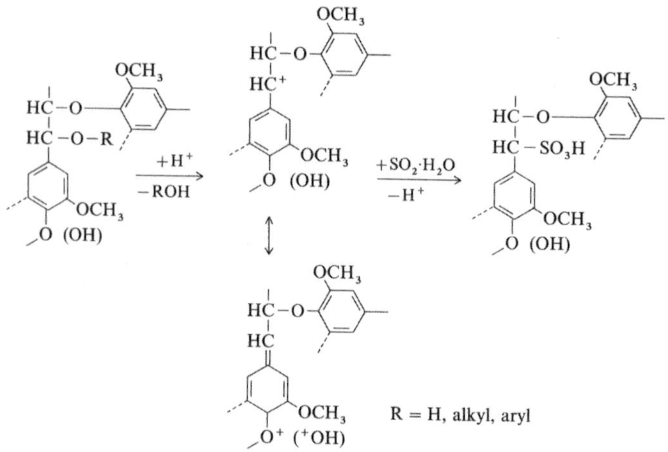

Figure 16. Reactions of lignin during acid sulfite pulping

combined SO$_2$ available at the end of digestion is not sufficient. The resulting "black digest" can be avoided by raising the bisulfite ion concentration, i.e., by increased addition of base. Phenolic extractives from the bark or heartwood constituents such as pinosylvin (3,5-dihydroxystilbene) and its methyl ether from pine heartwood, or taxifolin from the heartwood of larch and Douglas fir, readily condense with lignin and can greatly impede acid bisulfite pulping or bring it to a complete standstill.

In bisulfite pulping (pH ca. 4), these condensation reactions can be largely suppressed, and a more selective, but slower sulfonation is achieved by the bisulfite ions. The main disadvantage of bisulfite pulping is the lower degree of delignification that can be attained.

In the neutral sulfite process, the reactions of lignin are restricted to the phenolic structural units, which are converted to quinone methide structures, as in alkaline pulping. After sulfonation at the α-C atom, the neighboring β-aroxy group can be slowly cleaved sulfolytically [95]. The nonphenolic β-arylether units, however, are resistant to neutral sulfite pulping [29]. Since practically only phenolic lignin structures react, only semichemical pulps with a high residual lignin content can be produced.

In the alkaline sulfite process, which is carried out in the presence of NaOH (NH$_4$OH), the cleavage of α-ether bonds is also restricted to the phenolic lignin structures. Apart from the sulfolytic β-arylether cleavage described above for the neutral sulfite process, β-aryl ether (β-O-4) bonds of nonphenolic units are also cleaved to a certain extent by the action of alkali, analogous to the reaction in the soda process. Consequently, chemical pulps similar to kraft pulps can be produced by alkaline sulfite pulping.

The reactions of carbohydrates in sulfite pulping are also determined by the pH value. At high pH values, rapid deacetylation of hemicelluloses takes place and alkali-induced degradation reactions (see Section 1.3.1) occur. In the neutral pH range, carbohydrates are attacked to the lowest extent, but their susceptibility increases considerably with increasing acidity. The glucosidic bonds are readily cleaved by acid, especially under the conditions of acid bisulfite pulping. Hemicelluloses in particular are subject to hydrolytic degradation because their glycosidic bonds are more easily accessible and less resistant than the bonds of cellulose. With increasing degradation, highly depolymerized hemicelluloses dissolve and are hydrolyzed step by step until monomeric sugar building blocks are obtained. The depolymerization of cellulose leads essentially to a reduction in the degree of polymerization (DP value); only under drastic conditions are soluble cellulose fragments formed. In general, acid pulping of softwoods gives higher yields because galactoglucomannans, the dominant hemicelluloses, are relatively acid resistant, whereas the xylans occurring predominantly in hardwoods are degraded much faster [96–98].

The monosaccharides formed in pulping are not completely stable and are partially dehydrated and oxidized. Aldonic acids in particular are produced by bisulfite oxidation. The thiosulfate formed by the reduction of bisulfite can, in turn, react with reactive lignin structures and contribute to lignin condensation [99–101].

1.3.2.2. The Various Sulfite Processes

Acid Bisulfite Process. In the acid bisulfite process, pulping is performed with a large excess of free SO_2 at pH 1.2 – 1.8. All of the above-mentioned bases can be employed. No basic differences in their mode of action exist, but the pulping conditions considered to be optimal vary slightly. The course and the result of pulping are influenced by the base selected [102, 103]. For instance, magnesium bisulfite pulping results in the best yields and the highest pulp brightness. Use of the monovalent bases Na^+ and NH_4^+ results in stronger pulps and a lower shive (fiber bundle; see section 2.4.1) content. If ammonium is used as base, the pulping process is very fast, but the pulps are dark in color and cannot be used in the unbleached state. Pulps of the lowest strength are obtained with calcium as base, and this pulping process must be controlled more carefully because calcium precipitates readily.

The total SO_2 used in acid bisulfite pulping is in the range 18 – 25 %, depending on the type of wood used, degree of digestion desired etc., and the SO_2 to base ratio is adjusted to 4.5 – 6.0. During the digestion phase, the liquor to wood ratio is usually 3 – 4 : 1. The chips are first impregnated with a higher liquor ratio to achieve a homogeneous distribution of the pulping liquor. A part of the pulping liquor (side relief) and the SO_2 gas (top relief) is then removed after impregnation and returned to the accumulator to prepare fresh digestion liquor. Pulping temperatures of 125 –140 °C are used for the production of paper pulps. Unlike bisulfite pulping, the acid bisulfite process is also well suited to the production of dissolving pulps because the hemicelluloses can be dissolved to a large extent in the highly acidic pulping process. However, higher maximum temperatures of 145 – 150 °C are employed for this process. The digestion time is usually 2 – 5 h at this pulping temperature, but can be extended to more than 7 h at lower pulping temperatures. A long heating time is also necessary because the SO_2 penetrates into the wood much faster than the base due to the high vapor pressure of the excess free SO_2. Rapid heating can result in lignin condensations due to the high acidity within the chip centers. Since digester relief must also be effected slowly so that the excess SO_2 can be reused in the subsequent digestion, the result is a longer cover-to-cover cycle of 8 – 12 h.

From its first industrial application in 1874 until the end of World War II, the calcium bisulfite process was the principal process for wood pulping used worldwide because of the low costs and good availability of limestone and sulfur dioxide as basic materials for the production of the digestion acid. Since the pulping chemicals were inexpensive, it was not necessary to recover them before discharging the sulfite waste liquor into rivers or lakes. At the beginning of the 1950s, factories in North America started using ammonium as the base and it was possible to use the resultant waste liquor as fertilizer. Today, calcium bisulfite pulping, ammonium bisulfite pulping, and other sulfite pulping processes based on ammonium are of very little importance. If there is no possibility of utilizing the waste liquor, it is burnt for reasons of environmental protection and process energy production. However, base recovery is pursued neither in the calcium bisulfite process nor in the ammonium sulfite process.

The recovery of calcium fails because of the high temperatures of more than 1150 °C required to completely decompose the gypsum, formed during waste liquor combustion, to calcium oxide and sulfur dioxide. For ammonium sulfite waste liquor, semitechnical recovery processes based on ion exchange have been developed. However, these processes have not gained acceptance in industry because of high costs. Thus, recovery in both processes is limited to the portion of SO_2 that is not bound to the base, i.e., a maximum of 80 % of the sulfur dioxide charge is recoverable.

Therefore, the use of ammonium as base was only a temporary alternative. With the development of recovery processes for magnesium and sodium sulfite liquors the corresponding pulping processes have increasingly gained acceptance for economic and environmental reasons. Moreover, compared with the calcium bisulfite process, the possibility exists that improved paper pulp qualities and higher yields can be achieved and the utilizable range of raw materials can be extended by increasing the pH value of the pulping liquor.

Bisulfite Pulping. The pulping liquor used in the bisulfite process consists of approximately equal parts of free and combined SO_2, i.e., it contains no excess free SO_2. The pH value is between 3.5 and 5.5, depending on the temperature and the chemical concentration. The use of calcium as base in this pH range is not possible. Magnesium is usually employed as base (magnefite process) [104] because the recovery of chemicals is easy and inexpensive. Ammonium bisulfite [105] and sodium bisulfite pulping [106] (Arbiso process) have also found industrial applications in individual cases.

Bisulfite pulping is normally carried out with 4 – 5 % total SO_2 in the digestion liquor and a liquor to wood ratio of 4 : 1, i.e., the total SO_2 used, based on wood, is 16 – 20 %. The pulping temperature is ca. 155 °C for hardwoods and 165 °C for softwoods. The higher temperatures required, compared with the acid bisulfite process, compensate for the lower rate of delignification caused by the lower acidity of the pulping liquor. However, carbohydrates are also dissolved to a smaller extent during bisulfite pulping. Hence, these pulps have a high hemicellulose content, and yields can reach 60 %. In comparison with acid bisulfite pulping, shorter digestion times of 1 – 3 h are required because of the higher digestion temperatures. The heating period is also considerably shorter since the risk of lignin condensation is lower and, therefore, the digestion liquor can be pumped at higher temperature into the digester and heated more quickly to the maximum temperature. On the whole, the cover-to-cover cycle is much shorter (approx. 6 – 8 h), which increases the production capacity.

Since phenolic components of bark and heartwood interfere with bisulfite pulping only to a small extent, the raw material base is considerably wider than for the acid bisulfite process. In addition to spruce and fir, pine, Douglas fir, and larch can be used in the production of long-fiber pulps. Bisulfite pulps are brighter, stronger, and better suited to papermaking than the corresponding acid bisulfite pulps. However, these pulps are not suitable for further processing in the chemical industry because of their high hemicellulose contents. A serious disadvantage of the bisulfite process is that the extent of delignification attainable is lower than that obtained under highly acidic pulping conditions, since otherwise the losses of yield and strength that occur are excessively high. The higher residual lignin content of the pulp requires the use of more bleaching chemicals and hence produces a greater effluent charge. As a result of stricter environmental controls, many mills that had already switched from the acid bisulfite to the bisulfite process reversed their decision.

Neutral Sulfite Process. The NSSC (neutral sulfite semichemical) process is not suited for making chemical pulp because the degree of delignification attainable is too low. This process uses sodium as base and is widely applied in the production of semichemical pulps, especially in North America. These pulps are employed mainly for the middle layers of corrugated cardboard and in newsprint [107, 108]. Neutral sulfite pulp mills are sometimes used in combination with kraft mills, which allows the combined recovery of chemicals (cross recovery). The NSSC waste liquor can be used as make-up chemical to compensate for losses of chemicals in the kraft process.

The neutral sulfite process is a two-stage process with a chemical delignification step followed by mechanical defibration. In the chemical treatment, neutral to slightly alkaline sulfite solutions are used to dissolve hemicelluloses and lignins and, thus, weaken the bonds between the fibers. In this way, the mechanical effort required to expose fiber bundles and individual fibers is reduced considerably.

Hardwood is the raw material of choice; it can be digested easily and gives bright pulps with high stiffness and crush strength. Annual plants, such as straw and bagasse, and softwoods can also be used. Conifers produce pulps of good quality, but the chemicals requirement is higher and more energy is needed for defibration and subsequent beating.

In the neutral sulfite pulping of hardwood, between 8 and 18 % of sodium sulfite or ammonium sulfite is used, based on o.d. wood. To maintain the pH in the range 7 – 9, the digestion liquor must be buffered by the addition of 2.5 – 5 % of alkali ($NaHCO_3$, Na_2CO_3, or NaOH) because acids are released during digestion, particularly acetic acid by the cleavage of acetyl groups from hemicelluloses. The digestion time is 30 – 180 min at the maximum temperature (ca. 170 °C). The reaction proceeds very slowly and only 25 – 50 % of the lignin and about 30 – 45 % of the hemicelluloses are dissolved. Accordingly, high yields of about 65 – 75 % are obtained for deciduous woods and up to 85 % for conifers.

Alkaline Sulfite Pulping. Alkaline sulfite pulping is carried out with liquors containing sodium sulfite and sodium hydroxide (sodium carbonate) at initial pH values of 9 – 13. When, as in NSSC pulping, a low chemical charge, low alkalinity, and low temperature are applied, semichemical pulps can be prepared, especially from annual plants. The boundaries between the two processes are fluid. If the pulping liquor consists of approximately equal amounts of sodium sulfite and NaOH, and if strongly alkaline conditions (pH >13) are chosen, chemical pulps can be produced as well. This pulping exhibits characteristics typical of both the sulfite process (no odors due to formation of reduced sulfur compounds; good bleachability of pulps) and the kraft process (no raw material restrictions, high pulp strengths, low yields, and low brightness) [109]. However, the higher chemicals requirement, longer pulping times, higher residual lignin content, and the more complicated chemicals recovery, compared to the kraft process, prevented establishment of this process in industrial practice. However, the especially pronounced effect of anthraquinone in alkaline sulfate pulping may lead to renewed interest in this process.

1.3.2.3. Multistage Sulfite Processes

The various modifications that can be achieved in the sulfite pulping system by adjustment of the pH can be advantageously combined by changing the pH during pulping. In this way, it is possible to produce special pulps that are ideally suited to certain fields of application. Sodium, as a soluble base, is particularly suitable for multistage digestions of this type. Magnesium can be used as base within limits. In the two-stage systems, pulping is either carried out first at a higher pH and the pH is subsequently decreased or vice versa. The Stora process, applied for the first time in the Skutskär factory of Stora-Kopparsberg, has achieved commercial importance [110, 111]. In this process, a sodium bisulfite – sulfite solution at pH 6 – 8 is used in the impregnation and first digestion stages. Subsequently, excess pulping liquor is removed and the pH is adjusted to ca. 1.5 by introducing SO_2. This is followed by normal acid bisulfite pulping as the main delignification step. The reactive centers of lignin are sulfonated practically completely in the first stage, and thus condensation reactions with phenolic extractives under strongly acidic conditions are blocked, so that even pine heartwood, for instance, can be digested satisfactorily by this method. Furthermore, glucomannan stabilization takes place in the bisulfite – sulfite step such that the acetyl groups are cleaved first, allowing the glucomannans to attach themselves firmly to the cellulose fibrils by increased formation of hydrogen bonds [112–114]. The resistance to acid hydrolysis is thereby increased. However, this stabilization effect is restricted essentially to softwoods, with their high galactoglucomannan content. An increase in yield of over 5 % can be achieved in this manner. The very bright, easily bleachable pulps are specially suited to the production of greaseproof paper because of their high hemicellulose contents and their resultant fast beatability.

A corresponding process can be carried out with magnesium as base [115]. The term FB process results from the very fast beatability, a characteristic property of these pulps. Analogous to the process based on sodium, pretreatment in a slightly acidic to neutral medium is followed by pulping at a highly acidic pH. As a result of the limited solubility of magnesium sulfite, the pulping liquor is pumped into the digester at pH 5 for the first step. The required pH value of 5.5 – 6 is then adjusted in the digester by the direct addition of magnesium hydroxide. In the second pulping stage, the pH is lowered by addition of SO_2.

The second type of process involving an increase in the pH in the second pulping stage is the Sivola (Rauma) process [116–118]. In a two-stage procedure, a pH 3 – 4 sodium bisulfite solution is applied in the first stage, and the pH value is subsequently adjusted to 7 – 10 with sodium hydroxide. A three-stage variant of the Sivola process, with a somewhat higher pH in the first stage, an intermediate acid bisulfite stage, and a final neutral or slightly alkaline stage is especially suited to extractive-rich wood species, such as pine. This method can be used to produce dissolving pulps with a particularly high α-cellulose content [119, 120]. Dissolving pulps can also be produced by the two-step process, and it is also possible to make many different paper pulps by altering the pulping conditions. Hemicelluloses can be dissolved to a great extent by using a high pH in the second stage. The tear strength of the pulps thus obtained rivals that of kraft pulps. On the other hand, the hemicelluloses are largely retained in the pulp by second-stage digestion under neutral conditions, so that higher yields are achieved.

A corresponding two-stage process based on magnesium is called the HO process (high opacity). Impregnation with a bisulfite solution (pH 3.8 – 4) is carried out in the first stage. The pH is then adjusted to 6.5 – 7 by the injection of magnesium oxide milk and digestion is completed at a high pulping temperature (170 °C) [121]. Since dissolved organic acids apparently stabilize the pulping liquor, no precipitation occurs in spite of the high pH.

1.3.2.4. Anthraquinone-Catalyzed Sulfite Pulping

An exceptional feature of the sulfite pulping system is that the catalytic activity of anthraquinone is also exerted at slightly alkaline to neutral pH values, whereas effective acceleration of delignification in soda pulping requires a pH above 13 [122, 123]. This explains how sulfite pulping in a moderately alkaline pH range, which would otherwise permit only inadequate delignification, can be used for the production of chemical pulps if small amounts of AQ (0.1 – 0.2 %) are added [124–127, 73–78]. The pulping liquor used in this process consists predominantly of sodium sulfite. Sodium hydroxide solution and/or sodium carbonate is added as additional alkali. The alkali ratio (sodium sulfite/total alkali) is preferably 0.8 – 0.9. In comparison with the AS – AQ process with NaOH (pH 10 – 13.5), the term NS – (neutral sulfite) AQ process is frequently used when sodium carbonate is employed as the buffer below pH 10. The transition is, however, fluid and both process variants require alkaline conditions (final pH 7 – 10.5).

This pulping procedure is faster and more selective than alkaline sulfite pulping with approximately equal sodium sulfite and sodium hydroxide charges and without AQ. Better delignification is achieved in spite of the lower chemicals demand (20 – 25 % calculated as NaOH/o.d. wood) and shortened digestion times (150 – 240 minutes).

AQ-catalyzed sulfite pulping combines the advantages of the sulfite and kraft processes. First, there is no raw material restriction and high pulp strengths are achieved. Second, high yields are obtained, the pulps are easily bleachable and the emission of malodorous sulfur compounds is avoided. The main disadvantage of this process is the relatively high residual lignin content in the pulps, which have a high chemicals demand for bleaching and require the use of chlorine-containing bleaching agents.

1.3.3. Other Pulping Processes

Environmental problems and the high investment costs for pulp mills (€ 1500 – 2000 $t^{-1} a^{-1}$) have led to the development of new pulping processes. The aim is to develop a process that can produce high strength pulps with good yields from all lignocellulose-containing raw materials that are suited to fiber production. The lignin content of these pulps must be low and the residual lignin must be highly reactive to permit easy, high-brightness bleaching. An important criterion is also the bleachability with chlorine-free chemicals. This makes possible the treatment of the bleach effluent together with the waste digest liquor in the recovery stage and reduces the fresh water demand of a pulp mill to a minimum. Other aims are to reduce the cost of pulp production by reducing investment costs, e.g., in the fields of chemical recovery and utilization of useful byproducts from the dissolved wood components present in the black liquor.

1.3.3.1. Nitric Acid Pulping

The first investigations made by PAYEN (1838) showed that wood can be delignified with nitric acid. The breakdown of lignin proceeds via electrophilic substitution on the aromatic rings and subsequent oxidation reactions. Apart from the partial cleavage of methoxy groups and propane side chains, the ether bonds between the lignin units are also partially cleaved. As a result of the oxidation reaction, the nitrated lignin fragments possess mainly a quinoid structure and are highly colored. Alkali treatment is required to dissolve the partially water soluble nitrolignin.

Nitric acid pulping has achieved importance only to a modest extent in small pulp mills for the pulping of hardwoods and annual plants. The process was used for the first time on an industrial scale in the 1930s in Wolfen, Germany. Dissolving pulps with a high cellulose content (98 – 99 %) were produced from beech wood with a yield of ca. 40 %. The chips were first impregnated for 3 h under pressure at 40 – 45 °C with 15 % nitric acid. After the removal of excess acid and reduction of the acid concentration to 4 %, the chips were digested for 7 h at 70 – 95 °C. To dissolve the nitrated lignin, a two-step alkali extraction was then performed with 1 % NaOH for 60 min at 70 °C, followed by 3 % NaOH for 180 min at 80 °C.

Apart from this process, numerous other nitric acid pulping processes and modifications have been developed which differ primarily in the acid concentration, pulping temperature and time, and the extraction conditions [128–132].

Nitric acid pulping can be accelerated greatly by the addition of sulfuric acid or aluminum sulfate so that the digestion time for pressured digestion at 110 °C can be reduced to a few minutes. The alkali extraction step can be enhanced by the addition of oxygen, and ammonium hydroxide can be used instead of sodium hydroxide. The black liquor obtained in this way has a high nitrogen content and can be utilized as long-term fertilizer. Although the amount of nitric acid required has been reduced drastically from 550 kg/t, e.g., to 130 kg/t of pulp, by changing the original pulping procedure, this method may be considered economical only if cheap nitric acid is available and special pulps with a high market value can be made.

1.3.3.2. Pulping with Organic Solvents

Alcell Process. The use of low-boiling alcohols in wood pulping was proposed in the 1930s by KLEINERT, who applied mixtures of ethanol and water to dissolve lignin [133]. In the 1970s, this method [134–136], in a modified form, was adopted in many new pulping processes [137–159]. The Alcell process [137–141] comes closest to Kleinert's original idea: hardwoods are cooked with a 50:50 ethanol – water mixture. Three extractors in series are used in which the chips are delignified by successive exposure to different pulping liquors. Since the solution applied first has already been used previously in two extraction stages, it has the highest content of dissolved wood components. Fresh solution is employed in the third delignification stage. The exchange of pulping liquor is effected by displacement. The alcohol is recovered by condensation of the gases during relief of the digester after the third extraction stage. The alcohol remaining in the pulped material is recovered by introducing steam into the extractor.

The waste liquor from the first extraction step is processed, the alcohol is distilled off, and the lignin that then precipitates is separated from the carbohydrates remaining in the aqueous solution by centrifugation. In this process, lignin and low-molecular carbohydrates and their reaction products can be recovered in addition to pulp. Depending on the type of wood, 15 – 20 % lignin, 7 – 10 % xylose, and 1 – 3 % of other sugars based on o.d. wood are recovered from the black liquor.

At still acceptable pulp strength, the kappa number of the pulps is in the range 25 – 30. The yield for beechwood pulping is 50 % (based on o. d. wood) and thus, less than that given by kraft pulping. The relatively high lignin contents make it difficult to bleach the pulp to high brightness when exclusively chlorine-free bleaching agents are used. Alcell pulps subjected to conventional bleaching with chlorine-containing chemicals have a 10 % lower breaking length and at least 20 % lower tear strength than comparable kraft pulps. Until now, use of the Alcell process has been restricted to the pulping of hardwoods, which are easy to delignify. Due to lower costs for recovery of chemicals, it can be assumed that plant costs are 10 – 15 % lower than those of kraft pulp mills. The economy of this process can be further improved by alternative use of lignin and carbohydrates. However, this process cannot be regarded as a generally applicable alternative to the kraft process because the strength properties of the pulps are markedly lower than those of kraft pulps. Also, the restriction to hardwoods limits the applicability of the process.

Organocell Process. The limited capacity of aqueous alcohol solutions to dissolve lignin led to the development of organosolv processes, which involve the addition of conventional pulping chemicals. The Organocell process is a two-stage process [143–145]. In the first pulping stage, the chips are treated with a mixture of methanol and water (50 : 50) at 190 °C for 20 – 40 min. In this stage, equal amounts of lignin and carbohydrates, accounting for 10 – 20 % of the wood substance, are dissolved.

In the main pulping stage, the methanol content of the liquor is lowered to 35 vol % and ca. 20 % sodium hydroxide and 0.1 – 0.2 % anthraquinone based on o.d. wood are added. In this stage, the pulping time for softwoods is 40 – 60 min at ca. 170 °C. The yield and the residual lignin content of the pulp are comparable to the kraft process. Apart from pulp production, the utilization of substances in the black liquor is an objective of this process. Distillation is used to separate the alcohol from the pulping liquor from the first stage. Lignin precipitates in this process and can be removed by centrifugation. The carbohydrates remain in the aqueous solution. In this manner, about 6 % lignin and 5 % carbohydrates based on o.d. wood can be recovered in softwood pulping. Since the second stage requires sodium hydroxide, the carbohydrates dissolved in this stage are degraded extensively, and only lignin can be recovered from the black liquor. After methanol distillation, lignin precipitates after acid treatment or in the electrolytic recovery of sodium hydroxide. Since the recovery of sodium hydroxide by electrolysis consumes a lot of energy and the lignin recovered cannot be sold at a cost-covering price, the simplest solution is conventional black liquor treatment with evaporation, combustion, and recovery of chemicals. The recovery process, being free of sulfur, is unproblematic.

Organocell pulping can be used to process the same range of raw materials as the kraft process. However, it has disadvantages with regard to the pulp yields and the extent of delignification. The tear strengths of organocell pulps are comparable to those of kraft pulps, but the tensile strengths are much lower.

ASAM Process. Like the AS – AQ (NS – AQ) process, sodium sulfite is used as the main delignification agent in the alkaline sulfite process with the addition of anthraquinone and methanol (ASAM). Sodium hydroxide, sodium carbonate, or a mixture of the two can be added to increase the alkalinity of the pulping liquor. The optimal ratio of sodium sulfite to added alkali varies between 70 : 30 and 80 : 20, depending on the type of wood to be pulped and the desired pulping result. In addition to anthraquinone, used in amounts of 0.1 % based on o.d. wood, 10 – 25 vol % of methanol is added to the pulping liquor to intensify the delignification process.

The same range of raw materials as in the kraft process can be digested at 175 – 180 °C in 2 – 3 h. Both the yield and the pulp strength, especially the breaking length, are superior to those obtained from the kraft process. However, the essential advantages of the ASAM pulps are the lower residual lignin content (kappa numbers between 10 and 20), the relatively high brightness and the easy bleachability. Even completely chlorine-free bleaching sequences can produce high brightness levels [152–155].

1.3.3.3. Pulping with Organic Acids

Both formic acid and acetic acid, with or without the addition of peroxides, can extensively remove lignin from wood. In all cases, concentrated acids are required, and temperatures range from 100 to 190 °C. The use of lower temperatures has the advantage that pressureless systems can be employed. In this case, however, an acid catalyst (e.g., HCl, aluminum chloride) or the addition of peroxide is required. Very low kappa numbers can be achieved in acetic acid pulping. The pulps as well as the dissolved lignin and the hemicelluloses are acetylated. Dissolved carbohydrates and lignin components can be separated and processed into byproducts. Chlorine-free bleaching in the acetic acid system can follow directly if peroxides are used. Another bleaching chemical that has proved effective under these conditions is ozone. The pulps have fairly good tensile strengths; the tear strengths are similar to those of sulfite pulps [160–167].

1.4. Technology of Pulp Production

1.4.1. Pulping

Wood pulping can be carried out in a continuous or batch process. The steel digesters are protected internally against the corrosive digestion liquor. In older batch digesters, acid resistant tiles or carbon bricks were used for the digester lining.

A lining of this type reduces the digester volume by 7 – 15 %. In addition, the bricks and tiles must be inspected regularly because they can crack easily, especially at the joints, due to the high temperature differences during heating up and cooling down. Today, digesters are usually plated on the inside with stainless steel. Batch digesters (Fig. 17) have a volume in the range of 100 – 400 m³. Rotary digesters are older types with smaller volumes. They are rotated head over in order to thoroughly mix the chips with the digestion liquor, and are heated directly by the introduction of steam into the digester. In this way, the liquid volume increases and the concentration of chemicals is reduced.

In more modern pulp mills that use batch pulping systems, the digesters are arranged upright in batteries and each is equipped with a liquor circulation system. The liquor is removed at the bottom of the digester and pumped into the top of the digester. A heat exchanger in the circulation cycle is used for indirect heating. Digesters with liquor circulation systems and direct heating are very rare.

The investment costs for conventional batch pulping systems are lower than for continuous systems. However, it is more difficult to maintain the constant process conditions that are required for uniform pulping results. Moreover, the heat economy of batch pulping systems is unfavorable. To reduce or completely eliminate these disadvantages, energy saving systems have been developed for batch digestion systems, which use the heat content of the waste liquor to heat up the next digester. This can be achieved either by heat exchangers or by collecting the black liquor in thermally insulated collecting tanks and using it for heating chips. In this way, the residual pulping chemicals in the black liquor are also made available for pulping. Apart from improved heat economy, impregnation of the pulping material is also increased. The chemical dosing during the entire cook can be better adjusted to the pulping requirements, reducing side reactions, especially with the carbohydrates. Pulps made by such liquor displacement can be delignified to a greater extent. They have better yields and very good strengths.

Modern pulp technology is dominated by continuous digesters. The leading system worldwide was developed by Kamyr (Fig. 18). In this process, chips are continuously fed into upright digestion tubes from the top while the pulp is removed at the bottom of the digester. digestion systems of this type have a very wide capacity range of 200 – 2000 t/d. All conventional pulping processes can be conducted in this system, and many different types of pulp, ranging from semichemical pulp to dissolving pulp, can be produced. Accordingly, both the design and the process mode of this system can be modified.

The standard design functions as follows. Chips are introduced continuously via a low-pressure valve into a presteaming vessel, in which they are heated with low-pressure steam (150 kPa) to prepare them for subsequent impregnation. The chips are transferred to an impregnation tower through the chip shute together with the pulping liquor via a high-pressure valve. Complete impregnation of the chips takes place

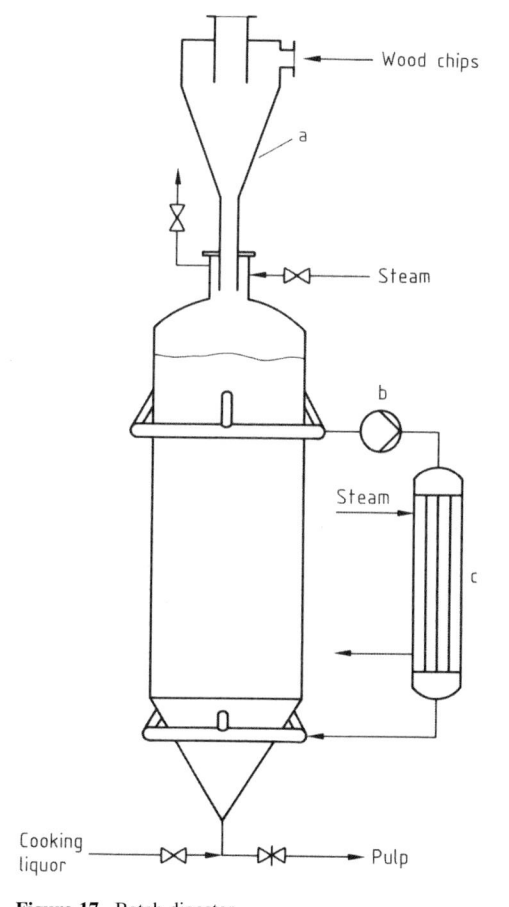

Figure 17. Batch digester
a) Wood chip collector; b) Circulation pump for digestion liquor; c) Heat exchanger

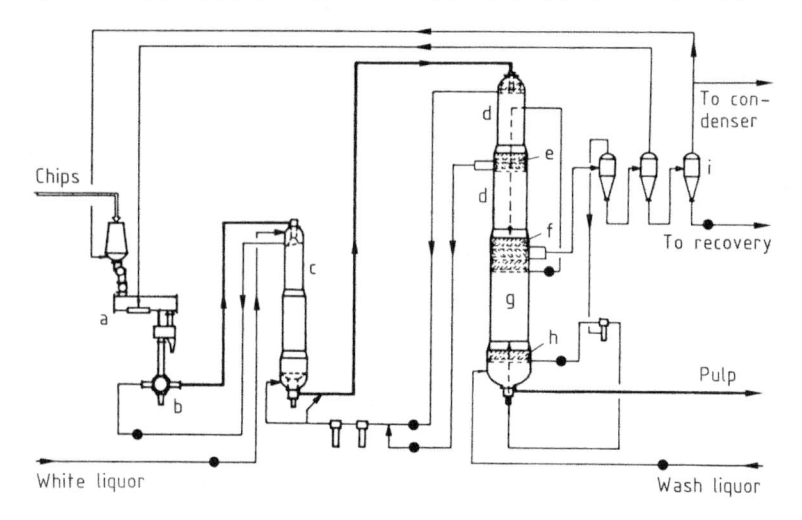

Figure 18. Continuous digester (Kamyr system)
a) Presteaming vessel; b) High pressure valve; c) Impregnation tower; d) Digester; e) Strainers for recirculation of digestion liquor; f) Strainers for waste black liquor; g) Washing zone; h) digestion liquor outlet; i) Heat exchangers

here. The chips are discharged from the bottom of the system and are conveyed to the digester together with the liquor. They are then heated to the maximum digestion temperature. The chips are transported from the top of the digester downwards and pulped in the process. The pulping occurs in the interchanging action of liquid and vapor phase digestion. Recirculation of the pulping liquor is achieved by means of a system of concentric tubes in the center of the digester and outlet screens set in the wall. Varying numbers of these screens are arranged at different levels in the digester. In each case, the inlet tubes end slightly above the outlet screens, and the pulping liquor is fed centrally into the digester. As a result of pressure differences, it flows from the center to the outlet screens in the wall. This liquid cycle causes the reaction products to be washed out of the chips and replaced by new pulping chemicals, which react in the vapor phase between the screens.

Wash liquor is added in the lower part of the digester and moves upwards countercurrently to the stream of chips. Thus the spent liquor present in this region is displaced, and the pulp is subjected to preliminary washing. In addition, this technique serves to reduce the fiber temperature, and only little fiber damage occurs during pulp discharge through a valve by rapid pressure relief. The wash liquor used comes from washers located behind the digester. The black liquor is withdrawn through the last screen ring and fed into the liquor evaporation system via a heat exchanger.

Apart from the Kamyr system, various companies have developed continuous pulping systems for the production of semichemical and chemical pulps based on both wood and annual plants. In addition to upright digesters, inclined and horizontal digesters are also employed, in which a conveyor system such as a screw, is used to transport chips and chemicals. All these systems are only partially suited to the production of chemical pulps from wood because liquid-phase pulping cannot be carried out due to the lack of liquor recycling systems. All the pulping chemicals must be homogeneously distributed throughout the raw material before pulping starts. Subsequently, pulping occurs in the vapor phase without exchange of chemicals.

1.4.2. Washing and Screening of Pulp

Brown-Stock Washing. In the production of chemical pulps, 45 – 55 % of the wood substance fed into the digester is dissolved. The remainder is present as fibers, largely still in the form of the original chips. In addition, ca. 200 – 500 kg of pulping chemicals per tonne of pulp are present, depending on the process and the extent of delignification. It is essential for economic and environmental reasons that the dissolved wood substances and the pulping chemicals are separated from the pulp as completely as possible with a minimum amount of wash water. The black liquor contains organic components of high heating value, whose combustion can cover all or most of the energy requirements of a pulp mill. For economic operation of the mill, the digestion chemical make-up demand should be as low as possible. The washing efficiency must be as high as possible to minimize the pollution load of the mill effluent stream. However, this objective must be achieved with a minimum of wash water because its evaporation before the subsequent black liquor combustion is an energy consuming process.

The washing process is influenced by many factors, e.g., the type of wood, the chip size, the packing density of the chips in the digester, the extent of delignification, the concentration of the black liquor, amount of wash water, flow rate, temperature, washing time, diffusion gradient, and the pH.

Pulp washing can be divided into two steps. The first stage involves the displacement of the free spent digest liquor. This occurs either with the wash water or by pressing the pulped material. The second step involves the removal of the black liquor from the chips or fibers. The washing process occurs largely by diffusion since the free flow of the wash water is impeded by the tissue structure of the wood. The washing process must be controlled such that both steps are carried out under optimal conditions. In other words, the free black liquor must first be displaced from the brown stock to establish a concentration gradient between the liquor inside the chips and the liquor surrounding the chips. Then optimal diffusion conditions must exist to ensure the fastest possible removal of the dissolved wood substances and the pulping chemicals from inside the chips and fibers. In practice, these two washing steps, displacement of free waste liquor and subsequent diffusion, cannot be clearly separated from the viewpoint of time and space. In the displacement of free black liquor, it is unavoidable that the black liquor mixes with the wash liquor. A region of low black liquor concentration is established where diffusion occurs again on account of the concentration gradient. The washing is influenced considerably by the accessibility of the chips and the individual fibers after pulping. The higher the delignification, the easier it is to wash the black liquor solids and the chemicals out of 'the chips. A very important factor for the washing efficiency is whether the process is dominated more by displacement washing or by diffusion. If the wood substance is still present as chips, the dissolved substances require a correspondingly longer time to find their way out of the chips and into the wash water. Diffusion is of importance primarily in the removal of dissolved substances from the cells. The rate of diffusion depends on the concentration gradient between the black liquor in the cell wall and in the wash water. Moreover, the washing temperature is of significance. At elevated temperatures, the solubility of the substances to be washed out increases and the viscosity of the wash water decreases. The diffusion rate can be doubled by increasing the temperature by 30 %, which favors high-temperature washing. However, this can be carried out only in closed systems. If the pulped material is still present as chips during washing, diffusion plays the dominant role and the time required to achieve the effective removal of black liquor is relatively high. In the case of defibrated material, displacement of black liquor by the wash water dominates.

Washing Systems. In the early days of pulping, washing was carried out in the digester or in the blow pit into which the digesters were emptied. The amount of wash water used did not have to be restricted because the waste liquor was not monitored; the dissolved wood substance and the pulping chemicals were simply discharged into the receiving stream. Today, a washing efficiency of 99 % must be attained with dilution factors of 2 (wash water/pulp) or less. This is possible only in closed systems in which the repeated use of the wash water takes place in a countercurrent washing process.

In the continuous pulping system described above, pulp is prewashed in the lower part of the digester with wash filtrate from a subsequent washing stage. The washing efficiency attainable at this stage is already above 80 %. The pulp is then fed into a single- or multistage diffusion washer (Fig. 19). Since this is a closed system, effective, high-temperature washing can be performed above 100 °C. The pulp enters the diffuser at the bottom with a consistency of 5 – 10 %. In the interior, concentric perforated outlet screens are to be found over the entire area. The

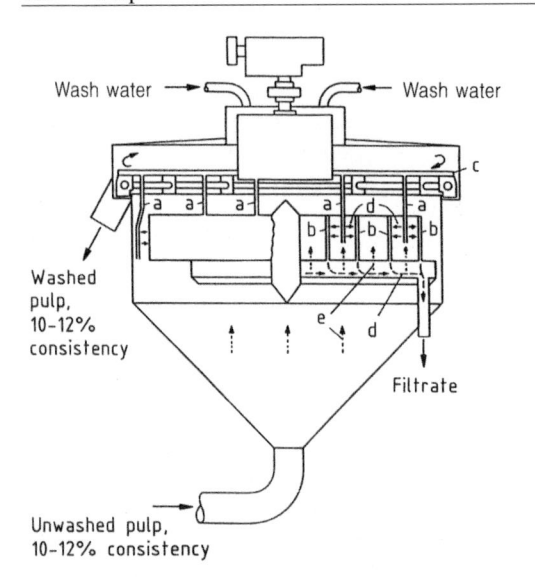

Figure 19. Diffusion washer
a) Rotating nozzle for wash water addition (attached to (c));
b) Concentric screen rings for removal of wash water;
c) Rotating pulp scraper; d) Wash water outlet; e) Upflowing pulp

wash water is injected into the system by rotating jets between the outlet screens and diffuses through the pulp to the outlet screens, where the waste liquor is withdrawn. Several of these washing units can be placed at different levels in one diffuser and operated in series, with fresh water being used only in the last step. The system is operated countercurrently and the pulp leaves the diffuser from the top. The digester washing and a subsequent two-step diffuser give a washing efficiency of more than 99 %.

The standard washing in a pulp mill is carried out in a washing line which operates with rotary filters (Fig. 20). In this process, drums are used that are closed at the sides and have a screen spanned over their shell surfaces. The drum dips into a chest containing the pulp suspension at a consistency of ca. 4 %. During rotation through the suspension, the fibers become attached to the filter surface, while the waste liquor enters the filter and is pumped off. The fiber mat on the filter surface is removed by a scraper. Modern filters are closed and can be operated above 100 °C at

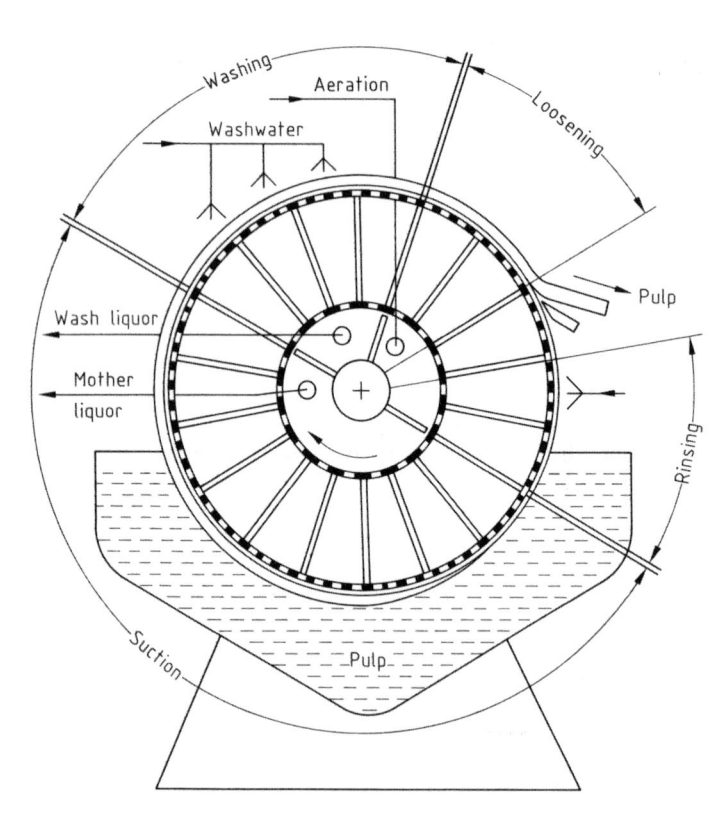

Figure 20. Rotary filter for washing pulp and recovering waste liquor

an overpressure. The washing effect is increased by spraying wash water onto the pulp web on the filter surface. In modern filter designs, several washing steps are installed in a single unit. A washing line consists of several filters operating in series. Fresh water is used only in the last filter, the preceding filters each operate with the filtrate from the following washing step. This countercurrent operation minimizes water consumption. The filtrate from the first washing stage, which is polluted to the greatest extent, is used to dilute the pulp in the digester.

Double-sieve presses and double-roll presses have also proved effective in pulp washing. In the former, the pulp is dewatered between two flat sieves which run on pressure rollers. The second system involves double presses with perforated surfaces. They dip into a pulp suspension and their surfaces are pressed together. While ca. 15 % solids content can be achieved in a pulp suspension by dewatering filters, the pulp consistency attainable with double sieves reaches 40 % and with double-roll presses >40 %.

Fiber Screening. After the removal of dissolved substances from the pulp, all solid impurities that are not part of the fibrous material must be removed. These include bark residues and undercooked wood particles such as knots and shives, as well as external impurities such as sand, stones, and even metal objects.

As a result of the varying size ranges and other sorting criteria, e.g., specific weight, screening and cleaning must be carried out with different separating aggregates in a multistage process.

Coarse screens or knotters, commonly of vibratory design, are used to separate large rejects (Fig. 21). The flat screens have perforations in the form of slots or round holes. The screen vibrates elliptically, and the pulp flows onto the screen with a consistency of 1 – 2 %. The accepts pass through the screen, while the rejects, owing to the vibration direction of the screen, are transported over its surface to the outlet. The spraying of water on the surface of the screen intensifies the screening process.

Vibrating screens can also be designed in cylindrical form. They consist of perforated, vibrating cylinders that have a longitudinal axis of rotation. Depending on the design of the screen surface, they can be used in both coarse and fine screening.

Centrifugal strainers have a stationary horizontal cylindrical screen. The pulp suspension enters this body with a stock consistency of less than 1 %. The rotating suspension is flung against the screen, causing the fibers to pass through it. To improve the screening process, water can be injected into the drum. Multistage screening in one aggregate is possible by using screen perforations of different sizes at different levels in the cylinder.

Today, pressure screens are most often used for pulp screening. They have a high output and produce excellent separation. These are vertical centrifugal screens with a vertical shaft and a fixed screen basket. The stock is introduced tangentially from the top. The centrifugal motion of the stock is intensified by the propeller blades rotating at the bottom of the screen. Very effective screening is possible because of the pressure pulsation that occurs due to the scraper closely passing the screen. Moreover, the blades keep the screen surface free. Since screening is accelerated by applying overpressure, the rejects must be removed batchwise. On the other hand, the accepts go into a chest or directly to the paper machine, thus maintaining the pressure in the system. Pressure screens are used almost entirely for fine screening.

The screens described above remove impurities primarily on the basis of particle size. Unwanted accompanying substances with the same dimensions as fibers, e.g., sand or bark residues, cannot be separated in this way. Centrifugal cleaners (hydrocyclones) are suitable for the separation of these contaminants. A cleaner of this type is shown in Figure 22. The stock enters tangentially so that a vortex flow is produced in the tube. The tangential, axial, and radial velocity components of the swirling motion of the suspension are dependent on the location. The suspension near the cyclone wall moves predominantly

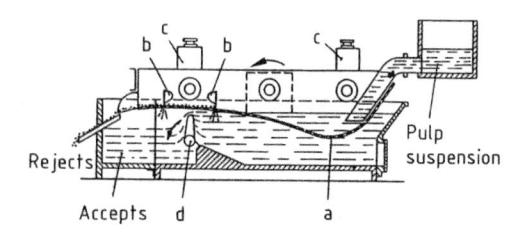

Figure 21. Jönsson shaker
a) Vibrating screen; b) Canned sprayer; c) Casing for spring; d) Overflow control

cifically heavy particles such as sand are removed from the stock flow relatively early. In the case of particles of the same specific gravity, e.g., fibers and some bark components, separation is based on the different ratio of projection area to volume. The larger the projection area of a particle, the higher the power of resistance of the liquid. As a result, even in an early phase of the separation process, large fibers with correspondingly large projection surfaces reach the fiber outflow in the center of the cleaner. In comparison, smaller particles are carried further down and then pass into the rising accepted stock flow. By using a particular entry velocity and dimensions of the centricleaner, particles with defined length to thickness ratios can be eliminated. Since fibers or fiber fragments and bark components can have the same dimensions, separation can be achieved only incompletely in a single step. For this reason, cleaners are arranged in several stages and the rejects are after-screened in each case. Cleaners normally operate with dilute suspensions with pulp consistencies of less than 1 %.

The arrangement and operation of the different screening systems in a pulp mill can vary considerably, and are determined largely by the type of pulp being made. Thus, in the prescreening of unbleached pulp, swinging perforated sieves and centrifugal sorting machines are used to remove knots and other coarse impurities. In final screening, however, swinging cylindrical sieves and centrifugal cleaners are used.

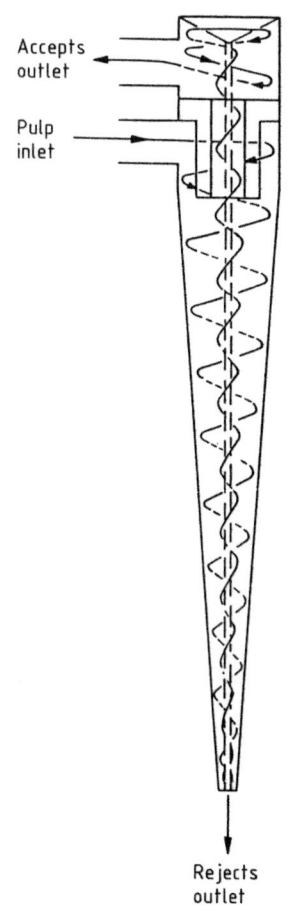

Figure 22. Schematic of a centrifugal cleaner (hydrocyclone)

towards the rejects outlet, while that in the interior moves towards the accepts outlet.

It can be assumed that in the cyclone, fibers and impurities move approximately with the tangential and axial velocity of the carrier liquid. In the radial direction, however, considerable differences in velocity occur which are responsible for the separation process. If a particle moves towards the cyclone wall as a result of the centrifugal force and the radial pressure gradient, this motion counteracts the resistance to fluid flow because the particle must move through the carrier liquid. The larger the projection area of a particle, the higher the resistance. In addition, the migration velocity of the particle depends on its specific weight. The higher the mass per unit projection area, the quicker it migrates to and along the wall to the rejects outlet. Hence, spe-

1.4.3. Pulp Drying

After being cleaned in the screening process, pulp is usually dried either in the bleached or unbleached state. This drying to a solids content of ca. 90 % is required in all cases for pulp that is to be marketed. Also, in the case of integrated pulp and paper mills, the pulp is sometimes dried because the fiber properties of interest for papermaking are changed by drying. Pulp that has been subjected to intermediate drying swells to a lower extent and can be drained more easily on the paper machine. The opacity of paper made from this pulp is higher. Intermediate storage in the dried state poses no problems.

Conventional dewatering and pulp drying is carried out on a Fourdrinier. As in the production of paper and board, the fiber suspension flows through a stock inlet onto the wire and is then drained. The single-layer mat of pulp thus formed is removed at the end of the wire, passed through presses onto conventional steam-heated drying cylinders where the final drying to a predetermined moisture content takes place. To increase drying efficiency, the final drying of the pulp, after web formation and pressing, is often completed in flakt or jet dryers. These are usually arranged in several levels one above the other. The pulp passes through them and is dried in a stream of hot air. After drying, the mat of pulp is cut into sheets, stacked, pressed, and packed in bales.

Fiber Flash Drying. In fiber flash drying, the moisture content of the pulp is first reduced to about 50 % in double-screen, disc, or double-roll presses. The predrained pulp is then flaked in a shredder. The pulp flakes are subsequently blown into a hot-gas channel and dried in two steps. The drying stages involve ventilators, high-performance suspension towers, and cyclone separators. The dried flakes are fed continuously into a bale press, which produces bales with a specific weight of about 0.5 g/cm^3. The bulk density is increased to at least 0.8 g/cm^3 by repressing the bales in a hydraulic bale press. As a result of the relatively severe drying conditions, the fibers can become horny, especially if the hemicellulose content of the pulps is high.

1.5. Waste Liquor Treatment

The waste liquor obtained from the production of chemical pulps contains at least 50 % of the wood substance and pulping chemicals. For environmental and economic reasons, this waste liquor must be utilized as far as possible. Indeed, its utilization to cover the energy requirement of pulp production is of primary importance, and in most processes, this is combined with the recovery of the pulping chemicals. In addition, part or all of the waste liquor can be utilized directly or after chemical or biological treatment.

1.5.1. Recovery of Useful Products from Waste Liquor

The composition of waste liquor depends on the raw material and pulping process used. For in-

Table 8. Composition of softwood sulfite spent liquor

Product	Quantity, kg/t pulp
Methanol	7–10
Acetic acid	30–90
Formic acid	0.5–1
Formaldehyde	2–6
Furfural	5–6
Sugar sulfonic acids and aldonic acids	150–250
Sugars	200–400
Cymenes	0.3–1
Lignosulfonates	600–800

stance, utilizable carbohydrates are found only in waste liquor from acid pulping processes, while resin acids and fatty acids occur only in black liquor from alkaline pulping processes that use resin-containing softwoods. The lignin structure is also influenced by the pulping process. Thus, water soluble lignosulfonates are found in the sulfite process, whereas alkali lignins are formed in the kraft or soda process. The composition of typical sulfite and kraft waste liquors from softwoods is shown in Tables 8 and 9.

Waste liquor from the sulfite process is better suited to the nonenergetical waste liquor utilization than kraft black liquor. Concentrated sulfite waste liquor is used both as a pelletizing agent in the feed industry as well as in the pelletizing of coal and ore. The carbohydrates obtained in the sulfite process can be utilized as shown in Figure 23. Hexoses are fermented to ethanol, while the pentoses are converted to yeast, or reduced to xylitol, or converted to furfural [168, 169].

The lignosulfonates in sulfite waste liquor can be used in many ways, e.g., they are employed as concrete additives to improve the viscosity. The

Table 9. Composition of softwood kraft black liquor

Product	Quantity, kg/t pulp
Methanol	5
Acetates	100–200
Carbohydrate degradation products (lactones, sugar acids, etc.)	350–400
Aliphatic sulfur compounds (methyl mercaptan, dimethyl sulfide, dimethyl disulfide)	1
Terpentine	8–10
Tall oil	20–100
Alkali lignin	400–600

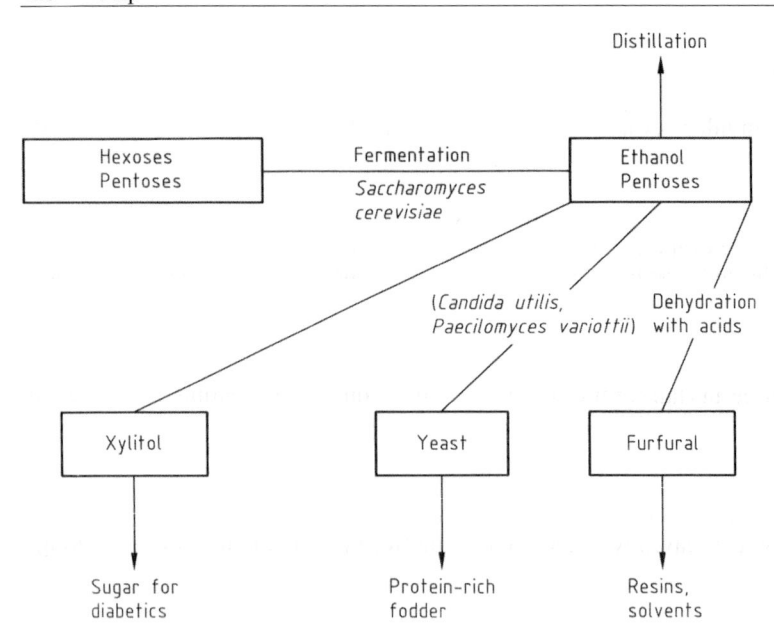

Figure 23. Processing of pentoses and hexoses

dispersing properties of lignosulfonates are exploited in the dispersion of pigments in dyes, as dispersing agents for water insoluble compounds in pesticides, as drilling agents, and as additives in the production of adhesives for the wood industry. The residual waste liquor components are present in amounts that are usually too small for cost-effective utilization. However, the elimination of these substances from waste liquor may be necessary for pollution control.

In the recovery of byproducts from kraft black liquor, only two groups of substances are of economic interest: alkali lignins and the resins and terpenes. The terpene fraction is steam volatile, so when the digester is relieved, this fraction is found in the gas phase and can be recovered by condensing the digester gases. The wood resins are dissolved in the black liquor. During concentration in black liquor evaporation plants, the resins can be skimmed off the waste liquor as sodium soaps with a solids content of ca. 28 %. They are salted out and separated by distillation.

Alkali lignins as such are of little interest economically. They are, however, subjected to fractional precipitation and sulfonation to give high-quality lignosulfonates. They can also be converted to polymers. The following reactions are of potential interest: lignin and formaldehyde to give phenolic plastics; lignin and epichlorohydrin to give epoxide resins; and lignin and epoxides to give polyalcohols, followed by reaction with diisocyanates to polyurethanes [170, 171]. However, these have not been used until now because products recovered from fossil raw materials can be produced in higher quality and often more cheaply. Lignin can be converted to a nitrogen-yielding fertilizer by reaction with ammonia and oxygen under pressure. A series of low molecular mass products based on lignin can also be produced. Vanillin and dimethyl sulfoxide are recovered on an industrial scale. Due to low demand, however, only a very small part of the waste liquor lignin obtained can be used for this purpose. Potential alternative uses include hydrogenolysis of lignin to give phenols, and high-temperature pyrolysis for the recovery of gaseous or liquid products such as carbon monoxide or ethane and methane. However, these processes are not used industrially because they are not cost effective.

1.5.2. Energy Generation from Waste Liquor

After pulp washing the black liquor has a solids concentration of 10 – 18 %. The solids content must be increased to 55 – 70 % before utilization for energy. Counterflow multistage vacuum systems are the most frequently used liquor evaporating systems. They consist of 4 to 7 evaporators arranged in series. The individual evaporators usually comprise a lower evaporating part and an upper vapor space. The dilute black liquor flows between the vapor space and the evaporating part into the evaporating body. The evaporating part consists of either a steam-heated tube bundle or falling-film evaporators. The waste liquor is heated in these systems, which operate as heat exchangers, and the steam produced is withdrawn through the vapor space and later removed from the system as vapor condensate. The waste liquor concentrated in this manner passes into the next evaporating body and is concentrated further with steam. It flows through the entire evaporating system in this way and leaves with the required final concentration. Because the steam is reused repeatedly, its temperature decreases. To maintain the evaporation capacity, this loss of temperature is compensated for by operating the evaporators under vacuum; the lower the steam temperature, the higher the vacuum employed. The higher the rise in concentration of waste liquor in the individual evaporating steps, the higher the evaporation temperature used. In this way a sufficiently high evaporating efficiency is ensured and, at high solids concentrations, the viscosity of the waste liquor is maintained at such a level that precipitation and deposition on the heat exchangers is avoided. A flow sheet of an evaporation plant is shown in Figure 24.

The risk of precipitation in the liquor evaporating system is especially high for more concentrated liquor. This problem can be combated by rinsing the evaporating system from time to time with vapor condensate. During this time, the evaporating body being rinsed cannot be used for evaporation. The evaporating system can be designed so that half of it is used for evaporation while vapor condensate flows through the other half. The two systems can be used alternatingly by changing from one to the other. Moreover, systems exist that operate without vacuum. In these systems, evaporation is also carried out in counterflow but at correspondingly higher temperatures. The lowest steam temperature of a system of this type used to concentrate the dilute black liquor flowing into the first stage must be

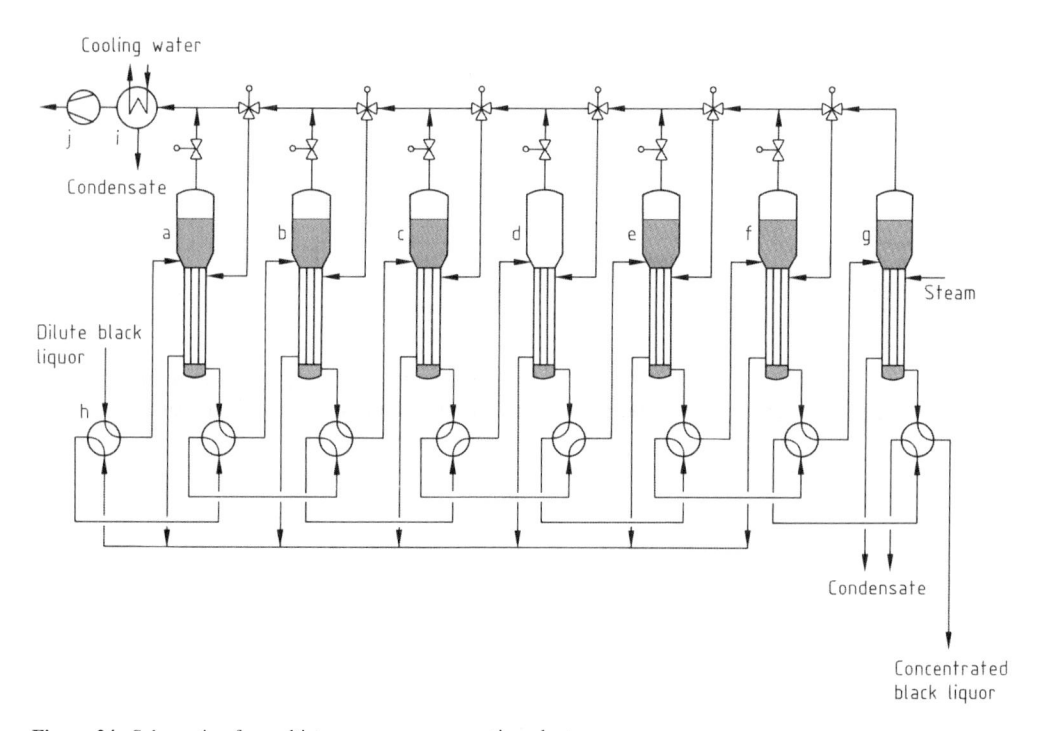

Figure 24. Schematic of a multistage vacuum evaporating plant
a) – g) Evaporators (Evaporator d is switched to washing); h) Reversing valve; i) Condenser; j) Compressor

well above 100 °C. Although the energy requirement in the individual evaporating stages is quite favorable in a system of this type, utilization of the waste steam is necessary if the total energy requirement is to be minimized. This often causes problems in a pulp mill, because most mill operations require higher steam temperatures.

In modern evaporating plants, liquor concentrations of up to ca. 80 % are attainable. However, the composition of the liquor plays a considerable part in concentration. The viscosity is very important as is the concentration-dependent solubility. Here, the molecular structure of lignin and its solubility in water and alkali are of significance. The concentration and composition of inorganic components in the waste liquor are also important. Silicates, which are present in large amounts particularly in annual plants, are extremely critical because they deposit as sodium silicates on heat exchanger surfaces during alkaline pulping. The concentration of problematic waste liquor of this type to a high solids content is possible only by direct evaporation. In this process, the liquor is dried directly in a hot-air drier, which is fed with waste gas from the black liquor combustion stage. This can lead to air pollution problems because reduced or oxidized gaseous sulfur compounds are liberated from the waste liquor. These must then be introduced into the liquor combustion system or the off-gas purification system.

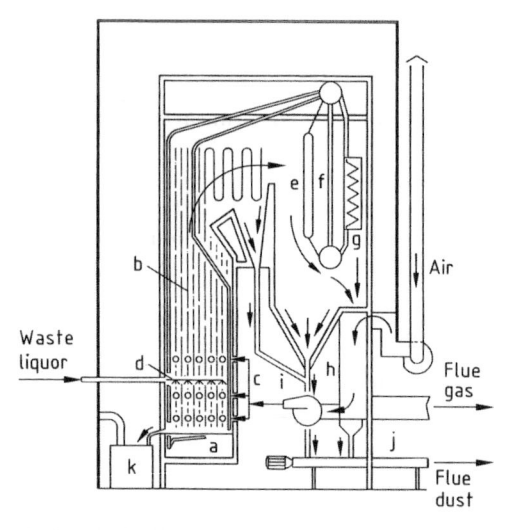

Figure 25. Black liquor combustion furnace
a) Melting zone; b) Boiling tubes; c) Nozzle for combustion air; d) Nozzle for black liquor; e) Superheater; f) Preevaporator; g) Fresh water preheater; h) Air preheater; i) Pipes for flue gas transport; j) Transport system; k) Smelt dissolver

Combustion of Concentrated Liquor. Liquor combustion combines several objectives. It is a means of disposal; serves the recovery of energy, making pulp production self-sufficient; and allows the recovery of the majority of the pulping chemicals charged. Combustion is carried out in a special combustion furnace which functions as a thermal power station and is equipped with a chemical recovery system. Both the combustion furnace and the following chemical regenerating system are adapted to a particular type of waste liquor. A furnace suited to the combustion of waste liquor based on sodium is shown in Figure 25.

The concentrated waste liquor is injected into the lower part of the furnace. In the combustion of waste liquor with a low content of organic components (e.g., waste liquor from the production of semichemical pulp), the simultaneous combustion of heavy fuel oil or coal dust is required. Injection occurs through oscillating nozzles to achieve uniform distribution on the furnace walls where the remaining water evaporates. The organic components of the liquor burn at ca. 1000 °C. The dosing of oxygen during the combustion process is of considerable importance. Sodium-containing waste liquor is normally burnt under reducing conditions in the furnace bed. This means that enough oxygen must be present in the combustion air, so that a complete conversion of organic substance to CO_2 and water occurs. However, the sulfur compounds in the waste liquor should be present in the reduced form, i.e., bound to sodium as sodium sulfide in the smelt. In contrast, in sulfite waste liquor based on magnesium, combustion conditions must be regulated so that magnesium oxide is present as ash, and SO_2 in the gas phase. Both compounds must be separated from the combustion gas. In the combustion of kraft black liquor, most of the pulping chemicals are present in the form of sodium sulfide and sodium carbonate in the smelt at the bottom of the furnace and are then transferred to a dissolving tank. The flue gases pass through heat exchangers in which high-pressure steam is produced. The gases contain a considerable amount of entrained dust and must therefore be purified in mechanical dust collectors and electrostatic filters. An additional gas purification step is required to meet environmental regulations. Sulfur-containing gaseous compounds, in particular, must be eliminated. This is achieved by wet cleaning in scrubbers. The solid and gaseous combustion products which contain pulping chemicals can either be reused directly in the production of pulping liquor or they must be first subjected to chemical conversion.

1.5.3. Chemicals Recovery

Kraft Process. A series of reactions take place in the combustion furnace during the combustion of kraft black liquor. The carbon in the organic material combines with oxygen to form carbon dioxide, which can react with free NaOH to give sodium carbonate. Since combustion partly takes place under reducing conditions, sodium compounds react with the sulfur to form sodium sulfide. The chemical losses that have occurred in the entire pulping process are compensated by addition of sodium sulfate as make-up before combustion. This sodium sulfate is reduced to sodium sulfide in the combustion furnace. For economic reasons, other sodium and sulfur-containing materials are often used to compensate for losses. In modern mills, the chemical losses, calculated as sodium sulfate, are 10 – 20 kg per tonne of pulp. This corresponds to >90 % recovery of the pulping chemicals used. The reactions occurring in the combustion furnace can be simplified as follows:

$C+O_2 \rightarrow CO_2$

$2\,NaOH+CO_2 \rightarrow Na_2CO_3+H_2O$

$Na_2O+CO_2 \rightarrow Na_2CO_3$

$Na_2SO_4+2\,C \rightarrow Na_2S+2\,CO_2$

Therefore, the smelt present at the bottom of the combustion furnace consists primarily of sodium carbonate and sodium sulfide. A solution of this smelt in water is referred to as green liquor. The greenish color is due to iron sulfides contained in the solution.

To recover the pulping chemicals used, sodium carbonate must be converted to sodium hydroxide. This takes place in the causticization step, in which sodium carbonate reacts with slaked lime to give sodium hydroxide and calcium carbonate:

$Na_2CO_3+CaO+H_2O \rightarrow 2\,NaOH+CaCO_3$

This is an equilibrium reaction which, under normal conditions, produces degrees of causticization of 85 to 90 %. The solubility of calcium carbonate increases with increasing OH^- content, and the reaction equilibrium is shifted to the right. Normally 2 – 3 % of calcium oxide remains in the sludge produced by causticization. Although the addition of more quicklime improves the degree of causticization slightly, a multistage causticization process is more effective. The calcium carbonate formed is separated from the white liquor and rewashed to reduce chemical losses. About 250 kg quicklime per tonne of pulp are required for causticization. Amounts of this magnitude are very expensive and the large capacity of kraft pulp mills makes this not only a procurement problem, but also a waste problem. Therefore, the lime sludge is reburnt in a rotary kiln in the plant, converting the calcium carbonate in an endothermic reaction to calcium oxide:

$CaCO_3 \rightarrow CaO+CO_2$

The lime kilns are a major source of air pollution because mercaptans are released from liquor residues in the sludge. The outlet air from the kilns can be returned to the liquor combustion furnace to reduce odors. A chemial recovery system for the kraft process is shown in Figure 26.

Calcium Bisulfite Process. Apart from calcium oxide and sulfur dioxide, calcium sulfate is formed predominantly in the combustion of calcium bisulfite liquors. The complete thermal decomposition of calcium sulfate occurs only above 1150 °C; temperatures above 1200 °C must be used to achieve an essentially quantitative decomposition. This causes problems because dead burning of lime occurs at these high temperatures, changing its crystal structure and reducing its reactivity.

Magnesium Bisulfite Process. In comparison with calcium sulfate, the magnesium sulfate

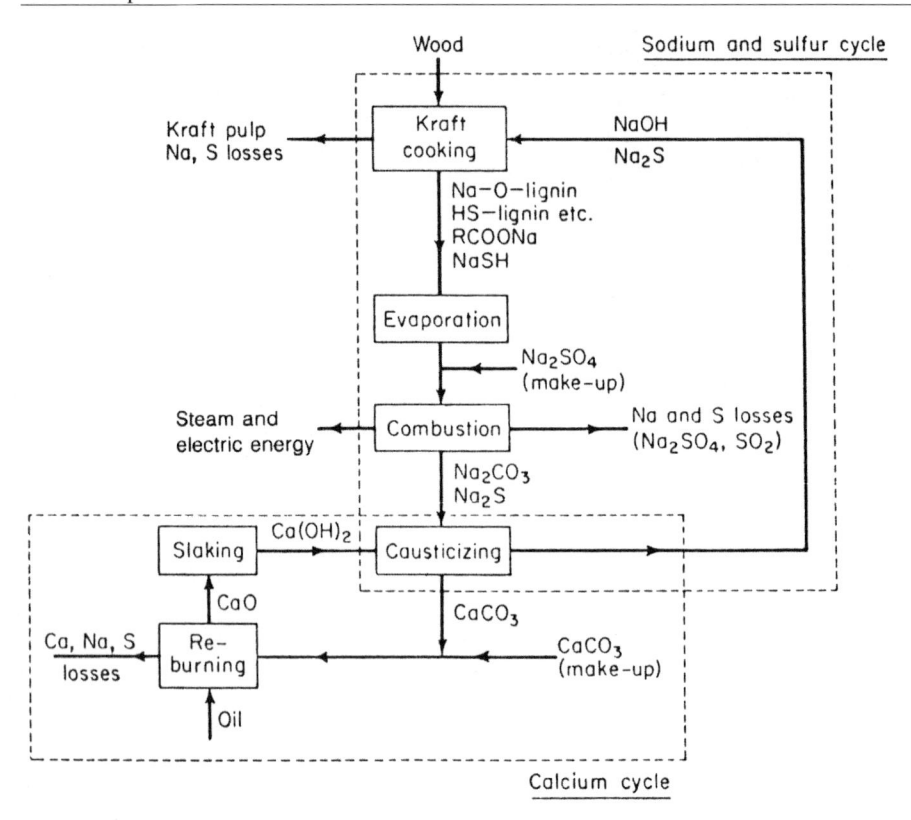

Figure 26. Recovery cycles for a kraft mill

obtained from the combustion of magnesium sulfite waste liquor can be decomposed thermally at normal combustion temperatures to give gaseous SO_2 and magnesium oxide. The latter can be separated from the combustion gases in powder form. In slaking tanks, the magnesium oxide is converted to magnesium hydroxide, which reacts with SO_2 to form a mixture of magnesium sulfite and magnesium bisulfite. In the fortification tower, this solution is treated with SO_2 so that either a pure magnesium bisulfite solution is formed which can then be used in the magnefite process, or a magnesium bisulfite solution with excess SO_2 for acid magnesium bisulfite pulping. A schematic of the process is shown in Figure 27.

Flue gas purification using electrostatic filters is not widely employed because chlorides, for example, are not removed in this process. Instead, the pulp particles are discharged from the gas flow in mechanical purification units or multiple cyclones. In the subsequent wet cleaning, water-soluble solid components of the flue gases are removed. The flue gases then again pass through wet dust separators, which precipitate further chlorides. Chloride discharging occurs in a retention tank for magnesium oxide. The chloride level of the chemical cycle can be controlled by regulating the overflow.

In sulfite mills, the spent liquor contains considerable amounts of acetic acid, furfural, and methanol, especially when hardwoods are used for pulping. For this reason, the vapor condensate obtained on evaporation is highly polluted. Therefore, the digestion spent liquor is usually neutralized before evaporation to convert the low-boiling, acidic components to salts and thus prevent their transfer to the vapor condensate. For waste liquor neutralization in magnesium bisulfite mills, the amounts of MgO required are of the same order of magnitude as the amounts of magnesium oxide used in digestion liquor preparation. The volatile compounds can also be separated by stripping the vapor condensates. Acetic acid can also be recovered by liquid – liquid extraction.

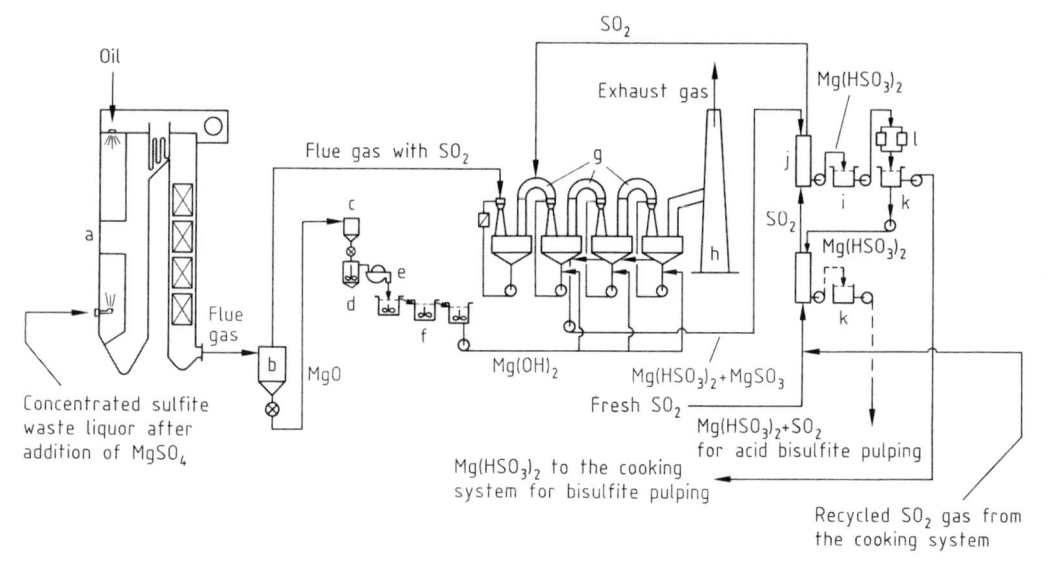

Figure 27. Recovery of the chemicals from the spent liquor from the magnesium bisulfite process
a) Waste liquor combustion furnace; b) Dust collector; c) MgO silo; d) Slurry tank; e) Wash filter; f) Hydration vessels; g) Flue gas scrubber; h) Stack; i) Sedimentation tank; j) Fortification tower; k) Storage tank; l) Filter

The degree of chemicals recovery in the magnesium bisulfite process is 80 – 90 % for both magnesium and sulfur. Therefore, the make-up requirement is 10 – 15 kg magnesium oxide and SO_2 per tonne of pulp.

Sodium-Based Sulfite Process. The recovery of chemicals in sodium-based sulfite processes is relatively complicated. For this reason, many different methods have been developed. Until now, only a few processes have proved useful in industrial practice. These are based on the principle of carbonation or pyrolysis. Direct sulfitation is a very interesting process, but it has never been used in a large-scale operation.

Carbonation. The concentrated liquor is burnt in a kraft-type furnace under reducing conditions. Sodium sulfide and sodium carbonate are present in the smelt. The sodium sulfide reacts with CO_2 from the flue gases or pure CO_2 or bicarbonate to yield sodium carbonate and hydrogen sulfide. The gaseous hydrogen sulfide is withdrawn and burnt to give SO_2. In an absorption tower, SO_2 reacts with sodium carbonate to give sodium sulfite or sodium bisulfite, depending on the stoichiometry. Carbonation processes in various modifications are still the only large-scale recovery processes for sodium sulfite liquor [172, 174].

Pyrolysis. Sodium and sulfur can be separated in a single step by pyrolysis of sodium sulfite waste liquor to produce sodium carbonate and solid carbon, while sulfur is obtained as gaseous hydrogen sulfide, which gives SO_2 on combustion. In comparison with other combustion processes, the pyrolysis temperatures are relatively low because no smelt should be formed in this process. Sodium carbonate is then dissolved in water and the carbon is filtered off and burnt in a separate step. Sodium carbonate and SO_2 are recombined to give the pulping liquor. In the industrial application of this process, however, substantial oxidation of sodium sulfur compounds occurs during pyrolysis, forming sodium sulfate. This must be carried through the process cycle as a chemical inert for pulping. Temperature control is also a problem in this process. If the temperature is too low, incomplete chemical conversion occurs. On the other hand, if the temperature is too high, a smelt is formed which contains sodium sulfides [175].

Direct Sulfitation. In the direct sulfitation, the smelt containing sodium sulfide and sodium carbonate, formed after spent liquor combustion under reducing conditions, is reacted directly with SO_2 to yield sodium sulfite. Sodium sulfide reacts to give sodium sulfite and H_2S, which is in turn converted to SO_2 by combustion. The industrial application of this process failed because thiosulfate is formed in considerable amounts from sodium sulfide, SO_2, and sodium sulfite in side reactions. Thiosulfate catalytically decomposes bisulfite during the digestion process.

Ammonium Sulfite Process. Today, ammonium sulfite solutions are seldom used for pulp production. The main reasons for this are the relatively high price of ammonia and the fact that base recovery is not cost effective. However, after combustion of the waste liquor the SO_2 can be recovered from the gas flow.

1.6. Pulp Bleaching

Lignin cannot be completely dissolved out of the wood in a pulping process because the rate of carbohydrate degradation in the final phase of delignification exceeds the rate of lignin dissolution. This leads to high yield losses and reduced pulp strengths due to the depolymerizing attack on the carbohydrates. For this reason, pulping must be stopped and the lignin that remains in the pulp must be removed by selective lignin dissolving chemicals. The residual lignin content of pulps depends on the type of raw material, the pulping process, and the required pulp properties. The residual lignin content of softwood pulps is normally higher than that of hardwood pulps, and sulfite processes can delignify more extensively than the kraft process. The normal residual lignin content of unbleached pulps is in the range 2 – 5 %. The elimination of this residual lignin is required primarily to brighten the pulps. However, mainly the structure and not the content of residual lignin is responsible for the color of unbleached pulp. Depending on the pulping reactions, more or less chromophoric groups are formed. Sulfite pulps always have a higher brightness than kraft pulps. Table 10 lists residual lignin contents and pulp brightnesses for various pulps.

Pulp bleaching is carried out in several stages, with the elimination of residual lignin as far as possible being the primary objective of the first stages. The subsequent bleaching stages are re-

Table 10. Residual lignin content and brightness of various pulps

Type of pulp	Residual lignin content (kappa number)	Brightness, % ISO
Softwood kraft	27–33	25–30
Hardwood kraft	17–20	25–35
Softwood sulfite	15–25	55–60
Hardwood sulfite	12–22	55–65
Semichemical pulp/neutral sulfite	>80	50–60
Semibleached pulps	1–5	70–85
Fully bleached pulps	<1	85–92

sponsible for brightening the pulp. Accordingly, bleaching is divided into prebleaching and final bleaching. The following abbreviations have been introduced for the bleaching stages:

C– chlorination
E– alkaline extraction
H– sodium or calcium hypochlorite bleaching
D– chlorine dioxide bleaching
O– oxygen bleaching in alkaline medium
P– peroxide bleaching
Z– ozone bleaching
Y– dithionite (hydrosulfite) bleaching

1.6.1. Conventional Bleaching

In conventional pulp bleaching, chlorine, hypochlorite, and chlorine dioxide are used with or without intermediate alkaline extraction steps. For purposes of comparison, the amounts of chlorine chemicals employed are expressed as active chlorine, the varying molecular weight and oxidation potential of the chlorine-containing bleaching agents being taken into account. Thus, the oxidation potential of chlorine in hypochlorite is twice that of chlorine and the oxidation potential of chlorine in chlorine dioxide is five times that of chlorine. Taking into consideration the different molecular masses of the various chlorine compounds, the chemical equivalence compared to molecular chlorine is 1.05 for sodium hypochlorite and 2.63 for chlorine dioxide.

The use of chlorine and chlorine-containing compounds in pulp bleaching is ecologically the most problematic step in the entire process of pulp production. The chlorine compounds react in various ways with the lignin of pulp. Substitution reactions on the phenol rings of lignin, which occur to a large extent especially in chlorination, are particularly critical with regard to environmental pollution. The effluents from the bleaching process contain chlorinated organic compounds and cannot be disposed of by combustion with the spent digestion liquor. It must be released into the receiving stream, possibly after being subjected to biological treatment. It pollutes this stream considerably with its oxygen demand and its toxic and genotoxic effects [176–179]. Among others, various dibenzofurans and dioxins and even the extremely toxic 3,4,7,8-tetrachlorodibenzo-p-dioxin have been detected in small concentrations [179–183]. The combustion of bleaching waste liquor of this type would cause corrosion problems because of the formation of hydrochloric acid and would lead to the production of other highly toxic chlorinated products.

Chlorination. The first step in conventional lignin-dissolving bleaching is usually chlorination [184]. Chlorine is an effective, selective, and inexpensive bleaching agent that reacts exceptionally quickly with lignin. However, instead of brightening the pulp it usually causes a decrease in the brightness. Chlorination is normally carried out at a low temperature (ca. 30 °C). A short reaction time is sufficient, usually 30 min for sulfite pulp and up to 60 min for kaft pulp. The stock density for chlorination is usually ca. 3 %. Due to the fast reaction of chlorine with pulp, this low stock consistency is required to guarantee quick chlorine mixing and, hence, a uniform reaction. In particular, overchlorination of parts of the pulp and the resultant fiber damage should be avoided. However, the development of high performance mixers has made chlorine bleaching possible even at higher pulp consistencies of up to 10 % [185].

Chlorine water is normally used for chlorination. As a result of the more favorable reaction conditions and the possibilities now available for the optimal mixing of chlorine and pulp, chlorine gas has also been used recently on an industrial scale.

The most important function of chlorination is the conversion of the residual lignin in the pulp to products that are soluble in water and alkali. Chlorine reacts primarily with the benzene rings of lignin in a nonradical reaction, in which both substitution and oxidation reactions take place.

At the same time, addition reactions occur on the propane side chains of the lignin monomers if double bonds are present. However, since this is very seldom the case, addition reactions play only a minor role.

Electrophilic substitution with the chloronium ions formed by heterolytic cleavage of the Cl–Cl bond results in the replacement of the hydrogen atoms of the benzene ring by chlorine atoms [36, 186–189]. The number of chlorine atoms and their position on the ring depend on the structure of the lignin molecule. The use of excess chlorine can result in the maximum replacement of the three ring hydrogens at positions 5, 6, and 2 of the aromatic nucleus by chlorine atoms; 50 – 70 % of the total chlorine requirement of a bleaching sequence is consumed in chlorination.

The amount of chlorine used for different pulps is normally in the range of 1.0 – 1.7 g active chlorine per gram residual lignin. Lower amounts are required for sulfite pulps, which contain more reactive residual lignin, than for kraft pulps. Independent of the pulping process, softwood pulps require more chlorine than hardwood pulps. Under industrial chlorination conditions, ca. 1 – 2 chlorine atoms per lignin unit are attached to the phenol ring in substitution reactions. If the phenolic OH group in *para*-position to the side chain is not etherified, the substitution reaction occurs preferentially at the 5-position of the aromatic ring. If there is an alkyl, aryl, or methoxy group at position 5 of the ring, substitution occurs only at positions 6 or 2. The substitution of chlorine on the guaiacyl ring has the following priority: $6 > 5 > 2$.

Electrophilic substitution can also occur at position 1 of the phenol ring if the α-C atom of the side chain bears an OH or OR group. In this reaction, the aliphatic moiety is cleaved from the aromatic ring of the lignin phenylpropane unit (side-chain elimination). In contrast to the substitution of hydrogen atoms on the aromatic ring, the lignin molecule is fragmented in this reaction. Apart from the substitution reaction, oxidation also occurs during chlorination [190]. The initial step in the oxidation reaction is again the electrophilic addition of chlorine. Subsequent demethylation gives rise to quinoid intermediates, which can be oxidized further to dicarboxylic acids upon ring opening. The reactions occurring on the aromatic rings during chlorination of lignin are shown in Figure 28 [36].

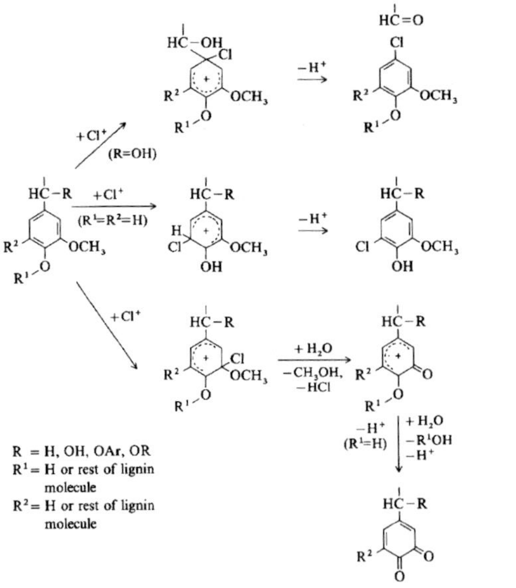

Figure 28. Substitution and dealkylation of phenolic and nonphenolic units by chloronium ions

Apart from the reactions described above, a large number of side reactions occur which give rise to many different chlorinated compounds, some of which are toxic, resistant to biological degradation, or potentially mutagenic or carcinogenic.

In industrial chlorination processes, the ratio of substitution to oxidation reactions is about 50 : 50. The chlorination conditions chosen previously favored the substitution reaction because carbohydrates are attacked thereby only to a minor extent. However, the chlorination process is now operated increasingly such that the oxidation reactions dominate, thus avoiding the aromatic chlorine compounds produced in the substitution reactions. This can be achieved by increasing the reaction temperature and pH.

Chlorine is a selective bleaching agent that reacts faster with lignin than with cellulose. Nevertheless, carbohydrates are also subject to attack by the chlorine radicals formed [184]. Carbohydrate degradation occurs via oxidation as well as hydrolysis. The positions 1, 2, 3 and 6 of anhydroglucose units are very sensitive. If oxidation occurs at position 1, depolymerization takes place by cleavage of the β-1,4-glycosidic bond.

In other oxidation reactions, primary and secondary alcohol groups can be converted to aldehyde and keto groups, which facilitates depolymerization of the cellulose chains in subsequent bleaching steps.

Alkaline Extraction. Chlorination makes lignin partially soluble in water and alkali. After chlorination, in fact, sulfite lignin is for the most part already soluble in water, whereas the majority of kraft lignin can be dissolved only in alkali [191]. Due to the action of alkali, chlorine atoms bound to the aromatic nuclei are partially replaced by hydroxyl groups, increasing lignin solubility. Apart from the dissolving of chlorolignin, the function of alkali extraction in the production of dissolving pulps is to remove the hemicelluloses from the pulp as extensively as possible. For this purpose, higher amounts of alkali and severer conditions are required. In the production of paper pulps, however, hemicelluloses should remain in the pulp as far as possible and the DP value of the cellulose should be maintained at the highest possible level. Therefore, alkali extraction should be carried out gently.

In the extraction step, 1 – 1.5 % NaOH based on o.d. pulp is used for sulfite pulps, and up to 3 % for kraft pulps. The extraction time is 60 – 90 min at 40 – 60 °C. For dissolving pulps, temperatures up to 90 °C and up to 15 % sodium hydroxide based on o.d. pulp are used if pulps with extremely high α-cellulose contents are to be made. The consistency in the extraction stage is generally ca. 10 %.

As a result of the intensive dissolution of lignins and carbohydrates in the chlorination and extraction stages of pulp bleaching, highly polluted wastewater is produced which accounts for up to 90 % of the COD of the total bleach and ca. 100 % of the color load. The chlorinated lignin degradation products have, in part, a potential toxic effect on living organisms as well as a mutagenic effect. In particular, chlorine compounds with molecular masses of less than 1000 are problematic because they are small enough to diffuse through the membranes of living cells and can accumulate in fat tissues. The effluents from the chlorination and extraction steps can be degraded only to a limited extent by biological treatment. In single-stage plants, rates of degradation of approximately 50 % can be achieved, carbohydrates being broken down preferentially.

Hypochlorite Bleaching. After chlorination and alkaline extraction, the delignification of pulp is almost complete. The residual lignin content of the pulp is ca. 0.5 – 0.8 %. In conventional bleaching, this lignin is dissolved with oxidizing bleaching agents. In the past, hypochlorite was the agent of choice, but today chlorine dioxide is used almost exclusively, especially for kraft pulps. If both bleaching agents are used, hypochlorite is usually used first, followed by chlorine dioxide. Whereas calcium hypochlorite was used predominantly in the past, especially in calcium bisulfite pulp mills, sodium hypochlorite is used almost exclusively today in pulp bleaching [192].

In the hypochlorite bleaching step, the pH is the most important influencing factor. It can be varied within a wide range by varying the ratio of alkali to chlorine. At pH values less than 2, only elemental chlorine is present in the bleaching liquor. However, HOCl is formed at higher pH values, and the dissociation of this acid produces OCl^- ions at still higher pH values. Above pH 9.5, only OCl^- ions are present. Chlorination reactions take place predominantly at acidic pH values, whereas oxidation reactions occur preferentially at higher pH values. Since the latter are preferred, the initial pH of the hypochlorite solution should be adjusted to 11 – 12. The pH drops during bleaching owing to the formation of acidic products such as carbonic acid and oxalic acid. The pH should not fall below 8 at the end of the reaction because cellulose is also attacked extensively at lower pH values.

Normally, 1 – 2 % hypochlorite based on o.d. pulp is used at 30 – 50 °C. Temperatures up to 70 °C are also acceptable if the pH does not fall below 11 during bleaching. The consistency in the hypochlorite stage is 10 %, and the bleaching time is 2 – 4 h, depending on the raw material used, the pulping process, and the pulp properties required.

The oxidation reactions occurring at high pH in hypochlorite bleaching attack the free phenolic hydroxyl groups or the phenolic ethers at positions α and β of the phenylpropane side chain. The products of lignin degradation from hypochlorite bleaching are methanol, organic acids, and carbon dioxide.

Hypochlorite is, however, not a selective oxidizing agent; it also reacts with carbohydrates. The following reactions can occur [192, 193]:

1. Oxidation of primary hydroxyl groups at C_6 to carbonyl and further oxidation to carboxyl groups
2. Formation of keto groups at atoms C_2 and C_3; ring cleavage to give dicarboxylic acids
3. Cleavage of the β-1,4-glycosidic bond and oxidation of the reducing end groups to carboxyl groups
4. Ring fission between C_1 and C_2 with formation of an aldehyde group at position 2 and a carboxylic ester at position 1; further oxidation to a carboxyl group at position 2 is possible

Depolymerization, which leads to poorer technological pulp properties, can be limited by using high pH values and lower reaction temperatures.

Chlorine Dioxide. Chlorine dioxide is an extremely effective and selective, but also expensive bleaching chemical. In the gas phase, chlorine dioxide is explosive at concentrations above ca. 12 %. It is, however, uncritical in dilute aqueous solution. The production of chlorine dioxide is carried out at the pulp mill by reduction of sodium chlorate. For a long time, it was not possible to use chlorine dioxide in pulp bleaching because of its toxicity, corrosivity, and the risk of explosion. It was the industrial application of chlorine dioxide shortly after World War II that made it possible to bleach even kraft pulps to high brightness, and led to the dominance of kraft pulping.

Chlorine dioxide has five oxidation equivalents. It can develop its full oxidation potential at a pH of 4 because all five oxidation equivalents can be used. In practice, the highest brightness is obtained when the pH at the beginning of bleaching is adjusted to approximately 6 and is around 4 at the end of the process [194].

The bleaching effect of chlorine dioxide is based on the oxidation of lignin. The initial step in lignin breakdown involves radical attack on structures with phenolic OH groups, forming phenoxy radicals, which undergo further reactions to give quinoid structures or, after ring cleavage, to form muconic acid derivatives [36, 195]. Ring chlorination reactions are caused by elemental chlorine, which is either contained in technical chlorine dioxide or formed by its decomposition.

Under normal bleaching conditions, chlorine dioxide hardly degrades carbohydrates. However, carbohydrate degradation can occur at temperatures above 80 °C, pH values above 7, and in the presence of excess chlorine dioxide, especially if the carbohydrates have already been attacked in the preceding bleaching stages.

The application of chlorine dioxide is a first response to the necessity of using environmentally harmless bleaching agents. The amount of bleaching agent used is small. Extensive degradation of lignin to aliphatic compounds occurs and chlorinated compounds are formed only to a very small extent.

Chlorine dioxide has an exceptionally positive effect on pulp brightness. It can degrade even largely unreactive lignin structures and is indispensible for the elimination of dirt specks in pulp.

The temperature during chlorine dioxide bleaching is between 70 and 80 °C and the bleaching time is 3 – 4 h. The pulp consistency is ca. 10 – 12 %. The amount of chemicals used can vary considerably depending on the position of the chlorine dioxide stage in the bleaching sequence and the type of pulp to be bleached. Normally, the amounts used are between 0.5 and 1.5 % active chlorine based on o.d. pulp.

Conventional Bleaching Sequences for Various Pulps. Depending on the type of wood, the pulping process used, the residual lignin content of the pulp, and the target brightness, different bleaching sequences must be used in the production of the various pulps. In pulps of all types, conventional bleaching starts with chlorination and extraction stages. In the case of readily bleachable sulfite pulps having low residual lignin contents, a hypochlorite stage follows CE prebleaching, and chlorine dioxide is used in the final bleaching. The bleaching of sulfite pulps to brightnesses of 90 % ISO and higher is possible with this CEHD standard sequence. Semibleached pulps having brightnesses of 80 % can be produced by using the sequence CEH.

In the conventional bleaching of kraft pulps, the sequence CEDED has become established. Brightnesses of 90 % can be achieved while maintaining good pulp quality. Bleaching yields of ca. 93 % are obtained for softwood kraft pulps with an initial kappa number of about 30 (residual lignin content ca. 4.5 %). Bleaching yields of 96 – 97 % are achieved for sulfite pulps with low unbleached kappa numbers in the range 12 – 15.

The properties of pulp change during bleaching in a characteristic manner. As a result of the removal of hemicelluloses, the bonding potential of the fibers is reduced, which, in turn, reduces the breaking length in particular. On the other hand, the tear strength can increase.

1.6.2. Chlorine-Reduced and Chlorine-Free Pulp Bleaching

In response to pollution linked to the use of chlorine in pulp bleaching, considerable efforts have been made to replace chlorine by an environmentally less harmful chemical. Moreover, efforts are also being made to replace as far as possible the other chlorine containing bleaching agents and to develop bleaching sequences that require no chlorine-containing compounds. Then, the requirements would be met not only for the combined evaporation and combustion of waste liquor from bleaching and digestion, but also for the recovery of the alkali used in the various bleaching stages and for carrying out pulp production with a greatly reduced water consumption.

At the beginning of this development, attempts were made to replace the highly polluting CE prebleaching by chlorine-reduced or even chlorine-free alternatives. The first steps in this direction were the partial or complete replacement of chlorine by chlorine dioxide and the introduction of oxygen delignification. Ozone has also been employed on a laboratory and pilot scale. Peroxide has proved useful in enhancing oxygen delignification and is also being used increasingly in final bleaching.

Substitution of Chlorine Dioxide for Chlorine in Prebleaching. Owing to the different modes of reaction of chlorine and chlorine dioxide, the formation of chloroorganic compounds decreases with increasing use of chlorine dioxide [196–201]. At the same time, the selectivity of the bleaching reaction is improved [202–208]. Even small amounts of chlorine dioxide added in a chlorination stage reduce carbohydrate degradation considerably because chlorine dioxide functions as a radical scavenger and suppresses the oxidation of alcoholic OH groups to carbonyl groups by chlorine radicals [209, 210].

In industrial prebleaching, up to 70 % of the chlorine is replaced by chlorine dioxide. Especially in the case of kraft pulps, the complete replacement of chlorine is impracticable because the delignifying effect of chlorine dioxide is much lower than that of chlorine. The use of mixtures of chlorine dioxide with chlorine has the following advantages:

1. The process can be carried out at higher temperatures (60 – 70 °C) and higher consistencies (10 – 15 %). This corresponds to the conditions used in the following bleaching stages. Thus, a reduction of the energy and water requirements is possible.
2. Since chlorine dioxide has five oxidation equivalents, the use of this agent, compared to chlorine, corresponds to a reduction of the chloride ion content to 20 %.
3. The oxidative attack of chlorine dioxide on lignin leads to water-soluble degradation products, so that the amount of alkali used in the subsequent extraction stage can be reduced.
4. The use of chlorine dioxide produces effluents that are less chlorinated, less colored, and contain smaller amounts of toxic and genotoxic substances.
5. Pulps bleached with chlorine dioxide are of a better quality. Their extractives content is lower, and the brightness, viscosity, brightness stability, and technological properties are improved. Yields increase since chlorine dioxide attacks the carbohydrates in pulp to a lesser extent.

Alkali/Oxygen Bleaching. The increased use of chlorine dioxide also has disadvantages in that the investment and chemical costs are higher. Because of the high oxidizing power of chlorine dioxide, bleaching equipment must be made of corrosion-resistant material.

The treatment of pulp with oxygen under alkaline conditions is a chlorine-free alternative to prebleaching [211–219]. Sodium hydroxide is usually used as alkali; magnesium hydroxide can be used in magnesium-based sulfite mills. The bleaching agent employed is gaseous oxygen, which is a very inexpensive chemical. However, oxygen is only very slightly soluble in water, especially at the high bleaching temperatures that must be used in the oxygen stage to achieve an adequate reaction rate. Therefore, the process must be carried out under pressure for a sufficient amount of oxygen to be available in the bleaching liquor. Oxygen in alkaline medium is, under these conditions, a relatively strong oxidizing agent. As a biradical, it can remove an electron from the phenolate ions present in the alkaline medium. The phenoxy radical thus formed is subject to further reactions. Hydroperoxides are produced which are further degraded by intramolecular nucleophilic attack of the peroxide anions [36, 218–221].

Oxygen does not, however, react selectively with lignin, but also attacks carbohydrates to a considerable extent [122, 123]. Above all, oxidations at atoms C 2 and C 3 of the anhydroglucose units occur. The unstable carbonyl compounds facilitate further degradation, which can result in the cleavage of the glycosidic bonds. To reduce carbohydrate degradation, the extent of delignification occurring in this stage must be limited. Inhibitors such as magnesium sulfate can be added to protect the carbohydrates. In industrial operation, delignification in the oxygen stage is restricted to about 50 % in spite of these additives. Consequently, oxygen bleaching is less effective than conventional prebleaching with chlorine.

Oxygen bleaching was originally developed as a high consistency bleaching process for pulp consistencies of 30 – 35 %. However, newly installed large-scale plants operate almost exclusively in the MC range (medium consistency), i.e., at consistencies of 10 – 15 % [224–227]. While high consistency bleaching requires temperatures above 100 °C and pressures up to 0.8 MPa, MC oxygen bleaching is carried out at temperatures below 100 °C and pressures of 0.2 – 0.4 MPa. The temperature can be reduced further to 70 °C in the bleaching of sulfite pulps. Modern MC oxygen stages usually operate under hydrostatic pressure. The oxygen is mixed in at the bottom of the bleaching tower under the hydrostatic pressure of the reactor. Thus, a reactor height of 20 m, for instance, results in a pressure of 0.2 MPa.

The reaction time in the oxygen stage is 30 – 90 min, and 20 – 40 kg alkali per tonne o.d. pulp and about 0.5 – 2.5 kg magnesium sulfate are charged. Depending on the delignification efficiency, 10 – 25 kg oxygen per tonne of pulp are consumed.

In order to improve delignification and to apply milder conditions, especially lower temperatures, hydrogen peroxide is added to the oxygen stage, particularly in sulfite pulp bleaching [228–231]. Sulfite pulps can be delignified by 50 – 60 % at 70 – 80 °C in the oxygen stage by adding 1 % peroxide. Moreover, the pulp properties profit from the milder reaction conditions.

Pulp Bleaching with Ozone. Since oxygen delignification is limited, other delignifying agents must additionally be used before final bleaching is performed. In practice, especially in kraft pulp bleaching, chlorine and chlorine dioxide are used today in varying proportions. Ozone has been tested for this purpose both on a laboratory and pilot-plant scale [232–240]. However, the use of ozone in pulp bleaching requires new technology. Like oxygen, ozone is only slightly soluble in water. Ozone is also very unstable and must be prepared at the mill by electrical discharge in air or oxygen. The ozone concentration in the air flow reaches 3 – 5 %, while concentrations of 6 – 10 % can be achieved with oxygen. In order that sufficient amounts of ozone reach the fibers, extremely low pulp consistencies of less than 2 % must be used; otherwise this process must be carried out in the gas phase at consistencies above 35 %. Pressurized medium consistency ozone bleaching is also possible. The water should be as free as possible of heavy metals to minimize catalytic decomposition of ozone.

The most important reaction of ozone with lignin is the cleavage of the β-O-4 bonds between the lignin units. Ozone can attack both the aryl and the alkyl moieties of the molecule. The attack on the aromatic rings can lead to ring cleavage between the C 3 and C 4 atoms. Double bonds in the aliphatic side chain are also attacked, forming carbonyl and peroxide structures [36, 241–244].

The intense attack on the carbohydrates is a disadvantage. The dominant reactions are cleavage of the glycosidic bonds and oxidations at the C 2 and C 3 atoms, which facilitate further chain breakdown in subsequent alkaline treatment [222, 245, 246].

The formation of hydroxyl and perhydroxyl radicals is of considerable importance in these reactions. The hydroxy radicals in particular react very unspecifically and lead to severe carbohydrate degradation. Since the formation of these radicals is favored by hydroxyl ions, the ozone treatment should be carried out under acidic conditions. For this reason, the pulp must be acidified, and SO_2 and sulfuric acid have proved effective for this purpose. Since ozone is extremely reactive, a very short reaction time can be used, provided that thorough mixing of ozone and pulp takes place. If this is the case, the entire duration of treatment including an appropriate after-reaction time can be less than 10 min. The temperature in the ozone step should be as low as possible to avoid harmful side reactions. This applies in particular to low-consistency bleaching. In high-consistency ozone bleaching with very thorough mixing, temperatures up to 50 °C can be used.

Low-consistency ozone bleaching is of advantage with regard to the selectivity of lignin degradation [247]. However, high-quality water and low temperatures are required. If the water cycle of the mill is largely closed, both requirements can be met only with difficulty. The stronger attack on carbohydrates in high consistency ozone bleaching must be restricted by limiting the amount of ozone charge. Amounts exceeding 1 % ozone based on o.d. pulp should not be applied. The pulp can be delignified by about 10 kappa numbers with this amount of ozone. Under these conditions, ozone can be used for pulps with a high lignin content as the second bleaching stage after oxygen prebleaching. Alternatively, it can be used as the sole delignifying agent for sulfite pulps with low initial kappa numbers. As a result of its higher delignifying capacity, ozone, unlike oxygen, is also well suited to the bleaching of low-residual-lignin pulps. The kappa number after the ozone stage should be in the range of 1.5 – 3.

An emerging technology is ozone bleaching at medium consistency. The MC ozone bleaching must be carried out under pressure to reduce the gas volume and to have a good mixing effect. On the other hand, the disadvantages of ozone bleaching at low or high consistency are greatly eliminated. First pilot plant results have been satisfactory; in particular, carbohydrate degradation was substantially reduced.

Until now, ozone has not been used for large-scale bleaching of pulp. However, the existing laboratory and pilot plant results and the necessity for chlorine-free bleaching suggest that the industrial application of this bleaching agent is in the offing.

Peroxide Bleaching. Hydrogen peroxide in alkaline medium is well suited to the final bleaching of chemical pulps [248]. The bleaching of mechanical pulp has shown that it destroys the chromophoric groups present in lignin by cleaving conjugated double bonds. At 70 – 80 °C, the highly nucleophilic perhydroxyl ion formed is capable of further degrading quinoid lignin structures, produced by the attack of electrophilic bleaching chemicals. This is generally accompanied by ring fission [36, 221, 249]. As mentioned above, peroxide is used in prebleaching as an intensifying agent in the oxygen stage. In final bleaching, it is often used as the concluding treatment. The amounts used are 1 – 2 % at consistencies of 10 % and temperatures of 70 – 80 °C. An advantage of the use of peroxide in final bleaching is the high brightness stability of peroxide-bleached pulps.

Chlorine-Free Bleaching Sequences. The choice of the bleaching sequence to be used, if chlorine-free bleaching agents are applied exclusively, depends on the raw materials used in pulping, the pulping process, the residual lignin content of the pulp, and the demands made on the quality of the bleached pulp. The sequence OZEP is suitable for sulfite pulps with high residual lignin contents. After the oxygen stage, the kappa number of the pulp should be 10. The kappa number can then be reduced to ca. 3 by the use of 0.8 % ozone based on o.d. stock. These pulps can be brightened to 90 % ISO by subsequent alkaline extraction and the addition of 1 % hydrogen peroxide in the final bleaching step.

At present, kraft pulps still cannot be bleached economically to high brightness levels in a chlorine-free sequence. In the case of softwood kraft pulps, strength losses of 20 – 25 % must be accepted in chlorine-free bleaching on reaching brightness levels of 85 % ISO.

On the other hand, several sulfite mills have switched to completely chlorine-free bleaching sequences; oxygen and hydrogen peroxide are employed. Prebleaching is achieved with alkali/oxygen, partially intensified by hydrogen peroxide. Peroxide is used in final bleaching, usually in a two-step process. Bleaching sequences of this type give brightnesses up to 85 % ISO. Compared with conventional bleaching and ozone bleaching, however, strength losses must be accepted.

1.6.3. Technology of Bleaching

Today, bleaching is always carried out in a closed system. The individual bleaching stages take place in bleach towers in which the material flows either from top to bottom or vice versa (Fig. 29). Washing units (filters or sieve presses) located between the bleach towers remove the used chemicals and the reaction products from the pulp. The washing efficiency is about 80 %. Countercurrent operation is employed as far as possible to minimize water consumption. In modern plants, the tendency is to use similar reaction conditions in all bleaching stages. The bleaching stages are preferentially performed at a medium consistency of 10 – 15 %. Bleaching towers with an upward flow can be used for this purpose without problems. The addition of the chemicals occurs before the pulp enters the tower. An exception is the chlorination step, which is performed at consistencies of 3 – 5 % and at the lowest possible temperatures. However, the temperature and the consistency can be adapted to those of the other bleaching stages by partially replacing chlorine by chlorine dioxide. Chlorination towers are usually rubber lined, while chlorine dioxide towers are made of high-alloy stainless steel.

High-consistency oxygen bleaching requires a technology which differs from conventional bleaching stages. Alkali and the degradation inhibitor (magnesium sulfate) are charged before the pulp enters the reactor. After mixing, the stock is adjusted to a consistency of at least 30 % and then fluffed at the top of the reactor. The tower is operated from the top downwards at a high oxygen pressure. The tower can be equipped with movable trays over which the pulp moves and is gassed with oxygen. Before being discharged, the pulp on the reactor floor is diluted to give a pumpable suspension and it then enters a washing stage.

On the other hand, the medium consistency oxygen stage used more frequently today uses an up-flow. Alkali and magnesium sulfate are added before the pulp enters the tower. Oxygen is added in an additional mixer. Bleaching occurs either at hydrostatic pressure or with oxygen overpressure. The stock is discharged continuously from the top of the tower and is transported to a washer. The MC oxygen bleaching technology is not problematic because no explosive mixtures of

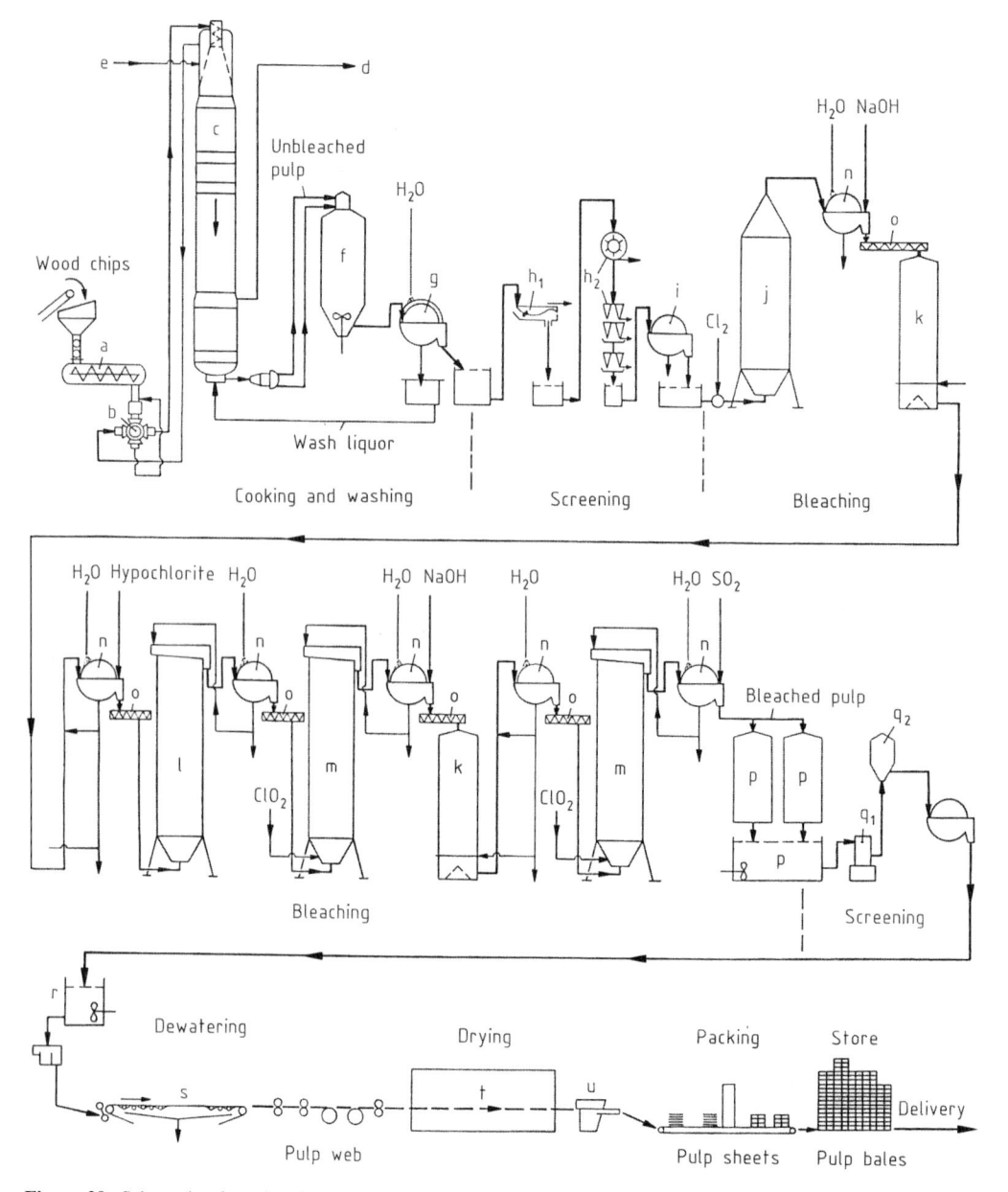

Figure 29. Schematic of a pulp mill
a) Steaming drum; b) High-pressure valve to the continuous digester; c) Continuous digester for kraft and bisulfite pulping; d) Waste liquor outlet; e) digestion liquor inlet; f) Blow tank; g) Filter for pulp washing; h_1) Coarse screening; h_2) Fine screening; i) Filter thickener; j) Chlorine bleaching tower; k) Bleaching tower for alkali treatment; l) Bleaching tower for hypochlorite treatment; m) Bleaching tower for chlorine dioxide treatment; n) Wash filters; o) Mixers; p) Storage tanks; q_1) Coarse postscreening; q_2) Fine postscreening; r) Machine chest; s) Dewatering machine; t) Tunnel dryer; u) Sheet cutter

carbon monoxide/oxygen can form. Increasing the consistency in the bleaching stages saves energy and reduces water consumption, but problems with the mixing of bleaching chemicals and the stock can occur at higher consistencies. The water and energy requirements in the medium-consistency range can be reduced by adapting the reaction conditions in the different bleaching steps and by using countercurrent washing. A system of this type is shown in Figure 30.

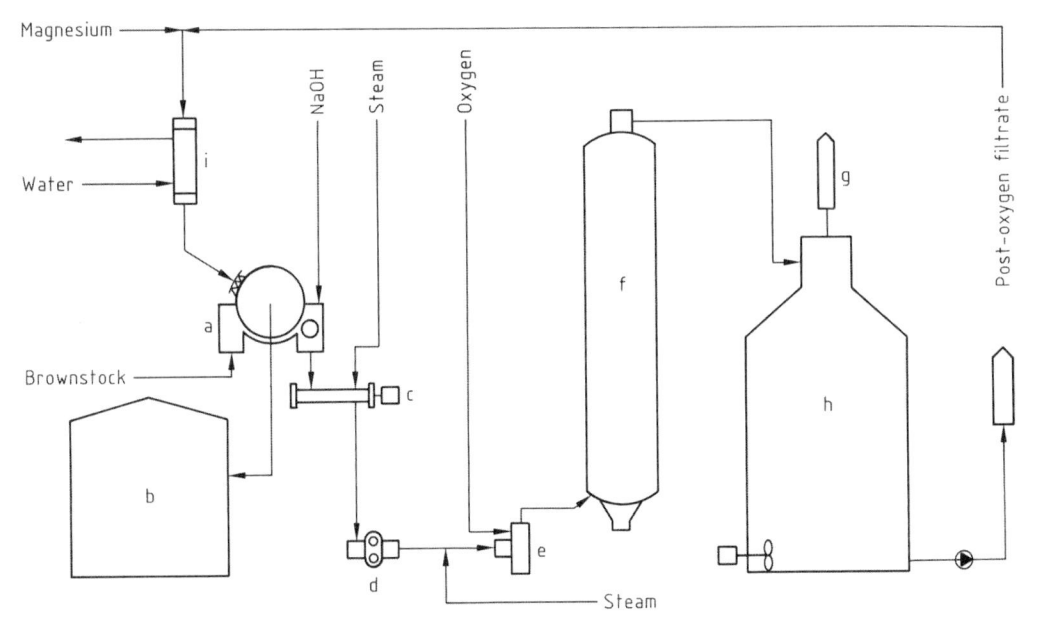

Figure 30. A medium consistency oxygen bleaching system
a) Brownstock washer; b) Seal tank; c) Steam mixer; d) Medium consistency pump; e) Oxygen mixer; f) Oxygen reactor; g) Vent; h) Oxygen blow tank; i) Post-oxygen washer

A high-consistency ozone stage can be operated with the technology of the high-consistency oxygen stage. However, a pressure reactor is not required. Ozone can be added in the fluffer. On the other hand, up-flow towers can be used for low and medium consistency ozone bleaching.

1.7. Environmental Aspects of Pulp Production

The production of pulp is associated with considerable pollution. The type and extent of pollution depend on the pulping and bleaching processes used and on the internal and external measures taken to reduce pollutants. If all the possible and economically justifiable measures are taken to reduce the environmental burden in the production of kraft pulp, pulpmaking with a wastewater COD below 50 kg/t of pulp appears to be possible. The specific water requirements of a mill of this type is ca. 50 m^3 per tonne of pulp. In the production of sulfite pulp, the COD burden is even lower. Values of 30 kg COD per tonne of pulp and water consumption corresponding to that of kraft pulp production are possible. Base losses of 1 – 1.5 % based on o.d. pulp and sulfur losses, measured as SO_2, that are only insignificantly higher are incurred by modern kraft and sulfite pulp mills. Today, combustion furnaces operate with dust emissions below 50 mg/m^3, NO_x emissions below 300 mg/m^3, and SO_2 emissions below 1.5 g/m^3.

Sources of Emission. In the pulping process, not only gaseous emissions, but also wastewater is produced. Today, the washing efficiency after the digestion process exceeds 99 %. The original load of the entire wood substance dissolved by pulping amounts to ca. 1500 kg COD per tonne of pulp and a final charge of ca. 15 kg COD is produced. Screening and liquor evaporation cause further pollution. In addition, pulp bleaching is a significant source of pollution. A rough assessment shows that about half of the COD of a pulp mill is caused by pulp production and the other half by pulp bleaching. In modern sulfite and kraft pulp mills, the gaseous sulfur compounds formed are withdrawn at the point of emission and charged into the chemical recovery system. The separation of gaseous sulfur compounds from the flue gases takes place in wet cleaners. The major sources of SO_2 in the sulfite process are the digester, the blow tank, the pulp

Table 11. AOX values from the bleaching of various pulp

Bleaching sequence	AOX, kg/t	
	Hardwood pulp	Softwood pulp
CEHDED	4–5	8–10
$O(C_{85}D_{15})$ EDED	2.5–5	4–5
$O(D_{70}/C_{30})$ (EO) DED	1–2	1.5–2.5

See Section 1.6 for abbreviations.

wash, and the chemical recovery system. In the kraft process, lime kilns are a further substantial source of emissions. Considerable amounts of chlorinated organic compounds are formed particularly in the chlorination and extraction stages of conventional bleaching. These compounds can be determined as organic halogens (AOX values) adsorbable on active charcoal. Ranges of AOX values for various hard- and softwood kraft pulps are listed in Table 11.

The chlorinated organic material can be divided into high and low molecular compounds, the dividing line being at a molecular mass of 1000. The high molecular mass fraction is not directly toxic. However, biological degradation of these compounds can give rise to low molecular toxic products. The low molecular portion of the chlorinated organic compounds can, in turn, be divided into a water soluble, an ether soluble, and a highly volatile fraction. The ether soluble substances, particularly the chlorinated phenolic compounds, are especially hazardous, some of them being highly toxic, chemically very stable, and able to accumulate in the fatty tissue of water organisms via the food cycle. This fraction can also contain polychlorinated dibenzofurans and dibenzodioxins.

Prevention Measures. Pollution caused by pulping can be reduced by both internal and external measures. Internal measures refer to changes in the production process that lead to a reduction of pollution. These include extended digestion to reduce the residual lignin content of unbleached pulp, increasing the black liquor recovery rate by recycling the wastewater into the process, and switching to chlorine-free bleaching.

External basic prevention measures include the elimination of pollutants formed. Wastewater treatment for the reduction of pollutants can be carried out by filtration, sedimentation, flotation, and removal of solids. These processes can be supplemented by physicochemical procedures such as precipitation of dissolved compounds or flocculation.

Dissolved organic compounds can be degraded biologically in subsequent process steps. Both aerobic as well as anaerobic wastewater treatment are used for this purpose. Biological treatment is indispensible in modern pulp mills. To achieve the highest possible rate of degradation, multistage processes are frequently used; the sequence anaerobic/aerobic has proved to be especially effective. Depending on the wastewater composition, the COD of pulp mill effluents can be reduced by 50 % in single-stage plants. In two- or three-stage treatment plants, a decrease in the COD of up to 80 % and a AOX reduction of 50 % can be achieved.

In pulp production, the atmospheric emissions can be greatly reduced by rigorous detection of all flows of exhaust air and their return to the chemical recovery system, and by the purification of the flue gases.

1.8. Pulp Properties and Applications

Pulps have a wide range of properties and uses. The technological properties of pulps depend largely on the raw material, the pulping process, the extent of delignification, and the bleaching sequences applied. Basic differences exist between paper and dissolving pulps. The latter must not only be absolutely free of lignin, but the content of extractives, inorganic impurities, and hemicelluloses must not exceed relatively narrow limits. In addition, the degree of polymerization of cellulose must fall within a definite range, depending on the intended use. Dissolving pulps are produced by using the prehydrolysis–kraft process or the acid sulfite process. In the prehydrolysis–kraft process, the hemicelluloses are prehydrolyzed to such an extent that pulp with a high α-cellulose content is formed in subsequent digestion. The bleaching conditions, especially in the extraction stage, are also characterized by relatively stringent conditions. This refers to both the alkali dosing as well as the bleaching temperatures. A survey of the requirements to be met by dissolving pulps for the production of various products is given in Table 12.

Table 12. Specifications of dissolving pulps for various products

Specifications	Viscose					Cellulose derivatives				
	Cordrayon	Modal fibers	Cellulose guts	Cellulose film	Acetyl cellulose	Cellulose[c] ethers	Cellulose[d] ethers	Cellulose nitrates	Methylcellulose	Carboxy-cellulose
Brightness, % ISO	>99	>96	>90	>90	96	89	93	>90	>88	>88
$R_{18}{}^a$	>98	>93[b]	>95	>93	94	93–95	92	>92	>92	>90
$R_{10}{}^a$			>93	>91		91–94	89	>90	>90	89
Extract in CH_2Cl_2, %	0.02[b]	0.02	<0.2	<0.2	0.1	0.1	0.4	<0.2	0.2	0.2
Ash	0.15	0.18		<0.1	0.1	<0.2	0.1	0.1	<0.2	0.2
$DP_W{}^d$	760–860	460–520	900–1700	900–1200				900–4500	500–4500	700–4500
Specific density, kg/m³					600	650–800	800	500–700	ca. 700	ca. 700

[a] Residues in 10 or 18 % sodium hydroxide at 20 °C.
[b] In ethanol.
[c] Softwoods.
[d] Hardwoods.
[e] Degree of polymerization.

For pulps used in the production of paper and board, the strength properties are of special importance. Kraft pulps are clearly superior to sulfite pulps in this respect, especially with regard to the tear strength. The pulp fibers form the supporting structure of paper and board, while mechanical pulp fibers improve the opacity and the volume of paper. Mineral fillers also increase the opacity and the surface smoothness of writing and printing papers. On the other hand, recycled fibers from wastepaper contributes little to paper strength. They are, however, inexpensive and their reutilization reduces waste problems. Indeed, wastepaper is used as a substitute for mechanical pulp in many cases. Today, the sulfite processes are of secondary importance in the making of paper pulp because their strength properties are inadequate for many applications. For instance, the tear strength of acid bisulfite pulps is only about two thirds that of kraft pulps. However, they have the advantage of easy bleachability and beatability. Semichemical pulps made by the bisulfite, neutral sulfite, or alkaline sulfite processes have good strength properties and, compared to competing pulps made by the kraft process, better brightness levels. A survey of pulps made by various processes, their properties, and possible uses is presented in Table 3.

2. Mechanical Pulp

2.1. Introduction

Mechanical pulp denotes the semifinished product, used for the production of paper and board, which is made solely by the use of mechanical energy to separate the wood into individual fibers and fiber fragments. According to present findings, the internal processes occurring in such nonchemical production processes are predominantly of a physical nature.

There are two main processes used for the production of mechanical pulp. In the stone groundwood (SGW) process, logs are pressed against a rotating grinder stone with simultaneous addition of water. Refiner mechanical pulps (RMP, TMP) are produced by defiberizing wood chips between refiner disks to give fibers and fiber fragments.

High yields are characteristic for these two processes, i.e., the losses occurring during the processes, mainly in the form of dissolved ligneous substances, are low.

The most serious disadvantages associated with all types of mechanical pulp are their low resistance to ageing, which results in loss of strength and the tendency to discolor. For this reason, the large scale use of this pulp is limited to "short-lived" paper such as newsprint.

In the Federal Republic of Germany mechanical pulp accounted for about 15 % of the total production of paper and board from 1986 to 1988 [250]. This percentage is also more or less valid for world production.

2.1.1. The Raw Material, Wood

For the production of mechanical pulps, which to a large extent are responsible for the structural stability of the paper produced, softwood is used worldwide. The most common woods used in Europe are spruce and fir, while in climatically more favorable countries, woods such as the fast growing *Pinus insignis, Pinus maritima,* hemlock, etc. are used. Softwoods have two predominant advantages over hardwoods. The fiber content of softwoods is ca. 90 % as compared to 50 % for hardwoods and, in addition, the fiber length of softwoods of 3 – 4 mm is considerably larger than the 1 – 2 mm of hardwoods. Under the same manufacturing conditions these two characteristics are always responsible for the higher strength of softwood mechanical pulps.

The logs required for the groundwood process are obtained by thinning out forests and are usually 10 – 20 cm in diameter. The bark has to be removed before the logs are processed. The wood chips used for the production of refiner mechanical pulps are predominantly obtained from sawmill residues. It is of great importance for the quality of mechanical pulp that the moisture content of the wood is as high as possible. This results in a higher flexibility of the fibers, which improves their fibrillation.

2.1.2. Characteristics Required of Mechanical Pulps

About 90 % of the mechanical pulp produced is used to make graphic papers, such as newsprint; uncoated and coated magazine papers, e.g., SC (super calendered), and LWC (light weight coated) papers [251]. The desired properties of mechanical pulps, used for graphic papers, are as follows:

1. The mechanical pulp should contribute substantially to the structural strength of the paper. Newsprint, for example, may contain more than 80 % mechanical pulp.
2. The printability of the paper is mainly determined by the quality of the mechanical pulp.
3. Since graphic paper is usually printed on both sides, a very low show-through and strike-through of the printing ink is very important. In offset printing, the surface of the paper should have a high picking resistance and be free of dust.
4. The drainage ability (freeness) of mechanical pulp has to be high to allow the paper machines to be operated at high speeds (ca. 1400 m/min). Furthermore, the paper has to be able to cope with the stresses within the printing machines.

The technological data of mechanical pulp is determined by means of standardized test methods. The United States and its neighbors follow the TAPPI regulations [252], the Scandinavian countries use the Scan methods [253], and Germany and its neighbors, the Zellcheming regulations [254].

Although the methods used vary and the results are thus not directly comparable, all standards require determination of the following mechanical pulp properties:

1. Drainage characteristics
2. Stability properties such as tensile strength tearing strength, and picking resistance
3. Optical properties such as brightness and opacity
4. The quantitative composition with respect to the portions of long fibers, fiber fragments, and fines as well as the size of the shives

Furthermore, proof printing and other tests provide information about the suitability of a mechanical pulp.

All in all, the test methods in question permit adequate assessment of the mechanical pulp and are essential for mechanical pulp production.

2.2. Groundwood

Until the 19th century, used textiles were virtually the only raw material used for papermaking. With the constantly increasing demand for paper, the supply of new raw materials was an urgent and widely known problem. In 1843 a weaver from Saxony, F. G. KELLER, shredded wood by pressing it against a hand-driven grinder stone adding water to the grinding process. From the resulting pulp, he then made a sheet of paper. By using wood, KELLER found a raw material which is still the most important in the paper industry. After this discovery was adopted, the first mechanical grinder was put into operation in 1852.

It was not until 1930 that the internal processes involved in grinding were systematically studied with the aim of obtaining a higher quality groundwood [255]. The investigations carried out in Canada in the early 1960s on an experimental grinder operating under pressure led to the pressure grinding process for the production of pressure groundwood (PGW) [256]. The Tampella company of Finland then developed this process and introduced it in 1978 [257]. The thermogrinding process for the production of thermogroundwood (TGW) was introduced by the Voith company in 1984 [258]. This grinding process is carried out at a normal pressure and with process temperatures that reach the feasible upper limit.

Theory of the Grinding Process. The purpose of the grinding process is to disintegrate the wood fiber bond system. The logs are pressed against the rotating grinder stone with the wood fibers parallel to the axis of the stone (Fig. 31) [259]. Based on the usual operating parameters, a stone particle requires about 0.02 s to pass the grinding zone, removing 0.03 mm of the wood substance. The total or partial separation of the fibers from the wood and the properties of the

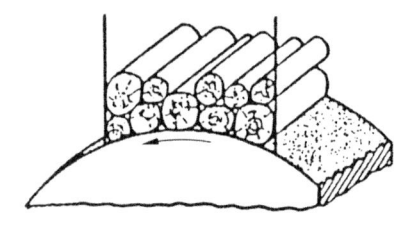

Figure 31. Arrangement of the wood on the grinding stone

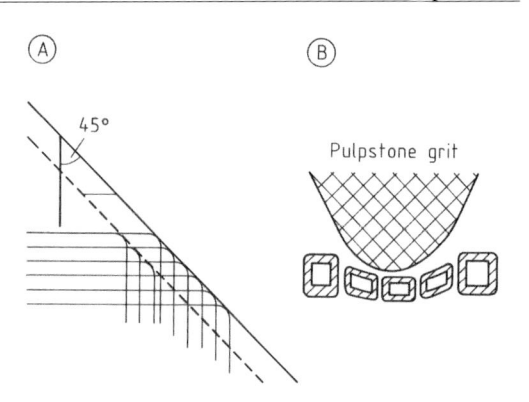

Figure 32. Processes occurring in the grinding zone
A) Angle of removal; B) Deformation of fibers by pulpstone grit

groundwood achieved with this process depend on a series of factors. These are, above all, the surface structure of the grinder stone, the nature of the abrasive particles (round or sharp-edged) and their depth of penetration into the wood. Fiber removal occurs as soon as the stone particles are in contact with the wood, and the availability of sufficient water is of basic importance for heat dissipation and friction reduction. When sliding along the wood surface, a stone particle not only causes deformation of the fibers, but simultaneously changes the direction of already separated fiber parts (Fig. 32). These deformations lead to the fracture of the fiber-to-fiber bonds and of the bonds within the fiber walls. The results of this effect, comparable to "brushing" of the deflected fiber parts, are the fibrillation of the fiber walls and the creation of fines [260].

It can be assumed that most of the energy put into the grinding process is transformed into heat; values of 90 to 99 % have been reported. The resulting increase in temperature is effective in the wood and in the grinding zone [261]. Due to the poor thermal conductivity of wood, temperatures of 180 °C or higher may develop within the wood directly above the grinding zone [262]. In the conventional grinding process at normal pressure, the average temperature in the grinding zone is below 100 °C. Evaporation may occur only in those areas where the amount of water present is insufficient to dissipate the heat [261].

During the development of grinding processes it was confirmed that the temperature is a decisive factor in the production of groundwood. The explanation for this was given by GORING,

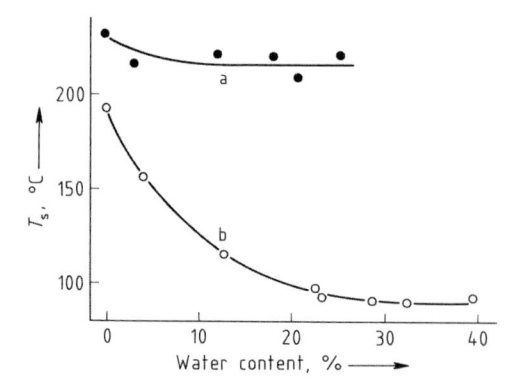

Figure 33. Variation of softening temperature T_s with water content
a) Sulfite pulp; b) Periodate lignin

Figure 34. Ceramic-bound grinder stone

who found that lignin with a water content of more than 30 % softens at temperatures as low as 90 °C (Fig. 33) [263]. Since the lignin-rich middle lamellae of the wood fibers contribute substantially to the mutual bonding of the fibers and to the structural stability of the wood, the softening of lignin must inevitably facilitate the exposure of the fibers.

A sharp-edged abrasive particle has an energy-saving cutting action on the wood structure, whereas round particles exert a predominantly fibrillating effect. The result is a groundwood with superior strength properties, but with the disadvantage of a lower production rate. The ratio of the motor energy to the amount of groundwood produced is called the specific energy requirement (kWh/t), and is an important process parameter. An increase in the specific energy requirement is always linked with better strength properties, but also with a lower drainage capacity. The magnitude of the specific energy requirement depends on the interaction of all grinding conditions and the surface characteristics of the grinder stone. The effect of the grinding conditions – such as the grinding surface area, the peripheral speed, the specific grinding pressure, etc. – on the grinding process and on the groundwood quality has been the subject of detailed studies, which have in turn contributed to the optimization of the process [255, 264–266].

2.2.1. The Grinder Stone

Apart from the cement stones, now of declining importance, the grinder stones used are exclusively of the ceramic-bound type (Fig. 34) with a diameter 1.5 – 1.9 m. The stone surface consists of segments fastened to the concrete core of the stone [267]. The abrasive particles used consist either of sharp-edged carborundum (SiC) or, increasingly, of the softer and less sharp-edged corundum (Al_2O_3). A pattern is impressed on the grinder stone surface (Fig. 35) by traversing sharpening rollers, a procedure which has to be repeated periodically depending on the wear of the pattern. The intervals at which the stone is resharpened vary greatly between 1 and 100 d. This sharpening process must be carried out with extreme care because at present there is no practically useful technique available for measuring the pattern [268, 269]. This sharpening pattern has two functions. It meets the requirements for the separations of the fibers from the wood substances, and it transports the water needed in the grinding zone within the grooves of the sharpening pattern.

Figure 35. Pattern on a grinder stone surface

2.2.2. Groundwood Production

For the production of groundwood two grinder systems are predominantly used. In press grinders the logs are transported automatically into chambers, where they are pressed against a grinder stone by means of hydraulic presses (Fig. 36 A). In continuous grinders, logs are pressed against the stone by means of specifically shaped chains (Fig. 36 B) [270]. The advantage of the latter system is that the continuous grinding process better utilizes the motor load. The two-press grinder, on the other hand, has a higher specific production due to its larger grinding surface area. A comparison of the two grinding systems is given in Table 13.

The grinders are equipped with control systems that keep the power input, the production rate, or the specific energy consumption constant.

Table 13. Comparison of the two types of grinder for wood of 1 m diameter

	Two-press	Continuous
Grinding surface, m^2	2.0	1.2
Motor output, kW	5000	3000
Motor output utilized, kW	4500	2700
Production (1250 kWh/t), t/d	78	50
Area required/unit, m^2	26	13
Space required/unit, m^3	100	80

Each of these control systems has its advantages and disadvantages. In fact, combinations of two of the above control systems are usually applied.

2.2.3. Newer Methods of Groundwood Production

The superior strength properties of refiner mechanical pulps, especially of TMP, forced further development of the conventional grinding process if it was to remain competitive. The findings on the influence of process temperature [256] led to the development of the pressure grinding process (PGW process). In this process, the Tampella pocket grinder is sealed so that the grinding process takes place at overpressure. Overpressures of up to 0.3 MPa, produced with compressed air, allow the process to be operated with white water temperatures of 95 °C and grinder pit temperatures of 125 °C. The more intensive softening of lignin associated with these temperatures results in improved groundwood qualities [271–275]. The strength of pressure groundwood is 30 – 35 % higher than that of conventional groundwoods produced with the same specific energy consumption. Although this process represents a tremendous improvement, the technical and financial requirements are considerably higher.

The manufacturers of continuous grinders were also compelled into improving the quality of their groundwood. This was achieved by reducing the heat losses [276] which result from evaporation in the grinding zone and by maximizing the process temperature, without using overpressure, by means of stability controllers [258]. The qualities of groundwood obtainable from the thermogrinding process (TGW process) are about 15 – 20 % higher than that of

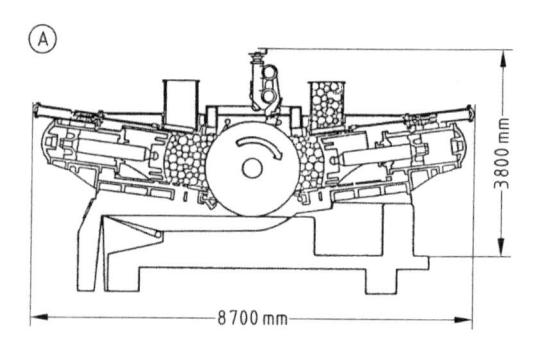

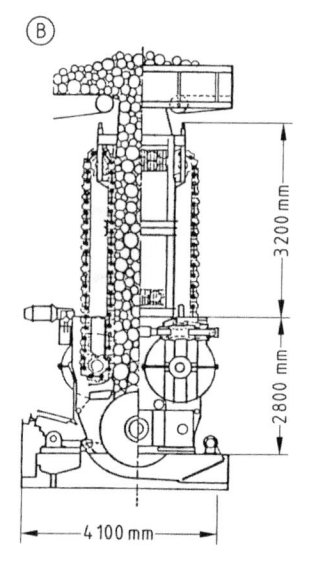

Figure 36. A) Tampella press grinder; B) Voith chain grinder

conventional groundwood and are comparable to lower-grade pressure groundwoods. A certain advantage is the fact that the conversion of existing grinders to the TGW process is relatively inexpensive.

2.3. Refiner Mechanical Pulps

2.3.1. Historical Survey

Refiners have been available since the thirties and were used particularly in the fiberboard industry. After 1955 they also found application in the paper industry for the production of mechanical pulps [277]. Sawmill residues are used as a cheap raw material by being transformed into wood chips. The refiner mechanical pulp (RMP) process, which is hardly used today, was the first stage within this development. After 1960, the advantages of higher process temperatures were exploited in the thermomechanical pulp (TMP) process. This process is still being improved and is used for the production of the majority of refiner mechanical pulps.

Since 1980, the chemothermomechanical pulp (CTMP) process is becoming increasingly important. Because wood chips are readily impregnated, small additions of chemicals give further improvements in the pulp properties. A further increase in the amount of chemicals used, as in the more recently developed CMP process, results in mechanical pulps comparable to semichemical pulps.

2.3.2. Theory of the Refiner Process

The objective of the refiner process is the same as that of groundwood production, i.e., the breaking loose and fibrillation of the wood fibers [278]. As a result of the large surface area of wood chips, the advantages of higher temperatures can be exploited to a much greater extent than in the grinding process. In the TMP process, the chips are subjected to thermal pretreatment, and the disintegration and defibration process is carried out in the refiner at overpressure. To obtain a large proportion of intact fibers, fiber separation should preferentially be a fiber-to-fiber action rather than an action between the refiner plates and the wood [279, 280]. Pulp consistencies in the refining zone of at least 40 % are suitable for this purpose. Nevertheless, the design of the refiner plate pattern exerts a marked influence on the quality of the wood pulp.

TMP with good strength properties is obtained as a result of the relatively small number of cutting actions. However, the specific energy consumption is 40 – 50 % higher than that of the grinding process. The temperatures resulting from the excess pressures of up to 5 bar in the refiners permit the recovery of the extra energy demand by means of heat exchangers [281, 282].

2.3.3. TMP Production

A schematic of the TMP process is shown in Figure 37 [283]. Since the pulping of wood chips to the desired pulp quality is usually not possible in a single refiner stage, two stages are used. The pulping of screen rejects is often carried out in the second refiner stage.

The capacity of a two-stage refiner system is much higher than that of a grinder; 300 o.d. t/d of TMP can be produced at an installed motor power of 15 000 kW per refiner. The specific energy consumption, including that of the auxiliary machines is ca. 2100 kWh/o.d. t. The arrangement of the refiner disks can be quite different (Fig. 38) and each of these models has its advantages [284].

The refiner disks consist of 6 – 12 segments and have a diameter of more than 1.5 m. The profile of the plates is designed such that the direction of disk rotation can be reversed, extending the plate life-time, which ranges between 1000 and 2000 h.

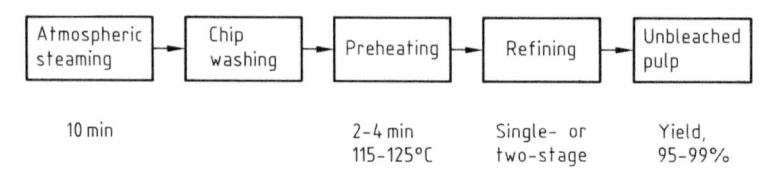

Atmospheric steaming	Chip washing	Preheating	Refining	Unbleached pulp
10 min		2-4 min 115-125°C	Single- or two-stage	Yield, 95-99%

Figure 37. Schematic of the TMP process

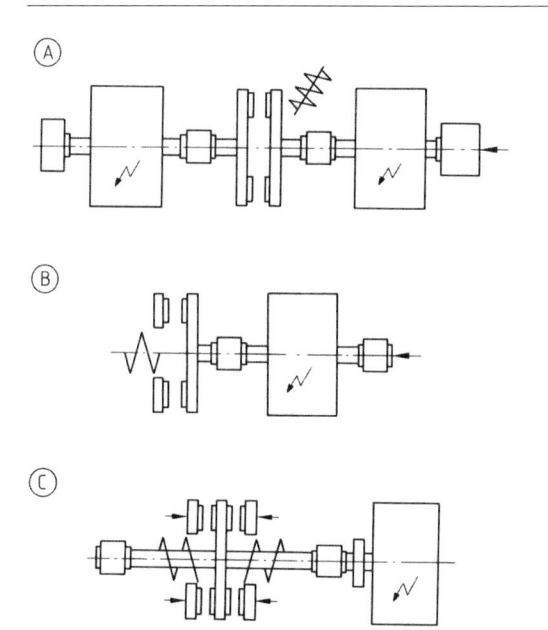

Figure 38. Refiner configurations
A) Double-disk refiner; B) Single-disk refiner; C) Twin refiner

The design of the refiner plates is also of great importance. This applies both to the optimum profile and to the alloy composition of the plate material [285]. The possibility of using elsewhere the heat generated in the process by means of heat exchangers has already been mentioned. In this way, 50 – 70 % of the motor energy can be saved [281, 282].

2.3.4. New Methods of Refiner Pulp Production

Although the addition of chemicals in the groundwood process did not prove worthwhile, their use in the refiner process has become increasingly important because wood chips can be impregnated very easily.

Sulfonation of the wood chips in combination with the TMP process has proved most effective. The properties of the CTMP can be varied to a great extent by changing the amount and the nature of the chemicals. CTMP has a wide field of application, ranging from tissue to LWC. A schematic of the process is given in Fig. 39 [286, 287].

Sulfonation results in permanent lignin softening which permits easy separation of the fibers. CTM pulps have a favorable dewaterability and are characterized by high strength, a large long fiber fraction, and a very low shive content [288].

A further development of the CTMP process is the CMP process. In this method, the process conditions are intensified, giving products comparable to semichemical pulps (Fig. 40) [286].

CTMP and CMP have the advantages of good strength properties and low shive content, but they also have the disadvantages of higher solvent losses and, therefore, environmental problems. Their tendency to yellowing is typical of all lignin-containing pulps [289]. For this reason, use of CTMP and CMP is limited to paper that is used for only a relatively short period of time.

2.4. Screening of Mechanical Pulps

2.4.1. Purpose of Screening

All mechanical pulps contain undesired components such as large, insufficiently pulped fragments (i.e., coarse rejects) and the shives, which consist of many fiber bundles (Fig. 41).

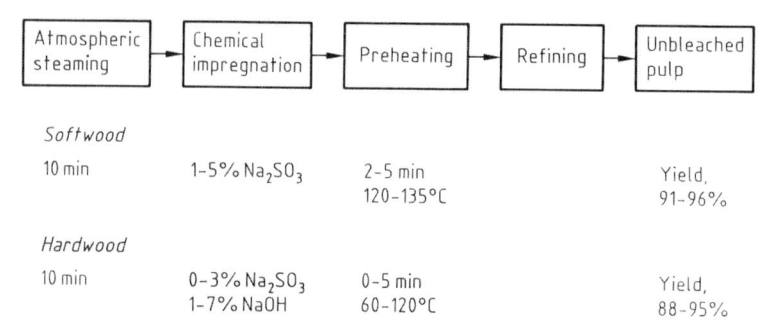

Atmospheric steaming	Chemical impregnation	Preheating	Refining	Unbleached pulp
Softwood				
10 min	1–5% Na_2SO_3	2–5 min 120–135 °C		Yield, 91–96%
Hardwood				
10 min	0–3% Na_2SO_3 1–7% NaOH	0–5 min 60–120 °C		Yield, 88–95%

Figure 39. Schematic of the CTMP process

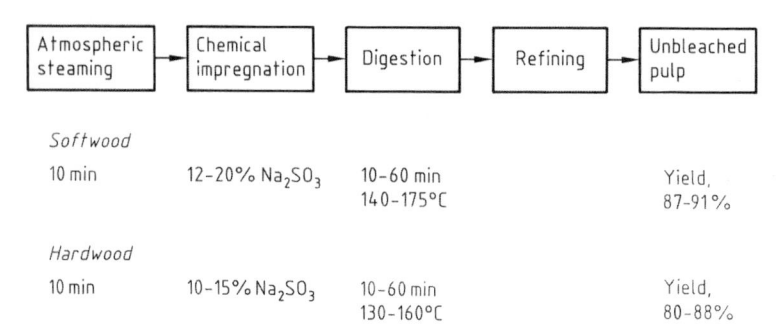

```
┌─────────────┐   ┌─────────────┐   ┌─────────┐   ┌─────────┐   ┌───────────┐
│ Atmospheric │──▶│ Chemical    │──▶│Digestion│──▶│Refining │──▶│ Unbleached│
│ steaming    │   │ impregnation│   │         │   │         │   │ pulp      │
└─────────────┘   └─────────────┘   └─────────┘   └─────────┘   └───────────┘
```

Softwood
10 min   12–20% Na_2SO_3 10–60 min   Yield,
                                    140–175 °C   87–91 %

Hardwood
10 min   10–15% Na_2SO_3 10–60 min   Yield,
                                    130–160 °C   80–88 %

Figure 40. Schematic of the CMP process

The shives have to be removed because they reduce the strength and the printing quality of paper [251]. The screening out of the coarse rejects is easy to perform, but the removal of the shives requires a more elaborate technique. The shive content of unscreened mechanical pulp may be as high as 5 %, depending on the process used. The purpose of screening is to remove the shives to a large extent and, at the same time, to keep the simultaneous screening out of valuable long fibers as low as possible. The extent of the required shive removal depends on the type of paper to be produced. With increasing printing quality, priority is given to shive removal over the prevention of fiber losses [290].

For the evaluation of the action of a screen or a screening system, it is not sufficient to evaluate the accepts alone; the quality of the inflow and of the rejects also has to be taken into consideration. The higher the efficiency of a screen, the higher the relative shive retention at the lowest possible overflow [291, 292].

2.4.2. Mechanical Pulp Screening

Coarse rejects are removed by vibrating flat screens (bull screens) equipped with perforated plates (3 – 10 mm). Hammer mills or pulpers are used for the reduction of the coarse rejects, also called "sauerkraut". The bull screen accepts mainly contain coarse shives and are usually added to the rejects of the fine screening system for further processing.

The screens used for the separation of shives usually have vertically arranged screen cylinders. The screening process is carried out at overpressure (Fig. 42). The size of the holes in the screen plates is chosen in such a manner that fibers and accepted stock pass through the screen, whereas shives above a certain size are retained. Depending on the type of paper required, screens with holes of 1.0 to 1.8 mm in diameter are employed, and the screening process is carried

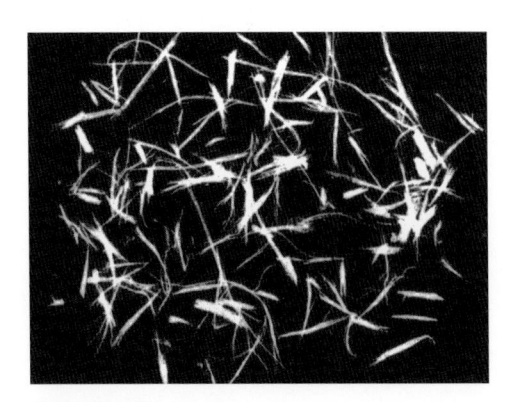

Figure 41. Fiber bundle or shives

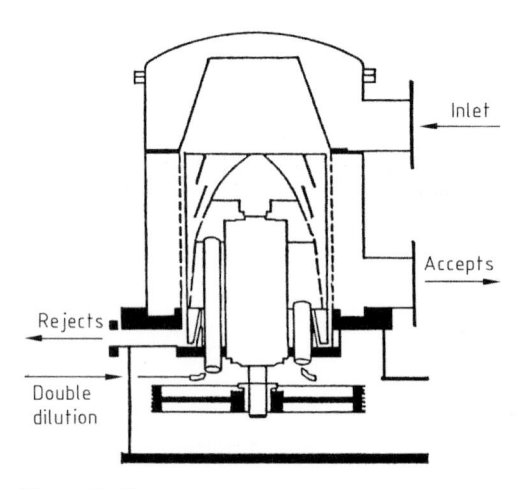

Figure 42. Hooper screen

out at high dilutions (pulp consistencies below 1 %) to promote the screening action.

Although older models are occasionally used, the new perforated plates equipped with slots (0.2 – 0.3 mm) instead of holes give an excellent screening effect [293, 294]. They relieve the cleaners, which are mainly used for the separation of cubelike particles, and are still an essential component of any multistage screening installation. The screening reject rate may be as high as 30 % of the inflow. A screening installation consists of various stages in order to reduce the strain on subsequent installations and to save most of the acceptable components contained in the rejects.

The rejects of the last stage, after thickening to ca. 20 % consistency, are processed in refiners and recycled to the screening process.

In the case of groundwood, the energy requirement for screening, for refining the screen rejects, and for subsequent thickening accounts for ca. 25 % of the total energy. The value for refiner mechanical pulp is ca. 29 % [295].

2.5. Bleaching of Mechanical Pulps

Mechanical pulps need to be brightened to suit their intended use. Since higher process temperatures are always accompanied by a loss of brightness, the amount of wood pulp that requires bleaching is constantly increasing. Two bleaching processes or a combination of the two are used to achieve the desired final brightness [296–298].

Reductive bleaching uses sodium dithionite ($Na_2S_2O_4$) and can lead to an increase in brightness of up to 12 % ISO. Depending on the required final brightness, the amount of $Na_2S_2O_4$ used is 0.1 to 1.2 %, based on o.d. pulp. A suitable pH value is 5.6 to 6.5, and a temperature up to 70 °C accelerates the bleaching process. However, a further increase in temperature is detrimental. The optimum pulp consistency is 3 – 5 % and the bleaching time is ca. 1 h. The bleaching tower operates with an overflow and is charged from the bottom upstream, due to the detrimental effect of the presence of free oxygen on the process [296, 297].

Oxidative bleaching is carried out in the pH range 10 – 10.5. Sodium peroxide (Na_2O_2) or hydrogen peroxide (H_2O_2) is used in amounts of 0.4 – 4 % H_2O_2 (50 %), based on o.d. pulp. The bleaching time is 1 – 2 h at 50 – 60 °C. High pulp consistencies of 15 – 20 % are favorable and the maximum increase in brightness is 14 % ISO. However, an increase in brightness of 16 – 20 points can be achieved by combining the two bleaching processes in the order peroxide–dithionite (Fig. 43) [297, 299]. Both bleaching processes result in lower brightness in the presence of heavy metal ions. For this reason chelating agents (0.3 – 0.5 %) are added before bleaching [296, 297, 300, 301].

The technological properties of mechanical pulps are not changed appreciably by bleaching. However, solvent losses vary substantially, being distinctly lower with reductive bleaching. Finally, it should be mentioned that the tendency of

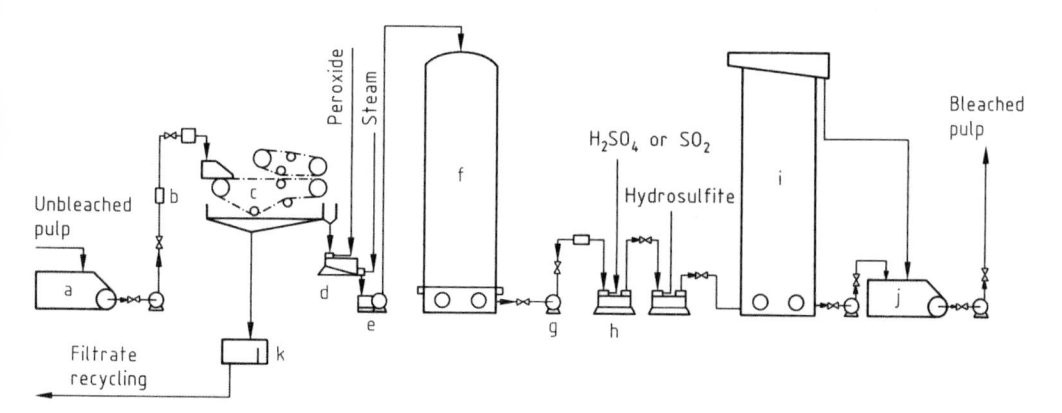

Figure 43. Schematic of a two-stage, continuous mechanical pulp bleaching process
a) Unbleached pulp chest; b) Pulp density regulator; c) Double wire press; d) High consistency mixer; e) High consistency pump; f) High consistency peroxide tower; g) Pump; h) Low consistency mixer; i) Low consistency hydrosulfite tower; j) Bleached pulp chest; k) Filtrate vessel

Table 14. Comparison of unbleached mechanical pulps (TAPPI standard)

	SGW	TGW	PGW	RMP	TMP	CTMP	CMP	Semibleached kraft pulp
Specific energy, kWh/t	1500	1530	1550	2100	2200	2100	1550	100
Process temperature, °C	75–80	93	115	90	120–140	120–140	150–160	
Yield, %	97–98	97–98	96–97	97–98	94–96	91–94	82–90	45
BOD, kg/t	7–15	9–16	10–17	ca. 15	18–25	30–40	50–85	200
Freeness (CSF), mL	100	100	100	100	110	110	300–400	350
Tensile index, Nm/g	30	34	37	36	42	47	54	65
Tear index, mN·m^2/g	3.4	3.9	4.7	3.9	7.3	7.3	9.0	14.0
Quality value (QV)*	98	112	131	114	188	193	234	345
Long fiber content (R 30), %	21	24	30	29	40	40		
Fines content (D 100), %	45	41	37	40	33	32		
Brightness, % ISO	63	62	60	58	56	60	52	

* QV = tensile index + 20 x tear index.

mechanical pulps to yellowing is not changed by bleaching.

2.6. Comparison of Mechanical Pulps

Each mechanical pulping method results in very specific properties of the pulps produced. Furthermore, the pulp properties can be influenced to a certain extent by varying the pulping conditions or by additional treatments. A decision in favor of one or the other process is always a compromise because of the great variety of often contrasting properties.

It is generally valid that increasing process temperatures result in increased solvent losses and reduced brightnesses. Although the use of chemicals (CTMP, CMP) is of a certain advantage (for example, allowing the pulping of hardwoods) it causes a further increase in solvent losses.

Table 14 lists some technological characteristics of the various processes after thickening and before bleaching. These values are intended only to indicate the trends at about the same freeness level (excluding CMP and halfbleached sulfate pulp), and it should be mentioned that the values for the specific energy requirement, the strength, and fiber classification have a margin of error of ca. ± 10%.

References

General References

1 J. P. Casey (ed.): *Pulp and Paper, Chemistry and Chemical Technology*, 3rd ed., Wiley Interscience, New York 1980.
2 D. Fengel, G. Wegener: *Wood, Chemistry, Ultrastructure, Reactions*, Walter de Gruyter, Berlin 1984.
3 M. J. Kocurek, O. V. Ingruber, A. Wong (eds.): "Sulfite Science and Technology," *Pulp and Paper Manufacture*, 3rd ed., vol. **4**, Joint Textbook Committee of the Paper Industry, Montreal – Atlanta 1985.
4 M. J. Kocurek, F. Hamilton, B. Leopold (eds.): "Secondary Fibers and Non-Wood Pulping," *Pulp and Paper Manufacture*, 3rd ed., vol. 3, Joint Textbook Committee of the Paper Industry, Montreal – Atlanta 1987.
5 M. J. Kocurek, T. M. Grace, E. Malcolm (eds.): "Alkaline Pulping," *Pulp and Paper Manufacture*, 3rd ed., vol. 5, Joint Textbook Committee of the Pulp and Paper Industry, Montreal – Atlanta 1989.
6 P. Lengyel, S. Morvay: *Chemie und Technologie der Zellstoffherstellung*, Güntter-Staib-Verlag, Biberach/Riß 1973.
7 A. Mimms, M. Kocurek, J. Pyatte, E. Wright (eds.): *Kraft Pulping*, Tappi Press, Atlanta 1990.
8 W. H. Rapson (ed.): "The Bleaching of Pulp," *Tappi Monogr. Ser.* **27** (1963).
9 S. A. Rydholm: *Pulping Processes*, Interscience Publ., New York 1965.
10 R. P. Singh (ed.): *The Bleaching of Pulp*, Tappi Press, Atlanta 1979.
11 E. Sjöström: *Wood Chemistry, Fundamentals and Applications*, Academic Press, New York 1981.
12 G. A. Smook: *Handbook of Pulp & Paper Technology*, Tappi Press, Atlanta 1990.
13 C. F. B. Stevens, M. J. Kocurek (eds.): "Properties of Fibrous Raw Materials and Their Preparation for Pulping," *Pulp and Paper Manufacture*, 3rd ed., vol. **1**, Joint Textbook Committee of the Paper Industry, Montreal – Atlanta 1983.
14 H. F. J. Wenzl: *Sulphite Pulping Technology*, Lockwood Trade Journal Co., New York 1965.
15 H. F. J. Wenzl: *Kraft Pulping*, Lockwood Publ. Co., New York 1967.

Specific References

16 J. E. Atchison, *Proc. Tech. Assoc. Pulp Pap. Ind.* 1988, 25 – 45.

17 B. Steenberg, *Sven. Papperstidn.* **64** (1961) 126 – 129.
18 L. Deslandes, *Paper (London)* **207** (1987) no. 5, 19 – 23.
19 B. Zobel, *Tappi J.* **72** (1988) no. 12, 42 – 46.
20 S. Sideaway, *Tappi J.* **72** (1988) no. 12, 47 – 51.
21 J. T. Jeyasingam, *Proc. Tech. Assoc. Pulp Pap. Ind.* 1988, 571 – 579.
22 A. M. Hurter, *Proc. Tech. Assoc. Pulp Pap. Ind.* 1988, 139 – 154.
23 S. Yllmer, B. Enström, *Sven. Papperstidn.* **59** (1956) 229 – 232.
24 S. Yllmer, B. Enström, *Sven. Papperstidn.* **60** (1957) 549 – 554.
25 I. Croon, B. F. Enström, *Tappi* **44** (1961) 870 – 874.
26 S. Axelsson, I. Croon, B. Enström, *Sven. Papperstidn.* **65** (1962) 693 – 697.
27 J. Gierer, I. Norén, *Acta Chem. Scand.* **16** (1962) 1713 – 1729.
28 J. Gierer, L. A. Smedman, *Acta Chem. Scand.* **19** (1965) 1103 – 1112.
29 J. Gierer, *Sven. Papperstidn.* **73** (1970) 571 – 596.
30 J. Marton in K. V. Sarkanen, C. H. Ludwig (eds.): *Lignins, Occurrence, Formation, Structure and Reactions*, Wiley-Interscience, New York 1971, pp. 639 – 694.
31 J. Gierer, I. Pettersson, L. A. Smedman, I. Wennberg, *Acta Chem. Scand.* **27** (1973) 2082 – 2094.
32 J. Gierer, S. Ljunggren, *Sven. Papperstidn.* **82** (1979) 503 – 508.
33 J. Gierer, S. Ljunggren, *Sven. Papperstidn.* **82** (1979) 71 – 81.
34 J. Gierer, *Wood Sci. Technol.* **14** (1980) 241 – 266.
35 J. Gierer, I. Norén, *Holzforschung* **34** (1981) 197 –200.
36 J. Gierer, *Holzforschung* **36** (1982) 43 – 51, 55–65.
37 M. H. Johansson, O. Samuelson, *Wood Sci. Technol.* **11** (1977) 251 – 263.
38 M. H. Johansson, O. Samuelson, *Sven. Papperstidn.* **80** (1977) 519 – 524.
39 B. Lindberg, *Sven. Papperstidn.* **59** (1956) 531 – 534.
40 D. B. Mutton, *Pulp Pap. Mag. Can.* **65** (1964) T 41 – T 51.
41 E. Sjöström, *Tappi* **60** (1977) no. 9, 151 – 154.
42 J. W. Green et al., *Tappi* **60** (1977) no. 10, 120 – 125.
43 O. Franzon, O. Samuelson, *Sven. Papperstidn.* **60** (1957) 872 – 877.
44 B. Alfredsson, L. Gedda, O. Samuelson, *Sven. Papperstidn.* **64** (1961) 694 – 698.
45 M. H. Johansson, O. Samuelson, *Carbohydr. Res.* **34** (1974) 33 – 41.
46 R. D. Brooks, N. S. Thompson, *Tappi* **49** (1966) 362 – 366.
47 C. H. Matthews, *Sven. Papperstidn.* **77** (1974) 629 –635.
48 Y. Z. Lai, *Ekman-Days 1981, Int. Symp. Wood Pulping Chem.* 1981, vol. 2, 26 – 33.
49 N. Hartler, *Tappi* **50** (1967) no. 3, 156 – 160.
50 P. J. Kleppe, *Tappi* **53** (1970) no. 1, 35 – 47.
51 A. R. Procter, R. H. Wiedenkamp, *J. Polym. Sci. Part C* **28** (1969) 1 – 13.
52 N. Hartler, L. A. Olsson, *Sven. Papperstidn.* **75** (1972) 559 – 565.
53 N. Hartler, *Sven. Papperstidn.* **62** (1959) 467 – 470.
54 R. Aurell, N. Hartler, *Tappi* **46** (1963) 209 – 215.
55 E. Hägglund, *Sven. Papperstidn.* **49** (1946) 204.
56 B. Alfredsson, O. Samuelson, B. Sandstig, *Sven. Papperstidn.* **66** (1963) 703 – 706.
57 N. Hartler, *Sven. Papperstidn.* **68** (1965) 369 – 377.
58 V. B. Lasmarias, R. C. Peterson, *Cellul. Chem. Technol.* **3** (1969) 479 – 496.
59 A. Teder, *Sven. Papperstidn.* **72** (1969) 294 – 303.
60 B. Bach, G. Fiehn, *Zellst. Pap. (Leipzig)* **21** (1972) 3 – 7.
61 H. H. Holton, *Pulp Pap. Can.* **78** (1977) T 218 – T 223.
62 H. H. Holton, F. L. Chapman, *Tappi* **60** (1977) no. 11, 121 – 125.
63 K. Goel, A. M. Ayroud, B. Branch, *Tappi* **63** (1980) no. 8, 83 – 85.
64 T. J. Blain, H. H. Holton, *Pulp Pap. Can.* **84** (1983) T 124 – T 129.
65 L. Löwendahl, O. Samuelson, *Sven. Papperstidn.* **80** (1977) 549 – 551.
66 L. Löwendahl, O. Samuelson, *Tappi* **61** (1978) no. 2, 19 – 21.
67 B. I. Fleming, G. J. Kubes, J. M. MacLeod, H. I. Bolker, *Tappi* **61** (1978) no. 6, 43 – 46.
68 J. R. Obst, L. L. Landucci, N. Sanyer, *Tappi* **62** (1970) no. 1, 55 – 59.
69 J. Gierer, O. Lindeberg, I. Norén, *Holzforschung* **33** (1979) 213 – 214.
70 G. J. Kubes, B. I. Fleming, J. M. MacLeod, H. I. Bolker, *Wood Sci. Technol.* **14** (1980) 207 – 228.
71 A. Wong, *CPPA Annual Meeting*, Montreal 1987, Prepr. 73 B, 249 – 254.
72 G. W. Kutney, *Pulp Pap.* **61** (1987) 73 – 75.
73 N. E. Virkola, J. Kettunen, I. Yrjälä, *Paperi ja Puu* **61** (1979) 685 – 700.
74 J. Kettunen, N. E. Virkola, J. Yrjälä, *Paperi ja Puu* **61** (1979) 686 – 700.
75 S. Raubenheimer, S. H. Eggers, *Papier (Darmstadt)* **34** (1980) no. 10 A, V 19 – V 23.
76 N. E. Virkola, R. Pusa, J. Kettunen, *Tappi* **64** (1981) no. 5, 103 – 107.
77 A. Wong, *PPI, Pulp Pap. Int.* **23** (1981) 55 – 59.
78 O. V. Ingruber, M. Stradal, J. A. Histed, *Pulp Pap. Can.* **83** (1982) T 342 – T 349.
79 R. E. Dörr, *Sven. Papperstidn.* **46** (1943) 361 – 369.
80 G. A. Richter, *Tappi* **39** (1956) 193 – 210.
81 M. V. Tuominen, *Paperi ja Puu* **50** (1968) 517 – 528.
82 L. A. Hiett, *Tappi J.* **68** (1985) no. 2, 42 – 48.
83 H. Crönert, *Papier (Darmstadt)* **40** (1986) 619 – 626.
84 H. Crönert, *Papier (Darmstadt)* **41** (1987) 63 – 71.
85 N. Hartler, L. A. Lindström, *Int. Dissolving Pulps Conf. (Prepr.)* 1987, 47 – 52.
86 D. K. Misra in J. P. Casey (ed.): *Pulp and Paper, Chemistry and Chemical Technology*, 3rd ed., vol. **1**, Wiley-Interscience, New York 1980, pp. 504 – 568.
87 Tappi, Non Wood Plant Fiber Pulping Progress Report Nr. 1 – 18.

88 M. F. Judt, *Proc. Tech. Assoc. Pulp Pap. Ind.* 1987, 451 – 459.
89 F. Wultsch, *Papier (Darmstadt)* 24 (1970) 394 – 408.
90 N. Hartler, L. Stockman, O. Sundberg, *Sven. Papperstidn.* **64** (1961) 33 – 49, 67 – 85.
91 N. H. Schön, L. Wannholt, *Sven. Papperstidn.* **72** (1969) 489 – 492.
92 G. Gellerstedt, J. Gierer, *Sven. Papperstidn.* **74** (1971) 117 – 127.
93 G. Gellerstedt, *Sven. Papperstidn.* **79** (1976) 537 –543.
94 D. W. Blennie in K. V. Sarkanen, C. H. Ludwig (eds.): *Lignins – Occurrence, Formation, Structure and Reaction*, Wiley-Interscience, New York 1971, pp. 597 – 637.
95 G. Gellerstedt, J. Gierer, *Acta Chem. Scand. B* 37 (1977) 729 – 730.
96 J. K. Hamilton, *Wood Chemistry Symp. Proc.* (1962) 197 – 217.
97 F. Shafizadeh, *Tappi* **46** (1963) 381 – 383.
98 J. F. Harris, *Appl. Polym. Symp.* 28 (1975) 131 – 144.
99 N. Hartler, L. Lind, L. Stockman, *Sven. Papperstidn.* **64** (1961) 160 – 166.
100 M. Goliath, B. O. Lindgren, *Sven. Papperstidn.* **64** (1961) 109 – 112, 469 – 471.
101 N. H. Schöön, *Sven. Papperstidn.* **61** (1964) 624 – 633.
102 N. E. Virkola, A. A. Alm, M. Mäkinen, R. Soila, *Paperi ja Puu* **42** (1960) 665 – 676.
103 F. M. Ernest, S. M. Harman, *Tappi* **50** (1967) no. 12, 110 A–116 A.
104 G. H. Tomlinson, G. H. Tomlinson II, J. R. G. Bryce, N. G. M. Tuck, *Pulp Pap. Mag. Can.* **59** (1958) 247 – 252.
105 T. W. Meloney, V. Gibbs, D. H. Andrews, *Pulp Pap. Mag. Can.* **74** (1973) T 378 – T 384.
106 R. M. Dorland, R. A. Leask, J. W. McKinney, *Pulp Pap. Mag. Can.* **59** (1958) 236 – 246.
107 W. W. Marteny in J. P. Casey (ed.): *Pulp and Paper. Chemistry and Chemical Technology*, 3rd ed., vol. 1, Wiley-Interscience, New York 1980, pp. 252 – 291.
108 H. E. Wörster, *Pulp Pap. Manuf.* **4** (1985) 130 –158.
109 O. V. Ingruber, G. A. Allard, *Pulp Pap. Mag. Can.* **74** (1973) T 354 – T 369.
110 S. Lagergren, B. Lundén, *Pulp Pap. Mag. Can.* **60** (1959) T 338 – T 341.
111 S. Lagergren, *Sven. Papperstidn.* **67** (1964) 238 –243.
112 G. E. Annergren, S. A. Rydholm, *Sven. Papperstidn.* **62** (1959) 737 – 746.
113 G. E. Annergren, L. Croon, B. F. Enström, S. A. Rydholm, *Sven. Papperstidn.* **64** (1961) 386 – 396.
114 G. E. Annergren, S. A. Rydholm, *Sven. Papperstidn.* **63** (1960) 591 – 605.
115 E. L. Bailey, *Tappi* **45** (1962) 689 – 691.
116 T. A. Pascoe, J. S. Buchanan, E. H. Kennedy, G. Sivola, *Tappi* **42** (1959) 264 – 281.
117 N. Sanyer, E. L. Keller, G. H. Chidester, *Tappi* **45** (1962) no. 2, 90 – 104.
118 J. E. Laine, J. Turumen, R. H. Rasanen, *Tappi* **62** (1979) no. 5, 65 – 68.
119 T. Ulmanen, R. Rasanen, M. Rantanen, *Tappi* **46** (1963) 151 A–153 A.
120 I. Hassinen, R. Räsänen, *Pulp Pap. Mag. Can.* **75** (1974) T 13 – T 15.
121 J. R. G. Bryce, G. H. Tomlinson, *Pulp Pap. Mag. Can.* **63** (1962) T 355 – T 361.
122 D. W. Cameron et al., *Appita* **35** (1982) 307 – 315.
123 I. D. Suckling, *Holzforschung* **43** (1989) 111 – 114.
124 A. Wong, *Tappi* **63** (1980) no. 4, 53 – 57.
125 E. Ojanen, J. Tulppala, J. Kettunen, *Paperi ja Puu* **64** (1982) 453 – 464.
126 O. V. Ingruber, *Proc. Tech. Assoc. Pulp Pap. Ind.* 1985, 461 – 469.
127 P. Tikka, N. E. Virkola, A. Wong, *Papier (Darmstadt)* **49** (1989) no. 10 A, V 1 – V 7.
128 R. P. Wither, H. A. Captein, *For. Prod. J.* **10** (1960) 174 – 177.
129 D. L. Brink, *Tappi* **44** (1961) 256 – 263.
130 D. L. Brink, J. Vlamis, M. M. Merriman, *Tappi* **44** (1961) 263 – 270.
131 D. L. Brink, M. M. Merriman, E. J. Schwerdtfeger, *Tappi* **45** (1962) 315 – 326.
132 J. H. Kalisch, *Tappi* **50** (1967) 44 A–51 A.
133 T. N. Kleinert, K. V. Tayenthal, *Angew. Chem.* **44** (1931) 788 – 791.
134 T. N. Kleinert, *Tappi* **57** (1974) no. 8, 99 – 102.
135 T. N. Kleinert, *Tappi* **58** (1975) 170 – 171.
136 T. N. Kleinert, *Papier (Darmstadt)* **30** (1976) no. 10 A, V 18 – V 24.
137 R. Katzen, R. E. Fredrickson, B. F. Brush, *Pulp Pap.* **54** (1980) no. 8, 144 – 149.
138 R. Katzen, R. E. Fredrickson, B. F. Brush, *Chem. Eng. Progr.* **76** (1980) no. 2, 62 – 67.
139 J. H. Lora, S. Aziz, *Tappi J.* **68** (1985) no. 8, 94 – 97.
140 E. K. Pye, *Pima* **69** (1987) no. 11, 21 – 23.
141 E. K. Pye, *Proc. Tech. Assoc. Pulp Pap. Ind.* 1990, 991 – 996.
142 M. Baumeister, E. Edel, *Papier (Darmstadt)* **34** (1980) no. 10 A, V 9 – V 18.
143 E. Edel, *DPW Dtsch. Papierwirtsch.* 1987, no. 1, T 39 – T 45.
144 E. Edel, *Papier (Darmstadt)* **41** (1989) no. 10 A, V 116 – V 123.
145 G. Dahlmann, M. C. Schröter, *Tappi J.* **73** (1990) no. 4, 237 – 240.
146 J. Nakano, H. Daima, S. Hosoya, A. Ishizu, *Ekman-Days 1981, Int. Symp. Wood Pulping Chem. Prepr.* 1981, vol. 2, 72 – 77.
147 R. Marton, S. Granzow, *Tappi* **65** (1982) no. 6, 103 – 106.
148 U. P. Gasche, *Papier (Darmstadt)* **39** (1985) no. 10 A, V 1 – V 7.
149 L. Paszner, N. C. Bahera, *Holzforschung* **39** (1985) 51 – 61.
150 L. Paszner, *Holzforschung* **43** (1989) 159 – 168.
151 L. Paszner, H. J. Cho, *Tappi J.* **72** (1989) no. 2, 135 – 142.
152 R. Patt, O. Kordsachia, *Papier (Darmstadt)* **40** (1986) no. 10 A, V 1 – V 8.

153 O. Kordsachia, R. Patt, *Holzforschung* **42** (1988) 203 – 209.
154 R. Patt, O. Kordsachia, K. Kopfmann, *Papier (Darmstadt)* **41** (1989) no. 10 A, V 108 – V 115.
155 O. Kordsachia, B. Reipschläger, R. Patt, *Paperi ja Puu* **72** (1990) 44 – 50.
156 J. Janson, R. Uvorisalo, *Paperi ja Puu* **68** (1986) 610 – 615.
157 B. Lönnberg, T. Laxén, R. Sjöholm, *Paperi ja Puu* **69** (1987) 757 – 762.
158 B. Lönnberg, T. Laxén, A. Bäcklund: *Paperi ja Puu* **69** (1987) 826 – 830.
159 T. Laxén, J. Aittamaa, B. Lönnberg, *Paperi ja Puu* **70** (1988) 891 – 894.
160 H. H. Nimz, R. Casten, *Holz Roh-Werkst.* **44** (1986) 207 – 212.
161 H. H. Nimz et al., *Papier (Darmstadt)* **41** (1989) no. 10 A, V 102 – V 108.
162 R. A. Young, E. B. Wiesmann, J. Davis, *Holzforschung* **40** (1986) 99 – 108.
163 R. Y. Young, *Tappi J.* **72** (1989) no. 4, 195 – 200.
164 K. Poppius et al., *Paperi ja Puu* **68** (1986) 87 – 92, 622 – 629.
165 J. Sundquist, *Paperi ja Puu* **68** (1986) 616 – 620.
166 J. Sundquist, L. Laamanen, K. Poppius, *Paperi ja Puu* **70** (1988) 143 – 148.
167 K. Poppius-Levlin, R. Mustonen, T. Huovila, J. Sundquist, *Paperi ja Puu* **73** (1991) 154 – 158.
168 S. A. Saponitzky: *Verwertung von Sulfitablauge*, VEB Fachbuchverlag, Leipzig 1963.
169 F. Melms, K. Schwenzon, *Verwertungsgebiete für Sulfitablauge*, VEB Deutscher Verlag für Grundstoffindustrie, Leipzig 1967.
170 S. Y. Lin, *Ekman-Days, Int. Symp. Wood Pulping Chem.* 1983, vol. 3, 108 – 113.
171 R. W. Coughlin, D. W. Sundstrom, H. E. Klei, E. Avni, "Conversion of Lignin to Useful Chemical Products," in D. L. Wise (ed.): *Bioconversion Systems*, C. R. C. Press, Boca Raton, Fla., 1984, 42 – 58.
172 O. V. Ingruber, *AIChE Symp. Ser.* **200** (1980) no. 76, T 196 – T 208.
173 T. Hauki, R. Reilama, *Pulp Pap. Can.* **85** (1984) T 120 – T 123.
174 R. Brännland, R. Gustafsson, B. Hultman, *Sven. Papperstidn.* **18** (1976) 591 – 594.
175 E. A. Horntvedt, *Tappi* **53** (1970) 2147 – 2152.
176 L. Stockman, L. Strömberg, F. de Sousa, *Cellul. Chem. Technol.* **14** (1980) 517 – 526.
177 I. Palenius, *Papier (Darmstadt)* **36** (1982) V 13 –V 19.
178 K. P. Kringstad, K. Lindström, *Environ. Sci. Technol.* **19** (1985) 1219 – 1224.
179 K. P. Kringstad, *Papier (Darmstadt)* **43** (1989) no. 10 A, V 12 – V 19.
180 H. Beck, K. Eckart, W. Mathar, R. Wittkowski, *Chemosphere* **17** (1988) 51 – 57.
181 S. E. Swanson, C. Rappe, J. Malmström, K. P. Kringstad, *Chemosphere* **17** (1988) 681 – 691.
182 J. J. Lindberg, *Paperi ja Puu* **80** (1988) 713 – 718.
183 C. Rappe et al., *Pulp Pap. Can.* **90** (1989) T 273 –T 278.
184 C. W. Dence, G. Annergren: "Chlorination," in R. P. Singh (ed.): *The Bleaching of Pulp*, Tappi Press, Atlanta 1979, pp. 29 – 80.
185 J. Gullichsen, *Tappi* **59** (1976) no. 11, 106 – 109.
186 C. W. Dence, K. V. Sarkanen, *Tappi* **43** (1960) no. 1, 87 – 96.
187 J. B. van Buren, C. W. Dence, *Tappi* **50** (1967) 553 –560.
188 J. B. van Buren, C. W. Dence, *Tappi* **53** (1970) 2246 – 2253.
189 C. W. Dence: "Halogenation and Nitration," in K. V. Sarkanen, C. H. Ludwig (eds.): *Lignins –Occurrence, Formation. Structure and Reactions*, Wiley Interscience, New York 1971, pp. 373 – 432.
190 J. Gierer, L. Sundholm, *Sven. Papperstidn.* **74** (1971) 345 – 351.
191 R. P. Singh, E. S. Atkinson: "The Alkaline Extraction," in R. P. Singh (ed.): *The Bleaching of Pulp*, Tappi Press, Atlanta 1979, pp. 81 – 100.
192 L. E. Larsen, H. deV. Partridge: "Bleaching with Hypochlorites," in R. P. Singh (ed.): *The Bleaching of Pulp*, Tappi Press, Atlanta 1979, pp. 101 – 112.
193 D. B. Mutton, *Pulp Pap. Mag. Can.* **63** (1964) T 41 – T 51.
194 W. H. Rapson, G. B. Strumila: "Chlorine Dioxide Bleaching," in R. P. Singh (ed.): *The Bleaching of Pulp*, Tappi Press, Atlanta 1979, pp. 113 – 157.
195 B. O. Lindgren, *Sven. Papperstidn.* **74** (1971) 57 – 63.
196 P. Axegård, *Pulp Pap. Sci.* **12** (1986) no. 3, J 67 – J 72.
197 P. Axegård, *Tappi J.* **69** (1986) no. 10, 54 – 59.
198 P. Axegård, *Proc. Tech. Assoc. Pulp Pap. Ind.* 1987, 105 – 111.
199 P. Axegård, *Proc. Tech. Assoc. Pulp Pap. Ind.* 1988, 69 – 76.
200 D. C. Pryke, *Tappi J.* **72** (1989) no. 10, 147 – 155.
201 D. C. Pryke, *Pulp Pap. Can.* **90** (1989) T 203 – T 207.
202 W. H. Rapson, C. B. Anderson, *Pulp Pap. Mag. Can.* **61** (1960) T 495 – T 504.
203 W. H. Rapson, C. B. Anderson, *Pulp Pap. Mag. Can.* **67** (1966) T 47 – T 55.
204 W. H. Rapson, C. B. Anderson, D. W. Reeve, *Pulp Pap. Can.* **78** (1977) T 137 – T 148.
205 D. W. Reeve, W. H. Rapson, *Tappi* **64** (1981) no. 9, 141 – 144.
206 J. V. Hatton, *Pulp Pap. Mag. Can.* **68** (1967) T 181 – T 190, T 204.
207 B. J. Fergus, *Tappi* **56** (1973) no. 1, 114 – 117.
208 U. Germgård, R.-M. Karlsson, *Paperi ja Puu* **66** (1984) 627 – 634.
209 P. S. Fredricks, B. O. Lindgren, O. Theander, *Tappi* **54** (1971) 87 – 90.
210 P. S. Fredricks, B. O. Lindgren, O. Theander, *Cellul. Chem. Technol.* **4** (1970) 533 – 547.
211 I. Croon, D. H. Andrews, *Tappi* **54** (1971) no. 11, 1813 – 1898.
212 R. P. Singh, B. C. Dillner: "Oxygen Bleaching," in R. P. Singh (ed.): *The Bleaching of Pulp*, Tappi Press, Atlanta 1979, pp. 159 – 209.

213 L. Almberg, I. Croon, A. Jamieson, *Tappi* **62** (1979) no. 6, 33 – 35.
214 G. L. Akim, *Paperi ja Puu* **63** (1981) 291 – 298.
215 P. J. Kleppe, *Papier (Darmstadt)* **39** (1985) no. 10 A, V 8 – V 13.
216 U. Germgård et al., *Sven. Papperstidn.* **88** (1985) R 113 – R 117.
217 T. J. McDonough, *Tappi J.* **69** (1986) no. 6, 46 – 52.
218 N. Liebergott, B. van Lierop, G. Teodorescu, G. J. Kubes, *Pulp Pap. Can.* **87** (1986) no. 5, 64 – 68.
219 L. Tench, S. Harper, *Tappi J.* **70** (1987) no. 11, 55 –61.
220 R. C. Eckert, H.-M. Chang, W. P. Tucker, *Tappi* **56** (1973) no. 6, 134 – 138.
221 J. Gierer, F. Imsgard, *Sven. Papperstidn.* **80** (1977) 510 – 518.
222 J. S. Gratzl, *Papier (Darmstadt)* **41** (1987) 120 –130.
223 O. Theander in J. S. Gratzl, J. Nakano, R. P. Singh (eds.): *Chemistry of Delignification with Oxygen, Ozone, and Peroxides,* Uni Publishers Co., Tokyo 1980, pp. 41 – 57.
224 L. Nasman, G. Annergren, *Tappi* **63** (1980) no. 4, 105 – 109.
225 E. F. Elton, V. L. Magnotta, L. D. Markham, C. E. Courchene, *Tappi* **63** (1980) no. 11, 79 – 82.
226 P. J. Kleppe, P. C. Knutsen, F. Jacobsen, *Tappi* **64** (1981) no. 6, 87 – 90.
227 K. Idner, *Tappi* **71** (1988) no. 2, 47 – 50.
228 H. Krüger, H. U. Süss, *Pulp Pap. Can.* **85** (1984) T 297 – T 299.
229 H. U. Süss, *Papier (Darmstadt)* **40** (1986) 10 – 15.
230 H. U. Süss, *Papier (Darmstadt)* **40** (1986) 150 –153.
231 D. Helmling, H. U. Süss, J. Meier, M. Berger, *Tappi J.* **72** (1989) no. 7, 55 – 61.
232 N. Soteland, *Pulp Pap. Mag. Can.* **75** (1974) T 153 – T 158.
233 N. Soteland, *Nor. Skogind.* **32** (1978) 199 – 204.
234 R. B. Secrist, R. P. Singh, *Tappi* **54** (1971) no. 4, 581 – 584.
235 R. P. Singh, *Tappi* **65** (1982) no. 2, 45 – 48.
236 N. Liebergott, B. van Lierop: "Oxygen, Ozone and Peroxide Bleaching," Tappi Seminar Notes, New Orleans 1978, pp. 90 – 106.
237 N. Liebergott, B. van Lierop, G. J. Kubes, *Ozone: Sci. Eng.* **4** (1982) 109 – 120.
238 N. Liebergott, B. van Lierop, G. J. Kubes, *Tappi J.* **67** (1984) no. 8, 76 – 80.
239 C.-A. Lindholm, *Paperi ja Puu* **69** (1987) 211 – 218.
240 R. Patt, O. Kordsachia, D. L.-K. Wang, *Papier (Darmstadt)* **42** (1988) no. 10 A, V 14 – V 23.
241 R. C. Eckert, R. P. Singh in J. S. Gratzl, J. Nakano, R. P. Singh (eds.): *Chemistry of Delignification with Oxygen, Ozone and Peroxides,* Uni Publishers Co., Tokyo 1980, pp. 229 – 243.
242 P. J. Balousek, T. J. Mc. Donough, R. D. McKelvey, D. C. Johnson, *Sven. Papperstidn.* **84** (1981) R 49 – R 54.
243 H. Kaneko, S. Hosoya, K. Iiyama, J. Nakano, *J. Wood Chem. Technol.* **3** (1983) 399 – 411.
244 T. Eriksson, J. Gierer, *Ekman-Days, Int. Symp. Wood Pulping Chem.* 1983, vol. 4, 94 – 98.
245 A. A. Katai, C. Schuerch, *J. Polym. Sci. Part A 1,* 1966, 2683 – 2703.
246 G. Pan, C.-L. Chen, H.-M. Chang, J. S. Gratzl, *Ekman-Days 1981, Int. Symp. Wood Pulping Chem.* 1981, vol. 2, 132 – 144.
247 C.-A. Lindholm, *Paperi ja Puu* **68** (1986) 283 – 290.
248 D. H. Andrews, R. P. Singh: "Peroxide Bleaching," in R. P. Singh (ed.): *The Bleaching of Pulp,* Tappi Press, Atlanta 1979, pp. 211 – 253.
249 G. Gellerstedt, I. Pettersson, S. Sundin, *Ekman-Days 1981, Int. Symp. Wood Pulping Chem.* 1981, vol. 2, 120 – 124.
250 VDP Papier 89: *Zahlenreihen zur Entwicklung der Zellstoff- und Papierindustrie in der BRD,* Verband deutscher Papierfabriken, Bonn 1989.
251 "Holzstoff – Halbstoff mit Zukunft," *Papier (Darmstadt)* **37** (1983) no. 10 A, V 11.
252 American National Standards, Tappi, Atlanta 1989.
253 *Scan Norm,* Testing Committee, Stockholm.
254 *Merkblatt VI/1/66,* Verein der Zellstoff- und Papier-Chemiker und -Ingenieure, Darmstadt.
255 W. Brecht: *Der Versuchsschleifer im Dienste der Holzschliffforschung,* Voith Forschung und Konstruktion, Heft 8, 1962.
256 F. G. Powell, F. Luhde, K. C. Logan: "Supergroundwood by Grinding," *Pulp Pap. Mag. Can.* **66** (1965) no. 8, T 399.
257 M. Aario, P. Haikkala, A. Lindahl: "Pressure Grinding – A New Method to Produce Mechanical Pulps," *CPPA Annual Meeting,* Montreal 1979.
258 A. Meinecke, R. Süttinger, K. Schmidt: "Das Thermoschleifen," *Wochenbl. Papierfabr.* **112** (1984) no. 9, 307.
259 *Ullmann,* 4th ed., **17,** 561.
260 A. Meinecke: "Über die Technologie und die Maschinen zur Holzstofferzeugung," *Wochenbl. Papierfabr.* **100** (1972) no. 9, 309.
261 R. Süttinger: "Ein Beitrag zur Kenntnis der Temperaturverhältnisse beim Schleifprozeß," *Wochenbl. Papierfabr.* **113** (1985) no 7, 247.
262 A. Atack, J. T. Pye: "The Measurement of Grinding Zone Temperature," *Pulp Pap. Mag. Can.* **65** (1964) Sept. T 363.
263 D. A. J. Goring: "Thermal Softening of Lignin, Hemicellulose and Cellulose," *Pulp Pap. Mag. Can.* **64** (1963) no. 12, T 517.
264 D. K. Alexander: "The Mechanism of Wood Grinding and Relationship between Energy and Fiber Quality," *7. International Mechanical Pulp Conference,* Ottawa, June 1971.
265 D. Atack, L. R. Heffell: "High Speed Grinding of Eastern Black Spruce," *Pulp. Pap. Mag. Can.* **73** (1972) no. 9, T 78.
266 R. Süttinger: *Die Technologie des Schleifprozesses,* Voith Forschung und Konstruktion, Heft 26, 1979.

267 K. H. Klemm: *Neuzeitliche Holzschlifferzeugung*, Sändig Verlag, Wiesbaden 1957, p. 106.
268 M. Loly: *Pulpstone Sharpening Manuel*, Norton International Inc., Worcester 1978.
269 G. Schweizer, N. Pancur, P. Hell: "Die Holzschliffqualität als Funktion des Steinprofils, *Wochenbl. Papierfabr.* **108** (1980), no. 11/12, 361.
270 K. H. Klemm: "Technik und Wirtschaftlichkeit der mechanischen Gewinnung von Faserstoffen aus Nadelholz," *Wochenbl. Papierfab.* **106** (1978) no. 11/12, 405.
271 A. Kärnä: "Pressure Groundwood, Status of Development and Use," *Eucepa Meet*, München Oct. 1980.
272 M. Aario, P. Haikkala: "Pressure Grinding is Proceeding," *Tappi* **63** (1980) no. 2, T 139.
273 D. Atack, J. Fontebasso, M. J. Stationvala: "Die Vorgänge beim Schleifen in einer Druckatmosphäre," *Papier (Darmstadt)* **35** (1981) no. 9, 397.
274 H. P. Sollinger: Die Temperatur und der Druck als Einflußgrößen auf die Schliffqualität beim Holzschleifverfahren, Dissertation, TH Aachen 1982.
275 J. Seehofer, G. Lippert: "Einjährige Erfahrungen mit einem Druckschleifer," *Wochenbl. Papierfabr.* **111** (1983) no. 17, 600.
276 R. Süttinger: *Die Weiterentwicklung des Holzschleifverfahrens*, Voith Forschung und Konstruktion, Heft 31, 1985, p. 9.
277 F. Luhde: "Der Aufschluß des Holzes beim Schleifverfahren und beim Hackschnitzel-Refinerverfahren," *Papier (Darmstadt)* **16** (1962) no. 11, 655.
278 D. Atack: "On the Characterization of Pressurized Refiner Mechanical Pulps," *Sven. Papperstidn.* **75** (1972) no. 3, T 89.
279 W. D. May: "A Theory of Chip Refining – The Origin of Fiber Length," *Pulp Pap. Mag. Can.* **74** (1973) no. 1, T 70.
280 D. Atack, W. D. May: *Mechanical Reduction of Chips by Double-Disk Refining*, Paper and Pulp Research Inst. Canada, Dec. 1962.
281 B. Ferrari, O. Danielsson: "Energierückgewinnung in TMP- und CTMP-Systemen," *Wochenbl. Papierfabr.* **114** (1986) no. 18, 739.
282 B. Engstrom, M. Jackson: *Heat Recovery Aspects of Sunds Defibrators TMP/CTMP Systems*, Sunds Defibrator AB, Sweden, 1984.
283 M. Jackson: *The Manufacture, Physical Properties and End-Uses of High-Yield Pulps*, Sunds Defibrator AB, Sweden 1988.
284 H. Münster: "Thermo-Mechanischer HolzstoffFaserstoff der Zukunft," *Wochenbl. Papierfabr.* **105** (1977) no. 5, 155.
285 B. R. Stein: *A Practical Approach to Refiner Plate Optimization*, vol. 1, nos. 1–3, J. & L. Plate, Optima, Wis., 1988.
286 G. äkerlund, M. Jackson: *Manufacture and End-Use Application of CTMP and CMP from Softwoods and Hardwoods*, Sunds Defibrator AB, Sweden, 1984.

287 J. V. Hatton, S. S. Johal: "Chemimechanical Pulps from Hardwood/Softwood Chip Mixtures," *Pulp Pap. Can.* **90** (1989) no. 3, T 91.
288 J. Kurdin: "Refinerbased Chemi-Thermo-Mechanical-Pulping," *Pulp Pap. Technol.* 1989.
289 H. U. Süss, W. Eul: "Hochgebleichter CTMP – ein Zellstoffersatz?" *Wochenbl. Papierfabr.* **114** (1986) no. 9, 320.
290 S. Nikula: "On Development of Mechanical Pulp Screening," *Paperi ja Puu* **71** (1989) no. 5, T 458.
291 E. Böttger, R. Rienecker: "Neue Entwicklungen bei der Herstellung von Holzschliff hoher Qualität," *Wochenbl. Papierfabr.* **108** (1980), no. 11/12, 429.
292 F. Stolze: *Beurteilungskriterien für Sortierschaltungen in Holzstofferzeugungsanlagen*, I. M. Voith, company publ. 1987, p. 2728.
293 K. Schmidt: *Holzschliffsortierung mit Schlitzsiebkörben*, I. M. Voith, company publ. 1987, p. 2743.
294 A. W. Hooper: "The Screening of Mechanical Pulp," Mechanical Pulping Course, CPPA Technical Section, S. W. Hooper Corp., Atlanta, Ga., 1987.
295 K. Süttinger: "Analyse des Energieverbrauchs in der Holzstofferzeugung," *Wochenbl. Papierfabr.* **107** (1979) no. 2, 40.
296 *Ullmann*, 4th ed., **17**, 565.
297 K. Schröter: *Die Holzschliffbleiche*, Güntter Staib Verlag, Biberach 1976.
298 M. J. Ducey: "Mechanical Pulp Bleaching Survey Explores New Equipment Technology," *Pulp Pap.* **63** (1989) no. 66, T 130.
299 R. Bott: "Moderne Verfahrenstechnik der Holzschliffbleiche," *Wochenbl. Papierfabr.* **100** (1972) no. 21, 791.
300 G. Schweizer: "Theorie und Praxis der Holzschliffbleiche," *Wochenbl. Papierfabr.* **100** (1972) no. 11/12, 433.
301 W. Auhorn, J. Melzer: "Untersuchung von Störsubstanzen in geschlossenen Kreislaufsystemen," *Wochenbl. Papierfabr.* **107** (1979) no. 13, 493. Chapter 3

Further Reading

M. Ash, I. Ash (eds.): *Handbook of Paper and Pulp Chemicals*, Synapse Information Resources, Endicott, NY 2000.
P. Bajpai: *Environmentally Benign Approaches for Pulp Bleaching*, Elsevier, Amsterdam 2005.
C. J. Biermann: *Handbook of Pulping and Papermaking*, 2nd ed., Academic Press, San Diego, CA 1996.
J. Brander, I. Thorn (eds.): *Surface Application of Paper Chemicals*, Blackie Academic & Professional, London 1997.
H. Holik (ed.): *Handbook of Paper and Board*, Wiley-VCH, Weinheim 2006.
M. A. Hubbe: *Paper*, "Kirk Othmer Encyclopedia of Chemical Technology", 5th edition, vol. 18, p. 89–132, John Wiley & Sons, Hoboken, NJ, 2005, online: DOI: 10.1002/0471238961.1601160512251405.a01.pub2

J. F. Kadla, Q. Dai: *Pulp*, Kirk Othmer Encyclopedia of Chemical Technology, 5th edition, vol. 21, p. 1–42, John Wiley & Sons, Hoboken, NJ, 2006, online: DOI: 10.1002/0471238961.1621121607051403.a01.pub2

J. C. Roberts (ed.): *Paper Chemistry*, 2nd ed., Blackie Academic & Professional, London 1997.

J. C. Roberts: *The Chemistry of Paper*, Royal Society of Chemistry, Cambridge 1996.

G. A. Smook: *Handbook for Pulp & Paper Technologists*, 3rd ed., Angus Wilde Publications, Vancouver, Bellingham 2002.

Polysaccharides

ALPHONS C. J. VORAGEN, Wageningen, Agricultural University, Department of Food Science, Wageningen, The Netherlands

CLAUS ROLIN, The Copenhagen Pectin Factory Ltd., Lille Skensved, Denmark

BEINTA U. MARR, The Copenhagen Pectin Factory Ltd., Lille Skensved, Denmark

IAN CHALLEN, NutraSweet Kelco, Waterfield, Tadworth, Surrey, United Kingdom

ABDELWAHAB RIAD, Setexam, Kenitra, Morocco

RACHID LEBBAR, Setexam, Kenitra, Morocco

SVEIN HALVOR KNUTSEN, Norwegian Food Research Institute, MATFORSK, Norway 1430 Ås

1.	Introduction	459	5.4.	Analysis	484
2.	Analysis and Characterization	460	5.5.	Properties	485
3.	Pectin	462	5.6.	Applications	487
3.1.	Occurrence and Structure	462	5.7.	Physiological Properties	488
3.2.	Pectolytic Enzymes	464	6.	Agar	488
3.3.	Production	464	6.1.	Production	488
3.4.	Properties	467	6.2.	Structure and Gelling Mechanism	489
3.4.1.	Physical Properties	467	6.3.	Quick Soluble Agar	490
3.4.2.	Gel Properties	468	7.	Gum Arabic	491
3.4.3.	Stability and Chemical Reactions	470	8.	Gum Tragacanth	493
3.5.	Analysis	472	9.	Gum Karaya	494
3.5.1.	Measurement and Standardization of Gel-Forming Capacity	472	10.	Gum Ghatti	495
			11.	Xanthan Gum	496
3.5.2.	Chemical Analysis	472	11.1.	Production	496
3.6.	Pharmaceutical and Nutritional Characteristics	473	11.2.	Structure and Properties	497
			11.3.	Analysis	500
3.7.	Application in the Food Industry	474	11.4.	Applications, Market	501
3.8.	Market	474	12.	Gellan Gum	501
4.	Alginates	474	13.	Galactomannans	503
4.1.	Occurrence	474	13.1.	Structure	504
4.2.	Production	475	13.2.	Production	504
4.3.	Structure	476	13.3.	Properties	505
4.4.	Properties	477	13.4.	Analysis and Composition of Commercial Preparations	505
4.5.	Propylene Glycol (Propane-1,2-diol) Alginate	479			
4.6.	Bacterial Alginates	480	13.5.	Derivatives	506
4.7.	Analysis	480	13.6.	Applications	506
4.8.	Applications	480	13.7.	Market	506
4.9.	Market	481	14.	Acknowledgement	506
5.	Carrageenan	481		References	507
5.1.	Structure	481			
5.2.	Sources and Raw Materials	483			
5.3.	Production	483			

1. Introduction [1–12]

For a general definition of the term polysaccharides, see → Carbohydrates: Occurrence, Structures and Chemistry.

Polysaccharides made up of only one type of neutral monosaccharide structural unit and with only one type of glycosidic linkage – as in cellulose or amylose – are denoted as *perfectly linear* polysaccharides. In branched polysaccharides the frequency of branching sites and the length of the side chains can vary greatly. Molecules with a long "backbone" chain and many short side chains are called *linearly branched* polysaccharides.

Polysaccharides are water soluble or swell in water, giving colloidal, highly viscous solutions or dispersions with plastic or pseudoplastic flow properties. Functional properties such as thickening, water holding and binding, stabilization of suspensions and emulsions, and gelling, are based on this behavior. Therefore, polysaccharides are often referred to as gelling or thickening agents, stabilizers, water binders, or fillers. A more generic name is gums. The terms hydrocolloids or, in the food industry, food colloids are also in use, but they include proteins with similar functional properties.

In this article the economically important polysaccharides, with the exception of starch (see → Starch) and cellulose (→ Cellulose) are described. Table 1 lists the polysaccharides, and gives an overview of their sources and composition. Their most important and various applications are summarized in Table 2. In only a few cases is their use specific. Examples are the use of pectin in commercial jam and marmalade production, where its heat stability coincides with its ability to gel in the typical pH range of these products, or the stabilization of drinking chocolate with carrageenan, which has a specific interaction with milk proteins and chocolate particles. In many cases, the manufacturer has various options and price plays an important role. For papermaking choice may be between, for example, alginates, galactomannans, and starch derivatives; for oil drilling, xanthan, starch derivatives, cellulose derivatives, galactomannans, or alginates can be used. Milk gels can be prepared with pectin, starch, alginates, or carrageenans. Often mixtures of polysaccharides are preferred to single polysaccharides because they combine various desirable properties (e.g., in ice cream or salad dressing). In a number of cases a synergistic interaction on the molecular level between polysaccharides can be exploited (e.g., improvement of the structure of carrageenan gels by the addition of guar gum or the preparation of gels by mixing xanthan and locust bean gum).

2. Analysis and Characterization

The analysis and characterization of polysaccharides can be aimed at the following:

Table 1. Origin and main sugar moieties of polysaccharides

Name	Origin	Main sugar moieties
Pectin	cell walls and middle lamella of higher land plants	D-galacturonic acid D-galacturonic acid methylester
Alginate	cells walls of brown seaweeds exopolysacchrides of *Azetobacter vinelandii*	D-mannuronic acid L-guluronic acid and their acetyl derivatives
Carrageenan	cell walls of red seaweeds	D-galactose 3,6-anhydro-D-galactose both sugars sulfated to higher or lower degree
Agar	cell walls of red seaweeds	D-galactose 3,6-anhydro-L-galactose few sulfate groups
Gum arabic	exudate of acacia species	L-arabinose, D-galactose L-rhamnose, (4-O-methyl) D-glucuronic acid
Gum tragacanth	exudate of astragalus species	L-arabinose, D-galactose D-galacturonic acid methylester D-xylose, L-rhamnose, L-fucose
Gum karaya	exudate of sterculia species	D-galacturonic acid, L-rhamnose, D-galactose D-glucuronic acid
Gum ghatti	exudate of *Anageissus latifolia*	L-arabinose, D-galactose D-mannose, D-xylose, D-glucuronic acid, L-rhamnose
Guar gum	endosperm of seeds of the guar plant (*Xyampsis tetragonolobus* L. Taub)	D-mannose D-galactose
Locust bean gum	endosperm of the seeds of the carob tree (*Ceratnia siliqua* L.)	D-mannose D-galactose
Tara gum	endosperm of the seeds of the tara tree (*Caesalpinia spinoza*)	D-mannose D-galactose
Xanthan gum	exopolysaccharide of *Xanthomonas campestris*	D-glucose, D-glucoronic acid, D-mannose substituted with acetate or pyruvate groups
Gellan gum	exopolysacchride *Pseudomonas elodea*	D-glucose, D-glucuronic acid, L-rhamnose

Table 2. Areas of application of polysaccharides*

Foods (→ Foods, 3. Food Additives)
1. Thickening and gelatinizing of fruit with a higher or lower content of sugar: preserves, marmalades, jellies, fruit for yogurt and ice cream, marmalade for cooking.
2. Gelatinizing of heat-reversible jelly and sugar glaze. Jellied sugar products, jellied sweets, dessert jellies (powder form).
3. Prevention of starch breakdown in bread and baked products, and in fruit-filled products based on starch (→ Bread and Other Baked Products).
4. Cloud stabilization in fruit juice. Providing body in refreshing powdered drinks (→ Beverages, Nonalcoholic).
5. Stabilization of powdered flavor emulsions. Microencapsulation of flavorings (→ Microencapsulation).
6. Gelatinization and thickening of milk: puddings and cream for hot and cold preparations; also as powdered instant products (→ Milk and Dairy Products).
7. Stabilization of chocolate milk and evaporated milk: prevention of sedimentation or separation of the cream fraction (→ Milk and Dairy Products).
8. Stabilization of ice cream: prevention of formation of ice crystals and dripping. Stabilization of the fat emulsion, imparting favorable slip properties for mold release (→ Milk and Dairy Products).
9. Water binding in cream cheese-type spreads, soft cheese, and cheese preparations.
10. Stabilization of fat emulsion in powdered coffee whiteners and low-fat margarine.
11. Stabilization of egg-white froth, whipped cream, and imitation whipped cream, meringues, and beer foam.
12. Stabilization and thickening of soups, sauces, dips, salad cream, mayonnaise, catsup, etc. Prevention of syneresis. Imparting freeze – thaw stability, and body in fat and starch-reduced products.
13. Binding agents in ground meat products (e.g., corned beef, sausages) and in canned dog and cat foods.
14. Gelatinized (jellied meat) or thickened stock from meat, fish, and vegetables.
15. Binding agents for (potato flour) snacks.

Cosmetics
16. Thickening agents in creams, ointments, lotions, and hair gels (→ Skin Cosmetics).

Household Goods and Industry
17. Solid air fresheners.
18. Stabilization of emulsified and suspended active agents in insecticides and herbicide sprays.
19. Stabilization of suspended abrasion materials in polishes and cleaning agents.
20. Glue.

Pharmaceuticals
21. Thickening agents in creams, lotions, ointments, and syrups.
22. Toothpaste (water-binding, consistency, → Dental Materials).
23. Denture adhesives.
24. Tablet-binding agents.
25. Dental impression materials (→ Dental Materials).

Industry
26. Binding agents in explosives.
27. Binding agents for layered coating of welding electrodes.
28. Different functions of the rinsing liquid used in deep drilling technology.
29. Paper binding agents. Paper coating agents (fat-resistant properties, improved printability, uniform distribution of pigment despite extremely high speeds of modern papermaking machines (→ Paper and Pulp, Production); additives to improve absorbency.
30. Sizing and finishing agents used in the textile industry (→ Textile Auxiliaries).
31. Thickening of textile printing pastes and coloring agents to ease application and prevent color bleeding.
32. Flocculation agents for water processing (→ Water; → Wastewater).
33. Pigment stabilization and paint consistency (thixotropy in emulsion paints).

*The numbering has been introduced merely to allow reference in the text and does not rank the applications in order of importance.

1. Estimation of their content in raw materials
2. Qualitative and quantitative identification of one or more polysaccharides in a preparation
3. Qualitative and quantitative identification of one or more polysaccharides in a (food) product
4. Estimation of specific properties such as gelling strength or viscosity
5. Estimation of specifications for purity (e.g., color, content of heavy metals or cations, moisture, ash, or microbial count)

The manufacturer is interested primarily in item 1); items 2) and 3) are of concern to the user and 3) for state inspection services; item 4) is of interest to the manufacturer in terms of standardization, and the user in terms of applications; item 5) is of particular interest to the consumer and to state inspection services in pharmaceutical and food applications.

For determination of items 2) and 3), several possiblities exist [13], [14]. In most cases, solutions or extracts must be prepared in which the polysaccharides are estimated qualitatively by precipitation reactions [15], or a special analytical procedure is used to separate anionic polysaccharides from neutral polysaccharides by precipitation with cetylpyridinium chloride [16], [17]. When the necessary equipment is available, quantitative analysis, based on the gas chromatographic determination of the constituent sugar moieties after hydrolysis and derivatization of the purified polysaccharide is often quicker and more reliable [29]. Recently, HPLC has been used more often in the analysis of constituent sugar moieties. Another new development is the use of immunological methods based on the development of antibodies against the polysaccharides. The requirements indicated under item 5) and the corresponding analytical methods are defined in pharmacopeias, food legislation, and specifications of national and international organizations [41, 62, 63]. The latter also describe identification reactions. Additional information on the analysis of individual polysaccharides is given in the corresponding sections.

3. Pectin

Pectin or pectic substances, also called galacturonans or rhamnogalacturonans in scientific literature, is a collective name for heteropolysaccharides that consist predominantly of partially methylated galacturonic acid residues [18, 74, 85, 96, 104–111]. The name pectin (Greek: *Pectos* = gelled) was coined by BRACONNOT [112] who first described this compound in detail in 1825 and indicated its primary use as a gelling agent. Native pectin plays an important role in the consistency of fruits and vegetables, and in textural changes during ripening and storage [85], [113]. The often desired cloud stability of fruit juices and fruit drinks is lost when enzymes endogenous to the fruit degrade pectin (tomatoes) or cause it to precipitate as calcium pectate after demethylation. On the other hand, enzymes may be added as processing aids to degrade native pectin, for example, to apple juice to facilitate clarification or to the pulp of berries to improve press and color yield [114–116]. Pectin is extracted on an industrial scale from the press residues in apple and citrus juice manufacture and used mainly as gelling and stabilizing agents in the food industry [117].

3.1. Occurrence and Structure

Occurrence. Pectins occur in virtually all higher plants, Zosteraceae seaweed, and certain freshwater algae. Pectins are major structural components of the primary cell wall and the middle lamella of young growing plant tissues (meristimatic and parenchymatic) but do not occur in more mature tissue. The composition of the cell wall therefore is of major importance in the texture of fruit and vegetables [19], [20]. The primary cell wall consists of 90 % polysaccharide and 10 % glycoprotein on a dry matter basis [21]. The polysaccharide composition is 30 – 60 % cellulose, 15 – 45 % pectic substances, and 15 – 25 % hemicellulose [22].

The biosynthesis of pectin takes place in the cell plate during cell division. Pectin is formed as polygalacturonic acid with UDP-D-galacturonic acid – arising from UDP-D-glucose by an epimerase- catalyzed reaction – as the most active glycosyl donor. Immediately after the galacturonan chain has been formed, methoxyl groups are formed with S-adenosylmethionine as the methyl group donor [120].

Pectic substances can be partially solubilized from plant tissues without degradation by using weakly acidic, aqueous solvents with or without calcium chelating agents. The pectin fraction that is not extractable with these extractants because of its attachment to other cell wall components by chemical, physical, or mechanical (enmeshment) bonds, is often designated as *protopectin*. Commercial pectin extraction must break down protopectin to a soluble, high molar mass pectin. This is achieved by acid hydrolysis in which, on the one hand, the molar mass of pectin molecules is lowered and, on the other hand, connections to the hemicellulose fractions are split. This transition from protopectin to soluble pectin also occurs during ripening of fruit or cooking of vegetables, resulting in textural changes [85], [113]. Pectin technology is therefore interested in the nature of protopectin and its fixation in the cell walls in an attempt to develop more efficient methods of extraction. FRY [119] has discussed cross-links in cell walls and agents used to cleave the individual bonds. Pure enzymes may also be used for extraction of firmly bound pectin [24], [25].

Structure. Pectin is composed of 1,4-linked α-D-galactopyranosyluronic acid units in the $^4C_1$

Figure 1. α-D-Galactopyranosyluronic acid in $^4C_1$ conformation (above); fragment of galacturonan chain, 40 % methylated (below)

conformation, with the glycosidic linkages arranged diaxially (Fig. 1). A proportion of the carboxyl groups is esterified with methanol. Commercial pectins are divided into low-ester and high-ester pectin. In *low-ester pectins (LM-pectins)* less than 50 % of the carboxyl groups are methylated (typical range is from 20 to 40 %) whereas in *high-ester pectins (HM-pectins)* more than 50 % are methylated (typical range is from 55 to 75 %). If less than 10 % of the carboxyl groups are methylated the polysaccharide is called *pectate* or *pectic acid*. Pectins prepared from pears, potatoes, sugar beets, and sunflower heads are acetylated to varying degrees at the secondary hydroxyl groups of the galacturonic acid residues [118].

The heteropolysaccharide nature of pectin derives from the fact that other sugars are incorporated in the pectin molecule. The most common ones being L-rhamnose (Fig. 2), occasionally inserted by α-1,2-linkages in the galacturonan backbone, providing "kinks" in the molecular chain. Other sugars are β-D-xylose, attached as single-unit side chains mainly to O-3 of the galactopyranosyluronic acid residues in the backbone; and D-galactose and L-arabinose, which occur in long side chains, only attached to rhamnopyranosyl residues (for projection formulae, see → Carbohydrates: Occurrence, Structures and Chemistry).

The frequency of rhamnose occurrence remains to be established, however it has been suggested that α-rhamnosyl units are concentrated in rhamnose-rich areas. In other words, the soluble pectin is built up of homogalacturonan-dominated areas, so-called *smooth regions* linked to rhamnogalacturonan areas rich in neutral sugars, so-called *hairy regions* (Fig. 3). The neutral sugars account for 10 – 15 % of the weight of the pectin. In the hairy regions, the neutral sugar chain length may be in the range 8 to 20 residues [26], [27]. By degradation with chemical β-elimination and endo-polygalacturonase, it was found that 90 % of the rhamnose units are found in the hairy region [28], [30].

By acid hydrolysis, i.e., splitting the acid labile glycosidic bonds between rhamnose and galacturonic acid, nearly pure homogalacturonic acids with molecular masses in the range 20 000 to 25 000 have been obtained [32], [33]. This corresponds to a chain length of 75 to 100 galacturonic acid residues. For comparison, a molecular mass of 90 000 has been quoted for intact pectin [34], but the molecular mass of pectin is somewhat uncertain (see Section 3.5.2).

Figure 2. 1,2-Linked L-rhamnopyranosyl unit in $^1C_4$ conformation

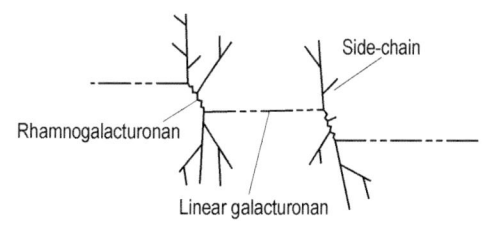

Figure 3. Schematic representation of pectin backbone, showing the "hairy" regions (rhamnogalacturonan and side-chains) and the "smooth" regions (linear galacturonan) [31]

Commercial pectins may also contain neutral polysaccharides that are not covalently attached to the pectic backbone, such as galactans, arabinans, arabinogalactans, and starch, often referred to as "ballast" compounds. Purified pectins prepared from apple pomace or citrus peel may contain 75 – 90 % anhydrogalacturonic acid on an ash- and moisture-free basis.

3.2. Pectolytic Enzymes

Pectins can be attacked by various enzymes [114], [115], [121]. The significance of native pectic substances in food technology can be evaluated properly only when the activity of these enzymes is taken into account.

Pectinesterase (PE, pectin methylesterase, pectase, pectin demethoxylase, pectin pectylhydrolase, EC 3.1.1.11) splits off the methoxyl groups and converts high-methoxyl pectins to low-methoxyl pectins. The latter are extremely sensitive to complex formation and precipitation with Ca^{2+} ions, particularly when a pectinesterase of plant origin is used. This type of enzyme does not saponify methyl esters in a random fashion as microbial PEs do, but acts along the galacturonan chain, creating blocks of free carboxyl groups. Pectinesterase occurs in many higher plants, particularly tomatoes, citrus, and other fruits; it is also produced by many fungi and bacteria.

Polygalacturonase (PG, pectinase, pectate hydrolase, poly-α-1,4-D-galacturonide glycanohydrolase, EC 3.2.1.15 and 3.2.1.67) preferentially hydrolyzes low-methoxyl pectins or pectic acid because these enzymes can cleave glycosidic linkages only next to free carboxyl groups. PGs can be divided into enzymes that degrade their substrate by an endo attack (splitting randomly in the backbone, *endoPG*), and enzymes that act from the nonreducing end removing mono- or digalacturonic acid (*exoPG*). PGs are produced by fungi and certain bacteria, and also occur in higher plants (tomatoes). The endoPGs, with their strong depolymerizing action, are of particular technological importance.

Pectate Lyase (PAL = pectic acid lyase, PATE = pectic acid transeliminase, LMPL = low-methoxyl pectin lyase, poly-α-1,4-D-galacturonide lyase, EC 4.2.2.2 and 4.2.2.9) also splits glycosidic linkages next to free carboxyl groups. In this group of enzymes, endo and exo enzymes also exist. The preferential substrates for *endo-PAL* are LM-pectins rather than pectic acid. Pectate lyases have an absolute requirement for Ca^{2+} ions. The glycosidic linkages are split by a *trans*-elimination reaction. PALs are produced predominantly by bacteria and are not important in fruit and vegetable processing because of their high optimum pH (8.5 – 9.5).

Pectin Lyase (PL, PTE = pectin transeliminase, pectinase, poly-α-1,4-D-methoxygalacturonide lyase, EC 4.2.2.10) splits glycosidic linkages between methoxylated galacturonide residues by a *trans*-elimination reaction. These enzymes therefore have a preference for HM-pectins. Pectin lyases are produced only by fungi.

A newer pectin-degrading enzyme acts in cooperation with a pectin acetylesterase only on highly branched regions of pectin to release oligosaccharides consisting of alternating sequences of α-1,2-linked L-rhamnosyl residues and α-1,4-linked D-galacturonosyl residues, with galactosyl residues β-1,4-linked to part of the rhamnosyl units. The nonreducing end is always a rhamnose unit [122], [123].

Commercially Available pectinases, used on an industrial scale in fruit and vegetable processing are of fungal origin and generally contain in addition to PE, PG, and PL, proteases and various hemicellulases and cellulases.

3.3. Production [106], [109], [117], [122], [124]

The production of pectin is summarized in Figure 4.

Raw Materials of importance to pectin manufacturing are currently various kinds of citrus peel, and apple pomace. Lemon and lime are the preferred citrus sources, and more pectin is produced from these than from apple or the less preferred citrus materials, orange and grapefruit. Some of the pectin producing companies which historically developed in connection to apple production now partly or wholly base their

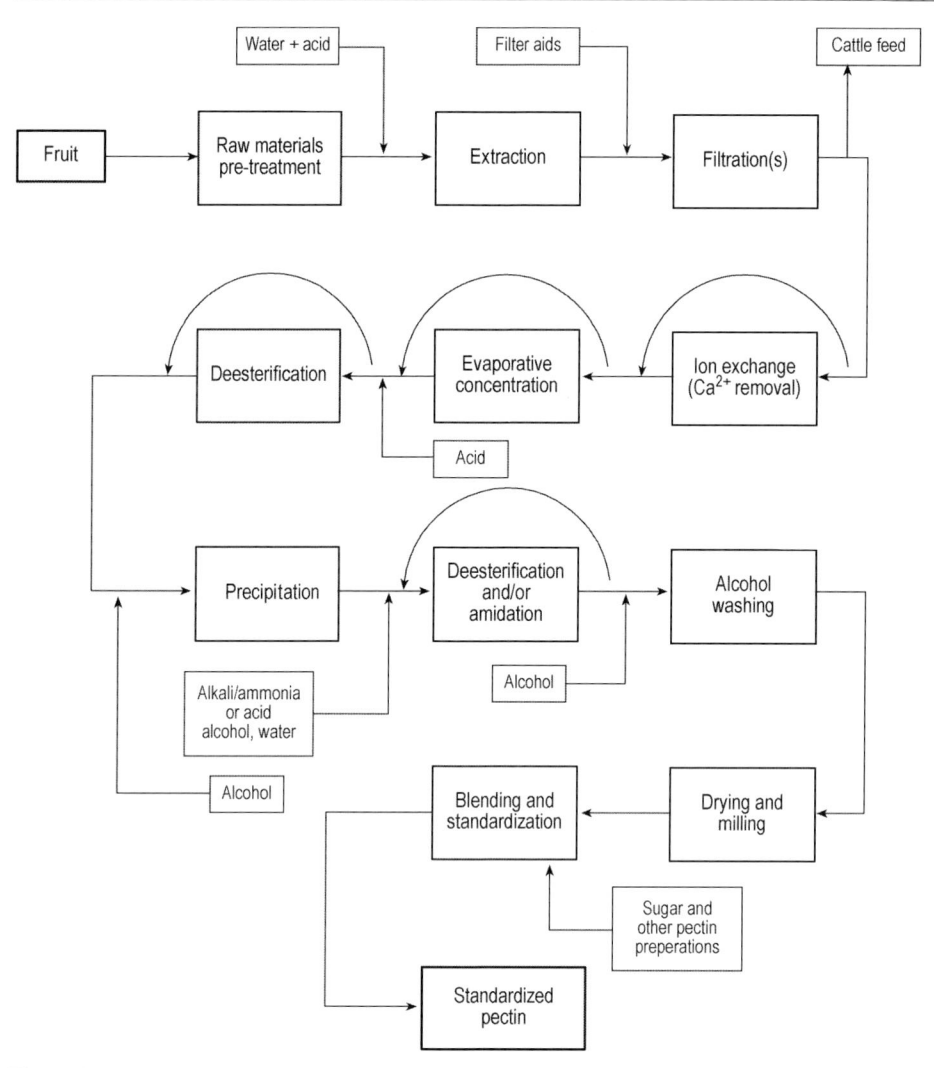

Figure 4. Flow chart of pectin production

production on imported citrus peel. Sugar beet was once used as a pectin source to some extent [35], but the pectin is inferior as a gelling agent compared to citrus or apple pectin. It has been reintroduced, and is currently being marketed as a stabilizer. Numerous other sources like mango [36], pea hulls [37], sunflower heads [38], [39], and pumpkin [40] have been suggested.

Citrus peel and apple pomace are available as byproducts from juice manufacturing. They are usually washed in water and dried before being used for pectin manufacturing, but some citrus material is used in pectin plants neighboring the juice production without previous drying. In either case processing of the raw materials has to commence immediately after juice production in order to prevent microbial degradation. The washing leaches out organic acids, sugar, and pigments and is thus one of the separation processes in the purification of pectin. Most importantly, it prevents discoloration either from browning of pigments or from caramelization during the raw materials drying.

Extraction. The pretreated raw material is extracted in water which has been acidified with e.g., hydrochloric or nitric acid. Typical conditions are: pH 1 to 3, temperature 50 to 90 °C, duration 3 to 12 h. During the extraction, limited depolymerization of the pectin and possibly of other connecting biopolymers takes place, and the pectin dissolves. The low pH further dissociates ionic linkages which hold the pectin in the plant tissue. In addition to hydrolyzing glycosidic bonds, the extraction conditions also hydrolyze ester linkages, more specifically the methyl ester at C-6, and the acetate to which pectin may be esterified by its hydroxyl groups. The extraction process thus causes a reduction in degree of polymerization as well as in degree of methyl and acetate esterification. The pectin yield increases with the acidity, the temperature, and the duration, but the product will lose too much in degree of polymerization if all these parameters are at their maximum. The combination of low pH and low temperature favors hydrolysis of ester linkages over hydrolysis of glycosidic bonds, and it is thus preferred for production of pectin with a relatively low degree of esterification.

Filtration in one or more stages separates the extract containing the solubilized pectin from the insoluble, but at this stage very soft and fragile, plant tissue. The rather difficult filtration requires reasonably low viscosity, and as a consequence the pectin concentration must be less than 0.6 to 1 %, depending upon the pectin type. Further, the solids must not have been comminuted by excessive mechanical treatment such as vigorous agitation. Water-insoluble materials like wood cellulose or diatomaceous earth may be added in order to improve the porosity and mechanical strength of filter cakes. Amylase may be added to remove starch from apple pectin extracts.

The spent plant raw material is typically used for cattle feed.

Isolation. Following filtration, the extract may optionally be passed through a column with cation-exchange resin and concentrated by evaporation. The pectin is then precipitated by mixing the extract with an appropriate alcohol, e.g., 2-propanol. Finally, the precipitate is separated from the spent alcohol, washed in more alcohol, pressed to drain as much liquid as possible, and then dried and milled. The powder is now ready for standardization, i.e., mixing with other pectin batches and/or sucrose in order to ensure uniformity. The alcohol is recovered by distillation. An alternative to alcohol precipitation is precipitation by adding appropriate metal salts to the extract. Pectin forms insoluble salts with, e.g., Cu^{2+} and Al^{3+}. The Al^{3+} precipitation [125] was previously used industrially. Removal of the metal ions from the precipitated pectin is done by washing in acidified aqueous alcohol.

Modification. Pectin derived from citrus or apple raw material as described above will normally have a degree of esterification between 55 and 75. A lower degree of esterification can be achieved by acidifying the extract and leaving it for some time before precipitation, or by treating precipitated (but not dried) pectin with acid or alkali during suspension in aqueous alcohol. Ammonia may convert methyl-esterified carboxylate groups of a pectin to primary amides [42], [43]. This is done industrially by suspending precipitated pectin in a mixture of alcohol and water with dissolved ammonia [127], [128]. Deesterification takes place concurrently. By choosing proper conditions with respect to ammonia concentration, water activity, and temperature, pectins with various proportions of amidated, methyl esterified and free carboxylate groups can be produced.

Standardization. The properties of botanical raw materials like those used for pectin fluctuate due to, e.g., weather conditions or sorts variation. Pectin as it appears directly from milling contains this variation. In order to maintain a constant quality, the pectin manufacturer may mix different batches. Further, pectin intended for food is typically diluted with sucrose in order to achieve a uniform grade (i.e., "strength," for a definition, see Section 3.5.1). Pectin without admixed sugar, e.g., for pharmaceutical purposes, is also available from the major manufacturers.

Due to the multitude of ways in which pectins may vary, it is not possible to ensure batch to batch consistency with respect to all possible attributes at the same time. Pectin is normally standardized with respect to a few properties which are measured in defined chemical systems which simulate the applications, e.g., breaking strength of a gel, gelation temperature, etc. The major pectin manufacturers have developed a great number of specialty types which are tailor-made for individual applications and which have each their set of standardization criteria (control methods). In fact, a pectin type is defined by its set of standardization criteria. When using pectin, it is obviously important to choose a type which has been standardized in a way that corresponds reasonably to the intended use.

3.4. Properties

A range of parameters – intrinsic and extrinsic – are important for the performance of pectin which is in most cases used to impart certain rheological properties, e.g., by forming a gel. The *intrinsic parameters* determine the nature of the gel and may include molecular mass, degree of esterification (DE), degree of acetylation (DA), neutral sugar content, and composition. The *extrinsic parameters* which determine the gelation process may include pectin concentration, pH, ionic strength, water activity, and temperature.

The most important properties of pectin preparations depend on their molecular mass and DE and DA. The proportion and nature of neutral sugars in the side chains as well as in the "ballast" are also of significance. Pectin with a high degree of polymerization is more viscous in solution compared to a pectin with a lower degree of polymerization. Further, the gel strength will typically increase with the degree of polymerization, i.e., less pectin is needed with a high molecular mass pectin [44–46]. The DE strongly influences the functional properties of pectin with the two main groups, HM- and LM-pectin being influenced differently. At the typical conditions in HM-pectin applications, high DE means high gelling temperature whereas at typical use conditions for LM-pectins, low DE means high gelling temperature. No experimental evidence is available to demonstrate the influence of the hairy regions (see Structure) on the functionality of pectin. It could be speculated that by being bulky and providing kinks in the molecular chain, these parts will prevent molecules from aligning throughout their entire length. This may contribute to preventing precipitation and reducing potential syneresis (spontaneous exudation of solvent from the gel).

Chemical Reactions. By treating pectin with ammonia under alkaline conditions in alcohol suspensions, ca. 20 % of the methyl ester groups are converted to acid amide groups, and amidated pectins are obtained [129]. Amidated pectins have a higher calcium reactivity than LM-pectins, and gels can be obtained with very few Ca^{2+} ions [134]. Carboxyl groups in pectin can be esterified easily with methanol [126], glycol, and glycerol but poorly with ethanol [135]. By using polyols, cross-linked, insoluble systems are obtained. Insoluble pectates can also be prepared by cross-linking with epichlorohydrin; these pectates have ion-exchange properties with a certain selectivity for calcium and heavy-metal ions. They are successfully used for the isolation of pectolytic enzymes [136], [137]. Pectins are readily degraded by oxidants [138] except for chlorite and chlorine dioxide, which can selectively oxidize aldehyde groups at the reducing chain end [139].

3.4.1. Physical Properties

Pectin is water-soluble, exhibiting an increased solubility with increasing DE and decreasing degree of polymerization. In order to ensure complete dissolution of pectin, it is necessary that it is properly dispersed without lumping. Once formed, lumps are extremely difficult to dissolve. Pectin, like any other gum or gelling agent, will not dissolve in media where gelling conditions exist. In order to add pectin to complex formulations such as food systems three alternative procedures are recommended: (1) dissolve the pectin in pure water and add the solution; (2) dry blend the pectin with five parts of sugar and add the mixture; (3) disperse the pectin in a liquid in which it is not soluble and add the dispersion.

Aqueous dilute pectin solutions, i.e., with a pectin content below 0.5 % are almost Newtonian whereas more concentrated pectin solutions exhibit pseudoplasticity, i.e., shear-thinning behavior. From dilute viscosity data, the intrinsic viscosity, $[\eta]$ (dL/g) may be determined. $[\eta]$ indicates the hydrodynamic volume of a polymer molecule and depends primarily on the molecular mass, however viscosity is also influenced in a complex manner by many other factors such as the DE, pH (dissociation), and ionic strength. $[\eta]$-values for pectins typically lie in the range of 1.0 to 6.0 dL/g [47–51]. The molecular mass of pectin is often estimated by intrinsic viscosity methods using the Mark – Houwink relationship $[\eta] = KM^\alpha$ (see Section 3.5). Originally an α-value of 1.34 [52] was suggested corresponding to a rigid rod-like molecular structure, however more recent findings estimate α to be in the range 0.7 – 0.8 indicating a random coil structure [47], [49], [53], [54].

Pectin is insoluble in most organic solvents such as alcohols and acetone. Pectin can also be precipitated from aqueous solutions by quaternary detergents, water-soluble cationic polymers including proteins, and multivalent cations. LM-pectins can be precipitated by calcium ions; pectates by alkali cations and by acid.

Pectin is a polycarboxylic acid. Dissolved pectin is negatively charged at neutral pH and approaches zero charge at low pH. Since pectin is a polyprotic acid it is not possible to determine an exact value of the apparent dissociation constant, pK_a. Rather, pK_a is different for varying carbohydrate concentrations and for varying degree of dissociation, α. The negative charge density, in turn, is dependent on DE which implies that pK_a-values are increasing with increasing content of unesterified galacturonic acid units. Typically, pK_a-values at 50 % dissociation, i.e., α = 0.5, lie in the range 3.5 – 4.5 [55–59]. The usual dependence of pK_a with polymer concentration and ionic strength is observed, i.e., pK_a is lowered with increasing concentration and ionic strength.

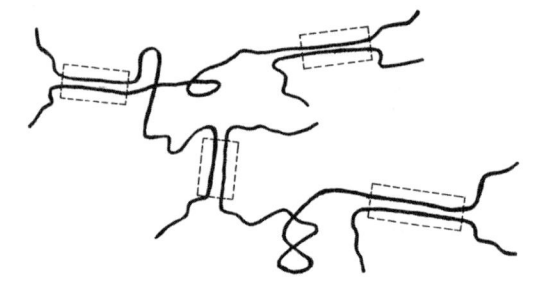

Figure 5. Network of a gel

3.4.2. Gel Properties

Pectin is used mainly as a gelling agent in industry; therefore its gelling properties are most important [4], [6], [106], [117]. The gel formation mechanism of pectin is similar to that of other gelling polysaccharides: Some regions of the polymer molecules associate in junction zones to form a three-dimensional network, which traps the solvent with cosolutes; free stretches of the molecules provide elasticity to the gel obtained [111], [140], [141] (Fig. 5). Irregularities in the pectin molecule, such as the distribution pattern of methyl ester or *O*-acetyl groups, rhamnosyl residues in the backbone, or the presence of side chains, limit the length of the junction zones and give shape to the free stretches of the macromolecules [142].

A comprehensive, coherent description of pectin gelation does not exist. Conventionally, a distinction is made between HM- and LM-pectin gelation, however, in reality this paradigm is too simple. It applies that different gelation mechanisms may act simultaneously. This is e.g., illustrated by the fact that gelation of LM-pectin, which is normally claimed to gel in the presence of certain metal ions, is further favored by a decrease in pH. If the gelation was solely determined by the formation of calcium-bridges between molecules, the opposite effect would be expected, an increase in pH leading to an increase in gel strength.

In order to adapt the conventional and still widely accepted theories for pectin gelation, the gelation phenomenon will, however, be treated as two distinct mechanisms, ie HM- and LM-pectin gelation. Gels used for jams and jellies are typically formed with HM-pectin at an acidic pH and require the presence of a high concentration of sugar. LM-pectin is typically used for yogurt fruit preparations; these gels can be formed without sugar over a wide pH range, however, the presence of a divalent cation is necessary. In most cases the cation, i.e., calcium, is inherently present in the fruit material. LM-pectin gels can be remelted whereas usually it is not possible to melt an HM-pectin gel, i.e., with HM-pectin gel preparations, the difference between the apparent temperatures of setting and melting is so large that the gels are said to be thermo-irreversible. Further, LM-pectin gels solidify almost immediately after gelling conditions have been introduced, while an HM-pectin gel will build up over time.

HM-Pectins form so-called *low-water-activity* or *pectin – sugar – acid gels* and are used in jam, jelly, and marmalade production. The basic galacturonan chain (smooth regions) of the pectin molecule apparently contains blocks with conformational regularity to provide opportunities to build up junction zones. The homogalacturonan part of the molecule is configured as helices with three anhydrogalacturonic acid units per turn, with the methyl ester groups protruding from the helix. According to OAKENFULL and SCOTT [143], junction zones are stabilized by different forces between pectin chain molecules: hydrogen bonds between undissociated carboxyl and secondary alcohol groups, and hydrophobic interactions between methoxyl groups (Fig. 6). Both types of forces are fortified by sucrose; the low pH suppresses the dissociation of carboxyl groups. Sugar reduces the water activity of the system and thereby influences hydrogen bonding by decreasing polymer – water interactions and increasing polymer – polymer interactions. To a certain extent, sugar and acid are interchangeable: at lower sugar concentration, lower pH is required, but at higher sugar concentration, higher pH values are possible and necessary to avoid setting during the boil. The lower limit for the sugar concentration is 55 %. At this concentration, the pH should not be higher than 2.8. At a sugar concentration of 80 % (jellies), the mass will also gel at pH 3.5. This means that within the gelling range, at the same sugar concentration, more acid will give a stronger gel and, at the same pH value, this is achieved by adding more sugar. Addition of urea to a gel cancels out hydrogen

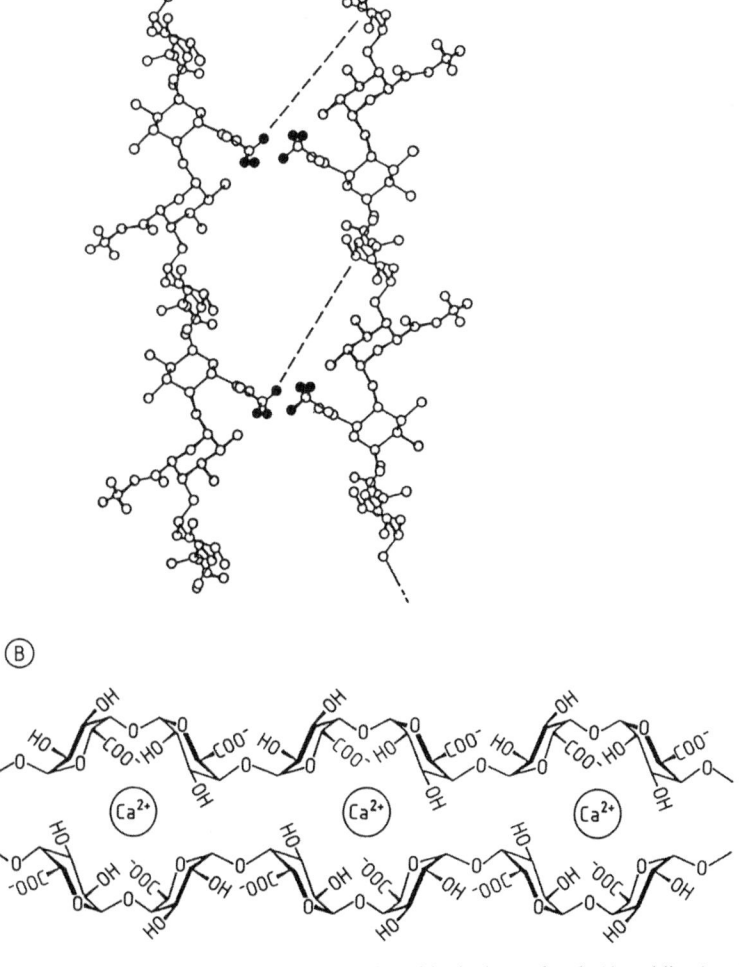

Figure 6. A) Junction zone in high ester pectin gel by hydrogen bonds (dotted lines) and hydrophobic interactions (filled circles); B) Junction zone in low ester pectin gel by dimerization of polygalacturonate (polyguluronate blocks) induced by their strong binding power for Ca^{2+} ions which fill the oxygen-lined cavities between the polysaccharide chains [110]

bonds and results in a weaker gel with a lower setting temperature [144].

In general, pectin – sugar – acid gels are prepared by a boiling process followed by cooling. At a certain temperature the system sets to a gel. The food technologist is interested in the rate of setting. At the same rate of cooling the rate of setting determines the setting temperature or the setting time. The parameters that contribute to stronger gels also accelerate the setting rate. Based on the pH limits of gelling as well as on the setting rate, HM-pectins can be subdivided into rapid-set (DM $> 70\,\%$) and slow-set (DM 60 – 65 %) pectins. At the same sugar concentration, rapid-set pectins have a higher pH limit for gelling because these very highly methyl esterifiedpectins have few carboxyl groups that must be protonated. A fully methyl esterified pectin gels with sugar alone and does not need acid. The concentration of pectin in the gel influences the rheological properties of the gel, not the gelling rate.

LM-Pectins form so-called *calcium pectate gels*. Theories about the chemical structure of calcium pectate gel junction zones were first developed in comparison with alginate gels (see Chapter 4). In calcium alginate gels, the junction zones are formed by α-1,4-L-polyguluronate blocks in which the diaxial configuration of the glycosidic linkages leads to a buckled ribbon with limited flexibility and a strong binding power for Ca^{2+} ions, which induce dimerization of alginate chains by filling the oxygen-lined cavities between them. This has evoked the picture of an eggbox, and the expression eggbox-type junction zones has become universally accepted [111], [141], [145], [146] (Fig. 6). In comparing the primary structures of poly-L-guluronate and poly-D-galacturonate (pectin), they are seen to be stereochemically analogous mirror images of each other, except at C-3. The two-fold helix, however, has not been observed with X-ray diffraction techniques, but molecular modeling calculations indicate that it can exist [60]. Circular dichroism data [61] suggest that a conformational change takes place when dissolved or Ca^{2+}-gelled pectate is dried and it is suggested that a transition takes place from a 2_1 ribbon-like to a 3_1 helical symmetry. Similar to the low-water-activity gels, junction zones in calcium pectate gels are terminated by rhamnosyl residues in the backbone, side chains attached to the backbone, or acetyl groups. The presence of some methoxyl groups does not inhibit formation of eggbox-type junction zones. With LM-pectins, acids or sugars are not so important. Gels of acidic fruit juices with LM-pectins can be made by addition of a calcium salt (low-sugar jams) or of milk with its neutral pH and calcium ions (desserts). The amount of calcium necessary for gelation depends on the degree of esterification, the way the LM-pectin has been prepared, and the types and amounts of other ingredients. Coagulation as a result of the addition of calcium salts must be absolutely avoided. Slow availability of Ca^{2+} for the pectin molecules is a prerequisite for obtaining a gel network. This can be accomplished in various ways: (1) an insoluble calcium salt (phosphate, citrate, tartrate) may be used resulting in a slow exchange of Ca^{2+} ions with the LM-pectin and formation of a gel; (2) use of calcium chelating agents such as diphosphates help to retard the availability of Ca^{2+} ions. The fact that calcium pectate gels and precipitates are often thermoreversible (i.e., they are soluble under conditions of gel formation at high temperature) can also be used to advantage. Soluble calcium salts such as calcium lactate or calcium chloride can therefore be added at boiling temperature, and gelling occurs on cooling. LM-pectins can be solubilized in milk by heating; the calcium caseinate of the milk provides the calcium necessary for the system to gel on cooling. Addition of sugars to such gels gives stronger gels; however, at higher concentrations the risk of coagulation increases. A solution to this problem is offered by *amidated pectins*. Gel formation with amidated pectins is also explained by chain associations via eggbox junction zones. Eggbox junction zones can accommodate amide groups, which, however, provide less drive for Ca^{2+} ion binding [146], [147]. On the one hand, they need fewer Ca^{2+} ions for gelation, and on the other hand in the presence of excess Ca^{2+} ions they are not as sensitive to coagulation [148]. In the United States, amidated LM-pectin is used for all applications of LM-pectins.

3.4.3. Stability and Chemical Reactions

Stability of pectin molecules in aqueous solution depends upon the temperature and the pH. Pectin has, in contrast to most other hydro-

tions after heating for 10 min at 90 °C [131]. As is evident from the figure, highly esterified pectin is vulnerable to high pH. Even at pH 5, depolymerization is considerable, in particular at elevated temperatures [133]. Consequently, it is difficult to raise the pH of pectin solutions without causing a decline in the average degree of polymerization, because when trying to mix an alkaline solution into a pectin solution, too high pH cannot be avoided locally. It is recommended to ensure good agitation and low temperature, and to avoid the use of hydroxides.

Low pH hydrolyzes ester bonds causing a decline in DE as well as in the content of *O*-acetyl groups, and it causes a decline in degree of polymerization by hydrolysis of glycosidic bonds [64], in particular at rhamnose insertions in the molecular backbone [65]. Very high acid concentration may degrade galacturonic acid to CO_2, furfural, reductic acid (2,3-dihydroxy-2-cyclopenten-1-one; $C_5H_6O_3$) and alginetin. Carbon dioxide production which is quantitative by boiling in 12 mol/L HCl has in the past been used for the quantitative determination of pectin [85], [132].

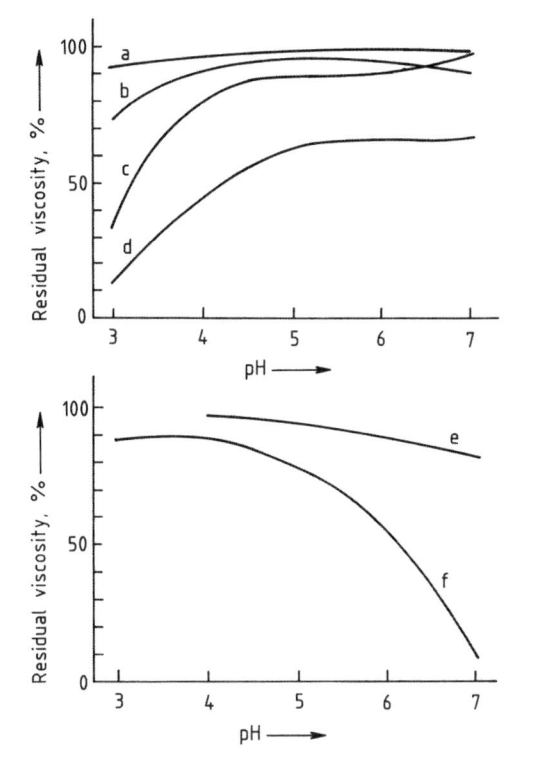

Figure 7. Stability of some polysaccharides at various pHs [131]: Residual viscosity after 10 min incubation at 90 °C
a) Carboxymethylcellulose – methyl cellulose; b) Locust bean gum; c) Agar; d) Carrageenan; e) Pectate; f) Pectin

colloids, optimal stability at pH 3.5 to 4. Figure 7 shows the stability of pectin and some other thickening agents at various pH values, expressed as residual viscosity of buffered solu-

High pH depolymerization is due to β-elimination [66] (Fig. 8), it requires the presence of a methyl ester group at the anhydrogalacturonic acid residue which has its 4-C attached to the bond being split. Since the presence of methyl esters is required for β-elimination, vulnerability to this degradation mechanism is related to the

* Enzymatically
X = OCH_3, Y = OCH_3 Pectin lyase (PL)
X = OH, Y = OH or OCH_3 Pectat lyase (PAL)

Chemically
X = OCH_3, Y = OCH_3 or OH   pH > 5

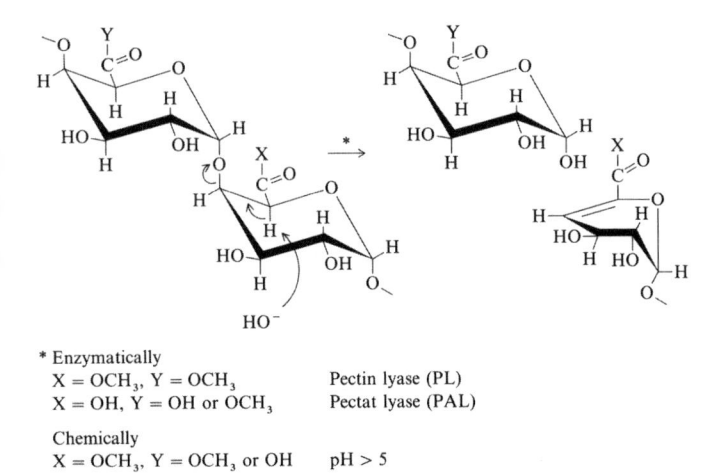

Figure 8. β-Eliminative depolymerization of a galacturonan chain by pectin lyase (PL) or pectate lyase (PAL) or chemically at pH ≥ 5

DE. High pH further reduces DE (whereby the β-elimination becomes incomplete) as well as the content of *O*-acetyl groups.

Derivatization. The low-pH hydrolysis of natural ester linkages may be reversed under conditions of low water activity, e.g., in mixtures of methanol and concentrated sulfuric acid [67], [68], [126], or, for introduction of *O*-acetyl groups, mixtures of concentrated sulfuric acid and acetic anhydride [69]. Further, the carboxyl group may readily be esterified with glycol or glycerol but poorly with ethanol [135]. By using polyols, cross-linked, insoluble systems are obtained.

At high pH, ammonia may convert methyl esterified carboxylate groups to amides [42], [129]. This is used industrially since amidated pectins are of commercial importance.

Insoluble pectates can be prepared by cross-linking with epichlorohydrin; these pectates have ion-exchange properties with a certain selectivity for calcium and heavy-metal ions. They are successfully used for the isolation of pectolytic enzymes [136], [137].

Pectins are readily degraded by oxidants [138] except for chlorite and chlorine dioxide which can selectively oxidize aldehyde groups at the reducing chain end [139].

3.5. Analysis

3.5.1. Measurement and Standardization of Gel-Forming Capacity

HM-Pectins are generally standardized to uniform strength at specified constant conditions. Expressing the sugar binding capacity of the pectin, the USA – SAG method suggested by the IFT Committee for Pectin Standardization, has been universally accepted for grading HM-pectins [149]. A standard gel is prepared in conical test glasses with the following conditions: soluble solids 65%, pH 2.20 – 2.40, gel strength 23.5%, SAG measured with a *ridgelimeter* [106], [153]. After 24 h at 25 ± 3 °C the gel is deposited on a glass plate, and the sagging of the gel under its own weight is measured after 2 min. From the SAG value and the pectin quantity, the grade can be calculated. Most commercial HM-pectins are standardized to 150 grade USA – SAG. (A gel strength of 150 grade SAG implies that 1 part of pectin is able to transform 150 parts of sucrose into a jelly with above standard properties.)

LM-Pectin may be standardized by closely analogous procedures, however, no universally accepted method exists. With LM-pectins it is difficult to set up a single universal test because the conditions under which LM-pectins are used may differ widely with respect to soluble solids, calcium content, and pH (see also Standardization).

3.5.2. Chemical Analysis [70]

In addition to gel-forming ability, the analysis of pectin preparations is concerned particularly with the *degree of of esterification, DE*. This is determined by converting the pectin to its acid form by passing it over an ion-exchange resin or washing it in an alcohol suspension, first with hydrochloric acid – alcohol and then with neutral alcohol. The acid and saponification equivalent is determined by titration, and from these values the anhydrogalacturonic acid content and the DE (in percent) can be calculated [63]. This principle for determination of DE has been adopted by the major legislative bodies [71–73]. Presence of *O*-acetyl groups in the pectin will result in an overestimation of DE as well as the anhydrogalacturonic acid content, therefore, it should be evaluated in a separate aliquot of the preparation. The pectin can also be precipitated with copper ions before and after saponification to determine the copper in the well-washed precipitate. From the amount of copper ions bound to the original pectin and the saponified pectin the anhydrogalacturonic acid and the DM can then be calculated without the interference of acetyl groups [151], [152]. Various methods exist for the separate determination of methyl ester and acetyl groups. Methanol is released on saponification of methyl ester and can be determined by GLC either directly [153] or in the headspace of a closed vial after conversion of the methanol to volatile methyl nitrite [154], [155]. *Acetyl groups* can be conveniently determined by an enzymatic spectrophotometric method (supplier Boehringer), by GLC [156], or by distillation and titration after alkaline saponification [157]. A convenient new method is to saponify the pectin preparation in alkaline alcohol, which is then analyzed by HPLC for methanol and acetic acid [158]. Methods measuring the relative content of carboxylate groups to total material, either using size exclusion chromatography with combined detection by conductivity and refractive index [75], or using capillary electrophoresis [76], have also been published. The DE can be inferred from calibration curves if it can be assumed that the anhydrogalacturonic acid content is the same in the samples compared. Previously, the analytical methods used by legislative bodies did not comprise a correction for *O*-acetyl groups since the content of those is small in citrus pectin and modest in apple pectin, but it has now been included in the latest version of the FAO/WHO specification for pectin [71].

In commerical pectins, up to 25 % of the carboxyl groups may be amidated. The *degree of amidation* is calculated from the amount of ammonia released on alkaline distillation [63], [159]. Amidated pectic acids undergo β-elimination reactions, whereas pectic acid does not, which permits quantitative analysis of mixtures of amidated and nonamidated pectins [160]. The anhydrogalacturonide content of pectins can also be determined from aqueous solutions by colorimetric methods with carbazole [161], the more specific *m*-hydroxydiphenyl [162], or sulfamate – *m*-hydroxydiphenyl [163]. Sometimes corrections must be made for interfering compounds (neutral sugars, amide groups, azide). The colorimetric methods can be automated easily and used for routine analyses of large series of samples [130].

With the above-mentioned methods, only the galacturonide residues in the backbone can be analyzed. The *neutral sugars* can be analyzed conveniently by gas chromatography after acid hydrolysis and conversion to volatile derivatives. By a preceding precipitation of pectins with copper ions the neutral sugars covalently attached to the galacturonan can be analyzed specifically. If starch is present it can be removed by enzymatic degradation [164].

To analyze the *pectin of plant material* the so-called alcohol-insoluble residue is usually prepared first by washing the plant material with refluxing alcohol. This inactivates endogenous enzymes and removes alcohol-soluble constituents. The pectin content is then determined in extracts of the alcohol-insoluble residue. The total pectin is determined in an alkaline extract or in the combined extracts of enzymatic and acid extraction. Another approach is the gradual extraction first with cold or hot water (HM-pectin), oxalate, ethylenediaminetetraacetic acid (EDTA), or cyclohexanediamine tetraacetate (CyDTA) (LM-pectin), and then acid or alkali (protopectin) [165], [166]. The total pectin content and the average degree of methoxylation can also be determined in the alcohol-insoluble residue when this is converted to the acid form before and after saponification. When treated with alcoholic calcium acetate solutions, the free carboxyl groups of pectin set free an equivalent amount of acetic acid that can be determined by titration. An alternative is determination of the bound copper ions from copper solutions [151].

Molecular Mass of pectin may be determined with viscosimetry [44], [77], membrane osmometry [78], size exclusion chromatography [47], [48], [54], light scattering [54], [77], [79], [80], ultracentrifugation [34], and analysis of reducing end-groups [80]. Quoted results vary, partly because the intermolecular distribution of molecular mass is broad, partly because pectin molecules aggregate and may contain slight amounts of insoluble material, and, of course, partly because samples are different. As an example, a weight average molecular mass of 90 000 ± 10 000 was reported by HARDING et al. [34]. Quoted values for the Mark – Houwink exponent, relating intrinsic viscosity to molecular mass, are generally in the vicinity of 0.8 [47], [49], [53], [54], suggesting a random coil molecular shape. Integrated systems combining high performance size exclusion chromatography, viscosity detection and light scattering detection are commercially available, including software with which data for molecular mass and Mark – Houwink parameters can be extracted.

3.6. Pharmaceutical and Nutritional Characteristics [167]

Pectin is not significantly degraded in the upper digestive tract of humans, and it can be recovered almost intact from the small intestine [81]. In the cecum and colon, it is fermented by microorganisms mainly to short-chain fatty acids, as can be concluded from in vitro studies [82], rat studies [83] and studies comparing the degradation patterns of humans and rats [84].

Since pectin is a dietary fiber, much attention has been paid to the possible health benefits of pectin which are: (1) reduced glycemic response [86–88], (2) prolonged gastric residence time [89–91], (3) reduced serum cholesterol level [86], and (4) effect against diarrhoea [92]. Pectic polysaccharides from certain botanical sources like ginseng root (*Panax ginseng*) [93], eel grass (*Zosteraceae*), and *Bupleurum falcatum* [94], [95] have shown healing effect on gastric and duodenal ulcers. Most studies, e.g., [97–99], conclude that pectin, in spite of its metal-binding ability, apparently does not inhibit the uptake of minerals from the diet. All of the above effects must be thought of as general tendencies in a vast amount of published findings which are not all mutually consistent. Some discrepancies may be explicable because it is attempted to generalize results achieved with different systems (in vitro, animal, or human) and with different pectins or pectin-rich plant materials. Publications often fail to specify important details about the pectin being used for the study, such as botanical origin, DE, etc.

3.7. Application in the Food Industry [168]

Indigenous manufacture of jams and marmalades, before commercial stabilizers were available, involved the use of pectin-rich fruit and in situ extraction of the pectin during prolonged cooking. Partly owing to this tradition, but mostly due to its superior stability and gelling ability at relatively low pH, pectin is the dominating gelling agent in modern production of jams as well as other products which are gelled, acidulous, and sweet. Examples are jelly fruits, and fruit preparations for industrial production of fruit-containing yogurt (Table 2 groups 1 and 2). Commercial pectins for these applications are tailor-made to yield specific gelation temperatures or gelling rates under specified conditions, and to exhibit specific functionalities such as heat reversibility, heat resistance, firmly gelled textures, pumpable semi-gelled textures, etc. HM pectin is used as a stabilizer in yogurt beverages and beverages in which milk proteins are heat-treated at relatively low pH during the production [100–102]. LM pectin finds use for thickening spoonable yogurt (Table 2 group 6). Particles of calcium pectinate are used as a substitute for fat in low-calorie foods [103] (Table 2 group 12). Other applications from Table 2 include groups 4, 8, 9, 10, and 11.

3.8. Market

Annual pectin production is estimated at 25 000 t (80 % citrus pectin), sold mostly in standardized form. Pectin production takes place in Brazil, Denmark, France, Germany, Mexico, Switzerland, and United Kingdom. Smaller amounts are produced in Eastern Europe and the former Soviet Union. Average prices on the world market are $10 – 11 per kilogram HM-pectin and $ 12 – 13 per kilogram LM-pectin.

4. Alginates

Alginate is a collective term for a family of linear 1,4-linked α-L-gulurono-β-D-mannuronans of widely varying composition and sequential structures [169]. Commercial preparations are usually designated as alginates and include alginic acid, its salts, and derivatives [2], [3], [5], [170–175].

4.1. Occurrence

Alginates form a group of polysaccharides that occur as structural components of the cell walls of brown seaweed (Phaeophyceae; Fig. 9) in which they make up to 40 % of the total dry matter and play an essential role in maintaining the structure of the algal tissue. They were isolated for the first time in 1880 by STANFORD [176], who also introduced the names algin for the soluble substances in an aqueous sodium carbonate extract and alginic acid for the material that could be precipitated from this extract by addition of acid.

The brown weeds occur in many parts of the world. For example, on the west coast of North America the giant kelp *Macrocystis pyrifera* is harvested mechanically by special ships, which cut the weeds at a depth of 1 m. The weeds grow so fast that they can be harvested several times a year. Other species are harvested semi-mechanically by fishing boats, collected by reaping

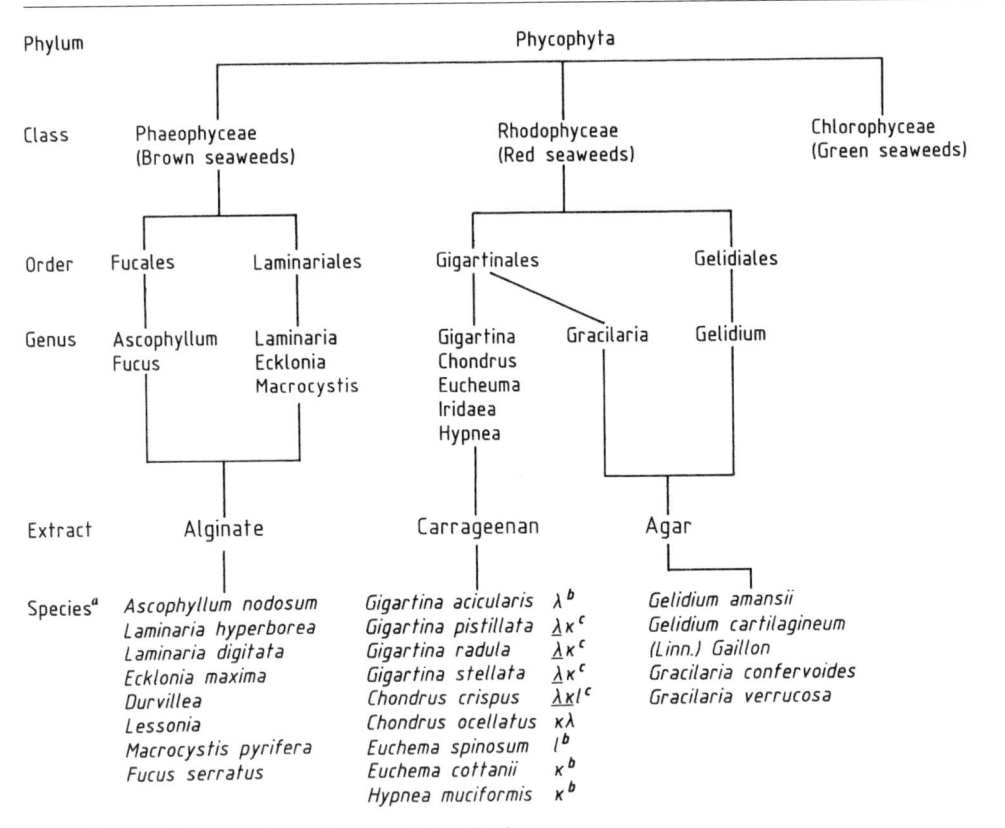

			Phycophyta			
Phylum						
Class	Phaeophyceae (Brown seaweeds)		Rhodophyceae (Red seaweeds)		Chlorophyceae (Green seaweeds)	
Order	Fucales	Laminariales	Gigartinales	Gelidiales		
Genus	Ascophyllum Fucus	Laminaria Ecklonia Macrocystis	Gigartina Chondrus Eucheuma Iridaea Hypnea	Gracilaria	Gelidium	
Extract	Alginate		Carrageenan	Agar		
Species[a]	Ascophyllum nodosum Laminaria hyperborea Laminaria digitata Ecklonia maxima Durvillea Lessonia Macrocystis pyrifera Fucus serratus		Gigartina acicularis λ[b] Gigartina pistillata λκ[c] Gigartina radula λκ[c] Gigartina stellata λκ[c] Chondrus crispus λκι[c] Chondrus ocellatus κλ Euchema spinosum ι[b] Euchema cottanii κ[b] Hypnea muciformis κ[b]	Gelidium amansii Gelidium cartilagineum (Linn.) Gaillon Gracilaria confervoides Gracilaria verrucosa		

Figure 9. Origin of seaweed extracts – general classification
[a] Species of economic significance; [b] Contains only component mentioned; [c] Contains predominantly underlined component

hooks, harvested from shore at low tide, or collected directly from the beach. Many regions have a tradition of harvesting or collecting brown seaweed, which in the past was used for the extraction of iodine and sodium carbonate.

4.2. Production

Various brown seaweed species are used as raw materials to obtain a range of specific properties (Table 3).

The production of alginates is shown schematically in Figure 10. Current processes are still based on patents that were filed in the 1930s by U.S. companies [185], [186]. Alginate production is based on a series of ion-exchange processes: The water-insoluble calcium alginate in the raw material is, for the purpose of extraction and purification, converted first to the soluble sodium

Table 3. Species of brown seaweeds (Phaeophyceae); alginate content and ratio between mannuronic and guluronic acid residues (data from various sources [283])

Species	% Alginate DM*	Man : Gul
Macrocystis pyrifera	13 – 24	1.4 – 1.8
Ascophyllum nodosum	20 – 30	1.4 – 1.9
Laminaria digitata	15 – 27	1.3 – 1.6
Laminaria hyperborea (leaf)	15 – 26	1.3 – 6.0
(stem)	27 – 33	0.4 – 1.0
Ecklonia maxima	30 – 38	1.4 – 1.8

	Mannuronan	Guluronan	Alternating
Macrocystis pyrifera	40.6 %	17.7 %	41.7 %
Ascophyllum nodosum	38.4 %	20.7 %	41.0 %
Laminaria hyperborea	12.7 %	60.5 %	26.8 %
Azetobacter vinelandii (acetylated)	17.8 %	0.5 %	81.7 %

*DM = Dry matter.

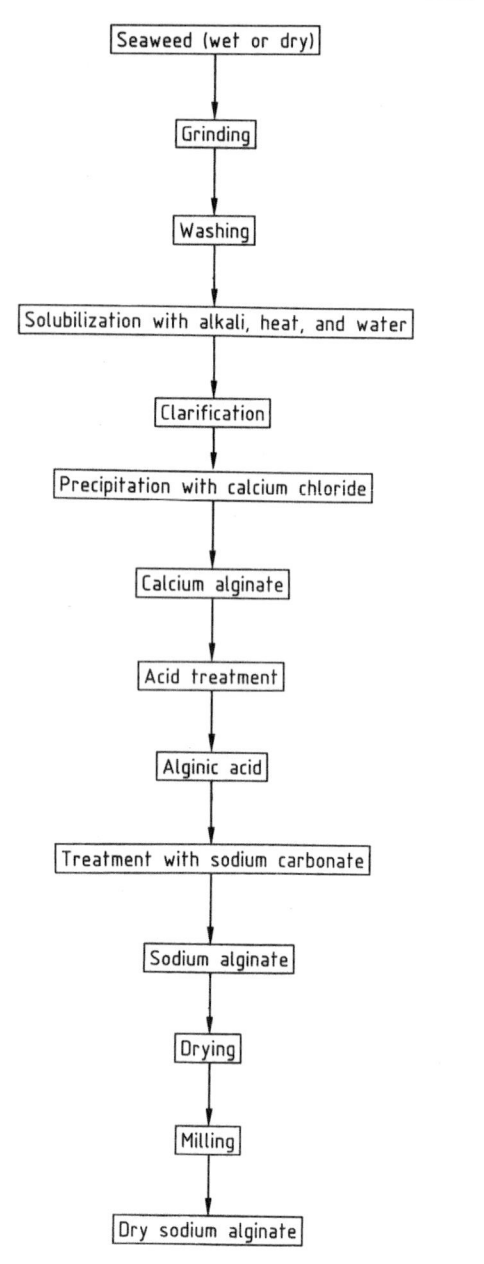

```
Seaweed (wet or dry)
        ↓
     Grinding
        ↓
     Washing
        ↓
Solubilization with alkali, heat, and water
        ↓
    Clarification
        ↓
Precipitation with calcium chloride
        ↓
   Calcium alginate
        ↓
   Acid treatment
        ↓
    Alginic acid
        ↓
Treatment with sodium carbonate
        ↓
   Sodium alginate
        ↓
      Drying
        ↓
      Milling
        ↓
  Dry sodium alginate
```

Figure 10. Flow sheet for the production of sodium alginate [171]

form and then, possibly via the calcium form, to insoluble alginic acid, which is neutralized to obtain the finished product.

While the *Macrocystis* is taken to the processing factory directly by boat other species are air dried before being transported for processing. The weeds are dried mechanically, ground, and then transported to factories. Salts, which might influence the solubility, and other impurities such as the high-polymeric laminarin (a β-1,3; β-1,6-D-glucan) are removed by washing. The alginate remains in an insoluble form as the calcium salt or as alginic acid. Extraction of the material with cold or hot soda follows, combined with a mechanical disintegration. A homogeneous mass with a pH of ca. 10 – 11 is obtained. After dilution with water, the slurry separates into a liquid phase (sodium alginate) and a solid phase consisting predominantly of cellulose. For removal of these solid impurities, a flotation step may precede the classical separation processes such as sieving and filtration. Alginic acid is obtained from purified sodium alginate by direct precipitation with acid or by precipitation with Ca^{2+} as calcium alginate and subsequent conversion to alginic acid by washing with acid. For the destruction of chloroplasts, either the sodium alginate solution or the calcium alginate precipitate is bleached with hypochlorite. Washed and mechanically dewatered alginic acid can be dried as such; in general, however, it is treated with sodium carbonate or other bases to produce alginate salts. The finished products are highly refined, odorless white powders, which are permitted as food additives. For industrial purposes, less purified preparations are often sufficient.

4.3. Structure

Until 1954, alginate was considered to be composed of polymannuronic acid. In 1955, FISCHER and DÖRFEL [177] discovered L-guluronic acid in addition to D-mannuronic acid, and since ca. 1964, alginic acid has been known to be a copolymer of these two residues. In principle, alginates are composed of three structural elements: the homopolysaccharides α-1,4-L-guluronan and β-1,4-D-mannuronan, and a heteropolysaccharide consisting of alternating 1,4-linked α-L-guluronic (G) and β-D-mannuronic acid (M) residues [178] (Fig. 11). The structure of alginate can therefore be represented schematically as

$$- M - G - M - (M - M)_n - M - G - (M - G)_q - M - G - (G - G)_p - G - M - G -$$

Alginates, like pectins, are linear polymers. The D-mannuronic acid residues are in the $^4C_1$

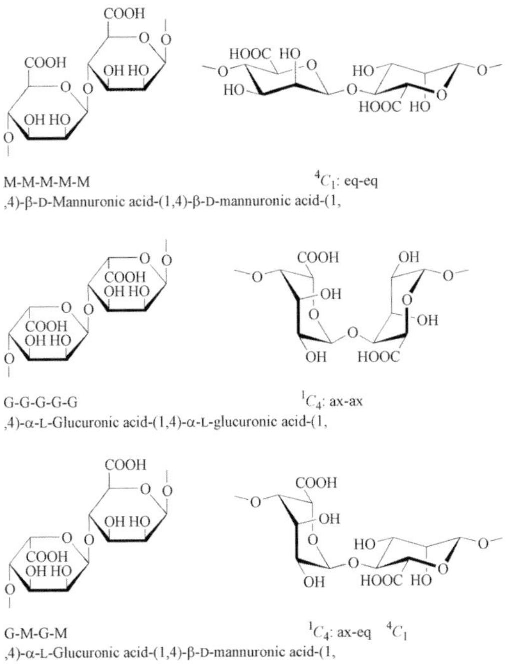

M-M-M-M-M $^4C_1$: eq-eq
,4)-β-D-Mannuronic acid-(1,4)-β-D-mannuronic acid-(1,

G-G-G-G-G $^1C_4$: ax-ax
,4)-α-L-Glucuronic acid-(1,4)-α-L-glucuronic acid-(1,

G-M-G-M $^1C_4$: ax-eq $^4C_1$
,4)-α-L-Glucuronic acid-(1,4)-β-D-mannuronic acid-(1,

Figure 11. Structural units in alginates

configuration, and therefore the glycosidic linkage between them is equatorial – equatorial. Since the L-guluronic acid residues are in the $^1C_4$ configuration, they are connected by axial – axial glycosidic linkages similar to the linkages between galacturonosyl residues in pectin. However, unlike pectin, alginates do not contain neutral sugar residues. The properties of alginates are determined largely by their molecular mass and by the ratio in which the three structural elements occur in the polymer. This ratio is determined by the variety of seaweed from which the alginate is extracted [172], [175], [179–182].

The *biosynthesis in plants* starts from D-mannose which, via guanosine diphosphate mannose, is oxidized to guanosine diphosphate mannuronic acid by a NAD dehydrogenase system and then polymerized to polymannuronic acid by a transferase enzyme. In homogenates of brown seaweed, guanosine diphosphate guluronic acid has also been identified, so at the end of the pathway copolymerization may also occur [173].

4.4. Properties [188]

Solubility. *Alginic acid* itself is insoluble in water and precipitates at pH < 3.5. The ammonium, alkali and magnesium salts, and alginate salts with organic bases are water soluble; calcium ions and other multivalent ions precipitate with alginates. *Alginates* are insoluble in organic solvents and can be precipitated with alcohols. They are, however, slightly miscible with simple alcohols, glycerol, and glycols, and these compounds can therefore be included to a certain content in alginate solutions. The dissociation constant pK of monomeric mannuronic acid is 3.38, and of guluronic acid 3.65. Alginic acid with a high proportion of guluronan has a pK of 3.74, when the proportion of mannuronan is high, this value is 3.42.

Degree of Polymerization. The properties of alginates depend largely on the degree of polymerization and the ratio of guluronan : mannuronan blocks in the molecules. The mannuronan regions in the molecules with their diequatorial glycosidic linkages are stretched and flat, while the diaxial glycosidic linkages in the guluronan regions give this part of the alginate molecule a buckled ribbon shape with limited flexibility.

Viscosity. The high viscosity of aqueous alginate solutions is a function of the degree of polymerization and is one of their most important properties (Table 4).

By mixing alginates of different sources, and by oxidative degradation, manufacturers can offer products with viscosities varying from 4 – 1000 mPa · s for a 1 % solution. Viscosity numbers alone are meaningless without the shear stress or shear rate because alginate solutions are Newtonian fluids only at very low shear rates and low concentrations. Under other conditions, they are pseudoplastic. Table 4 also shows the strong concentration dependence of viscosity. Cations have a significant effect on viscosity. Low concentrations of Ca^{2+} ions considerably increase the viscosity as a result of complex formation. In measuring the viscosity of alginate solutions a distinction is made between the "direct" viscosity and the "sequestered" viscosity. The latter is always lower because a sequestrant removes the Ca^{2+} ions.

Effect of Shear. In common with most other food hydrocolloids, sodium alginate solutions are pseudoplastic. In solution the alginate mole-

Table 4. Dynamic viscosities (in mPa · s) of polysaccharide solutions at different concentrations (data from various sources [283], [41], [62], [63])[*]

Concentration	Pectin	Alginate	Carrageenan	Agar	Gum arabic	Gum tragacanth	Gum karaya	Gum ghatti	Xanthan	Locust bean gum	Guar gum
1 %	50	214	57	4		54	3 000	6	2 000	59	3 025
2 %	200	3 760	397	25		906	8 500	10	7 000	1 114	25 060
3 %	550	29 400	4 411			10 605	20 000	40	11 500	8 260	111 150
4 %		39 660	25 356	400		440 265	30 000	60		39 660	302 500
5 %			51 425			7 111 000	45 000			121 000	510 000
10 %						17					
30 %						200					
50 %						4 163					

[*] Reported value give only general trends. Depending on preparation and measuring conditions differences may occur.

cule is long, thin, and rigid. When the solution is at rest, the alginate molecules interfere with each other forming loose entanglements. When shear is applied to the solution, the molecules are oriented in the direction of the shearing force. The association between the molecules is greatly reduced and there are no longer any entanglements. The result is an immediate lowering of the viscosity. When the shear is removed, the molecules entangle again resulting in the recovery of the viscosity.

Effect of pH. Between about pH 4.5 and 10, there is little change in solution viscosity. At about pH 3.8 the solution viscosity increases dramatically and, as the pH is reduced further the alginate is precipitated out of solution as alginic acid. At higher pHs viscosity is lost by a β-elimination process. The glycosidic linkages between mannuronic and guluronic acid residues are known to be less stable to acid hydrolysis than the linkages between two mannuronic or two guluronic acid residues [190]. The homopolymers are slightly more stable than heteropolymers.

Effect of Sequestrants. Sequestrants are used with alginates to soften the water so that the alginate can hydrate completely and thus modify the rate of the calcium – alginate reaction. They may also be used to sequester the calcium ions naturally present in the alginate itself. Different sequestrants bind different amounts of calcium ions at different pHs. Therefore, it is important to understand which sequestrant is being used and at what pH it will be used at.

Effect of Salts. The presence of *monovalent salts* reduce the viscosity of sodium alginate solutions. The viscosity drop occurs because alginates, in common with other polyelectrolytes, contract as the ionic strength increases.

As *polyvalent cations*, usually calcium ions, are gradually introduced into a sodium alginate solution, they cause the alginate polymer chains to start to align – this is seen as an increase in the solution viscosity. As more calcium is added the solution viscosity increases further to a point where there is evidence of some gel structure. At this stage, the solution shows thixotropic properties – the viscosity decreases when sheared, pumped, or shaken but the structure is regained when the shear is removed. Higher levels of calcium ions result in coherent gel formation. Typically, this occurs when more than about 70 % of the alginate has been converted into the calcium alginate form. The introduction of even more calcium ions increases the strength of the gel until, finally, all of the alginate is precipitated out of solution. At this stage all solution viscosity is lost.

Gel Properties. In food applications, sodium alginates are used as gelling agents, which in many aspects resemble LM-pectins and also form gels with Ca^{2+} ions. As pure polyuronic acids, alginates are, however, more sensitive to acid than pectins and are insoluble at pH < 3.5. Alginates also have a higher sensitivity to calcium than pectins, which is probably due to their higher degree of polymerization. For the preparation of acid gels, calcium salts are used that are soluble only under acidic conditions, together with compounds that release H^+ ions slowly, for instance, slowly soluble fumaric acid or slowly hydrolyzing glucono-1,5-lactone. The use of sequestrants has also been mentioned. The gel-forming ability of alginates is related mainly to the proportion of L-guluronan blocks. These blocks bind with Ca^{2+} ions according to a cooperative mechanism, giving rise to the formation of junction zones, which are referred to as the eggbox model (see Section 3.4.1). The junction zones in alginate gels are stronger than those in LM-pectin gels, and they are not thermally reversible even in the presence of sugar. Alginates rich in guluronan blocks form strong, brittle gels with a tendency to syneresis, whereas alginates rich in mannuronic acid form weaker more elastic gels that are less prone to syneresis [193]. Single mannuronic acid residues or mannuronan blocks interrupt the guluronan junction zones. *Propylene glycol alginates* form weaker gels. They can be prepared more easily because propylene glycol alginate is less sensitive to acid and Ca^{2+} ions and, therefore, less liable to coagulation. Alginates also gel in the presence of equimolar quantities of an HM-pectin. The strongest gels are obtained with HM-pectin and alginates with a high proportion of guluronic acid residues at low pH and in the absence of calcium ions. Sugar is not essential for gelation; but it does affect the gel properties [144], [194], [195].

Oxidation and Degradation. By using high acid concentrations and high temperatures, alginates and alginate esters can be quantitatively decarboxylated. Oxidants such as halogens or periodate, and redox systems such as polyphenols, ascorbic acid, or thiols, are able to depolymerize alginates. For production of low viscosity alginates, hydrogen peroxide is used. Bacterial alginate depolymerases have also been described [173], [191]. These enzymes have been used to study the distribution of L-guluronic and D-mannuronic residues along the backbone. They have no commercial significance; however, they condition the chemical conservation of alginate solutions. The degree of polymerization of commercial alginates is in the range of ca. 80 to ca. 800.

4.5. Propylene Glycol (Propane-1,2-diol) Alginate

Propylene glycol alginate (pga) is prepared by treating partially neutralized alginic acid with propylene oxide [187] (Fig. 12).

Pga's are completely soluble in both hot and cold water at neutral pH. Depending on the

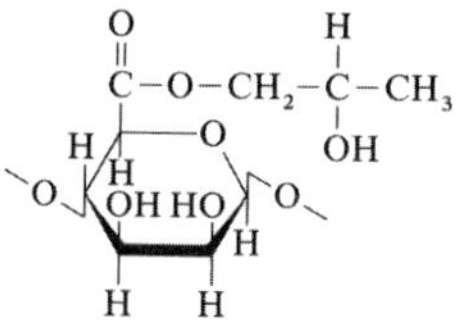

Figure 12. Propylene glycol mannuronate unit

degree of esterification, pga can be water soluble at a pH as low as 2.5. As with the soluble alginate salts, the viscosity of a pga solution is a function of the degree of polymerization. Typically, pgas are available in the viscosity range 25 to 500 mPa · s for a 1 % solution and degrees of esterification from about 50 % to about 90 %. Highly esterified products cannot be made with highly polymerized alginates due to steric hindrance.

The flow properties of pga solutions depend upon the gum concentration and the shear rate: at high concentrations (e.g., > 1 %) they are shear thinning over a wide range of shear rates. At low concentrations they exhibit long, smooth flowing characteristics.

With increasing temperature the viscosity – shear rate curves are approximately linear over a wide range of shear rates and show progressively less pseudoplasticity with increasing temperature. This effect is reversed on cooling.

The bulky propylene glycol groups are responsible for the other different properties of the molecule. They physically hinder the aggregation of the polymer chains and also afford protection to the β,1–4 linkages. As a consequence of esterification, fewer carboxyl groups are available for interaction with divalent metal cations such as calcium. Propylene glycol alginates with less than 60 % esterification will react with calcium ions to give enhanced viscosity – such solutions are thixotropic but there is no evidence of gel formation or precipitation. Highly esterified pgas are not affected by the presence of calcium ions.

Propylene glycol alginates are tolerant to ionic materials such as salt and acid. The property of salt tolerance is made use of in barbecue sauces, salad dressings, and other liquid products. The esterification process reduces the ionic character of the pga molecule, thereby increasing its tolerance to low pH. The higher the degree of esterification the more tolerant the pga is to lower pH, but in strong acid the alginate chains depolymerize, with a resultant loss of viscosity [171].

In acid solutions proteins behave as bases and can interact with pga. This leads to some thickening and a degree of solution thixotropy, i.e., the solution exhibits high viscosity at rest but low viscosity when shear is applied. This property is used in meringues, cakes, and in the stabilization of beer foam.

The lipophilic ester side chains interact with fatty substances. Propylene glycol alginates build a molecular film around oil droplets thereby enhancing emulsification by inhibiting coalescence and droplet growth. Thus pgas provide dual functionality – they act as emulsifiers and show other rheological benefits. This film-forming property is also reflected in pga's abiltiy to form stable foams in aqueous solution. Thus, they are useful in protein-containing systems such as marshmallows and the stabilization of beer foam.

Legal Aspects. The *safety* of pga was evaluated by the Joint FAD/WHO Expert Committee on Food Additives (JECFA) in 1971, when it was assigned an Acceptable Daily Intake (ADI) of 0.25 mg per kilogram body weight. Pga has been an EC-approved food additive for over 25 years. It is listed as E405, in the Emulsifiers, Stabilisers, Thickeners and Gelling Agents Directive, 74/329/EEC.

4.6. Bacterial Alginates

Alginate production has also been reported in bacteria (i.e., *Pseudomonas aeruginosa* (pathogenic), *P. mendocina*, *P. putida*, *P. fluorescens*, and *Azetobacter vinelandii*). These alginates are also formed of the same three structural elements and are further characterized by the presence of *O*-acetyl groups. For *A. vinelandii* it has been shown that polymannuronic acid is synthesized first; mannuronic acid residues can then be transformed to guluronic acid residues by a mannuronan C-5 epimerase. With this enzyme both bacterial and algal alginates can be epimerized. The activity of the enzyme depends strongly on the Ca^{2+} concentration [169], [183], [184].

4.7. Analysis

The alginate content in alginate preparations can be determined by titration as described for pec-tins. From this analysis the degree of esterification can be deduced if propylene glycol alginate is present. In mixtures or foods, the uronic acid content can be determined by the colorimetric methods used for pectins [162]. If necessary this analysis can be preceded by a purification step. Titration as well as colorimetric methods do not, however, distinguish between pectins and alginates. Pectins can be degraded in solution with specific enzymes whereupon they become alcohol soluble. Alginates can then be determined in the alcohol-insoluble residue. Polyuronides can also be precipitated from preparations, mixtures, or foods with Ca^{2+} ions and thus separated from other polysaccharides. After hydrolysis to monomeric uronic acid residues [181], the latter can be analyzed by gas chromatography [29] or HPLC [196]. This method also allows analysis of the uronic acid composition of alginates and, after partial hydrolysis, the proportions of mannuronan and guluronan blocks. The proportions of M and G residues in an alginate preparation and their distribution in M, G, and MG blocks can also be determined by ^{1}H-NMR [182], [197] and circular dichroism [198]. As a result of their wide range of applications, no internationally recognized standard methods exist to assess the quality of alginates. Manufacturers therefore usually specify the viscosity or gel strength of their products by methods agreed upon with clients.

4.8. Applications

Food Industry. The applications of *alginates* in the food industry are based on

1. Their potential to form highly viscous solutions with outstanding suspension-stabilizing properties
2. Their stability at high temperature and high pH
3. Their reactivity with Ca^{2+} ions, which enables them to gel
4. The thermal stability of these gels [168], [171]

As a result alginates are used in almost all groups listed in Table 2. For acid products, *propylene glycol alginates* are used. In all of these applications, lump formation must be considered when adding the alginate to an aqueous system, similar to pectins. When sugar is part of the ingredients, some may be admixed with the alginate. When the application allows the addition of alginate as a solution, the alginate can best be solubilized by dosing it gradually in the vortex of a powerful mixer. For laboratory use the alginate can be moistened with alcohol. For all listed applications, only a few grams of alginate are added per kilogram of product; for the stabilization of foams a few milligrams per kilogram suffice.

Other Applications. Twice as much alginate is used in applications outside the food industry. Except for groups 17, 20, 23, and 26, all applications in Table 2 have been described. Of greatest impact is group 31, which claims to cover ca. one-third of alginate sales. Groups 27 and 29 also represent a substantial market. One of the first applications of alginates was in the treatment of boiler water (group 32). The calcium is flocculated as calcium alginate and not deposited as scale. For this application, alginate paste is used, which does not have to be refined to the extent necessary for food applications. The strong swelling power of alginic acid makes it a good binding agent for tablets (group 24), and its calcium reactivity enables denture replicas to be made (group 25). The triethanolamine salt is often used as a thickening agent in cosmetic and pharmaceutical ointments. Another interesting application mentioned in the commercial literature is as a gel for embedding the roots of plants to protect them from mechanical damage and from dehydration after replanting. In the last decade alginates have been used in biotechnology research to immobilize bioactive cells and to purify enzymes [137].

4.9. Market

The alginate industry is limited to a small number of companies. Production plants are found in locations where suitable weed species are present (i.e., North America, Scotland, Brittany, and Japan). Alginate is also produced in the former Soviet Union. About 25 000 t of alginate is produced annually; ca. one-third of this is used in the food industry. The price of sodium alginate for food applications is ca. $ 9.0 – $ 15 per kilogram.

5. Carrageenan

Carrageenans belong to a family of polydisperse long chain galactans, which can be extracted from red seaweeds (Rhodophyta) [304], [305]. The major uses are as high quality ingredients in food and cosmetics with estimated worldwide sales approximately $ 263 \times 10^6$ in 1997 [306].

5.1. Structure

Carrageenans are build up by alternating 1,3-linked β-galactose (**G**-units) and 1,4-linked α-galactose (**D**-units) that can be partly esterified with sulfuric acid (**S**). In carrageenans with the ability to form a gel the major part of the 4-linked units are in the 3,6-anhydro form (**DA**). A short hand nomenclature system based on letters has been introduced [307] to simplify the old system based on Greek letters [308], [309].

Historically the three major commercial carrageenans types were named κ, ι, and λ with their corresponding structures and short names given in Table 5. Originally the two former (gelling family) were isolated based on their insolubility in KCl [310], whereas the latter formed the soluble fraction (nongelling family). These major types are representation in Figure 13. In the text the Greek nomenclature normally will be used for the three common commercial types only, whereas the letter system will be used to specify structural characteristics when appropriate.

There are several biological factors that complicate the structural picture of carrageenans.

In general the **DA** (in κ-carrageenan) or the **DA2S** (in ι-carrageenan) might be replaced with their 6-sulfated precursor residues, **D6S** (μ-carrageenan) and **D2S,6S** (ν-carrageenan), respectively [311]. The conversion of precursors to 3,6-anhydrogalactose (**DA** or **DA2S**) occurs during

Table 5. Most common carrageenan structures in commercial samples

4-linked unit	3-linked unit	Greek letter name
DA	G4S	kappa (κ)
DA2S-G4S	G4S	iota (ι)
D2S,6S-G2S	G2S	lambda (λ)
DA	G	beta (β)

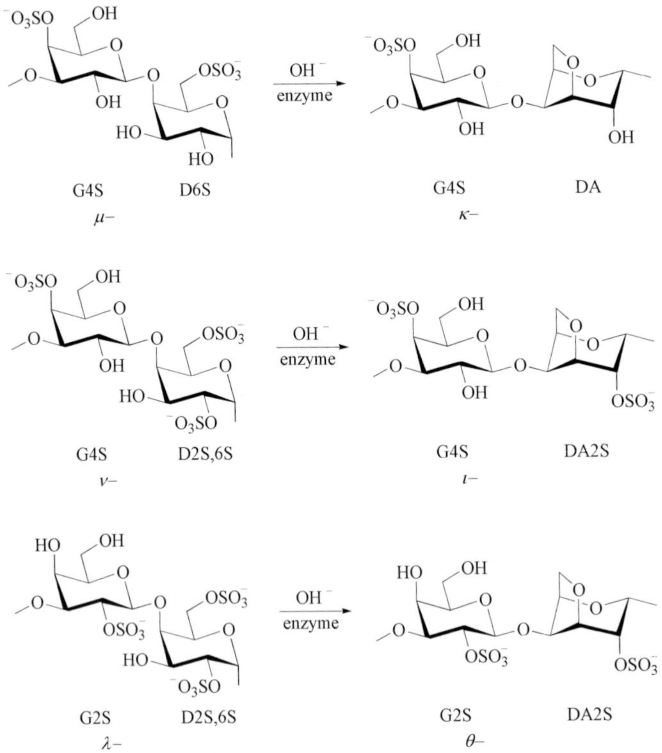

Figure 13. The repeating structures of the most common commercial carrageenans and the relation to their 6-sulfated analogues

biosynthesis in the algae [312], but can also be catalyzed by hot alkali during the commercial phycocolloid production. During the alkali treatment or the enzyme catalyzed elimination reaction the $^4C_1$ conformation of the 4-linked 6-sulfated precursor units (**D2S,6S** or **D6S**) is transformed into $^1C_4$ in 3,6-anhydrogalactose with the liberation of sulfate from position 6. The desulfatase enzyme has also been called "dekinkase" because it effects a change in the configuration of the transformed galactosyl unit and removes "kinks" present as equatorial-axial glycosidic linkages [204]. Some algae have a gametophyte generation producing carrageenans of the gelling family, and a sporophyte generation (both often identical in appearance), producing **D2S,6S-G2S** (λ) or structurally related carrageenans [313]. Such carrageenans have traditionally been ill defined due to difficulties in their characterization and are mostly obtained from *Chondrus* and *Gigartina* genera [306]. λ-Carrageenan can most probably only be transformed into a component containing a 3,6-anhydro bridge (**DA2S-G2S**, termed θ-carrageenan) during a hot alkaline treatment. Enzymes capable of catalyzing this reaction in the algae have not been detected and naturally occurring **DA2S-G2S** is rare [314].

In general the repeating sequence that dominates the structural character of a carrageenan is the basis for any nomenclature system. However, it should be stressed that the idealized structures most often are masked by precursors or other carrageenan elements. In fact by selecting the algal source carrageenans with wide **DA/DA2S** ratio can be obtained [315–317]. In addition to the carrageenan mentioned above in Table 5, which are the main constituents of commercial preparations, different carrageenan structures are constantly isolated from new sources [309], [314], [318] and/or found in small quantities in regular samples. Finally, algae containing molecules with a mixed carrageenan and agar (4-linked D-galactose replaced with 4-linked L-galactose) structure have been detected [319], [320], [321]. So far these extracts are not economically important.

Table 6. Examples of carrageenans that are present in neutral* or alkali-treated extracts. **n** is restricted to the gametophytic generation and **2n** is restricted to the sporophytic generation. The species considered as commercially most important are in bold phase.

Dominating structure(s)	Generic examples
DA-G4S	*Kappaphycus alvarezii*[a], *Hypnea musciformis*, **Chondrus crispus**(n), *Chondrus ocellatus*(n), *Gigartina tenella*(n)
DA2S-G4S	*Eucheuma denticulatum*[b], *Agardhiella tenera*, *Gymnogongrus concinna*,
DA2S-G4S and DA-G4S[c]	*Sarcothalia crispate*[d](n), *Gigartina skottsbergii*[e](n)
DA-G4S ≈ DA-G	*Furcellaria lumbricalis*[f]
DA-G > DA-G4S	*Eucheuma gelatinae*
D2S,6S-G2S*[g]	**Chondrus crispus**[h](2n), **Gigartina pistillata**[i](2n), *Sarcothalia crispate*(2n), *Gigartina skottsbergii*(2n)
DA2S-G2S[j]	*Sarcothalia crispate*(2n), *Gigartina skottsbergii*(2n)
DA-G6S	*Phyllophora sp, Risoella verruculosa*

[a] Trade name (Eucheuma) cottonii or cottoni.
[b] Trade name (Euchuma) spinosum.
[c] Copolymers referred to as kappa-2 with weaker and less milk reactivity than **DA-G4S**. For a full discussion of several species see [317].
[d] Sandpaper [324].
[e] Pigskin [324].
[f] Danish agar, Furcellaran or κ-furcellaran.
[g] May contain other structures such as pyruvated residues (**GP,2S**).
[h] Cultivated in Canada.
[i] Obtained from Morocco.
[j] Occur only in alkali treated sporophytic extracts. For a more detailed explanation of nomenclature see [305], [307]. Information mostly extracted from [306] and [317].

In addition to sulfate, substitution by methyl ethers and pyruvate or other sugars, such as xylose, glucose, and uronic acid, may be present in a carrageenan sample. Nitrogen containing components such as proteins and pigments are common impurities (0.1–1 wt %). Since the viscosity of several commercial carrageenan solution can be reduced by the addition of protein degrading enzymes, some possible linkage to the carrageenan chain is suspected [322]. Carrageenans may therefore be considered as proteoglycans [304].

To complete the description of the carrageenan structure, the mean number of constituent sugars, their weight average molecular mass M_w, and the distribution of the molecules (polydispersity factor = $<M_w>/<M_n>$ must be given. Representative values are in the range of 300 000–600 000 and 2.3–5.1, respectively [305].

5.2. Sources and Raw Materials

Carrageenan has been used for several hundred years in Europe and the Far East, and the name is linked to the Irish village Carraghen, where Irish Moss (*Chondrus crispus*) was harvested and utilized in milk products [323]. During the last years the harvesting of natural Irish Moss populations has been reduced and is being discussed as an environmental issue in several countries. Natural resources are mostly harvested in temperate regions (Canada, Chile, and France) and at present the largest consumption is based on cultivated tropical seaweeds such as *Kappapycus alvarezii* [306]. For the major commercial products and sources see Table 6.

5.3. Production

After harvesting the seaweeds material is washed to remove sand and stones and dried to a dry matter content of ca. 25 wt %. The carrageenan content can vary from 15–70 %, depending on the seaweed. The production of refined carrageenan is shown schematically in Figure 14. The dried algae are treated with alkali and ground to a paste. Alkaline conditions facilitate extraction of the macerated algae, retard acid-catalyzed depolymerization of the galactan and catalyze the conversion of C-6 sulfated precursor residues to 3,6-anhydrogalactopyranosyl residues, thus increasing the gelling potential of the product. The crude extracts are further purified by sieving and filtration by using a filter aid such as celite. If drum drying is anticipated, pigments are first removed with activated carbon. To prevent gelling, all of these operations must be carried out at higher temperatures.

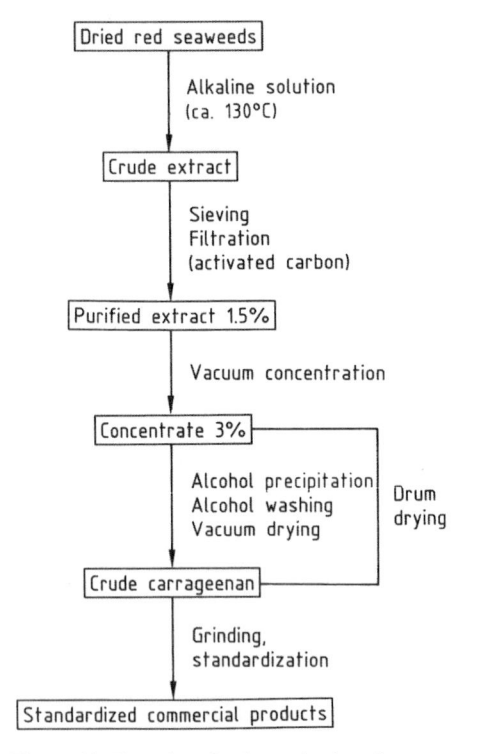

```
Dried red seaweeds
        │
        │ Alkaline solution
        │ (ca. 130°C)
        ▼
   Crude extract
        │
        │ Sieving
        │ Filtration
        │ (activated carbon)
        ▼
 Purified extract 1.5%
        │
        │ Vacuum concentration
        ▼
   Concentrate 3% ──────────┐
        │                   │
        │ Alcohol precipitation   Drum
        │ Alcohol washing        drying
        │ Vacuum drying
        ▼                   │
   Crude carrageenan ◄──────┘
        │
        │ Grinding,
        │ standardization
        ▼
 Standardized commercial products
```

Figure 14. Flow sheet for the production of carrageenans

Because as with pectin and alginates, dilute extracts must be processed to a dry product, the energy cost for dewatering and recovery of alcohol is an important factor. Isopropanol is used as the alcohol for precipitation of the carrageenan. An alternative process is available for κ-carrageenan by extruding the extract into a KCl solution, pressing the precipitate and removing water by freeze-thaw cycle. After drying and grinding several batches might be blended and/or standardized by the addition of salts and sugars. Furthermore, by selecting a milder alkaline extraction process than the conventional alkali modification, carrageenans with a remaining small fraction of precursor units and hence altered function might be obtained [324]. The most innovative processing procedures include enzymic tools based on molecular biology, mostly developed at the Goemar Laboratories, (Roscoff, France). Both stable enzyme preparations of glycosylhydrolases – to produce carrageenan oligosaccharides – as well as sulfohydrolases – to convert precursor residues into 3,6-anhydrogalactose on the polymer level have been cloned [325]. In the late 1990s [326] a semirefined carrageenan has been offered on the market. The product, Processed Euchuma Seaweed (PES), is regulated as E407b. Such carrageenan products are obtained by first treating the algae with hot KOH solutions, washing with water, and finally drying, bleaching, and grinding to a suitable particle size [327]. Under the conditions used the carrageenan is not extracted and the precursors, if present, are converted to 3,6-anhydrogalactose inside the algal material. PES will contain more acid insoluble matters and fiber components such as cellulose [328].

5.4. Analysis

Carrageenan preparations can be separated based on their solubility in aqueous KCl and further analyzed by several tools. The content of sulfate can be determined by gravimetry (FAO, 1992) and the content of 3,6-anhydrogalactose by the resorcinol reagent [329]. The content of both constituents can be determined by HPLC [330]. By a simple recording of the increase in the 3,6-anhydrogalactose content by the recorcinol reagent upon a hot alkaline treatment the level of precursors can be evaluated in a carrageenan sample.

The constituent sugars can be analyzed by *gas chromatography* after the introduction of a specialized hydrolysis and derivatization procedure to conserve the 3,6-anhydro units. This reductive hydrolysis in the presence of methylmorpholineborane [363] can even be applied directly to seaweed samples for evaluation of carrageenan structure and taxonomy [331]. The use of partial reductive hydrolysis after methylation can discriminate between agaran and carrageenan backbone and its substitution [332]. *IR-spectroscopy* can be applied to study the position of sulfation but has some limitation for quantitative analysis. This technique can be applied directly to carrageenan [333] or to dried and milled material by the use of FTIR diffuse reflectance spectroscopy [334].

By far the most powerful technique for carrageenan analysis is *NMR spectroscopy* [335]. ^{13}C-NMR spectroscopy is preferred for identifying the different residues in a carrageenan sample since most of the signals (resonances) from the residues and even some signal splitting due to sequence effects are well separated [336]. However, a low signal to noise ratio is encountered and components below 5 % may not be detected. For ^{1}H-NMR the sensitivity is better but a distinction between the different components is more difficult since the different signals may overlap in a quite narrow region of the spectra. For routine applications ^{1}H-NMR can be applied to 1 mg samples and analyzed within 15 min [316], [337], [338]. Spectroscopy can also be used for the estimation of carrageenan in foods [339]. Combined with the selective degradation with carrageenase [311] and chromatography the major as well as the minor constituents may be identified [340]. Apart from data for sugar composition and sulfate substitution an important parameter for carrageenans is their molecular mass. At present the common method is HPLC based *size exclusion chromatography* (HPSEC), coupled to a Multi-Angle Laser-Light Scattering (MALLS) detector, absolute mass and the molecular mass distribution can be obtained [341], [342]. Typical tests to demonstrate the presence of carrageenan are the formation of a blue coagulate with methylene blue, positive sulfate reaction after acid hydrolysis, and the formation of much stronger gels by the addition of potassium chloride. Recent progress in the analysis of carrageenans in food is reviewed elsewhere [343].

5.5. Properties

Gelation. All forms of λ-carrageenan as well as the sodium salts of κ- and ι-carrageenan are soluble in cold water. The potassium and calcium salts of κ- and ι-carrageenans, however, dissolve only at 70 °C and form gels or – depending on the ionic strength – viscous systems upon cooling (Table 4). κ-Carrageenan is more sensitive to potassium ions, whereas ι-Carrageenan is more sensitive to calcium ions. Once a highly potassium sensitive carrageenan resembling κ-carrageenan was produced from *Furcellaria*, i.e., furcellaran or Danish agar [344]. This carrageenan has a mixed **DA-G4S** and **DA-G** structure. However, since this free floating algae became extinct from Danish waters the holdfast variety is at present collected in Canada and processed together with *Chondrus crispus*. From a solution of the three major types of carrageenans, κ- and ι-carrageenan can be precipitated by potassium chloride. To obtain a gel the carrageenan molecules must undergo a transition from a random coil structure into helices that aggregate upon cooling. In general the ion induced network formation [345] is an intermolecular association that requires a minimum degree of polymerization of about 100 [346]. It is still under debate whether the fundamental ordered state is a single or a double-stranded helix. Results obtained by high-performance size exclusion chromatography-low angle light scattering (HPSEC-LALLS) (see Section 5.4) originally suggested a doubling of the molecular mass for the ordered conformation [341], [347] and hence double helices (Fig. 15). At present the common method is HPLC based Size Exclusion Chromatography, coupled to a Low- or a Multi-Angle Laser-Light Scattering (LALLS or MALLS, respectively) detector.

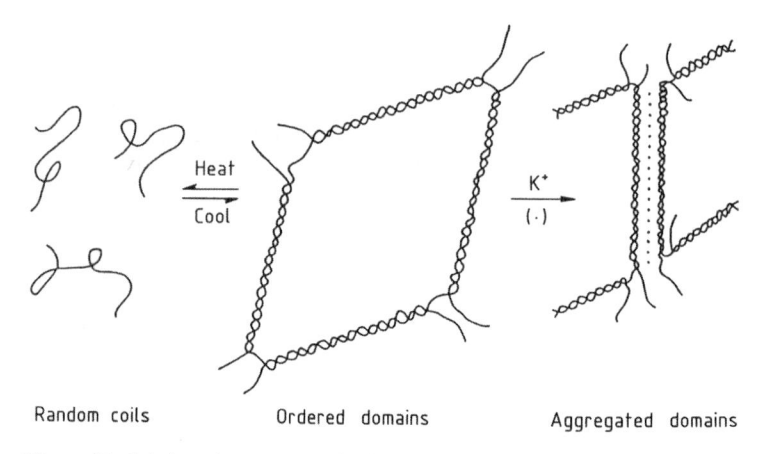

Figure 15. Gelation of carrageenans by formation of double helices and aggregation of helices [348]

However, the findings based on light scattering data have been questioned due to the preparation method of the solution and the concentrations used, and result have been presented in favor of single helixes of ι-carrageenan [349]. The carrageenan gels are thermoreversible and upon heating, the helices unfold, the molecules go in solution again as random coils and the gel melts. In the gel state the aggregation of helices may continue, the network contracts, and the gel becomes brittle (short) and shows syneresis. The inability of λ-carrageenan to form gels can be explained by the occurrence of **D2S,6S** units in its backbone. A solubilized preparation rich in κ- carrageenan gives a firm gel with a setting point of 40 °C and a melting point of 55 °C at a concentration of 0.5 %, in the presence of 0.2 % potassium salt. The gel strength increases strongly with increasing concentration of potassium ions; however, potassium ion concentrations > 0.2 % are unacceptable because of taste. Also, addition of sugar increases the gel strength. Both additions also increase the setting temperature, as well as the melting temperature, of the gels. The hysteresis, i.e. the difference between melting and setting temperature, remains small and is ca. 15 °C. The gels are brittle and have a tendency to become opaque and show syneresis. This can be prevented by adding ι-carrageenan. Gels of ι-*carrageenan* alone are transparent, they show no syneresis and little hysteresis. Due to the presence of **DA2S** residues, which may act as a wedging group and prevent tight aggregation of double helices, the gels are rather weak. Furthermore, the observation that gelation of a commercial ι-carrageenan (**DA2S**-units) showed a small specificity towards monovalent cations was interpreted as being due to the inclusion of a small proportion of κ-carrageenan (**DA**-units) [350].

Viscosity. In a solution with low ionic strength the carrageenan chains are extended due to the electrostatic repulsion of the negatively charged sulfate groups and the resulting solutions are highly viscous. The viscosity increases with concentration and decreases by addition of salts due to charge shielding, and will decrease exponential upon heating. Viscosity will be reasonable reversible upon cooling as long as the condition for maximum stability of the molecular mass is kept (pH 9). The viscosity of soluble carrageenan forms can be measured in 1 % solutions at room temperature, but because of gelation at intermediate temperature, viscosity is normally measured in 1.5 % solution at 75 °C. At this temperature values up to 800 mPa·s have been measured for a 1.5 % solution. In industrial applications measurements are performed with rotational viscometers. Since carrageenan solutions have pseudoplastic behavior the shear rate must also be reported. The molecular mass (M_w) dependence of the intrinsic viscosity [η] is given by the Mark-Houwink equation:

$$[\eta] = K (M_w)^a$$

The constants K and a are dependent of the carrageenan type and the solvent. A value for a close to unity indicates rigid rod-like structure [351].

Synergistic Effect with Other Gums. Of particular interest is the interaction of κ-carrageenan with locust bean gum galactomannan. By partially replacing the κ-carrageenan with locust bean gum, which does not gel on its own, a stronger gel with improved properties is obtained. It becomes more elastic, and has a lower tendency to syneresis and to become opaque. Similar observations were made with konjac glucomannans. The smooth regions of the mannan chain (i.e., regions with no galactose or glucose side groups) are thought to bind to the double helices of the κ-carrageenan, forming mixed junction zones [207], (Fig. 16). These mixed junction zones have a lower tendency to form tightly packed aggregates. The industry has placed preparations on the market that exhibit

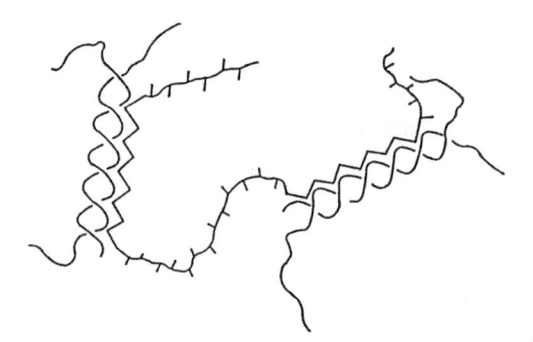

Figure 16. Synergistic gelation between helix-forming carrageenans and agar with galactomannans by direct association of smooth regions of the mannan chain with double helices in the seaweed polysaccharides [209]

constant gel strength and gel characteristics by blending different lots and adding potassium salts, sugar, and possibly locust bean gum. Because carrageenans may form lumps when added to water, premixing with sugar or adding the carrageenan in a powerful mixer is recommended. Carrageenans are also useful in altering the textural properties of a starch system. Adding ι-carrageenan (0.5 %) to a starch system increases the viscosity as much as 10 times, whereas no effect is obtained from k-carrageenan [352].

Stability. Carrageenans are not stable under acidic conditions, because the 3,6-anhydro ring and the 1→3 linkages are easily hydrolyzed. The substitution with 2-sulfate (**DA2S**) introduces some stability. Carrageenan solutions can be sterilized at pH > 4.5 but some reduction in molecular mass will occur [353]. In general decrease in molecular mass might be encountered during dialysis and due to autohydrolysis during freeze-drying or long storage. Gelled carrageenans are more stable and the common microorganisms found outside the marine environment do in general not degrade carrageenans.

Reactivity with Proteins. Carrageenans possess a strong anionic character because of their sulfate groups. These charges as well as associated ions (Na^+ versus K^+ and Ca^{2+}) and the conformation of the sugars in the chain determine the properties of carrageenans. Reactivity with proteins can be observed both with carrageenan of the gelling and nongelling family, but some chain regularity is important in different types of interactions [351]:

1. Below the pH of the isoelectric point (I-pH) of the protein the positively charged protein and the negatively carrageenan form a complex, which might result in a precipitate depending on the net charge ratio.
2. Above I-pH, when both polymers are negatively charged, the interactions are mediated by polyvalent cations such as Ca^{2+}.
3. Finally, interaction with a positively charged part of a molecule with a net negative charge may occur.

In milk systems a highly specific interaction between κ-casein and carrageenans has been established that causes κ-casein particles or casein micelles to attach to the carrageenan chain. When the molecular mass of the carrageenans is sufficient, helical regions can form and aggregate and a gel network is obtained. This can occur only with the gel forming κ- and ι-types. Addition of 0.2 % κ-carrageenan to milk at the pasteurization temperature and cooling to 43 °C, gives a firm gel that melts again at 60 °C.

Weak networks form at carrageenan concentrations as low as 0.02 %, which can fix casein particles. In chocolate milk, for instance, this network holds the cocoa suspension and, in creams the lipid globules. It is suggested that both type (2) and (3) are important in the gelation of milk and to stabilize chocolate milk. The ability to stabilize casein micelles is in the order **DA-G4S>DA2S-G2S>D2S,6S-G2S** [351]. The reaction between milk proteins and carrageenan may synergistically increase the gel strength about 10 times, and carrageenans forms milk gels such as flans at a concentration of 0.2 %. ι-Carrageenan forms elastic, κ forms brittle, and λ-carrageenan forms weak milk gels.

5.6. Applications

Medical Applications. Carrageenans are widely used in clinical tests for anti-inflammatory agents by their ability to induce so-called rat-paw edema. Furthermore, it has been shown that carrageenan may act as anticoagulant (heparin analogue) and by activating the Hagemann factor [354]. Carrageenans may also act as immunosuppressor agents [355] and show toxicity against macrophages [356]. It should be noted that some of the earliest studies may have been performed with quite ill defined carrageenans. Carrageenans have more recently been considered as potential antiviral agents [357], [358]. It has been shown that the activity is related to the sulfate substitution pattern and that **D2S,6S**-residues as found in precursor- and λ-carrageenan have the highest reactivity [359].

Other Applications. Carrageenan oligosaccharides have been investigated as fertilizers, stimulants of growth and reproduction and to stimulate defense mechanisms in plants [360].

Discussions on carrageenans for use as paint stabilizer, beverage clarifier, controlled release agent of drugs, in viscosifier in food and tooth paste and in a waste different other applications in the food, nonfood, and medicine sector are found elsewhere [328], [351], [361].

5.7. Physiological Properties

Carrageenan has been given GRAS status by the FDA, based on the assumptions that high molecular mass carrageenans are not absorbed in the intestine, and food grade carrageenans are not ulcerogenic in humans. However, it has been shown that commercial carrageenan samples have a quite wide molecular mass distribution and that carrageenan may be degraded under stomach juice simulated conditions [362]. Therefore the matter is constantly under debate.

6. Agar

Agar, also called agar-agar, kanten, or gelose, is the oldest known gel-forming polysaccharide.

Discovered in the 17th century in Japan and consumed for 200 years, agar is extracted from certain marine red algae of the class Rhodophyceae mainly from *Gelidium* and *Gracilaria* species, growing essentially along the coasts of Morocco, Spain, Portugal, Chile, Japan, and Korea [210].

The extreme inertness of agar led KOCH and PETRI in 1882 to use it as a medium in which to grow bacteria [211]. In his now classical preliminary note on the tubercle bacillus, published in 1882, KOCH formally announced agar as a new solid culture medium. Up to date, no better solidifying agent in microbiological media has been found. At present, agar has become an indispensable substance in microbiological, biotechnological, and public health laboratories, and an important colloid in other industries. Agar is also a permitted gelling, stabilizing, and thickening agent for food applications, authorized in all countries without limitations of daily intake [210]. It finds use in many food industries such as confectionery, bakery, pastry, beverage, sauces, wines, spreads, spices and condiments, meats and fishes, dairy, jams, etc.

Apart from its ability to gelify aqueous solutions and produce gel without the support of other agents, agar can also be used as a safe source of dietary fiber since it is not digestible by the human body.

Agar is insoluble in cold water, hardly soluble in hot water, and readily soluble in boiling water. In food or technical preparations of a high total solids content, agar should be solubilized separately before adding the other ingredients of the formulation. Usual agar is therefore not compatible with the continuous industrial processes.

Quick soluble agar (QSA) is a new agar version developed in 1993 by Setexam. It is obtained from a selected red seaweed and according to a patented manufacturing process without any chemical or genetic modifications.

Unlike traditional agar QSA has a great ability to dissolve in water or in milk at temperatures > 65 °C or during the pasteurization stage, without need to be cooked or heated to boiling, and is therefore compatible with most of the continuous industrial processes.

6.1. Production

Agar is insoluble in cold water but is colloidally dispersible in boiling water. Hence, most commercial agars are produced by hot water extraction. Other methods are possible, such as an extraction with glycerol, anhydrous ammonia, or other solvents [212], but the traditional process employing hot water is the most recommended for the agar destined to food applications.

The agar extraction procedures vary according to the treated seaweed variety. They generally follow three stages (see Fig. 17):

Extraction
Purification
Dehydration

First, the seaweeds are carefully washed with water in order to eliminate marine salt and foreign matter such as sand. The extraction is made by hot water under pressure or in open tank if the seaweeds have undergone an alkaline treatment. This latter is mainly applied to *Gracilaria* seaweed in order to lower the sulfate content in the resultant agar.

The extraction juice, composed of 99 % water and 1 % agar, is filtered at hot temperature, then cooled at room temperature. The resulting gel is dehydrated under mechanical pressure or by freezing-thawing.

1,4-linked 3,6-anhydro-α-L-galactose alternating with 1,3-linked β-D-galactose [214–216].

β-D-Galactose   3,6-Anhydro-α-L-galactose

Agaropectin is thought to have the same repeating unit as agarose, although some of the 3,6-anhydro-L-galactose residues can be replaced with L-galactose sulfate residues [217], [218] and the D-galactose residues are partially replaced with the pyruvic acid acetal 4,6-*O*-(1-carboxyethylidene)-D-galactose [217], [219]. Fractionation studies on DEAE Sephadex A-50 carried out by DUCKWORTH and YAPHE [217], [220] have indicated that agar is not made of one neutral and one charged polysaccharide but is composed of a complex series of related polysaccharides which range from a virtually neutral molecule to a highly charged galactan. These studies indicated that in agars containing 4,6-*O*-(1-carboxyethylidene)-D-galactose, this charged residue is always found in regions of the molecule with low sulfate content. Agars in which the D-galactose residues are partially replaced with 6-*O*-methyl-D-galactose contain this 6-*O*-methylated monosaccharide evenly distributed throughout the neutral and charged regions of the complex [217], [218].

Studies by GUISELEY in 1970 [221] indicated that in naturally occurring agars, an increase in gelling temperature occurred with increasing methoxyl group content. In this way, the gelling temperature of an agar from a *Gracilaria* species with high methoxyl content is higher than that of an agar from a *Gelidium* species with little or no methoxyl content.

Gelling Mechanism. The gelling properties of an agar are essentially effected by the agarose part [222], [223]. The rheological properties of the gel are strongly influenced by agaropectin.

Agar forms high strength gels which are completely reversible with a hysteresis cycle over a range of 40 °C (see Fig. 18): Upon dissolution in boiling water agar acquires a random coil configuration. The subsequent gel formation proceeds in two steps. The decrease in temperature of the hot solution, slightly above the geli-

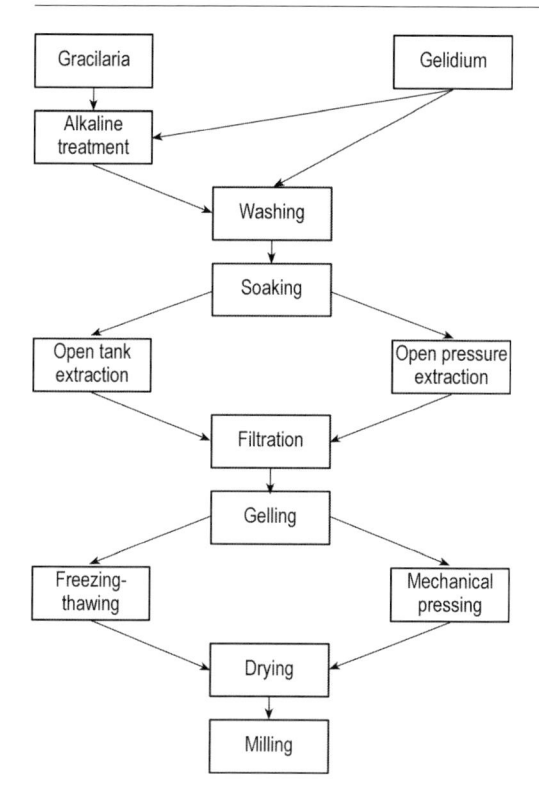

Figure 17. Flow sheet of traditional agar extraction

In *pressure dehydration* a pressure of 49 – 98 N/cm^2 is applied to press the gel through intercalated filters. Water passes the filters and the agar molecules are retained as filter cake.

Gel freezing is carried out at −18 °C. At that temperature the agar molecules are excluded selectively from the ice network and form strips. The thawing step melts the ice and agar strips are recovered selectively.

These two operations of dehydration taken individually remove about 70 % of the initial water content. A part of the residual moisture is then eliminated by drying with hot air. The dry agar flakes are ground to prepare a final product in powder form with different mesh sizes.

6.2. Structure and Gelling Mechanism

Structure. The elucidation of the complex chemical structure of agar by ARAKI had revealed the presence of two dominating polysaccharides: a virtually neutral polymer termed agarose, and a charged polymer termed agaropectin [213]. ARAKI deduced the structure of *agarose* to be

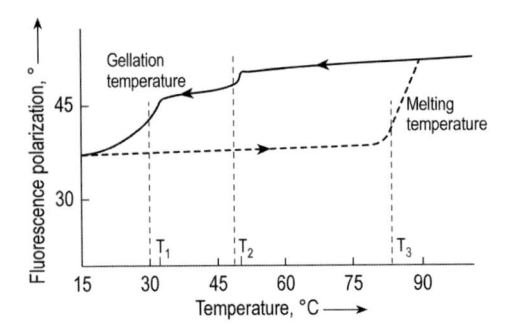

Figure 18. Hysteresis of 1.5 % agar gels

6.3. Quick Soluble Agar

Production. If the gelling mechanism is superposed to the extraction procedure, it can be concluded that agar powder is a dehydrated gel and the molecules adapt the aggregated double helix conformation. That is why a certain amount of heat energy is required to dissociate the hydrogen bonds and to dissolve the agar molecules in random coil form.

The amount of heat energy required can be considerably lowered if the molecules in agar powder are not obtained in aggregated double helix but in random coil configuration. Setexam managed to finalize a new procedure including neither a chemical transformation nor a genetic modification for the production of Quick Soluble Agar (see Fig. 20).

Properties. QSA has the same physical and chemical characteristics as traditional agar, i.e., gel strength, pH, color, total ash, etc., except for solubility. Contrary to traditional agar that dissolves only in boiling water, QSA dissolves instantly in water or milk at 65 °C or during pasteurization. This property of QSA simplifies the manufacturing of food or technical products, in batch and continuous processes. It eliminates all the disadvantages met with the traditional agar utilization such as the necessity of a separated dissolution stage, a large initial volume of water, blackening on cooking inner sides, degradation of the gelling power, etc.

fication temperature, provokes at first the formation of a double-helical conformation of the agarose chains [224–226]. The second stage is the self-associations by nucleation of these double helices to form aggregates [227], [228]. The obtained agar gel is completely heat-reversible since it remelts at above 85 °C (see Fig. 19).

The gelation process of agar is not linked with the cations present such as in the case of alginates, some pectins, or the carrageenans, but is only due to formation of hydrogen bonds which explains the thermoreversibility of agar gels. Three equatorial hydrogen atoms of the 3,6-anhydro-α-L-galactose residue are believed responsible for constraining the molecule so as to form a helix with a threefold screw axis [229]. Interaction of these helices causes gel formation.

Replacement of 3,6-anhydro-α-L-galactose residues with L-galactose sulfate causes kinks in the helix, and hence a polysaccharide of lower gel strength is formed [230].

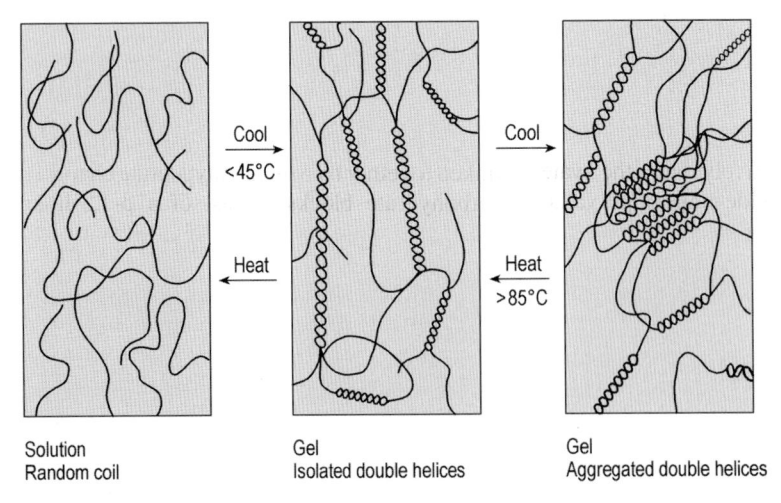

Figure 19. Gel formation mechanism in aqueous agar solutions

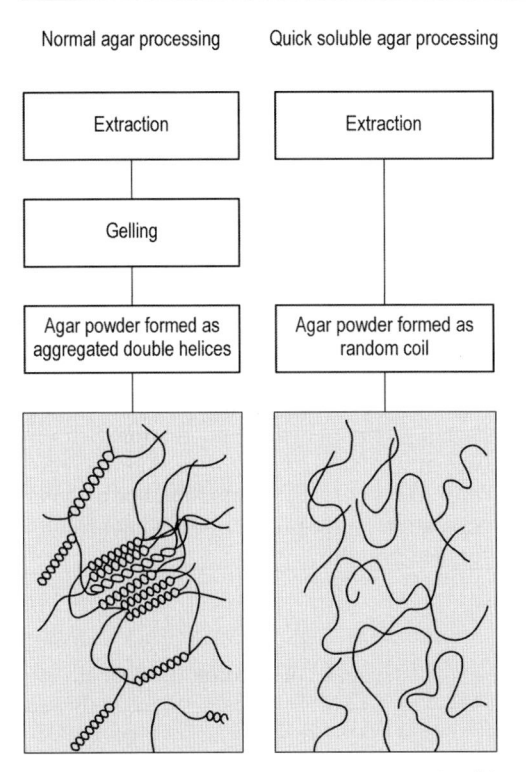

Figure 20. Comparison of production processes of traditional agar and QSA

Application of QSA is very easy, as it is introduced during the preliminary stages of any food or technical preparation and dissolves instantaneously during the heating, the pasteurization or the sterilization stage, which makes it compatible with most of the continuous industrial processes.

The thermoreversibility of QSA gels facilitates its handling and recycling; and its unmatched natural hysteresis offers a definite advantage particularly with regards to the shelf life of food or technical preparations.

The QSA reacts selectively with the water portion of any recipe in a wide range of pH values without interaction with neither the present cations nor with the present organic matters.

7. Gum Arabic

Gum arabic is a dried exudation obtained from the stem and branches of acacia trees, *Acacia senegal* L. or *Acaica seyal* (fam. Leguminosae). For commercial production the trees are tapped or drilled. After drying in air, the exudate tears are harvested by hand and transported to central collecting stations where they are sorted according to size, color, and contamination with impurities. Harvesting takes place during October to June. A young tree yields ca. 0.9 kg and an older tree up to 12 kg of gum annually. The trees grow in the tropical and subtropical areas of Africa. Exproting countries are Sudan, Tschad, Nigeria, Mali, Mauretania, Senegal, and Ethiopia. Acacia species can be found worldwide, but only the exudates of *A. senegal* and *A. seyal* may be traded as gum arabic. World production is around 40 000 t/a.

Raw gum is still sold in substantial quantities; however, a precleaning by air classification to remove sand, fines, and bark is required. The product can be further processed to a kibbled, granular, or powdered form. Importers in Western Europe and the United States produce speciality products for particular applications. For this purpose the gum is dissolved, and the gum solution is exposed to sieving or centrifugation and then drum or spray dried [1–4], [231].

Structure. Gum arabic consists of three principal fractions, an arabinogalactan fraction (AG) with very little protein, representing 88 % of the gum und having a molecular mass of 250 000; a high molecular mass (1.5×10^6) arabinogalactan – protein complex (AGP) representing about 10 % of the gum; and a glycoprotein (GI) with a molecular mass of 200 000, making up 1.2 % of the gum. By treatment with proteolytic enzymes the AGP fraction is degraded to fragments with a molar mass similar to AG. The wattle blossom model has been proposed to describe the structure of AGP: Carbohydrate blocks with molecular mass of ca. 250 000 are linked together by a main polypeptide chain. The carbohydrate blocks consist of a β-1,3-linked galactopyranose backbone chain, with numerous side chains linked through β-1,6-galactopyranose residues and containing arabinofuranose, arabinopyranose (Ara), rhamnopyranose (Rha), glucuronic acid (GlcA), and 4-*O*-methylglucuronic acid. The sugar moieties occur in the ratio Gal 36, Ara 31, Rha 13, and GlcA 18. The structure of a representative segment is shown in Figure 21 [232–235]; depending on growth conditions and age, some structural variations exist.

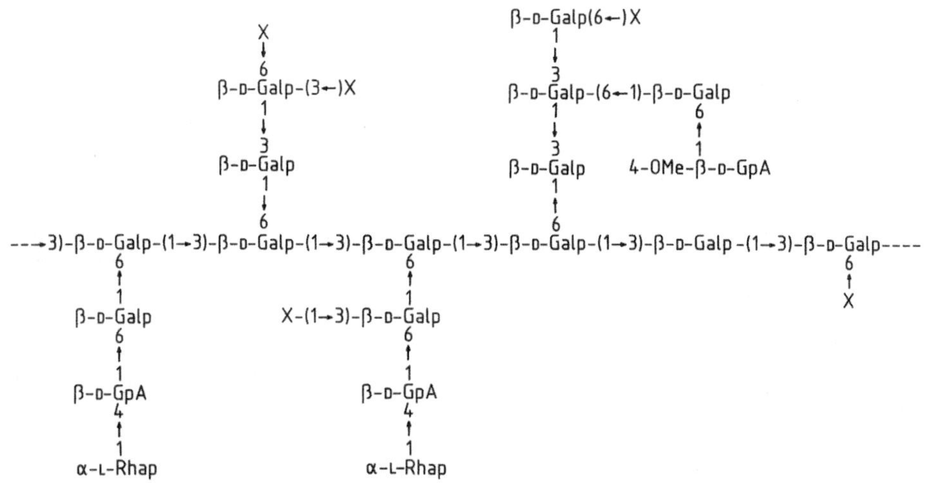

```
                              X                          β-D-Galp(6←)X
                              ↓                                1
                              1                                ↓
                              6                                3
                     β-D-Galp-(3←)X              β-D-Galp-(6←1)-β-D-Galp
                              1                        1                6
                              ↓                        ↓                ↑
                              3                        3                1
                         β-D-Galp                 β-D-Galp        4-OMe-β-D-GpA
                              1                        1
                              ↓                        ↓
                              6                        6
---→3)-β-D-Galp-(1→3)-β-D-Galp-(1→3)-β-D-Galp-(1→3)-β-D-Galp-(1→3)-β-D-Galp-(1→3)-β-D-Galp----
         6                             6                                           6
         ↑                             ↑                                           ↑
         1                             1                                           X
     β-D-Galp                 X-(1→3)-β-D-Galp
         6                             6
         ↑                             ↑
         1                             1
     β-D-GpA                      β-D-GpA
         4                             4
         ↑                             ↑
         1                             1
     α-L-Rhap                      α-L-Rhap
```

X = L-Araf-(1,
or α-D-Galp-(1,3)-L-Araf-(1,
or β-L-Arap-(1,3)-L-Araf-(1,
or L-Araf-(1,3)-L-Araf-(1,
or L-Araf-(1,3)-L-Araf-(1,3)-L-Araf-(1,
or β-L-Arap-(1,3)-L-Araf-(1,3)-L-Araf-(1,

β-D-Galp = β-D-Galactopyranose
β-D-GpA = β-D-Glucopyranosyluronic acid
4-OMe-β-D-GpA = β-D-Glucopyranosyluronic acid-4-methylether
α-L-Rhap = α-L-Rhamnopyranose
L-Araf = L-Arabinofuranose

Figure 21. Structural features of gum arabic

Properties. The molecular mass of gum arabic averages around 500 000; the glucuronic acid content (including 4-O-methylglucuronic acid) of ca. 18 % makes gum arabic an acidic polysaccharide. With a cation exchanger or by washing with acidified alcohol, gum arabic can be converted to arabinic acid, which exhibits the titration curve of a strong acid. A 5 % solution has an apparent dissociation constant of 10^{-3}. Solutions in water have a pH of 2.2 – 2.7.

The most striking feature of gum arabic is its *extreme solubility in water*. Due to its highly branched, compact structure, solutions containing up to 40 % gum arabic can be obtained, which still exhibit typical Newtonian flow behavior. Compared to other polysaccharides of similar molar mass, the viscosity of these solutions is very low (Table 4). Viscous solutions are obtained only at concentrations > 30 %, where effective molecular overlap begins to occur and solutions gradually assume pseudoplastic behavior. The *viscosity* of gum arabic solutions is highest in the pH range 4.5 – 5.5; in this range the carboxyl groups are ionized to a large degree. Addition of salts strongly decreases the viscosity due to suppression of the electrostatic charge. Gum arabic is compatible with most other hydrocolloids except for alginates and gelatin. It is also surface active and has been widely used to *stabilize oil-in-water emulsions*. The AGP fraction is responsible for its emulsifying ability. The polypeptide chain is believed to be at the periphery of the molecule and to adsorb onto hydrophobic surfaces. Gum arabic has been shown to be heat sensitive; proteinaceous material (AGP and GI) precipitates on prolonged heating, and as a consequence the solution viscosity drops considerably and the emulsifying ability is lost [231], [232], [235], [236].

Gum arabic can be *precipitated* with trivalent cations and by salts such as silicates, borates, and mercury nitrate. It is insoluble in oils and most organic solvents but soluble in aqueous ethanol up to ca. 60 % ethanol. In gum arabic preparations enzyme activities have often been demonstrated, for example, oxidases, peroxidases, amylases, and pectinases. For applications in foods, these activities are undesirable, and the enzymes must be inactivated by heat [2].

Applications, Market. Gum arabic was a commodity 4000 years ago in Egypt, and was used for the manufacture of cosmetics, dyes, and tinctures. Half of the amount imported by the United States is used in the food industry; this makes gum arabic one of the most useful plant hydrocolloids. The annual production has risen to > 50 000 t/a [231], [234]. Gum arabic is used in groups 2, 4, 8, and 11 (Table 2). Dried, powdered flavors (group 5) constitute a major application. High gum concentrations allow spray drying without loss of flavor. Its emulsifying and stabilizing ability, in combination with its use in high concentrations, makes gum arabic attractive for pharmaceutical and cosmetic applications (groups 16 and 21). In industrial applications such as an adhesive on gummed envelopes and stamps (group 20), and as a dispersive agent in dyes and tinctures, gum arabic is being replaced by starch and cellulose derivatives. Prices vary between $ 3 and $ 8.5 per kilogram.

8. Gum Tragacanth

Gum tragacanth is the exudate from breaks or wounds inflicted in the bark of shrubs of the *Astragalus* species, particularly *Astragalus gummifer* Labillardière (Leguminosae family), which are found in the mountainous regions of Iran, Syria, and Turkey. The gum was described by THEOPHRASTUS in the third century B.C. The name tragacanth is of Greek origin and means goat horn. It probably refers to the curved shape of the 2 – 3-mm-thick ribbons that represent the better quality gum. Gum tragacanth also exists as flakes, which are of poorer quality. The bushes are carved to stimulate gum production. Ribbons are obtained from April to September; flakes are formed in the following months until November. The translucent, whitish, elastic ribbons, which originate predominantly in Iran, are considered the best quality. The harvest is collected at central stations and is exported mainly to the United States and the United Kingdom. Importers process the raw materials to standardized powders [1–4], [232], [233], [237]. Prices range from $ 16 – $ 21 per kilogram; the annual harvest is ca. 3000 t/a.

Structure. Gum tragacanth is composed of tragacanthic acid, which is the major component, and a nearly neutral *arabinogalactan* that is water insoluble but swells to a gel. The constituent sugar residues of this arabinogalactan are: 75 % L-arabinose, 12 % D-galactose, 3 % D-galacturonic acid methyl ester, and L-rhamnose. *Tragacanthic acid* is water soluble and consists of a backbone of 1,4-linked α-D-galactopyranosyluronic acid residues (polygalacturonic acid chain as in pectins) with short side chains including single-unit β-1,3-linked D-xylopyranose, dimeric β-1,3-linked D-xylopyropyranosyl-1,2-α-L-fucopyranose, and dimeric β-1,3-linked D-xylopyranose-1,2-D-galactopyranose. The structure is shown schematically in Figure 22. The sugar moieties are present in the relative proportions

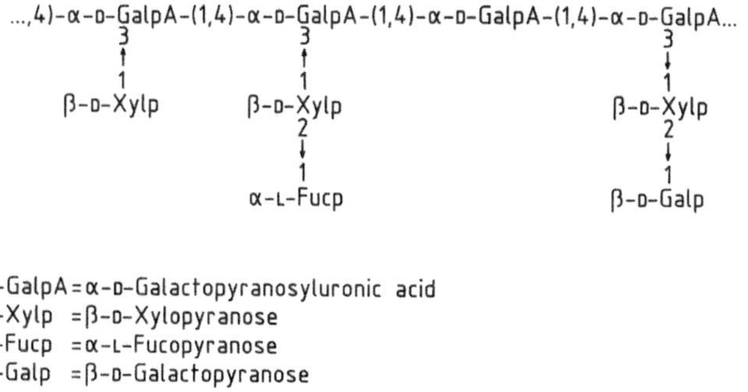

```
...,4)-α-D-GalpA-(1,4)-α-D-GalpA-(1,4)-α-D-GalpA-(1,4)-α-D-GalpA...
          3                 3                              3
          ↑                 ↑                              ↑
          1                 1                              1
       β-D-Xylp          β-D-Xylp                      β-D-Xylp
                            2                              2
                            ↑                              ↑
                            1                              1
                         α-L-Fucp                       β-D-Galp
```

α-D-GalpA = α-D-Galactopyranosyluronic acid
β-D-Xylp = β-D-Xylopyranose
α-L-Fucp = α-L-Fucopyranose
β-D-Galp = β-D-Galactopyranose

Figure 22. Structural features of gum tragacanth

D-galacturonic acid 43 %, D-xylose 40 %, L-fucose 10 %, and D-galactose 4 %. Tragacanthic acid has a high molecular mass (840 000) and gives accordingly highly viscous solutions (Table 4) that exhibit pseudoplastic behavior at concentrations > 0.5 %. A 1 wt % solution can have a viscosity of 3600 mPa · s. Accompanying Ca^{2+} and Mg^{2+} ions may contribute to the high viscosity and gelling properties of gum tragacanth solutions. The viscosity changes little in the pH range 2 – 10 and is stable during heating in a 1 % acetic acid solution [1–4], [232], [233], [237].

Applications, Market. The application of gum tragacanth in foods and acid products is based on its heat stability, and its stabilizing ability in emulsions and suspensions. Therefore, gum tragacanth is of particular interest for group 12 (Table 2). Because of the slimy consistency of gum tragacanth dispersions, this gum can also function as a fat replacer. Other applications are mentioned in groups 8 and 9. Outside the food industry gum tragacanth is also used in groups 16, 18, and 21. A special application is its use in spermicidal jelly.

9. Gum Karaya

Gum karaya is also named Sterculia gum, Kadaya gum, or Indian gum; it is the dried exudate from the bark of *Sterculia urens* or other *Sterculia* species (Sterculaceae family), trees occurring throughout northern and central India. The formation of exudate is artificially initiated by carving the bark. After a few days the exudate can be harvested as large irregularly shaped tears, which may weigh 1 kg or more. From April to June, 4.5 kg can be obtained per tree. The gum is sold to merchants at central trade stations and precleaned by removing pieces of bark; the tears are broken and graded according to purity and color. In consumer countries, the gum is then milled to a homogeneous powder. In the past, gum karaya was sold in a blend with gum tragacanth; in the meantime, however, it has found many applications because of its specific characteristics. The annual production is estimated at 5000 t, the larger part of which is used in the United States. Prices are ca. $ 4.8 –$ 6.4 per kilogram. Expansion of the production requires afforestation and takes at least 10 years [1–4], [232]. Gum karaya is a highly acetylated, acidic polysaccharide with a backbone chain of alternating 1,4-linked α-D-galacturonopyranosyluronic acid residues and 1,2-linked α-L-rhamnopyranosyl residues. Singleunit β-D-galactopyranosyl- and β-D-glucuronopyranosyluronic acid side chains are attached to O-2 and O-3 of the galacturonic acid residues, whereas half of the rhamnose residues have 1,4-linked β-D-galactopyranosyl residues at O-4 (Fig. 23). The molecule also has 13 wt % acetyl

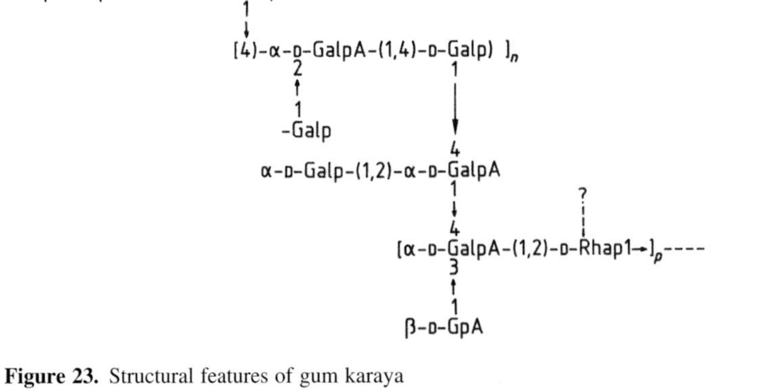

```
                    α-D-GalpA = α-D-Galactopyranosyluronic acid
                    β-D-GpA   = β-D-Glucopyranosyluronic acid
              ?     L-Rhap    = L-Rhamnopyranose
              ↓     α-D-Galp  = α-D-Galactopyranose
              4
----[4α-D-GalpA-(1,2)-L-Rhap]ₘ
    3                     1
    ↑                     ↓
    1                     4
  β-D-GpA              α-D-Galp
                          1
                          ↓
              [(4)-α-D-GalpA-(1,4)-D-Galp) ]ₙ
                    2                  1
                    ↑                  ↓
                    1                  4
                  -Galp         α-D-Galp-(1,2)-α-D-GalpA
                                              1
                                              ↓
                                              4        ?
                                        [α-D-GalpA-(1,2)-D-Rhap1→]ₚ----
                                              3
                                              ↑
                                              1
                                           β-D-GpA
```

Figure 23. Structural features of gum karaya

groups. Gum karaya is poorly water soluble; however, it swells readily in cold water to many times its original volume, particularly when finely ground (<200 mesh), giving a dispersion that appears homogeneous. By using an autoclave, solutions of up to 20 % can be obtained; their viscosity is, however, decreased as a result of degradation. Gum karaya solutions deviate from Newtonian behavior and exhibit viscoelasticity. Under alkaline conditions, deacetylation occurs and converts a gum karaya suspension into a ropy mucilage [232], [233]. The ability to swell and its resistance to enzymic and microbial degradation make powders of gum karaya good adhesives for dentures. Particles of ca. 0.6 mm swell to 60 – 100 times their volume and form a mucilage that allows drainage. Food applications are in groups 9, 11, 12, and 13 (Table 2). Gum karaya has also found use in the paper industry (group 29) and textile painting (group 30).

10. Gum Ghatti

Gum ghatti, also called India gum, is an exudate of *Anogeissus latifolia* (Combretaceae family), trees that grow in the deciduous forests of India and Sri Lanka. The Indian word gath means mountain pass and refers to the trade routes for this gum. The exudate usually has the shape of 1-cm-thick beads or tears, which may be colored by absorption of tannins from the bark. Also, dust and other particles may stick to the gum. Gum ghatti is harvested mainly in April and graded by traders according to color and impurities. The highest quality is pure and only lightly colored. Importers, particularly in the United States, purify and standardize the gum by grinding, sieving, and blending or by solubilization in water, filtration, and spray drying. It is of minor importance, with an annual production of ca. 1000 t [4].

Gum ghatti is composed of D-galactose, D-mannose, D-glucuronic acid, and L-rhamnose. A complex array of single-unit and oligomeric side is attached to a chain of alternating 1,4-β-D-glucuronic acid and 1,2-α-D-mannopyranose residues. The side chains comprise several 1,6-linked D-galactopyranose units, some terminated by glucuronic acid, joined to the mannopyranose residues at O-3 through 1,3-linked L-arabinopyranose residues. Single-unit L-arabinofuranose residues can also be found, some attached to the O-6 of mannopyranose residues, and other short sequences of L-arabinofuranose [233], [238]. O-acetyl groups have also been found. A schematic presentation of the structure is shown in Figure 24. In its acid form, gum ghatti has an equivalent mass of 1067; the gum occurs as the calcium or magnesium salt. The gum is not entirely soluble in cold water, but 90 % can be solubilized by maceration and heating at 90 °C. Removal of the divalent cations enhances solubility. The remaining 10 %, which is chemically identical to the soluble portion, is strongly swollen and also contributes to the viscosity [192]. Spray-dried preparations are completely soluble in water; they give, however, a lower viscosity of the aqueous solutions. The viscosity of gum

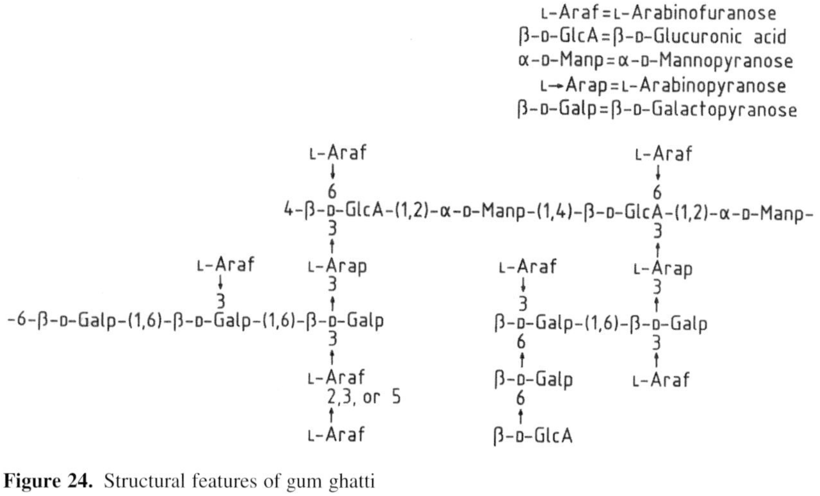

```
              L-Araf = L-Arabinofuranose
              β-D-GlcA = β-D-Glucuronic acid
              α-D-Manp = α-D-Mannopyranose
              L→Arap = L-Arabinopyranose
              β-D-Galp = β-D-Galactopyranose

                         L-Araf                       L-Araf
                            ↓                            ↓
                            6                            6
         4-β-D-GlcA-(1,2)-α-D-Manp-(1,4)-β-D-GlcA-(1,2)-α-D-Manp-
                            3                            3
                            ↑                            ↑
      L-Araf     L-Arap              L-Araf     L-Arap
         ↓          ↓                    ↓         ↓
         3          3                    3         3
 -6-β-D-Galp-(1,6)-β-D-Galp-(1,6)-β-D-Galp    β-D-Galp-(1,6)-β-D-Galp
                            3                    6         3
                            ↑                    ↑         ↑
                         L-Araf               β-D-Galp   L-Araf
                        2,3, or 5                6
                            ↑                    ↑
                         L-Araf               β-D-GlcA
```

Figure 24. Structural features of gum ghatti

ghatti solutions is higher than that of gum arabic solutions; viscosity increases quadratically with concentration. Like gum arabic, gum ghatti is an excellent stabilizer of emulsions and suspensions, and has found application in wax suspensions and pastes, oil emulsions, polishes, syrups, and as a dispersive agent in dried, vitamin-rich oil suspensions.

11. Xanthan Gum

Xanthan gum is the extracellular, high molecular mass heteropolysaccharide produced by the aerobic fermentation of *Xanthomonas campestris* NRRL-B1459 [1–5], [239]. The production of xanthan gum goes back to the investigations of JEANES and coworkers in the early 1960s at the Northern Regional Research Laboratory of the United States Department of Agriculture, Peoria, Illinois, in the framework of a large screening program for slime-producing microorganisms of industrial interest [240].

11.1. Production

Commercial production is carried out by a batchwise, submersed fermentation under strong aeration. During fermentation the bacterial cells are kept under a constant stress to direct their metabolism to metabolite production instead of growth. This is achieved by proper selection of medium composition. The medium cost forms a significant part (ca. 25 %) of the total production cost; the sugar concentration is adjusted so that it falls as low as possible at the end of the fermentation cycle, but depletion must be avoided to prevent the microorganisms from using the polysaccharide as a substrate [241]. For xanthan gum production, a typical medium contains a nitrogen source, phosphate and magnesium ions, trace elements, and glucose, which is kept at < 5 %. The pH must not be below 7. After 96 h at 30 °C, less than 0.1 % glucose is left in the medium, and a conversion to polysaccharide of more than 50 % can be obtained.

In the patent literature [242], much attention is devoted to propagation of the starter culture. In this stage, nitrogen is supplied by organic sources (e.g., soy peptone or corn steep liquor), which also contain some growth factors. During the main fermentation, nitrogen is supplied by mineral salts (e.g., ammonium nitrate). A typical composition of the medium at this stage is 0.06 % ammonium nitrate, 0.5 % potassium dihydrogenphosphate, 0.01 % magnesium sulfate heptahydrate, 2.25 % glucose, and 97.18 % water (Fig. 25). After inocculation of 5300 L of medium with 227 L of starter culture and vigorous aeration for 72 h at 28 – 31 °C, 77 kg of hydrocolloid can be recovered. The remaining medium contains < 0.1 % glucose. This means a conversion of 64.7 % and a final concentration of the polysaccharide of 1.45 %, which gives the medium a viscosity of 3000 mPa · s at 25 °C (Brookfield Viscometer, 60 rpm).

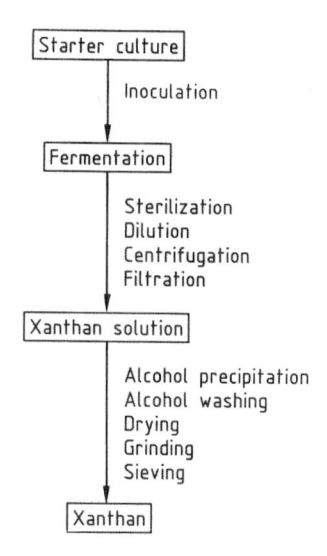

Figure 25. Flow sheet for the manufacture of xanthan gum

The usual precautions must be taken to preclude contamination of the batch fermentation through the culture medium, equipment and accessories, air, and neutralizer; thus it is doubtful if the batch operation can be replaced by a continuous process. Also the stability of the bacterial strain is a point of consideration. Since *Xanthomonas* bacteria do not form spores, the fermentation can be stopped by heat treatment in a heat exchanger. Further downstream processing involves centrifugation, filtration, and precipitation with ethanol or isopropanol as used for other hydrocolloids. The high viscosity of the fermentation liquor must be reduced by dilution with water or alcohol. The precipitate, dried by vacuum or with hot air, is processed to a marketable article by grinding and sieving. The powder is off-white to yellow in color.

11.2. Structure and Properties

Chemical Structure. Xanthan is a heteropolysaccharide composed of D-glucose, D-mannose, and D-glucuronic acid. It has a β-1,4-D-glucan (cellulosic) backbone substituted through C-3 on alternate glucose residues with a trisaccharide side chain consisting of β-D-mannose-(1,4)-β-D-glucuronic acid-(1,2)-α-D-mannose. The terminal mannose moiety may have pyruvate residues linked to the 4- and 6-positions, the internal mannose unit is acetylated at C-6 (Fig. 26) [243]. X-ray diffraction studies on oriented xanthan gum fibers [244] identified the molecular conformation as a right-handed five-fold helix with a pitch of 4.7 nm. In this conformation the trisaccharide side chain is aligned with the backbone and stabilizes the overall conformation by hydrogen bonding [245].

By modifying the biosynthetic pathway for xanthan production, the carbohydrate structure and the substitution pattern of the polymer can be genetically controlled. In this way, polysaccharides with quite different properties can be obtained [246].

Xanthan gum has a *molecular mass* of ca. 2×10^6. Since the structure and the molecular mass are genetically controlled, hydrocolloids with similar molecular mass and similar properties are obtained.

Solution Properties. Xanthan gum is completely soluble in hot or cold water. The gum can be hydrated in sugar solutions up to 60 %. The viscosity of the gum solution is a function of its concentration in the dispersion [247]. *Aqueous solutions* of xanthan gum are extremely viscous (Table 4) and pseudoplastic (i.e., they exhibit a reversible, highly shear-thinning behavior) [248] resulting from the conformation (rod-like) structure and the high molecular mass of the polymer [249]. A 0.25 % solution in water containing 1000 µg/g NaCl and 147 µg/g $CaCl_2 \cdot H_2O$ has a viscosity of ca. 1000 mPa · s at a shear rate of $1\ s^{-1}$ which decreases to ca. 50 mPa · s at a shear rate of $100\ s^{-1}$. The property of shear thinning is made use of in food processing and in improving the sensory qualities of food [250]. Equations relating apparent viscosity, concentration, shear rate, and temperature have been reported [251].

Xanthan gum solutions exhibit higher viscosities than other gums at the same low concentration. Shear thinning is more pronounced due to the semi-rigid conformation (compared to the

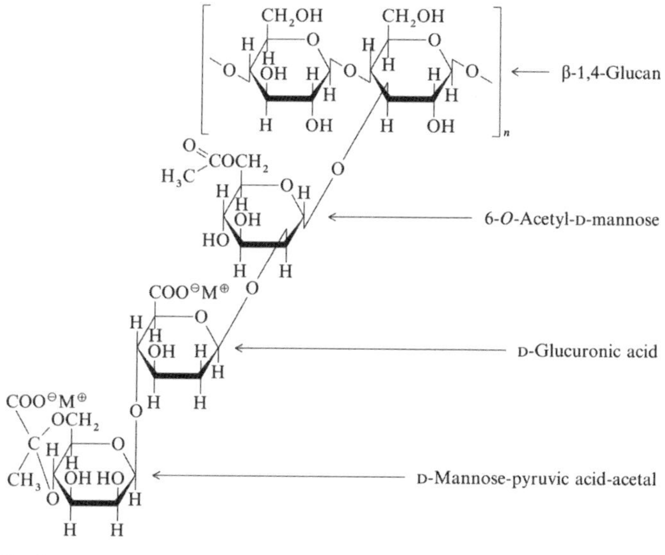

Figure 26. Structural features of xanthan gum

random coil conformation of the other gums). The formation of a weak gel network results in high yield point values. Theses interactions are shear-reversible. 1 % Xanthan gum in a 1 % KCl solution has a yield value of ca. 11 300 mPa compared to 4000 mPa for guar gum, 830 mPa for hydroxypropyl methyl cellulose (HPMC) and 360 mPa for locust bean gum (LBG). At a concentration of 0.3 % xanthan gum exhibits a yield value of 500 mPa while the other gums mentioned have no significant yield value. This quality is responsible for xanthan gum's ability to stabilize emulsions and dispersions [252].

Effect of Salts. In the presence of salts the solution viscosity is stabilized and maintained; polyvalent cations, however, can precipitate xanthan gum under alkaline conditions. Xanthan gum solutions are compatible with high concentrations of various salts [253].

Effect of Temperature. At low ionic strength xanthan gum solutions undergo a thermal transition (i.e., melting), as can be detected by a viscosity change [254]. Optical rotatory and circular dichroic transitions are coincident with the viscosity change, indicating a conformation transition. These data are consistent with the unwinding of an ordered structure into a random coil [255]. The transition temperature can be increased to > 90 °C when small amounts of salt are added [257].

The viscosity of a xanthan gum solution containing salt is almost unaffected over the temperature range 10 – 90 °C. Sterilized food products containing xanthan gum loose only ca. 10 % of their viscosity – much less than products stabilized with other hydrocolloids such as guar gum, alginate, and carboxymethyl cellulose (CMC).

In foods intended for use in microwave ovens xanthan gum eliminates moisture separation. At a gum level of 0.4 % the viscosity is not affected by electrolytes while at a 1 % gum level the viscosity increases significantly.

Effect of pH. The secondary structure (the side chains wrapped around the backbone) imparts xanthan gum stability against degradation by acids, bases, heat treatment, freeze – thaw cycles, enzymes and extended mixing processes [239].

Xanthan gum solutions are stable over a wide pH range (ca. < 2.5 to > 11) [247]. At pH $\simeq$ 9 and above the gum gradually deacetylates. Its stability depends on the gum concentration: the higher the concentration the more stable the solution.

Enzymes commonly present in food products or added to such products (amylase, pectinase, cellulase) do not degrade xanthan gum although it can be partly degraded by endo-β-1,4-glucannases after removal of calcium ions [256].

Interactions and precipitation can occur in acidic and/or heat-treated dairy-protein systems [239].

Oxidizing Agents. Xanthan gum solutions are degraded by high levels of strong oxidants such as hypochlorite, persulfate, and hydrogen peroxide [247], particularly at high temperature. The gum will dissolve in 8 % solutions of sulfuric, nitric and acetic acids, 10 % hydrochloric acid, and 25 % phosphoric acid [257]. Such solutions remain stable for several months.

Solvents. Up to 50 % ethanol and propylene glycol can be added to aqueous xanthan gum solutions without causing precipitation.

Other Hydrocolloids. Xanthan gum is highly compatible with alginates and starch exhibiting long term stability. It is completely compatible with most natural and synthetic thickeners.

Gel Formation. Two models have been proposed for weak gel formation in xanthan gum solutions; both models are based on the occurrence of xanthan molecules in helical structures (Fig. 27). In the model of NORTON et al., single helical regions associate to form a network [258]. The model of MORRIS [259] is based on a double-helical model for xanthan. The primary mechanism is considered to be end-to-end association into fibrous aggregates via double-helix formation.

Xanthan gum interacts with the galactomannans (locust bean, guar, and tara gum) and glucomannans synergistically, leading to enhanced viscosity or gel formation. The formation of thermoreversible gels by synergistic mixtures of xanthan gum with certain galacto- and glucomannans has been ascribed to intermolecular

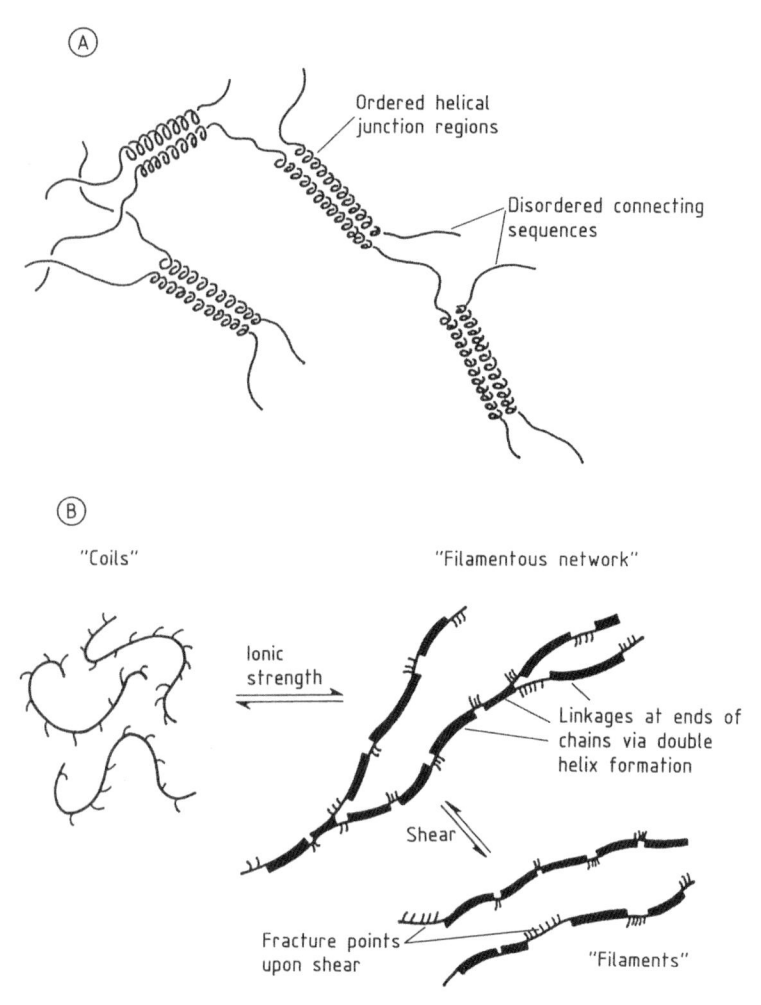

Figure 27. A) Weak gel formation in xanthan gum solutions by self-association of single helices and B) End-to-end association via double-helix formation into fibrous aggregates [260]

binding through cocrystallization of denatured xanthan chains within the mannan crystallite [260]. The cellulosic backbone of xanthan gum and the stereochemically similar mannan backbone both form ribbonlike structures. The mixed crystallites probably act as strong junction zones to consolidate the weak xanthan gum network.

Gel Formation with Galactonmannans. At all concentrations and blend ratios xanthan gum and *guar gum* interact to produce a synergistic viscosity increase. This interaction will occur in the cold although higher viscosities are generated if the system is heated and cooled.

It is presumed that the interaction occurs between the xanthan gum and the "smooth" regions of the galactomannans, thereby explaining the stronger interaction with locust bean gum than with guar gum [261–263]. Xanthan gum exhibits a similar reactivity with tara gum [35].

The xanthan gum – guar gum interaction leads to an increase in solution viscosity and elastic modulus (G'). Determinations of elastic modulus and apparent viscosity are important for the stabilization of emulsions and suspensions (e.g., salad dressing requires a high elastic modulus while sauces and soups need a high apparent viscosity). Maximum viscosity synergism is produced at total gum concentrations of up to 1%. The synergistic ratio varies with the total gum level; for example, at a total gum level of 0.5% the maximum viscosity is produced with a ratio of xanthan gum:guar gum of 30:70 while at a total gum concentration of 0.2% the ratio is nearer 50:50. The synergism decreases in the presence of ionic materials such as salt and acid. These synergistic blends have a lower elastic modulus which is important in some food applications. The presence of xanthan gum improves the heat stability of guar gum. These blends have a wide variety of applications due to the good sensory properties and lower cost than xanthan gum alone.

In *locust bean gum* the ratio of mannose (backbone) to galactose (side chains) is ca. 4:1 but, unlike guar gum, the distribution of the side chains is more irregular – there are regions that are highly substituted ("hairy") and unsubstituted ("smooth") regions. In xanthan gum – LBG blends at total gum levels of < 0.1% synergistic viscosity increases are produced; at gum levels > 0.1% strong cohesive thermoreversible gels exhibiting very little syneresis are obtained. The optimal gum ratios are xanthan gum:LBG of 40:60 to 60:40. The interaction, as demonstrated by gel strength, is stronger at neutral pH and decreases with decreasing pH. In food products this gel texture by itself is undesirable but it can be modified by the inclusion of starches and/or proteins [239].

Rheological and ultracentrifugation studies were conducted on heated and unheated mixtures of xanthan gum with whole LBG, and temperature fractions of the latter possessing different mannose:galactose ratios [264]. Results suggest that xanthan gum and LBG interact via two distinct mechanisms. One mechanism takes place at room temperature, gives weak elastic gels, and has little dependence upon the galactose content of the galactomannan, whereas the second mechanism requires significant heating of the polysaccharide mixture, gives stronger gels, and is highly dependent upon galactomannan composition [264].

Mixed gels of xanthan gum and galactomannans (guar gum and fractionated LBG) were compared by means of oscillatory shear measurements [265]. Whatever the proportion of LBG, maximal synergism was observed when the ratio of xanthan gum to galactomannan was 1:1. However, the magnitude of this maximum varied with LBG content in the mixture.

If guar gum is modified through partial removal of galactose substituents with an α-galactosidase, it also gels with xanthan gum [260]. The sedimentation rate in aqueous mixtures of xanthan gum and enzyme-modified galactomannans (the galactose content being decreased by the enzyme) was studied [266]. Results were related to rheological properties of the systems at large deformations. The shear modulus increased while the strain at yielding and sedimentation velocity decreased with decreasing galactose content of galactomannan. Properties of the xanthan gum enzyme-modified galactomannan system were compared with those of xanthan gum – LBG systems. LBG was suggested to be a better stabilizer than the enzyme-modified galactomannans studied [266].

Gel Formation with Glucomannans. Xanthan gum – konjac mannan (glucomannan) interactions are strong at total gum concentrations above 0.2% [267]. The optimum ratio is between 40:60 and 30:70. The presence of salt decreases the interaction. At temperatures > 50 °C, the system's G' is higher than its G'', indicating well defined elastic properties. At ca. 55 °C the gel structure collapses and there is a sharp decrease in the value of G'. At values > 55 °C the behavior is typical of a liquid which has G'' significantly higher than G'. At ca. 1.0% total gum concentration, maximum gel strength is achieved, with no further increase in this property with increasing gum concentration. The mechanical properties of xanthan gum – konjac mannan are also similar to those of xanthan gum – LBG.

Others. Small amounts of xanthan gum are sufficient to stabilize *starch solutions* during storage. Xanthan gum at 0.1 – 0.2% prevents starch solution retrogradation and increases its stability, especially at low pH (around 3). These small amounts of xanthan gum also improve starch stability during freeze – thaw cycles.

Reactions of xanthan gum with specific dextrins and incompatibility between xanthan gum and gum arabic at pH values below 5 have been reported. Xanthan gum also reacts with casein below its isoelectric point [268].

11.3. Analysis

The presence of xanthan gum can be indicated by the formation of a gel during cooling of a 1:1

mixture with carob gum in a 1% solution. Its presence can also be indicated from the reaction of pyruvate with 2,4-dinitrophenylhydrazone in a hydrolysate.

11.4. Applications, Market

The applications of xanthan gum are based especially on the viscous behavior and stabilizing properties of this hydrocolloid. In industry, xanthan gum is used in groups 18, 19, 26, 27, 28, 29, and 33 (Table 2). These applications enabled Kelco, the first producer of xanthan gum, to carry out a comprehensive toxicological evaluation that resulted, in 1969, in the acceptance of this hydrocolloid in foods, where it has found applications especially in groups 4, 5, 8, and 12.

Xanthan gum is produced by several companies. Present production is estimated at ca. 20 000 t/a. For food applications prices are ca. $ 11 – $ 13 per kilogram. Because of its specific properties the potential for this polysaccharide is great.

12. Gellan Gum

Gellan gum is produced by the bacterium *Pseudomonas elodea* (ATCC 31461). The bacterium was originally found growing in a lily pond in Pennsylvania, USA. Gellan gum has been approved for food uses [260], [269], [270]. It is a broad-spectrum, multipurpose gelling agent, which is fully compatible with many food systems and processes. Gellan gum is both cold setting and thermosetting; its melting point (thermoreversible and thermoirreversible) depends on ionic strength. Gelled products can be made to be stable or to melt when subjected to retorting and microwave cooking.

Production. The gum is produced commercially by inoculating a fermentation medium with the microorganisms. The medium contains a carbon source, phosphate, organic and inorganic nitrogen salts, and trace elements. The fermentation is carried out under sterile conditions with careful control of aeration, agitation, pH, and temperature [271]. When fermentation is complete, the viscous broth is pasteurized to kill viable bacteria before the gum is recovered.

Recovery can be achieved in several ways. Direct recovery yields the gum in its native or *high acyl form*. Treatment of the broth with alkali causes deacylation resulting in the *low acyl version*. The presence of the acyl substituents largely influences the resulting solution properties and gel characteristics [272].

Structure. Gellan gum is a linear anionic polysaccharide with a tetrasaccharide repeating unit:

-3)-β-D-Glc-(1,4)-β-D-GlcA-(1,4)-β-D-Glc-(1,4)-α-L-Rha-(1-

Light scattering and intrinsic viscosity measurements give a molecular mass of approximately 5×10^5 for the deacylated gum [273].

In the native polysaccharide two acyl substituents, L-glycerate and acetate, are present. Both substituents are located on the same 1,3-linked glucosyl residue. The glycerate is esterified on the hydroxyl group on C-2, the acetate on the hydroxyl group on C-6. On average, there is one glycerate per repeating unit and one acetate per every two repeating units [274].

Chemical analysis established that glycerate substitution predominates over that with acetate [275]. In addition, glycerate substitution dramatically influences gellan gum properties since this bulky substituent hinders chain association and accounts for the change in gel texture brought about by de-esterification [270].

X-ray diffraction studies of oriented fibers [275–277] shows that gellan gum exists as a three-fold, left handed, parallel double helix. The pair of molecules that comprise the helix is stabilized by hydrogen bonds at each carboxylate group. In the potassium salt of the deacylated version the potassium ion is coordinated to the carboxylate group, which, in turn, is involved in interchain hydrogen bonds to stabilize the duplex. In this structure the location of the potassium ions on both the inside and the outside of the helix provides a mechanism for both stabilization of the helix and the association of the helices that is necessary to produce a macromolecular gel structure to aggregate [278]. The calcium salt form fits the same model with the divalent calcium ion replacing the potassium – water – potassium bridge [279]. In these salt forms of the gel, helix aggregation is responsible for the brittle nature of the gel. In the native form the acyl substituents do not prevent helix formation.

However, the acetate groups interfere with aggregation, resulting in elastic rather than brittle gels.

Hydration. The pure sodium ion form of gellan gum hydrates at room temperature in deionized water. However, the commercial product is a mixed salt – predominantly in the potassium ion form but also containing other ions such as sodium, calcium, and magnesium. This form is not soluble in cold water since the low level of divalent ions is sufficient to hold the molecular chains together and, thereby, inhibit hydration. Hydration can be achieved by dispersing the gum in deionized water and heating to 75 °C. In tap water, the hydration temperature depends upon the concentration and types of ions in solution. For example, at around 180 ppm calcium carbonate hardness (0.0072 % calcium) the gum will not dissolve even in boiling water but it will hydrate at 90 °C in a 1 % sodium chloride solution [272].

In many systems, hydration of gellan gum by simply dispersing the gum in water and heating is not practical. In such systems hydration is effected by the addition of a sequestrant such as sodium citrate. The sequestrant removes the inhibitory divalent ions from both the gum and the water, but (if used at the appropriate amounts) does not introduce sufficient sodium ions into the system to prevent gum hydration. Depending upon the ions initially present and their concentrations, gellan gum can be hydrated at room temperature when a sequestrant is used [272].

Solution Properties. Highly viscous solutions of *low acyl gellan gum* can be prepared using a sequestrant. At a concentration of 1 wt % gellan gum solutions are less pseudoplastic than xanthan gum solutions but more pseudoplastic than solutions of high molecular mass sodium alginate [276]. A large drop in the solution viscosity of gellan gum is observed at 25 – 40 °C. This reflects a conformational change from some form of relatively ordered, nonaggregated double helix to a random coil. This change is reversible.

Native, or *high acyl gellan gum* dispersions in cold water have very high solution viscosities. Such solutions are very sensitive to salt concentration. In equally concentrated solutions of xanthan gum and gellan gum containing increasing levels of sodium chloride the viscosity of the native gellan gum is strongly dependent upon the salt concentration whereas that of the xanthan gum is not. High acyl gellan gum solutions are highly thixotropic; such behavior is the outcome of gel-like network formation. Similar thixotropic behavior is observed with solutions containing a blend of xanthan gum with locust bean gum.

Gel Formation. Unlike other gums gellan gum forms gels with many ions including hydrogen ion. However, its affinity for each ion is different. Divalent ions have a much stronger affinity than monovalent ones. Divalent ions like calcium and magnesium produce optimal gel strength at 3 % of the level required with sodium and potassium ions [280]. Since ions promote chain association in forming gels, the ion levels needed to cause gelation tend to prevent the gum from hydrating. Furthermore, in keeping with their respective affinities, divalent ions inhibit hydration much more strongly than monovalent ions.

Ions are needed to set up the gel network. The simplest way to make a gellan gum gel is to hydrate the gum in deionized water by heating the gum dispersion to above 75 °C, add ions such as calcium or sodium ions to the hot solution, and cool. When sequestrants are used for hydration, ions are already present in the system. Since the level of these ions is such that they do not prevent hydration, they are also not sufficiently abundant to form strong gels. In several applications such weak gels are quite acceptable but, to optimize gel strength, additional ions must be added. Again, these ions can be added to the hot gellan gum solution which will later gel when allowed to cool.

Gel Texture. The texture of gellan gum and other polysaccharide gels has been investigated using the technique of Texture Profile Analysis on an Instron (a testing machine produced by Instron). Free-standing gel disks were prepared by allowing hot solutions to cool in ring molds. The set, tempered, gels were compressed twice in succession to a given percentage of their original height at a given rate. The textural parameters (modulus, hardness, brittleness, elasticity and cohesiveness) were calculated from the resulting force/deformation curve [272]. The results show that gels made with low acyl gellan gum have a high modulus and are very brittle; in contrast high acyl gellan gum gels are very elastic, nonbrittle and cohesive [281], [282].

Low acyl gellan gum gels are clear, firm to the touch and do not melt, but they release liquid when eaten giving a melt-in-the-mouth sensation. This property is important to flavor release and can be useful in, for example, dessert products [283]. Other investigations have suggested that overall flavor increases as gel hardness decreases and gel brittleness increases [284].

Melting and Setting Points, which are important gel properties, depend upon ion concentration and type and, to a lesser extent, on gum concentration [269]. Gels differ in their setting temperature. For example, those containing calcium ions set between 25 and 40 °C, and those with sodium ions set at 40 – 50 °C. At lower ion levels, gels remelt upon heating, at higher ion levels they do not melt below 100 °C.

Applications. Gellan gum is used in many applications including microbiological media, tissue culture media, food and pet feeds, photographic emulsions, and microcapsules. It is also used as a model for studying gel swelling and gel shrinkage [269], [285], [286]. Gellan gum can be used as a fining agent for alcoholic beverages including beer, wines, and fortified wines [287].

Gellan gum is mainly used in confectioneries, jams and jellies, fabricated foods, water-based gels, pie fillings and puddings, icings and frostings, and dairy products. Gellan gum and polydextrose were used to produce low-calorie or sugar-free jelly sweets where the final solids content exceeded 80 %. In comparison to other gelling agents gellan gum does not appreciably add to the viscosity of the depositing mix and is, therefore, an ideal gelling agent for such systems [288].

Gellan gum is useful in cocoa and chocolate products [289] and in alcoholic and nonalcoholic beverages [290].

Gellan gum is used in a variety of Japanese food including Mitsumame jelly cubes, soft and hard red bean jelly and Tokoroten noodles [269]. It has been reported to stabilize oil-in-water dispersions [291] and can be used to prepare gels that are dried and subsequently rehydrated to their original shape [292].

Legal Aspects. Gellan gum has been shown to be completely safe for use in food. It has been given a "not specified" ADI by the Joint FAO/WHO Expert Committee on Food Additives [293] and by the EC's Scientific Committee for Food [294].

Gellan gum is approved in over 30 countries worldwide, including the United States, Canada, Australia, South Africa, Japan and most of South East Asia, e.g., Hong Kong, Philippines, Taiwan and Thailand, most of South America, and all 15 EU member states.

In the European Union gellan gum is one of the generally permitted additives listed (as E418) in Appendix I of the European Parliament and Council Directive 95/1/EU on food additives other than colors and sweeteners [295].

Specification and Purity. Kelcogel gellan gum complies with the specifications for gellan gum prepared by the Joint FAO/WHO Expert Committee on Food Additives (JECFA) [296] and the Food Chemicals Codex (FCC) [297].

Market. The commercially available products are K9A50, a nonclarified form of the gum for industrial applications; Kelcogel LT100, a nonclarified form and Kelcogel, a clarified form for food and industrial uses; and Gelrite, which is used for microbiological plating media. All gellan gum is produced by NutraSweet Kelco (a Monsanto company). Gellan gum is priced at about $ 53 per kg.

13. Galactomannans [2–6], [11], [298]

Galactomannans are plant reserve carbohydrates, like starch, that occur in the endosperm of the seeds of many Leguminoseae. During sprouting, the galactomannans are enzymatically degraded and used as nutrition. Of the many known galactomannans, up to now only three have been processed and used on an industrial scale:

1. **Locust bean gum (carob gum)** has been known for a long time. It is derived from seeds of the carob tree (*Ceratonia siliqua* L.) which grows around the Mediterranean Sea, in Spain, Portugal, and Morocco, where more attention is paid to its systematic cultivation. The annual production is estimated at 10 000 – 12 000 t.

2. **Guar gum** is at present the most important galactomannan. It is derived from the seeds of the guar plant (*Cyamopsis tetragonoloba* L. Taub.), which grows in India and Pakistan. Since 1944, guar plants have also been cultivated on a large scale in the southern United States. They have been introduced into Brazil, Africa, and Australia, where two crops a year can be obtained. The industrial production of guar gum began around 1940. The annual production is estimated at ca. 125 000 t.
3. **Tara gum** has been produced only since the 1970s, to a much lesser degree (1000 t/a), from the seeds of the tara tree (*Cesalpina spinosa* L.), which occurs mainly in Peru.

The galactomannans on the market are flours prepared from the endosperm of the corresponding seeds. The carob and tara kernels are particularly difficult to process because of their tough, hard hulls. Remnants of the hulls and seeds are, therefore, always present in small amounts. The pure polysaccharide extracted from the flour of locust bean gum is often designated carobin gum and the polysaccharide from guar flour, guaran.

13.1. Structure

Galactomannans have a backbone of β-(1,4)-glycosidically linked mannopyranosyl residues. This backbone is substituted with single-unit α-D-galactopyranosyl residues linked to the O-6 of certain mannose moieties. This substitution renders the mannans soluble. Within the various mannans a wide spectrum of chemical structures occurs. This diversity of structure includes not only a wide variation in their mannose: galactose ratio, but also significant differences in the distribution of galactose units along the mannan backbone (Fig. 28) [299], [300]. For *locust bean gum* a mannose: galactose ratio of 3.5 has been established, compared to a ratio of 3 for *tara gum* and 1.5 for *guar gum*. A measure of the differences in the distribution of galactose side groups has been obtained from an examination of the degree of hydrolysis and the characteristic array of oligosaccharides obtained by digestion with an endomannanase. In this way, DEA et al. [301] established that a *tara gum* has a more statistically random distribution of side chains while

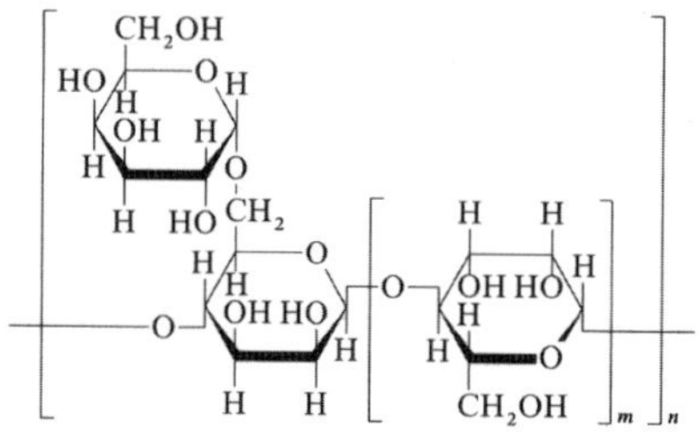

$m = 3$: Locust beam gum
$m = 1$: Guar gum
$m = 2$: Tara gum

Figure 28. Structural features of galactomannans

locust bean gum has a nonregular, nonstatistically random distribution with a certain proportion of unsubstituted blocks of intermediate length. *Guar gum* was found to have few, if any, regions that were unsubstituted (smooth regions) with galactose. Galactomannans differ further in their molar mass distribution; the highest values were established for guar gum. These differences in fine structure between the galactomannans, involving the extent and pattern of galactosyl substitution and molar mass distribution, account for their very different functional properties.

13.2. Production [4], [11], [298]

Guar Gum. For the separation of the germ and the endosperm halves, guar seeds are first screened to remove foreign matter and then fed to an attrition mill to split the seeds in two endosperm halves; finally, the germ material is sifted off. Despite their relatively low galactomannan content the remaining endosperm halves, covered with hull (splits), are suitable for many technical applications. For higher-quality demands, and particularly for food applications, the endosperm must be further purified and liberated from the adhering hull. Various processes are used for this; in general, the splits are treated with moist or dry, hot air, which results in the loosening of the hull as a result of different swelling properties. After removal of the hull by sifting, pure endosperm is obtained with a galactomannan content of 85 – 95 % based on dry matter. By suitable milling and screening techniques the endosperm can be worked up to the desired commercial products.

Locust Bean and Tara Gum. Since the hull of carob and tara gum is very hard and tough, these gums are difficult to process. For removal of the hulls, two different processes exist. In the *acid process* the hulls are carbonized by treating the kernels with moderately dilute sulfuric acid at elevated temperature. Similar to the guar gum process, the remaining hull fragments are then removed by washing and brushing operations. After a drying step the dehulled kernels are cracked and the germs sifted off from the endosperm. Commercial products are obtained by milling and screening.

In the vanishing *roasting process* the kernels are roasted in a rotating furnace, where the hulls pop. The endosperm is then obtained as described above. This process yields products of somewhat darker color. However, the use of sulfuric acid is avoided.

13.3. Properties

The most prominent properties of galactomannans are their high water-binding capacity, their potential to form highly viscous solutions even at low concentrations, and the interaction of locust bean gum with carrageenans, agar, and xanthan gum. In double logarithmic plots of zero shear viscosity versus concentration multiplied by intrinsic viscosity, galactomannans exhibit a different type of viscosity behavior from most polysaccharides for which nonspecific physical entanglement between fluctuating random coil molecules is the underlying mechanism. Galactomannans show an earlier onset of concentrated solution behavior and a substantially greater concentration dependence thereafter [299], [300]. This behavior is attributed to the occurrence of interchain associations inherent to β-1,4-D-mannan chain segments. Such self-association has also been observed for β-1,4-D-glucan and β-1,4-D-xylan structures.

Locust bean gum and guar gum are both *water soluble*; they differ, however, in their dissolution behavior. Whereas guar gum is readily soluble in cold water, locust bean gum can be solubilized only by boiling. Solutions of both galactomannans are, depending on their purity, somewhat cloudy because of the presence of some proteins and crude fiber.

Even 1 wt% solutions are highly *viscous* (2000 – 6000 mPa · s, measured with an RTV Brookfield viscometer at 20 rpm); 2 – 5 wt% swollen dispersions show gellike behavior. The viscosity of solutions is hardly influenced by ionic strength. Locust bean gum solutions can gel at high sodium hydroxide concentrations. Also boric acid and alkaline copper salt solutions cause flocculation or gelling of the galactomannans. Mixing of solutions of locust bean gum and xanthan gum results in an increase of the viscosity or even in gel formation (after boiling [209]). Such synergistic interactions have also been observed for combinations of locust bean gum with carrageenan or agar (Section 5.3). Dilute galactomannan solutions form elastic, transparent films upon dehydration.

Galactomannans are rapidly *degraded* in acidic, aqueous solutions at elevated temperature, resulting in a rapid loss of viscosity. They can also be degraded by oxidants and by microbial enzymes (mannanases); under alkaline conditions, they are rather stable [302]. When stored for longer periods of time, solutions of galactomannans must therefore be protected against microbial degradation by addition of preservatives.

13.4. Analysis and Composition of Commercial Preparations

Commercial galactomannan preparations still may contain remnants of germs and hull that are rich in protein and fiber, and reduce the galactomannan content. The galactomannan content is therefore an important factor in quality. This *gum content* is determined by analyzing the impurities and subtracting their total from 100%. Depending on the application, products with different galactomannan content are used.

Another quality factor is the *viscosity* of the aqueous solution, usually measured in 1 wt% systems. Locust bean gum must first be solubilized by heating.

The *presence of galactomannans* can be demonstrated by various precipitation reactions (e.g., gelation with borax). For determination of the *mannose : galactose ratio*, the preparation is hydrolyzed and the sugar moieties are analzyed by using gas chromatography [15], [29].

13.5. Derivatives

As with cellulose, galactomannans contain primary and secondary hydroxyl groups, which, in principle, can be derivatized and substituted like cellulose. Only hydroxyalkyl, carboxymethyl, and cationic derivatives, as well as partially depolymerized products, are of industrial significance, however.

Hydroxyalkyl Derivatives are obtained by treating galactomannans with ethylene or propylene oxide in an alkaline medium. These derivatives differ from the parent material by having a faster rate of hydration and being almost completely soluble.

Carboxymethylated Products are obtained by "etherification" of a certain portion of the hydroxyl groups with monochloric acid or sodium monochloroacetate. Unlike carboxymethyl celluloses, which are soluble only at degrees of substitution in the range of 0.7 – 1.0, galactomannans – if not already soluble prior to derivatization – become completely soluble at a degree of substitution of <0.05.

Cationic Derivatives are obtained by treatment of galactomannans with suitable reactive organic amines such as 2-hydroxy-3-chloropropyltrimethylammonium chloride or its reaction product 2,3-epoxypropyltrimethylammonium chloride, which forms in alkaline medium.

Partially Depolymerized Galactomannans with reduced viscosities are obtained by oxidative, hydrolytic (acid), or enzymatic degradation. Conversely, by reaction with bifunctional molecules such as epichlorohydrin, *cross-linked galactomannans* can be obtained, which – depending on their degree of cross-linking – are only partly water soluble or fully insoluble. Plant α-galactosidase that can remove galactose from galactomannans has been cloned in yeast and can be produced conveniently. This enzyme is active in systems containing up to 35 % galactomannan [303].

13.6. Applications

Guar Gum has found application in many industries; its use is still increasing. Most uses are based on its good solubility in water and the high viscosity of aqueous solutions. In many applications, its inertness against high concentrations of salts of monovalent cations, its ability to form gels with certain multivalent cations, and its flocculating action are also decisive. *Derivatives of guar gum* are used in the oil industry in secondary or tertiary oil recovery of nearly exhausted oil wells. For applications in the textile industry (Table 2, groups 30 and 31), a variety of native and derivatized types are produced. They are also used in the manufacture of paper (group 29), explosives (group 26), and cables. In the latter applications, guar gum is added in advance to immobilize the inevitably penetrating water. They are also used in the pharmaceutical industry and as flocculating agents in the processing of minerals and effluent purification (group 32). In the food industry, guar gums are used as efficient thickeners of aqueous systems, thereby controlling the mobilization of water. This influences the consistency, body, and shelf life of aqueous food systems, as well as the stability of oil – water and water – oil emulsions. They are used in groups 3, 5, 6, 8, 9, and 12, and have also found application in pet food (group 13) and cattle feed.

Due to its higher price, *locust bean gum* is used mainly in the food industry, in the same groups as the guar gums. Of special interest for locust bean gum is its synergistic action with other hydrocolloids such as carrageenans, agar, and xanthan gum.

13.7. Market

Of the 125 000 t of *guar gum* produced annually, ca. one-third is used in the food industry; the remainder has found technical application. *Locust bean gum* is used almost completely in the food and pet food industry. The prices on the world market average ca. $ 1.1 – $ 1.3 per kilogram for guar gum and $ 6.4 – $ 7.5 per kilogram for locust bean gum.

14. Acknowledgement

Updated from the corresponding chapter in Ullmann's 4th ed. written by WALTER PILNIK.

References

1. M. Glicksman: *Gum Technology in the Food Industry*, Academic Press, New York 1969.
2. R. L. Whistler, J. N. BeMiller (eds.): *Industrial Gums*, 2nd ed., Academic Press, New York 1973.
3. H. D. Graham (ed.): *Food Colloids*, AVI, Westport, Connecticut 1977.
4. H. Neukom, W. Pilnik: *Gelier- und Verdickungsmittel in Lebensmitteln*, Forster Verlag, Zürich 1980.
5. G. O. Aspinall: *The Polysaccharides*, vol. 2, Academic Press, New York 1983.
6. P. Harris: *Food Gels*, Elsevier Applied Science Publishers, London 1990.
7. J. R. Mitchell, D. A. Ledwàrd: *Functional Properties of Food Macromolecules*, Elsevier Applied Science Publishers, London 1986.
8. M. L. Fishman, J. J. Jen: "Chemistry and Functions of Pectin," *ACS Symp. Ser.* **310** (1986)
9. D. Oakenfull: "Gelling Agents," *CRC Crit. Rev. Food Sci. Nutr.* **26** (1987), 1 – 25.
10. M. Yalpani: "Industrial Polysaccharides; Genetic Engineering, Structures/Property Relations and Applications," *Prog. Biotechnol.* **3** (1987)
11. M. Glicksman: *Food Hydrocolloids*, CRC Press, Boca Raton Fla. 1986.
12. R. H. Walter: *The Chemistry and Technology of Pectin*, Academic Press, New York 1991.
13. W. Schmolck, E. Mergenthaler: "Beiträge zur Analytik von Polysacchariden, die als Lebensmittelzusatzstoffe verwendet werden," *Z. Lebensm. Unters. Forsch.* **152** (1973) 87 – 99.
14. M. E. Endean: The Separation and Identification of Polysaccharide Stabilizers, Thickeners and Gums from Foods, Technical circular no. 575, The British Food Manufacturing Industrie Research Association, Leatherhead, Surrey 1974.
15. *Schweizerisches Lebensmittelbuch*, vol. 3, chap. 40, "Gelier- und Verdickungsmittel," Eidgen. Drucksachen- und Materialzentrale, Bern 1989.
16. R. G. Morley, G. O. Phillips, D. M. Power, *Analyst (London)* **97** (1972) 315 – 319.
17. Scott, D. Glick (eds.): *Methods of Biochemical Analysis*, Interscience, New York 1960, pp. 145 –197.
18. C. Rolin in R. L. Whistler, J. N. BeMiller (eds.): *Industrial Gums*, 3rd ed., Academic Press, San Diego 1993, pp. 257 – 293.
19. R. Ilker, A. S. Szczezesniak: "Structural and chemical bases for texture of plant foodstuffs," *J. Text. Studies.* **21** (1990) 1.
20. J. P. Van Buren: "The Chemistry of Texture in Fruits and Vegetables," *J. Text. Studies.* **10** (1979) 1.
21. M. McNeil, A. G. Darvill, S. C. Fry, P. Albersheim, "Structure and Function of the primary Cell Walls of Plants," *Ann. Rev. Biochem.* **53** (1984) 625.
22. R. R. Selvendran: "The Chemistry of Plant Cell Walls", in G. G. Birch, J. Parker (eds.): *Dietary Fibre*, Applied Sci. Publ., London 1995.
23. M. C. McCann, K. Roberts: *The Cytoskeletal Basis of Plant Growth and Form*, Academic Press, Oxford 1991, pp. 109 – 129.
24. C. M. G. C. Renard, H. A. Schols, A. G. J. Voragen, J. F. Thibault, W. Pilnik: "Studies on Apple Protopectin. III. Characterization of the Material Extracted by Pure Polysaccharidases from Apple Cell Walls," *Carbohydr. Polym.* **15** (1991) 13.
25. C. M. G. C. Renard, A. G. J. Voragen, J. F. Thibault, W. Pilnik: "Comparison between enzymatically and chemically extracted pectins from apple cell walls," *Animal Feed Sci. Technol.* **32** (1991) 69.
26. J. A. De Vries, F. M. Rombouts, A. G. J. Voragen, W. Pilnik: "Enzymic Degradation of Apple Pectins," *Carbohydr. Polym.* **2** (1982) 25.
27. R. R. Selvendran: "Development in the Chemistry and Biochemistry of Pectic and Hemicellulosic Polymers," *J. Cell Sci. Suppl.* **2** (1985) 51.
28. T. P. Kravtchenko, A. G. J. Voragen, W. Pilnik: "Analytical comparison ot three industrial pectin preparations," *Carbohydr. Polym.* **18** (1992) 17.
29. H. Thier: "Identification and Quantification of Natural Polysaccharides in Food Stuffs," in G. O. Phillips, D. J. Wedlock, P. A. Williams (eds.): *Gums and Stabilisers for the Food Industry 2, Applications of Hydrocolloids*, Pergamon Press, Oxford 1984.
30. T. P. Kravtchenko, M. Penci, A. G. J. Voragen, W. Pilnik: "Enzymic and chemical degradation of some industrial pectins," *Carbohydr. Polym.* **20** (1993) 195.
31. R. H. Walter (ed.): *The Chemistry and Technology of Pectins*, Academic Press, New York 1991, p. 113.
32. J.- F. Thibault, C. M. G. C. Renard, M. A. V. Axelos, P. Roger, M.- J. Crepeau: "Studies of the length of homogalacturonic regions in pectins by acid hydrolysis," *Carbohydr. Res.* **238** (1993) 271.
33. J. A. de Vries: "Repeating units in the structure of pectin", in G. O. Phillips, P. A. Williams, D. J. Wedlock (eds.): *Gums and Stabilisers for the Food Industry*, vol. 4, IRL Press, Oxford 1988, p. 25.
34. S. E. Harding, G. Berth, A. Ball, J. R. Mitchell, J. Garcia de la Torre: "The Molecular Weight Distribution and Conformation of Citrus Pectins in Solution Studies by Hydrodynamics," *Carbohydr. Polym.* **16** (1991) 1.
35. Z. I. Kertesz: *The Pectic Substances*, Interscience Publishers, New York 1951.
36. O. P. Beerh, B. Raghuramaiah, G. V. Krisnamurthy: "Utilization of mango waste: peel as a source of pectin," *J. Food Sci. Technol.* **13** (1976) 96.
37. R. M. Weightman, C. M. G. C. Renard, J.- F. Thibault: "Structure and properties of the polysaccharides from pea hulls. Part 1: chemical extraction and fractionation of the polysaccharides," *Carbohydr. Polym.* **24** (1994) 139.
38. M. J. Y. Lin, E. S. Humbert: "Extraction of pectins from sunflowers heads," *Can. Inst. Food Sci. Technol. J.* **11** (1978) 75.

39 W. J. Kim, F. Sosulski, S. C. K. Lee: "Chemical and gelation characteristics of ammonia-demethylated sunflower pectins," *J. Food Sci.* **43** (1978) 1436.
40 N. M. Ptitchkina, I. A. Danilova, G. Doxastakis, S. Kasapis, E. R. Morris: "Pumpkin pectin: gel formation at unusually low concentration," *Carbohydr. Polym.* **23** (1994) 265.
41 EC Proposal for Counsil Directive on Food Additives Other than Colours and Sweeteners (III/3624/91/EM-REV3) 1992.
42 G. H. Joseph, A. H. Keiser, E. F. Bryant: "High-polymer ammonia-demethylated pectinates and their gelation," *Food Technol.* **3** (1949) 85.
43 J. C. E. Reitsma, J.-F. Thibault, W. Pilnik: "Properties of amidated pectins. I. Preparation and characterization of amidated pectins and amidated pectic acids," *Food Hydrocoll.* **1** (1986) 121.
44 P. E. Christensen: "Methods of Grading Pectin in Relation to the Molecular Weight (Intrinsic Viscosity) of Pectin," *Food Res.* **19** (1954) 163.
45 A. Kawabata: "Studies on chemical and physical properties of pectic substances from fruits," *Mem. Tokyo Univ. Agr.* **19** (1977) 115.
46 W. J. Kim, V. N. M. Rao, C. J. B. Smit: "Effect of chemical composition on compressive mechanical properties of low ester pectin gels," *J. Food Sci.* **43** (1978) 572.
47 H. A. Deckers, C. Olieman, F. M. Rombouts, W. Pilnik: "Calibration and Application of High-Performance Size Exclusion Columns for Molecular Weight Distribution of Pectins," *Carbohydr. Polym.* **6** (1986) 361.
48 M. L. Fishman, D. T. Gillespie, S. M. Sondey, Y. S. El-Atawy: "Intrinsic Viscosity and Molecular Weight of Pectin Components," *Carbohydr. Res.* **215** (1991) 91.
49 M. A. V. Axelos, J.- F. Thibault, J. Lefebvre: "Structure of Citrus Pectins and Viscometric Study of their Solution Properties," *Int. J. Biol. Macromol.* **11** (1989) 186.
50 J.-F. Thibault, M. Rinaudo: "Interactions of mono- and divalent counterions with alkali- and enzyme-deesterified pectins in salt-free solutions," *Biopolymers* **24** (1985) 2131.
51 C. Garnier, M. A. V. Axelos, J.- F. Thibault: "Phase Diagrams of Pectin-Calcium Systems: Influence of pH, Ionic Strenght, and Temperature on the Gelation of Pectins with Different Degrees of Methylation," *Carbohydr. Res.* **240** (1993) 219.
52 H. S. Owens, H. Lotzkar, T. H. Schultz, W. D. Maclay: "Shape and Size of Pectinic Acid Molecules Deduced from Viscometric Measurements," *J. Am. Chem. Soc.* **68** (1946) 1628.
53 G. Berth, H. Anger, F. Linow: "Streulichtphotometrische und Viskosimetrische Untersuchungen an Pektinen in wäßrigen Lösungen zur Molmassenbestimmung," *Nahrung* **21** (1977) 939.
54 H. Anger, G. Berth: "Gel Permeation Chromatography and the Mark Houwink Relation for Pectins with Different Degrees of Esterification," *Carbohydr. Polym.* **6** (1986) 193.

55 G. Ravanat, M. Rinaudo: "Investigation on Oligo- and Polygalacturonic Acids by Potentiometry and Circular Dichroism," *Biopolymers* **19** (1980) 2209.
56 F. Michel, J.- F. Thibault, J.- L. Doublier: "Viscometric and Potentiometric Study of High-Methoxyl Pectins in the Presence of Sucrose," *Carbohydr. Polym.* **4** (1984) 283.
57 E. Racapé, J.- F. Thibault, J. C. E. Reitsma, W. Pilnik: "Properties of Amidated Pectins. II. Polyelectrolyte Behavior and Calcium Binding of Amidated Pectins and Amidated Pectic Acids," *Biopolymers* **28** (1989) 1435.
58 I. G. Plaschina, E. E. Braudo, V. B. Tolstoguzov: "Circular-Dichroism Studies of Pectin Solutions," *Carbohydr. Res.* **60** (1978) 1.
59 A. Cesàro, F. Delben, S. Paoletti: "Thermodynamics of the Proton Dissociation of Natural Polyuronic Acids," *Int. J. Biol. Macromol.* **12** (1990) 170.
60 M. D. Walkinshaw, S. Arnott: "Conformations and interactions of pectins, I. X-ray diffraction analysis of sodium pectate in neutral and acidified forms," *J. Mol. Biol.* **153** (1981) 1055.
61 M. J. Gidley, E. R. Morris, E. J. Murray, D. A. Powell, D. A. Rees: "Spectroscopic and stoichiometric characterization of the calcium-mediated association of pectate chains in gels and in the solid state," *J. Chem. Soc. Chem. Commun.* **22** (1979) 990.
62 FAO Food and Nutrition Papers 34 and 37, Specifications for Identity and Purity of Certain Food Additives. Distribution and Sales Section, FAO, Rome 1986.
63 Food Chemicals Codex, 3rd ed., National Academy of Sciences, Washington, D.C. 1992.
64 J. N. BeMiller: "Acid-catalyzed hydrolysis of glycosides," *Adv. Carbohydr. Chem.* **22** (1967) 25.
65 D. A. Powell, E. R. Morris, M. J. Gidley, D. A. Rees: "Conformations and Interactions of Pectins, II. Influence of Residue Sequence on Chain Association in Calcium Pectate Gels," *J. Mol. Biol.* **155** (1982) 517.
66 P. Albersheim, H. Neukom, H. Deuel: "Splitting of pectin chain molecules in neutral solutions," *Arch. Biochem. Biophys.* **90** (1960) 46.
67 W. Heri, H. Neukom, H. Deuel, *Helv. Chim. Acta* **44** (1961) 1939.
68 R. Kohn, I. Furda: "Distribution of free carboxyl groups in the molecule of pectin after esterification of pectic and pectinic acid by methanol," *Collect. Czech. Chem. Commun.* **34** (1969) 641.
69 E. L. Pippen, R. M. McCready, H. S. Owens: "Gelation properties of partially acetylated pectins," *J. Am. Chem. Soc.* **72** (1950) 813.
70 L. W. Doner: "Analytical methods for determining pectin composition," *A.C.S. Symp. Ser.* **310** (1986) 13.
71 FAO, Food and Nutrition Paper 52, Addendum 1, Rome 1992, p. 87.
72 *Food Chemicals Codex*, FCC III Monographs, 3rd ed. (including supplements), National Academy Press, Washington D.C. 1981, p. 215.

73 EEC, Council Directive 78/663, *Off. J. EEC* 14 08 78 (plus Updates), 1978.
74 Z. I. Kertesz: *The Pectic Substances*, Interscience, New York 1951.
75 A. Plöger: "Conductivity detection of pectin: a rapid HPLC method to analyze degree of esterification," *J. Food Sci.* **57** (1992) 1185.
76 H.-J. Zhong, M. A. K. Williams, R. D. Keenan, D. M. Goodall, C. Rolin: *Carbohydr. Polym.* **32** (1997) 27.
77 G. Berth: "Studies on the Heterogeneity of Citrus Pectin by Gel Permeation Chromatography on Sepharose 2B/Sepharose 4B," *Carbohydr. Polym.* **8** (1988) 105.
78 R. C. Jordan, D. A. Brant: "An Investigation of Pectin and Pectic Acid in Dilute Aqueous Solution," *Biopolymers* **17** (1978) 2885.
79 S. Sawayama, A. Kawabata, H. Nakahara, T. Kamata: "A Light Scattering Study on the Effects of pH on Pectin Aggregation in Aqueous Solution," *Food Hydrocoll.* **2** (1988) 3.
80 R. H. Walter, H. L. Matias: "Pectin Aggregation Number by Light Scattering and Reducing End-Group Analysis," *Carbohydr. Polym.* **15** (1991) 33.
81 W. D. Holloway, C. Tasman-Jones, K. Maher: "Pectin digestion in humans," *Am. J. Clin. Nutr.* **37** (1983) 253.
82 E. C. Titgemeyer, L. D. Burquin, G. C. Fahey, Jr., K. A. Garleb: "Fermentability of various fiber sources by human fecal bacteria in vitro," *Am. J. Clin. Nutr.* **53** (1991) 1418.
83 M. Nyman, N.-G. Asp: "Fermentation of dietary fibre components in the rat intestinal tract," *Br. J. Nutr.* **47** (1982) 357.
84 M. Nyman, N.-G. Asp, J. Cummings, H. Wiggins: "Fermentation of dietary fiber in the intestinal tract: comparison between man and rat," *Brit. J. Nutr.* **55** (1986) 487.
85 J. J. Doesburg: "Pectic Substances in Fresh and Preserved Fruits and Vegetables," I.B.V.T. Communication no. 25, Sprenger Institute, Wageningen 1965.
86 R. A. Baker: "Potential Dietary Benefits of Citrus Pectin and Fiber," *Food Technol.* (1994) Nov., 133.
87 A. Siddhu, S. Sud, R. L. Bijlani, M. G. Karmarkar, U. Nayar: "Nutrient interaction in relation to glycaemic and insulaemic response," *Indian. J. Physiol. Phamacol.* **36** (1992) 21 – 28.
88 D. J. A. Jenkins, *et al.*: "Pectin and complications after gastric surgery: normalisation of postprandial glucose and endocrine responses," *Gut* **21** (1980) 574.
89 S. E. Schwartz, R. A. Levine, A. Singh, J. R. Scheidecker, N. S. Track: "Sustained Pectin Ingestion Delays Gastric Emptying," *Gastroenterology* **83** (1982) 812.
90 S. Satchithanandam, M. Vargofcak-Apker, R. J. Calvert, A. R. Leeds, M. M. Cassidy: "Alteration of gastrointestinal mucin by fiber feeding in rats," *J. Nutr.* **120** (1990) 1179.
91 C. D. Lorenzo, C. M. Williams, F. Hajnal, J. E. Valenzuela: "Pectin Delays Gastric Emptying and Increases Satiety in Obese Subjects," *Gastroenterology* **95** (1988) 1211.

92 D. M. Zimmaro, *et al.*: "Isotonic Tube Feeding Formula Induces Liquid Stool in Normal Subjects: Reversal by Pectin," *J. Parenteral and Enteral Nutr.* **13** (1989) 117.
93 X.-B. Sun, T. Matsumoto, H. Yamada: "Anti-Ulcer Activity and Mode of Action of the Polysaccharide Fraction from the Leaves of Panax ginseng," *Planta Med.* **58** (1993) 432.
94 H. Yamada, M. Hirano, H. Kiyohara: "Partial structure of an anti-ulcer pectic polysaccharide from the roots of Bupleurum falcatum L.," *Carbohydr. Res.* **219** (1991) 173.
95 X.-B. Sun, T. Matsumoto, H. Yamada: "Effects of a Polysaccharide Fraction from the Roots of Bupleurum falcatum L. on Experimental Gastric Ulcer Models in Rats and Mice," *J. Pharm. Pharmacol.* **43** (1991) 699.
96 A. Kawabata: *Studies on Chemical and Physical Properties of Pectic Substances from Fruits*, Memoirs of the Tokyo University of Agriculture XIX, 1977, pp. 115 – 200.
97 L. M. Drews, C. Kies, H. M. Fox: "Effect of dietary fiber on copper, zinc, and magnesium utilization by adolescent boys," *Am. J. Clin. Nutr.* **32** (1979) 1893.
98 K. Y. Lei, M. W. Davis, M. M. Fang, L. C. Young: "Effect of pectin on zinc, copper and iron balances in humans," *Nutr. Rep. Int.* **22** (1980) 459.
99 J. H. Cummings, *et al.*: "The digestion of pectin in the human gut and its effect on calcium absorption and large bowel function," *Br. J. Nutr.* **41** (1979) 477.
100 P.-E. Glahn: "Hydrocolloid stabilization of protein suspensions at low pH," *Prog. Fd. Nutr. Sci.* **6** (1982) 171.
101 A. Parker, P. Boulenguer, T. P Kravtchenko: "Effect of the addition of high methoxyl pectin on the rheology and colloidal stability of acid milk drinks", in K. Nishinari, E. Doi (eds.): *Food Hydrocolloids, Structures, Properties and Functions*, Plenum Press, New York 1994, p. 307.
102 P.-E. Glahn, C. Rolin: "Casein-pectin interaction in sour milk beverages," *Food Ingred. Eur. Conf. Proc.* 1994, 252.
103 B. U. Nielsen: "Fiber-Based Fat Mimetics: Pectin", in S. Roller, S. Jones (eds.): *Handbook of Fat Replacers*, CRC Press, Boca Raton FL, 1996, p. 161.
104 W. Pilnik, P. Zwiker: "Pektine," *Gordian* **70** (1970) 202 – 204, 252 – 257, 302 – 305, 343 – 346.
105 D. B. Nelson, C. J. B. Smit, R. R. Wiles in H. D. Graham (ed.): *Food Colloids*, Avi, Westport, Connecticut 1977, p. 418.
106 C. Rolin, J. De Vries in P. Harris (ed.): *Food Gels*, Elsevier Applied Science, London 1990, pp. 401 – 434.
107 G. O. Aspinall in J. Preiss (ed.): *The Biochemistry of Plants*, vol. **3**, Academic Press, New York 1980, p. 473.
108 J. N. BeMiller, *ACS Symp. Ser.* **310** (1986) 12.
109 J. K. Pederson in R. L. Davidson (ed.): *Handbook of Water Soluble Gums and Resins*, Chap. 15 – 1, McGraw-Hill, New York 1980.
110 S. Hojgaard Christensen: "Pectins" in M. Glicksman (ed.): *Food Hydrocolloids*, vol. 3, CRC Press, Boca Raton, Fla. 1986, p. 205.

111 W. Pilnik: "Pectin – a Many Splendoured Thing" in G. O. Phillips, D. J. Wedlock, P. A. Williams (eds.): *Gums and Stabilisers for the Food Industry 5*, IRL Press, Oxford 1990, pp. 209 – 221.
112 H. Braconnot, *Ann. Chim. Phys. Ser. 2* **28** (1825) 173 – 178.
113 W. Pilnik, A. G. J. Voragen: "Pectic Substances and Other Uronides," in A. C. Hulme (ed.): *The Biochemistry of Fruits and their Products*, vol. 1, Academic Press, London 1970, pp. 53 – 87.
114 F. M. Rombouts, W. Pilnik: "Pectic Enzymes" in A. H. Rose, (ed.): *Economic Microbiology*, vol. 5, "Enzymes and Enzymic Conversions," Academic Press, London 1980, pp. 225 – 280.
115 A. G. J. Voragen, W. Pilnik: "Pectin-Degrading Enzymes in Fruit and Vegetable Processing," *ACS Symp. Ser.* **389** (1989) 93 – 115.
116 W. Pilnik, A. G. J. Voragen: "The Significance of Endogenous and Exogenous Pectic Enzymes in Fruit and Vegetable Processing" in P. F. Fox (ed.): *Food Enzymology*, vol. 1, Elsevier Applied Science, London 1991, pp. 303 – 337.
117 C. D. May: "Industrial Pectins: Sources, Production and Application," *Carbohydr. Polymers* **12** (1990) 79 –90.
118 E. M. McComb, R. M. McCready, *Anal. Chem.* **29** (1957) 819 – 821.
119 S. C. Frey: "Cross-Linking of Matrix Polymers in the Growing Cell Wall of Angiosperms," *Ann. Rev. Plant. Physiol.* **37** (1986) 165 – 186.
120 D. H. Northcote, *ACS Symp. Ser.* **310** (1986) 134.
121 L. Rexova-Benkova, O. Markovic, *Adv. Carbohyd. Chem. Biochem.* **33** (1976) 101 – 130.
122 H. A. Schols *et al.*, *Carbohydr. Res.* **206** (1990) 105 –115.
123 M. J. F. Searle-van Leeuwen *et al.*, *Eur. J. Appl. Microbiol. Biotechnol.* **38** (1992) 347 – 349.
124 W. Pilnik: "Pektine" in R. Heiss (ed.): *Lebensmitteltechnologie*, Springer Verlag, Berlin 1988, pp. 228 – 234.
125 M. A. Joslyn, G. de Luca, *J. Colloid Sci.* **12** (1957) 108 – 130.
126 M. Manabe, *Nippon Nogei Kagaku Kaishi* **45** (1971) no. 4, 195 – 199.
127 Sunkist Growers, US 1 332 985, 1971.
128 California Fruit Growers Exchange, US 2 480 710, 1949.
129 W. J. Kim, C. J. B. Smit, V. N. M. Rao, *J. Food Sci.* **43** (1978) 74 – 78.
130 W. Pilnik, A. G. J. Voragen: "Gelling Agents (Pectins) from Plants for the Food Industry," *Adv. Plant Cell Biochem. Biotechnol.* **1** (1992) 219 – 270.
131 W. Pilnik, R. A. MacDonald, *Gordian* **68** (1968) 531 – 535.
132 M. Byland, A. Donutzhuber, *Sven. Papperstidn.* **15** (1968) 505 – 508.
133 T. P. Kravtchenko, I. Arnould, A. G. J. Voragen, W. Pilnik, *Carbohyd. Polym.* (in press).
134 G. H. Joseph, A. H. Kieser, E. F. Bryant, *Food. Technol (Chigago)* **3** (1949) 85 – 90.
135 M. Manabe, *Nippon Nogie Kogaku Kaishi* **45** (1971) no. 9, 417 – 422.
136 F. M. Rombouts, A. K. Wissenburg, W. Pilnik, *J. Chromatogr.* **168** (1979) 151 – 161.
137 W. Somers, J. Visser, F. M. Rombouts, K. van t'Riet: "Developments in Downstream Processing of (Poly) saccharide Converting Enzymes," *J. Biotechnol.* **11** (1989) 199 – 222.
138 H. Neukom, *Schweiz. Landwirtsch. Forsch.* **2** (1963) 112 – 122.
139 H. F. Launer, Y. Tomimatsu, *Anal. Chem.* **31** (1959) no. 9, 1569 – 1574.
140 D. A. Rees, *Adv. Carbohydr. Chem. Biochem.* **24** (1969) 267 – 332.
141 D. A. Rees, E. R. Morris, D. Thom, J. K. Madden in G. O. Aspinall (ed.): *Polysaccharides*, vol. 1, Academic Press, New York 1982, pp. 195 – 290.
142 D. A. Rees, A. W. Wight, *J. Chem. Soc. B* 1971, 1366.
143 D. Oakenfull, A. Scott: "New Approaches to the Study of Food Gels," in G. O. Philips, D. J. Wedlock, P. A. Williams (eds.): *Gums and Stabilisers for the Food Industry*, vol. 3, Elsevier Applied Science Publishers **1985**, London, pp. 465 – 475.
144 E. R. Morris *et al.*, *Int. J. Biol. Macromol.* **2** (1980) 237.
145 D. Thom, I. C. M. Dea, E. R. Morris, D. A. Powell, *Progr. Fd. Nutr. Sci.* **6** (1982) 87 – 108.
146 V. J. Morris in J. R. Mitchell, D. A. Ledward (eds.): *Functional Properties of Food Macromolecules*, Elsevier, Amsterdam 1986, pp. 121 – 169.
147 Racapé *et al.*, *Biopolymers* **28** (1989) 1435 – 1448.
148 B. Lockwood, *Food Process. Ind.* **41** (1972) 493, 47 – 51.
149 I. F. T. Pectin Standardization, *Food Technol. (Chicago)* **13** (1959) 496 – 500.
150 P. C. Crandall, L. Wicker: "Pectin Internal Gel Strength: Theory, Measurement and Methodology," *ACS Symp. Ser.* **310** (1986) 79 – 102.
151 M. J. H. Keijbets, W. Pilnik, *Potato Res.* **17** (1974) 169 – 177.
152 M. F. Katan, P. v.d. Bovenkamp, *Basic Clin. Nutr.* **3** (1981) 217 – 239.
153 R. F. McFeeters, S. A. Armstrong, *Anal. Biochem.* **139** (1984) 212 – 217.
154 L. G. Bartolome, J. E. Hoff, *J. Agric. Food Chem.* **20** (1972) 266 – 270.
155 M. A. Litchman, R. P. Upton, *Anal. Chem.* **44** (1972) 1495 – 1497.
156 G. C. Cochrane, *J. Chromatogr. Sci.* **13** (1975) 440 – 447.
157 E. L. Pippen, R. M. McCready, H. S. Owen, *Anal. Chem.* **22** (1950) 1457 – 1458.
158 A. G. J. Voragen, H. A. Schols, W. Pilnik, *Food Hydrocolloids* **1** (1986) 65 – 70.
159 FAO Food and Nutrition Paper 31/2 (Rome) 1984, p. 75.
160 J. C. E. Reitsma, W. Pilnik: "Analysis of Mixtures of Pectins and Amidated Pectins," *Carbohydr. Polym.* **10** (1989) 315 – 319.

161 T. Bitter, H. M. Muir, *Anal. Biochem.* **4** (1962) 330.
162 N. Blumenkranz, G. Asboe-Hansen, *Anal. Biochem.* **54** (1973) 484 – 489.
163 T. M. C. C. Filisetti-Cozzi, N. C. Carpita, *Anal. Biochem.* **197** (1991) 157 – 162.
164 T. P. Kravtchenko, A. G. J. Voragen, W. Pilnik, *Carbohydr. Polym.* **18** (1992) 17 – 25.
165 R. R. Selvendran, P. Ryden: "Isolation and Analysis of Plant Cell Walls," in P. M. Dey, J. B. Harborne (eds.): *Methods in Plant Biochemistry* **2** (1990) 549 –579.
166 C. M. G. C. Renard, A. G. J. Voragen, J. F. Thibault, W. Pilnik, *Carbohydr. Polym.* **12** (1990) 9 –25.
167 H. U. Endress: in [12], pp. 251 – 267.
168 P. A. Sanderson, J. Baird: "Industrial Utilization of Polysaccharides" in G. O. Aspinall (eds.): *The Polysaccharides*, vol. 2, Academic Press, London 1983, pp. 411 – 490.
169 T. J. Painter in G. O. Aspinall (ed.): *The Polysaccharides*, vol. 2, Academic Press, London 1983, pp. 257 – 275.
170 P. A. Sandford in R. R. Colwell, E. R. Pariser, R. J. Sinskey (eds.): *Biotechnology of Marine Polysaccharides* Hemisphere, Washington 1985, pp. 454 – 516.
171 W. J. Sime. Alignates in P. Harris (ed.): *Food Gels*, Elsevier Applied Science Press, New York 1990, pp. 53 – 78.
172 W. Pilnik, A. G. J. Voragen: "Pektine und Alginate" in H. Neukom and W. Pilnik (eds.): *Gelier- und Verdickungsmittel in Lebensmitteln*, Forster Verlag, Zürich 1980, pp. 67 – 94.
173 E. Percival, R. H. McDowell: "Algal Polysaccharides," *Methods in Plant Biochemistry* **2** (1990) 523 – 548.
174 V. J. Chapman: *Seaweeds and Their Uses*, 2nd ed., Methuen & Co. Ltd., London 1970.
175 A. Haug: "Composition and Properties of Alginates," Report no. 30, Norwegian Institute for Seaweed Research, Trondheim 1964.
176 E. C. C. Stanford, GB 142, 1881.
177 F. G. Fischer, H. Dörfel, *Hoppe Seyler's Z. Physiol. Chem.* **302** (1955) 186 – 203.
178 A. Haug, B. Larsen, O. Smidsrod, *Acta Chem. Scand.* **21** (1967) 691 – 704.
179 W. A. P. Black, *J. Mar. Biol. Assoc. U.K.* **29** (1950) 45 – 72.
180 H. A. Hoppe, O. J. Schmid, *Bot. Mar. Suppl.* **3** (1962) 16 – 66.
181 A. Haug, B. Larsen, *Acta Chem. Scand.* **16** (1962) 1908 – 1918.
182 A. Penman, G. R. Sanderson, *Carbohydr. Res.* **25** (1972) 273 – 285.
183 D. F. Pindar, C. Bucke, *Biochem. J.* **152** (1975) 617 – 622.
184 G. Skjåk-Break, B. Larsen, *Carbohydr. Res.* **139** (1985) 273 – 283.
185 Marine Colloids, US 2 128 551, 1938.
186 Kelco, US 2 036 934, 1936.
187 A. B. Steiner, W. H. McNeely, US 2 494 911, 1950.
188 P. A. Imeson in G. O. Phillips, D. J. Wedlock, P. A. Williams (eds.): *Gums and Stabilisers for the Food Industry 2*, Pergamon Press, Oxford 1984, p. 189.
189 A. Haug, B. Larsen, O. Smidsrod, *Acta Chem. Scand.* **20** (1966) 183 – 190.
190 A. Haug, B. Larsen, O. Smidsrod, *Acta Chem. Scand.* **21** (1967) 2859 – 2870.
191 J. Preiss, G. Ashwell, *J. Biol. Chem.* **237** (1962) 309 – 316.
192 M. Jefferies, G. Pass, G. O. Phillips, *J. Sci. Fd. Agric.* **28** (1977) 173 – 179.
193 O. Smidsrod, *Faraday Discuss. Chem. Soc.* **57** (1974) 263.
194 D. Thom, I. C. M. Dea, E. R. Morris, D. A. Powell, *Progr. Food Nutr. Sci.* **6** (1982) 97.
195 K. Taft, *Progr. Food Nutr. Sci.* **6** (1982) 89.
196 H. A. Schols, *Analyt. Biochem.* (in press).
197 H. Grasdalen, *Carbohydr. Res.* **118** (1983) 225 – 260.
198 J. S. Craigie, E. R. Morris, D. A. Rees, D. Thom, *Carbohydr. Polym.* **4** (1984) 237 – 252.
199 G. A. Towle in [281], pp. 237 – 252.
200 J. K. Pedersen in [283], pp. 113 – 133.
201 N. F. Stanley in [285], pp. 79 – 119.
202 Marine Colloids, US 3 094 517, 1963.
203 O. Smidsrod, B. Larsen, A. Pernas, A. Haug, *Acta Chem. Scand.* **21** (1967) 2585 – 2598.
204 C. J. Lawson, D. A. Rees, *Nature (London)* **227** (1970) 390 – 395.
205 E. R. Morris, D. A. Rees, G. R. Robinson, *J. Mol. Biol.* **138** (1980) 349.
206 O. Smidsrod, H. Grasdalen, *Carbohydr. Polym.* **2** (1982) 270.
207 I. C. M. Dea et al., *Carbohydr. Res.* **57** (1977) 249–272.
208 B. V. McCleary, *Carbohydr. Res.* **71** (1977) 249.
209 E. R. Morris, (1990) In *Food Gels*, (Ed P. Harris), Applied Science Publishers, London, pp. 291–359.
210 Marinalg International, world association of seaweed processors, Paris, www.marinalg.org.
211 E. Percival, R. H. Mc Dowel, "Algal Polysaccharides", in P. M. Dey, J. B. Harborne (eds.): *Plant Biochemistry*, Vol. **2**, Carbohydrates, Academic Press, New York 1990, pp. 536 – 537.
212 H. H. Selby, W. H. Wynne, "Agar in Industrial Gum.", in R. L. Whistler (ed.): *Polysaccharides and Their Derivatives*, 2nd ed., Academic Press, New York 1973, pp. 29 – 48.
213 C. Araki, *J. Chem. Japan* **58** (1973) 1338.
214 C. Araki, *Bull. Chem. Soc. Japan* **29** (1956) 543.
215 C. Araki, S. Hirase, *Bull. Chem. Soc. Japan* **33** (1960) 291.
216 C. Araki, K. Arai, *Bull. Chem. Soc. Japan* **30** (1957) 287.
217 M. Duckwork, K. C. Hong, W. Yaphe, *Carbohydr. Res.* **18** (1971) 1 – 9.
218 C. Araki, *Proc. 5th Intern. Seaweed Symp.*, Pergamon Press, Oxford 1966, p. 3.
219 S. Hirase, *Bull. Chem. Soc. Japan* **30** (1957) 68.
220 M. Duckwork, W. Yaphe, *Carbohydr. Res.* **16** (1971) 189.
221 K. B. Guisely, *Carbohydr. Res.* **13** (1970) 247 – 256.
222 D. A. Rees, *Chem. Ind.* (1972) 630 – 636.

223 M. Glicksman, Polysaccharides in Food, ed. J. M. V and Mitchell, J. R., In Blanshard (1979) 185 – 24, Butterworths, London.
224 A. Hayashi, *Polym. J.* **9** (1977) 219 – 225.
225 A. Hayashi, *Polym. J.* **10** (1978) 485 – 494.
226 A. Hayashi, *Polym. J.* **12** (1980) 447 – 453.
227 J. P. Flory, *Proc. R. Soc. London. Ser. A* **234** (1956) 73.
228 E. Pines, W. Prins *Macromolecules* **6** (1973) no. 6 888 – 895.
229 D. A. Rees, *Adv. Carbohyd. Chem. Biochem.* **24** (1969) 267.
230 D. A. Rees, *J. Chem. Soc.* (1961) 5168.
231 C. R. Williams: "The Processing of Gum Arabic to give Improved Functional Properties," in G. O. Phillips, D. J. Wedlock, P. A. Williams (eds.): *Gums and Stabilisers for the Food Industry*, 5th ed., IRL Press, Oxford 1990, pp. 37 – 40.
232 A. P. Imeson: "Exudate Gums," in A. Imseon (ed.): *Thickening and Gelling Agents for Food*, Blackie Academic and Professional, London 1992, pp. 66 – 97.
233 A. M. Stephen: "Structure and Properties of Exudate Gums," in G. O. Phillips, D. J. Wedlock, P. A. Williams (eds.): *Gums and Stabilisers for the Food Industry*, 5th ed., IRL Press, Oxford 1990, pp. 3 – 16.
234 S. Conolly, J. C. Fenyo, M. C. Vandevelde, *Food Hydrocolloids* **1** (1987) 477 – 480.
235 J. C. Fenyo, M. C. Vandevelde: "Physico-Chemical Properties of Gum Arabic in Relation to Structure," in G. O. Phillips, D. J. Wedlock, P. A. Williams (eds.): *Gums and Stabiliserer for the Food Industry*, 5th ed., IRL Press, Oxford 1990, pp. 17 – 23.
236 P. A. Williams, G. O. Phillips, R. C. Randall: "Structure-Relationships of Gum Arabic," in G. O. Phillips, D. J. Wedlock, P. A. Williams (eds.): *Gums and Stabilisers for the Food Industry*, 5th ed., IRL Press, Oxford 1990, pp. 25 – 36.
237 D. M. W. Anderson, *Food Hydrocolloids* **2** (1988) 417 – 423.
238 G. O. Aspinall, V. P. Bhavanadan, T. B. Christensen, *J. Chem. Soc.* (1965) 2677 – 2684.
239 B. Urlacher, B. Dalbe: "Xanthan Gum," in A. Imeson (ed.): *Thickening and Gelling Agents for Food*, Blackie Academic and Professional, London 1992, pp. 202 – 226.
240 A. R. Jeanes, J. E. Pittsley, F. R. Santi, *J. Appl. Polymer Sci.* **5** (1961) 519 – 526.
241 P. Delest: "Fermentation Technology of Microbial Polysaccharides," in G. O. Phillips, D. J. Wedlock, P. A. Williams (eds.): *Gums and Stabilisers for the Food Industry*, 5th ed., IRL Press, Oxford 1990, pp. 301 – 313.
242 A. A. Lawrence: *Edible Gums and Related Substances*, Noyes Data Corp., Park Ridge, N.J. 1973.
243 P. E. Janson, L. Keene, B. Lindberg, *Carbohydr. Res.* **45** (1975) 275 – 282.
244 R. Moorhouse, M. D. Walkinshaw, S. Anott: "Xanthan gum – molecular conformation and interactions", in P. A. Sandford (ed.): *Extracellular Microbial Polysaccharides. ACS Symposium Series*, No. 45.: American Chemical Society, Washington, D.C. 1977, pp. 90 – 102.
245 R. Moorhouse, in A. P. Imeson (ed.): *Thickening and Gelling Agents for Food*, "Chap. 9", Blackie A & O, Glasgow 1992, pp. 202 – 226.
246 M. R. Betlach et al.: "Genetic Engineering, Sturcture/Property Relations and Applications," in M. Yalpani (ed.): *Industrial Polysaccharides*, Elsevier, Amsterdam 1987, pp. 35 – 50.
247 D. J. Petit: "Xanthan gum", in M. Glicksman (ed.): *Food Hydrocolloids*, vol. 1, CRC Press, Boca Raton, FL 1982, pp. 127 – 149.
248 P. J. Whitcomb, C. W. Makosko, *J. Rheol.* **22** (1978) no. 5, 493 – 505.
249 Whitcomb, C. W. Makosko, "Rheology of xanthan gum solutions," *J P.J. Rheol.* **22** (1978) no. 5, 493 – 505.
250 Kelco Co.: "Xanthan gum Keltrol/Kelzan, in Natural Biopolysaccharides for Scientific Water Control," 2nd ed., Kelco Co., San Diego, CA 1975.
251 R. A. Speers, M. A. Tung: "Concentration and temperature dependence of flow behaviour of xanthan gum dispersions," *J. Food Sci.* **51** (1986) no. 1, 96 – 98, 103.
252 B. Launay, J. L. Doublier, G. Cuvelier, in J. R. Mitchell and D. A. Ledward (eds.): *Functional Properties of Macromolecules*, Elsevier Applied Science, London 1986, pp. 1 – 78.
253 W. H. McNeely, K. S. Kang in R. J. Whistler, N. J. BeMiller (eds): *Industrial Gums*, Academic Press, New York 1973, pp. 473 – 521.
254 A. R. Jeanes, J. E. Pittsley, F. R. Samtis: "Polysaccharide B-1459; a new hydrocolloid polyelectrolyte produced from glucose by bacterial fermentation," *J. Appl. Polym. Sci.* **5** (1961) 519 – 526.
255 E. R. Morris et al.: "Order – disorder transition for a bacterial polysaccharide in solution: a role for polysaccharide conformation in recognition between Xanthomonas pathogen and its plant host," *J. Mol. Biol.* **110** (1977) no. 1, 1 – 16.
256 M. Rinaudo, M. Milos, *Int. J. Biol. Macromol.* **2** (1980) 45 – 48.
257 G. A. R. Carnie: "Evaluation of a new polysaccharide gum," *Aust. J. Pharm.* **45** (1964) no. 19, 580 – 583.
258 I. T. Norton et al., *J. Mol. Biol.* **175** (1984) 371.
259 V. J. Morris, *Food Biotechnol.* **4** (1990) 45 – 57.
260 V. J. Morris: "Science, Structure and Applications of Microbial Polysaccharides," in G. O. Phillips, D. J. Wedlock, P. A. Williams (eds.): *Gums and Stabilisers for the Food Industry*, 5th ed., IRL Press, Oxford 1990, pp. 315 – 328.
261 A. W. Thomas, H. A. Murray: "Gum arabic," *J. Phys. Chem.* **32** (1928) no. 6, 676 – 697.
262 US 3 413 125, 1971 (H. R. Schuppner).
263 US 3 721 571, 1973 (M. Glicksman, H. E. Farks).
264 R. O. Mannion et al.: "Xanthan/locust bean gum interactions at room temperature," *Carbohydr. Polym.* **19** (1992) no. 2, 91 – 97.

265 P. B. Fernandez: "Influence of galactomannan on the structure and thermal behaviour of xanthan/galactomannan mixtures," *J. Food. Eng.* **2** (1995) 269 – 283.
266 H. Luyten, W. Kloek, T. Van Vliet: "Yielding behaviour of mixtures of xanthan and enzyme-modified galactomannans," *Food Hydrocolloids* **8** (1994) no. 5, 431 – 440.
267 B. Dalbe: "Interactions between xanthan gum and konjac mannan", in G. O. Phillips, D. J. Williams (eds.): *Gums and Stabilisers for the Food Industry*, vol. 6, Oxford University Press, Oxford 1992 pp. 201 – 208.
268 G. R. Sanderson: "The interaction of xanthan gum in food systems," *Prog. Food. Nut. Sci.* **6** (1982) 77 – 87.
269 G. R Sanderson: "Gellan Gum," in [285], pp. 201 – 232.
270 W. Gibson: "Gellan Gum," in A. Imeson (ed.): *Thickening and Gelling Agents for Food*, Blackie Academic and Professional, London 1992, pp. 227 – 249.
271 K. S. Lang et al.: "Agar-like polysaccharide produced by Pseudomonas species: production and basic properties," *Appl. Environ. Microbiol.* **43** (1982) 1086.
272 "Gellan Gum: Multifunctional Polysaccharide for Gelling and Texturising," NutraSweet Kelco publication, 1994.
273 H. Grasdalen, O. Smidsrod: "Gelation of gellan gum," *Carbohydr. Polym.* **7** (1987) 371.
274 M. S. Kuo, A. Dell, A. J. Mort: "Identification and location of L-glycerate, an unusual substituent in gellan gum," *Carbohydr. Res.* **156** (1968) 173.
275 C. Upstill, E. D. T. Atkins, P. T. Atwool: "Helical conformations of gellan gum," *Int. J. Biol. Macromol.* **8** (1986) 275.
276 R. Chandrasekaran, A. Radha, V. G. Thailambal: "Roles of potassium ions, acetyl and L-glyceryl groups in native gellan double helix: an x-ray study," *Carbohydr. Res.* **224** (1992) 1.
277 R. Chandrasekaran, R. P. Millane, S. Arnott: "Crystal structure of gellan," *Carbohydr. Res.* **175** (1988) 1.
278 R. Chandrasekaran, L. C. Puigjaner, K. L. Joyce, S. Arnott: "Cation interactions in gellan: an x-ray study of the potassium salt," *Carbohydr. Res.* **181** (1988) 23.
279 R. Chandrasekaran, V. G. Thilambal, *Carbohydr. Polym.* **12** (1990) no. 4, 431 – 442.
280 G. R. Sanderson, R. C. Clark: "Gellan Gum", *Food Technol.* **37** (1983) 63.
281 G. R. Sanderson, V. L. Bell, R. C. Clark, D. Ortega: "The texture of gellan gum gels", in G. O. Phillips, D. J. Wedlock, P. A. Williams (eds.): *Gums and Stabilisers for the Food Industry*, 4th ed., IRL Press, Oxford 1988, p. 219.
282 G. R. Sanderson, V. L. Bell, D. Ortega: "A comparison of gellan gum, agar, ϰ-carageenan and algin," *Cereal Foods World* **34** (1989) 991.
283 G. Owen: "Gellan-gum quick setting gelling systems", in G. O. Phillips, D. J. Wedlock, P. A. Williams (eds.): *Gums and Stabilisers for the Food Industry*, vol. 5, IRL Press, Oxford 1989, pp. 345 – 349.
284 R. C. Clark: "Flavour and texture factors in model gel systems", in A. Turner (ed.): *Food Technology International Europe*, Sterling Publications International, London 1990, pp. 272 – 277.
285 US 6027 930115, 1994 (W. F. Chalupa, G. R. Sanderson).
286 A. Nissinovitch, N. Peleg, E. Mey-Tal: "Continuous monitoring of changes in shrinking gels," *Lebensm. Wissen. Tech.* **28** (1995) no. 3, 347 – 349.
287 C. K Darty: "Applications of gellan gum as a fining agent in alcoholic beverages," Res. Disclosure, No. 348, 256 (1993).
288 W. Gibson et al.: "Production of low calorie (low joule) or sugar-free jelly sweets using polydextrose and gellan gum, where final total solids exceeds 80%," Res. Disclosure, No. 361, 276–7 (1994).
289 "Gellan Gum: a stabiliser of many means," *Prepared Foods* **160** (1995) no. 6, 125.
290 J. Giese: "Developments in beverage additives," *Food Technol.* **49** (1995) no. 5, 63 – 65, 68 – 70, 72.
291 EP-A 0 473 854 A1, 1992 (I. T. Norton).
292 V. L. Bell et al.: "Use of gellan gum to prepare gels that can be dried and subsequently rehydrated to their original shape," Res. Disclosure, No 345, 42 (1993).
293 Toxicological evaluation of certain food additives, WHO Food Additives Series 28, World Health Organisation, Geneva 1991.
294 Reports of the Scientific Committee for Food 26th Series, Commission of the European Committees, Report EUR 13913, Luxemburg 1992.
295 Official Journal of the European Committee, L61, April 18, 1995.
296 Compendium of Food Additive Specification, FAO Food and Nutrition Paper 52. Add 5., FAO, Rome 1997.
297 *Food Chemicals Codex*, 4th ed., National Academy Press, Washington D.C. 1996.
298 J. E. Fox: "Seed Gums," in A. Imeson (ed.) *Thickening and Gelling Agents for Food*, Blackie Academic and Professional, London 1992, pp. 153 – 170.
299 I. C. M. Dea: "Structure/Function Relationships of Galactomannans and Food Grade Cellulosics," in G. O. Phillips, D. J. Wedlock, P. A. Williams (eds.): *Gums and Stabilisers for the Food Industry*, 5th ed., IRL Press, Oxford 1990, pp. 373 – 382.
300 A. H. Clark, I. C. M. Dea, B. V. McCleary: "The Effect of the Galactomannan Fine Structure on Their Interaction Properties," in G. O. Phillips, D. J. Wedlock, P. A. Williams (eds.): *Gums and Stabilisers for the Food Industry*, 3rd ed., Elsevier Applied Science Publishers, London 1985, pp. 429 – 440.
301 I. C. M. Dea, A. H. Clark, B. V. McCleary, *Carbohyd. Res.* **142** (1986) 275 – 294.
302 P. M. Dey, *Adv. Carbohydr. Chem. Biochem.* **35** (1978) 341 – 376.
303 B. V. McCleary, *Int. J. Biol. Macromol.* **8** (1986) 349 – 354.
304 T. J. Painter, in G Aspinall (ed.): *The Polysaccharides*, Vol. 2, Academic Press, London, 1983, pp. 195– 285.

305 M. Lahaye, *J. Appl. Phycol.* **13** (2001) 173–184.
306 G. A. DeRuiter, B. Rudolph, *Trends Food Sci. Technol.* **8** (1997) 389–395.
307 S. H. Knutsen, D. E. Myslabodski, B. Larsen, A. I. Usov: "A modified system of nomenclature for red algal galactans", *Bot. Mar.* **37** (1994) 163–169.
308 D. A. Rees, *Adv. Carb. Biochem.* **24** (1969) 267–332.
309 E. Zablackis, G. A. Santos, *Botanica Marina* **29** (1986) 319–322.
310 D. N. Smidt, W. H. Cook, *Arch. Biochem. Biophys.* **45** (1953) 232–233.
311 C. Bellion et al., *Carbohydr. Res.* **119** (1983) 31–48.
312 K. F. Wong, J. S. Craigie, *Plant. Physiol.* **61** (1978) 663–666.
313 E. L. McCandless, J. S. Craigie, J. A. Walter, *Planta (Berl.)* **112** (1973) 201–212.
314 R. Falshaw, R. H. Furneaux, *Carbohydr. Res.* **276** (1995) 155–165.
315 R. Falshaw, H. J. Bixler, K. Johndro, *Food Hydrocoll.* **15** (2001) 441–452.
316 F. Vande Velde, H. A. Peppelman, H. S. Rollema, R. H. Tromp, *Carbohydr. Res.* **331** (2001) 271–283.
317 H. J. Bixler, K. Johndro, R. Falshaw, *Food Hydrocoll.* **15** (2001) 619–630.
318 A. Amimi et al., *Cabohydr. Res.* **333** (2001) 271–279.
319 A. I. Usov, V. V. Barbakadze, *Bioorg. Khim.* **4** (1978) 1107–1115.
320 M. Ciancia, M. C. Matulewicz, A. S. Cerezo, *Phytochemistry* **34** (1993) 1541–1543.
321 R. Takano et al., *Phytochemistry* **37** (1994) 1615–1619.
322 G. M. King, G. E. Lauterbach, *Botanica Marina* **30** (1987) 33–39.
323 C. K. Tseng, *Science* **101** (1945) 597–602.
324 H. J. Bixler, K. Johndro, R. Falshaw: "Kappa-2 carrageenan: structure and performance of commercial extracts II. Performance in two simulated dairy applications", *Food Hydrocolloids* **15** (2001) 619.
325 T. Barbeyron, D. Flament, G. Michel, P. Potin, B. Kloareg, *Cahiers De Biologie Marine* **42** (2001) 169.
326 H. J. Bixler, *Hydrobiologia* **327** (1996) 35.
327 R. A. Hoffmann, M. J. Gidley, D. Cooke, W. J. Frith, *Food Hydrocoll.* **9** (1995) 281–289.
328 A. P. Imeson, in G. O. Phillips, P. A. Williams (eds.): *Handbook of Hydrocolloids*, Woodhead Publishing Ltd., Cambridge, 2000, pp. 87–102.
329 W. Yaphe, G. P. Arsenault, *Anal. Biochem.* **13** (1965) 143–148.
330 C. N. Jol et al., *Anal. Biochem.* **268** (1999) 213–222.
331 A. I. Usov, N. G. Klochkova, *Botanica Marina* **35** (1992) 371–378.
332 R. Falshaw, R. H. Furneaux, *Carbohydr. Res.* **269** (1995) 183–189.
333 D. J. Stancioff, N. F. Stanley, *Proc. Int. Seaweed Symp.* **6** (1969) 595–609.
334 T. Chopin, E. Whalen, *Carbohydr. Res.* **246** (1993) 51–59.
335 A. I. Usov, *Botanica Marina* **27** (1984) 189–202.
336 M. Ciancia, M. C. Matulewicz, P. Finc, A. S. Cerezo, *Carbohydr. Res.* **238** (1993) 241–248.
337 D. Welti, *J. Chem. Res. (M)* (1977) 3566–3587.
338 S. H. Knutsen, D. E. Myslabodski, H. Grasdalen, *Carbohydr. Res.* **206** (1990) 367–372.
339 T. Turquois, S. Acquistapace, F. A. Vera, D. H. Welti, *Carbohydr. Polym.* **31** (1996) 269–278.
340 S. H. Knutsen, H. Grasdalen, *Carbohydr. Polym.* **19** (1992) 199–210.
341 T. Hjerde, O. Smidsrod, B. E. Christensen, *Biopolymers* **49** (1999) 71–80.
342 C. Viebke, P. A. Williams, *Food Hydrocoll.* **14** (2000) 265–270.
343 M. A. Roberts, B. Quemener, *Trends Food Sci. Technol.* **10** (1999) 169–181.
344 E. Bjerre-Petersen, J. Christensen, P. Hemmingsen, in R. L. Whistler and B. J. N. (eds.): *Industrial gums. Polysaccharides and their derivatives*, Academic Press, New York, 1973, pp. 123–136.
345 L. Piculell, in A. M. Stephen (ed.): *Food polysaccharides and their applications*, Marcel Dekker, Inc., New York, 1995, pp. 205–244.
346 T. Hjerde, O. Smidsrod, B. T. Stokke, B. E. Christensen, *Macromolecules* **31** (1998) 1842–1851.
347 C. Viebke, J. Borgstrom, L. Piculell, *Carbohydr. Polym.* **27** (1995) 145–154.
348 N. F. Stanley, in P. Harris (ed.): *Food gels*, Elsevier Applied Science, London, 1990, pp. 79–119.
349 K. Bongaerts et al. *Macromolecules* **33** (2000) 8709–8719.
350 L. Piculell, C. Hakansson, S. Nilsson, *Int. J. Biol. Macromol.* **9** (1987) 297–301.
351 K. B. Guiseley, N. F. Stanley, P. A. Whitehouse in R. L. Davidson (ed.): *Handbook of water soluble gums and resins*, McGraw-Hill, New York 1980.
352 FMC Biopolymer: "Not Just Products. Partners. (2000), p. 10 (www.fmcbiopolymer.com).
353 G. H. Therkelsen, in R. L. Whistler, J. N. BeMiller (eds.): *Industrial gums: Polysaccharides and their derivatives*, Academic Press, San Diego, 1993, pp. 145–180.
354 M. DiRosa, *J. Pharm. Pharmacol.* **24** (1972) 89–102.
355 L. Aschheim, S. Raffel, *Res. J. Reticuloendothel. Soc.* (1972) 253–262.
356 P. Cantanzaro, H. Schwartz, R. Graham, *Am. J. Pathol.* **64** (1971) 387–404.
357 D. J. Schaeffer, V. S. Krylov, *Ecotoxicol. Environ. Saf.* **45** (2000) 208–227.
358 C. Haslin, M. Lahaye, M. Pellegrini, J. C. Chermann, *Planta Medica* **67** (2001) 301–305.
359 M. J. Carlucci et al. *Antiviral Res.* **43** (1999) 93–102.
360 P. Potin, K. Bouarab, F. Kupper, B. Kloareg, *Curr. Opin. Microbiol.* **2** (1999) 276–283.
361 F. Vande Velde, G. A. DeRuiter, in S. De Baets, E. J. VanDamme, S. Steinbüchel (eds.): *Biopolymers*, vol. 6 Polysaccharides II, Wiley-VCH, Weinheim, Germany, 2002.
362 L. G. Ekstrøm, *Carbohydr. Res.* **135** (1985) 283–289.
363 T. T. Stevenson, R. H. Furneaux, *Carbohydr. Res.* **210** (1991) 277–298.

Further Reading

J. M. Aguilera, P. Lillford (eds.): *Food Materials Science*, Springer, New York, NY 2008.

G. L. Cote, J. A. Ahlgren: *Microbial Polysaccharides*, "Kirk Othmer Encyclopedia of Chemical Technology", 5th edition, John Wiley & Sons, Hoboken, NJ, online DOI: 10.1002/0471238961.1309031803152005.a01.

W. P. Edwards: *The Science of Bakery Products*, Royal Society of Chemistry, Cambridge, UK 2007.

H.-G. Elias: *Macromolecules*, Wiley-VCH, Weinheim 2008.

H.-J. Gabius (ed.): *The Sugar Code*, Wiley-VCH, Weinheim 2009.

H. G. Garg, M. K. Cowman, C. A. Hales: *Carbohydrate Chemistry, Biology and Medical Applications*, Elsevier, Amsterdam 2008.

T. Heinze (ed.): *Polysaccharides*, Springer, Berlin 2005.

T. Heinze, T. Liebert, A. Koschella: *Esterification of Polysaccharides*, Springer, Berlin 2006.

M. Rinaudo: *Polysaccharides*, "Kirk Othmer Encyclopedia of Chemical Technology", 5th edition, John Wiley & Sons, Hoboken, NJ, online DOI: 10.1002/0471238961.polyrina.a01.

Starch

JAMES N. BEMILLER, Whistler Center for Carbohydrate Research, Purdue University, West Lafayette, Indiana, USA

KERRY C. HUBER, School of Food Science, University of Idaho, Moscow, Idaho, USA

1.	Introduction	517
2.	Sources of Commercial Starches	518
3.	Worldwide Starch Production	518
3.1.	Starch Production and Consumption in the EU	518
3.2.	The USA Corn Wet-Milling Industry	519
3.3.	Tapioca Starch Production in Thailand	519
4.	Molecular Structures and Properties of Amylose and Amylopectin	519
5.	Structures and Properties of Starch Granules	521
5.1.	Granule Structure	521
5.2.	Physicochemical Properties	523
5.2.1.	Melting/Gelatinization	525
5.2.2.	Retrogradation	525
5.2.3.	Pasting and Viscoelastic Properties	526
6.	Minor Components of Starch Granules	528
7.	Starch Biosynthesis and Genetics	528
8.	Industrial Starch Production Processes	528
8.1.	Corn/Maize Wet Milling	528
8.1.1.	Grain Cleaning	528
8.1.2.	Kernel Steeping	528
8.1.3.	Kernel Milling and Fraction Separation	528
8.1.4.	Starch Drying	529
8.2.	Wheat Starch	529
8.3.	Potato Starch	529
8.4.	Tapioca/Cassava Starch	530
8.5.	Rice Starch	530
9.	Modified Starches	530
9.1.	Chemically Modified Starches	532
9.1.1.	Converted Starches and Hydrolyzates	532
9.1.1.1.	Thinned Products	532
9.1.1.2.	Starch Dextrins	533
9.1.1.3.	Dextrose Equivalency	533
9.1.1.4.	Maltodextrins	534
9.1.1.5.	Syrup Solids	534
9.1.1.6.	Syrups and Crystalline D-Glucose	534
9.1.1.7.	High-Fructose Syrups and Crystalline D-Fructose	535
9.1.1.8.	Other Starch Conversion Products	535
9.1.2.	Cross-linked Starches	535
9.1.3.	Stabilized Starches	536
9.1.4.	Cationic Starch	536
9.1.5.	Starch Graft Copolymers	537
9.2.	Thermally Modified Starches	537
9.2.1.	Instant Starches	537
9.2.2.	Annealed and Heat-moisture-treated Starches	538
9.2.3.	Dry Heating of Starches	538
9.2.4.	Destructurized and Thermoplastic Starch	538
9.3.	Genetically Modified Starches	539
9.4.	Multiple Modifications of Starches	539
10.	Examples of Uses of Starches and Products Derived from Starches	540
10.1.	Uses of Starches	540
10.2.	Uses of Products derived from Starches via Extensive Depolymerization	542
11.	Starch Digestibility	543
	References	543

1. Introduction

Starch [9005-25-8] is widely used in nonfood industrial applications, especially in the production of paper and paperboard products, as a textile sizing agent, and in fermentation processes to produce ethanol and other products, and in the processed food industry as a thickener/stabilizer, gelling agent, and a starting material for the production of sweeteners and polyols. It is also the principal source of dietary calories for the world's human population. The chemistry and technology of starch has been reviewed in several books [1–7].

Native starch occurs as discrete particles called granules. A starch from a specific biological (plant) source is unique among starches, i.e., starch granules from the various plant sources differ in appearance, particle size distribution, fine structure of the constituent polymer molecules, and physical properties. Most applications of starch are realized only after the starch is

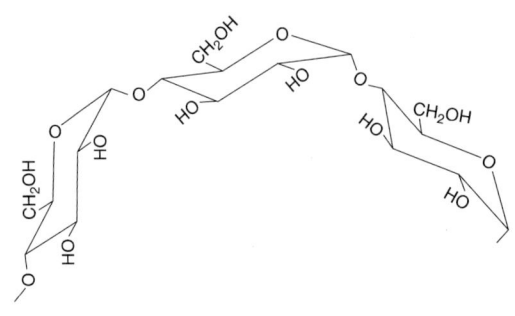

Figure 1. A three-glucosyl unit segment of an unbranched (i.e., linear) portion of an amylose or amylopectin molecule

heated (cooked) in the presence of water, a process that destroys its granular structure and releases its constituent polymer molecules.

Granular starch is generally composed of two types of molecules, amylose and amylopectin (Chap. 4). *Amylose* [9005-82-7] is a predominantly linear (1,4)-α-D-glucan (Fig. 1), although some amylose molecules are slightly branched. *Amylopectin* [9037-22-3] (Fig. 2) has a branch-on-branch structure consisting of mostly short chains of (1,4)-linked α-D-glucopyranosyl units (sometimes referred to as anhydroglucose units) linked to other short, linear chains via α-(1,6) branch points. The amylopectin fraction is composed of much larger molecules than those of amylose. Most normal starches contain 70–75% amylopectin molecules by weight. Different molecular structures and sizes, amylose/amylopectin ratios, and granular architectures give each type of starch its unique properties.

2. Sources of Commercial Starches

Major crops used for starch production include corn (maize), cassava, potato, and wheat. Corn used to produce starch includes hybridized genetic variants of corn (Chap. 4). Lesser amounts of starch are isolated industrially from arrowroot, mung bean, rice, sago palm, sweet potato, yam, yellow pea, and other plants.

3. Worldwide Starch Production

In 2006, it was reported that 99% of the global starch production of ca. 60×10^6 t originated from crops of corn/maize, cassava/tapioca, wheat and potatoes, with 73% being maize starch, 14% tapioca starch, 8.1% wheat starch, and 3.7% potato starch [8]. Starch is predominantly produced in the USA, the EU, Japan, and Thailand. The largest commercial starch producers worldwide are Cargill (9.2×10^6 t, 14.6%), CPI (5.2×10^6 t, 8.2%), ADM (5.2×10^6 t, 8.2%), National Starch and Chemical Co. (1.2×10^6 t, 1.9%), and Avebe (0.6×10^6 t, 1.0%). Approximately 33.6×10^6 t (53.1%) of starch is produced by small to medium-sized companies.

3.1. Starch Production and Consumption in the EU

The starch industry in Western Europe is comprised of 24 different companies with 68 plants, which in 2005, produced 9.6×10^6 t of starch distributed as follows: 46% corn/maize starch (4.4×10^6 t), 36% wheat starch (3.4×10^6 t), and 18% potato starch (1.7×10^6 t) [8]. Production of starch from yellow pea has also been instituted.

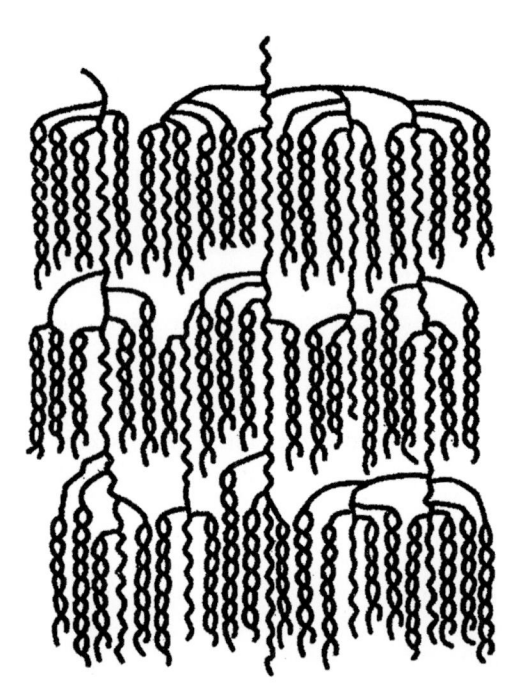

Figure 2. A representation of amylopectin molecules showing crystalline packing of double-helical pairs of branch chains
(Reprinted with permission from A. Imberty, A. Buléon, V. Tran, S. Perez, *Starch/Stärke* **43** (1991) 375.)

In 2005, 9.0×10^6 t of starch products were consumed in the EU for production of starch hydrolysates, including high-fructose syrups (5.1×10^6 t, 57%), native starches (2.1×10^6 t, 23%), and modified starches (1.8×10^6 t, 20%). Approx. 57% of the 9.0×10^6 t was used in the processed food industry and 43% in the nonfood sector. The breakdown by application areas/market sectors was as follows: sweets and drinks (confectionary, beverages, fruit processing) 30%, processed food (convenience food, bakery, food ingredients and food preparations, dairy products and ice cream) 27%, paper and corrugated board manufacture 28%, chemical, fermentation, and other industrial products 14% (Chap. 10).

3.2. The USA Corn Wet-Milling Industry

By far the majority of starch produced in the USA is isolated from corn/maize, with relatively small amounts of potato and wheat starch also being produced. Corn starch is produced by a wet-milling process (Section 8.1). Production figures for member companies of the Corn Refiner's Association (USA) for 2008 are given in Table 1 [9]. The production ratio of high-fructose syrups to glucose syrups was approx. 67:33. In addition, 4.5×10^9 L of fuel ethanol [64-17-5] was produced by corn wet millers, although most of the fuel ethanol made in the USA by fermentation was obtained from corn by a dry-grind process. The sale of coproducts (see Table 1) from the wet-milling process is important to the economic viability of the industry.

Table 1. Products from the corn wet-milling industry in the United States for 2008

Product[a]	Production, 10^6 t
Sweeteners[b]	5.2
Starches[c]	1.5
Coproducts[d]	5.9

[a] Dry weight.
[b] Also called conversion products. Includes high-fructose syrups (42% and 55% fructose), glucose syrups, crystalline glucose (dextrose), crystalline fructose, maltodextrins, and syrup solids.
[c] Includes native corn starches, modified corn starches, and dextrins.
[d] Includes corn oil, corn oil meal, corn gluten feed, corn gluten meal, and steepwater.

3.3. Tapioca Starch Production in Thailand

The greatest amount of tapioca starch is produced in Thailand. Lesser amounts are produced in Indonesia, Brazil, and China. In 2009, Thailand exported 1.8×10^6 t of native tapioca starch, 0.7×10^6 t of modified tapioca starch, and 0.02×10^6 t of tapioca pearls (total = 2.5×10^6 t) [10].

4. Molecular Structures and Properties of Amylose and Amylopectin

Amylopectin. The major polysaccharide of starch, amylopectin, is a very large, highly-branched molecule. The branch points constitute 4–5% of the total glycosidic linkages and occur in clusters. There are three general classes of chains. A chains are those that are connected to another chain via a (1, 6) linkage, but are themselves unbranched. B chains are those chains that are connected to another chain via a (1, 6) linkage and have one or more A chains or other B chains attached to them, i.e., they are further branched. B chains can be subdivided into chains of various lengths/sizes. The C chain is the one possessing the lone reducing end-unit of the amylopectin molecule. In starch granules, the A and short B chains occur as pairs of chains entwined around each other in double helices (Fig. 2), which pack together to give rise to the crystallinity of starch granules. Average molecular masses of amylopectin molecules are at least 10^7 and may be as large as 10^9.

The original cluster model for amylopectin has been refined several times, but has kept its basic form [11–14] until recently when a variation was proposed [15] that accounts for the superhelical nature of the amylopectin molecules proposed to be present in at least some starches.

Amylopectin is present in all known starches, constituting about three-fourths of most normal starches (Tables 2 and 3); indeed, some starches consist entirely of amylopectin (i.e., they lack amylose). Starches containing only amylopectin (all-amylopectin starches) are often referred to as *waxy starches*. Average fine structures, average molecular masses, molecular mass ranges, and

Table 2. General properties of granules and pastes of native corn/maize starches

	Normal corn starch	Waxy maize starch	High-amylose corn starch[a]
Granule size, μm	2–30	2–30	2–24
Amylose (approx.)[b], %	28	0	50–70
Gelatinization/pasting temperature[c], °C	74–81	66–71	66–71[d]
Relative viscosity	medium	medium-high	very low
Paste rheology (body)	short	long	short
Paste clarity	opaque	slightly cloudy	slightly opaque
Tendency to gel/retrograde	high	very low	very high
Gel consistency	firm	nongelling	very firm
Lipid, % ds[e]	0.8	0.2	–
Protein, % ds[e]	0.35	0.25	0.4
Starch-bound phosphorus, % ds[e]	0	0	0
X-ray diffraction pattern type	A	A	B

[a] Also known as amylomaize starch.
[b] The percentage of amylopectin is the difference between 100% and the percentage of amylose.
[c] From the initial temperature of gelatinization to complete cookout.
[d] Under ordinary cooking conditions, where the slurry is heated to 95–100 °C, high-amylose corn starch produces little viscosity. Gelatinization of high-amylose corn/maize starch with ca. 50% apparent amylose content may begin as low as 66 °C and end as high as 166 °C, depending on the moisture content. Gelatinization of high-amylose corn/maize starch with ca. 70% apparent amylose content may begin as low as 66 °C and end as high as 171 °C, depending on the moisture content.
[e] ds = dry solids.

perhaps, the shapes of amylopectin molecules vary from starch to starch and are important determinants of the physical properties of a starch, as well as foods and/or other products that contain starch.

Amylose, the other naturally occurring starch biopolymer, is an essentially linear chain of (1,4)-linked α-D-glucopyranosyl units (Fig. 1). Some amylose molecules are slightly branched, with branch points constituting 0.3–0.5% of the total glycosidic linkages. Because there are only a few branches, with the branches usually being either very long or very short chains, and because the branch points are usually far apart, amylose molecules behave as linear polymer molecules. The average molecular masses of amyloses from different commercial starches are in the 10^5–10^6 range, which means that they have average *degrees of polymerization* (DP, number of glucosyl units per molecule) of ca. 600–6000. The nature of the glycosidic linkages in amylose chains produces a natural right-handed helix (Fig. 1), which has consequences for its properties.

Most starches contain 25–30% amylose (as shown in Tables 2 and 3, where their general properties are outlined) and are known as normal starches. Starches containing more than this

Table 3. General properties of granules and pastes of potato, tapioca, and wheat starches

	Potato starch	Tapioca starch	Wheat starch
Granule size, μm	5–100	4–35	0.5–45[a]
Amylose (approx.)[b], %	21	17	28
Gelatinization/pasting temperature[c], °C	58–65	52–65	52–85
Relative viscosity	very high	high	medium-low
Paste rheology (body)	very long	long	short
Paste clarity	clear	clear	cloudy
Tendency to gel/retrograde	medium-low	low	high
Gel consistency	salve-like	soft	soft
Lipid, (% ds)[d]	< 0.1	< 0.1	0.9
Protein, % ds[d]	0.1	0.1	0.4
Starch-bound phosphorus, % ds[d]	ca. 0.08	0	0
X-ray diffraction pattern type	B	A	A

[a] Bimodal population.
[b] The percentage of amylopectin is the difference between 100% and the percentage of amylose.
[c] From the initial temperature of gelatinization to complete cookout.
[d] ds = dry solids.

amount of amylose are known as *high-amylose starches*. Two commercial corn starches, known as high-amylose corn starches or *amylomaize starches*, have apparent amylose contents of ca. 50% and >70%.

Several particular properties of amylose are important with regard to applications of starch. Heating of starches in water with shear forms what are called *pastes*. Pastes from amylose-containing starches are generally opaque and usually form firm gels upon cooling. Precipitation, rather than gelation, may occur in dilute pastes as they cool. The opacity and the formation of a gel or a precipitate results from amylose molecules associating with each other in crystalline order, a process that is called *retrogradation* (see Section 5.2.2), or when it is associated with determining paste properties, *setback* (see Section 5.2.3).

Helical amylose chains have hydrophobic interiors capable of forming inclusion complexes with linear hydrophobic portions of molecules that can fit within the lumen (i.e., the inner cavity) of the helix. When a hot, aqueous dispersion of starch is stirred with a slightly polar, slightly water-soluble organic compound, such as butan-1-ol, and then cooled, amylose complexes crystallize out and can be isolated by centrifugation. Complexation stabilizes the helix and converts the long, linear molecules into more uniform and more rod-like structures. Because only the amylose molecules crystallize, such complexation can be employed to isolate pure amylose.

Iodine (as I_3^-) complexes with amylose and amylopectin molecules, with complex formation also occurring within the hydrophobic interiors of helical segments of starch molecules. The long helical segments of amylose allow long chains of poly(I_3^-) to form and produce the blue color that is a diagnostic test for starch (more specifically, amylose). The amylose–iodine complex contains 19% iodine, and determination of the amount of iodine complexed is used to measure the amount of apparent amylose in a starch. Amylopectin forms reddish brown complexes with iodine because its branches are too short for the formation of long chains of poly(I_3^-).

Polar lipids (surfactants/emulsifiers and fatty acids) affect starch pastes and starch-based foods in one or more of three ways as a result of complex formation: (1) by affecting starch gelatinization (see Section 5.2.1) and pasting, (2) by modifying the rheological behavior of the resulting paste, and (3) by inhibiting (or in some cases accelerating) the crystallization of starch molecules associated with the retrogradation process. Specific changes to a paste that are observed upon the addition of a lipid depend on its structure and the starch employed, i.e., different lipids/surfactants affect the gelatinization and pasting behaviors of a given starch differently, and each lipid/surfactant affects starches from different botanical sources differently. Because complex formation with emulsifiers occurs much more readily with amylose and, as a result, has a much greater effect on amylose molecules than on amylopectin molecules, polar lipids have a much greater effect on normal starches than on waxy (all-amylopectin) starches. Addition of most lipids/surfactants/emulsifiers to starches containing amylose inhibits the processes associated with gelatinization and pasting, but some speed up these processes and/or cause them to occur at lower temperatures. Polar lipids inherent to native cereal starches generally inhibit retrogradation.

The ability of polar lipids to form complexes with amylose and amylopectin is associated with their chain length, their degree of unsaturation, and the nature of their hydrophilic group. In general, esters of myristic (C_{14}, saturated) and palmitic (C_{16}, saturated) acids are most effective.

5. Structures and Properties of Starch Granules

5.1. Granule Structure

Within plants, synthesized starch molecules are assembled in the form of semicrystalline aggregates, termed granules, which vary according to size (< 1–100 µm) and shape (spherical, ellipsoidal, polygonal, lenticular, etc.) depending on their botanical origin [16]. Starch granule biosynthesis occurs within an organelle called the *amyloplast* and originates at a point called the *hilum* (the approximate center of the granule), from which site starch molecules are deposited in a radial, spherocrystalline arrangement [17]. The precise architecture of starch granules is rather complex and consists of multiple levels of structural organization. At the most basic

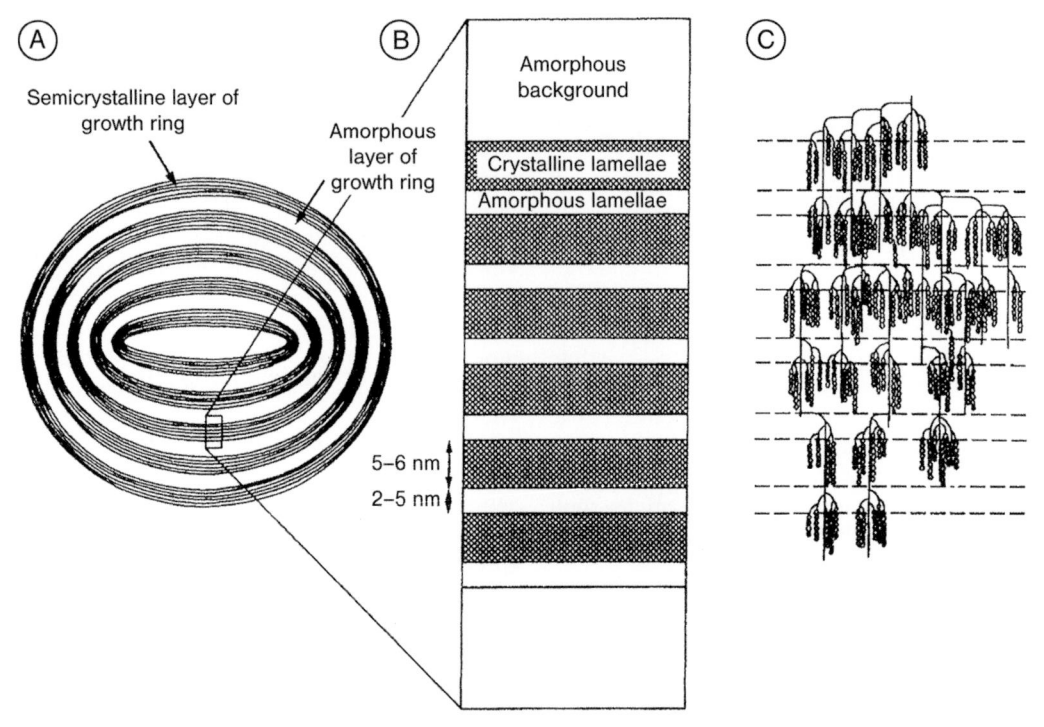

Figure 3. Schematic diagram of the starch granule
A) Structural relationship between the semicrystalline and amorphous shells (growth rings); B) Structural relationship between crystalline and amorphous lamellae; C) Amylopectin molecular structure
(Reprinted with permission from P.J. Jenkins, R.E. Cameron, A.M. Donald, W. Bras, G.E. Derbyshire, G.R. Mant, A.J. Ryan, *J. Polym. Sci., B-Polym. Phys.* **32** (1994) 1579.)

organizational level, the granule is comprised of alternating semicrystalline and less crystalline (frequently called amorphous) shells, commonly referred to as growth rings [17–19] (Fig. 3). On a higher organizational level, a semicrystalline shell or growth ring itself is comprised of alternating crystalline and amorphous lamellae, which correspond to the molecular features of amylopectin molecules. The branching regions of amylopectin molecules comprise amorphous areas within the crystalline regions of both shells, while the short, linear segments of amylopectin branch chains organized in double helices (Chap. 4) constitute the crystalline lamellae. Another level of organization, intermediate to those already described, involves aggregation of adjacent double-helical segments of amylopectin to form crystalline blocks [17, 18]. In short, amylopectin is primarily responsible for the organization of starch granules and accounts for their semicrystalline nature. In contrast, amylose is believed to be located in amorphous regions, where it is proposed to be randomly interspersed among amylopectin clusters and/or to be involved in complexes with native polar lipids [17].

The spherocrystalline radial arrangement of starch molecules in the granule is inferred from the birefringence (as evidenced by the Maltese cross) that is seen on microscopic examination of starch granules under plane-polarized light [17] (Fig. 4). As determined by X-ray diffraction, native starch granules generally exhibit crystallinities ranging from 15–45% [20, 21]. X-ray diffraction of native starches reveals three distinct patterns, A, B, and C (not to be confused with the A, B, and C chains of amylopectin molecules already described), which define the specific allomorphic packing arrangement of starch double helices (i.e., crystallites) present within granules (Fig. 5) [17, 20]. The specific type of crystallite packing observed in a starch is primarily a function of both the temperature condition within the plant during starch synthesis and the chain length. An average chain length in excess of DP 12 tend to favor the formation of the B allomorph over the A allomorph (average DP

Figure 4. Depiction of the Maltese cross observed within starch granules (potato) viewed under plane-polarized light

10–12) [17]. In general, tuber starches produce B crystallites (starch synthesized underground under cool conditions), while cereal starches possess an A crystalline arrangement (starch synthesized above ground under hot, dry conditions); legume starches exhibit the C allomorph, which is a mixture of A and B crystalline allomorphs [17, 20].

At the macrostructural level, granules of corn/maize and wheat starches possess pores on external granule surfaces [22–24], which features represent openings to channels leading into the granule interior [23–26]. Channels within granules of these starches have been demonstrated to impact granular patterns of amylolytic degradation and chemical modification by facilitating access of enzymes and chemical reagents to the granule interior [22, 27–29].

5.2. Physicochemical Properties

Native starch granules, which possess complex long-range and short-range molecular order, are insoluble in cold water, but are nevertheless capable of absorbing water (0.48–0.56 g/g dry starch) due to their semi-amorphous nature [30]. In hydrated granules, water has a plasticizing effect on amorphous regions, effectively lowering the overall glass transition temperature (T_g) and enhancing mobility of the biopolymers. Hydration (at 25 °C) produces increases in granule diameter (i.e., swelling) of approximately 10% and 25% for normal and waxy corn starch, respectively, which swelling is reversible in nature.

The most significant physicochemical properties of starch are associated with irreversible thermal transitions that result in increased starch granule swelling and dissolution of the starch biopolymers. Heating of starch granules in excess water (i.e., water-starch ratios > 1.5:1) brings about *gelatinization* (the irreversible loss

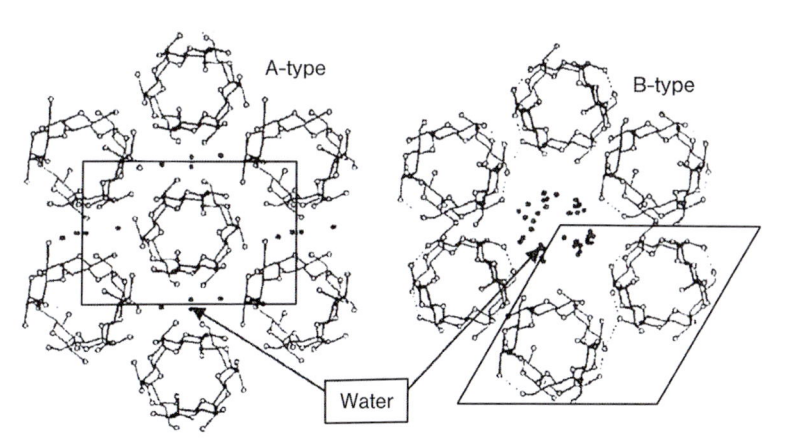

Figure 5. Depiction of A and B polymorphic structures for starch double helices (Reprinted with permission from H.F. Zobel, *Starch/Stärke* **40** (1988) 1.)

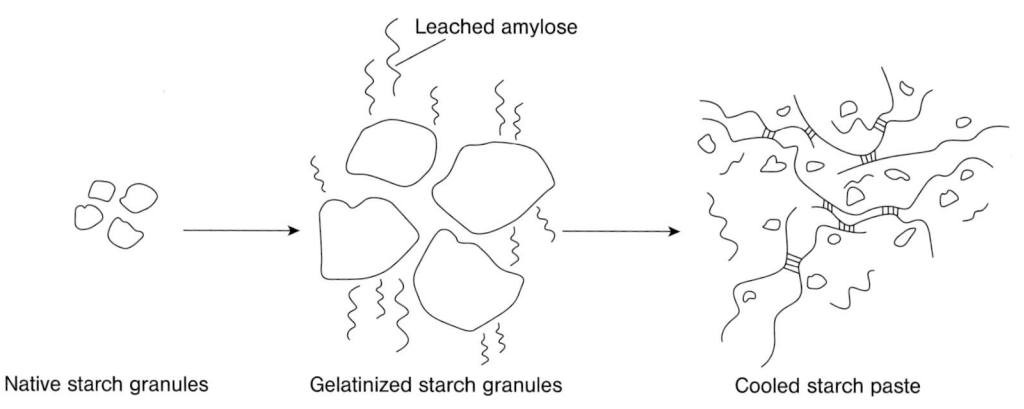

Figure 6. Schematic representation of the structural changes associated with the gelatinization and pasting of native starch granules. Gelatinization (loss of granular molecular order) is accompanied by granule swelling and leaching of soluble starch components (amylose) during heating in an aqueous system. With the application of shear, swollen granules undergo further disintegration to yield a paste, which is comprised of a continuous phase of solubilized starch and a dispersed phase of granule remnants. Upon cooling, amylose retrogradation (depicted by the cross-hatching between molecules within the paste) results in the formation of a gel network. (Reprinted with permission from K.C. Huber, A. McDonald, J.N. BeMiller: "Carbohydrate Chemistry", in Y. H. Hui, F. Sherkat (eds.): *Handbook of Food Science, Technology, and Engineering*, vol. 1, Taylor & Francis/CRC Press, Boca Raton, 2006.)

of granular/molecular order), which is accompanied by increased granule hydration, swelling, and leaching of soluble components (primarily amylose) (Fig. 6) [31]. Upon heating, hydrated molecules within granule amorphous regions begin to vibrate, inducing progressive disruption of hydrogen bonds to further increase hydration and plasticization of starch molecules within granules. Granule swelling becomes irreversible as the strain imposed within granule amorphous regions becomes sufficient to bring about assisted disruption of starch crystallites. Gelatinization is marked by the disappearance of birefringence and a complete loss of granule crystallinity [30]. When starch is present in sufficient concentration (generally 2–7%), swollen granules begin to press against one another as they occupy nearly the entire volume of the aqueous continuous phase, leading to greatly increased viscosity.

In the presence of shear, highly swollen granules are fragile and disintegrate to form a *paste*, which is comprised of hollow, swollen granules (ghosts) and broken granule remnants (ghost fragments) dispersed within a continuous phase of partially dissolved starch molecules. As a paste is cooled, amylose molecules reassociate or recrystallize in a process called *retrogradation*, which involves formation of double-helical structures that aggregate to form a three-dimensional gel network (Fig. 6) or precipitate [32]. Starch gels are a two-phase, mixed gel comprised of both amylose-rich and amylopectin-rich regions due to incompatibility of the two polymers [31]. Due to their branch-on-branch structure and relatively short branch chains, amylopectin molecules undergo very limited intermolecular associations and retrograde very slowly (although they may gradually crystallize somewhat with time) [32]. Thus, waxy starches, which consist primarily of amylopectin, lack the ability to form strong gel networks. Starch paste characteristics are defined by: (1) the rheology of the continuous phase, (2) the rigidity of granule remnants and the volume occupied by them, and (3) the extent and nature of interactions between the continuous and dispersed phases [30].

Short and long flow are terms that remain in use within the starch industry to describe the nature of a starch paste, although they have been largely replaced by more exact rheological characteristics (Section 5.2.3). Short and long flow refer to the draining behavior from a pipette or funnel. As the forming drop gets larger, it becomes heavier and the flow rate increases. If the weight of the stream causes it to break so that the fluid exits in small drops, the fluid is said to have a short flow. Such a fluid is pseudoplastic, i.e., shear-thinning. Fluids without shear-thinning behavior exit in long strings (long flow, slimy texture).

5.2.1. Melting/Gelatinization

Melting and gelatinization transitions of starches are endothermic in nature and may be both detected and quantified via differential scanning calorimetry (DSC). Due to the heterogeneous nature of crystallites present within granules, starch thermal transitions occur over a temperature range rather than at a defined temperature.

At low moisture levels (water/starch ratios < 1:1), the term "melting" is used to define the disappearance of starch crystallinity in response to heating [31]. Melting of native starch produces up to four endothermic transitions – the first two being associated with disordering of amylopectin double helices and the latter two reflecting order to disorder transitions of complexes of amylose with the native lipids of the starch granules [30]. Thermal transitions occurring under limited moisture conditions (as low as 11%) are generally shifted to elevated temperatures (as high as 180 °C), depending on the amount of water present for plasticization [30, 31]. Melting phenomena are likely to be encountered within high-temperature/low-moisture processes such as extrusion or baking of low-moisture products.

Gelatinization, which is characterized by the thermal disordering of starch (amylopectin) crystallites in excess water (water/starch ratios > 1.5:1) [31], generally occurs over a 10–15 °C temperature range [30]. Thus, a gelatinization endotherm obtained via DSC is defined by an onset (T_o), peak (T_p), and completion (T_c) temperature, as well as a gelatinization enthalpy (ΔH). For a native normal starch (i.e., amylose content in the range of 20–30%), the gelatinization temperature range generally falls between 55 and 80 °C, with the precise temperature range being specific to the botanical source of the starch [16, 30, 33]. High-amylose starches exhibit much higher and broader gelatinization temperature ranges (70–130 °C) relative to normal starches due to their abundance of long amylopectin branch chains [16, 33]. Aside from the events described for gelatinization, a second endothermic event (90–120 °C) may be observed for starches possessing amylose–lipid complexes (V-structures) (Chap. 4) [30, 31].

Starch gelatinization temperatures and ranges are impacted by the presence of other system constituents, particularly solutes. Sugars generally tend to increase the starch gelatinization temperature, but reduce the gelatinization temperature range [30]. Hypotheses proposed to explain this effect include competition of sugars with starch for water (solvent), physical sugar-starch interactions, and/or anti-plasticization effects of the sugar-water cosolvent [30, 31]. Low salt concentrations inherent to most foods may increase the starch gelatinization temperature; salt effects on gelatinization become much more varied and complex at high (salt) concentrations [30]. Monoacyl and other polar lipids, either native or incorporated, may increase starch gelatinization temperatures due to complexation with amylose chains [31]. At typical concentrations encountered in food systems, hydrocolloids exhibit minimal impact on starch gelatinization characteristics (as measured by DSC) [30], although they do significantly influence starch paste rheology (Section 5.2.3).

5.2.2. Retrogradation

Retrogradation involves reassociation (crystallization) of starch chains within a paste upon cooling (below T_m) to form either a viscoelastic gel network or a precipitate [31]. Although amylose and amylopectin chains are both capable of crystallization, the two polymers retrograde on vastly different timescales, with amylose crystallization being very rapid (minutes to hours) and amylopectin reassociation being very slow (days to weeks). For starch pastes comprised of both amylose and amylopectin, gelation involves both phase separation (formation of polymer-rich and polymer-deficient regions) and crystallization (primarily involving amylose chains), although the precise sequence of these two events is not fully resolved [30, 31]. Junction zones within normal starch gels are comprised almost exclusively of aggregates of amylose double-helical chain segments [30].

Starch crystallization is defined by three primary processes: nucleation (crystal initiation), propagation (crystal growth), and maturation (crystal perfection), of which nucleation is the rate-limiting step [30]. For starch pastes of less than 50% by weight concentration, nucleation is most favored at temperatures approaching T'_g (ca. −5 °C), whereas propagation (i.e., crystal growth) is enhanced at temperatures approaching the melting temperatures of the crystallites (T_m)

[30]. A hold or storage temperature of less than 60 °C, promotes formation of B crystallites (Section 5.1) [30]. The melting temperatures of amylose and amylopectin crystallites differ dramatically (140–160 °C and 45–60 °C, respectively), with amylose associations being much more thermoresistant [30]. Retrogradation may be monitored via a multitude of techniques including DSC, gel turbidity, light scattering, optical rotation, FTIR, NMR, and viscometry [30]. Within foods, excessive starch retrogradation often leads to product defects, such as syneresis, loss of viscosity, or staling of bakery products (amylopectin crystallization is thought to have a partial role here). Conversely, retrogradation may also be considered beneficial to various food products and processes (e.g., in the preparation of croutons, frozen potato products, and resistant starch) and may be promoted via defined cooling and/or temperature cycling schemes. Retrogradation is very much a phenomenon to be avoided in papermaking processes, particularly in sizing and coating operations.

The tendency for a starch to retrograde can be decreased through chemical modification (i.e., stabilization; Section 9.1.3). Starch retrogradation may also be greatly influenced by other system components, though the specific roles of particular classes of constituents are often varied and difficult to categorize in unified fashion. However, there is reasonable consensus for the effects of monacyl and polar lipids, which generally decrease the extent of retrogradation by promoting formation of amylose-lipid complexes (Chap. 4) [30, 31], although the reverse has also been reported. Effects of various low-molecular mass sugars and oligosaccharides on starch crystallization vary according to the stereochemistry of the solute. Generally, solutes that exhibit favorable hydrogen bonding interactions with the native water structure are suggested to reduce starch crystallization, whereas solutes that disrupt the water network tend to promote starch retrogradation [30]. Water-soluble polysaccharides can enhance the starch retrogradation rate (via phase separation and concentration of starch polymer microdomains), but may either increase or decrease gel rigidity depending on the specific hydrocolloid added [30, 31]. In general, nonstarch polysaccharides tend to enhance the water-holding capacities and freeze-thaw behaviors of starch gels [31, 34].

5.2.3. Pasting and Viscoelastic Properties

Upon cooking in excess water, a starch granule slurry experiences dramatic physical changes in response to swelling, gelatinization, pasting, and retrogradation events. A pasting profile provides a definitive, systematic fingerprint that embodies the flow behavior of a starch-containing material as it undergoes these defined events. While the *Brabender ViscoAmylograph* was traditionally used for generating pasting profiles, the *Rapid Visco Analyser (RVA)* has now emerged as the instrument of choice, due to a smaller sample requirement for analysis, reduced analysis times, and a defined unit of viscosity measure (cP [mPa·s] or Rapid Visco Analyser Units [RVU]). (A *Brabender Micro Visco-Amylo-Graph* with similar functions as the RVA is also available [30].) A pasting profile is generated by these instruments by monitoring the viscosity of an aqueous starch suspension over the course of a specified heating-cooling cycle, in the presence of shear, over a defined analysis time period. Heating and cooling rates, length of hold, rate of shear, and heating and cooling temperatures may all be varied by the operator, though traditional profiles generally heat to 95 °C, hold at 95 °C, and cool to 50 °C. Defined pasting attributes (i.e., pasting temperature, peak viscosity, etc.) may be obtained at various points of a pasting profile to yield specific information about starch swelling properties, as well as paste stability and retrogradation tendency (Fig. 7).

Pasting time represents the point in time (for a defined heating rate) at which sufficient starch granule swelling has occurred to register an initial rise in viscosity (above that of the original ungelatinized granule slurry). This value is often reported in regard to temperature (i.e., *pasting temperature*). However, temperature values provided by the RVA instrument reflect the instrument heating block temperature rather than that of the actual paste. Thus, there is a lag in the true paste temperature relative to that of the heating block, although the extent of lag can be mathematically corrected to estimate the true paste temperature for a given heating rate value [35]. Pasting temperature provides an estimate of the minimum temperature needed to cook a starch under given slurry conditions [36].

Peak viscosity results from a rapid swelling of granules upon gelatinization. Paste composition

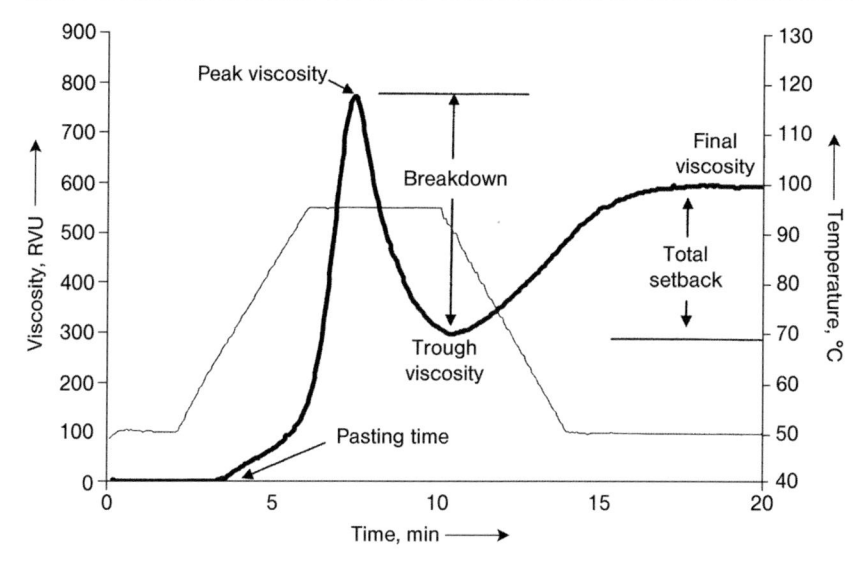

Figure 7. Example of a pasting profile depicting the viscosity of a wheat starch granule suspension over the course of a heating and cooling cycle. Specific pasting attributes obtained from the pasting profile are defined and labeled accordingly (bold line = starch viscosity profile; thin line = temperature profile)

at the peak is comprised of a combination of both highly swollen and ruptured granules within a continuous phase of solubilized starch. Peak viscosity is indicative of starch thickening power [30], and can be used to provide a rough estimate of the viscous load likely to be achieved in a cooking/mixing operation [36].

During the hold period at peak temperature (usually 95 °C), a starch paste will experience a loss in viscosity due to rupturing of the fragile, swollen granules until a *trough viscosity* (i.e., *hot paste viscosity*) is reached. (Trough viscosity represents the viscosity of the solubilized starch continuous phase containing dispersed granule remnants). *Breakdown* (peak viscosity minus trough viscosity) indicates the relative stability of a paste to high temperature and shear conditions. Chemical cross-linking of starch is conducted to stabilize granules against paste breakdown (Section 9.1.2).

Upon cooling a starch paste, amylose molecules begin to reassociate to form junction zones to achieve a *final viscosity*. *Setback* (final viscosity minus trough viscosity) reflects the extent of amylose recrystallization occurring within a starch paste, and is generally correlated positively with syneresis tendency [36]. Waxy starch pastes exhibit minimal setback, due to the lack of amylose. Normal starches, which generally exhibit a high propensity toward syneresis, may be chemically substituted to reduce the tendency of the starch to retrograde (Section 9.1.3).

Beyond pasting properties, more sophisticated analysis of starch dispersion and/or gel properties may be conducted under conditions of constant stress or constant strain using uniaxial compression testing (both large and small deformation tests), a rotary viscometer, or a dynamic oscillatory rheometer [30]. These analyses determine starch viscoelastic properties, defined as the degree of solid-like (storage modulus, G') versus liquid-like (loss modulus, G'') behavior present within a gel.

Nonstarch constituents present in a starch dispersion may have a significant effect on paste rheological behavior. High solute concentrations generally inhibit starch swelling, increase starch pasting temperature, and decrease paste peak viscosity [30]. Sugars may either increase or decrease gel strength, depending on their specific stereochemistry [30]; this effect was previously described in regard to retrogradation behavior (Section 5.2.2). Because of the essentially neutral character of starch, low concentrations of salts have little effect on gelation. Acids tend to decrease starch viscosity by promoting the hydrolysis of glycosidic linkages, effectively decreasing the molecular size of the starch polymers (Section 9.1.1). Polar lipids, through complexation with amylose chains (Chap. 4), restrict granule swelling, increase starch pasting temperature, and reduce overall paste viscosity [31]. Hydrocolloid addition may alter starch pasting properties (pasting temperature, peak viscosity, breakdown, setback, final viscosity), as well as both the storage and loss moduli of starch gels, although both the nature and degree of effects vary according to the specific polysaccharide [34]. To explain these effects, direct starch-hydrocolloid interactions and/or phase separation of the mixed polymer gel system have been proposed [31, 34].

6. Minor Components of Starch Granules

In addition to amylose and/or amylopectin molecules, starches contain other components, including lipid, protein, and ash. Only cereal starches contain significant amounts of lipids. Normal cereal starches contain ca. 0.6–1.2% total lipid by weight, consisting primarily of free fatty acids and lysophospholipids of differing ratios and compositions in different starches. Starches without amylose, such as waxy maize starch, contain only ca. 0.2% lipid. Noncereal starches, such as potato starch and tapioca starch, contain less than 0.1% lipid. Normal cereal starches contain ca. 0.35–0.40% protein by weight, while waxy maize starch contains ca. 0.25% protein and potato and tapioca starches contain ca. 0.1% protein. Potato starch contains monostarch phosphate ester groups (0.06%–0.1% P by weight). Starch ash contents usually range between 0.1 and 0.5% and are related to the lipid, protein, and phosphate ester contents.

7. Starch Biosynthesis and Genetics

The biosynthesis of amylose and amylopectin has been and continues to be an area of intense study [37] due to the complex natures of the molecules, especially the clustered branching pattern of amylopectin, and because of the desire to understand the impact of the biopolymer structures on the properties of a starch. The genetics of starch biosynthesis has also received intense study [38] because of the potential importance of altering such characteristics as the amylose/amylopectin ratio and the fine structure of amylopectin.

8. Industrial Starch Production Processes

8.1. Corn/Maize Wet Milling

Production of corn starch occurs primarily in the USA, China, EU, Japan, South Korea, Brazil, and Argentina. Corn wet milling consists of several steps: (1) cleaning the grain, (2) steeping of kernels, (3) milling of kernels and separation of fractions (i.e., starch and nonstarch components), and (4) drying of the isolated starch [39].

8.1.1. Grain Cleaning

Grain is first cleaned by screening and by magnetic means to remove foreign material including metal.

8.1.2. Kernel Steeping

Cleaned grain is steeped to soften the kernel and to facilitate its separation into the desired components. This process involves countercurrent steeping in water containing 0.10% sulfur dioxide at 48–52 °C for 30–40 h. Sulfur dioxide aids in the dissolution of the protein matrix to release the starch, combats growth of spoilage organisms, and maximizes the starch yield. The proper steeping temperature is critical for optimal growth of lactic acid bacteria, which lower the pH, restricting the growth of other organisms. The lactic acid bacteria also release proteolytic enzymes, which break down protein. Steepwater pH is typically buffered at ca. 4.0.

Steepwater may be further processed to remove phytic acid. In either form, it is then concentrated and sold as a nutrient solution for commercial fermentation to produce, for example, antibiotics, or more often dried by absorption onto corn fiber and sold as cattle feed.

8.1.3. Kernel Milling and Fraction Separation

After steeping, the grain is milled to obtain separation of components. Steeped grain is first put through an attrition mill to release the oil-containing germ, which is removed using hydroclones (liquid cyclones, hydrocyclones) and dried prior to extraction with hexane or pressed to remove the oil. Next, the residual grain is ground to release starch granules. The slurry is then screened to separate the corn hull or seed coat, which is recovered as a fibrous mass.

Starch is isolated from the remaining slurry using centrifuges or by passage through batteries of hydroclones that separate out not only the starch, but also the insoluble protein (called *corn gluten* in the industry). The process also washes the starch to lower the protein and ash contents. The protein (gluten) is less dense (1.1 g/cm^3) than starch (1.5 g/cm^3) and is easily separated by centrifugal forces. Starch thus produced contains 1–2% protein, which may be removed by a second washing with water to give a starch with a final protein content of $< 0.38\%$.

The final starch slurry is recovered by filtration or centrifugation, and the starch cake is dried in continuous driers (usually flash driers) (Section 8.1.4) and sold as unmodified (native) starch. Alternatively, the starch slurry can be pumped, cooked, and subjected to enzyme-catalyzed hydrolysis to produce syrups and other products (Section 9.1.1), pumped into tanks for chemical modification (Sections 9.1.2, 9.1.3, 9.1.4), or pumped onto hot rolls for production of pregelatinized starch (Section 9.2.1).

8.1.4. Starch Drying

Starch is usually dried by flash drying. In this method, the starch slurry is centrifuged or filtered and the moist starch cake is introduced at the bottom of a stream of rapidly moving hot air (93–127 °C). Drying is rapid and the starch is collected using cyclones. Drying parameters can be used to control both bulk density and particle size.

8.2. Wheat Starch

Wheat starch (along with starches of barley, rye, and triticale) is comprised of at least two distinct granule populations, commonly designated the A- and B-types. The respective A- and B-type granule populations possess different sizes (>10 μm vs. <10 μm), shapes (lenticular vs. spherical/polygonal), and starch characteristics and properties [40–42]. Of the two primary granule types, A-type granules are most industrially important, although small granule starches (high B-type granule contents) are also produced commercially [41].

In contrast to corn wet-milling, commercial processes for wheat starch isolation involve dry-milling of grain to flour (72% extraction or greater), which is the preferred starting material for starch isolation. Most commercial wheat starch isolation methods take advantage of the matrix forming properties of gluten proteins, combined with the insoluble nature of starch granules, to facilitate simultaneous separation of starch and gluten coproducts. Wheat starch is isolated on a commercial scale by one of several methods, including the Martin (dough ball), batter, Fesca, Raisio, hydrocyclone (dough-batter), or high pressure disintegration/tricanter processes [41]. The latter two processes are most commonly employed industrially, with advantages of reduced energy and water usage. In short, the process utilizing hydrocyclones (hydroclones) involves kneading and vigorous agitation of wheat flour-water mixtures to agglomerate gluten into thread-like pieces, after which hydrocyclones are used to separate starch and gluten streams. A series of hydrocyclones is further utilized to concentrate and purify the primary starch stream (yielding a wheat starch with a high concentration of A-type granules). Gluten, small-granule starch (predominantly B-type granules), and fiber (pentosans) may be recovered via additional processing. The high-pressure disintegration/tricanter process utilizes a combination of high-shear homogenization and decanter centrifugation processes to efficiently fractionate wheat flour batters into A-type granule starch, B-type granule starch, gluten, and fiber product streams. The primary producers of wheat starch are France, Germany, USA, and China.

8.3. Potato Starch

Potato starch is produced in largest quantities in Europe, especially in the Netherlands, Germany, and France [43]. Potato starch is also produced in China in substantial amounts. Industrial potatoes are washed with water to remove dirt and other foreign matter. They are then disintegrated in the presence of an antioxidant using drum rasps. Potato juice is removed using a decanter centrifuge. Sieving is used to remove fiber. After sieving, a disk-type, continuous centrifuge is used to concentrate the starch granules and remove any remaining small particles of fiber and protein. Although starch granules within potatoes occur in a wide range of sizes (see Table 3), the larger granules are primarily obtained in this process. Any remaining protein is removed by reslurrying the starch in water and concentrating the starch using a series of hydroclones in a countercurrent fashion. The starch is then collected by filtration and dried to ca. 20% moisture.

In the USA, potato starch is recovered commercially as a coproduct of potato processing operations, viz. slicing. Recovered starch is further washed and dried.

8.4. Tapioca/Cassava Starch

Tapioca starch is obtained from roots of the cassava plant, which is grown in tropical areas and is variously known as cassava, cassada, yuca, manioca, mandioca, and tapioca, depending on the region where it is grown [44]. Roots are harvested 10–12 months after planting and processed as quickly as possible. Processing consists of washing the roots, then converting them into a pulpy slurry that is passed through a series of screens to remove fiber. The resulting slurry of starch granules is concentrated and dewatered via centrifugation. Finally, the starch is dried. The yield of starch is ca. 26% (based on the fresh weight of roots of 60–70% moisture content).

8.5. Rice Starch

Most rice starch is prepared by steeping broken milled rice kernels in 0.3–0.5% NaOH solution at 25 °C to 50 °C for times of up to 24 h [45]. Steeping in an alkaline solution softens the kernels and dissolves the glutelin protein, which constitutes approx. 80% of the total protein in rice. Wet-milling of the steeped kernel produces a slurry of starch. An additional holding of the alkaline slurry for 10–24 h further solubilizes the proteins. Fiber is removed using screens or filters, and the starch is isolated from the slurry, washed, neutralized, and dried. Rice starch may also be produced by a mechanical process which involves physical separation of starch granules and protein bodies.

9. Modified Starches

The terminology related to modified starches, as with other terminology used in the starch industry, is not always precise. In this article, starches that are modified by reaction of their hydroxyl groups with a chemical reagent to convert them into esters, ethers, or other chemical derivatives, such as graft copolymers, are called *derivatized starches*. Those that are modified by some change in the basic polymer structure or molecular mass (by acid or enzyme treatment or oxidation) are called *converted* or *thinned starches* (Section 9.1.1). Starches that have undergone a physical (thermal) treatment are described by the nature of the product, most often being a type of *pregelatinized starch* (Section 9.2). Any starch treated in any way to alter one or more of its original physical or chemical properties is called a *modified starch*. A modified starch may be multiply-modified, e.g., derivatized, converted/thinned, and/or pregelatinized.

Native starches have low process tolerance and, in the case of food starches, generally produce undesirable textures. Their pastes often have poor stability. Conversion and/or derivatization produces higher quality products. Companies that produce and market starches make available a number of products that are the result of various combinations of starch type and modification. Converted, derivatized, and unmodified starches from different sources are often employed in the same general applications by different users.

Uses of native starches are limited because, as indicated above, their inherent physiochemical properties make them less than optimally desirable for most applications. Rather, the majority of starches used in various food and other industrial applications are modified starches. There are several reasons for modifying the properties of starches, but the two principal ones may be the following: (1) to prevent the formation of precipitates or microgel particles in a starch paste or gelation of the entire paste (Section 5.2.2), and (2) to reduce the viscosity of a starch paste (Section 5.2.3) so that higher solids solutions can be made. Both are needed for starch products used in the papermaking process. For starches used in processed food products, additional factors are important. In foods, starch ingredients provide bulk, body, and improved texture and mouthfeel, as well as other functionalities. For most applications, food processors prefer starches with better behavioral characteristics than those provided by native starches. Upon being cooked into slurries, native (i.e., unmodified), corn (maize) and waxy corn (waxy maize) starches produce weak-bodied, cohesive, rubbery pastes and undesirable gels. However, their functional properties can be improved dramatically by modification. In general, modifications are made to increase the ability of the paste produced by cooking to withstand the heat, shear, and low pH associated with processing conditions, to make the starch paste more stable, or to introduce specific functionalities.

Modifications can be accomplished by chemical (the majority) or physical treatments. Types of modifications include cross-linking of polymer chains (Section 9.1.2), non-cross-linking derivatization (sometimes called *stabilization* [Section 9.1.3] and depolymerization (Section 9.1.1) (all chemical modifications) and pregelatinization (a physical modification) (Section 9.2). Many improvements in the properties of a starch can be made with these modifications, singly or as combinations of them. Some specific property improvements that can be obtained are reduction in the energy required for cooking, modification of cooking characteristics, increased solubility, increased or decreased paste viscosity, increased freeze-thaw stability of pastes, enhancement of paste clarity, increased paste sheen, enhancement or reduction of gel formation and/or gel strength, reduction of gel syneresis, improvement of interaction with other substances, improvement of stabilizing properties, enhancement of film formation, improvement in water resistance of films, reduction in paste cohesiveness, and increased stability of the granules and pastes to lower pH, heat, and shear. Such imparted characteristics are itemized in Table 4.

The most important commercial derivatives of starch are those in which only a very few of the hydroxyl groups are derivatized. Normally, the hydroxyl groups of the starch polymer molecules are converted into ester or ether groups at very low *degree of substitution (DS)* values, the DS being the average number of hydroxyl groups per glucosyl unit that have been derivatized by ether or ester formation. DS values of modified food starches are most often < 0.1 and are generally in

Table 4. Characteristics imparted by primary starch modifications

Modification	Main Attributes
Hypochlorite-oxidized	Whiter
	Lowered gelatinization temperature
	Reduced maximum paste viscosity (usually)
	Softer, cleaner gels
	Greater adhesion of pastes
Cross-linked (food)	Increased stability to heat
	Increased cooking temperature (delayed pasting)
	Increased shear resistance
	Increased stability to low pH/acid
	Decreased setback of cooks
	Decreased or increased paste viscosity
	Increased body
Stabilized	Lowered gelatinization/pasting temperature (easier cooking)
	Improved tolerance of cold storage of products
	Improved freeze-thaw stability of products
	Decreased paste setback/retrogradation
	Reduced gelation
	More easily dispersed when pregelatinized
	Greater paste clarity
	Increased moisture control
Cross-linked and stabilized	Lowered gelatinization/pasting temperature, but increased paste viscosity
	Other attributes of cross-linking and stabilization
	A variety of textures and rheological properties
Octenylsuccinylated	Emulsifying properties
	Emulsion-stabilizing properties
	Ability to encapsulate hydrophobic materials
Thinned	Reduced hot-paste viscosity
	Lowered gelatinization/pasting temperature
	Increased solubility
	Increased gel strength
	Increased paste clarity or opacity
	Increased film-forming capability
Dextrinized	Imparts crispness (food)
	Increased stability (food)
	Reduced viscosity
	Improved film formation
	Emulsifying properties
	Increased browning (food)
	Increased adhesion of pastes
	Remoistenable adhesiveness
Pregelatinized[a]	Thickening without cooking

[a] Including cold-water swelling.

the range 0.002–0.2. Thus, generally, on average, there is one substituent group for every 5 to 500 D-glucopyranosyl units.

Starches are chemically modified as an aqueous slurry of granules. For esterification or etherification, a starch slurry of 30–40% solids is introduced into a stirred reaction tank. Sodium sulfate or sodium chloride is added to a 10–30% concentration to inhibit gelatinization. The pH is adjusted, generally with sodium hydroxide, to pH 8–12, the exact value depending on the reaction to be conducted, after which the chemical reagent is added to the starch slurry. The alkaline pH activates the starch for reaction by converting some of its hydroxyl groups to alkoxide ions for participation in nucleophilic substitution reactions. The reaction temperature is generally maintained at less than 60 °C, often to 49 °C, to prevent pasting of the granules so that the derivatized starch can be recovered in granule form. Following reaction to the desired degree of substitution, the starch is recovered by centrifugation and/or filtration, washed, and dried.

9.1. Chemically Modified Starches

9.1.1. Converted Starches and Hydrolyzates

In the starch industry, the term *conversion* refers to depolymerization. Products that are partially depolymerized are called *converted starches*. There are three basic processes for making converted starches: using an acid (with various moisture contents and degrees of heating), using an enzyme(s) [amylase(s)], using an oxidant in an alkaline system. Extensive conversion produces *maltodextrins, syrup solids, syrups, and dextrose*.

9.1.1.1. Thinned Products

To produce products variously known as *acid-modified, acid-converted, converted, thin-boiling, fluid,* or *fluidity* starches [46–48], which are the least converted of the products converted with an acid, either hydrochloric acid is sprayed onto well-mixed starch or stirred moist starch is treated with hydrogen chloride gas, and the mixture is heated until the desired degree of hydrolysis (which is very slight in any case) is obtained. The acid is then neutralized, and the product is recovered by filtration or centrifugation.

When to stop a conversion is usually determined empirically by monitoring the viscosity of a hot paste of the acid-modified starch. In the starch industry, this is known as determining *fluidity*. Acid conversion is practiced on both native starches and chemically modified starches, i.e., derivatized starch products (Sections 9.1.2, 9.1.3) may also be *thinned*, in which case, the pH of a stirred slurry of starch granules (36–40% solids) that has just been derivatized and held at a temperature below the gelatinization temperature of the derivatized starch (usually 40–60 °C) is lowered by the addition of a mineral acid, such as hydrochloric acid, until it is in the acid range. When the desired degree of limited hydrolysis (conversion) is reached, the suspension is neutralized and the granules are recovered by centrifugation and/or filtration, washed, and dried.

Compared to the parent starch, acid-converted starches undergo a greater degree of granule disintegration upon cooking and produce pastes with lower intrinsic viscosity values, reduced hot paste viscosity, increased gel strength, and better film-forming capabilities (Table 4). Specific characteristics are a function of the parent starch and the conversion conditions.

Many starch products are bleached with small amounts of an oxidant. Use of higher levels of an oxidant oxidizes the starch and often results in partial depolymerization. Conversion using an oxidant and an alkaline system occurs because oxidation of a hydroxyl group of a D-glucopyranosyl unit introduces a carbonyl group and leads to β-elimination and chain cleavage. Such a reaction is usually accomplished by treating a slurry of starch granules with a solution of sodium hypochlorite [49]. Use of hydrogen peroxide and cupric ions depolymerizes starch molecules without introducing carbonyl groups. Ammonium persulfate is used to depolymerize starches oxidatively for paper sizing and coating processes. The decision as to when to stop an oxidation is generally determined in the way previously noted for an acid-converted starch, i.e., by monitoring hot paste viscosity (fluidity).

Properties of starches converted by oxidative cleavage generally parallel those of starches converted by acid-catalyzed hydrolysis, but with two differences. Pastes of oxidized starches generally are clearer and more stable than those made from acid-modified starches, and there can be greater variability in products in oxidatively-thinned products due to choice of the parent starch, oxidizing agent, and conditions of oxidation. For example, starches treated with low levels of hypochlorite at an acidic pH (rather than the usual alkaline pH condition) result in products with a greater (rather than reduced) hot paste viscosity.

Thinning is primarily done so that higher solids solutions can be made without generating excessive viscosity.

9.1.1.2. Starch Dextrins

Starch dextrins [*9004-53-9*] are products resulting from a greater degree of conversion than are thinned products. *Dextrinization* is the process of making dextrins.

Starch dextrins (occasionally called pyrodextrins) are made by heating a starch with or without addition of an acid [46–50]. There are three general types of these dextrins: white dextrins, yellow dextrins, and British gums. The variables involved in making dextrins are the base starch used, the amount and type of acid used, the percentage of moisture in the starch, the temperature employed, and the time of conversion. In general, white dextrins are made using an acid catalyst, relatively low temperatures, and short conversion times. Yellow dextrins are products of higher temperatures and longer conversion times. British gums are also made using high temperatures and long times, but with less or no added acid. Hydrochloric acid is usually the acid used, particularly for food-grade dextrins. Starch at 10–22% moisture is either treated with gaseous hydrogen chloride or a solution of hydrochloric, sulfuric, or phosphoric acid is sprayed onto the starch as an atomized mist. The starch is then predried to a moisture content of 2–5% by heating it at a relatively low temperature, sometimes under reduced pressure, to keep hydrolysis to a minimum. During the actual dextrinization step (pyroconversion), the pre-dried starch is heated at 100–200 °C. The product is then cooled and rehumidified. Alternatively, the starch is first dried to a moisture content of 2–5% and is then treated with HCl (either as a gas or sprayed on in liquid form) before being heated.

Three types of reaction take place during dextrinization, the relative amounts of which are a function of the moisture content of the starch and the treatment temperature. (1) Hydrolysis of glycosidic bonds is the predominant reaction when the moisture content is high and is, therefore, the main reaction in the preparation of white dextrins with a low degree of conversion. Because of glycosidic bond cleavage and the resulting reduction in polymer molecular masses, final products generate lower solution viscosities. (2) Transglycosidation (transglycosylation) results in transfers of portions of glucan chains to hydroxyl groups on the same or different chains to create new glycosidic bonds. The result is more highly branched and more soluble macromolecules. (3) Repolymerization (reversion) occurs with catalytic amounts of acid at high temperatures and low moisture contents. Under these conditions, glycosidic bonds are formed from the reducing ends of the maltooligosaccharides (and any glucose) released during hydrolysis. The result is higher molecular mass and more highly branched macromolecules than were present before reversion occurred.

9.1.1.3. Dextrose Equivalency

Converted, i.e., depolymerized, products are classified by their *dextrose equivalency (DE)* which is defined as $100 \times DP$ (*degree of polymerization* or the average number of D-glucosyl units in the polymer or oligomer). The DE value is based on the facts that (1) a new reducing end-unit is generated with each hydrolytic cleavage of a starch polymer chain and (2) the reducing power of converted products is something that can be measured. Amylose and amylopectin have no measurable reducing power (DE = 0). The product of complete hydrolysis, D-glucose (dextrose) [*50-99-7*] (DP = 1), has by definition a DE of 100. Therefore, the DE of a product of starch hydrolysis is its reducing power as a percentage of the reducing power of dextrose and is inversely proportional to the average molecular mass of the product molecules. (Both DE and DP are average values for populations of molecules.)

9.1.1.4. Maltodextrins

Maltodextrins [*9050-36-6*] by regulatory definition (USA) are purified, nutritive mixtures of saccharide polymers obtained by partial hydrolysis of edible starch [51–53]. Most molecules in maltodextrin preparations are maltooligosaccharides of DP 2–20. Because the DE values of the products are mostly in the range 5–15, their average DPs are in the 7–20 range. Maltodextrins are produced by hydrolysis of the starch molecules within pastes using either a single amylase, several amylolytic enzymes, or an acid to achieve DE values of less than 20 (average DP >5). Maltodextrins made by acid-catalyzed hydrolysis alone possess a high quantity of linear chains that are able to retrograde. When it follows acid-catalyzed hydrolysis to DE 5–10, α-amylase-catalyzed hydrolysis produces maltodextrins of low hygroscopicity and high water solubility that generate less haze when dissolved. Maltodextrins are also produced either by use of an α-amylase alone or with a combination of an α-amylase and a debranching enzyme.

Maltodextrins that have been specially prepared so that they associate to form microcrystals that are about the size of oil or fat droplets may be used as fat mimetics/sparers/replacers. Another specially made maltodextrin, one in which digestible carbohydrate has been removed, is employed as soluble dietary fiber.

Maltodextrins are primarily used in processed food products; their collective attributes are listed below.

- Bland flavor
- Control moisture migration
- Crystallization inhibition
- Encapsulator of flavors and aromas
- Fat replacers, sparers, and mimetics (specially prepared maltodextrins in microcrystalline form)
- Film formers
- Good dispersibility in aqueous systems
- Good solubility in aqueous systems
- Low to no sweetness
- Provide body
- Provide a smooth texture
- Range of susceptibility to Maillard browning
- Readily digestible
- Spray dry easily

9.1.1.5. Syrup Solids

Hydrolysis to DE 20–35 (average DP 3–5), followed by drying, produces syrup solids.

9.1.1.6. Syrups and Crystalline D-Glucose

($\rightarrow$ Glucose and Glucose-Containing Syrups)

Continued hydrolysis of starch with an acid and/or enzymes produces mixtures of D-glucose, maltose, and other maltooligosaccharides in a concentrated form known as *glucose syrups* (usually called *corn syrups* in the USA) [54–57]. Most often, a glucose syrup is a purified, concentrated, aqueous solution of nutritive saccharides derived from food-grade starch with DE values of >20. (Syrups that are not required to be food grade are also produced for use in anti-freeze and deicing compositions.)

Glucose syrups are made by one of three processes: (1) acid conversion (pH ca. 2, temperature $>100\ °C$) (a minor process); (2) acid-enzyme conversion (an acid-converted hydrolyzate is treated with one or more hydrolyzing enzymes, the choice of which depends on the desired saccharide profile of the finished syrup); or (3) enzyme-enzyme conversion (the major process) (in a single step, a starch suspension is pasted and the paste is liquefied using a thermostable α-amylase. The product is then treated with one or more other enzymes depending upon the desired saccharide profile). Specifically, glucose syrups are most often made by passing an aqueous slurry of starch containing a thermally stable α-amylase through a jet cooker, where rapid gelatinization and enzyme-catalyzed hydrolysis (*liquefaction*) occur. After the solution is cooled to 55–60 °C, glucoamylase is added and hydrolysis is continued. When hydrolysis is complete, the syrup is clarified, decolorized, and concentrated. DE values of syrups may range from 25 to 95 with the lower-DE syrups having greater proportions of maltooligosaccharides and the higher-DE syrups having a greater proportion of D-glucose. Syrups are stable because the oligosaccharides present in them prevent crystallization and the high osmalality (ca. 70% solids) prevents growth of microorganisms. When crystalline D-glucose (dextrose) [*50-99-7*] or its monohydrate is desired, seed crystals are added to a syrup of DE >95

[54, 56]. *High-maltose syrups* are also produced, as well as being used as is; they are the starting materials for the production of maltitol [*585-88-6*] [57].

9.1.1.7. High-Fructose Syrups and Crystalline D-Fructose

For the production of D-fructose [*57-48-7*], a solution of D-glucose is passed through a column containing immobilized glucose isomerase, producing an equilibrium mixture of approximately 58% D-glucose and 42% D-fructose [56–58]. Higher concentrations of D-fructose are desired for many applications of *high-fructose syrups* (HFS), often called *high-fructose corn syrups* (HFCS) in the USA. To make HFS/HFCS with more than 42% D-fructose, the isomerized syrup is subjected to a continuous chromatographic separation process by passing it through a bed of cation-exchange resin in the calcium salt form that binds D-fructose; the resin is subsequently eluted to provide a fraction enriched in D-fructose. This fraction is added to a 42% fructose solution to produce syrups with higher concentrations of D-fructose. Two major types of high-fructose syrups are produced: HFS 42 (ca. 42% fructose) and HFS 55 (ca. 55% fructose). Although D-fructose is not easily crystallized, crystalline D-fructose can be obtained from the enriched solution [59].

9.1.1.8. Other Starch Conversion Products

Other commercial products of starch conversion include β-cyclodextrin [*7585-39-9*] [60] and α,α-trehalose [*99-20-7*] [57].

9.1.2. Cross-linked Starches

Cross-linking reactions utilize bi- or multifunctional reagents to induce formation of intramolecular and/or intermolecular cross-links between adjacent starch chains within starch granules. Reagents employed for the commercial production of cross-linked starches include phosphoryl chloride (phosphorus oxychloride, $POCl_3$), sodium trimetaphosphate (STMP), epichlorohydrin (EPI), and adipic–acetic mixed anhydride [61, 62]. Reaction conditions and schemes vary according to reagent type. For reaction with STMP, starch granules are impregnated with solutions containing both reagent and a catalyzing base, after which slurry moisture levels are reduced to less than 15% and granules are heated to 130 °C in the semi-dry state to drive the reaction [63]. For cross-linking with STMP, reaction is favored at pH values above 10. In contrast, $POCl_3$, EPI, and adipic–acetic mixed anhydride reagents are reacted with starch granules in aqueous slurries at pH values of 11–12, 8, and 10.5, respectively [62]. Products of cross-linking reactions are in the form Starch-O-X-O-Starch, where X represents $-PO_2^-$ — for reactions with $POCl_3$ and STMP, $-CH_2CH(OH)CH_2-$ for reactions with EPI, and $-COCH_2CH_2CH_2CH_2CO-$ for reactions with adipic–acetic mixed anhydride.

Starch cross-linking reactions are employed to strengthen the structure of swollen granules upon gelatinization (Section 5.2.1), imparting resistance to viscosity breakdown (Section 5.2.3) in the presence of mechanical shear, acid conditions, and/or high temperatures. Cross-linked starch rheology is ultimately defined by the nature and concentration of swollen granules within a paste, with the firmness of swollen granules increasing with increasing cross-linking levels (Section 5.2.3). Cross-linked starches typically possess one cross-link for every 100–3000 glucosyl units. Properties of the products vary with modification level. Very low levels of cross-linking usually stabilize the granule structure, allowing the modified starch to attain a higher degree of granule swelling during heating than would be observed for the unmodified native starch. In such cases, higher swelling powers and paste peak viscosities are generally observed (Section 5.2.3). Progressively higher levels of cross-linking generally lead to reduced granule swelling, solubility, extent of amylose leaching, paste clarity, and paste peak viscosity (see Table 4). Those products with significantly reduced peak viscosities are known as *inhibited starches*. Cross-linked starches generally possess improved paste texture (i.e., are less stringy), but do not generally exhibit stability to retrogradation or freeze-thawing without additional stabilization (i.e., dual modification) (Section 9.1.3).

9.1.3. Stabilized Starches

Stabilization (also called *substitution*) involves derivatization of a starch with a monofunctional reagent [64], converting starch hydroxyl groups to ester or ether groups depending on the reagent used. Both starch esters and ethers are produced under similar reaction conditions. Derivatization is most often conducted in the form of an aqueous granule slurry (30–40% solids) in the presence of a gelatinization-inhibiting salt (10–30% concentration; most often sodium sulfate, sometimes sodium chloride) between pH values of 8 and 12, the optimum pH being reagent specific. The temperature of the reaction medium is typically adjusted (most often to 49 °C) to further promote starch granule swelling and reactivity. Use of a gelatinization-inhibiting salt and a temperature below the gelatinization temperature of both the native starch and the modified product prevents pasting, allowing the starch to be recovered in granular form. Allowed DS/MS levels of products intended for use in foods vary from country to country (with those within the EU being uniform), but generally range from 0.002 to 0.2, with different maximum values for different reagents (as examples, maximum allowable derivatization levels for food ingredient products are noted in the ensuing discussion where appropriate).

Commercial monostarch esters include starch acetates (Starch-OCOCH$_3$), succinates (Starch- O-CO(CH$_2$)$_2$CO$_2^-$), octenylsuccinates (Starch-OCOCH[CH$_2$ CO$_2^-$]CH = CH(CH$_2$)$_5$ CH$_3$), and phosphates (Starch-OPO$_3^{2-}$). Acetylated starch (maximum allowable DS $\sim$ 0.09) is prepared using acetic anhydride or, in countries in which possession of acetic anhydride is prohibited, vinyl acetate [65]. Starch succinates are made by reaction with succinic anhydride, but are of limited commercial significance [66]. Reaction of starch with oct-1-enylsuccinic anhydride (OSA) produces the oct-1-enylsuccinic ester (OSA starch), for which the DS may not exceed 0.02 [66]. Monostarch phosphate esters are generated from reaction of starch with sodium orthophosphate or sodium tripolyphosphate (STTP) [67]. A DS maximum of 0.002 is permitted for monostarch phosphate esters. A large variety of other esters can be, and have been, made by selection of the proper reagent [65, 68]. These other esters include starch and amylose triacetate, which are water insoluble, thermoplastic materials (Section 9.2.4).

Manufactured monostarch ethers include hydroxylethyl (Starch-OCH$_2$CH$_2$OH), hydroxypropyl (Starch-O-CH$_2$CHOHCH$_3$), carboxymethyl (Starch-OCH$_2$CO$_2^-$) and cationic (discussed in Section 9.1.4) starches. Reaction with ethylene oxide produces hydroxylethyl starch [68, 69], reaction with propylene oxide produces hydroxypropyl derivatives (maximum MS $\leq$ 0.2) [68, 70], and reaction with sodium monochloroacetate produces sodium carboxymethyl starch [71], known as sodium starch glycolate in the pharmaceutical industry. A large variety of other ethers, including silyl ethers, can be made by selection of the appropriate reagent [68].

Esterified and etherified products generally exhibit similar physical properties, although ethers are stable to acids and bases, while esters are not. In general, *stabilized starches* are so named because they have a reduced tendency to undergo retrogradation (usually a very desirable attribute) due to the incorporation of bulky substituent groups along the starch chains, that block intermolecular associations. Some incorporated substituent groups also possess a negative charge (e.g., monostarch phosphate esters), further reducing interchain associations (through like-charge repulsion) to enhance paste stability. Thus, stabilized starches produce more stable starch pastes and gels with improved clarity, less syneresis, and greater freeze-thaw stability than their unmodified starch counterparts (see Table 4). Substitution (especially at high levels) can disrupt the native granule structure, resulting in reduced starch gelatinization and pasting temperatures, as well as increased starch swelling, paste peak viscosities, and solubilities. In the case of OSA starch, some hydrophobicity is also introduced to starch chains, giving the product emulsifying and emulsion-stabilizing properties (Table 4). Stabilization-type reactions are often conducted in combination with other types of chemical and/or physical modifications to yield multifunctional starches (Section 9.4).

9.1.4. Cationic Starch

Cationization reactions are carried out using monofunctional reagents that impart a positive charge to starch chains via derivatization with reagents that incorporate ether groups containing imino, amino, ammonium (quaternary amino), sulfonium, or phosphonium moieties [72]. The most industrially significant and commonly utilized reagent for starch cationization is 3-chloro-2-hydroxypropyltrimethylammonium chloride (CHPTMA) [72–74], although a multitude of potential reagents may be employed. Reaction with CHPTMA gives the following product: Starch-O-CH_2CHOH$CH_2$$N^+$$(CH_3)_3$.

Cationic starch may be prepared batchwise via granular slurry, semi-dry, and solubilized paste processes, and by continuous means such as reactive extrusion [73–75]. The first two processes yield a modified starch product with retained granule structure, while in the latter two processes starch is reacted in a gelatinized or molten state [75]. For the granular slurry process, reaction may be conducted in either an aqueous or aqueous alcohol reaction medium in the presence of NaOH at a temperature of 49 °C to achieve reaction efficiencies in the range of 80–88% [72–75]. Aqueous slurry reactions generally include sodium sulfate to prevent gelatinization of the starch granules during reaction, while use of an aqueous alcohol reaction medium eliminates the need for a gelatinization-inhibiting salt [75]. The semi-dry cationization process, which also maintains starch in a granular form, entails heating of dry granular starch (< 120 °C) in the presence of both catalyst (NaOH) and reagent [75], resulting in reaction efficiencies of 90–95% [74]. For cationization via the paste process, a starch slurry is adjusted to alkaline pH (ca. 11.3) and gelatinized via steam injection, prior to reaction at 50 °C [76]. The use of an extruder as a continuous reaction vessel (i.e., reactive extrusion) for derivatization of starch polymers in the molten state has been described for various types of modification [75]. For starch cationization, the reactive extrusion process is reported to achieve reaction efficiencies in the range of 90% [77–79]. However, the high temperature and shear conditions imposed by the extrusion process result in some thermomechanical degradation of starch molecules, the degree of which is dependent on the processing parameters [80, 81].

For cationic starches modified in granular form, relative starch crystallinities, gelatinization temperatures and enthalpies, and pasting temperatures decrease with increasing levels of cationization [72, 73]. Cationic starch derivatives also exhibit increased swelling powers, water sorption properties, peak and breakdown viscosities, stability of pastes and gels to syneresis, lipid-binding capacity, and emulsion-stabilizing properties relative to native (unmodified) starches [72, 73].

The primary applications of cationic starches are in the wet-end, surface-sizing, and coating processes in the paper industry, as well as industrial flocculants in sludge dewatering and mineral ore recovery operations [72–74]. Degree of substitution (DS) values for cationic starch derivatives used in papermaking applications generally range from 0.02–0.07, while those for flocculent applications are significantly higher (DS ca. 0.2–0.7) [73].

9.1.5. Starch Graft Copolymers

Starch graft copolymers are derivatives in which synthetic polymers are covalently bonded to starch molecules, most often through ether linkages. These products are generally obtained by generating free radicals from hydroxyl groups of native or hydroxyalkylated (generally hydroxyethylated) starch molecules, followed by reaction with an unsaturated monomer, such as an acrylic or vinyl monomer. A wide range of different products has been made using different methods for producing starch free radicals, different polymerizable monomers, and different starches and modified starches [82, 83].

9.2. Thermally Modified Starches

9.2.1. Instant Starches

Pregelatinized starches are precooked products that are cold-water soluble, i.e., they can be used without cooking/heating and will produce dispersions without lumps when correctly prepared [84, 85]. They may be made from either native or chemically modified (Section 9.1) starches by two general methods. In one method, a starch-water slurry is applied to a steam-heated roll or into the nip between two nearly touching and counterrotating steam-heated rolls, where the starch is rapidly gelatinized, pasted, and dried. The dry film is scraped from the roll and milled to a powder. The resulting product should contain no intact granules, except when the starting material is a highly cross-linked starch. The other way to make a pregelatinized product is to pass a starch and some moisture through an extruder, in which the heat and shear imparted by the process conditions disrupts the granules. In this process also, the dried product is ground to the desired mesh size.

Pregelatinized starch products are used in food and laundry products when (1) no heat for cooking is available, (2) no process step requires sufficient heat to cook the starch, or (3) heat cannot be applied because of the heat liability of one or more of the other ingredients. If modified starches are used in the preparation of the pregelatinized starch product, the properties of the paste introduced by any modification(s) are also exhibited by the pregelatinized product.

Products known as *cold-water-swelling (CWS) starches* contain granules that swell extensively in the presence of ambient temperature water [85]. They are made by two different processes. In one case, a slurry of starch granules in aqueous alcohol is heated to disrupt the internal granule structure, i.e., to gelatinize the granules, without much granule swelling and no dissolution of the polymers. In the other, an aqueous starch slurry is atomized via a special spray-drying nozzle into a stream of heated air. Granules in cold-water-swelling starches are gelatinized, but their granular form is maintained (as opposed to pregelatinized products). Both pregelatinized and CWS starches are categorized industrially as *instant starches*.

Starches that are neither pregelatinized nor cold-water-swelling are known as *cook-up starches*.

9.2.2. Annealed and Heat-moisture-treated Starches

While preparation of instant starches is widely practiced, less common are thermal treatments known as *heat-moisture treatments (HMT)* and *annealing* [86]. Annealing results from heating granular starch in excess water (> 40%) at a temperature above the glass transition temperature but below the onset temperature of gelatinization. The resulting changes in properties are a function of the parent starch, the treatment temperature, and the duration of heating, but the most common and consistent effects of annealing are increased onset and peak gelatinization temperatures, an increased enthalpy of gelatinization (not always observed), a narrowing of the gelatinization temperature range, and decreased digestibility (Section 5.2 and Chap. 11).

Heat-moisture treatment also involves heating of starch granules above their glass transition temperature, but differs in that it is conducted at reduced levels of moisture (< 35%) and at higher processing temperatures (80–140 °C). The variables in this process are the type of starch, the moisture content of the starch, the treatment temperature, the heat source, and the duration of the treatment. Common property changes are increases in onset, peak, and often conclusion gelatinization temperatures; reduced granule swelling and solubility; reduced leaching of amylose; decreased peak viscosity; increased paste stability; reduced enthalpy of gelatinization (though, in some cases, enthalpy may not change); and decreased digestibility (Section 5.2 and Chap. 11). However, due to the diverse range of treatment conditions and starches utilized for HMT, the resulting starch characteristics and properties can vary considerably.

9.2.3. Dry Heating of Starches

Commercial products are made by heating dry starch (< 15% moisture) at a temperature below that at which thermal degradation occurs (90–128 °C) [86]. Products generally exhibit greater stability to the conditions of heat and shear imposed by cooking processes. An increased paste viscosity and a creamy consistency for quick-cooking products are also claimed.

9.2.4. Destructurized and Thermoplastic Starch

Although destructurized and pregelatinized could be considered to be synonymous terms (because a pregelatinized starch is destructurized to mixtures of amorphous amylose and amylopectin), the term *destructurized starch* was originally (1987) used to describe starch of relatively low moisture content (10–25%) that was heated under pressure in an extruder to a temperature above its glass transition and melting temperatures to yield a thermoplastic mass that could be made into a molded article or film. The difference between a pregelatinized starch made by extrusion cooking and a thermoplastic destructurized starch is in the amount of moisture used as a plasticizer in the feed and, hence, the temperature needed to melt the starch [87, 88]. Addition of other plasticizers, such as glycerol, allows the starch to be converted into a thermoplastic material at lower moisture contents than are required for traditional starch pasting. Addition of a depolymerizing catalyst to reduce polymer molecular masses gives a product from which shaped articles with improved characteristics can be made.

Thermoplastic starch can be blended with natural or synthetic hydrophilic polymers. In some cases, the blends can be further mixed with hydrophobic polymers, such as polyethylene, for production of water-resistant, blown films [82]. Destructurized starch can also be mixed with plasticizers and other materials such as natural and synthetic polymers, graft copolymers, and lubricants to produce a large family of products also known as thermoplastic starch. Thermoplastic starch can also be made by esterifying or etherifying (e.g., hydroxypropylating) starch to high DS values (DS 2 or more) [82]. Thermoplastic starch materials can be subjected to various thermoforming processes such as spinning, extrusion, film blowing, injection molding, and foaming. The products are usually biodegradable.

9.3. Genetically Modified Starches

The characteristics of native starches can be modified genetically [89]. A major commercial starch that is the base starch for many modified food starch products is *waxy maize starch*, a maize starch devoid of amylose (Chap. 4). Waxy (all-amylopectin) and partial waxy (reduced amylose content) wheat starches, as well as all-amylopectin potato starch have been developed. Rice plants produce starches with a range of amylose contents from less than 1% to at least 33%. Commercial hybrids of corn, known as *high-amylose corn* or *amylomaize starches*, produce starches with ca. 55% and > 70% contents of amylose.

9.4. Multiple Modifications of Starches

Starches are modified for many and diverse specific applications (Chap. 10). Variables include choice of the parent starch, including a genetically modified starch (Section 9.3), type of chemical modification (Section 9.1), and any thermal treatment (Section 9.2). Many modified starch products are made by combinations of treatments. As examples, a coating binder starch for a papermaking process will likely be hydroxyethylated (stabilized) and thinned, and a food starch might be made by cross-linking, stabilizing, thinning, and pregelatinizing waxy maize starch (a genetic modification). Major commercial individual (combinations of these modification processes are commonly used) starch modification processes and products are listed below:

Chemical reactions

A. *Cross-linking*
1. Esterification (approved for food use)
    a. Distarch phosphates (approved for food use)
    b. Distarch adipates (approved for food use)
2. Etherification
    a. Starch cross-linked by reaction with epichlorohydrin [*106-89-8*] (not used in foods in the USA)

B. *Stabilization*
1. Etherification
    a. Hydroxyethylstarches
    b. Hydroxypropylstarches (approved for food use)
    c. Cationic starches
    d. Starch graft copolymers
2. Esterification
    a. Starch acetates (approved for food use)
    b. Starch oct-1-enylsuccinates (approved for food use)
    c. Monostarch phosphates (approved for food use)
3. Oxidation
    a. With hypochlorite (approved for food use)
    b. With hydrogen peroxide and Cu(II) ions (approved for food use)
    c. With ammonium persulfate

C. *Depolymerization*
1. Acid-catalyzed (approved for food use)

2. Oxidation, followed by alkaline pH (approved for food use)

D. *Transglycosylation plus depolymerization (dextrinization)* (approved for food use)

Physical/thermal transformations

A. *Pregelatinization* (approved for food use)
B. *Preparation of cold-water-swelling starch* (approved for food use)

Genetic control/plant breeding (approved for food use)

10. Examples of Uses of Starches and Products Derived from Starches

10.1. Uses of Starches

The primary uses of starches, which vary from country to country and region to region (see Table 5) [90], are to produce sweeteners (dextrose, glucose syrups, HFS) and related products (maltodextrins, syrup solids), in paper and paperboard product manufacture [91], in processed food preparations [92], and to produce fuel ethanol, although as already mentioned, most ethanol is produced by processes in which the starch is not isolated as a pure material. Some examples of uses of specific products follow, although essentially all products employed in paper manufacture and as a textile sizing agent have been thinned, and many of the products used in the food industry have been pregelatinized and/or thinned in addition to being both cross-linked and stabilized. Thus, relating a specific modified starch to a specific application is often not possible.

Selected examples of nonfood uses of starches, which term includes both native and modified (derivatized, oxidized, and acid-thinned products, starch-latex copolymers, dextrins, and pregelatinized products) starches are as follows:

- Absorbants
- Briquette binder
- Carpet and rug sizing
- Ceramics (clay binder)
- Fiberglass sizing
- Ink and dye thickeners
- Manufacture of paper, boxboard, corrugated board, fiberboard drums, spiral tubes, etc. (wet-end additive, sizing agent, coating binder, adhesives [remoistenable, case and carton sealing, laminating, tube winding, and bag adhesives])
- Ore refining (both flotation and electrolytic reduction processes)
- Tablet binder and excipient
- Textile finishing and sizing
- Wall-covering pastes

Selected examples of food uses of starches, which term includes both native starches and modified (derivatized [cross-linked and stabilized], oxidized, and acid-thinned products, dextrins, and pregelatinized) products are as follows:

- Baby foods
- Biscuits, crackers, cookies
- Batters and batter mixes
- Beverage emulsions
- Breadings and breading mixes
- Breakfast cereals
- Cake, pancake, doughnut, muffin, and waffle mixes
- Confections (gum, jelly, and panned products)
- Fillings for pies and other baked goods (cream and fruit)
- Glazes
- Gravies and gravy mixes
- Hot-filled products in jars and cans
- Noodles
- Pet foods
- Processed cheese products

Table 5. End-use demand for starches in some major markets [90]

	Approximate percent of total use		
	EU	USA	Japan
To produce glucose syrups, dextrose, and maltodextrins	28	12	27
To produce HFS[a]	5	31	27
In processed foods	18	3	21
To produce paper and paperboard	22	10	11
To produce fuel ethanol	2	42	5
Other industrial uses[b]	26	2	10

[a] High-fructose syrups.
[b] Includes textiles, personal care products, chemicals other than ethanol, detergents, pharmaceutical, and other industrial products.

- Processed meat products
- Pudding, pie filling, and other dessert mixes
- Retorted canned products
- Salad dressings (spoonable and pourable)
- Sauces and sauce mixes
- Snack and other extruded foods
- Soups and soup mixes
- Surimi
- Toppings and syrups
- Yogurt

Approximate utilization of corn/maize starch isolated by the wet-milling process in the USA: **As starch products (ca. 16%) [9]**

- In paper and paperboard manufacture (ca. 10%)
- In processed food products (ca. 3.0%)
- As a textile size (ca. 2.0%)

As the result of conversion (ca. 85%); excluding that used to produce ethanol [9]

- As high-fructose syrups (ca. 49%)
- As glucose syrups, dextrose, and maltodextrins (ca. 36%)

Uses of native (i.e., unmodified) starches are limited. In the food industry, they are used to dust molds for jelly-type confections, for moisture control (e.g., in salt), to help dissipate heat during grinding of sugar to produce powdered sugar, and for dusting of marshmallows and chewing gum. Nonfood uses include their use in papermaking, where the papermaker purchases a native starch, then subjects it to controlled oxidative (ammonium persulfate, potassium persulfate, sodium hypochlorite, or hydrogen peroxide) or enzyme-catalyzed (α-amylase) depolymerization or derivatization on site. Native starch is used to make porous ceramic filters (e.g., for diesel vehicle exhaust) and, with addition of other ingredients, as the adhesive for corrugated board manufacture [93]. Native starches are combined with plasticizers and other additives and pregelatinized/destructurized in an extruder to prepare water-soluble, loose-fill packaging material.

Acid-modified and other thinned products are used in the warp sizing of textiles, as adhesives, and in the manufacture of gypsum and corrugated board [93].

As mentioned, many derivatized starch products are also "thinned." Those starches that have been thinned, but otherwise unmodified (i.e., thin-boiling or fluidity native starches), are used to make gum candies (e.g., jelly beans, jujubes, orange slices, spearmint leaves) and processed cheese loaves.

Cross-linked products are mostly used in the food industry, and food processors use both cross-linked-only products and starches that have been both cross-linked and stabilized (i.e., dually modified). Cross-linked starches are used as thickeners in canned (retorted) and hot-filled products such as sauces, soups, gravies, pie fillings, puddings, spoonable salad dressings, and baby foods and in batter mixes. Cross-linked and stabilized starches are used in preparation acidic food products such as tomato-based sauces, salad dressings, and fruit pie fillings; retorted products such as sauces, soups, and gravies; aseptically-filled products such as puddings and cheese sauces; frozen foods such as pot pies, fruit pies, and gravies; and baby foods.

Stabilized-only starches are used in both food and nonfood applications. Starch acetates and hydroxypropylated starches are used as thickeners of food products when low-temperature stability is required (e.g., in chilled and frozen products) and in general when retrogradation/setback is undesirable. The most widely used stabilized starch for nonfood applications is hydroxyethyl starch, which is used in papermaking as a furnish additive, a surface sizing agent, and as a coating and pigment binder [91]. Hydroxyethyl starch also finds use in textile sizing [94]. Acetylated starches are used in surface sizing of certain paper products and for warp sizing of textiles.

Cationic starches are used in the papermaking process as a wet-end additive to balance fiber flocculation and retention in order to promote sheet dewatering and to attain good sheet formation and strength [91]. It is also used in surface sizing in certain paper products, and even less often as a coating binder. Cationic starch also finds use as a flocculant for anionic materials and in fabric softeners.

Monostarch phosphates find use in both food and nonfood applications. They can be used with alum as a wet-end additive in making paper at low pH. They are also used as flocculants and sedimentation aids for various cationic materials and in textile sizing and printing pastes.

The amphophilic *octenylsuccinylated starches* are special types of stabilized starch products. These starches are used in emulsion preparation and stabilization in products such as pourable salad dressings and flavored beverages. They are also used to encapsulate certain flavors, aromas and drugs and as fat sparers.

Carboxymethylated starches are used as sizing agents for textiles [94]. Higher-DS products are used as additives to control fluid loss and to modify the rheology of drilling muds in oil and natural gas production.

Uses and potential uses of starch graft copolymers have been reviewed [82, 95]. Graft copolymers of starch with a latex are useful as paper-coating materials. Some polyacrylate starch copolymers are employed as superabsorbants.

Pregelatinized starches are widely used. Nonfood uses include as a spray-on laundry starch, as an adhesive in foundry cores for metal casting, as a binder, and as an excipient in solid oral dosage forms of pharmaceuticals. Their use in foods is varied and widespread since cross-linked and/or stabilized starches can also be pregelatinized. A common use is in dry mixes for such products as instant puddings and soups. They are also used in pizza toppings, extruded snacks, soft cookies, bakery fillings, and low-fat salad dressings. Cold-water-swelling starches are used to thicken sugar solutions and glucose syrups in the preparation of gum candies, in certain desserts, and in muffin mixes.

Thermoplastic starch can be used to prepare food, lawn, and leaf bags for composting, food-service ware, pharmaceutical capsules, personal hygiene products, and packaging material and as fillers for tires.

10.2. Uses of Products derived from Starches via Extensive Depolymerization

The primary uses of dextrins are to prepare adhesives for paper and paperboard products [50], where they are used as bag-seam pastes (white dextrins, British gums), in tube winding (white and yellow dextrins, British gums), for case and carton sealing (white and yellow dextrins), in laminating (white and yellow dextrins, British gums), in preparation of remoistenable gummed tape and labels (white and yellow dextrins), and in envelope manufacture (white and yellow dextrins).

In the food industry, white dextrins are used in pan coating of confections, as a gloss for bakery products, and as carriers for flavorings, spices, and colorants.

Syrup solids and maltodextrins are utilized in foods in many of the same applications, which include use as agglomerating agents, as binders, as bulking agents, in coatings, as coffee whiteners, in frozen foods, in infant formulas, in meat products, in processed cheese products, and in whipped toppings. Maltodextrins are used to encapsulate flavors and aromas.

Glucose syrups are produced in large quantities and their use is widespread in food products. (Syrup viscosity is important as it relates to the handling characteristics of the syrup and the texture of the food product). They are used largely as humectants and to provide bulk and body with various degrees of sweetness. Products with low sweetness provide body to ice cream, canned fruits and vegetables, bakery products, and some confections. Products with moderate sweetness are used in foods requiring body but not high sweetness (e.g., in sauces, toppings, and some confections) and sweet bakery products. Glucose syrups are used to reduce water activity, lower the freezing point, increase the osmotic pressure, and increase the chewiness of a food product. They also inhibit crystallization of sucrose. In ice cream, they provide resistance to meltdown. They increase the shelf life of hard candies and peanut brittle by preventing crystallization and providing resistance to changes in moisture. They provide fermentable substrates and contribute to browning of bakery products, and are used in many other applications from bakery fillings to salad dressings.

Glucose syrups are the carbon source for a variety of fermentation processes that are used to produce a range of products from lactic acid to antibiotics, and as has already been mentioned, various salts are added to them to produce antifreeze and deicing solutions.

High-glucose syrups are the starting materials for the production of sorbitol [*50-70-4*], ascorbic acid (vitamin C) [*50-81-7*] via sorbitol, glucono-δ-lactone (D-glucono-1, 5-lactone) [*52153-09-0*], and polydextrose [68424-04-4], all of which are widely used in their own rights.

Crystalline D-glucose is used in many of the same products that utilize glucose syrups when a dry form of sugar is desired. In addition, when it is used as part of the sugar coating of doughnuts and in sandwich cookie and sugar wafer fillings, it imparts a desirable cooling sensation in the mouth because of its slight negative heat of solution.

Most often, hydrogenated starch hydrolysates (HSH) are used to make food products for diabetics. Some is also used to make sugarless candies and chewing gum.

High-fructose syrup containing ca. 42% fructose (HFS 42) is used as the sweetener in many fruit-flavored beverages, in doughs used in production of yeast-raised breads and rolls, and in batters used in production of cakes. HFS 42 is blended with sucrose and glucose syrups in fruit canning. Dairy products, namely, chocolate milk, yogurts, milk-based nutrition drinks, ice cream, frozen novelties, and frozen desserts, as well as nondairy products such as jams, fruit fillings, barbecue sauces, and cake frostings, may also contain HFS 42. Most cola drinks are made with HFS 55.

β-Cyclodextrin (→ Cyclodextrins) is primarily used to prepare powdered essential oils, which are thereby protected from oxygen and light, but which are readily released when the dry powder is added to an aqueous system. Such dry complexes are used in dry mixes for baked goods, beverages, and soups; flavored coffee and tea; savory snacks and crackers; breakfast cereals; chewing gum; tabletted candies; processed cheese products; gelatin desserts; and puddings. It is also used to protect other aromas, flavors, and certain pharmaceuticals from oxidation, volatilization, light-induced degradation, and thermally induced decomposition.

11. Starch Digestibility

Not all starch is equally digestible. In a food product containing starch, there will be fractions termed *rapidly digestible* (or *rapidly digesting*) starch (*RDS*), *slowly digestible* (or *slowly digesting*) starch (*SDS*), and *resistant starch* (*RS*). Rapidly digesting starch is the fraction which, in a laboratory analysis designed to mimic digestion in the stomach and small intestine, is converted into D-glucose within 20 min. Slowly digestible starch is the fraction that is converted into D-glucose in the time period between 20 and 120 min. Resistant starch is the portion of starch that does not undergo digestion in the small intestine (is not converted into D-glucose within 120 min in the laboratory test). It is fermented by the microflora in the large intestine (colon) and, therefore, falls under the definition of dietary fiber. There are four general classes of resistant starch: *RS1* (starch that is physically inaccessible to the digestive enzyme α-amylase because it is encased within plant cell walls, nondigested denatured protein, or something else that entraps it); *RS2* (uncooked, i.e., ungelatinized granular starch); *RS3* (retrograded starch in which starch polymer molecules [primarily amylose molecules] are highly associated and attacked by α-amylase only slowly); and *RS4* (granules that are resistant to the action of α-amylase because of chemical modification). The glycemic index of the fractions is RDS > SDS > RS.

References

1 R.L. Whistler, E.F. Paschall (eds.): *Starch: Chemistry and Technology*, Academic Press, New York, vol. I, 1965; vol. II, 1967.
2 R.L. Whistler, J.N. BeMiller, E.F. Paschall (eds.): *Starch: Chemistry and Technology*, 2nd ed., Academic Press, New York 1984.
3 P.J. Frazier, A.M. Donald, P. Richmond (eds.): *Starch Structure and Functionality*, The Royal Society of Chemistry, Cambridge 1997.
4 D.J. Thomas, W.A. Atwell: *Starches*, Eagen Press, St. Paul 1999.
5 J. BeMiller, R. Whistler (eds.): *Starch: Chemistry and Technology*, 3rd ed., Academic Press, Boston 2009.
6 A.C. Bertolini (ed.): *Starches: Characterization, Properties, and Applications*, CRC Press, Boca Raton 2010.
7 O.B. Wurzburg (ed.): *Modified Starches: Properties and Uses*, CRC Press, Boca Raton 1986.
8 Fachverband der Stärke-Industrie e.V.: Zahlen und Fakten zur Stärke Industrie Ausgabe 2006, Berlin 2006.
9 Corn Refiners Association: 2009 Corn Annual, Corn Refiners Association, Washington, DC 2009; http://www.corn.org (accessed 24 November 2010)
10 http://www.Thaitapiocastarch.org/export.asp (accessed 24 November 2010).
11 Z. Nikuni, *Chori Kagaku* **2** (1969) 6; *Depun Kagaku* **22** (1975) 78.
12 D. French, *Depun Kagaku* **19** (1972) 8.

13 K. Kainuma, *Chori Kagaku* **13** (1980) 83.
14 D. French, in [2], p. 211.
15 E. Bertoff, *Carbohydr. Polym.* **68** (2007) 433.
16 J.L. Jane, in [5], pp. 193–236.
17 S. Pérez, P.M. Baldwin, D.J. Gallant, in [5], pp. 149–192.
18 D.J. Gallant, B. Bouchet, P.M. Baldwin, *Carbohydr. Polym.* **32** (1997) 177.
19 P.J. Jenkins, R.E. Cameron, A.M. Donald, W. Bras, G.E. Derbyshire, G.R. Mant, A.J. Ryan, *J. Polym. Sci., B-Polym. Phys.* **32** (1994) 1579.
20 H.F. Zobel, *Starch/Stärke* **40** (1988) 1.
21 R.F. Tester, J. Karkalas, X. Qi, *J. Cereal Sci.* **39** (2004) 151.
22 J.E. Fannon, R.J. Hauber, J.N. BeMiller, *Cereal Chem.* **69** (1992) 284.
23 H.S. Kim, K.C. Huber, *J. Cereal Sci.* **48** (2008) 159.
24 M.A. Glaring, C.B. Koch, A. Blenow, *Biomacromol.* **7** (2006) 2310.
25 J.E. Fannon, J.M. Schull, J.N. BeMiller, *Cereal Chem.* **70** (1993) 611.
26 K.C. Huber, J.N. BeMiller, *Cereal Chem.* **74** (1997) 537.
27 K.C. Huber, J.N. BeMiller, *Carbohydr. Polym.* **41** (2000) 269.
28 K.C. Huber, J.N. BeMiller, *Cereal Chem.* **78** (2001) 173.
29 J.-H. Han, J.A. Gray, K.C. Huber, J.N. BeMiller, in M.L. Fishman, P.X. Qi, L. Wicker (eds.): *Advances in Biopolymers: Molecules, Clusters, Networks and Interactions*, American Chemical Society, Washington, DC, 2006.
30 C.G. Biliaderis, in [5] pp. 293–372.
31 P. Colona, A. Buleon, in [6] pp. 71–102.
32 K.C. Huber, A. McDonald, J.N. BeMiller, *Carbohydrate Chemistry*, in Y.H. Hui, F. Sherkat (eds.): *Handbook of Food Science, Technology, and Engineering*, vol. 1, Taylor and Francis/CRC Press, Boca-Raton 2006.
33 J. Jane, Y.Y. Chen, L.F. Lee, A.E. McPherson, K.S. Wong, M. Radosavljevic, T. Kasemuswan, *Cereal Chem.* **76** (1999) 629.
34 K.C. Huber, J.N. BeMiller, in [6], pp. 179–181.
35 J.L. Hazelton, C.E. Walker, *Cereal Chem.* **73** (1996) 284.
36 Anon, RVA-4 Series Instruction Manual, Newport Scientific Pty. Ltd., Warriewood, Australia.
37 J. Preiss, in [5], pp. 83–148 and references 1–14 therein.
38 J.C. Shannon, D.L. Garwood, C.D. Boyer, in [5], pp. 23–82.
39 S.R. Eckhoff, S.A. Watson, in [5], pp. 373–439.
40 B.P. Geera, J.E. Nelson, E. Souza, K.C. Huber, *Cereal Chem.* **83** (2006) 551.
41 C.C. Maningat, P.A. Seib, S.D. Bassi, K.S. Woo, G.D. Laster, in [5], pp. 441–510.
42 H.S. Kim, K.C. Huber, *J. Cereal Sci.* **51** (2010) 256.
43 H.E. Grommers, D. A. van der Krogt, in [5], pp. 511–539.
44 W.F. Breuninger, K. Piyachomkwan, K. Sriroth, in [5], pp. 541–568.
45 C.R. Mitchell, in [5], pp. 569–578.
46 R.G. Rohwer, R. E. Klem, in [2], pp. 529–541.
47 G.M.A. van Beynum, J.A. Joels (eds.): *Starch Conversion Technology*, Marcel Dekker, New York 1985.
48 O.B. Wurzburg, in [7], pp. 17–40.
49 K.C. Huber, J.N. BeMiller, in [6], pp. 156–157.
50 H.M. Kennedy, A.C. Fisher, Jr., in [2], pp. 593–610.
51 F.W. Schenck, R.E. Hebeda (eds.): *Starch Hydrolysis Products*, VCH Publishers, New York 1991.
52 R.J. Alexander, in [28], pp. 233–275.
53 I.S. Chronakis, *Crit. Rev. Food Sci.* **38** (1998) 599.
54 P.J. Mulvihill, in [28], pp. 121–176.
55 D. Howling, in [28], pp. 277–317.
56 L. Hobbs, in [5], pp. 797–832.
57 P. Hull: *Glucose Syrups: Technology and Applications*, Wiley-Blackwell, Chichester 2010.
58 J.S. White, in [28], pp. 177–199.
59 L.M. Hanover, in [28], pp. 201–231.
60 A. Hedges, in [5], pp. 833–851.
61 O.B. Wurzburg, in [7], pp. 41–53.
62 K.C. Huber, J.N. BeMiller, in [6], pp. 159–161.
63 S. Lim, P.A. Seib, *Cereal Chem.* **70** (1993) 137.
64 K.C. Huber, J.N. BeMiller, in [6], pp. 161–163.
65 W. Jarowenko, in [7], pp. 55–77.
66 P.C. Trubiano, in [7], pp. 131–147.
67 D.B. Solarek, in [7], pp. 97–112.
68 M.W. Rutenberg, D. Solarek, in [2], pp. 311–388.
69 K.B. Moser, in [7], pp. 79–88.
70 J.V. Tuschoff, in [7], pp. 89–96.
71 B.T. Hofreiter, in [7], pp. 179–196.
72 D.B. Solarek, in [7], pp. 113–129.
73 K.C. Huber, J.N. BeMiller, in [6], pp. 163–166.
74 C.-W. Chiu, D. Solarek, in [5], pp. 629–655.
75 K.C. Huber, J.N. BeMiller, in [6], pp. 177–179.
76 J. Radosta, W. Vorweg, A. Ebert, A. Haji Begli, D. Grülc, M. Wastyn, *Starch/Stärke* **56** (2004) 277.
77 M.E. Carr, *J. Appl. Polym. Sci.* **54** (1994) 1855.
78 N. Gimmler, F. Meuser, *Starch/Stärke* **46** (1995) 268.
79 F. Berzin, A. Tara, L. Tighzert, B. Vergnus, *Polym. Eng. Sci.* **47** (2007) 112.
80 A. Ayoub, F. Berzin, L. Tighzert, C. Bliard, *Starch/Stärke* **56** (2004) 513.
81 A. Ayoub, C. Bliard, *Starch/Stärke* **55** (2003) 297.
82 J.L. Willett, in [5], pp. 715–743.
83 K.C. Huber, J.N. BeMiller, in [6], p. 166.
84 E.L. Powell, in [1], vol. II, pp. 523–536.
85 K.C. Huber, J.N. BeMiller, in [6], pp. 166–168.
86 K.C. Huber, J.N. BeMiller, in [6], pp. 169–173.
87 R.M.S.M. Thiré, in [6], pp. 103–128.
88 L. Janssen, L. Moscicki (eds.): *Thermoplastic Starch*, Wiley-VCH, Weinheim 2009.
89 K.C. Huber, J.N. BeMiller, in [6], p. 181.
90 Corn Refiners Association: 2001 Corn Annual, Corn Refiners Association, Washington, DC, 2001.
91 H.W. Maurer, in [5], pp. 657–713.
92 W.R. Mason, in [5], pp. 745–795.
93 B.H. Williams, in [7], pp. 253–263.
94 K.W. Kirby, in [7], pp. 229–252.
95 G.F. Fanta, W. M. Doane, in [7], pp. 149–178.

Further Reading

J. N. BeMiller, R. L. Whistler (eds.): *Starch*, 3rd ed., Academic Press, London 2009.

A. C. Bertolini (ed.): *Starches*, CRC Press, Boca Raton 2010.

A.-C. Eliasson (ed.): *Starch in Food*, CRC Press, Boca Raton 2004.

L. P. B. M. Janssen, L. Moscicki (eds.): *Thermoplastic Starch*, Wiley-VCH, Weinheim 2009.

Wood

ULLMANN'S ENCYCLOPEDIA OF INDUSTRIAL CHEMISTRY

HORST H. NIMZ, Bundesforschungsanstalt für Forst- und Holzwirtschaft, Institut für Holzchemie und chemische Technologie des Holzes, Hamburg, Federal Republic of Germany

UWE SCHMITT, Bundesforschungsanstalt für Forst- und Holzwirtschaft, Institut für Holzbiologie und Holzschutz, Hamburg, Federal Republic of Germany

ECKART SCHWAB, Bundesforschungsanstalt für Forst- und Holzwirtschaft, Institut für Holzphysik und mechanische Technologie des Holzes, Hamburg, Federal Republic of Germany

OTTO WITTMANN, BASF Aktiengesellschaft, Ludwigshafen, Federal Republic of Germany

FRANZ WOLF, BASF Aktiengesellschaft, Ludwigshafen, Federal Republic of Germany

1.	Morphology and Properties	548
1.1.	Introduction	548
1.2.	Biology of Wood	549
1.2.1.	Anatomy	550
1.2.1.1.	Cell Types	551
1.2.1.2.	Softwoods	551
1.2.1.3.	Hardwoods	551
1.2.2.	Secondary Changes	553
1.2.3.	Cell Wall Formation and Architecture	554
1.3.	Chemistry of Wood	555
1.3.1.	Chemical Components of Wood	555
1.3.2.	Distribution of Components in the Cell Wall	559
1.3.3.	Macromolecular Structure of Cell Wall Components	559
1.4.	Physical Properties	560
2.	Wood-Based Materials	562
2.1.	Introduction	562
2.2.	Laminate Bonding	563
2.2.1.	Laminated Structural Timber (Glulam)	563
2.2.1.1.	Development	563
2.2.1.2.	Construction and Production	563
2.2.1.3.	Properties	564
2.2.2.	Veneers, Plywood, and Derived Products	564
2.2.2.1.	Development	564
2.2.2.2.	Production of Veneers	565
2.2.2.3.	Principal Types of Plywood	565
2.2.2.4.	Adhesives and Additives	566
2.2.2.5.	Production of Veneer Plywood	567
2.2.2.6.	Properties, Testing, and Uses	567
2.2.2.7.	Engineered Wood	569
2.3.	Particle Bonding	570
2.3.1.	Particle Board	570
2.3.1.1.	Development	570
2.3.1.2.	Types of Particle Board	570
2.3.1.3.	Raw Materials	571
2.3.1.4.	Adhesives and Additives	572
2.3.1.5.	Production	572
2.3.1.6.	Properties, Testing, and Uses	575
2.3.2.	Fiberboard	576
2.3.2.1.	Development	576
2.3.2.2.	Types of Fiberboard	578
2.3.2.3.	Raw Materials	579
2.3.2.4.	Binders and Additives	579
2.3.2.5.	Production Processes	580
2.3.2.6.	Properties, Testing, and Uses	584
2.3.3.	Wood-Based Materials with Mineral Binders	585
2.3.3.1.	Cement-Bonded Boards	585
2.3.3.2.	Gypsum-Bonded Board	586
2.3.3.3.	Magnesia-Bonded Boards	587
2.4.	Surface Treatment	587
2.4.1.	Veneers	587
2.4.2.	Liquid Paints and Coatings	587
2.4.3.	Coating with Sheets, Short-Cycle Films, and Decorative Laminates	588
2.4.3.1.	Coating with Foils	588
2.4.3.2.	Decorative Films (Short-Cycle Films)	589
2.4.3.3.	Decorative Laminates	590
2.4.4.	Properties, Testing, and Uses	591
2.5.	Environmental and Toxicological Aspects	592
2.5.1.	Wood Dust	592
2.5.2.	Emissions	592
2.5.2.1.	Emissions during Production of Wood-Based Materials	592
2.5.2.2.	Emissions from Finished Materials	593
2.5.3.	Disposal	594
2.6.	Polymer Wood	594
2.7.	Economic Aspects	595
	References	597

Abbreviations:

CPL: continuous pressed laminates
DP: degree of polymerization
DS: defibrator-seconds
F: formaldehyde
G: guaiacylpropane
GDP: guanosine diphosphate
H: hydroxyphenylpropane
HDF: high-density fiberboard
HPL: high-pressure laminate
LSL: laminated strand lumber
LVL: laminated veneer lumber
M: melamine
MDF: medium-density fiberboard
MDI: methylene diphenylene isocyanate
MF: melamine – formaldehyde
ML: middle lamella
MUF: melamine – urea – formaldehyde
MUPF: melamine – urea – phenol – formaldehyde
OSB: oriented structural board
P: primary wall; phenol
PF: phenol – formaldehyde
PMDI: polymeric methylene diphenylene isocyanate
PRF: phenol – resorcinol – formaldehyde
PSL: parallel strand lumber
S: syringyl
S1S: smooth one side
S2S: smooth two sides
U: urea
UF: urea – formaldehyde

1. Morphology and Properties

1.1. Introduction

Not only is wood the most important renewable natural resource in terms of quantity, but because of its composition and versatile properties, its use as a raw material for pulp, structural timber, sawn wood, panels, furniture and other purposes is also economically important. In industrialized countries, wood plays only a minor role as an energy carrier [19]. The total phytomass of the earth is estimated to be 1.24×10^{12} t, of which 80 % is attributed to wood [20]. The potentially utilizable annual wood growth is 1.1×10^{10} t [21].

Data from the Food and Agriculture Organization (FAO) of the UN on forested areas and roundwood removals worldwide are given in Table 1.

The forested area of Germany, for example, covers 10.8×10^6 ha. This is 30.4 % of the total area of Germany, of which 55.1 % is agricultural. Of the forested area, 32.8 % is occupied by spruce, 27.5 % by pine, 14 % by beech, and 8.5 % by oak [23]. According to these statistics, Germany is among the more highly forested countries of the world, but with regard to the forested area per capita, it is one of the less forested, with 0.14 ha per inhabitant.

Finland, in comparison, with 33.8×10^6 ha total land area and 20×10^6 ha forests, has about the same size but nearly twice as much forests as Germany. It has 4 ha forested area per inhabitant, consisting mainly of spruce and pine (40 % each) and birch (16 %). Similarly Sweden, with a total land area of 45×10^6 ha, of which 58 % is forested, has 3.1 ha forests per inhabitant. In the southern

Table 1. Forests and roundwood removals worldwide according to [22]

Area	Forested area (1990)			Roundwood removals (1991), 10^6 m^3				
	10^6 ha	% of land area	ha/inhabitant	Total	Coniferous wood	Deciduous wood	Wood for processing	Wood for combustion
North/Central America	530 744	25	1.24	737	494	243	593	144
South America	898 184	51	3.03	345	61	285	103	242
Africa	535 848	19	0.84	527	16	511	59	468
Europe	140 196	30	0.28	335	224	112	284	51
Former Soviet Union	754 958	34	2.61	355	291	65	274	81
Asia	463 221	18	0.15	1086	189	897	252	834
Pacific Rim	87 700	10	3.29	42	21	22	34	9
Total	3 410 851	26	0.64	3429	1295	2134	1599	1830

Table 2. Total wood balance (wood and wood products) in Germany in 10^6 m^3, according to [24]

Supply	1991	1992	Demands	1991	1992
Wood removal	32.1	27.7	Increase in stored wood	0	0
Waste paper (inland)	22.0	22.9	Export	50.9	51.0
Waste wood	0.3	0.4	Consumption	87.2	88.3
Import	77.3	80.3			
Decrease in stored wood	6.3	8.0			
Total supply	138.0	139.3	Total demands	138.1	139.3

part of Sweden spruce is predominant (51 %), followed by pine (29 %). This ratio is reversed in Northern Sweden with 48 % pine and 39 % spruce.

A quite different distribution of wood species is found in southern and tropical countries, where deciduous woods (hardwoods) often outweigh coniferous woods (softwoods).

Germany is a typical wood-importing country. In 1992, 80.3×10^6 m^3 of wood was imported and 27.7×10^6 m^3 of indigenous wood was cut. These figures take into account the different units used to quantify wood products, e.g., tonnes of pulp are expressed as raw wood equivalents in cubic meters (Table 2).

Finland, in contrast, as a typical wood exporting country, in 1992 had 55×10^6 m^3 roundwood removal from indigenous forests (cut wood) and additional 6×10^6 m^3 from Russia. Out of these 20×10^6 m^3 went into the domestic timber industry and 28×10^6 m^3 (plus 6×10^6 m^3 industrial wood residues) into the pulping industry. Finland exports some 1.5×10^6 t pulp and 6×10^6 t paper and cardboard, mainly to Western Europe, covering ca. 32 % of its national exports by wood products.

Important importing and exporting countries of wood, semifinished wood products, pulp, and paper are listed in Table 3.

The versatile technological properties of wood result from its anatomy and the chemical properties of its three main components, cellulose, lignin, and hemicelluloses, which, in turn, are affected in different ways by wood extractives occurring as minor components.

1.2. Biology of Wood

All woody plants reproduce by seeds and are part of the Spermatophyta. This systematic division is split into the Gymnospermae, to which the softwoods belong, and the Angiospermae, to which the hardwoods belong. The Monocotyledoneae are part of the Angiospermae; some of them (e.g.,

Table 3. Economics of forest products in the most important countries in 1993, according to [25]

	Production, 10^6 $	% of GDP*	Imports, 10^6 $	Exports, 10^6 $	% of trade	Consumption, 10^6 $
World	391 327	2	106 742	99 618	3	398 450
United States	93 189	2	16 873	13 401	3	96 661
China	34 659	2	4 648	1 121	1	38 186
Canada	30 655	6	2 082	19 295	13	13 442
Japan	21 828	1	16 767	1 648	0	36 912
Former Soviet Union	18 443		124	2 061		16 506
Brazil	17 437	5	308	1 995	5	15 751
India	16 543	4	262	17	0	16 788
Germany	14 481	1	9 502	5 751	2	18 232
Indonesia	13 519	6	556	5 158	14	8 917
Sweden	9 792	5	845	7 483	15	3 154
Finland	9 230	10	475	7 411	32	2 294
France	8 415	1	4 962	3 997	2	9 381
Nigeria	5 281	4	77	20	0	5 337
Malaysia	5 126	7	411	4 190	9	1 347
Italy	4 491	1	5 545	2 008	1	8 028
Korea, Rep.	4 097	2	3 126	570	1	6 652
United Kingdom	4 055	0	8 192	1 932	1	10 315
Spain	3 658	1	2 479	1 032	2	5 106

*GDP = Gross domestic product.

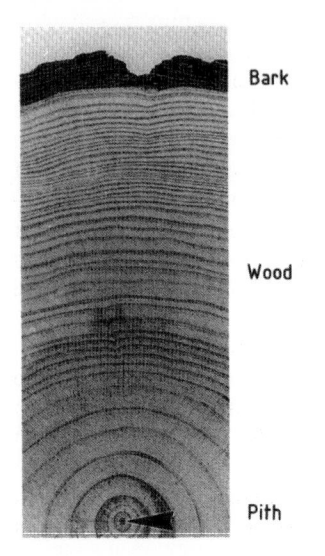

Figure 1. Macroscopic appearance of a *Pinus sylvestris* wood segment with wood and bark

bamboo and rattan) develop woody tissue but are for various reasons maintained as a separate entity.

Wood, except for the Monocotyledoneae, is defined as the inner tissue of stems (see Fig. 1), branches, and roots of perennial plants. It acts as a transport system for water and mineral solutes between the root tips and the parts above ground, serves as storage tissue for reserve material, and provides the plant with mechanical stability. Wood is surrounded by the *cambium* (Fig. 2), which consists of a few layers of meristematic cells, and the bark, which functions both as a transport system for assimilates and as protection against environmental influences. The cambial cells undergo frequent division during the vegetation period, producing wood cell initials in the adaxial direction and bark cells in the abaxial direction. With time, the wood cell initials differentiate and form the taxon-specific arrangement of functionally and morphologically different cell types.

1.2.1. Anatomy [26]

The histological term for wood as the tissue inside the cambium is *xylem*. Its formation is restricted to the annual vegetation period and therefore represents a periodic process. In the northern temperate zone, typical growth rings are laid down with subdivision into *earlywood* and *latewood* that determine the beginning and end of the growing season. A growth ring is strictly correlated with one year, whereas in the tropics and subtropics the growth increments may reflect alternating wet and dry seasons, not necessarily an annual cycle.

Monocotyledoneae do not possess a cambium and therefore lack the secondary thickening characteristic of gymnosperms and most angiosperms. Their woody tissue consists of numerous axially oriented vascular bundles, which are embedded in a parenchymatous ground tissue (Fig. 3). Within the vascular bundles, different

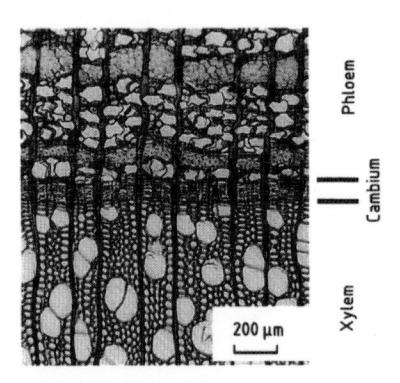

Figure 2. *Populus robusta*: Cross-section through the cambial area, adaxial xylem, and abaxial phloem
Light micrograph (LM)

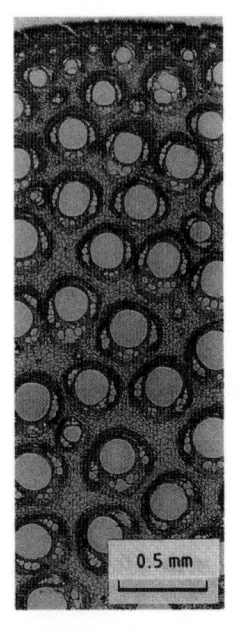

Figure 3. *Calamus caesius*: Cross-section through a rattan stem (LM)

cell types are associated with the main functions of water and assimilate transport, mechanical support, and to a certain extent, storage of reserve material. In the axial direction, the stem is subdivided into long internodes and short nodes.

The following sections describe wood formed by gymnosperms and angiosperms.

1.2.1.1. Cell Types

Wood in its unique structure is composed of vast numbers of cells. According to their morphology and function, four groups of cell types exist: parenchyma cells, tracheids, fibers, and vessels. *Parenchyma cells* in living wood contain cytoplasm (Fig. 4) and are involved in various processes such as metabolic pathways, short-distance transport, and storage; they form horizontal rays as well as axially oriented strands. Axial parenchyma is much more frequent in hardwoods, where it tends to form distinctive patterns, particularly in tropical timber species. *Tracheids* are the predominant axial cell element in softwoods that provide both water conduction and mechanical support. Tracheids also occur in some hardwoods. They are located adjacent to the vessels (vasicentric tracheids), or they resemble vessels (vascular tracheids) with the sole function of water transport. *Fibers* occur exclusively in hardwoods and represent specialized elements for mechanical support. Slight differences in morphology have given rise to a differentiation between "fiber-tracheids" and "libriform fibers." However, since continuous intergrading exists between them, the general term fiber is commonly used. *Vessels* are specialized elements for long-distance transport of water and mineral solutes in hardwoods; they are composed of individual vessel members that are associated to form a continuous channel with a length of 50 – 60 cm in some genera (*Populus, Fagus*) or greater (*Quercus*). During maturation of the vessels the end walls of the individual members abutting on other members above and below are removed partly or entirely by enzymatic action; this process results in a rather different appearance of the so-called vessel perforation plates.

Tracheids and vessels reach functionality only after the death of their cytoplasm. Fibers have commonly been considered dead cells too, but in some species, a few cells retain their cytoplasm over a long period after completion of wall thickening, in the extreme, up to heartwood formation. These fibers are termed living fibers and may store starch.

1.2.1.2. Softwoods

Softwoods (coniferous woods) exhibit a simpler structure than hardwoods (deciduous woods). They are composed mainly of tracheids as the principal cell type in the axial direction (Fig. 5 A). Tracheids of softwoods are extremely elongated elements with a length of 2.0 – 5.0 mm. Cell size, wall thickness, and lumen diameter vary greatly, with thin walls and wide lumina in earlywood tracheids but extreme wall thickening and narrow lumina in the latewood tracheids. The latter contribute mainly to mechanical stability, whereas the earlywood tracheids are conductive rather than supporting elements. Tracheids may also be horizontally oriented, forming lower and upper margins of the rays in some genera of Pinaceae. In some softwoods, axial parenchyma is absent (e.g., *Picea, Pinus*) or sparse (e.g., *Larix*); in others it is of regular occurrence (e.g., *Podocarpus, Juniperus*). Axial and radial resin ducts are characteristic features of some genera of Pinaceae (e.g., *Picea, Larix, Pinus, Pseudotsuga*) and are lined by parenchymatous epithelial cells responsible for the production and posterior secretion of resin into the ducts.

1.2.1.3. Hardwoods

The evolutionary advanced hardwoods possess a functionally and morphologically more diversified structure. The most distinctive feature is the

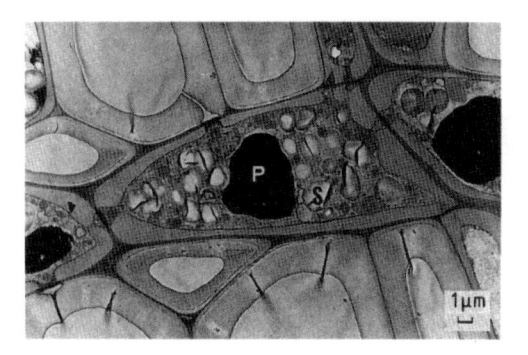

Figure 4. *Betula pendula*: Ray parenchyma cell with numerous starch grains (S); P = Phenolic body
Electron micrograph (EM)

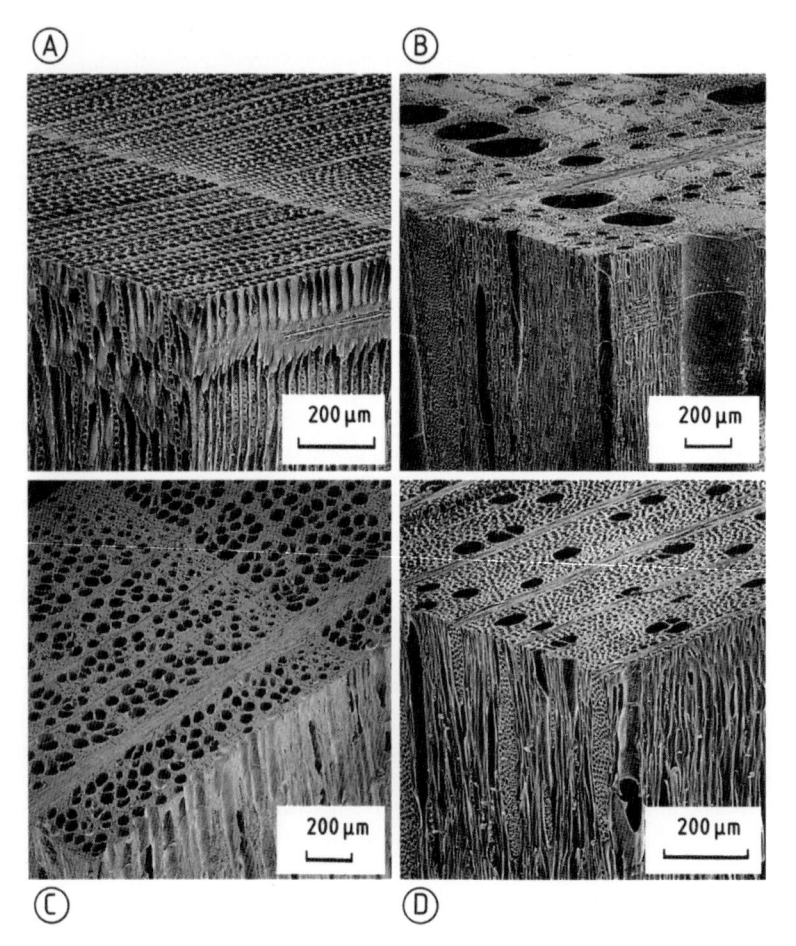

Figure 5. A) *Cedrus libani*: Arrangement of tracheids and parenchymal cells in softwoods (EM); B) *Quercus cerris*: Ring-porous hardwood showing large-diameter earlywood pores and narrow latewood pores (EM); C) *Fagus sylvatica*: Semi-ring-porous hardwood (EM); D) *Acer platanoides*: Diffuse-porous hardwood (EM)

occurrence of vessels, alternatively called "pores" when observed on transverse surfaces. Vessels vary in size and distribution, especially within a discernible growth increment. This diversity is called porosity and represents a very important feature for the identification of hardwoods attributed to three groups: *ring-porous*, *semi-ring-porou,*, and *diffuse-porous*. *Ring-porous woods* (Fig. 5 B) show distinct differences in the size of vessels within one growth increment; earlywood vessels are larger in diameter than latewood vessels, with an abrupt transition between both. Hence, in some cases the earlywood portion becomes clearly visible as a pore-ring even with the unaided eye. *Semi-ring-porous woods* (Fig. 5 C) are characterized by a gradual change from large-diameter earlywood vessels to small latewood vessels. *Diffuse-porous woods* (Fig. 5 D) display vessels of nearly the same diameter throughout one growth increment. The variation within a species, however, may range sometimes from ring porous to semi-ring porous or from semi-ring porous to diffuse porous. Vessels may be arranged in tangential bands, may exhibit a diagonal or radial pattern, or may reveal a branched appearance to form a dendritic pattern. Vessels may also occur solitarily, as groups in radial multiples of four or more, or as irregular clusters.

Axial parenchyma cells are much more abundant in hardwoods than in softwoods (Table 4). The amount and various distributional patterns display a marked diversity. The position of axial parenchyma cells in relation to vessels is used for

Table 4. Percentage of cell-type volume in some wood species

Species	Vessels	Fibers	Ray parenchyma	Axial parenchyma
Douglas fir (*Pseudotsuga menziesii*)		93	7	
European spruce (*Picea abies*)		95	5	
Pine (*Pinus sylvestris*)		93	6	
Aspen (*Populus tremula*)	26	61	13	
Birch (*Betula alba*)	25	65	8	2
Oak (*Quercus robur*)	8 – 40	44 – 58	16 – 29	sparse – 5
Ash (*Fraxinus excelsior*)	12	62	15	11
Beech (*Fagus sylvatica*)	39	40	16	5
Framiré (*Abbrevinalia ivorensis*)	20	53	17	10
Kosipo (*Entandrophragma candollei*)	12	45	25	18
Mahogany (African) (*Khaya ivorensis*)	18	57	21	4
Dark red meranti (*Shorea macrophylla*)	29	41	12	17
Teak (*Tectona grandis*)	12	66	16	6

a general subdivision into two types of parenchyma. The term *apotracheal parenchyma* reflects the isolation of vessels against axial parenchyma by other cells, whereas *paratracheal parenchyma* describes the association of axial parenchyma strands with vessels. Rays in hardwoods may be large or small in width and height, or both types may occur within one species.

1.2.2. Secondary Changes

With increasing age and secondary thickening of a woody plant the more centrally located xylem of many species undergoes natural changes. Consequently, the xylem must be subdivided into two parts. In a living tree the xylem portion close to the cambium serves as a pathway mainly for water and mineral solutes, but it also has a storage function; its parenchyma cells are physiologically active and involved in various metabolic processes. The more centrally located wood portion in mature trees, however, does not contain living parenchyma. In general, the terms *sapwood* for the outer, and *heartwood* for the central portions are used. Transition from sapwood to heartwood usually occurs within a narrow zone of a few cell rows, resulting in a slightly undulating sapwood – heartwood boundary. In most cases, the heartwood has a significantly darker color than the surrounding sapwood (Fig. 6), with certain variation of the width of the sapwood between and within species. Besides the obligatory formation of such a distinctly colored heartwood, some tree species develop less colored or nearly colorless heartwood; all of those species are therefore grouped into regular heartwoods.

Other tree species such as *Betula* or *Fraxinus* are not able to form heartwood under unaffected growth conditions. However, all woody plants develop *discolored wood* as a result of environmental influences (e.g., injury, attack by microorganisms, severe frost). The boundary between unaffected and discolored wood then often moves across a larger number of growth increments with an irregular periphery. Discolorations may also be induced in regular heartwood as a secondary process [27]. The misleading terms "false heartwood," "pathological heartwood,"

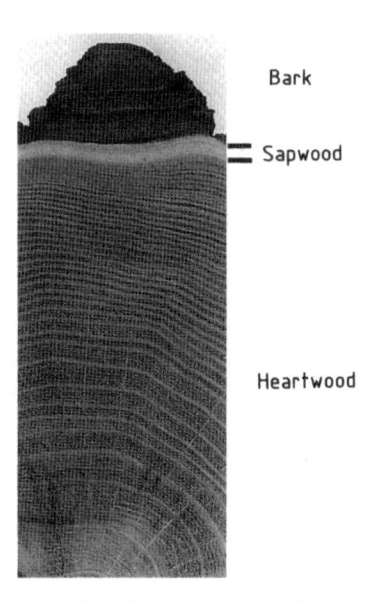

Figure 6. *Robinia pseudoacacia*: Division of the wood into light sapwood and dark heartwood

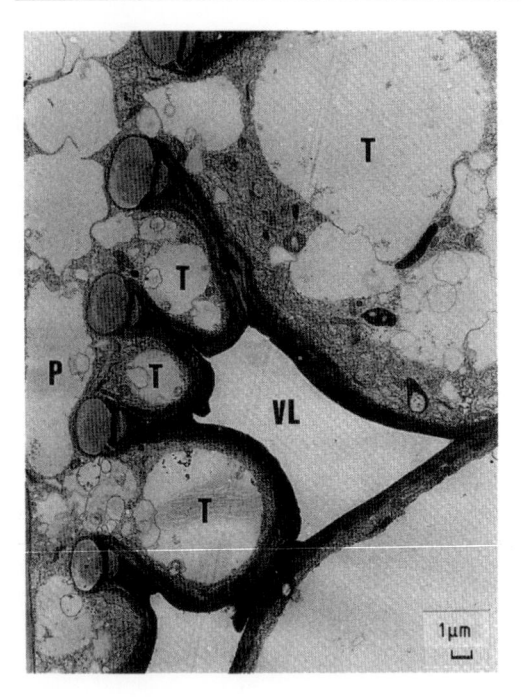

Figure 7. *Quercus robur*: Tylosis development – tylosis initials protrude from a parenchyma cell into a vessel (EM) T = Tylosis; P = Parenchyma cell; VL = Vessel lumen

and "frost heartwood" are often used for these environmentally initiated changes. The main criteria for the differentiation between heartwood and discolored wood are as follows: heartwood is characterized by increased natural durability and lower moisture content in comparison to sapwood; in discolored wood, however, the durability does not increase, and the moisture content may rise even above the sapwood level.

The change from sapwood to heartwood during natural heartwood formation in hardwoods is frequently accompanied by the occlusion of vessels. The lumina of some species are more or less filled with often dark-staining material secreted from adjacent parenchyma cells. Other species develop *tyloses*, representing balloon-like protrusions from parenchyma cells through the pits into adjacent vessels (Fig. 7).

1.2.3. Cell Wall Formation and Architecture

Walls typically envelop the cytoplasm of all plant cells to impart protection and firmness. Stabilization of a multicellular plant body is sustained by the continuum of adhering walls of all cells. The cell walls consist primarily of an amorphous polysaccharide matrix of pectins and hemicelluloses as well as crystalline cellulose – the basic structural polysaccharide, which is aggregated in the form of microfibrils. Nonpolysaccharide components, of which the most important is lignin, are finally incorporated. This combination of different chemical constituents determines the physical properties of cell walls. Details on the chemical composition of cell walls are given in Section 1.3.

Cell wall formation starts with the division of a cambial initial. A daughter cell becomes separated first at the equatorial plane by a thin, tangential wall, which consists mainly of pectic material. This initial layer, common to both adjacent cells, is termed the *middle lamella* (ML) and functions as a cementing substance between cells. At this early developmental stage the young wood cell begins to deposit new wall material against the ML. In contrast to the ML, this new layer contains cellulosic material embedded in a pectin- and hemicullose-containing matrix. This layer represents the first-formed cell wall layer and is called the *primary wall* (P). Here, the cellulose microfibrils show a dispersed texture. Both middle lamella and primary wall are extensible to allow cell expansion during differentiation. Cellulose fibrils and matrix material are continuously incorporated into the growing wall. As the developing wood cell approaches final size, deposition of the stiff secondary wall occurs. Secondary walls in most cases show a subdivision into three layers (S 1, S 2, S 3), with the narrow S 1 as the oldest layer next to the primary wall, the S 2 layer mostly representing the thickest portion, and the sometimes barely discernible or thin S 3 layer forming the interface with the cytoplasm in living cells or the cell lumen in dead cells. The main constituents of the developing secondary wall are hemicelluloses and, in a distinctly higher proportion than in the primary wall, cellulose. The microfibrils of all secondary wall layers are laid down in parallel arrays: in the S 1, the mean angle with the longitudinal cell axis is $50 - 75°$; microfibrils of the S 2 appear more closely packed, with an orientation of $10 - 30°$ to the cell axis; in the S 3 layer, microfibrils exhibit a mean angle of $60 - 80°$ with the cell axis (Fig. 8). Sculpturing

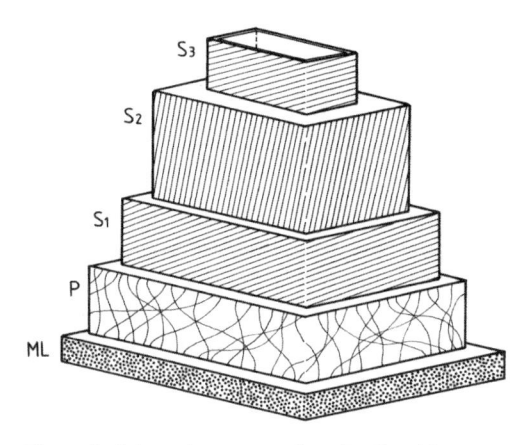

Figure 8. Schematic representation of wall architecture

on the lumen surface of the S 3 layer is sometimes observed (e.g., the "warty layer" in some softwoods and hardwoods). Incrustation with lignin completes wall development.

The formation of microfibrils during wall thickening takes place at the interface between plasma membrane and cell wall. Fine structural investigations on differentiating plant cells performed in the 1970s and 1980s revealed specific structures within the bilayered plasma membrane: terminal globules in the outer leaflet [28] and rosettes in the inner leaflet [29]. Both terminal globules and rosettes are probably part of the cellulose-synthesizing apparatus. The position of the terminal globules at ending imprints of microfibrils in the membrane implies that they may represent sites of cellulose synthases. The membrane rosettes, in most cases a complex of six subunits, are also believed to be structures with cellulose synthase activity; their frequent occurrence in certain phases and sites of wall synthesis may reflect involvement in microfibril synthesis. Future research must clarify the details of this putative structure – function relationship. Besides the obviously membrane-bound synthesis of cellulose microfibrils, their orientation also appears dependent on membrane-associated structures, the membrane-underlying microtubules. These cortical microtubules may act as guidelines for the synthase complexes within the plasma membrane during microfibril elongation.

At the onset of secondary wall thickening, the deposition of wall material is interdicted along some primary wall areas. These develop into *pits* that represent specific pathways for the conduction of water and mineral solutes. Pits therefore appear as well-defined areas connecting neighboring cells through a thin membrane composed of a middle lamella and a primary wall layer on each side. With rare exceptions, a conduit between two cells consists of a pair of pits. *Bordered pits* show the secondary wall overarching the membrane on both sides, thus reducing the pit aperture. This is true both for pits between the tracheids in softwoods (Fig. 9 A) and for those connecting hardwood tracheids, fibers, and vessels with each other. *Half-bordered pits* generally occur between those cell types and a parenchyma cell with the unbordered part at the parenchyma side (Fig. 9 B). *Simple pits* are without border and connect two adjacent parenchyma cells (Fig. 9 C). Direct cytoplasm contact between parenchyma cells is maintained by *plasmodesmata*, small strands of cytoplasm that extend from cell to cell through perforations in the pit membrane.

1.3. Chemistry of Wood

1.3.1. Chemical Components of Wood

The main components of the wood cell wall (wood substance) are cellulose ($\rightarrow$ Cellulose), lignin ($\rightarrow$ Lignin), and hemicelluloses. The proportion of cellulose in the most common European woods is 41 – 43 %, whereas the lignin content in coniferous woods (27 – 28 %) is higher than that in deciduous woods (22 – 24 %). Correspondingly, hemicelluloses occur in higher concentrations in deciduous woods (30 – 35 %) than in coniferous woods (25 – 30 %). Compression wood contains more lignin (ca. 40 %), and tension wood (e.g., poplar) more cellulose (> 50 %). Differences are also found in tropical or subtropical types of wood. Within one wood species, age, location, and growth conditions can also play a role.

Most European woods also contain 1 – 3 % extractives and 0.1 – 0.5 % mineral components, which can be determined as ash. Higher contents of extractives have been found in certain types of wood (up to 10 % resins in pines, and tannins in quebracho and eucalyptus); in others, larger quantities of inorganic salts are found. The chemical compositions of some important types of

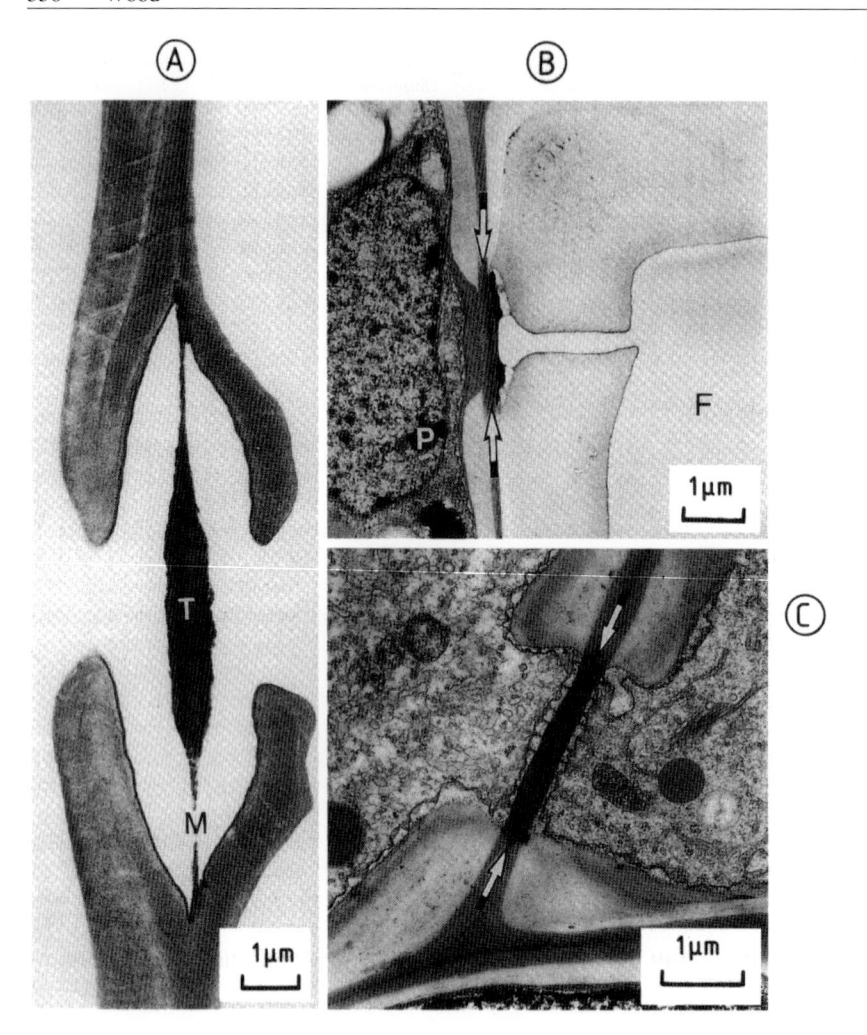

Figure 9. Electron micrographs of different types of pits
A) *Picea abies*: Bordered pit between tracheids; the pit membrane is divided into the central torus (T) and the peripheral margo (M); B) *Betula pendula*: Half-bordered pit between a parenchyma cell (P) and a fiber (F); arrows indicate the pit membrane; C) *Quercus robur*: Simple pit between parenchyma cells; arrows indicate the pit membrane.

wood are given in Table 5. Due to the lack of analytical methods that allow a clean separation of the wood components, the percentages given in Table 5 suffer from different isolation conditions used by different authors.

The *extractives* consist of a large number of low molecular mass compounds, which can be extracted from wood with organic solvents (terpenes, fats, waxes, and phenols) or hot water (tannins and inorganic salts). Depending on the quantity and the class of compound, which are specific to the tree type and genus to some extent, the extractives affect the chemical, biological, physical, and optical properties of the wood to varying degrees. As the swelling and biodegradability may be significantly reduced by lipophilic resins they improve the dimensional stability and durability of wood, while other physical and chemical properties, e.g., during mechanical and chemical pulping, sawing, or painting, are negatively affected.

Coniferous woods contain mainly *terpenes*, which can be obtained from pines by resinification of the living tree ($\rightarrow$ Resins, Natural, Section 5.2.1.) or as turpentine (monoterpenes) ($\rightarrow$ Turpentines) and tall oil (diterpenes, resin acids) from kraft (sulfate) pulping ($\rightarrow$ Tall Oil). *Long-chain fatty acids* are also present in tall oil and arise from fats and waxes in the wood. In other

Table 5. Chemical comparison of various wood species (% of dry wood weight) [30]

Species	Common name	Extractives*	Lignin	Cellulose	Glucomannan	Glucuronoxylan	Other polysaccharides
Softwoods							
Abies balsamea	balsam fir	2.7	29.1	38.8	17.4	8.4	2.7
Pseudotsuga menziesii	Douglas fir	5.3	29.3	38.8	17.5	5.4	3.4
Tsuga canadensis	eastern hemlock	3.4	30.5	37.7	18.5	6.5	2.9
Juniperus communis	common juniper	3.2	32.1	33.0	16.4	10.7	3.2
Pinus radiata	Monterey pine	1.8	27.2	37.4	20.4	8.5	4.3
Pinus sylvestris	Scots pine	3.5	27.7	40.0	16.0	8.9	3.6
Picea abies	Norway spruce	1.7	27.4	41.7	16.3	8.6	3.4
Picea glauca	white spruce	2.1	27.5	39.5	17.2	10.4	3.0
Larix sibirica	Siberian larch	1.8	26.8	41.4	14.1	6.8	8.7
Hardwoods							
Acer rubrum	red maple	3.2	25.4	42.0	3.1	22.1	3.7
Acer saccharum	sugar maple	2.5	25.2	40.7	3.7	23.6	3.5
Fagus sylvatica	common beech	1.2	24.8	39.4	1.3	27.8	4.2
Betula verrucosa	silver birch	3.2	22.0	41.0	2.3	27.5	2.6
Betula papyriferea	paper birch	2.6	21.4	39.4	1.4	29.7	3.4
Alnus incana	gray alder	4.6	24.8	38.3	2.8	25.8	2.3
Eucalyptus camaldulensis	river red gum	2.8	31.3	45.0	3.1	14.1	2.0
Eucalyptus globulus	blue gum	1.3	21.9	51.3	1.4	19.9	3.9
Gmelina arborea	yemane	4.6	26.1	47.3	3.2	15.4	2.5
Acacia mollissima	black wattle	1.8	20.8	42.9	2.6	28.2	2.8
Ochroma lagopus	balsa	2.0	21.5	47.7	3.0	21.7	2.9

*Extraction with CH_2Cl_2, followed by C_2H_5OH.

types of wood (oak, chestnut, eucalyptus, and quebracho), *tannins* and *flavonoids* predominate. *Phenolic compounds* (lignans and stilbenes) are found mainly in coniferous wood (spruce and pine), whereas *sterols* (mainly β-sitosterol) are found in smaller quantities in almost all types of wood. The *mineral components* are mainly carbonates or glucuronates of calcium (40 – 70 %), potassium (10 – 30 %), magnesium (5 – 10 %), iron (up to 10 %), and sodium. Other metals, such as manganese and aluminum, are also present as cations in smaller quantities, which are often dependent on the location. Oxalate, phosphate, and silicate anions can also be found.

Lignin is a cross-linked macromolecule made up of phenylpropane units (C_9), which are linked together by at least ten different C−C and C−O bonds. The irregular structure of lignin arises from its biosynthesis, in which the last step is a nonenzymatic, random recombination of phenoxy radicals of coniferyl, sinapinyl, and *p*-coumaryl alcohols. The dimeric and oligomeric intermediates, obtained in vitro in the dehydrogenative polymerization of coniferyl alcohol, formed the basis of the constitutional scheme of spruce lignin proposed by FREUDENBERG (→ Lignin) [31]. FREUDENBERG'S hypothesis of lignin biosynthesis was later confirmed for beech lignin by degradation with thioacetic acid and ^{13}C-NMR spectroscopy [32]. Variations in the chemical reactivity of lignin arise from the proportions of the three structural units. Whereas coniferous lignin consists mainly of guaiacylpropane (4-hydroxy-3-methoxyphenylpropane) units (G), deciduous wood lignins also contain up to 50 % syringyl (3,5-dimethoxy-4-hydroxyphenyl) groups (S), and compressive wood lignins up to 40 % *p*-hydroxyphenylpropane (H) units. Up to 30 % of the latter also occur in grass lignins.

On an industrial scale, lignin and hemicelluloses are removed from wood by pulping (→ Pulp, Chap. 3.) under partial degradation and isomerization. Lignin can be removed from wood more selectively under laboratory conditions by using chlorine, chlorine dioxide, or peracetic acid. From the residue known as holocellulose, hemicelluloses can be extracted either with 17.5 % sodium hydroxide – with α-cellulose as extraction residue – or with dimethyl sulfoxide.

Hemicelluloses are polysaccharides and differ from cellulose in that they consist of several sugar moieties, are mostly branched, and have lower molecular masses with a degree of polymerization (DP) of 50 – 200. The two main types of hemicelluloses are xylans and glucomannans. Structural and quantitative differences exist between hemicelluloses in deciduous and coniferous woods. *Deciduous woods* contain 10 – 35 % *O*-acetyl-4-*O*-methylglucuronoxylan and 3 – 5 % glucomannan, whereas *coniferous woods* contain 10 – 15 % arabino-4-*O*-methylglucuronoxylan and 15 – 20 % *O*-acetylgalactoglucomannan [33]. The upper part of Figure 10 shows a section of the structure of a coniferous wood galactoglucomannan. The molecule chain ("backbone") consists of β-(1,4)-glycosidically bonded glucose and mannose units in the ratio 1:3, with 0.1 – 1 α-(1,6)-bonded galactose moiety and 0.24 acetyl group per glucose unit as side groups. In the lower part of Figure 10, a deciduous wood xylan is depicted with β-(1,4)-glycosidically bonded xylose units in the chain, 0.1 α-(1,2)-4-*O*-methylglucuronic acid moiety, and 0.7 *O*-acetyl group per xylose unit as side groups.

Deciduous wood glucomannans contain only ca. 1.5 mannose units and no side groups per glucose unit; coniferous wood xylans have 0.2 α-(1,2)-4-*O*-methylglucuronic acid and 0.13α-(1,3)-L-arabinofuranose groups per xylose unit. Larch wood contains 10 – 25 % arabinogalactan, which consists of β-(1,3)-bonded galactose moieties in the main chain, with β-(1,6)-bonded short side chains of D-galactose and L-arabinose units. This compound is usually classified as a pectin. Pectins ($\rightarrow$ Polysaccharides) occur in small quantities in the primary walls of all wood cells.

The biogenesis of D-glucuronic acid, D-galacturonic acid, D-mannose, D-galactose, D-xylose, and L-arabinose starts from guanosine diphosphate(GDP) – glucose and involves epimerization, dehydrogenation, and decarboxylation.

Cellulose. In terms of quantity, cellulose is the most important wood component. It consists of unbranched long chains of β-(1,4)-glycosidically bonded D-glucose units, which exist as pyranose rings in the $^4C_1$ conformation with equatorial OH and CH_2OH groups. The DP of native cellulose is between 10 000 and 14 000. In industrial pulping however, cellulose is degraded to DP values of 1000 – 3000. Annually ca. 130×10^6 t of pulp is used for paper production worldwide, and only ca. 5×10^6 t for the production of viscose and cellulose esters (acetate, nitrate, and mixed esters) and ethers (methyl-, carboxymethyl-, and hydroxyethylcellulose). The byproducts lignin and hemicelluloses present in the waste liquor are incinerated to recover the pulping chemicals. They pollute the wastewater from pulping factories to a great degree.

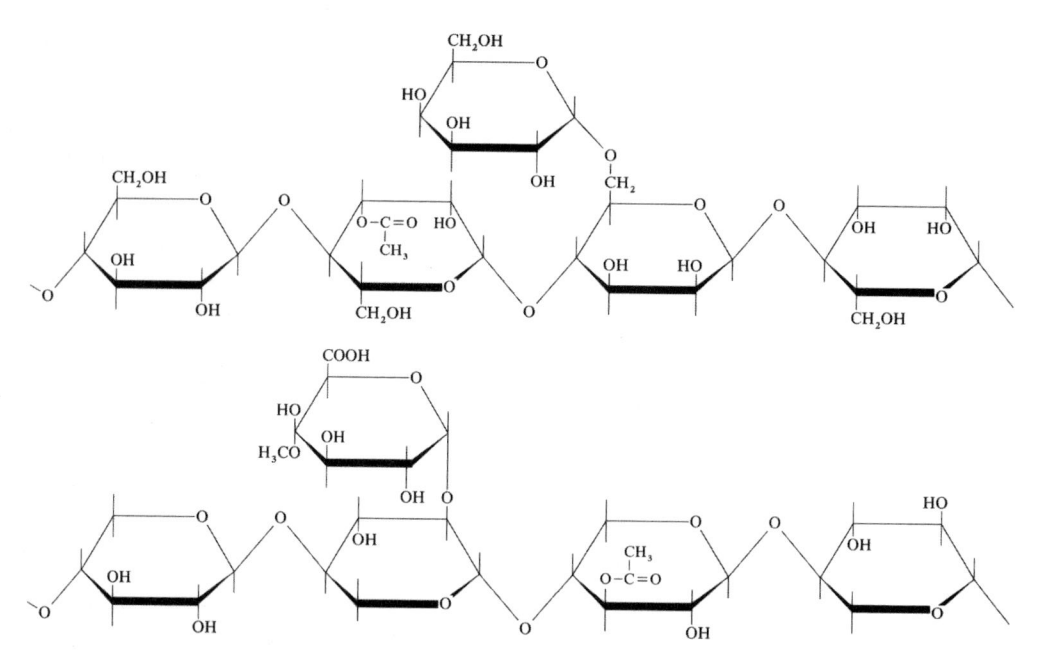

Figure 10. Sections of coniferous wood galactoglucomannan molecules (above) and deciduous wood 4-*O*-methylglucuronoxylan molecules (below)

1.3.2. Distribution of Components in the Cell Wall

The distribution of cellulose, lignin, and hemicelluloses varies among the individual layers of the wood cell walls. Lignin is formed at a distance of ca. 10 cell layers from the cambium only after formation of the cells themselves. It is deposited in the spaces between the cellulose fibrils in the secondary wall and between the cells in the middle lamella, so that its concentration in the middle lamella and in the corner sections is highest (70 – 90 %), while only 20 % is found in the secondary wall. Because of the larger volume of the secondary wall, the main quantity of lignin (63 – 74 %) is found in this layer, which also contains ca. 90 % of the cellulose and 70 – 80 % of the hemicelluloses, with the concentration decreasing toward the lumen. The lignin distribution in the wood cell wall can be determined from the absorption of UV light (240 nm) by very thin (0.1 μm) microtome sections under the microscope [34] (Fig. 11).

The proportions of lignin and tissue in the secondary wall, and the narrow and corner sections of the middle lamella in early and late spruce tracheids, are shown in Table 6. More lignin is found in the narrow and corner sections of the middle lamella in earlywood (28 %) than in latewood (19 %).

1.3.3. Macromolecular Structure of Cell Wall Components

Of the three cell wall components, only cellulose has a partially (ca. 70 %) ordered crystalline structure, whereas the hemicelluloses and lignin are amorphous. Because of regular hydrogen

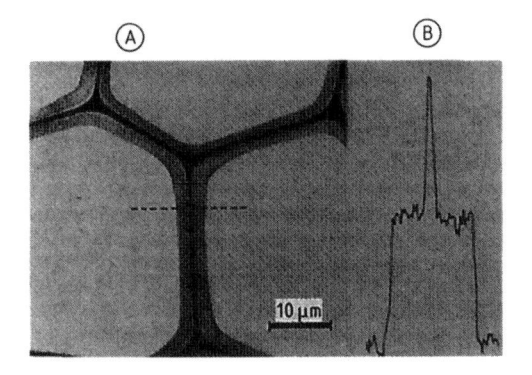

Figure 11. A) UV micrograph of microtome cross-section of an earlywood tracheid of black spruce (*Picea mariana*); B) Densitometric lignin distribution curve along the dotted line [34].

bonds, the cellulose chains are ordered to form elemental fibrils with a diameter of ca. 3.5 nm and alternating crystalline (micelles) and amorphous regions. These, in turn, are associated to form microfibrils, which are ca. 10 times thicker. The microfibrils form layers (Fig. 8) between which the lignin and the hemicelluloses are deposited. Chemical bonds are formed between the α-C atoms of lignin and the hydroxyl groups of the hemicelluloses (benzyl ethers), which prevent the leaching of lignin from the wood by neutral organic solvents. Irregular hydrogen bonds also exist between the hydroxyl groups of lignin, the hemicelluloses, and cellulose, through which a close, but irregular network of the three components is formed.

The cellulose microfibrils are first formed biogenetically as the framework of the cell wall, whose growth is finished before lignification begins. Lignification occurs in the secondary wall by filling the spaces between the cellulose microfibrils, whereby monomeric

Table 6. Distribution of lignin in black spruce tracheids (*Picea mariana*) [35]

Wood	Morphological differentiation	Relative absorbance	Tissue volume fraction, %	Lignin	
				% of tissue	% of total
Earlywood	secondary wall	1.0	87	22	72
	narrow sections of middle lamella	2.2*	9	50	16
	corner sections of middle lamella	3.8	4	85	12
Latewood	secondary wall	1.0	94	22	81
	narrow sections of middle lamella	2.7*	4	60	10
	corner sections of middle lamella	4.5	2	100	9

*Measured for the compound middle lamella. Measurements on oblique sections suggest absorbance values of 3.8 and 2.0 for the true middle lamella and primary wall sections, respectively.

phenoxy radicals add to already formed polymers (endwise polymerization).

1.4. Physical Properties

Wood is hygroscopic; i.e., it exchanges water vapor with its environment. In a uniform climate, the moisture in wood is in equilibrium with this climate. The *moisture content of wood* is defined as the mass of moisture contained in the wood relative to the mass of the absolutely dry wood substance and is given as a percentage. Figure 12 A shows the relation of the equilibrium moisture content to the relative humidity of the air at 20 °C. This sorption isotherm is approximately valid for most wood species. Only species that contain hydrophobic components (e.g., teak, afzelia, and movingui) have significantly lower sorption isotherms.

The following moisture levels are recommended for the use of wood:

Furniture, interior decoration	6 – 10 %
Parquet	7 – 11 %
Windows, outer doors	10 – 15 %
Covered, open buildings	12 – 18 %

Adherence to these standard moisture levels during the final working and incorporation of wood reduces later variations in moisture levels and dimensions to a minimum.

In the hygroscopic region (i.e., below fiber saturation; FS in Fig. 12), the dimensions of wood increase (swelling) with increasing moisture content and decrease (shrinkage) with a corresponding decrease in moisture content. Figure 12 B shows the variation with moisture levels in the *degree of swelling* of spruce in the three main directions. Wood can also store liquid water in the pore spaces above the FS level, but no further swelling takes place. Because of the usual surrounding climates, the range of wood moisture content is particularly important between 5 and 20 %. In this range, wood swells and shrinks by ca. 0.2 % in the radial and 0.3 % in the tangential direction on 1 % change in moisture content.

The density of the pure cell wall substance is close to 1.5 g/cm^3 for all wood species. However, the *density* of a piece of timber (mass of wood relative to volume including pore spaces) covers a wide range from ca. 0.17 (balsa) to 1.2 g/cm^3 (pock wood). This range results from the different proportion by volume occupied by the cell wall in a particular species – in balsa ca. 11 %, in pock wood ca. 80 %. The remaining volume is pore space, which is filled with air in

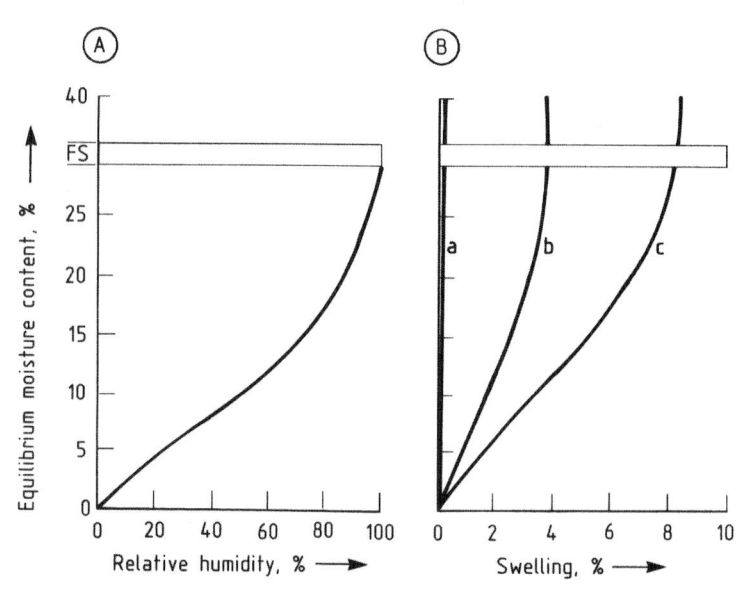

Figure 12. A) Sorption isotherm at 20 °C; B) Relation between degree of swelling and equilibrium moisture content in the three main directions of spruce
a) Longitudinal; b) Radial; c) Tangential

dry wood and water in wet wood. Density values quoted refer mostly to standard climate conditions (20 °C and 65 % relative humidity), i.e., a wood moisture level of ca. 12 %, as do the other moisture-dependent properties of wood. Since the density is a measure of the proportion of cell wall in the total wood volume, the following parameters tend to increase with increasing density: the modulus of elasticity and strength, abrasion resistance and hardness, and thermal conductivity.

Because of the structure of wood, all the properties mentioned, such as swelling and shrinkage, are anisotropic (i.e., direction dependent). The advantage of wood as a building material lies in the combination of high *strength* in the fiber direction and low thermal conductivity perpendicular to it. Wood is therefore particularly suitable for load-bearing and heat-insulating roof and wall components. Table 7 lists the average densities, moduli of elasticity, and strengths of a selection of wood species. It shows that for wood species in general, the compressive strength in the fiber direction is approximately half the value of the tensile strength. However, as the direction of the load deviates from that of the fibers, the tensile strength decreases more sharply than the compressive strength. Wood should therefore not be subjected to tension perpendicular to the fiber direction (cleavage cracks due to shrinkage). However, application of compression perpendicular to the fiber direction is possible and frequently leads only to densification, not breakage.

In each case, the elastic properties and strength in the hygroscopic region decrease with increasing moisture content. However, above fiber saturation (use of wood in marine construction work), these properties are no longer altered by the moisture level in the wood.

For wooden floors, *abrasion resistance* and hardness play an important role. The relation between *Brinell hardness* H_B and density ϱ_N is given by the equation

$$H_B = 76 \cdot \varrho_N - 24.5 \, (N/mm^2)$$

for the important density range of 0.5 – 1.0 g/cm³.

The following are values for the *thermal conductivity* perpendicular to the fiber direction:

Spruce, pine, and fir	0.13 W m⁻¹ K⁻¹
Beech and oak	0.20 W m⁻¹ K⁻¹
Azobe and greenheart	0.28 W m⁻¹ K⁻¹

Table 7. Average values for certain properties of a selection of wood species deabbrevined for small, defect-free test pieces at 12 % moisture content[*]

Wood species	Density, g/cm³	Modulus of elasticity, N/mm²	Strength, N/mm²		
			Tension	Bending	Compression
Softwoods					
Western red cedar	0.37	8 000	60	54	35
Spruce	0.47	10 000	80	68	40
Pine	0.52	11 000	100	80	45
Larch	0.59	12 000	105	93	48
Hardwoods					
Balsa	0.17	3 500	24	24	10
Abachi	0.40	6 800	79	60	40
Poplar	0.42	8 300	77	70	36
American Mahogany	0.54	9 500	100	80	45
Sipo Mahogany	0.59	11 000	110	100	58
Iroko	0.63	13 000	79	95	55
Oak	0.67	13 000	110	95	52
Beech	0.69	14 000	135	120	60
Dark red meranti	0.71	14 000	130	110	58
Angelique	0.76	14 000	130	120	70
Afzelia	0.79	13 500	120	115	70
Balau	0.94	20 000		142	76
Greenheart	1.00	22 000	220	180	100
Azobe	1.06	17 000	180	180	95

[*] The flexural strength is valid for loads perpendicular to the fiber direction, the tensile and compressive strengths for loads in the fiber direction of the wood.

2. Wood-Based Materials

2.1. Introduction

Wood has a number of favorable properties and has been a valuable material from time immemorial. It is relatively easy to obtain and can be worked readily. Wood has high strength at relatively low weight and has a decorative effect because of its color and graining. When used properly it often remains functional for hundreds of years even under adverse climatic conditions (e.g., outdoors).

Unfortunately, wood also possesses some less favorable properties. Since the arrangement of the elongated cells that form the bulk of the wood is predominantly parallel to the trunk (see Fig. 5), wood properties are anisotropic. Thus, the strength in the direction of the trunk is much greater than that perpendicular to it. Wood is also a hygroscopic material. Its moisture content is determined by the surrounding climate (see Section 1.4). Variations in the moisture level of wood in the hygroscopic region give rise to variations in its dimensions. These variations, in turn, differ depending on whether the anatomical direction is longitudinal, tangential, or radial. This phenomenon is known as warping. Wood is a typical natural material and is therefore inhomogeneous. It has growth irregularities and faults, such as knots, as a result of which the yield of usable material is often low. The lengths and particularly the widths available are limited, thus making the use of wood for certain applications difficult or impossible.

The primary aim of the development of wood-based construction materials is to eliminate or reduce the undesirable properties as much as possible. At the same time, attempts are made to retain the favorable characteristics and make them predominate.

Conversion of solid wood into wood-based materials always follows the same principle. Cordwood is cut into small pieces that are combined with a suitable binder (Fig. 13). Cutting into smaller pieces involves sawing, peeling, chopping, or defibration. Boards, veneers, particles of varying dimensions, and fibers are thus produced. As the size of the piece decreases, the properties of the wood-based panel or construction material differ more and more from those of wood. The wood-based material becomes more homogeneous, and the possibilities for size enlargement in two or three dimensions and molding increase. However, at a comparable density,

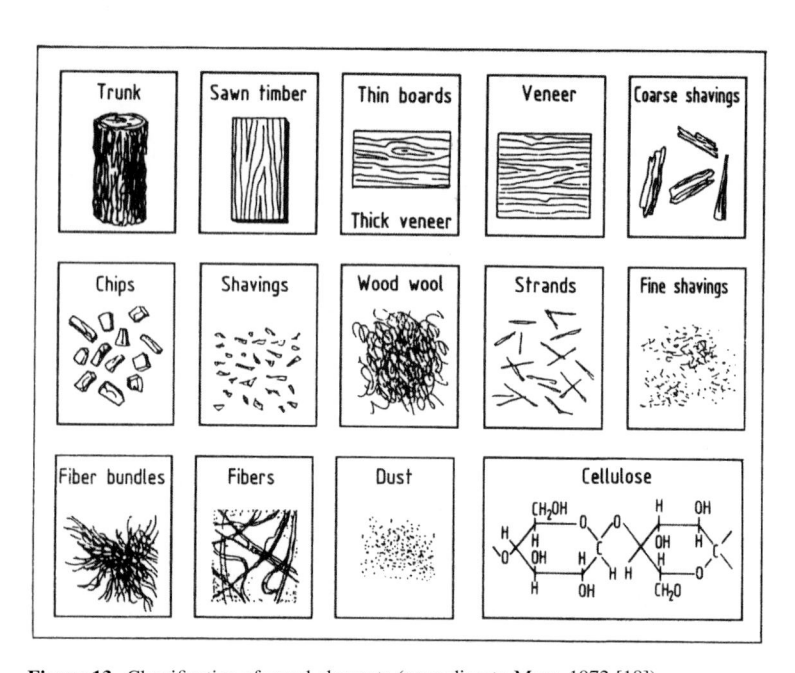

Figure 13. Classification of wood elements (according to MARA 1972 [18])

the strength decreases and the typical wood character is lost. This is exemplified by the changing properties in the series sawn timber, plywood, particle board, and medium-density fiberboard (MDF).

2.2. Laminate Bonding

2.2.1. Laminated Structural Timber (Glulam)

2.2.1.1. Development

The art of building with wood is many centuries old. Unfortunately, the quantity of trunk wood of sufficient dimensions available for making load-bearing building components with the spans desired is continuously decreasing. This consideration gave the German carpenter OTTO HETZER the idea in 1906 of using a completely new method. He cut the cordwood into boards and glued the graded and dried lamellar boards together with adhesive to form new cross-sections (glulam girders, i.e., glued laminated girders, or laminated beams).

The actual breakthrough of the process for using glulam girders in building occurred in 1930 with the development of synthetic adhesive resins. Large-scale industrial production opened up new areas of application for this method. Besides being used in tall buildings, glulam girders were used in airplanes (Mosquitoes in England and the Spruce Goose in the United States), sometimes with large wing spreads, and for mine casings because of their antimagnetic properties. Glulam girders are now used mainly for building in central and northern European countries and North America, where wood has traditionally been used widely for this purpose.

Glulam girders were traditionally produced in small, handcrafting workshops. These have been replaced in recent decades by efficient industrial plants, which use machine systems and automatic production lines and can offer quality assurance.

2.2.1.2. Construction and Production

Raw Materials. Laminated wood is made mostly from spruce, which is the easiest wood to process and fulfills the usual requirements concerning strength and durability.

Pine, larch, and Douglas fir are also used in special cases. Hardwoods are generally used only for building components that must support very heavy loads. For some hardwoods, the effect of wood components on the durability of adhesive compounds is not entirely clear.

Adhesives. In a normal indoor climate and for building components protected from weathering, glues based on urea resins are used. For building components that are exposed to strong climatic variations or weathering, weather-resistant adhesives based on phenol – resorcinol – formaldehyde (PRF) are employed. These PRF resins give a dark brown glue line, unlike adhesives based on amino resins. In the 1990s a special glue type for bonding laminated was developed based on using melamine – resorcinol – formaldehyde resins. These bondings are highly resistant to boiling water and give a pale glue line.

Production. Lamellar boards are sawed from cordwood trunks. The boards are predried to ca. 10 % residual moisture, graded, and then bonded in automatic plants to form continuous boards with dovetail joints. The moisture in the boards to be glued should be 2 – 3 % below the expected equilibrium moisture level. Variations in the moisture content of the different boards should not be greater than 2 %. The broad sides of these joined boards are planed until smooth and uniformly coated with adhesive. In constructing the cross-sections, particular attention is paid to the positions of the annual rings in the boards (Fig. 14). The rifts (vertical) and the flats (horizontal) should be distributed regularly.

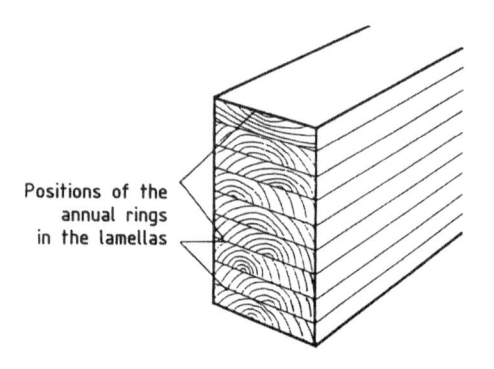

Figure 14. Lamellar construction of a glulam girder cross-section

Depending on how exactly the surfaces fit together, 400 – 600 g of adhesive liquor is applied per square meter. Glued boards are stacked and stored under pressure at room temperature or slightly higher (40 °C) so that the adhesive resin cures after a few hours and a composite beam is formed.

The geometry of the beams and their cross-sections can be varied widely and adapted to requirements of static calculations and aesthetics. Lamination allows bent beams with differing profile shapes and spans of 60 – 100 m to be produced.

Besides individually produced building components, the continuous production of laminated wood as a band was recommended [36]. Such standardization with regard to dimensions is expected to improve the economics of this process.

2.2.1.3. Properties

Laminated wood is clearly superior to squared logs in its performance. By eliminating the natural wood faults through careful sorting of the boards, a more homogeneous construction material is formed with regard to appearance and strength.

The behavior of wooden girders in the event of fire is especially favorable. Through the formation of a charcoal layer on the surface, the center of the girder is protected from further rapid attack by the fire and is thus insulated. Therefore, the desired extent of fire resistance can be determined only by the corresponding dimensioning of the load-bearing beam.

Unlike steel girders, glulam girders have high rigidity. Treatment with preservatives is recommended as a surface protection for building components exposed to weathering to prevent attack by microorganisms.

Testing and Standards. To test glulam girders, European countries have agreed on standards that stipulate classification, strength requirements, and test procedures. Because of the economic importance of glulam beams, bondings based on amino and phenolic resins are continually controlled according to EN 301 and EN 302. EN 302.1 – 4 describes the following tests:

1. Determination of shear strength
2. Determination of resistance to delamination
3. Determination of the effect of temperature and moisture cycles on tensile strength (damage by acid)
4. Determination of the effect of shrinkage on shear stength

The results of these short-term tests are used to predict the long-term behavior of glulam girders. The effects of climate change and of cold and boiling water are determined. The boiling-water test showed that short-term tests do not always reflect the behavior of a gluing in outdoor weathering [37]. Tests according to ASTM D 1101 – 59 and D 905 – 49 show comparable results in the delamination test. Contraction of the adhesives during setting should be as small as possible; otherwise cracks occur in thicker glue lines, which are formed when the sections to be glued fit less exactly.

Tests carried out over many years [38] on the aging of wooden building components gave, for example, unsatisfactory results for brittle pure urea resin adhesives in the case of thick glue lines. Broad application of urea resins in building with wood was made possible only by using suitable extenders and fillers [39].

2.2.2. Veneers, Plywood, and Derived Products

2.2.2.1. Development

Veneers are thin sheets of wood obtained by sawing, cutting, or peeling. They are among the longest-known wood-based construction materials. The oldest archaeological finds originate from the heyday of ancient Egypt (ca. 3000 B.C.). Everyday and cult objects were decorated with inlays of different types of wood, known as marquetry. Because of their texture and color, veneers were initially used mainly for decorative purposes. However, composite constructions, consisting of a core of plywood with a veneer glued on to it, were soon recognized to warp less than pure wood boards. Since beautifully grained wood was rare and difficult to obtain even in antiquity and in the Middle Ages, veneers also offered the possibility of using wood sparingly [40].

2.2.2.2. Production of Veneers

Veneers are produced from cordwood or from specially divided blocks by using peeling and cutting machines or, less commonly, by sawing (Fig. 15). Many types of wood – e.g., poplar (*Populus* sp.), birch (*Betula* sp.), or Douglas fir (*Pseudotsuga* sp.) – can be peeled or cut either fresh or saturated with moisture through storage in water. Other types of wood – e.g., beech (*Fagus silvatica*), oak (*Quercus* sp.), ash (*Fraxinus* sp.), walnut (*Juglans* sp.), or various tropical high-density deciduous woods – must first be softened by steaming or boiling to alow the wood to be cut more easily.

After peeling, the veneers have a moisture content of 30 – 110 % (based on oven dry weight). This must therefore be reduced by drying to 4 – 12 % for further processing. Special veneer dryers are used through which hot air is blown at 80 – 200 °C.

2.2.2.3. Principal Types of Plywood

By combining veneers of different types of wood, densities, numbers of layers (plies), and different core materials, a large number of materials can be produced with widely differing properties and areas of use [41]. Plywood can be classified according to different criteria, frequently according to its construction. EN 313 (see Section 2.2.2.6) differentiates among veneer, core, and composite plywood.

Veneer Plywood normally consists of at least three layers of wood glued to one another crosswise. The barrier technique (orientation of the grain of adjacent layers at an angle, usually 90°) requires a symmetry around the middle layer with regard to type of wood, grain direction, and thickness of individual layers. By forming a barrier, the contraction in width of layers with longitudinal grain direction depends extensively on the mechanical deformation of transverse layers. At the same time, the strength and deformation properties in the longitudinal and transverse directions of the plies are aligned with one another, thus decreasing the anisotropy.

Plywood boards are thus considerably more dimensionally stable and can be subjected to loads more uniformly than solid wood boards. The number of veneers is usually uneven, and the arrangements are symmetrical. Three-, five-, and seven-fold boards, etc., are produced (Fig. 16).

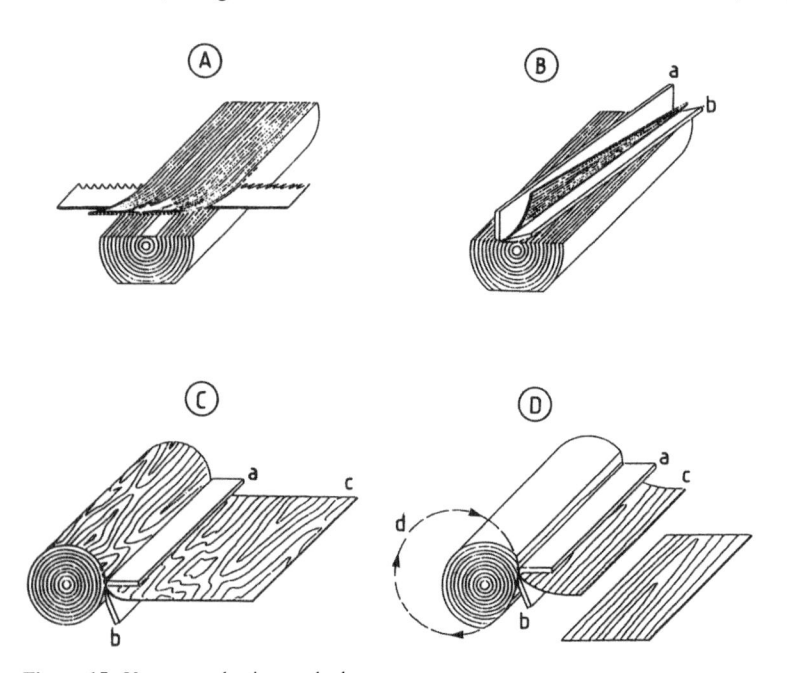

Figure 15. Veneer production methods
A) Sawn veneer; B) Cut veneer; C) Rotary cut veneer; D) Half-round cut veneer
a) Pressure pad; b) Blade; c) Veneer; d) Restraint

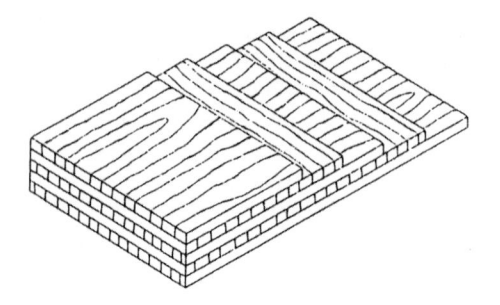

Figure 16. Five-layer veneer plywood

Core Plywood. In core plywood (blockboard), the core consists of battens or strips, usually of light coniferous wood that has been cut from a block of boards or veneers that have been glued together and then blocked on both sides with veneers (Fig. 17). Depending on the core, two types of sandwich board exist: *battenboard* and *laminboard*. Wood core plywood is used for high-grade joinery.

Composite Plywood. Two types of composite plywood exist. In the first type the core consists of a material, usually of low density, that is planked on both sides with veneers. Practically any material can be used for the core (e.g., extruded particle board, paper honeycomb, and for special applications, polyurethane foam). These types of boards are used, for example, for doors and partitions. In the reverse case, the core can be made of plywood and the covering of metal sheeting or glass-fiber-reinforced plastic. The possible combinations are endless. Composite materials with special properties, such as high strength, good insulating capacity, and high abrasion resistance, can be produced for particular applications (e.g., containers).

Particular Laminated Materials. *Glulam boards* consist of three or more layers of wood sheets that are glued together crosswise. The thickness of the layers is at least 8 mm. These boards are used as shuttering panels and to an increasing extent as furniture panels.

In *laminated veneer lumber*, the veneers are glued together in the direction of the fibers. This allows the high longitudinal tensile and flexural strengths of the solid wood to be retained. If the veneers used for producing laminated wood are impregnated with synthetic resins – mostly phenol – formaldehyde resins – and compressed under pressure, *densified laminated wood* with a density >1000 kg/m^3, is obtained. This is used mainly for technical purposes.

Molded plywood is formed when the plywood is molded by using appropriately shaped presses. It is used, for example, for chair backs and seats, furniture, wall paneling, and girder webs.

Wood-based construction materials that are used mainly in building are known in North America as *engineered wood* (see Section 2.2.2.7).

2.2.2.4. Adhesives and Additives

The natural adhesives used initially, such as protein adhesives (gluten, casein, soybean, and blood albumin adhesives), have disadvantages (low resistance to climate variations, sensitivity to attack by microorganisms, long pressing times, and sometimes highly complex processing). Since the 1930s they have been increasingly replaced by synthetic adhesives, and these are nowadays used almost exclusively. They include urea – formaldehyde (UF), melamine – formaldehyde (MF), and phenol – formaldehyde (PF) resins, as well as MUF or MUPF mixed condensates. The use envisaged and the cost determine the choice of adhesive. A detailed description of the chemistry and technology of wood adhesives can be found in [42].

Urea – Formaldehyde Resins. The most important adhesives for gluing wood are urea – formaldehyde resins. The use of UF resins to glue plywood was first described in 1929 [43]. Production of these wood adhesives has since developed enormously [44]. Urea – formaldehyde

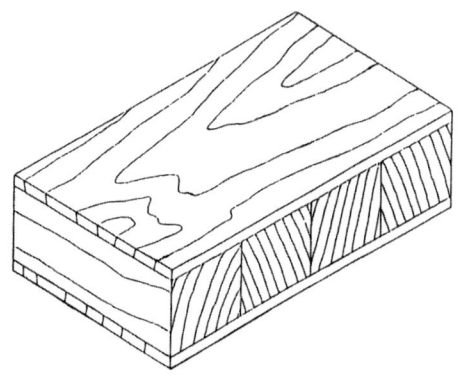

Figure 17. Core plywood

resins are used for adhesion of wood materials used in the interior and have limited resistance to moisture. They can be readily adapted to operational requirements through the nature and dosage of curing aids and additives. Ammonium salts (e.g., NH$_4$Cl) are generally used as curing aids. They lower the pH by reacting with formaldehyde. The rate of curing increases with increasing temperature and decreasing pH. The viscosity of the adhesive liquor is adapted to processing conditions by the addition of water, depending on the content of extenders and fillers. The pressing temperature is 80 – 120 °C.

Melamine – Formaldehyde Resins undergo a curing reaction analogous to that of UF resins. Melamine – formaldehyde resins are characterized by their high water resistance and colorless glue lines.

For cost reasons, however, pure MF resins are rarely employed in plywood production. Instead, MUF or MUPF mixed condensates are used when high weather resistance is required. Depending on the melamine content and the type of wood, adhesions for semiexterior or exterior applications are obtained.

Phenol – Formaldehyde Resins are generally cured by heating and also by a combination of heating and curing aids. Phenol – formaldehyde resins are less reactive than UF or MF resins. They therefore require higher pressing temperatures (115 – 140 °C). Curing of phenol – formaldehyde resins is more sensitive to wood moisture and to pressing conditions than curing with other adhesives. Their alkali content can have damaging effects, e.g., on subsequent surface treatment of veneered boards.

Poly(Vinyl Acetate) Adhesives are mainly used for bonding and veneer jointing.

Extenders and Fillers. In plywood production, extenders and fillers are used together with the adhesives to avoid glue penetration during hot pressing. *Extenders* are finely ground products of organic origin that are capable of swelling and gelatinizing and thus act as adhesives themselves (e.g., starches and cereal and vegetable flours). They are generally used with UF resins at levels of 15 – 30 % or higher (extender plus water, based on the adhesive resin) to reduce the brittleness of the glue line and resin consumption.

Fillers do not produce their own adhesion. They are finely powdered materials (e.g., nut shell or olive stone flours) or minerals, such as kaolin or pulverized gypsum (lenzin) that can be swelled only slightly or not at all. Fillers are normally used with PF resins.

The addition of extenders and fillers can decrease the resistance of the adhesion to water and microorganisms. Mineral fillers also increase the wear on tools.

2.2.2.5. Production of Veneer Plywood

The main operations involved in the production of veneer plywood are preparation, assembly, application of adhesive, stacking coated veneers (sometimes prepressing), hot pressing, conditioning, forming (sometimes sanding), and sometimes fault elimination.

The adhesive is generally applied by using two- or four-roll machines, by pouring using an adhesive curtain, or by melting an adhesive film. The coated veneers are stacked either manually or by machine. To make transportation easier and reduce stack height, coated veneer stacks are frequently precompressed in an unheated one-daylight press. Hot pressing occurs in one- or multi-daylight presses similar to those described in Section 2.3.1.5 but usually with a smaller size. Pressing conditions (temperature, time, and pressure) depend on the type of wood, the strength of the plywood, and the type of adhesive. Pressures of 0.6 – 2.5 N/mm^2 and temperatures of 80 – 140 °C are usually applied. After gluing and sometimes climatization, veneer plywood is sanded, cut to standard or ordered measurements, and graded according to quality.

2.2.2.6. Properties, Testing, and Uses

Standards. The properties and quality of the adhesion of plywood are laid down in most countries by standards and trademarks. The following standards are valid in Europe and the United States:

Europe
Plywood
 Classification and terminology — EN 313 (Parts 1 and 2) 1991
 Test methods and properties — EN 314 (Parts 1 and 2) 1991

United States
Hardwood and decorative plywood — PS 51–71, NBS-1972
Plywood for construction and industrial uses — PS 1–74, NBS-1974
APA grade – trademarks, available from American Plywood Assoc. P.O. Box 117 000, Tacoma, Wash. 98411–0700

Depending on the type of adhesive and the wood, a distinction is made between plywood for exterior and interior use; further classification is made for the following applications:

1. In dry areas, i.e., a normal indoor climate (interior)
2. In moist areas, i.e., with protection against direct weathering also suitable for outdoor climates (semiexterior)
3. For outdoor use, suitable for long-term weathering (exterior)

The combined tension and shear resistance is generally tested on appropriately shaped test objects. A further criterion for the quality of an adhesion is the nature of the fracture pattern (percentage of fracture in the wood). With good adhesion the break will occur not in the glue line but in the wood. Before the strength of adhesion is tested, the sample is pretreated depending on the type of bonding. Pretreatment normally involves storage under different conditions, e.g., in cold, warm, or boiling water with redrying, or cycle tests.

Properties. Veneer plywood is normally produced in thicknesses of 1 – 12 mm and as multiply boards wich thicknesses of 13 – 50 mm. Wood core plywood is available in thicknesses of 10 – 60 mm. The density of veneer plywood depends on the type of wood and the plywood construction (500 – 800 kg/m^3). Wood core plywood generally has a lower density (450 – 750 kg/m^3).

The *hygroscopic properties* of plywood correspond approximately to those of solid wood. For indoor uses, the equilibrium moisture level is between 8 and 12 %. The water absorption decreases as the number of veneer layers and glue lines increases.

Dimensional changes due to contraction and swelling are low in the surface direction because of the barrier technique. The changes correspond approximately to the values for solid wood in the longitudinal direction (0.005 – 0.01 % for changes in water content of 1 %). However, changes in thickness approach those of solid wood in the radial direction (0.012 – 0.019 % for changes in water content of 1 %) despite the adhesive layers.

The *strength properties* of veneer plywood depend mainly on those of the veneers lying parallel to the main line of stress and on the type of wood. Although in solid wood the modulus of elasticity at right angles to the fiber is only ca. 10 % of that in the fiber direction, the properties of veneer plywood approach isotropy as the number of layers increases. Star plywood has almost the same strength when stressed at any angle. The strength properties of a veneer plywood (DIN 68 705, part 3, 1968) are listed below:

Bulk density (at 12 % moisture content)	500 – 800 kg/m^3
Flexural modulus of elasticity	7000 – 12 000 N/mm^2
In the direction of the covering fibers (3 layers)	700 – 1200 N/mm^2
(5 layers or more)	3000 – 7000 N/mm^2

Uses. Most of the plywood produced is used in building. Douglas fir and pine are used predominantly worldwide. Geographically, the consumption of plywood varies widely and depends mainly on local building traditions. Countries such as Canada and the United States are the largest plywood consumers worldwide, with 105 and 77 m^3 per 1000 inhabitants, respectively. In Europe, the Netherlands is the leading consumer, with 28 m^3 per 1000 inhabitants; in Germany, consumption is only 12 m^3, and in Italy only 6 m^3 per 1000 inhabitants [45].

Typical uses for veneer plywood are as construction plywood for wall, floor, and ceiling panels and also as panels for concrete shuttering. Industrial plywood is used to build railway coaches (e.g., for high-speed trains). A large area of use is in containers, boxes, and packagings. Plywood with decorative covering veneers is frequently used in interior furnishings (e.g., wall panels or doors).

2.2.2.7. Engineered Wood

The term engineered wood is used predominantly in North America and refers to wood-based construction materials, made by gluing together small-sized woods, that are used in building. The reason for this development is the increasing scarcity of suitably sized tree trunks.

Particles of small-sized wood and wood residues, such as flakes, strands, or parts of veneers, are combined by oriented agglomeration to give new wood-based construction materials using a method similar to that of particle board production. The properties of these materials are very similar to those of solid wood. Phenol – formaldehyde resins, amino resins (UF/MF), and polymeric methylene diphenylene isocyanate (PMDI) are used to glue the oriented wood pieces, which are mostly Douglas fir and pine.

By using 4-cm-wide and 12-cm-long strands, building materials of the most varying types can be produced in this way. They are known in the building industry as *laminated strand lumber* (LSL). The material is produced in the form of boards with thicknesses up to 14 cm, from which the desired building components can be cut (Fig. 18).

If strands with a thickness of ca. 3 mm and a length of ca. 250 mm (e.g., from peeled veneers) are glued in a parallel orientation, *parallel strand lumber* (PSL) is obtained. These beams can have a cross-section of 28 × 49 cm and are produced in a batch process in special presses in lengths of 20 m. Parallel strand lumber can be worked with normal tools and can be used like laminated wood (Fig. 19).

Gluing together peeled veneers with the fibers parallel gives *laminated veneer lumber* (LVL). In principle, this corresponds to veneered laminated wood. LVL is produced in lengths of 20 m and thicknesses up to 9 cm. The thickness of the veneers used is 2 – 5 mm (Fig. 20).

Trade Names. These building materials are sold in North America under the commercial names Intrallam (LSL), Parallam (PSL), and Microllam by Trus Joist International Inc. and Mac-Millen Bloedel Ltd.

Figure 18. Laminated strand lumber

Figure 19. Parallel strand lumber

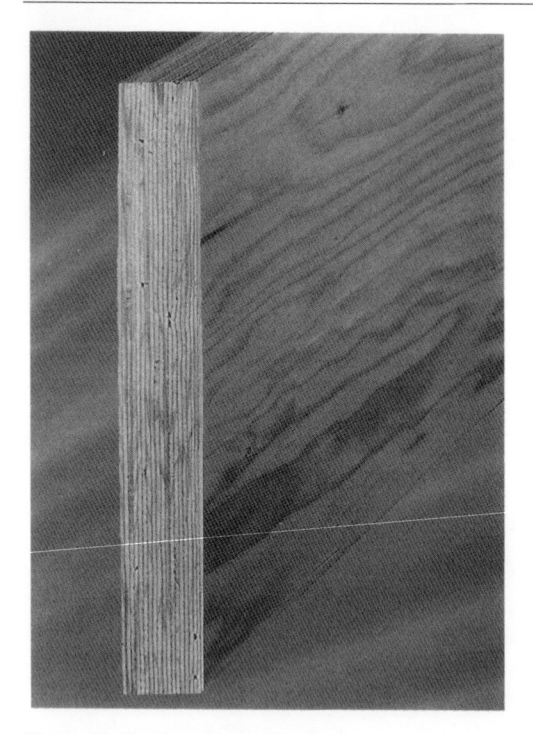

Figure 20. Laminated veneer lumber

2.3. Particle Bonding

To overcome the pronounced anisotropy of solid wood, it must be cut up into small pieces; then the pieces can be oriented to a desired fiber direction. The following wood-based particles have achieved economic importance in the production of wood-based construction materials:

1. Chips for defibration and grinding
2. Shavings from cutting, peeling, and sawing
3. Strands
4. Wafers or flakes
5. Fibers
6. Wood wool

In industrial production, these particles are formed in a certain distribution with regard to dimension and shape. In practice, therefore, they must be sieved or classified. After drying and application of adhesive, the properties of the wood-based construction material can be influenced by targeted orientation of the particles during mat forming and by molding.

2.3.1. Particle Board

2.3.1.1. Development

In 1902, HENRY WATSON of Indiana had the idea of making particle board from waste wood. Unfortunately, no adhesives available at this time could allow industrial realization of this idea.

Only when the chemical industry made a suitable adhesive system available with the development of amino resins, could the industrial production of particle board be begun in Germany in 1941. After the end of World War II, the development of particle board technology made rapid advances in Germany and subsequently in the whole of Europe. Important contributions came from the Swiss F. FAHRNI and the German W. KLAUDITZ. In 1991, particle board was the most important industrially produced wood-based construction material, with an annual worldwide production of ca. 50×10^6 m^3.

2.3.1.2. Types of Particle Board

Flat Pressed Particle Board. In flat pressed particle board, the particles are arranged predominantly parallel to the board surface and bonded. Most of the particle board produced worldwide belongs to this group. The wafer, flake, and strand boards and oriented structural boards (OSB) produced widely in North America are, in principle, flat pressed particle boards. Although *wafer* and *flake boards* are made up predominantly of (decorative) wood particles with large surfaces and a thickness up to 0.02 inch (ca. 0.05 cm), *strand boards* are made from strands with a high ratio of slenderness having a thickness up to 0.2 inch (ca. 0.5 cm) and a length of 3 – 6 inches (ca. 7.6 – 15 cm). The production, transport, gluing, and matforming of these particles require special machines.

Extruded Particle Board. These boards are produced by extrusion. The process was developed in the 1960s by O. KREIBAUM [46] and is still termed Okal process. Particles are arranged predominantly at right angles to the board surface. Short particles or granulated wood are mainly used. Adhesive-coated particles arrive in portions in the heated press, are compressed by using

Figure 21. Different texture of flat pressed (A) and extruded particle board (B)

a pressure piston driven by an eccentric flywheel, and are extruded through the hot zone of the press. A continuous band of board is formed as it leaves the press. Heating of the board is accelerated by heated rods that project into the press. The heated rods themselves form tube-shaped hollows in the board. Extruded particle board is generally used only in the planked or veneered state because of the low flexural strength resulting from the production process. The production of profiles constitutes a further variant of the extrusion process. Figure 21 shows the difference in texture between flat pressed and extruded particle board.

Molded Objects. The production of molded particle board was described in 1948 in a Swedish patent [47]. In the same year, work began on the Thermodyne process, in which natural wood constituents are activated as binders at pressing temperatures up to 290 °C and pressures up to 300 bar [48]. The process was not successful economically because of the extreme operating conditions.

The process patented by the Werz company (Werzalite process) in 1956 uses wood particles containing 20 – 30 % amino resins [49]. These are compressed to form a blank and then prehardened. The blank is then coated immediately on all sides with paper impregnated with melamine resin. Operating conditions are very similar to those employed in particle board production.

Production of molded objects using the Werzalite or similar processes was successful economically. Tabletops, window sills, paneling, palettes, and other molded objects are produced in this way.

2.3.1.3. Raw Materials

For particle board, low-value wood and industrial waste wood (e.g., remainders, slabs, splinters, and shavings from planing and sawing) are used. In countries with low indigenous wood production, the use of annuals such as cotton stalks, sugarcane bagasse, flax, rice husks, cereal straw, jute, esparto (alfa grass), and similar materials, instead of wood, is possible. These materials give a board that is very similar in properties to that made from wood [50].

In principle, all types of wood are suitable for making particle board. However, coniferous wood is used for most of the production in Europe and North America, which accounts for 75 % of the total worldwide particle board production. A combination of hard and soft types of wood is frequently used; the fact that these types of wood affect the properties of the board because of their differing sorption behavior is taken into account [51].

The time of wood harvest also affects sorption behavior and acidity, which in turn affect the curing of amino resins used to glue the particles. Some deciduous woods, such as oak or chestnut, have a high proportion of tannin and a low pH. Sometimes this impairs the strength produced by the gluing process. Wood bark used to be separated and incinerated or dumped, because the different structure of the bark generally required greater compression of the particles and the use of larger quantities of adhesive resins. Nowadays, however, the use of unpeeled wood has become common, even for the outer layer. Addition of 6 – 15 % bark to middle-layer particles does not affect board quality significantly [52]. Larger quantities of bark impair not only the appearance of the board, but also its strength and sorption properties. Excess bark in particle board factories is incinerated, and the waste gas is used for drying the particles [53].

Flat pressed particle board is now generally produced in several layers. The density of the individual layers and their construction determine the mechanical properties of the board. Light particle board has a density up to 500 kg/m^3.

Most of the board produced is, however, in the density range of 600 – 800 kg/m^3. The thickness of the board ranges from 3 to 50 mm.

2.3.1.4. Adhesives and Additives

Amino Resins (UF, MF, and MUF). Urea – formaldehyde adhesive resins are used as 55 – 72 % aqueous solutions or as powders. For gluing particles, the resins now used predominantly have a basic U : F molar ratio of 1:1.0 to 1:1.3 to limit formaldehyde emissions. Catalysts based on ammonium salts or acids are used to accelerate the curing process. The use of resins with increasingly less formaldehyde finally led to a lowering of the cross-link density and thus to impaired quality of adhesion. This was corrected by reinforcing the UF resins with melamine [54]. To adhere to the strict limiting values for formaldehyde emissions in many countries, a special process for low-emission boards was developed by using reactive formaldehyde-rich resins and formaldehyde scavengers [55]. Urea – formaldehyde is used in quantities of 6 – 12 wt % solid resin, based on absolutely dry wood. UF-glued particle board is produced mainly for interior use and for furniture.

Higher-quality adhesions are achieved with melamine – formaldehyde, phenol – formaldehyde, or polyisocyanate resins. Because of their chemical similarity to UF resins, MF resins are mixed or cocondensed with them. Melamine – urea – formaldehyde and phenol-containing MUPF resins with 15 – 25 % melamine and up to 5 % phenol are used for particle board that is exposed to particularly high moisture levels. The dosage of resin is increased to 12 – 15 %, based on the dry wood particles.

Phenol – Formaldehyde Resins are high-value products, particularly because of their resistance to hydrolysis. They are used as aqueous alkaline solutions with a P : F molar ratio of 1:1.5 to 1:2.5. The alkali content is 3 – 10 %, depending on the degree of condensation of the resin. Phenol – formaldehyde resins are generally less reactive than amino resins. This means that, in practice, longer pressing times must be used for processing PF resins compared with amino resins. Alkaline phenolic resins dye particle board light brown and increase its hygroscopic nature. PF-glued particle board produces no significant formaldehyde emissions. It is used for exterior applications. The usual dosage of resin is 8 – 12 %.

Polymeric Methylene Diphenylene Isocyanate. Since 1980 adhesives based on PMDI have increasingly been used. They consist of a mixture of monomeric, polymeric, and oligomeric diisocyanates, which are capable of reacting with the cellulose, lignin, and moisture in wood. The quality of wood adhesions using PMDI is comparable to that using PF resins. A significant disadvantage of isocyanate adhesives is their tendency to stick to metals. PMDI can therefore not be used without release agents.

Handling PMDI adhesives requires particular care because of their high reactivity toward water and bases. However, they are very economical. Adequate board strengths are achieved with 3 – 5 % adhesive.

Adhesives Based on Natural Materials. Adhesives made from waste sulfite pulping liquor and from wood and bark extracts are known, but have not achieved great economic importance.

Mineral Binders. Special boards for building purposes are also produced using inorganic binders. These include Portland cement, magnesia cement (Sorel cement), and gypsum (see Section 2.3.3).

Additives. For repelling water and delaying swelling during particle board production with synthetic resins, 0.1 – 1.5 % paraffin, based on dry wood, is generally added. Paraffin is mixed with the adhesive liquor as a dispersion or applied to the particles as a finely dispersed melt.

2.3.1.5. Production

Particle Production. The wood to be milled should have a moisture level of ca. 50 % and free of metallic or mineral substances. Low moisture levels increase the throughput of the size reduction machine, but also increase dust formation and fire hazards.

Chipping machines with very varying mechanisms are used for size reduction. For lengthwood (roundwood) machines with cutting tools fixed in a carrier disk or drum, together with spurs, are preferred. By pressing the wood onto the rotating blade carriers, uniformly thick particles are formed, which can be cut up further in a subsequent mill if desired. The most powerful chipping machines of this type are knife shaft machines with an operating width of 2 m. For making larger particles, knife ring machines are used, which have combined cutting and striking tools. To cut up the particles further, hammer, wing beater, and toothed disk mills are used. The target of the size reduction process is the production of a mixture of particles that are as uniform as possible in shape and thickness. The geometry of the particles affects both the further processing (distribution and dispersion of glue) and the mechanical properties of the board. Since particle board for use in surface coating is produced mainly in three or five layers, the aim of particle preparation is to produce different mixtures of particles. Fine particles for the outer and coarse ones for the middle layer are obtained by air classifying and screening. For particle board used in building, large proportions of fine particles and dust are undesirable.

Drying. To process the particles further, residual moisture levels of 1 – 3 % are necessary. High moisture levels, together with the water applied with the adhesive resin, generally cause steam cracking in the pressing process. Drying the particles is therefore indispensable.

Drying follows the rules of convection drying. In the first part of the drying process, wood moisture decreases almost linearly. The second part of the drying process is determined by vapor diffusion.

With particles of varying thickness and size, uniform drying is difficult to achieve. Separate drying of the coarse middle-layer and fine outer-layer materials is therefore necessary. To dry the particles the following powerful machines are used: (1) jet dryers, (2) tubular dryers, (3) stream dryers, and (4) drum dryers.

Glue Blending of the particles occurs continuously in flow mixers to achieve as uniform a distribution of the adhesive on the surface of the particles as possible. Dried particles are fed to the mixer via a belt weigher. The necessary quantity of adhesive is dosed volumetrically or gravimetrically with electronic control.

In the mixer, adhesive is rubbed on the surface of the particles or applied by using spray jets. The efficiency of the mixer depends on the design of

Figure 22. Particle glue-blending machine EK-AB 50 (Loedige, Paderborn, Germany)

the mixing drum and the conveying and braking tools. Because of the heat of friction produced, the mixing vessel must be cooled with water. Figure 22 shows a high-performance mixer from the Loedige company.

Forming Mats. Glue-blended particles are fed from the storage hopper into the dosing hopper of the forming station by using conveyor screws or belts.

By controlling the volume of the glued chips a uniform flow of particles is produced, which is deposited onto a moving forming belt via a toothed feed roller. The falling particles receive an impulse from a spreading roller, resulting in separation of the coarse and fine particles. Commonly, the mat is formed by using air classifying. In this process, air is blown against a curtain of falling particles so that fine particles are carried further with the air than coarse ones. Here also, classified particles fall onto a conveyor belt below. With two blowers set opposite one another, mats can be formed that have coarse particles in the center and finer particles on the outside. For reducing thickness, the mats are precompressed in a cold press and then passed to a hot press for activation of the adhesive.

Pressing and Pressing Techniques. In hot pressing the blank is compressed to the desired final thickness and the adhesive is irreversibly cured. Both continuously operating and batch presses are used.

Batch Presses. The output of presses for batch processing is limited by the pressing and

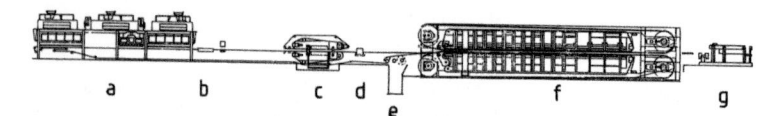

Figure 23. ContiRoll particle board press line (Siempelkamp, Germany)
a) Mat formers; b) Belt weighers; c) Prepress; d) Metal detector; e) Reject nose; f) ContiRoll press; g) Cut-to-size saw

cycle times. With *one-daylight* presses, high pressing temperatures (200 – 230 °C) are used to shorten the setting time of the adhesive. *Multi-daylight presses* have up to 20 daylights and are operated at < 200 °C. Automatic charging and emptying of a multi-daylight press require complex machinery. The tolerance in thickness of the boards produced by using the batch press is so great that calibration by sanding is indispensable.

Continuously Operating Presses. In larger particle board plants, continuously operating presses are mainly used. With this technique, boards are produced with significantly smaller variations in thickness than in batch production. Various continuously operating systems exist.

Roller Presses. In roller presses the application of pressure and transportation is effected by a bed of rods. Figure 23 shows a ContiRoll plant of the Siempelkamp company, as a typical example of this type of machine. Pressure and temperature control in the ContiRoll process is illustrated in Figure 24. Figure 25 shows how the pressure is conferred by means of the bed of rods. The operating temperature is 220 – 250 °C. Continuously operating roller presses are usually designed for capacities of more than 400 m^3 of board per day. The production costs are up to 8 %

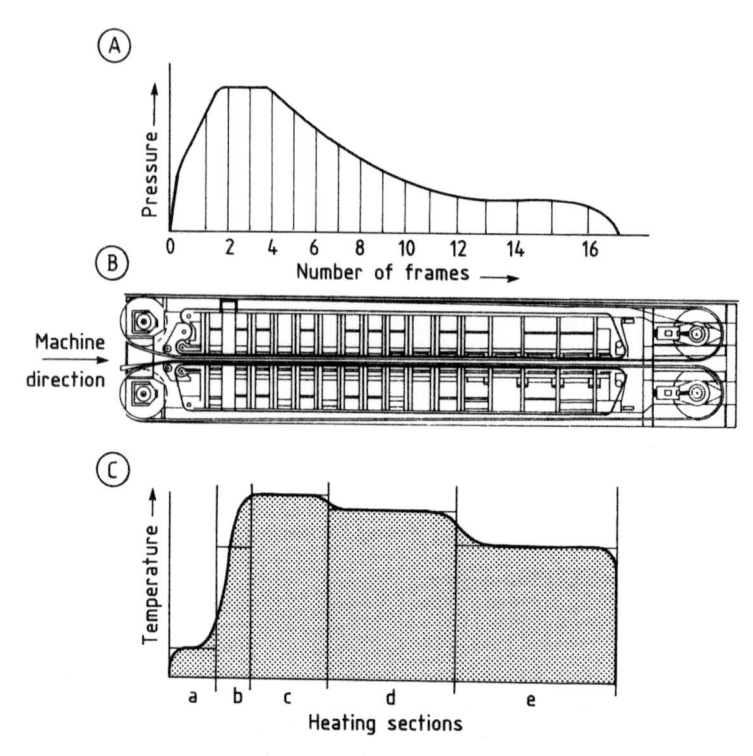

Figure 24. Pressure and temperature control in the ContiRoll press
A) Pressure program; B) Schematic of ContiRoll press; C) Temperature program

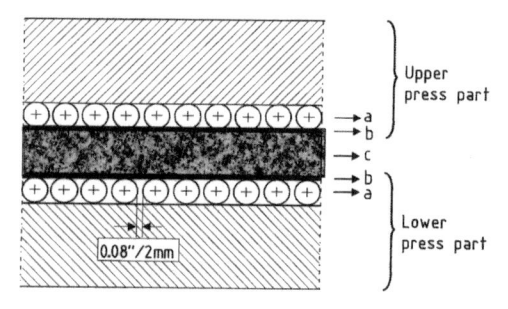

Figure 25. Moving rod system of the ContiRoll press
a) Bed of rolls; b) Steel belt; c) Mat

less than those for batch presses because of the lower energy consumption.

Gliding Film Presses. In gliding film presses the pressure is transmitted by an oil film (Hydro Dyn process of the Bison company) or by air cushions (Hymmen company).

Calender Presses. In calender presses pressure is applied to the blank by pressing it with steel bands on a heated cylinder (calender). Figure 26 shows a calender press that operates using the Mende process (producer Bison). The system is particularly suitable for producing thin particle board.

Extrusion Presses. In extrusion presses the particles are extruded continuously between heated plates or rods (Okal process, see Extruded Particle Board). Profiles and tubular boards are produced by this method.

Steam Injection Presses. Heating the mats can be accelerated by using steam injection presses. The steam is injected immediately before pressure is applied. The process is particularly suitable for thick boards, which are difficult to heat through. Since 1989, it is being developed further by Siempelkamp.

To *heat the presses* a heat-transfer oil is used predominantly as the energy carrier. Experience with the use of high-frequency energy is extensive. For cost reasons, it is now used only for warming the mats in prepresses.

The *pressing time* necessary is determined by the reactivity of the adhesive system and the pressing temperature. The specific pressure of 20 – 50 bar is decreased in steps as the spring back forces of the compressed mat decrease and the adhesive takes effect (Fig. 24).

Aftertreatment. After hot pressing, particle board is cooled and conditioned by storage for several days. This process leads to further setting of the adhesive so that the internal bonding of the board increases and the thickness swelling values decrease. The board is standardized by sanding.

2.3.1.6. Properties, Testing, and Uses

Properties. Even in glued particle board the hygroscopic properties of the basic material, wood, are noticeable. The state of equilibrium between the moisture in the wood and the relative air humidity causes ad- and desorption of water on climate change, which, in turn, lead to a change in volume. The properties of particle board are determined by the orientation of the particles, which depends on the process used. This effect is most obvious in the case of extruded and flat pressed particle boards (see Fig. 21). Even in flat pressed particle board the longitudinal and thickness-sectional swelling differ by a factor of 10. The swelling and shrinkage behavior of particle board is also determined by the intensity of compression of the particles (board

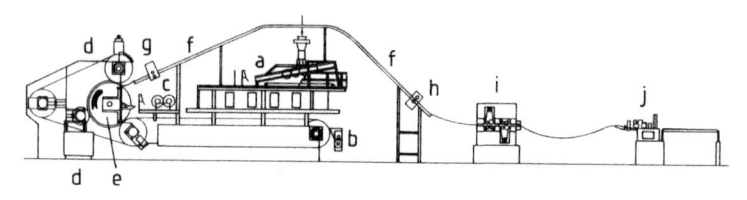

Figure 26. Calender press for thin particle board production (Bison, Germany)
a) Forming station with gamma-ray density control unit; b) Bin for misspreads; c) Paper unwinding device; d) Hydraulic for calender press; e) Calender press; f) Board transfer device; g) Extraction device; h) Braking device; i) Wide belt sander; j) Trimming and sizing saw

density), the nature and quantity of the adhesive, and the type and quality of the wood. Changing properties by altering wood quality is becoming less and less feasible because of the continually increasing use of residual and waste wood.

For mass production of boards in the furniture industry, dimensional stability is extremely important. Assessment criteria are the absorption of moisture and the swelling in moist air and on storage underwater. The sorption test in moist air and determination of the swelling after storage underwater do not always give the same results [56] since the final swelling is not achieved as rapidly in short-term storage underwater because of the delay in water absorption due to high board density or the use of a hydrophobing agent (paraffin).

Figure 27 shows the longitudinal and thickness swelling and the water absorption of flat pressed particle board based on spruce and glued with UF resin on storage in moist air (95 % R.H.).

Mechanical properties are adapted to the use envisaged as much as possible. The strength perpendicular to the board surface (*cross tensile strength*) is an indication of the quality of the bonding. It can be influenced by the nature and quantity of the adhesive, its distribution on the particles, and the compression of the board.

The *flexural strength* (i.e., bending strength) of the board is particularly important for use in building. It can be improved by orientation of the particles (OSB) in composite board. Generally, the outer layer is more highly compressed than the central layer so flexural strength can be influenced by the sandwich effect (low-density core- and high-density toplayers).

The strength properties depend on hygroscopic properties. Figure 28 shows the impairment of flexural and cross tensile strengths by moisture absorption on storage in water.

Testing. Many selective procedures are used to test the chemical, physical, and biological resistance of particle board [57].

Europe is the area with the highest particle board production. Here, efforts have been made to harmonize the standard requirements and tests. In EN 312, parts 1 – 7, the requirements are *classified* according to areas of use. In the United States, a differentiation is made between interior and exterior use in commercial standards CS 236. In the Japanese Industrial Standard JISA 5908, boards are classified according to their flexural strengths.

To assess *formaldehyde emissions* the desiccator test is used predominantly in the United States and Japan, while the perforator and gas analysis tests (EN 129 and EN 717) are used in Europe. Large test chambers are used in different countries to measure formaldehyde equilibrium concentrations, to give a reference value for the short-term emission test. For the corresponding European standard, operating conditions for a large testing chamber have not yet been laid down. In the United States, ASTM E 1333 is followed.

Resistance of the adhesions to *climate rotation* is determined by the short-term aging tests described in ASTM D 1037 and EN 321. By using a multistage cyclic test, involving storage in water, evaporation, action of frost, evaporation, and drying, the adhesions, in particular, are assumed to be exposed to aging. Also, boiling the test object in water for several hours according to DIN 68 763 (V 100) gives information on the resistance of adhesions.

Particle board for use in building often has high resistance to *biological decomposition* by microorganisms as well as high-quality adhesion (ASTM D 1413 and EN 335–3).

Widely differing test procedures are used to assess the *flammability* of particle board. Flame propagation, smoke development, and calorific value are determined according to ASTM E 84. In Germany, according to DIN 4102–8, in particular, the length of the test piece that remains undamaged is measured. British Standard BS 476–6 uses an index which is calculated from the temperature of the combustion gases.

2.3.2. Fiberboard

2.3.2.1. Development

Fiberboard consists, like paper, of lignocellulose-containing fibers. Both materials are related with regard to production and construction, and the history of fiberboard is closely associated with that of pulp and paper.

The first fiberboard was *softboard* (an insulating board). It was produced in the United Kingdom at the turn of the century – and somewhat later in the United States – from waste fibers that

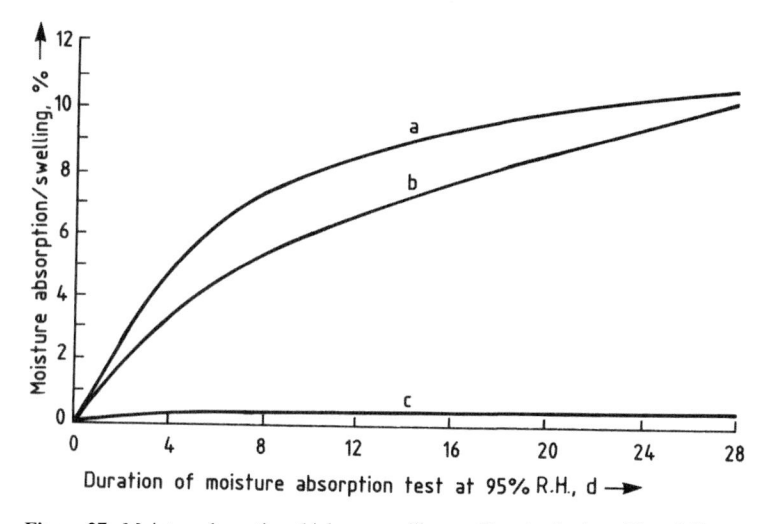

Figure 27. Moisture absorption, thickness swelling, and longitudinal swelling of 18-mm particle board (test object 10×10 cm, according to O. Liiri)
a) Moisture absorption; b) Thickness swelling; c) Longitudinal swelling

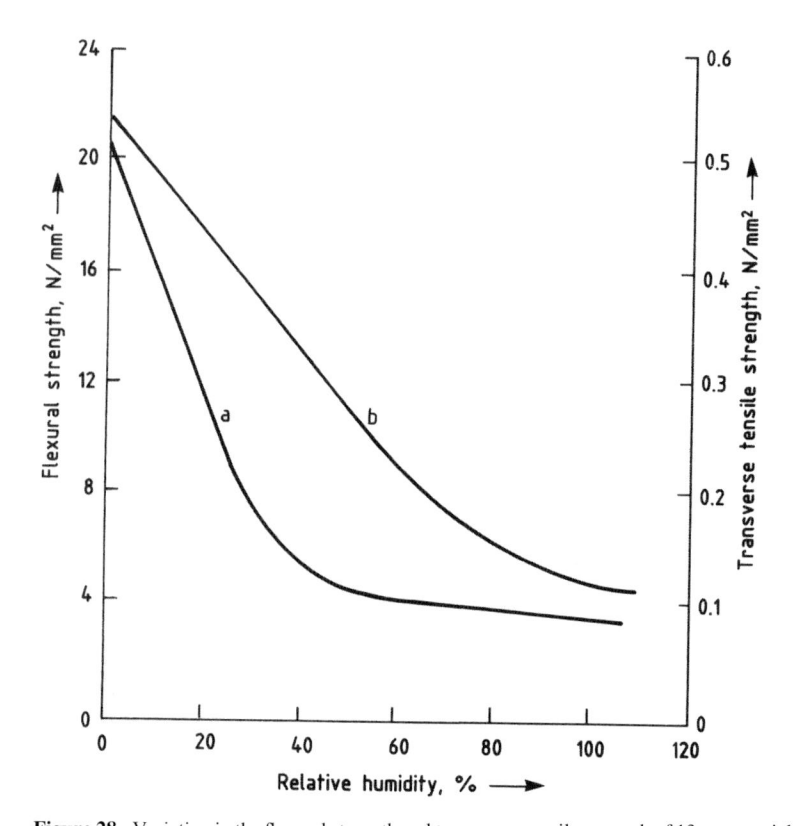

Figure 28. Variation in the flexural strength and transverse tensile strength of 13-mm particle board with moisture absorption on storage in water (according to O. Liiri)
a) Transverse tensile strength; b) Flexural strength

were unsuitable for paper production because they were too short. *Hardboard* was developed in the United States in the 1920s to utilize sawmill waste.

The production of fibers from chips is similar for all fiberboard production processes. The middle lamellae binding the fibers, which are rich in lignin, are softened by heat and moisture, and the fibers are then separated mechanically.

An important discovery was made by H. MASON, who developed the Mason cannon in 1926. In his process, the chips are first softened in a boiler under high pressure. This is followed by a sudden pressure release accompanied by a loud cracking sound, which indicates breakdown of the wood structure into fibers and bundles of fibers. These are pressed in a hot press to form hardboard without the addition of binders.

The Swede A. ASPLUND made the pioneering discovery of the *defibrator* in 1931. In the Asplund process chips that have been softened by steam are broken down into fibers by grinding between two disks.

The first hardboard was produced, like paper, by the *wet process* from a thin aqueous fiber suspension. The fiber mat transported to the hot press contained so much water that sieves were necessary for draining. The imprint of the sieve remains on the underside of the finished hardboard. Hence, this type of board is also known as *S1 S* (*smooth one side*).

The sieve imprint was undesirable for many applications. A board was subsequently developed that was smooth on both sides (*S2S, smooth two sides*). In this case, softboard is produced that is then compressed to hardboard between two smooth pressure plates.

Wet processes require enormous quantities of water. Pollutants, such as extractives, hemicelluloses, and lignin, are present in the wastewater. For environmental reasons the *dry process* was developed. Here, the fibers are dried, glue-blended, and pressed as in particle board production. This process gives an S2S board. For reasons of surface quality, gluing is carried out before drying. The *semidry* process lies between the wet and dry processes. Because fiber moisture is still present, sieves must be used and S1S board is produced.

2.3.2.2. Types of Fiberboard

The classification and nomenclature of the different types of fiberboard on the market seem confusing because the names either have a historic origin or represent current types of board or properties [58]. An overview is given in Table 8. Two important criteria used in classification are the production process and the board density.

Wet Process. The following types of board are produced by the wet process (S1S, density > 800 kg/m^3); tempered hardboard with higher strength and water resistance (density > 960 kg/m^3, thickness 2.5 – 5.0 mm); two types of mediumboard: low density (350 – 560 kg/m^3) and high density (560 – 800 kg/m^3); and softboard.

The density of softboard is < 350 kg/m^3 (thickness 8 – 16 mm). Impregnated softboard is treated with bitumen or other additives to

Table 8. Classification of types of fiberboard

Process	Board specification	Surface	Density, kg/m^3	Notes
Wet	standard hardboard	S1S	> 800	
	tempered hardboard	S1S	> 960	higher strength and water resistance
	low-density mediumboard	S1S	350 – 560	
	high-density mediumboard	S1S	560 – 800	
	softboard (insulationboard)		≤ 350	
	impregnated softboard		≤ 400	higher water resistance
Semidry	semidry process hardboard	S1S		
Dry	dry process hardboard	S2S	800 – 1000	adhesive PF
	medium-density fiberboard (MDF)	S2S	600 – 850	adhesive usually UF, occasionally PF or PMDI, thickness <4 mm
	low-density fiberboard	S2S	450 – 600	
	high-density fiberboard (HDF)	S2S	800 – 1000	
	MDF specialities			
	moisture resistant			
	exterior			
	flame and fire retardant			

increase its water resistance (density up to 400 kg/m^3).

Dry Process. Dry process hardboard is an S2S board that is normally produced by using PF resins (density 800 – 1100 kg/m^3). Medium-density fiberboard is generally produced by using UF resins (density 600 – 840 kg/m^3). Other types of fiberboard are, e.g., low-density board (450 – 600 kg/m^3); high-density fiberboard (800 –1000 kg/m^3, thickness < 4 mm); and moisture-resistant, exterior, and flame- and fire-retardant fiberboard.

Plasterboard. Besides the classical wood-based fiberboards, a fiberboard bonded with gypsum is becoming increasingly popular for use in interior furnishing for partitions and panels. Plasterboard can be produced from waste paper and gypsum from the desulfurization of exhaust gases. The production process is similar to that used for wood particle board using mineral binders (see Section 2.3.3.2).

2.3.2.3. Raw Materials

Fiberboard can be produced from all lignocellulose-containing materials with industrially utilizable fibers. However, for reasons of availability, wood is almost always used. To minimize variations in the process, only one type or a very limited number of types of wood are employed, the latter as a mixture of constant composition.

Deciduous and coniferous woods differ in their anatomical construction (see Section 1.2). Coniferous wood fibers are generally two to three times as long as those of deciduous wood (Fig. 29). The importance of the properties of the raw materials and fibers for the production process and for board quality is discussed in detail in [59]. Unlike paper, in fiberboard the fiber length is relatively unimportant. However, to a certain extent it determines the orientation of fibers in the board. Short fibers tend to be arranged vertically and longer ones horizontally. Fibers with thin walls nestle together better in the press than those with thick walls. The density of the wood used determines the thickness of the fiber mat. The lower the density, the thicker the mat, and consequently the higher are the degree of compression, the contact between fibers, and the resulting strength of the board. Dry processes

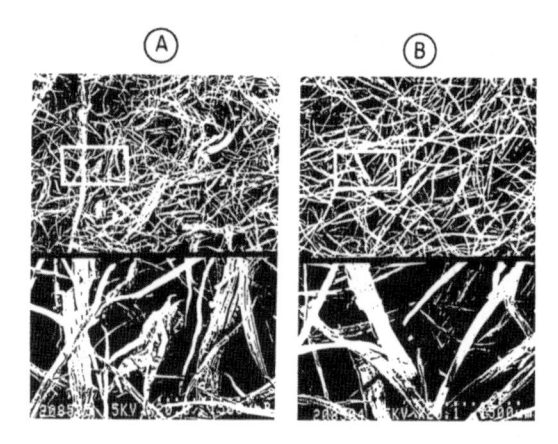

Figure 29. Wood fibers for MDF production
A) Deciduous wood; B) Coniferous wood

are generally less sensitive to the properties of wood than wet ones. In the wet process, deciduous woods give better hardboard and coniferous woods better softboard. Coniferous woods require longer steaming times and, therefore, more energy for fiber production than deciduous woods. Mixtures of deciduous woods are mostly available in groups classified according to density. The lower the density of the deciduous wood, the longer is the digestion time. For cost reasons, efforts are made to use waste wood instead of roundwood as far as possible. Debarked chips from sawmills give good fiberboard.

Sawdust is less suitable because of the irregular particle sizes and short fibers; it requires expensive preparation. Bark is not suitable because of its mineral components and dark color. Of the lignocellulose-containing annuals, only sugarcane bagasse is important for softboard production.

2.3.2.4. Binders and Additives

In the production of fiberboard by the wet process, no binders are required in principle because of the binding forces already present in the wood. However, adhesives are necessary for the dry process. Table 9 gives an overview of the main types of fiber bonding.

The main purpose of additives is to improve the mechanical properties and water resistance of fiberboard. In *softboard* the fibers are bonded predominantly by hydrogen bonds. For hydrophobing, molten paraffin (ca. 1% slack wax based on dry fiber) or asphalt emulsion is added.

Table 9. Types of fiber bonding and additives used in fiberboard production, according to [59]

Process	Board type	Main binder	Auxiliary binders
Wet	insulationboard	hydrogen bonds	asphalt, starch
	S1S	lignin	PF, drying oils
	S2S, MDF	lignin	rosin
Dry	S2S	PF	
	MDF	UF, MUF, PF, PMDI	

For *hardboard*, lignin, which is activated in the hot press, is the important binder.

S1S board is produced mainly with the addition of highly condensed, alkaline PF resins (0.5 – 2.0 % based on dry fiber), which are fixed on the fiber by the addition of $Al_2(SO_4)_3$. Paraffin (0.1 – 0.5 % based on dry fiber) and drying oils, such as linseed oil, are also added.

For *S2S board produced by the wet process*, phenolic resins cannot be used because the severe drying conditions would result in premature curing. Resins such as rosin are therefore employed. They are fixed on the fiber by lowering the pH with $Al_2(SO_4)_3$.

S2S dry process hardboard is generally produced by using PF resins (ca. 6 % based on dry fiber). For *MDF produced by the dry process*, UF resins, generally modified with melamine, are the most important binders (10 – 14 % solid resin based on dry fiber). To a small extent, mainly for special boards, MUF, PF, and PMDI are used. As with S2S boards, paraffin is used for hydrophobing.

2.3.2.5. Production Processes

Raw material processing and fiber production in the wet and dry processes are very similar, frequently even identical.

Raw Material Processing. The wood used is predominantly roundwood, also partly debarked chips from sawmills or whole-tree chips. Pieces of bark are generally undesirable. They contain minerals that shorten the service life of refiner disks and processing tools, and impair surface quality.

Roundwood is generally debarked before chip production. This can be carried out by using ring or drum debarkers.

Chips are most frequently produced by using disk chippers or cylinder chopping machines. The cutting disk of the disk chipper, arranged vertically or diagonally, is fitted with 4 – 16 blades, depending on the size. The positions of the blades determine the dimensions of the chips.

Fine chips (< 4-mm sieve mesh) and coarse chips (> 40-mm sieve mesh) are separated from normal chips by sieving on flat or drum sifters. The fine chips are incinerated, and the coarse ones are cut up further in hammer or wing beater mills. The optimum chip size is 20 – 25 mm × 20 mm × 5 mm.

To remove foreign material and mineral components, chips are washed before further processing. This is particularly important with whole-tree chips. Washing is carried out in specially designed plants with prewashers for separating coarse components and a screw drainer for separating fine mineral particles. The chips are then stored in reinforced concrete or metal silos. Storage acts as a buffer between chip and fiber production and at the same time allows the moisture distribution to become more uniform.

Fiber Production. The aim of defibration is to break up the previously chopped raw material as carefully as possible into individual fibers and bundles of fibers. This is carried out thermomechanically. The lignin-rich middle lamellae that bind the fibers are first softened by heat and steam. The fibers are then separated by mechanical abrasion. They are often broken in the process, giving fiber fragments. The anatomical nature of the fibers and their state after defibration are important for forming the fiber mat and for the properties of the board. This applies particularly to the wet process in which the freeness (draining behavior) and the type of matting are very important.

Thermal treatment of the chips before and during defibration causes part of the hemicelluloses to go into solution. Increasing the temperature and the length of the treatment period leads to increased softening of the interfiber bonding. In the wet process, this improves the natural adhesion between fibers during board production. At the same time, the sugar content of the wastewater and the energy consumption increase, so that a compromise between environmentally friendly and economic process design and board properties is usually aimed at. Fiber production is energy intensive; it consumes more than half of the total process energy. Table 10

Table 10. Processes for fiber production

Process (company)	Type of defibration	Conditions
Masonite (Masonite)	Mason cannon	1) 30 – 40 s steaming at 24 bar (220 °C) 2) 2 – 3 s steaming at 69 bar (280 °C) 3) explosive pressure release
Asplund (Sunds)	defibrator (one rotating grinding disk)	2 – 5 min steaming at 6 – 12 bar (160 – 195 °C) in preheater
Bauer (Sprout – Bauer)	Bauer refiner (two counterrotating grinding disks)	3 – 10 min steaming at 3 – 8 bar in the Bauer digester

gives an overview of important current fiber production processes.

In the *Masonite process*, both the softening of fiber bonds and the separation of fibers are effected by steam. This technique is used exclusively in the United States by the Masonite company (wet process).

Refiners. In all other processes defibration is carried out by using grinding disks. Some defibrators operate at atmospheric and others at elevated pressure. Also, machines exist with only one rotating grinding disk (Sunds) or with two counterrotatory disks (Sprout-Bauer). Refiners operating at atmospheric pressure have the disadvantages of high energy consumption and moderate quality of the fibers produced.

The refiner now used predominantly, which operates under pressure, gives better results. The Asplund defibrator (Sunds) is a prototype. This makes use of the fact that from ca. 170 °C, separation of fibers in the middle lamella is considerably easier and is associated with a significant lowering of the energy consumption and an improvement in fiber quality.

In the defibration process (see Fig. 30), washed chips are fed into the preheater through a conical opening using a screw feeder (c). The preheater (e) is usually a vertical, slightly conical pressure container, which is continuously charged with saturated steam at 150 – 180 °C. The steam heats the chips by condensation. Blowing steam out through the opening is hindered by the plug formed in it by the chips (d).

In 1 – 3 min the chips reach a conveyor screw (g) at the base of the preheater through which they are transferred to the grinding disks of the refiner (h), where defibration takes place under

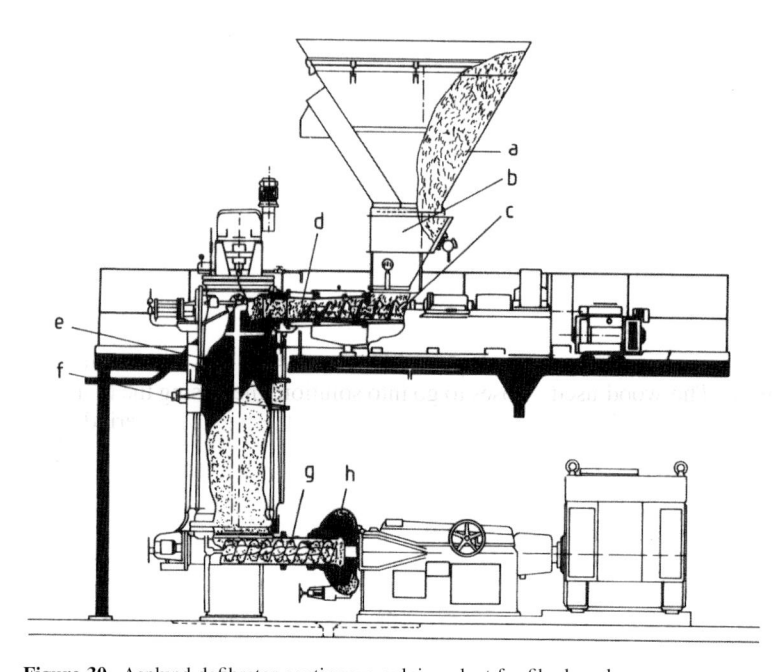

Figure 30. Asplund defibrator continuous pulping plant for fiberboard
a) Chip chute; b) Vibratory; c) Screw feeder; d) Chip plug; e) Preheater; f) Gamma-ray level controller; g) Conveyor screw; h) Asplund defibrator

steam pressure. The necessary grinding power is ca. 200 kW per tonne of dry fiber.

The important parameters affecting the quality of the fiber material and the energy consumption are the residence time and vapor pressure in the preheater, throughput quantity, defibrating pressure, distance between grinding disks (0.1 –0.4 mm), number of revolutions per minute, state of the grinding disk, nature of the raw material, and moisture level.

Defibration is tested by determining the degree of grinding or by fractionating the fiber material. The degree of grinding is a measure of the freeness of the fiber material and determined by means of the dewatering speed of a fiber suspension (standard freeness tester, defibrator-seconds, DS). In the wet process, higher degrees of grinding usually result in an improvement in board properties.

After defibration fibers are blown out through a pressure-regulated exit valve. From this stage onward the wet and dry processes differ and are described separately below.

Wet Process. A fiberboard production plant using the wet process is shown schematically in Figure 31. The fiber material is fed to silos (b) for mixing and adjusting concentration (4 – 6 %) or is further defibrated in disk refiners (f). For hard and semihard fiberboards a degree of grinding of 18 – 30 DS is aimed at, and for porous boards 42 – 70 DS [60].

A system of tanks and conveying devices is used for mixing, storing, and addition of adhesives and additives (see Section 2.3.2.4). A control panel connected to the system regulates concentration, temperature, and pH.

The fiber mat is formed on a sieve by drainage of water from the fiber suspension and subsequent sedimentation of the fibers. Mat formation was previously carried out as a batch process in tanks, but cylinder wet machines and conveyor sieve dewatering machines are predominantly used now. The main component of the latter is a continuously moving sieve. Fiber material is added through a headbox at a concentration of 1 – 2 % and is subsequently drained and compressed by means of a forming table, suction box, and wet pressing section.

To improve the surface properties of the smooth side of the fiberboard, fiber material of higher quality can be applied to the mat on the forming table through a second headbox. After the wet press the thickness of the mat is 20 – 25 mm, which gives a final thickness of 3 – 5 mm. The mat is then separated into individual sections corresponding to the pressed board format of the hot press.

In the production of *softboard* or *S2S hardboard* by the wet process, the drained fiber mats are dried in single- or multideck dryers at 150 – 170 °C from moisture levels of 100 – 120 % to 1 – 4 %.

To prevent dried boards from catching fire during subsequent storage, they are cooled immediately after drying by spraying with water and aspiration of ambient air.

S1S board is further dewatered in hydraulic multi-daylight presses by hot pressing using an established pressure – time program, compressed, and strengthened by curing the binders that are naturally present or have been added. For S1S, use of a draining sieve is necessary to remove the large quantities of water. For hardboard the pressing time is generally 2.0 – 3.5 min/mm at a pressing temperature of 180 – 200 °C. A typical pressing diagram consists of three phases, the compression and first high-pressure stage, drying, and the second high-pressure stage.

An improvement in the properties of fiberboard can be effected by thermal aftertreatment

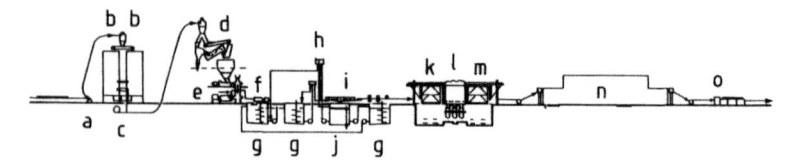

Figure 31. Schematic of fiberboard production by the wet process
a) Chopping machine; b) Silos; c) Chip sorting and disintegration; d) Chip washing; d) Defibrator; f) Refiner; g) Pulp vat; h) Quantity regulator; i) Screen; j) Wastewater vat; k) Pressure feed; l) Hot press; m) Press emptying; n) Thermal curing and climatization; o) Cutting on size

in chambers or continuously in tunnels or by impregnating with drying oils (see Section 2.3.2.4). Treatment results in a more uniform moisture distribution in the board as well as better dimensional stability.

Dry Process. The production of MDF is shown schematically in Figure 32. After defibration (g) the fibers pass down a blow line in which glue (see Section 2.3.2.4) is added and then reach the fiber dryer (i). Alternatively, they can be blown directly into the fiber dryer (i) by making use of the vapor pressure present in the refiner. In this case they are subsequently glue-blended in high-speed centrifugal machines, as in particle board production. In almost all plants, glue blending occurs before drying, thus avoiding spots of adhesive on the board surface that impair its quality. Paraffin added to improve hydrophobic properties is generally charged in the molten state to the lower conveyor screw of the refiner.

Tube dryers are now used exclusively for drying the fibers. They are mostly two-stage dryers with tube lengths of 80 – 100 m and diameters of 1.3 – 1.6 m. The air temperature at the entry is 140 – 170 °C. Because of the high air speed (ca. 30 m/s) and the small dimensions of the fibers, the drying time is only a few seconds [61]. The moisture level of the fibers is adjusted to 9 – 11 % by regulating the exit temperature of the dryer.

The fibers reach the forming station (m) via cyclones and fiber hoppers (j). The fiber mat is formed either mechanically by using a snowfall felter or pneumatically by using a pendistor on a vacuum screen conveyor.

The uniformity of distribution of the fiber material over the surface of the screen significantly affects fiberboard properties.

The mat coming out of the forming station has a very low bulk density and contains a large quantity of air, which must be removed before hot pressing. To decrease the size of the press opening, shorten the closing times of the hot press, and improve the transportability of the mat and the surface quality of the finished board, the mat is precompressed in continuously operating belt prepresses. Mat density is thus doubled or trebled.

Hot pressing of the fiber mat occurs either batchwise on multi-daylight presses with preliminary separation into mat sheets or, as is now increasingly common, continuously on double-belt presses or calenders for thin boards (up to 8-mm thickness). The presses are the same as those used for particle board production (see Section 2.3.1.5). To heat the fiber mat more rapidly and shorten the pressing time, the fiber mat is often subjected to high-frequency preheating before

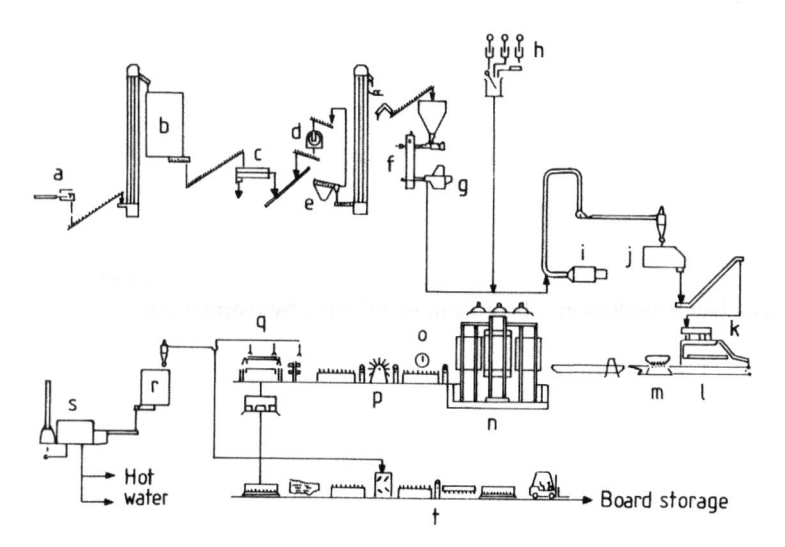

Figure 32. Schematic of MDF production
a) Chopper; b) Chip silo; c) Cleaning; d) Fine chopper; e) Sieve; f) Digester; g) Defibrator; h) Glue preparation; i) Dryer; j) Dry fiber hopper; k) Prepress; l) Spreading machine; m) Forming line; n) Hot press; o) Weigher; p) Cooler; q) Trimming; r) Dust silo; s) Heating boiler; t) Sanding machine

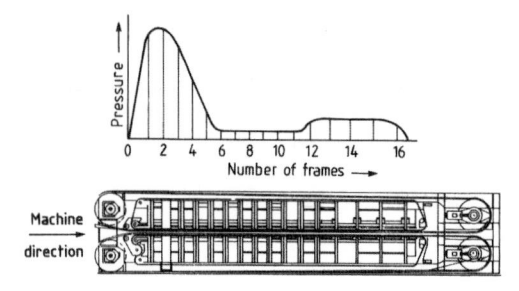

Figure 33. Pressure program for MDF production (Siempelkamp, Germany)

entering the hot press. Pressure control in MDF production differs from that in particle board production. Figure 33 gives an example of the pressure program.

The bulk density profile (perpendicular to the board thickness) and thus the properties of the board can be widely varied through the type of compression and the temperature control. High compression in the outer zones and as uniform a density as possible in the inner zones are aimed at (see Section 2.3.2.6).

2.3.2.6. Properties, Testing, and Uses

Properties. The main factor affecting almost all properties of fiberboard is its *bulk density*. This determines the flexural strength and modulus of elasticity, and affects the transverse tensile strength and swelling properties. Besides bulk density, the bulk density profile perpendicular to the board thickness is most important in the case of MDF. S1S hardboard produced by the wet process has a highly compressed upper side and a less dense underside (sieve side). S2S boards formed by the wet process have a symmetrical density profile with a minimum in the center of the board and maxima at a certain distance from the board surface. This gives the boards high flexural strength [59].

Figure 34 shows two different density profiles of MDF board. Highly compressed outer zones with a high density relative to the center, and the resulting high flexural strength and low transverse tensile strength, are achieved by closing the press rapidly. A more uniform density distribution with smaller density differences between the outer and middle layers, lower flexural strength, and higher transverse tensile strength is achieved

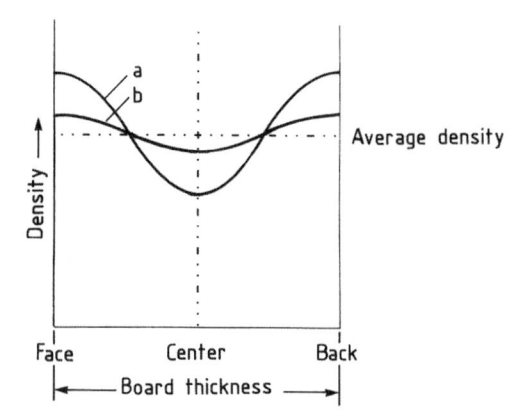

Figure 34. Examples of density profiles of MDF
a) High density contrast; b) Low density contrast at the same average density

by closing the press slowly [62]. In practice, a compromise between the two extremes is usually aimed at.

Quality Requirements and Test Methods for fiberboard are laid down in a large number of international standards. The standards for the United States and the EC are given below.

United States
Insulationboard	ANSI/AHA A 194.1 – 1985
Basic hardboard	ANSI/AHA A 135.4 – 1982
Prefinished hardboard	ANSI/AHA A 135.5 – 1988
Hardboard siding	ANSI/AHA A 135.6 – 1989
MDF	ANSI/AHA A 208.2 – 1990

EC
MDF, Industry Standard Euro MDF Board (EMB) (Gießen – 1990)
Fiberboards (dry process)EN 622 – 1993

Within the framework of production monitoring and quality assurance, the following fiberboard properties are generally tested:

1. Tensile strength perpendicular to the board surface, either dry or after the wet cyclic test
2. Flexural strength and flexural modulus of elasticity
3. Swelling after 24 h or after the wet cyclic test
4. Formaldehyde emission

The following properties are tested less frequently: screw withdrawal, dimensional stability, sand content, and surface absorption.

A comparison of the strength properties of hardboard and MDF with those of solid wood and particle board is given in Figure 35. The advantages and disadvantages of MDF compared with particle board, veneer plywood, and table board are given in Table 11.

Table 11 shows that MDF does not have any advantages over particle board with a fine covering layer with regard to finishing by veneering, laminating, and liquid coating. However, in all further processing in which the homogeneous fine fiber structure in the interior of the board is utilized, MDF has advantages over the other types of board. The edges can be milled, polished, and coated.

Uses. The main areas of use for fiberboard are as follows:

Insulationboard is used in interior and exterior applications, mostly as parts of walls, and for ceiling paneling, frequently in the form of acoustic panels.

Hardboard produced by the wet process generally consists of thin boards. It is employed mainly in buildings for inner and outer wall paneling. It is also used in the production of furniture or doors or molded for use in motor vehicle interiors.

MDF. Because of its excellent edge and surface properties and its good workability, MDF is used mainly for furniture production.

2.3.3. Wood-Based Materials with Mineral Binders

Wood-based materials with mineral binders consist of 30 – 90 % of a hardened inorganic binder and 10 – 70 % of wood particles. Because of the high binder content, these materials are heavy and particularly suitable for use in building. The most important mineral binders are cement, gypsum, and magnesia (magnesite) [64].

Mineral-bonded boards are characterized by low thickness swelling through the action of water. Cement- and magnesia-based boards are highly resistant to mold and termites [65]. Cement-bonded boards are used mainly for exterior applications because of their high resistance to weathering. All mineral-bonded wood-based construction materials are classified as building materials of low flammability because of their high binder content.

2.3.3.1. Cement-Bonded Boards

Cement was the first mineral binder used in the industrial production of boards. The originally long hardening times of this hydraulic binder have been shortened greatly by technical

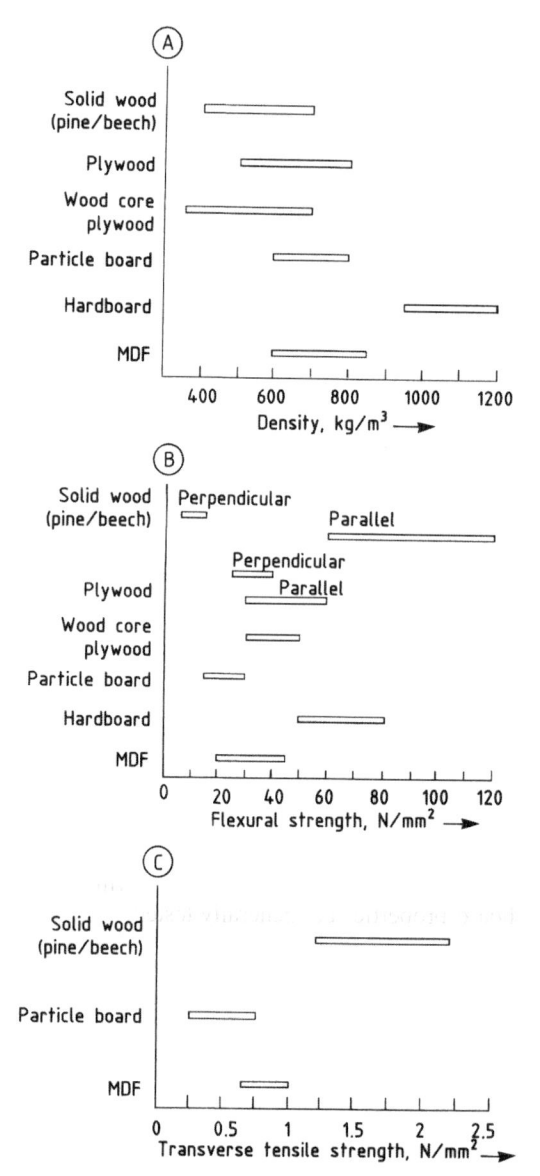

Figure 35. Comparison of density (A), Flexural strength (B) and transverse tensile strength (C) for wood-based construction materials

Table 11. Physicomechanical properties and ease of working of MDF in comparison with other wood-based construction materials [63]

	MDF in comparison with*		
	Particle board	Veneer plywood	Wood core plywood
Density	≈	≈	≈
Flexural strength	−	+	+
Transverse tensile strength	−	+	+
Modulus of elasticity	−	+	+
Linear expansion	+	+	+
Screw holding ability			
Surface	≈		+
Edge	−		+
Ease of molding			
Surface	−		−
Edge	−		−
Ease of varnishing			
Surface	≈	≈	≈
Edge	−		−
Ease of laminating	≈	−	−
Ease of veneering	≈	−	−

* − less favorable than MDF; ≈ same as MDF; + more favorable than MDF.

measures [66]. The setting time depends not only on the reactivity of a particular cement quality, but also on its compatibility with different types of wood. Wood components, such as free sugars, glucosides, and tannins, delay the hydration of cement and therefore impair the setting process [67]. Cement-bonded wood-based boards are produced predominantly by the wet or semidry processes.

The original production process developed by ELMENDORF is still used in different forms [68]. In the process, wood particles, cement, additives, and water are mixed, molded, and compressed. To shorten the hardening times, stacked boards are fed into a heated tunnel under pressure. The boards can generally be removed from their molds after 6 – 8 h and then conditioned for 12 – 18 d in storage until the final strength is reached.

Finished boards consist of ca. 60 wt % cement and additives, ca. 20 wt % wood, and ca. 20 wt % water. The ratio of total water to cement determines board properties such as compressive and flexural strength.

2.3.3.2. Gypsum-Bonded Board

For many years, gypsum has been an indispensable building material. Since the actual strength of pure gypsum is insufficient for the production of board material, it is reinforced with wood fibers, wood wool, or flakes. The suitability of gypsum as a binder is based on the de- and rehydration processes. The water of crystallization is partially removed from mined or industrially produced gypsum by calcination. In the setting process the water added becomes chemically bonded as water of crystallization. The quantity of water that is necessary for complete rehydration depends on the quality of the gypsum and is generally 14 – 21 %. Excess water is not bonded and must be removed from the system.

Wet Production Process. In the production of plasterboard by the wet process, a large excess of water is used. In a process used by the Siempelkamp company, excess water is first removed on a machine wire web during formation of the moldings and subsequently through drying of the boards [69]. The Knauf company in Germany carries out an interesting variant of the wet process in its production of multilayer boards by prefabricating the different layers separately. A particularly imaginative version involves forming and pressing fiberboard with uncalcined gypsum. The boards are calcined in an autoclave, and water is then added so that the desired strength can be obtained by the formation of gypsum hydrate [70].

Semidry Production Process. The semidry process is particularly suitable for the production of gypsum particle board. Only a small excess of water is used (ca. 30 %) so pourable particles can be processed. By decreasing the proportion of water, the energy balance of the gypsum particle board production is improved because less free water needs to be removed from the board by drying.

Industrially produced boards have a moisture content of 15 % after pressing. They are then dried to an equilibrium moisture level of 2 %. The bulk density of the board is ca. 1150 – 1250 kg/m^3. The pressing times for gypsum are generally shorter than those for cement. Further shortening of the setting time is possible [71].

2.3.3.3. Magnesia-Bonded Boards

Magnesia-based binders have been known as Sorel cement for more than 100 years. By calcination of magnesite or dolomite at ca. 800°C, magnesium carbonate ($MgCO_3$) is converted into magnesium oxide (MgO). The latter undergoes a highly exothermic reaction with water and magnesium chloride or sulfate to give an extremely hard material. This binder system has been used for a long time for the production of wood wool boards (trade name Heraklith). Because of the high solubility of magnesium salts, the water resistance of Sorel cement is not particularly great. However, the highly exothermic nature of the setting reaction is very advantageous. Short hardening and pressing times are thus possible at low pressing temperature (100 °C). For the production of magnesia-bonded particle board, ca. 70 % binder and water and ca. 30 % wood chips are used. The binder consists of 5 mol MgO, 1 mol $MgSO_4 \cdot 7 H_2O$, and 15 mol water. Normally, the water is applied to the wood chips; magnesium sulfate is added as a concentrated aqueous solution; and then magnesia is mixed in as a powder. At 100 – 150 °C the pressing time is ca. 10 min. The 18-mm-thick boards have a density of 1050 – 1150 kg/m^3, a transverse tensile strength of 0.65 N/mm^2, and a flexural strength of 13 N/mm^2. The thickness swelling of the boards on storage in water is < 2 %. The economic importance of magnesia-bonded board is currently low.

2.4. Surface Treatment

Wood-based panels are used mainly for interior furnishing and furniture production. Thus the appearance and decorative function of these materials are particularly important. Solid wood – both natural wood and veneers – has a very high aesthetic value. Surface treatment is aimed at increasing the decorative value as much as possible and improving properties of application.

Many techniques are known for producing certain effects in the surface treatment of solid wood or veneers. The nature of the surface can be altered mechanically by planing, grinding, polishing, or sanding, and chemically by bleaching, staining, and varnishing. Wood components in the surface are treated with bleaching or staining agents so that the natural color disappears completely or is rendered much more intense.

Materials based on glued wood particles are generally surface treated. Most are veneered, varnished, laquered, or coated with films, sheets, or laminates. The most important requirement for surface coating is that the surface of the material to be coated should be solid and smooth.

2.4.1. Veneers

The oldest process used for the finishing of wood-based construction materials is the application of veneers.

Today, peeled veneers, in particular, are used mainly for decorating furniture and panels. Chemical stains are often used to enhance natural effects and strengthen color contrasts.

The production of veneers and their processing to give decorative surfaces is described in Section 2.2.2.

2.4.2. Liquid Paints and Coatings

Coating is the most well known type of surface finishing. The coating technique used is determined by the nature of the surface. For glued particle materials such as particle board, MDF, hardboard, wafers, or OSB, which can have very absorbent surfaces, a thick coating is necessary. Besides absorbency, the surface structure of the wood-based construction materials must be taken into account. Here, MDF has advantages over boards made of coarser particles, such as particle board and OSB.

Technological parameters of the carrier board, such as the density of the outer layers, density profile, adhesive content, water absorption of the outer layers, surface hardness, face strength, surface macrostructure, porosity, and sanding quality, have a significant effect on the ease of varnishing. Tests on particle board show that for three-layer boards the outer layer should be made of fine particles so that the surface is as closed and smooth as possible [72]. The quantity of binder in the surface layers should be at least 12 %. The density of the surface layers should be 950 g/cm^3 (max.) and the face strength not less than 1 N/mm^2 [73].

In the industrial coating of wood-based construction materials, attempts are made to compensate for unavoidable deficiencies in the surface quality of the boards by pretreatment with primers or priming films (see Section 2.4.3.1).

On modern coating production lines, finely sanded particle board is cleaned and primed, and a colored undercoat is applied, followed by the actual coating paint. Transparent or pigmented priming or coating systems that can be cured by air or radiation are used. These include:

1. Photochemically curing coating systems (e.g., UV-curing systems),
2. Microwave or electron-beam curing systems, in particular for highly pigmented coatings
3. High-solid coatings with a solid material content of > 80 %
4. Application of prepregs based on unsaturated polyesters
5. Waterborne paints based on water-soluble or dispersed polymer systems

2.4.3. Coating with Sheets, Short-Cycle Films, and Decorative Laminates

Coating with foils, short-cycle films, and decorative laminates is extremely important in industrial finishing. In Western Europe, which has the highest wood-based construction material production worldwide, more than 40 % of the surfaces were coated with these materials in 1992 [74].

2.4.3.1. Coating with Foils

Foils are thin, preformed surfaces that are glued to a wood-based panel carrier. The following types of overlaying exist:

Thermoplastic Foils. The basic material consists of poly(vinyl chloride), polyolefins, or polyester resins, which can be mass dyed, subjected to decorative printing, and embossed on the surface.

Priming Foils consist of paper carrier web, which is impregnated by submersion in an aqueous amino resin solution and dried without touching in a suspension type dryer to a residual moisture level of 3 – 7 %; 60 – 120 g/m^2 paper is used. The resin content of the paper is relatively low – 60 % based on the weight of the paper.

Highly penetrating UF resins in aqueous solution are used for impregnation. These undergo extensive acid-catalyzed condensation during the drying process. Priming foils serve as a basic material for further surface treatment (e.g., printing or coating).

Decorative Coating with Foils. Decorative sheeting – an economically important surface material – also consists of paper that is impregnated with a mixture of amino resin and polyacrylate dispersions. For the production of thin sheeting, ca. 40 g/m^2 paper is used, and for thick sheeting, 80 – 120 g/m^2. The papers are usually printed before impregnation. After the paper web has been dried in a lay-on-air dryer, the surface is varnished by using doctor blades. The varnish layer of 10 – 20 g/m^2 is hardened immediately in a drying area. The sheeting is dispatched for further processing in the form of rolls.

The impregnating and drying processes place high demands on the wet tenacity of the paper. The absorption of water by the paper, its longitudinal swelling, and subsequent shrinkage in the drying tunnel are compensated for by electronic control of the web. Figure 36 shows a impregnating machine with an integrated varnish applicator (doctor blade) developed by the Vilts company in Germany.

The temperatures (100 – 180 °C) in individual drying areas can be controlled separately. The quantity of resin (UF resin – acrylate dispersion) applied to the paper is 40 – 70 % of the weight of the base paper. On applying the wood-graining decorative print, silicone-based wetting inhibitors can be added so that defects obtained during subsequent varnishing look like wood pores.

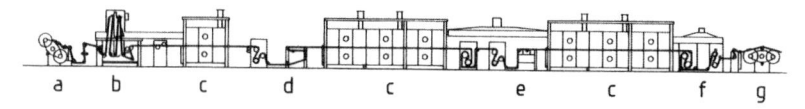

Figure 36. Schematic of an impregnating machine with contact-free drying and integrated coating applicator (Vilts, Germany) a) Paper roll exchange; b) Impregnating station; c) Drying tunnels; d) Applicator; e) Coating station; f) Sheet cooling; g) Sheet rolling and exchange

Edge Coating. In the furniture industry, edge coatings are required that fit in with the decorative surface sheeting. For edge sheeting, paper with a weight of 180 g/m² or more is used. The material itself is more rigid than that used for coating smooth surfaces, so that nonuniformity in the edges to be coated can be compensated for more easily. Paper-based edge sheeting is produced in various qualities. For less demanding applications, the printed paper is impregnated with an aqueous mixture of UF resin and polyacrylate dispersion. If greater resistance is required, etherified MF resins must be used. On impregnation, the paper absorbs 25 – 60 % of its weight of resin.

The resin is cured by using acid catalysts in drying tunnels heated with hot air. Paper-based edge sheeting is generally coated with varnish.

Application of this thermoset edge sheeting to a carrier board is often carried out with simultaneous forming of the edge material (soft forming). Edges with a purely thermoplastic basis, such as poly(vinyl chloride), polyolefin, or polyurethane, are also produced for these applications.

Sheeting and edge strips are fixed to the wood-based construction material with aqueous amino resins, poly(vinyl acetate) adhesives, melt adhesives, or solvent-containing adhesives. Melamine – formaldehyde resins and melt adhesives are frequently applied to the underside of the sheeting as a dry adhesive layer during sheet production. The sheeting can then be laminated onto the carrier board through the action of pressure and temperature in the press.

2.4.3.2. Decorative Films (Short-Cycle Films)

The application of quick-drying decorative films based on melamine – formaldehyde resins is the economically most important process for coating fiber- and particle board. In Europe in 1993, more than 1.3×10^9 m² of wood-based construction material surface was formed by this process.

An important feature of this coating process is the use of low pressure (1.2 – 2.5 N/mm²) at 140 – 230 °C and shorter pressing time (9 – 40 s). This allows direct coating of pressure-sensitive materials.

The high pressure and long pressing time necessary for conventional processing of resin-impregnated papers for laminate production (see Section 2.4.3.3) would lead to undesirable deformation and damage to the structure of the carrier in coating particle board.

The development of melamine – formaldehyde resins that have a sufficient melt flow at low pressure and can thus form a surface was an important prerequisite for the direct press coating of particle board. The company Th. Goldschmidt succeeded in modifying melamine – formaldehyde resins by incorporating elasticizing agents so that crack-free surfaces could be produced [75].

In principle, polyester resins or prepolymers based on diallyl phthalate are also suitable. These resins allow application of lower pressure (e.g., 1.0 N/mm²). However, for impregnating, they must be used in solvents such as acetone or methyl ethyl ketone. Also, the surface resistance and hardness of the much cheaper melamine – formaldehyde resins are not achieved when diallyl phthalate and polyester resins are used.

Highly filled papers – either of one color or with decorative printing – with a weight of 60 – 130 g/m² are used mainly for impregnation. To form a closed surface, 100 – 140 wt % solid melamine resin, based on the weight of paper, is required.

More strongly elasticized MF resins are now generally used for impregnating. In these resins the molar ratio of melamine to formaldehyde is 1:1.5 – 2. Polyols, p-toluenesulfonamide, caprolactam, or sugar, for example, is used as the elasticizing component. Small quantities of wetting agents, mold release agents, and catalysts are also added to the impregnating liquor. The aqueous liquor has a solids content of 48 – 55 %. The impregnation of paper is usually carried out in two stages to obtain a more uniform application of resin. This is achieved by predrying after the first stage. Applying the necessary quantity of resin to the paper from a 55 % liquor in a single step is difficult.

Two-stage impregnation also offers the possibility of replacing the relatively expensive MF resin partially or completely in the first stage by a more economical UF resin. After the first impregnation step, the papers are dried to a residual moisture level of 10 – 15 %. They are then coated once more on both sides with an MF-resin-based liquor using doctor blades or gravure rollers. Subsequent drying in a suspended dryer to a residual moisture level of 5.5 – 7 % gives films that are supplied for coating particle board in the form of rolls or stacked sheets. During drying the paper web is led through a tunnel of drying areas on a cushion of hot air introduced from above and below by a system of jets. Figure 37 shows the schematic arrangement of the jets and the progress of the paper web through a drying tunnel. Figure 38 shows a machine for impregnating and drying decorative films by the two-stage process (lay-on-air dryer).

The amino resin absorbed by the paper undergoes further condensation during drying, but not complete cross-linking. Final curing of the resin takes place only on coating the fiber- or particle board carrier.

Particle board used for coating must satisfy particular requirements. Generally, three-layer particle board is used, which has a highly compressed, fine outer layer with high face strength. The density of such boards is 700 kg/m^3, and the face strength is at least 1.2 N/cm^2. For more highly stressed surfaces, two films are sometimes pressed at the same time, so that possible non-uniformity of the carrier material can be compensated for more effectively.

In the *Unibord process*, decorative films impregnated with melamine resin are placed on the particle mat during actual particle board production so that a coated board can be formed in a single pressing process [76]. The process did not achieve economic success because the indispensable closed-circuit cooling process in the press requires too much energy and prolongs the cycle times of the plant.

In developing new processes for the continuous production of particle board, the idea of direct coating of boards during production is expected to be reinvestigated.

2.4.3.3. Decorative Laminates

Decorative laminates are produced by pressing several layers of paper impregnated with synthetic resin in heated, hydraulic presses at high

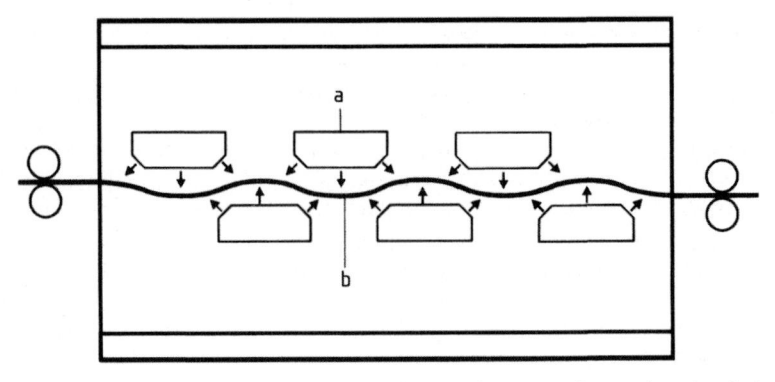

Figure 37. Schematic arrangement of the air jets and the course of paper through a drying tunnel
a) Air jets; b) Paper web

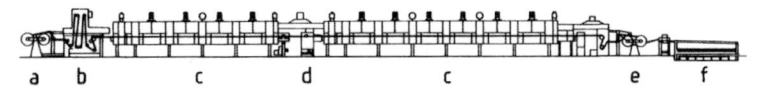

Figure 38. Schematic of a production plant for decorative film according to the two-stage process (Vilts, Germany)
a) Paper roll exchange; b) Impregnating station; c) Drying tunnel; d) Screen roller applicators; e) Paper rolling device; f) Transverse cutter and sheet deposition

pressure. The resin melts and hardens to give a composite material in which the cellulose fibers of the paper have a reinforcing and strengthening function. The number of resin-containing paper layers determines the laminate thickness.

The central layer of a laminate is generally formed from several sheets of kraft paper impregnated with phenolic resin. This central layer is coated with an underlay, a decorative, and an overlay paper; these three layers of paper are impregnated with MF resin. The use of MF resin has the advantage that any decorative coloration can be applied, because melamine – formaldehyde resins become absolutely colorless and transparent on pressing. Phenol – formaldehyde resins have a red-brown color and are thus not suitable.

High-Pressure Laminates.. Production of high-pressure laminates (HPL) is a batchwise process. The pressure in the hydrualic multi-daylight presses is $80 - 120$ bar, and the pressing temperature is $140 - 180$ °C. After pressing, the undersides of the layered sheets are sanded to facilitate subsequent gluing of these laminates to a carrier board. In the pressing process, several laminates are usually produced in one opening. Plates for modifying the structure allow a wide range of surface forms to be made.

To compensate for tension in the composite material and, in particular, to form a glossy surface, HPL is cooled under pressure to a temperature of ca. 80 °C before it is released from the mold. The pressing cycle times can be 30 min or more, depending on the thickness and number of laminates. Production of HPL is cost intensive because of the long pressing times required.

Continuous Pressed Laminates (CPL) are produced on double-belt continuous presses (Fig. 39) at pressures up to 50 bar with resin-impregnated papers from a roll.

Unlike the batch process, there is a continuous increase in pressure in the material after passing the entry of the continuous press. For technical reasons CPL can be produced economically only in a thickness range of $0.4 - 1.2$ mm.

The covering of wood carrier boards with the prefabricated laminate is effected by gluing under pressure, as in the production of veneers. Decorative laminates are rigid, very stable composite materials. They can also be glued to wood carrier boards with low requirements as regards surface quality. Amino resins, poly(vinyl acetate), or melt adhesives are used for gluing.

Although decorative laminates are regarded as thermoset surface materials, thermoformability can be achieved by employing certain measures. Postforming laminates are obtained by elasticizing the impregnating resin used and incomplete curing (cross-linking) of the resins during high-pressure laminating. Local heating using IR radiation or contact heat permits a single permanent forming even with a small radius of curvature and angles greater than 180°.

2.4.4. Properties, Testing, and Uses

Coating wood-based construction materials aims at both the production of a decorative effect and the improvement of performance characteristics. The coating protects these materials from chemical and physical attack.

Properties. The following properties are important for assessment of a surface:

1. Mechanical stability: scratch resistance, impact strength, abrasion resistance, and crack resistance
2. Resistance toward physicochemical attack by water and steam (loss of gloss), food and chemicals, weathering, heat, radiation (sunlight), and fire

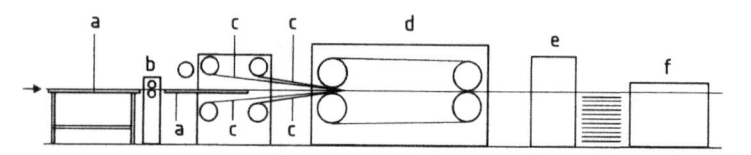

Figure 39. Schematic of a double-belt press for continuous lamination of particle board (Hymmen, Germany)
a) Boards; b) Cleaning; c) Saturated papers; d) Double-belt press; e) Saw; f) Double end tenoner (groove and tongue)

Test Standards. Besides standards for the assessment of painted surfaces, a number of national and international regulations exist for the classification, testing, and application of thermosets, which are widely used for coating wood-based materials. Classification/specification and determination of properties of HPL are listed in ISO 4586–1 and ISO 4586–2, respectively.

In Europe, particularly in Germany, coating of wood-based materials with short-cycle melamine films is widely used. Special standards were developed for assessment of these surfaces [77].

Uses. The surface coating of wood-based materials has been the major impetus for mass processing these materials. More than half of the fiber- and particle board produced worldwide is currently used in the mass production of furniture and building materials. Their application in wall and ceiling panels has increased considerably. Decorative laminates have revolutionized the entire kitchen furniture industry. Laminates that have been rendered highly abrasion resistant by incorporation of corundum, are being increasingly used as upper surfaces of floors. Plywood and particle board that are laminated with papers impregnated with phenol − formaldehyde resins can be repeatedly used for concrete shuttering. The use of short-cycle films in the processing of particle board moldings and profiles has also proved successful. Here, the coating and the compression molding are applied simultaneously (Werzalite process) [78].

2.5. Environmental and Toxicological Aspects

2.5.1. Wood Dust

In the production and processing of wood-based materials the formation of fine dust particles cannot be avoided. Suspicion has existed for a long time that prolonged occupational contact with wood dust increased the risk of development of adenocarcinomas in the inner nose. According to ACHESON workers in the wood industry have a thousandfold greater risk of developing nose cancer than other workers; he based his theory on statistical investigations in the wood industry at High Wycombe (United Kingdom) from 1956 to 1965 [79]. No clear explanation has been found as to why wood dust should be carcinogenic. Natural wood components, materials used in working and processing wood, metabolites of wood fungi, pyrolysis products from processing, or simple mechanical irritation could be responsible for the development of adenocarcinomas [80]. Wood components particularly those of hard deciduous woods, have been suggested to be mutagenically carcinogenic [81]. Besides a large number of mutagenicity tests with wood extracts, epidemiological investigations of nose cancer in various countries, in particular, have led to the formulation of preventive measures. In many countries, adenocarcinomas of the nose have been recognized as occupational diseases of workers in the wood industry.

2.5.2. Emissions

In the production of wood-based materials, processes are used in which the main components, such as wood, adhesives, varnishes, paints, and coatings and other additives, are heated. In these processes, substances are formed or liberated that constitute emissions from the production plant. Emissions from finished materials are also sometimes detected. These are a nuisance and impair the utility of the material.

2.5.2.1. Emissions during Production of Wood-Based Materials

Dust. In large wood processing plants such as fiber- and particle board factories, considerable quantities of dust are produced in cutting and sanding processes. The dust produced from sanding machines is also polluted with adhesives and additives. Removal of dust from waste gases is mandatory. Dust separation is carried out in filters, and the dust is incinerated to produce energy.

Emission of Wood Components. Emissions are also produced during the drying of wood, which depend mainly on the intensity of the drying process. Waste gases with intense odors are liberated from wood along with water vapor. These consist of wood components such as pinenes, terpenes, phenol derivatives, aldehydes, and alcohols. The quantity of these substances can be reduced by mild drying at lower temperature, or they can be eliminated by using gas scrubbers with up to 60 % efficiency.

Adhesives. For the production of glued wood-based construction materials, several million tonnes of adhesive resins are used per year.

Formaldehyde. Condensation products based on amino and phenol – formaldehyde resins are the most important adhesives economically. On hot pressing of glued particles, fibers, or veneers, formaldehyde is emitted. In the waste gas from a 12-daylight press in a particle board factory, 13 kg of formaldehyde per hour was measured for UF adhesives and only 0.6 kg/h for PF adhesives. The quantity of formaldehyde produced in the pressing process depends on the molar ratio of urea to formaldehyde in UF resins, pressing conditions, the moisture level of the molding, and the quantity of catalyst used [82].

Formaldehyde is a very reactive gas with a pungent odor, which can be detected even in very low concentrations. The odor threshold ranges from 0.06 to 1.2 mg per cubic meter of air (0.05 – 1 ppm) and varies from one individual to another.

The hazard potential of formaldehyde to humans is due to irritation of the eyes and mucous membranes of the respiratory passages and to skin allergies (→ Formaldehyde, Chap. 10.).

After two American animal experiments in 1980 showed that inhalation of high concentrations of formaldehyde over a long period can cause nose tumors in rats, extensive research was initiated worldwide into its possible carcinogenicity. The results of these and numerous epidemiological studies are now available. In animal experiments, varying results were obtained, from which carcinogenicity in humans could not be derived. In 1994 the UNO/WHO-IRAC had no information to support the assumption of carcinogenicity. As a precaution, emission and imission values for formaldehyde and maximum workplace concentrations have been officially regulated in almost all industrialized countries.

On using PF resins, a low emission of *phenol* or (hydroxymethyl)phenols must be expected. The more strongly precondensed the PF resin is, the lower are the emissions on hot pressing.

In the processing of PMDI monomeric components (MDI) can be emitted during pressing. The high reactivity of this compound with water usually leads to immediate, nontoxic polyurea formation.

During the drying of resin-impregnated paper for coating wood-based construction materials, formaldehyde, phenols, and sometimes additives are emitted. In modern plants, waste gases from the dryers are usually purified by incineration. Scrubbing these gases would be time consuming and expensive since 15 000 – 20 000 m^3 of waste gas are produced per hour per plant. If melamine – formaldehyde resins are used for impregnation, waste gases from the dryer contain 0.3 – 0.6 g formaldehyde per kilogram of solid resin. Under the same conditions, UF resins emit up to three times this quantity. Lamination of resin-impregnated and dried papers to the wood-based construction material generally results in low formaldehyde emissions.

2.5.2.2. Emissions from Finished Materials

The emission of formaldehyde from wood particle board was first reported in 1962 [83]. The emissions are caused by the adhesive used, whereby formaldehyde is liberated by hydrolysis of the condensation product [44]. In the meantime, a large number of publications on this subject have appeared [84], in which the causes are determined and possibilities for lowering emissions are described [55], [85]. Many methods of measurement have been developed to adhere to the limiting values stipulated by national authorities. In Europe the perforator method (EN 120) and the gas analysis method (EN 717–2) are used as process controls to monitor emissions from particle board.

Through international cooperation, conditions for a chamber test have been worked out. In this test, the formaldehyde concentration in ambient air is measured as an equilibrium concentration, and the emissions of the material are thus assessed.

In many countries, wood-based construction materials are divided into emission classes E1 – E3 according to their formaldehyde-emitting capacity. These classes represent certain ranges of values determined by the perforator method. For example, for boards in emission class E1, the consumer need not fear any emissions on proper use of the board.

No complaints of emissions from boards bonded with *phenolic resins* have been reported thus far. Values for emissions from *PMDI-bonded boards* have not yet been published.

2.5.3. Disposal

In the production of modern wood-based construction materials, considerable quantities of nonwood products are frequently used. The important materials remaining in the wood are adhesives, varnishes, paints, coating materials, dyes, wood preservatives, and flame retardants.

The waste material formed during production is mostly recycled to the process or incinerated to produce energy. Dumping is used in only a few exceptional cases, when allowed. Wood-based construction materials that are no longer utilizable and those that are at the end of their useful life are combusted. Here, their burning behavior must be taken into account, particularly in the case of nonwood substances. Incomplete combustion favors the emission of hazardous waste gas.

The emission of nitrogen oxides is determined mainly by the nitrogen content of the fuel [86]. For uncoated fiber- or particle board glued with amino resin, the nitrogen content is 3.5 – 5 %. For materials glued with phenolic resin, the alkali incorporated with the resin inhibits propagation of burning. Wood preservatives must be incinerated under controlled conditions and the ashes disposed of separately. Chloride-containing curing aids for amino resins or coatings based on poly(vinyl chloride) give rise to the hazard of formation of polychlorinated dibenzodioxins and dibenzofurans [87]. By controlled incineration, the behavior is improved and the risk of formation of these hazardous substances is decreased.

2.6. Polymer Wood

Polymer wood consists of solid wood or wood-based materials that have been impregnated with monomers or synthetic resin solutions. The plastic components are then cured thermocatalytically or by radiation.

Plastics are incorporated in the wood texture to give a composite material with improved physical and chemical properties. The increase in strength of the wood can be seen principally in improved hardness and transverse compressive strength. The abrasion resistance of polymer wood is determined by the hardness and chemical structure of the polymer components. The dimensional stability and sorption properties are also improved, the latter being due to a significant reduction in the rate of water absorption. The effect is limited only by the fact that most monomers are deposited in the cell lumens and capillaries, but not in the cell wall itself where water is active.

In the production of polymer wood, attention is paid to the differing impregnation capacities of various types of wood and of sap- and heartwood. The liquid monomers or prepolymer solutions are applied by the vacuum-pressure process to accelerate penetration. The proportion of polymer components (30 – 60 %) determines the properties of the composite material. Direct grafting of the monomer onto the cellulose of wood is desirable [88]. Chemical catalysts, thermal energy, or radiation is used to initiate polymerization. The use of higher temperatures has the disadvantage that considerable proportions of volatile monomers are lost from the system by evaporation. Also, some wood components may inhibit polymerization and cause incomplete conversion of the monomer. The use of high-energy γ-radiation (cobalt-60) is very effective to achieve polymerization in wood [89]. In practice, a radiation dose of ca. 4×10^4 Gy (4 Mrad) is used. The surface hardness can thus be improved by up to 25 % compared with purely thermal polymerization.

The following have been proposed as monomers or prepolymer solutions:

1. Condensation resins (e.g., UF, MF, and PF resins)
2. Polyester and epoxy resins
3. Diisocyanates (e.g., MDI monomer, prepolymer, and polymer)
4. Acrylic esters, acrylonitrile, vinyl chloride, styrene, and maleic anhydride

The degree to which these substances undergo direct chemical bonding with cellulose in wood is not known for many monomers. For example, no urethane groups could be detected when diisocyanates were used, which indicated that reaction with the hydroxyl groups in wood had occurred [90]. The NCO groups can therefore be assumed to have reacted with water in the cell wall to form polyurea.

On impregnating oak with UF resin, a resin absorption of 43 kg/m^3 is achieved, and with sapwood-rich pine, 90 kg/m^3. Water absorption decreased and the compressive strength of the impregnated wood increases with increasing absorption of resin. On prolonged storage in water, the absorption of water increases again probably due to leaching of components of the impregnating agent [91].

Some of the monomers proposed present problems in industrial utilization for toxicological reasons because both the handling of these substances and their quantitative conversion in wood are difficult. A large number of processes for polymer wood production have been worked out and patented [92].

Nevertheless, polymer wood has not yet been widely used industrially. Economic reasons are mainly responsible for the reluctant utilization of this material. Polymer wood production is expensive, and its subsequent working with tools is difficult. Its resistance to weathering is not particularly high. Polymer wood is used for floors (parquet) and also for sports apparatus, tool parts, rifle shafts, and highly stressed industrial hardware.

2.7. Economic Aspects

The worldwide economic development of the wood processing industry was very positive during the decade from 1981 to 1991. The main areas of use for wood-based boards are the furniture and building industries. From 1981 to 1991, worldwide production of wood-based boards increased from 100 to 122×10^6 m^3 or by ca. 22 % [74], [93]. All types of board have contributed to this growth. The highest growth rate, 33.8 %, has been in *fiberboard* production (MDF and hardboard). The production of *particle board* has increased by 27 % to 49.5×10^6 m^3, whereas that of plywood and veneer plywood has achieved a growth rate of 18 %, and light fiberboard (*softboard*) a growth rate of 6.5 %.

Worldwide production and consumption figures for wood products are given in Table 12.

The worldwide distribution of the production of wood-based construction materials gives an indication of the economic development of a region and the availability of corresponding raw material resources (Fig. 40).

Table 12. Production and consumption of wood products worldwide (in 1000 m^3)

Product	Year	
	1981	1991
Cordwood	2 926 027	3 429 426
Wood-based panels	100 363	122 425
Particle board	39 020	49 459
Plywood	40 329	47 636
Fiberboards (including softboard)	16 517	20 201

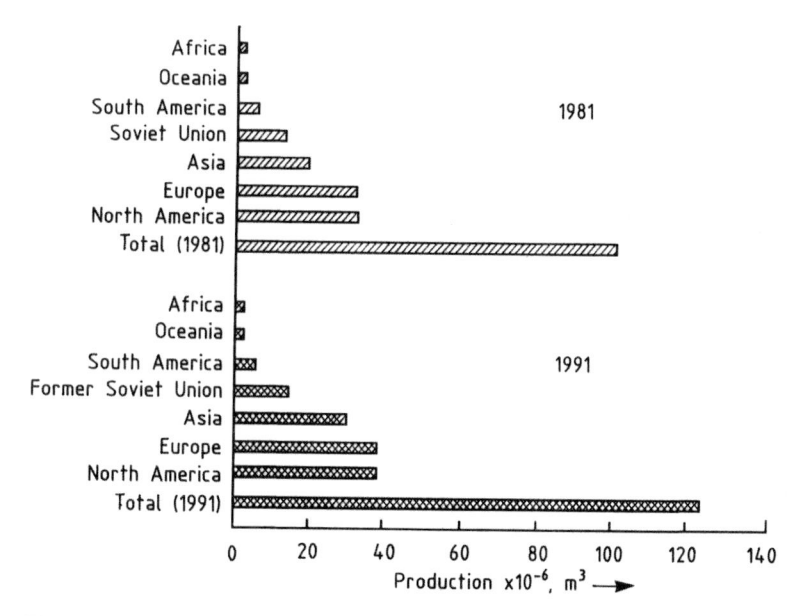

Figure 40. Worldwide production of wood-based panels in 1981 and 1991

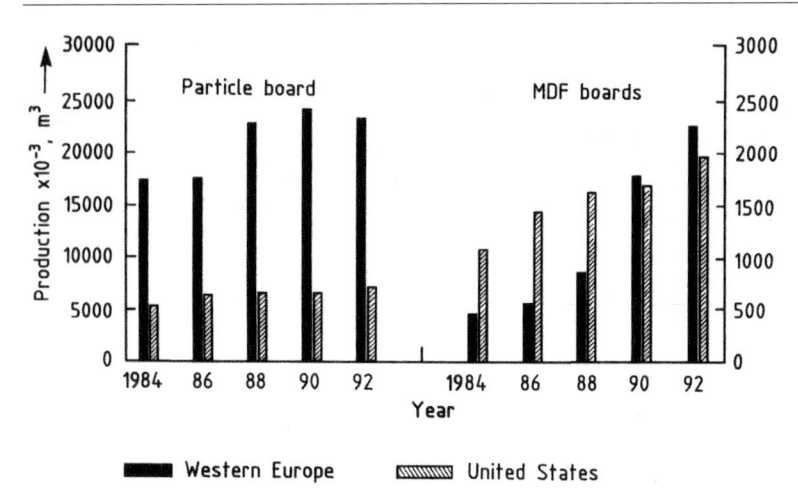

Figure 41. Development of particle board and MDF production in Western Europe and the United States from 1984 to 1992

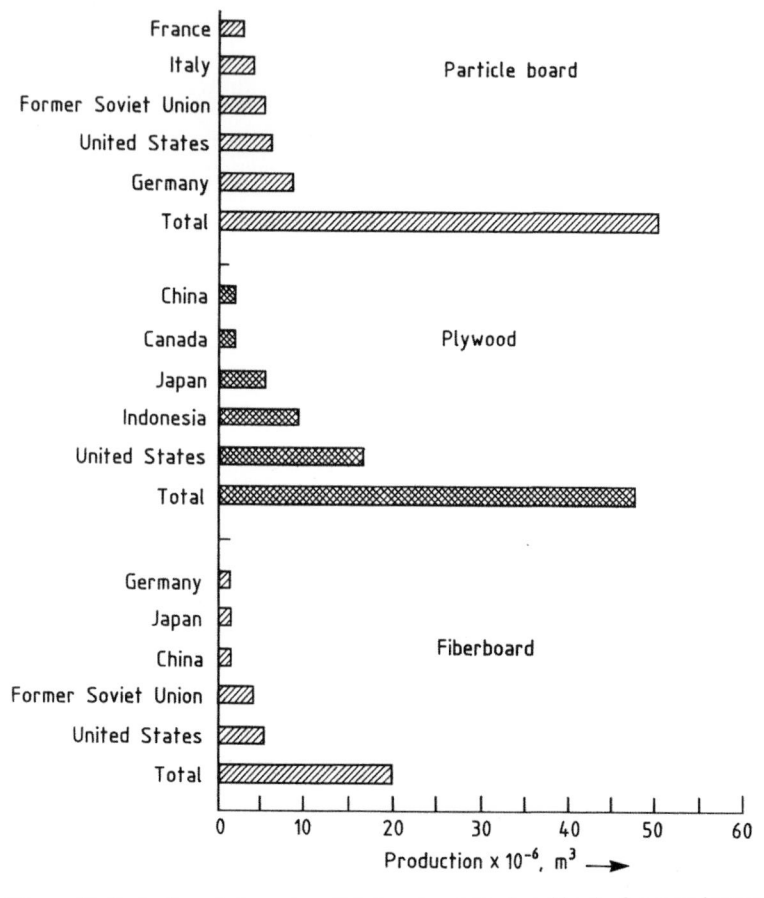

Figure 42. Production of plywood, particle board, and fiberboard by the five most important producer countries in 1991

If the three largest classes of wood-based construction materials – particle board, plywood, and fiberboard – are compared, since the 1960s particle board production has overtaken that of plywood, which was previously the leader in terms of quantity. A similar development is likely to occur with fiberboard (MDF). This can be seen, in particular, from a comparison between MDF production in Western Europe and in the United States from 1984 to 1992 (Fig. 41). In Figure 42 the quantities of particle board, plywood, and fiberboard produced by the five most important producer countries are compared.

Production of glued particle wood-based construction materials has increased markedly. The lack of round timber with large dimensions favors the use of lower-quality wood, mainly for fiber- and particle board. Improvement of the quality of glued particle wood-based construction materials, particularly in industrial processing, has also contributed to their reputation and use.

References

General References

1. H.-H. Bosshard: *Holzkunde,* vol. 2: Zur Biologie, Physik und Chemie des Holzes, Birkhäuser-Verlag, Basel 1984.
2. Deutsches Institut für Normung: *Normen über Holz,* DIN-Taschenbuch 31, Beuth-Verlag, Berlin 1992.
3. J. M. Dinnwoodie: *Timber – its Nature and Behaviour,* Van Nostrand Reinhold Co., New York 1981.
4. D. Fengel, G. Wegener: *Wood – Chemistry, Ultrastructure, Reactions,* 4th ed., De Gruyter, Berlin – New York 1989.
5. W. E. Hillis: "Heartwood and Tree Exudates," in T. E. Timell (ed.): *Springer Series in Wood Science,* Springer-Verlag, Berlin 1987.
6. E. A. Wheeler, P. Baas, P. E. Gasson (eds.) in IAWA (International Association of Wood Anatomists): "List of Microscopic Features for Hardwood Identification, " *IAWA Bull.* **10** (1989) 219 – 332.
7. F. W. Jane: *The Structure of Wood,* 2nd ed., Adam & Charles Black, London 1970.
8. W. Knigge, H. Schulz: *Grundriß der Forstbenutzung,* Paul Parey, Hamburg 1966.
9. F. Kollmann: *Technologie des Holzes und der Holzwerkstoffe,* 2nd ed., vol. 1, Springer-Verlag, Berlin 1951.
10. R. Mombächer (ed.): *Holz-Lexikon,* 3rd ed., vols. 1 and 2, DRW-Verlag, Stuttgart 1988.
11. P. Niemz: *Physik des Holzes und der Holzwerkstoffe,* DRW-Verlag, Stuttgart 1993.
12. K. V. Sarkanen, C. H. Ludwig: *Lignins, Occurrence, Formation, Structure and Reactions,* Wiley-Interscience, New York 1971.
13. E. Strasburger: *Lehrbuch der Botanik,* 33rd ed., G-Fischer-Verlag, Stuttgart – Jena – New York 1991.
14. R. Trendelenburg, H. Mayer-Wegelin: *Das Holz als Rohstoff,* 2nd ed., Hauser Verlag, München 1955.
15. R. Wagenführ, C. Scheiber: *Holzatlas,* 3rd ed., VEB-Fachbuchverlag, Leipzig 1989.
16. K. Wilson, D. J. B. White: *The Anatomy of Wood its Diversity and Variability,* Stobart & Son, London 1986.
17. H. Willeitner, E. Schwab (eds.): *Holz-Außenverwendung im Hochbau,* Verlagsanstalt Alexander Koch GmbH, Stuttgart 1981, p. 27.
18. M. Paulitsch: *Moderne Holzwerkstoffe,* Springer-Verlag, Berlin 1989, p. 59.

Specific References

19. K. Dreiner et al.: "Holz als umweltfreundlicher Energieträger – Eine Nutzen-Kosten-Untersuchung", *Schriftenr. Bundesmin. Ernähr. Landwirtsch. Forsten,* Reihe A: Angewandte Wissenschaft 432 (1994) 1.
20. L. Ajtay, P. Ketner, P. Durigneud: "Terrestrial Primary Production of Phytomass," in B. Berlin, E. T. Degens, S. Kunze, B. Ketner (eds.): *The Global Carbon Cycle,* J. Wiley and Sons, Chichester 1979, pp. 129 – 181.
21. E. F. Brünig, T. W. Schneider: "Verfügbarkeit forstlicher Rohstoffe," *Mitt. Bundesforschungsanst. Forst Holzwirtsch.* **130** (1980) 13 – 22.
22. FAO: *Yearbook of Forest Products 1961 – 1991,* Rome 1993.
23. Bundesministerium für Ernährung, Landwirtschaft und Forsten: *Nationaler Waldbericht der Bundesrepublik Deutschland,* 1994, pp. 611–617.
24. H. Ollmann: "Holzbilanzen 1991 und 1992 für die Bundesrepublik Deutschland", *Arbeitsbericht des Instituts für Ökonomie 93/3,* Bundesforschungsanstalt für Forst- und Holzwirtschaft, Hamburg, Aug. 1993.
25. FAO: *Yearbook of Forest Products 1945 – 1993,* Rome 1995.
26. R. Wagenführ: *Anatomie des Holzes,* 4th ed., VEB-Fachbuchverlag, Leipzig 1989.
27. J. Bauch: "Variation of Wood Structure due to Secondary Changes," in J. Bauch (ed.): Natural Variations of Wood Properties, *Mitt. Bundesforschungsanst. Forst Holzwirtsch.* **131** (1980) 69 – 97.
28. R. M. Brown, D. Montezinos: "Cellulose Microfibrils: Visualization of Biosynthetic and Orienting Complexes in Association with the Plasma Membrane, " *Proc. Natl. Acad. Sci. USA* **73** (1976) 143 –147.
29. W. Herth: "Plant Cell Wall Formation," in A. W. Robards (ed.): *Botanical Microscopy,* Oxford University Press, Oxford 1985, pp. 285 – 310.
30. E. Sjöström: *Wood Chemistry,* Fundamentals and Applications, 2nd ed., Academic Press, San Diego 1993, p. 249.
31. K. Freudenberg in K. Freudenberg, A. C. Neish (eds.): *Constitution and Biosynthesis of Lignin,* Springer-Verlag, Berlin 1968, pp. 47 – 116.
32. H. H. Nimz: "Das Lignin der Buche – Entwurf eines Konstitutionsschemas", *Angew. Chem.* **86** (1974) 336; *Angew. Chem. Int. Ed. Engl.* **13** (1974) 313.

33 T. E. Timell: "Recent Progress in the Chemistry of Wood Hemicelluloses, " *Wood Sci. Technol.* **1** (1967) 45 – 70.
34 B. J. Fergus, A. R. Procter, J. A. N. Scott, D. A. J. Goring: "The Distribution of Lignin in Sprucewood as Determined by Ultraviolet Microscopy, " *Wood Sci. Technol.* **3** (1969) 117 – 138.
35 J. A. N. Scott, A. R. Procter, B. J. Fergus, D. A. J. Goring in K. V. Sarkanen, C. H. Ludwig (eds.): *Lignin, Occurence Formation, Structure and Reactions*, Wiley-Interscience, New York 1971, p. 52.
36 N. Losberger: "Normbalken und Kundenwunsch, " *Holz Zentralbl.* **117** (1991) no. 42, 633.
37 R. Hinterwaldner: "Verleimung von tragenden Holzbauteilen, " *Holz Zentralbl.* **116** (1990) no. 30, 445 – 446.
38 K. Egner, H. Kolb: "Untersuchungen zum Alterungsverhalten für tragende Holzbauelemente, " *Holz Roh Werkst.* **24** (1966) no. 10, 439 – 442.
39 K. Egner: "Die Leimung tragender Holzbauteile, " *Holzbau* 1976, no. 4, 94 – 95.
40 F. Kollmann (ed.): *Furniere, Lagerhölzer und Tischlerplatten*, Springer-Verlag, Berlin 1962.
41 F. Kollmann, E. W. Kuenzi, A. J. Stamm: "Wood Based Materials", *Principles of Wood Science and Technology*, vol. II, Springer-Verlag, Berlin 1975, p. 223.
42 A. Pizzi (ed.): *Wood Adhesives*, Marcel Dekker, New York 1983.
43 IG Farbenindustrie (BASF), DE 550 647, 1929 (K. Vierling, M. Schmihing, H. Klingenberg).
44 B. Meyer: *Urea-Formaldehyde Resins*, Advanced Book Program, Addison-Wesley, Reading, Mass. 1979.
45 H. E. Edlund, E. Justuk, *Holz Zentralbl.* **107** (1981) no. 192, 1720 – 1721.
46 O. Kreibaum, *Holz Roh Werkst.* **15** (1957) 1.
47 Ji-Te Aktiebolag, SE 134 584, 1948 (H. Baumann).
48 R. Runkel, J. Jost, DE 841 055, 1948.
49 J. F. Werz KG, DE 1 055 125, 1956 (E. Munk).
50 C. Neumann, Ph. D. Thesis Universität Hamburg 1970.
51 A. A. Moslemi: *Particleboard volume 1, Materials; Particleboard volume 2, Technology*, Southern Illinois University Press, Carbondale – Edwardsville 1974.
52 K. Wishered, J. Wilson, *For. Prod. J.* **29** (1979) no. 2, 35 – 37.
53 H. J. Deppe, K. Ernst: *Taschenbuch Spanplattentechnik*, 3rd ed., DRW-Verlag Weinbrenner KG, Stuttgart 1991.
54 MCN Nederland, EP 0 062 389, 1982 (J. J. Hoetjer). Borden Inc., US 4 536 245, 1985 (D. W. Shiou, E. Smith).
55 VEB Leuna, EP 0 037 878, 1981 (J. Barse *et al.*). Statens Skogsindustrier, EP 0 027 583, 1980 (A. W. Westling). Fraunhofer-Gesellschaft, DE-OS 2 851 589, 1980 (E. Roffael, L. Mehlhorn).
56 A. Schneider, E. Roffael, H.-A. May, *Holz Roh Werkst.* **40** (1982) 339 – 344.
57 M. Paulitsch: *Methoden der Spanplattenuntersuchung*, Springer Verlag, Berlin 1986.
58 *Asian Pac. For. Ind.* **12** (1991) 41 – 42.
59 O. Suchsland, G. E. Woodson: *Fiberboard Manufacturing Practices in the United States*, Forest Products Research Society, Madison, 1990, p. 263.
60 H. Lampert: *Faserplatten*, VEB Fachbuchverlag, Leipzig 1966, 453 p.
61 G. Gran, *Holztechnologie* **28** (1987) 143 – 146.
62 O. Suchsland, G. E. Woodson: "Effect of Press Cycle Variables on Density Gradient of Medium-density Fiberboard," in T. M. Maloney (ed.): *Proc. 8th Particleboard Symposium, Pullmann*, Washington State University, Pullmann 1974, pp. 375 – 398.
63 Siempelkamp GmbH, *Holz Roh Werkst.* **40** (1982) 77 – 80.
64 F. Kollmann: *Technologie des Holzes und der Holzwerkstoffe*, Springer-Verlag, München 1951.
65 J. M. Dinwoodie, B. H. Paxton: "The Long-term Performance of Cement-bonded Wood Particle Board in A. A. Moslemi" (ed.): *Inorganic Bonded Wood and Fiber Composite Materials*, Forest Products Research Society, Madison, pp. 45 – 54.
66 Y. Aoki: "The Manufacture of Heat Hardening Cement Particleboard," in [65], pp. 35 – 44.
67 A. Kasim, M. H. Simatupang, *Holz Roh Werkst.* **47** (1989) 391 – 396.
68 U.S. Secretary of Agriculture, US 4 393 019, 1966 (R. L. Geimer).
69 Siempelkamp, DE 2 751 473, 1977.
70 A. Sattler, K. Lempfer in [47], pp. 19 – 25.
71 M. H. Simatupang, N. Seddig, DE 4 031 935, 1990.
72 C. v. Bismarck, *Holz Roh Werkst.* **37** (1979) 9 – 15.
73 EN 311 "Surface soundness of particleboard," 1992.
74 Fédération Europeén des Syndicats de Fabricants de Panneaux de Particule, FESYP Annual Report 1993, Brussels.
75 Goldschmidt AG, DE 1 954 383, 1969 (R. Mitgan, U. Hohlschmidt).
76 Uniboard AG, DE-OS 1 653 231, 1967 (R. Kunz).
77 DIN 68 861 1 – 8, Verhalten von Möbeloberflächen, 1981. DIN 68 751, Kunststoffbeschichtete dekorative Holzfaserplatten, 1987. DIN 53 799, Prüfung von Spanplatten mit dekorativen Oberflächen auf Aminoplastbasis, 1986. ISO 9352, Determination of resistance to wear by abrasive wheels, 1989.
78 Werz KG, DE 1 055 225, 1956 (E. Munk).
79 E. D. Acheson, R. H. Codwell, E. Hadfield, R. G. Macbeth, *Br. Med. J.* **2** (1968) 587 – 596.
80 H. Wolf *et al.*, *Arbeitsmed. Sozialmed. Präventivmed.* 1986, special issue 7.
81 J. B. Harborne, *Sci. Rep. Nr. 1*, Medical Research Council (MRC) Environmental Epidemiology Unit, Southampton 1982.
82 H. Petersen, W. Reuther, W. Eisele, O. Wittmann, *Holz Roh Werkst.* **30** (1972) 420 – 436; **31** (1973) 463 –469; **32** (1974) 402 – 410.
83 O. Wittmann, *Holz Roh Werkst.* **20** (1962) 221 – 224.
84 H. J. Deppe in K. Aurand, B. Seifert, J. Wegner (eds.): *Luftqualität in Innenräumen*, G.-Fischer-Verlag, Stuttgart 1982.

85 Fraunhofer-Gesellschaft, DE 2 829 021, 1980 (E. Roffael, L. Mehlhorn).
86 T. Nußbaumer, *Holz Roh Werkst.* **49** (1991) 445 – 450.
87 M. Strecker, R. Marutzky, *Holz Roh Werkst.* **52** (1994) 33 – 38.
88 R. Schandy, J. Wendrinsky, E. Proksch, *Holzforschung* **36** (1982) 197 – 206.
89 J. F. Siau, J. A. Meyer, *For. Prod. J.* **16** (1966) no. 8, 47 – 56.
90 A. Burmester, *Holz Roh Werkst.* **28** (1970) no. 5, 183 – 186.
91 K. Erler, D. Knospe, *Holz Roh Werkst.* **46** (1988) 327 – 329.
92 Dow Chemical Co., DE 2 707 368, 1977 (W. E. Broxtermann, F. L. Brown). Atlantic Richfield Co., DE 1 953 236, 1969 (B. V. Bell). Mitsui Toatsu Kagaku K. K., DE 2 403 953, 1974 (Y. Tadashi *et al.*).
93 *FAO Yearbook 1991*, Forest Products, Rome 1993.

Further Reading

J. R. Barnett, G. Jeronimidis: *Wood Quality and its Biological Basis*, Blackwell Publ., Oxford 2003.

C. A. S. Hill: *Wood Modification – Chemical, Thermal and Other Processes*, John Wiley & Sons, Chichester 2006.

T. Q. Hu (ed.): *Characterization of Lignocellulosic Materials*, Blackwell, Oxford 2008.

A. A. Klesov: *Wood-Plastic Composites*, Wiley-Interscience, Hoboken, NJ 2007.

D. Kretschmann et al.: *Wood*, "Kirk Othmer Encyclopedia of Chemical Technology", 5th edition, John Wiley & Sons, Hoboken, NJ, online DOI: 10.1002/0471238961.2315150404211802.a01.pub2.

W. F. Lehmann: *Wood-Based Composites and Laminates*, "Kirk Othmer Encyclopedia of Chemical Technology", 5th edition, John Wiley & Sons, Hoboken, NJ, online DOI: 10.1002/0471238961.2315150412050813.a01.

J. C. F. Walker: *Primary Wood Processing*, 2nd ed., Springer, Dordrecht 2006.

Author Index

A

Albrecht Wilhelm, Wuppertal, Federal Republic of Germany, *Cellulose*
Allen John M., Eastman Chemical Company, Kingsport, TN 37662, USA, *Cellulose Esters*
Astheimer Hans-Joachim, Rhodia AG, Freiburg, Federal Republic of Germany, *Cellulose Esters*

B

Babel Wilfried, Deutsche Gelatine-Fabriken Stoess AG, Eberbach, Federal Republic of Germany, *Gelatin*
Balser Klaus, Wolff Walsrode AG, Walsrode, Federal Republic of Germany, *Cellulose Esters*
BeMiller James N., Whistler Center for Carbohydrate Research, Purdue University, West Lafayette, Indiana, USA, *Starch*

C

Challen Ian, NutraSweet Kelco, Waterfield, Tadworth, Surrey, United Kingdom, *Polysaccharides*

D

Dick Eberhard, Deutsche Gelatine-Fabriken Stoess AG, Eberbach, Federal Republic of Germany, *Gelatin*

E

Eicher Theo, Stuttgart, Federal Republic of Germany, *Cellulose Esters*

G

Gareis Herbert, Deutsche Gelatine-Fabriken Stoess AG, Eberbach, Federal Republic of Germany, *Gelatin*
Giesen-Wiese Monika, Deutsche Gelatine-Fabriken Stoess AG, Eberbach, Federal Republic of Germany, *Gelatin*
Gil Luis, Instituto Nacional de Engenharia e Tecnologia Industrial, Lisboa, Portugal, *Cork*
Gruber Patrick R., Outlast Technologies Incorporated, Boulder, Colorado, USA, *Biorefineries – Industrial Processes and Products*

H

Hirano Shigehiro, Tottori University, Tottori, Japan, *Chitin and Chitosan*
Hoppe Lutz, Wolff Walsrode AG, Walsrode, Federal Republic of Germany, *Cellulose Esters*
Huber Kerry C., School of Food Science, University of Idaho, Moscow, Idaho, USA, *Starch*

K

Kamm Birgit, Research Institute Bioactive Polymer Systems (biopos e.V.), Teltow, Germany, *Biorefineries – Industrial Processes and Products*
Kamm Michael, Biorefinery.de GmbH, Potsdam, Germany, *Biorefineries – Industrial Processes and Products*
Knutsen Svein Halvor, Norwegian Food Research Institute, MATFORSK, Norway 1430 Ås, *Polysaccharides*
Kordsachia Othar, University of Hamburg, Hamburg, Germany, *Pulp*
Krässig Hans, Seewalchen, Austria, *Cellulose*

L

Lebbar Rachid, Setexam, Kenitra, Morocco, *Polysaccharides*
Lehnen Ralph, Institut für Holzchemie, Bundesforschungsanstalt für Forst- und Holzwirtschaft, Hamburg, Germany, *Lignin*
Lichtenthaler Frieder W., Clemens-Schöpf-Institut für Organische Chemie und Biochemie, Technische Universität Darmstadt, Darmstadt, Germany, *Carbohydrates: Occurrence, Structures and Chemistry*; *Carbohydrates as Organic Raw Materials*

M

Marr Beinta U., The Copenhagen Pectin Factory Ltd., Lille Skensved, Denmark, *Polysaccharides*
Mohring Marc, J. Rettenmaier & Söhne GmbH + Co, Rosenberg, Germany, *Cellulose*
Moiteiro Cristina, Instituto Nacional de Engenharia e Tecnologia Industrial, Lisboa, Portugal, *Cork*

N

Nimz Horst H., Bundesforschungsanstalt für Forst- und Holzwirtschaft, Institut für Holzchemie und chemische Technologie des Holzes, Hamburg, Federal Republic of Germany, *Wood*

P

Patt Rudolf, University of Hamburg, Hamburg, Germany, *Pulp*
Poth Ulrich, BASF Coatings AG, Münster, Federal Republic of Germany, *Drying Oils and Related Products*

R

Riad Abdelwahab, Setexam, Kenitra, Morocco, *Polysaccharides*
Rolin Claus, The Copenhagen Pectin Factory Ltd., Lille Skensved, Denmark, *Polysaccharides*

S

Saake Bodo, Institut für Holzchemie, Bundesforschungsanstalt für Forst- und Holzwirtschaft, Hamburg, Germany, *Lignin*
Schliefer Karl, Textilforschungsanstalt Krefeld, Krefeld, Federal Republic of Germany, *Cellulose*
Schlosser Harald, J. Rettenmaier & Söhne GmbH + Co, Rosenberg, Germany, *Cellulose*
Schmidt Marc, Wolff Cellulosics GmbH & Co. KG, Walsrode, Germany, *Cellulose Ethers*
Schmitt Uwe, Bundesforschungsanstalt für Forst- und Holzwirtschaft, Institut für Holzbiologie und Holzschutz, Hamburg, Federal Republic of Germany, *Wood*
Schott Annelore, Deutsche Gelatine-Fabriken Stoess AG, Eberbach, Federal Republic of Germany, *Gelatin*
Schrieber Reinhard, Deutsche Gelatine-Fabriken Stoess AG, Eberbach, Federal Republic of Germany, *Gelatin*
Schulz Dieter, Deutsche Gelatine-Fabriken Stoess AG, Eberbach, Federal Republic of Germany, *Gelatin*
Schurz Josef, University of Graz, Graz, Austria, *Cellulose*
Schwab Eckart, Bundesforschungsanstalt für Forst- und Holzwirtschaft, Institut für Holzphysik und mechanische Technologie des Holzes, Hamburg, Federal Republic of Germany, *Wood*
Seybold Uwe, Deutsche Gelatine-Fabriken Stoess AG, Eberbach, Federal Republic of Germany, *Gelatin*
Steadman Robert G., La Trobe University, Bundoora, Victoria, Australia, *Cellulose*
Stein Winfried, Fritz Häcker GmbH & Co., Vaihingen/Enz, Federal Republic of Germany, *Gelatin*
Steinmeier Hans, Rhodia Acetow AG, Freiburg, Federal Republic of Germany, *Cellulose Esters*
Süttinger Richard, Schongau, Germany, *Pulp*

T

Thielking Heiko, Wolff Cellulosics GmbH & Co. KG, Walsrode, Germany, *Cellulose Ethers*
Thomas Alfred, Unimills International, Hamburg, Federal Republic of Germany, *Fats and Fatty Oils*

V

Voragen Alphons C. J., Wageningen, Agricultural University, Department of Food Science, Wageningen, The Netherlands, *Polysaccharides*

W

Wandel Martin, Bayer AG, Leverkusen, Federal Republic of Germany, *Cellulose Esters*
Wittmann Otto, BASF Aktiengesellschaft, Ludwigshafen, Federal Republic of Germany, *Wood*
Wolf Franz, BASF Aktiengesellschaft, Ludwigshafen, Federal Republic of Germany, *Wood*

Subject Index

A

Abaca, 140, 149
Abbrevinalia ivorensis, 553
Abies balsamea, 557
Abrasives
    gelatin in, 375
Acacia mollissima, 557
Acacia senegal, 491
Acaica seyal, 491
Acer rubrum, 557
Acer saccharum, 557
Acetals
    synthesis, 82
2-Acetamido-2-deoxy-D-glucopyranose
    see N-Acetyl-D-glucosamine
Acetic acid
    recovery from cellulose acetate production, 204
Acetic anhydride
    recovery from cellulose acetate production, 204
Acetosolv process, 21
Acety, 218
N-Acetyl-β-d-glucosaminan
    see Chitin
N-Acetyl-D-glucosamine, 65
Acid bisulfite process, 413
Acidolysis
    of coconut oil, 327
Acid value
    of fats, 332
Activated charcoal
    for bleaching of oils, 320
S-Adenosylmethionine (SAM), 462
Adhesives
    gelatin in, 375
    in glulam, 563
    melamine–resorcinol–formaldehyde, 563
    phenol–resorcinol–formaldehyde
        (PRF), 563
Agar, 460, 488 460, 488
    extraction, 489
Agaropectin, 489
Albis, 218
Alcell process, 417
Alcohols
    producible by microbial fermentation, 107
    suitable as monomers, 113
Alcoholysis
    of fats, 327
Alcon extraction process, 313
Aldaric acid, 109
Aldehydes
    reaction with chitosan, 247
Alditols, 77
Aldoses
    reduction to alditols, 77
    structure and configuration, 61
Alfa–Laval Centriflow rendering plant, 315
Alfaprox procedure, 33
Alginate, 460, 474
    bacterial, 480
    solubility, 477
    structural units in, 477
Alginic acid
    solubility, 477
Alkylpolyglucosides
    production, 95
Almond oil
    fatty acid composition, 294
Alnus incana, 557
Aluminum
    production and environmental impact, 5

Amadori rearrangement, 76
Ameri-Bond, 389
Amino resin
    in particle board, 572
Ammonium
    sulfite pulping processes based on, 413–414, 432
Amylase–iodine complex, 521
Amylomaize starch
    see High-amylose corn starch
Amylopectin, 70
Amylopectin [9037-22-3], 518
    structures and properties, 519
Amyloplast, 521
Amylose [9005-82-7], 70, 518
    properties, 521
Anageissus latifolia, 460
Angiospermae, 549
3,6-Anhydrogalactose, 484
anhydroglucose unit, 229
Anisidine value
    of fats, 332
Anogeissus latifolia, 495
Anthrahydroquinone
    catalyst in lignin cleavage, 386
Anthraquinone
    anthraquinone/anthrahydroquinone redox system, 408
Antibiotics
    sugar-derived, 97
D-Apiose, 65
Arabinogalactan, 493
L-Arabinose, 60
Arachidic acid
    melting points of acid and its glycerides, 300
Arachidonic acid, 277
Arbiso process, 414
Artists' colors
    drying oils for, 282
ASAM process, 21, 418
Ascophyllum nodosum, 475
Ascophyllum vinelandii, 475
Ash, 553
Aspartic acid, 105
Aspen, 553
Asplund process
    in fiberboard production, 581
Astragalus gummifer Labillarière, 493
Avocado oil, 338
    fatty acid composition, 294
Azetobacter vinelandii, 460, 480

B

Babassu oil, 340
    composition and properties, 294
Bagasse pulp, 410
Ball drop viscosity, 191
Bamboo, 396
Bark, 400
Bast, 147
Baudouin reaction, 334
Bauer process
    in fiberboard production, 581
Beech, 553
Beech lignin, 382
Beef tallow, 354
    composition and properties, 294
    fatty acid composition, 294
Behenic acid
    melting points of acid and its glycerides, 300
Belt extractor, 313
Bergmann – Junk test, 191

Ullmann's Renewable Resources
© 2013 Wiley-VCH Verlag GmbH & Co. KGaA. Published 2013 by Wiley-VCH Verlag GmbH & Co. KGaA.
ISBN: 978-3-527-33369-1

Betula alba, 553
Betula papyrifera, 557
Betula verrucosa, 557
Bílik reaction, 74
Bioenergy, 37
Biofuel, 25, 37
Biogas, 26
Biomass
  definition, 36
  heat from, 26
  technology, 37
Biopol, 112
Biopolyesters
  synthetic, 110
Bioproducts, 37
Biorefinery, 29
  definition, 40
  design, 44
  generation I-, 40
  generation II-, 41
  generation III-, 40
  lignocellulosic feedstock, 44
  raw material biomass, 35
  research, 35
  two-platform concept, 49
  whole crop, 44, 46
Biosurfact, 389
Birch, 553
Björkman lignin, 389
Björkman procedure, 260
Black digest, 412
Blaw–Knox Rotocel extractor, 312
Bleaching
  of oils and fats, 319
Bleaching earths
  for refining of oils and fats, 320
Blown oils, 285
Blue gum, 396
Boiled oils, 285
Boleco oil, 277
Boleko oil, 352
Bone grease, 354
Borneo tallow, 342
Borresol, 389
Borresperse, 389
Brabender ViscoAmylograph, 526
Bradshaw process, 327
Brookfield viscometer, 505
Brown seaweed
  alginates in, 474
Bupleurum falcatum, 474
1,2,4-Butanetriol, 108
Butter
  composition and properties, 294

C

Caesalpinia spinoza, 460
Calcium
  sulfite pulping processes based on, 413–414, 429
Cambium, 550
Capric acid
  melting points of acid and its glycerides, 300
2-Capriodilaurin
  properties, 301
Caproic acid
  melting points of acid and its glycerides, 300
Caprylic acid
  melting points of acid and its glycerides, 300
Carbohydrates
  nomenclature, 72
  occurrence, structures and chemistry, 59
  as organic raw materials, 89
  reactions, 73
Carbon disulfide
  for xanthation of cellulose, 154
Carbonyl value
  of fats, 332
Carboxylic acids
  producible by microbial fermentation, 107
  suitable as monomers, 114
Carboxymethyl Cellulose (CMC)
  CMC, see carboxymethyl cellulose
Carboxymethyl cellulose acetate butyrate (CMCAB), 197, 209

Carboxymethyl starch, 536
Cardboard
  pulps, 414
Carob gum, 501, 503
  see also Locust bean gum
  production, 505
Carotene
  in fats, 298
ι-Carrageenan, 481
κ-Carrageenan, 481
λ-Carrageenan, 482
μ-Carrageenan, 481
ν-Carrageenan, 481
θ-Carrageenan, 482
Carrageenan, 460, 481
  analysis, 484
  gelation, 485
  production of, 484
  viscosity, 486
Castor oil, 352
  detection, 333
  as drying oil, 280
Catalysts
  for drying oil cross-linking, 283
CCL chipper, 402
Cellidor CP, B and S, 218
Cellit PR and BP, 208
Cellobiose, 66
Cellophane, 21
  see also Cellulose hydrate film
Celluloid, 193
Celluloid cotton, 184, 190
Cellulose, 68, 123
  acetylation reagents, 198
  alkalization, 152
  in cork, 260
  esterification, 201
  high-vacuum pyrolysis, 102
  production of microcrystalline, 170
  production of powdered, 170
  properties of microcrystalline, 173
  properties of powdered, 173
  as renewable resource, 21
  in wood, 558
  xanthation, 153
Cellulose acetate, 197–198
  analysis, 205
  catalysts in production of, 198
  manufacturers of, 203
  molding compounds from, 212
  precipitation and processing, 203
  solubility of, 205
Cellulose acetate butyrate (CAB), 197, 206
  in automobile metallic paints, 22
  characteristic data, 208
  films, 208
  molding compounds from, 212
Cellulose acetate fiber, 209
  see also Acetate fiber
Cellulose acetate phthalate, 197, 208
Cellulose acetate propionate (CAP), 197, 206
  characteristic data, 208
  films, 208
  molding compounds from, 212
Cellulose borates, 196
Cellulose butyrate, 206
Cellulose diacetate, 205
Cellulose ester, 136, 177
  industrial processes, 199
  mixed esters, 206
  polymer-modified mixed esters, 214
  as renewable resource, 22
  surface coatings, 209
Cellulose ester plastics, 211
  production, 217
  quality testing, 218
  toxicology, 220
Cellulose ether, 136, 225
  coagulation, 228
  flocculation point, 227
  molar mass distribution, 228
  as renewable resource, 22
  toxicology, 240

viscosity, 227
Cellulose fiber
  crystalline, 381
  from regenerated cellulose, 150
  man-made, 150
  natural, 140
Cellulose halogenides, 196
Cellulose nitrate, 180
  see also Nitrocellulose
Cellulose nitrate lacquers, 192
Cellulose nitrites, 196
Cellulose phosphates, 195
Cellulose phosphites, 196
Cellulose propionate, 206
Cellulose pulp
  from straw, 31
Cellulose sulfates, 194
Cellulose titanate, 196
Cellulose triacetate, 205
  hydrolysis, 202
Cellulose triacetate fibers, 210
Cellulose xanthate, 137, 197
Cement-bonded board, 585
Centrifish, 315
Cephalin, 297
Ceratonia siliqua, 460, 503
Cereal oils, 345
Cerebrosides, 297
Cesalpina spinosa, 504
Chaulmoogra oil, 352
Chemothermomechanical pulp (CTMP) process, 448
Chia seed oil, 347
Chicken fat, 355
  fatty acid composition, 294
Chitin, 68, 243
  content, in organisms, 245
  derivatives, chemical structure, 247
  toxicology, 250
Chitinase, 246
Chitosan, 243
  derivatives, chemical structure, 247
  enzymes for hydrolyses, 246
  toxicology, 250
Chlorine
  bleaching agent for pulp, 432
Chlorine dioxide
  bleaching agent for pulp, 435–436
3-Chloro-2-hydroxypropyltrimethylammonium chloride
  cationic starches with, 537
Chlorophyll
  commercial production from alfalfa, 33
Cholesterol
  in fats, 297
Chondrus crispus, 483
Cloud point
  of oil, 331
Clupanodonic acid, 277
Coal
  consumption worldwide, 14
  mining engineering, 12
  processing, 13
  as raw material, 8
  utilization, 13
Coal combustion
  CO_2, SO_2, and NO_x formed, 15
  reduction of emissions from, 15
  reduction of NO_x and SO_2 in FRG, 15
Coal liquefaction, 18
Cochius viscosity, 191
Cocoa butter, 341
  composition and properties, 294
Coconut oil, 339
  composition and properties, 294
  detection, 333
  fatty acid composition, 294
  transesterification, 327
Cod-liver oil, 356
Cohune oil, 340
Coir, 149
Coke production, 17
Cold-water-swelling starch, 538
Collex, 389
Collodium powder, 192

Combustion furnace
  for combustion of pulping waste liquor based on sodium, 428
Composite plywood, 566
D-Coniferin alcohol, 380
Conifers
  for production of wood pulp, 398
Coniferyl alcohol, 380
  dehydrogenation, 380
Continuous pressed laminates (CPL), 591
ContiRoll process
  in particle board production, 574
Cook-up starch, 538
Copolymers
  of drying oils, 286
Coreal, 193
Core plywood
  battenboard, 566
  laminboard, 566
Cork, 255
  composition, 267
  disks and stoppers, 266
  extraction, 265
  granulates and broken, 267
  recycling, 272
  toxicological aspects, 272
Corkboard
  insulation, 268
Cork quality council, 270
Corkrubber, 268
Cork standardization, 270
Corn
  raw material in biorefineries, 46
Corn germ oil
  composition and properties, 294
Corn oil, 344
Corn starch
  wet milling, 528
Corterra, 108
Cosmetics
  gelatin in, 373
Cotton, 140
  raw material for cellulose acetate, 199
  structural data, 161
Cotton fibers, 129
Cottonseed oil, 279, 343
  composition and properties, 294
  detection, 333
  fatty acid composition, 294
p-Coumaryl alcohol, 380
Cruciferous oils, 350
Crystallization
  fractional, of fats, 322
Cuoxam process, 165
  regenerated cellulose by, 22
Cuphea oil, 341
Cuprammonium rayon, 131
Cyamopsis tetragonoloba, 504
Cyclodextrins, 67
β-Cyclodextrin [7585-39-9]
  starch conversion product, 535
Cyclohexanediamine tetraacetate (CyDTA)
  gradual extraction of pectin by, 473

D

Dark red meranti, 553
Debarking, 400
Dehydration
  carbohydrates, 74
Dehydrogenation polymers, 380
DENOX process, 15
Densified laminated wood, 566
2-Deoxy-D-erythro-pentafuranose
  see 2-Deoxy-D-ribose
2-Deoxy-D-ribose, 65
6-Deoxy-L-galactopyranose
  see L-Fucose
6-Deoxy-L-mannopyranose
  see L-Rhamnose
De Smet extractor, 312
Detergents
  carboxymethyl cellulose, 231
Dextrans, 70
Dextrins

starch, 533
Dicarboxylic acid, 98
1,2-Dicapriolaurin
  properties, 301
1,3-Dicaprioolein
  properties, 301
1,2-Dicapriostearin
  properties, 301
1,3-Dicapriostearin
  properties, 301
1,2-Dilauroolein
  properties, 301
2,5-Dimethylfuran
  production, 98
1,3-Dipalmitoelaidin
  properties, 301
1,2-Dipalmitoolein
  properties, 301
1,3-Dipalmitoolein
  properties, 301
1,2-Distearoolein
  properties, 301
1,3-Distearoolein
  properties, 301
Difurfural diamine, 93
Dihydropyranones
  production, 99
Dika fat, 341
Dimethylolethylenurea
  for crease-resistant finishing of cellulose, 147
Direx process
  extraction, 313
Disk chipper, 402
Dissolving pulp, 442
Diwatex, 389
Douglas fir, 396, 553
Drum chipper, 402
Drying oils, 275
Dustex, 389
Dynasperse, 389

E
Earlywood, 550
E.C. 3.2.1.14
  see Chitinase
Ecklonia maxima, 475
E10-fuel, 34
Elaidic acid
  melting points of acid and its glycerides, 300
α-Eleostearic acid, 277
ELSOL, 94
Emission
  during production of wood-based materials, 592
Energy
  from biodiesel, 25
  from biogas, 25
  from biomass, 26
  from ethanol, 25
Engineered wood, 569
Entandrophragma candollei, 553
Epichlorohydrin
  in starch cross-linking, 535
Erucic acid
  melting points of acid and its glycerides, 300
Esters
  interesterification of fats, 306, 326
  reactions at the hydroxyl groups, 80
Ethanol (ethyl alcohol)
  biobased, 92
  in wood pulping, 417
Ethers
  reactions at the hydroxyl groups, 79
Ethylenediaminetetraacetic acid (EDTA)
  gradual extraction of pectin by, 473
Ethyl hydroxyethyl cellulose (EHEC)
  EHEC, *see Ethyl hydroxyethyl cellulose*
Eucalyptus camaldulensis, 557
Eucalyptus globulus, 557
Expelling
  of oil seeds, 309
Explosives
  from cellulose nitrate, 192
Extraction
  of oil seeds, 310
  selective, of fats, 324

F
Fachagentur Nachwachsende Rohstoffe, 19
Fagus sylvatica, 553, 557
Fats, 291
  synthetic, 315
Fats and fatty oils, 291
  fatty acid composition, 294
  as renewable resource, 19
  unsaturated, autoxidation, 307
  unsaturated, hydrogenation, 306
Fatty acids
  biosynthesis and biodegradation, 295
  mono–and diglycerides, 327
  in oils and fats, 293
  physical properties of tri–and monoglycerides, 300
  uses of oleochemical products, 20
Fatty acids, unsaturated
  in fats and oils, 295
  hydrogenation, 306
  isomerization and polymerization, 307
Fatty oils, 291
  see also Fats and fatty oils
FB process, 415
Fermentation
  guidelines in biorefineries, 41
Ferulic acid
  in cork, 258
Fiberboard
  binders and additives, 579
  classification of types, 578
  as hardboard, 585
  as insulationboard, 585
  production processes, 581
  properties and testing, 584
Fiber flash drying
  of pulp, 425
Fiber removal
  from pulp, 423
Fibers, man-made
  world production, 151
Fillers
  in plywood, 567
Film
  for wood-based material coating, 589
Filtration
  of oil–soil slurry, 320
Filtrex process
  extraction, 313
Fire point
  of fats and oils, 331
Fischer glycosidation, 75
Fischer projection formula, 61
Fish oil, 279, 356, 358
  composition and properties, 294
  isolation, 315
Fitelson test, 334
Flash point
  of fats and oils, 331
Flax, 140, 148
Floor coverings
  of cork, 267
Flue gas
  desulfurization with hydrated lime, 16
Food
  gelatin in, 371
Formaldehyde
  in wood-based materials, 593
Fractionation
  of fats, 322
Framiré, 553
Fraxinus excelsior, 553
D-Fructose, 60
D-Fructose 1,6-diphosphate, 80
β-D-Fructofuranosyl α-D-glucopyranoside
  see Sucrose
D-Fructose [57-48-7]
  in starch, 535
L-Fucose, 65
Fumaric acid
  production via glycolytic pathway, 108

Furan
 with a tetrahydroxybutyl side-chain, 99
Furanacrylic acid, 93
Furan-2,5-dicarboxylic acid, 109
Furanoses
 conformation of, 62
Furfural
 from carbohydrate sources, 93
 production, 31
Furfuryl alcohol, 93
Furfurylamine, 93
Furoic acid, 93
Furylidene ketone, 93

G
Galactolipids, 297
Galactomannan, 503
 derivatives, 506
 structural features of, 504
α-D-Galactopyranosyl-(1→6)-sucrose
 see Raffinose
α-D-galactopyranosyluronic acid
 in pectin, 462
β-D-Galactopyranosyl-(1→4)-D-fructopyranose
 see Lactulose
β-D-Galactopyranosyl-(1→4)-D-glucopyranose
 see Lactose
D-Galactose, 60
β-D-Glucopyranosyl-(1→4)-D-glucopyranose
 see Cellobiose
Galam butter, 342
Gas
 as raw material, 8
 from crude oil processing, 10
 transport, handling, and storage, 11
 uses, 12
Gasohol, 33
Gelatin, 363
 backing, 374
 baryta, 374
 gelling type [9000-70-8], 365
 nongelling type [68410-45-7], 365
 scrubbing, 374
 technical, 367
 top-coat, 374
Gellan gum, 460, 501
Girdler deodorizer, 322
Glucal, 101
α-D-Glucopyranosyl α-D-glucopyranoside
 see α,α-Trehalose
α-D-Glucopyranosyl-(1→6)-D-fructofuranose
 see Isomaltulose
α-D-Glucopyranosyl-(1→4)-D-glucopyranose
 see Maltose
D-Glucitol, 77
β-d-Glucosaminan
 see Chitosan
D-Glucose, 60
D-Glucose
 structure and configuration, 60
D-Glucose (dextrose) [50-99-7]
 dextrose equivalency, 533
 in starch, 533
 structure and configuration, 60
D-Glucose 1-phosphate, 80
D-Glucose 6-phosphate, 80
D-Glucuronic acid, 65
Glucose syrup, 534
Glulam
 adhesives in, 563
 boards, 566
 production, 563
 raw materials, 563
 testing and standards, 564
α-L-Gulurono-β-D-mannuronans, 474
Glutamic acid, 105
Gluten
 corn, 529
 wheat, 529
Glycerides
 of fatty acids, 293
Glycerol
 in cork, 258

Glycerol 1-phosphate, 297
Glycerophosphoric acid esters, 297
Glycolipids, 297
Glycosides
 reactions at the carbonyl group, 75
Glycosylamines
 reactions at the carbonyl group, 76
Glycosyl halides
 acylated, 81
Gmelina arborea, 557
Goitrin, 300
Goose fat, 355
 fatty acid composition, 294
Gorli oil, 353
Gossypol, 299
 in cottonseeds, 343
Graft copolymerization
 of cellulose, 137
Graft polymerization
 on chitin an chitosan, 248
Grape-seed oil, 348
 fatty acid composition, 294
Grass lignin, 379
Green Biorefinery, 40
 wet-fractionation, 48
Groundnut oil, 349
Groundwood, 445
see also Wood
 for papermaking, 445
 pressure groundwood (PGW), 445, 447
 thermogroundwood (TGW), 445, 447
Guaiacyl lignins, 382
Guaiacyl–syringyl lignins, 382
Guar gum, 460, 504
 production, 504
Guluronan, 477
L-Guluronic acid, 476
α-L-Gulurono-β-D-mannuronans, 474
Gum arabic, 460, 491
 structural features of, 492
Gum ghatti, 460, 495
 structural features of, 495
Gum karaya (Sterculia, kadaya gum), 460, 494
 structural features of, 494
Gum tragacanth, 460, 493
 structural features of, 493
Gun cotton, 178, 184, 190
Gun powder, 192
Gymnospermae, 549
Gypsum-bonded board, 586

H
Hagemann factor, 487
Halibut-liver oil, 356
Halphen reaction, 333
Hardening
 of oils and fats, 324
Hardwood, 396
see also Wood
 prehydrolysis kraft pulping
  conditions, 409
 for pulping and papermaking, 396
Hardwood lignin, 379
 composition of spent sulfite liquors from
  hardwood, 388
Headrige cutter, 402
Hemiacetals, cyclic, 61
Hemicelluloses
 in wood, 557
Hemp, 140, 149
Hempseed oil, 347
Henequen, 149
Heraklith, 587
Herring oil, 357
Heterooligosaccharides, 67
D-Hexulose (D-ketohexose)
 structure and configuration, 63
High-amylose corn starch, 521, 539
High-amylose starch, 521
High-fructose corn syrups, 535
High-fructose syrup, 535
High-maltose syrup, 535
High-pressure laminates (HPL), 591

High opacity (HO) process, 416
Hilum, 521
Horse fat, 355
  fatty acid composition, 294
Hottenroth index, 155
HWM fibers, 158
Hydnocarpus oils, 352
Hydrazones
  reactions at the carbonyl group, 76
Hydro Dyn process
  in particle board production, 575
Hydrogenation
  of unsaturated fatty acids, 324
  of unsaturated oils and fatty acids, 306, 324
Hydrogen peroxide
  bleaching agent for pulp, 438–439
Hydrolysis
  carbohydrates, 73
Hydroxyalkyl methyl celluloses (HAMC)
  HAMC see Hydroxyalkyl methyl cellulose
Hydroxyethyl Cellulose (HEC)
  HEC see Hydroxyethyl cellulose
Hydroxyglucal, 101
Hydroxyl ethyl starch, 536
Hydroxyl value
  of fats, 332
3-C-Hydroxymethyl-D-glycero-tetrose
  see D-Apiose
Hydroxypropyl Cellulose (HPC),
  HPC see Hydroxypropyl cellulose
Hydroxypropyl starch, 536
3-Hydroxybutyrolactone, 110
5-(Hydroxymethyl)furfural
  production, 97
5-Hydroxymethylfuroic acid, 98
3-Hydroxypropionic acid
  production, 106
Hypochlorites
  bleaching of pulp, 435

I

Illipé fat, 342
Imidazoles
  production, 102
Indian hemp, 149
Indulin, 389
INGEO, 94
Intrallam, 569
Intrinsic viscosity, 125
Inulin, 71
Iodine value
  of fats, 331
Isano oil, 277, 352
Isoderm, 193
Isomaltulose, 67
Isomerization
  carbohydrates, 74
Isomerized oils, 286
Isosorbide, 78
Itaconic acid
  production, 107

J

Jojoba oil, 356
Juniperus communis, 557
Jute, 140, 148

K

Kamyr system
  for pulping, 419
Kapok oil, 344
Kappaphycus alvarezii, 483
Karité butter, 342
Katiau fat, 342
Kelig, 389
Kenafseed oil, 344
Kennedy extractor, 313
D-Ketohexose
  see D-Hexulose
Ketones
  reaction with chitosan, 247
9-cis,11-trans,13-trans-4-Ketooctadecatrienoic acid
  see α-Licanic acid

Ketoses
  reduction to alditols, 77
Kevlar, 115
Khaya ivorensis, 553
Kiliani–Fischer extension, 77
Klason procedure, 389
Koenigs-Knorr reaction, 81
Kosipo, 553
Kraft black liquor
  composition of, 388
Kraft pulping (sulfate pulping), 21, 385, 405
  composition of spent Kraft liquors, 425
  lignin degradation by, 407
  prehydrolysis kraft process, 409
Krupp deodorizer, 322
Krupp Kontipress process, 314

L

Lacquer cotton, 184, 190
Lactic acid
  fermentation, 32
  production, 94
Lactose, 66
Lactulose, 67
Lallemantia oil, 347
Laminaria digitata, 475
Laminaria hyperborea, 475
Laminated strand lumber (LSL), 569
Laminated veneer lumber (LVL), 566
Lard, 353
  composition and properties, 294
  detection, 333
Larix sibirica, 557
Latewood, 550
Laurel oil, 341
Lauric acid oils, 339
1-Laurodimyristin
  properties, 301
1-Laurodipalmitin
  properties, 301
2-Laurodipalmitin
  properties, 301
1-Lauro-2-myristo-3-palmitin
  properties, 301
1-Lauro-2-myristo-3-stearin
  properties, 301
Lecithin (phosphatidylcholin), 297
  removal from soybean oil, 316, 349
Leguminous oils, 349
Levoglucosan, 100
Levoglucosenone, 100
Levulinic acid, 109
Levulinic acid
  production, 31
α-Licanic acid, 277
Lignan
  polymerization, 381
Lignin, 379
  commercial, 387
  components, 379
  detection methods, 389
  functional groups and elementary composition, 384
  molar mass, 384
  monomeric precursors, 380
  polymeric, 380
  structural units, 381
  thioacidolysis, 391
  toxicology, 392
Lignin
  chlorination reactions, 433
  in cork, 258
  vanillin from waste sulfite liquor, 426
  in wood, 557
Lignin carbohydrate complex, 382
LignoBond, 389
LignoSol, 389
Lignosulfonates, 384, 388
Linoleic acid, 277
  melting points of acid and its glycerides, 300
Linoleic acid, conjugated, 277
Linolenic acid, 277
  melting points of acid and its glycerides, 300
Linseed oil, 279, 346
  composition and properties, 294

fatty acid composition, 294
Linters, 146, 343
Liver oil
  isolation, 315
Lobry de Bruyn–van Ekenstein rearrangement, 74
Locust bean gum, 460
  see also Carob gum
Loop hydrogenation reactors, 325
Lovibond tintometer, 331
Lubricants
  noncirculating, 20
Lupine oil, 350
Lurgi deodorization plant
  for oils, 322

M
Macrocystis pyrifera, 474–475
Magnesia-bonded boards, 587
Mahogany (African), 553
Maillard reaction, 74
Maleinized oils, 286
Malic acid
  production via glycolytic pathway, 108
Maltitol [585-88-6], 535
Maltodextrins [9050-36-6], 534
  preparation and use, 534
Maltose, 66
Manila, 149
D-Mannitol, 78
D-Mannose, 60
Mannuronan, 477
D-Mannuronic acid, 476
Maraspers, 389
Marine oils, 314, 355
Mark–Houwink relationship, 467
Masonite process
  in fiberboard production, 581
Matches
  gelatin in, 374
Mäule test, 389
Melamine–formaldehyde resins
  in plywood, 567
Melon seed oil, 344
Mende process
  in particle board production, 575
Menhaden oil, 357
Mercerization, 146
Metals
  rate of recycling in FRG, 7
Methanol (methyl alcohol)
  in wood pulping, 418
N-Methyl-N-acyl-glucamides
  production, 95
Methyl cellulose (MC)
  MC, see Methyl cellulose
Methylene chloride
  recovery from cellulose acetate production, 204
Methylene diphenylene isocyanate (MDI)
  in particle board, 572
MIAG basket conveyor extractor, 312
Microllam, 569
Milled wood lignin, 384
Mineral
  classification, 3
Mineral resources
  classification of, 2
Mining
  underground, 12
Miscanthus reeds, 26
Mixed ethers hydroxyethyl methyl cellulose (HEMC)
  HEMC see Mixed ethers hydroxyethyl methyl cellulose
Modal fibers, 158
Molded plywood, 566
Monocotyledoneae, 549
Monosaccharides, 60
  structural variations of, 64
Mowrah fat, 342
Murumuru oil, 340
Mustard seed oil, 351
Mutton tallow, 355
  composition and properties, 294
Myristic acid
  melting points of acid and its glycerides, 300

N
Néou oil, 352
Neatsfoot oil, 354
Nef reaction, 77
Neutral sulfite semichemical (NSSC) process, 414
Newsprint, 444
  pulps, 414
Niger seed oil, 348
Nonuniformity factor, 126
Nonwoven fabrics (nonwovens)
  from cuprammonium fibers, 168
Norlig, 389
Nutmeg butter, 341

O
Oak, 553
Ochroma lagopus, 557
Oil
  as raw material, 8
  loading and storage, 10
  processing, 10
  production, 8–9
  transportation, 10
Oil seed
  expelling, 309
  extraction, 310
Oiticia oil, 279
Oiticica oil, 351
Okal process
  in particle board production, 575
Okra oil, 344
Oleic acid
  biosynthesis, 296
  melting points of acid and its glycerides, 300
Oleic–Linoleic acid oils, 345
Oligosaccharides, 65
  synthesis, 76
Olive kernel oil, 338
Olive oil, 338
  composition and properties, 294
  detection, 334
  fatty acid composition, 294
Ongoke oil, 352
Organocell process, 21, 417
Organometallic compounds
  from crude oil processing, 11
Osazones
  reactions at the carbonyl group, 76
Otoba fat, 341
Ouricuri oil, 341
Oxalates
  gradual extraction of pectin by, 473
Oxygen
  bleaching agent for pulp, 437, 439
Ozone
  bleaching agent for pulp, 418, 438

P
Paints and coatings
  for wood-based panels, 587
Palmitic acid
  melting points of acid and its glycerides, 300
Palmitic acid oils, 343
Palmitic–stearic acid oils, 341
2-Palmitodistearin
  properties, 301
1-Palmito-3-stearo-2-olein
  properties, 301
1,2-Propanediol, 108
1,3-Propanediol
  production, 108
1,2-Propylenediol alginate, 479
3-Pyridinols, 103
Palm kernel oil, 339
  composition and properties, 294
  fatty acid composition, 294
Palm oil, 336
  composition and properties, 294
  detection, 333
  fatty acid composition, 294
  interesterification, 327
Palm olein, 337

Palm stearin, 337
Panax ginseng, 474
Paper
  α-cellulose content, 139
  gelatin in, 375
  kraft pulps, 443
  sulfite pulps, 443
Parallam, 569
Parallel strand lumber (PSL), 569
Parenchyma
  apotracheal, 553
  cells, 551
  paratracheal, 553
Parinarium oils, 352
Particle board
  adhesives and additives in, 572
  extruded, 570
  flat pressed, 570
  molded objects, 571
  production, 572
  properties and testing, 575
Peanut butter, 350
Peanut oil, 349
  composition and properties, 294
  detection, 334
  fatty acid composition, 294
Pectate, 463
Pectate lyase (EC 4.2.2.2) (EC 4.2.2.9), 464
Pectin, 460, 462
  amidated pectins, 470
  analysis, 472
  as gelling agent, 468
  calcium pectate gels, 470
  degree of acetylation, 467
  degree of esterification, 467
  degree of of esterification (DE), 472
  high-ester (HM), 463
  low-ester (LM), 463
  low-water-activity, 468
Pectin, high-methoxyl, 464
Pectin, low-methoxyl, 464
Pectin lyase (EC 4.2.2.10), 464
  see also Pectin transeliminase
Pectin methylesterase (E.C. 3.1.1.11), 464
  see also Pectin glycosidase
Perchloric acid
  acetylation catalyst, 198
Perilla oil, 278, 347
Peroxide value
  of fats, 332
Pharmaceuticals
  gelatin in, 372
  sugar-derived, 97
Phellogenic acid
  in cork, 258
Phellonic acid
  in cork, 258
Phenol
  in wood-based materials, 593
Phenolic resins
  oil boiling with, 287
Phenol–formaldehyde resins
  in particle board, 572
  in plywood, 567
Phloionolic acid
  in cork, 258
Phloroglucinol
  biosynthesis, 104
Phosphatides
  in fatty oils, determination, 335
Phosphatidylethanolamine, 297
Phosphatidylinositol, 297
Phosphatidylserine, 297
Phospholipids, 297
Phosphoryl chloride (phosphorus oxychloride)
  in starch cross-linking, 535
Photography
  gelatin in, 373
Phulwara butter, 342
Picea abies, 553, 557
Picea glauca, 557
Pilchard oil, 357
Pine, 553

Pinus radiata, 557
Pinus sylvestris, 553, 557
Plasmalogens, 297
Plasterboard, 579
Plastiloid, 218
Plywood
  adhesives and additives in, 566
  extenders and fillers, 567
  properties and testing, 567
  types of, 565
Po-yoak oil, 352
Polyamides
  production, 112
Polydextrose [68424-04-4], 543
Polyesters, microbial, 112
Polyfon, 389
Polygalacturonase (PG) (E.C. 3.2.1.15), 464
  endo PG, 464
  exo PG, 464
Polyisoprene
  as renewable resource, 24
Polylactic acid
  production, 94
Polymers
  degree of polymerization, 125
  molar mass distribution and determination, 126
Polymer wood, 594
Polynosics, 158
Polysaccharides, 68, 459
  application, 461
  in cork, 260
  origin and main sugar moieties, 460
  as renewable resource, 24
Poly(vinyl acetate)
  in plywood, 567
Poppyseed oil, 348
Populus tremula, 553
Portland cement
  in particle board, 572
Potato starch
  processing, 529
  properties, 520
Powder cotton, 184, 190
Pressure grinding process (PGW process), 447
Prostaglandins (PG)
  precursors in biosynthesis, 296
Proteoglycan, 483
Protobind, 389
Protopectin, 462
Proxan procedure, 33
Pseudomonas aeruginosa, 480
Pseudomonas elodea, 460, 501
Pseudotsuga menziesii, 553, 557
Pulp
  for acetate fibers, 210
  anthraquinone as catalyst in the production, 408, 416
  bleaching, 432
  bleaching, non-polluting processes, 436
  bleaching, toxicology of chlorinated lignin degradation products, 434
  bleaching with chlorine and chlorine-containing compounds, 432
  carbohydrate degradation by pulping, 408, 413
  classification, 395
  dissolving pulp, 442–443
  hypochlorite bleaching, 435
  production and cellulose content, 138
  quality testing, 138
  residual lignin content and brightness of various pulps, 432
  specialty pulps, 409–410
Pulp bleaching
  bleaching chemicals for, 387
Pulp industry, 418
  waste liquor treatment, 425
Pulping, 403
  see also Pulp
  alkaline, 405–406
  nitric acid, 416
  pollution, 416, 441
  pulping with acetic acid, 418
  pulping with formic acid, 418
  soda pulping of annual plants, 409
  steam explosion or steam refining, 387
  waste liquor combustion for chemicals recovery, 429
Pulp, mechanical

bleaching, 451
pulp screening, 449
Pumpkin seed oil, 344
   fatty acid composition, 294
Pyranoses
   conformation of, 62
Pyrazoles
   production, 102
Pyrodextrin
   see Starch dextrins
Pyrones
   production, 99
Pyrroles
   production, 100

Q

Quercus robur, 553
Quick soluble agar (QSA), 488
Quinic acid, 104

R

Raffinose, 66
Ramie, 140
Rapeseed oil, 350
   detection, 334
   fatty acid composition, 294
Rapid Visco Analyser, 526
Raw materials
   availability of mineral materials, 5
   development of mineral deposits, 3
   formation of mineral deposits, 2
   mineral exploitation, 4
   mineral groups, 3
   primary, 1
   renewable resources, 18
   secondary, 6
Reax, 389
Reed pulp, 409
Refiner mechanical pulp (RMP) process, 448
Renewable resources, 18
   ecobalancing of, 26
   limits of use, 26
   uses, 19
Resins, natural
   as renewable resource, 24
L-Rhamnose, 65
L-Rhamnose
   in pectin, 463
Rhodophyta, 481
D-Ribose, 60
Rice bran oil, 345
   fatty acid composition, 294
Rice starch
   preparation, 530
Rice straw, 396
Ricinoleic acid
   melting points of acid and its glycerides, 300
Ridgelimeter, 472
Rosin
   oil boiling with, 287
Rotuba acetate, 218

S

Safety valve, subsurface, 11
Safflower oil, 279, 348
   composition and properties, 294
Sal butter, 342
Salt minerals, 5
Saponification value
   of fats, 332
Sardine oil, 357
Sawmill technique, 402
Scots pine, 396
Sesame oil, 346
   composition and properties, 294
   detection, 334
Setlithe, 218
Shea butter, 342
Sheets
   for wood-based material coating, 588
Sherwin–Williams extraction, 313
Shikimic acid, 380

metabolic engineering, 106
Shorea macrophylla, 553
Silver birch, 396
Silver fir, 396
Sinapinic acid
   in cork, 258
Sinapyl alcohol, 380
Sisal, 140, 149
β-Sitosterol
   in fats and oils, 297
Sivola (Rauma) process, 415
Smoke point
   of fats and oils, 331
Soda pulping, 388
Sodium cellulose xanthate, 197
Sodium chloride
   modified starch by, 532
Sodium sulfate
   from production of viscose, 157
   modified starch by, 532
Sodium trimetaphosphate
   in starch cross-linking, 535
Softwood, 396
see also Wood
   prehydrolysis kraft pulping conditions, 409
   for pulping and papermaking, 396
Softwood lignin, 379
   composition of spent sulfite liquors from softwood, 388
Solexol process
   for extraction of oil, 319
   for fractionation of fats, 324
Sorbitan esters, 95
Sorbitol production, 94
Sorel cement
   in particle board, 572
Sorona, 108
South American oil, 357
Soybean
   oil extraction from, 313
Soybean oil, 279, 349
   composition and properties, 294
   detection, 334
   fatty acid composition, 294
   removal of lecithin, 316
Spermatophyta, 549
Sperm oil, 356
Sphingolipids, 297
Sphingomyelins, 297
Sphingosine, 297
Spruce, 396, 553
Spruce lignin, 381
Spruce tracheids
   cell wall thickness, 398
Stack cooker
   for oil seeds, 309
Stand oils, 285
Starch
   as renewable resource, 23
   hydrolysis, 30
   industrial uses, 23
   occurence, 69
Starch acetates, 536
Starch dextrins, 533
Starch gels, 524
Starch graft copolymers, 537
Starch granules, 521
   breakdown, 527
   gelatinization, 523, 525
   melting, 525
   pasting time and temperature, 526
   peak viscosity, 526
   remnants, 524
   retrogradation, 525
   setback, 527
   structure, 523
   swelling, 523
   trough viscosity, 527
Starch [9005-25-8], 517
   annealing, 538
   cationic, 537
   converted, 532
   converted or thinned, 530

cross-linked, 535
crystallization, 525
degree of polymerization, 520, 533
degree of substitution, 531
derivatized, 530
destructurized, 538
dextrinization, 533
dextrose equivalency, 533
digestibility, 543
esterification or etherification, 532
heat-moisture treatments, 538
instant, 537
modification processes, 539
pregelatinized, 530, 537
stabilized, 536
thermoplastic, 539
thinned Products, 532
Starch octenylsuccinates, 536
Starch oct-1-enylsuccinic ester, 536
Starch phosphates, 536
Starch succinates, 536
Staudinger index, 125
Steam explosion
   production of fibers by, 387
Stearic acid
   melting points of acid and its glycerides, 300
Sterculia urens, 494
Sterols
   in fats and oils, 297
1-Stearodibehenin
   properties, 301
Stone groundwood (SGW) process, 443
Stora process, 415
Straw
   for chemicals and fuels, 51
   raw material in biorefineries, 46
Straw pulp, 410
Suberin
   in cork, 258
Succinic acid
   production via glycolytic pathway, 108
Sucrose, 65
Sucrose fatty acid monoesters
   production, 96
Sugar
   as renewable resource, 23
   production, 30
   ring forms of, 61
Sugar alcohols, 78
Sugar cane
   bagasse, 396
Sulfite process, 21
Sulfite pulping, 386, 405, 410
   alkaline, 415
   bisulfite pulping, 414
   compositions of spent sulfite liquors, 425
   hexoses and pentoses from, 425
   lignin reactions, 411
   lignosulfonates from, 425
Sulfur compounds
   from crude oil processing, 11
Sulfuric acid
   acetylation catalyst, 198
Sulfurized oils, 286
Sunflower oil, 279, 345
   composition and properties, 294
   fatty acid composition, 294
Surfactants
   renewable resources for, 20
   sugar-based, 95
Swiss Federal Office for Environment, Forests and Countryside, 6
Syrups
   glucose, 534
Syrup solids, 534

T

Tall oil
   as drying oil, 281
Tamiflu, 104
Tannin
   condensed, 262
   in cork, 262
   hydrolyzable, 262

Tapioca starch
   processing, 530
   properties, 520
Tara gum, 460
Teak, 553
Teaseed oil, 347
   detection, 334
Tectona grandis, 553
Tengkawang tallow, 342
Tenite acetate, 218
Tenite butyrate, 218
Tenite propionate, 218
Terpenes
   in fats, 298
   in wood, 556
Tetrahydrofuran, 93
Textile dyeing
   viscose fibers, 161
Textile finishing
   crease-resistant finishing of cellulose, 146
Thermogrinding (TGW) process, 447
Thermomechanical pulp (TMP) process, 448
Thioacetals
   reactions at the carbonyl group, 75
Thioglycosides
   reactions at the carbonyl group, 75
Tocopherols
   in fats and oils, 298
Totox value
   of fats, 332
Tracheids, 551
Tragacanthic acid, 493
Transesterification
   of drying oils, 287
α,α-Trehalose [99-20-7], 66
   starch conversion product, 535
Trembling aspen, 396
Trielaidin
   properties, 301
Trilaurin
   properties, 301
Trimyristin
   properties, 301
Triolein
   properties, 301
Tripalmitin
   properties, 301
Tristearin
   properties, 301
Tsuga canadensis, 557
Tucum oil, 337, 341
Tung oil, 279, 351
   detection, 334
Tyloses, 554

U

Ucuhuba fat, 341
Ufoxane, 389
Ulti-angle laser-light scattering (MALLS)
   for carrageenan analysis, 485
Ultrazine, 389
Unibord process
   application of films on panels, 590
Urea–formaldehyde resin
   in plywood, 566
Urethane oils, 288

V

Vanillin
   from lignin, 32
Veneer
   production methods, 565
Veneer plywood, 565
   production of, 567
Vernolic acid, 352
Vernonia oil, 352
Viscose fiber, 151
   grafted, 160
Vitamin A
   in fats and oils, 298
Vitamin E
   in fats and oils, 298

Index

Vitamins
  sugar-derived, 97
Voith chain grinder, 447

W

Wafex, 389
Walden-Reversion, 32
Waloran N, 193
Wanin, 389
Waste
  from cork, 272
Wax
  in fatty oils, 297
Waxes
  in cork, 262
Waxy maize starch, 539
Waxy starch, 519
Werzalite process
  in particle board production, 571
Whale oil, 355
  composition and properties, 294
  fatty acid composition, 294
  isolation, 314
Wheat-germ oil, 345
  fatty acid composition, 294
Wheat starch
  dry milling, 529
  properties, 520
Wheat straw, 396
White biotechnology, 34
Wiesner reaction, 389
Williamson ether synthesis, 229
Wood, 396, 445, 548
  see also Groundwood; Hardwood; Softwood
  abrasion resistance, 561
  bordered pits, 555
  Brinell hardness, 561
  cell types, 551
  cell wall formation, 554
  cell-type volume in some species, 553
  cellulose content, 138
  chemical comparison of wood species, 557
  chemical components, 555
  chips, production, 401
  debarking, 400
  degree of swelling, 560
  density, 560
  diffuse-porous, 552
  economics of forest products, 549
  fibers, 397
  for mechanical pulps, 444
  for pulp, 396
  forests and roundwood removals worldwide, 548
  grindler stones, 447
  half-bordered pits, 555
  hardwoods, 551
  heartwood, 553
  middle lamella, 554
  pits, 555
  plasmodesmata, 555
  primary wall, 554
  raw material for cellulose acetate, 199
  ring-porous, 552
  saccharification, 31
  sapwood, 553
  semi-ring-porous, 552
  simple pits, 555
  softwoods, 551
  springwood, 398
  storage damage, 399
  summerwood, 398
  thermal conductivity, 561
  total wood and wood products in FRG, 549
  toxicology, 592
Wood cell wall
  distribution of components in, 559
  structure of cell components, 559
Wood dust, 592
Wood fiber, 551
Wood-based materials, 562
  coating with sheets, 588
  decorative films for, 589
  decorative laminates for, 590
  disposal, 594
  emissions from finished materials, 593
  laminate bonding, 563
  with mineral binder, 585
  particle bonding, 570
  surface treatment, 587
Wood processing, 403

X

Xanthan
  manufacture of, 496
Xanthan gum, 460, 496
  manufacture of, 496
  structural features of, 497
Xanthomonas campestris, 460, 496
Xyampsis tetragonolobus, 460
Xylem, 550
Xylem tissue, 379
D-Xylose, 60

Z

Zenith process
  for neutralization of oil, 318
Zewa, 389